Pathology of Childhood and Adolescence

Consolato M. Sergi

Pathology of Childhood and Adolescence

An Illustrated Guide

Volume I

Consolato M. Sergi
University of Alberta
Edmonton, AB
Canada

ISBN 978-3-662-59167-3   ISBN 978-3-662-59169-7   (eBook)
https://doi.org/10.1007/978-3-662-59169-7

© Springer-Verlag GmbH Germany, part of Springer Nature 2020
This work is subject to copyright. All rights are reserved by the Publisher, whether the whole or part of the material is concerned, specifically the rights of translation, reprinting, reuse of illustrations, recitation, broadcasting, reproduction on microfilms or in any other physical way, and transmission or information storage and retrieval, electronic adaptation, computer software, or by similar or dissimilar methodology now known or hereafter developed.
The use of general descriptive names, registered names, trademarks, service marks, etc. in this publication does not imply, even in the absence of a specific statement, that such names are exempt from the relevant protective laws and regulations and therefore free for general use.
The publisher, the authors, and the editors are safe to assume that the advice and information in this book are believed to be true and accurate at the date of publication. Neither the publisher nor the authors or the editors give a warranty, express or implied, with respect to the material contained herein or for any errors or omissions that may have been made. The publisher remains neutral with regard to jurisdictional claims in published maps and institutional affiliations.

This Springer imprint is published by the registered company Springer-Verlag GmbH, DE part of Springer Nature.
The registered company address is: Heidelberger Platz 3, 14197 Berlin, Germany

To my patients and their families, to my family, and to my students and fellows

Foreword

University hospitals, major hospitals, and medical care centers must have a pathological institute, and this is a long-standing requirement. This facility is the prerequisite for the adequate diagnosis of most of the diseases that need to be treated. Since most major hospitals also have pediatric and pediatric and adolescent surgery, the affiliated pathological institute should either have a department for pediatric pathology or at least one or two, better three, designated pathologists of childhood pathology to meet the child-specific pathological diagnosis. It is well known that this is not always the case. There are very few institutes of pathology that have their department or division of pediatric pathology or at least have a pediatric pathologist. The well-known sentence "children are not small adults" is often not taken into account.

At present, at least in the German-speaking world, pediatric pathology is only a subdiscipline of pathology. It is carried out often by general pathologists who are particularly interested with this field, who receive their training on their initiative, by rotating in several renowned institutions and participating to congresses and meetings of the Society for Pediatric Pathology or Pediatric Pathology Society as well as national meetings of pediatric pathology.

What is the challenge for pediatric pathologists? The field of work comprises a vast field of diagnostic tasks, which implies two interlinked fields of knowledge: on the one hand, knowledge of physiological development processes, i.e., the continually changing processes for growth reasons, and typical findings of the various tissues of the human body, starting with the embryo and ending in adolescents and late youth, and on the other hand, knowledge of the possible deviating pathological findings for the respective age group. This means that the ever-changing morphology of healthy development represents an additional dynamic dimension in the evaluation of pathological changes that reduce the diagnostic effort of pediatric pathologists compared to that of general pathologists. An additional, not less critical, area of pediatric pathology is the fetal diagnosis, primarily because it can be used to clarify congenital malformations and intrauterine, perinatal, and postpartum deaths. Finally, placenta diagnosis should not be forgotten, as it can be used to detect relevant evidence not only for intra- and perinatal deaths or possible diseases in the newborn but also for pregnancy complications.

In summary, in addition to diagnostics, the participation of the pediatric pathologists in the context of rare or unusual findings with discussion in the setting of multidisciplinary team meetings with general and subspecialized pediatricians, pediatric surgeons, obstetricians, and human geneticists is very well welcomed. It is surprising that the children's pathology, although its diagnostic-scientific spectrum and its cooperation in the clinical implementation of its findings differ sizably from that of general pathology, is usually covered in very few textbooks. There is virtually no truly comprehensive, clinical, and molecular biology-oriented up-to-date manual that includes detailed information for all stakeholders.

Due to the far-reaching prerequisites mentioned, a great deal of courage is required to commit to the task of writing a handbook for children's pathology because, as mentioned above, children's pathology not only requires a comprehensive spectrum of knowledge per se but also requires solid knowledge of the physiological processes of the growing body and – in connection with diagnostics – also specific ideas regarding pediatric or pediatric surgical treatment options. These requirements make the creation of a manual an almost unmanageable challenge if all these criteria are to be considered efficiently. Consolato Sergi is in the field of pediatric pathology for more than 30 years collecting experiences and acquiring knowledge. He took up this challenge, as he told me about a year ago, after much consideration. His decision for this endeavor was born out of the idea of putting his extensive pediatric pathological knowledge on paper in order to fill the bibliographical gap that exists in this respect.

In this context, I would like to mention a few of my memories that connect me with the author of this book. Between 2004 and 2008, Sergi was employed as a pediatric pathologist at the Pathological Institute of the Medical University of Innsbruck, Austria. As director of the Department of Pediatrics and Adolescent Surgery, I had to deal with him very often, as we had to cope with a quite significant number of surgical patients. On the one hand, we repeatedly discussed not only critical problems of our shared patients (often not sparing any day of the week including Saturday and Sunday) but also scientific projects. On the other hand, we met fortnightly or weekly within the framework of multidisciplinary team meetings. The "case ideas" and discussions were very profitable and sustainable since Dr. Sergi, who is also pediatrician, already had, at that time, an extensive theoretical and clinical knowledge. Therefore, he was able to contribute a lot to the clinical discussions, which was generally well praised.

It was obvious that I remained in contact with him, who has proven himself to this day, after his departure from Innsbruck due to our excellent cooperation. I congratulated him on his decision to write up this textbook, which will complement the library of not only pathologists and pediatric pathologists but also pediatricians, pediatric surgeons, and human geneticists. Because of his knowledge and well-balanced scientific curriculum, I was sure he would be able to complete this project. In reading the table of the contents of his book, I saw my opinion confirmed. He had compiled a textbook covering all relevant pediatric pathological topics. What is important to mention in this context is the fact that not only had he brought his vast wealth of knowledge into text but also didactically presented his theoretical and practical knowledge with a lot of pictorial material so informative that it would benefit not only morphologists but also clinicians. Consequently, it represents for all those who are involved – directly – in the pathological diagnosis with all its facets and in the resulting treatment of sick children or adolescents who are involved in diagnostics and research in the field of pediatric pathology as well as trainees and fellows, an essential and comprehensive source of information.

Medical University of Innsbruck, Innsbruck, Austria   Josef Hager

Foreword

Pediatrics, the "Medicine of Children," is an area of medical study that is very broad, diverse, and challenging. Unlike different specialties and subspecialties, pediatrics does not deal with an organ, system, or specialty of function but with a developing organism, the human organism in its total entirety, genetics, embryonal, fetal, neonate to adolescent. The development through this continuum presents with a diversity of pathophysiology. The Department of Pediatrics, at the University of Alberta, has recognized the research and education opportunities to answer clinically important questions that make a difference in the health of children throughout the age spectrum. In collaboration with all our child health colleagues, including Dr. Sergi, the Department of Pediatrics is a leading recipient of national tri-council funding and through research truly making a clinical difference.

Dr. Sergi's book is a compilation of pathological illustrations and molecular biology data put together for the benefit of all who care for the newborn, child, and adolescent, including physicians, trainees, and allied Pediatric health providers. Clinical acumen depends predominantly on logical deduction but also on lateral interaction with disciplines that help to understand the biology of the developing human organism. Pediatric pathology is unique because it helps pediatricians to understand not only the pathological basis of disease but also inquire into molecular biological pathogenesis. Dr. Sergi's marvelous textbook in pediatric pathology is the crown of more than 30 years of the interaction of him with pediatricians. Joining the University of Alberta in 2008 as full professor of pathology and adjunct professor of pediatrics allowed me to interact with him very often as chair of the Department of Pediatrics at the University of Alberta, and importantly as a colleague in hepatology and liver transplantation.

For over a decade I have had the professional opportunity to interact clinically with Dr. Sergi. His clear explanation of concepts of pathophysiology to not only myself but to our multidisciplinary team, and most importantly our trainees, has enriched all of our intellect, and ongoing management of patients. I know that the value of such a compilation relies on extensive and thorough effort, as is witnessed in this textbook. The generosity of Dr. Sergi's dedication, and that of his family, for the advancement of pediatric medicine and pediatric pathology is immeasurable.

I wish all readers, my fellows and colleagues and myself will have the opportunity to interact more with Dr. Sergi and acquire the invaluable steps in progressing the knowledge in pediatrics.

Susan M. Gilmour, MD, MSc, FRCPC
Professor Pediatrics, Pediatric Gastroenterology/Hepatology
Department Chair 2009–2019
University of Alberta
Edmonton
Canada

Preface

The pediatric patient may be mistakenly considered an adult in miniature, but the precise definitions of diseases uniquely appearing in infancy, childhood, and adolescence make this field of medicine astoundingly rich in notions and knowledge ranging from the early intrauterine life to adolescence and youth. Pediatric pathology is one of the fastest growing subspecialties in medicine. Pediatric pathology is also unique because a few diseases that were earlier confined to adults are now occurring in childhood, adolescence, and youth. There may be several reasons, including the optimization of imaging technology, better surgical and clinical pathology criteria, but also some external factors are likely playing some role. The environment has become more and more impregnated with dangerous molecular compounds that are not only carcinogenic but also endocrine and metabolism disruptors. The pediatric pathology clubs, born in the United Kingdom and the United States, have seen growing number of participants in the last few years. The current pediatric pathology society (PPS) and societies of pediatric pathology (SPP) have reached an enormous interest not only in residents in pathology but also in pediatrics, obstetrics, gynecology, as well as other medical and surgical specialties. Long before the founding of the pediatric pathology clubs and their transformation into PPS and SPP, the German embryological schools played an unarguable and incontrovertible role in the expansion of our knowledge of perinatal medicine and congenital defects in the twentieth century.

Three figures have motivated me in collecting and producing this book. They are Guido Fanconi, Klaus Goerttler, and Harald P. Schmitt. Guido Fanconi is regarded as one of the founders of the modern pediatrics for his contributions to biochemistry and how biochemistry helped reshape modern pediatrics. His contributions to the pathology of pediatrics were numerous, and several diseases have been nominated after him. He was head of the *Kinderspital* in Zurich, Switzerland, for about 45 years and recognized Down syndrome years before the chromosomal identification. In 1934, the first patients affected with mucoviscidosis or cystic fibrosis of the pancreas were described in a doctoral thesis written under his direction. His contributions to pediatrics are countless, and some of these are highlighted in the present book. A few years ago, Stephan Lobitz, from the Department of Pediatric Oncology and Hematology, Charité–University Medicine Berlin, Berlin, Germany, and Eunike Velleuer, from the Department of Pediatric Oncology, Hematology and Immunology, University of Düsseldorf, Germany, wrote an outstanding contribution on Professor Fanconi. Most probably, one of the most paramount contributions for understanding the magnitude of Fanconi to pediatrics and the many mentees is a handwritten note kindly made available to Dr. Lobitz and Dr. Velleuer by Fanconi's son. Fanconi wrote "*Forschen: Aufdeckung und Deutung eines neuen Tatbestandes ist an eine Idee gebunden, der eine Unsumme fleissiger Arbeit folgt. Sie setzt einen ideenreichen Kopf und einen fleissigen, systematischen Arbeiter voraus. Ferner ist es wichtig, dass ein gründliches Wissen zugrunde liegt.*" ("Research: the discovery and interpretation of new facts are bound to an idea which is followed by an enormous amount of hard work. It presupposes a creative mind and a diligent, systematic worker. A sound knowledge base is also important.") This sentence has been a path to many pediatric pathologists who contributed to the self-determination, development, and autonomy of pediatric pathology starting about the second half of last century and continuing in this century. The stature of Professor Klaus Goerttler as top-ranking cardiac embryologist and pediatric pathologist has extensively contributed to the German embryology school, pediatric cardiology, and pediatric pathology. The German embryology school laid fundamental notions for the development of numerous concepts of knowledge in pediatric pathology. The clarification of the embryology of heart and its defects by the Heidelberg professor are milestones in the interpretation of the current pediatric cardiology and cardiac surgery as may be identified in his book *Normale und pathologische Entwicklung des menschlichen Herzens* (*Normal and Pathologic Development of the Human Heart*) of 1958. Hypoxia is a dreadful teratogen, causing disruption, particularly of neurulation if it interferes with early stages of embryonic development. Numerous experimental studies performed in amphibian and chick embryos showed that hypobaric-mediated hypoxia determines disruptions of the heart as well as head and brain. Most of the disruptions are induced at the beginning of gastrulation, i.e., before the onset of neurulation, when oxygen consumption is known to be exceptionally high. In these experiments, the underlying developmental mechanisms responsible for the malformations occurring with hypoxia were multiple. They include (1) the disturbance of the migration of blastema, or an alteration of the "*Topogenese*" (topogenesis" of the German embryological school according to Lehmann and of the "integrated cell and tissue movements" according to Gilbert), (2) the decreased induction of the altered blastema, and (3) the disturbance of the organ anlage or primordial organ to further differentiate. These studies continued the Spemann-Mangold experiments successfully on organizers and found that amphibian and chick investigations can also be extrapolated to humans. Professor Schwalbe highly influenced professor Goerttler and early pediatric pathologists worldwide. Ernst Theodor Karl Schwalbe (1871–1920) was a German pathologist, born in Berlin, who specialized in teratology. His medical study crossed three of the most prestigious universities (Strasbourg, Berlin, and Heidelberg). His habilitation (higher Ph.D. degree) was with a thesis on blood coagulation. In Heidelberg, Dr. Schwalbe worked as an assistant under Julius Arnold. As prosector and head of the pathology-bacteriology clinic, he worked at the city hospital in Karlsruhe in 1907/08. He worked for 12 years at the University of Rostock until 1920 when he was killed while serving as a volunteer during the Kapp Lüttwitz Putsch, which aimed to disengage the German Revolution of 1918–1919; overthrow the Weimar Republic, in governmental position after World War I; and establish an autocratic government in its place. In my opinion, Professor Schwalbe is probably one of the founders of pediatric pathology in general, and his studies have helped to shape the birth of several pediatric pathology clubs worldwide. Professor Horst P. Schmitt was my teacher in neuropathology during my residency in pathology. Professor Schmitt allowed me to revise several diseases in neuropathology and stimulated me in broadening my knowledge in developmental biology and pathology not only of humans but also of vertebrates.

The book *Pathology of Childhood and Adolescence* is not only a book of pathology for pathologists or surgeons (adult and pediatric pathologists) or residents in these disciplines but also contains a molecular biology approach to some of the very challenging diseases in pediatrics. This book came after 12 years work and tight cooperation with Springer Publisher. I am very grateful to Mr. Karthik Periyasamy, who looked at the production for several years; all the Springer team; to my colleague, Dr. Atilano Lacson; and my mentors (Francesco Callea, Walter J Hofmann, Herwart F. Otto, Axel von Herbay, Peter Sinn, Philipp Schnabel, Gregor Mikuz, Moorghen Morgan, Josef Hager, Lothar Bernd Zimmerhackl, Ulrich Schweigmann, and Brian Chiu, among others). I am also grateful to numerous adult and pediatric pathology colleagues and clinical colleagues who sent me consultation cases and advised me over the years. In this book, 18 chapters summarize and illustrate the pathology of childhood and youth comprehensively, including diseases of the cardiovascular system and respiratory tract, gastrointestinal tract, liver, pancreas and biliary tract, upper and lower urinary system, gynecological tract, breast, hematolymphoid system, endocrine system, soft tissues, arthro-skeletal system, central and peripheral nervous system, dermatological system, and placenta with pathology of the fetus and newborn. Pediatric pathology is not the search of the curiosities and does not involve the collection of rare specimens or the taking care of a cabinet of teratological cases. The pediatric pathologist tries to address specifically congenital defects in a context that may help to advance medicine and discover new cures. Most of the genes involved in morphogenesis may play a crucial role in carcinogenesis as identified most recently by pharmacology studies.

Apart from being a pathologist, several pediatric pathologists are engaged in various humanitarian missions. Several pediatric pathologists hold a second specialty in pediatrics or gynecology and membership in pediatric or gynecology societies. Our job as pathologist involves an impressive work for advocacy of children and families. Rudolf Virchow, who is probably one of the founders of cellular pathology, was particularly active in social medicine and public health. In 1902, the *British Medical Journal* suggested that Berlin was one of the most hygienic cities in Europe due to Professor Virchow's efforts on social milieu and social reforms in Germany. The increasingly authoritarian nature of his Imperial country moved Virchow to demand democracy and social welfare programs. Following the steps of Virchow's deep sense of humanitarianism, several pediatric pathologists are now engaged in middle- or low-income countries to improve the public health qualities in these geographic areas. Sir William Osler, first Baronet, FRS

FRCP (1849–1919), was a Canadian physician and one of the four founding fathers of Johns Hopkins Hospital in Maryland, United States. As a student, Sir William Osler spent some time in Berlin learning the art of the autopsy from Virchow and, on the pathologist's 70th birthday, emphasized that his consistent humanitarianism amplified Virchow's scientific achievements. Such Canadian esprit has permeated the twenty-first century as well. In 2006, Jean Vanier, a Canadian philanthropist, who recently passed away, gave a talk at Concordia University in the city of Montreal. He argued that the Western culture of the individualism which values beauty, money, and success also creates a gap between the healthy and the disabled.

In this century, healthcare is going to go through financial and technological challenges. It is imperative that the pediatric pathologist is prone to support social initiatives. Pathologists provide strong support to patients and physicians alike as they face the battle with cancer, and the Springer team and I hope that our pages provide the concepts and the foundation for the future of cancer prevention and therapy as we move to the era of personalized molecular oncology. I hope that our book will help the students to learn some pediatric pathology skills and that the physicians are motivated in continuing learning programs. We have employed a coded format that we consider is fruitful and powerful for achieving optimal learning and memorization (DEF, definition; AKA/SYN, also known as/synonyms; EPI, epidemiology; EPG, etio-pathogenesis; CLI, clinics; GRO, gross findings; CLM, conventional light microscopy; TEM, transmission electron microscopy, HSS, histochemical special stains; IHC, immunohistochemistry; FNA, fine-needle aspiration; CMB, cytogenetics and molecular biology; DDX, differential diagnosis; TRT, treatment; PGN, prognosis). We think that the reader will enjoy this book and forgive me if some areas have not been emphasized enough. I trust this book can be a good help for many colleagues in their daily practice. We are eager to share this book with all our health professional colleagues across the world honestly hoping the reader will find the content of pertinence and interest to the care of all children.

Edmonton, AB, Canada   Consolato M. Sergi

Coda
All health is global health, and all medicine is social medicine, and, in this setting, pediatrics plays a significant role for the well-being and health of all children no matter their origin, the skin color they have, the religion they profess, or the orientation their parents have. Pediatric pathology is at the basis of the diseases that can be discovered before birth and after birth. This book is a solid basis that can be useful not only for pediatricians and pathologists involved in the healthcare of developed countries but also for low- and middle-income developing countries. Thus, most of the profit of the author will go to charities involving children. Pediatric pathologists and pediatricians remain essential leaders in the advocacy for pediatric patients and their families. Many of us have been engaged in teaching and providing healthcare services in the setting of the World Health Organization integrated management solutions and nongovernmental organizations, and their efforts have paved the roads for suggesting improvements for better healthcare in low- and middle-income countries. We hope that the application of experience from resource-rich settings to resource-limited settings will be followed by voracious readers able to gather information and notes from this book for the continuing medical education in global health.

Disclaimer
Medicine is an evolving field of knowledge. Thus, please be advised of the following disclaimer. Here, I inform readers of this book that the views and opinions expressed in the text belong solely to the author. They do not necessarily belong to the author's employer, organization, committee, or other group or individual. Moreover, I do not make any warranties about the completeness, reliability, and accuracy of this information. Please verify your current standards of practice in your country of practice. Any action you take upon the information on this book is strictly at your own risk, and by reading and understanding this disclaimer, I will not be liable for any losses and damages in connection with the use of the text of this book.

Contents

Volume I

1 Cardiovascular System.. 1
    1.1 Developmental and Genetics 2
    1.2 Congenital Heart Disease............................. 6
           1.2.1   Sequential Segmental Cardio-Analysis 8
           1.2.2   Atrial Septal Defect (ASD), Ventricular Septal Defect (VSD), Atrioventricular Septal Defect (AVSD), Patent Ductus Arteriosus (PDA), and Ebstein's Tricuspid Anomaly (ETA).......... 11
           1.2.3   Tetralogy of Fallot (TOF) and Double Outlet Right Ventricle (DORV) 18
           1.2.4   Hypoplastic Right Heart Syndrome............... 24
           1.2.5   Transposition of the Great Arteries 26
           1.2.6   Common Trunk (Persistent Truncus Arteriosus).... 31
           1.2.7   Total Anomalous Pulmonary Venous Return....... 32
           1.2.8   Scimitar Syndrome............................ 32
           1.2.9   Hypoplastic Left Heart Syndrome and Coarctation of Aorta 33
           1.2.10 Heterotaxy Syndrome 41
           1.2.11 Common Genetic Syndromes with Congenital Heart Disease............................... 41
    1.3 Myocarditis and Nonischemic Cardiomyopathies......... 48
           1.3.1   Myocarditis 48
           1.3.2   Dilated Cardiomyopathy 48
           1.3.3   Hypertrophic Cardiomyopathy 66
           1.3.4   Infiltrative/Restrictive Cardiomyopathies 66
           1.3.5   Left Ventricular Non-compaction 78
           1.3.6   Arrhythmogenic Right Ventricular Dysplasia/Cardiomyopathy (ARVD/C) 79
           1.3.7   Lamin A/C Cardiomyopathy 81
           1.3.8   Mitochondrial Cardiomyopathies 81
           1.3.9   Cardiac Channelopathies and Sudden Cardiac Death............................... 83
    1.4 Ischemic Heart Disease, Myocardial Dysfunction, and Heart Failure 85
           1.4.1   Ischemic Heart Disease 85
           1.4.2   Myocardial Dysfunction 85
           1.4.3   Heart Failure 85

1.5	Valvular Heart Disease		87
	1.5.1	Endocarditis and Valvar Vegetation	87
	1.5.2	Mitral Valve Prolapse	91
	1.5.3	Calcific Mitral Annulus and Calcific Aortic Stenosis	91
	1.5.4	Rheumatic Heart Disease (RHD)	91
	1.5.5	Prosthetic Valves	92
1.6	Transplantation		92
	1.6.1	Allograft Cellular Rejection	92
	1.6.2	Allograft Humoral Rejection	95
	1.6.3	Cardiac Allograft Vasculopathy	97
	1.6.4	Extracardiac Posttransplant Lympho-proliferative Disorders (PTLD) Following HTX	100
1.7	Tumors		100
	1.7.1	Benign Tumors	102
	1.7.2	Malignant Tumors	109
	1.7.3	Metastatic Tumors	112
1.8	Pericardial Disease		115
	1.8.1	Non-neoplastic Pericardial Disease	115
	1.8.2	Neoplastic Pericardial Disease	116
1.9	Non-neoplastic Vascular Pathology		116
	1.9.1	Congenital Anomalies	116
	1.9.2	Fibromuscular Dysplasia	117
	1.9.3	Arteriosclerosis	117
	1.9.4	Large-Vessel Vasculitis	119
	1.9.5	Medium-Vessel Vasculitis	123
	1.9.6	Small-Vessel Vasculitis	125
Multiple Choice Questions and Answers			125
References and Recommended Readings			127

2 Lower Respiratory Tract 139

2.1	Development and Genetics		141
2.2	Dysmorphology and Perinatal Congenital Airway Diseases		145
	2.2.1	Foregut Cysts of Bronchogenic Type	146
	2.2.2	Agenesis, Aplasia, and Hypoplasia Pulmonis (Pulmonary Hypoplasia)	148
	2.2.3	CPAM and Congenital Lobar Emphysema	148
	2.2.4	Tracheal and Bronchial Anomalies	151
	2.2.5	Sequestrations and Vascular Anomalies	152
2.3	Infantile and Pediatric *NILD*		155
	2.3.1	Neonatal Respiratory Distress Syndrome	157
	2.3.2	"Old" and "New" Bronchopulmonary Dysplasia (OBPD and NBPD)	157
	2.3.3	Atelectasis	158
	2.3.4	Cystic Fibrosis	158
2.4	Infantile and Pediatric *ILD*		159
	2.4.1	Diffuse Lung Development-Associated ILD	159
	2.4.2	Trisomy 21 (Down Syndrome)-Associated ILD	161

		2.4.3	ILD Related to Other Chromosomal/Genomic Microdeletion Disorders	163
		2.4.4	Neuroendocrine Hyperplasia of Infancy (NEHI)	163
		2.4.5	Pulmonary Interstitial Glycogenosis (PIG)	164
		2.4.6	Surfactant Dysfunction Disorders	164
		2.4.7	Pulmonary Alveolar Proteinosis	166
	2.5	"Adult" ILD of the Youth		167
		2.5.1	Diffuse Alveolar Damage (DAD)	167
		2.5.2	Cryptogenic Organizing Pneumonia (COP)	170
		2.5.3	Usual Interstitial Pneumonia/Pneumonitis (UIP)	173
		2.5.4	Nonspecific Interstitial Pneumonia/Pneumonitis (NSIP)	175
		2.5.5	Desquamative Interstitial Pneumonia/Pneumonitis	176
		2.5.6	Respiratory Bronchiolitis ILD	177
		2.5.7	Pulmonary Langerhans Cell Histiocytosis	179
		2.5.8	Lymphoid Interstitial Pneumonia/Pneumonitis (LIP)	180
		2.5.9	Hypersensitivity Pneumonia/Pneumonitis (HSP)	181
		2.5.10	Acute Eosinophilic Pneumonia/Pneumonitis	182
		2.5.11	Progressive Massive Fibrosis	184
	2.6	Inflammation: Tracheobronchial		184
		2.6.1	Laryngotracheitis	184
		2.6.2	Acute Bronchitis	184
		2.6.3	Bronchial Asthma	184
		2.6.4	Pulmonary Hyperinflation	188
		2.6.5	Bronchiectasis	189
		2.6.6	Neoplasms	189
	2.7	Inflammation: Infectious (Pneumonia/Pneumonitis)		190
		2.7.1	Lobar Pneumonia	190
		2.7.2	Bronchopneumonia	191
		2.7.3	Interstitial Pneumonia	193
		2.7.4	Primary Atypical Inflammation (Pneumonia)	193
	2.8	Inflammation: Non-infectious (Pneumonia vs. Pneumonitis)		193
		2.8.1	Aspiration and Chemical Pneumonitis	193
		2.8.2	Lipoid Pneumonia	195
		2.8.3	Diffuse Pulmonary Hemorrhagic Syndromes	196
	2.9	Inflammation: Infectious/Non-infectious Granulomatous/Nongranulomatous		197
		2.9.1	Primary Pulmonary Tuberculosis	197
		2.9.2	Sarcoidosis	200
		2.9.3	Aspergillosis	201
		2.9.4	Others	202
	2.10	Chronic Obstructive Pulmonary Disease of the Youth		202
		2.10.1	Chronic Bronchitis	203
		2.10.2	Emphysema	203

		2.11	"Adult" Pneumoconiosis of the Youth	204
			2.11.1 Silicosis	205
			2.11.2 Asbestosis	206
			2.11.3 Non-silico Asbestosis-Related Pneumoconiosis	207
		2.12	Pulmonary Vascular Disorders	207
			2.12.1 Pulmonary Congestion and Edema	208
			2.12.2 Pulmonary Embolism/Infarction	208
			2.12.3 Pulmonary Arterial Hypertension (PAH)	210
		2.13	Transplantation-Related Disorders	212
			2.13.1 Acute Rejection	212
			2.13.2 Chronic Rejection	212
		2.14	Pediatric Tumors and Pseudotumors	215
			2.14.1 Pulmonary Teratoma	215
			2.14.2 Inflammatory Myofibroblastic Tumor	215
			2.14.3 Pulmonary Hamartoma	217
			2.14.4 Pulmonary Sclerosing Hemangioma	217
			2.14.5 Pulmonary Carcinoid	217
			2.14.6 Lymphangioma-(Leo)-Myomatosis (LAM)	217
			2.14.7 Kaposiform Lymphangiomatosis	219
			2.14.8 Pleuropulmonary Blastoma	220
			2.14.9 Metastatic Tumors	222
		2.15	"Adult"-Type Neoplasms in Childhood/Youth	222
			2.15.1 "Adult"-Type Preneoplastic Lesions	226
		2.16	Pleural Diseases	230
			2.16.1 Pleural Effusions and Pleuritis	230
			2.16.2 Pneumothorax	234
			2.16.3 Neoplastic Pleural Diseases	235
		2.17	Posttransplant Lymphoproliferative Disorders (PTLD)	237
		2.18	Non-thymic Mediastinal Pathology	240
			2.18.1 Anterior Mediastinal Pathology	241
			2.18.2 Middle Mediastinal Pathology	242
			2.18.3 Posterior Mediastinal Pathology	242
		Multiple Choice Questions and Answers		242
		References and Recommended Readings		244
3	**Gastrointestinal Tract**			255
	3.1	Development and Genetics		256
	3.2	Esophagus		258
		3.2.1	Esophageal Anomalies	258
		3.2.2	Esophageal Vascular Changes	263
		3.2.3	Esophageal Inflammatory Diseases	263
		3.2.4	Esophageal Tumors	274
	3.3	Stomach		277
		3.3.1	Gastric Anomalies	277
		3.3.2	Gastric Vascular Changes	280
		3.3.3	Gastric Inflammatory Diseases	281
		3.3.4	Tissue Continuity Damage-Related Gastric Degenerations	286
		3.3.5	Gastric Tumors	287

- 3.4 Small Intestine ... 304
  - 3.4.1 Small Intestinal Anomalies ... 304
  - 3.4.2 Abdominal Wall Defects (Median-Paramedian) ... 306
  - 3.4.3 Continuity Defects of the Intestinal Lumen ... 312
  - 3.4.4 Intestinal Muscular Wall Defects ... 316
  - 3.4.5 Small Intestinal Dystopias ... 317
  - 3.4.6 Composition Abnormalities of the Intestinal Wall ... 317
  - 3.4.7 Small Intestinal Vascular Changes ... 318
  - 3.4.8 Inflammation and Malabsorption ... 319
  - 3.4.9 Short Bowel Syndrome/Intestinal Failure ... 336
  - 3.4.10 Small Intestinal Transplantation ... 336
  - 3.4.11 Graft-Versus-Host Disease (GVHD) of the Gut ... 339
  - 3.4.12 Small Intestinal Neoplasms ... 340
- 3.5 Appendix ... 343
  - 3.5.1 Appendiceal Anomalies ... 343
  - 3.5.2 Appendiceal Vascular Changes ... 343
  - 3.5.3 Appendicitis ... 349
  - 3.5.4 Appendiceal Metabolic and Degenerative Changes ... 351
  - 3.5.5 Appendiceal Neoplasms ... 351
- 3.6 Large Intestine ... 353
  - 3.6.1 Large Intestinal Anomalies ... 353
  - 3.6.2 Large Intestinal Vascular Changes ... 373
  - 3.6.3 Large Intestinal Inflammatory Disorders ... 374
  - 3.6.4 Colon-Rectum Neoplasms ... 380
- 3.7 Anus ... 387
  - 3.7.1 Anal Anomalies ... 388
  - 3.7.2 Inflammatory Anal Diseases ... 389
  - 3.7.3 Benign Anal Tumors and Non-neoplastic Anal Lesions (Pseudotumors) ... 389
  - 3.7.4 Anal Pre- and Malignant Lesions ... 390
- 3.8 Peritoneum ... 393
  - 3.8.1 Cytology ... 393
  - 3.8.2 Non-neoplastic Peritoneal Pathology ... 394
  - 3.8.3 Neoplastic Peritoneal Pathology ... 394
- Multiple Choice Questions and Answers ... 399
- References and Recommended Readings ... 402

4 Parenchymal GI Glands: Liver ... 425
- 4.1 Development and Genetics ... 426
- 4.2 Hepatobiliary Anomalies ... 430
  - 4.2.1 Ductal Plate Malformation (DPM) ... 431
  - 4.2.2 Congenital Hepatic Fibrosis ... 435
  - 4.2.3 Biliary Hamartoma (von Meyenburg Complex) ... 435
  - 4.2.4 Caroli Disease/Syndrome ... 435
  - 4.2.5 ADPKD-Related Liver Cysts ... 435
- 4.3 Hyperbilirubinemia and Cholestasis ... 436
  - 4.3.1 Hyperbilirubinemia ... 436

	4.3.2	Decreased Bilirubin Conjugation and Unconjugated Hyperbilirubinemia	437
	4.3.3	Conjugated Hyperbilirubinemia	438
4.4	Infantile/Pediatric/Youth Cholangiopathies		439
	4.4.1	Biliary Atresia	440
	4.4.2	Non-BA Infantile Obstructive Cholangiopathies (NBAIOC)	443
	4.4.3	The Paucity of the Intrahepatic Biliary Ducts (PIBD)	444
	4.4.4	Neonatal Hepatitis Group (NAG)	444
	4.4.5	Primary Sclerosing Cholangitis (PSC)	456
	4.4.6	Primary Biliary Cirrhosis (PBC)	458
	4.4.7	Pregnancy-Related Liver Disease (PLD)	459
4.5	Genetic and Metabolic Liver Disease		460
	4.5.1	Endoplasmic Reticulum Storage Diseases (ERSDs)	461
	4.5.2	Congenital Dysregulation of Carbohydrate Metabolism	464
	4.5.3	Lipid/Glycolipid and Lipoprotein Metabolism Disorders	467
	4.5.4	Amino Acid Metabolism Disorders	471
	4.5.5	Mitochondrial Hepatopathies	471
	4.5.6	Peroxisomal Disorders	477
	4.5.7	Iron Metabolism Dysregulation	479
	4.5.8	Copper Metabolism Dysregulation	483
	4.5.9	Porphyria-Related Hepatopathies	485
	4.5.10	Shwachman-Diamond Syndrome (SDS)	485
	4.5.11	Chronic Granulomatous Disease (CGD)	485
	4.5.12	Albinism-Related Liver Diseases	486
4.6	Viral and AI Hepatitis, Chemical Injury, and Allograft Rejection		486
	4.6.1	Acute Viral Hepatitis	486
	4.6.2	Chronic Viral Hepatitis	488
	4.6.3	Autoimmune Hepatitis	489
	4.6.4	Drug-Induced Liver Disease (Chemical Injury) and TPN	490
	4.6.5	Granulomatous Liver Disease	495
	4.6.6	Alcoholic Liver Disease	496
	4.6.7	Non-alcoholic Steatohepatitis	498
	4.6.8	Acute and Chronic Rejection Post-Liver Transplantation	498
4.7	Hepatic Vascular Disorders		503
	4.7.1	Acute and Chronic Passive Liver Congestion	503
	4.7.2	Ischemic Hepatocellular Necrosis	505
	4.7.3	Shock-Related Cholestasis	505
4.8	Liver Failure and Liver Cirrhosis		505
4.9	Portal Hypertension		508
4.10	Bacterial and Parasitic Liver Infections		509

			4.10.1	Pyogenic Abscess.	509
			4.10.2	Helminthiasis	510
	4.11	Liver Tumors			510
			4.11.1	Benign Tumors.	510
			4.11.2	Malignant Tumors	517
			4.11.3	Metastatic Tumors	534
	Multiple Choice Questions and Answers				538
	References and Recommended Readings				540
5	**Parenchymal GI Glands (Gallbladder, Biliary Tract, and Pancreas)**				**551**
	5.1	Development and Genetics			551
	5.2	Biliary and Pancreatic Structural Anomalies			553
			5.2.1	Choledochal Cysts	553
			5.2.2	Gallbladder Congenital Abnormalities	553
			5.2.3	Pancreas Congenital Anomalies.	553
	5.3	Gallbladder and Extrahepatic Biliary Tract			556
			5.3.1	Cholesterolosis and Cholelithiasis	556
			5.3.2	Acute and Chronic Cholecystitis	557
			5.3.3	Gallbladder Proliferative Processes and Neoplasms.	558
			5.3.4	Extrahepatic Bile Duct Tumors and Cholangiocellular Carcinoma	561
	5.4	Pancreas Pathology			561
			5.4.1	Congenital Hyperinsulinism and Nesidioblastosis.	561
			5.4.2	Acute and Chronic Pancreatitis	563
			5.4.3	Pancreatoblastoma and Acinar Cell Carcinoma in Childhood and Youth.	566
			5.4.4	Cysts and Cystic Neoplastic Processes of the Pancreas.	567
			5.4.5	PanIN and Solid Pancreas Ductal Carcinoma	571
			5.4.6	Other Tumors	571
			5.4.7	Degenerative Changes and Transplant Pathology	571
	Multiple Choice Questions and Answers				574
	References and Recommended Readings				575
6	**Kidney, Pelvis, and Ureter**				**579**
	6.1	Development and Genetics			581
	6.2	Non-cystic Congenital Anomalies			589
			6.2.1	Disorders of Number	589
			6.2.2	Disorders of Rotation.	591
			6.2.3	Disorders of Position	591
			6.2.4	Disorders of Separation	591
	6.3	Cystic Renal Diseases			591
			6.3.1	Classifications	591
			6.3.2	Autosomal Dominant Polycystic Kidney Disease.	591
			6.3.3	Autosomal Recessive Polycystic Kidney Disease.	596
			6.3.4	Medullary Sponge Kidney	597

	6.3.5	Multicystic Dysplastic Kidney (MCDK)	597
	6.3.6	Hydronephrosis/Hydroureteronephrosis	597
	6.3.7	Simple Renal Cysts	598
	6.3.8	Acquired Cystic Kidney Disease (CRD-Related)	598
	6.3.9	Genetic Syndromes and Cystic Renal Disease	598
6.4	Primary Glomerular Diseases	599
	6.4.1	Hypersensitivity Reactions and Major Clinical Syndromes of Glomerular Disease	599
	6.4.2	Post-infectious Glomerulonephritis	600
	6.4.3	Rapidly Progressive Glomerulonephritis	602
	6.4.4	Minimal Change Disease	604
	6.4.5	(Diffuse) Mesangial Proliferative GN	604
	6.4.6	Focal and Segmental Glomerulosclerosis	604
	6.4.7	Membranous Glomerulonephritis	606
	6.4.8	Membranoproliferative (Membrane-Capillary) Glomerulonephritis	609
6.5	Secondary Glomerular Diseases	610
	6.5.1	SLE/Lupus Nephritis	610
	6.5.2	Henoch-Schönlein Purpura	611
	6.5.3	Amyloidosis	613
	6.5.4	Light Chain Disease	614
	6.5.5	Cryoglobulinemia	614
	6.5.6	Diabetic Nephropathy	614
6.6	Hereditary/Familial Nephropathies	615
	6.6.1	Fabry Nephropathy	615
	6.6.2	Alport Syndrome	615
	6.6.3	Nail-Patella Syndrome	617
	6.6.4	Congenital Nephrotic Syndrome	617
	6.6.5	Thin Glomerular Basement Membrane Nephropathy (TBMN)	617
6.7	Tubulointerstitial Diseases	618
	6.7.1	Acute Tubulointerstitial Nephritis (ATIN)	618
	6.7.2	Chronic Tubulointerstitial Nephritis (CTIN)	622
	6.7.3	Acute Tubular Necrosis	623
	6.7.4	Chronic Renal Failure (CRF)	623
	6.7.5	Nephrolithiasis and Nephrocalcinosis	624
	6.7.6	Osmotic Nephrosis and Hyaline Change	625
	6.7.7	Hypokalemic Nephropathy	625
	6.7.8	Urate Nephropathy	625
	6.7.9	Cholemic Nephropathy/Jaundice-Linked Acute Kidney Injury	625
	6.7.10	Myeloma Kidney	625
	6.7.11	Radiation Nephropathy	626
	6.7.12	Tubulointerstitial Nephritis and Uveitis (TINU)	626
6.8	Vascular Diseases	626
	6.8.1	Benign Nephrosclerosis	626
	6.8.2	Malignant Nephrosclerosis	626
	6.8.3	Renal Artery Stenosis	626

	6.8.4	Infarcts	627
	6.8.5	Vasculitis	627
	6.8.6	Hemolytic Uremic Syndrome	628
	6.8.7	Thrombotic Thrombocytopenic Purpura	630
6.9	Renal Transplantation		630
	6.9.1	Preservation Injury	630
	6.9.2	Hyperacute Rejection	630
	6.9.3	Acute Rejection	631
	6.9.4	Chronic Rejection	632
	6.9.5	Humoral (Acute/Chronic) Rejection	632
	6.9.6	Antirejection Drug Toxicity	633
	6.9.7	Recurrence of Primary Disorder	633
6.10	Hereditary Cancer Syndromes Associated with Renal Tumors		633
	6.10.1	Beckwith-Wiedemann Syndrome	633
	6.10.2	WAGR Syndrome	633
	6.10.3	Denys-Drash Syndrome	634
	6.10.4	Non-WT1/Non-WT2 Pediatric Syndromes	634
	6.10.5	Von Hippel Lindau Syndrome	634
	6.10.6	Tuberous Sclerosis Syndrome	634
	6.10.7	Hereditary Papillary Renal Carcinoma Syndrome	634
	6.10.8	Hereditary Leiomyoma Renal Carcinoma Syndrome	634
	6.10.9	Birt-Hogg-Dube Syndrome	634
6.11	Pediatric Tumors (Embryonal)		634
	6.11.1	Wilms Tumor (Nephroblastoma)	634
	6.11.2	Cystic Partially Differentiated Nephroblastoma (CPDN) and (Pediatric) Cystic Nephroma	645
	6.11.3	Congenital Mesoblastic Nephroma	646
	6.11.4	Clear Cell Sarcoma	646
	6.11.5	Rhabdoid Tumor	647
	6.11.6	Metanephric Tumors	647
	6.11.7	XP11 Translocation Carcinoma	647
	6.11.8	Ossifying Renal Tumor of Infancy	650
6.12	Non-embryonal Tumors of the Young		650
	6.12.1	Clear Cell Renal Cell Carcinoma	650
	6.12.2	Chromophobe Renal Cell Carcinoma	653
	6.12.3	Papillary Adenoma and Renal Cell Carcinoma	653
	6.12.4	Collecting Duct Carcinoma	656
	6.12.5	Renal Medullary Carcinoma	656
	6.12.6	Angiomyolipoma	656
	6.12.7	Oncocytoma	656
	6.12.8	Other Epithelial and Mesenchymal Tumors	659
6.13	Non-neoplastic Pathology of the Pelvis and Ureter		661
	6.13.1	Anatomy and Physiology Notes	661
	6.13.2	Congenital Pelvic-Ureteral Anomalies	661
	6.13.3	Congenital Ureteric Anomalies	661
	6.13.4	Lower Urinary Tract Abnormalities	662

	6.14	Tumors of the Pelvis and Ureter	666
		6.14.1 Neoplasms	666
		6.14.2 Genetic Syndromes	666
	Multiple Choice Questions and Answers		667
	References and Recommended Readings		668

7 Lower Urinary and Male Genital System ... 673

- 7.1 Development and Genetics ... 674
  - 7.1.1 Urinary Bladder and Ureter ... 674
  - 7.1.2 Testis, Prostate, and Penis ... 675
- 7.2 Lower Urinary and Genital System Anomalies ... 676
  - 7.2.1 Lower Urinary System Anomalies ... 676
  - 7.2.2 Male Genital System Anomalies ... 678
- 7.3 Urinary Tract Inflammatory and Degenerative Conditions ... 684
  - 7.3.1 Cystitis, Infectious ... 684
  - 7.3.2 Cystitis, Non-infectious ... 684
  - 7.3.3 Malacoplakia of the Young ... 686
  - 7.3.4 Urinary Tract Infections and Vesicoureteral Reflux ... 686
  - 7.3.5 Megacystis, Megaureter, Hydronephrosis, and Neurogenic Bladder ... 687
  - 7.3.6 Urinary Tract Endometriosis ... 688
  - 7.3.7 Tumorlike Lesions (Including Caruncles) ... 689
- 7.4 Preneoplastic and Neoplastic Conditions of the Urinary Tract ... 690
  - 7.4.1 Urothelial Hyperplasia (Flat and Papillary) ... 690
  - 7.4.2 Reactive Atypia, Atypia of Unknown Significance, Dysplasia, and Carcinoma In Situ (CIS) ... 691
  - 7.4.3 Noninvasive Papillary Urothelial Neoplasms ... 691
  - 7.4.4 Invasive Urothelial Neoplasms ... 693
  - 7.4.5 Non-urothelial Differentiated Urinary Tract Neoplasms ... 696
- 7.5 Male Infertility-Associated Disorders ... 696
  - 7.5.1 Spermiogram and Classification ... 696
- 7.6 Inflammatory Disorders of the Testis and Epididymis ... 696
  - 7.6.1 Acute Orchitis ... 697
  - 7.6.2 Epidermoid Cysts ... 697
  - 7.6.3 Hydrocele ... 700
  - 7.6.4 Spermatocele ... 700
  - 7.6.5 Varicocele ... 704
- 7.7 Testicular Tumors ... 704
  - 7.7.1 Germ Cell Tumors ... 704
  - 7.7.2 Tumors of Specialized Gonadal Stroma ... 720
  - 7.7.3 Rhabdomyosarcoma and Rhabdoid Tumor of the Testis ... 727
  - 7.7.4 Secondary Tumors ... 727

	7.8	Tumors of the Epididymis	727
		7.8.1 Adenomatoid Tumor	727
		7.8.2 Papillary Cystadenoma	728
		7.8.3 Rhabdomyosarcoma	728
		7.8.4 Mesothelioma	728
	7.9	Inflammatory Disorders of the Prostate Gland	728
		7.9.1 Acute Prostatitis	728
		7.9.2 Chronic Prostatitis	728
		7.9.3 Granulomatous Prostatitis	729
		7.9.4 Prostatic Malakoplakia of the Youth	729
	7.10	Prostate Gland Overgrowths	729
		7.10.1 Benign Nodular Hyperplasia of the Young and Fibromatosis	729
		7.10.2 Rhabdomyosarcoma, Leiomyosarcoma, and Other Sarcomas (e.g., Ewing Sarcoma)	729
		7.10.3 Prostatic Carcinoma Mimickers	730
		7.10.4 Prostatic Intraepithelial Neoplasia (PIN)	732
		7.10.5 Prostate Cancer of the Young	733
		7.10.6 Hematological Malignancies	739
		7.10.7 Secondary Tumors	739
	7.11	Inflammatory and Neoplastic Disorders of the Penis	739
		7.11.1 Infections	739
		7.11.2 Non-infectious Inflammatory Diseases	740
		7.11.3 Penile Cysts and Noninvasive Squamous Cell Lesions	741
		7.11.4 Penile Squamous Cell Carcinoma of the Youth	741
		7.11.5 Non-squamous Cell Carcinoma Neoplasms of the Penis	744
	Multiple Choice Questions and Answers		744
	References and Recommended Readings		746

Volume II

8	**Female Genital System**		**757**
	8.1	Development and Genetics	758
	8.2	Congenital Anomalies of the Female Genital System	760
	8.3	Ovarian Inflammatory and Degenerative Conditions	761
		8.3.1 Oophoritis	761
		8.3.2 Torsion	761
		8.3.3 Non-neoplastic Cystic Lesions	763
		8.3.4 Non-neoplastic Proliferations	763
	8.4	Infectious Diseases of the Female Genital System	764
		8.4.1 Viral Diseases	764
		8.4.2 Bacterial Diseases	765
		8.4.3 Fungal Diseases	767
		8.4.4 Parasitic Diseases	767
	8.5	Inflammatory, Reactive Changes and Degenerative Conditions of Vulva, Vagina, Cervix, Uterus, and Tuba	767
		8.5.1 Non- and Preneoplastic Vulvar and Vaginal Lesions	767
		8.5.2 Inflammation-Associated Cervical Lesions	769

		8.5.3	Reactive Changes of the Cervix.	770
		8.5.4	Cyst-Associated Cervical Lesions	770
		8.5.5	Ectopia-Associated Cervical Lesions	770
		8.5.6	Pregnancy-Associated Cervical Lesions	770
		8.5.7	Cervical Metaplasias	771
		8.5.8	Non-neoplastic and Preneoplastic Uterine Lesions	771
		8.5.9	Non-neoplastic Abnormalities of the Tuba	773
	8.6	Tumors of the Ovary		773
		8.6.1	Germ Cell Tumors	774
		8.6.2	Surface Epithelial Tumors	779
		8.6.3	Sex Cord-Stromal Tumors	790
		8.6.4	Other Primary Ovarian and Secondary Tumors	798
	8.7	Tumors of the Tuba		798
		8.7.1	Benign Neoplasms	798
		8.7.2	Malignant Neoplasms	798
	8.8	Tumors of the Uterus		798
		8.8.1	Type I/Type II EC	799
		8.8.2	Non-type I/Non-type II EC	799
		8.8.3	Uterine Leiomyoma (ULM) and Variants	799
		8.8.4	Uterine Leiomyosarcoma (ULMS)	799
		8.8.5	Uterine Adenomatoid Tumor	804
		8.8.6	Uterine Lymphangiomyomatosis (ULAM)	804
		8.8.7	Uterine Stroma Tumors	804
		8.8.8	Hematological Malignancies and Secondary Tumors.	804
	8.9	Tumors of the Cervix		805
		8.9.1	Benign Neoplasms	805
		8.9.2	Precancerous Conditions and Malignant Neoplasms	805
	8.10	Tumors of the Vagina		811
		8.10.1	Benign Tumors	811
		8.10.2	Precancerous Conditions and Malignant Tumors	813
	8.11	Tumors of the Vulva		816
		8.11.1	Benign Neoplasms	816
		8.11.2	Malignant Tumors	818
	Multiple Choice Questions and Answers			823
	References and Recommended Readings			825
9	**Breast**			833
	9.1	Development and Genetics		834
	9.2	Congenital Anomalies, Inflammatory, and Related Disorders		836
		9.2.1	Amastia, Atelia, Synmastia, Polymastia, and Politelia	836
		9.2.2	Asymmetry, Hypotrophy, and Hypertrophy	836
		9.2.3	Dysmaturity and Precocious Thelarche	838
		9.2.4	Acute Mastitis, Abscess, and Phlegmon	838

		9.2.5	Duct Ectasia, Periductal Mastitis, and Granulomatous Mastitis	839
		9.2.6	Necrosis, Calcifications, and Mondor Disease	839
	9.3	Pathology of the Female and Young Adult	840
		9.3.1	Fibrocystic Disease	840
		9.3.2	Soft Tissue Tumors and Hematological Malignancies	840
		9.3.3	Fibroadenoma	840
		9.3.4	Adenoma	841
		9.3.5	Genetic Background of Breast Cancer	844
		9.3.6	In Situ and Invasive Ductal Breast Carcinoma	847
		9.3.7	In Situ and Invasive Lobular Breast Carcinoma	849
		9.3.8	WHO Variants of the Infiltrating Ductal Carcinoma	849
		9.3.9	Sweat Gland-Type Tumors and Myoepithelial Tumors	850
		9.3.10	Phyllodes Tumor	851
	9.4	Cancer Mimickers	852
		9.4.1	Hyperplasia, Ductal and Lobular	852
		9.4.2	Adenosis	855
	9.5	Male Breast Disease	855
		9.5.1	Gynecomastia	855
		9.5.2	Breast Cancer of the Male	855
	Multiple Choice Questions and Answers			856
	References and Recommended Readings			857
10	**Hematolymphoid System**			**861**
	10.1	Development and Genetics		862
	10.2	Red Blood Cell Disorders		864
		10.2.1	Anemia	864
		10.2.2	Polycythemia	867
	10.3	Coagulation and Hemostasis Disorders		869
		10.3.1	Coagulation and Hemostasis	869
		10.3.2	Coagulation Disorders	869
		10.3.3	Platelet Disorders	870
	10.4	White Blood Cell Disorders		871
		10.4.1	Leukocytopenias and Leukocyte Dysfunctionalities	871
		10.4.2	Non-neoplastic Leukocytosis	871
		10.4.3	Leukemia (Neoplastic Leukocytosis) or *Virchow's "Weisses Blut"*	872
		10.4.4	Myelodysplastic Syndromes	878
		10.4.5	Hodgkin Lymphoma	878
		10.4.6	Non-Hodgkin Lymphomas	887
		10.4.7	Follicular Lymphoma	888
		10.4.8	Small Lymphocytic Lymphoma	890
		10.4.9	Mantle Cell Lymphoma (MCL)	891
		10.4.10	Marginal Cell Lymphoma	891

		10.4.11 Diffuse Large B-Cell Lymphoma (DLBCL)	892
		10.4.12 Lymphoblastic Lymphoma	894
		10.4.13 Burkitt Lymphoma	896
		10.4.14 Peripheral T-Cell Lymphoma	902
		10.4.15 Anaplastic Large Cell Lymphoma	903
		10.4.16 Adult T-Cell Leukemia/Lymphoma	903
		10.4.17 Cutaneous T-Cell Lymphoma (CTCL)	904
		10.4.18 Angiocentric Immunoproliferative Lesions	905
		10.4.19 Extranodal NK-/T-Cell Lymphoma	906
	10.5	Disorders of the Monocyte-Macrophage System and Mast Cells	907
		10.5.1 Hemophagocytic Syndrome	907
		10.5.2 Sinus Histiocytosis with Massive Lymphadenopathy (Rosai-Dorfman Disease)	908
		10.5.3 Langerhans Cell Histiocytosis	908
		10.5.4 Histiocytic Medullary Reticulosis	909
		10.5.5 True Histiocytic Lymphoma	910
		10.5.6 Systemic Mastocytosis	910
	10.6	Plasma Cell Disorders	910
		10.6.1 Multiple Myeloma	911
		10.6.2 Solitary Myeloma	912
		10.6.3 Plasma Cell Leukemia	912
		10.6.4 Waldenstrom's Macroglobulinemia	912
		10.6.5 Heavy Chain Disease	912
		10.6.6 Monoclonal Gammopathy of Undetermined Significance (MGUS)	913
	10.7	Benign Lymphadenopathies	913
		10.7.1 Follicular Hyperplasia	914
		10.7.2 Diffuse (Paracortical) Hyperplasia	918
		10.7.3 Sinus Pattern	919
		10.7.4 Predominant Granulomatous Pattern	920
		10.7.5 Other Myxoid Patterns	921
		10.7.6 Angioimmunoblastic Lymphadenopathy with Dysproteinemia (AILD)	921
	10.8	Disorders of the Spleen	921
		10.8.1 White Pulp Disorders of the Spleen	922
		10.8.2 Red Pulp Disorders of the Spleen	922
	10.9	Disorders of the Thymus	924
		10.9.1 Thymic Cysts and Thymolipoma	924
		10.9.2 True Thymic Hyperplasia and Thymic Follicular Hyperplasia	924
		10.9.3 HIV Changes	925
		10.9.4 Thymoma	925
	Multiple Choice Questions and Answers		927
	References and Recommended Readings		930
11	**Endocrine System**		**933**
	11.1	Development and Genetics	934
	11.2	Pituitary Gland Pathology	938

		11.2.1 Congenital Anomalies of the Pituitary Gland	939
		11.2.2 Vascular and Degenerative Changes	939
		11.2.3 Pituitary Adenomas and Hyperpituitarism	939
		11.2.4 Genetic Syndromes Associated with Pituitary Adenomas	941
		11.2.5 Hypopituitarism (Simmonds Disease)	942
		11.2.6 Empty Sella Syndrome (ESS)	943
		11.2.7 Neurohypophysopathies (Disorders of the Posterior Pituitary Gland)	943
	11.3	Thyroid Gland Pathology	943
		11.3.1 Congenital Anomalies, Hyperplasia, and Thyroiditis	946
		11.3.2 Congenital Anomalies, Goiter, and Dysfunctional Thyroid Gland	946
		11.3.3 Inflammatory and Immunologic Thyroiditis	948
		11.3.4 Epithelial Neoplasms of the Thyroid Glands	949
	11.4	Parathyroid Gland Pathology	965
		11.4.1 Congenital Anomalies of the Parathyroid Glands	965
		11.4.2 Parathyroid Gland Hyperplasia	965
		11.4.3 Parathyroid Gland Adenoma	966
		11.4.4 Parathyroid Gland Carcinoma	966
	11.5	Adrenal Gland Pathology	967
		11.5.1 Congenital Anomalies of the Adrenal Gland and Paraganglia	967
		11.5.2 Dysfunctional Adrenal Gland	969
		11.5.3 Adrenalitis	969
		11.5.4 Neoplasms of the Adrenal Gland and Paraganglia	974
		11.5.5 Syndromes Associated with Adrenal Cortex Abnormalities	996
	Multiple Choice Questions and Answers		997
	References and Recommended Readings		998
12	**Soft Tissue**		**1003**
	12.1	Development and Genetics	1004
	12.2	Vascular and Inflammatory Changes of Soft Tissue	1005
		12.2.1 Hyperemia	1005
		12.2.2 Necrotizing Fasciitis	1005
		12.2.3 Vasculitis-Associated Soft Tissue Changes	1005
		12.2.4 Miscellaneous	1005
	12.3	Soft Tissue Neoplasms: Scoring	1009
	12.4	Adipocytic Tumors	1010
		12.4.1 Lipoma	1011
		12.4.2 Lipoma Subtypes	1011
		12.4.3 Lipomatosis	1012
		12.4.4 Lipoblastoma	1012
		12.4.5 Hibernoma	1012

- 12.4.6 Locally Aggressive and Malignant Adipocytic Tumors... 1013
- 12.5 Fibroblastic/Myofibroblastic Tumors... 1017
  - 12.5.1 Fasciitis/Myositis Group ... 1019
  - 12.5.2 Fibroma Group... 1020
  - 12.5.3 Fibroblastoma Classic and Subtypes ... 1021
  - 12.5.4 Fibrous Hamartoma of Infancy (FHI) ... 1023
  - 12.5.5 Fibromatosis of Childhood ... 1023
  - 12.5.6 Infantile Myofibroma/Myofibromatosis... 1027
  - 12.5.7 Fibroblastic/Myofibroblastic Tumors with Intermediate Malignant Potential... 1027
  - 12.5.8 Malignant Fibroblastic/Myofibroblastic Tumors... 1033
- 12.6 Fibrohistiocytic Tumors... 1036
  - 12.6.1 Histiocytoma ... 1036
  - 12.6.2 Benign Fibrous Histiocytoma ... 1037
  - 12.6.3 Borderline Fibrous Histiocytoma... 1039
  - 12.6.4 Malignant Fibrous Histiocytoma ... 1039
- 12.7 Smooth Muscle Tumors... 1040
  - 12.7.1 Leiomyoma ... 1040
  - 12.7.2 EBV-Related Smooth Muscle Tumors... 1040
  - 12.7.3 Leiomyosarcoma ... 1040
- 12.8 Pericytic Tumors ... 1041
  - 12.8.1 Glomus Tumor... 1041
  - 12.8.2 Glomangiosarcoma ... 1042
  - 12.8.3 Myopericytoma ... 1042
- 12.9 Skeletal Muscle Tumors... 1044
  - 12.9.1 Rhabdomyomatous Mesenchymal Hamartoma (RMH) ... 1044
  - 12.9.2 Rhabdomyoma... 1046
  - 12.9.3 Rhabdomyosarcoma... 1046
  - 12.9.4 Pleomorphic RMS ... 1055
- 12.10 Vascular Tumors... 1059
  - 12.10.1 Benign Vascular Tumors ... 1059
  - 12.10.2 Vascular Tumors with Intermediate Malignant Potential ... 1062
  - 12.10.3 Malignant Vascular Tumors ... 1066
  - 12.10.4 Genetic Syndromes Associated with Vascular Tumors... 1068
- 12.11 Chondro-Osteoforming Tumors... 1069
  - 12.11.1 Extraskeletal Chondroma... 1069
  - 12.11.2 Extraskeletal Myxoid Chondrosarcoma... 1069
  - 12.11.3 Mesenchymal Chondrosarcoma... 1070
  - 12.11.4 Extraskeletal Aneurysmatic Bone Cyst ... 1070
  - 12.11.5 Extraskeletal Osteosarcoma (ESOS) ... 1070
  - 12.11.6 Extraskeletal Chordoma... 1070
- 12.12 Tumors of Uncertain Differentiation ... 1071
  - 12.12.1 Myxoma... 1071
  - 12.12.2 Myoepithelial Carcinoma... 1072

		12.12.3 Parachordoma 1072

```
              12.12.3   Parachordoma ..........................  1072
              12.12.4   Synovial Sarcoma .......................  1074
              12.12.5   Epithelioid Sarcoma.....................  1076
              12.12.6   Alveolar Soft Part Sarcoma .............  1076
              12.12.7   Clear Cell Sarcoma .....................  1079
              12.12.8   Extraskeletal Myxoid Chondrosarcoma
                        (ESMC) ................................  1080
              12.12.9   PNET/Extraskeletal Ewing Sarcoma (ESES) .  1081
              12.12.10  Desmoplastic Small Round Cell Tumor ....  1083
              12.12.11  Extrarenal Rhabdoid Tumor...............  1084
              12.12.12  Malignant Mesenchymoma .................  1084
              12.12.13  PEComa..................................  1084
              12.12.14  Extrarenal Wilms' Tumor ................  1085
              12.12.15  Sacrococcygeal Teratoma and Extragonadal
                        Germ Cell Tumor and Yolk Sac Tumor .....  1085
     Multiple Choice Questions and Answers .....................  1090
     References and Recommended Readings .......................  1091

13   Arthro-Skeletal System ....................................  1095
     13.1  Development and Genetics ............................  1096
     13.2  Osteochondrodysplasias...............................  1097
           13.2.1   Nosology and Nomenclature ..................  1097
           13.2.2   Groups of Genetic Skeletal Disorders .......  1098
     13.3  Metabolic Skeletal Diseases .........................  1103
           13.3.1   Rickets, and Osteomalacia...................  1103
           13.3.2   Osteoporosis of the Youth ..................  1104
           13.3.3   Paget Disease of the Bone ..................  1106
           13.3.4   Juvenile Paget Disease......................  1108
     13.4  Osteitis and Osteomyelitis ..........................  1109
           13.4.1   Osteomyelitis ..............................  1109
     13.5  Osteonecrosis........................................  1113
           13.5.1   Bony Infarct and Osteochondritis Dissecans..  1113
     13.6  Tumorlike Lesions and Bone/Osteoid-Forming Tumors ...  1115
           13.6.1   Myositis Ossificans ........................  1115
           13.6.2   Fibrous Dysplasia and Osteofibrous Dysplasia.  1116
           13.6.3   Non-ossifying Fibroma (NOF) ................  1118
           13.6.4   Bone Cysts..................................  1120
           13.6.5   Osteoma, Osteoid Osteoma, and Giant
                    Osteoid Osteoma ............................  1124
           13.6.6   Giant Cell Tumor ...........................  1127
           13.6.7   Osteosarcoma................................  1128
     13.7  Chondroid (Cartilage)-Forming Tumors ................  1136
           13.7.1   Osteochondroma .............................  1137
           13.7.2   Enchondroma.................................  1139
           13.7.3   Chondroblastoma.............................  1140
           13.7.4   Chondromyxoid Fibroma.......................  1143
           13.7.5   Chondrosarcoma .............................  1145
     13.8  Bone Ewing Sarcoma...................................  1147
```

13.9 Miscellaneous Bone Tumors 1149
    13.9.1 Chordoma 1149
    13.9.2 Adamantinoma 1150
    13.9.3 Langerhans Cell Histiocytosis 1151
    13.9.4 Vascular, Smooth Muscle, and Lipogenic Tumors 1153
    13.9.5 Hematologic Tumors 1153
13.10 Metastatic Bone Tumors 1153
13.11 Juvenile Rheumatoid Arthritis and Juvenile Arthropathies 1154
    13.11.1 Rheumatoid Arthritis and Juvenile Rheumatoid Arthritis 1154
    13.11.2 Infectious Arthritis 1156
    13.11.3 Gout, Early-Onset Juvenile Tophaceous Gout and Pseudogout 1157
    13.11.4 Bursitis, Baker Cyst, and Ganglion 1159
    13.11.5 Pigmented Villonodular Synovitis and Nodular Tenosynovitis 1159
Multiple Choice Questions and Answers 1161
References and Recommended Readings 1162

14 Head and Neck 1167
14.1 Development 1168
14.2 Nasal Cavity, Paranasal Sinuses, and Nasopharynx 1171
    14.2.1 Congenital Anomalies 1171
    14.2.2 Inflammatory Lesions 1173
    14.2.3 Tumors 1174
14.3 Larynx and Trachea 1183
    14.3.1 Congenital Anomalies 1183
    14.3.2 Cysts and Laryngoceles 1184
    14.3.3 Inflammatory Lesions and Non-neoplastic Lesions 1185
    14.3.4 Tumors 1186
14.4 Oral Cavity and Oropharynx 1189
    14.4.1 Congenital Anomalies 1189
    14.4.2 Branchial Cleft Cysts 1195
    14.4.3 Inflammatory Lesions 1197
    14.4.4 Tumors 1197
14.5 Salivary Glands 1203
    14.5.1 Congenital Anomalies 1203
    14.5.2 Inflammatory Lesions and Non-neoplastic Lesions 1203
    14.5.3 Tumors 1205
14.6 Mandible and Maxilla 1209
    14.6.1 Odontogenic Cysts 1210
    14.6.2 Odontogenic Tumors 1216
    14.6.3 Bone-Related Lesions 1218

	14.7	Ear 1218
		14.7.1 Congenital Anomalies 1218
		14.7.2 Inflammatory Lesions and Non-neoplastic Lesions 1218
		14.7.3 Tumors 1220
	14.8	Eye and Ocular Adnexa 1226
		14.8.1 Congenital Anomalies 1226
		14.8.2 Inflammatory Lesions and Non-neoplastic Lesions 1226
		14.8.3 Tumors 1226
	14.9	Skull 1230
	Multiple Choice Questions and Answers 1233	
	References and Recommended Readings 1235	
15	**Central Nervous System** 1243	
	15.1	Development: Genetics 1244
		15.1.1 Development and Genetics 1244
		15.1.2 Neuromeric Model of the Organization of the Embryonic Forebrain According to Puelles and Rubenstein 1246
	15.2	Congenital Abnormalities of the Central Nervous System 1247
		15.2.1 Ectopia 1247
		15.2.2 Neural Tube Defects (NTDs) 1250
		15.2.3 Prosencephalon Defects 1252
		15.2.4 Vesicular Forebrain (Pseudo-aprosencephaly) 1254
		15.2.5 Ventriculomegaly/Hydrocephalus 1257
		15.2.6 Agenesis of the *Corpus Callosum* (ACC) 1257
		15.2.7 Cerebellar Malformations 1258
		15.2.8 Agnathia Otocephaly Complex (AGOTC) 1259
		15.2.9 Telencephalosynapsis (Synencephaly) and Rhombencephalon Synapsis 1259
		15.2.10 CNS Defects in Acardia 1261
		15.2.11 CNS Defects in Chromosomal and Genetic Syndromes 1263
		15.2.12 Neuronal Migration Disorders 1264
		15.2.13 Phakomatoses 1264
	15.3	Vascular Disorders of the Central Nervous System 1267
		15.3.1 Intracranial Hemorrhage 1267
		15.3.2 Vascular Malformations 1272
		15.3.3 Aneurysms 1273
		15.3.4 Thrombosis of Venous Sinuses and Cerebral Veins 1274
		15.3.5 Pediatric and Inherited Neurovascular Diseases ... 1274
	15.4	Infections of the CNS 1275
		15.4.1 Suppurative Infections 1276
		15.4.2 Tuberculous (Lepto-)Meningitis 1279
		15.4.3 Neurosyphilis 1280
		15.4.4 Viral Infections 1280

		15.4.5 Toxoplasmosis 1282
		15.4.6 Fungal Infections 1282
	15.5	Metabolic Disorders Affecting the CNS 1282
		15.5.1 Pernicious Anemia 1283
		15.5.2 Wernicke Encephalopathy 1283
	15.6	Trauma to the Head and Spine 1284
	15.7	Head Injuries 1284
		15.7.1 Epidural Hematoma 1284
		15.7.2 Subdural Hematoma 1284
		15.7.3 Subarachnoidal Hemorrhage 1285
		15.7.4 Spinal Injuries 1285
		15.7.5 Intervertebral Disk Herniation 1286
	15.8	Demyelinating Diseases Involving the Central Nervous System 1286
		15.8.1 Multiple Sclerosis 1286
		15.8.2 Leukodystrophies 1287
		15.8.3 Amyotrophic Lateral Sclerosis 1288
		15.8.4 Werdnig-Hoffmann Disease 1288
		15.8.5 Syringomyelia 1289
		15.8.6 Parkinson Disease and Parkinson Disease-Associated, G-Protein-Coupled Receptor 37 (GPR37/PaelR)-Related Autism Spectrum Disorder 1289
		15.8.7 Creutzfeldt-Jakob Disease (sCJD or Sporadic), CJD-Familial and CJD-Variant 1290
		15.8.8 West Syndrome/Infantile Spasms, ACTH Therapy, and Sudden Death 1290
	15.9	Neoplasms of the Central Nervous System 1291
		15.9.1 Astrocyte-Derived Neoplasms 1291
		15.9.2 Ependymoma 1294
		15.9.3 Medulloblastoma 1295
		15.9.4 Meningioma 1299
		15.9.5 Hemangioblastoma and Filum Terminale Hamartoma 1299
		15.9.6 Schwannoma 1300
		15.9.7 Craniopharyngioma 1300
		15.9.8 Chordoma 1300
		15.9.9 Tumors of the Pineal Body 1302
		15.9.10 Hematological Malignancies 1302
		15.9.11 Other Tumors and Metastatic Tumors 1304
	Multiple Choice Questions and Answers 1309	
	References and Recommended Readings 1311	
16	**Peripheral Nervous System** 1321	
	16.1	Development 1321
	16.2	Disorders of the Peripheral Nervous System 1322
		16.2.1 Peripheral Neuropathy 1322
		16.2.2 Traumatic Neuropathy 1323
		16.2.3 Vascular Neuropathy 1323

		16.2.4 Intoxication-Related Neuropathy............... 1323

- 16.2.4 Intoxication-Related Neuropathy. 1323
- 16.2.5 Infiltration (e.g., Amyloid) Related Neuropathy . . . 1323
- 16.2.6 Neoplasms of the Peripheral Nervous System. 1323

16.3 Neuromuscular Disorders 1333
- 16.3.1 Muscle Biopsy Test 1333
- 16.3.2 Neurogenic Disorders 1335
- 16.3.3 Myopathic Disorders 1337
- 16.3.4 Glycogen Storage Diseases 1339
- 16.3.5 Mitochondrial Myopathies. 1339
- 16.3.6 Inflammatory Myopathies: Non-infectious 1341
- 16.3.7 Inflammatory Myopathies: Infectious 1341

Multiple Choice Questions and Answers 1341
References and Recommended Readings. 1342

17 Skin. 1345

17.1 Development, General Terminology, and Congenital Skin Defects 1347
- 17.1.1 Development 1347
- 17.1.2 General Terminology 1348
- 17.1.3 Lethal Congenital Contractural Syndromes. 1348

17.2 Spongiotic Dermatitis 1352
- 17.2.1 Conventional Spongiosis 1352
- 17.2.2 Eosinophilic Spongiosis. 1354
- 17.2.3 Follicular Spongiosis 1354
- 17.2.4 Miliarial Spongiosis. 1354

17.3 Interface Dermatitis 1355
- 17.3.1 Vacuolar Interface Dermatitis 1355
- 17.3.2 Lichenoid Interface Dermatitis 1360

17.4 Psoriasis and Psoriasiform Dermatitis 1360
- 17.4.1 Psoriasis. 1360
- 17.4.2 Psoriasiform Dermatitis 1360

17.5 Perivascular In Toto Dermatitis (PID) 1361
- 17.5.1 Urticaria 1361
- 17.5.2 Non-urticaria Superficial and Deep Perivascular Dermatitis 1362

17.6 Nodular and Diffuse Cutaneous Infiltrates. 1363
- 17.6.1 Granuloma Annulare 1363
- 17.6.2 Necrobiosis Lipoidica Diabeticorum (NLD). 1364
- 17.6.3 Rheumatoid Nodule. 1364
- 17.6.4 Sarcoid 1364

17.7 Intraepidermal Blistering Diseases 1364
- 17.7.1 Pemphigus Vulgaris, Pemphigus Foliaceus, and Pemphigus Paraneoplasticus. 1364
- 17.7.2 IgA Pemphigus and Impetigo 1366
- 17.7.3 Intraepidermal Blistering Diseases 1366

17.8 Subepidermal Blistering Diseases 1367
- 17.8.1 Bullous Pemphigoid and Epidermolysis Bullosa. 1367
- 17.8.2 Erythema Multiforme and Toxic Epidermal Necrolysis. 1368

	17.8.3 Hb-Related Porphyria Cutanea Tarda, Herpes Gestationis, and Dermatitis Herpetiformis 1368
	17.8.4 Lupus (Systemic Lupus Erythematodes)......... 1369

17.9 Vasculitis 1369
17.10 Cutaneous Appendages Disorders 1369
17.11 Panniculitis.. 1369
    17.11.1 Septal Panniculitis 1369
    17.11.2 Lobular Panniculitis........................ 1370
17.12 Cutaneous Adverse Drug Reactions.................... 1370
    17.12.1 Exanthematous CADR...................... 1371
17.13 Dyskeratotic, Non-/Pauci-Inflammatory Disorders 1372
17.14 Non-dyskeratotic, Non-/Pauci-Inflammatory Disorders ... 1372
17.15 Infections and Infestations........................... 1372
17.16 Cutaneous Cysts and Related Lesions 1372
17.17 Tumors of the Epidermis 1373
    17.17.1 Epidermal Nevi and Related Lesions........... 1373
    17.17.2 Pseudoepitheliomatous Hyperplasia (PEH)...... 1376
    17.17.3 Acanthoses/Acanthomas/Keratoses............ 1376
    17.17.4 Keratinocyte Dysplasia 1377
    17.17.5 Intraepidermal Carcinomas 1378
    17.17.6 Keratoacanthoma........................... 1378
    17.17.7 Malignant Tumors 1378
17.18 Melanocytic Lesions 1381
    17.18.1 Lentigines, Solar Lentigo, Lentigo Simplex, and Melanotic Macules (Box 17.8) 1381
    17.18.2 Melanocytic Nevi........................... 1381
    17.18.3 Variants of Melanocytic Nevi 1382
    17.18.4 Spitz Nevus and Variants 1384
    17.18.5 Atypical Melanocytic (Dysplastic) Nevi 1385
    17.18.6 Malignant Melanoma and Variants 1385
17.19 Sebaceous and Pilar Tumors 1387
    17.19.1 Sebaceous Hyperplasia 1387
    17.19.2 Nevus Sebaceous (of Jadassohn) (NSJ)......... 1387
    17.19.3 Sebaceous Adenoma, Sebaceoma, and Xanthoma 1388
    17.19.4 Sebaceous Carcinoma 1388
    17.19.5 Benign Hair Follicle Tumors 1388
    17.19.6 Malignant Hair Follicle Tumors............... 1390
17.20 Sweat Gland Tumors 1390
    17.20.1 Eccrine Gland Tumors...................... 1391
    17.20.2 Apocrine Gland Tumors..................... 1393
17.21 Fibrous and Fibrohistiocytic Tumors................... 1394
    17.21.1 Hypertrophic Scar and Keloid 1394
    17.21.2 Dermatofibroma........................... 1394
    17.21.3 Juvenile Xanthogranuloma 1394
    17.21.4 Dermatofibrosarcoma Protuberans............. 1396

	17.22	Vascular Tumors.	1396
	17.23	Tumors of Adipose Tissue, Muscle, Cartilage, and Bone	1396
	17.24	Neural and Neuroendocrine Tumors	1396
		17.24.1 Merkel Cell Carcinoma	1396
		17.24.2 Paraganglioma	1399
	17.25	Hematological Skin Infiltrates	1400
		17.25.1 Pseudolymphomas	1400
		17.25.2 Benign and Malignant Mastocytosis	1400
	17.26	Solid Tumor Metastases to the Skin.	1400
	Multiple Choice Questions and Answers		1403
	References and Recommended Readings		1405
18	**Placenta, Abnormal Conception, and Prematurity**		**1409**
	18.1	Development and Useful Pilot Concepts and Tables	1410
	18.2	Pathology of the Early Pregnancy	1422
		18.2.1 Disorders of the Placenta Formation	1423
		18.2.2 Disorders of the Placenta Maturation	1431
		18.2.3 Disorders of the Placenta Vascularization	1434
		18.2.4 Disorders of the Placenta Implantation Site.	1434
		18.2.5 Twin and Multiple Pregnancies	1442
	18.3	Pathology of the Late Pregnancy	1449
		18.3.1 Acute Diseases.	1450
		18.3.2 Subacute Diseases	1457
		18.3.3 Chronic Diseases	1468
		18.3.4 Fetal Growth Restriction	1482
	18.4	Non-neoplastic Trophoblastic Abnormalities	1485
		18.4.1 Placental Site Nodule.	1485
		18.4.2 Exaggerated Placental Site.	1485
	18.5	Gestational Trophoblastic Diseases, Pre- and Malignant	1486
		18.5.1 Invasive Mole.	1486
		18.5.2 Placental Site Trophoblastic Tumor.	1486
		18.5.3 Epithelioid Trophoblastic Tumor.	1486
		18.5.4 Choriocarcinoma	1486
	18.6	Birth Defects	1488
		18.6.1 Birth Defects: Taxonomy Principles	1490
		18.6.2 Birth Defects: Categories.	1493
		18.6.3 Birth Defects: Pathogenesis (Macro- and Micromechanisms).	1496
		18.6.4 Birth Defects: Etiology (Mendelian, Chromosomal, Multifactorial)	1515
	18.7	Infection in Pregnancy, Prom, and Dysmaturity	1533
		18.7.1 Infection in Pregnancy	1533
		18.7.2 Premature Rupture of Membranes (PROM)	1541
		18.7.3 Fetal Growth Restriction (FGR) and *Dys*maturity	1541
	18.8	IUFD and Placenta.	1546

  18.8.1 Fetal Death Syndrome (Intrauterine Fetal Demise, IUFD) 1546
  18.8.2 Step-by-Step Approach in the Examination of a Placenta 1548
 Multiple Choice Questions and Answers 1552
 References and Recommended Readings 1554

Index 1571

Cardiovascular System

Contents

1.1	**Developmental and Genetics**..	2
1.2	**Congenital Heart Disease**..	6
1.2.1	Sequential Segmental Cardio-Analysis..	8
1.2.2	Atrial Septal Defect (ASD), Ventricular Septal Defect (VSD), Atrioventricular Septal Defect (AVSD), Patent Ductus Arteriosus (PDA), and Ebstein's Tricuspid Anomaly (ETA)...	11
1.2.3	Tetralogy of Fallot (TOF) and Double Outlet Right Ventricle (DORV)..........	18
1.2.4	Hypoplastic Right Heart Syndrome..	24
1.2.5	Transposition of the Great Arteries...	26
1.2.6	Common Trunk (Persistent Truncus Arteriosus)...	31
1.2.7	Total Anomalous Pulmonary Venous Return..	32
1.2.8	Scimitar Syndrome..	32
1.2.9	Hypoplastic Left Heart Syndrome and Coarctation of Aorta............................	33
1.2.10	Heterotaxy Syndrome...	41
1.2.11	Common Genetic Syndromes with Congenital Heart Disease.........................	41
1.3	**Myocarditis and Nonischemic Cardiomyopathies**.......................................	48
1.3.1	Myocarditis...	48
1.3.2	Dilated Cardiomyopathy...	48
1.3.3	Hypertrophic Cardiomyopathy..	66
1.3.4	Infiltrative/Restrictive Cardiomyopathies...	66
1.3.5	Left Ventricular Non-compaction...	78
1.3.6	Arrhythmogenic Right Ventricular Dysplasia/Cardiomyopathy (ARVD/C)..	79
1.3.7	Lamin A/C Cardiomyopathy...	81
1.3.8	Mitochondrial Cardiomyopathies..	81
1.3.9	Cardiac Channelopathies and Sudden Cardiac Death.......................................	83
1.4	**Ischemic Heart Disease, Myocardial Dysfunction, and Heart Failure**......	85
1.4.1	Ischemic Heart Disease...	85
1.4.2	Myocardial Dysfunction..	85
1.4.3	Heart Failure...	85
1.5	**Valvular Heart Disease**...	87
1.5.1	Endocarditis and Valvar Vegetation..	87
1.5.2	Mitral Valve Prolapse..	91
1.5.3	Calcific Mitral Annulus and Calcific Aortic Stenosis......................................	91
1.5.4	Rheumatic Heart Disease (RHD)..	91
1.5.5	Prosthetic Valves..	92

© Springer-Verlag GmbH Germany, part of Springer Nature 2020
C. M. Sergi, *Pathology of Childhood and Adolescence*,
https://doi.org/10.1007/978-3-662-59169-7_1

1.6	**Transplantation**	92
1.6.1	Allograft Cellular Rejection	92
1.6.2	Allograft Humoral Rejection	95
1.6.3	Cardiac Allograft Vasculopathy	97
1.6.4	Extracardiac Posttransplant Lympho-proliferative Disorders (PTLD) Following HTX	100
1.7	**Tumors**	100
1.7.1	Benign Tumors	102
1.7.2	Malignant Tumors	109
1.7.3	Metastatic Tumors	112
1.8	**Pericardial Disease**	115
1.8.1	Non-neoplastic Pericardial Disease	115
1.8.2	Neoplastic Pericardial Disease	116
1.9	**Non-neoplastic Vascular Pathology**	116
1.9.1	Congenital Anomalies	116
1.9.2	Fibromuscular Dysplasia	117
1.9.3	Arteriosclerosis	117
1.9.4	Large-Vessel Vasculitis	119
1.9.5	Medium-Vessel Vasculitis	123
1.9.6	Small-Vessel Vasculitis	125

Multiple Choice Questions and Answers 125

References and Recommended Readings 127

1.1 Developmental and Genetics

The developing embryo has the cardiovascular system as the first system to function, and the blood begins to circulate by the end of the third week. Septation is key and is defined as a division between cavities. It evolves continuously and plays a major role in the construction as parts of an organism take progressively shape by partitions. The milestones of cardiac development are described in Box 1.1 (Fig. 1.1).

The *cardiac crescent* is indeed the first stage of the development of the heart before the end of the gastrulating embryo. Remarkably, it is characterized by the commitment of the mesodermal cells of the anterior lateral plate to the cardiogenic lineage. These cells are devoutly committed to migrate and organize in the structure, which is called the *cardiac crescent*. The commitment of these cells is under the influence of mostly BMP, FGF, and WNT proteins (BMP, bone morphogenetic proteins; FGF, fibroblast growth factor; WNT, portmanteau of Wingless and Int-1). In addition to these three cell signaling pathways, cardiac morphogenesis is also regulated by NOTCH, TGF, and VEGF. Specific knockout mice give origin to a particular phenotype that may be useful for the understanding of the multiplicity of congenital heart disease.

In particular, the *linear heart* is the second stage of development of the heart, which is the first functional organ in the embryo. In the elongated heart, there is migration and fusion along the ventral midline of the precursors of the cardiac crescent. *Cardia bifida* is the result of a failure of the fusion of these cell aggregates. One of the critical transcription factors playing a significant role in the second stage is GATA-4, which is synergistical with the SMAD proteins, several intracellular effectors of the TGF/BMP signaling pathway to activate NKX2.5 transcription.

Chamber formation occurs in the third and fourth stage of cardiac development and includes specification, septation, and trabeculation of the chambers. KO mouse technology is being pillared in identifying several pathogenetic pathways. It has allowed identifying several transcription factors that play a significant role during this stage. A secondary or anterior heart field (AHF), other than heart tube-derived cardiomyocytes, accompanies the development of the heart. AHF myo-

> **Box 1.1 Heart Development**
>
> *Day 15th*:
>
> - Lateral mesoderm-derived first heart field (TBX5, HAND1) develops in the left ventricle.
> - Second heart field (FGF-10, HAND2) aims to form the right ventricle outflow tract and atria.
>
> *Day 20th*:
>
> - Initial crescent-derived beating tube loops to the right side of the forming embryo.
>
> *Day 28th*:
>
> - Heart chambers steadily delineate.
> - Neural crest cells migrate with septation of the outflow tracts.
> - Aortic arches form.
> - Extracellular matrix swells up for the formation of the endocardial cushions.
>
> *Day 50th*:
>
> - Four-chambered heart forms.
>
> Notes: FGF-10, Fibroblast Growth Factor 10; Hand1, Heart And Neural Crest Derivatives Expressed 1; Hand2, Heart And Neural Crest Derivatives Expressed 2; TBX5, T-Box Transcription Factor 5.

cytes derive from the pharyngeal mesoderm and are incorporated in the developing heart tube during the looping stage at both the venous and arterial poles. The "cardiogenic area" of the splanchnic mesenchyme represents the area where cardiac development enters into the scene and precisely on day 16–19 of embryo development. The heart takes shape as two side-by-side longitudinal cell strands that are labeled cardiogenic cords. The canalization of the cardiogenic cords is key to form the endocardial tubes of the forming heart. These two tubular structures fuse together to form a single tubular heart or heart tube. On day 23–24 of embryo development, the bulboventricular loop progressively forms introducing asymmetry in the cardiac structure. The paired aortic arches are responsible for the development of the arterial system, while the venous end of the heart is responsible for the development of the venous system. In week 3 of the embryo development, the vascular system originates from mesodermal cells of the "blood islands." The peripheral cellular component of these structures is responsible for the formation of the endothelium, while the central portion gives rise to hematopoietic elements. Subsequently, pericytes from the surrounding mesenchyme will join the first blood vessels. They differentiate into the wall layers. The histology of the heart changes from pre- to postnatal state. In the fetal stage, the heart shows a significant amount of glycogen, while the heart in children and, mainly, adolescents is quite similar to young adults.

The cardiomyocyte in the adult mammalian organism is a differentiated postmitotic cell with no significant proliferative potential due to its inability to reactivate the cell cycle. It is known that the G_0 phase of a cell is a period of the cell cycle in which cell does exist in an inactive, non-cycling state. Although cells were previously thought to enter G_0 by default, this state is now considered as a way for cells to preserve essential and extremely precious cell physiology functions over time. In response to extrinsic stimuli, some of the G_0 cells may proliferate by reentering the cell cycle. In G_1 phase, there is a restriction point (R-point) with a dichotomic fate. Mammalian cells can enter the G_0 phase before the R-point but are committed to mitosis after the R-point. The lack of G_1 cyclin/cyclin-dependent kinases (Cdk), which are essential positive cell-cycle modulators, and the deficiency of high levels of cell-cycle inhibitors, including the retinoblastoma proteins pRb/p130 and the Cdk inhibitors p21/p27, limit the mitotic capacity of mature cardiomyocytes. Thus, the adult mammalian heart is functionally incapable of repairing itself after ischemic injury. In fact, cardiomyocytes that survive the insult suffer from cell enlargement, myofibrillar disarray, and re-expression of some fetal transcription factors and genes. Although the molecular mechanisms underlying heart failure are poorly understood, it is known that this process is a significant pathway. Cell hypertrophy with the disarray of the myofibrils and re-expression of some fetal transcription factors and genes are associated with con-

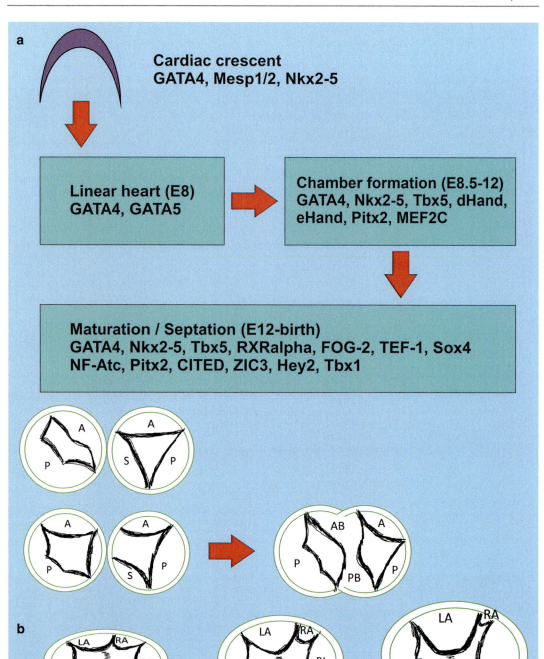

1.1 Developmental and Genetics

siderable morbidity and mortality and remains at long-term an inefficient mechanism to let differentiated postmitotic cells to survive. This process, which does not provide adequate or appropriate adjustment to the environment or situation with time, leads to the development of heart failure.

The most exciting result linking to the heart transcriptome is its regulation by the guardian of the human genome, the TP53 gene, which is located on the short arm of chromosome 17 (17p13.1). After several decades of research into TP53, there is substantial evidence that the tumor suppressor guardian of live, better known as p53, inhibits cell growth after acute stress by regulating the transcription of genes. It has been recently identified an intriguing and sensational role for p53 in the heart as a master regulator of the cardiac transcriptome. The TP53 gene spans 20 kB, with a noncoding exon 1 and a 10-kB-long first intron. The regulation of tumor-free survival is based on p53, which is a transcription factor that translates growth and survival signals into specific gene expression patterns. In cells that are in regular status, the E3 ubiquitin ligase Mdm2 keeps p53 expression at low levels by targeting p53 for proteasomal degradation. Mdm2 results inactivated in response to acute stress factors. The Mdm2 inactivation results in increased levels of p53 with the fate of cell division and apoptosis. There is a negative feedback that limits p53 by activating Mdm2 transcription using p53. This negative feedback loop defines p53 activity. Tumorigenesis is one of the consequences of having p53 mutated. In this setting, there is a proliferation of aberrant clones of genetically unstable cells, as shown in the p53 knockout mouse. Conversely, Mdm2 knockout mice die during embryogenesis through p53-induced apoptosis. This balance between p53 and Mdm2 is critical for the organism and tumorigenesis. In end-stage heart failure, high levels of p53 correlate with apoptosis of cardiomyocytes and cell hypertrophy. The normal by-products of 36-ATP-linked aerobic respiration are reactive oxygen species (ROS), which induce DNA damage and cellular defense systems. In fact, p53 plays a significant role in the cardiac transcriptome, because it acts as a pleiotropic regulator of the cardiac function as Mak et al. recently described.

The activation of process-specific transcription factors (Mef2a, Myocd, Pgc-1α, Tfam) results from p53 loss. The presence of Esrrγ, Gata4, and Mef2a in p53 immunoprecipitates from adult hearts of experimental animals may suggest complex formation between p53 and some process-specific transcription factors (TFs), and endogenous p53 protein specifically binds to these promoters in oligonucleotide precipitation experiments and PCR chromatin immunoprecipitation (ChIP) assays. This high degree of connectivity of p53 with TFs is probably a fascinating event and may represent a working platform for future studies. In animal models, some master transcription factors (Gata4, c-Myc, Nfat3, NF-κB) are linked to the development of HF. It has been demonstrated that physical exercise is crucial in increasing the expression of these specific gene sets. These very recent studies reinforce the postulation

Fig. 1.1 Cardiac crescent formation with transcription factor activation during early development and abnormalities of endocardial cushion development. (**a**) From the beginning of the cardiac crescent through the maturation/septation, there are a number of transcription factors that are activated. In (**b**) there is the representation of the endocardial cushion defects illustrating the variations of the Rastelli types (CITED, CREB-binding protein/p300-interacting transactivator with Asp/Glu-rich C-terminal domain; FOG-2 "FRIEND OF GATA2" or ZINC FINGER PROTEIN, MULTITYPE 2 / ZFPM2, GATA4 and GATA5, transcription factors characterized by their ability to bind to the DNA sequence "GATA"; TBX5, T-box transcription factor 5; dHAND and, eHNAD, basic helix-loop-helix transcription factors; HEY2, Hes Related Family BHLH Transcription Factor With YRPW Motif 2; NF-ATC, Nuclear Factor of Activated T Cells; PITX2, Pituitary homeobox 2; MEF2C, Myocyte-specific enhancer factor 2C, aka MADS box transcription Enhancer Factor 2; MESP 1/2, Mesoderm Posterior BHLH Transcription Factor 1 and 2; NKX 2-5, NK2 Homeobox 5; RXR-A, Retinoid X Receptor Alpha; SOX4, SRY-Box Transcription Factor 4; TBX1, T-Box Transcription Factor 1; TEF-1, Transcription-al enhancer factor 1; ZIC3, a member of the Zinc finger of the cerebellum (ZIC) protein family)

of a rheostat-like role for p53 with significant changes of the p53/Mdm2 circuitry. It seems evident that the circulatory benefit derived from physical exercise arises not only from muscular activity and efficiency of the pump function, but also from a well orchestrated cascade of cytokines.

In mouse and human, there are, probably, more than 2×10^3 TFs, which modulate the mRNA profiles corresponding to 23,000 genes. DNA-binding TFs are thought to be crucial in getting significant insights into the regulation of the cardiac transcriptome. Thus, we strongly need a systems biology based approach to magnificently complement the single-gene-focused biology of the last three decades.

1.2 Congenital Heart Disease

Congenital heart surgery has seen a rapid growth in the last three decades showing some new procedures emphasizing how cardiac pathology is a rapidly evolving field. Enhancement of patients' outcome has been witnessed not only in well-developed countries but also in developing countries. The introduction of life support devices such as the Berlin heart and the increase in knowledge in cardiac physiology integrated with new laboratory data is probably by this advancement. Pharmacologic intervention and hemodynamic support have petrified milestones in medicine. The extracorporeal membrane oxygenation provides better patient outcomes than decades ago. Congenital heart disease has become increasingly complex and affects nearly 1% of births per year in several well-developed countries. The prevalence of some congenital disabilities involving the heart, especially mild types, is increasing, while it seems that the prevalence of other types remains stable. The ventricular septal defect is the most common type of heart defect, which may be quite mild. However, 1/4 of newborns with a congenital heart disease present with a critical heart illness, and these infants usually need surgery or a palliative corrective procedure in their first year of life. The cause is multifactorial, although the increase of some environmental etiologies may be alarming. Right-to-left shunt with cyanotic heart disease is mostly seen in cardiovascular defects, such as the transposition of the great arteries (TGA) and the

Box 1.2 Congenital Heart Disease
1.2.1. Sequential Segmental Cardio-Analysis
1.2.2. Atrial Septal Defect (ASD), Ventricular Septal Defect (VSD), Atrioventricular Septal Defect (AVSD), Patent Ductus Arteriosus (PDA), and Ebstein's Tricuspid Anomaly (ETA)
1.2.3. Tetralogy of Fallot and Double Outlet Right Ventricle (DORV)
1.2.4. Hypoplastic Right Heart Syndrome
1.2.5. Transposition of Great Arteries (TGA)
1.2.6. Common Trunk ("Persistent Truncus Arteriosus")
1.2.7. Total Anomalous Pulmonary Venous Return (TAPVR)
1.2.8. Scimitar Syndrome
1.2.9. Hypoplastic Left Heart Syndrome and Coarctation of Aorta
1.2.10. Heterotaxy Syndrome
1.2.11. Common Genetic Syndromes with Congenital Heart Disease

tetralogy of Fallot (TOF). Left-to-right shunt is seen in atrial (ASD) or ventricular (VSD) septal defects and patent ductus arteriosus (PDA). These defects are also recognized as cyanotic heart diseases of late-type because the shunt induces pulmonary hypertension with consequent right ventricular hypertrophy and progressively inversion of the shunt from left to right into the right to left, a condition labeled Eisenmenger syndrome. Disorders contemplating the clinical morphology and the association of molecular genetic disorders became an essential part of the study of both a pediatrician and a pediatric pathologist. In this section, the critical anomalies in the congenital heart disease are highlighted following the presentation of the method to investigate such defects. Indeed, the investigative method and its principles are crucial to correctly identify these defects both *intra vitam* and at time of *post mortem* examination. The proposed program to orientate into congenital heart disease is shown in Box 1.2, although the reader should familiarize himself/herself with the normal configuration of the heart (Fig. 1.2).

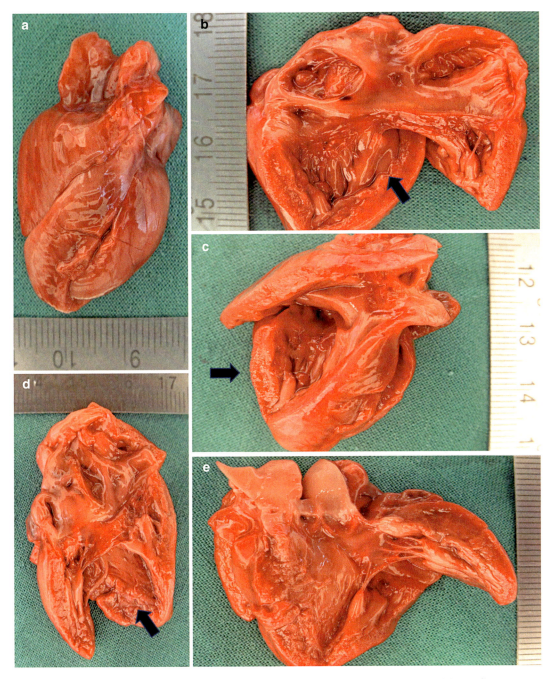

Fig. 1.2 View of the normal heart at fetal age with illustration of the right and left chambers. In (**a**) the anterosuperior face of the heart is seen with two cut opening the right and left chambers. In (**b**) the arrow points to the right ventricle with the high trabeculated septal wall, while in (**c**) the right ventricle (arrow) and its outflow tract are shown. In (**d**) the arrows points to the left ventricle showing the smooth parietal wall characterizing anatomically the left ventricle. In (**e**) the outflow tract of the left ventricle with the aortic valve are shown

1.2.1 Sequential Segmental Cardio-Analysis

Disorders involving an abnormal formation of the heart are better investigated if the cardiac specimen is studied schematically. Cardiac anomalies are one of the most frequent anomalies and may occur singly or in the setting of a multiple congenital anomaly syndrome. The recognition of a single cardiac anomaly, e.g., aortic stenosis, may represent a non-difficult task, but probably complex cardiac defects may pose a challenge for non-pediatric pathologists. Both the Bostonian school of segmental notation of single parts of the heart developed from the teaching of Professor Van Praagh, and the impulses and directions derived from the European school of sequential segmental analysis (SSA) with Professor Robert Anderson and Professor Becker have been fundamental in cardiac anatomy teaching. These approaches allow us to assess cardiac specimens that can be considered reproducible and of great value for gynecologists and cardiologists (Fig. 1.2).

Sequential segmental analysis (SSA) is a careful step-by-step approach to describe the cardiac anatomy analytically and is necessarily useful for all cardiac malformations leading to a systematic approval of its pathophysiology. SSA was developed between the late 1970s and early 1980s of the last century and is based on a method deriving from the separate analysis of three essential elements, including atria, ventricles, and great arteries. All three segments need to be described independently from their position, i.e., they should have supportive findings indicating if they are morphologically right or left and which great artery is. The anatomical features dominate the "crime scene," rather than their spatial orientation. The separation of the three segments needs the identification of two planes of partition, including the fibrous tissue plane of the atrioventricular (AV) junction of the valves separating the atria from the ventricles and the attachment of the arterial valves to the ventricles or ventriculoarterial (VA) connections separating the ventricular masses from the great arteries. First, the atrial situs is individualized by ascertaining the position of each atrium about each other within the chest cavity. Second, the AV and VA connections of the heart are determined. A segment can be right- or left-sided and, independently, morphologically right or left (Carvalho et al. 2005). Thus, the question is to identify the morphology of the single parts of the heart and is at the basis of the terminology used in SSA, which has been now used for decades by gynecologists, cardiologists, and pathologists. In SSCA, the analysis using SSA is complemented with transcriptomic analysis, which may be crucial in understanding pediatric versus adult failing hearts (Jana et al. 2018). SSCA may also be defined sequential segmental chemical analysis.

The atrial arrangement or situs is the initial approach describing the heart. In pediatric cardiology, the atrial situs is recognized by the method of the abdominal major blood vessels concerning the spinal column. In obstetrics, fetal laterality is the first category to ascertain. Situs is defined as the location that an organ occupies in a symmetric bilateral system. The morphologically right or left atrium is determined by the most constant anatomic feature, which is the atrial appendage. A broad-based triangular atrial appendage ("Snoopy's nose"-like) defines the atrial segment, which lodges the appendage, as for the right. Conversely, a narrow-based trapezoidal atrial appendage ("Snoopy's ear"-like) defines the atrial segment as left. There are three possible types of sites, including (1) situs solitus, (2) situs inversus, and (3) situs ambiguus. The *situs solitus* is the usual atrial arrangement, the *situs inversus* is the mirror-image arrangement, and the *situs ambiguus* presents with isomerism of right or left atrial appendages, i.e., in other words, two morphologically right or two morphologically left atrial appendages.

Isomerism means paired, mirror image sets of ordinarily single or nonidentical organ systems (atria, lungs, and other organs), frequently associated with other abnormalities (*vide infra*). In *right isomerism* both large vessels of the abdomen, i.e., aorta and IVC, are located on the same side of the spinal column (either left or right) with IVC on the front and aorta on the back. On the contrary, in *left isomerism*, the IVC shows an interruption. There is azygos continuation, which is represented as a venous structure posterior to the aorta. The hepatic venous flow only reaches, usually, the right atrium from below through IVC. In the azygos continuation, the IVC is interrupted distal to its passage through the liver, and blood flow reaches the right atrium through an azygos vein, which may be quite large, connecting the IVC to the SVC. In fact, an essential marker of left isomerism is an interrupted IVC with azygos continuation (Carvalho et al. 2005). The aorta is more centrally located, and the azygos vein is on either the right or left side. Usually, bilateral right-sidedness is peculiarly distinguished with the absence of left-sided structures, such as the spleen (asplenia), while bilateral left-sidedness with a surplus of left-sided structures, namely, more spleens or polysplenia. Paired morphologically correct structures identify right isomerism or asplenia syndrome: absence of the spleen, bilateral right bronchi, bilateral trilobed (right) lungs, two morphologic right atria, and multiple anomalies of systemic and pulmonary venous connections as well as other complex cardiac and non-cardiac abnormalities. Left isomerism or polysplenia syndrome is characterized by paired, morphologically left structures: multiple bilateral spleens, bilateral left bronchi, bilateral bilobed (left) lungs, midline liver, two morphologic left atria, and complex congenital cardiac and non-cardiac defects.

Thus, in autopsies restricted to heart and lungs, as often happens in post-cardiac surgical procedures or induced interruptions of pregnancy, a review of the *intra vitam* imaging is essential for a correct evaluation and pathology report. It is crucial to stress how important the relationship between the cardiologist and the pathologist is. Failure of communication or conflicts of interest at this level may have disastrous consequences. Moreover, the pathologist needs to rely on the support from the cardiologist in addressing more resources for the pathologist aiming to be an advocate for cardiac pathology in public health economics and politics. The description of the atria also foresees the correct identification of the coronary sinus, Koch's triangle, and Thebesian valve. These structures should also be shown to the cardiologist or the cardiac surgeon at the time of the presentation of the autopsy findings.

There are two possible options for *AV connections*, either *bi*ventricular AV connection with each atrium connecting to its ventricle or *uni*ventricular AV connection when each atrium is not related to a separate ventricular chamber. *Biventricular* AV connection also needs to be subdivided into concordant, discordant, or ambiguus. *Univentricular* AV connection needs to be further classified as absent right AV connection or absent left AV connection or double-inlet ventricle (DIV) with a ventricle that can be morphologically right or left. In biventricular AV connection, there can be concordance or discordance of the AV connection. *Concordant biventricular AV connection* is characterized by a morphologically right atrium connecting to a morphologically right ventricle with a morphologically left atrium connected to a morphologically left ventricle. On the other hand, *discordant biventricular AV connection* is characterized by a morphologically right atrium connected to a morphologically left ventricle and a morphologically left atrium connected to a morphologically right ventricle.

Three components characterize each ventricle, including an inlet, apical, and outlet components. All three need to be carefully (often magnification-lens-aided) evaluated before labeling a ventricle as right or left. The apical portion is the most constant part of a ventricle because the inlet and the outlet can be absent. The right sidedness of a ventricle is characterized by a tricuspid valve as AV junction with the tension apparatus attached to both the septal wall and to the papillary muscle. Moreover, coarse trabeculation at the apex and over the entire septal fence is found, and the arterial valve is supported by an infundibulum, which is entirely muscular and is separated from the AV valve by the muscle. Typically, there is also a muscular bar in the right ventricle, the moderator band, which connects the IVS to the anterior papillary muscle crossing the lower portion of the right ventricular chamber and acting as a primary conduction pathway to the free wall originating from the right bundle branch. The left-sides or laterality of a ventricle is characterized by an AV junction guarded by a bicuspid valve (mitral) with the attachment of the tension apparatus (chordae) to the papillary muscles. Moreover, fine, criss-crossing trabeculations at the apex and a smooth upper septal surface, as well as a fibrous continuity between arterial and AV valve, are seen. Concordant and discordant connections can only be seen in heart specimens with lateralization of the atria, while heart specimens with atrial isomerism, a biventricular connection, are labeled as ambiguus which underlines the absence of lateralization of the atria.

In this common congenital heart disease, the pulmonary venous circulation drains correctly into the left atrium and then into left-sided morphological right ventricle, but from this ventricle, blood is directed toward the systemic circulation through the aorta, which regarding SSA is indicated as *A-V-Discordance + V-Art-Discordance* or l-TGA. This type of TGA is different from d-TGA with *A-V-Concordance + V-Art-Discordance*, i.e., the ventricles are connected to the wrong, great artery and acronym for this defect. Hence, d-TGA is not to be confused with l-TGA or Levo-TGA, where there are both atrioventricular and ventriculoarterial discordance (AVD + VAD).

Univentricular AV connection is defined in case one or two atrial chambers connect to one ventricle, which is also called "main" or "dominant" ventricle. Double-inlet ventricle consists of both atria, no matter of their arrangement, connecting to the single ventricle. If one of the atria does not show any connection and only an atrium connects to the ventricle, the condition is called right or left absent AV connection. The other minor or absent ventricle is usually rudimentary and usually lacks the inlet component. The rudimentary right ventricle is located anterior and superior, if the dominant ventricle is of left ventricular morphology, and may acquire a right- or left-sided or directly anterior position. The rudimentary left ventricle has a posterior and inferior position if the dominant ventricle is of right ventricular morphology and may be either right- or left-sided. Very rarely, there may be an occasion of a ventricle of general morphology, but their existence has been a matter of debate.

An additional describing parameter in the SSA of heart specimens regards the fashion or mode of AV connection. There are four possibilities, including (1) two perforate valves, (2) one perforate and one without perforation, (3) a common AV valve, and (4) straddling and overriding valves. In the setting of an AVSD, there is neither a true mitral nor a correct tricuspid valve, but a 5-leaflet A-V valve guards a single atrioventricular orifice, in the severe form. A "right A-V valve" comprises the right anterosuperior leaflet, the right inferior leaflet, the right portions of the superior, and the inferior bridging leaflets. Conversely, a "left A-V valve" includes the left portions of the superior (anterior) and inferior (posterior) bridging leaflets and the left lateral leaflet. Cleft AV valve is characterized by a defect involving the left AV valve in AVSD designed by the unification of the superior and inferior bridging leaflets. Something similar but morphogenetically distinct entity may occur concerning the anterior or rarely posterior leaflet of the mitral valve in otherwise healthy hearts. *Overriding* is defined by the condition characterized by an AV valve overriding the interventricular septum above a VSD and emptying into both ventricles. The disease characterized by an AV valve with the anomalous insertion of valvar tensor apparatus into the contralateral ventricle is called *straddling* and is usually accompanied by a VSD.

The examination of the VA connections also belongs to SSA and aims to identify which great artery connects to which ventricle. An abnormal condition is *truncus arteriosus communis* or common arterial trunk with a single great artery arising from the ventricular chambers and giving rise to coronary arteries, pulmonary arteries, and head and neck blood vessels. VA connections can be of four types, including (1) concordant, (2) discordant, (3) double-outlet, and (4) single-outlet. Concordant VA connections are junctions between morphologically right ventricle connecting to the pulmonary artery and morphologically left ventricle connecting to the aorta. Discordant VA connections are characterized by morphologically right ventricle connecting to the aorta and morphologically left ventricle connecting to the pulmonary artery. Double-outlet VA connection occurs when both great arteries arise predominantly from a ventricle of right, left, or indeterminate morphology. In DORV, there is usually no fibrous continuity between the semilunar and the AV valves, and there is an associated VSD. If the VSD is located in subaortic position without RV outflow tract obstruction, the cardiac physiology simulates a pure VSD, while in case of RV outflow tract obstruction, the cardiac physiology simulates tetralogy of Fallot. A VSD located in the subpulmonary position (*Taussig-Bing anomaly*) determines cardiac physiology, affecting a complete transposition of the great arteries with VSD. Substantially, the Taussing-Bing anomaly includes a transposition of the aorta to the right ventricle in addition to a malposition of the pulmonary artery with an obvious subpulmonary VSD. A single-outlet is represented by either a common arterial trunk (persistent *truncus arteriosus*) or be associated with pulmonary or aortic atresia.

About the cardiac conduction system preparation, the sinus node is a small "cigar"-shaped structure located just subepicardial in the terminal groove laterally to the junction SVC and RA and is generally inferior to the crest of the appendage of the atrium. The AV node is located at the apex of the Koch's triangle. As indicated above, the sides of the triangle of Koch are the tendon of Todaro, which is the continuation of the *valvula eustachii* of the IVC (or its remnant) and the axis of the septal leaflet of the right AV valve. The coronary sinus and the sub-Thebesian sinus, which is also called *valvula sinus coronarii thebesii*, form the base of the triangle. Between the membranous and muscular components of the ventricular septum runs the AV bundle.

Finally, the SSA includes the investigation of associated anomalies of the atria, AV and VA connections, e.g., bilateral SVC, TAPVC, and coarctation of the aorta. In the chest, the cardiac position is independent of the chamber connections with three possibilities, including dextrocardia, levocardia, or mesocardia, if the heart is located in the right chest or left chest or centrally placed, respectively. In *situs solitus*, the center is in the left chest with the apex pointing to the left, while in *situs inversus*, the heart is in the right chest with the top usually leading to the right.

Numerous aspects of the delivery of care are probably trying to measure the cardiac anatomy comprehensively. Some may have an impact on the outcome. The sequential segmental analysis and the new approach associated with transcriptomics or sequential segmental cardio-analysis (chemical analysis) of heart specimens examined following transplantation or complete or limited autopsy are crucial for the progress of pediatric cardiology and the understanding of congenital heart disease in the adolescent and adulthood.

1.2.2 Atrial Septal Defect (ASD), Ventricular Septal Defect (VSD), Atrioventricular Septal Defect (AVSD), Patent Ductus Arteriosus (PDA), and Ebstein's Tricuspid Anomaly (ETA)

1.2.2.1 Atrial Septal Defect (ASD)

- *DEF*: Defect of separation of the atria with interatrial communication (four types), including *ostium primum* (5%) or low in septum, adjacent to AV valves; *ostium secundum* (90%), at the site of the foramen ovale; *sinus venosus* (5%), divided in superior (near SVC) and inferior (near IVC) *sinus venosus* defect; and *unroofed coronary sinus* (Fig. 1.3).
- *EPI*: 10% congenital cardiac defects, 1:1500 live births.
- *ETP*: Deficiency of the *septum interatriale cordis* due to different causes, including trisomy 21 syndrome (Down syndrome), Ebstein's anomaly, fetal alcohol syndrome, Holt-Oram syndrome, and Lutembacher's syndrome.
- *CLI*: Asymptomatic, mostly, but untreated defects can yield right atrial enlargement, cardiac arrhythmias, and heart failure as time progresses.
- *EKG*: Prolonged PR interval (a first-degree heart block) due to the enlargement of the atria and the increased distance due to the defect itself. Moreover, there is a left axis deviation of the QRS complex with a primum ASD, while there is a right axis deviation of the QRS complex with a secundum ASD and a left axis deviation of the P wave in case of a sinus venosus ASD. A common finding in the ECG is the presence of an incomplete right bundle branch block.
- *Echo*: Jet of blood from the left atrium to the right atrium.
- *TRT*: *Secundum*, *primum*, and coronary sinus defects with small shunts (Qp: Qs <1.5) do not require treatment. A corrective closure (percutaneous device closure more for minor errors and surgical closure for more substantial defects) is carried out usually at 2–4 years of age. It is mandatory to do surgery at an earlier

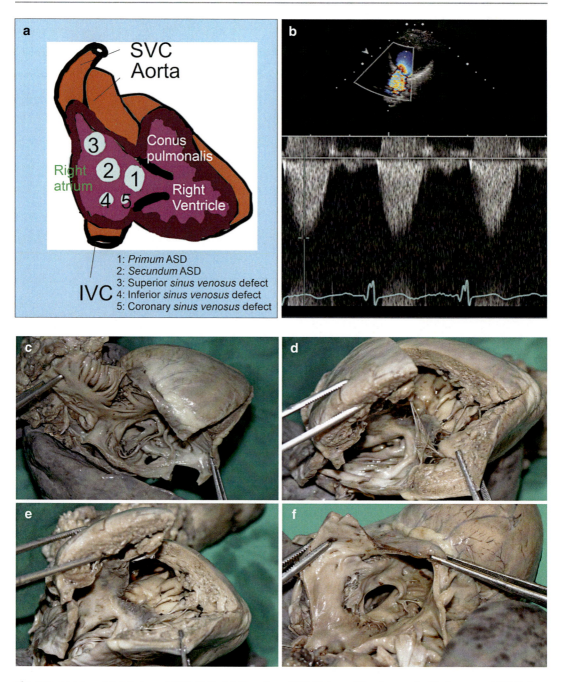

Fig. 1.3 Atrial septal defects and ASD II. In (**a**) there is the representation of the different atrial septal defects, while in (**b**) is shown the echocardiogram of a classic ASD-II type. The figures (**c–f**) show an ASD-II type (secundum type) observed from several perspectives

age if the shunt is more massive (Qp: Qs ≥1.5), there is a right atrial enlargement, or the patient has a *sinus venosus* defect.
- *PGN*: Good.

1.2.2.2 Ventricular Septal Defect (VSD)
- *DEF*: Defect of separation of the ventricles with interventricular communication, including either a deficiency occurring in the *pars membranacea septi interventricularis* or adjacent to the membranous septum (90%) or in the *pars muscularis septi interventricularis* (10%), usually congenital, but rarely acquired after myocardial infarction or trauma.
- *EPI*: 30% congenital cardiac defects. Increasing frequency in the last two decades.
- *ETP*: Deficiency of the *septum interventriculare cordis* in genetic syndromes, including tetralogy of Fallot. According to the Congenital Heart Surgery Nomenclature and Database Project, there are five types of VSD, including Type 1 or subaortic, Type 2 or perimembranous, cono-ventricular, membranous septal defect, and subaortic (70% of all kinds), Type 3 or inlet (or AV canal type), which is usually associated with AVSD (5%), Type 4 or muscular (trabecular), which is located in the muscular septum (20%), and Type Gerbode or left ventricular to right atrial communication.
- *CLI*: Significant left-to-right shunt determines pulmonary HTN, which, if left untreated, can progress to shunt reversal with cyanosis and the Eisenmenger syndrome.
- *EKG*: Normal sinus rhythm, PVCs, normal or mild ↑ PR interval, 1° AVB 10%, QRS axis: RAD (right axis deviation) with BVH, LAD (left axis deviation) 3–15%, Normal or rsr QRS Configuration, possible RBBB (right bundle branch block), possible RAE ± LAE (right vs left atrial enlargements), ventricular hypertrophy with BVH (bi-ventricular hypertrophy) 23–61% and RVH (right ventricular hypertrophy) with Eisenmenger.
- *Echo*: The flow is from the left ventricle to the right ventricle.
- *TRT*: Small shunts may close spontaneously in childhood and can be managed by observation only, but large shunts require mandatory surgical closure. Surgical indications include failure of congestive cardiac failure to respond to drugs, VSD with pulmonic stenosis, large VSD with pulmonary HTN, and VSD with aortic regurgitation.
- *PGN*: Good. Of note, the Gerbode defect is a very rare congenital anomaly with excellent outcome. The Gerbode defect is a rare defect representing less than 1% of CHD and consists in a left ventricular-right atrial communication due a structural abnormality of the central fibrous body in combination with arrested maturation of the membranous portion of the ventricular septum. The first successful closure of such a defect was reported by Kirby at the Hospital of the University of Pennsylvania in 1956, while the first successful series of patients operated on with a LV-RA shunt was reported by Frank Gerbode at Stanford University (Tidake et al. 2015).

1.2.2.3 Atrioventricular Septal Defect (AVSD)
- *DEF*: *Ostium primum* type ASD associated with a common AV valve, with or without an associated inlet (AV septal type) VSD (Figs. 1.4 and 1.5).
- *SYN*: Atrioventricular canal defect (AVCD), also known as "common atrioventricular canal" (CAVC) or "endocardial cushion defect" (ECD).
- *EPI*: 1:2000 newborns/year in the USA.
- *ETP*: Abnormal TBX2 (T-box transcription factor 2) expression during embryogenesis with maldevelopment of the endocardial cushions. Trisomy 21 (Down) syndrome and heterotaxy syndromes are often associated with AVCD. Other RFs include a CHD harboring parent, EtOH use during pregnancy, uncontrolled DM, as well as iatrogenic (specific drugs/medications during pregnancy). In the partial AVSD, there is an *ostium primum* defect with the inferior part of the atrial septum but no direct intraventricular communication, while the complete AVSD (CAVSD) refers to a large ventricular component beneath either or both the superior or inferior bridging leaflets of the AV valve. The valves are a common atrioventricular valve. The severity of the defect relies mostly on the sup-

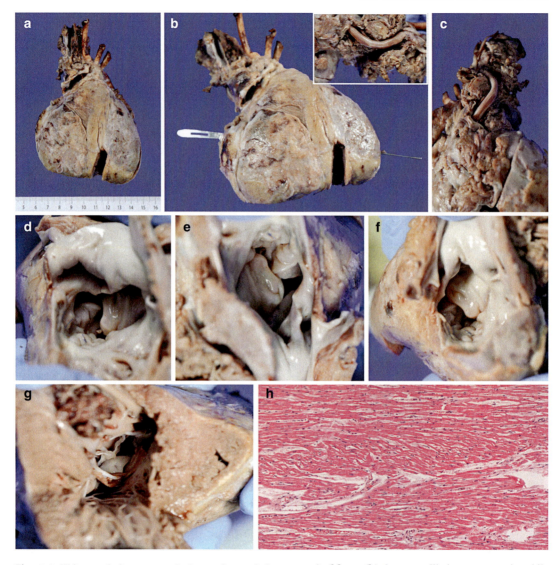

Fig. 1.4 This panel shows several views of an unbalanced atrioventricular septal defects in a patient with trisomy 21 syndrome (Down syndrome). The inset (upper corner) of figure (**b**) shows a palliative anastomosis, while (**h**) shows some of the non-characteristic aspects of interstitial fibrosis and signs of cardiomyocyte hypertrophy

porting attachments of the valve to the ventricles and whether the valve allows dominant flow ("unbalanced" flow) from the RA to RV and from LA to LV.
- *CLI*: In the absence of a VSD component or a small VSD and relatively good AV valve function, children may be asymptomatic, but in the presence of a large VSD component or significant AV valve regurgitation, children exhibit signs of heart failure (e.g., dyspnea with feeding, poor growth, tachypnea, and sweating or enhanced diaphoresis). There are also heart murmurs, and patients can suffer from tachypnea, tachycardia, and evidence hepatomegaly.
- *EKG*: Normal sinus rhythm, PVCs 30%; PR interval, 1° AVB >50%; moderate to extreme LAD of the QRS axis; usual with atypical QRS configuration, rSr′ or rsR′; atrial enlargement, possible LAE; and ventricular hypertrophy, uncommon in partial; BVH in complete; RVH with Eisenmenger syndrome.

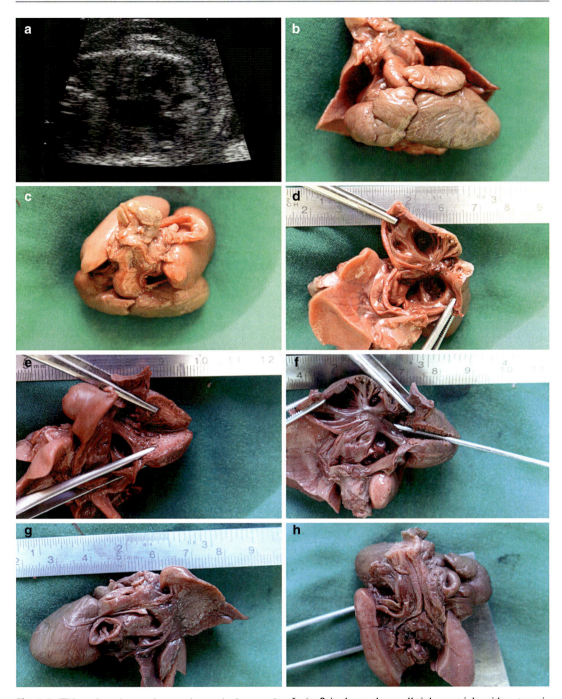

Fig. 1.5 This patient shows a large atrioventricular septal defect as identified first in the prenatal ultrasound (**a**). The figure (**b**) shows the anterosuperior face of the heart and the superior part of the congenital heart defect (**c**), while (**d**) demonstrates the large atrioventricular septal defect. In (**e**, **f**) is shown the small right ventricle with a stenosis of the pulmonary atresia. Figures (**g**, **h**) show the pulmonary venous return from the lungs to the left atrium and the trachea and its bifurcation, respectively

- *Echo*: Evidence of the defect using 2-dimensional echocardiography with color flow and Doppler studies.
- *TRT*: Surgical repair involving the closure of the atrial and ventricular septal defects and restoration of a competent left AV valve.
- *PGN*: Repair of the AVSD should be performed by age 2–4 months because of the heart failure and failure to thrive. In some patients, repair may be delayed up to 6 months, if they grow well, but not over this time to prevent the development of pulmonary vascular disease, specifically in patients with trisomy 21 syndrome (Down syndrome). Complications include a leaky mitral valve with consequent heart failure. In cases with proper treatment, most infants with AVSD grow up to have healthy lives. Endocarditis prophylaxis is required only for the first 6 months after repair or if there is a residual defect, which is located adjacently to a surgical patch.

1.2.2.4 Patent Ductus Arteriosus (PDA)

- *DEF*: Persistence of the unique blood vessel between the main pulmonary artery and the aorta in extrauterine life closing spontaneously by day 4 of age ($\uparrow O_2$ tension) ± VSD, CoA.
- *SYN*: *Ductus arteriosus Botalli apertus*
- *EPI*: 12% of CHDs; incidence: 2/10,000 and 6/100,000 of live newborns; ♂ < ♀.
- *RF*: Pregnancy rubella (first trimester), high altitudes (>10,000 feet), premature babies <1500 g.
- *ETP*: PDA in the preterm infant is due to the lack of normal closure mechanisms due to immaturity. Four fifths of preterm infants presenting with respiratory distress syndrome (RDS) also evidence a PDA, which may be due to the increased circulating prostaglandins (PGE2). Other RFs include high altitude at birth, genetic factors, and in utero exposure to rubella virus, although the pathogenesis may be more complex (Dice & Bhatia, 2007).
- *CLI*: Bounding pulses with widened pressure, S_2 narrowly split, rough "machinery" murmur.
- *CXR*: Enlargement of the left heart and prominent MPA and aorta.
- *EKG*: Normal to left ventricular hypertrophy (in case of RVOT obstruction, there is RVH).
- *Echo*: Enlargement of the left heart and prominent MPA and aorta.
- *TRT*: PGE maintains DA open, while $\uparrow O_2$ and indomethacin stimulate the closure.
- *PGN*: Patients with small shunt do quite well even without surgery, which may be required for large shunts and pulmonary HTN.

1.2.2.5 Ebstein's Tricuspid Anomaly (ETA)

- *DEF*: Downward displacement of the right AV or tricuspid valve with most of the valve being attached to the ventricular wall rather than to the fibrous ring of the valve. It leads to atrialization of the inlet portion of the RV with a variable degree of malformation and displacement of the anterior leaflet of the tricuspid valve (Fig. 1.6) and potentially associated with an ASD in 2/3 of cases with R → L shunt and cyanosis.
- *SYN*: Atrialized right ventricle, *caput mortuum*.
- *EPI*: 1/50,000–1/200,000 (prevalence), ♂ = ♀.
- *ETP*: Although it has been questioned, lithium exposure during the first trimester of pregnancy may lead to an increased risk of ETA. An increased risk has been observed in patients with Wolff-Parkinson-White syndrome (WPWS).
- *CLI*: Variable symptomatology is ranging from asymptomatic to SCD with the degree of severity depending upon the degree of malattachment of the AV valve and the associated anomalies. The Ebstein's tricuspid anomaly (ETA) is not a simple anatomical variant as often was present in older books. ETA has critical physiologic implications. During the pump function of the heart, in the setting of an ETA, an anomalous faulty of the hydraulic machine occurs. In fact, as the right ventricle contracts, some of the blood flowing from the right atrium into the right ventricle of the heart takes a backward flow, into the right atrium. In the association with an ASD ostium type II or I eventually, some of the blood will flow into the left ventricle bringing some unoxygenated blood into the systemic ejection chamber.

1.2 Congenital Heart Disease

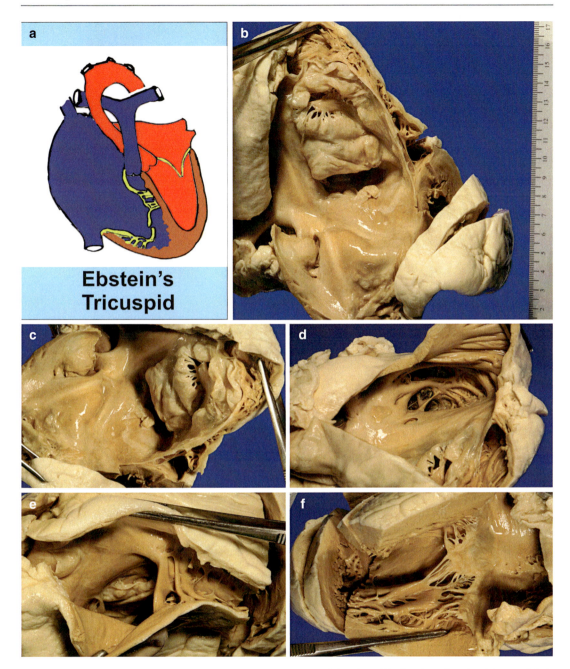

Fig. 1.6 In this panel the Ebstein's tricuspidal anomaly is represented using a cartoon (**a**) and an explanted heart (**b–f**). In the sequential gross photographs, the displacement of the tricuspid valve down into the right ventricle is shown. The atrialized right ventricle is particularly evident

- *EKG*: 50% ETA with WPWS, signs of right atrial enlargement or tall and broad "Himalayan" P waves' first degree AV block manifesting as a prolonged PR interval, low amplitude QRS complexes in the right precordial leads, atypical right BBB, T wave inversion in V1–V4, and Q waves in V1–V4 and II, III, and aVF.
- *Echo*: Carpentier et al. (1988) and Celermajer et al. (1994) Scoring systems for newborns with ETA are still crucial in clinical departments of cardiology.

- *TRT*: Annuloplasty aims to modify the level of the tricuspid orifice and decrease mitral insufficiency with arrhythmias as late complications.
- *PGN*: It depends on the severity of the atrialization of the RV (assessment of the "*caput mortuum*").

1.2.3 Tetralogy of Fallot (TOF) and Double Outlet Right Ventricle (DORV)

Conotruncal anomalies include various lesions such as TOF, absent pulmonary valve syndrome, DOV, malposition of the great arteries, and common arterial trunk. In all these cardiovascular defects, there is an abnormal conotruncal septum, and the tremendous variability explains the diversity of the cardiac malformations in the development of the sub-arterial conuses. The complete repair of all conotruncal anomalies includes, despite a wide anatomic variety, two operational steps. The first steps are to build an intracardiac tunnel to connect the left ventricle to one of the arterial orifices (either aortic or pulmonary), through the conoventricular VSD. The second step refers to a connection from the right ventricle to the pulmonary artery, either at the intracardiac or extracardiac level, according to the distance between the tricuspid valve and the pulmonary orifice. Some conotruncal anomalies are not ideal to biventricular repair, because of one or several of the following associated lesions, including hypoplasia of one of the ventricles, multiple VSDs making septation impossible, and severe anomalies of the AV valves (e.g., overriding, straddling, leaflet abnormalities) (Figs. 1.7 and 1.8).

In the reconstruction of the left ventricular outflow tract (LVOT), the construction of the intracardiac tunnel is often performed through three approaches. The first approach is the anchorage of a prosthetic pericardial patch to the inferior margin of the VSD through the right atrium. The second approach involves the conal septum, through the right ventricle, and is resected and the intracardiac tunnel constructed. Finally, in the third approach, the shaft is completed through the aortic orifice using interrupted mattress sutures without pledges. In cases of right ventricular outflow tract (RVOT), the aim is to develop a subpulmonary conus, and the distance between the tricuspid orifice and the pulmonary orifice may be crucial. In practice, when the tricuspid-to-pulmonary distance is long enough, i.e., $\geq$ aortic valve diameter, anatomic reconstruction is feasible. If the tricuspid-to-pulmonary distance is too short, i.e., < aortic valve diameter, IVR cannot be performed, because of the impossibility to build an LV-to-aorta tunnel without obstructing the native pulmonary orifice.

1.2.3.1 Tetralogy of Fallot (TOF)
- *DEF*: Complex congenital heart disease with stenosis of the pulmonary valve or RVTO, RV hypertrophy, VSD, and overriding of the VSD by the aortic root (Fig. 1.7). In 80% of TOF, the posteroinferior margin of the defect between the ventricles is formed by a fibrous continuousness between the leaflets of the aortic valve and tricuspid valve with the involvement of the remnant of the interventricular portion of the membranous septum. In 20% of TOF, the defect is muscular. There is muscular tissue between aortic and tricuspid valve, and the muscular part is represented by a continuum of the posteroinferior limb of the septomarginal trabeculation with the ventricular-infundibular fold. In almost all cases, the VSD is unrestrictive allowing for bidirectional shunting.
- *EPI*: 7–10% of all CHD, 3/10,000 live births.
- *ETP*: Multifactorial including maternal T1DM, PKU, retinoic acid, and chromosomal abnormalities (trisomies 13, 18, 21 and microdeletions of chromosome 22).
- *CLI*: Small and thin infants/children with cyanosis after the neonatal period, hypoxemic spells during infancy, and systolic ejection murmur at the upper left sternal border. Signs may include watch-crystal nails and drumstick fingers. The size of the shunt is by the degree of desaturation and the extent of cyanosis. The shunt is in turn dependent upon the resistance to outflow from the RV, the size of the VSD, and the systemic vascular resistance.
- *CXR*: "Boot"-shaped heart.

1.2 Congenital Heart Disease

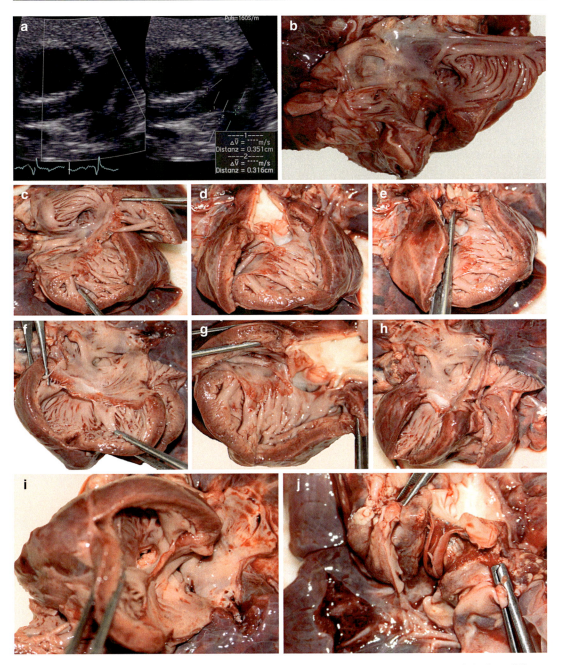

Fig. 1.7 This panel shows the echocardiogram and the gross photographs of a heart harboring a tetralogy of Fallot. In (**i–j**) a Blalock-Taussig shunt from the brachiocephalic artery to the right pulmonary artery is illustrated, while the gross photographs b through h show different gross views of the heart opened like a book using the cardiac sequential segmental analysis.

- *EKG*: Right deviation of the cardiac axis (+90°/+180°) with regular P waves or slight RAH.
- *Echo*: Thickening of the free RV wall, with overriding of the aorta, a VSD, obstruction at the level of the infundibulum and pulmonary valve, as well as the anatomy of the coronary arteries.
- *TRT*: Neonatal palliation (e.g., Blalock-Taussig [BT] shunt) and repair (VSD closure,

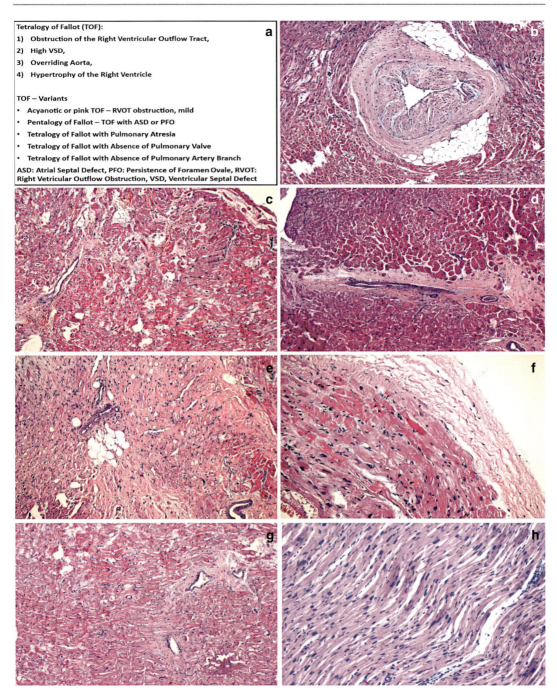

Fig. 1.8 In (**a**) the four components of the tetralogy of Fallot (TOF) and the TOF variants are shown. Figures (**b**–**h**) show some histologic aspects of the cardiac pathology in patients with TOF. Although not specific for TOF, they are important to be identified at the microscopic examination. In (**b**) there is a hypertrophy of the intima and media; in (**c**) and (**d**), an increase of some interstitial cells as well as perivascular fibrosis; in (**e**) an island of fatty cells is shown; in (**f**–**h**), there are several areas of cardiomyocyte hypertrophy with increase of the nuclear size more than three times of the normal size, and there are several bizarre shapes of nuclei

with patch channeling the left ventricle to the aortic root, excision of an excess of subpulmonic muscular tissue, and reconstruction of the pulmonary arteries, if it is necessary) at 5–6 months of age. The BT shunt involves the creation of an anastomosis between the SCA and the RPA, which is usually performed on the side opposite of the aortic arch. The Waterston-Cooley shunt between the ascending aorta and the MPA and the Potts shunt, an anastomosis between the aorta (descending portion) and the LPA, are other shunt procedures, which were developed for the same purpose, although they do not seem to be longer in use in cardiac surgery.

- *PGN*: Infants with the most severe degree of TOF lesions are commonly profoundly cyanotic at birth. Surgical correction is needed by the first decade of life. Those with moderate RVOTO do reasonably well, and cyanosis is not as severe as seen above. The second to third decades of life require surgical correction. The cause of death may rely on cerebrovascular accidents, brain abscess, subacute infective endocarditis, anoxia, or pulmonary hemorrhage. The patients with the mildest form of the disease seem to be relatively acyanotic. Since the degree of RVOTO increases as the patient gets older, death may occur by the third or fourth decade of life if no corrective surgery is performed.
- RR is 3%.

Aortopulmonary Collateral Arteries (APCAs)

Long considered a relatively minor element, APCAs have acquired tremendous importance for the survival of patients with TOF. APCAs represent the persistence of ventral segmental arteries and are an echocardiographic and angiographic finding in CHD (not only TOF). APCAs play a significant role in TOF accompanied with pulmonary atresia. However, APCA may occur in association with several different bronchopulmonary disorders (congenital and acquired type). APCA is common in VLBW infants and particularly in infants requiring prolonged positive pressure ventilation with secondary BPD.

Pulmonary Arterial Changes in VSD and Severe Pulmonary Hypertension

Large VSD may be a complicated pulmonary vascular disease with a significant rate of mortality. Morphometric studies of pulmonary arteries start with a good fixation of the lung biopsy and proceed with good slides and staining of the sections with elastic Van Gieson or Movat pentachrome staining. The external diameter is measured using a morphometric software, such as ImageJ software or a calibrated eyepiece. It is important to focus on the distance between an external elastic lamina and internal elastic lamina across the shortest axis. For arteries measuring 20–50 μm and 50–100 μm in thickness, the wall thickness is also measured determined from the internal elastic lamina to external elastic lamina along the shortest axis of the arterial blood vessel. It is essential to calculate the percentage as an index of the artery's outer diameter according to the following formula: (2 × wall thickness)/external diameter × 100. Haworth-Hislop reference values for infants aged 2 months and over of uninjected pulmonary arteries are 9.6% ± 3.4% in 25–50 μm arteries and 7.4% ± 2.5% of the ED in arteries 50–100 μm. Arterial changes of the vascular bed of the lung are more severe in patients with higher pressure or an increase in pulmonary resistance than in those with only an increase in pulmonary blood flow or a small rise in PA pressure. Using the Heath-Edwards grading system (HEGS), it may be identified that medial hypertrophy represents an early structural change of pulmonary arteries, which are exposed to increased pulmonary blood flow and pressure. Media thinning with or without dilatation lesions is, instead, a later and inconstant change. Cyclical patterns may take place when high blood flow increases wall shear stress and endothelial cell damage with consequent intimal proliferation and turbulence to flow and luminal occlusion, which in turns increases resistance and develops worsening structures. The best predictor of the early fall in pulmonary pressure following the closure of a VSD is the structure of the pulmonary bed, as determined by the HEGS, the density of peripheral arteries, and the degree of medial hypertrophy. In patients undergoing VSD closure, ≤1 grade III lesion according to HEGS

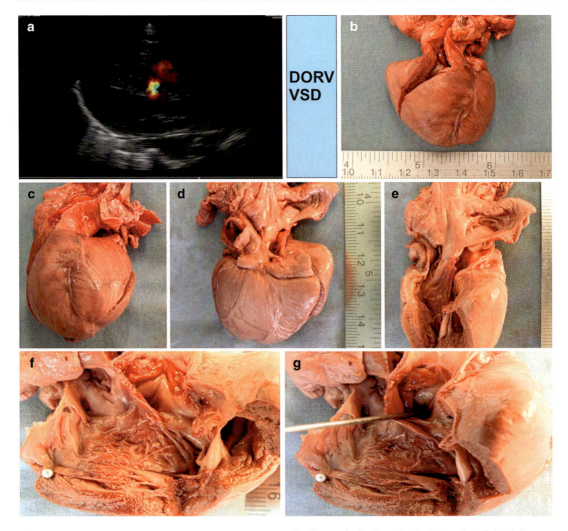

Fig. 1.9 In (**a–g**) there is the presentation (echocardiography and gross photographs) of a heart with double outlet right ventricle and ventricular septal defect. Specifically, the figures (**e**, **f**) show the double outlet of the right ventricle, while figure (**g**) illustrates the ventricular septal defect, which is indicated by the probe inside of the defect

and arterial thickness (≥0% age-related) are the best correlating parameters of the difference between pre- and early postoperative pulmonary arterial pressure.

Pentalogy of Fallot (POF): Variant of the more common TOF, comprising the standard four features with the addition of an ASD or PDA (Fig. 1.8).

1.2.3.2 Double Outlet Right Ventricle

- *DEF*: A right ventricle with two outlets. In terms of morphology, there are some diagnostic landmarks useful for the distinction of d-TGA from DORV with transposed aorta. The landmarks consist in the presence of a subaortic and subpulmonary *conus* that means noncontinuity between the mitral valve and pulmonary valve, alignment of the pulmonary root relative to the interventricular septum, and finally the coronary pattern (Figs. 1.9 and 1.10). DORV is subtyped based on the relationship of the always associated VSD (subaortic, doubly committed, subpulmonary, or noncommitted) with the great arteries.
- *ETP*: It is a deficiency of the cono-truncus with mutations of GDF1 (19p13.11) and

1.2 Congenital Heart Disease

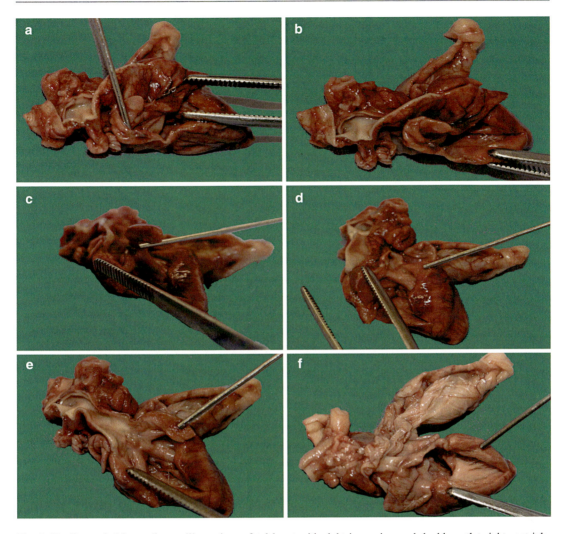

Fig. 1.10 Congenital heart disease illustrating a fetal heart with right isomerism and double outlet right ventricle. Figures (**a–f**) show the right ventricle in several views with a double outlet morphology

CFC1 (2q21.2), which control the cardiac progenitor cells. DORV is associated with chromosomal anomalies such as 22q11.2 deletion syndrome, trisomy 13 syndrome, trisomy 18 syndrome, and CHARGE syndrome.

- *EPI*: 1–3% of all CHD, 1/10,000 live births, a cardiac defect in which both the great arteries are in whole or in part connected to the right ventricular chamber with or without a VSD or TGA. Clinically, there is a variable presentation, which depends on how the anatomical defects affect the blood flow within the heart. If a subaortic VSD is present, the physiology is similar to TOF, while if a subpulmonic VSD (aka Taussig-Bing anomaly) is present, the physiology is similar to TGA. Other forms of DORV lead to cardiac physiology quite identical to a large VSD.
- *CLI*: Symptoms may start on the first day of life and include cyanosis, tachypnea, tachycardia, poor feeding, slow weight gain, and a heart murmur. Since DORV is always associated with a VSD, there is blood, which drains from the left ventricle. Other cardiac anomalies include pulmonary stenosis (DORV Fallot type, DORV with noncommitted VSD), various degrees of left and right ventricular hypoplasia, AVSD, and straddling AV valves,

while extracardiac anomalies include heterotaxy.
- *EKG*: If there is a subaortic VSD with no stenosis of the pulmonary valve, a superior QRS axis (−30° to −170°) is seen with either right ventricular hyperplasia (RVH) or biventricular hypertrophy and left atrial enlargement. Also, a first-degree AV block may be found. However, if there is subpulmonic VSD or in those with subaortic VSD and stenosis of the pulmonary valve, a right axis deviation, RVH with right atrial enlargement may be identified.
- *Echo*: Transthoracic echocardiography serves as the major imaging modality to diagnose DORV and surgical planning, while preoperative transesophageal echocardiography (TEE) is used to confirm the transthoracic findings, identify additional defects, and examine any interval anatomical or physiological changes. Post-repair TEE is performed to verify the adequacy of repair and assess heart function.
- *PGN*: Poor due to severe cyanosis, heart failure, or pulmonary hypertension, if no treatment is set up, but with biventricular repair, patients may have an average life expectancy.

1.2.4 Hypoplastic Right Heart Syndrome

- *DEF*: Underdevelopment of the atrioventricular and ventriculoarterial connections on the right side of the heart including the tricuspid valve, right ventricle, pulmonary valve, and pulmonary artery. HRH is commonly associated with ASD; pulmonary blood flow derived either from a PDA or collateral channels and ultimately hypoxygenated blood flow to the body and cyanosis (Fig. 1.11) (HRHS).
- *SYN*: Hypoplastic right heart.
- *TXN*: Pulmonary atresia with VSD (PSVSD/PAVSD), pulmonary stenosis/atresia with intact ventricular septum (PSVSD/PAIVS), and hypoplastic right heart with tricuspid atresia (HRHTA).
- *EPI*: 7% congenital heart diseases.
- *ETP*: Underdevelopment with stenosis of the valve cups but the underlying genetic mechanisms are unknown.

1.2.4.1 PAVSD
- *DEF*: Extreme form of TOF with complete atresia of the pulmonary valve and VSD with the pulmonary blood flow deriving from either a PDA or collateral channels.
- *CLI*: Phenotype depending upon the size of the PDA and collateral channels.
- *IMG*: Echo, cardiac catheterization, and angiocardiography are diagnostic.
- *TRT*: PGE_1 to dilate PDA and stabilize the patient until surgery is scheduled. The operation consists of a construction of a bypass for the RVOT obstruction and VSD closure.
- *PGN*: As recently indicated by Kaskinen et al. (2016), palliative surgery has a role in the therapy of PA and VSD because the size of pulmonary arteries increases after placement of systemic-pulmonary artery shunt. Also, subtotal repair by an RV-PA connection and septal fenestration improves the survival over extracardiac palliation.

1.2.4.2 PSIVS/PAIVS
- *DEF*: Valve-level RVOT obstruction due to PV stenosis/atresia and intact ventricular septum with the cusps of the pulmonary valve fused to form a membrane or diaphragm with a central hole (diameter: 0.2–1 cm) with commonly moderate to marked post-stenotic dilation of MPA and LPA and PDA. If the tricuspid valve remains competent, the right ventricle remains hypertrophic with a small chamber volume (Type 1), but if tricuspid incompetence occurs, the right ventricular chamber diameter is either normal or enlarged (Type II).
- *EPI*: 1–10% of all CHD.
- *ETP*: There are theories including viral or other infections resulting in local disruption of the pulmonary valve and proximal pulmonary artery embryogenesis with the destruction of the early vasculature.
- *CLI*: Asymptomatic with mild/moderate PSVSD, while cyanosis and right-sided CHF with severe PSVSD (PAIVS). Right ventricular lift with systolic ejection click at the pulmonary area with mild/moderate PSVSD and S_2 widely split with soft or silent P_2, grades I–VI obstructive systolic murmur. To maintain an adequate output across the PV, it

1.2 Congenital Heart Disease

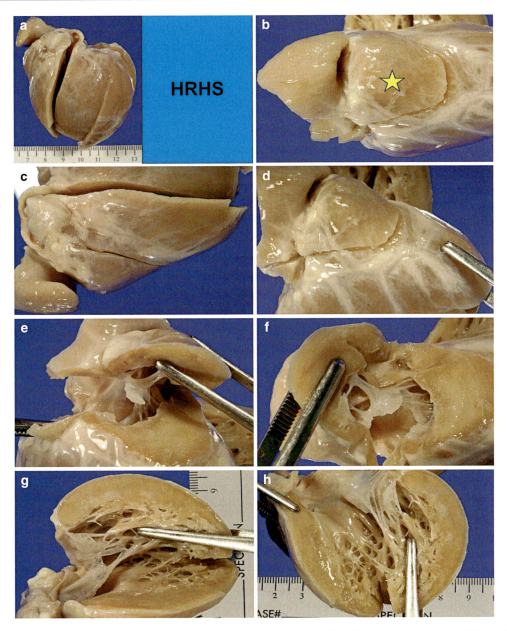

Fig. 1.11 Congenital heart disease with a hypoplastic right heart with figure (**a**) showing the anterior-superior wall, while (**b**) shows the rudimentary right ventricle (yellow star-labeling). Figures (**c, d**) show also the rudimentary right ventricle from the external point of view, while (**e, f**) show the right ventricle with a very tiny cavity internally. Figures (**g, h**) put in evidence the large left ventricle

increases pressure developed by the RV with pressures higher than systemic and considered life-threatening. The increased right-sided workload induces RVH and, eventually, right-sided CHD differently from TOF with a VSD. PDA allows an R-to-L shunt at the atrial level.

- *EKG*: RVH signs and RAD in moderate PSIVS, while RVH-signs, RAD, and RV strain pattern (deep inversion of the T wave) ± RAH in severe PSIVS.
- *CXR*: Dilated PA on posteroanterior CXR. In severe PSIVS, the pulmonary vascular markings are decreased.

- *Echo*: Atrial contraction and significant RV diastolic pressure causing early opening of the PV and echo-dense PV.
- *TRT*: Elective valvotomy for RV pressures >50 mm Hg or > 2/3 of systemic pressures or percutaneous balloon valvuloplasty for palliatively relieving the valvular obstruction.
- *PGN*: Mild PSIVS harboring patients have a normal lifespan, moderate PSIVS may present with easy fatigability, and dyspnea on exercise while severe PSIVS/PAIVS harboring patients manifest with severe cyanosis, and right-sided CHF in early life.

Other conditions that may be encountered in this setting of RVTO are infundibular pulmonary stenosis without VSD and distal pulmonary stenosis (supravalvar pulmonary stenosis, peripheral pulmonary stenosis, and absence of a pulmonary artery).

1.2.4.3 HRHTA
- *DEF*: Complete atresia of the tricuspid valve (right AV valve) with two subtypes depending upon the relationship of the great arteries (HRHTA I without TGA and HRHTA II with TGA).
- *EPI*: <1% of CHD.
- *CLI*: The absence of direct communication between RA and RV has as a consequence that the whole PVR flows through the atria using either a PFO or an ASD. The RV is hypoplastic severely according to the left-to-right shunt at the ventricular level, and the most severe form of HRH occurs when there is no VSD or the VSD is very small. Symptoms include neonatal cyanosis, failure to thrive, reduced feeding, tachypnea, dyspnea, anoxic spells, and right-sided CHF. Moreover, S_1 is a regular sound, S_2 is a single sound (aortic closure), and there is variable harsh blowing murmur (I–III/IV).
- *IMG*: Cardiac enlargement and small MPA and variable size of the RA and decreased pulmonary vascular markings (CHR) and TA (echo).
- *EKG*: Left axis deviation with tall and peaked P waves (RAH) + LVH.
- *TRT*: PGE_1 until surgery is performed. Fontan (RA connected to RV/PA).
- *PGN*: It depends upon the achievement of a balance of pulmonary blood flow allowing adequate oxygenation of the tissues.

1.2.5 Transposition of the Great Arteries
- *DEF*: A complete swap of the great vessels subdivided in d- and l-TGA with (complex) or without (simple TGA) other heart defects (60% PDA, 30% VSD) (Figs. 1.12, 1.13, 1.14, and 1.15).
- *d-TGA* (dextro-TGA): *A-V-Concordance + V-Art-Discordance* with hypoxygenated blood from the right heart is pumped through the aorta into the body, while oxygenated blood is pumped continuously back into the lungs through a patent pulmonary artery. Pulmonary and systemic circulations are in parallel (*circular TGA*).
- *l-TGA* (Levo-TGA): *A-V-Discordance + V-Art-Discordance* in which the morphological left ventricle is right-sided and the morphologically right ventricle is left-sided and their corresponding atrioventricular valves and primary arteries are also transposed with the aorta located anteriorly and to the left of the pulmonary artery (*congenitally corrected TGA*). Pulmonary and systemic circulations are connected. The left-sided but morphologically right ventricle works at high pressure against the high resistance residing in the systemic circulation.
- *EPI*: 4% congenital cardiac anomalies, 16% of all cyanotic CHD, and more common in infants of diabetic mothers; ♂ : ♀ = 3:1.
- *ETP*: Abnormal spiral division of the truncus arteriosus.
- *CLI*: Two groups and five subgroups, including TGA-1 with IVS, which is subdivided in 1a, without pulmonary stenosis, and 1b with pulmonary stenosis, and TGA-2 with VSD, which is divided in 2a, with pulmonary stenosis; 2b, with pulmonary vascular obstruction (PVO); and 2c, without PVO. Since the aorta arises directly from the RV, some interatrial communication is required for survival. TGA-1 with IAC and PDA, which may be small, experience severe cyanosis, while TGA-2 have decreasing levels of cyanosis (2a > 2b > 2c).

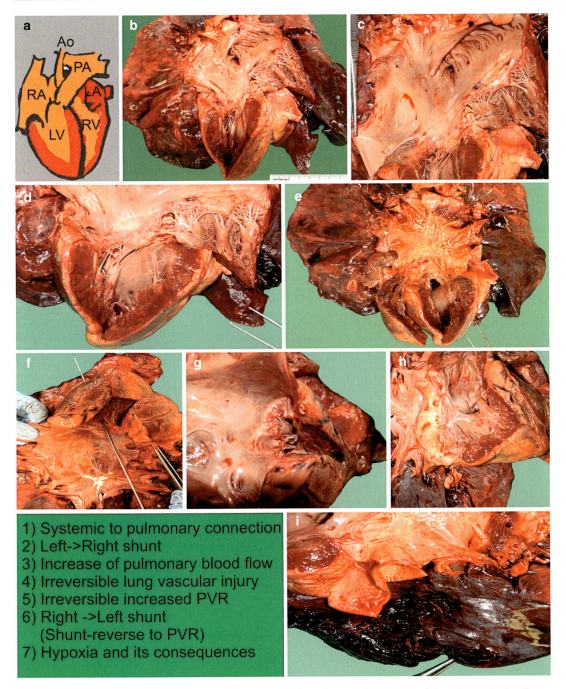

Fig. 1.12 Congenital heart disease showing a left (corrected) transposition of the great arteries with ventricular septal defect, pulmonary hypertension, and Eisenmenger syndrome. In (**a**) is a cartoon showing the corrected transposition of the great arteries with atrioventricular discordance and ventriculoarterial discordance. Figures (**b–h**) show the several segments of the atrioventricular discordance on both sides with a probe in (**f**) inserted in the ventricular septal defect. Figure (**i**) shows some additional lung hemorrhage that can be fatal as seen in our patient, and besides there is the physiopathology cascade of the L-TGA

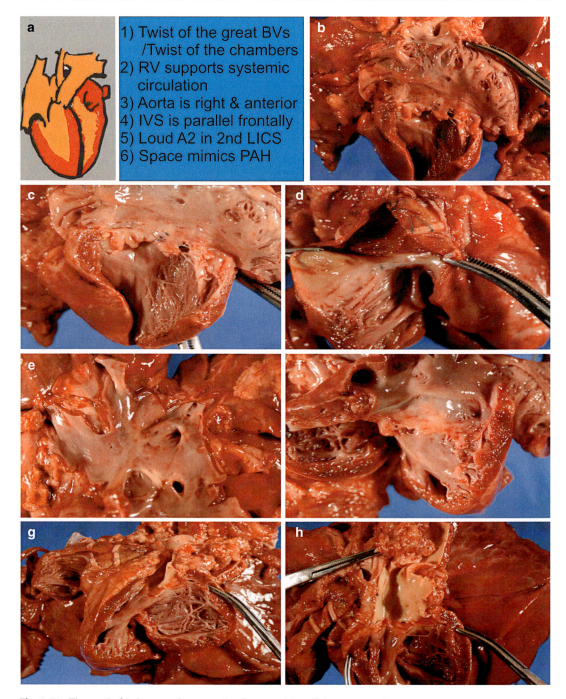

Fig. 1.13 Figures (**a–h**) show another example of transposition of the great arteries

1.2 Congenital Heart Disease

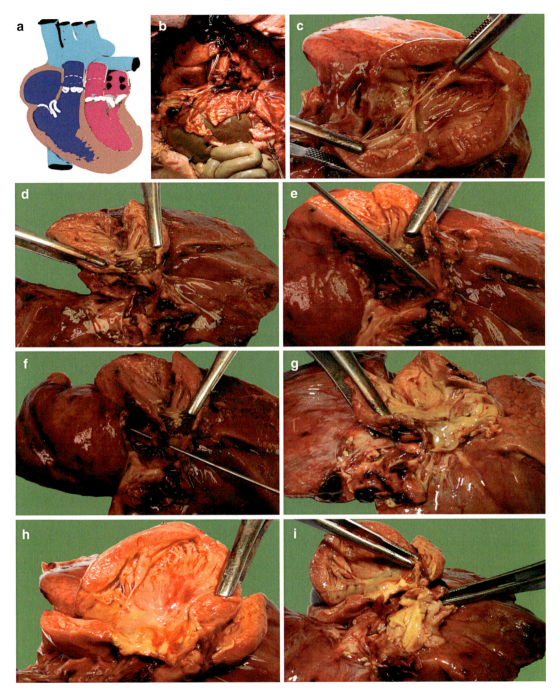

Fig. 1.14 Congenital heart disease with TGA. Figures (**a–h**) show another example of transposition of the great arteries. In this example, there is concordance between atria and ventricles (atrioventricular concordance)

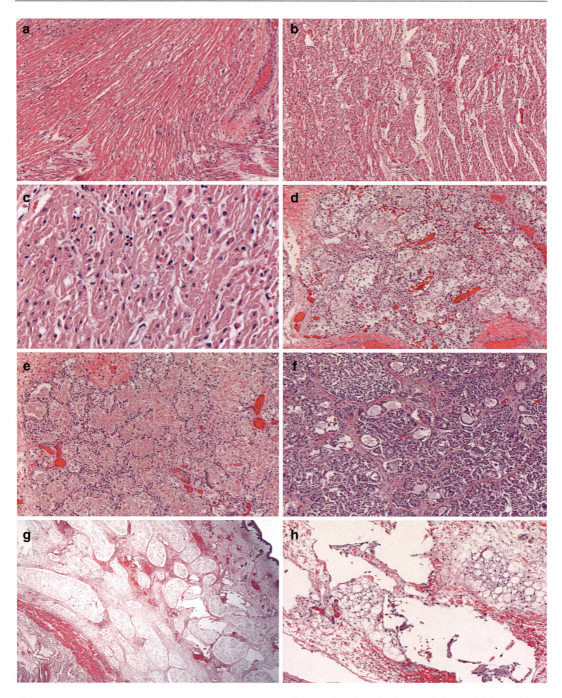

Fig. 1.15 Histopathology of transposition of the great arteries (TGA). Figures (**a–h**) show the histopathology associated with patients with TGA. Myocardial tissue is demonstrated in (**a–c**), while the remaining histologic photographs show the lung pathology with hemorrhage. In the heart, signs of cardiomyocyte hypertrophy accompany interstitial edema and fibrosis

- *PEX*: Tachypnea (hypoxia), tachycardia, murmur absent in 1a but present in 1b, 2a, and 2b due to the pulmonary stenosis and in 2c due to the VSD, palpable RV impulse and accentuated S_2.
- *CXR*: Mild cardiomegaly, mild ↑pulmonary vascular markings, and "egg-on-string/side" appearance to the cardiac silhouette.
- *EKG*: It is of little definite help, although the absence of positive findings of other defects, such as a left axis deviation of TA-HRH, may provide some information.
- *Echo*: Posterior great artery arises from LV and divides into RPA and LPA, while the aorta is anterior and appears from RV.
- *TRT*: Cardiac catheterization is performed to characterize the defect in preparation for corrective surgery further but also to perform a Rashkind septostomy (aka balloon atrial septostomy or BAS) once the cardiologist has identified the TGA with two well-developed ventricles. The BAS may be complicated by arrhythmias, inadequate or excessive size of the tear, and rupture of the left atrium. The Rashkind procedure is performed with PGE1 infusion to maintain PDA. Since 1958, the year when Senning introduced his surgical procedure for d-TGA or atrial switch operation, the prognosis for children born with TGA has changed dramatically. Subsequently, an alternative technique was proposed by Mustard. In the Senning procedure, the venous return is directed to the contralateral atrioventricular valve and ventricle using an atrial baffle or deflector made of the patient's septal tissue. In Mustard's technique, the atrial septum is excised, and a pericardial or synthetic baffle is created to direct the venous return. Senning procedure was popular in Europe, while Mustard technique was favorite in Canada and the USA. Late complications associated with these techniques include mostly atrial arrhythmias and right ventricular dysfunction. SCD remains a leading cause of death in many series, and the history of tachyarrhythmias is considered an independent predictive factor for mortality. Thus, the Arterial Switch Operation (ASO) or Jatene's procedure started to be used in the mid-1980s and is currently used for TGA. There is still considerable interest in the outcome of TGA patients operated with Senning or Mustard procedures, and both techniques may be used in cases in which atrial repair is not feasible. Rastelli repair for TGA is an operation for repair of complete TGA in association with a large VSD and pulmonary stenosis and consists of a neo-communication between the LV and the aorta via the VSD using a baffle within the RV, which is connected to the pulmonary artery using a valved conduit and finally by obliteration of the LV-PA connection. Thus, Rastelli repair allows the left ventricle to support the systemic circulation. Although atrial switch operation was performed worldwide is not done anymore, patients with heart corrected are still alive, and the permission for an autopsy may be granted. Thus, this knowledge is always essential to have. Rastelli can be performed with low mortality at an early stage, but later substantial rates of morbidity and mortality are associated with this procedure. Fatal consequences include conduit obstruction, left ventricular outflow tract obstruction, and arrhythmia. Thus, the first two structures need to be investigated by the pathologist. The current early and late results of the ASO are excellent. ASO has reduced the application of the Rastelli repair to an alternative when LVOTO is not prone to be relief at the time of ASO.
- *PGN*: Without surgery, congestive heart failure is common, because of the high cardiac output, inadequate oxygenation of the myocardium, and systemic pressure in both ventricles.

1.2.6 Common Trunk (Persistent Truncus Arteriosus)

- *DEF*: Developmental failure of separation of the aorta and MPA with a single large vessel. The Collett and Edwards classification divides the PTA into four types. Type I shows the main pulmonary artery arising from the truncus and then dividing into the right and left pulmonary arteries. Type II is identified by the right and left pulmonary arteries, which arise distinctly from the

posterior aspect of the trunk. Type III shows the right and left pulmonary arteries arising from the lateral parts of the root of the trunk. Finally, type IV, which is now reclassified as TOF with pulmonary atresia, shows both pulmonary arteries supplied by collateral vessels from the *aorta descendens*. Van Praagh updated this classification identifying a type A (*truncus arteriosus with VSD*) and a type B (*truncus arteriosus without VSD*). Van Praagh subdivided type A into four subtypes: A1 (the main pulmonary artery arises from the trunk and subsequently divides into right and left pulmonary arteries), A2 (the right and left pulmonary arteries to originate separately from the posterior part of the trunk), A3, and A4. In A3, one lung is supplied by a pulmonary artery branch that arises from the truncus, while a ductus-like collateral artery supplies the other lung. Finally, the subtype A4 is characterized by a truncus, which is a large pulmonary artery, and there is an aortic arch, which is interrupted or coarctation is existing.
- *EPI*: 1:10,000 newborns.
- *ETP*: Embryologically, the primitive truncus remains undivided into the pulmonary artery and aorta, i.e., there is a single, large, arterial trunk that overlies a large VSD, which is malaligned and perimembranous in localization.
- *CLI*: There is a mixture of oxygenated and non-oxygenated blood with resulting cyanosis and heart failure. The children show poor feeding, diaphoresis, and tachypnea. There is a typical first heart sound (S1) and a loud, single second heart sound (S2) and variable murmurs.
- *PGN*: Surgical management consists of complete repair. The VSD is closed so that the left ventricle ejects into the truncal root, and the continuity between the ventricle (right) and the confluence of the pulmonary arteries is permitted using a conduit with or without a valve.

1.2.7 Total Anomalous Pulmonary Venous Return

- *DEF*: Abnormal drainage of the oxygenated pulmonary blood flow directly or indirectly into the right atrium instead of the left atrium with mixing in the right atrium of systemic (hypoxygenated) and pulmonary (oxygenated) venous blood in association with either an ASD or a PDA and consequent right-to-left shunt and cyanosis with TAPVR: supracardiac (I), cardiac (II), infradiaphragmatic (III), or mixed (IV) types. The vascular connections are visualized in the Darling classification.
- *EPI*: 1.5% of all CHD, ♂:♀ = 3:1, isolated CHD in 2/3, but combined CHD in 1/3 of patients.
- *EPG*:
  - Persistence ≥1 veins that join to the right SVC left vertical vein/innominate (brachiocephalic) vein or umbilical/vitelline vein/portal vein is due to failure of the common pulmonary vein to connect with the pulmonary venous plexus.
  - Direct connection to the RA is due to either failure of the septum primum to usually form or abnormal septation of the *sinus venosus*.
- *CLI*: RA and RV enlargement due to the pulmonary venous return connecting to the systemic venous system, and in the case of pulmonary venous obstruction, RVH takes place. ASD or PDA are associated with defects that will allow life in maintaining a left ventricular outflow. Signs and symptoms include cyanosis, failure to thrive, tachypnea, and susceptibility to lower respiratory infections.
- *IMG*: RA and RV enlargement with lung edema and "8"/"Snowman"-CXR pattern and abnormal PPVVs connections using echo, angiography, and MRI.
- *EKG*: RVH signs.
- *TRT*: Surgery (reconstruction of the abnormal drainage RA → LA) and closure of ASD/PDA.
- *PGN*: CHF (RHF), arrhythmias, pneumonia/bronchopneumonia, and pulmonary HTN.

1.2.8 Scimitar Syndrome

- *DEF*: Hypoplastic lung syndrome due to abnormal venous return draining the right lower lobe of the ipsilateral lung entering the IVC (typically), the portal vein, the hepatic vein, or the right atrium instead of the left atrium and showing a small pulmonary artery

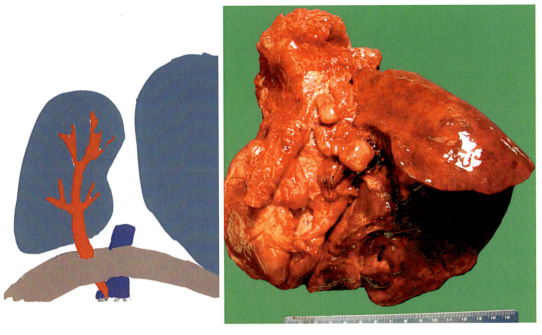
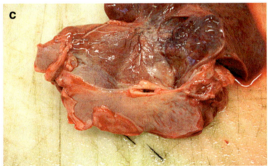
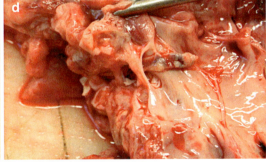

Fig. 1.16 In Scimitar syndrome, there is an association of congenital cardiopulmonary anomalies consisting of a partial anomalous pulmonary venous connection of the right lung to the inferior vena cava (**a**), right lung hypoplasia (**b**), dextroposition of the heart, and anomalous systemic arterial supply to the right lung. In this patient, the Scimitar syndrome was also accompanied by adhesion to the diaphragm (**c**) and thrombi (**d**) seen in the right and left pulmonary circulation

and potentially showing a less common lung systemic arterial supply (Fig. 1.16).
- *SYN*: Hypogenesis lung syndrome, hypogenetic lung, and pulmonary venolobar syndrome.
- *CXR/Echo*: "Scimitar shadow" (anomalous vein draining into the IVC).
- *DDX*: Pulmonary sequestration, which is a disconnected bronchopulmonary mass with systemic arterial supply (contralateral mediastinal shift, focal lung consolidation, more common systemic arterial supply, and less common abnormal venous drainage).

1.2.9 Hypoplastic Left Heart Syndrome and Coarctation of Aorta

1.2.9.1 Hypoplastic Left Heart Syndrome (HLHS)

- *DEF*: Life-threatening defect, which causes still 25% of infants' deaths due to multiple etiologies, including maternal, gestational, and familial/genetic/chromosomal conditions as well as fetal exposure to teratogens and active maternal infections. There is an abnormal development of the left-sided structures of the heart with blood flow obstruction at the level of the outflow tract of the left ventricular chamber. Although terminologically controversially debated, HLHS is probably a better-accepted term for this condition, including underdevelopment (hypoplasia/stenosis through atresia) of the mitral valve, aorta, and aortic arch. HLHS may show a considerable inconsistency in expressivity with variable degrees of left ventricular outflow tract obstruction (LVOTO), AV valve involvement, and aortic hypoplasia (Figs. 1.17, 1.18, 1.19, 1.20, and 1.21).

- *EPI*: 0.016–0.036% of all live births (≥ about 2:10,000).
- *CLI*: Full-term babies, apparently healthy (S_2 is loud and single) until the closure of arterial duct, when left atrial hypertension may develop quite quickly if there is a restriction of blood flow from the left to the right atrium due to an absent or inadequate atrial communication. There are also marked parasternal lift with poor peripheral perfusion, cyanosis, and missing or barely palpable pulses.
- *DGN*: CXR, EKG, and echo.
- *DDX*: Aortic stenosis, CoA, and IAA (structural cardiac diseases).
- *MNG*: Either heart TX or three-staged palliative procedures:
1. *Norwood* operation during the first week of life including the following steps:
   - The central PA is divided with the distal stump closed with a patch, and the DA is ligated.
   - One of the two alternatives to increase blood flow incrementing O_2 saturation is performed.
   - 1st: Right-sided modified Blalock-Taussig shunt (anastomosis between an SCA and RPA).
   - 2nd: Sano shunt (RVOT-PA conduit).
   - The atrial septum is distended, and the proximal MPA and hypoplastic aorta are joint together with an aortic or PA allograft/homograft (neo-aorta).

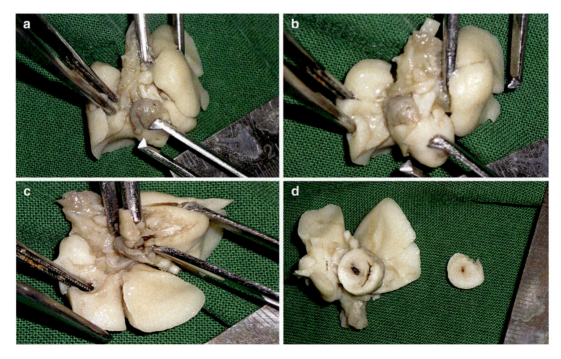

Fig. 1.17 Hypoplastic left heart syndrome (or better known as hypoplastic left heart). Figures (**a–d**) show a hypoplastic left heart in a fetus, and the rudimentary cavity is shown in (**d**)

1.2 Congenital Heart Disease

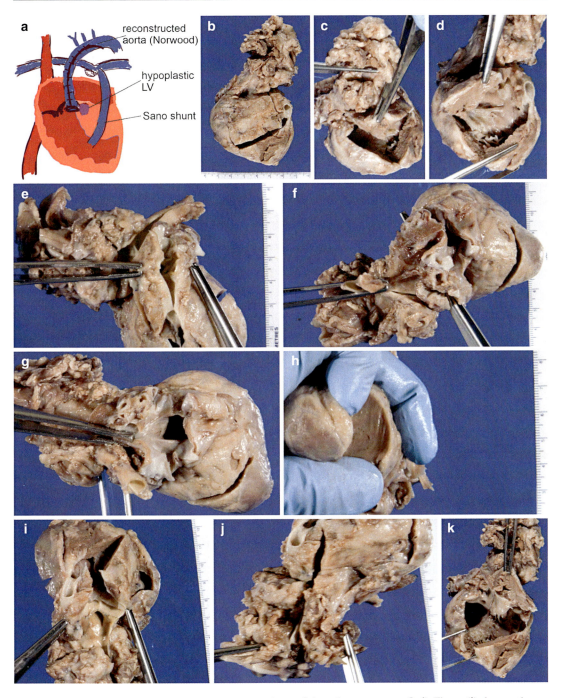

Fig. 1.18 This panel shows a cartoon (**a**) and several views of a hypoplastic left heart with status post Norwood and Glenn anastomosis between the superior caval vein and the pulmonary artery (**b–i**). Figure (**j**) shows a thrombus in the right coronary artery, while figure (**k**) shows the full right ventricular cavity

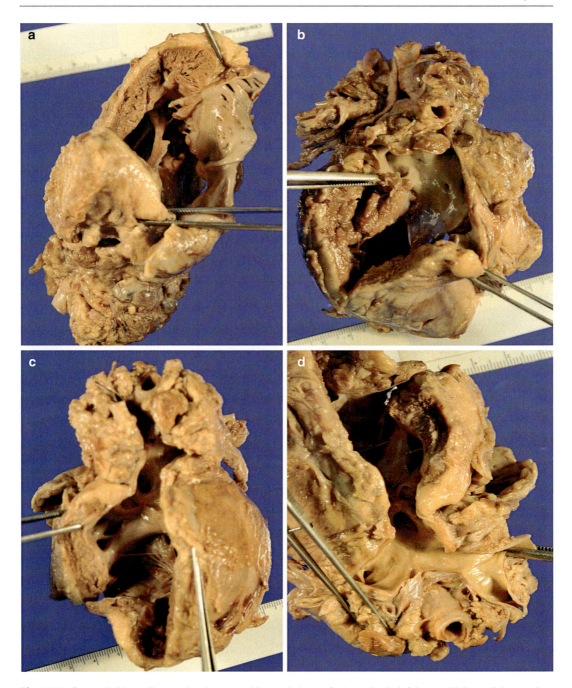

Fig. 1.19 Congenital heart disease showing several internal views of a hypoplastic left heart with large right ventricular cavity

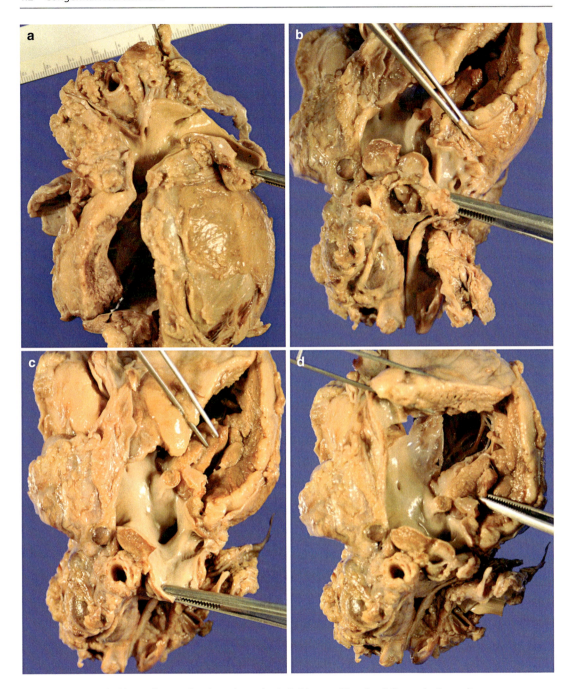

Fig. 1.20 Congenital heart disease showing a hypoplastic left heart with a tiny left ventricular cavity

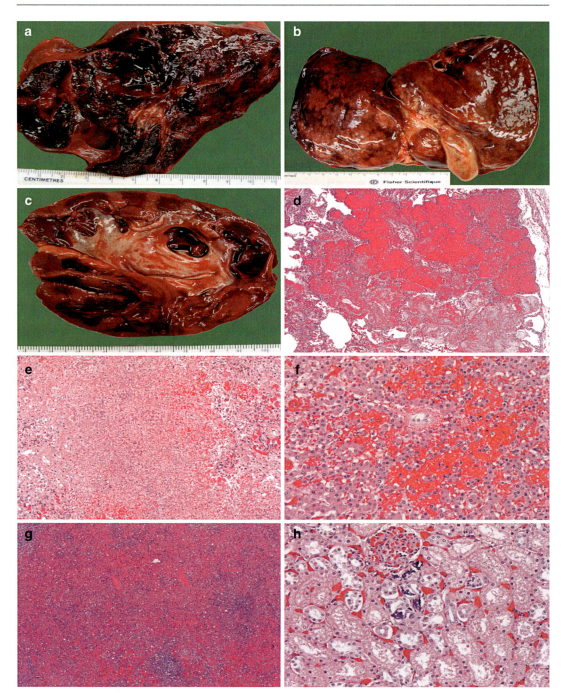

Fig. 1.21 Congenital heart disease with a hypoplastic left heart and pathologic changes identified in the organs. Several hemorrhages are shown in several organs from (**a**–**h**), including the lung (**a**), liver (**b**), kidney (**c**) on gross examination, and lung (**d**–**e**) with an hemorrhagic infarction, liver (**f**) with congestion of the sinusoids and rarefaction of the liver trabeculae, spleen (**g**) with congestion of the red pulp and relative reduction of the white pulp, and kidney (**h**) with hemorrhage and calcifications on histologic examination

2. *Bi-directional Glenn operation* (SVC to the RPA, end-to-side anastomosis) at 3–6 months.
3. *Fontan operation* (a baffle directs IVC flow within the RA into the lower part of the SVC that is divided or the RAA, which is connected to the PA; the upper portion of the SVC is jointed superiorly to PA) at 18–36 months.

- *PGN*: SR is 75% for stage 1, 95% for stage 2, and 90% for stage 3. Overall 5YSR: 70% after surgical correction. Factors: (1) low diastolic blood pressure in the aortic root is associated with decrease in coronary arterial perfusion (myocardial ischemia!), (2) bleeding with coagulopathy is a Norwood complication, and (3) neurodevelopmental disabilities (genetics or overt/occult CNS hypoperfusion due to thromboemboli). Recurrence risk in siblings is 0.5%, but accompanied by other forms of CHD, it raises to 13.5%. Genetic syndromes include Turner syndrome, Noonan syndrome, Smith-Lemli-Opitz syndrome, Holt-Oram syndrome, etc. Antenatal diagnosis is based on a fetal echo at 18–22 weeks of gestation.

1.2.9.2 Coarctation of the Aorta

- *DEF*: Narrowing of a section of the aorta most commonly between the left subclavian artery and the ductus arteriosus (*Isthmus aortae*) to an abnormal width. In the fetus and neonate, the aortic isthmus, the portion of the aorta between the left subclavian artery and the ductus arteriosus, usually is usually narrowed, and the lumen is about 2/3 of the ascending-descending parts of the developing aorta until 6–9 months of age when the narrowing indeed vanishes. Tubular hypoplasia is found in the aortic isthmus and is often referred to as preductal or infantile coarctation of the aorta. Interrupted aortic arch (IAA) is classified according to Celoria-Patton and can be subtyped in three types, A, B, and C. Type A involves interruptions distal to the left subclavian artery. Type B involves interruptions distal to the left common carotid artery, and, finally, type C involves interruptions distal to the innominate artery (proximal to the left common carotid arterial origin). Type A and type B have a similar frequency, despite some reports indicate that type B is the most common type, while type C is exceptionally uncommon. Each type is divided into three subtypes: 1: normal subclavian artery, 2: aberrant subclavian artery, and 3: isolated subclavian artery that arises from the *ductus arteriosus Botalli*. In CoA, there is a posterior eccentric fold (posterior shelf) inside the isthmus, opposite to the orifice of the arterial duct, which closing at birth or soon afterward, further narrows remarkably the aortic lumen. Bonnet classification: infantile or preductal (isthmic) form and adult or juxtaductal form (Fig. 1.22).
- *EPI*: 6% congenital cardiac anomalies, ♂:♀ = 3:1, ±Turner syndrome and accompanied by other defects in 50% of cases (PDA, bicuspid aortic valve, AS, ASD, VSD). Other associations include DiGeorge syndrome, *truncus arteriosus*, aortopulmonary septal "window", TGA, DORV, and functional single ventricle.
- *ETP*: The embryology of coarctation indicates that the fourth arch persists on the left side of the body to connect ventral aorta to dorsal aorta and form the aortic arch, while the sixth arch develops distally into *ductus arteriosus*. Coarctation of aorta results from the fourth and sixth arch. There are, commonly, two ETP hypotheses, including (1) the ductal tissue theory suggesting that the aortic narrowing is due to migration of ductal smooth muscle cell into the aorta and (2) the hemodynamic theory implying that the reduction is due to reduced flow through the aortic arch. The tubular hypoplasia of the aortic isthmus may be indeed related to blood flow paucity in the intrauterine life, which improves postnatally.
- *CLI*: From asymptomatic to congestive heart failure in early infancy with early symptoms of decreased exercise tolerance and fatigability in early childhood, pulse lag in lower extremities, ABP UEX > LEX (≥20 mmHg), and blowing systolic murmur in the interscapular area of the back (ABP, arterial blood pressure; UEX, upper estremities; LEX, lower extremities).
- *CXR*: Multiple signs including the "3" sign, rib notching of posterior third of ribs 3–8,

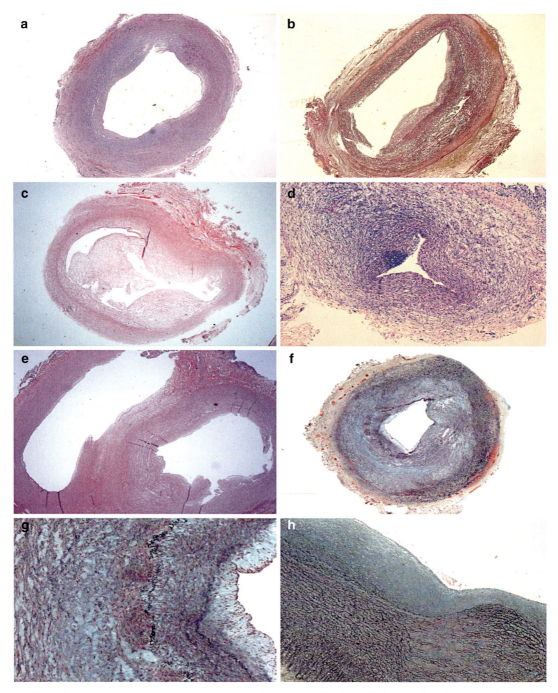

Fig. 1.22 Cross sections of several aortic specimens taken at several levels in different patients (**a–h**) showing marked fibrointimal hyperplasia and luminal stenosis

> **Box 1.3 Complications of the Coarctation of the Aorta**
> - Aortic Rupture
> - Bleeding of Extracardiac Type (e.g., Intracranial or Cerebrovascular Stroke)
> - Congestive Heart Failure
> - Disseminated (Sepsis) or Localized Infection (Endocarditis of Subacute Type)

"E"-shaped esophagus on barium swallow (the first arc of the E is due to dilatation of the aorta just proximal to the coarctation, the middle bar of the E is expected to coarctation itself, while the second arch of the E is due to post-stenotic dilatation of the aorta), infantile cardiomegaly with LV enlargement, and infantile pulmonary congestion.

- *EKG*: Normal to LVH signs ± RVH signs with or without congestive heart failure.
- *Echo*: Visualization of the CoA with data of pressure gradients, LVH, and anatomy details.
- *TRT*: Surgery but deferral of definitive correction may be opted in some cases using intravenous nitroprusside or propranolol to reduce the afterload.
- *PGN*: If not corrected, 60% die before age 40, often of aortic rupture (Box 1.3). The patients who survive the neonatal period without developing heart failure seem that they should do quite well throughout childhood and adolescence.

1.2.10 Heterotaxy Syndrome

- *DEF*: Abnormal arrangement of the internal organs in the chest and abdomen affecting the heart, lungs, liver, spleen, intestines, and other organs (έτερος, 'different' and τάξις, 'arrangement'). *Situs solitus* is the typical arrangement of the organs. The *situs inversus* is when the orientation of the internal organs is entirely flipped from right to left, and the *situs ambiguus* is when the direction of the internal organs is mixed. The affected organs may show three or two lung lobes bilaterally, two morphologically left/right chambers, right-located liver or located across the middle of the body, left/right stomach, *a*-splenia/*poly*-splenia, and malrotation of the intestines (Figs. 1.23, 1.24, and 1.25). Life is conditional on the organs involved. Signs and symptoms of heterotaxy syndrome can include cyanosis, breathing difficulties, an increased risk of infections, and food digesting difficulties. The most severe complications are generally determined by critical heart disease, a group of complex heart defects that are present from birth. Heterotaxy syndrome is often life-threatening in infancy or childhood, even with treatment, although its severity depends on the specific abnormalities involved.
- *SYN*: HTX, Ivemark syndrome, left/right isomerism, *situs ambiguus*, *situs ambiguus viscerum*, and visceral heterotaxy.
- *EPI*: 3% of all CHD, 1 in 10,000 people worldwide, and black/Hispanic > white.

1.2.11 Common Genetic Syndromes with Congenital Heart Disease

The development of the heart characterizes the period ranging from the third and the seventh week of gestation. In this time, complex folding and looping of the described initially as '*horseshoe*'-shaped tube give rise to the four-chambered cardiac pump; there are also some other organs that are forming and may be involved in phenotype leading to common congenital syndromes. In this period, both exogenous factors (e.g., rubella, CMV, alcohol, drugs, cannabis, cocaine, etc.) and intrinsic factors (genes and epigenetics) may play an etiopathogenetic role. In about 1/5 of all children born with a cardiac defect, there are extracardiac defects. A chromosomal, subchromosomal (microdeletion), or genetic disorder (monogenic), which is either inherited from the biologic parents or occurred spontaneously, may be the underlying common factor. Indeed, a genetic cause can be supposed if a cardiovascular defect or syndrome occurs frequently or recurrently in one family. Numerous reviews are highlighting congenital syndromes with congenital heart disease. Here, some of the

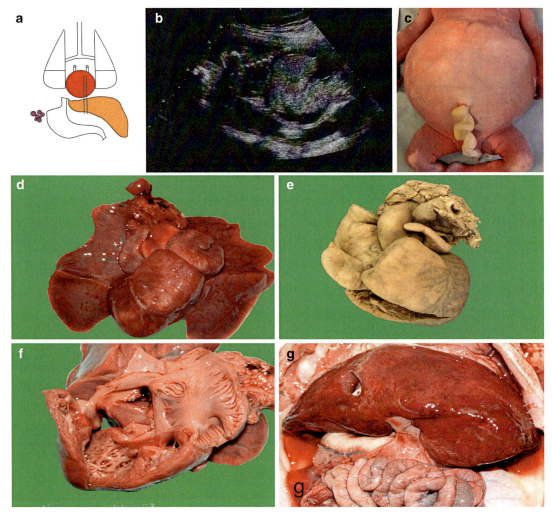

Fig. 1.23 Ivemark syndrome in a child with heterotaxy syndrome (**a**) showing a common atrium (**b**), left loop and unbalanced atrioventricular septal defect. There is malrotation or dislocation of the internal organs with prominent abdomen (**c**). The morphology of the heart is abnormal harboring a common atrium (**d–e**) as better shown on (**f**). In (**g**), the liver is located on the left side of the abdominal site

most critical congenital syndromes are summarized (Figs. 1.26 and 1.27). More congenital syndromes are also described in other chapters of this book.

- *Trisomy 13 syndrome* (Patau syndrome): 1:10,000 births, small for date, severe developmental defects of the brain, heart (80% of patients, mostly VSD, heterotaxy), urinary, and gastrointestinal systems.
- *Trisomy 18 syndrome* (Edwards syndrome): 1:3000/1:10,000, small for date, and severe developmental defects of the brain, eyes, palate, heart (>90% of patients have CHD, mostly VSD), urinary, and gastrointestinal systems.
- *Trisomy 21 syndrome* (Down syndrome): 1:650 births, appropriate for the delivery date, and mild to severe developmental defects of the heart (up to 50% of patients, mostly AVSD, ASD, VSD) and gastrointestinal system (atresia, aganglionosis, or Hirschsprung disease).
- *Monosomy X0 syndrome* (Ullrich-Turner syndrome): 1:2000 births (female, 45, X0) with

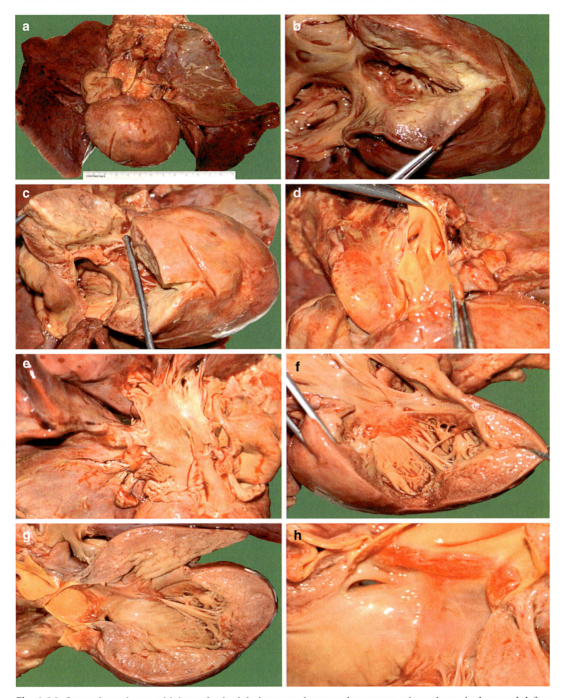

Fig. 1.24 Ivemark syndrome with hypoplastic right heart syndrome, pulmonary atresia, and ventricular septal defect as shown in several views of the heart dissection (**a**–**h**)

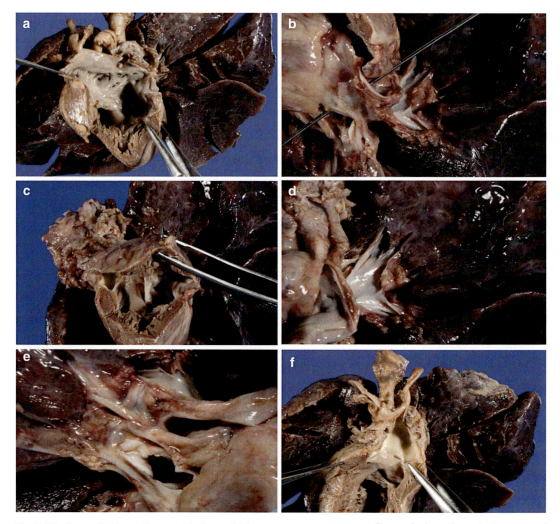

Fig. 1.25 Congenital heart disease with right atrial isomerism, double outlet right ventricle, balanced atrioventricular septal defect, and total anomalous pulmonary venous return as well as asplenia as shown in these several views of this heart dissection (**a–f**)

urogenital defects (underdeveloped "streak gonads") and cardiovascular defects (aortic isthmus stenosis).

- *1p36 deletion syndrome* (1p36 monosomy): 2/10,000 births, severe disruptive psychosocial behavior with cerebral seizures and self-destructive phenotype, general hypotonia, feeding problems, vision disturbances and hearing loss, and cardiovascular defects in 40% of the affected children (PDA, TOF, ASD, VSD, ECA), who may start or evolve into dilated cardiomyopathy (DCM).
- *22q11.2 microdeletion syndrome* (DiGeorge and Shprintzen syndrome): 1:4000–1:6000 births with cardiovascular defects other than somatic (85% RVOT defects, including the common trunk, VSD, TOF), thyroid, and parathyroid gland defects with resultant disorders of calcium metabolism, thymus agenesis, and psychological development.
- *4p- syndrome* (Wolf-Hirschhorn syndrome): 1:50,000 births, loss of a short arm of one of the chromosomes 4, 1:50,000 births, VLBW infants with microcephaly and typical facial

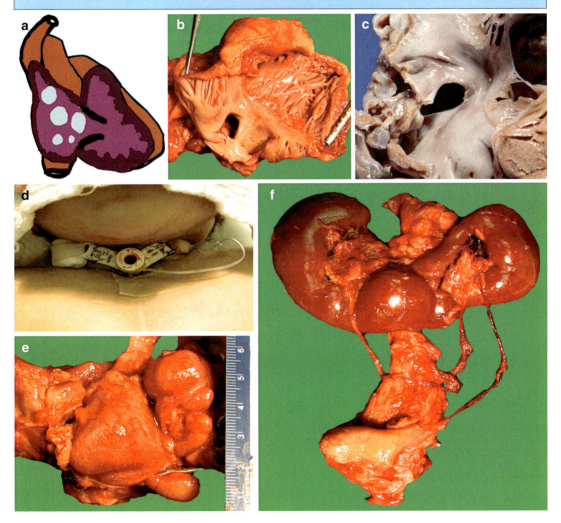

Fig. 1.26 VACTERL association is an association of multiple congenital anomalies including vertebral anomalies, anal atresia, cardiac defects, tracheoesophageal fistula, and/or esophageal atresia, as well as renal and radial anomalies and limb defects. In (**a**) are shown several types of atrial communication in a cartoon and two heart specimen preparations, fresh (**b**) and post-formalin fixation (**c**), showing an atrial septum defect of type II. In (**d**) is shown a tracheostomy, which is often necessary in these patients. Moreover, these children may develop thoracal lymphomegaly that may contribute to breathing difficulties (**e**). In (**f**) there is joining of the lower poles of both kidneys ("horseshoe kidney")

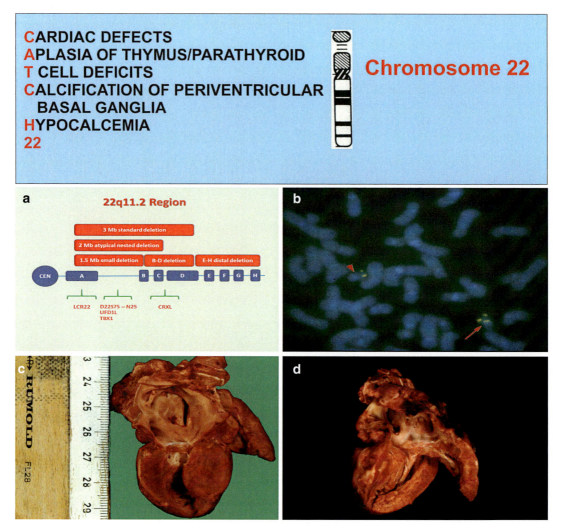

Fig. 1.27 Catch 22 syndrome is a genetic syndrome including cardiac defects, abnormal facial features with cleft palate, thymic hypoplasia with T-cell deficiencies, calcification of periventricular basal ganglia, and hypocalcemia. The second letter "C" of the acronym is also often used to designate the cleft palate. Catch 22 syndrome results from a deletion within chromosome 22q11 and includes DiGeorge syndrome, conotruncal anomaly face syndrome, and velocardiofacial syndrome. In (**a**) is shown the 22q11.2 region, while (**b**) shows a characteristic FISH of a patient harboring Catch 22 syndrome. The normal chromosome 22 shows a hybridization signal in band 22q11.2, the Catch 22 region, being marked by an arrow. No signal is observed in the other chromosome 22, indicating a microdeletion of the region 22q11.2 (arrow head). Both chromosomes 22 reveal signals for the control region. Hybridization on the distal long arm 22q13.3 was used as control. Karyotype (500 G- and Q-banding): 46, XX.ish del (22) (q11.2 q11.2) (D22S75-). In (**c, d**) there is the preparation of the heart showing tetralogy of Fallot, a frequent congenital heart defect of children affected with Catch 22 syndrome

features with hypertelorism, eyelids that slant down, broad nose, micrognathia, turned-down mouth, and ear tags or pits, mental retardation with seizures, failure to thrive. CHD includes ASD or VSD. 1-YSR < 30%.

- *5p- syndrome* (cri du chat syndrome, Lejeune syndrome): 1:50,000 births, failure to thrive, cardiac defects (50% of the patients with ASD, VSD, and complex CHD), microcephaly, and mental retardation.

- *CHARGE syndrome*: 0.1–1.2/10,000 births with *c*oloboma, *h*eart defects (80% with TOF, aortic arch defects, ASD, and VSD), *a*tresia of the nasal choanae, *r*etardation of growth and psychological development, *g*enital and urinary abnormalities, *e*ar abnormalities and deafness.
- *Ellis-van Creveld syndrome* (ECS) is a rare genetic disorder with very few cases reported affecting genes on the short arm of chromosome 4. Multiple defects include short ribs, polydactyly (hands and feet), cleft palate, short-limbed dwarfism, narrow chests, and CHD (60%) including a VSD. Life expectancy is directly linked to the respiratory difficulties caused by the tight chest and extension of the VSD.
- *Holt-Oram syndrome* (HOS) is a genetic disease (1:100,000 births) with spontaneous or hereditary mutations affecting the hand-heart axis (radius/thumb-heart), as well as of the heart. The HOS section of the gene is on chromosome 21. CHD (70%) includes ASD and VSD as well as cardiac arrhythmias. Life expectancy is directly linked to the extension of the involvement of CHD.
- *Jacobsen syndrome* is a rare chromosomal disorder (1:100,000 births) with deficiency of part of the long arm of chromosome 11 and includes typical features including a triangle-shaped head, retrognathia, hypertelorism, iris coloboma, and low-set ears as well as thrombocytopenia and mental retardation. CHD (55%) includes mostly HLH. Life expectancy is directly linked to the thrombocytopenia and extension of the VSD.
- *Marfan syndrome* is a genetic syndrome (1:5000–1:10,000 cases) with a mutated gene on the q arm of chromosome 15 affecting the development of fibrillin, which is a significant component of the microfibrils, which form a sheath around the elastin fibers of the connective tissue. There is an instability of the body's connective tissues with abnormally long limbs, joint flexibility, the curvature of the spine, as well as vision disorders. CHD is limited to aneurysms and tears of the aortic wall and cardiac valve defects. Life expectancy is directly linked to the regular monitoring and prompt treatment of CHD because these patients without therapy have a life expectancy of about 30 years.
- *Noonan syndrome* is the most frequent monogenetic cause of heart defects (1:500–1:2500) with mutations in the *PTPN11* gene causing about 50% of all cases (*SOS1* gene mutations cause an additional 15%, and *RAF1* and *RIT1* genes each account for about 5% of cases, mutations in other genes account for a small number of cases). The four genes provide instructions for making proteins that are important in the RAS/MAPK cell signaling pathway, which is required for cell division and proliferation, differentiation, and migration. Typical features include short stature, triangular face shape with hypertelorism, drooping of the eyelids, and a wide neck that accompany mental retardation. CHD (80%) embraces pulmonary stenosis or HCM. Life expectancy is directly linked to the extension of CHD.
- *VACTERL association* (1:5000 births) is a genetic disorder including V for vertebral anomalies, A for anal atresia, C for congenital heart disease, T for tracheoesophageal fistula, E for esophageal atresia, R for renal and lower urinary tract anomalies, and L for radial limb defect. CHD includes VSD. Life expectancy is directly linked to the extension of the abnormalities involved in this genetic association.
- *Williams-Beuren syndrome* is a genetic disorder (1:10,000 births) caused by the deletion of genetic material from the q arm of chromosome 7 (q11.23) when the gene coding for elastin is missing (*CLIP2*, *ELN*, *GTF2I*, *GTF2IRD1*, and *LIMK1* are the genes frequently deleted). There is reduced compliance of the blood vessels with CHD (80%) including supravalvular aortic stenosis or stenosis of the pulmonary artery as well as VSD. Typical features include a deep hoarse voice, noise sensitive, and vision defects, in addition to mental retardation. Life expectancy seems to be unaltered by CHD.
- *Alcoholic embryopathy* (1:1000 births) is due to alcoholic poisoning of the unborn

child using alcohol (any beverage) by the mother during gestation. The affected children are SGA and low-weight with microcephaly and damage to the brain and extracerebral defects. CHD (30%) includes VSD. Life expectancy depends on the deterioration of the organs.

- *Cannabis (marijuana)-associated CHD* is associated with the use of marijuana as identified by well-described case-control studies and is going to reach the levels of early embryopathy once the legalization is achieved in most of the countries. The Baltimore-Washington Infant Study reported an odds ratio of 1.36 (95% CI = 1.05–1.76) for isolated membranous VSD and attributable risk of 7.8% for TGA associated with the use of marijuana in fathers (not in mothers!) (Ewing et al. 1997).
- *Rubeola embryopathy* is due to rubella virus-infected pregnancy (1:10,000 newborns). The defects include damage to the inner ear, heart, eyes, and other organs, which frequently results in premature births and IUFD. CHD (50–80%) includes PDA, stenosis of the pulmonary or main arteries, ASD, and VSD. The vaccination before pregnancy does grant protection against rubeola embryopathy.

1.3 Myocarditis and Nonischemic Cardiomyopathies

1.3.1 Myocarditis

- *DEF*: It is a nonischemic inflammatory process of the myocardium, which is characterized by necrosis and/or degenerative changes of cardiomyocytes with accompanying inflammatory infiltrate. Some adapted criteria for myocarditis involve the recruitment of 5 lymphocytes/HPF (i.e., at 40×) in 20 fields (Figs. 1.28 and 1.29). However, variations of these criteria are present in the literature. Substantially, the Dallas criteria, although critically debated, are still valid. The immunohistochemistry is, currently, key and some proposed criteria suggest more than 14 leukocytes/mm^2 (Aretz, 1987; Caforio et al. 2013; Sobol et al. 2020). Noticeably, there is both inflammation (usually lymphocytic) and cardiomyocyte necrosis, and the consequence of it gives rise typically to a dilated cardiomyopathy.
- *ETP*: Coxsackie A and B, ECHO, polio, influenza A and B, HIV, *Trypanosoma cruzi* (Chagas disease), bacterial, fungal, parasitic, spirochetes (Lyme disease), *Chlamydia* spp., collagen vascular disease, radiation, heat stroke, sarcoidosis, and Kawasaki disease.

1.3.2 Dilated Cardiomyopathy

- *DEF*: Disorder involving the myocardium without a primary ischemic component evolving in a dilation of the heart (all four chambers usually enlarged). The affected individuals may have foci of apparent ischemic necrosis in myocardium due to catecholamine release and myocarditis is often considered the most common "overture" to dilated cardiomyopathy (DCM). Myocarditis as early clue for DCM may have four different clinico-pathologic patterns, including fulminant, acute, chronic-active, and chronic-persistent. Viral and bacterial infections, Lyme disease, *Trypanosoma cruzi* and non-infectious etiologies (i.e., cardiotoxic drugs, sarcoidosis, collagen vascular diseases, and rheumatic fever) may start the "inflammatory" process leading to dilation of the heart (DCM). (Figs. 1.30, 1.31, and 1.32).
- *EPI*: Despite recent advances in therapies in the twenty-first century, DCM remains a significant cause of morbidity and mortality and is a leading indication for cardiac transplantation (6:100,000 persons/year).
- *EPG*: EtOH, thiamin deficiency, cobalt toxicity, pheochromocytoma, and hemochromatosis. According to the WHO/ISFC criteria, cardiomyopathies are defined as diseases of the myocardium associated with cardiac dysfunction. They are classified as dilated, restrictive, and hypertro-

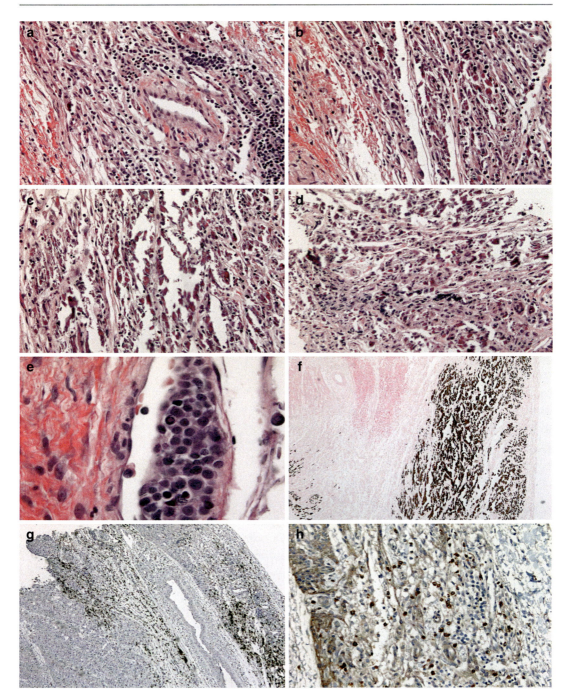

Fig. 1.28 The infiltration of the myocardium by lymphocytes with necrosis of the cardiomyocytes is shown in these microphotographs (**a–h**) (×200). Some nuclear changes with eosinophilic inclusion and peripheral location of chromatic are demonstrated in figure (**e**) at 630× as original magnification. A RT-PCR confirmed the histologic finding. The figure (**f**) illustrates some calcifications highlighted with the von Kossa special stain (50×), while macrophages (**g**) are highlighted by immunohistochemistry using an antibody against CD68 (×50) and lymphocytes (**h**) using an antibody against CD3 (×200)

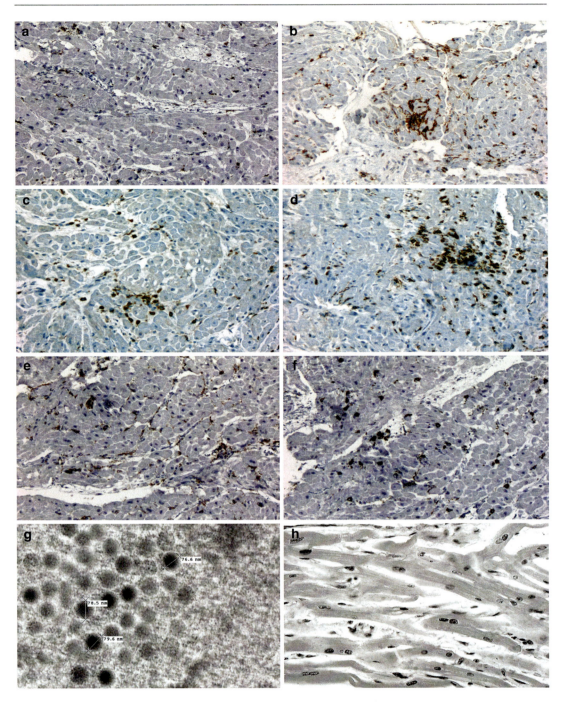

Fig. 1.29 This panel shows adenovirus myocarditis with some macrophages using antibodies against CD68 (**a**, ×200) and the recruitment of determined subpopulations of lymphocytes using CD3 (**b–d**, ×200), CD4 (**e**, ×200), CD8 (**f**, ×200), and the identification of the adenovirus by electron microscopy (**g**), marked intercellular edema with some disarray in myocarditis (**h**, ×400)

Lipoteichoic acids N-acetyl-B-D-glucosamine
=> Ficolin-2 and MBL => Complement activation
                                          and phagocytosis

Streptococcus pyogenes

Proteins **Genes**
High MBL levels MBL2
Low Ficolin-2 levels FCN2

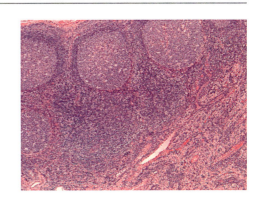

RHEUMATIC HEART DISEASE

due to the crossreactivity of the anti-streptococcal antibodies and T cells with the antigens in cardiomyocytes and valves

Streptococcal Pharyngitis

Pericardial Effusion

Endocarditis

Myocarditis with Ashoff bodies

Mitral valve
Aortic valve
Tricuspid valve

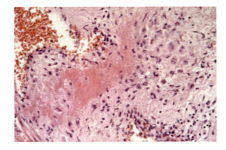

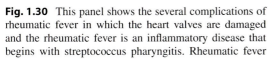

Fig. 1.30 This panel shows the several complications of rheumatic fever in which the heart valves are damaged and the rheumatic fever is an inflammatory disease that begins with streptococcus pharyngitis. Rheumatic fever can affect connective tissue throughout the body, especially in the heart, joints, brain, and skin. In the heart, all three layers are involved with presence of endocarditis, myocarditis, and pericarditis

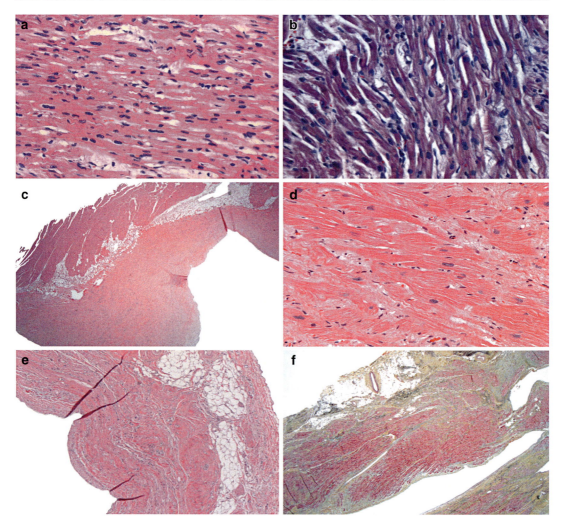

Fig. 1.31 There is interstitial fibrosis as well as cardiomyocyte hypertrophy with fibroadipose metaplasia, and the fibrosis is yellowish (old) on Movat staining. Myocyte loss with fibrotic and/or fibrofatty replacement is the pathologic hallmark of arrhythmogenic right ventricular (RV) dysplasia/cardiomyopathy (ARVD/C). However, the 2010 Task Force Criteria (TFC) do not support the use of either feature for the diagnosis of ARVD/C. Fibro-fatty infiltration assessment in the RV is highly subjective, lacking reproducibility and specificity. ARVD/C usually affects the RV only and manifests in ventricular arrhythmias and RV dysfunction. Diagnosis remains challenging because of the ARVD/C nonspecific findings. TFC were initially proposed in 1994 and subsequently revised in 2010

phic cardiomyopathy. Arrhythmogenic right ventricular cardiomyopathy has been added as an entity of its own. Although most DCM is sporadic, DCM may occur isolated or associated with additional conduction and muscular disorders. Autosomal recessive, mitochondrial, and X-linked DCM have also been described. In isolated autosomal dominant DCM, null and missense mutations were identified in eight different genes, including the α-cardiac actin (*ACTC*), desmin (*DES*), δ-sarcoglycan (*SGCD*), metavinculin (*VCL*), titin (*TTN*), troponin T (*TNNT2*), b-myosin heavy chain (*MYH7*), and a-tropomyosin 1 (*TPM1*). ACTC, DES, SGCD, and VCL are needed for the force transmission from the sarcomere to adjacent sarcomeres and the extra-

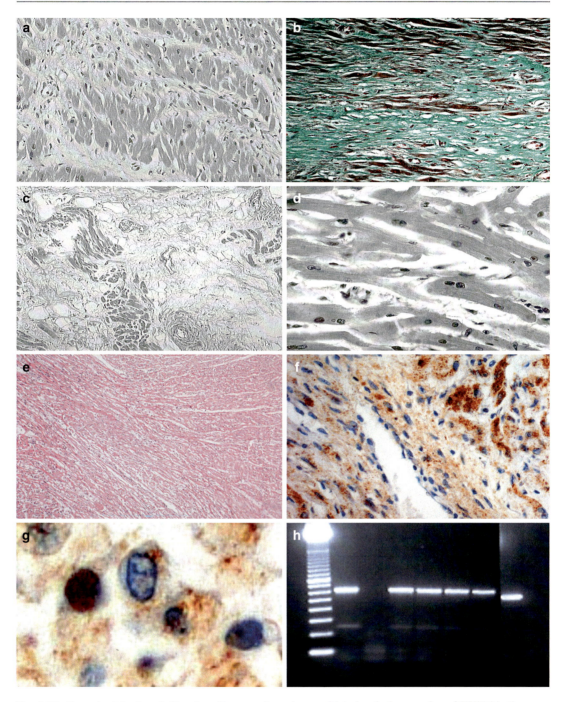

Fig. 1.32 Several etiologies of dilated cardiomyopathy (DCM). HIV cardiomyopathy is shown in figures (**a–d**) with interstial edema and mild to moderate hypertrophy of the cardiomyocytes (**a**, 160×; **b**, 160×; **c**, 100×; **d**, 320×). HHV-6 myocarditis/cardiomyopathy is shown in figures (**e–h**). Figures (**e, f**) show the dilative changes of the myocardium with remarkable interstitial edema and some cardiomyocyte hypertrophy. Figure (**f**) shows the immunohistochemical expression of HHV-6 in the myocardium. Figure (**g**) shows in high magnification the nuclear localization of the virus (×630). HHV-6 detection can be performed by in situ hybridization or by immunohistochemistry with an antibody against the virus proteins. Figure (**h**) shows the molecular biology confirmation using cardiac tissue specimens by amplifying viral sequences by polymerase chain reaction

cellular matrix, whereas TTN, TNNT2, MYH7, and TPM1 genes are essential for the force production generated by the sarcomere. In non-isolated DCM, lamin A/C (LMNA) gene mutations have been shown responsible for DCM associated with conduction system disease and muscular disorders. A familial origin may be identified in ~25% of cases. Various subtypes of familial DCM have been reported and characterized. Although not perfect, some clues for a familial DCM may be deducted by the younger age at diagnosis, higher ejection fraction, higher frequency of ST-T segment abnormalities on the EKG, and a higher frequency of AV conduction defects than sporadic DCM cases.

- *GRO*: Ventricular chamber enlargement, due to ventricular dilation with mild or minor hypertrophy, depressed myocardial contractility, and increased heart weight with the ventricular wall thickness generally standard because of the effect of dilation. Subendocardial and transmural scarring can follow even in the absence of thromboembolic obstruction of the blood supply through the coronary arteries, and, occasionally, interstitial, perivascular, and endocardial fibrosis are prominent.
- *CLM*: Cardiomyocyte degeneration, in particular sarcoplasmic degeneration, with irregularity in the hypertrophy of myofibers, and the occurrence of systolic contractile dysfunction, has been linked to these changes. The role of TEM in cardiac biopsies has been evaluated with etiologies elucidating cases of amyloid, granulomatous inflammation, and basophilic degeneration. TEM may have a potential to identify pathways and address therapies when other organs are involved clarifying glycogen, lipid, and mitochondrial accumulations and varying types of degeneration affecting myofibers. However, TEM, if used as a routinely ancillary technique with non-expert personnel, routing information, may be collected that may not be satisfactorily interpreted.
- *PGN*: DCM is also associated with high rates of SCD due to a ventricular arrhythmia (mainly VT) that may occur at any stage. 5-YSR: 30–50%.

Single nucleotide polymorphisms (SNPs) are DNA sequence variations that occur when a single nucleotide A (adenine), C (cytosine), T (thymine), or G (guanine) in the genome sequence is changed. Manual and automatic sequencing of the human genome can be used to identify single DNA variations or SNPs. Genome variations include mutations and polymorphisms that may be distinguished by frequency, in addition to the association with disease. For example, if a location in the genome where 96% of people have an adenine and the remaining 4% have thymine is defined as a *polymorphism*. If one of the potential sequences is present in less than 1% of the population (99.9% of people have a T, and 0.1% have an A), then the DNA variation is properly called a *mutation*. For a genetic variation to be considered an SNP, it must occur in at least 1% of the population. However, a word of caution should be spent in dealing with polymorphisms, because there are very conflicting data and one of the reasons relies on the heterogeneity of the sample of the population studied as well as the techniques used in identifying the polymorphisms (Corfield et al. 2010; Stambader et al. 2010).

1.3.2.1 Angiotensin I-Converting Enzyme (*ACE*) Gene

Angiotensin-converting enzyme (*ACE*) gene is one of the most frequently studied genes affecting idiopathic DCM. The ACE gene is localized on chromosome 17q23 and is characterized by a major insertion/deletion polymorphism resulting in three genotypes (DD, ID, and II), which affects serum and tissue ACE activity as well as other vasoactive substances. ACE is an ectoenzyme found on the external surface of the endothelial and epithelial cell membranes. It enhances the synthesis of angiotensin II, which is liberated from angiotensinogen (AGT) by the sequential action of renin and ACE. Angiotensin II promotes proliferation, migration, and hypertrophy of vascular smooth muscle cells. Moreover, the increased free radical generation by angiotensin II contributes to endothelial dysfunction (endothelial cell damage). A genetic polymorphism in the *ACE* gene has been found to have a strong association with higher risk for acute coronary events, SCD, vascular restenosis after angioplasty, and idiopathic CM and HCM. Moreover, the *RAS* gene is one of the important factors regulating blood pressure, as well as fluid and electrolyte balance. It may have a crucial role in the pathogenesis of HTN. The replication of epidemiologic-genetic studies and the creation of adequate experimental studies will help to definitively establish the pathogenetic role of the permanent increase in *ACE* expression associated with the deletion polymorphism genotype.

1.3.2.2 Angiotensin II Type 1 Receptor (*AGTR1*) Gene

The *AGTR1* was found selectively downregulated in failing left ventricle from patients with end-stage heart failure due to IDC, suggesting that the failing human heart is exposed to increased concentrations of angiotensin II at the cellular level. Genetic variation in the *AGTR1* gene might then influence susceptibility to CM or progression of the disease. The selection of two polymorphisms (A-153G and A + 39C) was based on the findings of susceptibility to myocardial infarction and essential HTN. However, others did not find such an association.

1.3.2.3 Angiotensinogen (*AGT*) Gene

Angiotensinogen (AGT) is the precursor of angiotensin I. Its concentration is rate-limiting for the generation of angiotensin II. A study revealed that a CYP11B2 gene variant but neither *ACE* nor *AGT* gene variants examined predicted improvement in LVEF measured after initiating medical therapy with furosemide, digoxin, and ACE inhibitors in patients of African ancestry with IDC. An association between the M235T and T174M polymorphisms of the *AGT* gene and essential HTN has been reported.

1.3.2.4 Tumor Necrosis Factor-α (*TNF-α*) Gene

An increase of TNF-α in the heart can determine lethal pump failure. The level of TNFα production is, in part, determined by promoter gene polymorphisms. There is no evidence to support an association between *TNF2* and *TNFα2* alleles and the development of Chagas disease, but it has been shown that patients, positive for *TNF2* or *TNFα2* alleles, have a shorter survival time compared to those carrying other alleles. Serum TNF-α level of patients with idiopathic DCM was found to be higher than those of the controls. These data indicate that the *TNFA2* allele has the potential to produce high levels of TNF-α under stimulated conditions. Although controversial data have been published, at least in the Japanese population, the *TNFA2* allele seems to be associated with the pathogenesis of idiopathic DCM, at least in some cohorts of patients.

1.3.2.5 Phospholamban (*PLN*) Gene

Phospholamban (PLN) is a protein that controls the sarcoplasmic reticulum Ca^{2+} pump and controls the size of the sarcoplasmic reticulum Ca^{2+} store during diastole. Abnormal homeostasis of the calcium ion is a prototypical mechanism for contractile dysfunction in failing cardiomyocytes. Depressed cycling of this ion in experimental and human heart failure reflects, at least in part, impaired calcium sequestration by the smooth reticulum. It has been evidenced that increases in the relative levels of PLN to Ca^{2+}-ATPase in failing hearts and resulting inhibition of Ca^{2+} sequestration during diastole impair contractility. The g.203A4C genetic variant in the human PLN promoter may contribute to depressed contractility and accelerate functional deterioration in heart failure. Estimates from previous reports evaluate the frequency of familial DCM causing mutations in PLN to 5%.

1.3.2.6 Bone Morphogenetic Protein-10 (*BMP10*) Gene

Bone morphogenetic protein-10 (BMP10) interacts with a protein called titin-cap (TCAP). BMP10 is localized on the cell surface and at the stretch-sensing Z disc of cardiomyocytes. There is an association between the variant of the human *BMP10* gene, Thr326Ile, with susceptibility to hypertensive DCM. The variant BMP10 showed decreased binding to Tcap and increased extracellular secretion of BMP10. BMP10 is an adherent component of the TGF-ß family. It has been shown that BMP10 possesses pro-hypertrophic activity and is upregulated in hypertensive cardiac hypertrophy.

1.3.2.7 Titin/Connectin Gene

Matsumoto et al. reported in an earlier study several mutations in the titin/connectin gene found in patients with DCM or HCM. A titin/connectin mutation (Arg740Leu) was found to increase the binding to actinin, while other titin/connectin mutations (Ala743Val and Val54Met) decreased the binding to actinin and Tcap/telethonin. Since the N2-B region expresses only in the heart, it was speculated that functional alterations due to the mutations cause cardiomyopathies. These findings indicate that N2-B region mutations may cause CM through dysregulation operated on the recruitment of metabolic enzymes.

1.3.2.8 Human Leukocyte Antigen (*HLA*) Gene

HLA-DQA1 0501 and DQB1 0303 seem to be related to the genetic susceptibility to IDC. DQA1 0201, DQB1 0502, and DQB1 0504 alleles confer protection from IDC. It has been shown that HLA-DQ polymorphisms may be useful as genetic markers for IDC and are probably involved in the regulation of immune-specific response to auto- or exterior anti-myocardium antibody. However, these results have been challenged by others being turned out quite variable, due to technical problems, a low number of samples, and lack of adequately matched controls. Ethnic variability is also a ponderable factor in Latin American countries where these studies have been initially carried out. The association of IDC with HLA-DRB1 1401 in nonfamilial Japanese patients seems to be stronger than the original studies.

1.3.2.9 Alpha2C-Adrenoceptor Gene

The 2C-adrenoceptor deletion may be a relatively novel, strong, and independent predictor of reduced event rates in DCM patients treated according to guidelines (99% ACEI, 76% b-blockers). DCM patients with the deletion variant Del322–325 in the a2C-adrenoceptor showed significantly decreased event rates. Genetic variation in the a2C-adrenoceptor gene (a2CDel322–325) is independently associated with survival and absence of events in cohorts with severe heart failure due to dilated DCM. Regitz-Zagrosek et al. suggest that the a2C-adrenoceptor gene (a2CDel322–325) polymorphism is an important and independent genetic factor that determines survival in patients with advanced DCM. Caucasian patients with the a2C322–325 deletions have a functional advantage in the presence of an equally impaired left ventricular function in comparison with the "wild types."

1.3.2.10 β1-Adrenoceptor Gene

The β1-adrenergic receptor (b1-AR) is a G protein-coupled receptor expressed in the heart and other tissues, acting as a receptor for catecholamines. Coding and promoter polymorphisms of this receptor have been identified in the general population. The variants β(1)Ser49, β(1)Arg389, and α(2c)Del322–325 were found in negative association with the susceptibility to risk factors for chronic heart failure due to DCM. However, there is a strong debate and probably evidence that α(2c)Del322–325 variant might be protective. Moreover, the Arg389 allele frequency in the Japanese population is similar to that of a Scandinavian group but a little lower than that in the French group.

1.3.2.11 β2-Adrenoceptor Gene

The β1- and β2-adrenergic receptors are G protein-coupled receptors for the catecholamines, epinephrine, and norepinephrine. β2-adrenoceptors (AR) play an essential role in the regulation of vascular and bronchial smooth muscle tone. They also exist in the human heart and contribute to the regulation of heart rate and contractility. In a study of 259 patients with chronic heart failure (CHF) due to ischemic or DCM, patients harboring the heterozygous status for Thr164Ile-β2AR exhibited a rapid and inexorable progression to death or heart transplantation (HTX).

1.3.2.12 Beta-Myosin Heavy Chain Gene

Four genes involved in familial DCM in a population of idiopathic DCM were investigated in 96 independent patients, of which seven mutations were found in *MYH7*, one in *TNNT2* and none in *PLN* or the *VCL* cardio-specific exon. In HCM patients, more than 80 different mutations have been found to date in *MYH7*. These findings confirm the genetic heterogeneity of FDCM, with a prominent role of *MYH7* in DCM with a 10% mutation frequency in familial forms and delayed onset of the disease in the studied population and a low occurrence of *PLN* and *VCL* mutations, a 2% frequency of *TNNT2* mutation (R141W is associated with high penetrance and early onset) (Stambader et al. 2010).

1.3.2.13 Myosin Binding Protein-C (*MYBPC3*) Gene

Myosin binding protein-C (MyBP-C or *MYBPC3*) is one of the sarcomeric proteins. Mutations in the *MYBP-C* gene, on chromosome 11, are a frequent cause of HCM. Konno et al. (2003) detected an Arg820Gln missense mutation in the *MYBP-C* gene in eight probands. This sequence variant was found in clinically affected patients and was absent in 100 normal controls. Konno et al. (2003) suggest that the Arg820Gln missense mutation in the *MYBP-C* gene may be associated with the disease. Elderly carriers with the Arg820Gln missense mutation may show LV systolic dysfunction and dilation. The Arg820Gln missense mutation in the *MYBP-C* gene is associated with variable clinical features, and the clinical expression of this mutation is often delayed until middle age. The screening of patients with DCM and HCM for this mutation is of vital importance.

1.3.2.14 Cardiac Troponin I (*TNNI3*) Gene

The main function of the troponin complex is to regulate muscle contraction and relaxation. This regulation is mediated via conformational changes of the I, T, and C complexes, which are induced by the variation of intracellular calcium ion concentration. Researchers, investigating the N-terminus of cTnI, have identified important interaction sites with both cTnC and cTnT. *TNNI3* is the first recessive disease gene identified in DCM. This impairment leads to diminished myocardial contractility and disease. The mutation in this gene could cause disease because this genetic variant was not found in 150 controls. *TNNI3* mutations in DCM are rare, but other recessive disease genes might play a role by the use of molecular genetic strategies suitable for the identification of homozygous sequence variations.

1.3.2.15 Cardiac Troponin T (*TNNT2*) Gene

Troponin T is a regulatory protein of the striated muscle. Mutations in the troponin T gene are the most frequent. Seven mutations in *MYH7*, one in *TNNT2*, and none in *PLN* or the *VCL* cardio-specific exon were found. *MYH7* seems to be the most frequently mutated gene in the familial DCM (FDCM), and mutation carriers present a delayed onset, in contrast to *TNNT2*. These findings confirm the genetic heterogeneity of FDCM with a 2% frequency of *TNNT2* mutation (R141W is associated with high penetrance and early onset), a low occurrence of *PLN* and *VCL* mutation, and a prominent role of *MYH7* in DCM (10% mutation frequency in familial forms and delayed onset of the disease).

1.3.2.16 Alpha-Cardiac Actin Gene

ACTC is one of the six actin genes in humans, some of which have been investigated in human disease. In cardiac myocytes, cardiac actin is the main component of the thin filament of the sarcomere. One end of the polarized actin filament builds cross bridges with myosin. The other end is immobilized, attached to a Z band or an intercalated disc. Thus, actin transmits force between adjacent sarcomeres and neighboring myocytes to effect coordinated contraction of the heart. The sarcomeric protein, actin, plays a central, dual role in cardiac myocytes. Actin generates contractile force by interacting with myosin and also transmitting force within and between cells. Two missense mutations in the ACTC postulated to impair force transmission have been associated with FDCM. Monserrat et al. (2007) reported that HCM, DCM, or left ventricular noncompaction (LVNC) and restrictive cardiomyopathy might appear as overlapping entities. The E101K mutation in the *ACTC* should be considered in the genetic diagnosis of LVNC, apical HCM, and septal defects. Olson et al. found that mutations in *ACTC* cause either HCM or DCM, depending on the functional domain of actin that is affected. Olson et al. identified two mutations (G867A and A1014G) in two DCM families of German or Swedish-Norwegian ancestry (Olson et al. 1998; Olson et al. 2000; Olson et al. 2002).

1.3.2.17 Endothelin-1, Endothelin-A (ETA), and Endothelin-B (ETB) Receptor Genes

Plasma endothelin-1 (ET-1) levels and endothelin-A (ETA) receptor densities are increased in patients with DCM. Several genetic polymorphisms in the genes encoding the endothelin system have been reported. In France, a common genetic polymorphism in the ETA receptor gene has been more often found in subjects with DCM, than in controls. Herrmann et al. identified that patients carrying the T allele of the ETA receptor gene polymorphism H323H show significantly worse cumulative survival compared to non-carriers.

1.3.2.18 Apolipoprotein E (ApoE) Gene

It has been shown that ApoE alleles are associated with both cardiovascular and Alzheimer diseases. Among alleles, the ε4 has been found to be associated with higher plasma cholesterol levels and is related to the risk of lipid disorder and coronary vasculopathy. ApoE is a major component of low-density and high-density lipoproteins. Three common alleles – ε2, ε3, and ε4 – encode for the three main isoforms, ApoE ε2, ApoE ε3, and ApoE ε4, circulating in the bloodstream. Genetically, the ε4 allele of the apolipoprotein E (*APOE*) gene is the strongest risk factor for late-onset Alzheimer disease, but also it has been identified as an important risk factor for coronary vasculopathy suggesting an association of *APOE* polymorphism with a severe form of DCM.

1.3.2.19 *SCN5A* Gene

Mutations in the *SCN5A* gene coding for the α-subunit of the cardiac Na$^+$ ion channel cause long QT syndrome, Brugada syndrome, idiopathic ventricular fibrillation, sick sinus node syndrome, progressive conduction disease, DCM, and atrial "stillstand." The identification of gene carriers is clinically important, particularly in sudden infant and adult death syndromes. *SCN5A* (MIM 600163) is the gene coding for the α-subunit of the cardiac depolarizing Na$^+$ ion channel Naþ channel (Nav1.5), which conducts a depolarizing inward Naþ current and is essentially responsible for the generation of the upstroke Phase 0 of the cardiac action potential.

1.3.2.20 Myotrophin Gene

The myotrophin gene is a gene, which consists of four exons separated by three introns. The myotrophin gene has been mapped and shown to be a novel gene localized in human chromosome 7q-33. Adhikary et al. (2005) found that this protein may be a common link initiating a different type of cardiac hypertrophy.

1.3.2.21 Vinculin and Metavinculin Genes

Vinculin and its isoform metavinculin are protein components of intercalated discs, structures that anchor thin filaments and transmit contractile force between cardiac myocytes. Vinculin is located on chromosome 10q22.1–q23. The smaller isoform, metavinculin, is ubiquitously expressed. Metavinculin contains an additional 68 amino acids and is expressed exclusively in cardiac and smooth muscle. In cardiac myocytes, vinculin and metavinculin are located at major sites of contractile force transmission. Investigations are suggesting a potential relationship between metavinculin and vinculin expression, intercalated disc abnormalities, and DCM. Olson et al. examined the possible heritable dysfunction of metavinculin in the pathogenesis of DCM and suggested that metavinculin plays an important role in the structural integrity and function of the heart. They also showed that inherited dysfunction of this protein is associated with altered actin filament organization in vitro, disrupted intercalated disc structure in situ, and DCM. These results agree with the hypothesis that defective contractile force transmission leads to DCM.

1.3.2.22 Lamin A/C Gene

Lamin contributes to the structural integrity of the nuclear envelope and provides mechanical support for the nucleus. Missense mutations of the lamin gene may alter interactions with some cytoplasmic proteins, particularly intermediate filaments of cytoskeleton (e.g., cytokeratins), but this has not yet been fully demonstrated. Lamins A and C are components of the nuclear envelope and are located in the lamina, a multimeric structure associated with the nucleoplasmic surfangiotensin converting enzyme of the inner nuclear membrane. This gene is also responsible for two skeletal myopathies: Emery-Dreifuss and limb-girdle muscular dystrophy. *LMNA* mutations have been linked to familial or sporadic DCM, with or without conduction system disease. The main findings of Hermida-Prieto et al. (2004) were the outstanding description of one novel and one recurrent mutation in the lamin A/C gene associated with severe forms of FDCM and the identification of isolated LVNC in a young carrier of the R190W lamin A/C mutation. Hermida-Prieto et al. found that the novel R349L mutation may contribute (but not definitely) to the disease. The R190W mutation has been associated with severe forms of FDCM with conduction system disease. The R190W mutation may contribute (but not definitely) to isolated LVNC in the patients studied. Arbustini et al. (2002) reported that *LMNA* gene mutations accounted for 33% of the cases of DCM with AVB, all of which were familial autosomal dominant DCM. The lamin A/C gene was found associated with the autosomal dominant form of DCM associated with a particular phenotype. In the literature, there have been reports of 11 families with DCM for mutations in the lamin A/C gene (Charron et al. 2012). Five families showed lamin A/C gene mutations. There were five missense mutations, including one in the lamin C domain (tail) and four in the α-helical domain (rod) of the lamin A/C gene. The results of Speckman et al. (2000) indicated that missense mutations in the lamin A/C gene cosegregate with familial partial lipodystrophy. However, it is not clear how the alterations described lead to adipocyte apoptosis or initiate loss of fat at puberty. Further studies are necessary to clarify why different alterations within lamin A/C are responsible for clinically distinct diseases.

1.3.2.23 Adenosine Monophosphate Deaminase-1 (AMPD1) Gene

The human adenosine monophosphate deaminase-1 (*AMPD1*) gene is located in the region p13–p211 of chromosome 1 and contains 16 exons. AMPD1 is an enzyme that catalyzes the deamination of AMP to inosine monophosphate as a part of purine catabolism. The AC-T transition at nucleotide 34 (codon 12 in exon 2) results in a nonsense mutation, predicting a severely truncated protein that loses its catalytic activity. The mutant AMPD1 cannot catalyze the deamination of AMP inosine monophosphate; thus, AMP turns into adenosine. The loss of the catalytic activity of the mutant AMPD1 increases the adenosine production in skeletal muscle. Adenosine, in turn, can attenuate the expression of TNF-α. This aspect suggests a TNF-α-related mechanism being responsible for a better clinical outcome, as observed in patients with CHF who have a mutant *AMPD1* allele. Gastmann et al. (2004) confirmed the result of a better survival associated with the mutant *AMPD1* allele.

1.3.2.24 Cypher/ZASP Gene

Cypher/ZASP encodes a Z-disc-associated protein (Z-band alternatively spliced PDZ motif). Myocardial function is associated with the activation or regulation of signal kinases, including protein kinase C (PKC). A modification of myocardial proteins by PKC plays a key role in the regulation of contractility and the growth of cardiomyocytes, whereas alterations in the expression, activity, or localization of PKC are associated with cardiac hypertrophy and failure. A gene for the PDZ and LIM domain-containing cytoskeletal protein, Cypher/ZASP, was identified in mouse (*Cypher*) and human (*ZASP*). Two assays demonstrated that the D626N mutation of *Cypher/ZASP* increased the affinity of the LIM domain for protein kinase C. This data suggests altered recruitment of molecules participating in intracellular signaling. LIM domains are protein structural domains, including two contiguous zinc finger domains, which are separated by a two-amino acid residue hydrophobic linker. Their name arises from the initial discovery in the proteins Lin11, Isl-1, and Mec-3 (LIM). Proteins containing LIM-domains play roles in cytoskeletal organisation, organ development, and tumorigenesis. LIM-domains are crucial because they mediate protein–protein interactions, which remains currently critical to cellular processes. ZASP is LIM domain binding 3 (LDB3), which is also known as Z-band alternatively spliced PDZ-motif (ZASP). ZASP remains a key in the sarcomeric protein interaction. ZASP belongs to the Enigma subfamily of proteins and stabilizes the sarcomeric units.

1.3.2.25 β-Sarcoglycan (SGCB) and δ-Sarcoglycan (SGCD) Genes

SGCB and *SGCD* genes may play a role in the etiopathogenesis of DCM. Both genes have a muscular-restricted expression profile, due to a molecular link between the extracellular matrix and sarcolemmal cytoskeletal proteins in the myocytes. Mutational data are linking mutations in these genes with genetically inherited muscular disorders, such as DCM and myopathy. The β-sarcoglycan (*SGCB*) and δ-sarcoglycan (*SGCD*) genes are strong candidates for a morbid role in DCM for several reasons. These genes are highly expressed in cardiac and skeletal muscle. They encode proteins involved in the cytoarchitecture of the cardiac cell as they are components of the dystrophin-associated sarcoglycan complex that forms a structural link between the F-actin cytoskeleton and the extracellular matrix. Moreover, the *SGCB* and *SGCD* genes have been implicated in limb-girdle muscular dystrophies or LGMD (LGMD2E and LGMD2F, respectively), and both disorders have been found associated with DCM. Currently, the δ-sarcoglycan is one of the four proteins (α, β, γ, and δ) in the sarcoglycan complex, which sequentially forms a part of the dystrophin-associated glycoprotein complex. This complex is located in the transmembrane region, and its function is to form a link between the intracellular and extracellular matrix. It has been suggested that the δ-sarcoglycan gene is unlikely to cause DCM in patients from eastern Finland (Kärkkäinen et al. 2002). Only two DCM-associated mutations in this gene have been previously reported. Tsubata et al. (2000) found two mutations in the δ-sarcoglycan gene. The Ser151Ala mutation was detected in three family members of one family and a 3-bp deletion in position 238 (del Lys238) in two sporadic cases without signs of skeletal muscle disease. The Ser151Ala and del Lys238 mutations caused a relatively severe form of DCM characterized by SCD and heart failure at a young age and a ineluctable need for HTX. The carriers of the Arg71Thr mutation (Kärkkäinen et al. 2003) have a relatively mild, late-onset disease and a good response to medication. Therefore the phenotype of the subjects carrying this mutation seems to be less severe than that of patients described by Tsubata et al. Sylvius et al. (2003) estimated the prevalence of *SGCD* gene mutations responsible for DCM at less than 1.5%. This finding underlines the fact that the SGCD gene is only marginally implicated in the disease. This aspect is in agreement with previous results obtained by mutation screening of other candidate genes such as *DES*, *ACTC* or *TNNT*, *MYH7*, and *TPM1*, which are also rarely mutated in DCM as none were over a 10% mutation frequency. Sylvius et al. found that the most frequently implicated gene in DCM appears to be the LMNA gene since ten different mutations responsible for DCM associated with conduction and muscular disorder have been reported. However, these clinically non-isolated forms of DCM represent only 10% of all familial DCM. As no major gene or locus have been identified in DCM and given the fact that morbid mutation identification concerns only a minor percentage of familial cases of DCM, it may be speculated that a large number of morbid genes remains to be identified. The δ-sarcoglycan gene was also demonstrated to be responsible for DCM. In one family with the autosomal dominant mode of inheritance, a Ser151Ala mutation was detected in three patients with isolated DCM at a young age. Intriguingly, Tsubata et al. suggested that δ-sarcoglycan is a disease-causing gene responsible for familial and idiopathic DCM and lend support to the "final common pathway" hypothesis that DCM is a "cytoskeleton pathology". The fact that mutations in δ-sarcoglycan and dystrophin, as well as mutations in *G4.5*, can also result in skeletal myopathy suggests that patients with DCM should be carefully evaluated for skeletal muscle weakness and that neurologists caring for patients with skeletal myopathies should be aware of the potential for associated cardiomyopathies in their patients. Mutations in the gene *G4.5*, originally associated with Barth syndrome, have been reported to generate multiple severe infantile X-linked cardiomyopathies.

1.3.2.26 Polyadenylate-Binding Protein 2 (*PABP2*) Gene

The weakness of distal limb muscle groups is a clinical hallmark of myopathies of distal type (MPDs). The first genetic locus for AD-inherited MPD was discovered in an Australian family, in which affected individuals developed particular weakness of foot and toe extensors, followed by progressive weakness of finger extensors and neck muscles. Distal myopathy (MPD) linked to chromosome 14q11–q13 (MPD1) is rare.

Genetically, MPDs are a heterogeneous group of muscle disorders. The coding sequence of *PABP2* (the polyadenylate-binding protein two genes) was evaluated by Hedera et al. in 2003. Haplotype analysis suggests that the gene causing MPD1 is located between polymorphic microsatellite markers D14S283 and D14S1034 on chromosome 14q11–q13. The MPD1 locus contains *PABP2*. Mastaglia et al. reported the exclusion of the *PABP2* gene in their family with MPD1. Hedera et al. (2003) also did not detect a coding change in this gene, thus excluding it as the cause of MPD1. Another important candidate gene within the MPD1 locus is *MYH7* (heavy chain cardiac ß-myosin). Mutations in this gene cause familial HCM, although these patients do not have signs of clinical myopathy. The history of idiopathic cardiomyopathy in some affected individuals may supply an important indication for additional candidate genes.

1.3.2.27 Genes Encoding the Four Major Components of the Heart Calcineurin Pathway, *PPP3CA*, *PPP3CB*, *GATA4*, and *NFATC4*

Despite the investigation of a large number of polymorphisms, only one among the four non-synonymous polymorphisms, NFATC4/G160A, was associated with the investigated phenotype. A Gly/Ala substitution of the NFATC4 protein (G160A) was associated with left ventricular mass and wall thickness.

1.3.2.28 Dystrophin Gene

Dystrophin is a large (427 kDa) cytoskeletal protein that localizes to the inner lamina of the plasma membrane or sarcolemma. Through the amino-terminal actin-binding domain, dystrophin is related with F-actin and the sarcomere. The carboxy-terminal domain is associated with a large transmembrane complex of glycoproteins, termed the dystrophin-associated glycoprotein (DAG) complex. In this manner, dystrophin is believed to play a critical role in establishing connections between the internal cytoskeleton and the sarcomeric structure and the external basement membrane. The absence of dystrophin leads to a disruption of the DAG complex, a loss of integrity of the plasmalemma and necrosis of cardiomyocyte fibers. Although no skeletal muscle involvement is evident (in contrast with Duchenne or Becker muscular dystrophy, where the dystrophin gene is also involved), plasma creatine kinase levels are usually high. DNA alterations involved in the dystrophin gene are located in the muscular promotor-first, muscular exon-first intron regions. These alterations consist of deletions or a point mutation in the splice consensus site of the first intron. They result in the absence of the protein in the cardiac muscle, whereas dystrophin expression is preserved or slightly reduced in skeletal muscle, in exons 2–7, 9, 45–49, 48–49, and 49–51, where other genetic alterations have been identified (deletions, duplication, and missense mutations). As dystrophin mutations may cause clinical or subclinical skeletal myopathy, it is possible that the muscular fatigue, seen chronically in many patients with DCM, could be due to primary skeletal muscle disease and not due primarily to chronic heart failure. Examining the dystrophin gene in DCM is this group's first application of Mendel leaping. More work is necessary to evaluate the general utility of this approach for selecting candidate genes for the complex disease. Point mutations seem to be associated with sporadic DCM without clinical evidence of skeletal myopathy. Moreover, Charron et al. have reported that the first gene responsible for DCM was identified in 1993, the dystrophin gene, as responsible for X-linked DCM. The genetically affected individuals usually develop a severe form of DCM at adolescence or in young adulthood.

1.3.2.29 Gene for Platelet-Activating Factor Acetylhydrolase (PAF-AH)

It was found that a variant allele (279Phe allele) of the gene for platelet-activating factor acetylhydrolase (PAF-AH), which has a reduced enzymatic activity as compared with the normal 279Val allele, is impressively associated with nonfamilial IDC in a Japanese population. Because PAF-AH is related to both inflammation and superoxide-induced tissue damage, this result might support the immune-related and oxygen stress-related etiologies of IDC. This association should be confirmed in other patients and control panels. The reported association of the PAF-AH polymorphism with nonfamilial IDC was examined in 106 nonfamilial IDC patients in the same publication in Human Genetics of 2000 by Arimura and colleagues. In contrast, no evidence was found to support the reported association, suggesting that the contribution of the PAF-AH variant in the susceptibility to nonfamilial IDC (if there is indeed any such contribution) may not be large enough to be confirmed by the analysis of 106 nonfamilial IDC patients.

1.3.2.30 Transforming Growth Factor β1 (*TGF-β1*) Gene

Transforming growth factor-β1 (*TGF-β1*), a regulatory cytokine produced by many cell types, has been studied about the pathogenesis of coronary artery disease. Both anti-atherogenic and pro-atherogenic activities of TGF-β1 have been reported. TGF-β1 inhibits the proliferation of many cells, including smooth muscle cells, endothelial cells, and epithelial cells. It could, therefore, inhibit the development of atherosclerosis. Furthermore, TGF-β1 has chemoattractant activities, enhances cell adhesion, and stimulates intracellular matrix deposition. The involvement of TGF-β1 in cardiomyopathy is still under intense investigation. Elevated *TGF-β1* gene expression was measured in ventricular biopsies from hypertrophic and dilated hearts, whereas others found decreased TGF-β1 plasma levels in patients with DCM. The contradictory findings in patients with ischemic heart disease (IHD) as well as in patients with CMP could be the result of different biologic activities of TGF-β1 during various stages of disease processes or be intimately connected to intraindividual variations in TGF-β1 protein production. Analysis of *TGF-β1* polymorphisms showed that the presence of the Arg25 allele is associated with increased blood pressure and with the development of vascular graft disease after cardiac transplantation, whereas the Pro25 allele seems to be associated with myocardial infarction. Another study, however, could not confirm a relation between these *TGF-β1* polymorphisms and coronary artery disease. Holweg et al. (2001) reported an association between the Leu10-Pro (codon 10) polymorphism in the *TGF-β1* gene with end-stage heart failure caused by DCM. Although other cytokines are involved, these observations suggest that TGF-β1 is implicated in the pathophysiology of DCM. The difference in *TGF-β1* gene polymorphism distribution in the patient group with DCM needs further investigation. This group is of interest, because a difference may exist between patients with hereditary DCM and patients with cardiomyopathy caused by toxic agents or viral infection.

1.3.2.31 *G4.5* Gene

Mutations have been continuously emphasized in the gene *G4.5* in patients with classic Barth syndrome (BTHS) as well as in patients with infantile DCM and isolated LVNC. In 2001, Ichida et al. have remarkably reported the identification of novel mutations in G4.5 in patients with isolated LVNC. In patients with LVNC associated with CHD, no mutations were found in *G4.5*. Instead, a mutation in the calcium-binding EF-hand domain of α-dystrobrevin was identified in one family. Several important points are becoming apparent about *G4.5* mutations. Many affected individuals develop severe infantile disease and succumb. The gene defect usually differs among families. However, there seems to be no distinct genotype:phenotype correlations that allow the differentiation of clinical course to be predicted. The cardiac phenotypes that occur as a result of *G4.5* mutations may vary significantly. The cardiac manifestations include DCM, endocardial fibroelastosis, LVNC, and HCM. Also, this phenotype can differ among family members and change over time, possibly in response to therapy. Finally, the systemic manifestations of BTHS are equally unpredictable. In some children, sudden death occurs. It is likely that modifier genes are involved in determining the phenotype and clinical severity. Ichida et al., in 2001, suggest that studies of patients with myocardial disorders having prominent systolic dysfunction should include evaluation of members of the cytoskeleton-sarcolemma complex, as well as *G4.5*, as candidate genes.

1.3.2.32 *HS426* (Nebulette) Gene

Abnormalities in genes for cytoskeletal proteins related to Z-disc function have been reported to cause idiopathic DCM. Therefore, the genomic organization of the gene for nebulette, a novel actin-binding Z-disc protein, may be an interesting investigational field. In 2000, Arimura et al. identified that the Asn654Lys variation was associated with nonfamilial idiopathic DCM only in the homozygous state and no increase in the frequency of heterozygotes was observed in the patients. Because the Asn654Lys variant of nebulette is the polymorphism present in the healthy population, the functional difference between the alleles of nebulette may not be so large as to cause any disease phenotypes in the heterozygous state.

1.3.2.33 Human Cardiotrophin-1 Gene (*CTF1*)

CT-1 (CTF1) belongs to the IL-6 family of cytokines and can stimulate cardiac myocyte growth *in vitro*, suggesting that the gp130 signaling pathway may play a role in cardiac hypertrophy. Erdmann et al. found genetic variants in the coding and promoter region of *CT-1* which might modify gene function and gene expression. Because of the low prevalence of this promoter polymorphism, further investigations are ongoing with very large patient and control groups to achieve a more precise estimate of the relative risk.

1.3.2.34 Desmin Gene

Desmin is the main intermediate filament of skeletal and cardiac muscle. It functions as a cytoskeletal protein linking Z bands to the plasma membrane and nuclear membrane, and it maintains the structural and functional integrity of the myofibrils. Desmin-related myopathy is a familial disorder of skeletal muscle characterized by the intracytoplasmic accumulation of desmin-reactive deposits in muscle cells and is often accompanied by cardiac involvement, such as conduction blocks and cardiomyopathy. Mutations in the desmin gene (*DES*) partly cause the myopathy. Nine disease-causing mutations have been identified: eight in the rod domain and one in the carboxy-terminal domain. All those in the rod domain were associated with skeletal muscle involvement, while those in the carboxy-terminal domain were not always associated. Miyamoto et al. (2001) reported that the mutation (Ile451Met) located in the carboxy-terminal domain caused familial DCM without clinically evident skeletal muscle abnormalities. The desmin gene was identified as responsible for the autosomal dominant inherited form of DCM. The *DES* gene was also identified as responsible for restrictive cardiomyopathy in several families. Miyamoto et al. showed that in a relatively large Japanese population affected with DCM, 3 out of 265 patients (1.1%) had the missense mutation (Ile451Met), previously reported as disease-causing. Another study in Europe showed that no mutation was detected in the population of 41 probands of DCM families and 22 sporadic cases. Another characteristic of cardiac involvement due to the mutations of the *DES* gene is conduction blocks. Seven mutations in the rod domain led to conduction blocks. Neither of the patients had any conduction blocks, similar to the cases due to the Ile451Met mutation reported by Miyamoto et al. (2001). As mentioned above the desmin gene is probably uncommon in DCM, since only one proband among 40 had a mutation in the study above and no mutations were found in 41 probands of FDCM and 22 sporadic cases (Tesson et al. 2000). The same group indicated that desmin and cardiac actin gene mutations are unlikely to cause DCM in the European population studied. Kärkkäinen et al. (2002) showed that the desmin and d-sarcoglycan genes are unlikely to cause DCM in patients originating from eastern Finland.

1.3.2.35 Endothelial Nitric Oxide Synthase (*NOS3*) Gene

In 2000, the CARDIGENE study was published by Tiret and colleagues. These authors studied eight candidate genes: the endothelial nitric oxide synthase (*NOS3*), the angiotensin I-converting enzyme (*ACE*), the angiotensin II type 1 receptor (*AGTR1*), the angiotensinogen (*AGT*), the aldosterone synthase (*CYP11B2*), the tumor necrosis factor-alpha (*TNF*), the transforming growth factor beta1 (*TGF β1*), and the brain natriuretic peptide (*BNP*) genes. Four hundred thirty-three patients with IDC and 401 controls were included. The controls were randomly sampled from the French population surveys carried out in Lille (northern France), Strasbourg (eastern France), and Toulouse (southern France) within the framework of the WHO MONICA (MONItoring in CArdiovascular diseases) project. Patients with heart failure show an enhanced basal production of nitric oxide, while an increased activity of inducible NOS has been found in cardiac tissue from patients with DCM. The CARDIGENE study is a large case-control study designed to investigate genetic factors involved in IDC. Tiret et al. (2000) could not find any association between the polymorphisms and the risk or the severity of IDC.

1.3.2.36 Skeletal Muscle Alpha-Actin Gene (*ACTA1*)

Muscle contraction results from the force generated between the thin filament protein actin and the thick filament protein myosin, which causes the thick and thin muscle filaments to slide past each other. There are skeletal muscle, cardiac muscle, smooth muscle, and non-muscle isoforms of both actin and myosin. Inherited diseases have been associated with deficiencies in cardiac actin (DCM and HCM), cardiac myosin (HCM), and non-muscle myosin. It has been reported that gene mutations in the human skeletal muscle α-actin gene (*ACTA1*) are associated with two different muscle diseases, congenital myopathy with excess of thin myofilaments (actin myopathy) and nemaline myopathy. Both diseases are categorized by structural abnormalities of the muscle fibers and variable degrees of muscle weakness. Fifteen different missense mutations resulting in 14 different amino acid changes were detected. Three genes mutated in numerous types of nemaline myopathy were detected: *TPM3* (encoding α-tropomyosin slow) in both dominant and recessive nemaline myopathy in which the nemaline bodies are restricted to slow, type I muscle fibers, *NEB* (encoding nebulin) in typical no- or slowly progressive congenital nemaline myopathy, and *ACTA1*. Mutations in *ACTA1* can also cause additional phenotypes. Nowak et al. (1999) identified several phenotypes associated with mutations in *ACTA1*.

1.3.2.37 Locus on Chromosome 6q12–q16 for Autosomal Dominant DCM

Haplotype reconstruction showed that all affected subjects, as well as all individuals with unknown status, shared a common haplotype on chromosome 6, between markers D6S1627 and D6S1716. The candidate interval corresponds to a 16.4 cM region localized on chromosome 6q12–16. The disease interval on chromosome 6q12–16 contains known genes encoding collagen IXa-1 polypeptide (COL9A1 [MIM 120210]), myosin VI (MYO6 [MIM 600970]), vascular endothelial growth factor (VEGF [MIM 192240]), malic enzyme cytoplasmic (ME1 [MIM 154250]), and several other genes encoding anonymously expressed sequence tags. Also, the genes encoding cardiac phospholamban (PLN [MIM 600133]) and laminin-α4 (LAMA4 [MIM 600133]), located near the disease interval, could also be considered as candidate genes.

1.3.2.38 A-Dystrobrevin (*DTNA*)

The α-dystrobrevin gene or Dystrophin-Related Protein 3 or DTNA is alternatively spliced, resulting in multiple isoforms of dystrobrevin (a, b, g), with different tissue distributions, of which only α-dystrobrevin is expressed in the heart. It is well known that α-dystrobrevin is a member of the dystrophin-associated protein complex (DAPC), which is composed of three subcomplexes, the dystroglycan complex, the sarcoglycan complex, and the cytoplasmic complex, which includes the syntrophins and dystrobrevins. The DAPC, which is located at the sarcolemma, connects the cysteine-rich and C-terminal domains of dystrophin with b-dystroglycan and the cytoplasmic complex, respectively. The protein known as β-dystroglycan is a transmembrane protein that binds to the laminin-binding protein α-dystroglycan in the extracellular matrix. At the N-terminus, dystrophin binds to actin. These interactions seem to effectively link the extracellular matrix to the dystrophin-based cytoskeleton of the muscle fiber at the C-terminus and the contractile apparatus at the N-terminus. Furthermore, α-dystrobrevin links the DAPC to the signaling protein neuronal nitric oxide synthase (nNOS). Disruption of these links results in severe muscle wasting or cardiac muscle pathology. For example, dystrophin mutations cause Duchenne muscular dystrophy or X-linked DCM. Ichida et al. reported the identification of novel mutations in *G4.5* in patients with isolated LVNC. In patients with LVNC associated with CHD, no mutations were found in *G4.5*. Instead, a mutation in the calcium-binding EF-hand domain of α-dystrobrevin was identified in one family. The α-dystrobrevin mutation described in this study resulted in a phenotype of cardiomyopathy with deep trabeculations, associated with congenital heart disease, consistent with the criteria for LVNC.

1.3.2.39 Manganese Superoxide Dismutase Gene (*SOD2*)

The *SOD2* gene encodes for manganese superoxide dismutase (MnSOD). MnSOD is an antioxidant enzyme localized in mitochondria to protect cells from oxidative damages and preferentially expressed in the heart, brain, kidney, and liver. Expression of MnSOD is induced by inflammatory cytokines such as IL1 and TNF, which are increased in sera of patients with idiopathic dilated cardiomyopathy or myocarditis, and overexpression of MnSOD promotes survival of cells damaged by these cytokines. Because the heart is rich in mitochondria as compared with the other organs, mitochondrial disorders frequently involve the cardiac tissue. It also has been suggested that MnSOD is involved in the pathogenic process of ischemic heart disease and adriamycin-induced cardiomyopathy, both of which often show the IDC-like phenotype. Hiroi et al. confirmed the association of idiopathic dilated cardiomyopathy with HLA-DRB1 1401 in nonfamilial Japanese patients and showed that the Val allele of the *SOD2* gene, especially in the homozygous state, was associated with nonfamilial idiopathic dilated cardiomyopathy in Japanese. They also showed that there was a difference in the mitochondrial processing efficiency of MnSOD (SOD2) leader signal depending on the Ala/Val polymorphism.

1.3.2.40 Mitochondrial DNA Abnormalities

Oxidative phosphorylation (OXPHOS) is one of the most important metabolic pathways. This pathway is essential for the cells, because they use enzymes to oxidize nutrients. Such oxidation releases the chemical energy of molecular oxygen. The oxygen is, then, used to produce adenosine triphosphate (ATP). In most eukaryotes, OXPHOS takes place inside mitochondria. The mtDNA is a double-stranded, circular DNA molecule. Because of the lack of histones, a repair system, and exposure to oxygen-free radicals, the mutation rate in mtDNA is more than ten times higher as in nuclear DNA. Also, mtDNA has no introns, so that a random mutation will usually strike a coding DNA sequence. The high mutation rate of the mtDNA causes a prevailing mtDNA variation rate in the human population and is a common cause for the mitochondrial disease. Mitochondrial DNA mutations have the potential to affect OXPHOS and in turn cellular death. The maternal inheritance and high mutation rate of the mtDNA mean that mtDNA alterations are common in human populations. Some pathogenic mtDNA mutations identified in patients with cardiomyopathy reside in tRNA genes of the mtDNA. These mutations have been shown to negatively affect protein synthesis of the mitochondria and specific respiratory enzyme activities. Two tRNA mutations were found in patients with DCM. The c.15924A > G tRNAThr mutation is connected with respiratory enzyme deficiency, mitochondrial myopathy, and cardiomyopathy. The tRNAArg—c.10458 T > C mutation is described as a known sequence polymorphism. In the D-loop region, several sequence alterations could be detected. The c.16189 T > C variant that is associated with susceptibility to DCM could be detected in the Caucasian cohort in 15.6% of patients with DCM and 9.7% of the control subjects. Some 17.2% of the c.16189 T > C variant in Caucasians with DCM versus 8.8% in controls have also been detected. A significant difference could be found in new mtDNA mutations in protein-coding genes of patients with DCM. Some of the mutations were found to be potentially pathogenic to DCM because they substituted evolutionarily conserved residues which might alter the function of the respective protein subunits. Mutations altering the function of the subunits of the respiratory chain can be responsible for the pathogenesis of DCM.

1.3.2.41 Dilated Cardiomyopathy Due to Chemotherapy-Related Cardiotoxicity

This form of DCM is likely to be secondary to some nutritional deficiency associated with the administration of adriamycin (doxorubicin). Histologically, there is patchy myocardial interstitial fibrosis with scattered vacuolated cardiomyocytes ("Adria cells"), fibroblastic proliferation, histiocytic infiltration (CD68+), partial/total loss of myofibrils, and vacuolar degeneration of the cells. The vacuolization of cardiomyocytes is due to dilation of the sarcotubular system as the earliest change, and the Adria cell shows loss of cross striations, homogeneous basophilic staining, and loss of myofilaments. The DCM due to chemotherapy-related cardiotoxicity is more often observed in young adults, either *intra vitam* (during life) by asymptomatic murmur on a routine physical exam or following sudden death during strenuous exercise.

In conclusion, DCM is a primary myocardial disease that causes considerable morbidity and mortality. Although this cardiomyopathy is clinically heterogeneous, genetic factors play an essential role in its etiology and pathogenesis. As demonstrated above, the association between gene polymorphisms and DCM has been investigated for various genes. The genetic contributions to heart failure can be broadly grouped into causative and modifier genes. Numerous components of the cytoskeletal system have been implicated as genes that cause DCM. In contrast, modifier genes become active after the disease is present and thus influence clinical course. The identification of genetic risk factors is essential for better understanding the pathogenesis of DCM. Up to now, several chromosomal loci and disease genes have been identified. But the data suggests that nearly all of the disease genes account for a relatively small proportion of all cases of DCM. As no significant gene or loci have been identified in DCM, it may be speculated that many morbid genes remain to be determined.

1.3.3 Hypertrophic Cardiomyopathy

- *DEF*: Multigenic/polygenic disorder mostly involving the interventricular septum but also an extension on some degree of left ventricular outflow tract obstruction (Fig. 1.33).
- *SYN*: Idiopathic hypertrophic subaortic stenosis (IHSS), asymmetric septal hypertrophy (ASH).
- *EPI*: ~0.2% (1:500) of young adults. Sarcomere protein mutations known to cause HCM are quite common in the general population.
- *ETP*: AD inheritance but also sporadic. Concentric (not asymmetric) left ventricular hypertrophy has been independently associated with an increase of the risk to have sudden cardiac death (SCD). Remarkably, the density of connexin 43 gap junctions at the intercalated disc in individuals with LVH + SCD and individuals with LVH + unnatural death is quite similar. An increase in the number of connexin 43-labeled gap junctions with an increase in the myocyte cross section may be found. These gap junctions may be at the basis of conduction of the electrical activity in the transverse orientation. The maintenance of gap junctions at the intercalated disc as an expected number and the longitudinal conduction properties in cardiomyocytes should remain intact, but the transverse conduction properties of the hypertrophied cardiomyocytes may be altered and be at the basis of SCD, at least in some cases.
- *CLI*: HCM remains a diagnosis of exclusion. Thus, secondary causes of left ventricular hypertrophy such as systemic hypertension, valvular/subvalvular aortic stenosis, and infiltrative cardiomyopathies must be ruled out. If echocardiography, computed tomography (CT), or cardiac magnetic resonance (CMR) in the absence of a secondary cause identify a wall thickness of ≥15 mm, this finding is consistent with HCM.
- *GRO*: Left ventricular hypertrophy.
- *CLM*: Myofiber hypertrophy, myofiber disarray, and interstitial fibrosis with intramyocardial located coronary vessels (~50%) and frequent endocardial fibrosis of the outflow tract are the essential characteristics for diagnosis.
- *TEM* reveals derangement of the cardiac cell anatomy at the subcellular level, including the loss of the normal alignment of myofibrils and Z disks.
- *PGN*: Disease progression is difficult to envisage, and genetic screening is jeopardized because there are differences in penetrance and expressivity.

1.3.4 Infiltrative/Restrictive Cardiomyopathies

Cardiomyopathies consisting of any disorder, which implicates a restriction of the ventricular filling phase of the cardiac pump, are grouped

1.3 Myocarditis and Nonischemic Cardiomyopathies

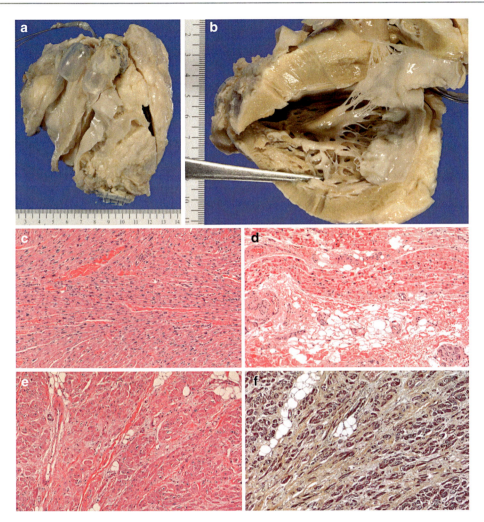

Fig. 1.33 The relationship between the heart and the brain is complex. Certain pathological conditions can interfere with the normal brain-heart regulatory mechanisms and result in impaired cardiovascular function and dilated cardiomyopathy. In neurogenic cardiomyopathy, there is dilatative form (**a–c**) with fibro-fatty infiltration of the myocardium at places (**d–f**)

under the label "infiltrative /restrictive cardiomyopathies." Although the main problem is the abnormal filling of the heart, the heart may not pump blood vigorously when the cardiovascular disease progresses. The abnormal cardiologic function can affect extracardiac organs. Restrictive cardiomyopathy may affect a single ventricle or both of the ventricles. Although restrictive cardiomyopathy is rare, the cardiologist and pathologist need to consider two of the most common causes, which are amyloidosis and glycogenosis. Other causes entail carcinoid heart disease, endomyocardial fibrosis, and Loeffler syndrome, iron overload (hemochromatosis or hemosiderosis), sarcoidosis, scarring after radiation or chemotherapy, scleroderma, and neoplasms of the heart (Figs. 1.34, 1.35, 1.36, 1.37, 1.38, 1.39, 1.40, 1.41, 1.42, 1.43, and 1.44). As part of systemic amyloidosis (AA type of amyloid) or isolated as in senile cardiac amyloidosis (AF type of amyloid [transthyretin]), amyloidosis may be an essential component of restrictive CM.

The result of the immunologic mechanisms involved in a multifactorial fashion with different types of amyloid. Substantially, it is a disorder of protein misfolding with at least 23 different pro-

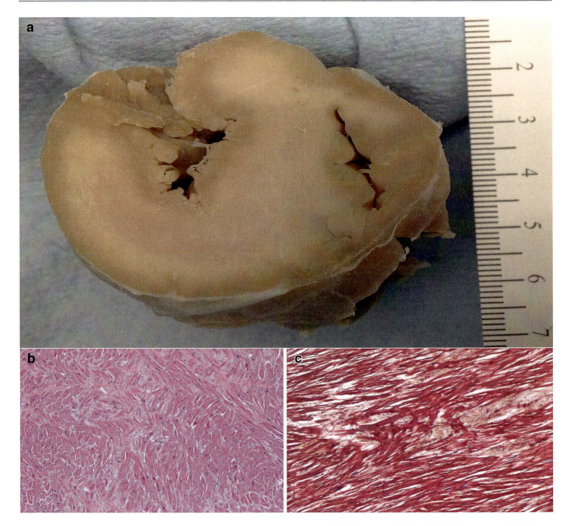

Fig. 1.34 In the hypertrophic cardiomyopathy, the ventricular septum is usually thicker than the left ventricular free wall (greater than 1.5 cm) (**a**). Histologically, there is myocardial disarray with interdigitating myocardial fibers. Although this finding is not pathognomonic for hypertrophic cardiomyopathy, it is indeed quite characteristic (**b, c**)

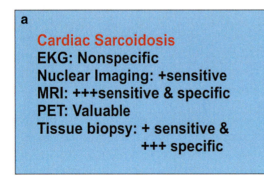
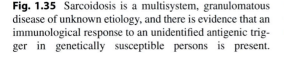
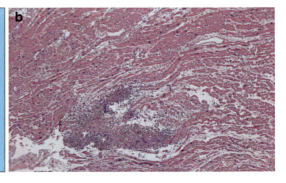

Fig. 1.35 Sarcoidosis is a multisystem, granulomatous disease of unknown etiology, and there is evidence that an immunological response to an unidentified antigenic trigger in genetically susceptible persons is present. Noncaseating granulomas are the histopathological hallmark (**b–g**), but the Schaumann bodies or asteroid bodies (**h**) are quite characteristic, although nonspecific

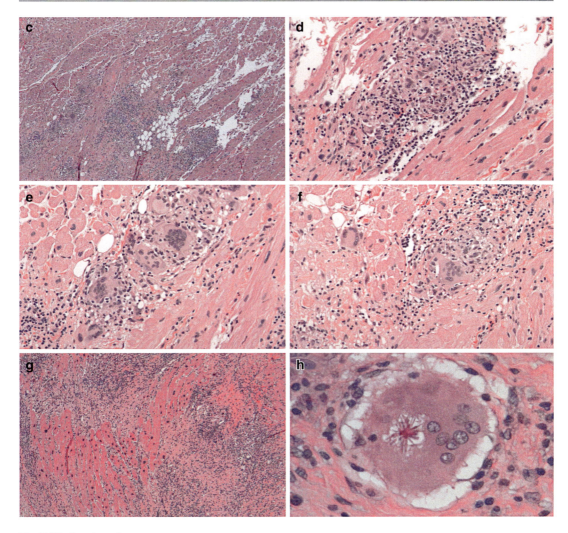

Fig. 1.35 (continued)

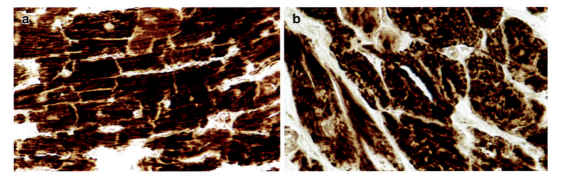

Fig. 1.36 Altered pattern of cytochrome c oxidase and succinate dehydrogenase (complex II) immunohistochemistry reveals a mosaic pattern of cytochrome c oxidase-deficient and cytochrome c oxidase-positive cardiomyocytes (top panel) with reduced succinate dehydrogenase staining (bottom panel) in the MELAS myocardium (right, top, and bottom) compared to a non-failing control (NFC) myocardium (left top and bottom). All the microphotographs have been taken at ×400 as original magnification

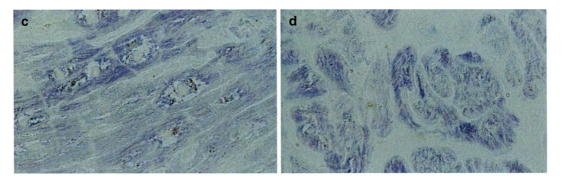

Fig. 1.36 (continued)

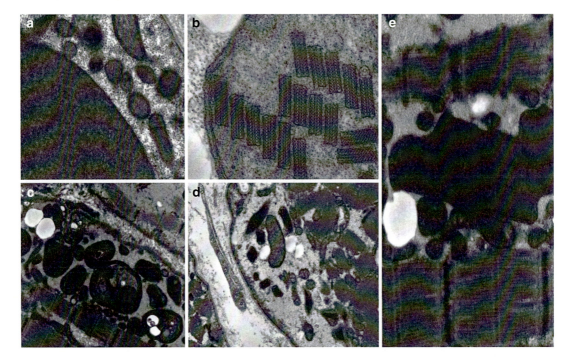

Fig. 1.37 Mitochondrial DNA cardiomyopathies with characteristic ultrastructural features on electron microscopy (EM). These features may range from large, subsarcolemmal accumulations of mitochondria to myofibers containing disordered myofibrils, degenerating mitochondria and vacuoles (**a**). One of the most frequent characteristic features of the abnormal mitochondria is the presence of very peculiar paracrystalline inclusions (**b**, **c**). The mitochondrial inclusions are sometimes described as having a "parking lot" appearance (**b**). Another characteristic feature is the presence of lipid droplets (**c**) together with these mitochondrial accumulations. Membranous whorled mitochondria can also be found. In the relaxed state, the wide gap between two Z lines is about 2.2 μm and is the functional basic unit of muscle, which is also referred to as sarcomere. In isotonic, i.e., associated with a muscle shortening, contraction shortens the length of sarcomeres of a muscle fiber at the same time. Nemaline inclusion myopathy is a congenital myopathy characterized by gross hypotonia in early infantile cases with cardiopulmonary insufficiency. The predominant feature is the presence of nemaline rods. Nemaline rods are identified within the cytoplasm and can also be demonstrated with a modified Gomori trichrome staining using histochemistry. Nemaline rods can be present as dense, irregular bodies in a random orientation just beneath the sarcolemma. On EM, they present as thickenings of the Z-lines within type I fibers. The nemaline rods are derived from Z-band material and have also been characterized immunohistochemically with an antibody anti-alpha-actinin. At higher magnification, nemaline rods show variable rod length in the longitudinal section. Thin filaments extending from each end interdigitate with thicker filaments. Inclusion body myositis contains a variety of inclusions (**f**, **g**) ranging from abnormal filaments to myelin figures (membranous whorls). Filaments can be identified within both the nucleus (**l**) and cytoplasm (**g**) of muscle fibers. In the cytoplasmic body myopathy (**i**), there is a sharply defined dense structure (**i**), which characteristically shows radiation of thin filaments from the center of the inclusion

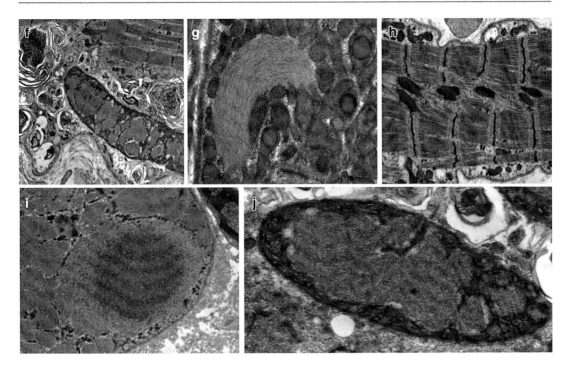

Fig. 1.37 (continued)

teins that aggregate forming fibrils with the appearance of amyloid. The red-green dichroism or birefringence is due to the dye Congo red, which binds to the amyloid-forming fibrils. Congo red is a water-soluble organic compound and its use in pathology is still *en vogue*.

The endomyocardial fibrosis of children and young adults, which often occurs in Africa, is characterized by endocardial fibrosis extending from apex into outflow tracts of the left and right ventricle. The scarring may extend into inner 1/3 of myocardium and may involve tricuspid and mitral valves. Microscopically, varying amounts of inflammation, including eosinophils, often seen at the edge of scarring can be demonstrated. Loeffler endocarditis may be seen as early as 2-year-old infants and is similar to EMF spectrum with three stages: acute myocarditis (necrotic), organizing thrombus, and endomyocardial fibrosis. Eosinophilic infiltration of other organs may be observed, and the outcome may reveal a rapid downhill course to death. Grossly, endocardial fibroelastosis shows pearly-white thickening of the endocardium due to an increase in collagen and elastic fibers on the surface, mostly in the left ventricle.

1.3.4.1 Sarcoidosis

Cardiac involvement of this idiopathic multisystemic granulomatous disease is seen in ~ ¼ of patients with systemic disease (mostly autopsy). Sarcoidosis involves the lymph nodes, lungs, eyes, kidneys, central nervous system and skin, and occasionally heart. Cardiac symptoms occur in less than 5% of patients with sarcoidosis and include papillary muscle dysfunction, congestive heart failure, pericarditis, and conduction abnormalities, and SCD as well as occasionally sarcoid myocarditis, which can mimic ARVC. The differential diagnosis of granulomas in the heart includes tuberculosis, toxoplasmosis, rheumatic fever, fungal, and idiopathic giant cell myocarditis, among others. Special stains are mandatory in these cases and include PAS, PASD, Giemsa, GMS, ZN, auramine-rhodamine, and Fite staining. IHC is also important and the pathologist should have in his repertoire antibodies against *T. gondii*.

Thalassemia major is the most dreadful form of β-thalassemia, which is a hematologic disorder that reduces the production of hemoglobin. The β-thalassemia is classified into two types depending on the severity of symptoms, including thal-

Fig. 1.38 Glycogenosis cardiomyopathy is often found in Pompe's disease. There is vacuolation of the cardiomyocytes with (**a**) or without (**b**) signs of cardiomyocyte hypertrophy and increase of the nuclear size of the cardiomyocytes. The ultrastructural features of glycogenosis are an accumulation in the lysosomes (Pompe's disease) or in the cytoplasm of the cardiomyocytes (**c**)

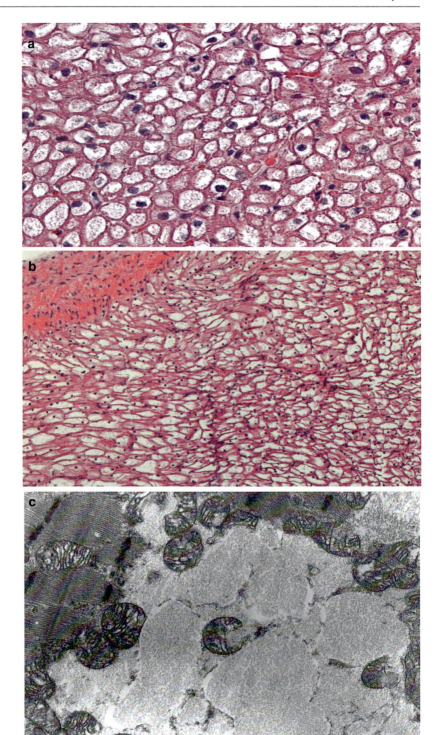

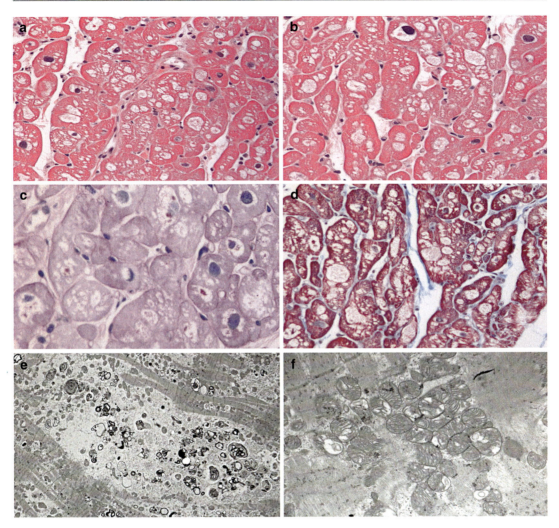

Fig. 1.39 Fabry disease is a lysosomal storage disease which involves dysfunctional metabolism of sphingolipids (sphingolipidosis). Cardiomyocytes show multiple vacuolated cells or perinuclear vacuoles on hematoxylin and eosin (H&E) staining (**a**, **b**), periodic acid Shiff (PAS) negativity (**c**), interstitial fibrosis on Movat pentachrome staining or trichromic stain (**d**), and ultrathin slide sections showing enlarged secondary lysosomes packed with storage material, which has a somewhat heterogeneous appearance and displacement of the mitochondria, which may show vacuolation and loss of cristae (**e**, **f**)

assemia major (or Cooley anemia, aka "Mediterranean anemia") and thalassemia intermedia. Thalassemia minor is also widely known as β-thalassemia trait. The β-thalassemia is caused by mutations in the *HBB* gene, which is located on *11p15.5*. *HBB* provides instructions for making a protein called beta-globin (β-globin). Beta thalassemia occurs most often in Mediterranean countries, North Africa, Middle East, India, Central Asia, and Southeast Asia. One of the significant complications of thalassemia-driven microcytic anemia with the early destruction of the abnormal erythrocytes in these patients is excessive siderosis. This complication, if untreated or poorly treated, can determine iron deposition in major organs, including the liver and heart. Cardiac iron accumulation is the precise cause of intractable cardiac failure if no treatment is started. However, despite iron chelating agents, patient survival may be reduced due to the intrinsic damage related to the deposition of iron in the cardiac muscle, causing sec-

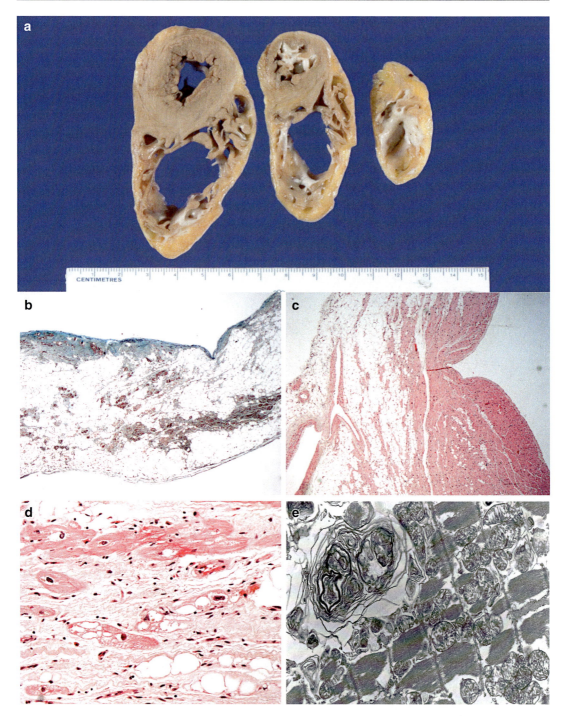

Fig. 1.40 Sanfilippo cardiomyopathy is a rare mucopolysaccharidosis. Accumulation of partially degraded GAGs progressively affects multiple organ systems including the heart. There is dilatation of the right chambers and hypertrophy of the left ventricle (**a**). On histology, there is infiltration of the myocardium of cells containing foamy cells (**b–d**) that illustrate concentric multilamellar bodies on electron microscopy (**e**)

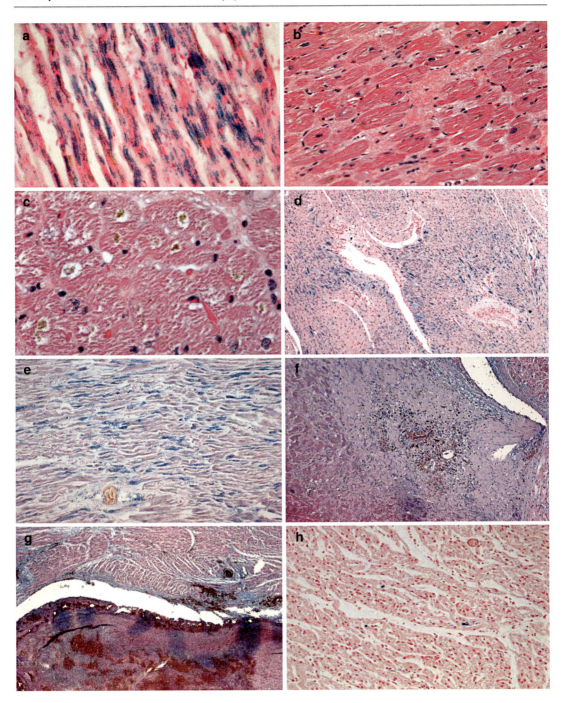

Fig. 1.41 Unusual combination of patient with restrictive cardiomyopathy due to siderosis-related myelodysplastic syndrome and Caroli disease. The cardiomyocytes show perinuclear and accumulation of hemosiderin (Fe) and interstitial fibrosis (**a**–**h**) at places or diffusely

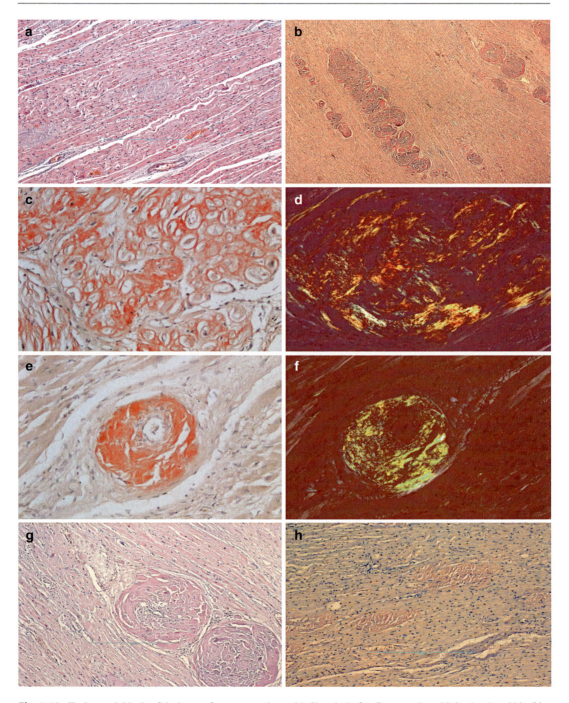

Fig. 1.42 Early amyloidosis of the heart of a young patient with fibrosis (**a**, **b**), Congo red positivity (**c**, **e**) and birefringence (**d**, **f**). At places, the amyloidosis is quite marked at level of the blood vessels (**g**, **h**)

ondary cytologic myocardial cell dysfunction. Thus, myocardial biopsies may be taken to investigate iron deposition and quantify hypersiderosis using either histochemical assessment-based or X-ray microanalysis-standardized scores. Cardiac biopsies should be analyzed by routine histology, transmission electron microscopy, and Perls Prussian blue (PPB) histochemical stain for detecting iron deposits in the cardiac tissue. PPB is a commonly used method in histology, histopathology, and clinical pathology to identify the presence of Fe in tissue

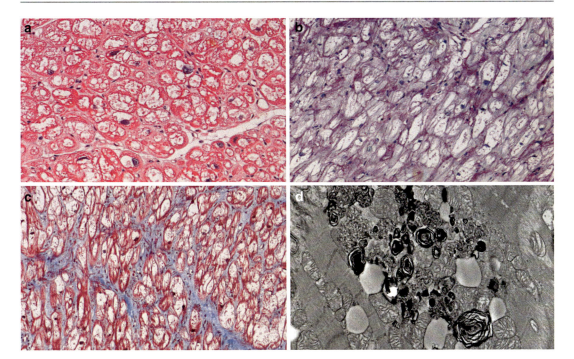

Fig. 1.43 Chloroquine toxicity of the cardiomyocytes in a young patient with restrictive cardiomyopathy showing pathology images of endomyocardial biopsy specimens. Light microscopy using hematoxylin and eosin stain demonstrates vacuolization of cardiomyocytes at ×200 magnification (**a**); light microscopy using periodic acid-Schiff stain at ×200 magnification shows that the vacuoles are periodic acid-Schiff-negative at ×200 magnification (**b**); there is an impressive interstitial fibrosis of the myocardium as highlighted using Movat pentachrome staining at ×200 magnification (**c**); electron microscopy at ×25,500 magnification demonstrates myeloid and curvilinear myeloid bodies, as well as vacuolated cytoplasm characteristic of chloroquine-associated cardiomyocyte damage (**d**). Myeloid and curvilinear myeloid bodies, as well as vacuolated cytoplasm characteristic of chloroquine-associated cardiomyocyte damage

or cell samples. This last method is the most used and cheaper method for follow-up patients. Masson trichrome histochemical stain is also useful for evaluating the extent of fibrosis, which may be related to early heart failure. The Buja and Roberts schema is highly reproducible and generally used in the routine diagnostics to quantify the visualized iron deposits in the cardiomyocytes. It has been suggested that an immunohistochemistry procedure may be used to highlight the iron-laden macrophages using antibodies against the CD68 antigen of the macrophages, but obviously histochemistry is cheaper than immunohistochemistry. Moreover, cardiomyocytes can be detected using antihuman desmin antibodies. Electron microscopy may also be very helpful revealing the presence of electron-dense cytoplasmic pleomorphic granules within cardiomyocytes, which correspond to small primary and large secondary lysosomes distributed randomly in the cytoplasm of the cardiac muscle cell. The dysfunctional cardiomyocytes often show discontinuities in the structure of sarcomeres and loss of myofilaments. There is also the loss of the sarcomeric arrangement by many siderosomes that disrupt the cardiac muscle cell and damage of both thick and thin filaments and disorganization of the Z-bands with the disappearance of the intercalated discs at places. Mitochondria show swelling and very few cristae. It does not seem that electron-dense iron-containing particles are found within other cytoplasmic organelles, such as Golgi apparatus. Nuclear changes are also observed and include elongation, an increase of the electron density, and increase of heterochromatin but no iron-containing particles. Basement membranes and cell membranes appear to be intact. X-ray microanalysis may be used to confirm the presence of iron inside of the primary lysosomes.

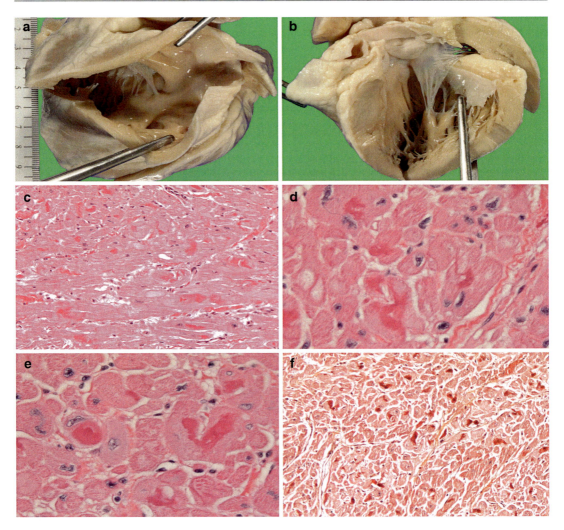

Fig. 1.44 Restrictive cardiomyopathy due to myofibrillar myopathy. In (**a**, **b**) dilatation of the heart, while (**c–f**) show the characteristic features of the cardiomyocytes exhibiting abnormal intracellular protein inclusions, which may be single or multiple. The inclusions are eosinophilic and may be congophilic (**c–f**), which may poses some differential diagnosis in older patients

1.3.5 Left Ventricular Non-compaction

- *DEF*: Spongy degeneration of the cardiac musculature with weak functionality and predisposing to heart failure due to an impaired ability to pump blood (Figs. 1.45 and 1.46).
- *EPI*: 8–12 per one million individuals per year.
- *ETP*: Mutations in the *MYH7* gene and *MYBPC3* gene account for about 1/3 of cases. LVNC can also be inherited as an X-linked inherited recessive disease.
- *CLI*: Symptoms vary considerably according to the degree of the involvement of the heart. In consideration of the malfunction, symptoms frequently overlap with those associated with dilated, hypertrophic, or restrictive cardiomyopathy.
- *GRO*: The cross sections of a cardiac specimen with LVNC show deep recesses extending to the inner half of the ventricle. These

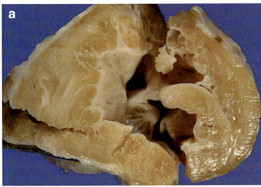
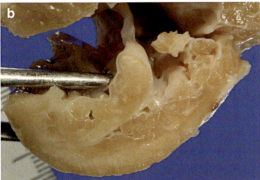
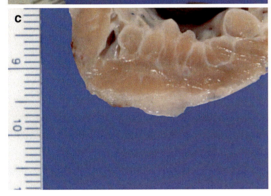

Fig. 1.45 Left-ventricle non-compaction (LVNC) with gross features characteristic of this syndrome, which is a relatively rare congenital disorder characterized by prominent trabeculations and intertrabecular recesses. There is the potential for thromboembolism, cardiac arrhythmias, and sudden cardiac death as adverse early or late effects

recesses are most conspicuously in the midventricle and toward the *apex cordis*. The gross appearance may vary from anastomosing trabecula to a quite smooth endocardial surface with narrow beginnings of the recesses to the cavity of the ventricle. The patterns of recesses may be divided into three types, including (1) anastomosing broad trabecula, (2) coarse trabecula like many papillary muscles, and (3) interlacing smaller muscle bundles or compressed invaginations, which are identified chiefly microscopically.

- *CLM*: Two patterns of myocardial structure in the superficial noncompacted layer are recognized, but in most hearts, the patterns overlap. The first pattern is characterized by anastomosing muscle bundles forming irregularly branching endocardial recesses with a staghorn appearance, while the second pattern reminds many small papillary muscles, creating an irregular appearance to the surface.
- *PGN*: It depends on the severity of the condition, but cardiac transplantation seems to be the only option when the heart fails.

1.3.6 Arrhythmogenic Right Ventricular Dysplasia/Cardiomyopathy (ARVD/C)

- *DEF*: Inherited, genetically determined CM characterized by structural and functional abnormalities, predominantly of the right ventricle resulting from progressive, fibrofatty replacement of the myocardium starting from the epicardium with an AD pattern and variable expressivity (Fig. 1.46).
- *SYN*: Uhl anomaly.
- *EPI*: 1:1000–1:5000 in the general population but accounts for up to 17% of all SCD of the youth.
- *ETP*: Point mutations in genes encoding for desmosomal proteins are the underlying genetic dysregulation. A triad of ARVC, palmoplantar keratosis, and woolly hair characterize an AR-inherited subtype (Naxos disease).
- *CLI*: RV dilatation, arrhythmias, and SCD are the characteristic triad. CT angiography is particularly useful for the detection of akinetic or dyskinetic bulging in the triangle of dysplasia, but MRI with gadolinium enhancement is superb for its potential in detecting regional and diastolic ventricular dysfunctions, which can be the early signs of ARVD/C. Two primary criteria or one major plus two minor criteria or four

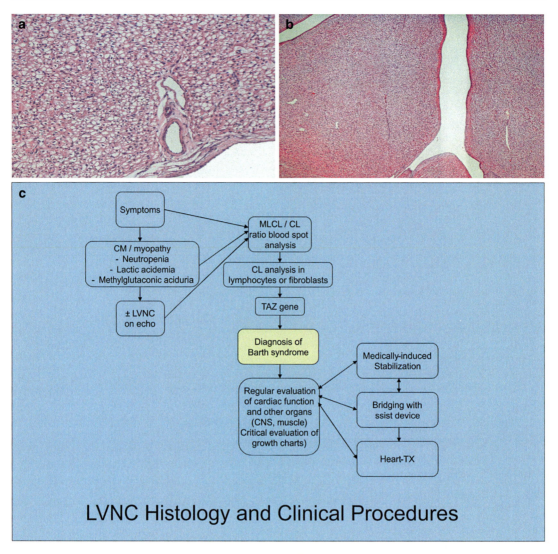

Fig. 1.46 Histology of LVNV showing the recesses and trabeculations using hematoxylin-eosin and connective tissue stain (**a**, **b**). In (**c**) are presented the clinical procedures. In the early stages, a vacuolation may be prominent. It has been proved that early metabolic stabilization and regular cardiac evaluation are a must for the young affected infants with LVNC. A surveillance plan for patients harboring Barth syndrome (a rare X-linked inherited disorder characterized by cardioskeletal myopathy, peripheral neutropenia, psychomotor and growth retardation, and 3-methylglutaconic aciduria) before heart transplantation has been suggested

minor criteria are essential for the diagnosis of ARVD/C. Six categories include global and regional dysfunction and structural alterations (I), wall morphology (II), re-/depolarization, conduction disturbances, arrhythmias (III–V), and family history (VI) (Marcus et al., 2010). The significant rules of category I include severe dilatation and reduction of RV ejection fraction with no/mild LV dysfunction, localized RV aneurysms, and severe segmental RV dilatation. The minor criteria of category I entail mild global RV dilatation and ejection fraction reduction with normal LV function, mild segmental dilatation of the RV, and regional RV hypokinesia. Category II includes fibrofatty replacement of the myocardium (only one major criterion), while categories III–V involve one primary criterion of ε-waves or localized prolongation of the QRS complex in the right precordial leads and three minor criteria, including inversion of

T-waves in the right precordial leads, late potentials, left bundle branch reentrant ventricular tachycardia, and recurrent ventricular extrasystoles. Finally, category IV relies on the family history confirmed at autopsy or surgery (primary) and family history of premature SCD in a young (<35 years) with suspicion of ARVD/C and family history (two minor).

- *GRO*: Fibrofatty replacement of the right ventricular myocardium and occasionally of the left ventricular myocardium leading to wall thinning and aneurysm formation is the hallmark. These changes are typically located at the inferior-apical and infundibular walls. However, limited diagnostic specificity should be determined in the fibrofatty replacement of the right ventricle, because it may be a common finding in obese and elderly individuals.
- *CLM*: Fibrofatty degeneration of the right ventricular myocardium (Fig. 1.47). Morphologically, there is focal thinning to the absence of the right ventricular myocardium due to the replacement of muscular by adipose and fibrous tissue. The diagnosis is based on well-standardized major and minor criteria detected by EKG, echo, angiography, MRI, or radionuclide scintigraphy.
- *DDX*: See Box 1.4 for VT for ARVC.
- *PGN*: An accurate postmortem and correct diagnosis of ARVD/C is critical for any surviving members, because this disease is hereditary. It is important to emphasize that SCD can be prevented and has obvious implications for relatives of individuals affected with ARVD/C.

1.3.7 Lamin A/C Cardiomyopathy

Lamins are type V intermediate filament proteins. They are known to provide as major structural components of the nucleus. It is known that there are two major types of lamins, A type and B type. A-type lamins include lamins A and C. They are alternative splice variants arising from a single gene LMNA. Also, B-type lamins include lamins B1 and B2, products of two separate genes, LMNB1 and LMNB2. The human LMNA gene contains 12 exons located on chromosome 1q21.2–21.3. LMNA encodes A-type lamins: A, AD10, C, and C2 through alternative splicing. The human LMNA gene was first identified in 1986, but it was not until 1999 that the LMNA mutation was found responsible for Emery-Dreifuss muscular dystrophy (EDMD), muscular dystrophy that involves humans. Currently, laminopathies encompass four groups, depending on the affected tissue: (1) striated muscles (skeletal muscle, heart), (2) adipose tissue (lipodystrophy), (3) nervous system, and (4) aging syndrome of accelerated type. LMNA mutations are known to explain for 6–8% of DCM with conduction defect and foreshadow a poor prognosis and are associated with a high risk for SCD (Fig. 1.48).

> **Box 1.4 Ventricular Tachycardia for ARVD/C–DDX**
>
> Congenital heart disease (repaired TOF, Ebstein's tricuspid anomaly, ARVD/C, ASD, PPVR) and acquired heart disease (tricuspid valve disease, pulmonary hypertension, right ventricular myocardial infarction, bundle-branch reentry tachycardia) as well as different cases (pre-excited AV reentry tachycardia, idiopathic right ventricular outflow tract tachycardia).
>
> Notes: ARVD/C, arrhythmogenic right ventricular dysplasia/cardiomyopathy; ASD, atrial septum defect; AV atrio-ventricular; PPVR, partial anomalous pulmonary venous return; TOF, tetralogy of Fallot.

1.3.8 Mitochondrial Cardiomyopathies

The heart is the second most common visceral organ to be affected by *mitochondriopathies* (MCPs) (Fig. 1.37). MCPs with cardiac abnormalities include mitochondrial encephalopathy, lactic acidosis, and stroke-like episodes (*MELAS*), myoclonic epilepsy and ragged red fibers (*MERRF*), Kearns-Sayre syndrome (*KSS*), chronic progressive external ophthalmoplegia (*CPEO*), Leber hereditary optic neuropathy

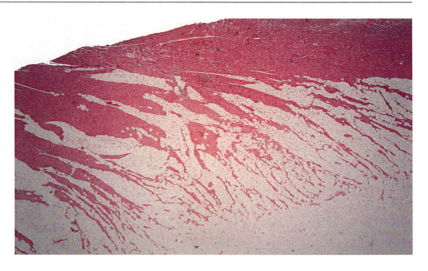

Fig. 1.47 Arrhythmogenic right ventricular cardiomyopathy (ARVC) is characterized by the fibro-adipose substitution of the myocardium as highlighted in this hematoxylin-eosin-stained slide. Although considered pathognomonic for ARVC, there a few other conditions that may exhibit the fibro-fatty replacement of the myocardium

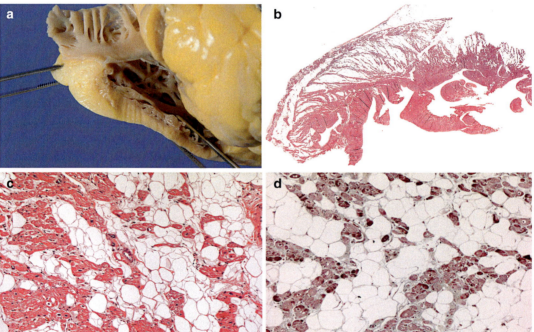

Fig. 1.48 Lamin-A/lamin-C cardiomyopathy is mimicker of arrhythmogenic cardiomyopathy. Gross pathological (**a**) and histological (**b–d**) characterization of the heart shows substantial fatty change and replacement of the myocardium. In (**b**, **c**), the right ventricle shows remarkable fatty infiltration of the right ventricular myocardium (hematoxylin-eosin staining, 1.5× original magnification). At places, the ventricular septum of the left ventricle shows some cardiomyocyte disarray in some areas and replacement fibrosis (hematoxylin-eosin staining and Masson trichrome staining, ×100 original magnification) (**d**)

(*LHON*), neurogenic muscle weakness, ataxia, retinitis pigmentosa (*NARP*), and *Leigh syndromes*. Manifestations of the cardiac involvement include impulse anomalies (generation and conduction disturbances), HCM, DCM, and left ventricular non-compaction (LVNC). The EKG often shows atrial fibrillation, AV block, Wolff-Parkinson-White (WPW) syndrome, bundle branch block, QT prolongation, or ST- and T-wave abnormalities. Occasionally, there have been reports of Takotsubo phenomenon with apical ballooning of the left ventricle associated with MCPs. LVNC manifested as a meshwork of interwoven myocardial fibers and cords distally to the papillary muscles and lined with the endocardium. Both cardiologic and neurologic manifestations may be subclinical. It means that every patient with MCP presenting with minimal cardiologic anomalies should undergo a complete neurological investigation. MCPs are also often associated with either arterial hypertension or orthostatic hypotension.

Mitochondrial DNA (mtDNA) defects, which can occur as both deletions and tRNA point mutations, have been associated with CMP. Ultrastructural changes of mitochondria of patients with idiopathic CMP and pathological mtDNA mutations include ring mitochondria, giant mitochondria with membrane fusion, and circular, concentric, whorled, and undulated cristae. Cristolysis, mitochondrial swelling, and variable size of mitochondria should be considered as nonspecific changes. Approximately 1/4 of patients with idiopathic DCM and ultrastructural changes of mitochondria show pathological mtDNA mutations. All pathological variations are usually of heteroplasmic type and should be of clinical significance if the mutated DNA is higher in the heart than in peripheral blood. Interestingly, these mutations carry the same features that identify pathogenicity in mitochondrial encephalopathies and are associated with a decrease of cytochrome C oxidase (Cox) activity as well.

1.3.8.1 mtDNA Depletion Syndrome

Clinically heterogeneous disorders are characterized by a severe quantitative reduction of total mitochondrial DNA and usually inherited with an autosomal recessive pattern with a 25% recurrence risk. There are three clinical phenotypes, which are quite distinct and include *hepatocerebral*, *myopathic*, and *cardiomyopathic*. Hepatocerebral mtDNA depletion syndrome has an onset in the first 6 months of age with hepatocellular disease and early progressive liver failure with lactic acidosis and hypoglycemia. Fatality of these cases is often the rule, even after liver transplantation. Intrauterine growth restriction has been described with mtDNA depletion disorders as well as cardiomyopathy, atrial, and ventricular septal defects and occasionally with complex congenital heart disease, such as the hypoplastic left heart syndrome.

1.3.9 Cardiac Channelopathies and Sudden Cardiac Death

Cardiac channelopathies are a growing and evolving group of clinical syndromes that affect the cardiovascular electrical system, particularly the myocardial ion channels, which include Na^+, K^+, and Ca^{2+} channels. Channelopathies have also been addressed remarkable interest for the association of sudden death, specifically SCD of the young and sudden infant death syndrome. Cardiac channelopathies occur when one of the proteins forming the channels is malfunctioning due to a germ line or altered somatic functionality. The result arising from these changes points to abnormalities of the electrical properties of the heart. *Pro*-arrhythmic (or sometimes called also arrhythmogenic) events are the consequence. These syndromes are included in Box 1.5.

In the USA, approximately 335,000 people each year succumb to death, specifically SCD. In some of them, cardiac channelopathies are the etiologic factor, although the estimate is unknown. Recently, some ethnic differences have been delineated using meta-analyses and observational studies (Kong et al. 2017; Sergi 2019).

Long QT Syndrome (LQTS)

- *DEF*: Cardiac channelopathy, in which there is a delay of repolarization showing that the cardiomyocytes take longer than usual to recover electrically after each beat, mostly due

---

Box 1.5 Cardiac Channelopathies
- Arrhythmogenic right ventricular dysplasia/cardiomyopathy (ARVD/C)
- Brugada syndrome (BrS)
- Catecholaminergic polymorphic ventricular tachycardia (CPVT)
- Long QT syndrome (LQTS)
- Short QT syndrome (SQTS)

to alterations in the sodium or potassium channels (Na^+/K^+ channels). There are 14 types of LQTS with several mutations.
- *EPI*: 1:2000–1:3000 in the USA/Canada population.
- *CLI*: Sudden, temporary loss of consciousness (syncope). Individuals with LQTS are at risk of developing an abnormally rapid heart rhythm ("*Torsades de Pointes*").
- *DGN*: The normal QTc is 0.35–0.46 s with 95% of people having a QTc between 0.38–0.44 s, but in 2.5% of the average population, prolonged QT interval are found, and 10–15% of patients affected with LQTS have a standard QT interval. The LQTS "diagnostic score" is used for the diagnosis of LQTS. Various criteria are used to calculate the score, and the LQTS diagnosis relies on ≥4 points (high probability), while with LQTS score ≤1, there is a low probability of having LQTS.

Short QT syndrome (SQTS) is an inherited cardiac channelopathy characterized by a short QT interval, which harbors an increased risk of arrhythmias. Diagnosis is founded on the evaluation of symptoms, including cardiac arrest and palpitations, a collection of detailed patient's family history, and 12-lead ECG. This investigation can at times be challenging due to the variable range of QT intervals identifiable in healthy subjects.

Brugada syndrome (BrS) is a cardiac channelopathy of genetic origin, in which the cardiomyocytes have altered electrical recovery, which occurs after each beat. This syndrome is due to genetic alterations in cardiac ionic channels, and the diagnosis relies specific morphology of the electrocardiogram (EKG): a "coved-type" (smooth, peaking) ST-segment elevation of ≥2 mm in precordial leads V1–V3 with inverted T waves and complete or incomplete right bundle branch block (electrical conduction delay to the right side of the heart). Affected individuals may develop loss of consciousness, tachycardia, and SCD. Another EKG pattern showing "saddleback-type" ST segment morphology (also quite universally known as type 2) may also be present. Genetically, BrS is AD inherited (1:1000 in the general population), and the central gene associated with this disease is *SCN5A*, which controls the cardiac Na^+ current. If there is a family history of unexplained syncope or SCD, including death in an otherwise healthy person or upon recognition of a BrS pattern on any EKG, the general practitioner or primary physician should seek further evaluation and refer family members for necessary screening in a third-level cardiac center.

Catecholaminergic polymorphic ventricular tachycardia (CPVT) is an inherited cardiac channelopathy (1:10000 in the general population), in which the cardiomyocytes are electrically unstable when under stress and individuals with this condition are prone to develop ventricular tachycardia. CPVT involves one of two identified abnormal genes (*RYR2* or ryanodine receptor gene or *CASQ2* or calsequestrin 2) that control the proteins that are responsible for Ca^{2+} handling and storage within the cardiomyocytes at the sarcoplasmic reticulum. This syndrome is often undetected before death and is not recognized as the cause of death. CPVT is readily treatable, and most deaths may be preventable.

Electronic medical devices are part of routine medical practice in many countries worldwide and have downloadable memories in most of the cases. Electronic medical devices with downloadable memories include implantable cardiac pacemakers, defibrillators, drug pumps, insulin pumps, and glucose monitors. Weitzman summarized in a very educational way the electronic medical devices that may be of importance for the pathologist, particularly at a time of autopsy but also at the time of surgical pathology when a surgical specimen is delivered to the pathology department with such a device. Weitzman suggested guidelines for the examination and reporting of electronic medical devices received at departments of laboratory medicine and pathology. With regard to the autopsy, in addition to the above-listed guidelines, it may be essential to record date and time of autopsy, type of autopsy (hospital, forensic/medicolegal), activity, and location at death or at onset of terminal episode involving electronic medical devices that lead to death (e.g., sleeping at home, sitting in office, walking near home, working in the backyard, driving), traumatic injury (e.g., gunshot wound, stab wound, blunt force injuries, car accident), exposure to electromagnetic sources (e.g., magnetic resonance imaging, radiation therapy, metal detector, electronic antitheft surveillance systems, lithotripsy, electrocautery, high-voltage power lines, cellular telephone, aircraft communication systems), anatomic sites (anatomic location of generator and leads, evidence of twiddler's syndrome, other), abandoned edges (fibrosis, thrombus, adherence), complications (e.g., infection, lead migration, perforation of viscera, other data), and device memory temporal (date and time) interrogation with relevant copies of downloaded information, evidence of device malfunction, and battery status.

1.4 Ischemic Heart Disease, Myocardial Dysfunction, and Heart Failure

1.4.1 Ischemic Heart Disease

Ischemic disorders of the myocardium seem, at first glance, a topic of the elderly, but myocardial infarction can occur in newborns, children, and youth representing a challenge for both clinician and pathologist (Fig. 1.49). By far, in more than 90% of cases of the general population, the most common cause of SCD is a decrease of the coronary blood flow secondary to atherosclerosis, thrombosis, vasospasm, arteritis, emboli, and increased right atrial pressure. However, these data are not relevant for pediatric age, apart from some thromboembolic disease or cardiac aneurysms of the coronary arteries as observed in conditions such as Kawasaki disease. Ischemic heart disease is satisfactorily described in adult pathology books.

1.4.2 Myocardial Dysfunction

Group B streptococcal (GBS) infection is a significant cause of neonatal morbidity and mortality, and systemic hypotension characterizes the severe early-onset form of GBS infection, decreased cardiac output, and hypoxemia, which lead to multi-organ failure (MOF). GBS β-hemolysin/cytolysin seems to be at the basis of the molecular effectors of GBS-induced cardiac dysfunction. GBS sepsis induces an impairment of the cardiomyocyte viability and function and promotes cardiomyocyte death upon expression of its β-hemolysin/cytolysin. This aspect has enormous relevance for neonatologists and perinatal pathologists who are required to perform perinatal autopsies. Meningococcal sepsis represents another catastrophic event. The systemic meningococcal disease is probably the most common cause of death in children in several developed countries of the Western world, and mortality is approximately 10%. Meningococcal septicemia and septic shock are responsible for mortality ranging between 20% and 40%, despite intensive care. In fatal cases of meningococcal sepsis, autopsy findings are highlighted by a constellation, including hypovolemic shock changes, disseminated intravascular coagulation, and MOF. The mechanisms responsible for inducing shock in a setting of meningococcal sepsis include an increase in vascular permeability leading to a "capillary leak syndrome," myocardial dysfunction, and diffuse intravascular thrombosis with the development of *purpura fulminans* and MOF. The problem occurs when patients become refractory to inotropic stimulation, which leads to low-output and failure of multiple organs. Cardiac troponin I, which is released from injured cardiomyocytes, is a sensitive and specific marker of myocardial cell death and seems to be related to the severity of disease and degree of cardiac dysfunction. There is evidence for cardiomyocyte death in meningococcal sepsis, although myocardial depression of septic shock is often reversible in some survivors. This aspect is particularly relevant for children because the pediatric myocardium, which is still in a growth phase, may have more reserves and has an essential capacity for restoration of function following injury.

1.4.3 Heart Failure

It is a condition that usually develops in adults, but it may be present in childhood and even in infancy, and the etio-pathogenesis is variable. Among several causes, iatrogenic and

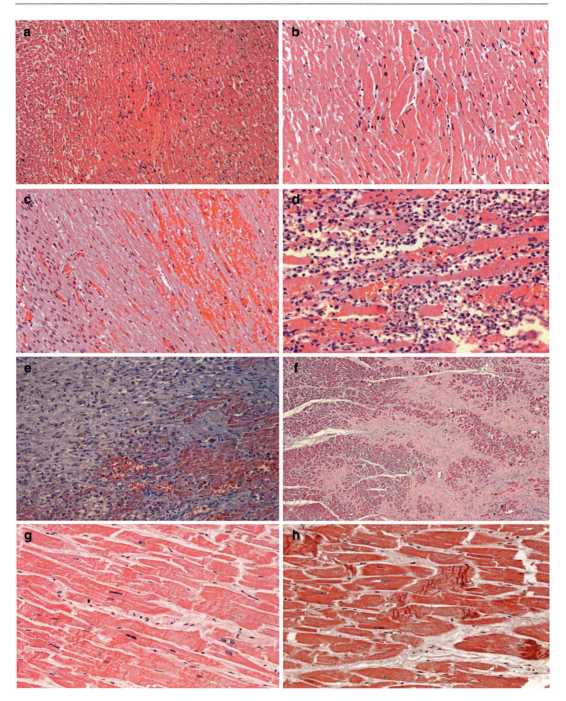

Fig. 1.49 Several areas of cardiomyocyte necroses at different ages of ischemia with hypereosinophilia, neutrophilic granulocytic recruitment, and scar formation (**a–f**). Contraction bands may be identified on both H&E and Movat pentachrome staining (**g, h**). They represent early ischemic changes of the cardiomyocytes

septic cardiomyopathy may be two reasons for this condition in childhood (Buschmann et al. 2017; Sergi et al. 2017; Turillazzi et al. 2016). Following the damage of the heart, it becomes weakened. Heart failure is defined to occur when the pumping action of the heart is not strong enough to move blood forward and unable to lodge the flow of blood back from the lungs with

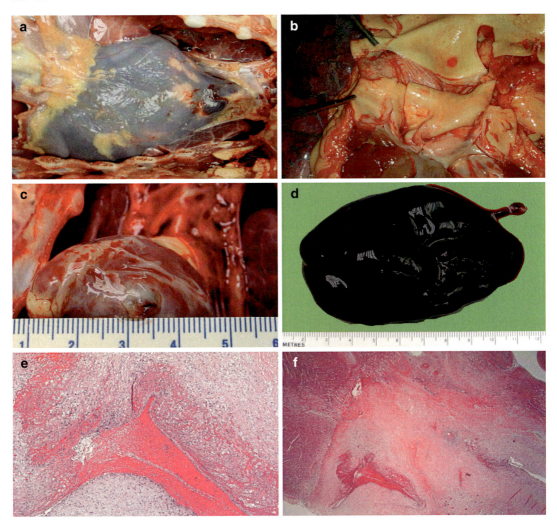

Fig. 1.50 Hemopericardium (**a**) with filling of the pericardial sac with blood and identification of the aortic dissection as cause of the hemopericardium in a young patient with Marfan syndrome. However, rupture of the heart at the free wall of the left ventricle with hemopericardium may be iatrogenic (e.g., surgical ablation for cardiac arrhythmias) with hemopericardium and blood clot recovered from the pericardial sac (**c–f**)

congestion in several peripheral parts of the body and centrally into the lungs with a potential life-threatening condition (acute pulmonary edema) (Figs. 1.50 and 1.51). In septic cardiomyopathy, several mechanisms are under investigation, including the calcium desensitization of the contractile apparatus, disruption of contractile units, presence of myocardial edema unable to be dislodged, and abnormal pro-fibrotic mechanisms of remodeling. Dietary fats contribute early to aortic and carotid intima-media thickness and distensibility (Laitinen et al. 2020).

1.5 Valvular Heart Disease

1.5.1 Endocarditis and Valvar Vegetation

Infective endocarditis colonization or invasion of the valves by infective organisms which lead to the formation of bulky, friable vegetations on flow side near the free edge with acute and subacute forms, depending on the virulence of the organism creating turbulent flow predispose: septal defects, valvular stenosis/reflux, and prosthetic valves. In 65% of cases, endocarditis (acute) is

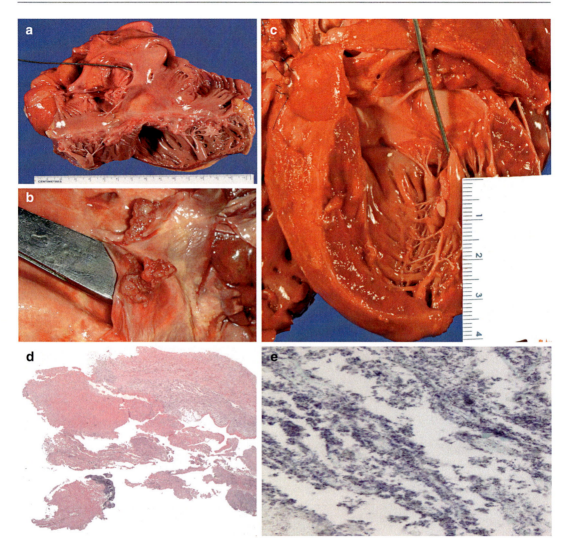

Fig. 1.51 Gross photographs (**a-c**) and microscopic photographs (**d-e**) of infective endocarditis (d, hematoxylin and eosin staining, original magnification x12.5; e, Gram staining, original magnification x630). This young adult suffered from a bacterial endocarditis and developed septic cardiomyopathy despite hesitant treatment. The force-generating capacity of the myocardium are significantly reduced in patients with infective endocarditis. No matter the age, these patients experience that an increased resting force is required to achieve a given force amplitude. Young adults are not immune from atherosclerosis, which is probably mediated by dyslipidaemia, dysregulation of glucose metabolism, and lipodystrophy at first, and accompanied by inflammation, endothelial dysfunction, and a prothrombotic state, secondly

caused by *Streptococcus* (*viridans*, *bovis*, *fecalis*, etc.), a low virulent type, while the principal cause of subacute endocarditis (20–30%) is caused by *Staphylococcus aureus*, a more virulent type. Other microorganisms include *E coli*, *N. gonorrhoeae*, *Candida* spp., and *Aspergillus* spp. Nonbacterial thrombotic endocarditis (marantic endocarditis) is the precipitation of small sterile masses of fibrin on valve leaflets on the line of closure, loosely attached and commonly involving multiple valves and often in chronic diseases (e.g., malignant tumors). Endocarditis of SLE (Libman-Sacks disease, systemic lupus erythematosus) mitral and tricuspid valvulitis is occasionally seen in SLE with numerous small vegetations of the leaflets, usually on the back side of the valves, mainly the posterior mitral leaflet with potential extension to the endocardium. Lambl excrescences are small, heaped-up lesions on flow side of valves characterized by a collection of fibrins at sites of endothelial damage (Figs. 1.52, 1.53, and 1.54).

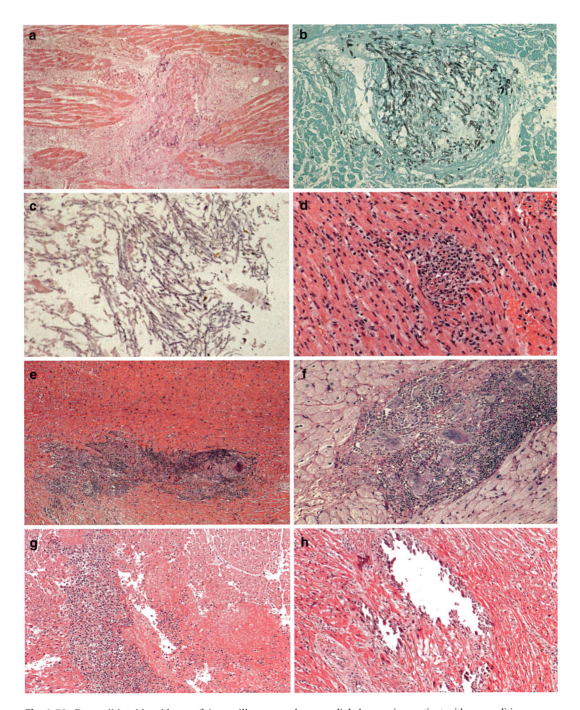

Fig. 1.52 Pancarditis with evidence of *Aspergillus* spp. and myocardial changes in a patient with pancarditis

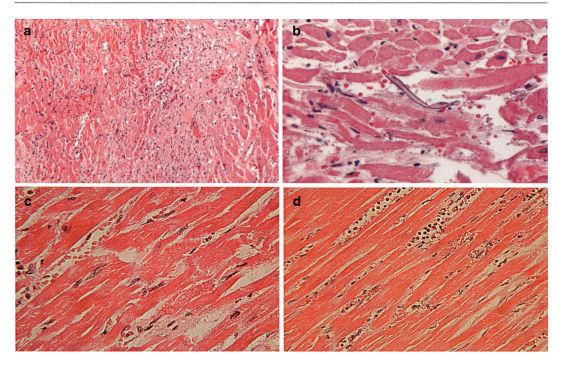

Fig. 1.53 Sepsis-carditis may show some necrosis of the cardiomyocytes (spotty), the presence of microorganisms, and the recruitment of neutrophils in the interstitium

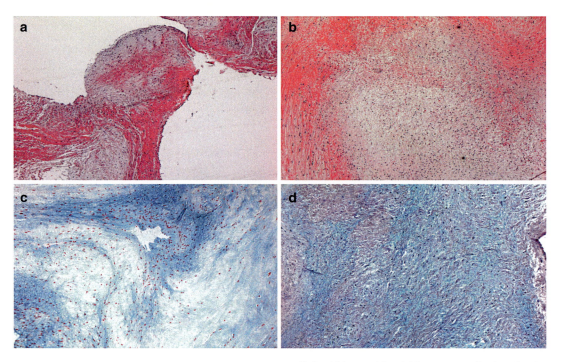

Fig. 1.54 Valve dysplasia with the valves being graded according to Becker's criteria, which are morphometrically validated. In assessing valvar dysplasia, the differentiation and detachment of the valve from the ventricular wall should be considered. The concept of valvar dysplasia is supported by substantial morphologic and morphometric measurements with change of the mucopolysaccharide content of the valves, among others

1.5.2 Mitral Valve Prolapse

- *DEF*: It is a degeneration of the mitral valve and is also called "floppy mitral."
- *EPI*: Widespread (5–7% general population), ♂:♀ = 2:3, usually 20s–40s.
- *ETP*: ± Marfan syndrome.
- *CLI*: Usually asymptomatic but mid-systolic click; if valve leaks, also late systolic murmur, intercordal (between the chords) ballooning (hooding), elongated *chordae tendineae*.
- *CLM*: Myxoid degeneration of *zona fibrosa* with thickening of *zona spongiosa* (standard ratio 1:1).
- *PGN*: Risk of infective endocarditis, mitral insufficiency, arrhythmias, or SCD.

1.5.3 Calcific Mitral Annulus and Calcific Aortic Stenosis

The *calcific mitral annulus* is the presentation of stony hard beading behind valve leaflets at the base, without accompanying inflammation, rarely affecting valve function. It is usually seen in elderly patients, often in association with ischemic heart disease, but young patients with TGA surgically operated with palliative procedures may present some aspects of such anomaly.

Calcific aortic stenosis is an idiopathic, male-predominant calcific degeneration of the aortic valve, which usually occurs with a predisposing congenital *uni*cuspid valve with children that become symptomatic before the age of 15 years or a *bi*cuspid valve later in life (generally after the 50s). In more rare occasions, it may also occur with a standard (tricuspid) valve, but it is usually seen in the elderly. Grossly, there are calcified masses within the valve leaflets and visible protruding from both sides of the cusps. The consequence is a valve with practically immobile leaflets with stenosis, insufficiency, or both with final left ventricular concentric hypertrophy and insufficiency. Interestingly, the degeneration starts at free edges in bicuspid valves, while the beginning is at the bases in tricuspid valves. Two lines of therapy are considered, including either valve replacement or balloon valvuloplasty.

1.5.4 Rheumatic Heart Disease (RHD)

It is a major problematic issue in the healthcare of developing countries. RHD is the cause of most of the cardiovascular mortality in young individuals (5–15 years), leading worldwide to approximately 250,000 deaths per year (Fig. 1.30).

- *ETP*: Abnormal autoimmune response to group A streptococci infection in a genetically susceptible host. RF occurs 1–5 weeks following a streptococcus group A pharyngitis (streptococcal infections elsewhere are not rheumatogenic). Antibodies cross-react with streptococcal antigens or autoantibodies triggered by streptococcal antigens and localize to the sarcolemmal membranes.
- *CLI*: Fever, migratory polyarthritis (in adults), carditis, subcutaneous nodules (giant Aschoff bodies), *erythema marginatum* of the skin (targetoid lesions), Sydenham chorea.
- *CLM*: Edema, collagen fragmentation, and fibrinoid change dominate the early phase (2–4 weeks after pharyngitis), while Aschoff bodies, i.e., subendocardial or perivascular foci of fibrinoid necrosis with lymphocytes, later with macrophages which can become epithelioid (granulomatous) and caterpillar cells (large multinucleated giant cells with chromatin clumping lengthwise in nucleus, appearing owl-eyed on cross section) dominate the subsequent phase. MacCallum plaques are map-like thickening of endocardium over lesions, typically located in the left atrium.
- *PGN*: Chronic rheumatic heart disease develops in only a small number of patients with acute rheumatic fever (ARF) ≥10 years (thickening of valve leaflets, fibrous bridging across commissures, calcifications, "fish-mouth" deformity, thickening and shortening of *chordae tendineae*) with mitral involvement alone in 65–70% and mitral and aortic participation in 25%, rarely other valves.

1.5.5 Prosthetic Valves

The pathology of the prosthetic valves includes calcifications, dysfunction (e.g., occlusion) due to tissue overgrowth, infective endocarditis, most commonly *Staphylococcus*, paravalvular leaks, tears in leaflets/structural breakdown of the valve, and thrombosis and thromboembolism.

Pulmonary homografts are often used in cardiac surgery, and their primary use is the reconstruction of the right ventricular outflow tract due to obstruction in children and for reconstruction in right ventricular outflow tract in patients undergoing the Ross procedure for the valvular disease of the aorta. Cryopreserved homografts may be susceptible to late stenosis, and an underlying immune mechanism may be the cause. Relatively poor results have been recorded for the reconstruction of the left ventricular outflow tract.

1.6 Transplantation

1.6.1 Allograft Cellular Rejection

Cardiac transplantation is currently one of the most frequent transplantations, particularly in some developed countries. In the Western of Canada, cardiac transplantation is leading in many statistics for quality assurance following other organs. In the working formulation of the International Society for Heart and Lung Transplantation (ISHLT) of the 1990s, the grading of allograft rejection was mainly based on the amount of inflammatory infiltrate and the cardiomyocyte damaging evidence. Kappa factor was highly variable among institutions and pathologists following its publication. Pathologists, cardiologists, and cardiac surgeons met a few years later in 2001 to discuss their experiences using the 1990s' classification. Currently, the 2001 revised classification is used (Figs. 1.55, 1.56, and 1.57).

Most probably, one of the significant discordances in evaluating endomyocardial biopsies for cellular rejection is the correct identification of cardiomyocyte damage. In the revised ISHLT classification, it is described as "clearing of the sarcoplasm and nuclei with nuclear enlargement and occasionally prominent nucleoli." However, this definition may be quite restrictive, and some more features may be indicative of cardiomyocyte damage in the setting of cardiac allograft rejection. Cardiac pathologists list some myocyte injuries in the setting of a cell itself or at intercellular levels. Thus, myocyte injuries include vacuolization, the irregular border of the cardiomyocyte, ruffling of the cardiomyocyte sarcoplasmic membrane, perinuclear halo, and at intercellular level with splitting or branching of cardiomyocytes and encroachment of the cardiomyocytes with their partial disruption. The presence of nuclear pyknosis and hypereosinophilia would suggest necrosis of the cardiomyocytes. A lymphocyte is surrounding or accompanying the myocyte damage mean rejection. However, it is important not to make a mistake looking at a previous biopsy site as indicative of rejection or lesions called "Quilty lesions," i.e., dense circumscribed subendocardial or intramyocardial lymphocytic infiltrates, seen in 15% of posttransplant biopsies, often composed mainly of B-lymphocytes, unlike rejection which is predominantly constituted by an involvement of T lymphocytes. Quilty lesions seem to occur exclusively in the endocardium of cardiac allografts. They have been explained as effects of cyclosporine-based immunosuppression, personal responses to cyclosporine A, and simultaneous infection with Epstein-Barr virus. The Quilty lesions have an extracellular matrix, i.e., collagen, between the infiltrating lymphocytes and some capillaries in the middle of the infiltrate. The Quilty type B lesion or infiltrating lymphocytes similar to Quilty of subendocardial location or Quilty type A of the 1990s' classification should be considered as foci of cellular rejection, but immunohistochemistry and intradepartmental/interdepartmental consultations may be needed.

Other than the definition of cardiomyocyte damage and avoiding calling rejection a Quilty lesion, it is important to stress about the adequacy and proper handling of the endomyocardial biopsy specimens. Both competence and processing of tissue biopsies are critical in any ser-

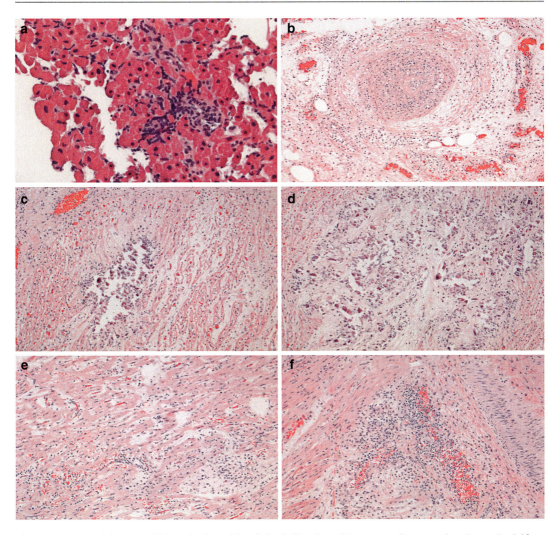

Fig. 1.55 Heart with acute cellular rejection with cellular infiltration of the myocardium, arteriopathy, and calcifications (**a–f**)

vice of cardiac transplantation service. An endomyocardial biopsy is the most sensitive indicator of rejection, but appropriate processing and an adequate number of these are essential. We need at least four pieces of tissue with the myocardium to be a proper sampling. However, six pieces of tissue, depending on the size of the bioptome used, are necessarily required for a correct histopathologic interpretation. Some biopsies less than four should be considered suboptimal because if three pieces taken show no rejection, there is 5% chance of missing a mild cardiac allograft cellular rejection. If four pieces are taken, the false-negative rate is reduced to 2%, i.e., from 1 in 20 transplanted patients to 1 in 50 transplanted patients. The 2004 revised working formulation recommends at least three biopsies, but in many cardiac transplantation centers, this number is considered suboptimal. Each biopsy piece has to contain 50% or more of myocardium and no evidence of previous biopsy site or scar. It is essential that endomyocardial biopsies that do not meet these criteria should be diagnosed as "inadequate biopsy." The ideal fixative is 10% phosphate-buffered formalin at 25 °C avoiding using cold fixatives that may form contraction bands. The use of unbuffered formalin solution has a pH of 3–4 and is unsuitable for good laboratory practices. The acidity can react with hemoglobin in the myocardial tissue to pro-

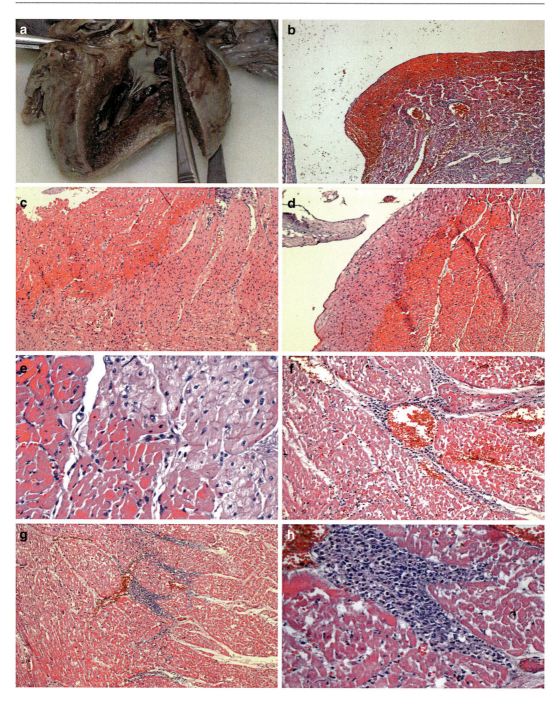

Fig. 1.56 Heart with hyperacute cellular rejection (ACR3R) with remarkable gross and microscopic photographs showing extensive necroses not only at subendocardial level (**a-h**). The recruitment of inflammatory cells can be seen particularly evident in (**f–h**)

1.6 Transplantation

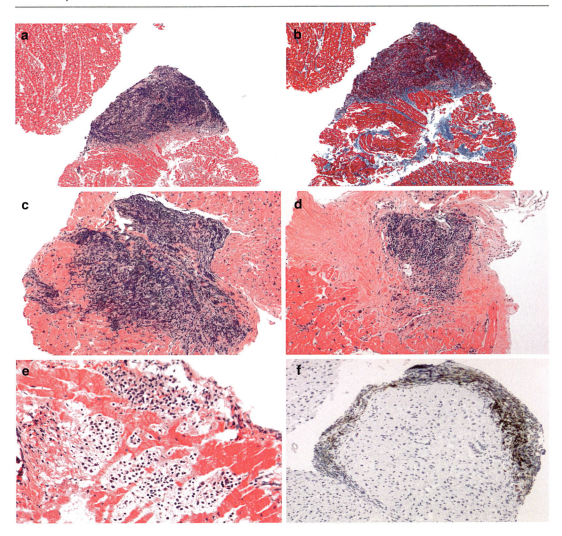

Fig. 1.57 Quilty lesions with microphotographs (**a**–**e**) from hematoxylin-eosin-stained tissue sections showing subendocardial and interstitial clusters of lymphocytes without evidence of myocardiocytic cell damage. In (**f**), there is the immunohistochemical demonstration that these cells are B lymphocytes (anti-CD20, avidin-biotin-complex immunostaining, 100× original magnification)

duce dark brown acid formaldehyde hematin, which precipitates and complicate histological slide interpretation. It is key to adjust the 10% formalin solution to a neutral pH using sodium phosphate. The frequency of biopsy surveillance following heart transplantation varies among institutions worldwide. It has been suggested that for the first month, biopsies are taken once weekly and every 2 weeks for the second month. Between the third and the end of the first year, one biopsy every 6–8 weeks should be the rule. This frequency can be decreased quarterly, biannually, or annually after the first year.

1.6.2 Allograft Humoral Rejection

In the setting of an immunopathogenic process of activation of the complement system, the allograft and the patient can experience a humoral immune response. The allograft humoral rejection can be elicited simultaneously with the cellular rejection. C4d is the standard marker performed on the frozen tissue (immunofluorescence) to identify antibody-mediated rejection (AMR) (Fig. 1.58). In the absence of frozen tissue, immunohistochemistry is performed. Immunologic evidence of AMR includes the deposition of complement

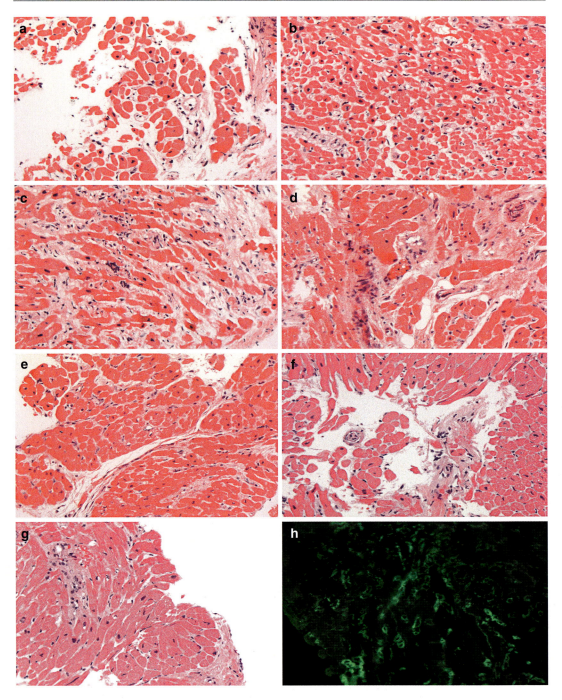

Fig. 1.58 The heart in the setting of acute humoral rejection may show capillaritis, thrombosis, interstitial fibrosis (**a–g**), and circumferential C4d deposition as highlighted by immunofluorescence (**h**)

and immunoglobulins (IgG, IgA, and IgM) in capillary walls, intraluminal aggregation of macrophages within capillaries using an anti-CD68 monoclonal antibody or double immunohistochemistry using a marker for the endothelium (anti-CD31 or anti-CD34), and circumferential C4d deposition within the capillary walls by paraffin immunohistochemistry. It is also important to invite clinicians to assess the patients for circulating antibodies to HLA class I or II as well as non-HLA donor antigens. Histologic clues for an AMR include worrisome features, such as interstitial hemorrhage, capillary fragmentation, mixed inflammatory infiltrate, pyknosis of the endothelial cells, karyorrhexis, and marked edema (Fig. 1.58).

Risk factors for eliciting an AMR include pregnancy, previous transplantation, blood transfusions, sensitization by OKT3 immunosuppressive regimen, use of ventricular assist device (VAD), the occurrence of B-lymphocyte flow cytometry cross-match, and increased panel-reactive antibodies. Rarely, but most often fatally, *hyperacute rejection* is encountered. This kind of rejection is a graft injury by preformed antibodies and occurs quickly following implantation of the graft. The time variable is within minutes to very few hours. Experimental discordant xenografts have been essential to understanding the phenomenon of hyperacute rejection, which is also called vascular rejection or early humoral rejection in the old literature. It has been postulated that preformed antibodies to epitopes of the ABO and HLA systems and vascular endothelial cells are probably dramatic predisposing factors. Other predisposing factors include previous pregnancies, multiple surgeries with the use of blood products, and earlier cardiac or other organ transplants. Currently, the pathogenesis of hyperacute rejection is an antibody-mediated activation of the complement cascade with platelet activation followed by the coagulation cascade and thrombosis. It is debated which blood vessel is involved first, although most of the authors favor the cardiac venules. Gross examination of the heart facing a hyperacute rejection shows a swollen dusky specimen with dilation of both ventricles and scattered hemorrhages mostly located at the subendocardium. The histopathologic examination shows swelling of the endothelial cells, vascular thrombosis, red blood cell extravasation, prominent interstitial edema, clusters of neutrophilic granulocytes, and cardiomyocyte necrosis (hyperacute rejection). Deposits of IgM, IgG, and complement and fibrin deposits are seen in the blood vessel walls by immunohistochemistry.

1.6.3 Cardiac Allograft Vasculopathy

Mixed acute cellular rejection and antibody-mediated rejection are quite rare; it has been reported and carries a high risk of mortality. However, in the long term, patients suffering from AMR can develop a cardiac allograft vasculopathy (CAV) quickly (Fig. 1.59). In most transplanted hearts, CAV may produce at a variable rate with early occurrence dating 3 months after transplantation. Angiographic studies of the ISHLT registries evidence that about half of adults with heart transplantation show some degree of CAV at 9.5 years post-grafting and one-quarter of children with heart transplantation show some degree of CAV at 7 years post-grafting. Early CAV is defined when it occurs within 3 years following transplantation and late CAV when it happens to start from the fourth year and later. Early CAV-associated risk factors include donor hypertension, infection within 2 weeks post-grafting, and rejection during the first year. Late CAV-associated risk factors include donor history of diabetes as an underlying disorder and intracranial hemorrhage as donor cause of death. Independent risk factors that are continuous for both early and late CAV include male gender, donor age, recipient age with an inverse relationship, the volume of the center, and the recipient pretransplant body mass index. The mechanisms of CAV are complex and include immunologic and non-immunologic pathways. Endothelial cells express ruling histocompatibility complex class I and class II antigens and are the target of cell-mediated and humoral immunologic response. The activation of T-lymphocytes promotes the secretion of cytokines (interleukins, interferons, and tumor necrosis factors) encouraging the proliferation of alloreactive T cells, the activation of monocytes and macrophages, and the expression of specific adhesion molecules by endothelial cells. The humoral immune response is indicated by the production of antibodies against HLA and antigens of the endothelial cells (Fig. 1.60). The complexity of the environment surrounding the endothelial cells promotes vascular thrombosis, vasoconstriction, and proliferation of vascular smooth muscle cells. There are

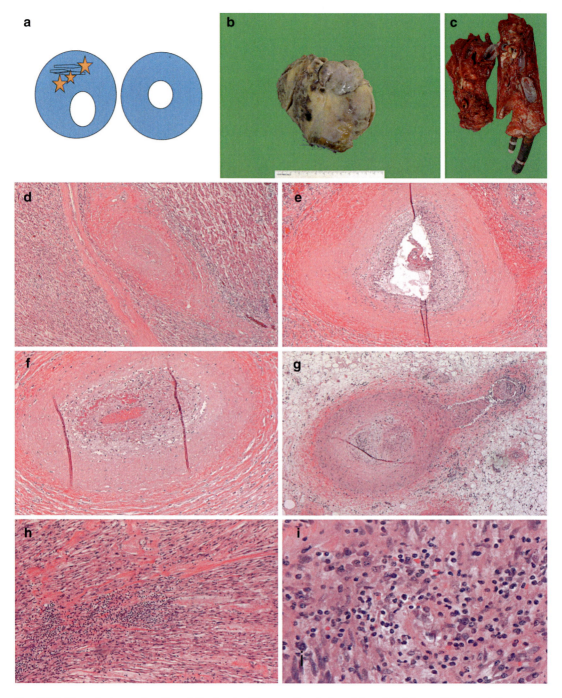

Fig. 1.59 Transplant vasculopathy of the heart with a cartoon showing the multifocal involvement (**a**), gross photograph (**b**) with extracorporeal membrane oxygenation devices (**c**), and histology of thromb-inflammatory vasculopathy (**d**–**g**) and infiltration of the myocardium (**h**–**i**)

also some nonimmune factors that have been associated with CAV and its progression. The non-immunologic factors associated with cardiac allograft vasculopathy include arteritis/vasculitis, cytomegalovirus infection, donor-transmitted coronary atherosclerosis, dishormonal status, fibrinolysis cascade deficiency, immunosuppression, abnormal lipid profile, and myocardial ischemia.

It is important to stress that the endomyocardial biopsy has limited or no sensitivity in recognition of CAV because it samples only the smallest branches of the intramyocardial arteries and arterioles. These blood vessels usually do not

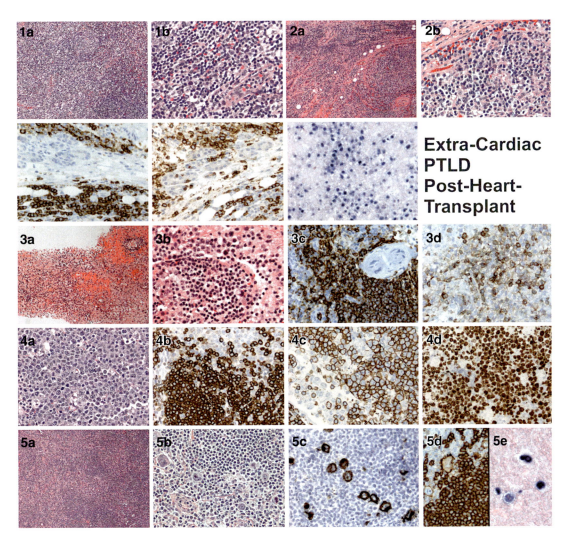

Fig. 1.60 Several aspects of extra-cardiac posttransplant lymphoproliferative disorders (PTLDs) following heart transplantation. **1a, b**: Early PTLD with effacement and expansion of parafollicular zones with persistent germinal centers and mixed population of large and small lymphoid cells and many plasma cells. (H&E stains: **1a** 100×, **1b** 400×); **2a–e**: Polymorphic PTLD in parotid gland with mixed B and T cell population with EBV status. (H&E stain: **2a** 100×, **2b** 400×, **2c** IHC-CD20 400× (B cell), **2d** IHC-CD3 400× (T cell), **2e** EBV-ISH 400× atypical lymphoid cells); **3a–d**: polymorphic PTLD in brain with perivascular infiltrate of mixed population of plasma and small atypical lymphoid cells. (H&E stain: **3a** 100×, **3b** 400×, **3c** IHC-CD20 (B cell), **3d** IHC-CD3 (T cell); **4a–d**: diffuse large B cell lymphoma PTLD) (H&E stain: **4a** 100×, **4b** IHC-bcl-2400×, **4c** IHC-CD20 400× (B cell), **4d** IHC-MIB1 400×); **5a–e**: Hodgkin PTLD with Reed Sternberg cells. (H&E stain: **5a** 100×, **5b** 400×, **5c** IHC-CD20 400× (B cell), **5d** IHC-CD15 400× (malignant cells), **5e** IHC-EBV-ISH 400× atypical lymphoid cells)

show histologic features of CAV. Using an elastica staining, it is easy to identify an intact or only focally disrupted internal elastic lamina of the epicardial branches, while the vasculitis present in the intramyocardial branches evidences itself in the adventitia extending to the medial layer with the destruction of the external elastic lamina. The classic features of CAV in epicardial vs. intramyocardial branches (IMB) include calcification and pools of extracellular lipid in association with atheromatous plaques; diffuse concentric narrowing with luminal vascular stenosis (proliferation of smooth muscle cells and myofibroblasts) (epicardial > IMB); eccentric atheromatous plaques with the superimposed intimal proliferation; fibrinoid necrosis of the media (IMB); T lymphocytes, macrophages, and foamy cells (epicardial > IMB); vascular thrombosis (IMB); and vasculitis, endothelialitis, and ACR (IMB).

The examination of the myocardium often shows an ischemic injury on both sides of the heart with inherent patchiness. This acute and healing ischemic injury is at the basis of the concept that intramyocardial branches are entirely or subtotal occluded before the large epicardial branches are stenosed. The pathology of CAV does not differ substantially between children and adults, although the underlying lipid profile may play an aggravating role often observed in adults.

1.6.4 Extracardiac Posttransplant Lympho-proliferative Disorders (PTLD) Following HTX

Heart transplantation (HTX) is associated with a burden of immunosuppression, and there is the risk of posttransplant lymphoproliferative disorders (PTLDs). In Fig. 1.60, some extracardiac posttransplant lymphoproliferative disorders have been collected. PTLD occurs between 1.2% and 9% of cardiac transplant patients across ages and gender. The first year of transplantation is the most vulnerable time to develop PTLDs. Risk factors for developing PTLDs are the infection with Epstein-Barr virus (EBV) and some immunosuppressive regimens, such as OKT3. EBV is a ubiquitous γ-herpesvirus able to establish lifelong persistent infection and can be intermittently shed in the saliva. EBV has a characteristic dual tropism for both B lymphocytes and epithelial cells. EBV exhibits a biphasic life cycle. Although the infection is often benign, EBV is associated with some lymphomas and carcinomas that arise in several anatomical sites. It has been questioned if EBV is a carrier or a driver for the oncogenesis. Currently, EBV is considered a carcinogenetic cofactor in cellular environments due to its ability to immortalize B lymphocytes and its occurrence in a subgroup of tumors considering that EBV may act more as an agent of tumor progression rather than tumor initiation. The majority of PTLDs are malignant B-cell lymphomas (Fig. 1.60). In a previous study, our group could identify heterogeneity of PTLD after heart transplantation. Our study confirmed the heterogeneity of PTLD and the involvement of multiple anatomic sites. The significant manifestation of PTLD in the aerodigestive tract is impressive and may be related to the immunologic reactions in mucosa-associated lymphoid tissues (MALT) in these sites. In many transplantation centers, we consider EBV to be a crucial etiologic agent in the development of PTLD. The emerging EBV-negative PTLD cases may warrant an investigation into other causal agents with different mechanisms of malignant transformation.

1.7 Tumors

Cardiac tumors are rare, although the exact incidence seems to be unknown. Several autopsy studies indicate a rate ranging from 0.0017% to 0.33%. In 90% of cardiac tumors in adulthood, myxomas and sarcomas play a contributive role, while in childhood, rhabdomyomas and fibromas comprise a vast majority of cardiac neoplasms (Figs. 1.61–1.63). Symptomatology of cardiac tumors is observed in four settings, including

1. *Intracavitary growth* causing either partial or occlusive obstruction of the blood flow

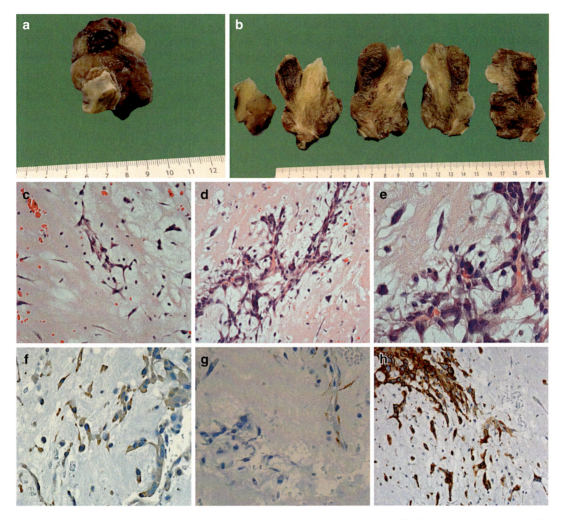

Fig. 1.61 Cardiac myxoma in a patient with Carney complex. Carney complex includes spotty skin pigmentation, myxomas, endocrine overactivity, and other conditions. Carney complex is inherited as an autosomal dominant trait and is caused by mutation of the PRKAR1A gene in about two-thirds of the cases. Grossly, cardiac myxomas are pedunculated nodules or masses, often attached to left atrial wall of the heart by a pedicle (**a**, **b**). Histologically, cardiac myxomas are endothelial cells in fibromyxoid matrix (**c**–**e**). Immunohistochemically, the tumor cells are mostly positive for S100 (**f**, **g**, ×200 as original magnification) and HLA-DR (**h**, ×100 as original magnification)

obstruction or interference with valvar functionality with different symptomatologies if the valves involved are atrioventricular or ventriculoarterial
2. *Intramyocardial growth* causing refractory arrhythmias, sudden cardiac death (SCD), and pericardial effusions up to tamponade
3. *Embolization* with transient ischemic attacks (TIA), strokes, and scotomas
4. *Systemic or constitutional symptoms*, which may be quite variable

Primary cardiac tumors may be detected as an abnormal finding of a chest x-ray or another imaging test performed for an unrelated reason. The more detailed anatomy and pathology of cardiac tumors are carried out using echocardiography, magnetic resonance imaging, and computed tomography.

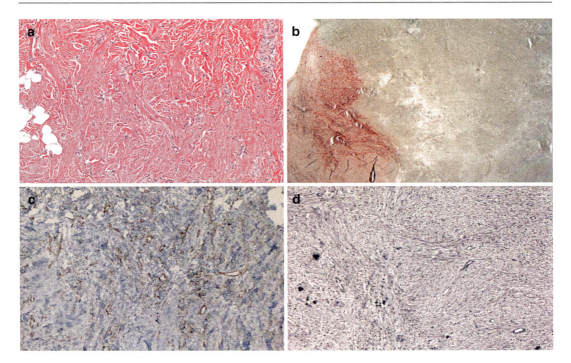

Fig. 1.62 Heart with fibrous hamartoma (**a**, hematoxylin and eosin staining, ×100 original magnification), fibroma (**b**, Movat pentachrome staining, 12.5 original magnification), and immunohistochemical evidence of actin (**c**, antiactin immunostaining, avidin-biotin-complex, ×100), and evidence of elastic fibers (**d**, Elastic-Van Gieson, ×100)

1.7.1 Benign Tumors

Cardiac disorders with neoplastic significance but also including tumorlike conditions with variable presentation and prognosis (e.g., recurrence) depending on size, localization, invasiveness, friability, growth rate, and histologic variety. Benign tumors of the heart can be classified as either cardiac specific or non-cardiac specific. The latter category comprises tumors that have a pattern similar to their counterparts in the body and will only be listed in this chapter. Some conditions need to be carefully distinguished from each other because morphology can remarkably be quite similar. An important differential diagnosis is the correct identification of fatty-appearing tumors of the heart, which include lipoma, fibrolipoma, lipomatous hypertrophy of the IAS, hemangioma, and rhabdomyoma. The differential diagnosis is crucial, and it is important to keep in mind the most frequent malignancies, including angiosarcoma, liposarcoma, synovial sarcoma, and the metastasis to the heart from malignancies, which are different in the lifespan from infancy to adolescence and adulthood. Moreover, some entities are also linked to genetic syndromes, and appropriate genetic counseling may need to be addressed. In infancy and childhood, rhabdomyoma, fibroma, and histiocytoid cardiomyopathy (also called Purkinje cell hamartoma) comprise a vast majority of cardiac tumors. In adulthood, ¾ of primary cardiac tumors are benign, and approximately half of these are myxomas, while the rest consists of lipomas, papillary fibroelastoma, and rhabdomyoma.

1.7.1.1 Myxoma
- *DEF*: Most common primary cardiac tumor (~50% of all primary cardiac neoplasms) presenting with heart failure, stroke, and constitutional symptoms and harboring chromosomal abnormalities, variable DNA content, and microsatellite instability supporting the concept of neoplasm rather than a reactive process (Fig. 1.61).
- *EPI*: Sporadic (~95%) and familial types (~5%): *Carney or myxoma complex* (skin pigmentation

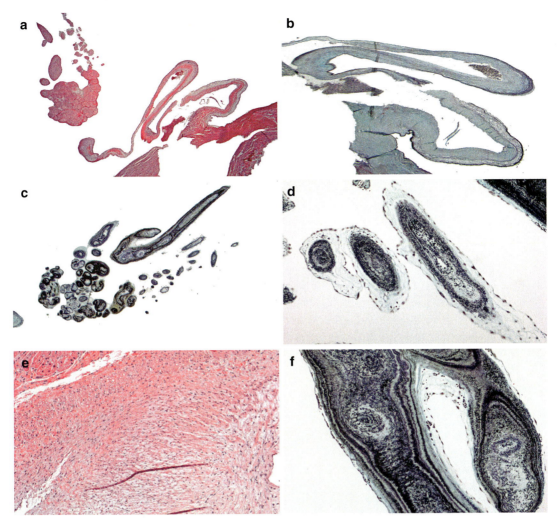

Fig. 1.63 Papillary fibroelastoma with the characteristic avascular branching papillae (**a**, **e**, hematoxylin and eosin staining; **b**, **c**, **d**, **f**, Masson's trichromic stain)

with lentigines and blue nevi, calcifying Sertoli-Leydig testicular tumors, cutaneous myxomas, psammomatous melanotic schwannoma, myxoid fibroadenomas of the breast, pigmented adrenal cortical hyperplasia, pituitary hyperactivity, and neoplasms of the thyroid gland) caused by *PRKAR1A* (protein kinase cAMP-dependent regulatory type 1α) gene mutations. Sporadic (single, 40–60 years, ♂ < ♀, left atrium in ~9/10 cases, rare recurrence) vs. familial (multicentric, ~20 years, ♂ > ♀, left atrium in ~2/3, recur in 1/3).

- *GRO*: Pedunculated tumor with irregular, nonhomogeneous lucencies on echo and divided into two types – solid (smooth and lobular) and papillary (friable and irregular with organized thrombi) and heterogeneity (cut surface) with cysts, calcifications, necrosis, and hemorrhage.
- *CLM*: Lepidic (Gr. λεπίς, scaly) cells (stellate cells with ovoid nucleus and ± distinct nucleoli and bulky eosinophilic cytoplasm with indistinct cell borders) singly, in cords or in pseudo vascular arrangement embedded in a gel-like mucopolysaccharide-rich stroma ± BV, extravasated RBC, recent and old hemorrhage, hemosiderin, calcifications (e.g., Gamna-Gandy bodies), and inflammatory cells ± heterologous component (columnar

epithelium with or without glands, extramedullary hematopoiesis, and thymic tissue).
- *IHC*: (±) Calretinin and vimentin.
- *DDX*: Organizing thrombus (lepidic cells in cords or ring structures forming vessels) and myxosarcoma (invasion of the tumor stalk and substantial areas of hypercellularity, necrosis, and cellular atypia).
- *PGN*: After surgery, possible recurrence (e.g., multicentric) and embolization risk and genetic counseling in the familial type.

1.7.1.2 Lipoma, Valve Fibrolipoma, and Lipomatous Hypertrophy of the Interatrial Septum (LHIAS)

- *Lipoma*: Well-localized, encapsulated, yellow, spherical/elliptical, subendo-/subepi-/subpericardial or intramyocardial, anywhere, with cardiomegaly and dyspnea, made of *mature fat cells* ± fat necrosis and histiocytes (DDX: WD-lipoma-like LPS showing lipoblasts!).
- *Valve Fibrolipoma*: Lipomatous "hamartoma" of the AV valve (properly neoplasm, because fat is not a standard component of the valve) with an admixture of *mature fat cells*, *fibrosis*, and *thin-walled BV*.
- *LHIAS*: Unencapsulated dumbbell mass of the IAS ± cardiac hypertrophy and arrhythmias and constituted by *mature and brown fat cells* and *degenerated entrapped cardiomyocytes* that typically spares the *fossa ovalis*.

1.7.1.3 Papillary Fibroelastoma (PFE) and Fibroma

Papillary Fibroelastoma
- *DEF*: First most common valvular and second most common primary benign cardiac tumor, which presents as small, stalky, homogenous valvular mass on echo causing blood flow turbulence. It has been suggested to be an exaggerated form of Lambl excrescence.
- *EPI*: Any age (~60 years), ♂ = ♀, with embolization risk (tumor or thrombus), usually ventricular surface of valves and the atrial surface of AV valves (R > L in children and R < L in adults).
- *GRO*: "pom-pom" or "sea anemone"-like a cluster of hairlike projections 2–5 mm in length on the valve (aortic>mitral>tricuspid>pulmonary) or endocardium.
- *CLM*: Three layers constituted by central fibroelastic stroma (+ EvG), outer myxoid, proteoglycan-rich matrix, and overlying hyperplastic endothelium (+ CD34/CD31/FVIIIRA).
- *DDX*: Lambl excrescences (along with the lines of closure – nodules of Arantii – of cardiac valves), vegetations, and valvular myxomas (-EvG).
- *PGN*: Conservatively (anticoagulants)/surgery with leaflet repair or valve replacement.

Fibroma
- *DEF*: The fibroma is the second most common pediatric primary cardiac tumor after rhabdomyoma with possible association with Gorlin syndrome or NBCCS (*PTCH1* on 9q22.3, multiple BCC of the skin, jaw cysts, and bifid ribs, frequently discovered in utero or infancy) (Fig. 1.63).
- *CLI*: Large, noncontractile, solid mass with heterogeneity on echo mimicking HCM in a setting of heart failure, arrhythmias, SCD or incidentally.
- *GRO*: Small-to-large, firm, white, well-circumscribed (nonencapsulated), or infiltrative mass of the free walls of the ventricles or IVS.
- *CLM*: Fibromatosis-like appearance with spindle cells harboring a bland-looking morphology embedded in collagen (children: cellularity > collagen deposition; adults: cellularity < collagen deposition) ± scattered inflammatory cells and central dystrophic calcification.
- *HSS/IHC*: (+) EvG, VIM, SMA, but B-cat (−).
- *DDX*: fibromatosis (B-cat +), fibrosarcoma ((+) VIM, P53, Coll. Type 1, MIB1), synovial sarcoma ((+) VIM, TLE1, CKs, EMA, Bcl2, β-Cat, CD99, CD56, CD57, CALR, SYT, ZO-1, Claudin-1, and occludin). Immunostaining for SYT and TLE1 proteins discriminates synovial sarcoma from other soft tissue tumors, because recent studies evidence that more tumors may have TLE1 positivity. Glandular component of

most biphasic cases are also usually positive for ZO-1, Claudin-1, and occludin.
- *PGN*: Surgery and no recurrence in partial resection; cardiac TX for unresectable lesions.

1.7.1.4 Rhabdomyoma
- *DEF*: Most common pediatric primary cardiac tumor, which is associated with tuberous sclerosis (*TSC1* on 9q34/*TSC2* on 16p13), discovered in utero (e.g., stillbirth) or in the first year of life, often multiple, ± arrhythmias/valvular anomalies/LVOTO, but the tendency to spontaneous regression.
- *GRO*: Small, firm, gray-white, well-circumscribed myocardial nodules/masses are protruding into ventricular chamber.
- *CLM*: Mixture of large polygonal cells with glycogen-rich cytoplasmic vacuoles separated by strands of cytoplasm radiating from cell center or nucleus (spider cells) in children prompting to the direction of a hamartoma but + in adults indicating a benign tumor.
- *IHC*: (+)PAS; ACT, DES, MGB, VIM; apoptosis-associated ubiquitin; HMB-45; hamartin, tuberin; remarkably MIB1/Ki67 −.
- *PGN*: Spontaneous regression or surgery if + symptoms (arrhythmias, valvular obstruction).

1.7.1.5 Hemangioma
- *DEF*: Benign vascular neoplasm, which is usually localized at the right atrium or ventricular septum, ± association with other cutaneous or visceral angiomas (diffuse angiomatosis) and may present with arrhythmias, pericardial effusion, heart failure, and sudden cardiac death.
- *EPI*: Any age, although more often seen in adults, ♂ > ♀.
- *GRO*: 2–4 cm sessile or polypoid, red-purple nodule.
- *CLM*: Capillary, cavernous, arteriovenous or intramuscular pattern, which may also have fat and fibrous tissue (Fig. 1.64).
- *DDX*: Epithelioid hemangioma/epithelioid hemangioendothelioma, angiosarcoma, (atypia, necrosis, mitoses, pericardial extension), lipoma, fibrolipoma.
- *PGN*: Usually good with surgery but also spontaneous regression and regression after steroids.

1.7.1.6 Inflammatory Myofibroblastic Tumor
- *DEF*: Rare tumor with obscure etiology (reactive, autoimmunity, neoplastic), which presents as a polypoid mass with a broad base, attached to the endocardium and covered with fibrin as well as constitutional symptoms mostly in children and adolescents.
- *GRO*: Circumscribed, not encapsulated whitish mass with whorled fleshy or myxoid cut surface and occasional foci of hemorrhage, necrosis, or calcification.
- *CLM*: There is an admixture of myofibroblasts, fibroblasts, and inflammatory cells in a myxoid or poorly collagenous stroma without significant pleomorphism and atypia.
- *IHC*: (+) SMA, VIM, calponin, but (−) CD34 and ALK1 (differently from non-cardiac IMT).
- *DDX*: Fibroma (non-endocardial, calcification, whorls of bland-appearing homogeneous fibroblast), fibrosarcoma (pleomorphism and atypical mitoses), myxoma (no multinucleated syncytia or perivascular rings, CD34+).
- *PGN*: Usually good with surgery but also spontaneous regression and regression after steroids.

1.7.1.7 Paraganglioma
- *DEF*: Rare extra-adrenal pheochromocytoma, which originates from paraganglia cells and typically found within the atrial walls (Fig. 1.65).
- *EPI*: Third to fourth decade of life.
- *CLI*: Mass in the roof of the right atrium ± symptoms hyper-catecholaminemia (HTN, headache, sweating, palpitations, and dyspnea) and positive ^{131}I-MIBG scintigraphy (MIBG, metaiodobenzylguanidine).
- *GRO*: Soft, fleshy, tan-to-brown, poorly circumscribed mass with a broad base.
- *CLM*: 'Zellballen' pattern constituted by nests of large polygonal epithelioid chromaffin cells

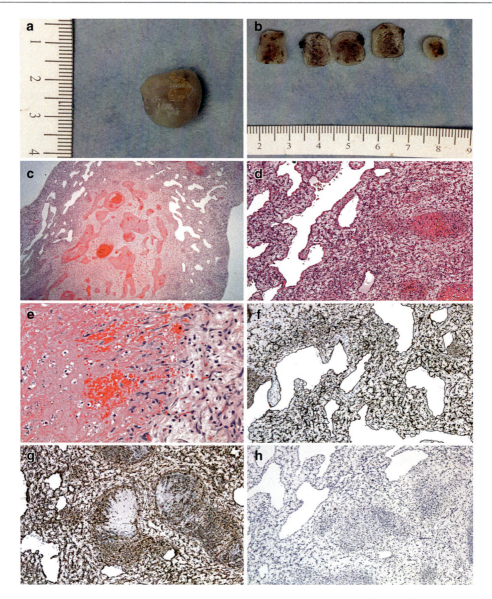

Fig. 1.64 Cardiac angioma with a semisolid vascular mass (**a**, **b**) and histological evidence of proliferating blood vessels (**c–h**). In (**f**) is shown the immunohistochemical expression of CD31 (×50, original magnification). In (**g**) is shown the immunohistochemical evidence of actin (×50, original magnification), and in (**h**) is shown the negative staining with antibody against calretinin, a mesothelial marker (×50, original magnification)

(+NSE, CGA, SYN) separated by sustentacular cells (+S100).
- *DDX*: Other extracardiac locations of paragangliomas/pheochromocytoma
- *PGN*: Usually good with surgery but potential local invasion and metastasis → malignancy.

1.7.1.8 Cardiac Hamartoma
- *DEF*: Rare, disorganized overgrowth of mature cardiomyocytes with abnormal development of embryonic cellularity.
- *EPI*: Any age, ♂ > ♀.
- *CLI*: Arrhythmias or incidentally.

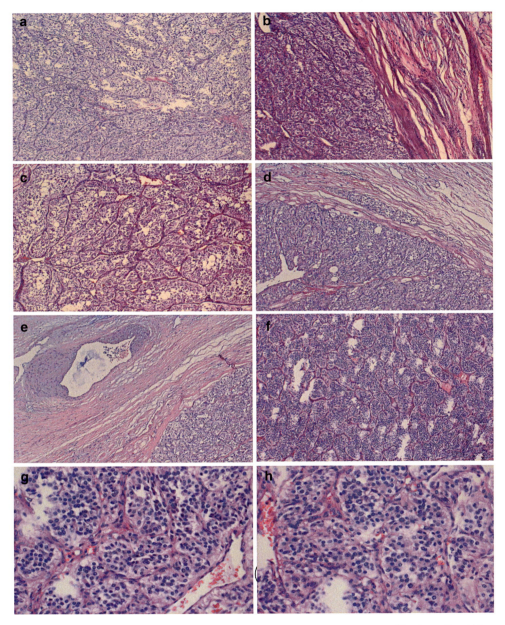

Fig. 1.65 Chemodectoma at the glomus aorticus showing (all figures are hematoxylin-eosin-stained). Chemodectoma has an appearance similar to the normal paraganglia with the tendency to form nests of 15–20 chief cells, which are arranged haphazardly within a moderately to very rich vascular stroma (**a–h**). Ultrastructural features include the presence of neurosecretory-like granules in the chief cells

- *GRO*: Small-to-large firm, pale white, poorly demarcated mass.
- *CLM*: Haphazardly or herringbone/whorling organization of mature hypertrophic cardiomyocytes ± thick-walled BV, fibrosis, and fatty tissue.
- *DDX*. Myocardial infarction (healing or scars), hypertrophic CM (diffuse process), fibroma (fibroblasts and collagen), and rhabdomyoma (no spider cells).
- *PGN*: Surgery with no recurrence.

1.7.1.9 Histiocytoid CM (HICM)/ Purkinje Cell Hamartoma

- *DEF*: It is a heterogeneous genetic *mitochondrial* disorder of cardiomyocytes. Although some controversy exists about the origin of HCM, it is being considered a first genetic CM according to AHA and not as a developmental anomaly of the atrioventricular conduction system, a multifocal tumor of Purkinje cells, or as a developmental arrest of cardiomyocytes.
- *EPI*: Mostly infant (< 2 years) girls (♂ < ♀).
- *CLI*: Cardiomegaly, arrhythmias, LVNC, SCD, and extracardiac manifestations, including abnormalities of the CNS (hydrocephalus), eyes (corneal opacities, retinal hypoplasia, microphthalmia), endocrine system (oncocytic change), and liver steatosis.
- *GRO*: Yellow-to-tan subendocardial/subepicardial/valvular nodular lesions.
- *CLM*: Homogeneous clusters of large, pale, polygonal (rounded to oval-shaped) cells with a central hyperchromatic nucleus are often showing apoptosis and large, finely granular/foamy cytoplasm (aka "transformed Purkinje cells," mild PAS+).
- *IHC*: (+) MSA, DES, MGB (myoglobin, MSN (myosin), and (−) CD68.
- *EM*: Abundance of mitochondria, which may show broken *cristae* and dense amorphous inclusions (flocculent densities).
- *PGN*: Poor, if no pacemaker is implanted.

1.7.1.10 Congenital Polycystic Tumor of Atrioventricular Node (AVN)

- *DEF*: Developmental embryonic rest of endodermal origin (1-layer: hamartoma; ultimobranchial heterotopic element, similar to solid cell nests of the thyroid gland), because the AVN region is an area of embryologic fusion with potential foreign tissue trapping.
- *EPI*: Third to fourth decades (but also in newborns with CHD, such as TGA or midline developmental defects), ♂ < ♀, incidental or SCD (smallest causing lesion).
- *GRO*: Inapparent to small cysts (2–20 mm).
- *CLM*: Multicystic lesion without smooth muscle but showing benign epithelium with several patterns of differentiation (squamous, cuboidal, transitional, sebaceous).
- *IHC*: (+) CEA, EMA, B72.3, CK and (−) calretinin, CD34/CD31/FVIIIRA.
- *DDX*: Bronchogenic cyst, mesothelial cyst, teratoma.
- *PGN*: Surgery and permanent pacemaker implantation.

1.7.1.11 Calcified Amorphous Tumor

- *DEF*: Unusual non-neoplastic tumor presenting as a pedunculated highly echogenic mass anywhere and embolism and associated with renal insufficiency.
- *GRO*: Firm, yellow-white lesion.
- *CLM*: Nodular calcium cluster in a background of hyalinization, degenerated blood elements, and chronic inflammatory cells.
- *DDX*. Myxomas, thrombi, infections (e.g., hydatid cysts), and valve vegetations.
- *PGN*: Surgery and possibility of recurrence.

1.7.1.12 Mesothelial/Monocytic Incidental Excrescences (MICE)

- *DEF*: Reactive or artifactual cardiac lesions.
- *CLI*: Usually incidentally identified.
- *EPI*: Anywhere without clear clinical significance, age or sex predilection.
- *GRO*: Pale thrombus-like excrescence or soft clot-like fragment.
- *CLM*: Mixture of benign mesothelial cells arranged in strips and glandular-like formations and histiocytes as well as fat vacuoles or adipocytes, inflammatory cells, and RBC.
- *IHC*: (+) CALR, D2–40, CK5/6, WT1, CD141, as well as (+) CD68.
- *DDX*: Metastases from carcinoma (lack of atypia and different stroma) need to be ruled out.
- *PGN*: No specific treatment.

1.7.1.13 Mural Thrombi

Tumorlike condition that can be intracavitary, intramural, or within the pericardial sac and tendency to thromboembolism in the setting of mitral valve disease or atrial fibrillation (atrial thrombi), myocardial infarction or dilated CMP (ventricular thrombi), chronic anabolic steroid abuse, and different diseases (Behçet syndrome, Loeffler endocarditis, Churg-Strauss syndrome, coagulopathies). Thrombi are characterized by three layers, including luminal (RBC and fibrin), intermediate (dense fibrin network), and abluminal (acellular-degraded fibrin).

- *DDX*. Myxoma, hemangioma/lymphangioma, papillary fibroelastoma, and hydatid cyst.
- *PGN*: Anticoagulants.

1.7.1.14 Rare Non-cardiac Benign Tumors

Tumors similar to the counterpart found in other localization of the body (adenomatoid tumor, angiomatosis, angiomyolipoma, benign fibrous histiocytoma, granular cell tumor, lymphangioma, xanthogranuloma, neurofibroma, schwannoma, and teratoma). Neurofibromas and schwannomas may occur on the epicardium and have been described in patients affected with neurofibromatosis and following radiation therapy.

1.7.2 Malignant Tumors

Malignancies are rare in both children and adolescents but may occur in young adults and later. Of all malignant tumors, almost all of them are sarcomas arising in the fourth decade of life without gender bias. The right atrium is often involved, and clinical presentation includes therapy-refractory heart failure, arrhythmias, and myocardial ischemia/infarction as well as bloody pericardial effusion or pericarditis (Figs. 1.66 and 1.67).

1.7.2.1 Epithelioid Hemangioendothelioma

- *DEF*: Low-grade malignancy with vascular differentiation. Rarely found in the heart (Fig. 1.68).

1.7.2.2 Angiosarcoma

It is the most common adult malignant cardiac tumor, which is localized typically in the right atrium, and, in case of a large mass, it may present as an intracavitary extension with apparent infiltration of the myocardium and right-sided heart failure or tamponade. Imaging may show areas of increased T1 signal intensity either focally ("cauliflower" appearance) or linearly along the pericardium ("sunray" appearance). The histology is similar to angiosarcoma detected elsewhere (Fig. 1.69). Briefly, hemorrhagic masses constituted by blood vessels with atypical epithelium, mitoses, necrosis, and invasiveness are found.

1.7.2.3 Rhabdomyosarcoma

- *DEF*: A most common pediatric malignant cardiac tumor arising from either ventricle with often multicentric intramyocardial localization (>10 cm) and frank invasiveness with possible extension to valve leaflets and pericardium. Embryonal rhabdomyosarcoma is usually found in children and adolescents, while the pleomorphic type is characteristically found in adults. There is a slightly male preference and are gelatinous and friable, firm, and fleshy with cystic or necrotic degeneration, grossly. Imaging shows characteristically a smooth or irregular low-attenuation mass localized in either ventricle.

1.7.2.4 Synovial Sarcoma

Synovial sarcoma is a tumor of uncertain cell origin unrelated to the synovium, despite the name (Fig. 1.70). It can involve any site and the histology includes an epithelial and and a mesenchymal component in different proportions. It is subdivided in monophasic, when only one component is present, and biphasic, when both components are found. Immunohistochemistry is positive for keratin and epithelial membrane antigen (EMA), CD99, S-100 protein, and BCL2 apoptosis regulator. Transducin-like enhancer of split 1 (TLE-1) is the most sensitive and specific marker for synovial sarcoma. TLE-1 is crucial in distinguishing this entity from fibrosarcoma, malignant peripheral nerve sheath tumor, carcinosarcoma, and mesothelioma. Cytogenetically, the synovial sarcoma exhibits the t(X;18) (p11;q11) translocation, which involves one of the synovial sarcoma-X (SSX) genes and the SS18 gene (aka SYT), respectively. Duran-Moreno et al. (2019) could find a total of 37 cases of pericardial synovial sarcoma (including their own patient). This tumor at this location has a

Fig. 1.66 Heart with a teratoma with yolk sac tumor component

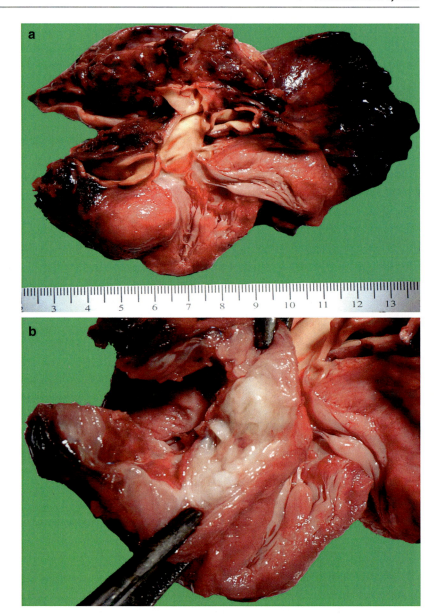

male predilection, with a male:female ratio of 4.28:1, and an age range of 13-67 years. The median age of 34 years has a kurtosis value of −1.168 and a skewness value of 0.331, highlighting an incidence slightly greater in those patients younger than 30 years. In the heart, the left atrium is the favorite site with 3/4 of cases having the tumor arising from the upper part of the pericardium. Dyspnea and cough are the most frequent symptoms. Echocardiogram is usually the first diagnostic test performed and needs to complemented with computerized tomography and magnetic resonance imaging with or without positron-emission tomography before any surgery. Chemotherapy and radiotherapy may provide some benefit in survival and tyrosine-kinase inhibitors may be attempted (e.g., pazopanib) (Duran-Moreno et al. 2019).

1.7.2.5 Others

Kaposi sarcoma, leiomyosarcoma (usually located in the posterior wall of the left atrium

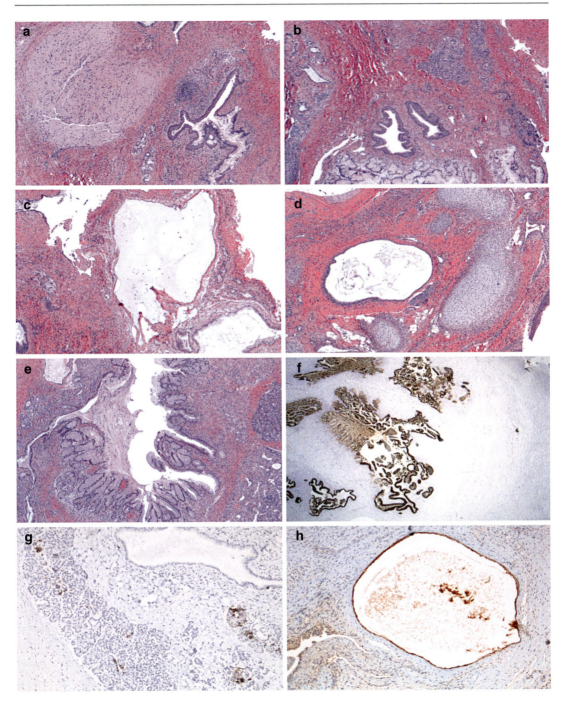

Fig. 1.67 The histology of the teratoma is apparent with structures of the three germ cell layers (**a–g**). In (**f**) and (**g**), there is immunohistochemical evidence of cytokeratin 7 and chromogranin A expression (avidin-biotin complex; **f**, ×12.5 original magnification; **g**, ×100 original magnification). A yolk sac tumor component with angular cystic pattern stained with an antibody against alpha-fetoprotein (avidin-biotin complex) is seen (**h**, ×100 original magnification)

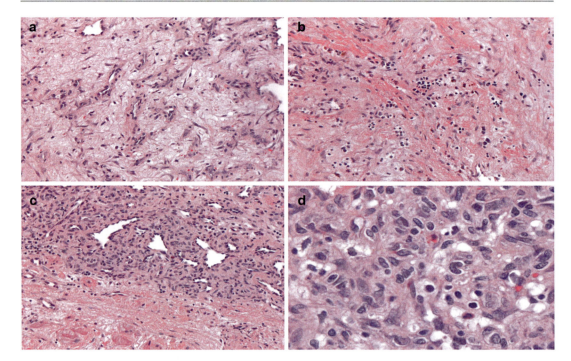

Fig. 1.68 Cardiac epithelioid hemangioendothelioma with a proliferation of large epithelioid tumor cells presenting at places atypical features and a mitotic index of 2–3 mitoses per 10 high power fields. Immunohistochemically, antibodies against CD 34 and CD 31 show a strong stained cytoplasmic vascular lumen (not shown)

with a tendency to invade the pulmonary veins), osteosarcoma, liposarcoma, pleomorphic sarcoma, and undifferentiated sarcoma are extremely rare entities. Metastatic osteosarcoma is generally found in the right atrium, while primary osteosarcoma is typically more often attached to the wall of the left atrium. The liposarcoma of the left or right atrium is also a very rare neoplasm. Hematopoietic tumors should also be considered and include leukemias, Hodgkin lymphoma, and NHL. Lymphomas are rare, more often in the right atrium, and their incidence increased simultaneously to the increasing rate of lymphoproliferative disorders related to EBV infection in immunosuppressed patients. Posttransplant lymphoproliferative disorder (PTLD) is a well-known complication of transplantation due to the use of potent immunosuppressive therapy and is subdivided in early lesions, polymorphic PTLD, and monomorphic PTLD according to the World Health Organization classification. Before giving some notes on the metastatic cardiac tumor, paraneoplastic syndromes involving the heart should also be considered, and the most frequent paraneoplastic syndrome is carcinoid syndrome secondary to carcinoid tumors. This kind of tumors can produce and release vasoactive substances, including serotonin, histamine, tachynins, etc. Symptoms include cutaneous flushing, gastrointestinal hypermotility, bronchospasm, severe local and distant fibrosis, mainly endocardial plaques, and fibrotic degeneration of the leaflets and subsequent valvar dysfunction (right > left).

1.7.3 Metastatic Tumors

Cardiac metastases have been reported to be up to 100-fold more common than primaries. The heart may be reached by four different pathways, including direct extension from mediastinal neoplasms (*per contiguitatem*), hematological spread, intra-

Fig. 1.69 Cardiac angiosarcoma with gross features of invasion (**a**) and histopathologic features of malignant cells and proliferation of blood vessels (**b–c**). There is a high rate of mitotic figures

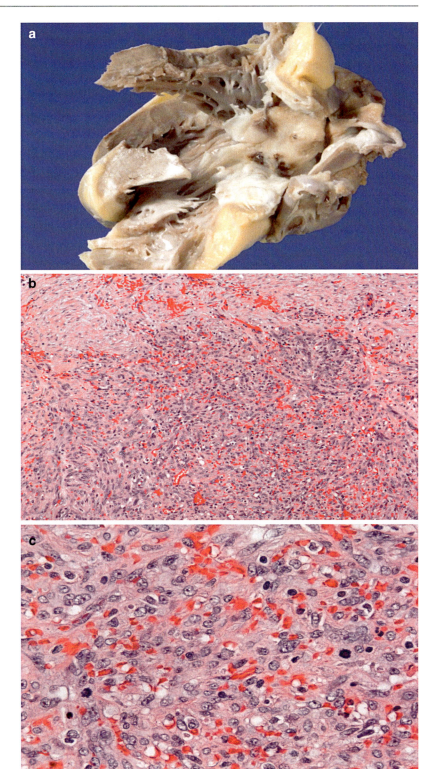

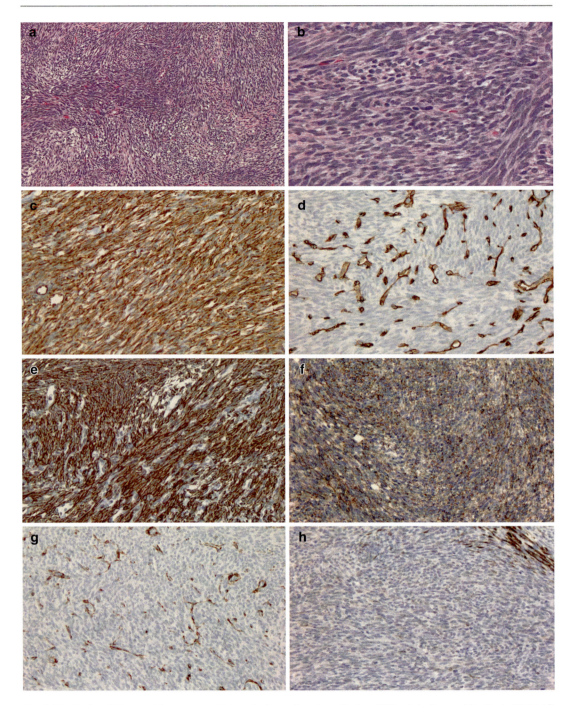

Fig. 1.70 Pericardial synovial sarcoma with interlacing tumor cells. There is one fibrous type, spindle or sarcomatous cell, with relatively small and uniform morphology and one epithelial in appearance with a classical synovial sarcoma of biphasic appearance (**a–b**). Immunohistochemically, the tumor cells show positivity for vimentin (**c**, ×200 original magnification), CD34 (**d**, ×200 original magnification), Bcl-2 (**e**, ×200 original magnification), CD99 (**f**, ×200 original magnification), actin (**g**, ×200 original magnification), but negativity for calretinin (**h**, ×200 original magnification)

1.8 Pericardial Disease

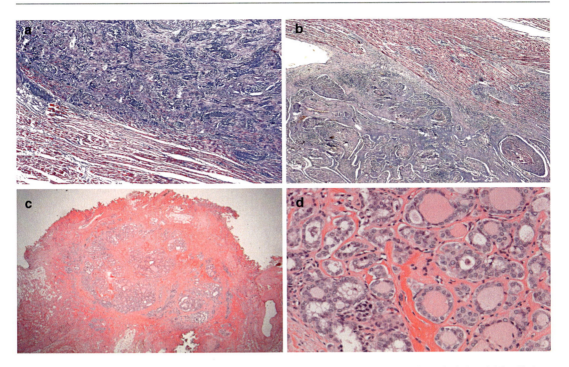

Fig. 1.71 Supraaortic metastasis of a follicular thyroid carcinoma with infiltration of atypical thyroidal cells in a stroma

cavitary expansion from the IVC, and through lymphatics. The bronchial carcinoma is the most frequent secondary tumor of the heart. In order of frequency following lung carcinomas, there is carcinoma of breast, melanoma, lymphoma, leukemia, renal cell carcinoma, and choriocarcinoma. A secondary tumor may often involve the pericardium, be multifocal, or produce an intracavitary lesion. In children and adolescents, secondary tumors are extremely rare (Fig. 1.71).

1.8 Pericardial Disease

1.8.1 Non-neoplastic Pericardial Disease

1.8.1.1 Pericardial Sac Fluid Storage Through Frank Accumulation Through Tamponade

30–50 cc thin, transparent, straw-colored fluid ordinarily present in the pericardial sac with epicardium lining the myocardium of the heart and the proper pericardial layer on the heart.

Pericardial effusion can be:

- *Serous*: congestive heart failure, hypoproteinemia
- *Serosanguinous*: blunt chest trauma, CPR
- *Chylous*: lymphatic obstruction (benign or malignant)
- *Cholesterol-rich*: myxedema

Hemopericardium rupture of heart or intrapericardial aorta following MI, dissecting aneurysm, or acute chest trauma. The factors underlying a hemopericardium in children and adolescents include chest wall PNET (Askin tumor), dermatomyositis, dissecting aortic aneurysm (thoracic), heart attack (acute MI), heart failure, heart surgery (post-surgery), heart tumors, heart wounds, iatrogenic (e.g., placement of central lines), renal failure, leukemia/lymphoma, pericarditis (following bacterial or viral infections), radiation therapy to the chest, recent invasive heart procedures (nonsurgical), systemic lupus erythematosus (SLE), and dysfunctional (underactive) thyroid gland.

Tamponade occurs with the simultaneous or independent occurrence of either fluid accumulation > pericardial sac capacity and when the collection does take place rapidly. Tamponade can cause cardiac compression and SCD. Pericardial effusion gives to the heart space, which is less than usual. Thus, the heart is unable to expand causing pump dysfunction. The fluid-filled pericardial sac can range from a few hundred ml to liter. Conversely, tamponade is filling with liquid in the sac but faster than effusion and more than emanation. Few hundred ml filled suddenly would induce a maladaptation of the heart, which decreases in contractility. In the end, the body goes into shock.

1.8.1.2 Pericarditis

Currently, the most important and practical question about the classification of pericardial disease is the distinction between viral pericarditis and autoreactive pericarditis. While the autoreactive forms benefit of systemic and intrapericardial corticosteroid treatment, their use in viral forms remains currently contraindicated.

Anatomic-pathologic classification has minor significance indeed. This classification includes *fibrinous* pericarditis, seen in the post-myocardial infarct, rheumatic fever, trauma, uremia, radiation, SLE, and severe pneumonia characterized histologically by an intertwined mass of thin eosinophilic fibers or sizeable amorphous mass and clinically evidenced by a loud friction rub. It may resolve or become organized to produce adhesive pericarditis (*cor villosum* → "*Panzerherz*" of the German literature). *Serous* pericarditis is encountered in rheumatic fever, SLE, systemic scleroderma, tumors, uremia, virus, or TB. Histologically, there is mild inflammation of epicardial and pericardial surfaces with both acute and chronic inflammatory cells. In the suppurative (purulent) inflammation, bacteria, fungi, or parasites are the most common etiologic factors. The pathogenesis follows direct extension, hematogenous seeding, lymphatic spread, and iatrogenic seeding with potential fatal mediastinitis. A sequel of the purulent pericarditis is the constrictive pericarditis.

- *Hemorrhagic* pericarditis: Iatrogenic (post-open heart surgery) or non-iatrogenic (TB or malignancy from the lung or breast).
- *Caseous* pericarditis: TB.
- *Constrictive* pericarditis: Idiopathic as a sequela of previous serous, fibrinous, or caseous pericarditis with eliminated pericardial space and substituted by a thick scar tissue of up to 10 mm in thickness with potential dystrophic calcification, causing encasement of the heart pump with restrictive mal-functionality (*concretio cordis*). The adhesion from the pericardial sac into the mediastinum is also a common event. The mal-functionality of the heart pump may induce hypertrophy and dilation of the heart according to the remaining free space and if the heart is not tightly adherent to the pericardium.

1.8.2 Neoplastic Pericardial Disease

It is mostly associated with metastasis by small round blue cell tumors in childhood and youth.

1.9 Non-neoplastic Vascular Pathology

1.9.1 Congenital Anomalies

1.9.1.1 Arteriovenous Malformations

An abnormal communication between arterial and venous systems (without passing through capillaries) that may be encountered in infancy, childhood, and youth. If the arterio-venous malformation (AVM) is large, it can induce heart failure. Apart from the natural forms, acquired forms include AVM as a result of penetrating injury, inflammatory necrosis, and aneurysmal rupture.

1.9.1.2 Berry Aneurysm

The most common vascular malformation of cerebral vessels is at the polygon of Willis at the basis of the brain. They are called a berry aneurysm due to the shape and account for 95% of cerebral aneurysms. The rupture of berry aneurysms may cause SCD. Such a breach is a significant component of the etiologies of SCD and needs to be ruled out with a complete autopsy.

Typically, berry aneurysm occurs at the bifurcation of major arteries.

1.9.1.3 Media Dysplasia

Ehlers-Danlos (EDS) and Marfan syndrome (MFS) are multisystemic diseases. These diseases are genetically inherited and mainly affect the soft connective tissues. EDS is heterogeneous and characterized by hyperextensibility of the skin, atrophic scar, joint hypermobility, and tissue fragility. The etiology relies on mutations in genes encoding fibrillar collagens or collagen-modifying enzymes. MFS is an AD-inherited disorder that affects the cardiovascular, ocular, and skeletal system. There are aortic root dissection, *ectopia lentis*, and bone overgrowth. Mutations in the *FBN1* gene cause MFS, encoding the microfibrillar protein fibrillin-1 (Figs. 1.72, and 1.73).

1.9.2 Fibromuscular Dysplasia

- *DEF*: Non-atherosclerotic, noninflammatory vasculopathy, which shows abnormal growth within the wall of an artery, mainly the renal and carotid arteries. Its pathogenesis is unknown. There is no necrosis, calcification, inflammation, and arteriosclerosis. Typically, FMD develops by 20–30 years of age and involves large- and medium-sized muscular arteries: renal, carotid, axillary, and mesenteric BVs. Microscopically, there is an unusual arrangement of cellular and extracellular elements of the wall, particularly media, with disorderly proliferation and distortion of the vascular lumen (Fig. 1.74). Six types are distinct, including:
  1. *Intimal fibroplasia*, which is blurry from the proliferative stage of atherosclerosis but no discernible lipid deposition
  2. *Medial fibroplasia*, characterized by "string of beads," i.e., alternating stenosis (intimal fibrosis and medial thickening) and building of mural aneurysms
  3. *Medial hyperplasia* characterized by an increase in cells of the *tunica media*
  4. *Perimedial fibroplasia*, which involves fibrosis of outer half of the tunica media, being the inner portion regular – with circumferential and uniform thickening of vessel and narrowing of the lumen of the blood vessel
  5. *Medial dissection* with medial fibrosis and dissecting aneurysms
  6. *Periarterial fibroplasia*, which includes perivascular fibrosis and inflammation

1.9.3 Arteriosclerosis

This process is not dissimilar from the adult pathology, but it does not seem to play any major role in pediatrics. The Mönckeberg arteriosclerosis refers to a medial calcific sclerosis with calcifications in media of medium to small arteries with no associated inflammation and potential ossification and rare in adolescents and youth. The cystic medial necrosis refers to an accumulation of amorphous material in the media often forming cysts or mucoid pools. We have a disruption of the structure of the tunica media by the appearance of small clefts filled with a slightly basophilic ground substance. The cystic medial necrosis is observed in Marfan syndrome and predisposes to vascular dissections. Pregnancy significantly augments the risk of vascular events. There is a several-fold higher risk of venous thromboembolism, myocardial infarction, and stroke than nonpregnant women of childbearing age. These risks are not minimal but extend for several months into the postpartum period and need to be known by the general practitioners. Aortic rupture and complications in pregnancy have been described in Marfan syndrome, Loeys-Dietz syndrome, Ehlers-Danlos syndrome, Turner syndrome, a bicuspid aortic valve (Fig. 1.75). On the other hand, aortic complications have also been reported in pregnant women without any other known risk factors and the risk of aortic dissection or rupture in pregnancy has been recently investigated (Kamel et al. 2016). These authors found that there is an absolute increase in the risk of aortic dissection or rupture attributable to pregnancy in ≈4 per million pregnancies. This data are helpful when counseling patients about the risks of pregnancy and when evaluating thoracal symptoms in pregnant or postpartum patients.

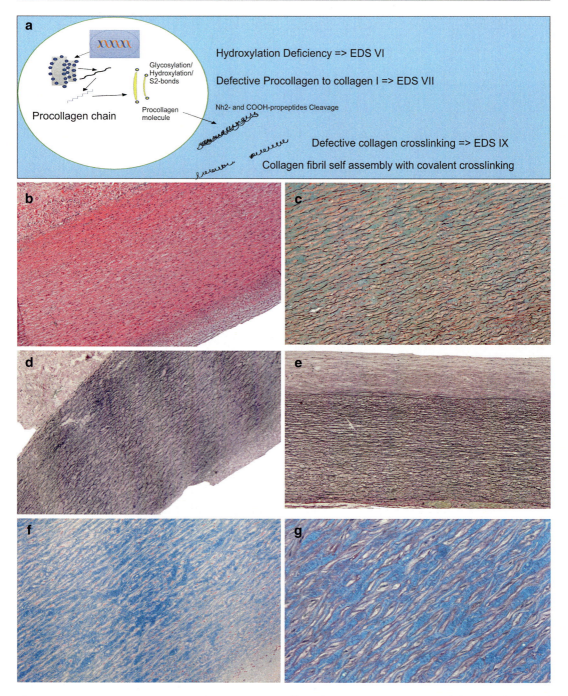

Fig. 1.72 Genetics and histopathology of Ehlers-Danlos syndrome (EDS) with myxoid degeneration and fibrillar degeneration (**a–b**). EDS is a group of genetic connective tissue disorders with symptoms including loose joints, stretchy skin, and abnormal scar formation and complications including aortic dissection, joint dislocations, and scoliosis/osteoarthritis. Figure (**c**) shows the rupture of the fibers by Movat pentachrome staining (×100, original magnification). Figures (**d**, **e**) show the rupture of the elastic fibers by Vierhoff staining (×50, original magnification). Figures (**f**, **g**) show the myxoid degeneration of the aorta using the Alcian Blue-periodic acid-Schiff special stain (**f**, ×100 original magnification; **g**, ×200 original magnification)

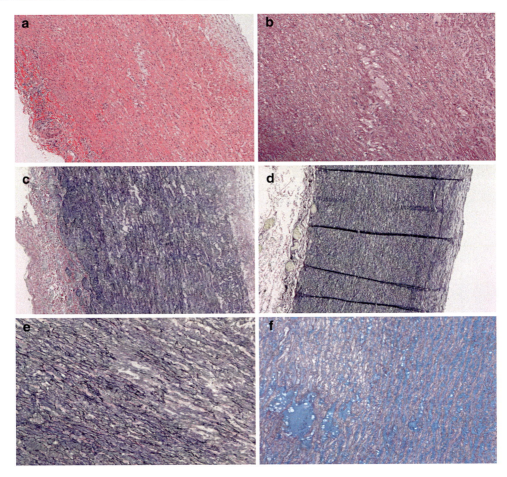

Fig. 1.73 Histopathology of Marfan syndrome with lacunae in the wall of the aorta (**a**, Hematoxylin and Eosin staining, ×50 and **b**, periodic acid-Schiff staining, ×100). An Ellis Van Gieson staining can show the differences between an aorta of a patient with Marfan syndrome (**c**, Ellis Van Gieson ×50 original magnification) and a normal age-matched control aorta (**d**, Ellis Van Gieson ×50 original magnification). Figure (**e**) shows a high magnification of the rupture of the elastic fibers (Ellis Van Gieson ×100 original magnification). Figure (**f**) shows the myxoid degeneration with accumulation of mucopolysaccharides (Alcian Blue-periodic acid-Schiff staining, ×100 original magnification)

1.9.3.1 Vasculitides

The vasculitides are presented here (Box 1.6).

1.9.4 Large-Vessel Vasculitis

1.9.4.1 Takayasu Arteritis
- *DEF*: Chronic granulomatous arteritis involving media and adventitia accompanied with intimal fibrosis and fibrous thickening of the aortic arch and marked focal narrowing depending from the location from the origins

---

Box 1.6 Vasculitides

1.9.4. *Large-Vessel Vasculitis* (Takayasu Arteritis, Horton Arteritis)

1.9.5. *Medium-Vessel Vasculitis* (Kawasaki Disease, Polyarteritis Nodosa)

1.9.6. *Small-Vessel Vasculitis* (Wegener Granulomatosis, EGPA or Churg-Strauss Syndrome, Microscopic Polyangiitis)

Note: EGPA, Eosinophilic Granulomatosis with Polyangiitis.

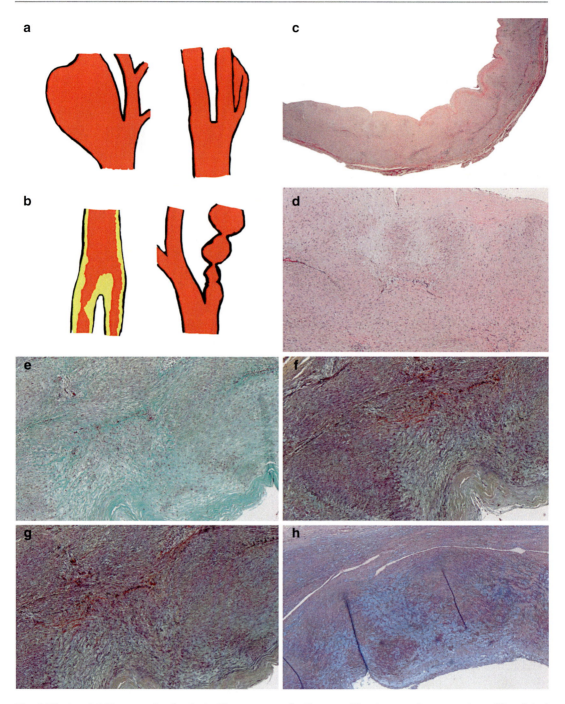

Fig. 1.74 Arterial fibromuscular dysplasia. The cartoon shows arterial fibromuscular dysplasia (**a**, **b**). Figures (**c**–**e**) show the fibromuscular dysplasia histologically (**c**, Hematoxylin and Eosin stain, ×12.5 original magnification; **d**, Hematoxylin and Eosin stain, ×50 original magnification; **e**, Movat pentachrome stain, ×50 original magnification; **f**, Movat pentachrome stain, ×50 original magnification; **g**, Movat pentachrome stain, ×50 original magnification; **h**, Alcian Blue periodic acid-Schiff stain, ×50 original magnification). See text for details

1.9 Non-neoplastic Vascular Pathology

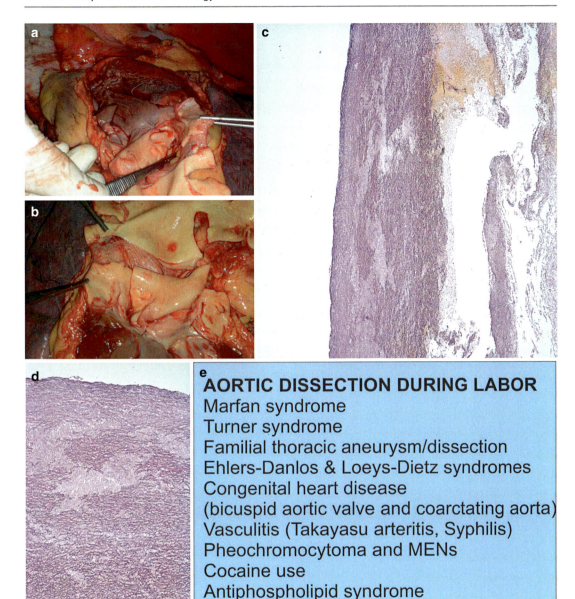

Fig. 1.75 Dissecting aortic aneurysms in labor is a tragedy that occurs rarely. In **a** and **b**, the gross photographs of such a dissection are presented. Histologic microphotographs (**c**, **d**) show the dissecting aneurysms. Figure (**e**) shows the underlying causes of a dissecting aneurysms in labor

of the arch vessels to the distal aorta (Fig. 1.76).
- *EPI*: Asian>Western, 15–45 years-old, ♂:♀ = 1:3, HLA-DR4 + (often).
- *EPG*: Autoimmune.
- *CLI*: Ocular disturbances, weakened pulses in upper limbs ("pulseless disease"), subclavian bruits, hypertension, syncope, hemiplegia
- *IMG*: Angiographically, the type I involves only the branches of the aortic arch, while the type IIa involves the ascending aorta, aortic arch, and its branches. The type IIb involves the ascending aorta, aortic arch, and its branches as well as the descending thoracic aorta. Type III affects the descending thoracic aorta, the abdominal aorta, and/or the renal

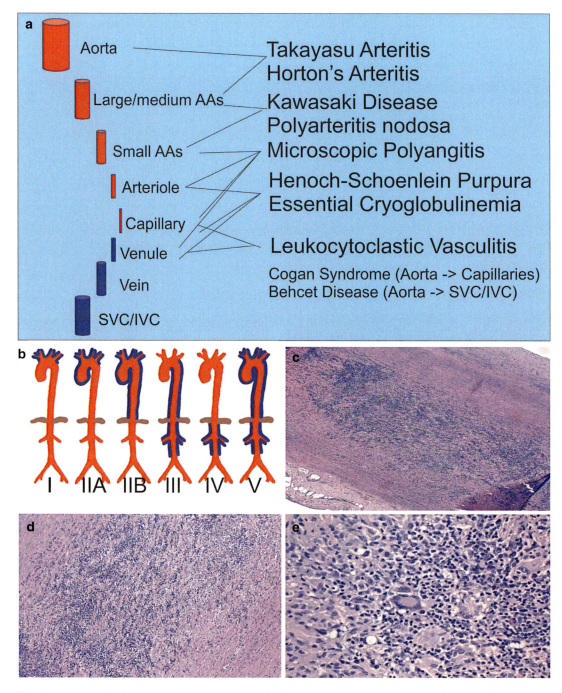

Fig. 1.76 Vasculitis schema (**a**, **b**) and Takayasu aortitis shown with microphotographs (**c–e**: Hematoxylin and Eosin stain, ×40, ×100, and ×400 original magnification, respectively)

arteries, while type IV involves only the abdominal aorta and/or the renal arteries, and, finally, the type V, which is the most common type, has combined features of both types IIb and IV. Moreover, the optional involvement of the coronary and pulmonary arteries is indicated as C+ and P+, respectively.

- *TRT*: Glucocorticosteroids starting at a high dose and tapered over weeks and months and cytotoxic drugs (e.g., methotrexate and azathio-

prine) as well as antitumor necrosis factor drugs such as infliximab. Surgery may be necessary to relieve blocked arteries or repair aneurysms.
- *PGN*: Mortality rate is 5%, but complications include stenosis of blood vessels, myocarditis, pericarditis, endocarditis, heart failure, an aneurysm in the aorta, arterial hypertension, ischemic stroke or transient ischemic attack (TIA), myocardial infarct, and pulmonary artery dysfunction.

1.9.4.2 Horton Arteritis (Cranial Arteritis, Giant Cell Arteritis)

- *DEF*: Chronic granulomatous arteritis with optional necrosis of medium and small arteries of the head and neck region with the destruction of the internal and external elastic laminae.
- *EPI*: Western>Asian, elderly, ♂:♀ = 1:3, HLA-DR4 + (often).
- *EPG*: Autoimmune.
- *CLI*: Headache, scalp tenderness, jaw pain, visual loss, ↑ESR.
- *GRO*: Hardening of the arterial segment in a temporal biopsy in 60% of patients.
- *CLM*: Chronic granulomatous arteritis with giant cells (see definition).
- *TRT*: Steroids.
- *PGN*: Rapid remission with steroid therapy.

1.9.5 Medium-Vessel Vasculitis

1.9.5.1 Kawasaki Disease

- *DEF*: Acute systemic vasculitis of unknown cause (Fig. 1.77).
- *SYN*: Mucocutaneous lymph node syndrome.
- *EPI*: In the USA, 19:100,000 children <5 years are hospitalized with Kawasaki disease annually. According to the USA and Japanese guidelines, Kawasaki disease is a clinical diagnosis. Asian and black American children are 2.5 and 1.5 times more likely to develop KD than white children, respectively, suggesting a genetic link.
- *EPG*: Acute, systemic vasculitis of small- and medium-sized arteries that predominantly affects patients younger than 5 years.
- *CLI*: DGN relies on a fever ≥5 days, accompanied by 4/5 findings: (1) bilateral conjunctival injection, (2) oral changes such as cracked and erythematous lips and strawberry tongue, (3) cervical lymphadenopathy, (4) limb changes such as erythema or palm and sole desquamation, and (5) polymorphous rash.
- *DDX*: Drug hypersensitivity, juvenile idiopathic arthritis, staphylococcal scalded skin syndrome, Stevens-Johnson syndrome, streptococcal scarlet fever, toxic shock syndrome, and viral infection.
- *IMG*: Transthoracic echocardiography is the imaging modality of choice to detect coronary aneurysms and other cardiac artery abnormalities.
- *TRT*: Corticosteroids and IVIG as initial treatment significantly reduced the risk of coronary abnormalities compared with IVIG alone.
- *PGN*: Coronary aneurysms. The corticosteroid-IVIG combined protocol reduces the development of coronary artery abnormalities from approximately 25% to less than 5% and the development of giant aneurysms to 1%.

1.9.5.2 Polyarteritis Nodosa (PAN)

- *DEF*: Necrotizing systemic medium vessel vasculitis, which is usually ANCA negative.
- *EPI*: Rare disease, young adults, ♂:♀ = 2–3:1.
- *EPG*: With the decline of HBV worldwide, mainly in Western countries, and the evolving definitions of vasculitis, PAN is becoming a rare disease, but variants of PAN include single-organ illness and cutaneous PAN. PAN can be idiopathic or secondary to other causes (e.g., HCV, HIV, parvovirus B19, and hairy cell leukemia). The most common knowledge postulates that viral antigens trigger the complement cascade, resulting in a neutrophil-rich and lymphocyte-rich inflammatory infiltrate within the arterial media. This inflammation will eventually lead to fibrosis, thrombosis, and aneurysmal degeneration.
- *CLI*: FUO, hematuria, albuminuria, hypertension, abdominal pain, and melena, muscular aches, and peripheral motor neuritis (FUO, *febris e causa ignota*, fever of unknown origin).
- *GRO/CLM*: Nodular lesions of small- or medium-sized muscular arteries [especially

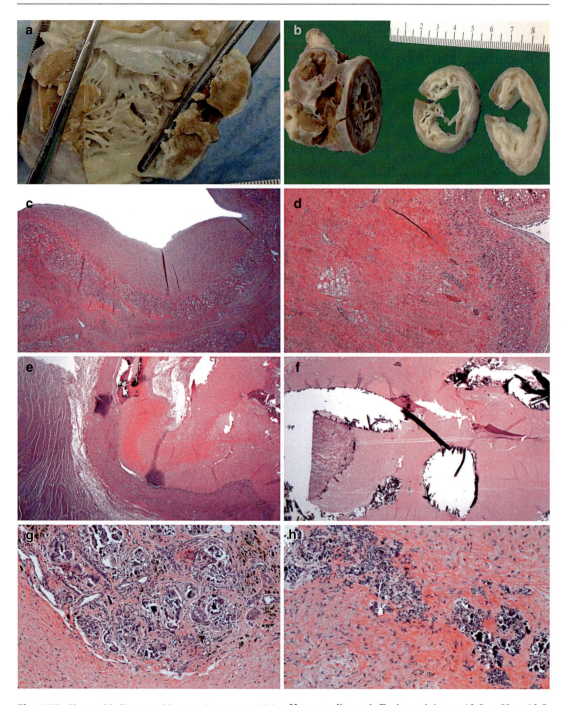

Fig. 1.77 Kawasaki disease with extensive myocardial necroses (**a**, **b**) and microphotographic documentation with necrosis and dystrophic calcifications (**c–h**, Hematoxylin and Eosin staining, ×12.5, ×50, ×12.5, ×12.5, ×100, and ×100, as original magnifications, respectively)

renal (85%), coronary (75%), hepatic (65%), and GI tract (50%); pulmonary circulation spared] with sharp demarcations (segmental lesions); most commonly at branch points and injuries frequently scattered and of various ages.
- *DDX*: The formal distinction between PAN and microscopic polyangiits (MPA) was made at the Chapel Hill Consensus Conference (CHCC) in 1994. PAN is limited to necrotizing inflammation of medium-sized and small arteries without the involvement of smaller vessels (i.e., arterioles, venules, or capillaries). Conversely, MPA is described as a pauci-immune (few or no immune deposits) necrotizing vasculitis affecting small vessels, with or without the involvement of medium-sized arteries.
- *TRT*: If PAN is not associated with a viral syndrome, treatment relies mostly on corticosteroid therapy, but immunosuppressive treatment has been proposed for a stronger response to treatment.
- *PGN*: Frequently leads to aneurysms (especially in intestinal branches) and thrombosis with infarction of supplied areas.

1.9.6 Small-Vessel Vasculitis

Wegener Granulomatosis
It is an acute necrotizing disease with granulomas of upper and lower respiratory tract (with giant cells and leukocytes), accompanying necrotizing vasculitis of the pulmonary vessels and glomeruli, leading to persistent pneumonitis, chronic sinusitis, nasopharyngeal ulcerations, and renal disease. It is better described in other chapters. Serum antineutrophil cytoplasmic antibodies serve as a marker for disease activity (typically C-ANCA or lysosomal enzymes).

Eosinophilic granulomatosis with polyangiitis
(EGPA) (formerly Churg-Strauss syndrome) is a systemic inflammatory necrotizing vasculitis with multiple etiological factors and a suspected vital role of eosinophils in its pathophysiology.

Microscopic polyangiitis
Is an autoimmune disease in which antineutrophil cytoplasmic autoantibodies (ANCA), particularly those reacting with MPO, are pathogenic and multiple organs can be affected (e.g., kidneys and lungs). Variable as presentation, MPA is part of the ANCA-associated vasculitides, which are characterized by necrotizing vasculitis of small vessels. The involvement is systemic and this disease predominantly affects small-caliber blood vessels.

Multiple Choice Questions and Answers

- CVS-1 Inductive signaling from adjacent endoderm and overlying ectoderm together with inhibitory signaling refine the sites in which the cardiomyogenic transcriptional program is initially activated. These signals include bone morphogenetic protein (BMP), fibroblast growth factor (FGF), and WNT signals that result in the activation of critical upstream transcriptional regulators of the cardiac phenotype. Which of the following triplet is the correct one that ectopically activated is enough to drive cardiomyogenesis in noncardiogenic regions of the embryo?
  (a) SMARCD3, GATA-4, and TBX5
  (b) FGF, WNT, and GATA-4
  (c) FGF, SMARCD3, and TBX5
  (d) NGF, EGF, and TBX5
- CVS-2 A baby born is noted to be cyanotic in the first hours after birth. Although the child has no respiratory distress, a single second heart sound is recorded by the physician. By echocardiography, a D-transposition is identified. Which of the following statements is true?
  (a) In D-transposition the pressure of the right ventricle is decreased.
  (b) A Rashkind (balloon atrial septostomy) should be performed to address the condition improving systemic oxygenation palliatively.
  (c) The aortic valve is anterior to the pulmonary valve.

- (d) The pulmonary valve is posterior to the aortic valve.
- (e) A Sano procedure should be performed as the final step to address the condition improving systemic oxygenation palliatively.

- CVS-3 The DKS procedure is named for three physicians – Paul Damus, Michael Kaye, and H. C. Stansel. The DKS procedure has been independently reported in the literature by these cardiac surgeons in the 1970s. In which condition is the DKS indicated?
  - (a) D-Transposition
  - (b) L-Transposition
  - (c) Tetralogy of Fallot
  - (d) Pulmonary atresia with intact ventricular septum
  - (e) TGA in combination with a small left ventricle

- CVS-4 Both clinical presentation and microscopic evaluation are variables in mitochondrial cardiomyopathy leading to a broad differential diagnosis including storage, metabolic, genetic, and environmental disorders. Which of the following statement is correct?
  - (a) Mitochondrial cardiomyopathies are mostly autosomal recessive inherited.
  - (b) In mitochondrial cardiomyopathy, there is fibro-fatty infiltration of the right ventricle.
  - (c) Although rarely seen in children, ragged red fibers are a typical feature of mitochondrial cardiomyopathy.
  - (d) DNA depletion syndrome is an autosomal dominant inherited mitochondrial cardiomyopathy.
  - (e) Dilated cardiomyopathy is the most common presentation of a mitochondrial cardiomyopathy.

- CVS-5 A 6-month-old child is operated, and a cardiac tumor is resected. The diagnosis provided by the pathologist is cardiac rhabdomyoma. Which of the following statements of cardiac rhabdomyoma is TRUE?
  - (a) Cardiac rhabdomyoma is less than 20% of primary heart tumors in children.
  - (b) It is usually discovered in patients older than 5 years of age.
  - (c) It is considered a hamartoma due to a mutation in *TSC1* and *TSC2* genes.
  - (d) Almost all patients have tuberous sclerosis (Bourneville disease).
  - (e) All cardiac tumors require surgery.

CVS-6 In acute rheumatic carditis, the vegetations occur most often in which location?
  - (a) The insertion of the *chordae tendineae*
  - (b) The annulus of the mitral valve
  - (c) The Valsalva sinuses
  - (d) The free margins of the mitral valve
  - (e) The coronary ostia

- CVS-7 A 12-year-old child presented to the family doctor following the discovery of an abnormal rhythm at the local training soccer camp. Echocardiography shows a mass at the cardiac apex in the right ventricle. No other relevant findings were observed. The resection of the tumor identifies a cardiac fibroma. Which of the following reports is NOT correct regarding the cardiac fibroma?
  - (a) Cardiac fibromas can present with heart failure, arrhythmia, cyanosis, chest pain, or sudden death.
  - (b) In children, cardiac fibroma is the first most common benign cardiac tumor, followed by the rhabdomyoma.
  - (c) In 10% of cardiac fibromas, Gorlin syndrome (nevoid basal cell carcinoma syndrome) is diagnosed.
  - (d) Grossly, the tumor presents usually as a solitary discrete bulging mass composed, microscopically, by bland fibroblasts and collagen, often with calcification, and an infiltrative margin.
  - (e) The differential diagnosis should include a scar, inflammatory myofibroblastic tumor, benign fibrous histiocytoma, and fibrosarcoma.

- CVS-8 According to the 2013 ISHLT classification for humoral rejection, which criterion should not be included in the correct definition of pAMR 3 or severe pathologic AMR?
  - (a) Interstitial hemorrhage
  - (b) Capillary fragmentation
  - (c) Capillary proliferation
  - (d) Mixed inflammatory infiltrates
  - (e) Endothelial cell pyknosis and/or karyorrhexis

CVS-9 Which of the next elements does NOT belong to the pentalogy of Fallot?
(a) Ventricular septal defect
(b) Right ventricular hypertrophy
(c) Dextroposition of the aorta
(d) Pulmonary valve stenosis
(e) Atrial septal defect
(f) Patent ductus arteriosus

- CVS-10 Which of the following criteria is NOT one criterion defining the major criteria for diagnosing Kawasaki disease in addition to the presence of ≥5 days of fever according to the most updated guidelines?
(a) Bilateral non-exudative conjunctivitis
(b) Erythema of the lips and oral mucosa (e.g., strawberry tongue, cracked lips)
(c) Changes in the extremities (swelling, redness or desquamation of the palms, desquamation in the subacute phase)
(d) Polymorphous exanthema
(e) Cervical lymphadenopathy
(f) Coronary arterial aneurysms

References and Recommended Readings

Abraham MR, Olson LJ, Joyner MJ, Turner ST, Beck KC, Johnson BD. Angiotensin-converting enzyme genotype modulates pulmonary function and exercise capacity in treated patients with congestive stable heart failure. Circulation. 2002;106(14):1794–9.

ACHA Q and A. Transposition of the great arteries after mustard/Senning repair. The Adult Congenital Heart Association. http://heart.ucla.edu/workfiles/Adult_Congenital/tgamustardweb.pdf

Adhikary G, Gupta S, Sil P, Saad Y, Sen S. Characterization and functional significance of myotrophin: a gene with multiple transcripts. Gene. 2005;353(1):31–40.

Alikasifoglu M, Tokgözoglu L, Acil T, Atalar E, Ali Oto M, Sirri Kes S, Tuncbilek E. Tumor necrosis factor-a polymorphism in Turkish patients with dilated cardiomyopathy. Eur J Heart Fail. 2003;5(2):161–3.

Allegue C, Gil R, Blanco-Verea A, Santori M, Rodríguez-Calvo M, Concheiro L, Carracedo A, Brion M. Prevalence of HCM and long QT syndrome mutations in young sudden cardiac death-related cases. Int J Legal Med. 2011;125(4):565–72. https://doi.org/10.1007/s00414-011-0572-7. PubMed PMID:21499742

Anderson RH, Becker AE, Freedom RM, Macartney FJ, Quero-Jimenez M, Shinebourne EA, Wilkinson JL, Tynan M. Sequential segmental analysis of congenital heart disease. Pediatr Cardiol. 1984;5:281–7.

Araoz PA, Eklund HE, Welch TJ, Breen JF. CT and MR imaging of primary cardiac malignancies. Radiographics. 1999;19(6):1421–34. Review. PubMed PMID:10555666

Arbustini E, Diegoli M, Fasani R, Grasso M, Morbini P, Banchieri N, Bellini O, Dal Bello B, Pilotto A, Magrini G, Campana C, Fortina P, Gavazzi A, Narula J, Viganò M. Mitochondrial DNA mutations and mitochondrial abnormalities in dilated cardiomyopathy. Am J Pathol. 1998;153(5):1501–10. PubMed PMID:9811342; PubMed Central PMCID: PMC1853408

Arbustini E, Pilotto A, Repetto A, Grasso M, Negri A, Diegoli M, Campana C, Scelsi L, Baldini E, Gavazzi A, Tavazzi L. Autosomal dominant dilated cardiomyopathy with atrioventricular block: a lamin A/C defect related disease. J Am Coll Cardiol. 2002;39(6):981–90.

Ardizzone N, Cappello F, Di Felice V, Rappa F, Minervini F, Marasà S, Marasà L, Rabl W, Zummo G, Sergi C. Atrial natriuretic peptide and CD34 overexpression in human idiopathic dilated cardiomyopathies. APMIS. 2007;115(11):1227–33. PubMed PMID:18092954

Aretz HT. Myocarditis: the Dallas criteria. Hum Pathol. 1987;18(6):619–24. PubMed PMID:3297992

Arimura T, Hayashi T, Terada H, Lee SY, Zhou Q, Takahashi M, Ueda K, Nouchi T, Hohda S, Shibutani M, Hirose M, Chen J, Park JE, Yasunami M, Hayashi H, Kimura A. A Cypher/ZASP mutation associated with dilated cardiomyopathy alters the binding affinity to protein kinase C. J Biol Chem. 2004;279(8):6746–52.

Arimura T, Nakamura T, Hiroi S, Satoh M, Takahashi M, Ohbuchi N, Ueda K, Nouchi T, Yamaguchi N, Akai J, Matsumori A, Sasayama S, Kimura A. Characterization of the human nebulette gene: a polymorphism in an actin-binding motif is associated with nonfamilial idiopathic dilated cardiomyopathy. Hum Genet. 2000;107(5):440–51.

Aryana A, Williams MA. Marijuana as a trigger of cardiovascular events: speculation or scientific certainty? Int J Cardiol. 2007;118(2):141–4. Review. PubMed PMID:17005273

Bachinski LL, Abchee A, Durand JB, Roberts R, Krahe R, Hobson GM. Polymorphic trinucleotide repeat in the MEF2A gene at 15q26 is not expanded in familial cardiomyopathies. Mol Cell Probes. 1997;11(1):55–8.

Bijlsma FJ, Bruggink AH, Hartman M, FH G-M, Tilanus MG, de Jonge N, de Weger RA. No association between IL-10 promoter gene polymorphism and heart failure or rejection following cardiac transplantation. Tissue Antigens. 2001;57(2):151–3.

Bijlsma FJ, vanKuik J, Tilanus MG, deJonge N, Rozemuller EH, van den Tweel JG, Gmelig-Meyling FH, deWeger RA. Donor interleukin-4 promoter gene polymorphism influences allograft rejection after heart transplantation. J Heart Lung Transplant. 2002;21(3):340–6.

Bons LR, Roos-Hesselink JW. Aortic disease and pregnancy. Curr Opin Cardiol. 2016;31(6):611–7. Review. PubMed PMID:27652809

Bossone E, Pyeritz RE, O'Gara P, Harris KM, Braverman AC, Pape L, Russo MJ, Hughes GC, Tsai TT, Montgomery DG, Nienaber CA, Isselbacher EM,

Eagle KA. International Registry of Acute Aortic Dissection (IRAD) Investigators. Acute aortic dissection in blacks: insights from the International Registry of Acute Aortic Dissection. Am J Med. 2013;126(10):909–15. https://doi.org/10.1016/j.amjmed.2013.04.020. PubMed PMID:23953874

Botenga AS. The significance of broncho-pulmonary anastomoses in pulmonary anomalies: a selective angiographic study. Radiol Clin Biol. 1969;38(5):309–28. PubMed PMID:5371133

Bowles KR, Gajarski R, Porter P, Goytia V, Bachinski L, Roberts R, Pignatelli R, Towbin JA. Gene mapping of familial autosomal dominant dilated cardiomyopathy to chromosome 10q21-23. J Clin Invest. 1996;98(6):1355–60.

Bowles NE, Javier Fuentes-Garcia F, Makar KA, Li H, Gibson J, Soto F, Schwimmbeck PL, Schultheiss HP, Pauschinger M. Analysis of the coxsackievirus B-adenovirus receptor gene in patients with myocarditis or dilated cardiomyopathy. Mol Genet Metab. 2002;77(3):257–9.

Brion M, Allegue C, Santori M, Gil R. Blanco-Verea A, Haas C, Bartsch C, Poster S, Madea B, Campuzano O, Brugada R, Carracedo A. Sarcomeric gene mutations in sudden infant death syndrome (SIDS). Forensic Sci Int. 2012;219(1–3):278–81. https://doi.org/10.1016/j.forsciint.2012.01.018.. PubMed PMID:22361390

Bruggink AH, van Oosterhout MF, De Jonge N, Gmelig-Meyling FH, De Weger RA. TNFalpha in patients with end-stage heart failure on medical therapy or supported by a left ventricular assist device. Transpl Immunol. 2008;19(1):64–8.

Buja LM, Roberts WC. Iron in the heart. Etiology and clinical significance. Am J Med. 1971;51(2):209–21. PubMed PMID:5095527

Caforio AL, Pankuweit S, Arbustini E, Basso C, Gimeno-Blanes J, Felix SB, et al. European Society of Cardiology Working Group on Myocardial and Pericardial Diseases. Current state of knowledge on aetiology, diagnosis, management, and therapy of myocarditis: a position statement of the European Society of Cardiology Working Group on Myocardial and Pericardial Diseases. Eur Heart J. 2013;34(33):2636–48. https://doi.org/10.1093/eurheartj/eht210. Epub 2013 Jul 3. PubMed PMID:23824828

Candy GP, Skudicky D, Mueller UK, Woodiwiss AJ, Sliwa K, Luker F, Esser J, Sareli P, Norton GR. Association of left ventricular systolic performance and cavity size with angiotensin-converting enzyme genotype in idiopathic dilated cardiomyopathy. Am J Cardiol. 1999;83(5):740–4.

Carpentier A, Chauvaud S, Macé L, Relland J, Mihaileanu S, Marino JP, Abry B, Guibourt P. A new reconstructive operation for Ebstein's anomaly of the tricuspid valve. J Thorac Cardiovasc Surg. 1988;96(1):92–101. PubMed PMID:3386297

Carvalho JS, Ho SY, Shinebourne EA. Sequential segmental analysis in complex fetal cardiac abnormalities: a logical approach to diagnosis. Ultrasound Obstet Gynecol. 2005;26(2):105–11. Review. PubMed PMID:16041685

Castillo JG, Silvay G. Characterization and management of cardiac tumors. Semin Cardiothorac Vasc Anesth. 2010;14(1):6–20. Review. PubMed PMID:20472615

Cave D, Ross DB, Bahitham W, Chan A, Sergi C, Adatia I. Mitochondrial DNA Depletion Syndrome-An Unusual Reason for Interstage Attrition after the Modified Stage 1 Norwood Operation. Congenit Heart Dis. 2011; https://doi.org/10.1111/j.1747-0803.2011.00569.x. [Epub ahead of print] PubMed PMID:22011012

Cave D, Ross DB, Bahitham W, Chan A, Sergi C, Adatia I. Mitochondrial DNA depletion syndrome-an unusual reason for interstage attrition after the modified stage 1 Norwood operation. Congenit Heart Dis. 2013;8(1):E20–3. https://doi.org/10.1111/j.1747-0803.2011.00569.x. PubMed PMID:22011012

Celermajer DS, Bull C, Till JA, Cullen S, Vassillikos VP, Sullivan ID, Allan L, Nihoyannopoulos P, Somerville J, Deanfield JE. Ebstein's anomaly: presentation and outcome from fetus to adult. J Am Coll Cardiol. 1994;23(1):170–6. PubMed PMID:8277076

Celoria GC, Patton RB. Congenital absence of the aortic arch. Am Heart J. 1959;58:407–13.

Chang AN, Potter JD. Sarcomeric protein mutations in dilated cardiomyopathy. Heart Fail Rev. 2005;10(3):225–35.

Chang RA, Rossi NF. Intermittent cocaine use associated with recurrent dissection of the thoracic and abdominal aorta. Chest. 1995;108(6):1758–62. PubMed PMID:7497801

Charron P, Arbustini E, Bonne G. What Should the Cardiologist know about Lamin Disease? Arrhythm Electrophysiol Rev. 2012;1(1):22–8. https://doi.org/10.15420/aer.2012.1.22. PMID: 26835025; PMCID: PMC4711561

Charron P, Komajda M. Are we ready for pharmacogenomics in heart failure? Eur J Pharmacol. 2001;417(1–2):1–9.

Chen PL, Chang HH, Chen IM, Lai ST, Shih CC, Weng ZC, Hsieh YC, Yang AH. Malignancy after heart transplantation. J Chin Med Assoc. 2009;72(11):588–93. PubMed PMID:19948436

Chiu B, Jantuan E, Shen F, Chiu B, Sergi C. Autophagy-inflammasome interplay in heart failure: a systematic review on basics, pathways, and therapeutic perspectives. Ann Clin Lab Sci. 2017;47(3):243–52. Review. PubMed PMID:28667023

Chiu B, Sergi C. Dilated cardiomyopathy: etio-morphologic investigation. Front Biosci (Schol Ed). 2010;2:112–6. Review. PubMed PMID:20036933

Connor JA, Thiagarajan R. Hypoplastic left heart syndrome. Orphanet J Rare Dis. 2007;2:23. Review. PubMed PMID:17498282; PubMed Central PMCID: PMC1877799

Corfield A, Meyer P, Kassam S, Mikuz G, Sergi CSNP. At the origins of the databases of an innovative biotechnology tool. Front Biosci (Schol Ed). 2010;2:1–4. Review. PubMed PMID:20036923

Corone S, Iliou MC, Pierre B, Feige JM, Odjinkem D, Farrokhi T, Bechraoui F, Hardy S, Meurin P. Cardiac Rehabilitation working Group of the French Society of Cardiology. French registry of cases of type I acute aortic dissection admitted to a cardiac rehabilitation center after surgery. Eur J Cardiovasc Prev Rehabil. 2009;16(1):91–5. https://doi.org/10.1097/HJR.0b013e32831fd6c8. PubMed PMID:19237998

Crawford JD, Hsieh CM, Schenning RC, Slater MS, Landry GJ, Moneta GL, Mitchell EL. Genetics, Pregnancy, and Aortic Degeneration. Ann Vasc Surg. 2016;30:158.e5–9. https://doi.org/10.1016/j.avsg.2015.06.100. PubMed PMID:26381327

Cuoco MA, Pereira AC, de Freitas HF. de Fátima Alves da Mota G, Fukushima JT, Krieger JE, Mansur AJ. Angiotensin-converting enzyme gene deletion polymorphism modulation of onset of symptoms and survival rate of patients with heart failure. Int J Cardiol. 2005;99(1):97–103.

Dacey LJ. Pulmonary homografts: current status. Curr Opin Cardiol. 2000;15(2):86–90. Review. PubMed PMID:10963144

Dazert P, Meissner K, Vogelgesang S, Heydrich B, Eckel L, Böhm M, Warzok R, Kerb R, Brinkmann U, Schaeffeler E, Schwab M, Cascorbi I, Jedlitschky G, Kroemer HK. Expression and localization of the multidrug resistance protein 5 (MRP5/ABCC5), a cellular export pump for cyclic nucleotides, in human heart. Am J Pathol. 2003;163(4):1567–77.

De Groot AC, Anderson RH. Straddling and overriding atrioventricular valves: morphology and classification. Am J Cardiol. 1979;44(6):1122–34. PubMed PMID:495507

De Ruiter MC, Gittenberger-de Groot AC, Rammos S, Poelmann RE. The special status of the pulmonary arch artery in the branchial arch system of the rat. Anat Embryol (Berl). 1989;179(4):319–25. PubMed PMID:2735526

De Sousa F, Gómez JA, Grandi C. Teratogenesis caused by marihuana (Cannabis sativa) in various stages of development of the chick embryo. Rev Med Panama. 1982;7(3):223–237. Spanish.. PubMed PMID:7146501.

Deng H, Xia H, Deng S. Genetic basis of human left-right asymmetry disorders. Expert Rev Mol Med. 2015;16:e19. https://doi.org/10.1017/erm.2014.22. Review

Dewar K, Nolan S. Chronic hypertension, recreational cocaine use and a subsequent acute aortic dissection in a young adult. BMJ Case Rep. 2017;2017:pii: bcr-2016-218235. https://doi.org/10.1136/bcr-2016-218235.. PubMed PMID:29079671; PubMed Central PMCID: PMC5665273.

Dice JE, Bhatia J. Patent ductus arteriosus: an overview. J Pediatr Pharmacol Ther. 2007;12(3):138–46. https://doi.org/10.5863/1551-6776-12.3.138. PMID: 23055849; PMCID: PMC3462096

Dos L, Teruel L, Ferreira IJ, Rodriguez-Larrea J, Miro L, Girona J, Albert DC, Gonçalves A, Murtra M, Casaldaliga J. Late outcome of Senning and Mustard procedures for correction of transposition of the great arteries. Heart. 2005;91(5):652–6. PubMed PMID:15831655; PubMed Central PMCID: PMC1768896

Drigo SA, Cunha-Neto E, Ianni B, Cardoso MR, Braga PE, Faé KC, Nunes VL, Buck P, Mady C, Kalil J, Goldberg AC. TNF gene polymorphisms are associated with reduced survival in severe Chagas' disease cardiomyopathy patients. Microbes Infect. 2006;8(3):598–603.

Drigo SA, Cunha-Neto E, Ianni B, Mady C, Faé KC, Buck P, Kalil J, Goldberg AC. Lack of association of tumor necrosis factor-alpha polymorphisms with Chagas disease in Brazilian patients. Immunol Lett. 2007;108(1):109–11.

Duran-Moreno J, Kampoli K, Kapetanakis EI, Mademli M, Koufopoulos N, Foukas PG, et al. Pericardial Synovial Sarcoma: Case Report, Literature Review and Pooled Analysis. In Vivo. 2019;33(5):1531-1538. https://doi.org/10.21873/invivo.11633.. Review. PubMed PMID:31471401; PubMed Central PMCID:PMC6754991.

Erdmann J, Hassfeld S, Kallisch H, Fleck E, Regitz-Zagrose V. Genetic variants in the promoter (g983G>T) and coding region (A92T) of the human cardiotrophin-1 gene (CTF1) in patients with dilated cardiomyopathy. Hum Mutat. 2000;16(5):448.

Ewing CK, Loffredo CA, Beaty TH. Paternal risk factors for isolated membranous ventricular septal defects. Am J Med Genet. 1997a;71(1):42–6. PubMed PMID:9215767

Ewing CK, Loffredo CA, Beaty TH. Paternal risk factors for isolated membranous ventricular septal defects. Am J Med Genet. 1997b;71(1):42–6. PubMed PMID:9215767

Faé KC, Drigo SA, Cunha-Neto E, Ianni B, Mady C, Kalil J, Goldberg AC. HLA and beta-myosin heavy chain do not influence susceptibility to Chagas disease cardiomyopathy. Microbes Infect. 2000;2(7):745–51.

Familiari A, Morlando M, Khalil A, Sonesson SE, Scala C, Rizzo G, et al. Risk Factors for Coarctation of the Aorta on Prenatal Ultrasound: A Systematic Review and Meta-Analysis. Circulation. 2017;135(8):772–85. https://doi.org/10.1161/CIRCULATIONAHA.116.024068. Epub 2016 Dec 29. Review. PubMed PMID:28034902

Fantel AG, Person RE. Involvement of mitochondria and other free radical sources in normal and abnormal fetal development. Ann N Y Acad Sci. 2002;959:424–33. Review. PubMed PMID:11976215

Feinstein JA, Benson DW, Dubin AM, Cohen MS, Maxey DM, Mahle WT, Pahl E, Villafañe J, Bhatt AB, Peng LF, Johnson BA, Marsden AL, Daniels CJ, Rudd NA, Caldarone CA, Mussatto KA, Morales DL, Ivy DD, Gaynor JW, Tweddell JS, Deal BJ, Furck AK, Rosenthal GL, Ohye RG, Ghanayem NS, Cheatham JP, Tworetzky W, Martin GR. Hypoplastic left heart syndrome: current considerations and expectations. J Am Coll Cardiol. 2012;59(1 Suppl):S1–42. Review. Erratum in: J Am Coll Cardiol. 2012 Jan 31;59(5):544. PubMed PMID:22192720

Finsterer J. Histiocytoid cardiomyopathy: a mitochondrial disorder. Clin Cardiol. 2008;31(5):225–7. Review. PubMed PMID:18473377

Finsterer J. Overview on visceral manifestations of mitochondrial disorders. Neth J Med. 2006;64(3):61–71.

Fisher A, Holroyd BR. Cocaine-associated dissection of the thoracic aorta. J Emerg Med. 1992;10(6):723–7. PubMed PMID:1491155

Forbess L, Bannykh S. Polyarteritis Nodosa. Rheum Dis Clin N Am. 2015;41(1):33–46. https://doi.org/10.1016/j.rdc.2014.09.005. Review. PubMed PMID:25399938

Fox CH. Cocaine use in pregnancy. J Am Board Fam Pract. 1994;7(3):225–8. Review. PubMed PMID:8059626

Fried R, Falkovsky G, Newburger J, Gorchakova AI, Rabinovitch M, Gordonova MI, Fyler D, Reid L, Burakovsky V. Pulmonary arterial changes in patients with ventricular septal defects and severe pulmonary hypertension. Pediatr Cardiol. 1986;7(3):147–54. PubMed PMID:3808993

Gastmann A, Sigusch HH, Henke A, Reinhardt D, Surber R, Gastmann O, Figulla HR. Role of adenosine monophosphate deaminase-1 gene polymorphism in patients with congestive heart failure (influence on tumor necrosis factor-alpha level and outcome). Am J Cardiol. 2004;93(10):1260–4.

Gedeon AK, Wilson MJ, Colley AC, Sillence DO, Mulley JC. X linked fatal infantile cardiomyopathy maps to Xq28 and is possibly allelic to Barth syndrome. J Med Genet. 1995;32(5):383–8.

Gerbode F, Hultgren H, Melrose D, Osborn J. Syndrome of left ventricular-right atrial shunt successful surgical repair of defect in five cases, with observation of bradycardia on closure. Annals of Surgery. 1958;148(3):433–46. [PMC free article] [PubMed] [Google Scholar]

Gilbert EF, Hodach RJ, Cheung MO, Bruyère HJ Jr. The effects of cocaine in the production of cardiovascular anomalies in beta-adrenoreceptor stimulated chick embryos. Experientia. 1976;32(8):1026–7. PubMed PMID:8328

Glueckert R, Rask-Andersen H, Sergi C, Schmutzhard J, Mueller B, Beckmann F, Rittinger O, Hoefsloot LH, Schrott-Fischer A, Janecke AR. Histology and synchrotron radiation-based microtomography of the inner ear in a molecularly confirmed case of CHARGE syndrome. Am J Med Genet A. 2010;152A(3):665–73. https://doi.org/10.1002/ajmg.a.33321. PubMed PMID:20186814

Greco A, De Virgilio A, Rizzo MI, Gallo A, Magliulo G, Fusconi M, Ruoppolo G, Tombolini M, Turchetta R, de Vincentiis M. Microscopic polyangiitis: advances in diagnostic and therapeutic approaches. Autoimmun Rev. 2015;14(9):837–44. https://doi.org/10.1016/j.autrev.2015.05.005. Review. PubMed PMID:25992801

Guan DW, Zhao R. Postmortem genetic testing in sudden cardiac death due to ion channelopathies. Fa Yi Xue Za Zhi. 2010;26(2):120–7. Review. Chinese. PubMed PMID:20653139

Haberer K, Buffo-Sequeira I, Chudley AE, Spriggs E, Sergi C. A case of an infant with compound heterozygous mutations for hypertrophic cardiomyopathy producing a phenotype of left ventricular noncompaction. Can J Cardiol. 2014;30(10) 1249.e1–3 https://doi.org/10.1016/j.cjca.2014.05.021. PubMed PMID:25262865

Haghighi K, Chen G, Sato Y, Fan GC, He S, Kolokathis F, Pater L, Paraskevaidis I, Jones WK, Dorn GW 2nd, Kremastinos DT, Kranias EG. A human phospholamban promoter polymorphism in dilated cardiomyopathy alters transcriptional regulation by glucocorticoids. Hum Mutat. 2008;29(5):640–647.

Haworth SG, Hislop AA. Pulmonary vascular development: normal values of peripheral vascular structure. Am J Cardiol. 1983;52(5):578–83. PubMed PMID:6613881

He R, Patel RM, Alkan S, Hammadeh R, Weiss SW, Goldblum JR, et al. Immunostaining for SYT protein discriminates synovial sarcoma from other soft tissue tumors: analysis of 146 cases. Mod Pathol. 2007;20(5):522–8. Epub 2007 Mar 2. PubMed PMID: 17334346

Hedera P, Petty EM, Bui MR, Blaivas M, Fink JK. The second kindred with autosomal dominant distal myopathy linked to chromosome 14q: genetic and clinical analysis. Arch Neurol. 2003;60(9):1321–5.

Hensler ME, Miyamoto S, Nizet V. Group B streptococcal beta-hemolysin/cytolysin directly impairs cardiomyocyte viability and function. PLoS One. 2008;3(6):e2446. PubMed PMID:18560574; PubMed Central PMCID: PMC2409074

Hermida-Prieto M, Monserrat L, Castro-Beiras A, Laredo R, Soler R, Peteiro J, Rodríguez E, Bouzas B, Alvarez N, Muñiz J, Crespo-Leiro M. Familial dilated cardiomyopathy and isolated left ventricular noncompaction associated with lamin A/C gene mutations. Am J Cardiol. 2004;94(1):50–4.

Herrmann S, Schmidt-Petersen K, Pfeifer J, Perrot A, Bit-Avragim N, Eichhorn C, Dietz R, Kreutz R, Paul M, Osterziel KJ. A polymorphism in the endothelin-A receptor gene predicts survival in patients with idiopathic dilated cardiomyopathy. Eur Heart J. 2001;22(20):1948–53.

Hiroi S, Harada H, Nishi H, Satoh M, Nagai R, Kimura A. Polymorphisms in the SOD2 and HLA-DRB1 genes are associated with non-familial idiopathic dilated cardiomyopathy in Japanese. Biochem Biophys Res Commun. 1999a;261(2):332–9.

Hiroi S, Harada H, Nishi H, Satoh M, Nagai R, Kimura A. Polymorphisms in the SOD2 and HLA-DRB1 genes are associated with nonfamilial idiopathic dilated cardiomyopathy in Japanese. Biochem Biophys Res Commun. 1999b;261(2):332–9.

Hofman-Bang J, Behr ER, Hedley P, Tfelt-Hansen J, Kanters JK, Haunsøe S, McKenna WJ, Christiansen M. High-efficiency multiplex capillary electrophoresis single strand conformation polymorphism (multi-CE-SSCP) mutation screening of SCN5A: a rapid genetic approach to cardiac arrhythmia. Clin Genet. 2006;69(6):504–11.

Holweg CT, Baan CC, Niesters HG, Vantrimpont PJ, Mulder PG, Maat AP, Weimar W, Balk AH. TGF-beta1 gene polymorphisms in patients with end-stage heart failure. J Heart Lung Transplant. 2001;20(9):979–84.

Holzer R, Ladusans E, Malaiya N. Aortopulmonary collateral arteries in a child with trisomy 21. Cardiol Young. 2002;12(1):75–7. PubMed PMID:11922447

Hsu YH, Yogasundaram H, Parajuli N, Valtuille L, Sergi C, Oudit GY. MELAS syndrome and cardiomyopathy: linking mitochondrial function to heart failure pathogenesis. Heart Fail Rev. 2016;21(1):103–16. https://doi.org/10.1007/s10741-015-9524-5. Review. PubMed PMID:26712328. http://www.isachd.org/proglossary/index.php?&ob=a#atrio-ventricular%20valve

Hulot JS, Jouven X, Empana JP, Frank R, Fontaine G. Natural history and risk stratification of arrhythmogenic right ventricular dysplasia/cardiomyopathy. Circulation. 2004;110(14):1879–84. Review. PubMed PMID:15451782

Hwa J, Ward C, Nunn G, Cooper S, Lau KC, Sholler G. Primary intraventricular cardiac tumors in children: contemporary diagnostic and management options. Pediatr Cardiol. 1994;15(5):233–7. PubMed PMID:7997428

Ichida F, Tsubata S, Bowles KR, Haneda N, Uese K, Miyawaki T, Dreyer WJ, Messina J, Li H, Bowles NE, Towbin JA. Novel gene mutations in patients with left ventricular noncompaction or Barth syndrome. Circulation. 2001;103(9):1256–63.

Ichida F. Left ventricular noncompaction. Circ J. 2009;73(1):19–26. Epub 2008 Dec 4. Review. PubMed PMID:19057090

Ito M, Takahashi H, Fuse K, Hirono S, Washizuka T, Kato K, Yamazaki F, Inano K, Furukawa T, Komada M, Aizawa Y. Polymorphisms of tumor necrosis factor-alpha and interleukin-10 genes in Japanese patients with idiopathic dilated cardiomyopathy. Jpn Heart J. 2000;41(2):183–91.

Iwai C, Akita H, Shiga N, Takai E, Miyamoto Y, Shimizu M, Kawai H, Takarada A, Kajiya T, Yokoyama M. Suppressive effect of the Gly389 allele of the beta1-adrenergic receptor gene on the occurrence of ventricular tachycardia in dilated cardiomyopathy. Circ J. 2002;66(8):723–8.

Jacobs JP, Anderson RH, Weinberg PM, Walters HL 3rd, Tchervenkov CI, Del Duca D, Franklin RC, Aiello VD, Béland MJ, Colan SD, Gaynor JW, Krogmann ON, Kurosawa H, Maruszewski B, Stellin G, Elliott MJ. The nomenclature, definition and classification of cardiac structures in the setting of heterotaxy. Cardiol Young. 2007;17(Suppl 2):1–28. https://doi.org/10.1017/S1047951107001138. Review

Jain D, Maleszewski JJ, Halushka MK. Benign cardiac tumors and tumorlike conditions. Ann Diagn Pathol. 2010;14(3):215–30. Review. PubMed PMID:20471569

Jana S, Zhang H, Lopaschuk GD, Freed DH, Sergi C, Kantor PF, et al. Disparate Remodeling of the Extracellular Matrix and Proteoglycans in Failing Pediatric Versus Adult Hearts. J Am Heart Assoc. 2018;7(19):e010427. https://doi.org/10.1161/JAHA.118.010427. PubMed PMID: 30371322; PubMed Central PMCID:PMC6404896

Joffe AR, Lequier L, Robertson CM. Pediatric outcomes after extracorporeal membrane oxygenation for cardiac disease and for cardiac arrest: a review. ASAIO J. 2012;58(4):297–310. PubMed PMID:22643323

Jurkovicova D, Goncalvesova E, Sedlakova B, Hudecova S, Fabian J, Krizanova O. Is the ApoE polymorphism associated with dilated cardiomyopathy? Gen Physiol Biophys. 2006;25(1):3–10.

Kain ZN, Kain TS, Scarpelli EM. Cocaine exposure in utero: perinatal development and neonatal manifestations – review. J Toxicol Clin Toxicol. 1992;30(4):607–36. Review. PubMed PMID:1433431

Kajander OA, Kupari M, Perola M, Pajarinen J, Savolainen V, Penttilä A, Karhunen PJ. Testing genetic susceptibility loci for alcoholic heart muscle disease. Alcohol Clin Exp Res. 2001;25(10):1409–13.

Kamel H, Roman MJ, Pitcher A, Devereux RB. Pregnancy and the Risk of Aortic Dissection or Rupture: A Cohort-Crossover Analysis. Circulation. 2016a;134(7):527–33. https://doi.org/10.1161/CIRCULATIONAHA.116.021594. Epub 2016 Aug 4. PubMed PMID: 27492904; PubMed Central PMCID:PMC4987245

Kamel H, Roman MJ, Pitcher A, Devereux RB. Pregnancy and the risk of aortic dissection or rupture: a cohort-crossover analysis. Circulation. 2016b;134(7):527–33. https://doi.org/10.1161/CIRCULATIONAHA.116.021594. PubMed PMID:27492904; PubMed Central PMCID: PMC4987245

Karch SB. Cocaine cardiovascular toxicity. South Med J. 2005;98(8):794–9. Review. PubMed PMID:16144174

Kärkkäinen S, Miettinen R, Tuomainen P, Kärkkäinen P, Heliö T, Reissell E, Kaartinen M, Toivonen L, Nieminen MS, Kuusisto J, Laakso M, Peuhkurinen K. A novel mutation, Arg71Thr, in the delta-sarcoglycan gene is associated with dilated cardiomyopathy. J Mol Med. 2003;81(12):795–800.

Kärkkäinen S, Peuhkurinen K, Jääskeläinen P, Miettinen R, Kärkkäinen P, Kuusisto J, Laakso M. No variants in the cardiac actin gene in Finnish patients with dilated or hypertrophic cardiomyopathy. Am Heart J. 2002;143(6):E6.

Kasai T, Suzuki T, Kodama M, Okuno Y. Isthmus aortae--its definition and significance. Okajimas Folia Anat Jpn. 1987;64(2-3):141–6. PubMed PMID:3431791

Kaskinen AK, Happonen JM, Mattila IP, Pitkänen OM. Long-term outcome after treatment of pulmonary atresia with ventricular septal defect: nationwide study of 109 patients born in 1970-2007. Eur J Cardiothorac Surg. 2016a;49(5):1411–8. https://doi.org/10.1093/ejcts/ezv404. Epub 2015 Nov 29. PubMed PMID:26620210

Kaskinen AK, Happonen JM, Mattila IP, Pitkänen OM. Long-term outcome after treatment of pulmonary atresia with ventricular septal defect: nationwide study of 109 patients born in 1970-2007. Eur J Cardiothorac Surg. 2016b;49(5):1411–8. https://

doi.org/10.1093/ejcts/ezv404. Epub 2015 Nov 29. PubMed PMID:26620210

Kauferstein S, Kiehne N, Jenewein T, Biel S, Kopp M, König R, Erkapic D, Rothschild M, Neumann T. Genetic analysis of sudden unexplained death: a multidisciplinary approach. Forensic Sci Int. 2013a;229(1–3):122–7. https://doi.org/10.1016/j.forsciint.2013.03.050. PubMed PMID:23683917

Kauferstein S, Kiehne N, Neumann T, Pitschner HF, Bratzke H. Cardiac gene defects can cause sudden cardiac death in young people. Dtsch Arztebl Int. 2009;106(4):41–7. https://doi.org/10.3238/arztebl.2009.0041. Review. PubMed PMID:19564966; PubMed Central PMCID: PMC2695303

Kauferstein S, Kiehne N, Peigneur S, Tytgat J, Bratzke H. Cardiac channelopathy causing sudden death as revealed by molecular autopsy. Int J Legal Med. 2013b;127(1):145–51. https://doi.org/10.1007/s00414-012-0679-5. PubMed PMID:22370996

Khanna AD, Burkhart HM, Manduch M, Feldman AL, Inwards DJ, Connolly HM. Composite hodgkin and non-hodgkin lymphoma of the mitral and aortic valves. J Am Soc Echocardiogr. 2010;23(10) 1113. e5–7. PubMed PMID:20451348

Kim SJ. Heterotaxy syndrome. Korean Circ J. 2011;41(5):227–32. https://doi.org/10.4070/kcj.2011.41.5.227.

Kirby C, Johnson J, Zinsser H. Successful Closure of a Left Ventricular-Right Atrial Shunt. Annals of Surgery. 1957;145(3):392–4. [PMC free article] [PubMed] [Google Scholar]

Komajda M, Charron P, Tesson F. Genetic aspects of heart failure. Eur J Heart Fail. 1999;1(2):121–6. Review

Kong T, Feulefack J, Ruether K, Shen F, Zheng W, Chen XZ, et al. Ethnic Differences in Genetic Ion Channelopathies Associated with Sudden Cardiac Death: A Systematic Review and Meta-Analysis. Ann Clin Lab Sci. 2017;47(4):481–90. Review. PubMed PMID:28801377

Konno T, Shimizu M, Ino H, Matsuyama T, Yamaguchi M, Terai H, Hayashi K, Mabuchi T, Kiyama M, Sakata K, Hayashi T, Inoue M, Kaneda T, Mabuchi H. A novel missense mutation in the myosin binding protein-c gene is responsible for hypertrophic cardiomyopathy with left ventricular dysfunction and dilation in elderly patients. J Am Coll Cardiol. 2003;41(5):781–6.

Konstantinov IE, Alexi-Meskishvili VV, Williams WG, Freedom RM, Van Praagh R. Atrial switch operation: past, present, and future. Ann Thorac Surg. 2004;77(6):2250–8. PubMed PMID:15172322

Kosemehmetoglu K, Vrana JA, Folpe AL. TLE1 expression is not specific for synovial sarcoma: a whole section study of 163 soft tissue and bone neoplasms. Mod Pathol. 2009;22(7):872–8. https://doi.org/10.1038/modpathol.2009.47. Epub 2009 Apr 10. PubMed PMID:19363472

Kreutzer C, De Vive J, Oppido G, Kreutzer J, Gauvreau K, Freed M, Mayer JE Jr, Jonas R, del Nido PJ. Twenty-five-year experience with rastelli repair for transposition of the great arteries. J Thorac Cardiovasc Surg. 2000;120(2):211–23. PubMed PMID:10917934

Küçükarabaci B, Birdane A, Güneş HV, Ata N, Değirmenci I, Başaran A, Timuralp B. Association between angiotensin converting enzyme (ACE) gene I/D polymorphism frequency and plasma ACE concentration in patients with idiopathic dilated cardiomyopathy. Anadolu Kardiyol Derg. 2008;8(1):65–6.

Kuehl KS, Loffredo C. Risk factors for heart disease associated with abnormal sidedness. Teratology. 2002;66(5):242–8. PubMed PMID:12397632

Kyriacou K, Michaelides Y, Senkus R, Simamonian K, Pavlides N, Antoniades L, Zambartas C. Ultrastructural pathology of the heart in patients with beta-thalassaemia major. Ultrastruct Pathol. 2000;24(2):75–81. PubMed PMID:10808552

Laitinen TT, Nuotio J, Rovio SP, Niinikoski H, Juonala M, Magnussen CG, et al. Dietary Fats and Atherosclerosis From Childhood to Adulthood. Pediatrics. 2020;145(4) pii: e20192786 https://doi.org/10.1542/peds.2019-2786. Epub 2020 Mar 24. PubMed PMID:32209700

Lee SE, Kim HY, Jung SE, Lee SC, Park KW, Kim WK. Situs anomalies and gastrointestinal abnormalities. J Pediatr Surg. 2006;41(7):1237–42. PubMed PMID:16818055

Leineweber K, Tenderich G, Wolf C, Wagner S, Zittermann A, Elter-Schulz M, Moog R, Müller N, Jakob HG, Körfer R, Philipp T, Heusch G, Brodde OE. Is there a role of the Thr164Ile-beta(2)-adrenoceptor polymorphism for the outcome of chronic heart failure? Basic Res Cardiol. 2006;101(6):479–84.

Li Y, Ji C, Zhang J, Han Y. The effect of ambient temperature on the onset of acute Stanford type B aortic dissection. Vasa. 2016;45(5):395–401. https://doi.org/10.1024/0301-1526/a000555. PubMed PMID:27351414

Lin A, Yan WH, Xu HH, Tang LJ, Chen XF, Zhu M, Zhou MY. 14 bp deletion polymorphism in the HLA-G gene is a risk factor for idiopathic dilated cardiomyopathy in a Chinese Han population. Tissue Antigens. 2007;70(5):427–31.

Lin AE, Krikov S, Riehle-Colarusso T, Frías JL, Belmont J, Anderka M, Geva T, Getz KD, Botto LD. National Birth Defects Prevention Study. Laterality defects in the national birth defects prevention study (1998–2007): birth prevalence and descriptive epidemiology. Am J Med Genet A. 2014;164A(10):2581–91. https://doi.org/10.1002/ajmg.a.36695.

Little BB, Snell LM, Klein VR, Gilstrap LC 3rd. Cocaine abuse during pregnancy: maternal and fetal implications. Obstet Gynecol. 1989;73(2):157–60. PubMed PMID:2911419

Liu M, Lu L, Sun R, Zheng Y, Zhang P. Rheumatic heart disease: causes, symptoms, and treatments. Cell Biochem Biophys. 2015;72(3):861–3. https://doi.org/10.1007/s12013-015-0552-5. Review. PubMed PMID:25638346

Liu W, Li W, Sun N. Association of HLA-DQ with idiopathic dilated cardiomyopathy in a northern Chinese Han population. Cell Mol Immunol. 2004a;1(4):311–4.

Liu W, Li WM, Sun NL. Polymorphism of the second exon of human leukocyte antigen-DQA1, -DQB1 gene and genetic susceptibility to idiopathic dilated cardiomyopathy in people of the Han nationality in northern China. Chin Med J. 2005;118(3):238–41.

Liu W, Li WM, Sun NL. Relationship between HLA-DQA1 polymorphism and genetic susceptibility to idiopathic dilated cardiomyopathy. Chin Med J. 2004b;117(10):1449–52.

Lopez KN, Marengo LK, Canfield MA, Belmont JW, Dickerson HA. Racial disparities in heterotaxy syndrome. Birth Defects Res A Clin Mol Teratol. 2015;103(11):941–50. https://doi.org/10.1002/bdra.23416. Epub 2015 Sep 2

Macarie C, Stoian I, Dermengiu D, Barbarii L, Piser IT, Chioncel O, Carp A, Stoian I. The electrocardiographic abnormalities in highly trained athletes compared to the genetic study related to causes of unexpected sudden cardiac death. J Med Life. 2009;2(4):361–72. PubMed PMID:20108749; PubMed Central PMCID: PMC3019018

Mak TW, Hauck L, Grothe D, Billia F. p53 regulates the cardiac transcriptome. Proc Natl Acad Sci U S A. 2017;114(9):2331–6. https://doi.org/10.1073/pnas.1621436114. PubMed PMID:28193895

Malhotra V, Ferrans VJ, Virmani R. Infantile histiocytoid cardiomyopathy: three cases and literature review. Am Heart J. 1994;128(5):1009–21. Review. PubMed PMID:7942464

Malik FS, Lavie CJ, Mehra MR, Milani RV, Re RN. Renin-angiotensin system: genes to bedside. Am Heart J. 1997;134(3):514–26.

Mangin L, Charron P, Tesson F, Mallet A, Dubourg O, Desnos M, Benaïsche A, Gayet C, Gibelin P, Davy JM, Bonnet J, Sidi D, Schwartz K, Komajda M. Familial dilated cardiomyopathy: clinical features in French families. Eur J Heart Fail. 1999;1(4):353–61.

Maron BJ, Towbin JA, Thiene G, Antzelevitch C, Corrado D, Arnett D, Moss AJ, Seidman CE, Young JB, American Heart Association; Council on Clinical Cardiology, Heart Failure and Transplantation Committee; Quality of Care and Outcomes Research and Functional Genomics and Translational Biology Interdisciplinary Working Groups; Council on Epidemiology and Prevention. Contemporary definitions and classification of the cardiomyopathies: an American Heart Association. Scientific Statement from the Council on Clinical Cardiology, Heart Failure and Transplantation Committee; Quality of Care and Outcomes Research and Functional Genomics and Translational Biology Interdisciplinary Working Groups; and Council on Epidemiology and Prevention. Circulation. 2006;113(14):1807–16. PubMed PMID:16567565

Marshall JD, Hinman EG, Collin GB, Beck S, Cerqueira R, Maffei P, Milan G, Zhang W, Wilson DI, Hearn T, Tavares P, Vettor R, Veronese C, Martin M, So WV, Nishina PM, Naggert JK. Spectrum of ALMS1 variants and evaluation of genotype-phenotype correlations in Alström syndrome. Hum Mutat. 2007;28(11):1114–23.

Mata M, Wharton M, Geisinger K, Pugh JE. Myocardial rhabdomyosarcoma in multiple neurofibromatosis. Neurology. 1981;31(12):1549–51. PubMed PMID:6796904

Matsumoto Y, Hayashi T, Inagaki N, Takahashi M, Hiroi S, Nakamura T, Arimura T, Nakamura K, Ashizawa N, Yasunami M, Ohe T, Yano K, Kimura A. Functional analysis of titin/connectin N2-B mutations found in cardiomyopathy. J Muscle Res Cell Motil. 2005;26(6–8):367–74.

McKenna WJ, Thiene G, Nava A, Fontaliran F, Blomstrom-Lundqvist C, Fontaine G, Camerini F. Diagnosis of arrhythmogenic right ventricular dysplasia/cardiomyopathy. Task Force of the Working Group Myocardial and Pericardial Disease of the European Society of Cardiology and of the Scientific Council on Cardiomyopathies of the International Society and Federation of Cardiology. Br Heart J. 1994;71(3):215–8. PubMed PMID:8142187; PubMed Central PMCID: PMC483655

Mestroni L, Rocco C, Gregori D, Sinagra G, Di Lenarda A, Miocic S, Vatta M, Pinamonti B, Muntoni F, Caforio AL, McKenna WJ, Falaschi A, Giacca M. Camerini. Familial dilated cardiomyopathy: evidence for genetic and phenotypic heterogeneity. Heart Muscle Disease Study Group. J Am Coll Cardiol. 1999;34(1):181–90.

Michaud K, Grabherr S, Jackowski C, Bollmann MD, Doenz F, Mangin P. Postmortem imaging of sudden cardiac death. Int J Legal Med. 2014;128(1):127–37. https://doi.org/10.1007/s00414-013-0819-6. PubMed PMID:23322013

Michaud K, Lesta Mdel M, Fellmann F, Mangin P. Molecular autopsy of sudden cardiac death: from postmortem to clinical approach. Rev Med Suisse. 2008;4(164):1590–3. French. PubMed PMID:18711970

Milo S, Ho SY, Macartney FJ, Wilkinson JL, Becker AE, Wenink AC. Gittenberger de Groot AC, Anderson RH. Straddling and overriding atrioventricular valves: morphology and classification. Am J Cardiol. 1979;44(6):1122–34. PubMed PMID:495507

Milting H, Lukas N, Klauke B, Körfer R, Perrot A, Osterziel KJ, Vogt J, Peters S, Thieleczek R, Varsányi M. Composite polymorphisms in the ryanodine receptor 2 gene associated with arrhythmogenic right ventricular cardiomyopathy. Cardiovasc Res. 2006;71(3):496–505.

Miyamoto Y, Akita H, Shiga N, Takai E, Iwai C, Mizutani K, Kawai H, Takarada A, Yokoyama M. Frequency and clinical characteristics of dilated cardiomyopathy caused by desmin gene mutation in a Japanese population. Eur Heart J. 2001;22(24):2284–9.

Mochizuki Y, Zhang M, Golestaneh L, Thananart S, Coco M. Acute aortic thrombosis and renal infarction in acute cocaine intoxication: a case report and review of literature. Clin Nephrol. 2003;60(2):130–3. Review. PubMed PMID:12940616

Mohammed S, Bahitham W, Chan A, Chiu B, Bamforth F, Sergi C. Mitochondrial DNA related cardiomyopathies. Front Biosci (Elite Ed). 2012;4:1706–16. Review. PubMed PMID:22201986

Monserrat L, Hermida-Prieto M, Fernandez X, Rodríguez I, Dumont C, Cazón L, Cuesta MG, Gonzalez-Juanatey

C, Peteiro J, Alvarez N, Penas-Lado M, Castro-Beiras A. Mutation in the alpha-cardiac actin gene associated with apical hypertrophic cardiomyopathy, left ventricular non-compaction, and septal defects. Eur Heart J. 2007;28(16):1953–61.

Montgomery HE, Keeling PJ, Goldman JH, Humphries SE, Talmud PJ, McKenna WJ. Lack of association between the insertion/deletion polymorphism of the angiotensin-converting enzyme gene and idiopathic dilated cardiomyopathy. J Am Coll Cardiol. 1995;25(7):1627–31.

Moric-Janiszewska E, Markiewicz-Łoskot G, Loskot M, Weglarz L, Hollek A, Szydlowski L. Challenges of diagnosis of long-QT syndrome in children. Pacing Clin Electrophysiol. 2007;30(9):1168–70. https://doi.org/10.1111/j.1540-8159.2007.00832.x. PMID 17725765

Mudhar HS, Wagner BE, Suvarna SK. Electron microscopy of myocardial tissue. A nine year review. J Clin Pathol. 2001;54(4):321–5. PubMed PMID:11304852; PubMed Central PMCID: PMC1731405

Muntoni F, Cau M, Ganau A, Congiu R, Arvedi G, Mateddu A, Marrosu MG, Cianchetti C, Realdi G, Cao A, Melis MA. Brief report: deletion of the dystrophin muscle-promoter region associated with X-linked dilated cardiomyopathy. N Engl J Med. 1993;329(13):921–5.

Murphy RT, Mogensen J, Shaw A, Kubo T, Hughes S, McKenna WJ. Novel mutation in cardiac troponin I in recessive idiopathic dilated cardiomyopathy. Lancet. 2004;363(9406):371–2.

Nakano N, Hori H, Abe M, Shibata H, Arimura T, Sasaoka T, Sawabe M, Chida K, Arai T, Nakahara K, Kubo T, Sugimoto K, Katsuya T, Ogihara T, Doi Y, Izumi T, Kimura A. Interaction of BMP10 with Tcap may modulate the course of hypertensive cardiac hypertrophy. Am J Physiol Heart Circ Physiol. 2007;293(6):H3396–403.

Nathan BM, Sockalosky J, Nelson L, Lai S, Sergi C, Petryk A. The use of hormonal therapy in pediatric heart disease. Front Biosci (Schol Ed). 2009;1:358–75. Review. PubMed PMID:19482707

Nemer M. Genetic insights into normal and abnormal heart development. Cardiovasc Pathol. 2008;17(1):48–54. Review. PubMed PMID:18160060

Nienaber CA. The role of imaging in acute aortic syndromes. Eur Heart J Cardiovasc Imaging. 2013;14(1):15–23. https://doi.org/10.1093/ehjci/jes215. Review. PubMed PMID:23109648

Nonen S, Okamoto H, Akino M, Matsui Y, Fujio Y, Yoshiyama M, Takemoto Y, Yoshikawa J, Azuma J, Kitabatake A. No positive association between adrenergic receptor variants of alpha2cDel322-325, beta1Ser49, beta1Arg389 and the risk for heart failure in the Japanese population. Br J Clin Pharmacol. 2005;60(4):414–7.

Norris MK, Hill CS. Assessing congenital heart defects in the cocaine-exposed neonate. Dimens Crit Care Nurs. 1992;11(1):6–12. PubMed PMID:1740090

Nowak KJ, Wattanasirichaigoon D, Goebel HH, Wilce M, Pelin K, Donner K, Jacob RL, Hübner C, Oexle K, Anderson JR, Verity CM, North KN, Iannaccone ST, Müller CR, Nürnberg P, Muntoni F, Sewry C, Hughes I, Sutphen R, Lacson AG, Swoboda KJ, Vigneron J, Wallgren-Pettersson C, Beggs AH, Laing NG. Mutations in the skeletal muscle alpha-actin gene in patients with actin myopathy and nemaline myopathy. Nat Genet. 1999;23(2):208–12.

Olson TM, Doan TP, Kishimoto NY, Whitby FG, Ackerman MJ, Fananapazir L. Inherited and de novo mutations in the cardiac actin gene cause hypertrophic cardiomyopathy. J Mol Cell Cardiol. 2000;32(9):1687–94.

Olson TM, Illenberger S, Kishimoto NY, Huttelmaier S, Keating MT, Jockusch BM. Metavinculin mutations alter actin interaction in dilated cardiomyopathy. Circulation. 2002;105(4):431–7.

Olson TM, Michels VV, Thibodeau SN, Tai YS, Keating MT. Actin mutations in dilated cardiomyopathy, a heritable form of heart failure. Science. 1998;280(5364):750–2.

Pachler C, Knez I, Petru E, Dörfler O, Vicenzi M, Toller W. Peripartum aortic dissection. Anaesthesist. 2013;62(4):293–5. https://doi.org/10.1007/s00101-013-2153-2. German. PubMed PMID:23494023

Parajuli N, Valtuille L, Basu R, Famulski KS, Halloran PF, Sergi C, Oudit GY. Determinants of ventricular arrhythmias in human explanted hearts with dilated cardiomyopathy. Eur J Clin Investig. 2015;45(12):1286–96. https://doi.org/10.1111/eci.12549. PubMed PMID:26444674

Parlakgumus HA, Haydardedeoglu B, Alkan O. Aortic dissection accompanied by preeclampsia and preterm labor. J Obstet Gynaecol Res. 2010;36(5):1121–4. https://doi.org/10.1111/j.1447-0756.2010.01263.x. PubMed PMID:20846256

Pasquini L, Parness IA, Colan SD, Wernovsky G, Mayer JE, Sanders SP. Diagnosis of intramural coronary artery in transposition of the great arteries using two-dimensional echocardiography. Circulation. 1993a;88(3):1136–41. PubMed PMID:8353875

Pasquini L, Sanders SP, Parness IA, Colan SD, Van Praagh S, Mayer JE Jr, Van Praagh R. Conal anatomy in 119 patients with d-loop transposition of the great arteries and ventricular septal defect: an echocardiographic and pathologic study. J Am Coll Cardiol. 1993b;21(7):1712–21. PubMed PMID:8496542

Patel DD, Kapoor A, Ayyagari A, Dhole TN. Development of a simple restriction fragment length polymorphism assay for subtyping of coxsackie B viruses. J Virol Methods. 2004;120(2):167–72.

Pelletier JS, LaBossiere J, Dicken B, Gill RS, Sergi C, Tahbaz N, Bigam D, Cheung PY. Low-dose vasopressin improves cardiac function in newborn piglets with acute hypoxia-reoxygenation. Shock. 2013;40(4):320–6. https://doi.org/10.1097/SHK.0b013e3182a4284e. PubMed PMID:23856923

Planas S, Ferreres JC, Balcells J, Garrido M, Ramón Y, Cajal S, Torán N. Association of ventricular non-compaction and histiocytoid cardiomyopathy. Case report and review of the literature.

Pediatr Dev Pathol. 2012;15(5):397–402. PubMed PMID:22758650

Poirier O, Nicaud V, McDonagh T, Dargie HJ, Desnos M, Dorent R, Roizès G, Schwartz K, Tiret L, Komajda M, Cambien F. Polymorphisms of genes of the cardiac calcineurin pathway and cardiac hypertrophy. Eur J Hum Genet. 2003;11(9):659–64.

Putko BN, Wen K, Thompson RB, Mullen J, Shanks M, Yogasundaram H, Sergi C, Oudit GY. Anderson-Fabry cardiomyopathy: prevalence, pathophysiology, diagnosis and treatment. Heart Fail Rev. 2015;20(2):179–91. https://doi.org/10.1007/s10741-014-9452-9. Review. PubMed PMID:25030479

Raaf HN, Raaf JH. Sarcomas related to the heart and vasculature. Semin Surg Oncol. 1994;10(5):374–82. Review. PubMed PMID:7997732

Raffray L, Guillevin L. Treatment of eosinophilic granulomatosis with polyangiitis: a review. Drugs. 2018;78(8):809–21. https://doi.org/10.1007/s40265-018-0920-8. PubMed PMID:29766394

Raisky O, Vouhé PR. Pitfalls in repair of conotruncal anomalies. Semin Thorac Cardiovasc Surg Pediatr Card Surg Annu. 2013;16(1):7–12. https://doi.org/10.1053/j.pcsu.2013.02.001. Review. PubMed PMID:23561812

Regitz-Zagrosek V, Hocher B, Bettmann M, Brede M, Hadamek K, Gerstner C, Lehmkuhl HB, Hetzer R, Hein L. Alpha2C-adrenoceptor polymorphism is associated with improved event-free survival in patients with dilated cardiomyopathy. Eur Heart J. 2006;27(4):454–9.

Rheeder P, Simson IW, Mentis H, Karussett VO. Cardiac rhabdomyosarcoma in a renal transplant patient. Transplantation. 1995;60(2):204–5. PubMed PMID:7624965

Risch N, Merikangas K. The future of genetic studies of complex human diseases. Science. 1996;273(5281):1516–7.

Rodríguez-Calvo MS, Brion M, Allegue C, Concheiro L, Carracedo A. Molecular genetics of sudden cardiac death. Forensic Sci Int. 2008;182(1–3):1–12. https://doi.org/10.1016/j.forsciint.2008.09.013. Review. PubMed PMID:18992999

Ruíz-Sánchez R, León MP, Matta V, Reyes PA, López R, Jay D, Monteón VM. Trypanosoma cruzi isolates from Mexican and Guatemalan acute and chronic chagasic cardiopathy patients belong to Trypanosoma cruzi I. Mem Inst Oswaldo Cruz. 2005;100(3):281–3.

Ruppert V, Nolte D, Aschenbrenner T, Pankuweit S, Funck R, Maisch B. Novel point mutations in the mitochondrial DNA detected in patients with dilated cardiomyopathy by screening the whole mitochondrial genome. Biochem Biophys Res Commun. 2004;318(2):535–43.

Saguil A, Fargo M, Grogan S. Diagnosis and management of Kawasaki disease. Am Fam Physician. 2015;91(6):365–71. Review. PubMed PMID:25822554

Sakamuri SS, Takawale A, Basu R, Fedak PW, Freed D, Sergi C, Oudit GY, Kassiri Z. Differential impact of mechanical unloading on structural and non-structural components of the extracellular matrix in advanced human heart failure. Transl Res. 2016a;172:30–44. https://doi.org/10.1016/j.trsl.2016.02.006. PubMed PMID:26963743

Sakamuri SS, Takawale A, Basu R, Fedak PW, Freed D, Sergi C, Oudit GY, Kassiri Z. Differential impact of mechanical unloading on structural and nonstructural components of the extracellular matrix in advanced human heart failure. Transl Res. 2016b;172:30–44. https://doi.org/10.1016/j.trsl.2016.02.006. PubMed PMID:26963743

Sanderson JE, Young RP, Yu CM, Chan S, Critchley JA, Woo KS. Lack of association between insertion/deletion polymorphism of the angiotensin-converting enzyme gene and end-stage heart failure due to ischemic or idiopathic dilate cardiomyopathy in the Chinese. Am J Cardiol. 1996;77(11):1008–10.

Sano S, Ishino K, Kawada M, Arai S, Kasahara S, Asai T, Masuda Z, Takeuchi M, Ohtsuki S. Right ventricle-pulmonary artery shunt in first-stage palliation of hypoplastic left heart syndrome. J Thorac Cardiovasc Surg. 126(2):504–9; discussion 509–10. https://doi.org/10.1016/s0022-5223(02)73575-7. PMID 12928651

Sawada A, Inoue M, Kawa K. How we treat chronic active Epstein-Barr virus infection. Int J Hematol. 2017;105(4):406–18. https://doi.org/10.1007/s12185-017-2192-6. Review. PubMed PMID:28210942

Schmidt A, Kiener HP, Barnas U, Arias I, Illievich A, Auinger M, Graninger W, Kaider A, Mayer G. Angiotensin-converting enzyme polymorphism in patients with terminal renal failure. J Am Soc Nephrol. 1996;7(2):314–7.

Schwartz PJ, Moss AJ, Vincent GM, Crampton RS. Diagnostic criteria for the long QT syndrome. An update. Circulation. 1993;88(2):782–4. https://doi.org/10.1161/01.CIR.88.2.782. PMID 8339437

Sclair M, Nassar H, Bar-Ziv Y, Putterman C. Dissecting aortic aneurysm in systemic lupus erythematosus. Lupus. 1995;4(1):71–4. PubMed PMID:7767343

Sergi C, Böhler T, Schönrich G, Sieverts H, Roth SU, Debatin KM, Otto HF. Occult thyroid pathology in a child with acquired immunodeficiency syndrome. Case report and review of the drug-related pathology in pediatric acquired immunodeficiency syndrome. Pathol Oncol Res. 2000a;6(3):227–232. PubMed PMID:11033465.

Sergi C, Daum E, Pedal I, Hauröder B, Schnitzler P. Fatal circumstances of human herpesvirus 6 infection: transcriptosome data analysis suggests caution in implicating HHV-6 in the cause of death. J Clin Pathol. 2007;60(10):1173–7. PubMed PMID:17545558; PubMed Central PMCID: PMC2014822

Sergi C, Himbert U, Weinhardt F, Heilmann W, Meyer P, Beedgen B, Zilow E, Hofmann WJ, Linderkamp O, Otto HF. Hepatic failure with neonatal tissue siderosis of hemochromatotic type in an infant presenting with meconium ileus. Case report and differential diagnosis of the perinatal iron storage disorders. Pathol Res Pract. 2001;197(10):699–709. discussion 711–3. PubMed PMID:11700892

Sergi C, Magener A, Ehemann V, De Villiers EM, Sinn HP. Stage IIa cervix carcinoma with metastasis to the heart: report of a case with immunohistochemistry, flow cytometry, and virology findings. Gynecol Oncol. 2000b;76(1):133–8. PubMed PMID:10620458

Sergi C, Schiesser M, Adam S, Otto HF. Analysis of the spectrum of malformations in human fetuses of the second and third trimester of pregnancy with human triploidy. Pathologica. 2000c;92(4):257–63. PubMed PMID:11029886

Sergi C, Schmitt HP. Central nervous system in twin reversed arterial perfusion sequence with special reference to examination of the brain in acardiusanceps. Teratology. 2000;61(4):284–90. PubMed PMID:10716747

Sergi C, Serpi M, Müller-Navia J, Schnabel PA, Hagl S, Otto HF, Ulmer HE. CATCH 22 syndrome: report of 7 infants with follow-up data and review of the recent advancements in the genetic knowledge of the locus 22q11. Pathologica. 1999;91(3):166–72. Review. PubMed PMID:10536461

Sergi C, Shen F, Lim DW, Liu W, Zhang M, Chiu B, Anand V, Sun Z. Cardiovascular dysfunction in sepsis at the dawn of emerging mediators. Biomed Pharmacother. 2017a;95:153–60. https://doi.org/10.1016/j.biopha.2017.08.066. Review. PubMed PMID:28841455

Sergi C, Shen F, Lim DW, Liu W, Zhang M, Chiu B, et al. Cardiovascular dysfunction in sepsis at the dawn of emerging mediators. Biomed Pharmacother. 2017b;95:153–60. https://doi.org/10.1016/j.biopha.2017.08.066. Epub 2017 Sep 12. Review. PubMed PMID:28841455

Sergi C, Weitz J, Hofmann WJ, Sinn P, Eckart A, Otto G, Schnabel PA, Otto HF. Aspergillus endocarditis, myocarditis and pericarditis complicating necrotizing fasciitis. Case report and subject review. Virchows Arch. 1996;429(2–3):177–80. Review. PubMed PMID:8917720

Sergi CM. Sudden cardiac death and ethnicity. CMAJ. 2019;191(45):E1254. https://doi.org/10.1503/cmaj.73297. PubMed PMID: 31712362; PubMed Central PMCID: PMC6861156

Shapiro AJ, Davis SD, Ferkol T, Dell SD, Rosenfeld M, Olivier KN, Sagel SD, Milla C, Zariwala MA, Wolf W, Carson JL, Hazucha MJ, Burns K, Robinson B, Knowles MR, Leigh MW. Laterality defects other than situs inversus totalis in primary ciliary dyskinesia: insights into situs ambiguus and heterotaxy. Chest. 2014;146(5):1176–86. https://doi.org/10.1378/chest.13-1704

Shaw GM, Malcoe LH, Lammer EJ, Swan SH. Maternal use of cocaine during pregnancy and congenital cardiac anomalies. J Pediatr. 1991;118(1):167–8. PubMed PMID:1986094

Shichi D, Kikkawa EF, Ota M, Katsuyama Y, Kimura A, Matsumori A, Kulski JK, Naruse TK, Inoko H. The haplotype block, NFKBIL1-ATP6V1G2-BAT1-MICB-MICA, within the class III – class I boundary region of the human major histocompatibility complex may control susceptibility to hepatitis C virus-associated dilated cardiomyopathy. Tissue Antigens. 2005;66(3):200–8.

Shim YH, Kim HS, Sohn S, Hong YM. Insertion/deletion polymorphism of angiotensin converting enzyme gene in Kawasaki disease. J Korean Med Sci. 2006;21(2):208–11.

Shinebourne EA, Macartney FJ, Anderson RH. Sequential chamber localization – logical approach to diagnosis in congenital heart disease. Br Heart J. 1976;38:327–40.

Sobol I, Chen CL, Mahmood SS, Borczuk AC. Histopathologic Characterization of Myocarditis Associated With Immune Checkpoint Inhibitor Therapy. Arch Pathol Lab Med. 2020; https://doi.org/10.5858/arpa.2019-0447-OA. [Epub ahead of print] PubMed PMID:32150459

Speckman RA, Garg A, Du F, Bennett L, Veile R, Arioglu E, Taylor SI, Lovett M, Bowcock AM. Mutational and haplotype analyses of families with familial partial lipodystrophy (Dunnigan variety) reveal recurrent missense mutations in the globular C-terminal domain of lamin A/C. Am J Hum Genet. 2000;66(4):1192–8.

Stambader JD, Dorn L, Mikuz G, Sergi C. Genetic polymorphisms in dilated cardiomyopathy, Front Biosci (Schol Ed). 2010a;2:653–76. Review. PubMed PMID:20036975

Stambader JD, Dorn L, Mikuz G, Sergi C. Genetic polymorphisms in dilated cardiomyopathy. Front Biosci (Schol Ed). 2010b;2:653–76. Review. PubMed PMID:20036975

Steffensen TS, Barness EG. Cardiac conduction disorders in children. Front Biosci (Elite Ed). 2009;1:519–27. Review. PubMed PMID:19482666

Steger CM, Hager T, Antretter H, Hoyer HX, Altenberger J, Pölzl G, Müller L, Höfer D. Cardiac sarcoidosis mimicking arrhythmogenic right ventricular dysplasia. BMJ Case Rep. 2009; https://doi.org/10.1136/bcr.08.2009.2204. Epub 2009 Nov 18.. PubMed PMID:22096465; PubMed Central PMCID: PMC3027594.

Steger CM, Hager T, Ruttmann E. Primary cardiac tumours: a single-center 41-year experience. ISRN Cardiol. 2012;2012:906109. Epub 2012 Jun 27. PubMed PMID:22792486; PubMed Central PMCID: PMC3391967.

Strano-Rossi S, Chiarotti M, Fiori A, Auriti C, Seganti G. Cocaine abuse in pregnancy: its evaluation through hair analysis of pathological new-borns. Life Sci. 1996;59(22):1909–15. PubMed PMID:8950288

Sutherland MJ, Ware SM. Disorders of left-right asymmetry: heterotaxy and situs inversus. Am J Med Genet C Semin Med Genet. 2009;151C(4):307–17. https://doi.org/10.1002/ajmg.c.30228. Review

Sutliff RL, Cai G, Gurdal H, Snyder DL, Roberts J, Johnson MD. Cardiovascular hypertrophy and increased vascular contractile responsiveness following repeated cocaine administration in rabbits. Life Sci. 1996;58(8):675–82. PubMed PMID:8594317

Swalwell CI, Davis GG. Methamphetamine as a risk factor for acute aortic dissection. J Forensic Sci. 1999;44(1):23–6. PubMed PMID:9987866

Sylvius N, Duboscq-Bidot L, Bouchier C, Charron P, Benaiche A, Sébillon P, Komajda M, Villard

E. Mutational analysis of the beta- and delta-sarcoglycan genes in a large number of patients with familial and sporadic dilated cardiomyopathy. Am J Med Genet A. 2003;120A(1):8–12.

Sylvius N, Tesson F, Gayet C, Charron P, Bénaïche A, Peuchmaurd M, Duboscq-Bidot L, Feingold J, Beckmann JS, Bouchier C, Komajda M. A new locus for autosomal dominant dilated cardiomyopathy identified on chromosome 6q12-q16. Am J Hum Genet. 2001;68(1):241–6.

Takahashi H, Kadowaki T, Maruo A, Yutaka O, Oshima Y. Mid-term results of mitral valve repair with autologous pericardium in pediatric patients. J Heart Valve Dis. 2014;23(3):302–9. PubMed PMID:25296453

Takai E, Akita H, Kanazawa K, Shiga N, Terashima M, Matsuda Y, Iwai C, Miyamoto Y, Kawai H, Takarada A, Yokoyama M. Association between aldosterone synthase (CYP11B2) gene polymorphism and left ventricular volume in patients with dilated cardiomyopathy. Heart. 2002;88(6):649–50.

Takai E, Akita H, Shiga N, Kanazawa K, Yamada S, Terashima M, Matsuda Y, Iwai C, Kawai K, Yokota Y, Yokoyama M. Mutational analysis of the cardiac actin gene in familial and sporadic dilated cardiomyopathy. Am J Med Genet. 1999;86(4):325–7.

Tamarappoo BK, John BT, Reinier K, Teodorescu C, Uy-Evanado A, Gunson K, Jui J, Chugh SS. Vulnerable myocardial interstitium in patients with isolated left ventricular hypertrophy and sudden cardiac death: a postmortem histological evaluation. J Am Heart Assoc. 2012;1(3):e001511. https://doi.org/10.1161/JAHA.112.001511. Epub 2012 Jun 22. PubMed PMID:23130141; PubMed Central PMCID: PMC3487319

Tan CD, Baldwin WM 3rd, Rodriguez ER. Update on cardiac transplantation pathology. Arch Pathol Lab Med. 2007;131(8):1169–91. Review. PubMed PMID:17683180

Tavora F, Burke A. Pathology of cardiac tumors and tumour-like conditions. Diagn Histopathol. 2009;16:1–9.

Teele SA, Jacobs JP, Border WL, Chanani NK. Heterotaxy syndrome: proceedings from the 10th international PCICS meeting. World J Pediatr Congenit Heart Surg. 2015;6(4):616–29. https://doi.org/10.1177/2150135115604470. Review

Tesson F, Sylvius N, Pilotto A, Dubosq-Bidot L, Peuchmaurd M, Bouchier C, Benaiche A, Mangin L, Charron P, Gavazzi A, Tavazzi L, Arbustini E, Komajda M. Epidemiology of desmin and cardiac actin gene mutations in a European population of dilated cardiomyopathy. Eur Heart J. 2000;21(22):1872–6.

Tester DJ, Ackerman MJ. The molecular autopsy: should the evaluation continue after the funeral? Pediatr Cardiol. 2012;33(3):461–70. https://doi.org/10.1007/s00246-012-0160-8. Review. PubMed PMID:22307399; PubMedCentral PMCID: PMC3332537

Tester DJ, Ackerman MJ. The role of molecular autopsy in unexplained sudden cardiac death. Curr Opin Cardiol. 2006;21(3):166–72. Review. PubMed PMID:16601452

Thiru Y, Pathan N, Bignall S, Habibi P, Levin M. A myocardial cytotoxic process is involved in the cardiac dysfunction of meningococcal septic shock. Crit Care Med. 2000;28(8):2979–83. PubMed PMID:10966282

Thomas-de-Montpréville V, Nottin R, Dulmet E, Serraf A. Heart tumors in children and adults: clinicopathological study of 59 patients from a surgical center. Cardiovasc Pathol. 2007;16(1):22–8. PubMed PMID:17218211

Thude H, Gerlach K, Richartz B, Krack A, Brenke B, Pethig K, Figulla HR, Barz D. No association between transmembrane protein-tyrosine phosphatase receptor type C (CD45) exon A point mutation (77C>G) and idiopathic dilated cardiomyopathy. Hum Immunol. 2005;66(9):1008–12.

Tiago AD, Badenhorst D, Skudicky D, Woodiwiss AJ, Candy GP, Brooksbank R, Sliwa K, Sareli P, Norton GR. An aldosterone synthase gene variant is associated with improvement in left ventricular ejection fraction in dilated cardiomyopathy. Cardiovasc Res. 2002;54(3):584–9.

Tidake A, Gangurde P, Mahajan A. Gerbode Defect-A Rare Defect of Atrioventricular Septum and Tricuspid Valve. J Clin Diagn Res. 2015;9(9):OD06–8. https://doi.org/10.7860/JCDR/2015/14259.6531. Epub 2015 Sep 1. PMID: 26500939; PMCID:PMC4606268

Tiret L, Mallet C, Poirier O, Nicaud V, Millaire A, Bouhour JB, Roizès G, Desnos M, Dorent R, Schwartz K, Cambien F, Komajda M. Lack of association between polymorphisms of eight candidate genes and idiopathic dilated cardiomyopathy: the CARDIGENE study. J Am Coll Cardiol. 2000;35(1):29–35.

Tsubata S, Bowles KR, Vatta M, Zintz C, Titus J, Muhonen L, Bowles NE, Towbin JA. Mutations in the human delta-sarcoglycan gene in familial and sporadic dilated cardiomyopathy. J Clin Invest. 2000;106(5):655–62. PubMed PMID:10974018; PubMed Central PMCID: PMC381284

Turillazzi E, Fineschi V, Palmiere C, Sergi C. Cardiovascular Involvement in Sepsis. Mediators Inflamm. 2016;2016:8584793. https://doi.org/10.1155/2016/8584793. Epub 2016 May 18. PubMed PMID: 27293322; PubMed Central PMCID:PMC4887644

Turillazzi E, La Rocca G, Anzalone R, Corrao S, Neri M, Pomara C, Riezzo I, Karch SB, Fineschi V. Heterozygous nonsense SCN5A mutation W822X explains a simultaneous sudden infant death syndrome. Virchows Arch. 2008;453(2):209–16. https://doi.org/10.1007/s00428-008-0632-7. PubMed PMID:18551308

Tynan D, Alphonse J, Henry A, Welsh AW. The Aortic Isthmus: A Significant yet Underexplored Watershed of the Fetal Circulation. Fetal Diagn Ther. 2016;40(2):81–93. https://doi.org/10.1159/000446942. Epub 2016 Jul 6. Review. PubMed PMID:27379710

Tynan MJ, Becker AE, Macartney FJ, Jimenez MQ, Shinebourne EA, Anderson RH. Nomenclature and classification of congenital heart disease. Br Heart J. 1979;41:544–53.

Valtuille L, Paterson I, Kim DH, Mullen J, Sergi C, Oudit GY. A case of lamin A/C mutation cardiomyopathy with overlap features of ARVC: a critical role of genetic testing. Int J Cardiol. 2013;168(4):4325–7. https://doi.org/10.1016/j.ijcard.2013.04.177. PubMed PMID:23684604

Van Praagh R, Bernhard WF, Rosenthal A, Parisi LF, Fyler DC. Interrupted aortic arch: surgical treatment. Am J Cardiol. 1971;27:200–11.

Van Praagh R, David I, Van Praagh S. What is a ventricle? The single-ventricle trap. Pediatr Cardiol. 1982;2:79–84.

Van Praagh R. Terminology of congenital heart disease. Glossary and commentary. Circulation. 1977;56:139–43.

Van Praagh R. The segmental approach to diagnosis in congenital heart disease. Birth Defects Orig Artic Ser. 1972;8:4–23.

Vancura V, Hubácek J, Málek I, Gebauerová M, Pitha J, Dorazilová Z, Langová M, Zelízko M, Poledne R. Does angiotensin-converting enzyme polymorphism influence the clinical manifestation and progression of heart failure in patients with dilated cardiomyopathy? Am J Cardiol. 1999;83(3):461–2. A10

Villard E, Duboscq-Bidot L, Charron P, Benaiche A, Conraads V, Sylvius N, Komajda M. Mutation screening in dilated cardiomyopathy: prominent role of the beta myosin heavy chain gene. Eur Heart J. 2005;26(8):794–803.

Virmani R, Kolodgie FD, Burke AP, Farb A, Schwartz SM. Lessons from sudden coronary death: a comprehensive morphological classification scheme for atherosclerotic lesions. Arterioscler Thromb Vasc Biol. 2000;20(5):1262–75.

Vranes M, Velinovic M, Kovacevic-Kostic N, Savic D, Nikolic D, Karan R. Pregnancy-related aortic aneurysm and dissection in patients with Marfan's syndrome: medical and surgical management during pregnancy and after delivery. Medicina (Kaunas). 2011;47(11):604–6. PubMed PMID:22286575

Wagoner LE, Craft LL. Singh B, Suresh DP, Zengel PW, McGuire N, Abraham WT, Chenier TC, Dorn GW 2nd, Liggett SB. Polymorphisms of the beta(2)-adrenergic receptor determine exercise capacity in patients with heart failure. Circ Res. 2000;86(8):834–40.

Wang D, Shah KR, Um SY, Eng LS, Zhou B, Lin Y, Mitchell AA, Nicaj L, Prinz M, McDonald TV, Sampson BA, Tang Y. Cardiac channelopathy testing in 274 ethnically diverse sudden unexplained deaths. Forensic Sci Int. 2014;237:90–9. https://doi.org/10.1016/j.forsciint.2014.01.014. PubMed PMID:24631775

Wang DG, Fan JB, Siao CJ, Berno A, Young P, Sapolsky R, Ghandour G, Perkins N, Winchester E, Spencer J, Kruglyak L, Stein L, Hsie L, Topaloglou T, Hubbell E, Robinson E, Mittmann M, Morris MS, Shen N, Kilburn D, Rioux J, Nusbaum C, Rozen S, Hudson TJ, Lipshutz R, Chee M, Lander ES. Large-scale identification, mapping, and genotyping of single-nucleotide polymorphisms in the human genome. Science. 1998;280(5366):1077–82.

Wang ZJ, Reddy GP, Gotway MB, Yeh BM, Hetts SW, Higgins CB. CT and MR imaging of pericardial disease. Radiographics. 2003;23(Spec):S167–80. Review. PubMed PMID:14557510

Webster WS, Abela D. The effect of hypoxia in development. Birth Defects Res C Embryo Today. 2007;81(3):215–28. Review. PubMed PMID:17963271

Weigang E, Görgen C, Kallenbach K, Dapunt O, Karck M, GERAADA Study Group. German Registry for Acute Aortic Dissection Type A (GERAADA) – new software design, parameters and their definitions. Thorac Cardiovasc Surg. 2011;59(2):69–77. https://doi.org/10.1055/s-0030-1250748. Review. PubMed PMID:21384302

Weitzman JB. Electronic medical devices: a primer for pathologists. Arch Pathol Lab Med. 2003;127(7):814–25. Review. PubMed PMID:12823035

Wenzel K, Felix SB, Bauer D, Heere P, Flachmeier C, Podlowski S, Köpke K, Hoehe MR. Novel variants in 3 kb of 5′UTR of the beta 1-adrenergic receptor gene (−93C>T, −210C>T, and −2146T>C): −2146C homozygotes present in patients with idiopathic dilated cardiomyopathy and coronary heart disease. Hum Mutat. 2000;16(6):534.

Williams J, Wasserberger J. Crack cocaine causing fatal vasoconstriction of the aorta. J Emerg Med. 2006;31(2):181–4. PubMed PMID:17044582

Wilson PD, Loffredo CA, Correa-Villaseñor A, Ferencz C. Attributable fraction for cardiac malformations. Am J Epidemiol. 1998;148(5):414–23. PubMed PMID:9737553

Witsch T, Stephan A, Hederer P, Busch HJ, Witsch J. Aortic Dissection Presenting as "Hysteria". J Emerg Med. 2015;49(5):627–9. https://doi.org/10.1016/j.jemermed.2015.05.039. PubMed PMID:26272546

Wright C. Cardiac surgery 2002: staged repair of hypoplastic left heart syndrome. Crit Care Nurs Q. 2002;25(3):72–8. PubMed PMID:12450161

Yogasundaram H, Hung W, Paterson ID, Sergi C, Oudit GY. Chloroquine-induced cardiomyopathy: a reversible cause of heart failure. ESC Heart Fail. 2018;5(3):372–5. https://doi.org/10.1002/ehf2.12276. PubMed PMID:29460476; PubMed Central PMCID: PMC5933951

Yogasundaram H, Paterson ID, Graham M, Sergi C, Oudit GY. Glycogen storage disease because of a PRKAG2 mutation causing severe biventricular hypertrophy and high-grade atrio-ventricular block. Circ Heart Fail. 2016;9(8) pii: e003367. https://doi.org/10.1161/CIRCHEARTFAILURE.116.003367. PubMed PMID:27496753

Yuan SM, Jing H. Palliative procedures for congenital heart defects. Arch Cardiovasc Dis. 2009;102(6–7):549–57. https://doi.org/10.1016/j.acvd.2009.04.011. PMID 19664575. Retrieved 2010-02-27

Zeman M, Sepši M, Vojtíšek T, Sindler M. Suddenly deceased young individuals autopsied at the Department of forensic medicine, Brno – analysis, Soud Lek. 2012;57(3):44–7. PubMed PMID:23057440

Lower Respiratory Tract 2

Contents

2.1	**Development and Genetics**...	141
2.2	**Dysmorphology and Perinatal Congenital Airway Diseases**....................	145
2.2.1	Foregut Cysts of Bronchogenic Type..	146
2.2.2	Agenesis, Aplasia, and Hypoplasia Pulmonis (Pulmonary Hypoplasia)........	148
2.2.3	CPAM and Congenital Lobar Emphysema..	148
2.2.4	Tracheal and Bronchial Anomalies..	151
2.2.5	Sequestrations and Vascular Anomalies..	152
2.3	**Infantile and Pediatric *NILD***...	155
2.3.1	Neonatal Respiratory Distress Syndrome..	157
2.3.2	"Old" and "New" Bronchopulmonary Dysplasia (OBPD and NBPD).........	157
2.3.3	Atelectasis...	158
2.3.4	Cystic Fibrosis..	158
2.4	**Infantile and Pediatric *ILD***...	159
2.4.1	Diffuse Lung Development-Associated ILD...	159
2.4.2	Trisomy 21 (Down Syndrome)-Associated ILD..	161
2.4.3	ILD Related to Other Chromosomal/Genomic Microdeletion Disorders....	163
2.4.4	Neuroendocrine Hyperplasia of Infancy (NEHI).......................................	163
2.4.5	Pulmonary Interstitial Glycogenosis (PIG)..	164
2.4.6	Surfactant Dysfunction Disorders...	164
2.4.7	Pulmonary Alveolar Proteinosis..	166
2.5	**"Adult" ILD of the Youth**...	167
2.5.1	Diffuse Alveolar Damage (DAD)..	167
2.5.2	Cryptogenic Organizing Pneumonia (COP)..	170
2.5.3	Usual Interstitial Pneumonia/Pneumonitis (UIP)......................................	173
2.5.4	Nonspecific Interstitial Pneumonia/Pneumonitis (NSIP)..........................	175
2.5.5	Desquamative Interstitial Pneumonia/Pneumonitis...................................	176
2.5.6	Respiratory Bronchiolitis ILD...	177
2.5.7	Pulmonary Langerhans Cell Histiocytosis..	179
2.5.8	Lymphoid Interstitial Pneumonia/Pneumonitis (LIP)...............................	180
2.5.9	Hypersensitivity Pneumonia/Pneumonitis (HSP).....................................	181
2.5.10	Acute Eosinophilic Pneumonia/Pneumonitis..	182
2.5.11	Progressive Massive Fibrosis..	184

© Springer-Verlag GmbH Germany, part of Springer Nature 2020
C. M. Sergi, *Pathology of Childhood and Adolescence*,
https://doi.org/10.1007/978-3-662-59169-7_2

2.6	**Inflammation: Tracheobronchial**	184
2.6.1	Laryngotracheitis	184
2.6.2	Acute Bronchitis	184
2.6.3	Bronchial Asthma	184
2.6.4	Pulmonary Hyperinflation	188
2.6.5	Bronchiectasis	189
2.6.6	Neoplasms	189
2.7	**Inflammation: Infectious (Pneumonia/Pneumonitis)**	190
2.7.1	Lobar Pneumonia	190
2.7.2	Bronchopneumonia	191
2.7.3	Interstitial Pneumonia	193
2.7.4	Primary Atypical Inflammation (Pneumonia)	193
2.8	**Inflammation: Non-infectious (Pneumonia vs. Pneumonitis)**	193
2.8.1	Aspiration and Chemical Pneumonitis	193
2.8.2	Lipoid Pneumonia	195
2.8.3	Diffuse Pulmonary Hemorrhagic Syndromes	196
2.9	**Inflammation: Infectious/Non-infectious Granulomatous/Nongranulomatous**	197
2.9.1	Primary Pulmonary Tuberculosis	197
2.9.2	Sarcoidosis	200
2.9.3	Aspergillosis	201
2.9.4	Others	202
2.10	**Chronic Obstructive Pulmonary Disease of the Youth**	202
2.10.1	Chronic Bronchitis	203
2.10.2	Emphysema	203
2.11	**"Adult" Pneumoconiosis of the Youth**	204
2.11.1	Silicosis	205
2.11.2	Asbestosis	206
2.11.3	Non-silico Asbestosis-Related Pneumoconiosis	207
2.12	**Pulmonary Vascular Disorders**	207
2.12.1	Pulmonary Congestion and Edema	208
2.12.2	Pulmonary Embolism/Infarction	208
2.12.3	Pulmonary Arterial Hypertension (PAH)	210
2.13	**Transplantation-Related Disorders**	212
2.13.1	Acute Rejection	212
2.13.2	Chronic Rejection	212
2.14	**Pediatric Tumors and Pseudotumors**	215
2.14.1	Pulmonary Teratoma	215
2.14.2	Inflammatory Myofibroblastic Tumor	215
2.14.3	Pulmonary Hamartoma	217
2.14.4	Pulmonary Sclerosing Hemangioma	217
2.14.5	Pulmonary Carcinoid	217
2.14.6	Lymphangioma-(Leo)-Myomatosis (LAM)	217
2.14.7	Kaposiform Lymphangiomatosis	219
2.14.8	Pleuropulmonary Blastoma	220
2.14.9	Metastatic Tumors	222
2.15	**"Adult"-Type Neoplasms in Childhood/Youth**	222
2.15.1	"Adult"-Type Preneoplastic Lesions	226
2.16	**Pleural Diseases**	230
2.16.1	Pleural Effusions and Pleuritis	230
2.16.2	Pneumothorax	234
2.16.3	Neoplastic Pleural Diseases	235

2.17	Posttransplant Lymphoproliferative Disorders (PTLD)	237
2.18	**Non-thymic Mediastinal Pathology**	**240**
2.18.1	Anterior Mediastinal Pathology	241
2.18.2	Middle Mediastinal Pathology	242
2.18.3	Posterior Mediastinal Pathology	242

Multiple Choice Questions and Answers 242

References and Recommended Readings 244

2.1 Development and Genetics

Lung development as endodermal derivate is central in the pathology of children and youth because the lung maturation is completed only after birth and precisely runs all through infancy until the age of 8 years. The acinus is the terminal respiratory unit. Parameters of lung maturity are the Wigglesworth coefficient and the radial alveolar count (Box 2.1). To study the lung maturity in open lung biopsies or babies who died during the pregnancy may be challenging. The collapsed lung specimen may jeopardize the proper evaluation. Moreover, many parameters may not be in line with the somatic growth. Effective and efficient criteria for the diagnosis of hypoplastic lungs have been searched for a long time. Wigglesworth proposed a coefficient considering his tremendous previous work in perinatal pathology. Wigglesworth focused on the ratio of lung weight to body weight and stated that this procedure is the most practical means of overcoming the intrauterine challenges. An LW/BW ratio ("lung weight/body weight") ratio of ≤ 0.012 is generally associated with a reduction in alveolar number as indicated by the radial alveolar count (RAC) (*vide infra*). Wigglesworth and Desai considered appropriate to define lung hypoplasia in infants or fetuses with a gestational age <28 weeks as half the mean LW/BW ratio of the 20–27-week group (i.e., <0.015) rather than taking the value of 0.012 used for infants with gestational age ≥ 28 weeks. These authors found that babies with renal agenesis, severe renal cystic dysplasia, or total urinary outflow obstruction had a lung DNA content at 34–40 weeks' gestation equivalent to that found in the normal fetal lung at 20–22 weeks' gestational age. In normal fetuses, i.e., fetuses with normal karyotype and without external or internal congenital anomalies, the graph of total lung DNA increase shows that the lung cell population doubles between 17 and 20 weeks' gestation. The RAC is calculated according to Emery and Mithal's method. RAC <2.0 is usually found in the lungs of fetuses aged less than 18 weeks. A considerable interindividual variation is present between 18 and 25 weeks. In fetuses with a gestational age between 25 and 30 weeks, an increase in RAC occurs. RAC >6 occurs only in fetuses at near full-term birth. Since the estimation of RAC overcomes the effects of varying degrees of alveolar collapse, RAC is useful also for the determination of the fetal age in cases of advanced putrefaction (Figs. 2.1 and 2.2).

Some procedures should be in line to ensure reproducibility in the RAC calculation. The chILD Pathology Co-operative Group proposed some extremely useful guidelines (Langston et al. 2006) (here slightly modified), which include the proper handling of a sterile lung specimen received unfixed from the pediatric surgeon or nurse directly from the pediatric operating room. We advise to sample tissue for microbiology directly in the operating room avoiding the potential contamination of transportation and pathology laboratory. Specifically, 10% of tissue is destined to the microbiology laboratory for microbiology cultures to identify potential viral,

---

Box 2.1 Parameters of Lung Maturity
- Wigglesworth coefficient: *Lung weight/ body weight ratio* (LW/BW)
- Emery and Mithal's *radial alveolar counting* (RAC)

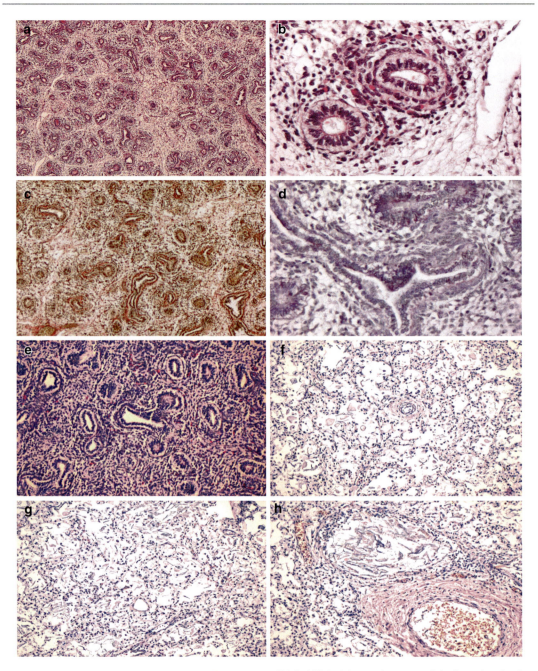

Fig. 2.1 Lung development: (**a**) lung architecture at 14th weeks' gestation without chromosomal abnormalities or karyotypic abnormalities (50×, H&E) showing a prominent interstitium rich in interstitial fluid; (**b**) high magnification at 14th weeks' gestation with cylindrical appearance of the future alveoli, which at this stage are labeled pseudoglands (H&E staining, 400×); (**c**) EvG stain highlights the addensation of elastic fibers around the pseudoglands (EvG, 100×); (**d**) the Periodic Acid Schiff staining, which is specific for glycogen, highlights the glycogen-rich content of the lung from a 14th weeks' gestation old fetus (PAS, 400×); (**e**) saccular stage of the lung showing less edema-rich interstitium and abnormal ovoidal shape ("sacculi") of the lung architecture. The epithelium is still cylindrical but at places may become cuboidal; (**f**), (**g**), and (**h**) come from a mature newborn with alveolar stage of development but suffering from birth asphyxia with presence of squamous lamellae in some of the alveolar structure. The blood vessels have a physiologic hypertrophy of the media at this age. No hyaline membranes can be recognized (H&E staining, 200× magnification for all three photographs)

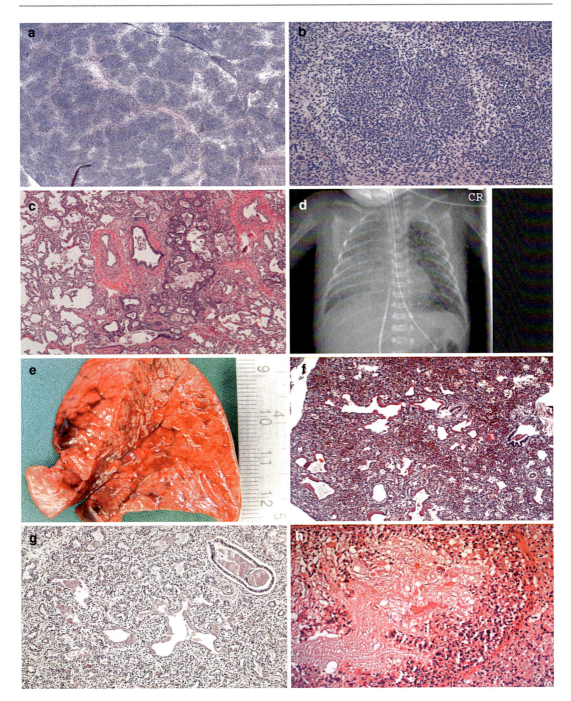

Fig. 2.2 Lung maldevelopment, immaturity: (**a**), (**b**) lung development arrest at 13 weeks (H&E staining, 40× and 200×); (**c**) lung development disorder in 1-year-old infant with alveolar enlargement and simplification, lobular remodeling with increased interstitial smooth muscle and mild fibrosis, interstitial inflammation, patchy, mild, peribronchiolar metaplasia and mucus stasis, pulmonary arteriopathy with mild to moderate muscularization, and myointimal proliferation (H&E staining, 50×); (**d**) Chest X-ray of an immature baby showing a granular-looking lung pattern with non-areated areas ("salt" granules) and aerated areas ("pepper" granules). (**e**) 30 weeks' gestation premature baby with GBS-Sepsis, SPH and normal heart; (**f**) 30 weeks' gestation premature baby with hyline membrane disease (H&E staining, 50×); (**g**) PAS-positive hyaline membranes in a premature baby (PAS staining, 50×); (**h**) obliteration of the alveolar structures by epithelial squames and neutrophils in a newborn with meconium aspiration syndrome (H&E staining, 50×)

bacterial, and fungal microorganisms, 10% of tissue is snap frozen (–70°C) for molecular biology studies (e.g., surfactant genes mutations), 10% of tissue is snap frozen (–70°C) in cryomatrix embedding resin for immunofluoresence, laser capture microdissection, or other studies requiring frozen tissue sections securing and supporting tissue with a firm bond with the specimen holder, <5% (usually three specimens of 1 mm^3) fixed in glutaraldehyde for transmission electron microscopy (multiple sites), 5% for flow cytometry for clonality and lymphocytic cell populations (e.g., post-transplant lymphoproliferative disorder) using Tissue Transport Medium or TTM (if TTM is not available, normal saline is the best choice), 0% touch preparations (imprints) for cytologic examination or rapid identification of organisms (two slides air dried and two alcohol fixed stained with Papanicolau staining and a Romanowski-based stain, e.g., Diff-Quik), 60% formalin inflation using a syringe and applying a gently infusion fixed for light microscopy. It is important to recall that the vials of glutaraldehyde 1%–2% solution are usually in the fridge (4 °C). On the other hand, glutaraldehyde seem to show no change in concentration after one year of storage at 25°C, but some glutaraldehyde products have shown a longer shelf life when they are placed in the 4 °C. TTM needs to be stored at 4°C for up to 1 month from the date of preparation.

TTM vials should have an orange-pink appearance and a color change to bright pink indicates a pH change. At this point, TTM should be discarded. When the respiratory bronchiole is symmetrical, the geometric center of the lung specimen (histological slide) should be used as a starting point. In case the bronchiole is irregular, the starting point should be in the middle of the way between the most proximal part (i.e., the wall lined by conducting epithelium) and the first branch point. No count should be made if the respiratory bronchiole is nearer to the edge of the slide than to the most adjacent connective tissue septum. In fact, cutting issues may conceal an artifact. A closer septum might have been aloof in taking the paraffin block. Finally, all alveoli traversed by the perpendicular should be counted. In other terms, a perpendicular line is drawn from the midpoint of the most distal respiratory bronchiole

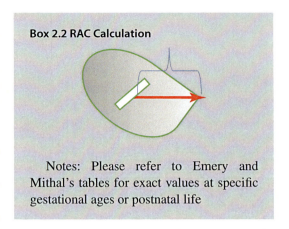

Box 2.2 RAC Calculation

Notes: Please refer to Emery and Mithal's tables for exact values at specific gestational ages or postnatal life

(transition point from respiratory to alveolar epithelium) to the nearest connective tissue septum, and each saccule is counted along this line.

In Box 2.2, the RAC calculation is shown.

Lobation: Right lung is trilobed, while the left lung is bilobed (the lingula of the left lung should be considered the equivalent of the right middle lobe). The lingula is often taken for sampling in case of suspicious interstitial lung disease. About the tracheal bifurcation, it occurs at the level of the 4th–5th ribs with the right main stem bronchus positioned more vertical than the left main stem bronchus. Lung aspiration is more frequent on the right side than on the left side.

Eparterial Versus Hyparterial Broncho-Arterial Crossing

About the concept of isomerism as seen in the previous chapter of the cardiovascular system, it is important to remember some important concepts of eparterial and hyparterial crossing connections of the bronchial ways. This is a useful criterion in the sequential segmental analysis. The right pulmonary artery is anterior to the right main stem bronchus, and in on occasion, it does cross over the bronchus because the bronchus is slightly superior (eparterial bronchus). Conversely, the left main stem bronchus is longer than the right, and the left pulmonary artery crosses the left main stem bronchus (hyparterial bronchus) superiorly. In the pathology of the isomerism, which involves an abnormal lobation and broncho-arterial crossing other than congenital heart disease and other visceral organs, we distinguish between right- and left-sidedness. Right-sidedness is associated with bilateral trilobed (right) lungs (~4/5 cases), asplenia, and an intact inferior vena cava (IVC), whereas left-sidedness is related to bilateral bilobed (left) lungs (~2/3 cases), polysplenia, and an interrupted IVC.

Moreover, some more words need to be spent on the trachea and bronchial system. The trachea and significant bronchi have C-shaped cartilage rings, which is of an obstacle to a full occlusion of the lumen by bronchoconstriction in comparison with the most distal airway structures that can be fully occluded. Bronchi have discontinuous plates of cartilage and are recognized explicitly by submucosal glands. The lungs have a dual arterial system: pulmonary and bronchial. Hemorrhagic infarction is possible if other than occlusion also stasis/left ventricular insufficiency is present. There is a single venous system: pulmonary. The pseudostratified ciliated columnar epithelium is from the larynx to bronchioles (except the vocal cords: squamous) with type I pneumocytes covering ~95% alveolar surface while type II pneumocytes covering the rest and contain lamellar bodies of surfactant.

> **Box 2.3 Anatomic and Immunohistochemical Characteristics of Pleura**
>
> - Pleurae: 2, i.e., (1) visceral, coating the lung, and (2) parietal, adherent to the thoracic cage
> - Lining: Mesothelium constituted by mesothelial cells of mesodermal origin
> - Mesothelium distinctive features: tight apical junctions, desmosomes, long surface microvilli, and thick glycocalyx (pericellular matrix)
> - Non-neoplastic mesothelium reactivity: P53 (−) ≠ from mesothelioma (P53+)
> - Non-neoplastic mesothelium reactivity: EMA (dim) ≠ from mesothelioma (spikes)
> - Non-neoplastic mesothelium reactivity: DES (+) ≠ from mesothelioma (−)
>
> Notes: ≠ different; EMA, epithelial membrane antigen; DES, desmin

The Fate of Inhaled Particles

Inhaled particles can reach the terminal stations of the acinus according to the size. If the particle size is >5 μm, it does usually impact high in airways and is cleared, and <1 μm typically remain suspended and are exhaled. Conversely, particles with a 1< Ø <5 μm reach the terminal airways and alveoli and deposit there eliciting no tissue reaction or a tissue reaction according to the chemical or physical composition of the particle. The pulmonary defense mechanisms are nasal riddance, a mucociliary clearance, and the alveolar macrophages (MΦ).

The anatomic and immunohistochemical characteristics of the pleura are summarized in Box 2.3.

Physiologically, the pleura play essential roles, including the crucial one that takes place when the individual breaths. The pleural cavity serves as optimal functioning of the lungs during breathing and contains pleural fluid, which acts as a "lubricant" allowing the pleurae to slide effortlessly against each other during respiratory movements. Pleural pathology is either autonomous in origin or secondary to some underlying pulmonary pathology (most of the cases), being co-involved due to the adhesion to the lung parenchyma.

2.2 Dysmorphology and Perinatal Congenital Airway Diseases

The mnemonic word *FACTS* can be used to remember significant dysmorphology and perinatal airway diseases (Box 2.4). It includes *F*oregut cysts and bronchogenic cysts, *A*genesis or hypoplasia (partial or total), *C*PAM (congenital pulmonary airway malformation) and *C*ongenital emphysema (lobar overinflation), *T*racheal and bronchial anomalies (atresia, stenosis, tracheoesophageal fistula), as well as *S*equestrations and vascular abnormalities. It is important to remember that bronchogenic cysts and sequestrations have a systemic arterial supply and do not communicate with the tracheobronchial tree, while CPAMs have regular pulmonary arterial supply and do communicate with the tracheobronchial tree.

> **Box 2.4 Dysmorphology and Perinatal Congenital Airway Diseases**
> 2.2.1. *F*oregut Cysts and Cysts of Bronchogenic Type
> 2.2.2. *A*-/Dys-genesis (Gr. ἀ-/δυσ- + γενησία): Agenesis, Aplasia, and Hypoplasia (sive Pulmonis (total or partial development of the lung) ≠ Agenesis (lack of lung anlage)
> 2.2.3. CPAM/CCAM and Congenital Lobar Emphysema
> 2.2.4. *T*racheal and Bronchial Anomalies (Atresia, Stenosis, and Tracheoesophageal Fistula)
> 2.2.5. *S*equestrations and Vascular Anomalies
>
> Notes: Agenesis derives from ἀγένητος "uncreated", while hypoplasia originates from ὑπο- "under" + πλάσις "formation"; TEF, Tracheoesophageal

2.2.1 Foregut Cysts of Bronchogenic Type

Foregut cysts may be of a bronchogenic, esophageal, or enteric type. The bronchogenic cyst is usually a mucinous cyst or an air containing extrapulmonary cavity, anywhere in the chest (adjacent to bronchi or bronchioles). It may be single, ranging from microscopic size to an effective Ø of ≥ 5 cm. Ciliated epithelium lines foregut cysts of bronchogenic type. The underlying connective tissue usually contains cartilage and submucous glands. Complications include mostly infections. If infection supervenes, progressive metaplasia of lining or total necrosis of the wall can ensue. Subsequently, this inflammatory process may lead to lung abscess (*vide infra*). Other complications may include rupture of the foregut cysts into either bronchi or pleural cavity (Fig. 2.3).

Cutaneous bronchogenic cysts may also occur and need to be differentiated by the branchial derivatives and thyroglossal duct cysts. The embryology and the lung development help in explaining the origin of bronchogenic cysts in the extrathoracic subcutaneous tissue. In the 5th week of gestation, the separation occurs of the primitive foregut. The laryngotracheal groove starts and determines the termination of the primitive foregut into two well specific but yet not well delineated structures, i.e., the dorsal and ventral structures. Two weeks later, at the 17th week of gestation, the laryngotracheal groove is complete determining two well-distinct embryonic structures, i.e., the ventral component forming the foregut and the dorsal element forming the lung buds. This period is crucial, and an abnormal process in the lung bud formation during this time implicates the origin of bronchogenic cysts. In the subcutaneous tissue of the chest (anterior side), bronchogenic cysts occur either by anterior migration of an intrathoracic bronchogenic cyst or by a sequester of the fusing sternal bars on the developing lung parenchyma. Conversely, extrathoracic cysts located in other sites (e.g., neck, shoulder, and chin) can be interpreted by the migration of these sequestered structures at their location in the developing embryo. They should not be construed as neoformations. A ciliated pseudostratified columnar epithelium covers the cutaneous bronchogenic cysts. This epithelium is commonly invariably interspersed with goblet cells, which may be recognized using special stains (alcian PAS or alcian pH 2.5). The cutaneous bronchogenic cysts also show other components, including cartilage, mucous glands, and smooth muscle fibers. The most common element is the smooth muscle layer. The differential diagnosis of a cutaneous bronchogenic cysts includes cutaneous ciliated cyst (lower extremities, female), thyroglossal duct cyst (midline, respiratory/squamous epithelium lining, TG (+), and TTF (+) thyroid follicles), epidermal inclusion cyst (stratified squamous epithelium), branchial cyst (stratified squamous epithelium and surrounding composed of lymphoid tissue), dermoid cyst (epidermis lining with several epidermal appendages), and trichilemmal cyst (squamous epithelium lining without an intervening granular cell layer). Although trichilemmal or

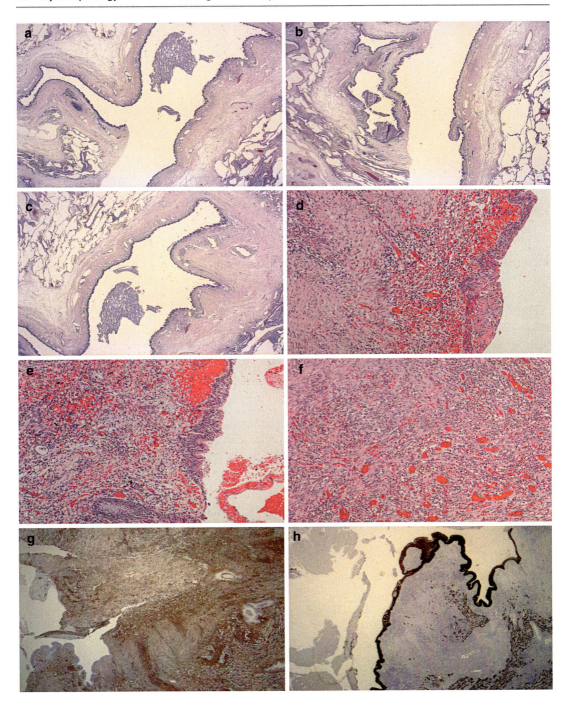

Fig. 2.3 Bronchogenic cyst showing a ciliated epithelium as lining with focal squamous metaplasia. In the wall, there is smooth muscle, connective tissue, and respiratory-type mucous glands with cartilaginous islands (**a**) H&E, 50×, (**b**) H&E, 50×, (**c**) H&E, 50×, (**d**) H&E, 100×, (**e**) H&E, 100×, (**f**) H&E, 100×, (**g**) Vimentin staining by immunohistochemistry, 12.5×, and (**h**) AE1-3 staining by immunohistochemistry, 12.5×

pilar cysts arise mostly in body ares of high hair follicle concentrations (e.g., scalp), they can also be found everywhere. In one third of cases, trichilemmal cysts are solitary, but in two thirds they are multiple. The histology is paramount in giving the definitive diagnosis to the pediatric surgeon.

2.2.2 Agenesis, Aplasia, and Hypoplasia Pulmonis (Pulmonary Hypoplasia)

Agenesis of the lung is particularly rare and has been observed in very few cases from the literature. In lung agenesis, there is a total absence of any lung tissue. There is no foregut budding *ab initio*. In lung aplasia, there is an extreme imperfect development of the organ with some minimal primordial mass (the lung anlage is present). Lung aplasia is a rare congenital pathology in which there is the unilateral or bilateral absence of proper lung tissue. It is arguably distinguished from lung agenesis, the main difference being that there is a short blind-ending bronchus in aplasia. However, said that agenesis and aplasia are difficult to distinguish from each other both clinically and pathologically, in the end, they are quite often used in the same sense. Lung hypoplasia may be part of the oligohydramnios sequence (aka "Potter sequence") but also part of multiple congenital anomalies (MCA)-associated syndromes. Partial or total hypoplasia may be encountered more often with or without congenital heart disease. The Askenazi-Perlman criteria may be beneficial for the diagnosis of lung hypoplasia. These authors suggest the following guidelines for the determination of pulmonary hypoplasia (LH). Lung weight (LW)/body weight (BW) ratio <0.009, LH very likely, and the radial alveolar count (RAC) are not obligatory to perform. If LW/BW ratio is 0.010–0.012, LH is probable, and RAC is indicated for confirmation of diagnosis. If LW/BW ratio is 0.013–0.017, LH is possible, and RAC is judicious and prudent to perform. If LW/BW ratio >0.018, LH is unlikely. In this setting, RAC may not contribute. Where RAC is estimated, a value <75% of the mean standard value (equivalent to 43% of cubed value) for that laboratory probably indicates LH.

2.2.3 CPAM and Congenital Lobar Emphysema

The acronym CCAM for congenital cystic adenomatoid malformation was in use for a while. Ultimately, it has been changed with CPAM for congenital pulmonary adenomatoid malformation, but CCAM is often still in use in the medical literature. CCAM/CPAM has been described as a hamartoma, and, in 1977, Stocker grossly classified CCAM into three types based mostly on cyst size. Class I included multiple large cysts (>2 cm in diameter) or a single large cyst surrounded by smaller cysts and was by far the most common type of CCAM/CPAM. Type I is associated with an excellent outcome. Multiple small cysts characterized type II and accompanied by other congenital anomalies that may affect the outcome (e.g., renal agenesis), while type III included numerous microcysts ($\varnothing \leq 0.5$ cm). Subsequently, in 1993, Adzick reported his group system of CPAM lesions, subclassifying microcystic structures and macrocystic structures. Microcystic structures (cysts with $\varnothing \leq 0.5$ cm) are associated with fetal hydrops and harbor a poor prognosis, while macrocystic structures (i.e., cysts with $\varnothing > 0.5$ cm) are usually not associated with fetal hydrops and harbor a relatively good prognosis. Stocker reviewed the topic, and currently, five types of CCAM/CPAM exist. CCAM/CPAM is substantially a maturation defect of the lung development which is a significant cause of perinatal morbidity. It often requires the baby to stay for some time in the Neonatal Intensive Care Unit (NICU). There are five known types of CPAM. These types are divided according to the most updated Stocker type, their relative incidence, genetics, gross features, light microscopy findings, and some outcome details. It is essential to differentiate the types of CPAM, because of the possibly associated congenital defects and potential complications, including malignancy (Figs. 2.4 and 2.5). Stocker CPAM/CCAM type 0 (~5%) shows small and solid cystic change with acinar dysplasia

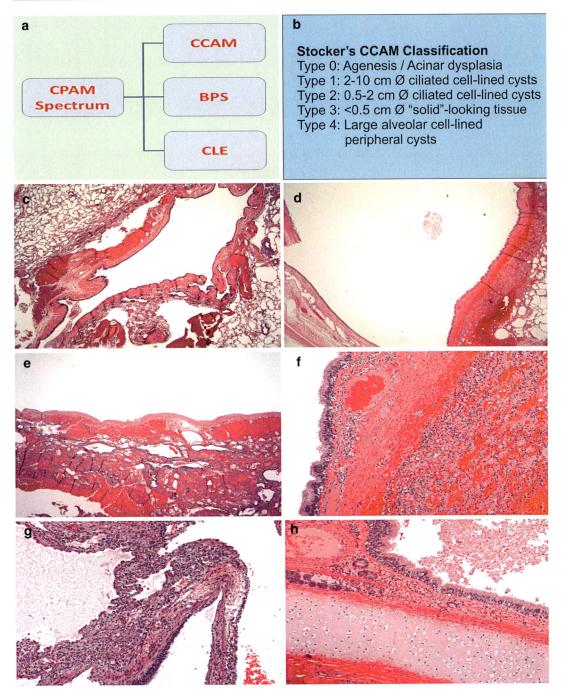

Fig. 2.4 (**a**) CPAM Spectrum, (**b**) CPAM classification, (**c**, **d**, **e**, **f**, **g**, **h**) show several features of CPAM type I (H&E staining for all slides, **c**–**e** at 12.5×, **f**–**h** at 100×). The microphotographs show large cysts lined by pseudostratified ciliated cells interspersed with mucus cells

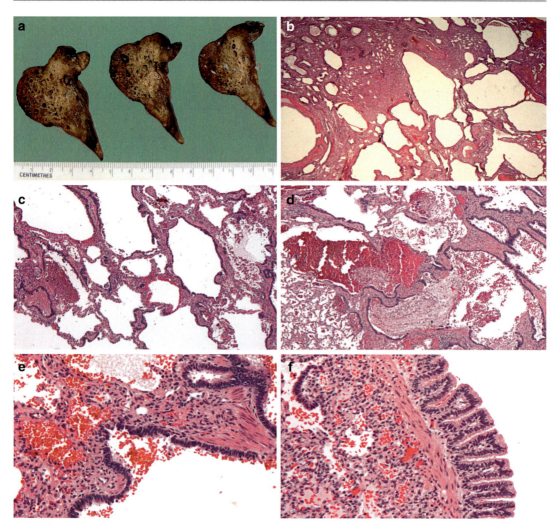

Fig. 2.5 Gross photograph (**a**) and microphotographs (**a–f**) showing the features of CPAM-II (**a** shows the gross photograph, while the other microphitographs are taken at 12.5×, 50×, 50×, 200×, and 200×, respectively, from hematoxylin and eosin stained histological slides). CPAM-II shows small cysts up to 2 cm, resembling dilated bronchioles separated by normal alveoli, and very little or no cartilage

(Ø ≤ 0.5 cm), respiratory lining, cartilage, but no normal alveoli and evolves to respiratory insufficiency. The type 1 (~60%) shows a localized mass grossly, and microscopically there are 2–10 cm cysts, respiratory epithelium, mucus cell rows, polypoidal infolds, and intervening fibromuscular tissue. The type 1 may evolve to bronchioloalveolar carcinoma or sarcoma if not resected. The type 2 (~15%) involves one lobe and shows characteristically 0.5–2 cm cysts, mono-columnar cells, and interspersed alveolar ducts, with or without striated muscle and/or accompanying urinary tract anomalies. There is a risk of rhabdomyomatous dysplasia and cardiovascular anomalies. The type 3 (~5%) resembles fetal lung grossly and is characterized microscopically by small cysts lined by mono-cuboidal cells. This type can evolve to respiratory insufficiency. Finally, the type 4 (~15%) is constituted by cysts located at the periphery of the lung. Microscopically, there are mesenchymal cysts, alveolar lining, and neither respiratory epithelium nor cartilage. There is a risk to develop pleuropulmonary blastoma (PPB) by harboring the type 4 of CCAM/CPAM.

Apart from types II and probably IV, CCAM/CPAM have a terrible prognosis, and except for type II, CCAM/CPAM present in newborns. The differential diagnosis of these lesions from other neonatal lesions is crucial for the correct management of these cystic lesions. Simple foregut cysts need to be differentiated from CPAM I, and the latter has polypoidal infolds and rows of mucous cells. A bronchogenic cyst needs also to be distinguished from CPAM I, and the latter has no cartilage and mural mucus glands that can be found in bronchogenic cysts. CPAM IV needs to be separated from PPB particularly the well-differentiated PPB type or type I. This differential diagnosis may also be impossible or may require multidisciplinary team meetings or clinicopathologic conferences. The cystic mesenchymal hamartoma has a mesenchymal cambium layer, which is not present in CPAM. Finally, except type 0, CPAM has a pulmonary arterial blood supply, bronchial tree communication, and presence of normal alveoli, all features important in the differential diagnosis with sequestration.

Congenital Lobar Emphysema (CLE) is designated as sudden progressive RDS (respiratory distress syndrome) in young children. CLE usually affects the upper lobe or right middle lobe and there is massive overdistention of airways without tissue destruction (DDX: Lung emphysema of the adult). CLE is part of the CPAM spectrum including CCAM/CPAM and BPS other than CLE.

2.2.4 Tracheal and Bronchial Anomalies

Lung development may be affected by upper respiratory tract malformations of the trachea and primary bronchi. Tracheal and bronchial anomalies (atresia, stenosis, and tracheoesophageal fistula) are significant to diagnose either during the intrauterine life or after birth, because they may be life-threatening conditions. Atresia (Greek ἀ- "not" and τρῆσις "hole") is the medical condition in which an aperture (*orificium*) or transition (channel, tube) in the body is absent or closed, while stenosis (Greek στένωσις derives from στενός "narrow") is an abnormal narrowing or contraction of a body orifice or passage.

2.2.4.1 Laryngeal Web

The laryngeal web is an incomplete recanalization of the larynx that should take place during the 10th week of embryogenesis. A membranous network may be visualized at the level of the vocal cords with consequent sub-obstruction of the airways.

2.2.4.2 Tracheoesophageal Fistula

Tracheoesophageal fistula (TEF) represents an incomplete division of the first gastrointestinal tract or foregut into digestive and respiratory portions. There are five different types of TEF with differently associated pathology. *TEF I* is focal esophageal atresia consisting of proximal esophagus ending as a blind pouch and evidencing dilation and distal esophagus arising from trachea just above bifurcation (85%). Both esophageal segments characterize *TEF II* and are blind pouches (8%). Both complete trachea and esophagus characterizes *TEF III* but remains present a singular connection at the level of tracheal bifurcation (4%). *TEF IV* is categorized by a proximal esophagus, which connects to the trachea at the level of bifurcation, while the distal portion of the esophagus arises as a blind pouch. Finally, *TEF V* shows that both proximal and distal esophageal segments communicate with the trachea.

The communication of the trachea with the esophagus is a condition that may allow some gastric juice to violate into airways with consequent aspiration pneumonia, which may remain silent for a while. The use of bronchoalveolar lavage with mandatory Oil Red O (ORO) staining of the cellularity may disclose alveolar macrophages filled with red vacuoles and suggest that aspiration pneumonia did occur. About the ORO staining, vide infra.

2.2.4.3 Tracheal Stenosis/Atresia

TEF may usually be associated with some tracheal stenosis or tracheal atresia, although they are quite rare.

2.2.5 Sequestrations and Vascular Anomalies

Sequestration means segregate or seizure something by law, and pulmonary sequestration is partial or complete segregation or separation of a portion of a lobe from the "surrounding" lung where nonfunctioning pulmonary tissue is separated from the rest of the parenchyma and supplied with blood from often an artery from systemic circulation due to an abnormal budding of the primitive foregut (Fig. 2.6). Another medical condition in which we use the sequestration terminology is sequestrum, which is defined as a fragment of bone or other dead tissue that has separated from the "mother" organ during necrosis,

- *DEF*: Congenital respiratory anomaly consisting of a portion of nonfunctioning normal lung tissue being excluded by the pulmonary arterial blood supply but receiving arterial blood supply from the systemic circulation and diagnosable in utero by prenatal ultrasound.

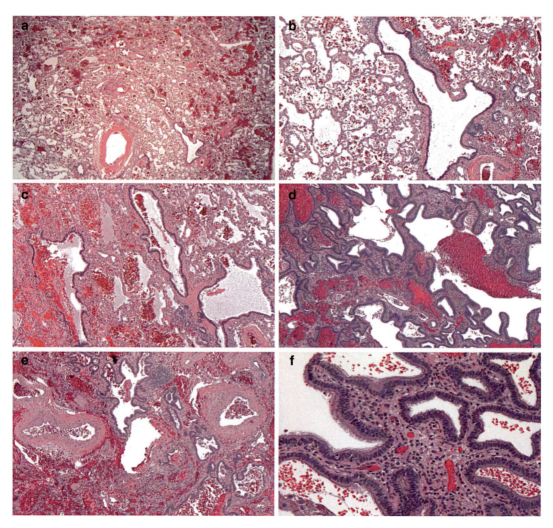

Fig. 2.6 Microphotographs (**a–c**) show intralobar pulmonary sequestration hematoxylin and eosin staining, 12.5×, 50×, and 100× magnifications), (**d–f**) show extralobar pulmonary sequestration (hematoxylin and eosin staining at 50×, 50×, and 200× magnification, respectively). Fibrosis, chronic inflammation, vascular sclerosis, and cystic changes replace the pulmonary parenchyma in intralobar sequestration (internal to lung)

- *SYN*: Bronchopulmonary sequestrum.
- *EPI*: 1–5% of all congenital pulmonary manifestations.
- *CLI*: Asymptomatic→persistent dry cough.

All sequestrations have systemic blood supply from the aorta and not from pulmonary arteries and no connection to tracheobronchial tree or airway system. In 1/5 cases there is CPAM-like histology. The sequestration is subclassified as *intralobar* if the sequestered portion is enclosed within the pleura of the rest of the lung and *extralobar* if the sequestered portion is separated from the pleura of the rest of the lung. The blood supply of 75% of pulmonary sequestrations arises from the aorta (thoracic or abdominal), while the remaining 25% of lung sequestrations receive their blood from different sources (subclavian, intercostal, pulmonary, pericardiophrenic, innominate, internal mammary, celiac, splenic, or renal arteries). Intralobar sequestrations may be acquired or congenital, but extralobar is only innate. The usual location of the ILS is the posterior segment of the left lower lobe. ELS-associated congenital anomalies include congenital cystic adenomatoid malformation, congenital diaphragmatic hernia (defect), vertebral anomalies, lung hypoplasia, and colonic duplication (LLL, left lower lobe; RLL, right lower lobe).

Generally, the ILS has the arterial blood supply from the lower thoracic aorta or upper abdominal aorta and the venous drainage to the left atrium via pulmonary veins establishing a left to right shunt. In the medical literature, abnormal connections to the vena cava, azygous vein, or right atrium have been described as well. The ILS may become infected when microorganisms migrate through the "Pores of Kohn" or if the sequestration is incomplete. Commonly, the ELS arterial supply is an aberrant blood vessel of the aorta, while its venous drainage is directed to the right atrium, vena cava, or azygous systems. The ELS rarely gets infected because it is enveloped in its pleural sac. Imaging was associated with an invasive arteriogram showing the systemic blood supply, but new noninvasive techniques are now preferred (e.g., CT scan or MRI). Chest radiograph shows a uniformly dense mass within the thoracic cavity or pulmonary parenchyma. Recurrent pulmonary infections may lead to having cystic change or cystic areas within the mass. Air-fluid levels due to bronchial communication can also be observed. Ultrasound displays an echogenic homogeneous mass with regular or irregular borders and a more solid or more cystic appearance, and Doppler studies may demonstrate the characteristic aberrant systemic artery. CT scan can show a solid mass that may be homogeneous or heterogeneous with or without cystic change. Usually, there are emphysematous changes at the margin of the lesion, and the CT scan has an overall 90% accuracy to identify pulmonary sequestrations. CT scans may be performed with multiplanar and 3D reconstructions. Contrast-enhanced MRI and T1-weighed spin-echo images may help in the diagnosis of lung sequestrations, although sharper images may be reached better with CT than MRI. There are eight significant complications for sequestrations, including aspergillosis, bleeding, bronchiectasis, chronic infection, left-right shunt, neoplastic degeneration of epithelial type (squamous cell carcinoma of the lung), neoplastic degeneration of neuroendocrine type (bronchial carcinoid), and tuberculosis.

Treatment of the lung sequestration is surgical with the removal of the lung tissue that may show pneumonia. It may be essential to take some swabs from the resection site or at the time the pathologist receives the specimen for the identification of the microorganism. If the fetus is diagnosed with lung sequestration, the Harrison catheter shunt may be used in case drainage of the fluid collection is mandatory. Fetal laser ablation procedure may also be used in some settings. In Scimitar syndrome, cysts in the hypoplastic lung are associated with the absent pulmonary blood vessel. The congenital pulmonary venolobar syndrome (aka Scimitar syndrome), originally described by Catherine Neill in 1960, is a rare congenital heart defect individualized by an anomalous venous return from the right lung draining into the systemic veins, rather than directly to the left atrium. This abnormal drainage creates a curved shadow on chest x-ray resembling a scimitar (Neill et al.

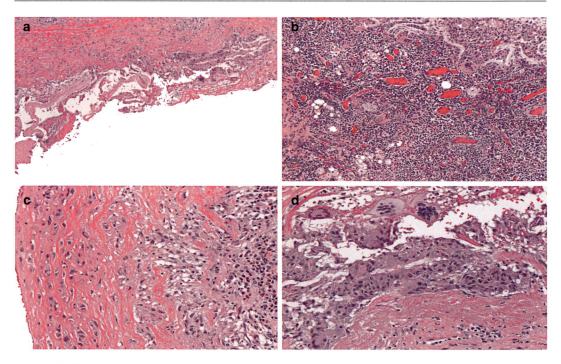

Fig. 2.7 Echinococcal cyst showing characteristic changes of the lung parenchyma with part of the chitinous membrane, scolices (not apparent), and multinucleated giant cells (**a–d**: H&E, **a** 50×, **b** 100×, **c**, **d** 200×)

1960). Hydatid's *Echinococcus granulosus* (Echinococcosis) (Fig. 2.7) is an important differential diagnosis in pediatrics.

In 1777, Huber provides the first description of an aberrant systemic artery supplying the lung publishing his findings in a Swiss medical journal (Huber, 1777). In 1946, the concept of "sequestration" was described by Pryce, but this lexicon could not outline all variants of sequestration, and the concept of a "sequestration spectrum" came forward properly in (Sade et al. 1974). Ten years later, the "sequestration concept" was reviewed, and the importance of the venous drainage was emphasized. This concept led to the Clements' "wheel theory," of whom came forth the term "pulmonary malinosculation" (*mal-* in *osculum*, a Latin word for "small mouth", ōs "mouth" + -culum, neuter form of -culus, which is the suffix forming a diminutive form of a word) (Clements and Warner, 1987). Malinosculation is the presence of abnormal communications by small opening or anastomoses. In 2007, Lee et al. proposed the "haphazard theory" to better substantiate this concept describing seven types of pulmonary malinosculation. Typically, pulmonary AV malformations are congenital lesions caused by abnormal development of capillaries. Their etiology is not completely understood, and some genetic/epigenetic factors are thought to play a role in the formation of disease. Lee et al. based their observation on whether the abnormal communication involved the bronchi, arteries of the lungs, or veins of the lungs in isolation or different combinations. Lee et al. schema consists of dividing bronchopulmonary vascular malformations systematically into seven subcategories (types A–G) using the "malinosculation theory." Different surgeries have been reported, including ligation of the anomalous artery and lobectomy, surgical ligation of the anomalous artery only without lobectomy, and disconnection of the anomalous artery from the aorta and anastomosing it to the pulmonary artery. Pulmonary malinosculation is a very intriguing investigative field, which is based on the malinosculation term. Stedman's Medical Dictionary defines malinosculation as the creation of abnormal communications using openings or anastomoses, applied particularly to the establishment of such communications between already existing blood vessels or other tubular structures that come into contact. The proposal of random branching patterns, which involve two independent systems to support the concept of bronchopulmonary malinosculation, may facilitate clinical classification and implication of these malformations. Shield et al. conceived three main categories, including the anomalies originating from the pulmonary arteries, those arising from the pulmonary veins, and anomalies arising from lymphatic vessels. In 1974, Dines et al. divided arteriovenous malformations into two groups as those with idiopathic AVMs and Rendu-Osler-Weber disease (ROWD), which is also known as hereditary hemorrhagic telangiectasia, HHT, and those with idiopathic AVMs. The cases with pulmonary AVMs are generally congenital and are observed with ROWD at a rate of 70%. Congenital PAVMs, which are incomplete and transmitted and inherited by a dominant gene, are more frequently seen in female individuals. PAVMs usually involve the lower lobe, are solitary, and localize superficially. Multiple lesions are seen in up to 50% of cases, and HHT often accompanies them. PAVMs accompanied by HHT progress rapidly and inexorably, and their complication rate is quite high. Congenital pulmonary anomalies can occasionally be encountered with malformations of the vascular supply. A coexistence of Scimitar syndrome and pulmonary sequestration has been designated as "venolobar syndrome."

2.3 Infantile and Pediatric NILD

There are lung diseases of infantile and/or pediatric onset with surface epithelial involvement and no direct interstitial involvement, but "adult" interstitial lung diseases may also occur in the adolescent and youth and need to be considered when evaluating a patient with respiratory problems according to age, physical examination, family history, and environmental issues (Box 2.5).

It is crucial to spend some words on surfactant, before dealing with the neonatal respira-

> **Box 2.5 Infantile and Pediatric NILD**
> 2.3.1. *Neonatal Respiratory Distress Syndrome* (NRDS) (Fig. 2.8)
> 2.3.2. *"Old" and "New" Bronchopulmonary Dysplasia* (OBPD and NBPD) (Fig. 2.8)
> 2.3.3. *Atelectasis* (focal and diffuse)
> 2.3.4. *Cystic Fibrosis* (mucoviscidosis)

tory distress syndrome (NRDS). The surfactant consists of a combination of proteins and lipids, mostly phosphatidylcholine, which are secreted by type II pneumocytes. The four main proteinaceous components of the surfactant are SP-A, SP-B, SP-C, and SP-D. The corresponding genes are *SFTPA*, *SFTPB*, *SFTPC*, and *SFTPD*. Type II pneumocytes and alveolar macrophages are responsible for clearing the surfactant. SP-B and SP-C are hydrophobic proteins and have been seen playing a critical role in determining the right balance for surfactant tension activity and cellular homeostasis. SP-A and SP-D, encoding for collecting-like molecules, have not associated with pediatric ILDs. Biallelic mutations cause SP-B deficiency with AR-inherited transmission and have been associated with RDS in term neonates. SP-C, which is monoallelic and transmitted in half of the cases as AD and half of the affected patients being sporadic, has been associated with quite a broad spectrum of clinical manifestations, including neonatal RDS, ILD of infancy and childhood, familial pulmonary fibrosis of adulthood, recurrent lung infections, and bronchiectasis. ABCA3 is a lipid transporter associated with lamellar bodies in the pneumocyte type II. Biallelic homozygous mutations or double heterozygous mutations of the *ABCA3* gene, cause a deficiency of the protein, which is responsible for an early-onset RDS similar to SP-B deficiency but also observed not too rarely in adolescents with ILDs.

TTF-1 is a nuclear factor, which plays an essential role in lung, thyroid, and CNS maturation. TTF-1 plays a significant role in the expression of *SP-B*, *SP-C*, and *ABCA3*. Mutations or deletions that cause haploinsufficiency of the

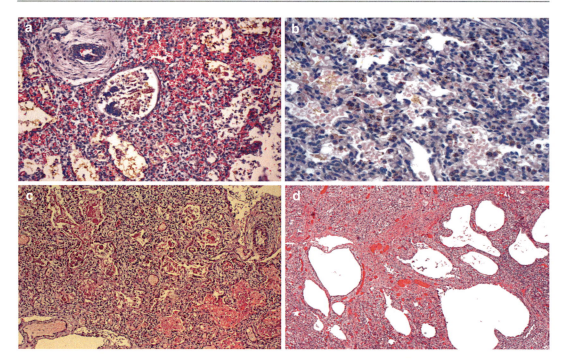

Fig. 2.8 Lung microphotographs (**a**, **b**) show meconium aspiration syndrome in an infant (H&E staining and PAS staining, 200× and 400×); Microphotograph (**c**) shows the lung histology of a classic bronchopulmonary dysplasia showing airway epithelial lesions, smooth muscular hyperplasia, extensive peribronchiolar and interstitial fibrosis, focal hemorrhage, areas of overdistension and atelectasis, and hypertensive vascular disease (H&E staining, 100×). In the microphotograph (**d**) is shown the lung histology of a 27 weeks' gestation premature baby with bronchopulmonary dysplasia type II or variant with significant remodeling of the alveolar architecture with focal overinflation (H&E staining 50×). Interstitial fibrosis and histologically mild pulmonary arterial hypertension (masculinization of small arteries) are also seen

gene encoding TTF-1, i.e., NKX2.1, have been found leading to variable combinations of respiratory failure, hypothyroidism, and brain-related disturbances, such as hypopituitarism and choreoathetosis, which is an extrapyramidal disorder characterized by combined involuntary movements of chorea (irregular contractions of migrating type) and athetosis (twisting and squirming). Haploinsufficiency is a genetic condition in which one copy of a gene is deleted or inactivated, while the remaining functional copy of the gene is present but remains inadequate to translate a sufficient gene product unarguably needed to preserve the normal function of the gene. The function of the surfactant is dual, including the function (1) to decrease the alveolar surface tension preventing the end-expiratory alveolar collapse and the function (2) to increase the armamentarium for an efficient anti-infectious defense. Gene mutations of *SFTPA*, *SFTPB*, *SFTPC*, and *SFTPD* are responsible for quantitative and qualitative alterations of intra-alveolar surfactant, ultimately leading to ineffective surfactant and abnormal clearance. The toxicity of the non-secreted proteins targets cells involved in its production. Granulocyte-macrophage colony-stimulating factor (GM-CSF) is a critical cytokine in the pathophysiology of abnormal surfactant clearance. GM-CSF has two subunits, alpha (α) and beta (β), which are coded by the respective genes, *CSF2RA* and *CSF2RB*. Mutations in these genes prevent GM-CSF signaling. GM-CSF is a growth factor for granulocytes and monocytes, stimulating differentiation, proliferation, and cell survival of myeloid cells, monocytes, neutrophils, and dendritic cells. Moreover, antibodies against GM-CSF can impair surfactant clearance due to hypoactive (hypofunctional) macrophages. The pulmonary surfactant composition can be easily determined and results of ~80% phospholipids, of which dipalmitoyl-phosphatidylcholine (DPPC) is ~60% and phosphatidylglycerol/ethanolamine/inositol is ~20%. Moreover, there are 10% neutral lipids (e.g., cholesterol, which is the most represented neutral lipid compound), 10% surfactant-associated proteins (SP-A, SP-B, SP-C, and SP-D), and <1% infitesimal traces of other substances.

SP-A (SFTPA) and SP-D (SFTPD) are hydrophilic proteins with CHO recognition domains that allow them to coat bacteria and viruses promoting phagocytosis by macrophages, while SP-B (SFTPB) and SP-C (SFTPC) are hydrophobic proteins that increase the rate the surfactant can spread over the surface. Surfactant synthesis is modulated by a variety of hormones and growth factors, including cortisol, insulin, prolactin, thyroxine, and TGF-β. The knowledge of factors and clinical scenarios potentially impairing surfactant synthesis is vital for both pediatrician and pathologist.

2.3.1 Neonatal Respiratory Distress Syndrome

- *DEF*: Multifactorial syndrome at neonatal onset with lower respiratory tract involvement due to absolute or relative, mostly global, surfactant deficiency and guarded prognosis if no treatment is undertaken (Fig. 2.8).
- *EPG*: (Causal/concausal) – hyaline membrane disease (HMD) of the preterm and appropriate for gestational age (AGA), intrauterine hypoxia (e.g., intrauterine pericervical cord coiling), aspiration of blood/amniotic fluid, fetal head (skull) injury during delivery, and excessive or inappropriate sedation of the mother.
- HMD: Preterm/AGA, ♂ > ♀, maternal diabetes, C-section mode of delivery, deficiency of pulmonary surfactant (lecithin or DPPC, phosphatidylglycerol, hydrophilic SP-A and SP-D, hydrophobic SP-B and SP-C).
- *DDX*: The differential diagnosis of RDS is meconium/amniotic fluid aspiration syndrome (MAS); HMD; bronchopulmonary dysplasia (BPD); congenital heart disease (CHD), especially cyanotic; congenital diaphragmatic defect (CDD); congenital pneumonia; pulmonary immaturity; positive end-expiratory pressure (PEEP)-related RD ("respiratory therapy-related); and transient tachypnea of the newborn (TTN).

The examination of the lungs is critical.

- *GRO*: Airless, and red-purple solid lungs prone to sink in water.
- *CLM*: Atelectasis with poorly developed alveoli + hyaline membranes (fibrin (PAS+) and cell debris chiefly from necrotic type II pneumocytes.
- *TEM*: The precise histological structure of the alveolar wall in HMD has been studied by Lauweryns (1965) using the electron microscope. The alveolo-capillary membrane is made up of three continuous and distinct layers: (1) external, the alveolar cells, (2) the interstitium, and (3) the capillary endothelium (flat endothelial cells). The hyaline membranes as seen by conventional microscopy lie directly upon the reticular fibers of the alveoli and alveolar ducts. This finding confirms the necrosis of the epithelial portion of the wall, and this hypothesis is reinforced by the remarkable and incontrovertible discovery in the same pulmonary tissue sections of areas with hyaline membranes and absent epithelium, along with normal alveoli with an epithelial lining but no hyaline membranes

O_2 *toxicity*: Retrolental fibroplasia (or retinopathy of prematurity) due to VEGF changes (initially decreased in hyper-O_2 status and then increased in hypo-O_2 status as room air ventilation) and bronchopulmonary dysplasia (*vide infra*).

2.3.2 "Old" and "New" Bronchopulmonary Dysplasia (OBPD and NBPD)

"Old"/classical type is defined by airway epithelial hyperplasia, squamous metaplasia, alveolar wall thickening, and peribronchial as well as interstitial fibrosis, while the "new" type shows a decrease of alveolar septation (i.e., large alveolar structures, which seem "simplified") and dysmorphic capillary configuration. New BPD infants are less than 30 weeks' gestation (usually ≤28 weeks) with BW <1250 g and an ♂ > ♀ predominance. The lower the weeks' gestational age is, the higher the risk is to develop the new BPD. Currently, this pathology occurs more than before and the reason may also lie to the increased survival in very low birth weight (VLBW) babies. Worldwide the prevalence of BPD is also increased (Fig. 2.8). Substantially, the severity of the new BPD is less than the old BPD.

2.3.3 Atelectasis

Atelectasis arises from two Greek words combined (ἀτελής, incomplete, and ἔκτασις, extension) and is a complete or partial collapse of a lung or lung lobe developing when the alveoli within the lung become deflated.

- *Neonatal atelectasis*: Incomplete expansion of the lung
- *Acquired atelectasis*: The collapse of the lung structure of the previously inflated organ

Acquired atelectasis may occur for several causes. In the case of resorption (obstruction), it is due to complete blockage of an airway (bronchial asthma, chronic bronchitis, bronchiectasis, post-op surgical immobilization, and foreign body aspiration). In the case of compression, the atelectasis is due to a partial/complete filling of the pleural cavity (exudate, tumor, blood, air). In the case of a contraction, it is expected to have local/generalized fibrotic changes that occur in the lung/pleura preventing full expansion. Some authors prefer using the term "*dys*telectasis" instead of the most commonly used term of *a*-telectasis for acquired atelectasis that has not been present at the time of birth, but this word seems to be more diffuse in some European countries than worldwide. Substantially, resorption is a process of lung volume decrease, which is linked to obstruction of the full type of an airway directing to a lung section and may be due to bronchial asthma, chronic bronchitis, bronchiectasis, post-surgery, or foreign body aspiration (FBA)-associated bronchial ways obstruction. Compression is a process of lung volume decrease due to partial or complete filling of the pleural cavity by solid (e.g., tumor), liquid (e.g., exudate or blood), or air (e.g., pneumothorax). Finally, contraction is a process of lung volume decrease due to local or generalized fibrotic changes of the lung parenchyma or of the pleura preventing full expansion of the lung. These concepts are essential to understand the radiograph or CT of the lung and later to understand the basis for the chronic obstructive pulmonary disease of the adult.

2.3.4 Cystic Fibrosis

Cystic fibrosis is a quite frequent disease with multivisceral involvement that both the medical student and the physician come across often both during the years of medical studies since biology and during the medical practice across several specializations.

- *DEF*: Progressive, a genetic disease that causes persistent lung infections with progressive respiratory failure.
- *SYN*: Mucoviscidosis.
- *GEN*: AR, carrier: 1/20 (white Caucasians), CFTR gene (7q22–7q31) with variable penetrance.
- *EPG*: Gene mutations (mostly Delta Phe508 or ΔP508) hampering the correct function of the gene product, which is a Cl^- transporter resulting in defective anion (X^-) transport, particularly Cl^- and HCO_3^-, with subsequent abnormal H_2O movement through the plasmatic membrane.
- *CLI*:
  1. Lung: ↓ Cl^- transport into lumen results in viscous secretions, which plug airways and determine progressively lung disturbed function aiming to usual interstitial pneumonia (UIP)
  2. Pancreas: Pancreas insufficiency
  3. Liver: Cholestatic liver disease
  4. Testis: Infertility

> **Box 2.6 Infantile and Pediatric NILD**
> 2.4. *Pediatric Interstitial Lung Diseases (ILD)*
>   2.4.1. *Diffuse Lung Development-Associated ILD*
>   2.4.1.1. Congenital Acinar Dysgenesis/Dysplasia
>   2.4.1.2. Congenital Alveolar Dysgenesis/Dysplasia
>   2.4.1.3. Alveolar Capillary Dysplasia With/Without Misalignment of the Pulmonary Veins
>   2.4.2. *Trisomy 21 (Down Syndrome)-Associated ILD*
>   2.4.3. *ILD Related to Other Chromosomal/Genomic Microdeletion Disorders*
>   2.4.4. *Neuroendocrine Hyperplasia of Infancy (NEHI)*
>   2.4.5. *Pulmonary Interstitial Glycogenosis (PIG)*
>   2.4.6. *Surfactant Dysfunction Disorders (SDDs)*
>   2.4.6.1. Dysfunctional SP-B-Related ILD
>   2.4.6.2. Dysfunctional SP-C-Related ILD
>   2.4.6.3. Dysfunctional ABCA3-Related ILD
>   2.4.6.4. Lysinuric Protein Intolerance
>   2.4.7. *Pulmonary Alveolar Proteinosis*

5. Skin: Oversaturated (hypersaline) sweat due to absent (somewhat rare) or ↓Cl⁻ resorption from secretions – sweat high in NaCl!
- *TRT*: Symptomatic until lung transplantation.
- *PGN*: Currently, CF remains one of the main reasons for lung transplantation in the young population. There is also an increased risk for lung infections by *Staphylococcus* and *Pseudomonas* spp. These infections are not little things but may be often fatal.

2.4 Infantile and Pediatric *ILD*

There are lung diseases at infantile and pediatric onset with surface epithelial involvement and direct interstitial involvement. Also, adult interstitial lung diseases may also occur in the adolescent and youth. They need to be engaged into account according to the family history and environmental connection and exposure (Box 2.6).

2.4.1 Diffuse Lung Development-Associated ILD

Interstitial lung disease associated with a diffuse lung development may have different patterns that need to be recognized by light microscopy. However, immunology plays a significant role in pulmonology. Thus, concepts of innate versus adaptive immunity should be evident in mind investigating a pulmonary pathology. Central lymphoid organs constitute tissues of the immune system (e.g., thymus), peripheral lymphoid organs (e.g., BALT, i.e., bronchial mucosa-associated lymphatic tissue), and lymphocytes of the circulating blood or fluids. HLA (human leukocyte antigens) and disease association are well known (e.g., B27, DR3, DR4, etc.). Cytokines of innate immunity are TNF, IL-1, IL-12, type I IFNs, IFN-γ, and chemokines, while the cytokines of adaptive immunity are IL-2, IL-4, IL5, IL-17, and IFN-γ. Subsets of T_H cells comprise T_{H1}, which produce IFN-γ. IFN-γ is responsible for MΦ activation and IgG production. Another subset comprise T_{H2}, which secrete IL-4, IL-5, and IL-13 and are accountable for the recruitment of eosinophils with IgE involvement in the process. Finally, the T_{H17} subset produce IL-17, IL-22, and chemokines, which are responsible for PMN and monocytes/MΦ recruitment.

T_{H1} is vital for host defense against intracellular microbes, while T_{H2} is central for host defense against helminths, and T_{H17} is crucial for fighting extracellular bacteria and fungi. Humoral immunity is played by naive B lymphocytes recognizing antigens and under the influence of T_H cell subsets as well as other stimuli. During the humoral immunity, the B lymphocytes are activated to proliferate and differentiate into Ab-secreting cells, i.e., plasma cells. Some of the activated B lymphocyte undergo heavy chain class switching and affinity maturation, and some become the cell population subset, known as memory cells. Hypersensitivity reactions may be life-threatening. They include allergen-associated, cellular-associated, Ig-associated, and delayed-associated reactions.

In a children's hospital clinic-pathologic conferences or multiteam disciplinary meetings, by the discussion of a child with respiratory disorders of specific gravity, it is often discussed the necessity to gather some lung tissue for histology and molecular studies. It is usually a discussion if it is convenient or appropriate to perform an open lung biopsy or a transbronchial lung biopsy. About the lung biopsy, timing is crucial (e.g., immunosuppression), and late-stage biopsy may be inconclusive. The decision is ultimately individualized. Transbronchial biopsy is feasible in the setting of eosinophilic pneumonia, lymph-angioleiomyomatosis (LAM), pulmonary Langerhans cell histiocytosis, sarcoidosis, vasculitis, malignancy, and lung transplantation. The lung biopsy interpretation relies on the pathology of the architecture (airspace, interstitium, mixed), distribution of the process (diffuse, mono-/bilaterality, severity variation), pattern (specificity, randomness), location (bronchiolar/bronchocentric), and lymphovascular involvement.

Desquamative interstitial pneumonia (DIP) is characterized by a diffuse alveolar presence of macrophages, monotonous, uniform pattern, limited thickening alveolar walls, and no alveolar debris/fibrin. Nonspecific interstitial pneumonia (NSIP) is a diffuse interstitial process, uniform in the stage of pathologic changes with mild inflammatory and fibrotic interstitial changes, no intra-alveolar exudate, and remarkably better prognosis than usual interstitial pneumonia/pneumonitis (UIP). Indeed, UIP is characterized by the heterogeneity of the pathologic changes with often confounding hyaline membranes in the early stages, diffuse alveolitis (lymphocytes, macrophages) and lymphoid aggregates, thickening of alveolar walls (↑ fibroblasts, smooth muscle, and collagen), and fibroblastic foci in the interstitium in all stages. The acute interstitial pneumonia (AIP) is categorized as diffuse alveolar damage, and the pathologic changes are uniform, focal hyaline membranes, mild alveolitis (lymphocytes, macrophages), some thickening of alveolar walls (diffuse proliferation of fibroblasts), but no obvious collagen fibrosis. Pathologic changes leading to end-stage lung fibrosis (honeycomb lung) are interstitial pneumonia, diffuse alveolar damage (DAD), inorganic dust exposure, interstitial granulomatous disease (infections, extrinsic allergic alveolitis, sarcoidosis, berylliosis), Langerhans cell histiocytosis, and prolonged or untreated severe GERD. In Box 2.7 are presented the salient features of the ILD of adult type.

2.4.1.1 Congenital Acinar Dysgenesis/Dysplasia

- *DEF*: Primary developmental lung abnormality with a severe form of hypoplasia and abnormal lung architecture with acini with incomplete formation of airspaces. It repre-

Box 2.7 Salient Features of the ILD of Adult Type

Feature	AIP	DIP/RBILD	NSIP	UIP
Course	Uniform	Uniform	Uniform	"Capricious"
Architecture change	–	–	–	"Honeycomb"
Alveolar MΦ	–	++ (peribronchiolar)	–/+	–/+
Interstitial pneumonitis	–/+	–/+	+++	–/+
Fibrosis	–	~ (variable)	~	~ (irregular)
Fibroblast foci	–	–	– (habitually)	+++
Organizing pattern	–	–	– (habitually)	–
HMD	–/+	–	–	–

sents an arrest of growth of the lung bud with architecture like ("stopped to") the pseudoglandular stage of development.
- *EPI*: Very rare, ♂ < ♀.
- *EPG*: Disturbance of the epithelial-mesenchymal interaction with genetic linkage including TBX4 or T-box transcription factor mutations, de novo missense mutation p.E86Q (c.256G > C) in DNA binding, homozygous *FGFR2* mutations in ectrodactyly-pulmonary acinar dysplasia, of which a missense mutation may be found in the IgIII domain (D3) (p.R255Q).
- *CLI*: Associated congenital defects are not typical (e.g., right-sided aortic arch, renal dysplasia) ± pregnancy complicated by chronic placental abruption with fetal adrenal hemorrhage and cerebral infarcts of the baby.
- *PGN*: Uniformly lethal.

2.4.1.2 Congenital Alveolar Dysgenesis/Dysplasia

- *DEF*: Primary developmental lung abnormality with the incomplete acinar development of the pulmonary architecture with an arrest of growth of the lung bud at the level of the saccular stage of lung development.
- *EPI*: Rare disease, usually term infants.
- *EPG*: Genetic etiology unknown.
- *PGN*: Longer survival than acinar dysplasia.

2.4.1.3 Alveolar Capillary Dysplasia With/Without Misalignment of Pulmonary Veins (ACD/MPV)

- *DEF*: Disorder affecting the development of both lungs and its vascular supply characterized by a drastic and dramatic reduction of capillaries with an abnormal location of the capillaries within the walls of the alveoli and consequent deficiency of the O_2 exchange. It may be associated with MPV (Misalignment of Pulmonary Veins) when the pulmonary veins are improperly associated with the pulmonary arteries showing the adventitial arterial layer encasing the pulmonary veins (Fig. 2.9).
- *CLI*: Cause of persistent pulmonary hypertension in the newborn (PPHN) with infants at term or near term presenting with respiratory distress in the first few days of life. ACD/MPV can present with associated defects, including gastrointestinal (GI) malrotation limited to the large bowel, other GI abnormalities, cardiovascular, and genitourinary abnormalities.
- *EPG*: Two genetic links, including (1) mutations of the *FOXF1* gene (transcription factor), essential for the development of the lungs and blood vascular supply. *FOXF1* is also involved in the development of the GI tract and (2) chromosome deletion on 16q24.1.
- *CLM*: Deficient numbers of capillaries, centrally placed in widened alveolar walls, alveolar maldevelopment, muscularized pulmonary arterioles, malposition of veins and venules within lobules, and variable lymphangiectasia.

2.4.2 Trisomy 21 (Down Syndrome)-Associated ILD

- *DEF*: Multifactorial syndrome at neonatal onset with lower respiratory tract involvement and considered to be reduced formation of airspaces and alveoli at the periphery of the lung. Lung pathology in Down syndrome is typified by small size with ↓ of lung volume and ↓ rate of airway generation, RACs, total number of alveoli, and alveolar surface area and ↑ mean linear intercept, large alveoli and alveolar ducts with a double capillary layer of the alveolar wall ("double alveolar capillary network"), alveolar hypoplasia with deficiency of the elastic fibers in the alveolar walls, and subpleural cysts (Ø: 0.1–0.2 cm) with some fibrosis in the intercystic spaces.

Moreover, infants with Down syndrome and structural CHD may develop earlier and progress more rapidly to irreversible pulmonary vascular changes than infants without Down syndrome. The more rapid development of pulmonary hypertension in infants with Down syndrome is probably due to some specific reasons, including chronic upper airway obstruction (e.g., tracheomalacia and tracheal bronchus "malacia"), recurrent pulmonary infection, and alveolar hypoventilation. The development of

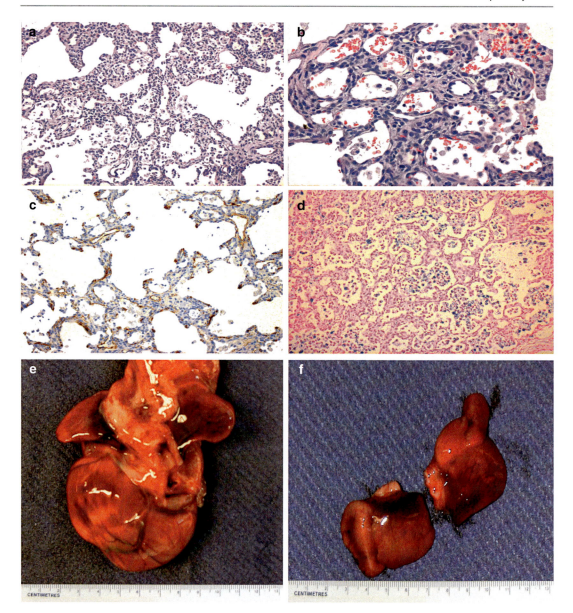

Fig. 2.9 Lung microphotographs (**a–d**) show capillary alveolar dysplasia with abnormal blood vessels in the lungs that do not allow proper blood oxygenation (**a**, H&E staining, 200×; **b**, H&E staining, 400×; **c**, immunohistochemistry with antibody against smooth muscle actin, 200×; Perls' Prussian blue for Fe, 200×); Gross photographs (**e**) and (**f**) show bilateral lung hypoplasia in a patient with congenital diaphragmatic hernia due to a congenital diaphragmatic defect. Note the hypoplasia of both lungs that have similar size

progressive pulmonary hypertension (PPHN) in Down syndrome is associated with failure of neonatal pulmonary vascular remodeling in such infants. Criteria of PPHN include hypoxemia refractory to O_2 therapy or lung recruitment strategies, which are achieved by conventional ventilation and high positive end-expiratory pressure (PEEP) initially and now replaced by high-frequency oscillatory ventilation later, as well as the presence of extrapulmonary shunt at atrial or ductal level in the lack of severe pulmonary parenchymal disease. Early development of the pulmonary vascular disease is seen in 1/10 to almost all children with Down syndrome and affected with structural CHD. RACs are usually evaluated by dropping a perpendicular line from the respiratory bronchiolar geometrical center to the nearest pleura or septum and counting the number of alveolar walls intersected according to Emery-Mithal instructions ("Emery-Mithal counts"). Schloo et al. evaluated lungs of patients with Down syndrome. These authors found five being 143–162% above expected (of which four to above adult values), five were as expected, and one was below expected (82%). In addition to abnormal RACs, Schloo et al. also found disturbances in alveolar multiplication and conclude that the reduction in airway branching in the lungs of patients with Down syndrome suggests interference with development before birth.

2.4.3 ILD Related to Other Chromosomal/Genomic Microdeletion Disorders

Other than trisomy 21 syndrome, there are some disorders with either chromosomal aberration or microdeletions of the genome that may show interstitial lung disease in infancy and childhood. Microdeletions with ILD include (but are not limited) thrombocytopenia-absent radius syndrome and 1q21.1 deletion, Prader-Willi syndrome due to chromosome 15q11.2-14 deletion, and the proximal chromosome 14q microdeletion syndrome.

2.4.4 Neuroendocrine Hyperplasia of Infancy (NEHI)

- *DEF*: Interstitial lung disease with chronic hypoxemia, concomitant persistent tachypnea, and characteristic appearance on CT imaging plus hyperplasia of neuroendocrine cells (pathology) in the lungs of infants (Fig. 2.10).
- *EPI*: Pulmonary disorder affecting infants less than 2 years of age but unknown detailed epidemiologic data.
- *EPG*: Neuroendocrine hyperplasia of unknown etiology with familial cases of NEHI suggesting genetic susceptibility. However, thus far no specific genetic defect has been identified except for a single example of

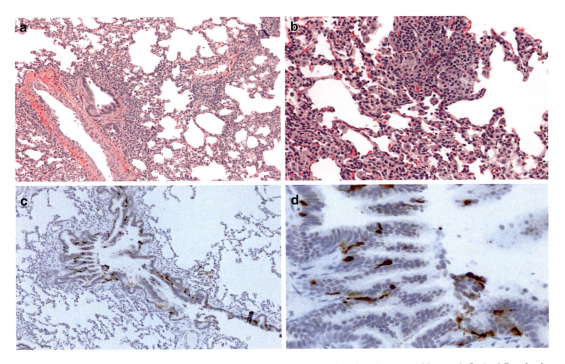

Fig. 2.10 Neuroendocrine hyperplasia of infancy (NEHI) showing NEHI with peribronchiolitis-bronchiolitis (**a**, H&E at 100×; **b**, H&E at 200×; **c**, Anti-Bombesin Immunohistochemistry at 100×; and **d**, Anti-Bombesin Immunohistochemistry at 400×) (Avidin-Biotin Complex)

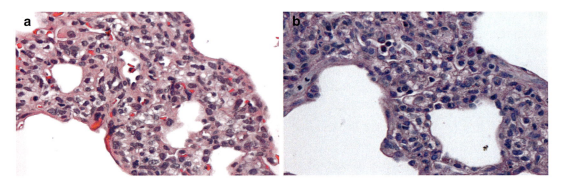

Fig. 2.11 Pulmonary interstitial glycogenosis (PIG) showing a thickening of the alveolar septs (**a**, H&E, 400×), which is also partly Periodic Acid Schiff-positive (**b**, PAS, 400×)

heterozygous *NKX2.1* mutation identified in a proband with a diagnosis of NEHI.
- *CLI*: Children with NEHI have normal KL-6 levels, a marker of regenerating airway and alveolar epithelial cells and lack pro-inflammatory cytokine profile in BAL samples, thus distinguishing this condition from other forms of pediatric ILD.
- *CLM*: ↑# of bombesin/gastrin-releasing peptide (GRP) and 5-HT (serotonin) immunopositive primary NE cells without other abnormalities and limited or absent inflammation.
- *IMG*: High-resolution chest CT documents patchy areas of ground-glass opacity with remarkable mosaic regions in air trapping.
- *TRT*: Mostly O_2 to maintain adequate O_2 saturation, with some receiving bronchodilators and systemic steroids, although the choice is quite individual.
- *PGN*: All patients with NEHI reported thus far in the literature are apparently alive and well at 8–10 years of age, although some still presenting mild residual symptoms. The long-term prognosis is, however, unknown. The possibility of evolution of NEHI into an adult form of diffuse idiopathic pulmonary neuroendocrine cell hyperplasia (DIPNECH) has been suggested.

2.4.5 Pulmonary Interstitial Glycogenosis (PIG)

The glycogen is a component frequently encountered in lung tissue of fetal age. Pulmonary interstitial glycogenosis (PIG) is not a life-threatening disorder but started to play a significant role in PICU in the last couple of decades.

- *DEF*: Congenital respiratory syndrome occurring at neonatal age with lower respiratory tract involvement due to absolute or relative, mostly global, deficiency of enzymes related to glycogen metabolism in interstitial fibroblasts and variably associated with congenital heart disease and PPHN.
- *CLI*: Variable clinical pictures ranging from mild disturbances of the lung function through respiratory failure.
- *CLM*: Presence of glycogen in the interstitium better highlighted using periodic acid Schiff (PAS) staining (Fig. 2.11).
- *TEM*: Primitive interstitial cells with numerous monoparticulate glycogen and few organelles, but no abnormal glycogen.
- *TRT*: Corticosteroid-responsive, due to the induction and acceleration of the interstitial cell maturation.
- *PGN*: Good.

2.4.6 Surfactant Dysfunction Disorders

Multifactorial syndrome at neonatal onset with lower respiratory tract involvement due to absolute or relative, mostly diffuse, deficiency of factors playing a critical role in surfactant tension dynamic properties. The *SFTPB* gene indel g.1549C > GAA (121ins2) is responsible for about 60% of the mutant alleles underlying complete surfactant protein B deficiency, which shows AR inheritance. Somatic recombination and about 50% spontaneous mutation account for the monogenic disorder, which is autosomal dominantly inherited. ATP-binding cassette, subfamily A, member 3 (*ABCA3*) is characterized by AR inheritance and uniparental disomy (UPD), while the NK2 homeobox 1 (*NKX2-1*) monogenic cause of surfactant deficiency shows an AD inheritance and harbors about 50% spontaneous mutations. In UPD, both copies of a chromosome pair or portions of a chromosome pair are inherited from a single parent with heterodisomy (inheritance of two homologous chromosomes from the same parent) or isodisomy (inheritance of two copies of the same chromosome from the same parent).

2.4.6.1 Dysfunctional SP-B-Related ILD

- *DEF*: Surfactant dysfunction disorder encompassing a variety of mutations involving the gene that encodes surfactant protein B (*SFTPB*).
- *CLM*: There is widened interstitial space with fibroblast proliferation, the arrest of the acinar development with a remarkable decrease of the number of alveoli, cuboid metaplasia of the alveolar epithelium, and alveolar lumen filled with eosinophilic proteinaceous material, which contains macrophages and desquamated epithelial cells (DIP-like or PAP-like pattern according to different authors who have intensely studied this entity; DIP, desquamative interstitial pneumonia/pneumonitis; PAP, pulomnary alveolar proteinosis).
- *TEM*: Large cytoplasmic inclusions with irregularly arranged lamellae associated with membranous and vesicular structures.

2.4.6.2 Dysfunctional SP-C-Related ILD

- *DEF*: Surfactant dysfunction disorder encompassing a variety of mutations involving the gene that encodes surfactant protein C (*SFTPC*) with the phenotype caused by aberrant surfactant protein folding, decreased endogenous SP-C secretion, ER stress, and apoptosis of alveolar epithelial cells.
- *CLI*: The lung disease associated with *SFTPC* mutations is characterized by marked phenotypic heterogeneity with a range of severe respiratory distress in infants to NSIP and UIP/idiopathic pulmonary fibrosis (IPF) in older adults.
- *CLM*: There is DAD, interstitial thickening with fibroblast proliferation and collagen deposition, numerous foamy macrophages and granular lipoprotein material in the alveolar lumina, and type II pneumocyte hyperplasia (AEC2) corresponding to DIP-like or PAP-like patterns according to the authors. Multiple fibrosis foci have also been encountered.
- *TEM*: ER stress and apoptosis of alveolar epithelial cells.

2.4.6.3 Dysfunctional ABCA3-Related ILD

- *DEF*: Surfactant dysfunction disorder encompassing a variety of mutations involving the gene that encodes ATP-binding cassette transporter A3 (*ABCA3*).
- *CLM*: Patterns with DIP- or PAP-like appearance like either SP-B or SP-C deficiencies.
- *TEM*: Abnormal lamellar body (LB) with tightly packed concentric membranes and a cluster of electron-dense material ("fried eggs").

2.4.6.4 Lysinuric Protein Intolerance

- *DEF*: AR inherited disease caused by an intrinsic defect of cationic amino acid transport. LIP is etiologically associated with mutations of the *SLC7A7* gene, the protein of which is located at the membrane of epithelial cells in the intestine and kidney. The expression of *SLC7A7* is a target of GM-CSF that may explain the reduced activities of the alveolar macrophages and peripheral blood monocytes of patients harboring this gene defect.
- *CLI*: Most patients present in infancy with failure to thrive, growth retardation, protein aversion, muscular hypotonia, hepatosplenomegaly, and osteoporosis. Lysinuric protein intolerance is diagnosed either by gene sequencing or by the presence of excessive

amounts of dibasic amino acids (Arginine, Lysine, Ornithine) in the urine, particularly following protein ingestion (e.g., red meat ingestion-based test).
- *CT/MRI*: Patients who developed a fatal respiratory insufficiency are usually children younger than 15 years of age, and adult patients with ILD on CT scans of the chest show inter- and intralobular septal thickening and subpleural cysts, but only a few are symptomatic.

A combined genetic and morphologic approach to SF deficiency disorders includes intradepartmental and extra-institutional consultation, whole blood or a buccal swab from affected infant ± parents for further studies, molecular biology investigation targeting the hotspot mutations and genes, sequencing of the entire exomes, and the lung biopsy. The lung tissue biopsy received at the Department of Pathology should be processed according to the ChILD guidelines. Medical consideration in infants with surfactant deficiency includes genetic diagnosis essential to inform the choice of medical interventions, evaluation of extrapulmonary organ dysfunction, unavoidable CNS-hypoxia changes and CNS anomalies, recurrence risk in siblings, and ability to be transferred to a lung transplantation center. Family considerations include progression of lung failure, adequate provision of information about the diagnosis and prognosis, and discussion of compassionate care in selected cases unsuitable for lung transplantation (e.g., numerous CNS-hypoxia-related changes, multiple congenital anomalies).

2.4.7 Pulmonary Alveolar Proteinosis

- *DEF*: Lung disease with a restrictive ventilatory defect (↓ lung volume and capacity, hypoxemia with ↑P(A-a) O_2) characterized by the accumulation of surfactant-like (lipo-)proteinaceous material in the alveoli due to abnormal homeostasis or clearance by the alveolar MΦ.
- *SYN*: Pulmonary alveolar lipoproteinosis.
- *EPI*: Rare, 20–50 years, ♂:♀ = 2:1.
- *EPG*: Abnormal surfactant homeostasis due to defective chemotaxis, phagocytosis, and phagolysosomal fusion of alveolar MΦ with broad pathogenic categories.
- *CLI*: In children, the presentation may be pulmonary (e.g., cyanosis) or extrapulmonary with diarrhea, vomiting, and failure to thrive, while in adults slowly progressive pulmonary infiltrates, dyspnea, cough, sputum, and fever are more typical, but 1/3 of patients may be asymptomatic.
- There is ↑ P(A-a) O_2 with shunting of blood from right to left through an entire pulmonary capillary bed (*sine qua non conditio*), differently from poorly ventilated alveoli filled with material. If PAP presents in infants <1 year, there is an association with *thymic alymphoplasia*.
- *CXR*: Airspace consolidation ("*bat's wings pattern*" = bilateral central symmetrical opacities with costophrenic and apical sparing) to diffuse poorly delimited small acinar nodules with ground-glass opacification ("*military pattern*"). The "*bat's wings pattern*" is typical in adults, while the "*military pattern*" with diffuse opacities is mostly common in children.
- *HCT*: Airspace consolidation ("*crazy paving pattern*") ± thickening of interlobular septs (ILS).
- *LAB*: ↑GM-CSF, ↑LDH, ↑KL6, ↑SP-A, ↑SP-D ± ↑tumor markers, depending on the type.
- *DGN*: BAL + HCT (± TEM on BAL) ± TBB/OLB. If either a transbronchial biopsy (TBB) or open lung biopsy (OLB) is performed, findings may be inconclusive, because of the patchiness of the alveolar process.
- *GRO*: Patchy areas of yellow parenchymal consolidation with some oily material exuding following cutting the lung (post-transplantation, autopsy) or OLB, which is rarely required.
- *CLM*: (1) ± (2) features, including:
  1. *Alveolar space filling with PAS(+)/DPAS(+)/AB(−)-(lipo)-proteinaceous*

material with normal interstitial pulmonary architecture ≠ mucus, which is PAS(+)/DPAS(−)/AB(+)
2. *Septal thickening* (pauci-cellular = edema, cellular = lymphocytic infiltration)
- *BAL*: "Milky Fluid", due to a basophilic granular extracellular deposition with some enlarged foamy MΦ and cellular debris on cytology.
- *TEM*: Abundant cellular debris and LBs ± sparse tubular myelin structures.
- *DDX*: Airspace consolidation ("*crazy paving pattern*") ± thickening of ILS may raise the differential of lipid pneumonia or bronchoalveolar carcinoma (BAC) or other ILDs.
- *TRT*: Whole lung lavage, GM-CSF, rituximab (a chimeric monoclonal antibody against the protein CD20, which is primarily found on the surface of B lymphocytes), plasmapheresis, targeting the underlying disorder, e.g., cancer.
- *PGN*: Excellent (25%, spontaneous resolution), but it depends exquisitely on the pathogenic type.

PAP categories include a congenital type (~2%, neonates or also known as "chronic pneumonitis of infancy" with inherent SF-abnormality), a secondary type (~8%, children, and adults, due to anti-granulocyte-macrophage colony-stimulating factor (anti-GM-CSF) antibodies or other precipitating diseases, including hematological neoplasms, immunodeficiency, immunosuppression, and pneumoconiosis), and an idiopathic type (~90%, adults/youth or acquired).

PAP has been reported following inhalation of both mineral particles (silica, talc, cement, kaolin), metals (aluminum, titanium, indium), and organic particles (fibers of cellulose). The massive inhalation of silica is also called silicoproteinosis. In summary, genetic PAP includes mutations in the genes, *SFTB*, *SFTC*, ATP-binding cassette 3 (*ABCA3*), NK2 homeobox 1 (*NKX2-1*), and *CSFRA* and *CSFRB*. Lung histology of patients with mutations of *SFTB*, *SFTC*, *ABCA3*, *NKX2-1* is remarkably similar showing varying degrees of thickening of the alveolar interstitium and remodeling of the alveolar epithelium. There is type II cell hyperplasia in addition to the intra-alveolar accumulation of eosinophilic, PAS (+) lipo-proteinaceous granular material, while in patients with *CSFRA* and *CSFRB* mutations, there is the relative integrity of alveolar architecture with only the accumulation being identifiable. *CSFRA* mutations have been described in children, while *CSFRB* has been described in both children and adults. As mentioned above, SP-A and SP-D bind to bacteria, viruses, and fungi facilitating the uptake of these microorganisms by the alveolar MΦ. The binding of SP-A and SP-D to the collecting receptor on alveolar MΦ determines an oxidative burst and determines an enhancement of phagocytosis. The lipo-proteinaceous material also acts as a scavenger for free radicals and hereafter impairs the oxidative burst that occurs in the alveolar MΦ during the phagocytosis. Of note, asphyxia may occur in a patient during general anesthesia with alveolar flooding. The PAP-associated risks are asphyxia, emphysema, fibrosis, infection, and pneumothorax. If the patient is scheduled for lung transplantation, a checklist should be filled including number and duration (written documentation) of meetings with parents/guardian about 5-YSR/outcome, no extrapulmonary organ dysfunction, optimization of enteral nutritional status, patient stable enough to be transported to a pediatric lung transplantation center, and the written documentation of the use of HFOV (high-frequency oscillatory ventilation).

2.5 "Adult" ILD of the Youth

As indicated above "adult" interstitial lung disease may be present in the adolescent and young male or female and need to be taken into account according to age, physical examination, family history, and environmental issues.

2.5.1 Diffuse Alveolar Damage (DAD)

- *DEF*: A multifactorial syndrome with common grossly and microscopic sequelae. It is linked to severe acute lung injury (ALI) with

acute onset of dyspnea following a known etiologic agent (in the absence of cardiac failure) and potential progression to adult respiratory distress syndrome (ARDS) with severe hypoxemia, a decrease of lung compliance, and a high mortality rate (50%) (Fig. 2.12).
- *SYN*: Acute lung injury (ALI), acute interstitial pneumonia (AIP).
- *EPG*: *ARDS* mnemonic for:
  - *A*spiration (gastric content), *A*ir embolism, and *A*mniotic fluid embolism
  - *R*adiation-Trauma-Burns (physics)
  - *D*IC, *D*rugs (chemotherapy, heroin), near *D*rowning, *D*ialysis, *D*ysregulation of metabolic pathways (pancreatitis, uremia, and ingestion of paraquat)
  - *S*hock, *S*epsis, and *S*moke inhalation

In determining the etiology of DAD/AIP, clinical history and radiographic imaging are paramount (Box 2.8). DAD is an abnormal (pathologic) respiratory pattern with a multi-etiologic background. In the event of an idiopathic etiology/setting DAD is termed acute interstitial pneumonia, although both names remain often interchangeable, at least for some extent and for most authors and pulmonary specialists. There is a combined endothelial and epithelial injury with bilateral and diffuse lung involvement. Pulmonary capillary wedge pressure (PCWP) represents the measurement of this pressure and is quite diffuse in the medical literature, although balloon occlusion occurs in the pulmonary arteries. Thus, the value obtained at time of the procedure is not indeed equal to the pulmonary capillary pressure in nonoccluded areas. Since the term PCWP is misleading, the term "pulmonary artery occlusion pressure". PCWP ≤18 mmHg excludes cardiogenic causes of lung edema. Nuclear factor kappa-light-chain-enhancer of activated B cells, labeled as NF-κB, is the key molecule. NF-κB is a protein complex found in almost all cell types of animals and is basilar to control DNA transcription. NF-κB is mainly involved in cellular responses to diverse stimuli such as general cellular stress, some cytokines, free radicals, UV irradiation, oxidized LDL, and several antigens of bacteria and viruses. Also, NF-κB plays a key role in regulating the immune response to infection being light chains of kappa-type critical regulators for the formation of Ig. NF-κB dysregulation has been linked to abnormal cellular responses leading to cancer, chronic inflammation, autoimmunity, and septic shock.

↑ NF-κB ⇒ increase of IL1, IL8, and TNF-α ⇒ endothelial cells and PMN activation ⇒ microvascular sequelae with an increase of leukotrienes, platelet activating factor, and other proteases.

- *CLI*: Abrupt onset of severe, unresponsive *arterial hypoxemia + CXR* variable from almost normal (beginning) to diffuse bilateral infiltrates (advanced stage).
- *HCT*: In the early phase, lung consolidation developing into bilateral diffuse ground-glass opacities and in the late stage there is cyst formation and honeycombing pattern of diffuse type.
- *GRO*: Heavy, congested solid lungs.
- *CLM*: Essentially, two stages:
  - *Exudative* (1st week, interstitial and intraalveolar edema, hemorrhage, fibrin deposition, and microthrombi in small blood vessels and capillaries, hyaline membranes (from the 3rd day after insult) and sloughing of alveolar cells with BM denudation.
  - *Proliferative* (2nd week, interstitial inflammation and fibroblast proliferation and interstitial fibrosis with organization/phagocytosis of the hyaline membranes, type II alveolar cells hyperplasia, and reactive atypical squamous metaplasia following bronchiolar damage). Some fragments of the hyaline membranes may remain in the alveolar lumina and are incorporated into the interstitium, and the alveoli are lined by reactive alveolar lining cells with hobnail phenotype. At this time, mitoses as well as marked reactive atypia with enlarged nuclei and prominent nucleoli as well as squamous metaplasia of bronchiolar and adjacent airspaces can be observed.

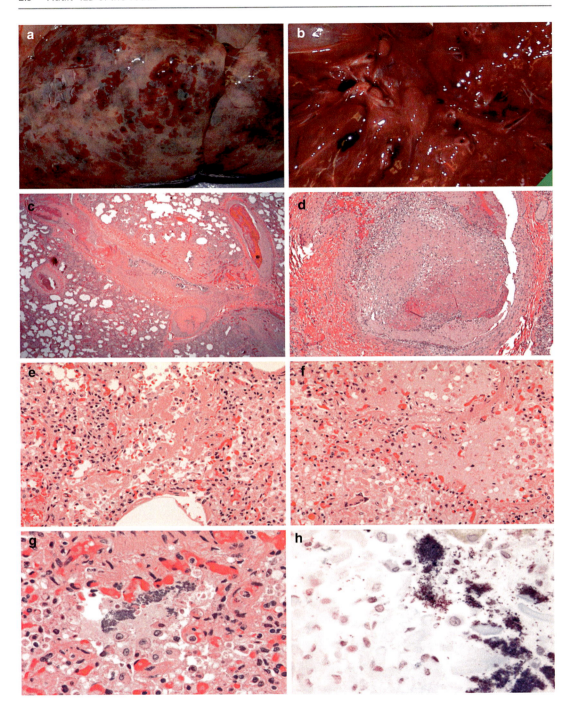

Fig. 2.12 Diffuse alveolar damage (DAD) and infections in a patient with *Candida spp.* and *Pseudomonas spp.* infection. The areas of discoloration with punctate hemorrhages are present in both gross photographs (**a**) and (**b**). Microphotographs (**c**–**h**) show thrombi as well as diffuse alveolar damage and Gram-positive and Gram-negative colonization of the thrombi (**c**, H&E, 12.5; **d**, H&E, 50×; **e** and **f**, H&E at 200×; **g**, H&E at 400×, and **h**, Gram staining at 630×)

Box 2.8 AIP Versus UIP Versus NSIP Versus COP

	DAD/AIP	UIP	NSIP	COP
Acute onset	+	−	−	+
Uniformity	+	−	+	−
Diffuseness	+	− (patchy)	+	− (patchy)
Honeycombing	−	+	−	−
HM	+	−	−	−
Eosinophils	−/+	−/+	−/+	−/+

Notes: *DAD/AIP* diffuse alveolar damage, *UIP* usual interstitial pneumonia, *NSIP* nonspecific interstitial pneumonia, *COP* cryptogenic organizing pneumonia, *AEP* acute eosinophilic pneumonia, *HM* hyaline membranes. AEP is characterized by numerous eosinophils

- *PGN*: Etiology-dependent, lung interstitial fibrosis-dependent, and variable survival rates (6-MSR: ~50%).
- *TRT*: Mechanical ventilation and hemodynamic support are the mainstay of DAD. Critical is the recognition that the fluid is entrapped in the lung by a diffuse fibrin gel that cannot be mobilized by diuresis, and perfusion pressures should be kept with near-normal wedge pressures. Moreover, overdiuresis, which can theoretically deplete intravascular volume and decrease right heart filling pressures, needs to be absolutely avoided.

Hamman-Rich disease: Rapidly progressive, idiopathic form of ALI of young adults with bad PGN (2-MSR: ~50%) that may follow flu or flu-like illness and be characterized by diffuse temporally synchronized and hypocollagenous fibroblast proliferation, proteinaceous exudate, and hyaline membrane deposition.

2.5.2 Cryptogenic Organizing Pneumonia (COP)

- *DEF*: Multifactorial *diffuse ILD*, *temporally uniform* and *bronchiolo-centric* distributed, which is characterized by intra-airspace plugs of granulation tissue and showing filling proliferative, branching or serpiginous, polypoid *fibroblastic plugs* (Masson bodies) with mild *interstitial inflammation*, and normal adjacent parenchyma (Figs. 2.13 and 2.14).
- *SYN*: Bronchiolitis obliterans organizing pneumonia (BOOP). The term "BOOP" was previously known in the literature but it is important to remember that an obliterating bronchiolitis can or cannot be present.
- *EPI*: 6–7 cases per 10^5 hospital admissions at a major teaching hospital in Canada and 1.1 per 10^5 in Iceland, any age, but pediatric cases are quite rare, ♂ = ♀.
- *EPG*: Specific causes of OP need to rule out and differentiated from the cryptogenic form of OP, and medical history/diagnostic tasks will include questions and laboratory tests to exclude infection, toxins, drugs, radiation, and inflammatory bowel disease. When an etiology cannot be identified, the adjective "cryptogenic" is used.
- *CLI*: Subacute onset of dyspnea and cough with a prodromal infectious event of the upper respiratory tract.
- *HCT*: Patchy airspace consolidation and nodules.
- *CLM*: There are airspaces filled with proliferative polypoid fibroblastic plugs (branching or serpiginous) or Masson bodies with lightly stained fibrosis with a myxoid pale-staining matrix rich in mucopolysaccharides and mild interstitial inflammation (lymphocytes, macrophages, plasma cells, and neutrophils) of terminal bronchioles, alveolar ducts, and spaces. Normal adjacent parenchyma but obstructive pneumonia may be seen distally to the fibroblastic plugs. COP may occur as acute exacerbation of UIP, and clinico-pathological correlation is recommended.

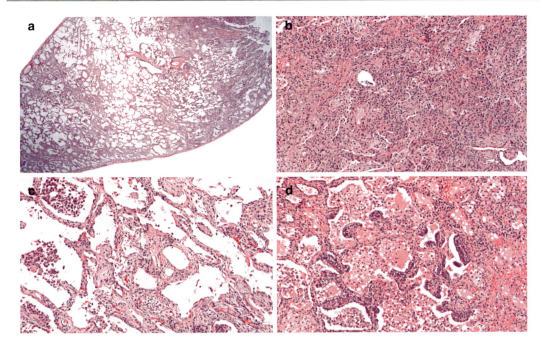

Fig. 2.13 Cryptogenic organizing pneumonia (COP) showing the distal airways and adjacent alveoli filled by fibromyxoid plugs of granulation tissue with temporal uniformity (all plugs appear to be at the same stage) with alveolar ducts and alveoli mainly involved and adjacent lung parenchyma normal. There is a mild interstitial infiltrate of plasma cells and lymphocytes and scant fibrin in airspaces and at places alveoli filled with lipid-laden foamy macrophages. There is no remodeling or honeycomb change as seen in UIP (**a**–**d** H&E, 12.5×, 100×, 100×, and 100×, respectively)

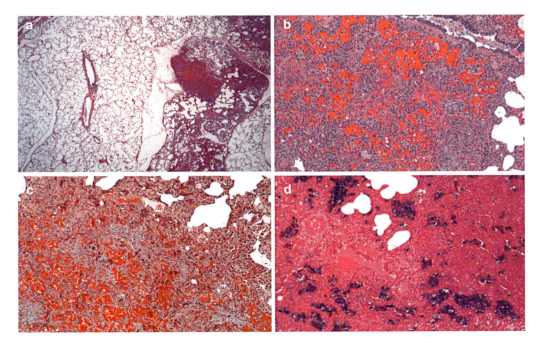

Fig. 2.14 Cryptogenic organizing pneumonia (COP) with hypersensitivity pneumonia (HSP) showing airway-centered (peribronchiolar) change with interstitial cellular infiltration and poorly formed non-necrotizing granulomas or interstitial giant cells with barely visible cholesterol clefts (**a**–**d**) (**a**, H&E, 12.5×; **b**, H&E, 100×; **c**, Movat pentachrome stain, 100×; **d**, Perls' Prussian Blue (PPB) for iron, 100×). In PPB there is evidence of hemosiderin-lade macrophages

- *TEM*: Type I pneumocyte necrosis (early phase).
- *DDX*: UIP, AIP, AEP.
- *UIP*: Spatiotemporal heterogeneity, random subpleural scarring, interstitial fibrosis, "honeycomb", fibroblastic foci adjacent to mature collagen with mostly collagenous fibrosis.
- *AIP*: Diffuse hyaline membranes in early stage and potential honeycomb change later.
- *AEP*: Acute inflammation with predominant eosinophils (AEP, acute eosinophilic pneumonia/pneumonitis).
- *TNT*: Steroid-responsive.
- *PGN*: Good (recovery usually in a few weeks) but wrong in the setting of a restriction to bronchioles (pure BOOP) in patients with BMT (bone marrow transplantation) and H/L-Tx (heart/lung transplantation) and rheumatoid arthritis.

A rare form of acute fibrinous and organizing pneumonia (AFOP) may also occur in childhood. It is characterized by bilateral basilar lung infiltrates on CXR. Histologically, there is organizing pneumonia and intra-alveolar fibrin in the form of "fibrin balls" (Fig. 2.15).

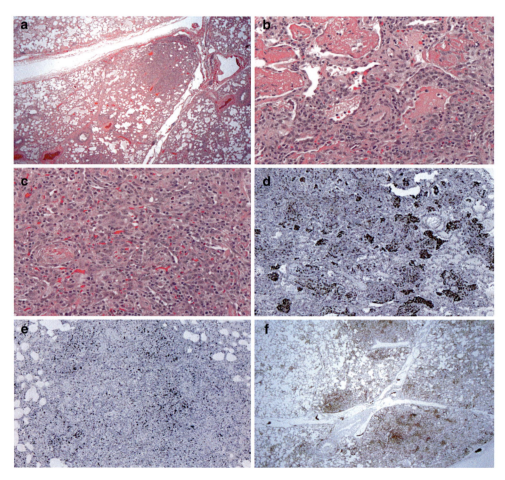

Fig. 2.15 Acute fibrinous pneumonia in a patient post-T-cell leukemia showing aggregates of inflammatory cells, including macrophages, lymphocytes, and fibrin (**a–c**, H&E, 12.5×, 200×, and 200×) supported by an immunohistochemical study with antibodies against CD68 (macrophages) (**d**, anti-CD68, 50×; **e**, anti-CD7, 50×; and **f**, anti-CD3, 12.5×). CD7 is a transmembrane glycoprotein, which is also a member of the immunoglobulin supergene family. CD7 does play a key role in T-cell and T-cell/B-cell interactions during the early stages of lymphoid development, while the CD3 T cell co-receptor is crucial in activating both the cytotoxic T cell and also T helper cells.

2.5.3 Usual Interstitial Pneumonia/Pneumonitis (UIP)

UIP is the most common type of chronic interstitial pneumonia in adults, although in at least half of the affected individuals, UIP is idiopathic. UIP is characterized by a *spatiotemporal diversity of architectural distortion* (old honeycomb fibrosis and active fibroblast foci) and chronic progression to lung failure (4–5 years) requiring H/L-Tx.

- *DEF*: Idiopathic/specific T_{H2}-*response*-driven *fibrosis* with ↑*TGF-β1* (Mϕ activation and innocent bystander tissue destruction) of the lung bilaterally with insidious onset (Fig. 2.16).
- *SYN*: Sclerosing/fibrosing alveolitis.
- *EPI*: Worldwide, ♂ > ♀, >40 years, apart from some children with *ABCA3* gene mutations.
- *EPG*: UIP is based on epithelial activation/injury – iterative processes (the so-called "UIP cycles") by some non-identified agents promoting
    1. ⇒ *TGF-β-1* (released from injured type I alveolar cells)
    2. ⇒ *TCF* effects, including:
        - ⇒ ↓ *T*elomerase (increase of senescence and apoptosis of alveolar cells)
        - ⇒ *C*aveolin-1 inhibition (fibroblast-directed inhibition)
        - ⇒ *F*ibroblasts proliferation with myofibroblasts transformation (*fibrogenesis*)
- *CLI*: Nl. FEV1/FVC or Tiffeneau-Pinelli ratio, including the forces expired volume in the first second and forced vital capacity with the normal value for the FEV1/FVC ratio being 70%, bilateral infiltrates (CXR), and BAL and CT also very useful for PGN.
- *GRO*: Nodular pleural surface with scarring and retraction.
- *CLM*: Patchy interstitial inflammation (lymphocytes, histiocytes [tight intra-alveolar clusters of macrophages], plasma cells, and germinal centers) with foci of heterogeneity ranging from old lesions of cystic changes and honeycomb pattern to active fibroblast foci and narrowed airspaces, subsequent pulmonary hypertensive changes, and smooth muscle hyperplasia.
- *TNT*: Steroid response in <20% of cases ⇒ lung and/or H/L-Tx.
- *PGN*: Lung transplantation as *ultima ratio*.

In pediatric lung biopsies, UIP gross and histology are excessively rare and is likely to remain an exceedingly rare occurrence. However, Young et al. reported UIP in a patient harboring ATP-binding cassette A3 (*ABCA3*) gene mutations. In a multicenter-based systematic review, no UIP was found over a 5-year period reviewing over 400 children. In our opinion and according to most of the pediatric pathologists, it may be worthy to follow-up more closely patients with *ABCA3* gene mutations. ABCA3 protein is located in the limiting membrane of lamellar bodies in alveolar type II cells and both in vitro, and now murine data indicate that ABCA3 functions in the transport of surfactant lipids into lamellar bodies. Thus, it is required to maintain pulmonary surfactant phospholipid homeostasis. Loss of ABCA3 function is responsible for lung disease, and one missense mutation, E292V, was identified in multiple unrelated children who were compound heterozygotes for *ABCA3* gene mutations. Respiratory viral infections have been temporally related to the onset of disease in patients harboring *SFTPC* gene mutations. Clinical genetic testing for mutations in *SFTPC* and *ABCA3* genes may be considered in children and younger adults with lung fibrosis. The progression of UIP is relentless toward idiopathic pulmonary fibrosis (IPF) and is quite worrying, particularly for the more frequent use of tobacco smoking at a younger age. The widespread diffusion of vaping and e-cigarette as well as cannabis may worse irreversibly the lung function of children and youth. The primary challenge in the field of ILDs is the early and accurate diagnosis of this distressing form. IPF has a worse prognosis than that of numerous cancers and has been less tackled by research funding agencies.

The rates of prevalence and incidence of IPF are challenging to estimate because numerous clinical conditions evolved toward this entity in the last decades. Allergic asthma may be an essential and common underlying disorder, but cystic lung dis-

a **P**atchy, paraseptal, subpleural, with ST-Heterogeneity
Interstitial involvement
Lympho-histiocytic inflammatory character
Scarring development with active fibroblastic foci
Normal-looking spared lung tissue
Elimination pattern of resolution
Relentless in the evolution

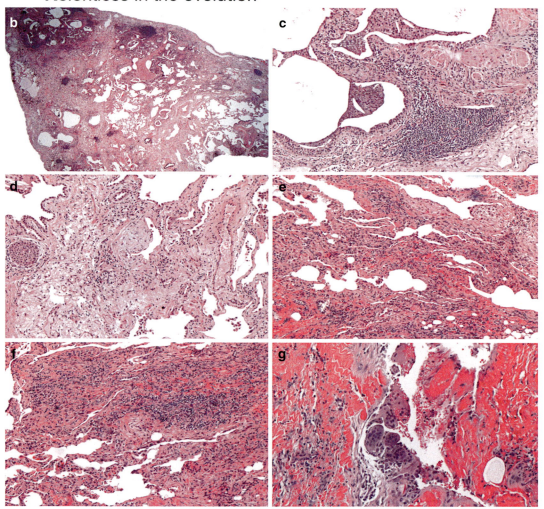

Fig. 2.16 Usual interstitial pneumonia – lung fibrosis highlighting in (**a**) the characteristic features of this interstitial lung disease. These features are demonstrated in the microphotographs **b**–**g** (**b**–**g**, H&E stain only with **b** 12.5× and **c**–**g**, 100×)

ease, scleroderma, radiation, and chemotherapy-related fibrosis, granulomatous lung disease, sarcoidosis, and environmental causes, as well as smoking-related COPD or the new vaping-related lung disease, are also listed behind the diagnosis of pulmonary fibrosis. The fibrotic stages of these conditions are often fatal because a median survival time has been estimated to be 3–5 years following the diagnosis of pulmonary fibrosis. Although not all pulmonary fibrotic conditions follow the simple paradigm of wound healing, this model may be the underlying pathogenic mechanism for most of these conditions. Immunologically, it is essential to classify the immune response as type 1, type 2, or type 17. Class 1, which is generally considered pro-inflammatory, is characterized by Th_1 cells, IFN-γ, TNF-α, and IgG2 antibody response; class 2 shows Th_2 cells, IL-4, IL-5, IL-13, and IgE and is usually considered a wound healing response; finally, class 17 is characterized by Th_{17} cells, which have been associated with pro-inflammatory conditions and based upon the T-helper cell-dominant cytokine responses.

Despite new progression in the immunological research of the last 20 years, the pathogenesis of pulmonary fibrosis is far to be fully understood. It is well known, one or more insults cause the lung to exhibit fibrosis ("*Gewebefibrosierung*" or fibrotization), and these insults or singular insult may be occasionally known, although they are unknown most of the time. The insult will lead to alveolar epithelial cell injury and subsequent repair, which may be or, actually, is dysregulated in pulmonary fibrosis very often. There is excessive extracellular matrix (ECM) and loss of typical architecture of the lung parenchyma and, subsequently, loss of lung function. In pulmonary fibrosis, fibroblasts that produce ECM components differentiate into myofibroblasts, which are considered the hallmark cells in the full development and establishment of the full picture of pulmonary fibrosis. Multiple pathways are implicated in the pathogenesis of this disease. Among others, coagulation, apoptosis, and oxidative stress are involved in the pathogenesis of pulmonary fibrosis.

Four distinct stages have been identified in wound healing. There is a clotting/coagulation phase (1), an inflammatory cell-associated migration phase (2), a fibroblast-associated migration/proliferation/activation phase (3), and a tissue remodeling and resolution phase (*restitutio ad integrum*) (4). Following a lung injury, there is a release of inflammatory mediators by epithelial cells that initiate an anti-fibrinolytic coagulation cascade, which triggers platelet activation and blood clot formation. After this initial phase, leukocytes (e.g., neutrophils, macrophages, and T cells) enter the sequence, and the recruited leukocytes secrete pro-fibrotic cytokines such as IL-1β, TNF, IL-13, and TGF-β. Activated macrophages and neutrophils then progressively remove dead cells and invading microorganisms. Later, fibrocytes from the bone marrow and orthotopic resident fibroblasts ("autochthonous cells") proliferate and differentiate into myofibroblasts, which release several components of the ECM. Epithelial-mesenchymal transition (EMT) should also be considered to be at the basis of the transformation of epithelial cells into fibroblasts and myofibroblasts. Finally, tissue remodeling and resolution phase take place, and activated myofibroblasts promote wound repair, leading to wound contraction and restoration of vascular nutrition. Fibrosis can develop if a dysregulation can arise at any stage in the tissue repair program or when the lung-damaging stimulus persists. Myofibroblasts, which may be considered dysregulated, originate from a resident pool of tissue fibroblasts in response to tissue injury. Interestingly, these cells organize into small areas of progressive fibrosis, which are called fibroblastic foci. These cells in these areas are the active source of an excessive assembly of ECM proteins.

2.5.4 Nonspecific Interstitial Pneumonia/Pneumonitis (NSIP)

- *DEF*: Idiopathic or specific (e.g., collagen vascular disease, drug toxicity, hypersensitivity, post-allogeneic hematopoietic stem cell transplantation) with the *temporal homogeneity* of cellular interstitial inflammation (e.g., plasma cells) and a *variable amount of fibrosis* (no or very few fibroblastic foci → inactive, or mostly inactive) (Fig. 2.17). Children with NSIP have been reported (Van Zuylen et al. 2019; Meignin et al. 2018; Hauber et al. 2011; Pediatric Diffuse Parenchymal Lung Disease/Pediatric Interstitial Lung Disease Cooperative Group, The Subspecialty Group of Respiratory Diseases, The Society of Pediatrics, Chinese Medical Association, 2011).
- *EPI*: Youth/middle age (20–50 years), usually younger than UIP patients.
- *CLI*: Dyspnea and cough over several months and bilateral interstitial infiltrate on CXR, steroid-responsive, and good PGN; two NSIP types:

a **U**niformity spatial and temporally of interstitial involvement
Lymphocytic (lymphoid clusters) = Cellular pattern
Fibrosing pattern with diffuse or patchy interstitial fibrosis, focal OP, subpleural fibrosis, and thickening of vascular media, but lack of ST-heterogeneity, fibroblast foci or aspects reminiscent of honeycombing.

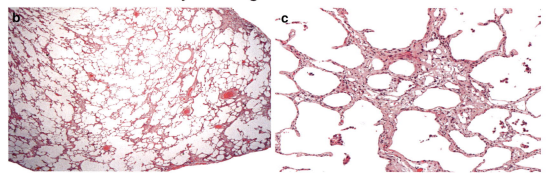

Fig. 2.17 Nonspecific interstitial pneumonia (NSIP) showing the key characteristics described in the text and grouped in **a**. The microphotographs **b** and **c** have been taken from an adolescent boy suffering from NSIP (**b**, H&E stain, 12.5×; **c**, 100: H&E stain, 100×)

1. *Cellular* (interstitial chronic inflammation with lymphocytes and plasma cell and preservation of the lung architecture)
2. *Fibrosing* (patchy/diffuse, but temporally uniform interstitial fibrosis)

Patients with NSIP, fibrosing pattern are older than patients with NSIP, cellular model.

- *HCT*: Mostly ground-glass opacities, bilateral, basal and peripheral as well as reticular, linear patterns in 1/2 of patients.
- *CLM*: Lung architecture-preserving spatiotemporal uniformity with mild-moderate inflammation ± lymphoid follicles, type II pneumocyte hyperplasia, and scant to dense interstitial fibrosis and "traction" bronchiolitis. No granulomas, no MNGC, no bronchiolocentric distribution, no acute lung injury pattern, no hyaline membranes, no prominent eosinophils, and no evidence of infection. NSIP mnemonic word is ULF ("uniform lymphoid follicles").
- *DDX*: UIP (spatiotemporal heterogeneity with fibroblastic foci, collagenous deposition, and honeycomb changes). It is mandatory to rule out the following diagnoses in every case: collagen vascular diseases, drug reaction, chronic hypersensitivity pneumonitis, immunodeficiency-related changes, infection, and slowly resolving acute interstitial pneumonia.
- *TRT*: Steroid-responsive.
- *PGN*: 5-YSR: 80–90%.

2.5.5 Desquamative Interstitial Pneumonia/Pneumonitis

- *DEF*: DILD, often associated with cigarette smoking (potentially also vaping), relatively good steroid-responsive, but occasionally relentless with significant mortality, and characterized by a uniform (spatially and temporally) intra-alveolar involvement of the lung more than interstitial inflammation (Fig. 2.18).
- *EPI*: Smokers, 30–60 years, depending on the length of confinement in dwellings with poor air quality (urban vs. rural).
- *EPG*: Spectrum with RB-ILD with mild pneumocyte hyperplasia vs. diffuse involvement of

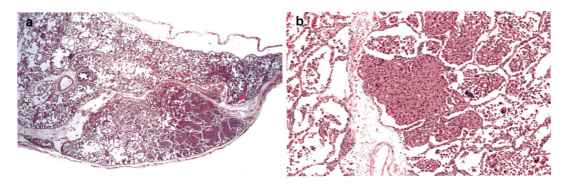

Fig. 2.18 Desquamative interstitial pneumonia (DIP) showing diffuse and massive accumulation of intra-alveolar macrophages (**a**, H&E stain, 12.5×; **b**, H&E stain, 100×)

airspaces by MΦ and prominent pneumocyte hyperplasia.
- *CLI*: Subacute dyspnea, cough.
- *CTS*: Bilateral ground-glass opacities, lower lung fields.
- *GRO*: COPD-like appearance of lungs (chronic bronchitis and emphysema).
- *CLM*: Numerous *diffusely, loosely aggregated, evenly dispersed* large collections of *MΦ* (granular golden-brown pigment-laden MΦ, Fe+) in airspaces (alveoli and distal airways) without tendency to distend lumina and admixed with *type II* pneumocytes, and loss of type I pneumocytes (+), *blue bodies* (PAS+, laminated iron containing microstructures within or surrounding MΦ), and architecture preservation (no scarring or honeycomb change). Moreover, no/mild *interstitial chronic inflammatory infiltrate (lymphocytic infiltrate) fibrosis and no/mild interstitial fibrosis.*
- *DDX*: LCH, giant cell interstitial pneumonia, UIP, AEP, and post-obstructive pneumonia.
- *TRT*: (Passive/active) smoking/vaping cessation ± steroids.
- *PGN*: 10-YSR: 70% (better than PGN of UIP!).

2.5.6 Respiratory Bronchiolitis ILD

- *DEF*: ILD with similarities to DIP, but harboring the tendency to patchiness, limit to the involvement of respiratory bronchioles (bronchiolocentric) ± alveolar ducts, and tendency to distend lumina (bronchiolar and peribronchiolar MΦ accumulation with standard large areas of intervening lung parenchyma) (Fig. 2.19).
- *EPI*: Early smokers, 30–60 years, depending on the amount and length of smoking and living conditions (urban vs. rural). The role of vaping is under intense investigation, and this phenomenon seems to be quite diffuse in teenagers. In several countries the regulation of the sell and use of these products may take place at late stages because the current use of vaping among children and teenagers or children and adolescents exposed to vaping in social events has fearfully increased in the last couple of years with rates ranging from 40% to 60%.
- *CLI*: Chronic dyspnea, cough with mild restrictive or mixed restrictive/obstructive pulmonary function tests.
- *IMG*: Centrilobular nodules or ground-glass opacities, patchy.
- *GRO*: Smoking-associated changes related to COPD (chronic bronchitis and emphysema).
- *CLM*: Patchy bronchiolo-centric process with intraluminal, Fe (+)-MΦ-accumulation (PPB+) centered on respiratory bronchioles with mild fibrosis surrounding the bronchioles and involving adjacent alveolar septa and preservation of the pulmonary architecture.
- *DDX*: The following features should not be present, such as diffuse filling of alveoli by MΦ, fibroblastic foci, honeycomb changes, NSIP pattern of diffuse involvement, hyaline membranes, bronchiolitis obliterans/organizing

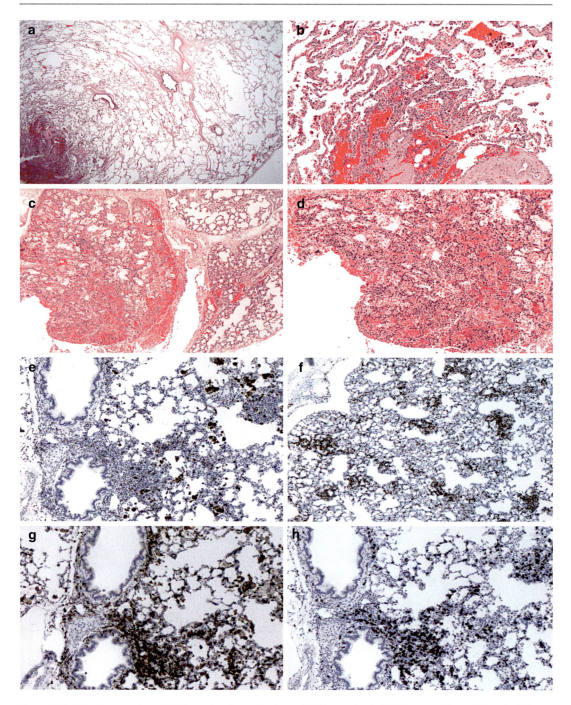

Fig. 2.19 Respiratory bronchiolitis showing focal chronic inflammation of terminal bronchioles and alveolar ducts with adjacent focal interstitial inflammation and mild fibrosis (**a–d**, H&E stain with **a**, 12.5×; **b**, 100×; **c**, 50×; **d**, 100×; **e**, Anti-CD68, 100×; **f**, anti-CD3, 50×; **g**, anti-CD4, 100×; **h**, Anti-CD8, 100×)

pneumonia, granulomas, MNGC, a foreign material other than pigment of anthracosis, numerous eosinophils, and neutrophils.
- *TRT*: Passive/active smoking cessation, because passive cigarette smoke exposure may prevent improvement or lead to recurrence of the disease. Some patients benefit from corticosteroids.
- *PGN*: The natural clinical course of the disease is unknown, but the prognosis is good with the avoidance of passive or active smoking exposure.

2.5.7 Pulmonary Langerhans Cell Histiocytosis

- *DEF*: ILD with multinodular infiltrates of Langerhans cells with the centrilobular involvement of respiratory bronchioles (peribronchiolar) and with large common areas of intervening lung parenchyma (Fig. 2.20). In adults, it is smoking-associated with "smokers' macrophages" (Prussian blue +), while in children and adolescents, it is infrequently associated with passive smoke exposure. The pediatric form may be characterized by multifocal involvement (10–15%) with mainly bone and systemic disease.
- *SYN*: Langerhans cell granulomatosis, histiocytosis X.
- *EPI*: Adult PLCH (smokers, 30–60 years, depending on the amount and length of smoking, and living conditions) and pediatric PLCH (rare, insufficient data).
- *CLI*: Dyspnea, ± fever, and pneumothorax (PNX), which represents air-collection between parietal and visceral pleura with consequent collapsed lung developing as a complication of the trapped air.
- *IMG*: Centrilobular/peribronchiolar reticulonodular infiltrates, diffusely scattered, mainly upper and mid-lung fields.
- *GRO*: Smoking-associated changes related to COPD (chronic bronchitis and emphysema).
- *CLM*: Patchy, nodular interstitial lesion with centrilobular distribution with stellate architectural involvement/scarring due to extension along alveolar septa with numerous histiocytes with variable numbers of eosinophils, plasma cells, lymphocytes, and MΦ and often nodular cavitation leading to cyst formation.
- *IHC*: (+) S100, CD1a, Langerin.
- *TEM*: Birbeck granules ("tennis-racket" cytoplasmic organelles of Langerhans cells; Birbeck et al. 1961). Birbeck granules' function is controversially discussed with some authors suggesting that these granules migrate to the periphery of the cells releasing their contents into the ECM, while other authors being main proponents behind the theory that Birbeck granules' are receptor-mediated endocytosis units, similar to clathrin-coated pits. Michael SC Birbeck (1925–2005), a British scientist and electron microscopist reported these granules in 1961 following intensive work at the Chester Beatty Laboratory of the Institute of Cancer Research, London, United Kingdom.

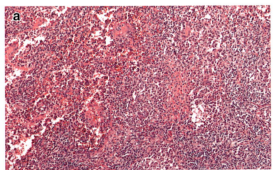 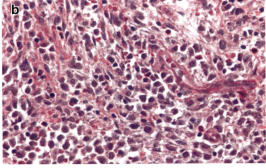

Fig. 2.20 Langerhans cell histiocytosis showing infiltration of Langerhans cells and eosinophils in a male newborn with respiratory distress following an uncomplicated pregnancy and term delivery. The neoplastic infiltrate with Langerhans cells was positive for S100 and CD1a (not shown). Birbeck granules were identified ultrastructurally using transmission electron microscopy (not shown). (**a**, H&E, 100×; **b**, H&E, 400×)

- *DDX*: DIP, RB-ILD, UIP, HSP, CEP (chronic eosinophilic pneumonia), and eosinophilic pleuritis.
  - DIP: Diffuse, temporally uniform, ± eosinophils, no cysts, no scarring, no peribronchiolar nodules of Langerhans cells, CD1a/Langerin negative.
  - UIP: Irregular subpleural scarring, honeycomb, no peribronchiolar nodules of Langerhans cells, CD1a/Langerin negative.
  - RB-ILD: Intra-alveolar process, ± eosinophils, no cysts, no scarring, no peribronchiolar nodules of Langerhans cells, CD1a/Langerin negative.
  - HSP: ± Exposure Hx. Granulomas and MNGC, generally present, no peribronchiolar nodules of Langerhans cells, CD1a/Langerin negative.
  - CEP: Mostly, intra-alveolar process, pools of eosinophils, no cysts, no scarring, no peribronchiolar nodules of Langerhans cells, CD1a/Langerin negative.
  - Eosinophilic pleuritis: Pleural procedure, mostly eosinophils with proliferating mesothelial cells, ± eosinophils, no cysts, no scarring, no peribronchiolar nodules of Langerhans cells, CD1a/Langerin negative.
- *PGN*: Variable outcome, including spontaneous remission, remission following smoking cessation, and progression to end-stage pulmonary fibrosis. In adults, hematolymphoid malignancies and carcinoma have been reported (synchronous or metachronous malignancy).
- *TRT*: If pediatric and multisystemic, chemotherapy is necessary. Otherwise, it is often very responsive to smoke exposure cessation.

2.5.8 Lymphoid Interstitial Pneumonia/Pneumonitis (LIP)

- *DEF*: ILD with a mixed cellular interstitial infiltrate of lymphocytes, plasma cells, epithelioid histiocytes, germinal centers, and granulomas with a variable degree of symptomatology and imaging features (Fig. 2.21). Pulmonary lymphoproliferative

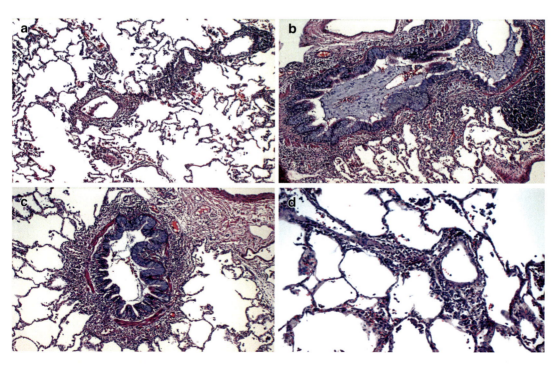

Fig. 2.21 Lymphoid interstitial pneumonia (LIP) showing interstitial lung disease with diffuse and dense lymphocytic infiltration and generally CD4+ or CD8+ T cells or B cells predominate (**a–d**, H&E stain; **a**, 100×; **b**, 100×; **c**, 100×; **d**, 200×)

disorders may also be suggested as LIP-associated disorders of the lung tissue. In fact, LIP has also been considered an early lymphoproliferative lesion of pulmonary MALT (mucosa-associated lymphatic tissue).
- *EPI*: Any age.
- *EPG*: Autoimmunity (e.g., Sjögren's syndrome), viruses (e.g., HIV), dysproteinemia have been associated to LIP.
- *CLI*: Gradual onset of dyspnea, cough (>6 months) ± immune-related disease (e.g., collagen vascular diseases, multicentric Castleman disease, immunodeficiency, e.g., AIDS, SCID) ± polyclonal γ-GB-changes (↑/↓-γ-GB-emia).
- *CTI*: Ground-glass opacities, perivascular cysts ± perivascular honeycombing (late stage).
- *CLM*: Spectrum of peribronchiolar/peribronchial polyclonal lymphoplasma cellular proliferations that include follicular bronchitis/bronchiolitis (BALT hyperplasia) with scattered reactive lymphoid follicles (Ø ≥ 1 mm) around bronchioles and bronchi, nodular lymphoid hyperplasia, and, in case of monoclonality, low-grade B-cell lymphoma (MALT-oma). The lymphoid follicles can compress bronchiolar lumina, and distal changes of obstructive pneumonia can be observed, and a mild infiltrate can also be seen into adjacent alveolar septs and when a congruous interstitial infiltrate is present there may be an overlap between follicular bronchiolitis and LIP. In follicular bronchiolitis as well as LIP, there is a mixed polytypic infiltrate of B and T cells, although T lymphocytes predominate in LIP, while B lymphocytes dominate in follicular bronchiolitis (FOB).
- *DDX*: It includes follicular bronchiolitis, ENMZ-BCL, nodular lymphoid hyperplasia, and hypersensitivity pneumonitis. Moreover, the marked interstitial lymphoplasmacytic infiltration, which expands the alveolar septa, also needs to be carefully distinguished from cellular NSIP, i.e., the fibrosing subtype of NSIP with collagenous fibrosis, which is diffuse but lacking structural remodeling.
- *PGN*: Variable depending on the stage.
- *TRT*: Steroid-responsive with regard to the follicular bronchiolitis/bronchitis.

An entity that seems to be controversially discussed in books and papers is lymphomatoid granulomatosis (LYG). Although not necessarily associated with LIP, it may be necessary to address LYG here for the predominance of lymphocytic infiltrates seen in both conditions. Angiocentric and angio-destructive pathologic disease process that more often affects the lung as bilateral nodular infiltrates constituted of a mixed population of lymphoreticular cells lacking specific granulomatous features.

LYG affects both sexes (♂ > ♀), the fifth decade of life with lung>CNS>skin>liver>kidney, but adolescents and youth may also been affected. It is debated how much is the lymphoproliferative process active in LYG. However, the current WHO criteria indicate that LYG is a distinct form of T-cell-rich large B-cell lymphoma. Diagnostic criteria include (1) mixed mononuclear cell infiltrate composed of an angiocentric and angio-destructive formation of large and small lymphoid cells with often plasma cells and histiocytes and the fate to replace the lung parenchyma implementing a clear-cut vascular infiltration and (2) variable cell population of T-cell-rich B-cell infiltration, which are CD20-(+) large B cells in a background of CD-3–(+) small lymphocytes. There are optional findings, although supportive, not often seen. They include necrosis of the cellular infiltrate with positivity for EBER (in situ hybridization for Epstein Barr virus or EBV RNA), multiple lung nodularity, skin involvement and CNS involvement. The WHO recommendation is that LYG is graded as 1, 2, or 3, according to the number of EBV-positive large B cells, being grade 1 containing <5 EBV+/HPF, grade 3 > 50 EBV+/HPF, while grade 2 between 5 and 50 EBV (+)/HPF. Treatment of LYG includes IFN α-2b for both classes 1 and 2, while DLBCL therapy is indicated for class 3 LYG.

2.5.9 Hypersensitivity Pneumonia/Pneumonitis (HSP)

- *DEF*: ILD with a specific immunologic reaction to inhaled agents (e.g., fungus, mold, animal proteins), often labeled according to the job of affected people (e.g., farmer's lung, maple bark stripper's disease, pigeon breeder's disease), or to the etiologic agent (e.g., byssinosis, actinomycosis).

- *SYN*: Extrinsic allergic alveolitis.
- *EPG*: Combination of *type III* (immune complex)-associated hypersensitivity responsible for the acute inflammation features and *type IV* (cell-mediated)-driven hypersensitivity reactions accountable for the formation of granulomas.
- *CLI*: Essentially, two forms.
- *Acute* (substantial exposure): Severe dyspnea, cough, fever about 6 h after exposure and usually resolving in 12–18 h (self-limited), although continued exposure can lead to permanent damage (chronic ⇒ DILD – chronic hypersensitivity pneumonia/itis).
- *Chronic*: Prolonged exposure to small doses and insidious onset of dyspnea and dry cough (restrictive lung disease) and characterized by "BIG":
  1. *B*ronchiolo-centric *I*nterstitial pneumonia/pneumonitis
  2. (patchy, peribronchiolar, temporally synchronized, with uninvolved lung parenchyma in between, with CD4 < CD8 T-lymphocytes, plasma cells, histiocytes, and rare eosinophils/neutrophils)
  3. Non-necrotizing epithelioid *G*ranulomas with giant cells, ±
  4. BOOP-like changes (2/3 cases)
- *TRT*: Steroids.
- *PGN*: Steroid-responsive, elimination of the etiologic agent, restrictive lung disease.

Mnemonic word: BIG (*B*ronchiolo-centric *I*nterstitial pneumonia/pneumonitis and non-necrotizing epithelioid *G*ranulomas with giant cells).

Actinomycosis, which may have some different clinical forms (cervicofacial, thorax, and abdomen), may enter in the differential diagnosis. Actinomycosis is a subacute/chronic bacterial infection triggered by filamentous, Gram (+), non-acid-fast anaerobic to microaerophilic bacteria. It has a contiguous spread ± "sulfur discharge" and suppurative and granulomatous inflammation ± multiple abscesses and fistulas. The clinical forms of actinomycosis include the cervicofacial, thorax, and abdomen (pelvic actinomycosis).

2.5.10 Acute Eosinophilic Pneumonia/Pneumonitis

Eosinophilic pneumonia is a sophisticated group of diseases accommodated in a heterogeneous collection characterized by the accumulation of eosinophils in the lung parenchyma with or without peripheral blood eosinophilia. Eosinophilic pneumonias are usually divided according to their etiology (idiopathic or secondary), clinical presentation, or morphology.

- *DEF*: Eosinophilic pneumonia characterized by acute onset (<7 days), the presence of fever, severe hypoxemia, radiological pulmonary infiltrates ± PB eosinophilia (Fig. 2.22).
- *EPI*: Unknown incidence, all ages, ♂ = ♀.
- *EPG*: New-onset smokers, fireworks smoke, heroin and crack cocaine smoking. A sentinel case was reported in a young firefighter after inhaling World Trade Center dust following the September 11, 2001 terroristic attack (Rom et al. 2002). The etiology in childhood and adolescence is somewhat heterogeneous.
- *CLI*: Rapid onset (<7 days) with fever, dyspnea, hypoxemia, BAL >25% Eos, ↑ SP-A and SP-D and patchy bilateral, mostly peripheral pulmonary infiltrates. Other symptoms may include myalgias and pleuritic chest pain. On auscultation, patients show bibasilar crepitation without wheezing (Box 2.9).
- *IMG*: Thickening of bronchovascular bundles (random distributed) and pleural effusions.
- *GRO*: DAD-like appearance but depends on the extension of the DAD.
- *CLM*: Alveoli filled with eosinophils and large MΦ or giant cells, DAD, necrotic debris, and palisading histiocytes with accompanying interstitial infiltrate of eosinophils and other inflammatory cells, mild non-necrotizing vasculitis, and ± bronchiolitis obliterans.
- *DDX*: The laboratory workup should include a

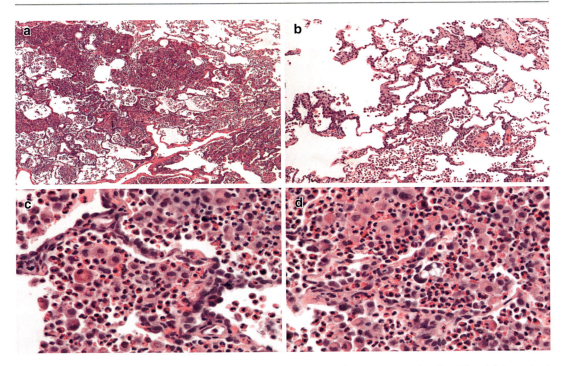

Fig. 2.22 Eosinophilic pneumonia of a male adolescent showing patchy aggregates of eosinophils with intermixed some macrophages (H&E with 50×, 100×, 400×, and 400× as original magnifications for **a**–**d** microphotographs)

Box 2.9 Classification of Eosinophilic Pneumonia/Pneumonitis
1. *Chronic* eosinophilic pneumonia (CEP) (chronic asthma, eosinophilia, and ↑ IgE)
2. *Acute* eosinophilic pneumonia (AEP) (*vide supra*)
3. *Simple* eosinophilic pneumonia (Loeffler syndrome) (mild, self-limited, PB eosinophilia, and interstitial infiltrate of eosinophils, but no vasculitis, fibrosis, or necrosis)
4. *Secondary* eosinophilic pneumonia (ABPA-linked and bronchocentric granulomatosis)
5. *Tropical* eosinophilic pneumonia (microfilariae, *Wuchereria bancrofti* and *Brugia malayi* within pulmonary capillaries)

Notes: The mnemonic word *CAST* serves to memorize the five types, keeping in mind that the letter S is used twice; *PB* peripheral blood

complete blood count with differential, bronchoalveolar lavage, analysis of stool for ova and parasites, serum immunoglobulin levels, ESR, and cultures for bacteria and fungi.
- *TRT*: Steroids and respiratory support.
- *PGN*: Variable, but usually good if there is a prompt response to corticosteroids.

If AEP is compared to CEP, there are a few aspects that need to be considered. The severity of damage to the basal lamina and the amount of following intraluminal fibrosis in CEP is more than AEP. In fact, in AEP, intraluminal fibrosis is virtually absent. Some drugs have been associated with AEP, including acetaminophen, bleomycin, carbamazepine, chlorpropamide, cromolyn sodium, dapsone, daptomycin, glafenine, imipramine, Maloprim, minocycline, naproxen, oxymetazoline, procarbazine, sulfanilamide-containing vaginal cream, sulfasalazine, trazodone, and tetracycline. An accurate medical and pharmacological history is paramount for both pediatric and adult cases. In Churg-Strauss syndrome, tissue eosinophilia is accompanied by vasculitis (PAN-like) and necrobiotic palisaded granulomas. The list of diagnoses of the thickening of the bronchovascular bundles on chest CT imaging other than the pulmonary interstitial edema includes acute eosinophilic pneumonia, lymphangitis carcinomatosis, lymphoid interstitial pneumonia, microscopic polyangiitis, mycoplasma pneumonia, and sarcoidosis.

2.5.11 Progressive Massive Fibrosis

Progressive massive fibrosis (PMF) is an extreme complication of *silicosis*, *asbestosis*, and *coal worker's pneumoconiosis* with virtually sparing of children and youth. PMF needs to be explained to youth affected with UIP and the relapse and persistence of some environmental issues. Clinically, dyspnea, even at rest, associated with localized chest pain is common, and right ventricular hypertrophic heart may be the reason for H-L/TX in some affected individuals. Microscopically, there are large amorphous masses of fibrous tissue. There is a tendency to obliterate and contract the lung parenchyma. It occurs preferentially in the upper lobes, where silicosis is frequent. Two life-threatening complications are cavitation, due to necrobiosis or TB and Caplan syndrome, which is characterized by multiple firm tan nodules with central necrosis (aka rheumatoid nodules) in the setting of PMF (Chiavegatto et al. 2010).

2.6 Inflammation: Tracheobronchial

Inflammation of the tracheobronchial system is persistent in pediatric and young adult population. Primary concerns are the increased level of air pollution in several countries and the habit to tobacco smoking and vaping, which seems to be increased in a very young population, at least in some countries. Although it should be illegal selling cigarettes or other tobacco products to individuals younger than 18 years old, the reality is that 6.2% Canadians (~1/20) aged 15–24 indicated using e-cigarettes in the past month, while 23.9% (~1/4) indicated having ever tried e-cigarettes (Mehra et al. 2019). (Box 2.10).

> **Box 2.10 Inflammation: Tracheobronchial**
> 2.6.1. *Laryngotracheitis*
> 2.6.2. *Acute Bronchitis*
> 2.6.3. *Bronchial Asthma*
> 2.6.4. *Pulmonary Hyperinflation*
> 2.6.5. *Bronchiectasis*
> 2.6.6. *Neoplasms*

2.6.1 Laryngotracheitis

Laryngotracheitis may herald a lung disease and may be associated to several drugs that may be ingested or substances that can be inhaled. There are a few, but essential, drugs that are important to remember and that can cause lung disease in children and youth. The pathology of laryngotracheitis mirrors the adult pathology (Fig. 2.23). A textbook of clinical pharmacology and/or respiratory medicine are essential to complement the information of this chapter.

2.6.2 Acute Bronchitis

Inflammation of the bronchial system with critical neutrophils in both lumen and invading the surface epithelium. This pathology mirrors that of adulthood.

2.6.3 Bronchial Asthma

- *DEF*: Increased responsiveness of the tracheobronchial tree to various stimuli (extrinsic or intrinsic), leading to paroxysmal airway constriction, which is potentially life-threatening (Fig. 2.24).
- *EPI*: The prevalence of asthma rose progressively from the 1980s until the late 1990s, when it reached a plateau. In 2007, about 10% of children 0–17 years of age had asthma, according to data from the National Health

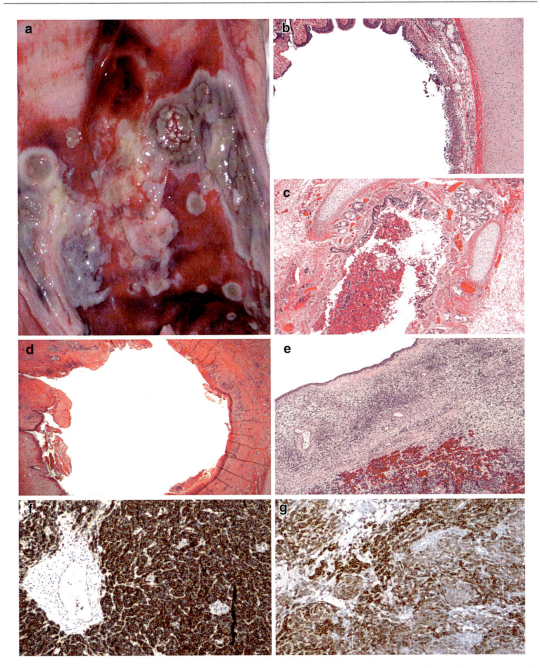

Fig. 2.23 Figure (**a**) shows the gross image of a *Candida* tracheitis with numerous circumscribed punched out erosions. The microphotograph (**b**) shows an erosive tracheitis without evidence of fungi (hematoxylin and eosin staining, ×50 original magnification). An erosive tracheitis with or without fungal colonization may be found following long-term intubation periods in both children and adolescents. The microphotograph (**c**) shows evidence of tracheal hemorrhage in the setting of a hemorrhagic diathesis due to cardiac septicemia (hematoxylin and eosin staining, ×50 original magnification). A tracheal stenosis is seen in microphotograph (**d**) with circumferential sclerosis in a patient who underwent a resection for a congenital tracheal pathology (hematoxylin and eosin staining, ×12.5 original magnification). The trachea can also rarely be involved by a neoplastic process as seen in the microphotograph (**e**) from a child with primitive neuroectodermal tumor (PNET) or extraskeletal Ewing sarcoma (hematoxylin and eosin staining, ×50 original magnification). The immunohistochemical analysis of the PNET revealed a positivity for CD99 (**f**) and CD56 (**g**) and confirmed by a translocation between chromosomes 11 and 22, which fuses the EWS gene of chromosome 22 to the FLI1 gene of chromosome 11 (immunostaining with avidin-biotin complex, ×100 original magnification for both images)

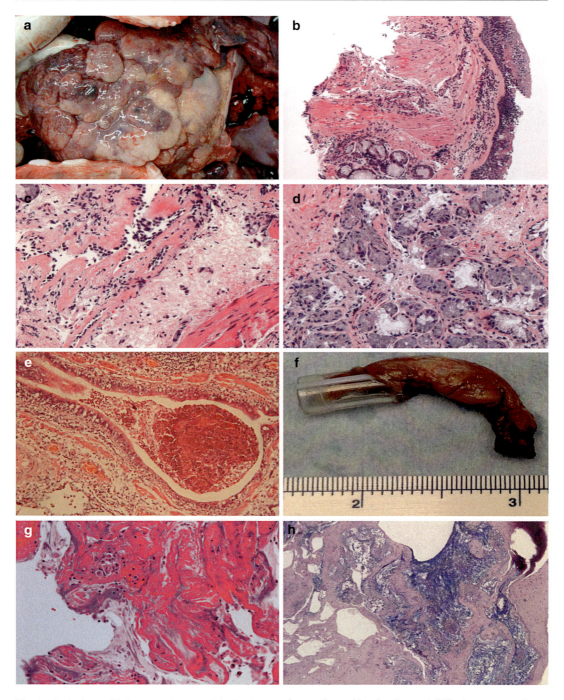

Fig. 2.24 Asthma with the gross photograph in (**a**) showing the classic features of *volumen pulmonum auctum* or acute exacerbation in case of an acute, untreatable attack of bronchial asthma, while the microphotographs **b**–**d** show the increase of the thickness of the subepithelial space with increase of periluminal glands, smooth muscle hypertrophy, and a thick subepithelial eosinophilic band (**b**). In (**e**) there is a plug of neutrophils and erythrocytes in a patient with relapsing and difficult to manage bronchial asthma. In (**f**) is shown a cast of plastic bronchitis, while (**g**) and (**h**) show the microphotographs of a cast of plastic bronchitis with mucus and positivity for alcian blue special stain (**b**, H&E stain, 100×; **c**, H&E stain, 200×; **d**, H&E stain, 200×; **e**, H&E stain, 100; **g**, H&E stain, 200×; **h**, alcian blue-PAS stain, 50×)

Interview Survey of the USA, and the lifetime prevalence of asthma in children remains about 10%.
- *EPG*: Three types – extrinsic (atopic, allergic), most common, and intrinsic (idiosyncratic); now recognized a mixed form. In the extrinsic type, there is the initial sensitization affecting the T_{H2}, which release IL4/5, which promote IgE release by B cells, mast cells, and eosinophils. The re-exposure to allergen leads to mediator release from mast cells that reside in the bronchial mucosa.
- *CLI*: Unremitting attacks (*status asthmaticus*) can be fatal (pulmonary hyperinflation).
- *IMG*: Typically, CXR is not significant, because history and physical exam are the cornerstones for the diagnosis; CXR can disclose some complications, including pneumomediastinum, pneumonia, or a foreign body in the airway.
- *GRO*: Overdistension of the lungs with areas of atelectasis and mucus plugs in proximal bronchi. The condition may reach a full overdistension of both lungs, which is also called "*Volumen pulmonum auctum*" corresponding to the German word "Lungenüberblähung" or, literally, enlarged volume of the lungs.
- *CLM*: Bronchial lumen plugging by thick mucus plugs containing eosinophils, whorls of shed epithelium (Curschmann spirals), and Charcot-Leyden crystals (eosinophilic membrane protein) and directly communicating distal airspaces over distended with thickening of the subepithelial basement membrane, edema, and inflammation in bronchial walls with prominence of eosinophils and hypertrophy of bronchial wall muscle (Box 2.11).
- *DDX*: Foreign body aspiration is the first differential diagnosis that needs to be ruled out in an emergency department.
- *TRT*: Therapeutic agents are aimed at increasing cAMP levels either by increasing production (β2-agonists, e.g., epinephrine) or decreasing degradation (methylxanthines, e.g., theophylline). Cromolyn sodium prevents mast cell degranulation
- *PGN*: It may be potentially fatal. There may be a tracheobronchial disease with a series of

> **Box 2.11 Morphologic Characteristics of Bronchial Asthma ("GILST")**
> - *G*oblet cell metaplasia in the airways epithelium
> - *I*nfiltrate of inflammatory cells
> - *L*ymphoid tissue hyperplasia
> - *S*mooth muscle hypertrophy and hyperplasia
> - *T*hickening of the basement membrane and "thicken" mucus (mucostasis)
>
> Notes: ~ variable; #, number; Curschmann's spirals are microscopic spiral-shaped mucus plugs found in the sputum and arising from subepithelial mucous gland ducts of bronchi. Of note is that in patients with quiescent asthma or history of asthma, the airways may appear entirely normal or otherwise show minimal changes of inflammatory or fibrotic type (Fig. 2.24)

possible complications, including ABPA, right heart-related congestive heart syndrome (*cor pulmonale chronicum*), acute hyperinflation without expiration possibilities, and *volumen pulmonum auctum* as indicated above (*auctus* derives from the Latin and means enlarged).

Right Middle Lobe Syndrome (RMLS)
- *DEF*: Atelectasis in the right middle lobe of the lung caused by various etiologies without a consistent clinical definition.
- *EPI*: The precise incidence in children is unknown. RMLS is widely underdiagnosed and frequently unrecognized. RMLS is more often to see in girls than in boys and particularly in children with a history of asthma or atopy.
- *EPG*: Although the mechanism by which asthma leads to lobar atelectasis is unknown, inflammation, bronchospasm, and secretions that cause mucus plugging are likely significant contributors. Specific anatomical features make the RML susceptible to transient obstruction as a result of frank inflammation

or edema. The slender Ø of the lobar bronchus and the acute takeoff angle create poor conditions for drainage. Bronchial obstruction can result from extrinsic compression as in hilar lymphadenopathy or tumor of neoplastic origin; however, atelectasis in children usually occurs from a process such as bronchial asthma-associated edema and inflammation. Foreign body aspiration into the RML orifice can also predispose to collapse of the lobe.
- *GRO*: Wedge-shaped density that extends anteriorly and inferiorly from the hilum of the lung.
- *CLM*: Histologically, chronic bronchitis/bronchiolitis with lymphoid hyperplasia, patchy organizing pneumonia, atelectasis, granulomatous inflammation, and an abscess can be seen. In patients with granulomatous inflammation, an associated atypical mycobacterial infection can also be observed (Fig. 2.25).
- *TRT*: Cause-specific management. Many RMLS patients respond to medical therapy alone.
- *PGN*: The chance of reinflation once atelectasis occurs is due to both the relative anatomical isolation of the RML and poor collateral ventilation decrease. Severity in children ranges from mild atelectasis and scarring of no consequence to severe bronchiectasis requiring surgical resection.

2.6.4 Pulmonary Hyperinflation

- *DEF*: Life-threatening acute/hyperacute increased air content of the lung.
- *SYN*: *Volumen pulmonum auctum* (*vide supra*).
- *EPI*: Case fatality from asthma in the USA is estimated at 5.2 per 10^5, with wide variations across Europe (e.g., 1.6 per 100,000 in Finland and 9.3 per 100,000 in Denmark).
- *ETP*: In obstructive airways, the expiratory airflow is more affected than the inspiratory airflow; at the end of an exhalation, there is ↑air in the lung (acute pulmonary obstruction). Although the ↑ of the intrathoracic gas volume

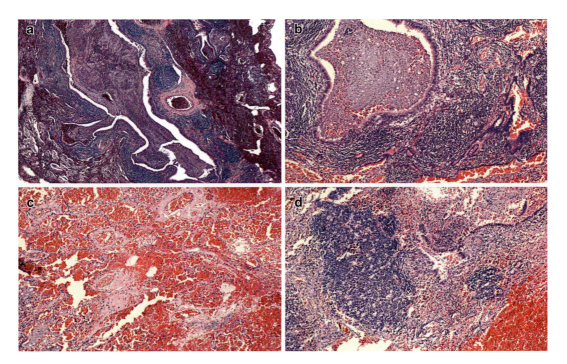

Fig. 2.25 Right middle lobe syndrome showing atelectasis, bronchiectasis, infection, and hemorrhage (**a–d**, H&E, 40× for **a**, and 100× for **b–d** as original magnifications)

is first reversible, chronic pulmonary hyperinflation results in irreversible pulmonary emphysema as a result of structural changes (e.g., asthma in teenagers and COPD in adults). In summary, two forms have been delineated, including an *absolute* hyperinflation or static (anatomically fixed pulmonary hyperinflation) with ↑ in total lung capacity (TLC) and, a probably enigmatic *relative* hyperinflation or dynamic with air trapping and ↑ residual volume (RV) at the expenses of the vital function. The use of e-cigarettes, vaporizers, and inhalers is being intensely investigated worldwide, but it seems that the damage from these forms of recreational activity may increase the rate of respiratory insufficiency of young adults in the nearest future.

- *CLI*: Life-threatening emergency with shortness of breath.
- *CXR*: Hypertransparency of the lungs (diminished density of the lung tissue), diaphragm level with flattened diaphragm couplings, drop heart – heart drawn with the heart downward, with the heart tip pointing downward as a result of the diaphragmatic lobe, and "Barrel chest" ("*Fassthorax*" of the German literature), broadened sagittal diameter with horizontally arranged ribs and extended intercostal spaces.
- *GRO*: Hyperinflation of both lungs with apical bullae.
- *CLM*: Acute marked distension of the alveolar spaces.
- *DDX*: It includes mostly a foreign body aspiration with occlusion of the bronchial airways.
- *TRT*: Some protocols are available for acute exacerbation of bronchial asthma and are available in internal medicine and pediatric respiratory books.
- *PGN*: Life-threatening condition with immediate medical intervention.

2.6.5 Bronchiectasis

- *DEF*: Permanent abnormal dilation of bronchi and bronchioles, usually associated with mucus plugging of the draining bronchial system and chronic, necrotizing inflammation.
- *EPI*: It may be quite variable, and CF, CVID, and ABPA may influence the correct rates.
- *CLI*: Fever, cough, and foul-smelling sputum.
- *GRO*: More often seen in left lung and more common in lower lobes.
- *CLM*: Bronchial lumen dilation + mucus plugs + chronic necrotizing inflammation.
- *DDX*: It includes some conditions including alpha-1-antitrypsin deficiency, congenital cartilage deficiency, RMLS, primary cilia dyskinesia (immotile cilia, Kartagener syndrome), pulmonary sequestration, tracheobronchomegaly, and "unilateral hyperlucent lung" (Swyer-James syndrome). Middle lobe syndrome is recurrent/permanent atelectasis of the right middle lobe or lingual with chronic inflammation associated with lymphadenopathy and/or malignancy (Gudbjartsson and Gudmundsson, 2012).
- *TRT*: It includes the prevention of exacerbations with regular vaccinations and measures to help clear airway secretions, avoidance of passive smoke exposure, the use of bronchodilators and inhaled corticosteroids, and antibiotics and bronchodilators for acute exacerbations. In rare cases, some patients will benefit from surgical resection for localized disease with intractable symptoms or persistent bleeding.
- *PGN*: Quite variable, but patients' experience ↓ in FEV1 is ~50–55 mL/year (normal ↓ in healthy individuals – 20–30 mL/year), and underlying disorders, such as CF, have the poorest prognosis, with a median survival of about 36 years with steadily intermittent exacerbations.

2.6.6 Neoplasms

There are very few literature data on tracheal neoplasms in childhood. The most frequent diagnoses of tracheal neoplasms include the chondroid hamartoma, tracheal chondroma, inflammatory myofibroblastic tumor, mucoepidermoid carcinoma, and rare occurrences of primary small round blue cell tumor (e.g., Wilms' tumor or rhabdomyosarcoma) (Fig. 2.23). Some of these entities will be discussed in detail later in the neoplasms of the lower bronchial system in this chapter.

2.7 Inflammation: Infectious (Pneumonia/Pneumonitis)

Infections of the bronchial and lung system are usually different among children, adolescents, young adults, older adults, and immunocompromised patients. Bacterial pneumonia is generally seen in adults, while viral or mycoplasma pneumonia is seen in children and youth. Opportunistic organisms (e.g., *Candida*, *Histoplasma*, *Pneumocystis jiroveci*) are at the origin of pneumonia in patients with altered host defense mechanisms such as prolonged antibiotic therapy, cancer chemotherapy, immunosuppressant drugs, and immunodeficiencies. Bacteria tend to cause pneumonia, which shows an intra-alveolar exudate leading to tissue consolidation, while viruses, *Chlamydia* spp., and mycoplasmas are prone to induce pneumonitis with a mainly interstitial infiltrate. Pneumonia may also be subdivided in segmental or lobular if the inflammatory process is patchy involving only a segment of a lobe and lobar if an entire lobe is involved. The term bronchopneumonia is used, if both alveoli and bronchi are affected. Patients affected with complicated pneumonia and pneumonitis may show congestive heart failure accompanied by poor ventilation and reduced vascular perfusion (Box 2.12) (Figs. 2.26, 2.27, and 2.28).

Box 2.12 Inflammation: Infectious (Pneumonia/Pneumonitis)
2.7.1. *Lobar Pneumonia*
2.7.2. *Bronchopneumonia*
2.7.3. *Interstitial Pneumonia/Pneumonitis*
2.7.4. *Primary Atypical Inflammation (Pneumonia)*

2.7.1 Lobar Pneumonia

- *DEF*: Multifactorial syndrome at neonatal onset with lower respiratory tract involvement due to absolute or relative, mostly diffuse, inflammatory alteration of the lung parenchyma. In more than 90% of cases, the agent is *Streptococcus pneumoniae*, most commonly types 1,3,7, and 2. Other agents are *Klebsiella pneumoniae, Staphylococcus aureus*, and *Haemophilus influenzae*.

Classically, four stages are seen, including:

- *Congestion* (~24 h): Vascular engorgement, intra-alveolar fluid, and bacteria, few/no PMN
- *Red hepatization*: Extravasated RBCs, fibrin, and increasing numbers of PMN
- *Gray hepatization*: Remarkable amount of fibrin and PMN as well as RBCs lysis
- *Resolution* with *restitutio ad integrum*

Pneumococcus (*Streptococcus pneumoniae*) is a Gram-positive, α- or β-hemolytic, facultative anaerobic member of the genus Streptococcus and characteristically the etiologic agent in alcoholic, debilitated, or malnourished individuals. During the four stages, there is initially an exudate of RBCs, PMNs, and fibrin that is progressively removed by the intra-alveolar entry of MΦ and lymphocytes, which characterize the late stages of lobar pneumonia with eventual resolution. It is important to remember that bacterial lobar pneumonia may complicate lung parenchyma following trauma, hemorrhage, infarction, neoplastic infiltration, and pneumonitis due to aspiration of gastric contents or inhalation of chemicals or toxins.

- *IMG*: In bacterial pneumonia, there are alveolar opacities, while interstitial opacities characterize atypical pneumonia. Lobar pneumonia: alveolar opacity, homogeneous, with pan-lobular extension and scissural demarcation and (+) air bronchogram (DDX – lobar atelectasis). *Staphylococcus* pneumonia: bronchopneumonia with disseminated foci with a lobular character with a tendency to an

2.7 Inflammation: Infectious (Pneumonia/Pneumonitis)

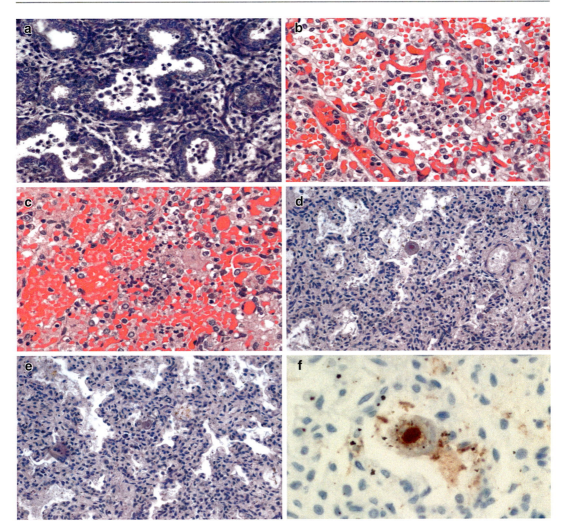

Fig. 2.26 Congenital pneumonia with microphotographs **a–c** showing intra-alveolar neutrophils in partly immature lung architecture from different premature babies, while microphotographs **d–f** show CMV pneumonia with nuclear inclusions (**a–e**, H&E at 400× original magnification. Microphotograph (**f**) shows the positive result of an immunohistochemical investigation with anti-cytomegalovirus (CMV) at 400× original magnification)

abscess formation (thin and regular walls quite different from the TB caverns) and pleural extension. *K. pneumoniae* pneumonia: bronchopneumonia with multiple and confluent foci. Bronchopneumonia: basal opacities, lobular, multiple, confluent with a cloudy aspect. Atypical pneumonia: ground-glass opacities, hilar enlargement, and characteristic conjunction striae. Air bronchogram: air-filled bronchi (dark) with surrounding alveolar opacification (gray/white) suggesting pulmonary consolidation, edema, nonobstructive atelectasis, ILD, bronchioloalveolar carcinoma, pulmonary lymphoma, infarct, and hemorrhage, other than normal expiration.

2.7.2 Bronchopneumonia

Multifactorial syndrome with lower respiratory tract involvement due to primary infection from *Staphylococcus*, *Pseudomonas*, *Proteus*, *Klebsiella*, and other Gram-negative coliform bacteria, which may often be considered hospital-

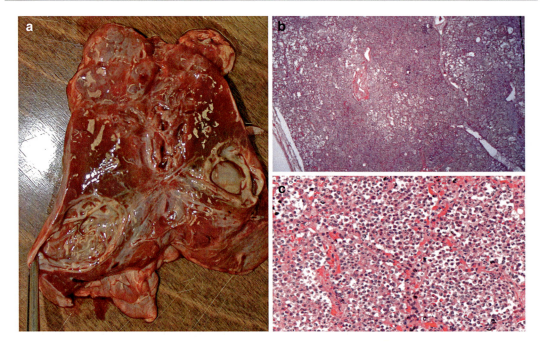

Fig. 2.27 Abscess pneumonia (**a**) showing an abscess in the lower lobe, while (**b**) and (**c**) show lobar pneumonia with packed neutrophils in the alveolar spaces (H&E for both microphotographs with (**b**) taken at 12.5× and (**c**) at 200× as original magnification)

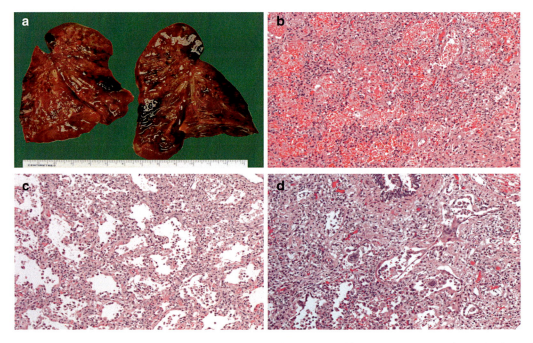

Fig. 2.28 Herpes simplex virus (HSV)-necrotizing pneumonitis showing a lung with bosselated surface and variegated color (**a**). The microphotographs (**b–f**) highlight the characteristics of HSV pneumonitis with hemorrhagic necrosis of the lung parenchyma (**b**), infiltration of the alveolar septs by inflammatory cells in a nonspecific interstitial pneumonitis (NSIP) pattern (**c**), intra-alveolar and interstitial multinucleated giant cells (**d** and **e**), and positivity for HSV antigens using monoclonal antibodies specific for HSV by immunohistochemistry (**f**) (**b–d**, H&E 100×; **e**, H&E, 400×; anti-HSV immunohistochemistry, 100×)

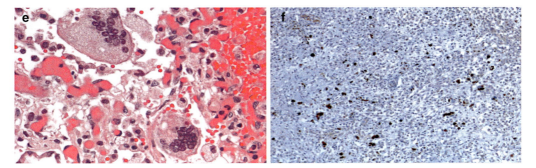

Fig. 2.28 (continued)

acquired infection. Bronchopneumonia has similar changes to lobar pneumonia, but a bronchial inflammation is also seen. Usually, right lower lobe, then RUL, then LLL, and significant destruction (out of proportion to inflammation) with foreign body multinucleated giant cells (FBGC or foreign body MNGC) are observed. Most commonly basal bronchopneumonia may evolve to abscesses, empyema, pneumatocele, and pyo-PNX, mainly when coagulase-positive staphylococci are the etiologic agent (Figs. 2.26 and 2.27).

2.7.3 Interstitial Pneumonia

It is usually virally based pneumonia with CMV, RSV, and measles virus as well as ortho- and paramyxoviruses as critical etiologic factors, particularly in immunocompromised patients. Viruses have a cytopathic effect directly against the epithelial cells of the lower respiratory tract with subsequent cell necrosis, necrotizing bronchitis, and terminal bronchiolitis, particularly adenovirus. Microscopically, there is atelectasis of the central lobule, prominent and edematous septs, and an inflammation consisting of lymphocytes, plasma cells, and histiocytes involving the interstitium. Hyaline membranes may also be seen in severe cases due to diffuse alveolar damage.

2.7.4 Primary Atypical Inflammation (Pneumonia)

It is infectious pneumonia, which is often seen in children and young adults and due mainly to *M. pneumoniae*, *Legionella pneumophila*, and *Chlamydia trachomatis*. The term atypical refers to the absence of the typical exudates and consolidation of lobar pneumonia. Microscopically, there are interstitial inflammation with bronchiolitis, erosion of the bronchial epithelium, and peribronchial lymphoplasmacytic infiltrates. A rising titer of cold agglutinins is seen in primary atypical pneumonia.

2.8 Inflammation: Non-infectious (Pneumonia vs. Pneumonitis)

Non-infectious forms of lung inflammation or non-infectious pneumonia include aspiration and chemical pneumonitis, lipoid pneumonia/pneumonitis, hemorrhagic pneumonitis, eosinophilic pneumonia, and hypersensitivity pneumonitis (Box 2.13).

2.8.1 Aspiration and Chemical Pneumonitis

Aspiration and chemical pneumonitis are defined as the passage of material from the oropharynx

> **Box 2.13. Inflammation: Non-infectious (Pneumonia Versus Pneumonitis)**
> 2.8.1. Aspiration/Chemical Pneumonitis
> 2.8.2. Lipoid Pneumonitis
> 2.8.3. Hemorrhagic Pneumonitis

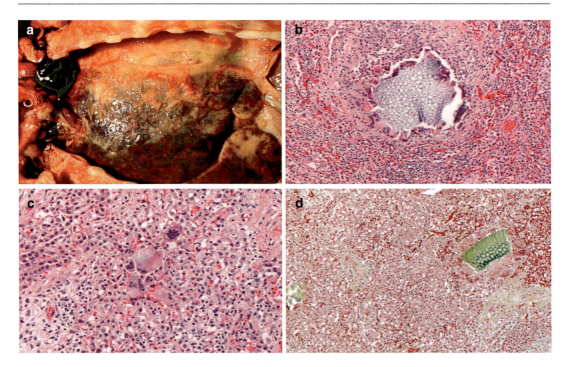

Fig. 2.29 Aspiration pneumonia showing an induration of the lung (gross photograph in **a**) and microscopic findings of foreign bodies and multinucleated giant cell reaction of the lung architecture (**b–d** with **b**, H&E at 100×; **c**, H&E at 200×; and **d**, Movat pentachrome stain at 100×)

into the tracheobronchial tree (Fig. 2.29). The predisposing conditions leading to tracheobronchial aspiration include neurological disease; esophageal diverticula, such as Zenker's diverticulum, gastroesophageal reflux, and swallowing disorders; intoxication/drug overdose/anesthesia (recent); and seizures (including febrile fits associated with a significant ↑ in body °C).

Symptoms depend on the type of material aspirated, the volume aspirated, and state of consciousness of the subject. The fate of aspiration depends on what is aspirated into the lungs. Thus, un-neutralized gastric acid produces a chemical pneumonitis, aka Mendelson syndrome. This syndrome can produce almost instantaneously airspace disease. If the volume of the aspiration is massive, the pattern is that of pulmonary edema. Mortality is up to 60% of the cases. Conversely, water or neutralized gastric acid may produce an ephemeral disease that can resolve in hours if aspiration does occur. Aspiration of bacteria can produce pneumonia. Community-acquired infections tend to be caused by *Streptococcus pneumoniae*, *Staphylococcus aureus*, *Haemophilus influenzae*, and *Enterobacteriaceae*, while hospital-acquired infections tend to be etiologically associated with *Pseudomonas* and other Gram-negative organisms. Typically, aspiration occurs in the gravity-dependent portions of the lung, such as the superior segments of the right-sided lower lobes (on the right, because of the larger caliber and straighter course of the right main bronchus) and the posterior parts of the upper portions. In a prone patient, aspiration is mostly in the right upper lobe and middle lobe or the lingual. Chronic aspiration pneumonia/pneumonitis is most frequently seen in patients with esophageal diverticula, achalasia, TEF, neuromuscular diseases, and GERD. Lipoid material causing lipoid pneumonia/pneumonitis may be mineral oil (used as a laxative) or oily nose drops. Clinically, there are choking or swallowing, acute respiratory distress, productive cough, and low-grade fever. Conventional CXR shows a fleeting infiltrate if bland and non-infected, while lobar consolidation is seen with the aspiration of anaerobic organisms or un-neutralized HCl. Grossly, the lung is firm and the lung parenchyma appears flesh-like. An important task requested to the pathologist, particularly to pediatric pathologist receiving a bronchoalveolar lavage (BAL) from a patient with a history of aspiration or suspicion of aspiration pneumonitis, is to perform the semiquantitative evaluation of the Oil Red O (ORO)-stained macrophages. ORO special stain needs some precautions, including avoiding alcohol-containing fixatives (propylene glycol is used instead) and frozen section (10 microns in thickness) or cytospin air-dried for at least 15 min. The adrenal gland or fatty liver should be used as control tissue. Neutral lipid stains in red, whereas nuclei are counterstained in blue using Mayer's hematoxylin. Differential cell counting is helpful to identify how many macrophages are present, but ORO stain is a useful tool for the quantification of lipid-laden macrophages, a finding identified as early as 1928. Continued aspiration leads to the accumulation of lipid-laden macrophages in the alveolar spaces. Repeated aspiration of gastric contents may be the consequence of recurrent pneumonia, asthma, bronchitis, and atelectasis. It has been proposed that the mere presence of lipid-laden macrophages in the lower respiratory tract is a marker of pulmonary disease related to aspiration. However, intra- and interobserver variability is a significant problem in evaluating this parameter in the BAL. In 1985, Corwin and Irwin found an increased number of lipid-laden macrophages in the BAL specimens of chronic aspirators and used these cells to create a semiquantitative test of aspiration called the *lipid-laden macrophage index* (LLMI).

Subsequently, Colombo and Hallberg modified the LLMI using extensive studies in rabbits, and the test is now commonly used to evaluate aspiration in children. Langston and Pappin use descriptive terms in their reports such as none, rare, moderate, or numerous lipid-laden macrophages rather than the numerical values. Collins et al. have proved to use a 4-grade system to score the number of lipid-laden macrophages in a semiquantitative way (grade 0, none; 1, 1–25; 2, 26–50; 3, >50 lipid-laden macrophages per slide). Yang et al. have suggested counting 100 macrophages but to use a 2-grade system (grade 1, if macrophages have less than 50% cytoplasmic lipid and grade 2 if macrophages show more than 50% lipid). However, several studies have raised some limitations for this score or any other scoring systems, and it is not considered a gold standard yet, although in many institutions is used. LLMI limitations come out when healthy non-aspirating children and adults have elevated LLMIs. Possible causes are parenteral nutrition with intravascular intralipid in infants, subclinical aspiration in otherwise healthy children, and tissue breakdown of surfactant. LLMI may also be elevated in ILD, lung neoplasms, obesity, COPD, and smoking. Any quantitative measurement tool is subject to inter- and intraobserver variability, but the degree of variability should be minimal if test results are to be considered reliable and reproducible. A k-value of ≥ 0.75 has been suggested as an optimal cutoff. The lack of such characteristics allows questioning the utility and validity of such test. Clinicians and (cyto-)/pathologists comparably should be aware of this pitfall and interpret LLMI accordingly.

2.8.2 Lipoid Pneumonia

- *DEF*: An explicit form of pneumonia that develops when lipids enter the bronchial tree.
- *SYN*: "Golden pneumonia."
- *EPG*: (1) Endogenous lipoid pneumonia due to the tumor, LN, abscess ⇒ lipid-derived from degenerating type II pneumocytes, cholesterol clefts, and MNGC. It is paramount to remember that exogenous lipoid pneumonia due to aspiration of lipid material more often occurs into the right lung than left and this phenomenon takes place because of the non symmetrical bronchial airways anatomy.
- *GRO*: Well circumscribed, firm lung or part of it.
- *CLM*: Lipid accumulation within foamy MΦ accompanied by inflammation, proliferating type II pneumocytes, cholesterol clefts, MNGC ± reactive endarteritis.

> **Box 2.14 Diffuse Pulmonary Hemorrhagic Syndromes (DPHS)**
> 2.8.3.1. Goodpasture Syndrome
> 2.8.3.2. Idiopathic Pulmonary Hemosiderosis
> 2.8.3.3. Vasculitis-Associated Hemorrhage (Necrotizing Capillaritis)
>
> Notes: Necrotizing capillaritis is substantially the basis that is present in microscopic polyangiitis (MPA), Wegener granulomatosis (WG), systemic lupus erythematosus (SLE), but also Churg-Strauss syndrome/eosinophilic granulomatosis with polyangiitis (CSS/EGPA), although it is less commonly seen in this entity.

2.8.3 Diffuse Pulmonary Hemorrhagic Syndromes

Diffuse pulmonary hemorrhagic syndromes is a heterogeneous group of syndromes with diffuse lung involvement and characterized by intrapulmonary bleedings (Box 2.14).

2.8.3.1 Goodpasture Syndrome
- *DEF*: Autoimmune pulmonic-renal disease characterized by circulating autoantibodies against the non-collagenous 1 (NC1) domain of an α3 chain of collagen IV
- *SYN*: Anti-basement membrane (BM) disease.
- *EPI*: Worldwide, ♂:♀ = 2:1, children/youth.
- *EPG*: HS-II with the initial inflammatory-based destruction of both the BM of pulmonary alveoli giving rise to a necrotizing hemorrhagic interstitial pneumonitis and the BM of renal glomeruli with a secondary proliferative and rapidly progressive glomerulonephritis (HS-II, hypersensitivity type II).
- *CLI*: (1) Lung symptomatology with hemoptysis, pulmonary infiltrates (CXR), and anemia-related respiratory symptoms and (2) kidney symptomatology with the acute nephritic syndrome (*vide infra*).
- *LAB*: (+) Serum for an anti-NC1 α3 chain of collagen IV antibodies, anemia, ↑ of nephritic syndrome-related serology markers.
- *GRO*: Heavy lungs ("pulmōnēs gravēs") with focal areas of necrosis.
- *CLM*: (1) Necrotizing hemorrhagic interstitial pneumonitis with RBCs accumulation and hemosiderin deposition in the alveoli, ↑ alveolar Mφ and pneumocytes type II, and nonspecific thickening of the alveoli but no vasculitis (usually) and (2) crescentic glomerulonephritis.
- *IFM*: Diffuse linear staining for IgG/IgM/IgA along alveolar septa.
- *TEM*: Breaks of capillary BM and gaps are widening between endothelial cells.
- *TNT*: Aggressive therapy is needed (life-threatening disease), which requires plasmapheresis.
- *PGN*: Renal involvement-dependent.

2.8.3.2 Idiopathic Pulmonary Hemosiderosis
- *DEF*: Probably autoimmune pulmonic-restricted disease characterized by diffuse alveolar hemorrhages without the renal involvement and no identifiable circulating autoantibodies.
- *EPI*: Worldwide, ♂ = ♀, children and youth.
- *EPG*: Although it is not fully understood, IPH is probably an autoimmune disorder with initial damage to the lung capillaries of the alveoli and subsequent bleeding with iron accumulation and progressive pulmonary fibrosis.
- *CLI*: Lung symptomatology with hemoptysis, pulmonary infiltrates (CXR), and anemia-related pulmonary symptoms.
- *LAB*: Serology (−) for an anti-NC1 α3 chain of collagen IV antibodies, in addition to anemia and lack of nephritic syndrome-related serology, as seen in Goodpasture syndrome.
- *GRO*: Brown heavy lung parenchyma.
- *CLM*: Necrotizing hemorrhagic interstitial pneumonitis with RBCs accumulation and hemosiderin deposition in the alveoli, ↑ alveolar Mφ and pneumocytes type II, and nonspecific thickening of the alveoli but no vasculitis (usually).
- *PGN*: Often self-limiting, improving without therapy, but may be fatal in some cases.

2.8.3.3 Vasculitis-Associated Hemorrhage (Necrotizing Capillaritis)

Necrotizing capillaritis is the basis that is present in microscopic polyangiitis (MPA), Wegener granulomatosis (WG), systemic lupus erythematodes (SLE), but also Churg-Strauss syndrome, although it is less common in this entity.

MPA is a necrotizing capillaritis, while *WG* is a T-cell-mediated hypersensitivity reaction, probably to inhaled infective agents or some environmental agents with subsequent development of a necrotizing capillaritis and granulomas of the upper and lower respiratory tract as well as focal or diffuse necrotizing glomerulonephritis with the crescent formation of the top urinary system.

Granulomatosis with polyangiitis, idiopathic pulmonary hemosiderosis, and Goodpasture syndrome may be the major players of the pulmonary hemorrhagic syndromes.

Granulomatosis with Polyangiitis (Wegener Granulomatosis)

- *DEF*: Disorder that causes inflammation of the blood vessels in the nose, sinuses, throat, lungs, and kidneys.
- *SYN*: Wegener granulomatosis.
- *CLI*: Upper and lower respiratory tract (URT and LRT) mucosal ulceration + c-ANCA (+) (70%).
- *CLM*: Extravascular necrotizing granulomas (URT and LRT) + generalized vasculitis + GN (necrotizing FSGN/crescentic, pauci-immune): serpiginous/geographic necrosis with nuclear debris ("dirty" necrosis) surrounded by loose/palisaded granulomatous inflammatory ± MNGC.
- Variants: Eosinophilic, bronchocentric, BOOP-like.
- *DDX*: Lymphomatoid granulomatosis (atypical lymphocytes), TB (caseating granulomas, ZN (+), ARC (+)), sarcoidosis with necrotizing angiitis (confluent granulomas), ABPA (GMS (+)).
- *PGN*: Better prognosis is the limited form of the lung when compared to the multivisceral involvement, but the lung can also heal with moderate/severe fibrosis ⇒ UIP.

2.9 Inflammation: Infectious/Non-infectious Granulomatous/Nongranulomatous

In this category, there are varieties of gross and microscopic patterns, and it is essential to keep in mind that an infectious granulomatous inflammation can mimic any of the non-infectious granulomatous diseases (Box 2.15).

Thermally dimorphic fungi have hyphae that produce spores at environmental temperatures and grow as yeasts (spherules/ellipses) at body temperature (intrapulmonary). Characteristically, in Ohio and Mississippi rivers as well as the Caribbean, the *Histoplasma* is ubiquitous. In Central and Southeastern USA, there is more often *Blastomyces* as fungus, while the *Coccidioides* is prevalent in Southwest and Far West of the USA and Mexico.

2.9.1 Primary Pulmonary Tuberculosis

- *DEF*: Chronic granulomatous inflammation of the lung and other organs (multivisceral forms) caused by the infection with *Mycobacterium tuberculosis* (Fig. 2.30).
- *SYN*: Koch bacillus infection.
- *EPI*: The epidemiology of tuberculosis (TB) in high-income countries is increasingly linked with immigration, and the influence of this trend on pediatric TB is far to be correctly defined. Foreign-born children or North America-born children of foreign-born parents account for an increasingly large proportion of total cases currently.

> **Box 2.15 Inflammation: Infectious/Non-infectious Granulomatous/Nongranulomatous**
> 2.9.1. *Primary Pulmonary Tuberculosis* (Fig. 2.30)
> 2.9.2. *Sarcoidosis*
> 2.9.3. *Aspergillosis*
> 2.9.4. *Others*

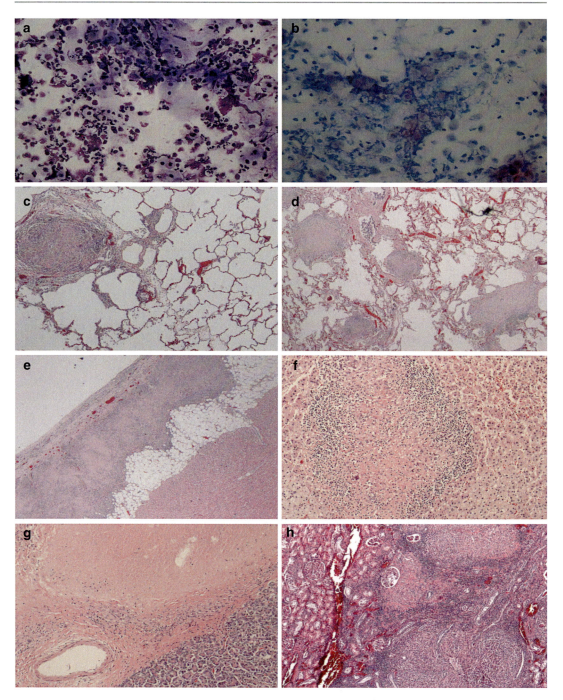

Fig. 2.30 Miliary tuberculosis showing BAL positive for mycobacteria (**a–b**) and several microphotographs (**c–h**) showing multiple granulomas of caseous necrosis type in the lung and in extrapulmonary sites (liver, **f**; pancreas **g**; and kidney, **h**). Miliary tuberculosis is named as such because of the innumerable tiny spots that form in the lungs are the size of millet, which is the small round seed in bird food (mean, 2 mm; range, 1–5 mm)

- *EPG*: *M. tuberculosis*. Other mycobacteria may present with similar findings. Initially, *M. tuberculosis* was on the spot because a transmission by inhalation and targeting humans was determined, but *M. bovis* has more recently been found to cause GI infections from infected milk and be an important cause of tuberculosis dissemination. In fact, according to recent epidemiological and microbiological investigations, *M. bovis* has been found as the cause of pulmonary tuberculosis as well.
- *CLI*: Signs and symptoms of active TB include >3 weeks lasting coughing, bloody expectorate, chest pain, or pain with breathing or coughing, failure to thrive, weight loss, fatigue, fever, night sweats, and chills.
- *GRO*: Consolidation of the focus at the pulmonary areas where there is the most significant volume of air flow.
- *CLM*: Slightly wavy or beaded bacteria, which are Gram+, ZN+, SS+, and AFB+ and located in the caseating granuloma, which is the hallmark lesion of tuberculosis ("soft tubercle"). The "hard tubercle" represents granulomas with a cellular center. *Ghon focus*: 1–1.5 cm gray-white inflammatory nodule seen at the periphery of the upper part of a lower lobe or lower part of upper lobe which becomes granulomatous and then centrally necrotic by the 2nd week and usually clinically silent. Conversely, the *Ghon complex* is a combination of primary lung lesion, draining lymphatics, and ipsilateral lymph node involvement. Occasionally, the primary focus may enlarge, erode into bronchi, giving rise to the satellite with bloodstream (miliary tuberculosis). The term "miliary" arises from the similarity to innumerable "millet seeds" observed on CXR in cases of tuberculosis dissemination. The term *Ranke complex* is reserved to a Ghon complex undergoing fibrosis and calcification.
- *DDX*: Other granulomatous infections of the lung.
- *TRT*: The most common medications used to treat tuberculosis include isoniazid, rifampin, ethambutol, and pyrazinamide.
- *PGN*: Although rare in high-income countries, miliary dissemination may become a reality, particularly in immunodeficient children. The increasing burden of pediatric TB in both foreign-born children and North America-born children of foreign-born parents suggests for more timely diagnosis of source cases. More effective and efficient screening of latent TB infection should be one of the primary tasks of the future.

Typical mycobacteria (e.g., *M. tuberculosis*, *M. bovis*, *M. leprae*) are 3–5-µm-long bacteria, which may be slightly wavy or beaded. Conversely, atypical mycobacteria (e.g., *Mycobacterium avium intracellulare*) are somewhat longer and thicker. The identification of mycobacteria can be very time-consuming. It has been advised to use thick sections, screen for 20 min at 40× and focusing up and down. Secondary TB usually arises from reactivation of old primary lesions creating progressive pulmonary TB, tuberculous empyema, intestinal TB, in case of swallowing of aspirated material, or miliary seeding. Isolated distant organ involvement includes extrathoracic lymph nodes, meninges, kidneys, adrenal glands, bones, fallopian tubes, testis, and epididymis. Secondary TB is usually accompanied by fever, night sweats, weakness, failure to thrive, fatigability, and loss of appetite. As historical notes, Heinrich Hermann Robert Koch (1843–1910) was a German physician and microbiologist. He is at the basis not only of the tuberculosis, but he is still recognized for paving the basis of clinical microbiology using his four postulates, including 1) the need of the microorganism to be always present in every case of the disease; 2) the need of the isolation of the microorganism from a host containing the disease and ability to grow in pure culture; 3) the need to prove a causative role for the microorganism taken from pure culture ro induce disease after inoculation into a healthy animal; 4) the need to have the same disease and same microorganism in the inoculated animal as first isolated from the primary diseased host. The date of March 24, 1882 is recorded as a milestone in clinical microbiology because that day Robert Koch announced to the Society of Physiology in Berlin Physiological Society that he had discovered the cause of tuberculosis. His presentation was followed by a scientific paper a few weeks later. His name is behind tuberculosis, cholera, and anthrax and received the Nobel Prize in Physiology or Medicine in 1905 for his research on tuberculosis.

2.9.2 Sarcoidosis

- *DEF*: Multisystem-involving disease of unknown etiology (probably or possibly infectious) characterized by a frequent involvement of the lower respiratory tract with non-necrotizing granulomatous inflammation showing numerous clusters of epithelioid histiocytes, ↑ of B lymphocytes, polyclonal hypergammaglobulinemia, and CD4 > CD8 rates.
- *SYN*: Besnier-Boeck-Schaumann disease.
- *EPI*: Children and youth (15–35 years), ♂ < ♀, Scandinavia, US blacks.
- *EPG*: Unknown agent, but the characteristic immunologic pattern includes
  1. Abnormal ↓ T-cell-mediated immunity (CD8) with ↓ T cells in PB and *"anergy"* due to HS-type IV failure using antigen injected in intradermal skin tests (PPD, Purified Protein Derivative)
  2. ↑ T_{H1}–response inducing:
     ⇒ ↑ IL2 ⇒ T-cell exp.
     ⇒ ↑ IFN-γ ⇒ Mφ activation
     ⇒ ↑IL-8, TNF-α, MIP1α ⇒ epitheliod cell-based granulomas (mainly TNF-α) (MIP, Macrophage Inflammatory Protein)
  3. ↑ T-cell activity of helper class (CD4), ↑ humoral immunity (↑ B cells, ↑ Ig-polyclonal) ⇒ BAL showing CD4 > CD8 (5:1/15:1) on flow cytometry (BAL, bronchoalveolar lavage; FC, flow cytometry).
- *CLI*: Several and different organs and tissues may be involved. Lungs are frequently involved. The involvement of the lower respiratory system may be the only manifestation in >1/2 of cases, CXR+ (interstitial infiltrates with hilar adenopathy), and TBB+ (non-necrotizing granulomatous inflammation) (Box 2.16) (TBB, transbronchial biopsy). The term "potato nodes" is sometimes used in this setting because of the striking enlargement of the lymph nodes observed on CXR or CT scans.
- *LAB*: ↓ T-cell-mediated immunity (CD8), ↑ T-cell activity of helper class (CD4), ↑ humoral immunity (↑ B cells, ↑ Ig-polyclonal) ⇒ BAL showing CD4 > CD8 (5:1/15:1) on FC.
- *Serum*: ↑ACE, hypercalcemia, hypercalciuria, (+) Kveim test. The Kveim test, Nickerson-Kveim or aka Kveim-Siltzbach test is not usually performed because of the risk that certain infections, such as bovine spongiform encephalopathy, could be transferred during this test.
- *GRO*: Consolidation without caseous necrosis.
- *CLM*: Non-necrotizing granulomatous inflammation composed of tight clusters of epithelioid histiocytes, MNGC of Langerhans type, few lymphocytes, and tendency to locate in interstitium vs. airspaces, along lymphatics, ± granulomatous vasculitis, and tendency to progressively hyalinize but no caseous necrosis, which is typically seen in tuberculosis.

Nodular sarcoidosis: Coalescence of large areas of hyalinized granulomas.

- *DDX*: Infection (special stains and culture for fungi), TB (special stains and culture),

---

Box 2.16 Sarcoidosis: Clinical Presentation and Tissue/Organ Involvement
1. *C*ardio involvement
2. *R*enal involvement
3. *I*ntestinal involvement
4. *L*ung involvement (bilateral coalescent granulomatous areas with ↑ hilar LNs)
5. *L*iver/spleen involvement
6. *L*N involvement (non-necrotizing granulomatous lymphadenitis)
7. *E*yes involvement: uveitis
8. *S*kin involvement: *erythema nodosum*
9. *S*keletal involvement: sarcoid arthritis

Notes: The mnemonic word CRI-LESS is useful to remember the various organs involved in this multisystem disease

Mycobacterium avium intracellulare (special stains and culture), reactions to metals/minerals (clinical history, polarization field), hypersensitivity pneumonitis (loose granulomas with more cellular inflation), local sarcoid-like responses (as seen adjacent to neoplasms and in draining lymph nodes), Wegener granulomatosis (suppurative "dirty" necrosis with neutrophils and necrosis surrounded by palisaded histiocytes).
- *TRT*: In Nathan's study patient data were analyzed after a follow-up of 18 months and 4–5 years. In the <10 years group, 7 of the 14 patients recovered completely and were free of health-related problems from sarcoidosis, while three patients exhibited improved clinical manifestations. In 1/3 of the cohort, relapses were seen leading to an increase in steroid treatment and for three of them the introduction of immunosuppressive therapy (mycophenolate mofetil, hydroxychloroquine, infliximab, methotrexate, azathioprine, cyclophosphamide). In the 4–5-year follow-up, in the <10 years group, only 1/3 of the patients experienced transient relapses. In the ≥10 years group, declines were seen in about half of the patients with current treatment regimens including steroids (oral or IV), of which seven received immunosuppressive drugs.
- *PGN*: 2/3 hilar sarcoidosis ⇒ spontaneous remission, but if lung disease is (+) ⇒ chronicity!

2.9.3 Aspergillosis

- *DEF*: Fungal infection due to *Aspergillus* spp. with three types of infection (Fig. 2.31).
- *EPI*: Ubiquitous fungus.
- *EPG*: Aspergillus 2nd most common infective fungus, especially in hospitals with 45° angle branching ("Y"-branching) septate hyphae with standard thickness (~4 μm) (dichotomous branching) (≠ mucormycotic zygomycosis due to *Rhizopus* spp.) (*vide infra*). *Aspergillus fumigatus* is the most common class identified in all pulmonary syndromes. *Aspergillus flavus* is more often involved in several forms of allergic rhinosinusitis, postoperative aspergillosis, and fungal keratitis, while *Aspergillus terreus*, which is amphotericin B resistant, is often observed in IA (invasive aspergillosis) in some institutions. Occasionally, *Aspergillus niger* is seen in IA or aspergillus bronchitis.
- *CLI*: Several forms mainly according to the immune system of the host, including invasive aspergillosis (IA), chronic pulmonary aspergillosis (CPA), allergic bronchopulmonary aspergillosis (ABPA) and aspergilloma as a late manifestation of CPA.
- *CLM*: Basophilic hyphae (viable) and eosinophilic hyphae (necrotic) and conidia (spores) and conidiophores (fruiting bodies) can form and be observed in cavities.

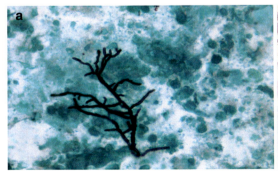
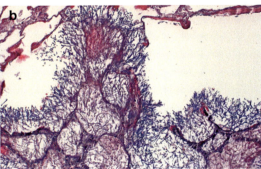

Fig. 2.31 Aspergillosis of the lung showing the characteristic hyphae (45 degrees branching and septation) in the bronchoalveolar lavage (BAL) of a patient positive for *Aspergillus spp.* (**a**, Grocott methenamine silver, GMS, 400×), a histologic preparation with diffuse Aspergillus proliferation in the alveoli (**b**, H&E at 100×), two gross photographs of the hemorrhagic necrosis with two levels of intensity of pulmonary necrosis (**c, d**), and microphotographs of the necrotic areas showing poor preservation of the tissue following the infection with Aspergillus in this patient (**e**, H&E, 12.5×; **f**, H&E, 400×)

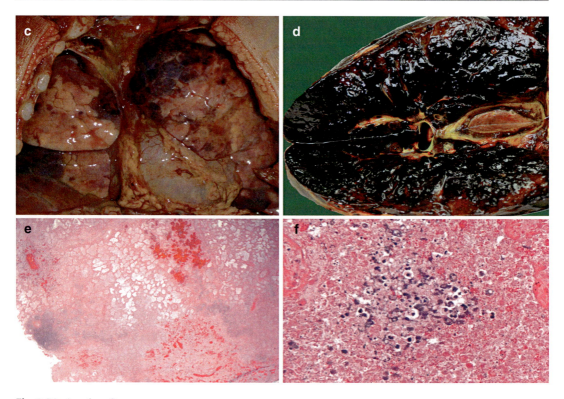

Fig. 2.31 (continued)

- *DDX*: *Mucormycotic zygomycosis* due to *Rhizopus* spp. shows irregularly branching aseptate/pauci septate hyphae of variable thickness (up to 20 μm) and random orientation with 90° angle branching and thinner than the stem (nondichotomous branching). Other two differential diagnoses are *Pseudallescheria* spp. (conidia differentiation) and fusariosis (90° branching).
- *TRT*: Voriconazole is the antifungal agent of choice for treatment of IA.
- *PGN*: It depends on the clinical form and the immune system of the host.

IA mainly affects patients with profound life-threatening defects in immune function (e.g., hematopoietic stem cell TX or solid organ TX recipients, patients undergoing CHT or steroid treatment). The more indolent form of aspergillosis is the *CPA*, which affects patients with underlying lung disease (e.g., cystic fibrosis), but with no or only minimal generalized immune compromise. The third form, *ABPA*, implicates an allergic response to inhaled *Aspergillus* spp. and results in severe asthma with fungal sensitization and asthmatic crises. The *aspergilloma* is rounded cluster of fungal hyphae, mucus, fibrin, and cellular debris that originates in pre-set lung cavities that have become colonized with *Aspergillus*.

2.9.4 Others

These lesions include the following etiologic categories: histoplasmosis, coccidioidomycosis, cryptococcosis, blastomycosis, dirofilariasis, pneumocystosis, mucormycosis, cytomegalovirosis, and adenovirosis.

2.10 Chronic Obstructive Pulmonary Disease of the Youth

In this session, chronic bronchitis and emphysema of the young are presented. Both are quite rare in most of the countries with children with respiratory disorders. However, the early start of smoking, the global use of vaping, the setting of urban dwelling, and the chronicity of pathology have influenced to bring this characteristic of adult pathology to children.

2.10.1 Chronic Bronchitis

- *DEF*: Inflammation of the bronchial tree characterized by a persistent cough on most days for ≥3 months for ≥2 consecutive years.
- *ETP*: ± Evidence of airway obstruction, 10–25% of urban dwelling adults, and active and passive smoking are the most critical causes with hypersecretion of mucus.
- *CLI*: There is a persistent cough on most days for ≥3 months for ≥2 consecutive years.
- *MIC*: Hypersecretion of mucous with ↑ # goblet cells in both small airways and large airways and ↑ size of submucosal glands in large airways (abnormal Reid index) + peribronchiolar chronic inflammation (lymphocytes) (Box 2.17).
- *TRT*: Stage-dependent, but always treat acute lung infections, adapt to low-flow O_2 therapy, postural drainage, percussion, nebulized corticosteroids, elimination of active/passive smoking.
- *PGN*: Progressive disease (poor PGN markers include severe airflow obstruction, inadequate capacity of physical activity, severely short of breath, significant over/underweight, respiratory failure, *cor pulmonale*, active smoking or continuous exposure to passive smoking, frequent acute exacerbations).

The Reid mean value in healthy individuals is ~0.25 (~1/4 of the thickness of the wall between the surface epithelial basement membrane and the external perichondrium). In patients with chronic bronchitis as well as in patients with cystic fibrosis who may develop chronic bronchitis, Reid index is ~0.6.

2.10.2 Emphysema

- *DEF*: Abnormal permanent enlargement of the distal airspaces due to the destruction of the alveolar walls with rarefaction of the alveolar septs and definitive loss of respiratory tissue
- *EPG*: There is a lack of elastic recoil, and the most common cause is smoking, which pro-

Box 2.17 Reid Index
It represents the ratio of the thickness of mucous glands (MGT: mucosal glandular thickness) to a depth of the wall (WT: whole thickness) between the basement membrane of the surface epithelium and perichondrium.

Note: the blue bar is the cartilage, while the yellow circle is the cluster of the subepithelial mucous gland.

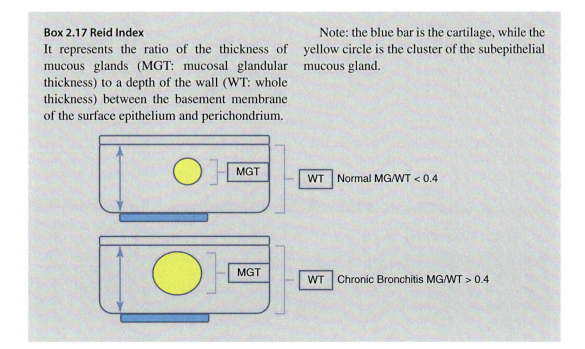

duces a combination of emphysema and chronic inflammation, while alpha-1-antitrypsin deficiency (AATD) produces almost pure emphysema. The use of vaping and electronic cigarette (e-cigarette) is increasing at a very alarming rate. Vaping is becoming more popular with youth in middle and high school. No or poor regulation and diffuse pseudo-knowledge about safety have worsened the situation in the last few years. Vaping includes the term of e-cigarette or other vaping device. E-cigarettes are the battery powered devices that deliver a nicotine-containing liquid, which changes into a vapor when turning on the devices. Chronic e-cigarette use does increase the neutrophil elastase and matrix metalloprotease levels in the lung (Ghosh et al. 2019).

- CLI: 2 Major Types
  - *Centroacinar* (*centrilobular*) emphysema: Proximal parts of the acini (respiratory bronchioles) with sparing of the distal alveoli and more severe involvement in upper lobes, especially apical segments, observed in smoking and secondary to coal dust.
    Blue bloater-type: Severe effects due to shunting not compensated ⇒ ↑ CO_2
    Hypercapnic respiratory failure (e.g., pulmonary HTN, cor pulmonale)
  - *Panacinar* (*panlobar*) emphysema: Uniform enlargement of all acini in a lobule without sparing of any portion and more severe involvement of the lower lobes, observed in AATD
  - *Pink puffer-type*: Milder effects with compensation by ↑ ventilation ⇒ ↓ CO_2

Other types include the *paraseptal* variety (distal acinar involvement with normal proximal acinus and most prominent adjacent to pleura and along the lobular connective tissue septs as observed in adolescents and youth with spontaneous PNX). Moreover, there are also the *bullous* type (any form of emphysema which produces large subpleural bullae with Ø > 1 cm), *interstitial* type (air penetration into the connective tissue stroma of the lung, mediastinum, or subcutaneous tissue), and the *compensatory* type (pseudo-emphysematous dilatation of alveoli without destruction of septal walls in response to loss of lung tissue). The *senile* type (pseudo-emphysematous change in lung geometry with larger alveolar ducts and smaller alveoli without loss of lung tissue) does not play any role in childhood or youth unless we consider the progeria syndrome and progeroid conditions.

2.11 "Adult" Pneumoconiosis of the Youth

The concept that pneumoconiosis is diseases of the adult is probably not valid anymore. There are still vast geographical areas, where asbestos is still part of the production, and children of poor-developed countries may suffer from the exposure to asbestos due to the contact with dresses and wearing used from parents and relatives. This aspect may also be real for other mineral or organic dusts as emphasized in most recent reviews (Box 2.18).

Typically, silicosis is a disease of long latency affecting mostly older workers, but silicosis deaths in young adults (aged 15–44 years) implicate acute or accelerated disease. In 2017, CDC (Centers for Disease Control and Prevention) published a report which analyzed the underlying and contributing causes of death using multiple cause-of-death data (1999–2015) and industry and occupation information abstracted from death certificates (1999–2013). CDC identified 55 pneumoconiosis deaths of young adults during 1999–2015. They used the International Classification of Diseases, 10th Revision (ICD-10) code J62 (pneumoconiosis due to dust containing silica). Of the 55 pneumoconiosis deaths in youth, 38 had pneumoconiosis due to other soil containing silica, and 17 had pneumoconiosis due to talc dust. Decedents worked in the manufacturing and construction industries

Box 2.18 Adult Pneumoconiosis of the Youth
2.11.1. *Silicosis*
2.11.2. *Asbestosis*
2.11.3. *Non-silico-Asbestosis-Related Pneumoconiosis*

and production occupations where silica exposure occurs. The 17 deceased adolescents and young people had an underlying or a contributing cause of death listed that indicated multiple drug use or drug overdose, but it does not seem that their talc-related death was occupational because none was working or was exposed in talc exposure-associated jobs. Mesothelioma is rare cancer most commonly diagnosed in people in their 60s and 70s, but over 300 cases of mesothelioma have been reported worldwide in young adults, children, and even infants. There is plenty of literature that in most cases of mesothelioma diagnosed in childhood and youth, there is no clear-cut family or personal history of exposure to asbestos differently from the adult cohorts, but the publications are often missing a proper environmental incidental report. In fact, most exposures to asbestos may come from the environment or exposure from fathers or mothers working in asbestos exposure-related job companies and bringing the working dress home. Secondhand exposure from a parent who worked with asbestos is a very critical cause in mesotheliomas of childhood and youth. Moreover, asbestos is still present in schools, and environmental asbestos is still a significant burden today because children can breathe the air or play in or eat contaminated soil. Asbestos in toys such as chalk, crayons, and modeling clay may be an additional factor (Fig. 2.32).

2.11.1 Silicosis

- Silica associated lung disease: SiO_2 (e.g., quartz).

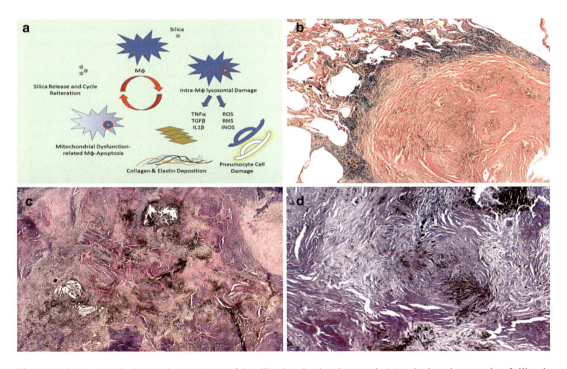

Fig. 2.32 Pneumoconiosis showing a schema of the silica involved pathogenesis (**a**) and microphotographs of silicosis of the youth, which is similar to the pneumoconiosis of advanced age. There is fibrous collagenous silicosis with hemosiderin-laden macrophages (**b**: Perls' Prussian Blue stain, 40×; **c**: H&E, 20×; **d**: Periodic acid Schiff stain, 100×). The main steps of the iterative cycle of pneumoconiosis include direct cellular toxicity, production of reactive oxygen species (ROS), reactive nitrogen species (RNS), and inducible nitric oxide synthase, secretion of inflammatory and fibrotic mediators, lung remodeling through collagen and elastin deposition, and programmed cell death with release of indigestible silica. Freshly fractured elements of silica stimulate mitogen-activated protein kinase (MAPK) family members, as was proved by Ding et al. (1999) evidencing the phosphorylation of p38 MAPK and extracellular signal-regulated protein kinases

- *Acute silicosis* ("accelerated silicosis"): Heavy exposure over 1–3 years with a pulmonary alveolar proteinosis-like appearance with variable amounts of interstitial fibrosis.
- *Chronic silicosis*: Pneumoconiosis following 20–40 years exposure to dust containing up to 30% quartz. Activities with mining, sandblasting, metal grinding, tunneling, and ceramics are at risk. There is a highly debated frame of progression with regards to the years after exposure and subsequent proliferation of hyalinized nodules and the reality is that there are individual and environmental factors that may play an important role. Chronic silicosis usually prefers the peripheral areas at the upper lobes. There may be variable amounts of black pigment (concomitant exposure to coal dust), well-delimited, concentric lamellated hyaline, with rare MNGC or granulomas and rare fields with necrosis. Finally, an expansion of the nodules with obliteration of the small airways does eventually occur.
- *CLM*: Whorled laminar collagen with doubly birefringent refractile round particles (1–2 μm) and peripheral plasma cells ⇒ silicotic nodules ⇒ conglomeration (obliteration of the small airways), which is often responsible for the final lung insufficiency.
- *PGN*: Complications – TB (in average 20× increased risk to contract this infection; 10–30× relative risk as a common range in numerous institutions involving individuals of different age).

In childhood, silicosis due to secondary exposure can occur and particularly among the children of mining communities in India. Unfortunately, malnourishment, tuberculosis, and silicosis are co-morbid conditions, especially in vulnerable populations like children who live close to stone mines. Important works of the American Academy of Pediatrics (AAP), Section on International Child Health (SOICH), in advocating more heath for children worldwide are key. Children are especially susceptible to silicosis from nonoccupational exposure in stone-mining areas where (a) communities are impoverished and have little access to modern medical facilities, (b) families live close to the workplace, and casual child labor is common, and (c) mine owners do not take preventive measures. Although mine owners do not comply with legislation to protect the health and safety of workers, particularly in some geographical areas, the state continues ignoring this situation as the stone-mining industry generates much foreign exchange, and there is limited legislative protection for families from subsequent exposure.

2.11.2 Asbestosis

Clinicopathologic diagnosis requiring histological confirmation with (1) diffuse interstitial fibrosis of the asbestosis type and (2) evidence of asbestos as the etiology (asbestos bodies).

1. Interstitial fibrosis of at least grade 3 or grades 1–2 with a typical distribution.
   Asbestosis typically manifests in the lower lobes with irregular "reticular" markings on CXR toward the lung periphery and involves the costophrenic angles with subpleural accentuation.
2. Asbestos bodies (AsBds):
   - Identifiable when ferruginated as golden foreign bodies using PPB (Pearls Prussian Blue histochemical special stain)
   - Located in the walls of the respiratory bronchioles and associated with intraalveolar Mφ
   - Significance level when >1 AsBd in ≥1–5-μm-thick PPB-stained lung tissue section from ≥1 several FFPE lung tissue blocks (standard: 1 AsBd/100 sections; FFPE, formalin-fixed and paraffin-embedded) (it has been proposed to have 30-μm-thick sections to reduce the time to find asbestos bodies, but it is not feasible at all times)
   - Quantification also in BAL and TEM
   - TEM evaluation: The reference values are roughly $1-2 \times 10^6$ fibers·g^{-1} dry lung for total amphibole fibers and 0.1×10^6 fibers·g^{-1} for amphibole fibers longer than 5 μm.
   - Asbestos-related pleural plaques (macronodular and microscopic "basket-weave" collagen)

- Compensation is foreseen for asbestosis, diffuse pleural fibrosis, or asbestos-related neoplasm (malignant mesothelioma and any lung carcinoma).

- *DDX* of asbestos-related pleural fibrosis include:
  - Sarcomatoid mesothelioma
  - Tuberculosis
  - Connective tissue disorders
  - Lung adenocarcinoma

Interstitial fibrosis similar to UIP, beginning as peribronchiolar fibrosis, preferentially lower lobes, with pleural fibrosis/calcification, parietal pleural plaques. The grading of the severity is related to the fibrosis.

- *PGN*: Progressive massive fibrosis and "new" Caplan syndrome. There is a ↑ risk of bronchogenic carcinoma – risk relative to unexposed is 5× (relative risk for smokers, 11×; relative risk for smokers exposed to asbestos, 55×). Asbestos exposure increases unavoidably the risk for mesothelioma (both pleural and peritoneal), although the non-bronchogenic mesothelioma is still more rare than bronchogenic mesothelioma.

2.11.3 Non-silico Asbestosis-Related Pneumoconiosis

Coal Workers Pneumoconiosis (CWP)
Simple CWP does present after years of exposure to coal. Two definitions are useful. *Dust macule* is defined as interstitial dust-filled MΦ surrounding dilated respiratory bronchioles and in the interlobular septa, usually affecting the upper portions of lobes, with minimal fibrosis, while *coal nodule* is defined as discrete palpable <1 cm (Ø) lesion with central hyalinized collagen and pigment-laden MΦ.

Other Pneumoconiosis
Pulmonary mycotoxicosis or organic dust syndrome is caused by massive inhalation – probably toxic vs. immune bronchiolitis with intra-alveolar and interstitial neutrophils. Siderosis occurs following exposure to inert metallic iron or its oxides. There are macules and perivascular dust-like CWP but with coarse brown-black particles of Fe and gold-brown hemosiderin. Berylliosis in the acute form produces DAD, while the chronic forms a systemic disease with a latent period up to 15 years. Most often, there is interstitial fibrosis with non-caseating subpleural, peribronchiolar, and perivascular granulomas which may have Schaumann bodies or asteroid bodies like sarcoidosis (*vide supra*). Other agents include talc, aluminum powder, hard metal pneumoconiosis, alginate powder, polyvinyl chloride, and fibrous glass.

2.12 Pulmonary Vascular Disorders

In this category, we list the vascular disorders as shown in Box 2.19.

Box 2.19 Pulmonary Vascular Disorders
2.12.1. *Pulmonary Congestion and Edema*
2.12.2. *Pulmonary Embolism/Infarction* (Fig. 2.33)
2.12.3. *Pulmonary Hypertension* (Figs. 2.34 and 2.35)
2.12.4. *Diffuse Pulmonary Hemorrhagic Syndromes*
    2.12.4.1. Goodpasture Syndrome
    2.12.4.2. Idiopathic Pulmonary Hemosiderosis
    2.12.4.3. Vasculitis-Associated Hemorrhage (Necrotizing Capillaritis)

Notes: Necrotizing capillaritis is substantially the basis that is present in microscopic polyangiitis (MPA), Wegener granulomatosis (WG), systemic lupus erythematodes (SLE), but also CSS/EGPA although it is less frequently observed in this entity (Churg-Strauss Syndrome/Eosinophilic Granulomatosis with Poly-Angiitis).

2.12.1 Pulmonary Congestion and Edema

Pulmonary Congestion

Pulmonary *congestion* is a hemodynamic increase of hydrostatic pressure (e.g., Starling law), as seen in congestive heart failure showing blood-filled blood vessels and capillaries full of blood. Ernest Frankling is probably the most influential personality in cardiac physiology studying the hemodynamics of heart failure (Patterson and Starling, 1914). The harbinger, Frank Starling, tectonically demonstrated that stroke volume increases proportionally with end diastolic volume (preload). This phenomenon is epitomized as "Frank-Starling Law". This principle allows the heart to actively compensate for impaired contractile function and can be a forewarning of the clinical syndrome of heart failure. Pulmonary edema is the extravasation of fluid into interstitium (*interstitial edema*) and, then, into alveoli (*alveolar edema*) due to either from hemodynamic disturbances (hemodynamic or cardiogenic pulmonary edema, usually left-sided congestive heart failure for mitral insufficiency or hypoplastic left heart, HLH) or direct increase in capillary permeability. Grossly, wet and heavy lungs are seen with fluid coming out on the dissecting table at the time of the cutting of the lungs in two halves. The intra-alveolar fluid occurs only after lymphatic drainage has increased by about tenfold and is more prominent in the lower lobes and potentially accompanied by alveolar micro-hemorrhages and hemosiderin-laden Mφ ("heart failure cells" or "*Herzfehlerzellen*" of the German literature). In situations when a long-standing process of lung congestion occurs, the wet and heavy lungs become firm and assume a brown coloration, a process called "brown induration" of the lungs. It should be pondered that all these changes of the hemodynamics in the lung predispose to infection, other than impair normal respiratory function. This is quite often the case of the increased risk of contamination of children with HLH treated with Norwood procedure, bidirectional Glenn anastomosis, and hemi-Fontan procedure before a TX organ may become available. The Western Canadian Heart Transplantation Program is leading in research and clinics in North America and worldwide with institutions and laboratories located at the Mazankowski Institute and University of Alberta, Edmonton, AB, Canada. An increased number of hemosiderin-laden Mφ occur in infants with sudden infant death syndrome (SIDS) and sudden unexpected death in infancy (SUDI), although markedly increased hemosiderin-laden Mφ has been associated to a nonnatural death. Microvascular injury may occur by inflammation, infection, toxins, and shock, and when diffuse, the acronym ARDS for acute respiratory distress syndrome is used as a diagnosis. Regarding acute lung injury (ALI) is summarized as a spectrum of pulmonary lesions played on an endothelial and epithelial level that can be initiated by some factors. ALI may present as congestion, edema, surfactant disruption, and atelectasis. ALI may be transitory or progress to ARDS or acute interstitial pneumonia/-itis.

2.12.2 Pulmonary Embolism/Infarction

Rudolf Virchow's triad is still actual and part of one of the bases of teaching. It consists of vascular damage, slowdown of the circulation (*rouleaux*), and abnormality of the coagulation which can induce thrombosis.

Pulmonary Embolism/Infarction

It occurs as occlusion of pulmonary arteries by a blood clot, which is almost always embolic in the source. In 95% of cases, thrombi from deep veins of the legs may cause death not only in the elderly but also in adolescents and even children. Large emboli (e.g., saddle embolus) often cause sudden death. If no sudden death is recorded, the clinical picture may resemble myocardial infarction. If the patient survives, the clot may retract and eventually become lysed, leaving only small membranous webs. Infarction most commonly involves the lower lobe (75%), with a wedge-shaped pleural-based lesion, hemorrhagic at first, but then becomes pale as red cells lyse and fibrous replacement begins (at the edges) (Fig. 2.33).

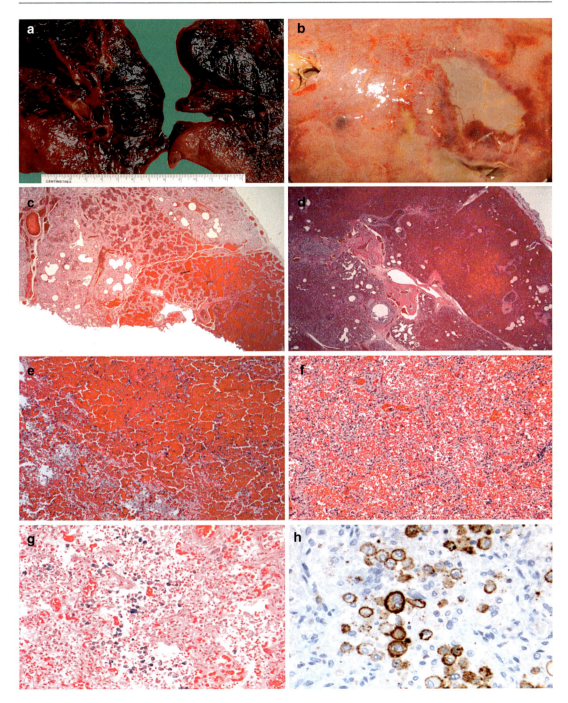

Fig. 2.33 Lung infarction with gross photographs at early and late event (**a**) and (**b**). Microphotographs (**c–f**) show H&E staining of the hemorrhagic infarction with filling of the alveoli by packed red blood cells (12.5× for (**b**) and (**c**), 100× for (**e**) and (**f**)). Microphotographs (**g**) and (**h**) show hemosiderin-laden macrophages (Fe special stain and 200×) and megakaryocytic recruitment (anti-CD43 immunohistochemistry and 400×)

2.12.3 Pulmonary Arterial Hypertension (PAH)

Pulmonary arterial pressures are, usually, only 1/8 of systemic arterial pressures. Atherosclerotic changes may be seen in pulmonary arteries in case of long-standing permanent hypertension and are generally indicative of pulmonary hypertension. The etiology is protean. Most probably, the most systematic and straightforward approach is to proceed along the cardiopulmonary pulmonary circulation, differentiating the causes that are found at each site. In 2003, the 3rd World Symposium on primary PAH updated the classification system in Venice, Italy (Fig. 2.34).

The normal conditions reveal that the pulmonary arterial pressure is 25 mmHg at rest and 30 mmHg during exercise. The arteries have an internal and external elastic lamina, and the muscle wall thickness is up to 7% of the external diameter. The veins characteristically travel in interlobular septs and have indistinct muscle and elastic laminae, while arterioles and venules have a single elastic lamina and lack muscle. Collateral arteries branch directly from larger arteries into the respiratory parenchyma and represent the typical site of dilatative and plexiform lesions. To increase the reproducibility of studies on PAH, pulmonary BVs need to be measured with an internal caliper properly. Using an electronic caliper to measure the inner diameter (ID), the external diameter (ED), and the adventitial diameter (AD), vascular morphometric parameters including relative adventitial thickness (%AT), relative media thickness (%MT), and wall thickness (WT) are calculated congruously. The arterial wall thickness (WT) is calculated as WT = (ED − ID)/2. The relative medial thickness (%MT) is defined as %MT = (ED − ID)/ED∗100 and relative adventitial thickness (%AT) as %AT = (AD − ED)/ED∗100 in consideration that we talk about percentage change. If V_1 represents the old value and V_2 the new one, percentage change = $\Delta V/V_1 = V_2 - V_1/V_1 *100$ and some devices directly support this calculation via a %CH or Δ% function (Box 2.20). When the variable in the query is a percentage itself, it is better to readdress its change by using percentage points, to avoid confusion in the reader between relative difference and absolute difference. In every lung, RAC also needs to be determined and added to the studies of PAH because a degree of lung maturation will help to understand the substrate for pulmonary hypertensive changes. In PAH investigation, it is advisable that RAC is usually measured in 20 different places to include the entire surface of the evaluated lung specimen. Vascular measurements are performed on 10 peripheral pulmonary arterioles. All these measurements need to be combined to calculate means and standard deviations, and t-test and ANOVA statistically compare the results.

Primary arterial hypertension is characterized by the occurrence in young females, involvement of liver disease, connective tissue disorders, infections (e.g., HIV), drug exposure, eosinophilia, parasitosis, and medial and intimal thickening. Conversely, primary venous hypertension occurs in a setting of bone marrow transplantation, chemotherapy, and radiotherapy. It may be secondary to arterial hypertension without plexogenic pulmonary vasculopathy. Primary venous hypertension is characterized by thrombosis and recanalization of septal veins with local congestion and intimal fibrosis, septal fibrosis, capillary hemangiomatosis, and granulomas to Fe/Ca deposits in the elastic laminae.

Secondary arterial hypertension is often due to ILD and chronic hypoxia with thrombotic and thromboembolic characters. Secondary venous hypertension is due to acquired CVD, mediastinal disease, sarcoid with venous medial hypertrophy and arterialization, and hemosiderosis are the main etiologic factors. Lymphangiectasis, septal edema and fibrosis can occur. The Heath-Edwards grading is an evaluation of the pulmonary vascular alterations, introduced by Heath and Edwards (1958). It has been widely used for assessment of the severity of the hypertensive disease of the pulmonary blood vessels. However, a lot of criticisms has been raised pointing to the increasing awareness of the complexity of the vascular lesions, and Wagenvoort (1981) suggested that this grading system no longer fulfills the requirements of unambiguous evaluation of the severity of vascular disease. Wagenvoort (1981) indicated that the degree and extent of the various lesions, the different types of intimal fibrosis, and the

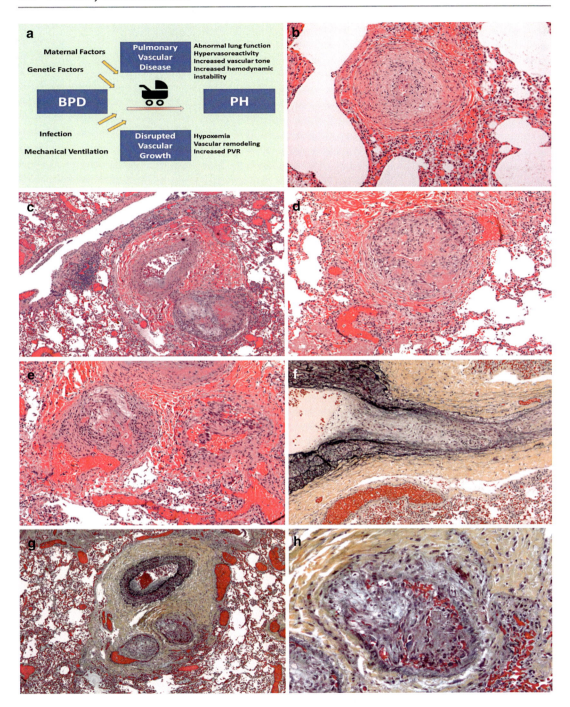

Fig. 2.34 Pulmonary hypertension schema with maternal factors, genetic factors, infective factors, and mechanical factors able to produce a change in the vascular growth of the lungs with prematurity and BPD. In (**b**) through (**h**), several degrees of vascular changes in case of pulmonary hypertension. (**b**), H&E, 100×; (**c**), H&E, 50×; (**d**), H&E, 100×; (**e**), H&E, 100×; (**f**), Movat pentachrome stain, 100×; (**g**), Movat pentachrome stain, 50×; (**h**), Movat pentachrome stain, 200×. (All photographs are illustrated in their original magnification)

Box 2.20 Formulas for Studies of PAH

Wall thickness (WT)	WT = (ED − ID)/2
Relative medial thickness (%MT)	%MT = (ED − ID)/ED∗100
Relative adventitial thickness (%AT)	%AT = (AD − ED)/ED∗100

Notes: PAH, *pulmonary hypertension*; AD, *adventitial diameter*; ED, *external diameter*; ID, inner diameter.

eventual decrease in number and size of vessels should all be assessed for all blood vessels, not only for the arteries. However, the Heath-Edwards grading is still a good standard in PICUs and cardiology departments because it is easy to perform and has a good ratio of inter- and intraindividual variability. In the Heath-Edwards grading, grades I–III are considered potentially reversible, while categories IV–VI are usually regarded as irreversible about the prognosis of patients harboring hypertensive pulmonary vascular disease. The plexiform lesions are vascular endothelial growth factor (VEGF)-1+ and show cellular recanalization-like plugs but with proximal local cellular intimal thickening, distally dilatative injuries, occurring at supernumerary arterioles and need to be differentiated from thrombotic or embolic recanalization sites, which show no dilatative injuries and usually affect larger blood vessels.

Veno-occlusive disease is depicted in Fig. 2.35.

2.13 Transplantation-Related Disorders

In the case of bone marrow transplantation, pulmonary complications may occur in ~50% of patients. These are usually of infectious origin (bacteria) with lobar pneumonia or interstitial pneumonia/-itis but may also involve the bronchioles (obliterative bronchiolitis) or the bronchi (lymphocytic bronchitis). In the setting of heart-lung transplantation, 50% develop obliterative bronchiolitis after a mean interval of 10 months, and there is a 50% mortality – probably due to chronic reject.

2.13.1 Acute Rejection

The 2007 Working Formulation for the Standardization of Nomenclature in the diagnosis of lung rejection is followed (Fig. 2.36). Grade 0 has no mononuclear inflammation, hemorrhage, or necrosis. Grade 1 (minimal) shows infrequent (scattered) perivascular 2–3-cell-thick mononuclear infiltrates in the alveoli of the lung parenchyma, but no eosinophils and no endothelialitis. In grade 2 (mild), there are frequent perivascular 2+ cell-thick mononuclear infiltrates (lymphocytes, plasmacytoid lymphocytes, MΦ, and eosinophils) and endothelialitis, but no apparent infiltration into the adjacent alveolar spaces/airspaces. In grade 3 (moderate), there are readily detectable perivascular 2+ cell-thick mononuclear infiltrates (lymphocytes, plasmacytoid lymphocytes, MΦ, and eosinophils) and endothelialitis, with apparent infiltration into the adjacent alveolar spaces/airspaces. Finally, grade 4 (Severe) shows an architecture with diffuse perivascular, interstitial, and intra-alveolar mononuclear infiltrates (lymphocytes, plasmacytoid lymphocytes, MΦ, and eosinophils) with pneumocyte damage, endothelialitis ± intra-alveolar necrotic cells, MΦ, hyaline membranes, bleeding, and PMNs and with or without parenchymal necrosis, infarction, or necrotizing vasculitis.

2.13.2 Chronic Rejection

There are morphological features of chronic rejection that are common to all solid organ allografts. These features include (a) patchy organized interstitial inflammation, (b) patchy interstitial fibrosis and associated atrophy of the parenchyma, (c) graft vascular disease with initial fibrointimal hyperplasia of arteries, (d) destruction of epithelial-lined ducts, and (e) destruction and atrophy of organ-associated lymphoid tissue and lymphatic channels. The immunological injury seems to play a primary role in the initiation and progression of these lesions. The final phenotypic expression of chronic rejection is dependent on an interaction between immunologic and physiologic factors that results in the prevalence of one

2.13 Transplantation-Related Disorders

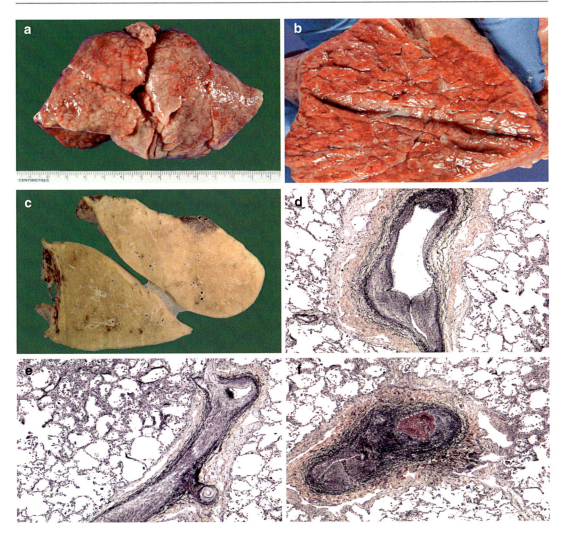

Fig. 2.35 Veno-occlusive disease of the lung with gross photographs showing an induration or consolidation of the pulmonary parenchyma (**a–c**), while **d–f** show the myointimal and media changes in occluded or semi-occluded veins (Movat pentachrome stain and 50× as original magnification for all three microphotographs)

or another of these features. The role of environmental factors is still under evaluation.

IBAFIS or ImageJ-Based Automated Fibrosis Index Score

The interpretation of fibrosis in histologic glass slides as well as the digital file may be quite challenging and harbor quite a considerable interindividual and intraindividual variability. Thus, the use of objective methods may be quite desirable. IBAFIS is the acronym for a procedure performed in our laboratories at the University of Alberta Hospital. The following data may explain the system in detail. First, the histologic slide is stained with Masson's trichrome, then the stained slide is scanned with some software (e.g., Aperio™), or three fields are identified by random numbers using a random number generator coupled with a positional field determining machine. The following equations are considered crucial for the determination of fibrosis: area (total) = area (cells) + area ICM dense + area ICM loose + area (empty space). Area ICM dense is the area corresponding to fibrosis. Thus, it results from subtracting the cellular area, the free ICM area, an area constituted by space from the total area. A

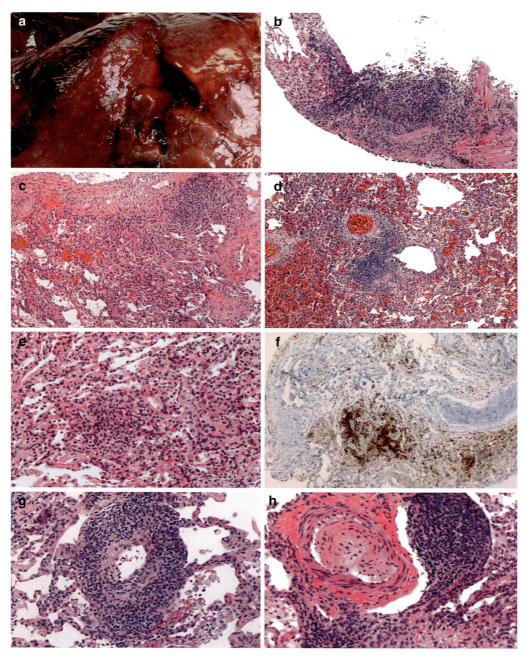

Fig. 2.36 In this panel, figure (**a**) shows the gross image of a "brown induration" of the lung following hemosiderosis following lung transplantation in an adolescent. There is not only a sign of hemorrhage on the left side of the photograph but the change of the color of the lung in other fields of the gross photograph. Microphotograph (**b**) shows an acute rejection A1 (hematoxylin and eosin staining, ×100 original magnification). Figure (**c**) shows an A2 acute rejection (hematoxylin and eosin staining, ×100 original magnification). Figure (**d**) shows another aspect of the lung rejection with a lymphocytic infiltration, which is not only interstitial but also perivascular (hematoxylin and eosin staining, ×100 original magnification). Figure (**e**) another example of lung rejection (A2 cellular rejection) (hematoxylin and eosin staining, ×200 original magnification). The microphotograph (**f**) is an immunohistochemical detection of A2 cellular rejection performed using an antibody against CD3 lymphocytes highlighting the infiltrating lymphocytic population (anti-CD3 immunostaining, avidin-biotin complex, ×100 original magnification). Figures (**g**) and (**h**) show the histology of severe rejection (A3 cellular rejection) (hematoxylin and eosin staining, ×200 original magnification) (see text for details)

variant of the IBAFIS is the IBANIS, which is the ImageJ-based automated nuclear index score. This last score may be necessary to identify the increase of the nuclear size of the cardiomyocytes in case of hypertrophic cardiomyopathy.

2.14 Pediatric Tumors and Pseudotumors

In this category, only very few pathologies may be encountered in childhood or youth. It is indeed challenging sometimes to distinguish between benign and malignant neoplasms because some benign tumors can behave very aggressively in the postsurgical course or follow an open lung biopsy and some tumors classified as lethal may have a better outcome if resection was performed entirely and staging does not show any metastatic colonization. In Box 2.21, the most frequent pediatric tumors and pseudotumors are listed and will be part of the topic of this chapter. In childhood, pleuropulmonary blastoma and carcinoid (NE, neuroendocrine) tumors play the significant role, but small cell carcinoma, adenocarcinoma, squamous cell carcinoma, as well as mucoepidermoid carcinoma can be encountered but will be discussed in the "adult"-type malignant primary neoplasms of the lung (Box 2.21).

Some pathways involved in lung carcinogenesis are depicted in Fig. 2.37a.

2.14.1 Pulmonary Teratoma

Intrathoracic teratomas are typically seen in the mediastinum, but occasionally they arise in the pulmonary parenchyma as intrapulmonary teratomas (Fig. 2.37b, c). The criteria for the diagnosis of an intrapulmonary teratoma are the exclusive origin of the tumor from the lung and exclusion of a gonadal site or other extragonadal primary sites. Histologically, PUT is similar to other benign cystic teratomas that have endodermal, mesodermal, and ectodermal tissue elements of the primitive three germ cell layers. At places, a malignant component can be seen, including primitive neuroectodermal tumor, Wilms' tumor,

> **Box 2.21 List of the Described Pediatric Tumors and Pseudotumors**
> 2.14 Pediatric Tumors and Pseudotumors
>   2.14.1 Pulmonary Teratoma (PUT)
>   2.14.2 Inflammatory Myofibroblastic Tumor (IMF)
>   2.14.3 Pulmonary Hamartoma (PUH)
>   2.14.4 Pulmonary Sclerosing Hemangioma (PSH)
>   2.14.5 Pulmonary Carcinoid (PUC)
>   2.14.6 Lymphangioleiomyomatosis (LLAM)
>   2.14.7 Kaposiform Lymph-Angiomatosis (KLAM)
>   2.14.8 Pleuropulmonary Blastoma (PPB)
>   2.14.9 Metastatic Tumors

neuroblastoma, or carcinoma that upstage the tumor. The presence of squamous differentiation and mature cartilage is observed in numerous cases. The presence of neuroectodermal tissue portends to the diagnosis of immature teratoma. Since PUT can have pancreas tissue and other endodermal elements, digestive or proteolytic enzymes can be secreted by the tumor that make the teratoma prone to rupture.

2.14.2 Inflammatory Myofibroblastic Tumor

In childhood, solid lung masses are typically caused by underlying inflammatory, infectious, or reactive processes. A plasma cellular infiltration may occur in the setting of chronic inflammation but also in the environment of a posttransplant lymphoproliferative disorder (PTLD) (Fig. 2.37d, e) (*vide infra*).

Inflammatory "Pseudotumors"
The plasma cell granuloma (aka fibroxanthoma and histiocytoma) is often an incidental finding on CXR and occurs in childhood and youth. It shows mature plasma cells and lymphocytes in

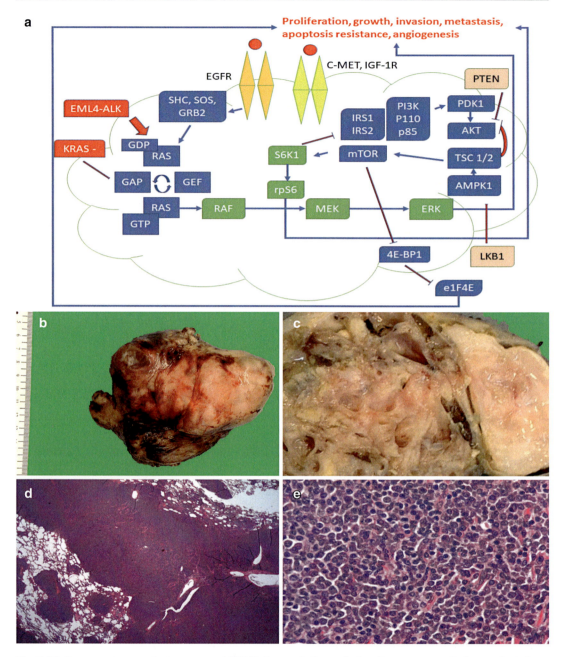

Fig. 2.37 Lung carcinogenesis, teratoma, and PTLD. In the cartoon shown in (**a**) markers of lung carcinogenesis are highlighted. They target the proliferation, growth, invasion, metastasis, apoptosis resistance, and angiogenesis collectively or at selective steps for a series of lung neoplasms. In (**b**) and (**c**) is shown a primary lung teratoma (gross photographs only), which showed the differentiation of the three germ cell layers (ectoderm, mesoderm, endoderm) (not shown). In (**d**) and (**e**) is shown the histology of a posttransplant lymphoproliferative disorder of monomorphic type with diffuse infiltration of plasma cells and blasts (**d**, H&E stain, 12.5; **e**, H&E stain, 400×)

a background of fibrosis and granulation tissue. Another pseudotumor is the pseudolymphoma, which is similar to LIP. It is a residuum of a healing inflammatory lesion. There is a nodule, which replaces a portion of the lung parenchyma and shows scarring in the center with densely packed collagen, fibroblasts, and lymphocytes but no necrosis at the periphery.

2.14.3 Pulmonary Hamartoma

In childhood and youth, intrapulmonary solid lung masses constituted by a heterotopic tissue are defined as hamartomas. In teratoma (*vide supra*), a germ cell layer differentiation different from autochthonous proliferation is seen.

2.14.4 Pulmonary Sclerosing Hemangioma

- *DEF*: Benign tumor of the lung showing a papillary pattern or islands filled with bland polygonal cells and copious eosinophilic cytoplasm and indistinct cell borders, frequently with sclerotic stroma and blood lakes.
- *SYN*: Sclerosing pneumocytoma.
- *EPI*: Rare, ♂ < ♀, youth-middle age.
- *GRO*: Small (~5 cm), well-circumscribed intraparenchymal (unrelated to bronchus).
- *CLM*: Papillary, angiomatous, and solid growth patterns with bland round/polygonal cells without nucleoli and no mitoses or necrosis. Focally, foamy Mφ, fibrosis, and angiomatous areas.
- *IHC*: (+) Surfactant, VIM, EMA, (±) AE1-3, (±) NE markers (NE, neuroendocrine markers).
- *TRT*: Limited resection.
- *PGN*: Almost always benign, but 2–4% have LN (+) – not affecting the outcome!

2.14.5 Pulmonary Carcinoid

- *DEF*: Well-differentiated NE tumor.
- *EPI*: Children, not associated with smoking.
- *CLI*: Paraneoplastic syndrome – carcinoid syndrome, Cushing syndrome, acromegaly.
- *GRO*: Central tumors are trabecular while peripheral insular.
- *CLM*: NE architecture with finely granular ± slight clumping with small inconspicuous nucleoli and thin-walled and variably dilated blood vessels. Variants – adenopapillary architecture, biphasic patterns (S100+, sustentacular cells), CK+. The term of pseudoparagangliomas has been used because paragangliomas are CK-. Other options of PUC include oncocytic, melanin-producing, mucin- producing, cartilage- and bone-rich as well as amyloid-secreting.
- *IHC*: (+) CGA, SYN and (±) TTF-1.
- *DDX*: Paraganglioma, adenocarcinoma, cell sugar tumor, glomus tumor, and SmCC (carcinoids show crush artifacts and dot-like CK, but not the Azzopardi phenomenon, chromatin texture, Ki-67 (if <20% it is unlikely to be SmCC, while if >50% it is doubtful to carcinoid and stromal vessel features). Moreover, the peripheral location favors carcinoid over SmCC.

Atypical carcinoid (moderately differentiated NE tumors) is defined by an MI of 2–10/10 HPF (Ki67/MIB1, 10–48%) ± necrosis (criteria, ≥2 criteria of MI ≥ 5/10HPF, focal necrosis, some pleomorphism, and some focal loss of the architecture). Clinically, the atypical carcinoid lacks the SmCC chemosensitivity (5-YSR: 60%), and in about 2/3 of cases, metastases are found. Histologically, the nuclei show a more coarsely granular/vesicular chromatin pattern and variable nucleoli. It has been suggested that peripheral tumors with SmCC-like cytology are likely to be atypical carcinoids. The first differential diagnosis is SmCC (Fig. 2.38).

2.14.6 Lymphangioma-(Leo)-Myomatosis (LAM)

- *DEF*: A neoplastic disease that affects lungs, kidneys (angiomyolipoma), and the lymphatic system, characterized by haphazard proliferation of smooth muscle cells ("LAM cells") throughout the interstitial space involving the

walls of blood vessels, lymphatics, bronchioles, and septa (Fig. 2.39).
- *EPG*: 2 forms – (1) inherited form, which is associated with tuberous sclerosis complex (*TSC1* and *TSC2* genes) and (2) isolated or sporadic form.
- *EPI*: ♂ < ♀ (often women of childbearing age).
- *CLI*: Dyspnea, chest pain, cough, hemoptysis, spontaneous PNX, and chylous effusion.
- *CLM*: Overgrowth of abnormal smooth muscle-like cells ("LAM cells") of endothelial-lined structures with the tendency to give rise to the cystic formation and the consequent destruction of healthy lung tissue and chylothorax.
- *IHC*: αSMA (+), HMB45 (+) (PEComa), S100 (−).
- *DDX*: UIP (end stage), which can show smooth muscle proliferation.
- *TRT*: Progestagens.
- *PGN*: Progressive disease ("neoplastic") unless therapy (or menopause) is started. LAM cells are cells that may spread between tissues and recur after lung TX.

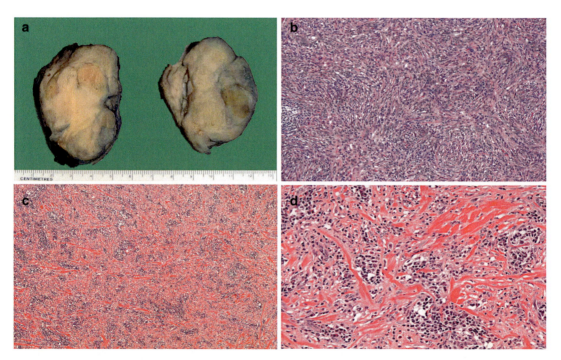

Fig. 2.38 This panel shows an inflammatory myofibroblastic tumor, which is an extremely rare type of pseudotumor that occurs most commonly in children and young individuals. Figure (**a**) shows the gross image of the tumor. The tumor is firm, gray-white, and quite homogeneous (**a**). The microscopic examination reveals proliferation of spindle cells arrayed in fascicles and admixed with lymphocytes, plasma cells, and eosinophils (**b**, hematoxylin and eosin staining, ×100 original magnification; **c**, hematoxylin and eosin staining, ×50 original magnification; **d**, hematoxylin and eosin staining, ×200 original magnification). The four immunohistochemical microphotographs (**e–h**) show the positivity of the tumor for CD138 or syndecan-1 (a plasma cell marker), actin (cytoskeleton antigen), Bcl2 (anti-apoptotic cell marker), and Ki67 (cell proliferation antigen) (**e**, anti-CD138 immunostaining, ×50 original magnification; **f**, anti-actin immunostaining, ×200 original magnification; **g**, anti-Bcl2 immunostaining, ×200 original magnification; **h**, anti-Ki67 immunostaining, ×100 original magnification). All immunostainings have been performed using the avidin-biotin complex. This tumor may be considered a true neoplasm because of the presence of the fusion gene anaplastic lymphoma kinase (ALK) 1 observed in the myofibroblastic component. The ALK gene, a tyrosine kinase oncogene, which is located on chromosome 2p23, was initially found to be arranged in anaplastic large cell lymphomas. This gene fusion leads to constitutive overexpression of the ALK determining cell proliferation in over half but not all inflammatory myofibroblastic tumors

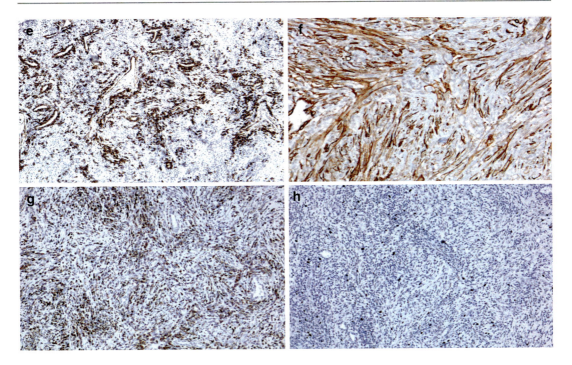

Fig. 2.38 (continued)

2.14.7 Kaposiform Lymphangiomatosis

- *DEF*: It is a diffuse proliferation of abnormal, complex lymphatic channels that may involve a lobe, part of a lung lobe, or the entire lung and often accompanied by chylous pleural effusions with tendency to unstoppable progression and dramatic fatality from respiratory failure.
- *EPI*: Rare, children and young adults, ♂ = ♀.
- *EPG*: Congenital, possibly from abnormal lymphatic development.
- *CLI*: Dyspnea, wheezing, and cough, which may be misinterpreted as bronchial asthma, but the diagnosis may be made if imaging (bilateral ground-glass opacities with smooth thickening of interlobular septa and bronchovascular bundles) is characteristic and bone is involved.
- *GRO*: Tumor with subpleural and septal thickening.
- *CLM*: There is a diffuse proliferation of complex, anastomotic lymphatic channels without significant dilatation, lined by benign appearing, flattened endothelial cells, as well as the prominence of lymphatic channels in visceral pleura, interlobular septa, and bronchovascular bundles with surrounding bundles of spindle cells, interspersed collagen, and blood vessels. Acellular, eosinophilic, proteinaceous material may be present in lymphatic channels. A hemorrhagic "kaposiform" component, with compressed vascular channels, plump spindle cells, and hemosiderin deposition, may be present.
- *IHC*: (+) CD31, D2–40, and F-VIII in endothelial cells and VIM, ACT, and DES in spindle cells while (−) HMB45, AE1/3, and HHV8.
- *DDX*: Kaposiform hemangioendothelioma, Kaposi's sarcoma (+) HHV8, lymphangiectasis, with dilation of existing lymphatics in a normal distribution, and lymphangioleiomyomatosis (+) HMB45 positive and more smooth muscle cells, pulmonary capillary hemangiomatosis.
- *TRT*: Supportive, including observation and medical management ± surgical resection for

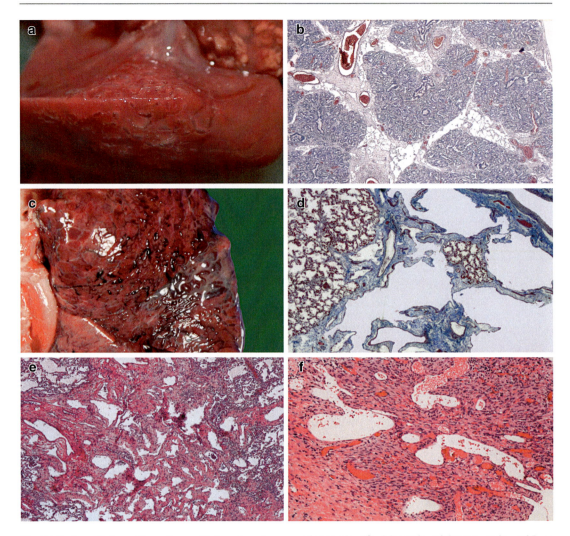

Fig. 2.39 Lymphatic malformations with lymphangiectasia (gross in **a**, microphotograph in **b** hematoxylin and eosin staining at 40× as original magnification), hemorrhagic lymphangiectasia (gross photograph in **c**), lymphangiomatosis (**d**: Masson's trichrome stain, ×25 as original magnification), and kaposiform lymphangiomatosis (**e–f**: H&E staining and 50× and 100× as original magnifications) (see text for details)

localized disease ± pleurocentesis, pleurodesis, pleurodectomy, and ligation of thoracic duct ± successful lung TX.
- *PGN*: Worse prognosis for children than adults and pleural involvement is an adverse prognostic factor.

2.14.8 Pleuropulmonary Blastoma

- *DEF*: Rare, primitive primary neoplasm of the thorax of young children, frequently associated with cystic lung lesions arising in pulmonary parenchyma, mediastinum, and pleura with three subtypes of PPB (I, II, III) (Fig. 2.40).
- *SYN*: Pneumoblastoma, mesenchymal cystic hamartoma, cystic mesenchymal hamartoma, pulmonary rhabdomyosarcoma (misnomer), rhabdomyosarcoma in lung cyst (misnomer), and pediatric pulmonary blastoma.
- *EPI*: Childhood, mostly 1st infancy (type I), while types II and III also older than 2 years of age.

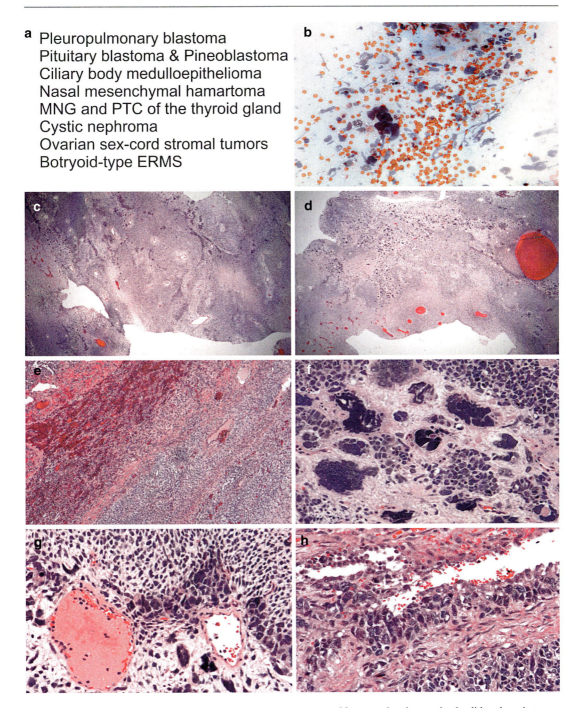

Fig. 2.40 Pleuropulmonary blastoma listing in (**a**) a series of different entities, probably correlated with the pleuropulmonary blastoma according to most sources of the scientific literature. In (**b**) is shown the cytologic preparation of a pleuropulmonary blastoma with large and hyperchromatic cells and bizarre nuclei (TP-Giemsa, 200×). The microphotographs illustrate the histology of the pleuropulmonary blastoma showing a mixed solid and cystic tumor with an apparently heterogeneous tumor composed of primitive blastema-like small cells with hyperchromatic nuclei, high nucleus to cytoplasma ratio, numerous mitoses, spindle cells, and ovoid cells embedded in a myxoid stroma as well as nodules of immature chondroid or myxoid elements (**c–h**, H&E stain; **c, d**, 12.5×; **e**, 50×; **f–h**, 200×)

- *EPG*: The tumor arises from the primitive mesenchymal cell, likely a stem cell, in patients with PPB family tumor and dysplasia syndrome (1/3 of patients) with most children harboring a mutation of the *DICER1* gene.
- *CLI*: Dyspnoea or RDS including persistent coughing ± atelectasis. Importantly, 1/10 of PPB cases harbor multilocular cystic nephroma and rarely Wilms' tumor.
- *IMG*: Right-sided (often), pleural-based, peripherally located the tumor.
- *GRO*: Situated peripherally, multicystic and thin-walled tumor (type I), mixed solid and cystic tumor characterized by variable thickened or nodule-like areas (type II), and well circumscribed, mucoid, white-tan solid mass attached to the pleura, involving a lobe or a lung (type III) ± necrosis and hemorrhage in friable areas.
- *CLM*: Peripherally located, multicystic and thin-walled architecture in type I but mixed solid and cystic tumor with variable thickened or nodule-like areas in type II. Conversely, type III is characterized by heterogeneity. The tumor is composed of ≥1 of the following components – primitive blastema-like small cells with nuclear hyperchromasia, high N/C ratio, high MI, spindle cells/ovoid cells embedded in a myxoid stroma, nodules of immature, dysplastic or malignant chondroid elements, and isolated or clusters of large anaplastic cells with pleomorphic nuclei, atypical mitoses, or eosinophilic hyaline bodies.
- *IHC*: (+) VIM, CD117 (focally), AAT (focally), CD99 (weakly), AE1-3 (surface epithelium), MSA (rhabdomyoblasts and primitive cells), and DES (rhabdomyoblasts and primitive cells) while (−) EMA, myogenin, S100, GFAP, NSE, TTF1, AFP, CGA, and SYN.
- *DDX*: ERMS (type III), large bronchogenic cyst/lung cyst (type I), FLIT (fetal lung interstitial tumor), MPNST, malignant teratoma, yolk sac tumor, mesothelioma, monophasic synovial sarcoma, PNET, and undifferentiated sarcoma.
- *TRT*: Multimodal, including surgery, chemotherapy, and radiation therapy.
- *PGN*: Poor, because of frequent relapses and distant metastases (brain and bone), especially types II and III.

2.14.9 Metastatic Tumors

In childhood, solid lung masses are typically caused by Ewing sarcoma, osteosarcoma, or hepatoblastoma (Figs. 2.41, 2.42, and 2.43). These neoplasms are described in the soft tissue, arthro-skeletal, and liver chapter, respectively. Finally, the condition of lymphangiosis carcinomatosa should be mentioned. Lymphangitis carcinomatosa is extremely rare in childhood. It is an inflammation of the lymph vessels (lymphangitis) caused by an epithelial or mesenchymal malignancy. In adults, breast, lung, stomach, pancreas, and prostate cancers are the most common neoplasms that may result in lymphangitis carcinomatosa. However, Chiang et al. (1991) reported on an 18-year-old adolescent male, who was admitted due to acute respiratory failure. Bilateral fine reticular densities and right massive pleural effusion were seen in his CXR. Autopsy findings included pulmonary lymphangitis carcinomatosa arising from an adenocarcinoma of choledochal cyst with global dissemination. Originally, lymphangitis carcinomatosa was first described in 1829 by Gabriel Andral, who was a pathologist investigating a patient with uterine cancer.

2.15 "Adult"-Type Neoplasms in Childhood/Youth

In this paragraph, we focus on a quite problematic issue, the onset of adult-type lung neoplasms in childhood and youth. These neoplasms are very rare, but the modern therapy and new chemotherapy protocols and immunosuppressive drugs have suggested pediatric oncologists and pediatric pathologists to discuss on this category in pediatric and pediatric pathology textbooks. The issue is problematic because there is the risk of misdiagnosing some of these entities because these neoplasms are non-familiar to pediatricians and pediatric pathologists. Thus, the review of the literature is essential first to highlight the rise of this concern and second to look at the new classification criteria that have been consolidated in the last couple of years in pediatric respiratory medicine (Box 2.22) (Fig. 2.44).

The majority of pediatric patients with a primary malignant pulmonary tumor present with carcinoid tumor or mucoepidermoid carcinoma

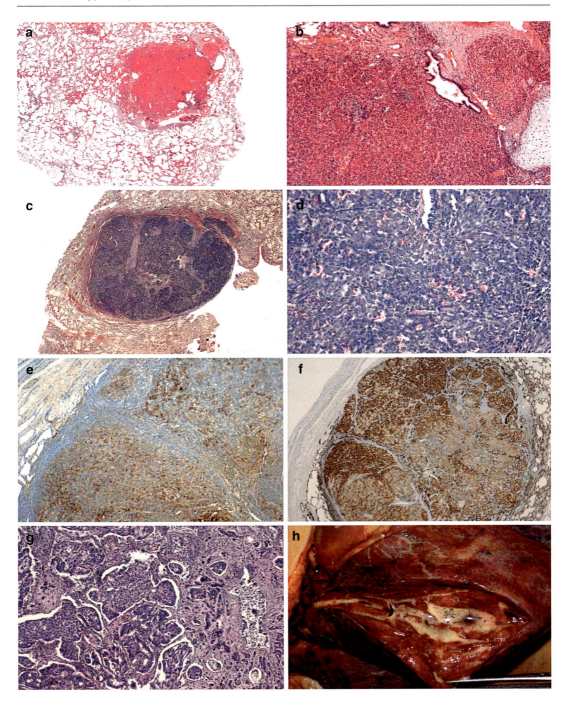

Fig. 2.41 This panel illustrates some metastatic tumor into the lung. Figures a-d show metastatic pulmonary hepatoblastoma of fetal (**a**, **b**) and embryonal type (**c**, **d**) (**a**, hematoxylin and eosin staining, ×20 original magnification; **b**, hematoxylin and eosin staining, ×100 original magnification; **c**, hematoxylin and eosin staining, ×20 original magnification; **d**, hematoxylin and eosin staining, ×200 original magnification). The hepatoblastoma shows an immunohistochemical positivity for alpha-feto-protein (**e**, anti-alpha-feto-protein immunostaining, ×100 original magnification) and cytokeratin 19 (**f**, anti-cytokeratin 19 immunostaining, ×40 original magnification). All immunostainings have been performed using the avidin-biotin complex. Microphotograph (**g**) shows the pulmonary metastasis of a hepatocellular carcinoma (hematoxylin and eosin staining, ×100 original magnification) in an adolescent. Figure (**h**) is the gross photograph of a diffuse pulmonary metastasis of a non-Hodgkin lymphoma of B type also in an adolescent

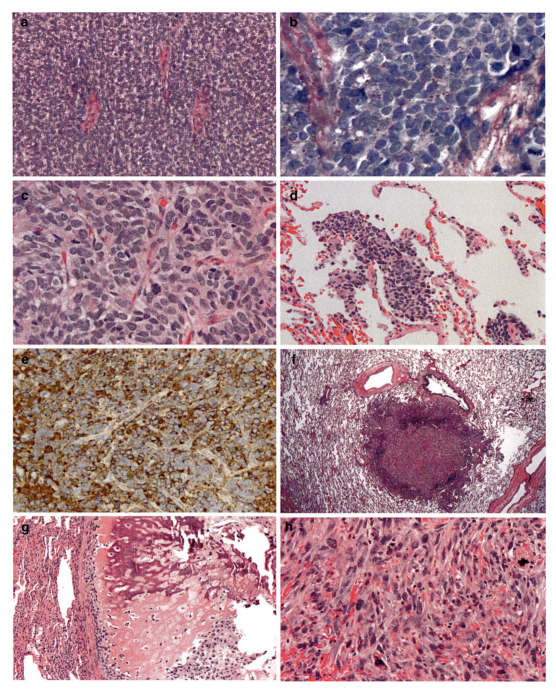

Fig. 2.42 This panel of metastatic lung tumor shows a metastatic Ewing sarcoma or primitive neuroectodermal tumor (PNET) of the soft tissue (**a**, hematoxylin and eosin staining, ×200 original magnification; **b**, Periodic acid Schiff staining, ×630 original magnification). It is easy to recognize the particulate positivity of the tumor cells containing intracytoplasmic glycogen. Microphotographs (**c**) and (**d**) show the metastasis of a neuroectodermal tumor at high power and low power (**c**, hematoxylin and eosin staining, ×400 original magnification; **d**, hematoxylin and eosin staining, ×200 original magnification), while figure e shows the immunohistochemical confirmation of the neuroendocrine tumor using an antibody against chromogranin A and the avidin-biotin complex (anti-chromogranin A immunostaining, ×200 original magnification). Figures (**f–h**) show several aspects of lung metastasis of osteosarcoma of bone with the evidence of malignant osteoid with anaplastic cells (**f**, hematoxylin and eosin staining, ×12.5 original magnification; **g**, hematoxylin and eosin staining, ×100 original magnification; **h**, hematoxylin and eosin staining, ×200 original magnification)

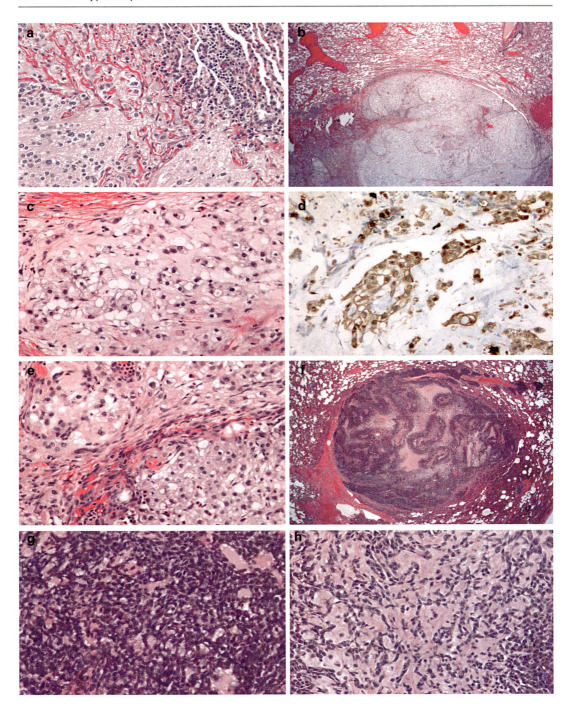

Fig. 2.43 This panel of metastatic lung tumor illustrates an intrapulmonary metastasis of a neuroblastoma with neuroblasts and neuropil (**a**, hematoxylin and eosin staining, ×200 original magnification), a chordoma (**b**, hematoxylin and eosin staining, ×12.5 original magnification; **c**, hematoxylin and eosin staining, ×200 original magnification; **d**, anti-S100 immunostaining with avidin-biotin complex, ×200 original magnification; **e**, hematoxylin and eosin staining, ×200 original magnification). Microphotographs (**f–h**) show the pulmonary metastasis of a myoepithelial carcinoma (**f**, hematoxylin and eosin staining, ×12.5 original magnification; **g**, hematoxylin and eosin staining, ×200 original magnification; **h**, hematoxylin and eosin staining, ×200 original magnification). The metastasis of a myoepithelial carcinoma recapitulates the morphology of the originary tumor with cords of epithelioid, plasmacytoid cells in hyalinized stroma

(MEC). Typically, they have a quite favorable prognosis. Lung cancers which are common in adults, but rare in children, have a worse prognosis in this age group. The histologic types of tumors encountered are similar to lung tumors occurring in adults, although the frequency of the various kinds differs from series to series. Carcinoid tumors are more frequent than bronchogenic carcinoma of small cell type. Carcinoids are also prevalent in the pediatric age group. Similar to the adult population, the prognosis of these tumors is dependent on histology and stage. Patients with carcinoid tumors seem to have the best outcome, followed by adenocarcinoma. The highly aggressive basaloid carcinoma has the worst sequel.

> **Box 2.22 List of the "Adult"-Type Neoplasms in Childhood/Youth**
> 2.15. *"Adult"-Type Neoplasms in Childhood/Youth*
>   2.15.1. Bronchogenic Carcinoma
>   2.15.2. Carcinoid/Neuroendocrine Tumor of the Lung
>   2.15.3. Pulmonary Mesenchymal Neoplasms

2.15.1 "Adult"-Type Preneoplastic Lesions

Preinvasive Neoplastic Lesions of Lung Tumors

In Box 2.23 are listed the preinvasive neoplastic lesions of lung tumors.

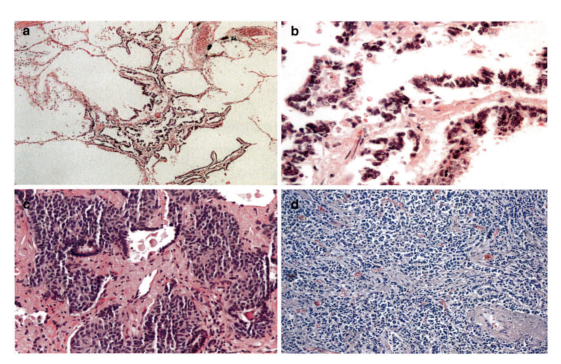

Fig. 2.44 This panel of non-pediatric lung tumors would offer the reader the possibility to consider the diagnosis of adult lung tumors even in a non-characteristic setting. Microphotographs (**a**) and (**b**) show the histology of an atypical adenomatous hyperplasia (**a**, hematoxylin and eosin staining, ×50 original magnification; **b**, hematoxylin and eosin staining, ×400 original magnification). Figure (**c**) shows a diffuse idiopathic pulmonary neuroendocrine cell hyperplasia (DIPNECH) (**c**, hematoxylin and eosin staining, ×200 original magnification). DIPNECH is a rare lung disorder, which is characterized by diffuse hyperplasia of bronchiolar and bronchial pulmonary neuroendocrine cells. Figure (**d**) shows a small cell lung carcinoma (**d**, hematoxylin and eosin staining, ×200 original magnification). Figures (**e**, **f**) show a lung adenocarcinoma (**e**, hematoxylin and eosin staining; ×100 original magnification; **f**, anti-carcinoembryonic antigen immunostaining with avidin-biotin complex, ×100 original magnification). Figures (**g**) and (**h**) show a squamous cell carcinoma and *pleuritis carcinomatosa* (**g**, hematoxylin and eosin staining; ×100 original magnification; **h**, hematoxylin and eosin staining; ×40 original magnification)

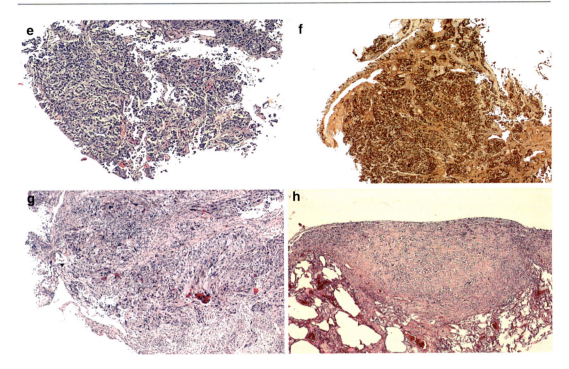

Fig. 2.44 (continued)

Box 2.23 Histologic Classification of Lung Cancer: Preinvasive Lesions
- *AAH* (atypical adenomatoid hyperplasia)
- *DIPNECH* (diffuse idiopathic neuroendocrine cell hyperplasia)
- *AIS* (adenocarcinoma in situ)
- *CIS*/squamous dysplasia

The preinvasive lesions may develop early in the youth, particularly in patients with genomic mutations of tumor suppressor genes or in young adults who started smoking cigarettes early in life, as seen in some countries. It is uncertain the role of the diffuse use of cannabis in the young generations about pulmonary oncogenesis. Smoked marijuana has deleterious effects on nerve cells, but most probably has also a role in inducing/promoting the carcinogenesis of lung tumors as identified in several studies eliminating confounding factors, such as tobacco and alcohol.

About the outcome and the therapeutic protocols, it is now known that some recommendations have become compulsory in some countries. This aspect may be the case for the youth living with an adult-type bronchogenic carcinoma as well. Pathologists, oncologists, pulmonologists, methodologists, and laboratory scientists, including molecular biologists as well as patient representatives, collaborate constantly to develop new guidelines. In addition to the 2013 recommendations of the College of the American Pathologists, new genes have been suggested to be investigated in case of an NSCLC regardless of the age of the patient. The following tests are now recommended:

1. *EGFR* and *ALK* mutation testing
2. *ROS1* mutations

EGFR T790 M mutational testing is essential to perform in patients who have progressed after tyrosine kinase inhibitor (TKI) treatment with a 1st-/2nd-generation EGFR TKI therapy, while *ROS1* mutation is a new and strongly recommended investigation to be carried out for all lung cancer patients independently of clinical characteristics. Moreover, ALK immunohistochemistry is now considered as equivalent to fluorescence in situ hybridization (FISH). In Fig. 2.37a are depicted some critical interactions. Tyrosine kinase is an enzyme able to phosphorylate, i.e., to transfer a phosphate group, in a protein, while TKI are tyrosine kinase inhibitors. Thus, EGFR is an important predictive marker, being part of the group of clinical, cellular, or molecular markers that predict the response of a tumor to treatment, which is assessed by tumor shrinkage or survival benefit from treatment. A molecular test should be part of the standard diagnostic workup because EGFR is a significant predictive marker. HER1-4 exhibit similar molecular structures with an extracellular, Cys-rich ligand-binding domain, a single α-helix transmembrane domain, a cytoplasmic TK domain (except HER3), and a carboxyl-terminal signaling domain. Downstream EGFR signaling promotes increased proliferation, angiogenesis, metastasis, and decreased apoptosis. The TK activity may be dysregulated by several mechanisms, including *EGFR* mutations, an increase of gene copy number, and EGFR protein expression. Inhibition of tumor cell apoptosis and ultimately tumor progression occurs in case of an improper activation of EGFR TK (gain-of-function or activating mutations). Moreover, EGFR interacts with the integrin pathway and activates matrix metalloproteinases to modify cellular adhesion, motility, and invasion into the surrounding environment. The activation of EGFR-dependent intracellular signaling is not the only mechanism, but several pathways have been identified in pulmonary carcinogenesis. HER 2 or sometimes called ERBB-2 is a transmembrane tyrosine kinase receptor. *HER2* gene mutations occur in around 1–5% of NSCLC, and gene variations involve most often in-frame insertions in exon 20, which codes part of the kinase domain. As *EGFR* mutation, the frequency of *HER2* mutations has been demonstrated to be increased in adenocarcinoma, never smokers, women, and Asian patients. ALK is a critical member of the insulin superfamily of receptor TK and is usually expressed only in the CNS, small bowel, and testis. Initially, the *ALK* gene translocation, t(2;5) (p23;q35), was found in a subset of anaplastic large cell lymphomas in the mid-1990s, while fusions of *ALK* with another upstream partner, *EML4*, were found in NSCLC more than 10 years later. *EML4-ALK* fusions result in protein oligomerization, and constitutive activation of the kinase causes aberrant activation of downstream oncogenic signaling pathways such as MAP kinase, PI3-kinase, and signal transducers and activators of transcription (STAT). *EML4-ALK* fusion is unusual, occurring in ~3–7% of patients with NSCLC and is more prevalent in individuals who never smoked or in light smokers and patients with adenocarcinoma and is commonly mutually exclusive with *EGFR* or *KRAS* mutations and associated with resistance to EGFR TKIs. *KRAS* has a key role in the EGFR signaling network and appears activated through a mutation in ~5–30% NSCLC. The exon 2 and codon 12 or 13 occur frequently (NSCLC, 97%). These missense mutations impair the functionality of RAS GTPase, locking the RAS signaling in active mode. Mutations of *RAS* occur predominantly in "smoking adenocarcinoma" patients, but 15% of never smokers with adenocarcinoma also have transition mutations [G → A (purine for purine)]. An activating mutation of *KRAS* is present in ~25–35% of TKI nonresponsive cases; somatic mutations of the *KRAS* oncogene are highly specific negative predictors of response to single-agent EGFR TKIs in advanced NSCLC, mostly adenocarcinomas. *EGFR* and *KRAS* mutations are hardly detected in the same tumor. The *BRAF* gene encodes a protein (a cytoplasmic serine/threonine kinase) that has a key role downstream of KRAS in cell signaling pathway. B-RAF is one of the three crucial members of the RAF kinase family (A-RAF, B-RAF, and RAF-1/c-RAF). *BRAF* gene mutations are associated with ↑ kinase activity that leads to activation of the downstream pathway. *BRAF* gene mutations are found in 1–3% of lung cancers (adenocarcinomas, usually) and appear in a mutually exclusive pattern with *KRAS* and *EGFR* gene mutations. *PIK3CA* mutations are seen in ~1–5% of NSCLC and appear more often in squamous cells. *PIK3CA* mutations most frequently affect residues Glu542 and Glu545 in exon 9 and 20 encoding the helical and catalytic domain.

PIK3CA copy number gains occur in approximately 20–30% of lung cancers, with higher frequency in SqCCs. The *AKT1* gene encodes protein kinase B, which helps to mediate PI3K signaling. The *AKT1* E17K mutation was found in 1% of SqCC of the lung but not in pulmonary adenocarcinoma. The phosphatase and tensin homolog (*PTEN*, located on 10q23), a tumor suppressor, encodes a lipid phosphatase that negatively regulates the phosphatidylinositol 3-kinase/AKT pathway. It dephosphorylates PIP3 into phosphatidylinositol-3,4-bisphosphate (PIP2), thereby inhibiting PI3K–AKT signaling. The loss of PTEN activity leads to hyperactivation of the PI3K–AKT pathway. *PTEN* inactivation, which can occur at the genomic level or the protein level, becomes apparent typically through epigenetic mechanisms. Promoter methylation is found in about 1/3 of PTEN-negative NSCLC, both adenocarcinoma and squamous cell. It has been proved that neoplasms with PTEN loss may be more sensitive to inhibitors of the PI3K pathway. *MET* is a proto-oncogene located on chromosome 7q21, which encodes a transmembrane tyrosine kinase receptor (hepatocyte growth factor receptor). MET signals via RAS, PI3K/Akt, and stat pathways. Both copy number gain and true amplification of MET have been identified in NSCLC. *MET* amplification has been reported in about 1/5 of tumors in patients with *EGFR* mutation and acquired resistance to EGFR TKI. *LKB1* (also called STK11) (located on 19p13) is a tumor suppressor gene that encodes for a serine-threonine kinase that phosphorylates *AMPK*, which plays a role in the cellular energy status. LKB1 regulates the cell polarity. Other tumor-suppressing properties of LKB1 may be mediated by inhibition of mTOR, inhibition of cell cycle, and activation of p53. *LKB1* inactivation in lung cancer occurs mainly as a result of deletions or insertions or nonsense mutations leading to a complete absence of LKB1 protein, and *LKB1* is the third most commonly altered gene in lung adenocarcinomas (5% of lung SqCC and 23% of adenocarcinomas), after *TP53* and *KRAS*. *FGFR1* is a critical member of the FGFR family of receptor tyrosine kinases. *FGFR1* activation leads to downstream signaling via RAS/RAF-mitogen-activated protein kinase (MAPK), phosphoinositide 3-kinase (PI3K)-AKT, STAT, and phospholipase Cγ. High-level amplification of *FGFR1* had been reported in about 1/4 of SqCC. The amplification sensitizes the tumors to FGFR1 inhibition. Furthermore, *FGFR1* amplification induced a strong FGFR1 dependency that could be exploited therapeutically, resulting in induction of apoptosis. Thus, *FGFR1* amplification represents an opportunity for targeted therapy in SqCC. On 15q26, there is the insulin-like growth factor 1 (*IGF-1*) receptor (IGF-1R), which is an emerging target for cancer treatment. This aspect is due to its overexpression in many cancers, including NSCLC. *IGF1R* activation triggers downstream pathways, including the RAS/RAF/MAPK pathway and the PI3K pathway, leading to cell proliferation and inhibition of programmed cell death. In lung cancer, overexpression of *IGF1R* is more commonly seen in squamous cells. The discoidin domain receptor 2 (*DDR2*, located on 1q23) is a receptor tyrosine kinase that binds collagen. It has been shown to promote cell migration, proliferation, and survival when activated by ligand binding and phosphorylation. *DDR2* mutations in lung cancer were identified specifically in about 1/20 of SqCC. *SOX2* (3q26) is a transcription factor that plays a key role in the development of lung epithelium. SOX2 is one of the most frequent chromosome gain seen in a squamous cell with amplification in ~20% of SqCC. It had been shown that RNA interference knockdown of *SOX2* reduced cellular proliferation; however, SOX2 alone does not appear to be transforming, and *SOX2* amplification may represent an important event that requires other additional events to be transforming. Recently, lung adenocarcinomas were reported to harbor a novel gene fusion involving the *RET* tyrosine kinase gene partnered with *KIF5B* (the kinesin family 5B gene) and others like *CCDC6*. *RET* fusion occurs in 1.4% of NSCLC, and 1.7% of lung adenocarcinoma, and the *KIF5B-RET* fusion induces an aberrant activation of RET kinase and is considered to be a potential new driver of lung adenocarcinoma. There are individuals with a specific phenotype (women, Asian, never smokers, adenocarcinoma type of bronchogenic carcinoma) and a particular genotype of tumor (*EGFR* gene amplification and/or somatic mutations in the kinase domain of the *EGFR* gene) that experience good response following treatment of the lung neoplasm with tyrosine kinase inhibitors (TKIs). Up to 1/5 of non-small cell lung cancer contains *EGFR* gene changes, and these patients do indeed benefit from TKIs. PCR-based *EGFR* mutational analysis and FISH amplification are the methods used to predict responsiveness with TKIs. The PCR method targets the detection of the most frequent *EGFR* mutations (exon 19 deletions and exon 21 L858R substitutions) accounting for about 9/10 of reported *EGFR* mutations. *EGFR* gene is amplified from either fresh-frozen tissue or formalin-fixed and paraffin-embedded tissue and targeting the gene sequence between exons 18 and exon 21 of the tyrosine kinase domain of the *EGFR* gene. The amplified sequences are then sequenced in a bidirectional fashion to identify mutations. Repeated sequencing of the tumor sample usually establishes a confirmatory method. FISH is used to detect *EGFR* gene amplification that may be both gene amplification (≥ 2 copies of *EGFR* comparing with the signal of a centromeric chromosome 7 control probe) and high polysomy (≥ 4 copies of *EGFR* per nucleus in >40% of cores). Conversely, patients with specific habit (smoking history) and mutations in the K-ras gene (*KRAS*) have a lack of response to EGFR inhibitors. *KRAS* gene mutations, mostly affecting codon 12 and codon 13, are present in about 1/4 of bronchogenic carcinoma of adenocarcinoma type. Pathology work is essential not only to determine the right diagnosis of lung cancer but also to identify the most suitable tumor tissue block for molecular studies. Diagnostic accuracy is necessary for pathology for correct treatment. In many reviews, the diagnostic accuracy of lung cancer is around 90%, which may be right from the statistical point of view, but is deleterious for the single patient. Striving for diagnostic accuracy is crucial in oncology as well as for non-oncologic diseases. Diagnostic accuracy may be improved by an awareness of the several and, frequently, multiple pitfalls that may lead to a diagnostic error of lung cancer. These pitfalls include crush artifacts, lymphoid tissue, non-small cell neuroendocrine tumors, basaloid variants of squamous cell carcinoma, etc. Thyroid transcription factor (TTF1) is, in the vast majority of reviews, the most useful marker of immunohistochemistry to detect an adenocarcinoma subtype in NSCLC-NOS cases. In fact, in the context of bronchial biopsy samples, TTF1 has a positive predictive value of 90% and a predictive accuracy of ~85%.

2.16 Pleural Diseases

In this category, both inflammatory and neoplastic diseases play a significant role. One of the most frequent requisitions for consultation with a department of pediatric pathology or pathology for an adolescent or young adult is resection of part of pleura for chronic pleuritis due to spontaneous PNX. Mesothelial cells undergo hyperplastic response to injury or persistent irritation, which can be difficult to distinguish from well-differentiated mesothelioma. The tumor suppressor gene, *TP53*, particularly its expression has been noted to be elevated in 70% of mesotheliomas but is normal in reactive mesothelium. Thus, this marker may be used in the immunohistochemical investigation of several pleural lesions (Fig. 2.45, 2.46, and 2.47).

2.16.1 Pleural Effusions and Pleuritis

Pleural effusion is defined as increase >15 ml of the fluid between visceral and parietal pleurae. The etiology of serous pleural effusion includes cardiac failure, renal failure, and liver cirrhosis. A serofibrinous effusion may list pneumonia, TB, lung infarcts, lung abscess, bronchiectasis, rheumatoid arthritis, SLE, and postradiation. A suppurative (empyema) etiology includes bacteria or mycotic seeding of the pleural cavity, while a hemorrhagic effusion should point to malignancy, hemorrhagic diathesis, and rickettsial infection. A hemothorax may be due to rupture of the aorta and a chylothorax (milky) due to pulmonary lymphatic blockage.

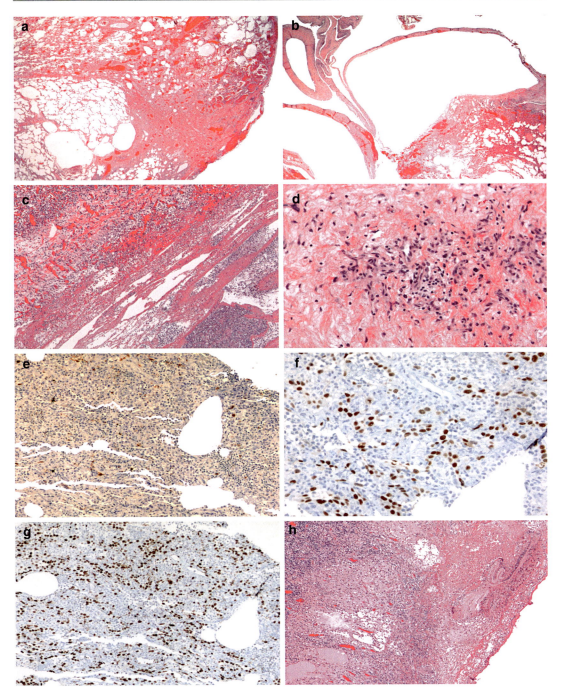

Fig. 2.45 This panel of pleural non-neoplastic diseases include examples of spontaneous pneumothorax due to pleural bullae (**a**, **b**, hematoxylin and eosin staining, ×12.5 original magnification), pleuritis from florid status (**c**) to granulation tissue, and fibrous tissue (**d**) (**c**, hematoxylin and eosin staining, ×50 original magnification; **d**, hematoxylin and eosin staining, ×200 original magnification). In figure (**e**) there is pneumothorax-related type pneumocyte hyperplasia highlighted by S100 positivity (anti- immunostaining, ×100 original magnification), which is also supported by the positive immunostaining for thyroid transcription factor (TTF) (**f**: anti-TTF immunostaining, ×200 original magnification) and proliferative activity (**g**: anti-Ki67 immunostaining, ×100 original magnification). All immunostainings procedures have been carried out using the avidin-biotin complex. Echinococcosis of the pleura is shown in Figure (**h**) (hematoxylin and eosin staining, ×50 original magnification)

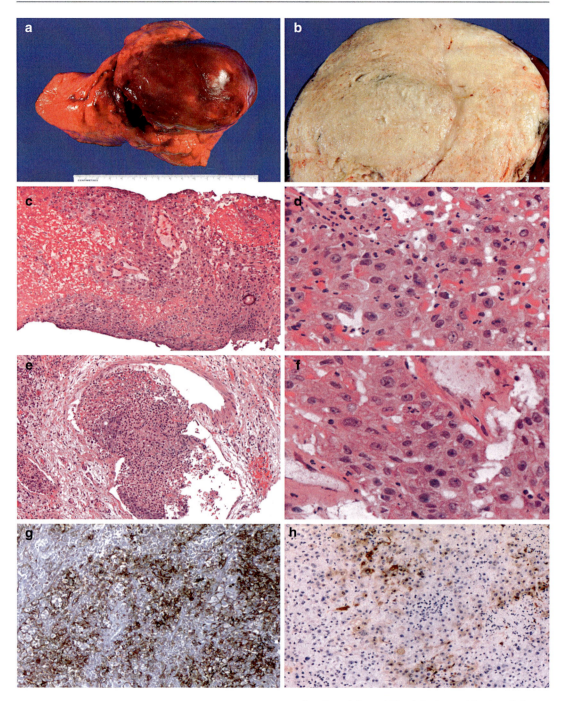

Fig. 2.46 This panel of pleural neoplastic diseases includes the rare occurrence of a pleural mesothelioma with a mass bulging from the pleura in the pleural cavity (**a**). The cut surface of this tumor shows a quite homogeneous appearance with gray yellowish color (**b**). The histologic examination of this tumor reveals solid sheets of polygonal cells with focal tubulopapillary growth pattern. The tumor cells have round nuclei, eosinophilic cytoplasm, and conspicuous nucleoli (**c–f**) (**c**, hematoxylin and eosin staining, ×100 original magnification; **d**, hematoxylin and eosin staining, ×400 original magnification; **e**, hematoxylin and eosin staining, ×100 original magnification; **f**, hematoxylin and eosin staining, ×400 original magnification). The images of immunohistochemistry show podoplanin positivity (**g**, anti-podoplanin – D2-40 immunostaining, ×100 original magnification) and calretinin positivity (**h**, anti-calretinin immunostaining, ×100 original magnification)

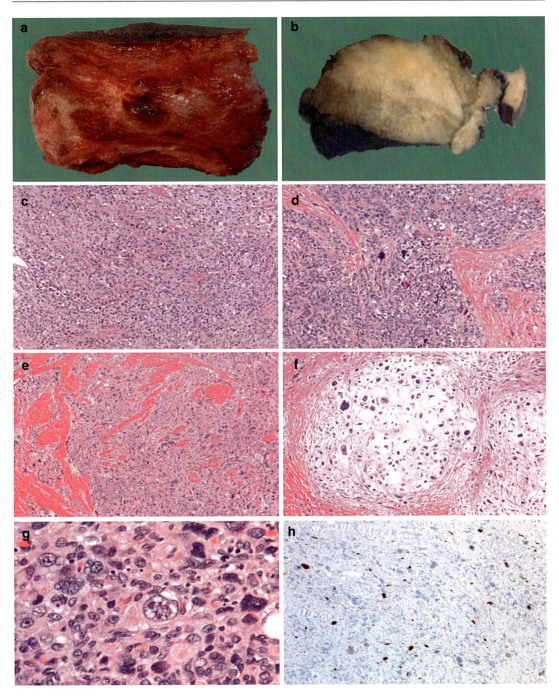

Fig. 2.47 This panel of pleural neoplastic diseases includes two examples of anaplastic rhabdomyosarcoma (**a**, **b**). The histologic examination reveals rhabdomyosarcoma with anaplastic features (**c–g**) (**c**, hematoxylin and eosin staining, ×100 original magnification; **d**, hematoxylin and eosin staining, ×100 original magnification; **e**, hematoxylin and eosin staining, ×50 original magnification; **f**, hematoxylin and eosin staining, ×200 original magnification; **g**, hematoxylin and eosin staining, ×400 original magnification). The immunostaining for MYF4 reveals positivity in the nuclei of some tumor cells (**h**: anti-MYF4 immunostaining, ×50 original magnification)

Pleural plaques: Hyalinized fibrous tissue usually but not always associated with asbestos exposure, which are characteristically parietal or diaphragmatic. Pleural plaques may calcify.

2.16.2 Pneumothorax

- *DEF*: Air/gas detectable pleural pathology located between the peripheral margin of the lung and the chest wall, mediastinum, or diaphragm.
- *EPG*: The etiology of PNX includes peri-pulmonic pneumonic breakthrough causes. These causes include external trauma through a perforation of the chest wall and parietal pleura, an internal trauma through an esophageal fistula, and an internal trauma through a perforated viscus (e.g., gastric ulcer).Visceral pleural breakthrough, non-syndromic subtype entails lung emphysema, Langerhans cell histiocytosis, lymphangioleiomyomatosis (LAM), CF, thoracic endometriosis, LIP, necrotizing pneumonia (e.g., *Pneumocystis jirovcii*), and light-chain deposition disease. The visceral pleural breakthrough, syndromic subtype comprises Marfan syndrome, Ehler-Danlos syndrome,and Birt-Hogg-Dube syndrome.
- *CLI*: PNX is usually unilateral and occurs abruptly as chest pain, most often during physical exertion and is followed by shortness of breath, particularly in adolescents with (idiopathic) spontaneous PNX of tall males or in patients with CF. In some cases of spontaneous PNX, a thoracotomy is performed on a focus of fibrosis (so-called "apical cap").
- *GRO*: 1+ bulla/ae surmount PNX up to 1 cm in Ø. A hole may be observed in the pleural surface adjoining the bullae. Alternatively, a tear may be seen in the wall of one of the bullae.
- *CLM*: There is a focus of fibrosis, and the overlying visceral pleura may contain air pockets. These findings correspond to interstitial emphysema and are termed blebs and are distinct from bullae, which represent enlarged alveolar spaces. There is an alveolar collapse, an accumulation of chronic inflammatory cells, and reactive mesothelial hyperplasia. The inflammatory infiltrate is often rich in eosinophils, and the term "reactive eosinophilic pleuritis" has been proposed in the scientific literature. A fourth histologic component is represented by the reactive type II pneumocyte hyperplasia, which may be quite brisk sparking suspicion of adenocarcinoma in some cases.
- *DDX*: Eosinophilic granuloma of the lung (pulmonary Langerhans cell histiocytosis) and eosinophilic pneumonia, which show deep infiltrates (not superficial), and vasculitis are the main differential diagnoses. Some small BV may be involved, but the process is limited to the subpleural region in the spontaneous PNX and does not involve deeper territories of the lung as seen in true vasculitis.
- *TRT*: In cases, repeated PNX occurs, a procedure called "pleurodesis" may be proposed, which can obliterate the pleural cavity. The obliteration may be carried out surgically, by stripping the pleura, or medically, by the installation of sclerosing compounds such as talc or tetracycline. Pleurodesis may be contraindicated in patients, who will eventually necessitate lung TX because it will make the lung explant procedure challenging to perform. The identification of a fibrotic pleura with many birefringent crystals indicates that pleurodesis was undertaken in the past. Talc pleurodesis-associated neoplasia is rare but also needs to be taken into account before performing such a procedure.
- *PGN*: Mesothelial hyperplasia may evolve into squamous metaplastic change if gas is maintained in the pleural cavity for many months, as was a frequent line of therapy for pulmonary TB. Squamous metaplasia may represent metaplasia of the mesothelium, progressive extension of metaplastic bronchial epithelium through a bronchopleural fistula or skin implantation. This metaplastic change has been noted to be an underlying disorder for progression to malignancy. In patients with CF, focal replacement of the mesothelium by ciliated columnar or squamous epithelium has been found. This aspect is likely due to a rupture of bronchiectatic abscesses contributing to the pathogenesis of PNX in this disease.

2.16 Pleural Diseases

Empyema-Based Bacteria-Producing Gas

In some cases, the label of hemopneumothorax is used, when the local blood vessels are also damaged. Iatrogenic forms of PNX occur when the air is deliberately introduced into the pleural sac for the treatment of chronic TB allowing collapsing and resting the lung (i.e., artificial PNX). In thoracic surgery, the pleural sac is often opened and specifically approved to be filled with air as well. PNX is also divided in the "open" and "closed" subtype being the former a PNX with aperture remaining patent and air pass freely into and out of the sac, while the latter shows progressive resorption of the air. Importantly, the gradual adsorption of air ensues over a few weeks, O_2 first and nitrogen later. The valve PNX may sometimes occur and consists of a PNX that allows air to enter, but not to escape. This clinical situation, which is called pressure or tension PNX, constitutes a medical emergency because the increased pressure displaces the structures of the mediastinum progressively to the opposite side similarly to a diaphragmatic defect.

The most striking differential diagnoses are bullous emphysema, LCH, and cystic LAM.

2.16.3 Neoplastic Pleural Diseases

There are very few tumors that have dominated the epidemiological platform in a specific age group in the past, but currently, mesothelioma is being diagnosed at an earlier age than before, and an increase to asbestos fibers exposure is overwhelming particularly in regions where asbestos fibers are still used or manufactured. Among tumors, a solitary fibrous tumor of pleura, benign mesothelioma, and malignant mesothelioma will be considered.

2.16.3.1 Solitary Fibrous Tumor of Pleura

- *DEF*: Benign mesenchymal tumor of pleura, usually visceral pleura, often pedunculated, and of interstitial dendritic cell origin, without apparent nexus to asbestos exposure.
- *SYN*: Solitary fibrous mesothelioma.
- *EPI*: Uncommon in adults but rare in children, ♂:♀ = 1:1. Very few pediatric case reports have been reported, including in the pleura of a 7-year-old boy, in the mesentery of a 6-year-old boy, or in the orbit of a 15-year-old child (Penel et al. 2012).
- *GRO*: Well circumscribed, ±encapsulated, firm, lobulated, neoplasm harboring a whirled cut surface.
- *CLI*: (1) IGF-related hypoglycemia (paraneoplastic hypoglycemia)-manifesting SFT, (2) hypertrophic pulmonary osteoarthropathy (HPO)-manifesting SFT (digital clubbing), (3) mass-related symptoms (cough, chest pain, dyspnea, effusion), and (4) asymptomatic SFT.
- *CLM*: Four microscopic aspects characterize this tumor, including:
  1. "Patternless" pattern, described by a tangled network of fibroblast-like cells with deposition of copious reticulin and collagen fibers ("ropey" collagen) and cellularity varies from densely cellular to edematous to densely collagenous with some keloidal appearance
  2. Wavy neuroid pattern
  3. Storiform/herringbone
  4. Hemangiopericytomatous
- *IHC*: (+) CD99, (+) BCL2, (+) CD34 (positive triad) and (+) VIM as well as (−) S100, CKs, SMA, DES, FXIIIa, and Coll-IV.
- *DDX*: Synovial sarcoma, which is (+) CD99, (+) BCL2, but (−) CD34.
- *TEM*: Non-mesothelial architecture.
- *TRT*: Usually cured by simple excision and paraneoplastic hypoglycemia will vanish.
- *PGN*: Recurrence in up to 1/5 of cases with "aggressive" behavior showing repeated recurrences. Malignancy has been identified in SFT showing high expression of *TP53* and high proliferation rate detected by MIB1 (Ki67) (Box 2.24).

2.16.3.2 Benign Mesothelioma

Benign mesothelial solitary tumor (aka pleural fibroma) seems to be relatively rare in pleural cavity, differently from the peritoneal cavity, where it is more common. The benign mesothelioma manifests usually as soft friable, well-circumscribed, mottled, gray/pink/yellow mass with papillary processes lined by one or several lay-

> **Box 2.24 "Aggressive" Signs of SFT**
> - Size: Ø > 5 cm
> - Lack of pedicle ("non-pedunculated SFT")
> - Hypercellularity accompanied by cellular pleomorphism and/or giant cells
> - Hyperdivision of the cells (high mitotic index with MI > 4/10 HPF)
> - Necrosis
>
> Notes: *MI*, mitotic index.

ers of cuboidal mesothelial cells without atypia histologically. It is important to emphasize here that the amphiboles are more carcinogenic than chrysotile asbestos properly in consideration of the next paragraph.

2.16.3.3 Malignant Mesothelioma

- *DEF*: Malignant pleural mesothelial tumor (MPM).
- *EPI*: Rare, <5% of presenting cases are children, and less than 300 cases of such have been published in the literature (Scharf et al. 2015).
- *EPG*: ~2/3 of malignant mesotheliomas are related to asbestos exposure, and chronic asbestos exposure carries a 2–3% risk of mesothelioma – risk, which is not increased by tobacco smoking. Long latency between exposure and tumor onset, 25–45 years, and asbestos bodies more commonly found in the lung (≠ ferruginous bodies). In the pediatric cases, there is no apparent association with asbestos or radiation exposure. An example of pediatric MPM has been reported after *in utero* isoniazid exposure took place. Moreover, malignant mesothelioma has been found in a child with ataxia-telangiectasia syndrome, which is a hereditary disorder characterized in part by susceptibility to malignancy.
- *GEN*: Deletion of chromosomes 1, 3, 6, 9, and 22; *TP53* and *P16* genes changes.
- *CLI*: Pulmonary symptoms are related to the extension of the tumor in the pleura opposing a normal movement of the pleura.
- *GRO*: Multiple gray or white ill-defined nodules are diffusely thickening of pleura, which may grow extensively to fill pleural space, producing effusion and encasing lung.
- *CLM*: (1) *Epithelial* type, (2) spindle or *sarcomatoid* type, and (3) *mixed* type.
  1. Epithelial type, showing papillae or pseudoacini or even solid nests of cuboidal, columnar, or flattened cells (tubular-papillary, glandular-acinar, sheets/solid, and mixed).
  2. Spindle or sarcomatoid type with more nodularity and less plaque-like, high cellularity, nuclear atypia, high MI, interweaving bundles of spindle cells, and common areas of hemorrhage, necrosis, and cystic change. In case of a significant amount of collagen, the sarcomatoid-type of malignant mesothelioma has been designated "desmoplastic mesothelioma."
  3. Mixed type (biphasic).
- *IHC*: (+) CALR, D2-40, K5/6, WT-1, thrombomodulin (CD141).

Calretinin is stronger in both cytoplasm and nuclei in the sarcomatoid type (+++) while is less strong (+) in the epithelial type of malignant mesothelioma. Other markers are much useful in differentiating between the two types of malignant mesothelioma.

- *TEM*: Long slender microvilli of the neoplastic mesothelial cells (≥15 times than Ø).
- *DDX*: Bronchogenic adenocarcinoma, metastasis, synovial sarcoma, etc.
- *TRT*: Chemotherapy alone is the first choice for patients with stages I–IV disease who are not candidates for surgery and for patients with sarcomatoid histology, which does not usually occur in youth.
- *PGN*: 1-YSR: ~50%; sarcomatoid type has a worse prognosis than epithelial type.

Distinguishing malignant mesothelioma from adenocarcinoma may be very challenging. However, some features are more distinctive of one entity, and some other features are more prone to suggest the other entity. Mesothelioma produces hyaluronic acid (intracellular and extracellular) which can be stained with alcian blue in a hyaluronidase sensitive manner, while mucicarmine or PAS positive droplets in cells may strongly suggest adenocarcinoma. Moreover, mesothelial cells are keratin, VIM, EMA, S100 positive, and almost always CEA, Leu-M1 (CD15), Ber-EP4 (anti-EpCAM) negative, while adenocarcinoma cells are steadily keratin, EMA, CEA, Leu-M1, and Ber-EP4 positive. HMFG-2 stains only cell membranes of mesothelial cells but shows cytoplasmic staining for adenocarcinoma. Mesothelial cells are also negative for p63, CD117, K20, and MUC4. In Box 2.25 there is a useful diagnostic panel to distinguish malignant mesothelioma from adenocarcinoma.

The criteria of malignancy are branching glands, extensive storiform pattern, lack of zonation, large cell groupings of neoplastic cells, a marked cellular atypia/pleomorphism, the presence of necrosis, the identification of papillae, and the obvious stromal invasion. Overall, the features favoring mesothelioma vs. carcinoma are (+) acid mucopolysaccharide, which is inhibited by hyaluronidase, (+) CALR, D2-40, WT-1, K5/6, and CD141, (−) TTF-1, CEA, B72.3 and BerEp4, p63, CD117, K20, MUC4, CD15, perinuclear rather than peripheral staining for keratins, and long and slender microvilli and abundant tonofilaments but absent microvillous rootle and lamellar bodies on TEM.

Box 2.25 Useful Diagnostic Panel for Pleural Malignant Mesothelioma (PMM) Versus Adenocarcinoma (ACA)

	CALR	D2-40	WT-1	K5/6	TTF-1	B72.3	Napsin
PMM	+	+	+	+	−	−	−
ACA	−	−	−	−	+	+	+

Notes: CALR, calretinin; WT-1, Wilms tumor-1; K5/6, keratin 5/6; TTF-1, thyroid transcription factor, CEA, carcinoembryonic antigen; B72.3, monoclonal antibody that recognizes tumor-associated glycoprotein 72 (TAG-72), which is a mucin-like carbohydrate and protein complex on the surface of numerous cancer cells, Napsin A is a functional aspartic proteinase playing an important role in identifying primary lung ACA. The acronym "Caldinek" may be used for the positive markers in PMM (Calretinin, D2-40, nephroblastoma antigen, and keratins)

2.17 Posttransplant Lymphoproliferative Disorders (PTLD)

PTLD are uncommon complications of transplantation and represent a heterogeneous group of diseases that can lead to significant morbidity and mortality. We investigated the frequency of anatomic site occurrence, the subtypes of PTLD, and the Epstein-Barr virus (EBV) status of the malignant cells. The EBV, also called human herpesvirus 4, is one of eight known human herpesvirus types in the herpes family. EBV is one of the most diffuse viruses in humans. EBV is infamously best known as the cause of infectious mononucleosis and the etiologic agent of the "first kissing" disease of the adolescents or "the kissing disease" as first described in the 1920s by Thomas Peck Sprunt and Frank Alexander Evans, who coined the term "infectious mononucleosis" (Sprunt and Evans, 1920). We reviewed the surgical pathology of 53 cases of PTLD in 47 consecutive patients with solid organ transplantation at the University of Alberta, Edmonton, AB, Canada (Yu et al. 2010). The anatomic occurrence of the lesions, subtype of PTLD, and the EBV status of the malignant cells were studied. In the 47 patients (11 children, 36 adults, age mean = 41, range 2–71, ♂:♀ = 3:2) with solid organ transplants (19 kidneys, 10 livers, and 9 each of hearts and lungs), 2 (4%) patients developed early PTLD with plasmacyte hyperplasia (E-PTLD) and 19 (36%) patients with polymorphic (PM) PTLD and 28 (53%) patients had monomorphic (MM) subtype of predominantly diffuse large B-cell lymphoma and 4 (7%) patients of Hodgkin disease (HD). The aerodigestive tract (ADG) represented the largest involved group, n = 19 (36%) (GI = 11, respiratory tract = 8), with 8 cases (42%) of PM and 11 (58%) MM-PTLD. Six patients had multiple organ involvement. In 35 patients tested by EBV-ISH, 26 (74%) patients were found positive (17 MM-HD, 9 PM), and 9 (26%) patients were EBV-negative (7 MM, 2 PM). The EBV status was compared to PTLD subtype but no statistical significance was observed. In an EBV-negative case, polyomavirus was detected. This study confirmed the heterogeneity of PTLD and the involvement of multiple anatomic sites. The significant manifestation of PTLD in the aerodigestive tract is crucial and may be related to the immunologic reactions in mucosa-associated lymphoid tissues (MALT) of these sites. EBV continues to be a critical etiologic agent in the development of PTLD. The emerging EBV-negative PTLD cases may warrant an investigation into other causal agents with different mechanisms of malignant transformation.

Fine needle aspiration cytology plays a major in the diagnostics of pulmonary disorders (Figs. 2.48 and 2.49). Contaminants of percutaneous FNA include cutaneous squamous cells, mesothelial cells, skeletal muscles, fibroblasts/fibroconnective tissue, and hepatocytes. Mesothelial cells are characterized by flatness, cohesiveness, uniformity, and "windows"-ness. The uniformity is explicit in showing round to oval nuclei with small nucleoli, a moderate amount of cytoplasm. Conversely, reserve cell hyperplasia is characterized by the cohesiveness of packed cells, small cell size with scant cytoplasm, nuclear molding with smudged dark chromatin, but no mitosis or necrosis. Some structures need to be kept in mind and recognized cytologically. Creola body is a cluster of reactive bronchial cells – ciliated columnar cells (usual nuclear features of bronchial cells, cilia of bronchial cells, ±eosinophils) found in the sputum of some asthmatic patients. Curschmann spiral is a microscopic finding in the sputum of asthmatic patients. Charcot-Leyden crystal is the by-product of eosinophilic degeneration. Curschmann's spiral is found in lavage fluid

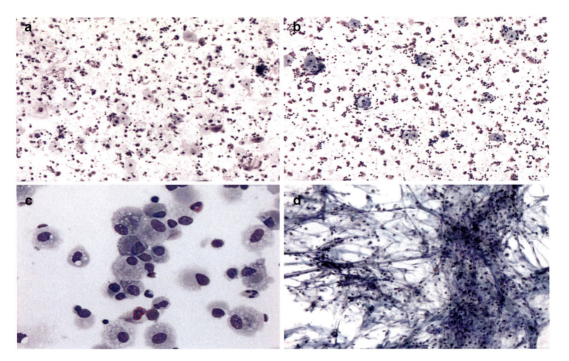

Fig. 2.48 Lung cytology panel showing reactive inflammatory cells in bronchoalveolar lavage (BAL) (**a**, Giemsa staining, ×100 original magnification). Figure (**b**) shows several acute inflammatory cells in BAL (Giemsa staining, ×100 original magnification). Figure (**c**) shows a BAL with nonmalignant cells and rare eosinophils (**a**, Diff-Quik staining, ×400 original magnification), while figure (**d**) discloses a BAL with numerous macrophages and reactive inflammatory cells in a child with active tuberculosis (Diff-Quik staining, ×200 original magnification). The same BAL was also processed with Ziehl-Neelsen showing acid fast bacilli (**e**, Ziehl-Neelsen staining, ×630 original magnification). Figure (**f**) shows a BAL with evidence of *Candida* infection (Grocott methenamine silver staining, ×400 original magnification). Figure (**g**) shows two scores for the evaluation of lipid-rich macrophages. Figure (**h**) shows an Oil Red O (ORO)-stained cytology specimen from BAL with lipid-rich macrophages (×200 original magnification). The macrophages are highlighted in the inset of Figure (**h**)

2.17 Posttransplant Lymphoproliferative Disorders (PTLD)

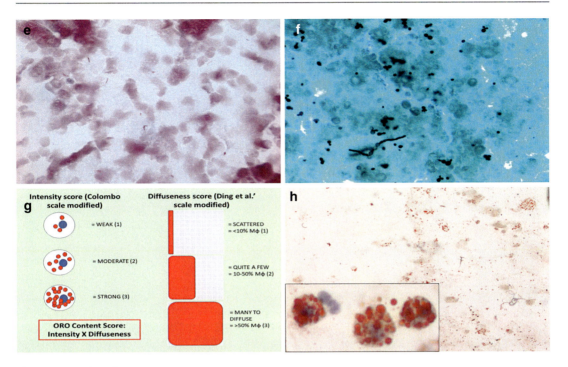

Fig. 2.48 (continued)

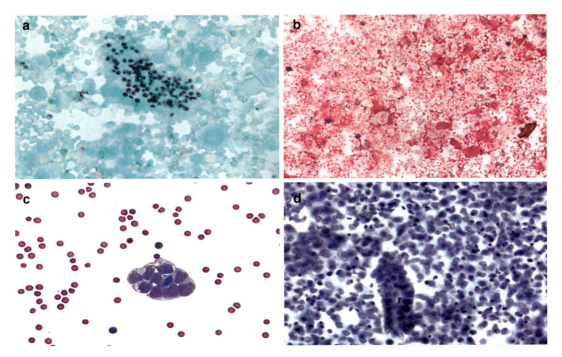

Fig. 2.49 Lung cytology panel showing *Pneumocystis jiroveci* in Figure (**a**) (Grocott methenamine silver staining, ×400 original magnification); a pulmonary siderosis view in Figure (**b**) (Perls' Prussian Blue staining, ×100 original magnification); a neuroblastoma in the pleural fluid in Figures (**c**) and (**d**) (**c**, Diff-Quik staining, ×400 original magnification; **d**, Toluidine blue staining, ×400 original magnification); metastasis of a fibrolamellar carcinoma of the liver in the pleural fluid in Figure (**e**) (Papanicolaou staining, ×630 original magnification); pleural effusion in a patient with posttransplant lymphoproliferative disorder in Figure (**f**) (hematoxylin and eosin staining on cell block, ×200 original magnification); malignant mesothelioma in Figures (**g**) and (**h**) (Diff-Quik staining, ×400 original magnification)

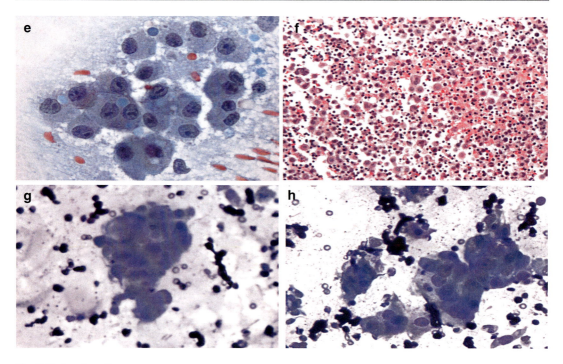

Fig. 2.49 (continued)

of asthmatic patients, and the Papanicolaou staining will highlight a reddish core with a fuzzy margin. To this margin, numerous cells are attached secondarily. The spiral can reach 1 mm in length. It is not uncommon to see contamination with vegetable cells in some FNAs. Sarcoidosis features include aggregates of epithelioid histiocytes, MNGC, and lymphocytes.

2.18 Non-thymic Mediastinal Pathology

The mediastinum is located in the thoracic cavity in the center between the two lungs. It is surrounded by loose connective tissue, as an undelimited region that contains a group of structures within the thoracic cavity. The mediastinum contains the heart, the cardiac outflow tract-derived blood vessels, the esophagus, the trachea, the phrenic and cardiac nerves, the thoracic duct, the thymus, and the lymph nodes of the central chest. Anatomists and surgeons divide the mediastinum into three compartments. The mediastinum is split into three compartments, including anterior, middle, and posterior. However, such a system of compartmentalizing may have deficiencies. The assessment performed on a lateral radiograph of the chest does not have real and definite fascial planes. Also, several disorders that may present as mediastinal masses cross boundaries or arise in more than one compartment. On the other hand, the practice of assigning a mediastinal mass to a specific chamber is still convenient because such a method enables the physician to formulate a manageable differential diagnosis. It effectively directs further imaging workup, and it yields valuable information, particularly for surgical planning. Numerous methods for classifying the mediastinal compartments have been published. The most typical of these methods include the Shields' classification schema, the Fraser and Pare, Felson, Heitzman, Zylak, and Whitten models, which may have more a radiologic practice than clinical practice. These schemata and classifications can be retrieved in specific textbooks. Several aspects contribute to make the terminology of this field typical of Babelic confusion.

There is quite a lot of confusion because there is considerable overlap. Terminological differences and diverse methodologies have resulted in confusion among physicians. In the following paragraphs, we are going to illustrate to the best of our knowledge the pathology occurring in the anterior mediastinum, middle mediastinum, and posterior mediastinum.

2.18.1 Anterior Mediastinal Pathology

The anterior mediastinum can have thymomas that are discussed in the hematologic chapters and non-thymomas anterior mediastinal pathology, which includes thyroid tissue masses and lymphadenopathy. Thyroid masses in the anterior mediastinum can arise either from ectopic thyroid tissue or via an extension of the orthotopic thyroid gland tissue into the anterior mediastinum. The most common pathology is a cervical thyroid goiter which extends into the substernal location. Overall, the lymphadenopathy is the most common cause of mediastinal masses. The characteristics of the patient (age, sex, medical history, etc.) and the clinical setting usually determine the roots of the lymphadenopathy. The differential diagnosis should always include metastatic sarcoma (rhabdomyosarcoma)/carcinoma, malignant lymphoma of Hodgkin type and non-Hodgkin type, and infectious/inflammatory conditions, especially granulomatous processes such as tuberculosis, histoplasmosis, and sarcoidosis. Most common types of lymphomas that can involve the mediastinum are the nodular sclerosis Hodgkin lymphoma, the lymphoblastic lymphoma, and the diffuse large cell lymphoma. Chiefly in pediatrics, some other small round blue cell tumors can mimic malignant lymphomas, including neuroblastoma, embryonal rhabdomyosarcoma, primitive neuroectodermal tumor/Ewing sarcoma, seminoma, and metastatic small cell carcinoma. The question to dispute is mainly a morphological issue if the biopsy is small or has been crushed. The classical Hodgkin lymphoma, nodular sclerosis variant shows marked fibrosis at low power, which is typical of the lesion. The use of the 20× ocular usually allows the identification of the neoplastic Reed-Sternberg cells, Reed-Sternberg cell variants, and Hodgkin cells. The Reed-Sternberg cells, which will be discussed in more detail in the hematologic chapter, have a distinctive multilobed nucleus and prominent nucleoli, while the background shows a mixture of benign inflammatory cells. The low power examination is also useful in identifying the sarcoidosis with the effaced lymph nodes showing numerous confluent non-necrotizing epithelioid granulomas with surrounding fibrous tissue. Histoplasmosis needs to be ruled out using a silver staining (histochemical special stain) to highlight multiple black yeast forms of *Histoplasma capsulatum*. In the anterior mediastinum, lymphangiomas and hemangiomas may also occur in addition to the relatively rarer occurrences of lipomas and sarcomas.

Germ cell tumors arise almost exclusively from primitive, undifferentiated cells generally found in the sexual organs but which can get sprinkled in various other sites of the body during embryogenesis, especially along the midline (retroperitoneum, skull base other than mediastinum). Germ cell tumors occur at any age, but usually, adolescents and young men are most often involved. Histologically, germ cell tumors are classified in seminomas ("pure seminoma") and non-seminomatous germ cell tumors (NSGCTs). The latter entity includes teratoma, choriocarcinoma, yolk sac tumor, embryonal carcinoma, and mixed germ cell tumors. The histology of the seminoma and NSGCTs is not dissimilar from the GCTs of the testis and ovary and will be highlighted in the chapter in detail. Typically, GCTs spread via lymphatics but can metastasize anywhere especially the choriocarcinoma type. Since testicular germ cell tumors can metastasize to the mediastinum, it is mandatory to study entirely male adolescents and young men with mediastinal masses for a testicular imaging investigation. Serum alpha-fetoprotein (AFP), which increases notably in case of yolk sac tumor, and human chorionic gonadotropin (HCG), which increases with choriocarcinoma, also need to be investigated both serologically and by immunohistochemistry. Moreover, thymic cysts and cysts of the parathyroid gland can also occur and need to be taken into the differential diagnosis.

2.18.2 Middle Mediastinal Pathology

In this category, mediastinal lymphadenopathy as illustrated above and pericardial and bronchogenic cysts play a significant role.

2.18.3 Posterior Mediastinal Pathology

In this part the mediastinum gastroenteric cysts, which are a maldevelopment defect, and neurogenic tumors play a key role. Neurogenic tumors are a typical cause of a posterior mediastinal mass, and they rarely arise elsewhere in the mediastinum. Most of these tumors are benign, with the malignant ones generally occurring in younger patients. They are classified as tumors of the sympathetic nervous system (neuroblastoma, ganglioneuroblastoma, and ganglioneuroma) and tumors of peripheral nerve and nerve sheath (schwannoma, neurofibroma, and malignant schwannoma or malignant nerve sheath tumors). Both tumors of the sympathetic nervous system and tumors of the peripheral nervous system will be illustrated in detail in other chapters (soft tissue and nervous system). Briefly, the neuroblastoma is an undifferentiated tumor composed of small cells with a high nucleus: cytoplasm ratio, variable Schwannian stroma, and a variable number of mitoses and karyorrhexis. The schwannoma is, conversely, a lobulated, encapsulated mass with a right capsule and homogeneous white-yellowish cut surface and constituted of normal spindled cells with no significant mitotic activity.

Multiple Choice Questions and Answers

- RSP-1 Which sequence corresponds to the right progression of the different phases of intrauterine lung development?
  (a) Canalicular phase → saccular phase → alveolar phase
  (b) Pseudoglandular phase → saccular phase → alveolar phase
  (c) Pseudoglandular phase → canalicular phase → saccular phase → alveolar phase
  (d) Canalicular phase → saccular phase → alveolar phase
- RSP-2 The mother brings a 2-year-old boy to the pediatrician with a history of recurrent pneumonia. Mother does not live with the father and has multiple partners. A few animals live in the house including a cat, a dog, and a parrot. On physical examination, wheezing and crackles are heard. Moreover, digital clubbing is recognized. Blood eosinophilia is not noted, and no signs of child abuse are evident. What is the most likely diagnosis?
  (a) Cystic fibrosis
  (b) Psittacosis
  (c) Human immunodeficiency virus (HIV) infection
  (d) Lung sequestration
  (e) Allergic bronchopulmonary aspergillosis
- RSP-3 A 3-year-old girl, ex premature baby, who received some oxygen at birth, is brought by the parents to the pediatrician with a history of relapsing pneumonia of the left lung. Imaging and bronchography show that the area involved does not fill contrasting with the surrounding parenchyma. What is the most likely diagnosis?
  (a) Bronchopulmonary dysplasia
  (b) Cystic fibrosis
  (c) Kartagener syndrome
  (d) Pulmonary sequestration
  (e) Postprimary tuberculosis of the lung (tuberculous cavern)
- RSP-4 Pneumocyte of type II possesses all the following features with the exclusion of one of the following statements. Which one?
  (a) These cells produce lung surfactant.
  (b) Ultrastructural examination of these cells reveals osmiophilic lamellar bodies.
  (c) They constitute more than 50% of the alveolar surface in healthy conditions.
  (d) Microvilli are identified on ultrastructural examination.
  (e) Defects in type II pneumocytes contribute to infantile and adult respiratory distress syndrome.

- RSP-5 Adult respiratory distress syndrome (ARDS) can present in both childhood and adulthood. Which of the following morphologic features are NOT present in ARDS?
  - (a) Hyaline membranes located in the alveoli
  - (b) Decreased permeability of the capillary endothelial cells of the lungs
  - (c) Hyperplasia/proliferation of type II pneumocytes
  - (d) Alveolar wall damage
  - (e) Pulmonary edema
- RSP-6 Which of the following statements on human parainfluenza viruses (HPIVs) is NOT correct?
  - (a) HPIVs are single-stranded, enveloped DNA viruses of the *Paramyxoviridae* family.
  - (b) Four serotypes are known, and they cause respiratory illnesses in children and adults.
  - (c) HPIVs bind and replicate in the ciliated epithelial cells of the upper and lower respiratory tract.
  - (d) Seasonal HPIV epidemics account for 40% of pediatric hospitalizations for lower respiratory tract illnesses and 75% of croup cases.
  - (e) HPIVs are associated with a broad spectrum of illnesses which include otitis media, pharyngitis, conjunctivitis, croup, tracheobronchitis, and pneumonia.
  - (f) Immunity resulting from the disease in childhood is incomplete, and reinfection with HPIV accounts for 15% of respiratory illnesses in adults.
- RSP-7 Which of the following statements on primary tuberculosis is NOT correct?
  - (a) An incubation period of 2–10 weeks follows the initial infection.
  - (b) Presenting symptoms include slightly elevated but persistent temperature, weight loss, fatigue, irritability, and malaise.
  - (c) Chest X-ray shows diffuse lobar infiltration of the right middle lobe.
  - (d) Mediastinal lymph node involvement is common.
  - (e) Tuberculous pleurisy is a late complication.
- RSP-8 Pulmonary hemosiderosis shows hemosiderin-laden macrophages histologically in alveolar lumina. There is also shedding and hyperplasia of alveolar epithelial cells. Marked alveolar capillary congestion can be seen. Pulmonary hemosiderosis is evidenced by which special stain?
  - (a) Gram stain
  - (b) Von Kossa stain
  - (c) Perls' Prussian blue stain
  - (d) Picro-sirius
  - (e) Masson Goldner stain
- RSP-9 The pathology of alveolar capillary dysplasia (ACD) with misalignment of the pulmonary veins has specific features in addition to the misalignment of the pulmonary veins. Generally, the pulmonary veins accompany pulmonary arteries within the same adventitial sheath and become parts of the bronchovascular bundles within the centers of the lobules. In ACD with misalignment of the pulmonary veins, the pulmonary veins reside within the interlobular septa at the peripheries of the lobules. Which of the following statements does NOT belong to ACD?
  - (a) A general reduction of alveolar capillaries
  - (b) Prominent thickening of the alveolar septa, containing fibroblasts and myofibroblasts and extracellular matrix and centrally placed alveolar capillaries that are at least several cell widths detached from the nearest alveolar spaces;
  - (c) Intra-alveolar hyaline membranes
  - (d) Striking pulmonary vascular changes
  - (e) Variable degrees of alveolar underdevelopment
- RSP-10 An 8-year-old male presented to his family physician complaining of dyspnea, weight loss, and lethargy. He was referred to a pediatric surgeon of a tertiary hospital with expertise in neoplastic lung disease. A chest radiograph and CT scan demonstrated a large right-sided chest mass that appeared to be arising from the upper lobe. There was no known exposure to asbestos or radiation. Neither intra- nor extrapulmonary sequestration were diagnosed in his infancy. A needle biopsy was performed, and the hematoxylin and eosin staining of the histology is shown here:

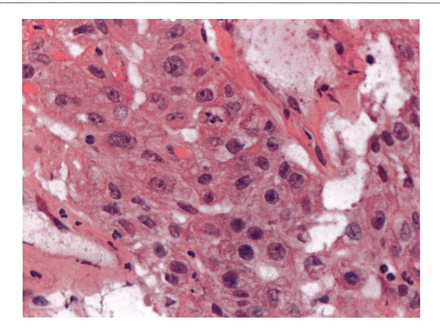

- What is the most likely diagnosis in consideration that the tumor cells were positive for vimentin, epithelial membrane antigen, WT1, calretinin and showed numerous long, narrow microvilli by transmission electron microscopy?
  (a) Metastasis of a rhabdomyosarcoma
  (b) Germ cell tumor of the mediastinum
  (c) Diffuse large B-cell lymphoma
  (d) Pleural malignant mesothelioma
  (e) Lung adenocarcinoma

References and Recommended Readings

Adzick NS. Fetal thoracic lesions. Semin Pediatr Surg. 1993;2(2):103–8. Review. PubMed PMID: 8062026.

Adzick NS, Harrison MR. Management of the fetus with a cystic adenomatoid malformation. World J Surg. 1993;17(3):342–9. Review. PubMed PMID:8337881.

Adzick NS, Harrison MR, Flake AW, Howell LJ, Golbus MS, Filly RA. Fetal surgery for cystic adenomatoid malformation of the lung. J Pediatr Surg. 1993;28(6):806–12. PubMed PMID: 8331508.

Alasaly K, Muller N, Ostrow DN, Champion P, FitzGerald JM. Cryptogenic organizing pneumonia. A report of 25 cases and a review of the literature. Medicine (Baltimore). 1995;74(4):201–11. Review. PubMed PMID: 7623655.

Allen JN, Pacht ER, Gadek JE, Davis WB. Acute eosinophilic pneumonia as a reversible cause of noninfectious respiratory failure. N Engl J Med. 1989;321(9):569–74. PubMed PMID: 2761601.

Aryal S, Nathan SD. Contemporary optimized practice in the management of pulmonary sarcoidosis. Ther Adv Respir Dis. 2019;13:1753466619868935. https://doi.org/10.1177/1753466619868935. PubMed PMID: 31409257; PubMed Central PMCID: PMC6696842.

Askenazi SS, Perlman M. Pulmonary hypoplasia: lung weight and radial alveolar count as criteria of diagnosis. Arch Dis Child. 1979;54(8):614–8. PubMed PMID: 507916; PubMed Central PMCID: PMC1545796.

Bakdounes K, Jhala N, Jhala D. Diagnostic usefulness and challenges in the diagnosis of mesothelioma by endoscopic ultrasound guided fine needle aspiration. Diagn Cytopathol. 2008;36(7):503–7. https://doi.org/10.1002/dc.20811. PubMed PMID: 18528879.

Ban N, Matsumura Y, Sakai H, Takanezawa Y, Sasaki M, Arai H, Inagaki N. ABCA3 as a lipid transporter in pulmonary surfactant biogenesis. J Biol Chem. 2007;282(13):9628–34. Epub 2007 Jan 30. PubMed PMID: 17267394.

Barberis I, Bragazzi NL, Galluzzo L, Martini M. The history of tuberculosis: from the first historical records to the isolation of Koch's bacillus. J Prev Med Hyg. 2017;58(1):E9–E12. PMID: 28515626; PMCID: PMC5432783.

Besnier E. Lupus pernio de la face; synovites fongueuses (scrofulo-tuberculeuses) symétriques des extrémités superieures. Annales de dermatologie et de syphilographie, Paris. 1889;2nd series; 10:333–6.

Betz P, Nerlich A, Bussler J, Hausmann R, Eisenmenger W. Radial alveolar count as a tool for the estimation of fetal age. Int J Legal Med. 1997;110(2):52–4. PubMed PMID: 9168319.

Biehl JR, Burnham EL. Cannabis smoking in 2015: a concern for lung health? Chest. 2015;148(3):596–606. https://doi.org/10.1378/chest.15-0447. Review. PubMed PMID: 25996274; PubMed Central PMCID: PMC4556119.

Birbeck MS, Breathnach AS, Everall JD. An electron microscope study of basal melanocytes and high-level clear cells (Langerhans cells) in vitiligo. J Invest Dermatol. 1961;37(1):51–64.

Boag S. The pathology of mediastinal masses. Queen's University at Kingston. http://www.thymic.org/uploads/mainpdf/PH2MEDIA_pixels.pdf

Boeck CPM. Multipelt benignt hud-sarcoi. Med 4 tavler. (En hidtil ikke beskreven hudaffektion). Norsk Magazin for Lægevidenskaben. Kristiania (Oslo). 1899;60:1321–34.

Boettger MB, Sergi C, Meyer P. BRCA1/2 mutation screening and LOH analysis of lung adenocarcinoma tissue in a multiple-cancer patient with a strong family history of breast cancer. J Carcinog. 2003;2(1):5. PubMed PMID: 14583096; PubMed Central PMCID: PMC239936.

Borie R, Danel C, Debray MP, Taille C, Dombret MC, Aubier M, Epaud R, Crestani B. Pulmonary alveolar proteinosis. Eur Respir Rev. 2011;20(120):98–107. https://doi.org/10.1183/09059180.00001311. Review. PubMed PMID: 21632797.

Braun L, Kisting S. Asbestos-related disease in South Africa: the social production of an invisible epidemic. Am J Public Health. 2006;96(8):1386–96. Epub 2006 Jun 29. PubMed PMID: 16809596; PubMed Central PMCID: PMC1522094.

Brock TD. Robert Koch: a life in medicine and bacteriology. Washington, DC: ASM Press; 1988. p. 296. ISBN 9781555811433.

Brunner J, Freund M, Prelog M, Binder E, Sailer-Hoeck M, Jungraithmayr T, Huemer C, Sergi C, Zimmerhackl LB. Successful treatment of severe juvenile microscopic polyangiitis with rituximab. Clin Rheumatol. 2009;28(8):997–9. https://doi.org/10.1007/s10067-009-1177-0. Epub 2009 Apr 24. PubMed PMID: 19390907.

Bunn PA Jr. Worldwide overview of the current status of lung cancer diagnosis and treatment. Arch Pathol Lab Med. 2012;136(12):1478–81. https://doi.org/10.5858/arpa.2012-0295-SA. Review. PubMed PMID: 23194039.

Cagle PT, Chirieac LR. Advances in treatment of lung cancer with targeted therapy. Arch Pathol Lab Med. 2012;136(5):504–9. https://doi.org/10.5858/arpa.2011-0618-RA. Review. PubMed PMID: 22540298.

Cagle PT, Myers J. Precision medicine for lung cancer: role of the surgical pathologist. Arch Pathol Lab Med. 2012;136(10):1186–9. PubMed PMID: 23020720.

Cagle PT, Olsen RJ. Verifying the role of surgical pathologists in the precision medicine of lung cancer. Arch Pathol Lab Med. 2013;137(9):1176–8. https://doi.org/10.5858/arpa.2012-0659-ED. Epub 2012 Dec 31. PubMed PMID: 23276149.

Cancer Genome Atlas Research Network. Comprehensive molecular profiling of lung adenocarcinoma. Nature. 2014;511(7511):543–50. https://doi.org/10.1038/nature13385. Epub 2014 Jul 9. Erratum in: Nature. 2014 Oct 9;514(7521):262. Rogers, K [corrected to Rodgers, K]. PubMed PMID: 25079552; PubMed Central PMCID: PMC4231481.

Carter R. Pulmonary sequestration. Ann Thorac Surg. 1969;7(1):68–88. Review. PubMed PMID: 4883836.

Centers for Disease Control (CDC). Centennial: Koch's discovery of the tubercle bacillus. MMWR Morb Mortal Wkly Rep. 1982;31(10):121–3. PubMed PMID: 6817044.

Chakraborty RK, Sharma S. Pulmonary sequestration. 2019 May 14. StatPearls [Internet]. Treasure Island (FL): StatPearls Publishing; 2019 Jan. Available from http://www-ncbi-nlm-nih-gov.login.ezproxy.library.ualberta.ca/books/NBK532314/. PubMed PMID: 30335347.

Cheong N, Zhang H, Madesh M, Zhao M, Yu K, Dodia C, Fisher AB, Savani RC, Shuman H. ABCA3 is critical for lamellar body biogenesis in vivo. J Biol Chem. 2007;282(33):23811–7. Epub 2007 May 31. PubMed PMID: 17540762.

Chiang YC, Lee CH, Lin PY. Pulmonary lymphangitis carcinomatosa due to adenocarcinoma arising from choledochal cyst: report of an autopsy case. J Formos Med Assoc. 1991;90(9):860–2. PubMed PMID: 1683388.

Chiavegatto CV, Carneiro AP, Dias EC, Nascimento MS. Diagnosis of severe silicosis in young adults working in stone polishing and mining in Minas Gerais, Brazil. Int J Occup Environ Health. 2010;16(2):147–50. PubMed PMID: 20465059.

Ciardiello F, Tortora G. EGFR antagonists in cancer treatment. N Engl J Med. 2008;358(11):1160–74. https://doi.org/10.1056/NEJMra0707704. Review. Erratum in: N Engl J Med. 2009 Apr 9;360(15):1579. PubMed PMID: 18337605.

Citti A, Peca D, Petrini S, Cutrera R, Biban P, Haass C, Boldrini R, Danhaive O. Ultrastructural characterization of genetic diffuse lung diseases in infants and children: a cohort study and review. Ultrastruct Pathol. 2013;37(5):356–65. https://doi.org/10.3109/01913123.2013.811454. PubMed PMID: 24047351.

Ciuleanu T, Brodowicz T, Zielinski C, Kim JH, Krzakowski M, Laack E, Wu YL, Bover I, Begbie S, Tzekova V, Cucevic B, Pereira JR, Yang SH, Madhavan J, Sugarman KP, Peterson P, John WJ, Krejcy K, Belani CP. Maintenance pemetrexed plus best supportive care versus placebo plus best supportive care for non-small-cell lung cancer: a randomised, double-blind, phase 3 study. Lancet. 2009;374(9699):1432–40. https://doi.org/10.1016/S0140-6736(09)61497-5. Epub 2009 Sep 18. PubMed PMID: 19767093.

Clements BS, Warner JO. Pulmonary sequestration and related congenital bronchopulmonary-vascular malformations: nomenclature and classification based on anatomical and embryological considerations. Thorax. 1987;42:401–8. [PMC free article] [PubMed].

Colby TV. Current histological diagnosis of lymphomatoid granulomatosis. Mod Pathol. 2012;25(Suppl 1):S39–42. https://doi.org/10.1038/modpathol.2011.149. Review. PubMed PMID: 22214969.

Collins KA, Geisinger KR, Wagner PH, Blackburn KS, Washburn LK, Block SM. The cytologic evaluation of lipid-laden alveolar macrophages as an indicator of aspiration pneumonia in young children. Arch Pathol Lab Med. 1995;119:229–31.

Colombo JL, Hallberg TK. Recurrent aspiration in children: lipid-laden alveolar macrophage quantitation. Pediatr Pulmonol. 1987;3(2):86–9. PubMed PMID: 3588061.

Colombo JL, Hallberg TK. Pulmonary aspiration and lipid-laden macrophages: in search of gold (standards). Pediatr Pulmonol. 1999;28(2):79–82. PubMed PMID: 10423305.

Colombo JL, Hallberg TK. Airway reactivity following repeated milk aspiration in rabbits. Pediatr Pulmonol. 2000;29(2):113–9. PubMed PMID: 10639201.

Colombo JL, Hallberg TK. Aspiration: a common event and a clinical challenge. Pediatr Pulmonol. 2012;47(4):317–20. https://doi.org/10.1002/ppul.21560. Epub 2011 Oct 17. PubMed PMID: 22006650.

Colombo JL, Hallberg TK, Sammut PH. Time course of lipid-laden pulmonary macrophages with acute and recurrent milk aspiration in rabbits. Pediatr Pulmonol. 1992;12(2):95–8. PubMed PMID: 1570193.

Contreras AL, Rossi C, Schwartz AM. Acute eosinophilic pneumonia. Pathol Case Rev. 2007;12:100–4.

Cooney TP, Wentworth PJ, Thurlbeck WM. Diminished radial count is found only postnatally in Down's syndrome. Pediatr Pulmonol. 1988;5(4):204–9. PubMed PMID: 2976929.

Corwin RW, Irwin RS. The lipid-laden alveolar macrophage as a marker of aspiration in parenchymal lung disease. Am Rev Respir Dis. 1985;132(3):576–81. PubMed PMID: 4037530.

Cristallo Lacalamita M, Fau S, Bornand A, Vidal I, Martino A, Eperon I, Toso S, Rougemont AL, Hanquinet S. Tracheal agenesis: optimization of computed tomography diagnosis by airway ventilation. Pediatr Radiol. 2018;48(3):427–32.

Cutz E. Hyperplasia of pulmonary neuroendocrine cells in infancy and childhood. Semin Diagn Pathol. 2015;32(6):420–37. https://doi.org/10.1053/j.semdp.2015.08.001. Epub 2015 Aug 29. Review. PubMed PMID: 26584876.

Da Costa AC. Les formations vesiculeuses dans les glandes endocrines [The vescicular formations in the endocrine glands]. C R Assoc Anat. 1928;23:69–75.

Da Cunha SG, Shepherd FA, Tsao MS. EGFR mutations and lung cancer. Annu Rev Pathol. 2011;6:49–69. https://doi.org/10.1146/annurev-pathol-011110-130206. Review. PubMed PMID: 20887192.

Dacic S. Molecular diagnostics of lung carcinomas. Arch Pathol Lab Med. 2011;135(5):622–9. https://doi.org/10.1043/2010-0625-RAIR.1. Review. PubMed PMID: 21526960.

Dalli A, Selimoglu Sen H, Coskunsel M, Komek H, Abakay O, Sergi C, Cetin TA. Diagnostic value of PET/CT in differentiating benign from malignant solitary pulmonary nodules. J BUON. 2013;18(4):935–41. PubMed PMID: 24344020.

de Sa DJ. Stress response and its relationship to cystic (pseudofollicular) change in the definitive cortex of the adrenal gland in stillborn infants. Arch Dis Child. 1978;53(10):769–76. PubMed PMID: 727790; PubMed Central PMCID: PMC1545415.

De Vuyst P, Karjalainen A, Dumortier P, Pairon JC, Monsó E, Brochard P, Teschler H, Tossavainen A, Gibbs A. Guidelines for mineral fibre analyses in biological samples: report of the ERS working group. European Respiratory Society. Eur Respir J. 1998;11(6):1416–26. PubMed PMID: 9657589.

Deepak D, Shah A. Thoracic sarcoidosis: the spectrum of roentgenologic appearances. Indian J Radiol Imaging. 2001;11:191–8.

Deutsch GH, Young LR, Deterding RR, Fan LL, Dell SD, Bean JA, Brody AS, Nogee LM, Trapnell BC, Langston C, Pathology Cooperative Group, Albright EA, Askin FB, Baker P, Chou PM, Cool CM, Coventry SC, Cutz E, Davis MM, Dishop MK, Galambos C, Patterson K, Travis WD, Wert SE, White FV, ChILD Research Co-operative. Diffuse lung disease in young children: application of a novel classification scheme. Am J Respir Crit Care Med. 2007;176(11):1120–8. Epub 2007 Sep 20. PubMed PMID: 17885266; PubMed Central PMCID: PMC2176101.

Devine MS, Garcia CK. Genetic interstitial lung disease. Clin Chest Med. 2012;33(1):95–110. https://doi.org/10.1016/j.ccm.2011.11.001. Epub 2011 Dec 6. Review. PubMed PMID: 22365249; PubMed Central PMCID: PMC3292740.

Dhawan V, Bown J, Lau A, Langlois-Klassen D, Kunimoto D, Bhargava R, Chui L, Collin SM, Long R. Towards the elimination of paediatric tuberculosis in high-income, immigrant-receiving countries: a 25-year conventional and molecular epidemiological case study. ERJ Open Res. 2018;4(2) https://doi.org/10.1183/23120541.00131-2017. pii: 00131-2017. eCollection 2018 Apr. PubMed PMID: 29750144; PubMed Central PMCID: PMC5938491.

Dheda K, Gumbo T, Maartens G, Dooley KE, McNerney R, Murray M, Furin J, Nardell EA, London L, Lessem E, Theron G, van Helden P, Niemann S, Merker M, Dowdy D, Van Rie A, Siu GK, Pasipanodya JG, Rodrigues C, Clark TG, Sirgel FA, Esmail A, Lin HH, Atre SR, Schaaf HS, Chang KC, Lange C, Nahid P, Udwadia ZF, Horsburgh CR Jr, Churchyard GJ, Menzies D, Hesseling AC, Nuermberger E, McIlleron H, Fennelly KP, Goemaere E, Jaramillo E, Low M, Jara CM, Padayatchi N, Warren RM. The epidemiology, pathogenesis, transmission, diagnosis, and management of multidrug-resistant, extensively drug-resistant, and incurable tuberculosis. Lancet Respir Med. 2017; https://doi.org/10.1016/S2213-

2600(17)30079-6. pii: S2213-2600(17)30079-6. [Epub ahead of print] Review. PubMed PMID: 28344011.

Diaz-Frias J, Widrich J. Scimitar syndrome. 2019 Sep 4. StatPearls [Internet]. Treasure Island (FL): StatPearls Publishing; 2019 Jan. Available from http://www-ncbi-nlm-nih-gov.login.ezproxy.library.ualberta.ca/books/NBK546602/. PubMed PMID: 31536209.

Dines DE, Arms RA, Bernatz PE, Gomes MR. Pulmonary arteriovenous fistulas. Mayo Clin Proc. 1974;49(7):460–5. PubMed PMID: 4834927.

Ding M, Shi X, Dong Z, Chen F, Lu Y, Castranova V, Vallyathan V. Freshly fractured crystalline silica induces activator protein-1 activation through ERKs and p38 MAPK. J Biol Chem. 1999;274(43):30611–6. PubMed PMID: 10521445.

Dishop MK, Askin FB, Galambos C, for the chILD Network. Classification of diffuse lung disease in older children and adolescents: a multi-institutional study of the Children's interstitial lung disease (chILD) pathology working group. San Diego: Society of Pediatric Pathology; 2007.

Donato L, Mai Hong Tran T, Ghori UK, Musani AI. Pediatric interventional pulmonology. Clin Chest Med. 2018;39(1):229–38. https://doi.org/10.1016/j.ccm.2017.11.017. Review. PubMed PMID: 29433718.

Dougherty RH, Fahy JV. Acute exacerbations of asthma: epidemiology, biology and the exacerbation-prone phenotype. Clin Exp Allergy. 2009;39(2):193–202. https://doi.org/10.1111/j.1365-2222.2008.03157.x. Review. PubMed PMID: 19187331; PubMed Central PMCID: PMC2730743.

Douillard JY, Pirker R, O'Byrne KJ, Kerr KM, Störkel S, von Heydebreck A, Grote HJ, Celik I, Shepherd FA. Relationship between EGFR expression, EGFR mutation status, and the efficacy of chemotherapy plus cetuximab in FLEX study patients with advanced non-small-cell lung cancer. J Thorac Oncol. 2014;9(5):717–24. https://doi.org/10.1097/JTO.0000000000000141. PubMed PMID: 24662454.

Doyle L. Gabriel Andral (1797–1876) and the first reports of lymphangitis carcinomatosa. J R Soc Med. 1989;82(8):491–3. PubMed PMID: 2674433; PubMed Central PMCID: PMC1292257.

El Azbaoui S, Sabri A, Ouraini S, Hassani A, Asermouh A, Agadr A, Abilkassem R, Dini N, Kmari M, Akhaddar A, Laktati Z, Aieche S, El Hafidi N, Ben Brahim F, Bousfiha AA, Ailal F, Deswarte C, Schurr E, Amar L, Bustamante J, Boisson-Dupuis S, Casanova JL, Abel L, El Baghdadi J. Utility of the QuantiFERON-TB gold in-tube assay for the diagnosis of tuberculosis in Moroccan children. Int J Tuberc Lung Dis. 2016;20(12):1639–46. PubMed PMID: 27931340.

Elias JA, Tanoue LT. Systemic sarcoidosis. In: Baum GL, Crapo JD, Celli BR, Karlinsky JB, editors. Textbook of pulmonary diseases. 6th ed. Philadelphia: Lippincott-Raven; 1998. p. 407–30.

Emery JL, Mithal A. The number of alveoli in the terminal respiratory unit of man during late intrauterine life and childhood. Arch Dis Child. 1960;35:544–7. PubMed PMID: 13726619; PubMed Central PMCID: PMC2012643.

Emery JL, Mithal A. In Emery J, editors. The anatomy of the developing lung. London: Heinemann Medical; 1969. p. 203–5.

Evans AS. The history of infectious mononucleosis. Am J Med Sci. 1974;267(3):189–95. PubMed PMID: 4363554.

Fabre OH, Porte HL, Godart FR, Rey C, Wurtz AJ. Long-term cardiovascular consequences of undiagnosed intralobar pulmonary sequestration. Ann Thorac Surg. 1998;65(4):1144–6. PubMed PMID: 9564949.

Faddoul D. Childhood tuberculosis: an overview. Adv Pediatr Infect Dis. 2015;62(1):59–90. https://doi.org/10.1016/j.yapd.2015.04.001. Review. PubMed PMID: 26205109.

Fan LL, Dishop MK, Galambos C, Askin FB, White FV, Langston C, Liptzin DR, Kroehl ME, Deutsch GH, Young LR, Kurland G, Hagood J, Dell S, Trapnell BC, Deterding RR. Children's interstitial and diffuse lung disease research network (chILDRN). Diffuse lung disease in biopsied children 2 to 18 years of age. Application of the chILD classification scheme. Ann Am Thorac Soc. 2015;12(10):1498–505. https://doi.org/10.1513/AnnalsATS.201501-064OC. PMID: 26291470; PMCID: PMC4627419.

Felson B. Chest roentgenology. Philadelphia: WB Saunders; 1973.

Fitzgerald ML, Xavier R, Haley KJ, et al. ABCA3 inactivation in mice causes respiratory failure, loss of pulmonary surfactant, and depletion of lung phosphatidylglycerol. J Lipid Res. 2007;48:621–32.

Fraser RG, Paré JA. The normal chest. Diagnosis of diseases of the chest. 2nd ed. Philadelphia: WB Saunders; 1977. p. 1–183.

Fraser RS, Müller NL, Colman N, Paré PD. The mediastinum. Fraser and Paré's diagnosis of diseases of the chest. 4th ed. Philadelphia: WB Saunders; 1999. p. 196–234.

Fregonese F, Ahuja SD, Akkerman OW, Arakaki-Sanchez D, Ayakaka I, Baghaei P, Bang D, Bastos M, Benedetti A, Bonnet M, Cattamanchi A, Cegielski P, Chien JY, Cox H, Dedicoat M, Erkens C, Escalante P, Falzon D, Garcia-Prats AJ, Gegia M, Gillespie SH, Glynn JR, Goldberg S, Griffith D, Jacobson KR, Johnston JC, Jones-López EC, Khan A, Koh WJ, Kritski A, Lan ZY, Lee JH, Li PZ, Maciel EL, Galliez RM, Merle CSC, Munang M, Narendran G, Nguyen VN, Nunn A, Ohkado A, Park JS, Phillips PPJ, Ponnuraja C, Reves R, Romanowski K, Seung K, Schaaf HS, Skrahina A, Soolingen DV, Tabarsi P, Trajman A, Trieu L, Banurekha VV, Viiklepp P, Wang JY, Yoshiyama T, Menzies D. Comparison of different treatments for isoniazid-resistant tuberculosis: an individual patient data meta-analysis. Lancet Respir Med. 2018;6(4):265–75. https://doi.org/10.1016/S2213-2600(18)30078-X. Erratum in: Lancet Respir Med. 2018 Apr 18; PubMed PMID: 29595509.

Garmany TH, Moxley MA, White FV, Dean M, Hull WM, Whitsett JA, Nogee LM, Hamvas A. Surfactant composition and function in patients with ABCA3 mutations. Pediatr Res. 2006;59(6):801–5. Epub 2006 Apr 26. PubMed PMID: 16641205.

Geiger R, Treml B, Pinna A, Barnickel L, Prossliner H, Reinstadler H, Pilch M, Hauer M, Walther C, Steiner HJ, Giese T, Wemhöner A, Scholl-Bürgi S, Gottardi W, Arnitz R, Sergi C, Nagl M, Löckinger A. Tolerability of inhaled N-chlorotaurine in the pig model. BMC Pulm Med. 2009;9:33. https://doi.org/10.1186/1471-2466-9-33. PubMed PMID: 19602222; PubMed Central PMCID: PMC2722574.

Ghon A. Der primäre Lungenherd bei der Tuberkulose der Kinder. Berlin/Wien: Urbach & Schwarzenberg; 1912.

Ghosh A, Coakley RD, Ghio AJ, Muhlebach MS, Esther CR Jr, Alexis NE, Tarran R. Chronic E-cigarette use increases neutrophil elastase and matrix metalloprotease levels in the lung. Am J Respir Crit Care Med. 2019; https://doi.org/10.1164/rccm.201903-0615OC. [Epub ahead of print] PubMed PMID: 31390877.

Giordano P, Cecinati V, Grassi M, Giordani L, De Mattia D, Santoro N. Langerhans cell histiocytosis in a pediatric patient with thrombocytopenia-absent radius syndrome and 1q21.1 deletion: case report and proposal of a rapid molecular diagnosis of 1q21.1 deletion. Immunopharmacol Immunotoxicol. 2011;33(4):754–8. https://doi.org/10.3109/08923973.2011.557077. Epub 2011 Mar 23. PubMed PMID: 21428712.

Gomes VC, Silva MC, Maia Filho JH, Daltro P, Ramos SG, Brody AS, Marchiori E. Diagnostic criteria and follow-up in neuroendocrine cell hyperplasia of infancy: a case series. J Bras Pneumol. 2013;39(5):569–78. https://doi.org/10.1590/S1806-37132013000500007. English, Portuguese. PubMed PMID: 24310630; PubMed Central PMCID: PMC4075883.

Gonzales SK, Goudy S, Prickett K, Ellis J. EXIT (ex utero intrapartum treatment) in a growth restricted fetus with tracheal atresia. Int J Pediatr Otorhinolaryngol. 2018;105:72–4. https://doi.org/10.1016/j.ijporl.2017.12.010. Epub 2017 Dec 9. PubMed PMID: 29447823. Mar;48(3):427-432. https://doi.org/10.1007/s00247-017-4024-5. Epub 2017 Nov 17. Review. PubMed PMID: 29147912.

Graser F. Hundert Jahre Pfeiffersches Drusenfieber [100 years of Pfeiffer's glandular fever]. Klin Padiatr. 1991;203(3):187–90. German. PubMed PMID: 1857056.

Griffin N, Devaraj A, Goldstraw P, Bush A, Nicholson AG, Padley S. CT and histopathological correlation of congenital cystic pulmonary lesions: a common pathogenesis? Clin Radiol. 2008;63(9):995–1005. https://doi.org/10.1016/j.crad.2008.02.011. Epub 2008 May 13. PubMed PMID: 18718229.

Grunewald J, Grutters JC, Arkema EV, Saketkoo LA, Moller DR, Müller-Quernheim J. Sarcoidosis. Nat Rev Dis Primers. 2019;5(1):45. https://doi.org/10.1038/s41572-019-0096-x. Review. Erratum in: Nat Rev Dis Primers. 2019 Jul 16;5(1):49. PubMed PMID: 31273209.

Guastadisegni MC, Roberto R, L'Abbate A, Palumbo O, Carella M, Giordani L, Cecinati V, Giordano P, Storlazzi CT. Thrombocytopenia-absent-radius syndrome in a child showing a larger 1q21.1 deletion than the one in his healthy mother, and a significant downregulation of the commonly deleted genes. Eur J Med Genet. 2012;55(2):120–3. https://doi.org/10.1016/j.ejmg.2011.11.007. Epub 2011 Dec 8. PubMed PMID: 22201559.

Gudbjartsson T, Gudmundsson G. Middle lobe syndrome: a review of clinicopathological features, diagnosis and treatment. Respiration. 2012;84(1):80–6. https://doi.org/10.1159/000336238. Epub 2012 Mar 1. Review. PubMed PMID: 22377566.

Hanley A, Hubbard RB, Navaratnam V. Mortality trends in asbestosis, extrinsic allergic alveolitis and sarcoidosis in England and Wales. Respir Med. 2011;105(9):1373–9. https://doi.org/10.1016/j.rmed.2011.05.008. Epub 2011 Jun 24. PubMed PMID: 21704503.

Harbut MR, Endress C, Graff JJ, Weis C, Pass H. Clinical presentation of asbestosis with intractable pleural pain in the adult child of a taconite miner and radiographic demonstration of the probable pathology causing the pain. Int J Occup Environ Health. 2009;15(3):269–73. PubMed PMID: 19650581.

Hauber HP, Bittmann I, Kirsten D. Nicht spezifische interstitielle Pneumonie [non-specific interstitial pneumonia (NSIP)]. Pneumologie. 2011;65(8):477–83. https://doi.org/10.1055/s-0030-1256284. Epub 2011 Mar 24. Review. German. PubMed PMID: 21437858.

Heath D, Edwards JE. The pathology of hypertensive pulmonary vascular disease; a description of six grades of structural changes in the pulmonary arteries with special reference to congenital cardiac septal defects. Circulation. 1958;18(4 Part 1):533–47. PubMed PMID: 13573570.

Heitzman ER. The mediastinum. 2nd ed. New York: Springer; 1988.

Hill DA, Dehner LP. A cautionary note about congenital cystic adenomatoid malformation (CCAM) type 4. Am J Surg Pathol. 2004;28(4):554–5; author reply 555. PubMed PMID: 15087677.

Hirai T, Ohtake Y, Mutoh S, Noguchi M, Yamanaka A. Anomalous systemic arterial supply to normal basal segments of the left lower lobe: a report of two cases. Chest. 1996;109:286–9. [PubMed].

Hoeper MM, Bogaard HJ, Condliffe R, Frantz R, Khanna D, Kurzyna M, Langleben D, Manes A, Satoh T, Torres F, Wilkins MR, Badesch DB. Definitions and diagnosis of pulmonary hypertension. J Am Coll Cardiol. 2013;62(25, Suppl D). 2013 by the American College of Cardiology Foundation ISSN 0735-1097. https://doi.org/10.1016/j.jacc.2013.10.032.

Horgan AM, Yang B, Azad AK, Amir E, John T, Cescon DW, Wheatley-Price P, Hung RJ, Shepherd FA, Liu G. Pharmacogenetic and germline prognostic markers of lung cancer. J Thorac Oncol. 2011;6(2):296–304. https://doi.org/10.1097/JTO.0b013e3181ffe909. PubMed PMID: 21206385.

Huber JJ. Observationes aliquot de arteria singulari pulmoni concessa. Acta Helvet. 1777;8:85.

Karger B, Fracasso T, Brinkmann B, Bajanowski T. Evaluation of the Reid index in infants and cases of SIDS. Int J Legal Med. 2004;118(4):221–3. Epub 2004 Jan 16. PubMed PMID: 14727122.

Kerr KM. Personalized medicine for lung cancer: new challenges for pathology. Histopathology. 2012;60(4):531–46. https://doi.org/10.1111/j.1365-2559.2011.03854.x. Epub 2011 Sep 14. Review. PubMed PMID: 21916947.

Koch R. Die Ätiologie der Tuberkulose. Berlin Klin Wochenschrift. 1882;15:221–30.

Kosmidis C, Denning DW. The clinical spectrum of pulmonary aspergillosis. Thorax. 2015;70(3):270–7. https://doi.org/10.1136/thoraxjnl-2014-206291. Epub 2014 Oct 29. Review. PubMed PMID: 25354514.

Kveim MA. En ny og spesifikk kutan-reaksjon ved Boecks sarcoid. En foreløpig meddelelse. Nordisk Medicin. 1941;9:169–72.

LaCasce AS. Post-transplant lymphoproliferative disorders. Oncologist. 2006;11:674–80.

Lal DR, Clark I, Shalkow J, Downey RJ, Shorter NA, Klimstra DS, La Quaglia MP. Primary epithelial lung malignancies in the pediatric population. Pediatr Blood Cancer. 2005;45(5):683–6. PubMed PMID: 15714450.

Langston C, Dishop MK. Diffuse lung disease in infancy: a proposed classification applied to 259 diagnostic biopsies. Pediatr Dev Pathol. 2009;12(6):421–37. https://doi.org/10.2350/08-11-0559.1. PubMed PMID: 19323600.

Langston C, Pappin A. Lipid-laden alveolar macrophages as an indicator of aspiration pneumonia. Arch Pathol Lab Med. 1996;120(4):326–7. PubMed PMID: 8619741.

Langston C, Patterson K, Dishop MK, chILD Pathology Co-operative Group, Askin F, Baker P, Chou P, Cool C, Coventry S, Cutz E, Davis M, Deutsch G, Galambos C, Pugh J, Wert S, White F. A protocol for the handling of tissue obtained by operative lung biopsy: recommendations of the chILD pathology co-operative group. Pediatr Dev Pathol. 2006;9(3):173–80. PubMed PMID: 16944976.

Lanza LL, Wang L, Simon TA, Irish WD. Epidemiologic critique of literature on post-transplant neoplasms in solid organ transplantation. Clin Transpl. 2009;23:582–8.

Lauweryns JM. Hyaline membrane disease: a pathological study of 55 infants. Arch Dis Child. 1965;40(214):618–25. PubMed PMID: 5891757; PubMed Central PMCID: PMC2019490.

Lee ML, Tsao LY, Chaou WT, Yang AD, Yeh KT, Wang JK, Wu MH, Lue HC, Chiu IS, Chang CI. Revisit on congenital bronchopulmonary vascular malformations: a haphazard branching theory of malinosculations and its clinical classification and implication. Pediatr Pulmonol. 2002;33(1):1–11. PubMed PMID: 11747254.

Lee ML, Yang SC, Yang AD. Transcatheter occlusion of the isolated scimitar vein anomaly camouflaged under dual pulmonary venous drainage of the right lung by the Amplatzer ductal Occluder. Int J Cardiol. 2007;115(2):e90–3. Epub 2006 Nov 28. PubMed PMID: 17126429.

Lee ML, Lue HC, Chiu IS, Chiu HY, Tsao LY, Cheng CY, Yang AD. A systematic classification of the congenital bronchopulmonary vascular malformations: dysmorphogeneses of the primitive foregut system and the primitive aortic arch system. Yonsei Med J. 2008;49(1):90–102. https://doi.org/10.3349/ymj.2008.49.1.90. PubMed PMID: 18306475; PubMed Central PMCID: PMC2615272.

Lelii M, Patria MF, Pinzani R, Tenconi R, Mori A, Bonelli N, Principi N, Esposito S. Role of high-resolution chest computed tomography in a child with persistent tachypnoea and intercostal retractions: a case report of neuroendocrine cell hyperplasia. Int J Environ Res Public Health. 2017;14(10). pii: E1113. https://doi.org/10.3390/ijerph14101113. PubMed PMID: 28946688; PubMed Central PMCID: PMC5664614

Lim C, Tsao MS, Le LW, Shepherd FA, Feld R, Burkes RL, Liu G, Kamel-Reid S, Hwang D, Tanguay J, da Cunha SG, Leighl NB. Biomarker testing and time to treatment decision in patients with advanced nonsmall-cell lung cancer. Ann Oncol. 2015;26(7):1415–21. https://doi.org/10.1093/annonc/mdv208. Epub 2015 Apr 28. PubMed PMID: 25922063.

Liptay MJ, Ujiki MB, Locicero J. Congenital vascular lesions of the lungs. In: Shields WT, editor. General thoracic surgery. 6th ed. Philadelphia: Lippincott Williams & Wilkins; 2004. p. 1144–52.

Liu AP, Tang WF, Lau ET, Chan KY, Kan AS, Wong KY, Tso WW, Jalal K, Lee SL, Chau CS, Chung BH. Expanded Prader-Willi syndrome due to chromosome 15q11.2-14 deletion: report and a review of literature. Am J Med Genet A. 2013;161A(6):1309–18. https://doi.org/10.1002/ajmg.a.35909. Epub 2013 Apr 30. Review. PubMed PMID: 23633107.

MacSweeney F, Papagiannopoulos K, Goldstraw P, Sheppard MN, Corrin B, Nicholson AG. An assessment of the expanded classification of congenital cystic adenomatoid malformations and their relationship to malignant transformation. Am J Surg Pathol. 2003;27(8):1139–46. PubMed PMID: 12883247.

Marabese M, Ganzinelli M, Garassino MC, Shepherd FA, Piva S, Caiola E, Macerelli M, Bettini A, Lauricella C, Floriani I, Farina G, Longo F, Bonomi L, Fabbri MA, Veronese S, Marsoni S, Broggini M, Rulli E. KRAS mutations affect prognosis of non-small-cell lung cancer patients treated with first-line platinum containing chemotherapy. Oncotarget. 2015;6(32):34014–22. https://doi.org/10.18632/oncotarget.5607. PubMed PMID: 26416458.

Margaritopoulos GA, Romagnoli M, Poletti V, Siafakas NM, Wells AU, Antoniou KM. Recent advances in the pathogenesis and clinical evaluation of pulmonary fibrosis. Eur Respir Rev. 2012;21(123):48–56. https://doi.org/10.1183/09059180.00007611. Review. PubMed PMID: 22379174.

Martin P, Leighl NB, Tsao MS, Shepherd FA. KRAS mutations as prognostic and predictive markers in non-small cell lung cancer. J Thorac Oncol. 2013;8(5):530–42. https://doi.org/10.1097/JTO.0b013e318283d958. Review. PubMed PMID:23524403.

Mazurek JM, Wood JM, Schleiff PL, Weissman DN. Surveillance for silicosis deaths among persons aged 15–44 years – United States, 1999–2015. MMWR Morb Mortal Wkly Rep. 2017;66(28):747–52. https://doi.org/10.15585/mmwr.mm6628a2. PubMed PMID: 28727677; PubMed Central PMCID: PMC5657940.

Mehra VM, Keethakumar A, Bohr YM, Abdullah P, Tamim H. The association between alcohol, marijuana, illegal drug use and current use of E-cigarette among youth and young adults in Canada: results from Canadian tobacco, alcohol and drugs survey 2017. BMC Public Health. 2019;19(1):1208. https://doi.org/10.1186/s12889-019-7546-y. PubMed PMID: 31477067; PubMed Central PMCID: PMC6721192.

Meignin V, Thivolet-Bejui F, Kambouchner M, Hussenet C, Bondeelle L, Mitchell A, Chagnon K, Begueret H, Segers V, Cottin V, Tazi A, Chevret S, Danel C, Bergeron A. Lung histopathology of non-infectious pulmonary complications after allogeneic haematopoietic stem cell transplantation. Histopathology. 2018;73(5):832–42. https://doi.org/10.1111/his.13697. Epub 2018 Aug 19. PubMed PMID: 29953629.

Mendelson CL. The aspiration of stomach contents into the lungs during obstetric anesthesia. Am J Obstet Gynecol. 1946;52:191–205. PubMed PMID: 20993766.

Moller DR. Systemic sarcoidosis. In: Fishman AP, Elias JA, Fishman JA, Grippi MA, Kaiser LR, Senior RM, editors. Fishman's pulmonary diseases and disorders. 3rd ed. New York: McGraw-Hill; 1998. p. 1055–68.

Murai Y. Malignant mesothelioma in Japan: analysis of registered autopsy cases. Arch Environ Health. 2001;56(1):84–8. PubMed PMID: 11256861.

Murlidhar V. An 11-year-old boy with silico-tuberculosis attributable to secondary exposure to sandstone mining in Central India. BMJ Case Rep. 2015;2015 https://doi.org/10.1136/bcr-2015-209315. pii: bcr2015209315. PubMed PMID: 26106174; PubMed Central PMCID: PMC4480122.

Nacaroğlu HT, Ünsal-Karkıner CŞ, Bahçeci-Erdem S, Özdemir R, Karkıner A, Alper H, Can D. Pulmonary vascular anomalies: a review of clinical and radiological findings of cases presenting with different complaints in childhood. Turk J Pediatr. 2016;58(3):337–42. PubMed PMID: 28266205.

Nathan N, Taam RA, Epaud R, Delacourt C, Deschildre A, Reix P, Chiron R, de Pontbriand U, Brouard J, Fayon M, Dubus JC, Giovannini-Chami L, Bremont F, Bessaci K, Schweitzer C, Dalphin ML, Marguet C, Houdouin V, Troussier F, Sardet A, Hullo E, Gibertini I, Mahloul M, Michon D, Priouzeau A, Galeron L, Vibert JF, Thouvenin G, Corvol H, Deblic J, Clement A, French RespiRare® Group. A national internet-linked based database for pediatric interstitial lung diseases: the French network. Orphanet J Rare Dis. 2012;7:40. PubMed PMID: 22704798; PubMed Central PMCID: PMC3458912.

Nathan N, Marcelo P, Houdouin V, Epaud R, de Blic J, Valeyre D, Houzel A, Busson PF, Corvol H, Deschildre A, Clement A. RespiRare and the French sarcoidosis groups. Lung sarcoidosis in children: update on disease expression and management. Thorax. 2015;70(6):537–42. https://doi.org/10.1136/thoraxjnl-2015-206825. Epub 2015 Apr 8. PubMed PMID: 25855608.

Nathan N, Sileo C, Calender A, Pacheco Y, Rosental PA, Cavalin C, Macchi O, Valeyre D, Clement A, French Sarcoidosis Group (GSF); Silicosis Research Group. Paediatric sarcoidosis. Paediatr Respir Rev. 2019;29:53–9. https://doi.org/10.1016/j.prrv.2018.05.003. Epub 2018 May 19. Review. PubMed PMID: 30917882.

Naylor B. Curschmann's spirals in pleural and peritoneal effusions. Acta Cytol. 1990;34:474–8.

Neill CA, Ferencz C, Sabiston DC, Sheldon H. The familial occurrence of hypoplastic right lung with systemic arterial supply and venous drainage "scimitar syndrome". Bull Johns Hopkins Hosp. 1960;107:1–21. PubMed PMID: 14426379.

Niu H, Wang F, Liu W, Wang Y, Chen Z, Gao Q, Yi P, Li L, Zeng R. 儿童肺部病变215例临床病理学分析 [Pediatric lung lesions: a clinicopathological study of 215 cases]. Zhonghua Bing Li Xue Za Zhi. 2015a;44(9):648–52. Chinese. PubMed PMID: 26705281.

Niu H, Wang F, Liu W, Wang Y, Chen Z, Gao Q, Yi P, Li L, Zeng R. Pediatric lung lesions: a clinicopathological study of 215 cases. Zhonghua Bing Li Xue Za Zhi. 2015b;44(9):648–52. Chinese. PubMed PMID: 26705281.

Oton AB, Wang H, Leleu X, Melhem MF, George D, Lacasce A, Foon K, Ghobrial IM. Clinical and pathological prognostic markers for survival in adult patients with post-transplant lymphoproliferative disorders in solid transplant. Leuk Lymphoma. 2008;49(9):1738–44. https://doi.org/10.1080/10428190802239162. PubMed PMID: 18798108.

Oviedo SP, Cagle PT. Diffuse malignant mesothelioma. Arch Pathol Lab Med. 2012;136(8):882–8. https://doi.org/10.5858/arpa.2012-0142-CR. Review. PubMed PMID: 22849735.

Papagiannopoulos K, Hughes S, Nicholson AG, Goldstraw P. Cystic lung lesions in the pediatric and adult population: surgical experience at the Brompton hospital. Ann Thorac Surg. 2002;73(5):1594–8. PubMed PMID: 12022556.

Park HS, Son HJ, Kang MJ. Cutaneous bronchogenic cyst over the sternum. Korean J Pathol. 2004;38:333–6.

Patterson SW, Starling EH. On the mechanical factors which determine the output of the ventricles. J Physiol. 1914;48(5):357–79.

Pediatric Diffuse Parenchymal Lung Disease/Pediatric Interstitial Lung Disease Cooperative Group, The Subspecialty Group of Respiratory Diseases, The Society of Pediatrics, Chinese Medical Association. 小儿间质性肺疾病14例临床-影像-病理诊断分析 [clinical, radiologic, pathological features and diagnosis of 14 cases with interstitial lung disease in children]. Zhonghua Er Ke Za Zhi. 2011;49(2):92–7. Chinese. PubMed PMID: 21426684.

Pedra F, Tambellini AT, Pereira Bde B, da Costa AC, de Castro HA. Mesothelioma mortality in Brazil, 1980–2003. Int J Occup Environ Health. 2008;14(3):170–5. Erratum in: Int J Occup Environ Health. 2009;15(4):391. PubMed PMID: 18686716.

Penel N, Amela EY, Decanter G, Robin YM, Marec-Berard P. Solitary fibrous tumors and so-called hemangiopericytoma. Sarcoma. 2012;2012:690251. https://doi.org/10.1155/2012/690251. Epub 2012 Apr 8. PubMed PMID: 22566753; PubMed Central PMCID: PMC3337510.

Peruzzi B, Bottaro DP. Targeting the c-Met signaling pathway in cancer. Clin Cancer Res. 2006;12(12):3657–60. Review. PubMed PMID: 16778093.

Pinkerton H. Reaction of oils and fats in the lung. Arch Pathol. 1928;5:380–401.

Planchard D. Identification of driver mutations in lung cancer: first step in personalized cancer. Target Oncol. 2013;8(1):3–14. https://doi.org/10.1007/s11523-013-0263-z. Epub 2013 Feb 1. PubMed PMID: 23371030.

Pryce DM. Lower accessory pulmonary artery with intralobar sequestration of lung; a report of seven cases. J Pathol Bacteriol. 1946;58(3):457–67. PubMed PMID: 20283082.

Ram G, Chinen J. Infections and immunodeficiency in down syndrome. Clin Exp Immunol. 2011;164(1):9–16. https://doi.org/10.1111/j.1365-2249.2011.04335.x. Epub 2011 Feb 24. Review. PubMed PMID: 21352207; PubMed Central PMCID: PMC3074212.

Ramazzini C. Asbestos is still with us: repeat call for a universal ban. Int J Occup Med Environ Health. 2010a;23(2):201–7. https://doi.org/10.2478/v10001-010-0017-4. PubMed PMID: 20682491.

Ramazzini C. Asbestos is still with us: repeat call for a universal ban. Odontology. 2010b;98(2):97–101. https://doi.org/10.1007/s10266-010-0132-5. Epub 2010 Jul 23. PubMed PMID: 20652786.

Reid L. Measurement of the bronchial mucous gland layer: a diagnostic yardstick in chronic bronchitis. Thorax. 1960;15:132–41. PubMed PMID: 14437095; PubMed Central PMCID: PMC1018549.

Reid-Nicholson M, Kulkarni R, Adeagbo B, Looney S, Crosby J. Interobserver and intraobserver variability in the calculation of the lipid-laden macrophage index: implications for its use in the evaluation of aspiration in children. Diagn Cytopathol. 2010;38(12):861–5. https://doi.org/10.1002/dc.21298. PubMed PMID: 20049966.

Resheidat A, Kelly T, Mossad E. Incidental diagnosis of congenital tracheal stenosis in children with congenital heart disease presenting for cardiac surgery. J Cardiothorac Vasc Anesth. 2018; https://doi.org/10.1053/j.jvca.2018.04.027. pii: S1053-0770(18)30258-1. [Epub ahead of print] PubMed PMID: 29753667.

Reungwetwattana T, Weroha SJ, Molina JR. Oncogenic pathways, molecularly targeted therapies, and highlighted clinical trials in non-small-cell lung cancer (NSCLC). Clin Lung Cancer. 2012;13(4):252–66. https://doi.org/10.1016/j.cllc.2011.09.004. Epub 2011 Dec 8. Review. PubMed PMID: 22154278.

Rojas Y, Shi YX, Zhang W, Beierle EA, Doski JJ, Goldfarb M, Goldin AB, Gow KW, Langer M, Vasudevan SA, Nuchtern JG. Primary malignant pulmonary tumors in children: a review of the national cancer data base. J Pediatr Surg. 2015;50(6):1004–8. https://doi.org/10.1016/j.jpedsurg.2015.03.032. Epub 2015 Mar 14. PubMed PMID: 25812444.

Rom WN, Weiden M, Garcia R, Yie TA, Vathesatogkit P, Tse DB, McGuinness G, Roggli V, Prezant D. Acute eosinophilic pneumonia in a New York City firefighter exposed to world trade center dust. Am J Respir Crit Care Med. 2002;166(6):797–800. PubMed PMID: 12231487.

Rubin EM, Garcia H, Horowitz MD, Guerra JJ Jr. Fatal massive hemoptysis secondary to intralobar sequestration. Chest. 1994;106(3):954–5. Review. PubMed PMID: 8082388.

Sade RM, Clouse M, Ellis FH Jr. The spectrum of pulmonary sequestration. Ann Thorac Surg. 1974;18(6):644–58. Review. PubMed PMID: 4611367.

Savic B, Birtel FJ, Tholen W, Funke HD, Knoche R. Lung sequestration: report of seven cases and review of 540 published cases. Thorax. 1979;34(1):96–101. PubMed PMID: 442005; PubMed Central PMCID: PMC471015.

Scharf JB, Lees GM, Sergi CM. Malignant pleural mesothelioma in a child. J Pediatr Surg Case Rep. 2015;3(10):440–3.

Schaumann J. Étude sur le lupus pernio et ses rapports avec les sarcoides et la tuberculose. Annales de dermatologie et de syphilographie, Paris. 1916–1917;6:357–73.

Scheier M, Ramoni A, Alge A, Brezinka C, Reiter G, Sergi C, Hager J, Marth C. Congenital fibrosarcoma as cause for fetal anemia: prenatal diagnosis and in utero treatment. Fetal Diagn Ther. 2008;24(4):434–6. https://doi.org/10.1159/000173370. Epub 2008 Nov 19. PubMed PMID: 19018145.

Schloo BL, Vawter GF, Reid LM. Down syndrome: patterns of disturbed lung growth. Hum Pathol. 1991;22(9):919–23. PubMed PMID: 1833304.

Schwienbacher M, Treml B, Pinna A, Geiger R, Reinstadler H, Pircher I, Schmidl E, Willomitzer C, Neumeister J, Pilch M, Hauer M, Hager T, Sergi C, Scholl-Bürgi S, Giese T, Löckinger A, Nagl M. Tolerability of inhaled N-chlorotaurine in an acute pig streptococcal lower airway inflammation model. BMC Infect Dis. 2011;11:231. https://doi.org/10.1186/1471-2334-11-231. PubMed PMID: 21875435; PubMed Central PMCID: PMC3178512.

Sergi C, Willig F, Thomsen M, Otto HF, Krempien B. Bronchopneumonia disguising lung metastases of a painless central chondrosarcoma of pubis. Pathol Oncol Res. 1997;3(3):211–4. https://doi.org/10.1007/BF02899923. PubMed PMID: 18470732.

Sergi C, Roth SU, Adam S, Otto HF. Mapping a method for systematically reviewing the medical literature:

a helpful checklist postmortem protocol of human immunodeficiency virus (HIV)-related pathology in childhood. Pathologica. 2001;93(3):201–7. Review. PubMed PMID: 11433613.

Sergi C, Daum E, Pedal I, Hauröder B, Schnitzler P. Fatal circumstances of human herpesvirus 6 infection: transcriptosome data analysis suggests caution in implicating HHV-6 in the cause of death. J Clin Pathol. 2007;60(10):1173–7. Epub 2007 Jun 1. PubMed PMID: 17545558; PubMed Central PMCID: PMC2014822.

Sergi C, Gekas J, Kamnasaran D. Holoprosencephaly-polydactyly (pseudotrisomy 13) syndrome: case report and diagnostic criteria. Fetal Pediatr Pathol. 2012;31(5):315–8. https://doi.org/10.3109/15513815.2012.659390. Epub 2012 Mar 20. PubMed PMID: 22432933.

Shah PS, Hellmann J, Adatia I. Clinical characteristics and follow up of down syndrome infants without congenital heart disease who presented with persistent pulmonary hypertension of newborn. J Perinat Med. 2004;32(2):168–70. PubMed PMID: 15085894.

Shepherd FA, Bunn PA, Paz-Ares L. Lung cancer in 2013: state of the art therapy for metastatic disease. Am Soc Clin Oncol Educ Book. 2013:339–46. https://doi.org/10.1200/EdBook_AM.2013.33.339. PubMed PMID: 23714542.

Sheth PR, Hays JL, Elferink LA, Watowich SJ. Biochemical basis for the functional switch that regulates hepatocyte growth factor receptor tyrosine kinase activation. Biochemistry. 2008;47(13):4028–38. https://doi.org/10.1021/bi701892f. Epub 2008 Mar 7. PubMed PMID: 18324780; PubMed Central PMCID: PMC2729649.

Shields TW. Congenital vascular lesions of the lungs. In: Shields TW, Cicero JL, Ponn RB, editors. General thoracic surgery. 5th ed. Philadelphia: Lippincott Williams & Wilkins; 2000. p. 975–87.

Shields TW, Shields TW. Primary tumors and cysts of the mediastinum. In: General thoracic surgery. Philadelphia: Lea & Febiger; 1983. p. 927–54.

Sprunt TPV, Evans FA. Mononuclear leukocytosis in reaction to acute infection (infectious mononucleosis). Bull Johns Hopkins Hosp. 1920;31:410–7.

Stocker JT. Cystic lung disease in infants and children. Fetal Pediatr Pathol. 2009;28(4):155–84. PubMed PMID: 19842869.

Stocker JT, Madewell JE, Drake RM. Congenital cystic adenomatoid malformation of the lung. Classification and morphologic spectrum. Hum Pathol. 1977;8(2):155–71. PubMed PMID: 856714.

Suri HS, Yi ES, Nowakowski GS, Vassallo R. Pulmonary langerhans cell histiocytosis. Orphanet J Rare Dis. 2012;7:16. https://doi.org/10.1186/1750-1172-7-16. Review. PubMed PMID: 22429393; PubMed Central PMCID: PMC3342091.

Torgyekes E, Shanske AL, Anyane-Yeboa K, Nahum O, Pirzadeh S, Blumfield E, Jobanputra V, Warburton D, Levy B. The proximal chromosome 14q microdeletion syndrome: delineation of the phenotype using high resolution SNP oligonucleotide microarray analysis (SOMA) and review of the literature. Am J Med Genet A. 2011;155A(8):1884–96. https://doi.org/10.1002/ajmg.a.34090. Epub 2011 Jul 8. Review. PubMed PMID: 21744488.

Travis WD, Brambilla E, Noguchi M, Nicholson AG, Geisinger KR, Yatabe Y, Beer DG, Powell CA, Riely GJ, Van Schil PE, Garg K, Austin JH, Asamura H, Rusch VW, Hirsch FR, Scagliotti G, Mitsudomi T, Huber RM, Ishikawa Y, Jett J, Sanchez-Cespedes M, Sculier JP, Takahashi T, Tsuboi M, Vansteenkiste J, Wistuba I, Yang PC, Aberle D, Brambilla C, Flieder D, Franklin W, Gazdar A, Gould M, Hasleton P, Henderson D, Johnson B, Johnson D, Kerr K, Kuriyama K, Lee JS, Miller VA, Petersen I, Roggli V, Rosell R, Saijo N, Thunnissen E, Tsao M, Yankelewitz D. International association for the study of lung cancer/American thoracic society/European respiratory society international multidisciplinary classification of lung adenocarcinoma. J Thorac Oncol. 2011;6(2):244–85. https://doi.org/10.1097/JTO.0b013e318206a221. Review. PubMed PMID: 21252716; PubMed Central PMCID: PMC4513953.

Travis WD, Brambilla E, Nicholson AG, Yatabe Y, Austin JH, Beasley MB, Chirieac LR, Dacic S, Duhig E, Flieder DB, Geisinger K, Hirsch FR, Ishikawa Y, Kerr KM, Noguchi M, Pelosi G, Powell CA, Tsao MS, Wistuba I, WHO Panel. The 2015 World Health Organization classification of lung tumors: impact of genetic, clinical and radiologic advances since the 2004 classification. J Thorac Oncol. 2015;10(9):1243–60. https://doi.org/10.1097/JTO.0000000000000630. PubMed PMID: 26291008.

Truitt AK, Carr SR, Cassese J, Kurkchubasche AG, Tracy TF Jr, Luks FI. Perinatal management of congenital cystic lung lesions in the age of minimally invasive surgery. J Pediatr Surg. 2006;41(5):893–6. PubMed PMID: 16677877.

Turner BM, Cagle PT, Sainz IM, Fukuoka J, Shen SS, Jagirdar J. Napsin A, a new marker for lung adenocarcinoma, is complementary and more sensitive and specific than thyroid transcription factor 1 in the differential diagnosis of primary pulmonary carcinoma: evaluation of 1674 cases by tissue microarray. Arch Pathol Lab Med. 2012;136(2):163–71. https://doi.org/10.5858/arpa.2011-0320-OA. PubMed PMID: 22288963.

Van Zuylen C, Thimmesch M, Lewin M, Dome F, Piérart F. Pneumopathie interstitielle non spécifique: une entité rare chez l'adolescent [Non-specific interstitial pneumonia: a rare clinical entity in adolescents]. Rev Med Liege. 2019;74(4):197–203. French. PubMed PMID: 30997969.

Varela P, Torre M, Schweiger C, Nakamura H. Congenital tracheal malformations. Pediatr Surg Int. 2018;34(7):701–13. https://doi.org/10.1007/s00383-018-4291-8. Epub 2018 May 30. Review. PubMed PMID: 29846792.

Wagenvoort CA. Grading of pulmonary vascular lesions – a reappraisal. Histopathology. 1981;5(6):595–8. PubMed PMID: 7319479.

Whitten CR, Khan S, Munneke GJ, Grubnic S. A diagnostic approach to mediastinal abnormalities. Radiographics. 2007;27:657–71.

Wigglesworth JS, Desai R. Use of DNA estimation for growth assessment in normal and hypoplastic fetal lungs. Arch Dis Child. 1981;56(8):601–5. PubMed PMID: 7271300; PubMed Central PMCID: PMC1627277.

Wigglesworth JS, Desai R, Guerrini P. Fetal lung hypoplasia: biochemical and structural variations and their possible significance. Arch Dis Child. 1981;56(8):606–15. PubMed PMID: 7023390; PubMed Central PMCID: PMC1627249.

Wilson MS, Wynn TA. Pulmonary fibrosis: pathogenesis, etiology and regulation. Mucosal Immunol. 2009;2(2):103–21. https://doi.org/10.1038/mi.2008.85. Epub 2009 Jan 7. Review. PubMed PMID: 19129758; PubMed Central PMCID: PMC2675823.

Wu X, Liu X, Koul S, Lee CY, Zhang Z, Halmos B. AXL kinase as a novel target for cancer therapy. Oncotarget. 2014;5(20):9546–63. Review. PubMed PMID: 25337673; PubMed Central PMCID: PMC4259419.

Wynn TA. Integrating mechanisms of pulmonary fibrosis. J Exp Med. 2011;208(7):1339–50. https://doi.org/10.1084/jem.20110551. Review. PubMed PMID: 21727191; PubMed Central PMCID: PMC3136685.

Yamanaka A, Hirai T, Fujimoto T, Hase M, Noguchi M, Konishi F. Anomalous systemic arterial supply to normal basal segments of the left lower lobe. Ann Thorac Surg. 1999;68:332–8. [PubMed].

Yancheva SG, Velani A, Rice A, Montero A, Hansell DM, Koo S, Thia L, Bush A, Nicholson AG. Bombesin staining in neuroendocrine cell hyperplasia of infancy (NEHI) and other childhood interstitial lung diseases (chILD). Histopathology. 2015;67(4):501–8. https://doi.org/10.1111/his.12672. Epub 2015 Apr 13. PubMed PMID: 25684686.

Yang YJ, Steele CT, Anbar RD, Sinacori JT, Powers CN. Quantitation of lipid-laden macrophages in evaluation of lower airway cytology specimens from pediatric patients. Diagn Cytopathol. 2001;24:98–103.

You WK, McDonald DM. The hepatocyte growth factor/c-Met signaling pathway as a therapeutic target to inhibit angiogenesis. BMB Rep. 2008;41(12):833–9. Review. PubMed PMID: 19123972; PubMed Central PMCID: PMC4417610.

Young LR, Nogee LM, Barnett B, Panos RJ, Colby TV, Deutsch GH. Usual interstitial pneumonia in an adolescent with ABCA3 mutations. Chest. 2008;134(1):192–5. PubMed PMID: 18628224.

Yu D, Mohammed S, Girgis S, Bacani J, Sergi C, Chiu BK. Posttransplantation lymphoproliferative disorders (PTLD): clinicopathologic and morphologic studies. Presentation at the 2010 Annual General Meeting of the Canadian Association of Pathologists/Association Canadienne des Pathologistes, Montreal, QC, Canada.

Zhang Z, Stiegler AL, Boggon TJ, Kobayashi S, Halmos B. EGFR-mutated lung cancer: a paradigm of molecular oncology. Oncotarget. 2010;1(7):497–514. Review. PubMed PMID: 21165163; PubMed Central PMCID: PMC3001953.

Zhang M, Shen F, Petryk A, Tang J, Chen X, Sergi C. "English disease": historical notes on rickets, the bone-lung link and child neglect issues. Nutrients. 2016;8(11). pii: E722. Review. PubMed PMID: 27854286; PubMed Central PMCID: PMC5133108.

Zylak CJ, Pallie W, Jackson R. Correlative anatomy and computed tomography: a module on the mediastinum. Radiographics. 1982;2:555–92.

Gastrointestinal Tract

Contents

3.1	**Development and Genetics**	256
3.2	**Esophagus**	258
3.2.1	Esophageal Anomalies	258
3.2.2	Esophageal Vascular Changes	263
3.2.3	Esophageal Inflammatory Diseases	263
3.2.4	Esophageal Tumors	274
3.3	**Stomach**	277
3.3.1	Gastric Anomalies	277
3.3.2	Gastric Vascular Changes	280
3.3.3	Gastric Inflammatory Diseases	281
3.3.4	Tissue Continuity Damage-Related Gastric Degenerations	286
3.3.5	Gastric Tumors	287
3.4	**Small Intestine**	304
3.4.1	Small Intestinal Anomalies	304
3.4.2	Abdominal Wall Defects (Median-Paramedian)	306
3.4.3	Continuity Defects of the Intestinal Lumen	312
3.4.4	Intestinal Muscular Wall Defects	316
3.4.5	Small Intestinal Dystopias	317
3.4.6	Composition Abnormalities of the Intestinal Wall	317
3.4.7	Small Intestinal Vascular Changes	318
3.4.8	Inflammation and Malabsorption	319
3.4.9	Short Bowel Syndrome/Intestinal Failure	336
3.4.10	Small Intestinal Transplantation	336
3.4.11	Graft-Versus-Host Disease (GVHD) of the Gut	339
3.4.12	Small Intestinal Neoplasms	340
3.5	**Appendix**	343
3.5.1	Appendiceal Anomalies	343
3.5.2	Appendiceal Vascular Changes	343
3.5.3	Appendicitis	349
3.5.4	Appendiceal Metabolic and Degenerative Changes	351
3.5.5	Appendiceal Neoplasms	351
3.6	**Large Intestine**	353
3.6.1	Large Intestinal Anomalies	353
3.6.2	Large Intestinal Vascular Changes	373

© Springer-Verlag GmbH Germany, part of Springer Nature 2020
C. M. Sergi, *Pathology of Childhood and Adolescence*,
https://doi.org/10.1007/978-3-662-59169-7_3

3.6.3	Large Intestinal Inflammatory Disorders	374
3.6.4	Colon-Rectum Neoplasms	380
3.7	**Anus**	387
3.7.1	Anal Anomalies	388
3.7.2	Inflammatory Anal Diseases	389
3.7.3	Benign Anal Tumors and Non-neoplastic Anal Lesions (Pseudotumors)	389
3.7.4	Anal Pre- and Malignant Lesions	390
3.8	**Peritoneum**	393
3.8.1	Cytology	393
3.8.2	Non-neoplastic Peritoneal Pathology	394
3.8.3	Neoplastic Peritoneal Pathology	394

Multiple Choice Questions and Answers 399

References and Recommended Readings 402

3.1 Development and Genetics

The *esophagus* is a retro-tracheal structure, located in the posterior mediastinum with C6 to T11/T12 extension. Nonkeratinized squamous epithelium lines its mucosa with papillae not extending more than 1/3 of the epithelial thickness. Usually, zero to two intraepithelial lymphocytes per high-power field (40×) may represent a typical finding, but no intraepithelial eosinophils. Moreover, there are submucosa, *muscularis propria* (the outer longitudinal muscle layer is striated for the first 6–8 cm), and *tunica adventitia*. The absence of serosal coating in the esophagus is the reason why neoplastic lesions can easily spread into the mediastinum. The esophagus length is 10–11 cm at neonatal age while 25 cm when the child becomes adult. It is paramount to consider that adult gastroenterologists performing an endoscopy identify the gastroesophageal junction (GEJ) at 38–41 cm measuring from the incisor line. The esophagus has three points of luminal narrowing, which are localized at cricoid cartilage, at the crossing with the left mainstem bronchus, and at the level of the diaphragmatic muscle. There are no morphological landmarks for the upper and lower sphincters, which are defined by manometry only. The development of the esophagus begins as part of the foregut tube. By the 10th week, the esophagus is lined by ciliated epithelial cells. At 4 months, the ciliated epithelium is replaced by squamous epithelium, and at both ends of the esophagus the esophageal glands start. The upper esophagus is derived from branchial arches 4 through 6, but the derivation of the lower esophagus is unknown. The full development of the esophageal wall requires coordination of a variety of genes and growth factors. In the first trimester, the esophageal peristalsis begins, and gastroesophageal reflux is demonstrated in the second trimester. The *stomach* layers are mucosa, submucosa, *muscularis propria*, and serosa. The surface cells are mucous and line both mucosal surface and gastric pits. Neck mucous cells are the progenitor cells responsible for both the glandular epithelium and the pit as well as the surface epithelium. The neck mucous cells are the cells where apoptotic bodies need to be assessed in the case of graft-versus-host disease. Glands of the cardia and antrum regions are similar and show the lining of neck mucous cells. The antrum contains endocrine cells, aka enteroendocrine, enterochromaffin, Kulchitsky, or APUD (amine precursor uptake and decarboxylase) cells. In the fundus, the glands contain the parietal (HCl and intrinsic factor, which is responsible for the absorption of vitamin B_{12}) and chief (pepsinogen) cells. The regional blood vascular supply is essential to correlate with the changes that occur in the case of liver cirrhosis or other conditions raising the pressure in the portal system. Lymphatics and regional lymph nodes are essential in the case of gastrectomy for cancer because they affect TNM staging. The epithelium regenerates every 72–96 h (3–4 days). The cell types are constituted by undifferentiated, Paneth, goblet, and endocrine cells. The embryological development of the gut starts at the 4th week. The primordial gut, which forms as a flat, three-layered embryonic disc, goes through the process of median and horizontal plane infolding and is closed by the oropharyngeal and cloacal membranes. The *stomodeum* is the depression between the brain and the pericardium. It is considered as the precursor organ of the mouth. Subsequent phases include the formation of the stomach by asymmetric enlargement and rotation of the distal part of the foregut and the structure of the *duodenum* from both distal most foregut and proximal midgut pulling along with a rotation of the stomach. Then, the configurations of the *jejunum*, *ileum*, and most of the *colon* derive from midgut, physiologic herniation into the embryo-proximal portion of the umbilical cord from the 6th through the 10th week, and 180° counterclockwise rotation. The hindgut gives origin to the distal colon and rectum, which is separated by the urinary tract by the urorectal septum. The *colon* is retroperitoneal along most of its length with rectum ~6 inches (~15 cm) long having a proximal portion in the peritoneal cavity, while the distal portion is extraperitoneal with peritoneal reflection forming the pouch of Douglas (a common site for tumor implantation) and essential landmark at the time of grossing. Vascular supply is vital for diagnostic and management directions and includes the superior mesenteric artery supplying the cecum, right colon, and transverse colon; the inferior mesenteric artery supplying the descending (left) colon, sigmoid colon, and proximal rectum; and the hemorrhoidal branches of an internal iliac artery supplying the distal rectum. Moreover, watershed areas are splenic flexure, between superior and inferior mesenteric arteries, at the rectum between inferior mesenteric and hemorrhoidal arteries.

The importance of collateral blood supply from the posterior abdominal makes transmural infarction uncommon. The superior hemorrhoidal veins (drain to the portal system) anastomose tightly with the inferior hemorrhoidal veins (drain to inferior vena cava, IVC). The superior hemorrhoidal veins become dilated in portal hypertension. Finally, nervous supply comprises Auerbach plexus (within muscularis propria) and Meissner plexus (within the submucosa) formed from neuroblasts (neural crest cells), which migrate in a cephalo-caudal direction during development (ontogenesis). The neural crest cells usually reach the rectum by 12 weeks of gestational age. The anal region is defined as the canal between the *anorectal ring*, which is considered the upper or proximal limit, and the *anal verge*, which is the lower or distal limit. This extension has been accepted by both the *American Joint Committee on Cancer* (AJCC) and the *Union Internationale Contre le Cancer* (UICC). The anorectal ring is a muscular bundle formed by the junction of the upper part of the internal sphincter muscle, the distal part of the longitudinal muscle, the puborectalis muscle, and the deep portion of the external sphincter muscle. In the anal canal, a fundamental structure is a *dentate line*, which is the not sharp boundary between the columnar epithelium of the upper portion and the stratified squamous epithelium of the lower part and represents the area where the anal glands empty. The dentate line divides the anal canal into two zones with anatomical and pathological significances. Thus, we have three histologic types, including glandular (proximal), transitional (aka intermediate, cloacogenic), and keratinized or nonkeratinized squamous (distal). The *anal transitional zone (ATZ)*, which is 0.3–1.1 cm long, has a wrinkled glistening appearance grossly and is mostly composed of transitional epithelium. It resembles urothelium with four to nine cell layers, has minimal mucin production, and may have features of squamous epithelium and focal glands in the submucosa, as well as endocrine cells, and rare melanocytes and is +CK7 and −CK20 (different from CRC, which is CK7−/CK20+). Moreover, ganglion cells are usually absent 1–2 cm above the dentate line. The squamocolumnar junction is anatomically and histologically very similar to the squamocolumnar of the cervix, and both may have squamous metaplasia areas that are particularly susceptible to the oncogenic effects of human papillomavirus (HPV). There is little or no doubt that high-grade intraepithelial neoplasia is the precursor lesion for anal cancer. Moreover, similar to high-grade cervical intraepithelial neoplasia it is much less likely to regress than low-grade lesions.

3.2 Esophagus

3.2.1 Esophageal Anomalies

Esophageal congenital and acquired anomalies are summarized in Box 3.1 (Fig. 3.1).

> **Box 3.1 Esophageal Congenital and Acquired Anomalies**
>
> 3.2.1.1. Esophageal Atresia With/Without Tracheoesophageal Fistula
> 3.2.1.2. Duplications (Cysts, Diverticula, Tubular Malformation)
> 3.2.1.3. Ectopia/Heterotopia (Gastric, Pancreatic, Salivary Gland)
> 3.2.1.4. Obstructive Disorders (Schatzki Rings, Webs, Dysphagia Lusoria)
> 3.2.1.5. Acquired Anomalies with Low/Eumotility
> 3.2.1.6. Acquired Anomalies with Dysmotility
>
> Note: Mnemonics acronym: *ADEO* (Latin, adeo means "I approach, I visit, I undertake, etc." from the infinitive (adire) = to approach, etc. It may also mean "To such a point, precisely, exactly, etc." as an adverb. Acquired anomalies with low or *eu*motility (normal motility) comprise diverticula, hiatal hernia, lacerations, and varices. Acquired anomalies with *dys*motility include achalasia, Plummer-Vinson syndrome, progressive systemic sclerosis, idiopathic muscular hypertrophy, and leiomyomatosis. The term "dysphagia lusoria" (Bayford-Autenrieth dysphagia) includes a Greek word (δυσ- "difficult, hard" and -φαγος from ἔφαγον "I ate") and a Latin word (lūsōrius, from lūsor "player" and represents the impairment of swallowing (dysphagia) due to compression arising from an aberrant right subclavian artery (arteria lusoria). Importantly, the term originates from the phrase lusus naturae, which means "caprice of nature" referring specifically to the anomalous course of the right subclavian artery (Bayford 1794).

3.2.1.1 Esophageal Atresia ± Tracheoesophageal Fistula (EA-TEF)

- *DEF*: Group of relatively common congenital anomalies that result from failure of the foregut to divide appropriately into the trachea and esophagus during the 4th week of embryonic development (foregut malformation of continuity interruption) when the median pharyngeal groove develops in the ventral side of foregut at day 22 of gestation (Box 3.2) (Fig. 3.1). TEF typing includes type "A," i.e., blind proximal esophagus with no fistula (8%); type "B," i.e., proximal fistula with completely interrupted distal esophagus (<1%); type "C," i.e., blind proximal esophagus with distal TEF (88%); type "D," i.e., esophagus with proximal and distal TEF (<1%); and type "E," i.e., TEF without EA or H-type TEF (3%).
- *EPI*: 1:2,400–1:4,500 live births.
- RF: Male, prematurity/VLBW (very low birth weight) infants, trisomy 21 syndrome, trisomy 18 syndrome, partial 13 trisomy syndrome, DiGeorge syndrome, Pierre-Robin sequence, VACTERL (vertebral abnormalities, anal atresia, cardiac anomalies, EA-TEF, renal agenesis and/or dysplasia, and limb defects) association, and CHARGE (coloboma, heart anomalies, atresia choanae/arum, retardation, genital anomalies, and ear anomalies) syndrome.
- *CLI*: Food regurgitation, drooling, aspiration, and recurrent pneumonia, particularly in patients harboring the H shape of TEF, which may be overlooked and diagnosed later in older children. Isolated EA is associated with other anomalies in 50–70% of the cases, while EA-TEF is associated with other anomalies in approximately 1/4 of the cases. Congenital defects are cardiac (35%), genitourinary (25%), gastrointestinal (25%), skeletal (8%), and abnormalities of the CNS (7%).

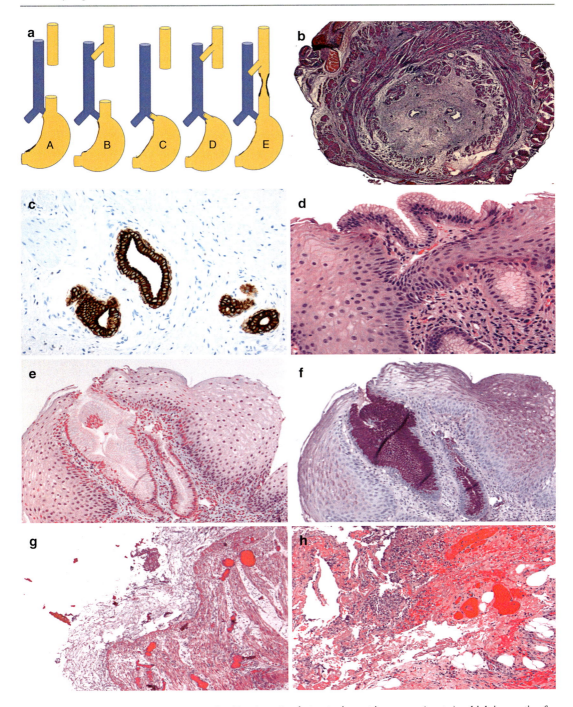

Fig. 3.1 In (**a**) is presented the current classification of tracheoesophageal fistula associated with esophageal atresia. The microphotographs (**b**) (H&E stain, 40×) and (**c**) (anti-AE1-3, 200×) show an example of a distal esophageal atresia with minimal lumina. The microphotographs (**d**) (H&E stain, 200×), (**e**) (Alcian blue pH 2.5 stain, 100×), and (**f**) (Alcian blue PAS stain, 100×) represent heterotopic gastric mucosa (center), which is negative for intestinal mucins but positive for gastric mucins. The microphotographs (**g**) (H&E stain, 50×) and (**h**) (H&E stain, 100×) represent a lesion of the surface following a trauma (post-traumatic mucosal necrosis), which is accompanied by fibrins and scattered neutrophils

> **Box 3.2 Dys-embryogenesis**
> - If the abnormal epithelial-lined link between the two separating tubes occurs, then it is TEF.
> - If excess mesenchymal growth is present instead, then it is EA.
>
> Note: It is common to have both defects, and two classifications are in use (Gross and Vogt classification systems). The excess tissue growth may lead to incorporation of part of the esophagus into the posterior wall of the tracheal tube

- *IMG*: AP and lateral X-ray films show the catheter in the upper esophageal blind pouch and the distal esophagus and stomach are filled with air when TE fistula exists. Air and catheter may be trapped in the esophagus. Air trapping occurs in the stomach in the case of congenital pyloric stenosis, and congenital esophageal stenosis may mimic atresia.
- *GRO*: Two classification systems are in use, including the Gross and Vogt classifications. Probably, the first one is more widely used than the second (Fig. 3.1).
- *CLM*: Microscopically, four structural abnormalities may be seen, including tracheobronchial remnants, such as ciliated clusters of seromucous glands and cartilage, disorganized muscular layers, and a modification/reduction of the Auerbach plexus of the esophagus/stomach.

Congenital esophagus stenosis has been classified as Group I, tracheobronchial rests (cartilage, respiratory mucus glands, ciliated epithelium); Group II, membranous diaphragm; and Group III, fibromuscular stenosis.

In early infancy, patients may present with aspiration and recurrent pneumonia. Dysphagia and regurgitation of solid food are symptoms that are observed in late infancy when more solid nutrients are introduced to the child's diet. If the esophageal stenosis is not severe, its diagnosis may be postponed until the child becomes adolescent or adult.

- *PGN*: Excellent prognosis is seen if early diagnosis and surgical correction are performed, but the presence of associated (extra-gastrointestinal) anomalies or abnormal karyotypes may change the outcome. The trachea, especially at the site of the previous EA-TEF, remains U-shaped with a full, membranous portion, rather than C-shaped (short, membranous portion). This anomaly is at the basis of tracheomalacia of varying degrees of severity. Moreover, loss of the ciliated respiratory epithelium, as well as squamous metaplasia, can occur. The surgical management usually starts in early infancy with the division of the TEF and primary esophageal anastomosis. The most common esophageal atresia with distal (lower) TEF is surgically managed by taking off the lower esophagus, the trachea, and joining to the upper/mid-esophagus. Some severe cases may be treated with a delay of the anastomotic procedure, with placement instead of a cervical esophagostomy, until enough esophageal growth takes place to allow a proper anastomosis. The complications of EA-TEF include anastomotic leak, anastomotic stricture, gastrointestinal symptoms (e.g., dysphagia due to abnormal esophageal peristalsis, postsurgical esophageal stricture, gastro-esophageal reflux disease or GERD), low percentiles of weight and height (sometimes controversially reported in the scientific literature), recurrent EA, recurrent infections, and recurrent TEF. Postoperative esophageal stricture occurs in up to 40% of patients with EA-TEF.

3.2.1.2 Esophageal Duplication
- *DEF*: Intra- or extramural duplication is located in the lower 1/3 of the esophagus bulging into the posterior mediastinum and caused by the failure of normal development around the 5th–8th week of embryonic development (<8th week mucosal lining is columnar epithelium *not* squamous epithelium). They can manifest as cysts, diverticula, or tubular malformations. Bronchogenic cysts protrude from the anterior mediastinum, while (neuro)enteric cysts are located in the posterior mediastinum and have an hourglass configuration.

- *EPI*: 1:8,000 autopsies, 10–20% of all GI duplications, ♂=♀.
- *CLI*: Early feeding difficulties and respiratory distress are observed.
- *IMG*: Intramural defects in the lower 1/3 of the esophagus in the case of duplications, mass effect when protruding into the anterior mediastinum in the case of bronchogenic cyst, and association with vertebral anomalies in the case of neuroenteric cysts.
- *GRO*: See imaging details (*vide supra*).
- *CLM*: Duplications are located within the esophageal wall and show intact muscle layers with an overlying epithelium of squamous, cuboidal, or ciliated epithelium. Bronchogenic cysts may contain cartilage in their walls. Conversely, neuroenteric cysts are usually lined with intestinal epithelium or gastric mucosa.
- *DDX*: When we see diverticula, they can be either congenital or acquired. Thus, the DDX includes acquired diverticula, which are outpouchings occurring in adults and showing communication with the esophageal lumen. The lining is squamous epithelium and may show erosions and ulcerations.
- *PGN:* There is a good outcome following adequate surgical resection.

3.2.1.3 Ectopia/Heterotopia

- *Gastric mucosa* (distal to cricoid cartilage with imaging showing a filling defect and lined as mucin-secreting cells with or without the chief and parietal cells and prone to ulcerate and transform toward neoplasms, e.g., gastric adenocarcinoma)
- *Pancreatic mucosa* (usually in trisomies 13 and 18 chromosomal aneuploidies and as metaplasia in GERD)
- *Salivary gland tissue* (middle or distal esophagus and composed of sebaceous glands)

3.2.1.4 Obstructive Disorders

These congenital disorders include *intramural* obstructive diseases, such as *Schatzki rings* and *webs*, and *extramural* obstructive diseases, such as *dysphagia lusoria*. Webs (networks) and rings are a fold of mucosa or annular thickening of muscle *above* (in the case of a web) or *below* the aortic arch (in the case of a circle) and may be either asymptomatic or present with dysphagia. Webs are more often in women and associated with Plummer-Vinson syndrome (sideropenic anemia, glossitis, cheilosis) with an increased risk to develop carcinoma of squamous cell type of the upper esophagus, oropharynx, or tongue (Gr. γλῶσσα "tongue" for glossitis and χεῖλος "lip" for cheilosis/cheilitis). *Dysphagia lusoria* is defined as an aberrant right subclavian artery originating on the left side with a retroesophageal course. A similar mechanism can be observed, and consequently, dysphagia can occur in the presence of persistence of the right aortic arch (*dysphagia aortica*).

3.2.1.5 Acquired Anomalies of Low or Eumotility

This section comprises diverticula, hiatal hernia, lacerations, and varices. The acquired diverticula can be subdivided into Zenker (pulsion) diverticulum, traction diverticulum, and epiphrenic diverticulum. The Zenker (pulsion) diverticulum is a pharyngoesophageal outpouching (false diverticulum or pseudo-diverticulum) just above the UES (cricopharyngeal muscle) with an increased risk to develop a squamous cell-based malignancy. The mid-portion (traction) diverticulum is a true diverticulum (*diverticulum verum*) near the hilum of lungs and due to scarring of mediastinal or pulmonary lymph nodes infected with *M. tuberculosis*. The epiphrenic diverticulum is an outpouching due to dysfunction of the LES (e.g., achalasia).

Hiatal hernia is a sac-like dilatation of the stomach present above the diaphragmatic muscle. Two types of hernias are observed: sliding hernia and rolling hernia. *A sliding hernia* (90%) is a *short esophagus*, which can be congenital or acquired from esophageal scarring, with the extent of herniation and degree of symptoms accentuated by swallowing and, ultimately, predisposes to GERD. *Rolling (paraesophageal) hernia* (10%) is a typically long esophagus with a portion of cardia protruding through the diaphragm into the thorax alongside the esophagus. This type of hernia is prone to strangulation and infarction.

Lacerations (aka Mallory-Weiss tears) are linear irregular lacerations oriented longitudinally usually at the GEJ. Well known in alcoholics which can occur following fits of vomiting, they can rarely be seen in childhood and adolescence in the setting of liver cirrhosis. Tears may involve from only mucosal lining to the full wall and constitute up to 10% of patients presenting with hematemesis.

Varices represent a dilatation of vascular channels (coronary veins) in the distal 1/3 of the esophagus able to divert flow out of the portal system in patients with portal hypertension, often secondary to liver cirrhosis. Although endoscopically not difficult to identify, they can become inapparent postmortem, because varices-prone blood vessels collapse. Hematemesis can complicate the clinical picture, and varices harbor a mortality rate of 40% in adults (Stanley et al. 2017; Poddar 2019). The worldwide mortality rate for upper gastrointestinal bleeding (UGIB) in children can range from 5% to 15% according to a recent Italian study by Romano et al. (2017).

3.2.1.6 Acquired Anomalies with Dysmotility

These lesions include lesions, which are rarely seen in childhood, although occasional cases have been observed in young adults. *Esophageal dysmotility* is a multi-etiologic syndrome with primary and secondary (and miscellaneous) motility disorders. *Central motility disorders* include achalasia, diffuse esophageal spasm (DES), "nutcracker esophagus," hypertensive LES, and nonspecific esophageal motility dysfunction (NEMD). *Secondary motility disorders* also affect the esophagus and consist of connective tissue diseases, diabetes mellitus, Chagas disease, chronic idiopathic intestinal pseudo-obstruction, and neuromuscular disorders of striated muscle. GERD may also be linked to motility disturbances of the esophagus. In *achalasia*, motor dysfunction of the esophagus due to abnormality or destruction of the myenteric plexus results in decreased peristalsis, incomplete relaxation of LES, and increased basal LES tone. Characteristically, there is a progressive dilatation of the esophagus above the LES. A key etiology comprises Chagas disease or trypanosomiasis, which shows the destruction of myenteric plexus of the esophagus, duodenum, colon, and ureter and harbors a risk of neoplastic development with esophageal carcinoma in ~5% of cases. Alterations in heart, skeletal, and smooth muscles develop 10–20 years after the infection with *T. cruzi*. In addition to contact with feces infected with parasites, the Chagas disease can also be caused by blood transfusion, organ transplantation, mother-to-baby transmission, professional exposure in the lab, and eating raw food contaminated with feces with microorganisms (infected bugs). *Non-achalasia-acquired anomalies* play a significant role in childhood, particularly with esophageal dysmotility including *pre- and post-fundoplication motility disorder*. Nissen fundoplication (NFP) is indeed a widely used surgical option for the treatment of severe GERD. Both NFP and gastro-jejunal feeding tubes are often employed to reduce GER, but there is no significant difference between these two surgical approaches in preventing aspiration pneumonia or improving overall survival in children. NFP is designed to avoid GER by correcting *hiatus* herniation, lengthening the intra-abdominal portion of the esophagus, tightening the crura, and increasing the pressure at the level of the LES. Thus, the gastroesophageal anatomy and function are permanently altered, which means that it can lead to some complications. Post-fundoplication problems are more often observed in children with neurologic or respiratory diseases, esophageal atresia, or a generalized motility disorder. Esophageal dysmotility has also been associated with *Plummer-Vinson syndrome*, *progressive systemic sclerosis*, *idiopathic muscular hypertrophy*, and *leiomyomatosis of the esophagus*.

Plummer-Vinson syndrome is characterized by the triad "sideropenic anemia, atrophic gastritis, and dysphagia" and usually affects women, and there is an association with esophageal webs and a risk to develop squamous cell carcinoma of the upper esophagus, oropharynx, or tongue. Progressive systemic sclerosis with esophageal involvement is part of CREST syndrome (calcinosis, Reynaud's, esophageal dysmotility, sclerodactyly, telangiectasias), which also shows vasculitis with muscle wall degeneration. Other disorders include idiopathic muscular hypertrophy and leiomyomatosis of the esophagus, which is a rare hamartomatous disorder with various presentations, characterized by proliferation of smooth muscle cells with thickening of the esophagus described mostly in children. This condition may occur sporadically or be inherited in AD pattern and has been associated with leiomyomata of other portions of the gastrointestinal tract, visceral leiomyomatosis, tracheobronchial lesions, genital lesions, and Alport syndrome.

3.2.2 Esophageal Vascular Changes

The endoscopically red points or strips of the esophageal mucosa correspond to vascular changes that may be observed in addition to superficial acute inflammatory changes, basal zone hyperplasia, and the proximity of the stromal papillae to the epithelial surface. These vascular phenomena probably represent the first sign of esophagitis recognizable both endoscopically and histologically.

3.2.3 Esophageal Inflammatory Diseases

Inflammatory diseases of the esophagus are grouped in Box 3.3.

3.2.3.1 Reflux Esophagitis

Gastroesophageal reflux (GER) is not uncommon in healthy infants and toddlers and even in children, adolescents, and young adults. GER is defined as the migration of gastric contents into the esophagus and is a normal physiologic process but may cause discomfort to patients and caregivers. Conversely, GERD is defined as the passage of gastric contents into the esophagus that results in *symptoms* (heartburn, epigastric pain, and regurgitation) considered inconvenient, annoying, or alarming and *complications*, such as reflux esophagitis, are common. The commonly observed regurgitation of infants, or "spitting up," is the passage of gastric contents into the pharynx or mouth, while vomiting is the forceful expulsion of the gastric contents. Both conditions need to be differentiated by rumination, which is defined as a voluntary, habitual, and effortless regurgitation of recently ingested food, which can be finally expulsed from the mouth or re-swallowed. About half of infants younger than 3 months of age and about 2/3 of infants at 4 months of age show at least one episode of GER daily. However, only 1 in 20 children shows regurgitation by 12 months of age. GERD symptoms affect up to 7–8% of children and adolescents, and several clinical settings seem to be a significant risk to develop GERD. These children may have been *premature* and harboring *EA-TEF* and may present with *neurologic impairment* (*e.g., West syndrome*), *obesity*, and *lung disease* (specifically cystic fibrosis). The early infantile epileptic encephalopathy (EIEE), the early myoclonic encephalopathy (EME), and early onset West syndrome are well distinct in early infancy. EIEE and West syndrome are considered to be part of a spectrum of epileptic encephalopathies, although some specific features are recognized in each syndrome with numerous numbers of pre-, peri- and postnatal damages identified in West syndrome. GERD is produced by recurrent or prolonged reflux and promoted by contact of the high acidity of the gastric content with the mucus-free squamous mucosa of the esophagus and disordered esophageal motility. GERD is also associated with sliding hiatal hernia, Zollinger-Ellison syndrome, and scleroderma. Microscopy changes are *basal cell hyperplasia* (>15–20% of the epithelial thickness); *elongation of the vascular papillae* of the lamina propria to >1/2–2/3 mucosal height; *intraepithelial eosinophils*, few (IEE < 6/HPF); and *intraepithelial lymphocytes*, numerous (IEL > 10/HPF) (Fig. 3.2).

> **Box 3.3 Esophageal Inflammatory Diseases**
> 3.2.3.1. Reflux Esophagitis
> 3.2.3.2. Eosinophilic Esophagitis
> 3.2.3.3. Infectious Esophagitis (Bacterial, Viral, Fungal)
> 3.2.3.4. Injury-Associated Esophagitis (Irritant Ingestion, Prolonged Gastric Intubation, Radiation/Chemotherapy)
> 3.2.3.5. Esophagitis in Systemic Disorders
> 3.2.3.6. Graft-Versus-Host Disease

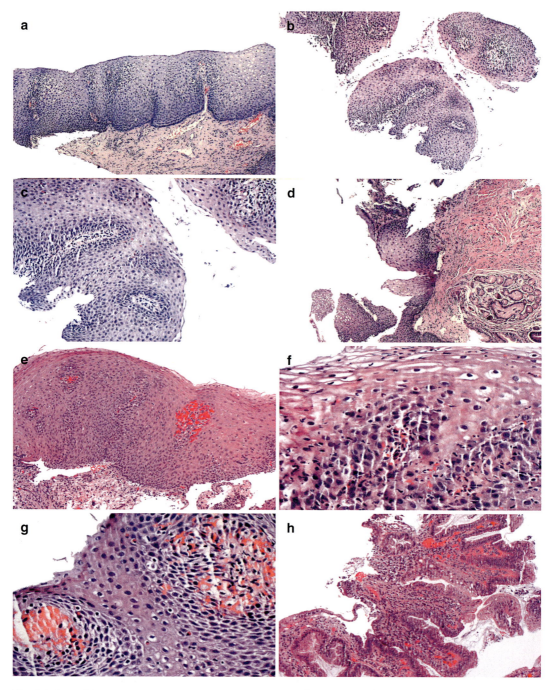

Fig. 3.2 In (**a–h**) are shown several features of GERD with basal cell hyperplasia, elongation of the papillae or rete pegs, scattered lymphocytes and a few eosinophils, as well as an inflammatory polyp (**h**). All microphotographs are H&E stained and are 100× as original magnification with the exception of (**c**) at 200× and (**f**) and (**g**) both at 400×

It is important to remember that GER without GERD histologic features of inflammation may show some ectasia of the vascular papillae only and should not be diagnosed as esophagitis. If nececessry, the pathological report can indicate "reactive changes" with a comment for the physicians that clinical and pathological correlations are needed. A systematic search of the biomedical literature allowed developing an international consensus on the definition of *GERD* in childhood based on the Montreal (QC, Canada) definition of *GERD* in adults as a framework.

Three critical consequences and caveats to keep in mind are:

1. Toddlers or young children when verbal communication, description of pain and intensity, localization, and severity may be unreliable until 8–10 years in most of the children.
2. The primary role of histology is to rule out other conditions.
3. Intestinal metaplasia solely as a definition for Barrett esophagus (BE).

There are some conditions predisposing children or young adults to develop severe, chronic GERD. These include neurologic conditions, such as cerebral palsy, Cornelia de Lange syndrome (CdLS), and Down syndrome; congenital esophageal abnormalities, such as EA and congenital diaphragmatic hernia, cystic fibrosis, hiatus hernia, and obesity/overweight; and family history positive for severe GERD, BE or esophageal malignancy.

In older children and adolescents, GER is the most common cause of heartburn, which is defined as a perceived uncomfortable burning sensation behind the sternum that can potentially reach a painful intensity. Moreover, from neonatal age to young adulthood, GERD may be associated with sleep disturbances. There are indeed some variables, which influence the validity of histology as an indispensable tool for diagnosing GERD. These variables include lack of standardization on number and localization of biopsies (e.g., sampling error) and laboratory variability in the choice of fixative, histology techniques of mounting, plane of orientation angle, and cutting thickness and accuracy of tissue samples as well as uniformity in interpreting morphometric parameters (e.g., hyperplasia of the basal cell layer, elongation of rete pegs, eosinophilic leukocytes intactness and granularity, spongiosis, and lymphocytes).

Eosinophilic leukocytes should not be present in the otherwise normal biopsies from asymptomatic children and adults, although one to two occasional eosinophils only without other histologic features has been variably observed in diet/therapy-reduced inflammation but this finding is strongly debated and not universally accepted. There may also be some overlap between GERD, EE, and food protein sensitivity, and eosinophilic leukocytes may be found in Crohn disease and infections, such as *Cytomegalovirus* (CMV), *herpes simplex virus* (HSV) types I and II, and *Candida albicans*.

Endoscopically suspected esophageal metaplasia or ESEM is a relatively new term, not yet widely used, that describes endoscopic findings consistent with BE that await histological examination and precise confirmation. To accurately identify ESEM, critical endoscopic landmarks of the GEJ should be documented in centimeters from the teeth and photographically documented to be handed to the pathologist, and multiple four-quadrant circumferential biopsies may represent the most sensible approach. The presence of specialized or intestinal metaplasia is the *sine qua non* condition for the diagnosis of BE, and the goblet cells stain for acid mucin with Alcian blue at pH 2.5 is particularly useful. Sandifer syndrome (*torticollis*) is a peculiar manifestation of GERD in children. There is insufficient data that GERD may cause or worsen sinusitis, pharyngitis, serous otitis media, and even lung fibrosis in children. Conversely, chronic cough, chronic laryngitis, hoarseness, asthma, infantile bronchopulmonary dysplasia (BPD), and dental erosions have been satisfactorily associated with GERD. A particularly intriguing phenomenon and its significance is represented by the finding of acid mucin-positive nongoblet columnar cells (NGCC) in the distal esophagus and GEJ. Importantly, GI mucins are classified into two broad categories, neutral and acidic. *Neutral* mucin is the type of mucin, which is generally detected in the gastric epithelium (surface-foveolar and glandular), while *acidic* mucins are located in the small and large intestines. Acidic mucins are subclassified into *sialo*-mucins (either N- or O-acetylated) and *sulfo*-mucins. Sialomucins are usually the mucins identified in the goblet cells of the small intestine (a small amount of such mucins has been identified in the gland neck-regenerative zone of the proximal stomach), whilst sulfomucins are observed exclusively in the goblet cells of the distal small and large intestine.

Alcian blue pH 2.5/PAS stain is usually used to detect *neutral* (magenta-stained) and *acidic* (blue-stained) mucins and high iron diamine (HID)/AB pH 2.5 stain is used to differentiate *sulfated* (sulfur) from *nonsulfated* (sialo) *acidic* mucins. HID stains the sulfomucins imparting an intense brown-colored staining, whereas AB pH 2.5 stains strikingly the sialomucins blue. HID/AB pH 2.5 may be of particular help for the detection of surface epithelial sulfomucins in tissue biopsies obtained from patients suspected to have BE. The presence of HID/AB pH 2.5 should direct the gastroenterologist to repeat endoscopy and biopsy in more areas to look for goblet cells with accuracy. The development of endoscopically performed ablative therapies, including photodynamic therapy, radiofrequency ablation, and cryotherapy, is changing the management of BE radically. The total eradication of BE is essential not only for known lesions but also for synchronous and metachronous of the epithelium at risk. However, detectable injuries in the setting of high-grade dysplasia or intramucosal carcinoma should be treated with a modality aiming at the acquisition of tissue for microscopic confirmation important for both staging and therapeutic goals.

GERD in Children with Severe Neurological Impairment

Infants and children with severe neurological impairment suffer most often from GER, which has been reported to be between 1/3 and 3/4 of all subjects. Untreated GER evolves to inadequate nutrition, esophagitis, and aspiration pneumonitis. The use of a medical approach is often ineffective, and Nissen fundoplication (NFP) is used to control symptoms, although guidelines have been slightly differently applied between North-America, Europe, and Australia/New Zealand. It is also important to emphasize that fundoplication in this well specific patient population may recur and may indicate at least at some point failure of the surgical procedure. It has been estimated that the failure rate in neurologically impaired children may vary between 5% and 25%. One of the causes of failure is wrap herniation. Some more important factors that may influence the failure of a NFP in children with severe neurologic impairment are aerophagia, constipation with abdominal-thoracal pressure changes, dysmotility of the GIT, and some seizure activity of different origins, which may be difficult to manage medically.

The pediatric surgeons' desire to develop a "once-and-for-all" effective therapy did let explore the use of total esophagogastric disconnection (EGD) to treat GER. In EGD and Roux-en-Y esophagojejunostomy, the esophagogastric junction is divided and the gastric end part is stapled. The surgeon, then, would bring up a 20–30 cm jejunal limb behind the transverse colon to the distal esophagus. An end-to-end esophago-jejunal and end-to-side jejuno-jejunal anastomoses are the next steps. Finally, the Heineke-Mikulicz pyloroplasty is performed, and the gastrostomy is preserved. Bianchi proposed EGD with Roux-en-Y esophagojejunostomy in 1997 as a permanent procedure of rescue type in children with severe neurological impairment and NFP failure (Bianchi 1997; De Lagausie et al. 2005). However, some complications well described by Buratti et al. (2004) have been recorded since its application in pediatric surgery. Early complications (≤4 weeks) include anastomosis leak, anastomosis stricture, enterocolitis, enterocutaneous fistula, hernia of paraesophageal type, necrosis, pneumonia, sepsis, slow emptying of the stomach, small bowel obstruction, urinary infection, and wound infection. Late complications (>4 weeks) include bile reflux and gastric irritation, enterocolitis, hernia of paraesophageal type, jejuno-esophageal bile reflux, pancreatitis, pneumonia, and small bowel obstruction.

Thus, Buratti et al. (2004) do not recommend EGD with Roux-en-Y esophagojejunostomy as primary management of GER instead of NFP in any case. These authors underline that this technique may be considered in the setting of significant complications such as repeat wrap failure, adhesions with small bowel obstruction, tube intussusception, and continuous manipulations for tube migration, fracture, dislodgement, and blockage.

Achalasia is usually an adult condition well known in some parts of South America where the infection with the protozoan parasite *Trypanosoma cruzii* is prevalent. Chagas disease is a major endemic parasitic disease with 16 million patients living throughout South America. Acute infection is mostly asymptomatic, while about 1/3 of the infected population develops chronic Chagas disease cardiomyopathy, an inflammatory and dilated type of cardiomyopathy, as late as 20–30 years after the initial infection. In pediatrics, non-parasite associated achalasia, which can also occur independently from Chagas cardiomyopathy, is rare, but may be seen in some genetic syndromes, such as the triple A syndrome (*Triple-A syndrome*) or AAA syndrome (Achalasia-Addisonianism-Alacrima) or aka Allgrove syndrome. Triple-A syndrome is a rare AR-inherited disease characterized by achalasia, adrenal insufficiency of primary type, and insufficiency of tears. Achalasia of the LES with failure to relax at the cardia has as a consequence, a slow progression of food going through the esophagus to the stomach with dilation of the thoracic esophagus and food impaction. Triple-A syndrome may be life-threatening, and there is some mental retardation associated with this syndrome, but individuals, adequately trained, may reach an average lifespan. Mutations in the *AAAS* gene, which encodes a protein known as ALADIN (ALacrima Achalasia aDrenal Insufficiency Neurologic disorder), have been found.

3.2.3.2 Eosinophilic Esophagitis

- *DEF*: Clinical diagnosis characterized by an isolated and severe eosinophilic infiltration of the esophagus in subjects with symptoms like GERD and characteristic histological markers. Early childhood symptoms of eosinophilic esophagitis (EoE) include dysphagia (95%), food impaction (67%), heartburn (25%), vomiting, regurgitation, nausea, thoracic pain/discomfort, and epigastralgia. Late childhood/youth EoE symptoms include dysphagia, food impaction, vomiting, and thoracic pain/discomfort.

In adults, there is a peak in the 3rd or 4th decade, while the peak of incidence in children is between 6 and 8 years of age. In 50–80% of the patients, there is a positive history of allergies, and a family history of atopic disease is found in 3/4 of cases. In 1/3 of cases, there is peripheral blood eosinophilia, and serum IgE levels are increased in about half of the patients. The mechanism of eosinophilic recruitment involves numerous mediators, including pro-inflammatory cytokines (e.g., IL-3, IL-4, IL-5, GM-CSF), specific chemokines such as RANTES, and proteins, which are pro-macrophagic such as eotaxins 1, 2, and 3. However, it seems that only IL-5 and the eotaxins play a significant role in orchestrating the eosinophilic inflammatory process of the esophagus. Both IL-5 and IL-13 cause release of eotaxin, which in turn attracts eosinophils. There is wide variability of blood eosinophilia in these patients (2–60%), but a family history positive for allergy, asthma, rhinitis, and ectopic eczema is often found, and about 80% of patients with EoE have another allergic pathology compared with 1/3 only seen in patients with GERD. The eotaxin-3 expression is a peripheral marker that may be useful in distinguishing EoE from other causes of esophagitis. Endoscopic findings, which are abnormal in more than 90% of the patients, usually include mucosal fragility (~60%), rings or corrugated lining (~50%), and strictures (~40%). The hallmark of EoE is a normal corrugated esophagus showing beautiful concentric mucosal rings due probably to acetylcholine release, which contracts muscle fibers in the *muscularis mucosae* resulting in the formation of concentric rings. These changes may be reminiscent of airway remodeling in asthma, and EoE endoscopy has been called "asthma of the esophagus." Two classifications are used in endoscopy. The first is the *Savary-Miller classification* used mainly in children, while the second is the *Los Angeles classification*, which is used in young adults.

Histologically, four aspects are frequently encountered in the biopsies of patients with EoE: (1) peak eosinophilic count ≥15 cells/HPF (40× objective × 10× = 400× resolution); (2) simultaneous infiltration of both proximal and distal esophagus by eosinophils; (3) clusters, mainly superficial, of eosinophils (eosinophilic microabscesses); and (4) persistence of the eosinophilic infiltrate following 2 months of PPI therapy. There is a histologic grading, although it does not correlate perfectly with the endoscopic staging or grading. At least 1 HPF must contain at least 15 intraepithelial eosinophils (Fig. 3.3), and the standard number of eosinophils in the GI tract needs to be considered. In the esophagus, there are nil eosinophils, while in the stomach, small bowel, and large bowel,

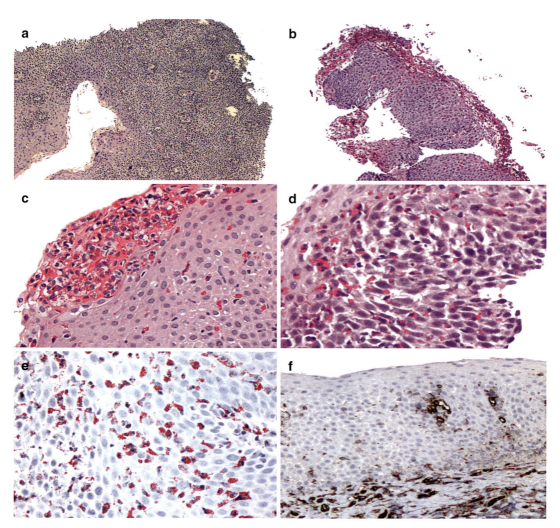

Fig. 3.3 In (**a–f**) are shown the histologic features of the eosinophilic esophagitis with numerous eosinophils, layering of eosinophils (**b**), eosinophilic abscesses (**c**), acanthosis (**d**), highlighted eosinophils by Luna staining, and increased expression of vascular endothelial growth factor (**f**). The microphotographs (**a–d**) are H&E stained. The magnification of the microphotographs (**a–f**) is 100×, 100×, 400×, 400×, 400×, and 200×, respectively

there are 2–3, 3–8, and 8–12 eosinophils per HPF, respectively. Symptoms and histological severity do not correlate well in patients with longstanding EE, but a modest correlation may be found between symptomatology and histological severity in newly diagnosed, untreated patients. Eosinophils increase in number as we descend examining the GI tract portions, although a consensus is not present worldwide. Moreover, some data may be variable for different ethnics and associated or underlying co-morbidities. Box 3.4 illustrates some differences between GERD and EoE. In the case of neutrophils, the pathologist needs to clarify an infective etiology (e.g., Candida, HSV, CMV) or a traumatic origin using histochemical stains (e.g., PAS, D-PAS, GMS), immunohistochemical stains (anti-HSV and anti-CMV), and the accurate use of polarizing lens.

The treatment of EoE is a topic of discussion at many pediatric GI meetings. Systemic therapy includes corticosteroids and antagonists of the leukotrienes. Topical treatment comprises corticosteroids and chromones, while dietary advice consists of an elimination diet that may be efficacious. Chromones are derivative of benzopyran with a substituted keto group on the pyran ring. Derivatives of chromone are collectively known as chromones, and most, though not all, chromones are also phenylpropanoids. Chromones inhibit histamine and the release of eicosanoids. Topical corticosteroids, e.g., fluticasone or budesonide, are started at 440–880 µg/day for children and 880–1760 µg/day for adolescents for 6–8 weeks. A metered dose inhaler without a spacer is administered by inserting it into the mouth and sprayed with lips sealed around the device, the powder is swallowed, and patients should not eat or drink for at least 30 min. Monitoring involves an EGD with biopsy every 4–6 months until resolution of the esophageal eosinophilia. Esophageal metaplasia (Barrett intestinal metaplasia or cardia-type metaplasia) has not been described in EoE, a quite different setting identified in patients who suffer from GERD.

3.2.3.3 Infectious Esophagitis

Bacteria, viruses, and fungi may colonize the squamous mucosa, cause recruitment of inflammatory cells, and induce erosions. CMV shows normal submucosal cells with inclusions, while HSV infection manifests histologically with epithelial cells harboring ground-glass nuclei and often coalesced in form of multinucleated giant cells (MNGCs). Budding blastospores and pseudohyphae of Candida and hyphae of mucormycosis (aseptate and dividing at a right angle) or aspergillosis (septate and splitting at 45°) may also be etiologic factors of infectious esophagitis (Fig. 3.4).

3.2.3.4 Injury-Associated Esophagitis

Ingestion of irritants (e.g., drug-/caustic-exposed esophagitis), prolonged gastric intubation (aka traumatic esophagitis), radiation, and anticancer chemotherapy may be the underlying disorder for some esophagitis. Liaising with the clinician and discussion of findings at multidisciplinary team meetings is essential. Appropriate information in the paper-based or electronic requisition form sent to the pathologist with the endoscopic specimen is also courteous and collegial (Fig. 3.4).

Box 3.4 Comparison of GERD and EoE

	GERD	EoE
Basal cell hyperplasia	+/++ (mild)	++/+++ (marked)
↑ Height of papillae	+/+++ (variable)	++ (moderate)
Spongiosis	+/++	++/+++ (marked)
Erosions/ulcerations	−/+/++	−/+
Intraepithelial eosinophilic infiltrates	−/+ (few)	++/+++ (numerous)
Intraepithelial lymphocytic infiltrates	+/++/+++ (variable)	−/+ (nil/few)
Intraepithelial neutrophilic infiltrates	−/+ (nil/few)	−/+ (nil/few)
Papillary telangiectasias	+	−

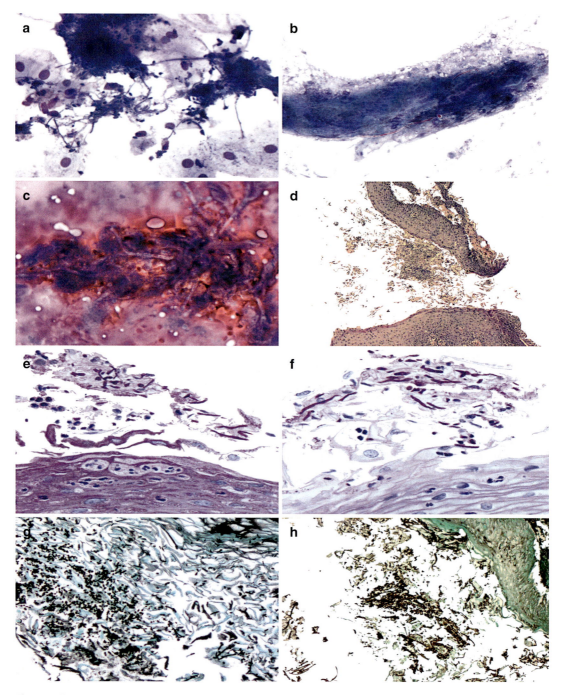

Fig. 3.4 In (**a–c**) are esophageal brushing for cytology showing *Candida* hyphae and spore (cytology), while the same pattern of hyphae and spore is illustrated in the histology in the microphotographs (**d–h**) (**a**, DQ stain at 400×; **b**, PAP stain at 400×; **c**, DQ stain at 400×; **d**, H&E stain at 100×; **e**, PAS stain at 400×; **f**, PASD stain at 400×; **g**, GMS stain at 200×; **h**, GMS stain at 200×)

3.2.3.5 Esophagitis in Systemic Disorders

Esophagitis may also be part of systemic disorders, including Crohn disease, dermatologic conditions (*pemphigus vulgaris*, bullous pemphigoid, erythema multiforme, TEN/Stevens-Johnson syndrome, and lichen planus), and, even, chronic renal insufficiency or uremia (Fig. 3.5).

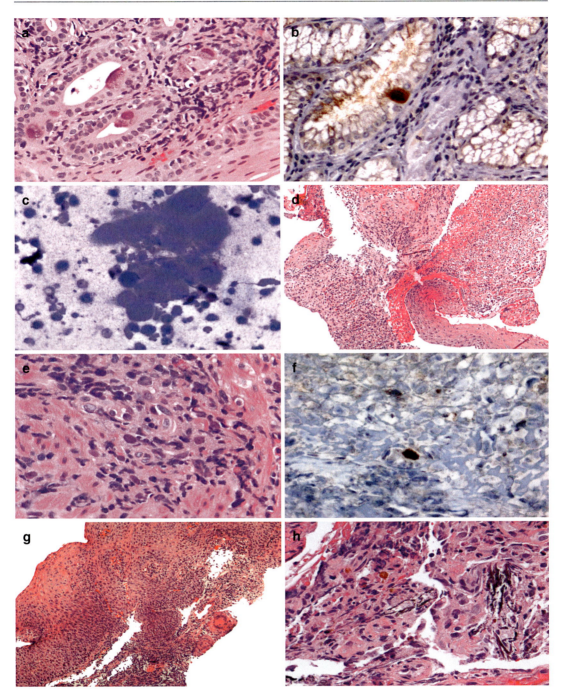

Fig. 3.5 The microphotographs (**c**–**f**) are an example of Herpes simplex virus infection with positive Tzanck smear (**c**, Tzanck smear, 400×), esophagitis on H&E (**d**, 100×; **e**, 400×) with Cowdry A inclusion, which is positive using an antibody against HSV by immunohistochemistry (630×). The microphotograph (**g**) illustrates a case of Crohn esophagitis with granulomas (H&E stain, 100×), while (**h**) is a case of foreign body esophagitis following ingestion of an unspecified object in an infant (H&E stain, 400×). Swallowing of foreign bodies is often observed in children aged between 6 months and 6 years and the most common objects are coins and nonmetallic sharp objects (NMSO), e.g. glass. In older children and adults fish bones and bones constutte the majority of foreign body objects

3.2.3.6 Graft-Versus-Host Disease

Graft-versus-host disease (GVHD) may show few or numerous intraepithelial inflammatory cells with few or many apoptotic keratinocytes. A recent study in adults indicates that the majority of cases of acute GVHD shows diffuse upper and lower GI involvement with esophageal, gastric, duodenal, sigmoid, and rectal involvement. The histologic findings of GVHD in the esophagus demonstrated increased apoptosis of squamous epithelial cells in the absence of an inflammatory infiltrate in almost all cases. Thus, in suspicion of GVHD, it is advisable to have multiple biopsies of the upper and lower GIT. There are some histologic criteria to recognize an intraepithelial apoptotic body, including clearing of cytoplasm and fragmentation and condensation of the nucleus. The lower 1/3 of the esophageal squamous mucosa is a proliferative compartment, like duodenal crypts and gastric glands, and typically has negligible apoptosis. Thus, it is an excellent location to be sampled during the endoscopy for GVHD grading. On the other hand, some *mimickers* may also show some GVHD-like features and need to be taken into consideration. Remarkably, mycophenolic acid is an immunosuppressant drug commonly used in patients undergoing solid organ transplant, and mycophenolate mofetil-associated injury of the upper GI tract, like that in the lower GI tract, is characterized by prominent apoptosis similar to that of mild or grade I graft-versus-host disease injury.

Endoscopic and Histological Criteria

Endoscopic findings may be staged according to the classification proposed by Cruz-Correa et al. (2002). Histological criteria may be phased according to McDonald and Sales (1984). In grade I, apoptosis should be investigated on the medium power of lens magnification (e.g., 10× or 20× as objectives). Grade II consists of crypt/glandular abscesses, epithelial flattening, and glandular/crypt dilation.

3.2.3.7 Inflammation-Associated Lining Changes

The prolonged exposure of the high acidic gastric content to the mucus-free squamous epithelium of the esophagus may start a metaplastic process to change the squamous epithelium to gastric and goblet cellularity. It is already irrefutable that controversy exists in defining *Barrett metaplasia*. According to several authors, the term "Barrett esophagus" should be discarded because it is misleading and does not differentiate between metaplasia and dysplasia, which is a precancerous condition. The use of inflammation or inflammatory process associated with esophageal lining changes may be encouraged because it is neuter. The controversial debate exists because there is some puzzle in distinguishing between gastric metaplasia and gastric mucosa of a short esophagus risking increasing the rate of Barrett metaplasia inappropriately. This aspect may be changed if the length and photographic material are provided to the pathologist. In the USA, Canada, and most parts of the world, Barrett metaplasia is defined by specialized type (intestinal or colonic), while in the UK and some other countries, Barrett is also described as gastric/cardiac or junctional type. Goblet cell change in the gastroesophageal junction (GEJ) may be recognized on H&E or better using Alcian blue (pH 2.5)/PAS staining, which allows the pathologist to discern between goblet cells stained as purple (Alcian blue, pH 2.5) and mucin cells of the cardiac mucosa, which are stained as pink (PAS) (Fig. 3.6) (*vide supra*).

Barrett metaplasia is defined as conversion of the stratified squamous epithelium to columnar epithelium in the lower portion of the esophagus. It may be distinguished between type I and type II. In Barrett type I, there are no residual squamous islands, while in Barrett type II, squamous islands are still present as scattered foci among the columnar mucosa. Also, different types of metaplasia may be recognized as briefly indicated in the previous paragraphs.

- *Specialized Type* (intestinal or colonic): Goblet cells and columnar cells (goblet cells contain sialomucin showing a positive staining with Alcian blue at pH 2.5) and most commonly seen in young and middle-aged adults (incomplete, gastric foveolar cells intermixed with intestinal goblet, absorptive, Paneth, and endocrine cells, and complete, characterized by only intestinal-type cells).

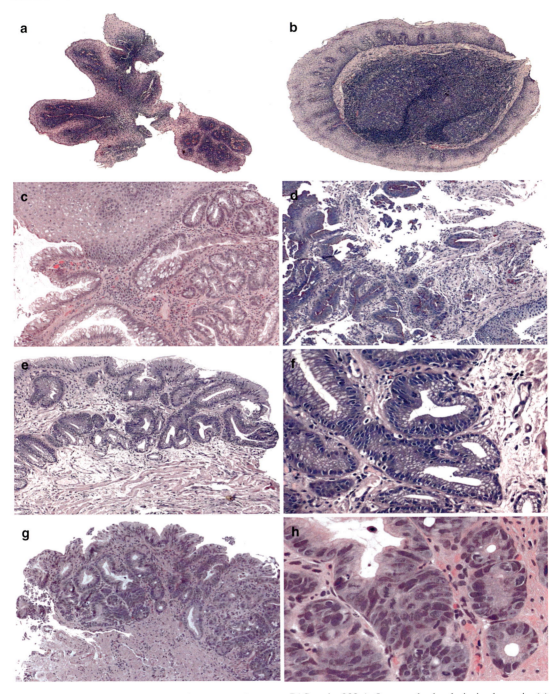

Fig. 3.6 The microphotographs (**a**, **b**) are two Barrett mimickers showing a papilloma (H&E stain, 40×) and a hyperplastic lymphoid follicle (H&E stain, 40×), both from the distal esophagus. Barrett metaplasia (intestinal metaplasia) is shown in **c** (H&E stain, 100×) and **d** (AB PAS stain 200×). Low-grade dysplasia is shown in (**e**) (H&E stain, 100×) and (**f**) (H&E stain, 400×). High-grade dysplasia is shown in (**g**) from a young adult with a long-history of untreated GERD (H&E stain, 100×) and (**h**) (H&E stain, 400×)

- *Cardiac or Junctional Type*: Almost entirely mucin cells, but generally, there is the criterion of biopsy ≥3 cm above the GEJ (i.e., <35 m) before calling metaplasia (unlike "irregular Z-line"). However, there is some controversy exists between Europe and North America on this topic as indicated above.
- *Atrophic Fundal Type*: characterized by few parietal and chief cells and may be observed in childhood. Barrett metaplasia may be complicated with ulceration, stricture, and neoplastic transformation (e.g., adenocarcinoma).

Barrett dysplasia is classified as a low and high grade in oncological terms. There are four possible diagnoses that may be encountered by the pathologist. There is the diagnosis of "negative for dysplasia" (NFD), which is displayed by the regularity of glands with mild crowded and hyperchromatic nuclei restricted to the basal layer and normal basal-superficial squamous maturation. Low-grade dysplasia (LGD) is characterized by mild architectural distortion with cytologic changes, including nuclear enlargement, nuclear crowding, and nuclear hyperchromatic pattern extending up to the surface epithelium, but no loss of nuclear polarity. High-grade dysplasia (HGD) is defined by marked architectural distortion (marked crowding and villiform/cribriform and back-to-back glandular patterns ± necrotic debris in dilated glands) with marked cytologic changes, including nucleomegaly, crowding, pleomorphism, and hyperchromatic pattern extending to the surface epithelium with loss of nuclear polarity. The infamous "indefinite for dysplasia" (IFD) entails (1) mild cytological and architectural changes with focal surface involvement, (2) marked cytoarchitectural changes, but the surface is not assessable, and/or (3) the impression that changes fall short of LGD. Some pearls and pitfalls need to be taken into consideration during the evaluation of an esophageal biopsy to evaluate the risk of adenocarcinoma (Box 3.5).

- AB 2.5-PAS staining purposes:
  1. To define the PAS (−) (glycogen-free) basal layer of the esophagus
  2. To detect dark-blue intestinal mucus in globoid goblet cells (when IM is present)
  3. To detect fungi (also GMS +)
  4. To identify the signet ring cell adenocarcinoma (as screening stain)
- Immunohistochemistry (IHC) role in LGD/HGD:
  – CK7 and CK20 sequential (not combined) IHC is useful to differentiate gastric cardia IM from esophageal IM (i.e., Barrett metaplasia). In the esophageal IM, there is a diffuse, superficial, and deep CK7 staining and a band-like, superficial CK20 staining.
  – P53 and MIB1 (Ki67) sequential (not combined) IHC is useful to improve intra- and interobserver agreement in diagnosing the processes of dysplasia and LGD and reactive/regenerative changes (stomach/esophagus): P53−, Cyclin-D1−, and MIB1+ in the central areas of glands only, while HGD is characterized by P53+, CyclinD1+, and MIB1+ with extension to the surface epithelium.
- Substantially, IFD should be used only when:
  1. There is a genuine difficulty in distinguishing between reactive changes from LGD.
  2. There are worrisome atypical features, but they are focal only.
  3. There is limited amount of supporting evidence for LGD and no evidence of HGD.
  4. There is a poor quality of tissue processing (from biopsy collection to lab processing).

Box 3.5 Pearls and Pitfalls
- In the nuclear polarity assessment, the long nuclear axis needs to be perpendicular to the basement membrane.

3.2.4 Esophageal Tumors

Esophageal cancer is rare in childhood, probably ≤$2/10^6$ children (prevalence), and the rarity of this kind of tumor probably relies on the extended period of exposure to promoting agents (e.g., chronic irritation from a wide variety of known environmental carcinogens and gastric contents that may be present in chronic reflux). However, risk factors have been identified in numerous studies targeting the young adults and need to be kept in mind. These include caustic ingestion, smoking, severe GERD, Barrett metaplasia, HPV infection, positive family history of esophageal cancer, inherited bone marrow failure syndromes (Fanconi anemia, dyskeratosis congenita), and *TP53* gene germinal mutation. Since some vaping constituents have been classified possibly or probably carcinogenic by the IARC/WHO, it is strongly suggested that this element in the patient history is accurately screened.

3.2.4.1 Benign Esophageal Tumors

Inflammatory fibrous polyp (aka fibrovascular polyp or inflammatory pseudotumor) is uncommon, asymptomatic and an incidental finding often during the endoscopic investigation for screening or other pathologies. It is usually pedunculated and solitary, and more than 3/4 of these polyps are in the *upper* 1/3 of the esophagus. Histologic sections show submucosal proliferation of vascularized, inflamed fibrous stroma with tissue eosinophilia similarly to the same lesion that can be found in the stomach.

Squamous papilloma is also uncommon and an incidental finding but may be multiple and is usually located in the *lower* 1/3 of the esophagus. It may be HPV-6- or HPV-11-associated and not-HPV-associated. It may be part of the *recurrent respiratory papillomatosis*, which is an HPV-associated chronic disease that occurs in both children and adults with the tendency to recur and spread throughout the respiratory tract.

Adenoma may be found but arises in Barrett mucosa and is a lesion usually seen in adults.

Other benign lesions include *leiomyoma*; *granular cell tumor*, which is PAS (+) and S-100 (+); and the *amyloidosis* of the localized type that are lesions of adults typically.

3.2.4.2 Malignant Esophageal Tumors

Squamous Cell Carcinoma

- *DEF*: Malignant epithelial tumor with squamoid cell differentiation (Fig. 3.7).
- *EPI*: Sporadic in pediatric age, but it is ~80% of all esophageal carcinomas and 10% of all carcinomas of the GI tract. There is a worrying high prevalence of Barrett metaplasia and esophageal squamous cell carcinoma (SqCC) after repair of esophageal atresia with Barrett metaplasia 4-fold higher in youth with esophageal atresia, and the prevalence of esophageal SqCC being 108-fold higher than in the general population (Vergouwe et al. 2018).
- *RFs*: Lye strictures; chronic alcohol consumption; tobacco use; consumption of foods rich in *Aspergillus*, nitrites, or nitrosamines; chronic vitamin deficiency; achalasia; esophagitis and Barrett metaplasia; Plummer-Vinson syndrome (sideropenic anemia, esophageal web, and atrophic glossitis); and tylosis.
- *CLI*: Dysphagia, hemorrhage, sepsis, and TEF, and sites involved are 20% upper 1/3, 50% middle 1/3, and 30% lower 1/3.
- *GRO*: Fungating polypoid lesion > necrotic deeply ulcerating lesion > diffusely infiltrating wall with thickening, rigidity, and luminal narrowing.
- *CLM*: Conventional *squamous cell carcinoma* (SqCC): tumor cells showing cytological or tissue architectural features of squamous cell differentiation (e.g., keratin, tonofilament bundles, or desmosomes). The differentiation of the tumor toward squamous epithelium could be difficult to ascertain on routine histologic sections, and sensitive, specific markers of squamous differentiation may be required: p63 (nuclear) and CK5/6!
- *PGN*: Extensive circumferential and/or longitudinal spread common due to rich lymphatic supply, but the prognosis is not bad if it is confined to the wall and 1/2 of the cases are prone to surgical excision. 5-YSR: 5–10% (overall). The high prevalence of Barrett metaplasia and esophageal SqCC in patients who underwent esophageal repair because of atresia has impor-

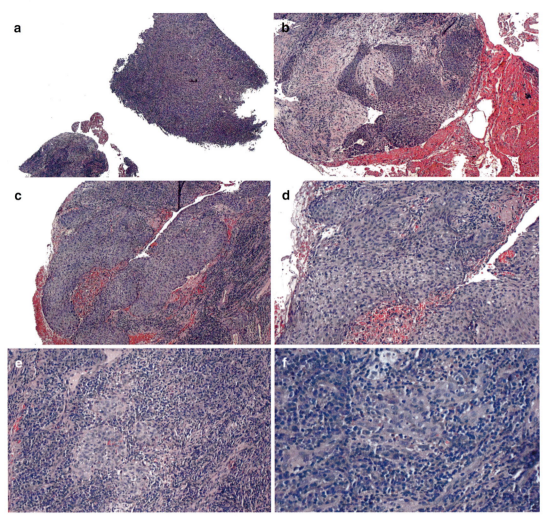

Fig. 3.7 An extremely rare occurrence of an esophageal carcinoma in a young male is shown with the histologic features of poorly squamoid differentiation. All microphotographs are stained with H&E and have 40×, 100×, 100×, 200×, 200×, and 400× in (**a–f**), respectively, as original magnification. There is a high prevalence of Barrett esophagus (Barrett metaplasia) and esophageal squamous cell carcinoma after repair of esophageal atresia (see text)

tant implications for transition of youth from pediatric care to adult gastroenterology departments. These patients need to receive life-long endoscopic follow-up evaluation to pinpoint early diagnosis of preneoplastic lesions.

Four major variants of squamous cell carcinoma of the esophagus are known. *Verrucous carcinoma* is a predominantly exophytic growth with spires of well-differentiated epithelium with minimal cytologic atypia, inconspicuous kerato-hyaline granules, and chronic inflammation, pushing margins and slow growth. It is a rare type of carcinoma, which rarely metastasizes. *Spindle cell carcinoma* (aka epidermoid carcinoma with spindle cells, pseudosarcoma, carcinosarcoma) is another variant with large size, polypoid aspect, and relatively better prognosis than traditional SqCC and constituted by a predominant spindle cell stroma with similarity to MFH with possible heterologous differentiation (e.g., cartilage, bone, or skeletal muscle differentiation) and minimal epithelial component. IHC markers, including p63 and CK5/6 as well as pan-CK, are handy, particularly in children, to rule out regional soft tissue sarcoma, such as mesenchymal chondrosarcoma, osteogenic sarcoma, or rhabdomyosarcoma, which are more often observed. *Basaloid squamous cell carcinoma* is a variant with highly aggressive behavior. Histologic sections show solid nests of large, hyperchromatic pleomorphic nuclei with scant cytoplasm and nuclear palisading at the periphery and central necrosis. *Small cell carcinoma* is a highly malignant variant, which is characteristically fungating in its gross appearance. Microscopically, there are solid sheets, nests, or ribbons of small round-oval hyperchromatic cells with speckled chromatin, indistinct nucleoli, nuclear molding, scant cytoplasm, and a diffuse infiltrative growth pattern.

Adenocarcinoma
Approximately 10% of esophageal carcinomas, mostly in white ethnics, ♂:♀ = 5:1, young-middle aged adults, 2/3 lower esophagus. Most of the cases arise in the setting of Barrett metaplasia and present as mass or nodule in the otherwise intact mucosa. Histologically, three types are recognized and the acronym "IDA" may be used for mnemonic purposes:

- *Intestinal*: as seen in the stomach or small/large bowel
- *Diffuse*: diffuse infiltration of mucin-producing cells (signet ring cell-like)
- *Adenosquamous*: a mixture of squamous cell and adeno-Ca portions

Esophagectomy indications in children and youth are Barrett esophagus with high-grade dysplasia, caustic-related stenosis/occlusion, and esophageal carcinoma. In some scientific literature HGD is a very controversial indication and a vivid debate exists.

Other Tumors of the Esophagus
Other malignant tumors include adenosquamous carcinoma, mucoepidermoid carcinoma, leiomyosarcoma, malignant melanoma, malignant lymphoma, and plasmacytoma. All of them have not been reported in detail or represent a rarity in childhood, although their occurrence in young adults has been occasionally observed.

3.3 Stomach

Box 3.6 lists the categories of disorders we are going to deal in this chapter.

> **Box 3.6 Disorders of the Stomach**
> 3.3.1. Defects (Congenital and Acquired)
> 3.3.2. Vascular Changes
> 3.3.3. Inflammation
> 3.3.4. Metabolic and Degenerative Diseases
> 3.3.5. Tumors

3.3.1 Gastric Anomalies

Box 3.7 summarizes the innate and acquired anomalies of the stomach (Figs. 3.8 and 3.9).

> **Box 3.7 Congenital and Acquired Anomalies of the Stomach**
> 3.3.3.1. Agenesis and Dystopias
> 3.3.3.2. Duplications (Cysts, Diverticula, Tubular Malformation)
> 3.3.3.3. Ectopia/Heterotopia (Gastric, Pancreatic, Salivary Gland)
> 3.3.3.4. Obstructive Disorders (Congenital and Acquired Hypertrophic Pyloric Stenosis)
> 3.3.3.5. Gastric Antral Vascular Ectasia
> 3.3.3.6. Acquired Anomalies with Dysmotility

3.3.1.1 Gastric Agenesis and Dystopias
Gastric agenesis or gastric atresia (often antrum atresia) is an interruption of the continuity of the upper gastrointestinal tract that is extremely rare. Developmental defects with final obstruction or disruption of the continuity of the upper digestive tract include congenital pyloric stenosis, acquired pyloric stenosis, antral web, and pancreas heterotopia in the differential diagnosis.

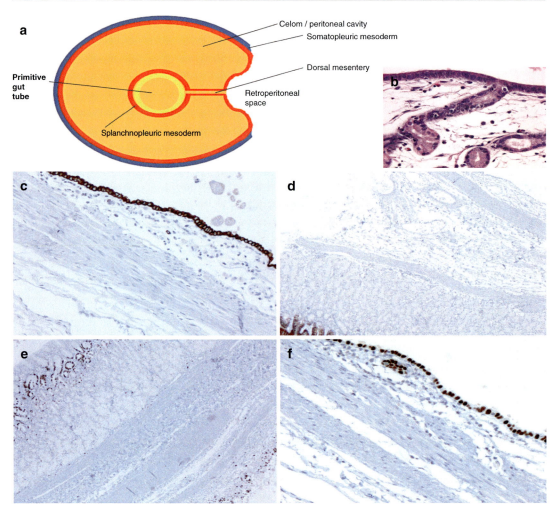

Fig. 3.8 In (**a**) is a cartoon showing the elements that combined will form the diaphragma and the primitive gut. In (**b**) a cyst lining is shown (H&E stain, ×400), which is CK7 positive (**c**) (×200), CK20 negative (**d**) (×50), low proliferative (Ki67) (**e**) (×50), and TTF, a marker of lung epithelium, positive (**f**) (×200)

3.3.1.2 Gastric Duplications

Gastric duplications account for about 7% of gastrointestinal tract duplications. In a comprehensive review, most of the gastric duplications have been characterized as noncommunicating, spheric, or ovoid closed cysts. In the majority of cases, the most common site is the greater curvature. About the mucosal lining, this is usually gastric mucosa. However, both pseudostratified respiratory epithelium and pancreatic tissue have been reported. Bremer theory (1944) focuses on persistent vacuoles within the primitive foregut epithelium to explain the small gastric cysts, but McLetchie et al. (1954) may be right indicating that large duplications may be the result of faulty endoderm and notochord separation early in the human development. Clinically, most duplications present within early childhood with symptoms, including vomiting and abdominal pain. However, infants are frequently asymptomatic and are discovered to harbor such disease later in life.

3.3.1.3 Ectopia/Heterotopia

Ectopic/heterotopic tissue may mean gastric antral mucosa in other regions of the stomach, which do not usually have antral mucosa (e.g., body), pancreatic tissue, and salivary gland tissue. Heterotopic pancreas is quite common and observed in 1–2%

3.3 Stomach

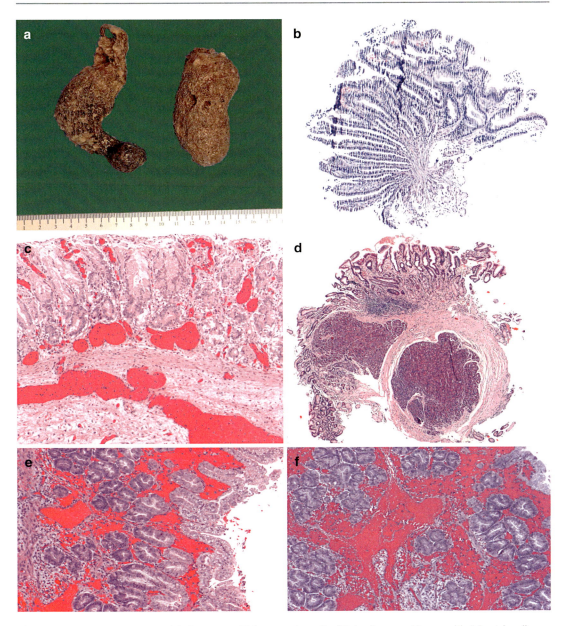

Fig. 3.9 In (**a**) is shown two trichobezoars, which are intraluminal accretions of ingested hair. While trichobezoars are a rare entity, they occur in children and psychiatric patients. They may present with significant complications, such as obstructions and perforations. In consideration of the infection risk and considerable size of many of these bezoars, an open removal is a safe procedure. In (**b**) is shown a biopsy with Menetrier disease (H&E stain, ×100). In (**c**) (H&E stain, ×100) hyperemia is seen with intact mucosa and no necrosis, which conversely present in (**e**, **f**) showing hemorrhagic necrosis (H&E stain, ×100). In (**d**) is a singular case of pancreas heterotopia in a stomach (H&E stain, ×40). Pancreas heterotopia is a rare cause of gastrointestinal bleeding in children

of the general population. Grossly, there is a dome-shaped mass (Ø: 1–2 cm), nipple-like projection, or symmetric cone. Microscopically, the heterotopic pancreatic tissue is found in 75–85% of the cases in the submucosa and 15–25% in the muscularis propria with about 2/3 located in the antrum and about 1/4 of cases in the pylorus. It is constituted by exocrine tissue only in about 2/3 of cases, while endocrine pancreas (islets) are present in the remaining 1/3 of cases.

3.3.1.4 Obstructive Disorders

Obstructive disorders include both congenital and acquired hypertrophic pyloric stenosis. *Congenital hypertrophic pyloric stenosis* (CHPS) is a familial malformation, which is seen in 1/300–900 live births with a male preponderance (♂:♀ = 4:1) and most commonly observed in firstborn infants. CHPS is characterized by hypertrophy and hyperplasia of the circular muscle of the muscularis propria of the pylorus region. Infants with CHPS present with regurgitation and vomiting in the 2nd or 3rd week of life. In contrast, acquired pyloric stenosis is a long-term complication of chronic antral gastritis and/or peptic ulcer disease and may be seen in subjects with disabilities and abnormal development of the nervous system. The acquired form of pyloric stenosis can also be seen in carcinomas of the pylorus, head of the pancreas, or malignant lymphomas, particularly Burkitt lymphoma. An unusual situation is a condition determined by bezoars, trichobezoar, and Rapunzel syndrome. Infants and children may acquire the practice of swallowing foreign material, which if it persists may lead to bezoars in the gastrointestinal tract, usually the stomach and rarely small intestine. This habit may mainly be present in infants and children with the psychosocial delay or mentally disturbed. The syndrome was named after the maiden in the Grimm brothers' fairy tale of 1812. Rapunzel's long hair allowed her to flee out of prison being rescued by her prince. The underlying psychiatric disorder in young females is trichophagia and is more frequent than in children or infants (Gr. θρίξ "hair" and -φαγία from φαγεῖν "to eat").

There is often a history of trichotillomania, which is characterized by recurrent pulling out of their hair, increasing sense of tension before pulling out, and relief or pleasure following the action (obsessive-compulsive behavioral disorder). The site is usually scalp, but also eyelashes, eyebrows, pubic area, or other parts of the body. Trichobezoars make up to 55% of all bezoars, but the foreign material may be vegetable or any other substance. In all situations of trichobezoars, strands of swallowed hair extend as a tail through the duodenum beyond the stomach. The typical presentation is an obstruction of the gastrointestinal tract with nausea and vomiting, gut perforation, acute pancreatitis with potentially life-threatening necrosis of the pancreatic parenchyma, obstructive jaundice, hypochromic anemia, vitamin B_{12} deficiency, weight loss, or an abdominal mass causing pain or discomfort. The abdominal mass is usually well defined and exquisitely mobile; in the majority of cases, it may calcify at the edges and may be indented ("Lamerton sign"). Laparoscopic removals are adequate for small bezoars, but open surgery is the first choice for the removal of large trichobezoars. Psychological support may be productive. Some patients may respond well to fluoxetine or other serotonin reuptake inhibitors, and the prognosis remains excellent if trichophagia can be controlled.

3.3.1.5 Acquired Anomalies with Dysmotility

Dysmotility may be the primary symptom of acquired anomalies, and studies of physiology should be carried out promptly to address the patient correctly.

3.3.2 Gastric Vascular Changes

Vascular processes include a short number of conditions that are summarized below (Box 3.8).

> **Box 3.8 Abnormal Vascular Conditions of the Stomach**
> 3.3.2.1. Hyperemia
> 3.3.2.2. Gastric Antral Vascular Ectasia

3.3.2.1 Hyperemia

It refers to the passive or active congestion of the blood vessels of the stomach, and one of the most frequent conditions is passive congestion due to backflow from the supragastric organs.

3.3.2.2 Gastric Antral Vascular Ectasia

Gastric antral vascular ectasia is very rare in youth, but genetic syndromes may be a predisposing factor.

- *DEF*: Distinctive pattern of gastric changes with an associate distinctive endoscopic appearance.
- *SYN*: Watermelon stomach (from the endoscopic appearance).
- *CLI*: >40 years, but younger patients have been reported, ♂<♀, Fe deficiency anemia, and there is an association with autoimmune (AI) disorders and connective tissue disorders (CTD) (autoimmune atrophic gastritis, hypothyroidism, primary biliary cirrhosis, Reynaud phenomenon, sclerodactyly, and CREST syndrome).
- *IMG*: Parallel longitudinal red stripes in antrum are converging on pylorus ("Watermelon stomach") on endoscopy.
- *CLM*: Mucosal vascular capillary ectasia ± tortuous blood vessels in the submucosa and fibrin thrombi, CD62 positive, in the capillaries without evidence of a vascular malformation. There is also edema, hemorrhage, and fibromuscular hyperplasia of lamina propria with a tendency to hyalinize and scattered chronic inflammatory cells. As a consequence of it, atrophic gastritis with intestinal metaplasia and endocrine cell hyperplasia may be present.
- *TRT*: Ablation of blood vessels may help.

3.3.3 Gastric Inflammatory Diseases

Gastritis is the inflammation of the stomach. It can be acute or chronic as well as having some specific characteristics (specific gastritis) (Box 3.9) (Figs. 3.10 and 3.11).

> **Box 3.9 Gastric Inflammation**
> 3.3.3.1. Acute Gastritis
> 3.3.3.2. Chronic Gastritis
> 3.3.3.3. Specific Gastritides

3.3.3.1 Acute Gastritis
- *DEF*: Acute mucosal inflammation, usually transient, with a frequent continuum including acute gastritis, acute hemorrhagic gastritis, acute erosive gastritis, and severe stress erosions. The inflammation is typically limited to the mucosa, unless a specific underlying disorder is recognized (e.g., Crohn disease).
- *EPG*: PICU (pediatric intensive care unit) stress determinants (e.g., shock), prolonged corticosteroids, excessive alcohol (EtOH) consumption ("18th birthday or initiation parties"), chronic use/abuse of NSAIDs or aspirin, rarely smoking (e.g., tobacco, marijuana) and vaping.
- *CLM*: The process involves typically only partial erosion of the mucosa, and there is no penetration of the *muscularis mucosae*, but there is often focal hemorrhage into the mucosa, which is highlighted by erythrocytes' extravasates. Severe forms (acute hemorrhagic erosive gastritis) are characterized by sloughing of the mucosa and extensive bleeding in the *lamina propria*.
- *PGN*: If severe and/or cofactors are associated (screen PICU stress determinants), mortality can easily exceed 50% in both childhood and youth.

3.3.3.2 Chronic Gastritis
- *DEF*: Infiltration of lamina propria by lymphocytes and plasma cells combined with varying degrees of gastric atrophy and intestinal metaplasia. Many classification schemes are used, but the most common is the following listed in Box 3.10. Overall, the ABC Classification is the most used worldwide. Although rare in pediatric patients, pernicious anemia caused by vitamin B_{12} deficiency may be encountered.

In the Sydney classification, ≥2 biopsies each from the fundus and antrum need to be taken from the endoscopist. The diagnosis relies on:

- *Etiology*: (if known), e.g., autoimmune, bacterial, and chemical (A/B/C)
- *Type*: acute, chronic, and special (granulomatous, eosinophilic)
- *Distribution*: antral, fundic, or pangastritis
- *Degree of variables*:
  - *Inflammation*, activity (mild = 1/3 pits, moderate = 1/3–2/3 pits, severe > 2/3 pits)
  - *Metaplasia* (present/absent)
  - *Atrophy* (present/absent)
  - *Helicobacter pylori* (*Hp*) (present/absent) and density (+/++/+++)

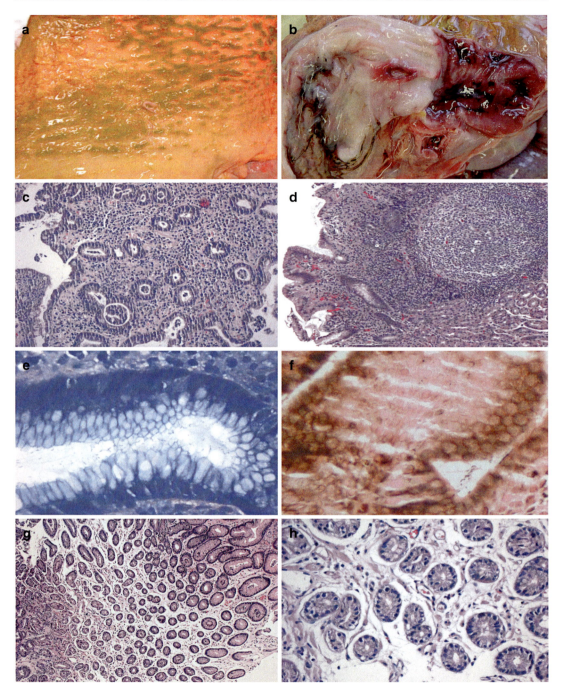

Fig. 3.10 In (**a**) is shown septic shock erosive gastritis. In (**b**) is shown a stress-ulcerative gastroduodenitis in a 1-year-old patient affected with hypoplastic left heart syndrome who underwent a series of palliative cardiac surgical procedures. In (**c**, **d**), there is evidence of a chronic, active *Helicobacter pylori* gastritis with lymphofollicular hyperplasia (**c**, H&E stain, ×200; **d**, H&E stain, ×100) with identification of the bacteria using Giemsa stain (**e**) and Warthin-Starry stain (**f**) at ×630 as original magnification. In (**g**, **h**) are two microphotographs of a C (chemical)-gastritis with impressive edema of the interstitium and retronuclear vacuoles in a young patient with NSAID exposure (**g**, H&E stain, ×100; **h**, H&E stain, ×400)

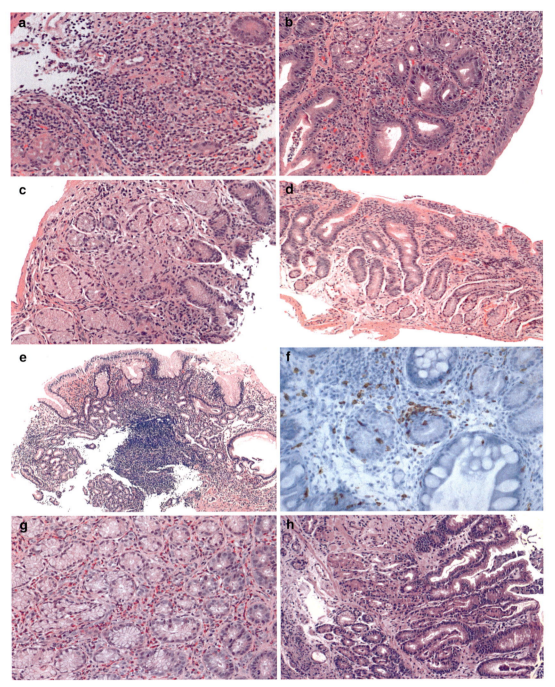

Fig. 3.11 In (**a**) is a focally enhanced gastritis (FEG) in a patient with Crohn disease, while (**b**) shows a erosive Crohn gastritis. FEG is more frequently found in patients with Crohn disease than patients with ulcerative colitis. In (**c**) is shown a granulomatous gastritis in a patient with Crohn disease. In (**d**) is shown a lymphocytic gastritis. In (**e**) is shown an autoimmune gastritis with CD8 attack of the glandular epithelium (**f**) (Anti-CD8 immunohistochemistry). In (**g**) is shown an eosinophilic gastritis, while in (**h**) there is a gastric ulcer with regenerative hyperplasia (H&E stain for all apart **f**, ×200, ×200, ×200, ×100, ×100, ×400, ×200, ×200 as original magnification)

Box 3.10 Classification of Chronic Gastritis
Type A, aka autoimmune gastritis, which involves mainly fundic mucosa with usually sparing of the antrum and patients harboring circulating antibodies to parietal cells and/or intrinsic factor, common hypo-/achlorhydria, loss of parietal cells due to the autoimmune destruction, and hypergastrinemia. In 10% of adults, it may develop pernicious anemia after several years. A-Gastritis is associated with other autoimmune disorders (Hashimoto thyroiditis, Addison disease).

Type B, aka nonimmune gastritis, which is associated with *Helicobacter pylori* infection and does not have associated parietal cell antibodies or pernicious anemia. Intestinal metaplasia may be present but does not show any bacterium on the surface epithelium.

- Hypersecretory subtype: Antrum-restricted, minimal fundic atrophy, hyperacidity with often duodenal ulcers, and normal gastrin levels.
- Environmental subtype: Fundus and antrum-involving inflammation with a strong association with atrophy, metaplasia, and carcinoma.

Type C, aka chemical gastritis, caused by bile reflux has a characteristic appearance like NSAIDs or alcohol-related gastritis. Bile reflux should be probably considered a promoter of gastric carcinogenesis in the postsurgical stomach and an appropriate follow-up is required.

3.3.3.3 Specific Gastritides

Lymphocytic Gastritis
- *DEF*: Morphologic endpoint of gastritis with several and different etiologies characterized by an increase of *IELs > 25/100 epithelial cells* in the surface and upper foveolar epithelium (consecutive fields read using a 400× magnification: ocular ×10 and objective ×40) without taking into consideration any increase of inflammatory cells of the lamina propria, which should be assessed as nonspecific.
- *CLI*: Endoscopically, the stomach may exhibit varioliform appearance, nodularity, hypertrophic folds, and aphthous erosions.
- *EPG*: Infection (e.g., *Hp*), celiac disease, autoimmune.
- *CLM*: IELs are not difficult to count because their nuclei are smaller than the nuclei of the epithelial cells and scant cytoplasm is seen. The inflammatory glandular involvement, i.e., glands containing inflammatory cells, should also be counted of a fixed total number of superficial glands and calculated as percent rate of the involved glands (optional). Moreover, the distribution and number of lymphoid aggregates differentiating between the presence and not germinal centers in the antrum should be provided in the final pathologic report.
- *IHC*: Monoclonal antibodies against CD8 (T lymphocytes, mostly suppressor/cytotoxic) and CD20 (B lymphocytes) are useful to evaluate the cellular and humoral immune response, respectively. CD8+ cells are characteristically higher in GSE-associated LG than Hp gastritis. Additional optional staining may include monoclonal antibodies against CD4 (T lymphocytes, mostly helper). CD57 (NK lymphocytes) and CNA42 (follicular dendritic cells), although they are not particularly useful, because NK activity is negligible in both Hp- and GSE-associated gastritis and the process of antigen presentation occurs in the lymph node.

It is essential to remind in this setting some pearls and pitfalls. In particular, the dominant cytokine response to Hp and GSE has a T-helper 1 (Th-1)-type profile. The pathogenetic relationship between three conditions associated with inflammation of the stomach, including Hp gastritis, the gastric involvement of celiac disease, and lymphocytic gastritis, is still not completely clarified. Gastric Hp presence does attenuate duodenal lesions of celiac disease patients. In pediatric patients, LG has CD3, CD7, and CD8 (+) IELs, while Hp gastritis has CD3, CD4 (+)

IEL, and prominent lymphoid aggregates with an increased number of mononuclear cells (B and T lymphocytes, plasma cells, MΦ, and mast cells) in the lamina propria. B-cell proliferation is driven by activated T lymphocytes (CD4+) that progressively activate mononuclear phagocytes. In young patients with detection of Hp and suspicion of atypical celiac disease (negative serology with increased IELs and normal villous-to-crypt ratio – Marsh I) with a differential between gastric involvement of celiac disease and Hp-caused gastritis, two factors may play a significant role, including the number of *CD8+ IELs* in the antrum and the number of *lymphoid follicles*. The first would point to a gluten-sensitive enteropathy with gastric involvement, while the latter would direct toward the infective etiology.

Granulomatous Gastritis

Gastritis with granulomas is usually observed in Crohn disease but also in sarcoidosis or as an isolated disorder or following an infectious disease, such as TB and mycosis. The inflammatory involvement may extend beyond the borders of the *muscularis mucosae*.

Eosinophilic Gastritis

Gastritis with infiltration of the mucosa and sometimes submucosa by numerous eosinophils usually targeting the distal stomach and proximal duodenum with or without necrotizing vasculitis and probably allergic reaction to the ingested material. Currently, an average density of ≥ 127 eosinophils/mm^2 in ≥ 5 separate fields (≥ 30 peak eosinophils/400× HPF) in gastric biopsies with no known causes of eosinophilia has been proposed for the definition of histological eosinophilic gastritis. In case of small biopsies, three separate fields may suffice to the diagnosis of eosinophilic gastritis. Also, an abnormal eosinophil localization in the surface or foveolar epithelium, muscularis mucosae, or submucosa in the presence of (1) mucosal damage (e.g., foveolar hyperplasia), (2) architectural distortion, and (3) significant chronic (lymphocytic/lympho-plasmacellular) or active (neutrophilic) inflammation has also been advocated as essential clues for the diagnosis of eosinophilic gastritis (Caldwell et al. 2014).

Helicobacter heilmannii Gastritis

It is a rare form of bacterial gastritis caused by *H. heilmannii* (previously known as *Gastrospirillium hominis*), which can be found in primates, cats, pigs, and carnivorous mammals. A transmission to humans usually occurs from domestic animals (Box 3.11).

Box 3.11 *Helicobacter heilmannii* Versus *Helicobacter pylori*

H. heilmannii: long and straight (4–10 μm × 0.5–0.9 μm) bacterium with a corkscrew morphology usually not attached to the surface epithelium

H. pylori: short and narrow (3.5 μm × 0.5–1.0 μm) bacterium with a curvilinear, "seagull" morphology, usually attached to the surface epithelium

---

- *EPI*: Rare in children (0.3–0.4 of children undergoing esophago-gastro-duodenoscopy (EGD) ≠ *H. pylori* gastritis, which is up to 9% using such procedure).
- *CLI*: Symptoms are similar to those of *H. pylori* gastritis and include epigastric abdominal pain, with or without nausea and vomiting, heartburn, dysphagia, and acute upper GI bleeding. Endoscopy shows antral erythema, thickened folds, gastric erosions, gastric and duodenal ulcers, esophagitis, and micronodular transformation of the antrum.
- *CLM*: Mild chronic gastritis, conversely from *H. pylori*, only occasionally causes peptic ulceration, gastric carcinoma, or MALT lymphoma. There is chronic gastritis with lympho-plasmacytic infiltrate, foveolar hyperplasia, vasodilation, edema of the lamina propria, an increase of the intracytoplasmic mucin of the epithelium, and occasionally foci of active inflammation with neutrophils. The microorganism may be detected by histochemical special stains (HSS) other than the routine hematoxylin and eosin (H&E) staining. HSS include Giemsa (for bacteria), Warthin-Starry or other silver stains, and cresyl violet stains. Immunohistochemistry can also be used and

several centers use the same polyclonal antibody used to detect *H. pylori* (cross-reactivity), which is linked to a rapid urease test (+). Finally, sub-specific characterization using FISH and partial 16S ribosomal DNA sequencing has been used.
- *TRT*: H2 blockers or PPI and two antibiotics of choice (clarithromycin, amoxicillin, and metronidazole) or bismuth compounds.

3.3.4 Tissue Continuity Damage-Related Gastric Degenerations

The lesions recognized under the umbrella of the degenerations related to the tissue continuity damage are described in Box 3.12.

3.3.4.1 Gastric Ulcers

Stress Ulcers
- *DEF*: Multiple small lesions, mainly gastric (sometimes duodenum) with shedding of the superficial epithelium to full-thickness ulceration occurring in the setting of "severe stress," e.g., shock, sepsis, severe trauma, etc.
- *ETP*: There are three typical ulcers, including:
  – Cushing ulcers: ↑ intracranial pressure with ↑ vagal tone and ensuing ↑ acid secretion
  – Steroid ulcer: following steroid use
  – Curling ulcers: following extensive burns

Peptic Ulcers
- *DEF*: Chronic, usually solitary lesion (80%) occurring at any level of GI tract exposed to acid-peptic juices with preferred sites: first portion duodenum (~80%), antrum of the stomach (~20%), but also Barrett metaplasia and Meckel diverticulum.
- *EPI*: ♂:♀ = 3:1 (duodenal); 1.5–2:1 (gastric).
- *CLI*: Nearly all patients experience pain after eating, concurrent chronic antral gastritis (e.g., *Helicobacter pylori* gastritis), ↑ of both the basal and stimulated level of acid secretion, and fast gastric emptying.
- *GRO*: Oval, sharply delimited defect with a tendency for overhanging mucosal margins with minimal if any heaping up of margins (≠ in carcinoma) with gastric folds radiating out from an ulcer.
- *PGN*: Remitting and relapsing course. Fatal in 5% (e.g., perforation → bleeding → hemorrhagic shock → death). 1–3% will develop gastric cancer (adenocarcinoma).

3.3.4.2 Gastric Hyperplasia

It is a condition, which is neither hypertrophic nor inflammatory but hyperplastic. It corresponds to giant, cerebriform enlargement of the rugal folds, and radiographically, it can be easily confused with malignant lymphoma.

Menetrier Disease
- *DEF*: Hyperplasia of the surface mucous (foveolar) cells with tortuous, corkscrew cystic dilatation extending to the base of the glands with two different gastric patterns, including:
  1. *Polyadenomes polypeux* (probably equivalent to multiple hyperplastic polyps)
  2. *Polyadenomes en nappe*, which genuinely corresponds to Menetrier disease
- *SYN*: Hypertrophic/hyperplastic gastropathy, giant hypertrophic gastritis, giant hypertrophy of gastric rugae.
- *ETP*: Unknown, idiopathic.
- *CLI*: Hypochlorhydria (chronic and severe) and chronic protein loss ⇒ hypoproteinemia. It is usually progressive with gastric antrum typically uninvolved and presenting with nausea, anorexia, epigastric pain, and diarrhea.
- *GRO*: Marked proliferative change with zenith along the greater curvature and characterized by markedly hypertrophic, *gyri-like rugae*

Box 3.12 Tissue Continuity Damage-Related Degeneration
3.3.4.1. Ulceration
3.3.4.2. Hyperplasias
3.3.4.3. Metaplasias
3.3.4.4. Metabolic/Degenerative (Amyloidosis, Hemochromatosis)
3.3.4.5. Dysplasias

with almost pathognomonic lack of antral involvement.
- *CLM*: Histological triad including:
  1. Foveolar hyperplasia, with tortuous pits, some degree of cystic dilatation, and extension into the base of the glands or occasionally even beyond the muscularis mucosae
  2. Atrophy of the underlying secretory (glandular) component (mild to moderate)
  3. Edema and inflammation of the surrounding stroma
- *DDX*: Malignant lymphoma, carcinoma.
- *PGN*: It harbors a slightly increased risk of carcinoma.

Zollinger-Ellison Syndrome
- *DEF*: Hyperplasia of the glandular cells with parietal cells tending to overwhelm the chief cells and potential proliferation of the enterochromaffin-like cells secondary to gastrin-secreting tumor due to MEN-I (1/5 of cases) (gland lumen size is average; no cyst formation).
- *SYN*: Hypertrophic hypersecretory gastropathy.
- *CLI*: Hyperchlorhydria ± hypergastrinemia with two variants ("protein-losing" mimicking Menetrier disease and non-protein-losing with a more characteristic histology pattern as described above).

3.3.4.3 Metaplasias
- *Pyloric Metaplasia*: Age-dependent fundic gland mass ↓ and replacement by pyloric glands.
- *Ciliated Cell Metaplasia*: Replacement by mucosa with ciliated columnar cells.
- *Intestinal Metaplasia*: Replacement by mucosa with goblet cells, absorptive cells, Paneth cells, etc. Both may be either complete or incomplete:
  - Complete (type I): Sialomucin predominates (goblet cells).
  - Incomplete (type II): No absorptive cells, but gastric foveolar cells retained.
- Type IIA: Foveolar cells contain neutral mucins.
- Type IIB: Foveolar cells contain sulfomucins.
- If IM is of incomplete type, special stains are needed to detect the different staining properties of the mucins (e.g., Alcian PAS and Alcian 2.5).

3.3.5 Gastric Tumors

Gastrointestinal (GI) tract malignancies are an infrequent diagnosis in childhood, and less than 5% of pediatric cancers account for this kind of malignancy in the USA and Canada as well as in the most industrialized countries (Box 3.13) (Figs. 3.12, 3.13, 3.14, and 3.15).

3.3.5.1 Benign Epithelial Neoplasms
Polyp: Any lesion that projects into the gastric lumen above the plane of the mucosa. Gastric polyps are identified in ~5% of upper endoscopies of combined children and adults, although the finding of gastric polyps in children is relatively low. Gastric polyps can be classified histogenetically according to the cell of origin in epithelial, mesenchymal, lymphoid, and miscellaneous. Epithelial polyps may be subclassified in fundic gland polyp, hyperplastic polyp, adenoma, neuroendocrine tumor, polyp or polypoid carcinoma, and polyp or polypoid metastatic carcinoma. Some tumors are increasingly identified in children and young adults, because of the screening due to genetic or familial diseases to detect dysplasia at an early stage. From the statistical point of view, hyperplastic polyps are the

---

Box 3.13 Gastric Tumors
3.3.5.1. Benign Epithelial Neoplasms
3.3.5.2. Malignant Epithelial Neoplasms
3.3.5.3. Neuroendocrine Tumors
3.3.5.4. Lymphoepithelial Lesions
3.3.5.5. Stroma Tumors and Smooth Muscle Differentiating
3.3.5.6. Nonstroma/Smooth Muscle Differentiating Mesenchymal Tumors
3.3.5.7. Secondary Tumors

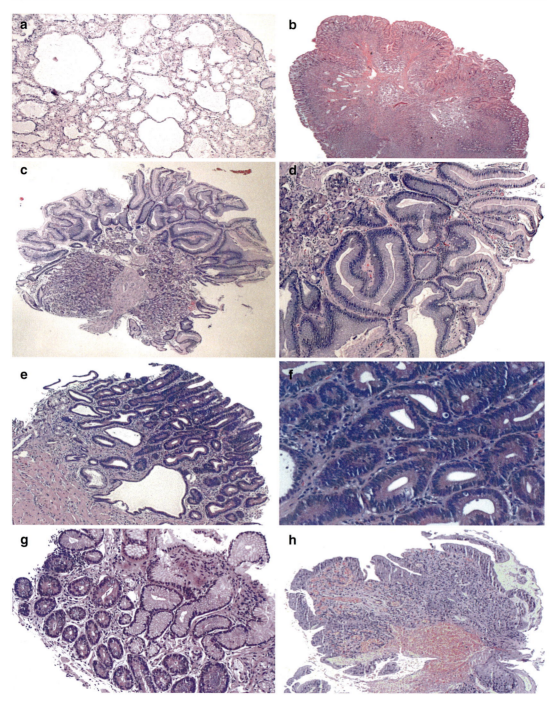

Fig. 3.12 In (**a**) is shown a fundic gland polyp. In (**b**) is shown a hamartomatous polyp. In (**c**, **d**) are shown two microphotographs of a hyperplasiogenic polyp of the stomach, while (**e**–**g**) show low-grade dysplasia of a tubular adenoma of the stomach with hyperchromasia of the nuclei, pseudostratification of the nuclei, and decrease of mucus cells. In (**h**) is shown an inflammatory cardia polyp of the stomach

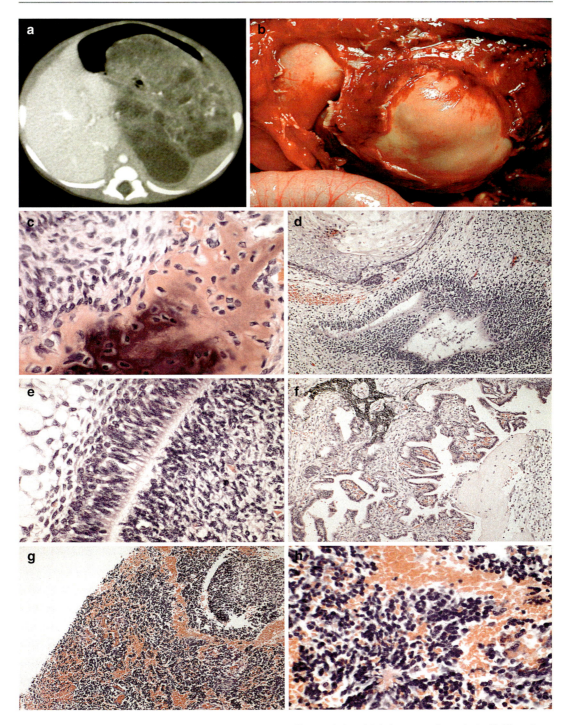

Fig. 3.13 This panel illustrates a gastric teratoma with the presence of neuroblastic tissue (**a**, imaging; **b**, gross photograph). In (**c**) is shown a site of membranous ossification, while (**d**) shows neuroepithelium of a primitive tube (immature teratoma). Pseudostratified ciliary epithelium and choroidal plexus are shown in (**e**, **f**). The microphotographs in (**g**, **h**) show neuroectodermal tissue portend to a diagnosis of neuroblastoma, which was synaptophysin positive and chromogranin A positive other than S100 and other neural markers (not shown)

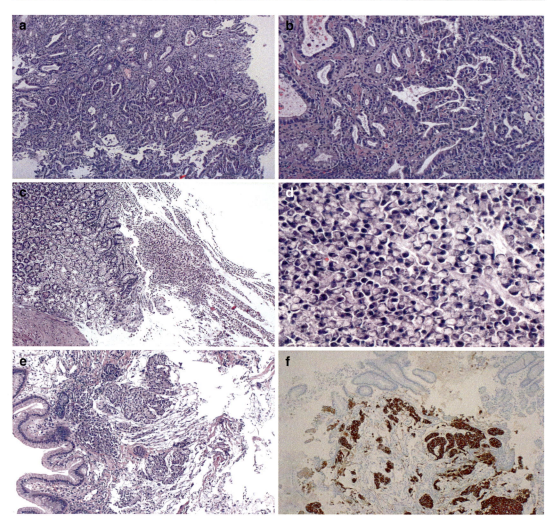

Fig. 3.14 In (**a**, **b**) are shown microphotographs of an intestinal-type gastric carcinoma, while in (**c**, **d**) are shown microphotographs of a diffuse-type gastric carcinoma using the Lauren classification. In the gastric carcinoma of intestinal type, detectable glands that range from well-differentiated to moderately differentiated tumors are seen, while in the diffuse type, poorly cohesive cells diffusely infiltrating the gastric wall with no or little glandular formation are seen. The signet ring cell type of the WHO classification corresponds to the diffuse type. In (**e**, **f**) are shown microphotographs of a carcinoid of the stomach with chromogranin A immunostain (**f**) and synaptophysin immunostain (not shown)

most often recognized polyp followed by the adenomatous polyp and others.

Hyperplastic (85%) (aka inflammatory, regenerative) polyps represent an exaggerated regenerative response to injury and are often multiple, sessile, <1 cm, and randomly distributed above the plane of the gastric mucosa. Microscopic features comprise elongated and distorted glands, tubules, and microcysts with a single layer of healthy cells, located predominately in the foveolar region. Polyp stroma shows some edema and inflammation and possibly patchy fibrosis. It is essential to rule out an early gastric carcinoma in these patients.

Adenomatous (10%) is a neoplastic lesion similar to that observed in other localizations of the GI tract. It is usually single, located in the antrum, pedunculated, and up to 3–4 cm. Microscopically, these polyps are characterized by the adenomatous change of the epithelium,

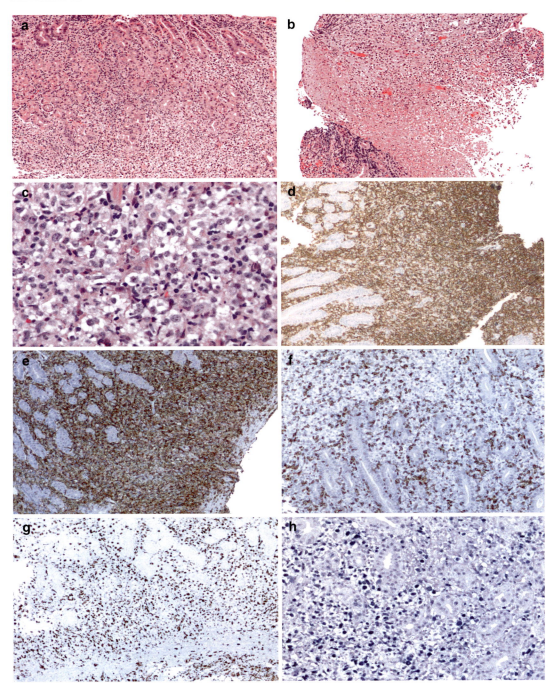

Fig. 3.15 Posttransplant lymphoproliferative disorder (PTLD) can complicate organ transplantation and can occur in several organs; among them the stomach may be the target. In (**a**–**c**), a monomorphic PTLD with positivity for CD20, a B-cell marker (**d**), CD43 (**e**), but virtual negativity for CD3 (**f**), high proliferation rate with anti-Ki67 (MIB1) (**g**), and positivity for EBV (EBER in situ hybridization) (**h**). In (**f**), the positive cells are some local T-lymphocytes identified by the antibody against the CD3 epitope. In (**h**), the dark blue dots indicate the RNA sequences hybridized in situ of EBV

including hyperchromasia of the cells, loss of mucin content, and architecture of closely packed gland-like tubular structures or villi, and may also arise in regions with intestinal metaplasia. There are tubular, villous, and tubule-villous varieties, as seen in the small and large intestine. There is a variable risk of synchronous and metachronous gastric adenocarcinomas described in the literature among different studies: 4–50% and 3–75%, respectively. Adenomatous polyps are dysplastic and represent a type of gastric dysplasia, which is subclassified in adenomatous, foveolar, pyloric gland adenoma, and tubule neck dysplasia.

Other benign epithelial polyps include fundic gland polyp, inflammatory fibroid polyp, hamartomatous, and polypoid mucosal prolapse. *Fundic gland polyps* are multiple small polyps with cystically dilated oxyntic glands forming microcysts lined by foveolar-type fundic epithelium. They can be sporadic and be associated with β-cat gene mutations or FAP associated with APC gene mutations. They have also been observed following PPI drug use.

Inflammatory fibroid polyps, as seen in the esophagus, can occur anywhere in the GI tract, and 3/4 of them are located in the antrum of the stomach. According to the size, a presentation with pain and/or obstruction has been reported. Histologically, there is a proliferation of loose connective tissue in the submucosa with small blood vessels surrounded by hypocellular stroma and variable inflammatory cells underneath a mucosa, which may be "stretched" with possible superficial ulceration. IFPs (CD34+, c-kit/CD117−) should be kept in the differential diagnosis with GIST (CD34+, c-kit/CD117+).

Hamartomatous polyps are subclassified as Peutz-Jeghers polyps or *STK11* gene-mutated polyps, Cowden/Ruvalcaba or *PTEN* gene-mutated polyps, and juvenile (retention) polyps or *SMAD4/BMPR1A* gene-mutated polyps.

Finally, further consideration should be paid to the condition called *polypoid mucosal prolapse*, which may occur at the gastroenterostomy stoma and has similar histologic features as hyperplastic polyps. Box 3.14 shows a recapitulating classification of nonepithelial gastric polyps.

Box 3.14 Nonepithelial Gastric Polyps

Mesenchymal gastric polyps

- Inflammatory fibroid polyp
- Gastrointestinal stromal tumor
- Inflammatory myofibroblastic tumor
- Smooth muscle tumors (e.g., leiomyoma)
- Glomus tumor
- Neural cell tumors (schwannoma, ganglioneuroma, granular cell tumor)
- Lipidic tumors
- Vascular tumors

Lymphoid polyps

- Polypoid gastritis
- Lymphoid hyperplasia
- Lymphoma

Miscellaneous

- Xanthoma
- Pancreatic heterotopia
- Brunner gland adenoma
- Granulomas
- Amyloid
- Langerhans cell histiocytosis
- Nonepithelial metastatic tumor (e.g., PNET)

3.3.5.2 Malignant Epithelial Neoplasms

Hereditary cancer syndromes may play a significant role in the occurrence of carcinoma of the stomach in both childhood and youth. As we will see below, the diffuse type is more frequent in the young than the intestinal form, which is more characteristic of the middle-aged and older population. The WHO classification subdivides the malignant epithelial neoplasms in tubular, papillary, mucinous (>50% extracellular mucin pools), and signet ring cell type (>50% signet ring tumor cells). In hereditary diffuse gastric carcinoma (HDGC), multiple foci of intramural signet ring cell carcinoma show a preference for the body/antral transition zone, which needs to be carefully sampled. The overall incidence in most industrialized countries has been steadily dropping over past several years, but the rate of carcinoma in the cardia region of the stomach is conversely increasing with about half of gastric cancers in white men being localized in the cardia. In 1993, McGills et al. reported 17 cases of primary gastric adenocarcinoma in patients under 21 years of age and mainly single case reports had been published since. In children, the majority of gastric carcinomas occur in the antrum, the esophagogastric junction, and the greater or lesser curve. Overall, the histological diagnosis is adenocarcinoma, linitis plastica, or poorly differentiated signet ring cell carcinoma. Ataxia-telangiectasia, IgA deficiency, and thymic dysplasia (lack of Hassal bodies) ± cerebellar atrophy are the pathologies that can occur in combination with pediatric gastric carcinoma.

Adenocarcinoma
- *DEF*: Malignant epithelial tumor-forming autonomous atypical glandular proliferations associated or not to some relevant genes/syndromes (Box 3.15) (Haller JA (1987), Lauren (1965), Asherson (1979), Lyle HH. VIII (1911), Agnes (2017), CEBM, Centre for Evidence-Based Medicine).
- *EPI*: In general terms, it represents 85–90% of gastric malignancies with increased incidence unusually high in Japan, Chile, Scotland, and Finland, while it is lower in the USA, the UK, Canada, and Greece. RF: dietary/environmental factors (e.g., nitrogen compounds), chronic active gastritis with intestinal metaplasia, pernicious anemia, gastric adenomatous polyps, post-gastrectomy, gastro-stump, Menetrier disease, and Epstein-Barr Virus (EBV) (cardiac carcinoma).

> **Box 3.15 Most Relevant Genes/Syndromes Associated with Gastric Cancer**
> *Intestinal*-type adenocarcinoma: HNPCC and FAP
> *Diffuse*-type adenocarcinoma: CDH1 and BRCA2
>
> Note: *HNPCC*, hereditary non-polyposis colon carcinoma; *FAP*, familial adenomatous polyposis; *CDH1*, E-cadherin gene; *BRCA2*, Breast Cancer 2

- *CLI*: Hypochlorhydria accompanies gastric adenocarcinoma in 85–90% cases, and chronic atrophic gastritis is usually present, and the lesser curvature is involved three times more frequently than greater curvature.
- *CLM: Lauren Classification*
  - *Intestinal* Type (aka *expanding carcinoma*): Relatively cohesive nodular, polypoid, or ulcerated, well-circumscribed mass of tumor cells with characteristic pushing margins.

    Microscopically, tumor cells may arise from metaplastic epithelium and show foveolar cells, intestinal columnar cells, and/or goblet cells. Its incidence is relatively decreasing when compared to the diffuse type. It is more prone to give liver metastases than diffuse type.
  - *Diffuse* Type (aka *infiltrative carcinoma*): Diffuse infiltration, which offers a hard aspect to the stomach (*linitis plastica*) and may be challenging to recognize grossly. Microscopically, the individual infiltrating tumor cells show intracellular mucin vacuoles, often forming signet ring cells with a stroma characterized by severe inflammation and desmoplasia. Its incidence is slightly increased due to screening programs for families harboring genetic defects. The term "linitis plastica" was coined by William Brinton (1823–1867) in 1854 and derives from two Latin words (*linitis* recalling the similar visual aspect of the hypertrophic submucosa resembling fibers of linen., while *plastica* meaning inelastic).

    Metastases are often into the peritoneum, lung, adrenal gland, and ovary (Krukenberg tumor).
  - *Mixed* Type: A combination of intestinal and diffuse type.
- *PGN*: Often present in advanced stage, gastric adenocarcinoma may present with metastases with only about half the tumors easy to resect at diagnosis, although it depends on the age, being youth most frequently harboring a worrisome prognosis. In about 80–90%, there are local LN metastases at the time of diagnosis. Peculiarly, Virchow node is the isolated metastasis to left

supraclavicular LN (also called Trousseau sign). The overall 5-YSR is still 5–15%, although several approaches have been established. The early form of gastric adenocarcinoma (superficial spreading carcinoma) represents 10–35% of gastric carcinomas and is limited to mucosa and submucosa with less than 1/5 harboring LN metastases and with 5-YSR of 80–95%. Some features may help in identifying good and bad behavior in gastric tumors (Box 3.16). In addition to the four mentioned poor prognostic features described in Box 3.16, some authorities also suggest other characteristics, including young age, deep invasion, positive margins, LN invasion, and vascular invasion.

Other Carcinomas

Another histologic type of carcinomas is infrequent and seems to not play any role in childhood. They are also rare in adults constituting less than 1% of gastric cancer and include adenosquamous carcinoma, squamous cell carcinoma, mucinous carcinoma, hepatoid adenocarcinoma, and parietal gland carcinoma. The mucinous gastric carcinoma (MGC) seems to have a worse prognosis at first glance. The prediction of MGC patients is worse compared to that of non-mucinous gastric carcinoma (NMGC) patients, as the former group consists of more advanced-stage cases. When patients with similar stages of disease are compared, the incidence of peritoneal metastasis is significantly higher among MGC patients, but hepatic metastasis is significantly more often observed in NMGC patients. Although MGC is rarer and mostly detected at an advanced stage, the diagnosis of the mucinous histological subtype is not an independent prognostic factor. Another subtype which is gastric carcinoma with neuroendocrine differentiation is described in the next paragraph. Medullary carcinoma is a lymphoepithelial-like carcinoma, poorly differentiated with prominent lymphoid stroma, and pushing borders and composed of cells arranged in a syncytial fashion with vesicular nuclei, EBV (+), and harboring a better outcome than other subtypes. It is usually identified as a neoplasm of the older individuals, of intestinal type, and is characterized by an expanding growth pattern and low TNM at the time of oncological staging.

3.3.5.3 Neuroendocrine Tumors

This term replaces the term carcinoid tumor, which has repeatedly been criticized in the scientific literature. The term carcinoid should not be included in the pathology report as diagnosis, but apparently can be inserted in the "comment." Neuroendocrine tumors (NETs) and neuroendocrine carcinomas (NEC) can arise in many organs of the body, but most of the research and studies have been conducted on neoplasms of the gastrointestinal tract, pancreas, and lung. In this chapter, NETs and NECs, as well as neoplasms, may be interchangeable terms, being mostly used by different corporation bodies or institutions across Europe and worldwide. However, the WHO classification of the tumors of the digestive tract (2010) reserves the term NET for the first two grades only and NEC for the highest grade. Although originally thought tumors are derived from the embryonic neural crest, it has been proved that NETs recapitulate cells of *endodermal origin* and have been classified as *endocrine*. However, the simultaneous expression of neural markers and epithelial markers let retain the term neuroendocrine, which has been emphasized in the WHO classification of the digestive system (2010). In the case of metastasis of unknown origin, the organ of derivation may be identified using specific markers, such as TTF1 for lung, CDX2 for intestines or pancreas, and PDX1 or

---

Box 3.16 Behavior Characteristics of Gastric Tumors

Characteristics that portend a *good* outcome

- Small size, distal gastric involvement, pushing margin, and intestinal type

Characteristics that portend a *poor* outcome

- Large size, proximal gastric involvement, infiltrative margin, and diffuse type

ISL1 for the pancreas. The term *differentiation* indicates how the neoplastic cells look like their non-neoplastic counterparts. In well-differentiated NETs, there is an organoid arrangement of the tumor cells with characteristic patterns, including nests, trabecular, or gyriform. The neoplastic cells are characterized by relative uniformity with the production of neurosecretory granules, which can be visualized as membrane-bound granules with a dense core and peripheral halo on TEM and expressing chromogranin A and synaptophysin by immunohistochemistry. Poorly differentiated NETs show less similarity to the cells of origin, and the growth pattern is less organoid but more sheetlike or diffuse. Both TEM and IHC would probably work hard in some cases to demonstrate neuroendocrine differentiation. The differentiation between argyrophilia and argentaffinity is paramount in the study of neuroendocrine tumors (Box 3.17).

> **Box 3.17 Argyro-philic vs. Argent-affinic**
> - *Argyro-philic*: Tumor able to take silver (Ag) up, but unable to reduce it.
> - *Argent-affinic*: Tumor able to take silver (Ag) up and reduce it.
> - *Argyro-philic* stain: Grimelius.
> - *Argent-affinic* stain: Fontana-Masson.
> - *Foregut* NETs are argyrophilic-positive, but argentaffin-negative.
> - *Midgut* NETs are argyrophilic-positive and argentaffin-positive.
> - *Hindgut* NETs are often argyrophilic-positive, but argentaffin-negative.
> - *Foregut*: (+) NSE, CGA, low 5-HT, multihormonal, ± carcinoid syndrome, + bony MTX.
> - *Midgut*: (+) NSE, CGA, high 5-HT, multihormonal, ++ carcinoid syndrome, ± bony MTX.
> - *Hindgut*: (+) NSE, CGA, (±) 5-HT, multihormonal, ± carcinoid syndrome, + bony MTX.
>
> Note: NETs, neuroendocrine tumors; NSE, neuron-specific enolase; CGA, chromogranin A; 5-HT, 5-hydroxytryptamine

The chromogranins (A, B, and C) are acidic polypeptides that are the primary component of the secretory granules, and neuron-specific enolase (NSE) is a γ-γ dimer of the glycolytic enzyme enolase, while synaptophysin (SYN) is a calcium-binding vesicle membrane glycoprotein that is singularly expressed independently of other neuroendocrine proteins. In foregut NETs, cytoplasmic granules are 180-nm (Ø) round and variable density. In midgut NETs, cytoplasmic granules are 230 nm (Ø), are pleomorphic, but have uniform thickness. In hindgut NETs, there are cytoplasmic granules with 190 nm in Ø, and they are round and show variable density.

The diagnosis of NET needs to be formulated using grade and differentiation, which are essential parameters to address the therapy correctly. The use of a *Minimum Pathology DataSet* or *CAP* (College of American Pathologists) *Tumor Checklist* is paramount information to be entirely included in pathology reports on NETs. Conversely, the term *grade* is linked to the biologic aggressiveness of the neoplasm, i.e., low-grade NETs have an indolent course, and high-grade NETs show a more aggressive outcome. As a general measure of understanding, well-differentiated NETs may be of low or intermediate grade, while poorly differentiated NETs are usually of high rank.

The current WHO classification identifies four categories for NETs with slightly different terminology in different medical journals. Grade: Low (G1) means neuroendocrine tumor (NET) I G1 and corresponds to the traditional entity known as "carcinoid"; Intermediate meaning NET II G2 corresponds to "atypical carcinoid tumor"; and neuro-intestinal carcinomas (NEC) or G3 are subdivided into a small cell and large cell carcinomas. Mitotic count and Ki67/MIB1 index are used to classify the grade of a NET and should be included in the pathology report supporting the diagnosis. In particular, the number of mitoses per *10 HPF* or *2 mm^2* should be determined using manual or computer-based counting in *50 HPF* of the most mitotically active areas or the highest labeling Ki67/MIB1 density. The mostly used grading system for the NET of the GIT is as follows. According to the WHO NET categories, grading is low if MI <2/10HPF AND <3% Ki67 index, intermediate if MI 2–20/10HPF OR 3–20% Ki67 index, and high if MI >20/10HPF OR >20% Ki67 index. NET G1, NET G2, and NEC of the stomach correspond to what was formerly called carcinoid tumor (also named well-differentiated endocrine tumor), atypical carcinoid tumor, and endocrine cell carcinoma (also named small cell carcinoma), respectively.

Slightly different grading systems exist for the pancreas and lung and will be highlighted in the respective chapters. In the *Minimum Pathology DataSet* or *CAP Tumor Checklist* for resection of primary tumors, some data need to be collected in the synoptic pathology report:

- Anatomic site of cancer, diagnosis, and size (three dimensions);
- Multicentricity;
- Histologic and cellular features (e.g., oncocytic, clear cell, gland-forming);
- Grade with mitotic count value (per 10 HPF in ≥50 HPF) or mean percentage of the highest labeling density (ki67/MIB1) as % on 500–2000 cells (optional immunostaining is that proposed by Volante et al. (2007) using somatostatin receptor type 2A (SSTR2A) score);
- Nonischemic tumor necrosis or other pathologies (e.g., non-neuroendocrine components);
- The extent of the depth of invasion;
- Vascular invasion;
- Perineural invasion;
- Immunohistochemical findings (CGA, SYN, MIB1);
- Lymph node metastases (positive nodes/total number of nodes examined);
- TNM staging;
- Resection margins indicating positivity, negativity, and proximity and proliferation changes or other abnormalities in non-neoplastic neuroendocrine cells.

The use of immunohistochemistry tools including chromogranin A (CGA) and synaptophysin (SYN) helps both to verify neuroendocrine differentiation and to classify endocrine neoplasms as hypergastrinemia-related or as sporadic gastrinoma. Adjacent mucosa in hypergastrinemia-related carcinomas can show changes of *linear hyperplasia* (≥5 endocrine cells in a linear arrangement), *nodular hyperplasia* (cluster of ≥5 endocrine cells with Ø < 150 μm), *endocrine cell dysplasia* (cluster of ≥5 endocrine cells with 150 μm < Ø ≤ 500 μm), and *endocrine cell neoplasia* or *carcinoid* (cluster of ≥5 endocrine cells with Ø > 500 μm).

Although considered uniform, NET may still contain some degree of variability in its architecture. The histologic patterns of NET are organoid subdivided into solid/insular, trabecular with ribbons or festoons, glandular (acinar, rosette-like, or pseudorosette), and mixed. Optional data may include peptide immunohistochemical results, if a specific clinical setting suggests a link to a functional syndrome, immunohistochemical enhancement of the endothelial cells in assessing lymphovascular invasion (e.g., CD31, D2-40), and distance of the tumor from resection margin if the cancer is within 0.5 cm. Specific data sets exist for biopsy of primary tumors, resection of metastatic tumors, and biopsy of metastatic tumors, although all of them are pretty similar.

NETs make up ~1% of all neoplastic lesions; of these, gastric NETs are <10%, and an estimated incidence rate is ~0.33/100,000. Gastric NET is typically derived from the histamine-secreting enterochromaffin-like (ECL) cell but may occasionally have a different phenotype indicating an origin from other cell types, such as serotonin-secreting EC cells, somatostatin analogs, or ghrelin cells. Gastrin/CCK2 receptor is expressed in all kind, although ECL cell tumors are classified as either gastrin dependent or independent. Gastrin-dependent tumors are associated with clinical situations inducing hypergastrinemia (e.g., CAG) and subdivided into *NET type I* (70–80%) and *NET type II* or gastrinomas/ZES (5–6%). Conversely, gastrin-independent tumors or *NET type III* (~20%) can be found with normal gastrin levels.

NET I is a characteristic of women of the 5th to the 6th decade with CAH and pernicious anemia.

NET II is a ZES-MEN-1-associated hypergastrinemia-driven tumor that may occur in the young. There are, generally, small multiple lesions, which can be heterogeneous in a determined population, show local infiltration, and harbor metastases in ~1/10 of cases. Menin is the gene product of *MEN-1*, a tumor suppressor gene, whose disruption by germline mutations and/or LOH induces a loss of proliferative inhibition, probably using retinoblastoma pathway, inactivation of cell cycle inhibitors (e.g., P16INK4A), or changes in the AKT/mTOR cellular signaling pathway. Approximately 1/3 of individuals with MEN-I develop *NET* II lesions, which may also occur in different endocrine disorders, such as hypothyroidism, diabetes mellitus, and Addison disease.

Finally, *NET III* lesions are aggressive neoplasms associated with normo-gastrinemia and occur more frequently in men of the fifth to sixth decade harboring a worrisome prognosis (5-YSR < 50%). Recently, AMACR, or alpha-methylacyl-coenzyme A racemase, which plays an essential role in the β-oxidation of branched-chain fatty acid and its derivatives and is a useful marker to detect prostatic adenocarcinoma and high-grade PIN (prostatic intraepithelial neoplasia), was found to be significantly higher in NEC (90%) and NET G2 (67%) than in NET G1 (0%). AMACR correlates with the Ki67/MIB1 index. To remember is that the mitotic rate should be reported as number of mitoses per 2 mm^2. This can be accomplished by evaluating at least 10 mm^2 in the most mitotically active part of the neoplasm. For example, if using a microscope with a field diameter of 0.55 mm, it is paramount to count 42 high power fields (10 mm^2) and divide the resulting number of mitoses by 5 to establish the number of mitoses per 2 mm^2, which is needed for the correct designation of the tumor grade. YF476 is a specific gastrin receptor (CCK2) antagonist, which inhibits acid secretion and enterochromaffin-like cell proliferation and reduces type I and possibly type II NET lesions in both size and number. In recapitulating the endocrine lesions of the stomach considering the tumor-like lesions as well, we need to consider the tumor-like lesions (hyperplasia and dysplasia); the *well-differentiated NET (carcinoid)*, including ECL cell carcinoid (types I/II/III);

EC-cell serotonin-producing carcinoid, and G-cell gastrin-producing tumor; the s*mall cell carcinoma or poorly differentiated NET*; and the *large cell carcinoma* (variant of the previous neoplastic lesion). Type III tumors can present either as a mass lesion, nonfunctioning, and clinically with gastric bleeding, obstruction, or metastasis or with endocrine symptoms characteristic of atypical carcinoid syndrome with red cutaneous flushes and lack of diarrhea, but usually coupled with liver metastases (5-HT-producing cells).

It may also be important to recall that the background mucosa is different according to the types: type I, chronic active gastritis and intestinal metaplasia; type II, hypertrophic oxyntic glands and hyperplastic parietal cells; and type III, normal.

The outcome of the tumor relies on several factors. Characteristics portend a benign behavior: mucosa (M)/submucosa (S-M)-confined tumor, Ø < 1 cm, chronic active gastritis (CAG) or multiple endocrine neoplasia (MEN)1/ZES (Zollinger-Ellison syndrome), nonfunctioning, and non-AI (angio-invasion). Conversely, features which portend a destructive behavior are muscularis propria (MP)-invading tumor (or beyond), Ø > 1 cm, sporadic, functioning, and AI. The concluding diagnosis should be made according to current WHO categories, grading according to pTNM and ENETS, and staging according to pTNM or ENETS. Moreover, the definition of the predominant cell type should include at least insulin B, gastrin G, and serotonin EC immunophenotyping as producing cells and suggestion for the primary tumor in liver and or lymph node metastases when this information is known.

3.3.5.4 Lymphoproliferative Lesions

Gastrointestinal (GI) tract lymphomas in children are rare and can be quite variable. The most common malignancy in pediatric tropical African setting remains a lymphoma, particularly Burkitt lymphoma, and often, there is intestinal involvement. In children, primary GI tract lymphomas tend to occur in the small and large bowel, unlike adults who have more frequently a gastric involvement. The GIT is the most common site of extranodal primary lymphoma, and non-Hodgkin lymphoma (NHL) is the principal diagnosis, usually involving the stomach and small intestine, as primary colonic lymphoma is relatively rare. Primary intestinal Hodgkin lymphoma is relatively rare. There are specific clinical settings that predispose to develop lymphomas, including immunologic disorders, such as *collagen vascular disease*, *congenital immune deficiency syndromes*, *acquired immunodeficiency syndrome*, and *organ-transplanted patients using immunosuppressive agents*. In general, patients with T-NHL with intestinal involvement behave, in several studies, worse than those harboring a B-NHL with intestinal localization. Data to gather for the staging of lymphomas and COG/SIOP/UKCCSG trials include tumor histology, extraintestinal lymph node involvement, the extent of metastasis, and the size of the tumor. Poor prognostic factors include advanced stage, aggressive tumor growth, tumor size >7 cm, invasion through the intestinal wall, unresectability, and *BCL-2* oncogene rearrangement. The 2-year survival rate is overall good in many series with approximately 80% for the early-stage lymphomas, but in the advanced stage, the survival rate decreases to around 30%. Thus, early diagnosis is crucial. The most frequent extranodal GI lymphomas are B-cell lymphomas. T-cell NHL includes the enteropathy-associated T-cell lymphoma, NK/T cell, nasal type, γ/δ, and anaplastic large T-cell lymphoma (Box 3.18). A pregerminal center neoplasm is the mantle cell lymphoma. Germinal center and post-germinal center lymphomas according to the WHO definition and classification are grouped in Box 3.18.

One of the most common EN GI lymphomas is diffuse large B-cell lymphoma (DLBCL). To better investigate pediatric gastrointestinal tract lymphomas, Kassira et al. have utilized the Surveillance, Epidemiology, and End Result (SEER) database, which is an outstanding program of the National Cancer Institute, U.S.A. (https://www.cancer.gov/research/areas/publichealth/what-is-seer-infographic). SEER includes 17 population-based registries, comparable to 1/4 of the population. This extensive database has been used for multiple pediatric malignancies. Kassira et al.'s object of the investigation was 265 children to evaluate the efficacy of surgical resection or radiation therapy on patient overall and disease-specific survival (DSS). In this study having as target pediatric GI malignancies, lymphomas widely predominate with similar distribution among diverse age groups. There is an overwhelming majority of boys, although there is no difference in survivals between boys and girls. Burkitt lymphoma is the predominant histological variant in their cohort of patients. An independent predictor of survival was tumor location with tumors of the small and large intestine associated with >80% of established 10-YSR, while gastric malignancies had a 10-YSR of about 60%. In the rectum and anus, malignancies exhibited the longest 10-YSR of 100%. About surgery, the role and extent of surgical resection is still a matter of controversial discussion, at least in some countries. Although Magrath promotes operative debulking for pediatric GI lymphomas before chemotherapy, Morsi et al. advocate a crucial role of surgery only in localized NHL. Secondary malignancies are common in patients treated for NHL, and appropriate follow-up about duration and frequency remains mandatory for both solid and hematologic malignancies.

Gastrointestinal B-NHL

Gastrointestinal NHL of B-cell type can be of the pre-germinal center type, germinal center type, or post-germinal center type. There are several

---

Box 3.18 GC and Post-GC EN B-Cell Lymphomas According to the Origin of the Germinal Center of the WHO Definitions

GC neoplasms	*FL*, *BL*, *HL* (nodular sclerosis)
	DLBCL (a proportion of this large group)
Post-GC neoplasms	*DLBCL* (a proportion of this large group), *MZL*, *MALT*

Note: *EN* extranodal, *GC* germinal center
Strikethrough are entities that do not appear as gastrointestinal lymphomas or have not been reported yet. See text for the abbreviations

IHC markers, pan-B type, and GC type, but three tags are most essential in distinguishing most of the usual variants, including CD5, CD23, and CD43. *CD5* is a T-cell marker expressed in small lymphocytic lymphoma and specifically in the MCL. *CD43* is another T-cell marker (pan-T-cell marker) important in T-cell and neutrophilic adhesion to endothelium, which is expressed in small lymphocytic lymphoma and mantle cell lymphoma, but also in some marginal zone B-cell lymphomas and Burkitt lymphoma. *CD23* is a crucial molecule for B-cell activation and growth, which is expressed in small lymphocytic lymphoma and some follicular lymphomas. Classic B symptoms are less prevalent than in B-NHL with nodal involvement.

Mantle Cell Lymphoma

- *DEF*: B-cell, pre-GC neoplasm made up of uniform - *monomorphic - small- to medium-sized lymphoid cells* with irregular nuclear contour and scant cytoplasm.
- *SYN*: Multiple lymphomatous polyposis.
- *EPI/CLI*: It accounts for approximately 5–10% of all GI lymphomas and occurs as polypoid mass(es) in both the stomach and small intestine. Rare in children and young individuals, MCL usually presents in middle-aged, ♂>♀, with weight loss, fatigue, diarrhea, abdominal pain, and often Fe deficiency. Most cases have sIgM, and 50% of patients have sIgD and are characterized by an *aggressive* clinical course.
- *CLM*: *Naked benign germinal centers surrounded by malignant cells* is the major histologic distinguishing feature with the absence of neoplastically transformed cells (centroblasts), para-immunoblasts, and proliferation centers. *"Cancerous" lymphoid cells usually straddle the muscularis mucosae, without involving the mucosa.* There are neither lymphoepithelial lesion nor neoplastic germinal centers, which are two important features for the differential diagnosis with other lymphomas.
- *DDX*: MALT lymphoma and other GC neoplasms, such as follicular lymphoma, which may also present as lymphomatoid polyposis (Grade 1 lesions composed of centrocytes without residual germinal centers). MCL usually does not transform to large cell lymphomas. There are four variants, including the small cell, marginal zone-like, blastoid, and pleomorphic variants. Approximately 1/4 of patients with MCL of the nodal type have GI involvement, although the neoplastic GI involvement is more often observed in the lower GI tract than the upper GI tract.
- IHC phenotype:
  - (+) CD19, CD20 (B-cell) and (+) CD5, CD43 (T-cell) and (+) Cyclin D1
  - (−) CD23 and CD10, Bcl6 (but Bcl2+)
  - t(11;14)(q13;q32) ⇒ *IGH/CCND1* ⇒ Upregulation of CCND1 gene ⇒ Cyclin D1 overcoming cell cycle suppressors (RB1, p27Kip1) (the t(11;14)(q13;q32) is a hallmark of MCL, but it has been found less often in other lymphoproliferative disorders, such as B-prolymphocytic leukemia, plasma cell leukemia, chronic lymphocytic leukemia, and multiple myeloma)
  - In CYCLIN-D1-negative MCL, the overexpression of the transcription factor SOX11 allows the differentiation of MCL from other indolent NHL.
- *TRT*: The combination of chemotherapy + rituximab/early autologous stem cell transplant (ASCT) in suitable candidates represent the mainstay of therapy for MCL.
- *PGN*: Median survival of 3–4 years is characteristic for MCL. The blastoid and pleomorphic variants have a more aggressive clinical course, and the Ki67/MIB1 index has PGN association and should be part of the pathology report.

Follicular Lymphoma

Germinal center (GC) neoplasm with extranodal GI involvement of the small intestine, particularly the *duodenum*, usually in individuals of the fifth to sixth decade (rarely younger patients), ♂=♀, may occur as multiple small polyps to lymphomatoid polyposis with usually a *grade 1 morphology-based (centrocytic) without residual germinal centers*, unlike MCL. Distinction from reactive follicular hyperplasia is paramount and includes effacement of the architecture, minimal variation in size/shape and even distribution of the follicles, poorly defined follicle borders, no polarization (CC-CB), no mantle, monomorphism, no TBM (tingible body macrophages), minimal phagocytosis, and atypical cells in interfollicular regions. TBMs are a type of macrophages often found in germinal centers, containing several to numerous phagocytized, apoptotic cells in various states of degradation, which are referred to as tingible bodies.

- IHC phenotype:
  - (+) CD19, CD20, CD22, CD79a (B cells) and (−) CD5, CD43 (T-cell markers) and (−) Cyclin D1
  - (+) CD10 and Bcl6 and CD23 ±
  - t(14;18)(q32;q21) ⇒ *IGH/BCL2* ⇒ Upregulation of BCL2 oncogene

The extranodal follicular lymphoma differs from the nodal type, because of the expression IgA as well as α4β7 mucosal homing receptor suggesting an origin from local B cells, expression of *VH4* gene, and no activation-induced cytidine deaminase expression.

- *PGN*: Indolent course with a low stage and 5-YSR ~65% with a usual "watch and wait" approach, because they do not develop extraintestinal spread. In rare cases with advanced disease or grade 3B, chemotherapy + rituximab or involved field radiotherapy (IFRT) to localized disease has been suggested. Maintenance with rituximab following first- or second-line treatment has also been indicated.

Burkitt Lymphoma
- *DEF*: GC B-cell neoplasm, mostly EBV associated with the aggressive clinical course and three presenting subtypes, including endemic (EBV+: 95%), sporadic (EBV+: 20%), and HIV-associated (EBV+: 30%).
- *CLI/EPI*: The endemic BL occurs mostly in the jaw, orbit, kidney, with highly adrenal gland, and ovaries (JAKOO as mnemonic word for sites) of children with a mean age of 10 years in equatorial Africa, while the sporadic BL occurs in the ileum at an age relatively older. The HIV-associated BL does usually not belong to childhood and youth.
- *GRO*: Bulky, fleshy tumors with or without necrosis.
- *CLM*: *Monotonous small round cells with a high mitotic rate* approaching 100% (Very Important Point!!!) and several *prominent basophilic nucleoli* and scant cytoplasm with numerous *intracytoplasmic fat vacuoles with* ORO positivity on touch preparation giving rise to the classic *"starry-sky pattern"* of the classic hematology books.
- IHC phenotype:
  - (+) CD19, CD20 (Pan-B markers) and (+) EBER (EBV) and (−) Cyclin D1
  - (+) CD10 and Bcl6, and CD23−, CD43+
  - t(8;14) ⇒ c-myc from 8q24 and Ig loci from 14q32
  - Other translocations t(8;2) and t(8;22).
- *TRT*: Resection and chemotherapy.

Diffuse Large B-Cell Lymphoma
- *DEF*: Heterogeneous B-NHL with GC and post-GC (activated B-cell) subtypes with highly prognostic relevance.
- *CLM*: There is a diffuse growth of cells lacking cohesiveness and showing cells with approximately double the size of normal small lymphocytes. It is important to recall that small lymphocytes are 7 to 10 μm in diameter, while large lymphocytes are 14 to 20 μm in diameter. These cells show open vesicular chromatin pattern with nucleoli adjacent to the cell membrane (in the case of centroblasts) or centrally located. Nucleoli can be eosinophilic in the case of immunoblasts.
- *DDX*: Poorly differentiated carcinoma, which is positive for mucin and keratin staining, negative B-cell markers, the presence of syncytial cell aggregates or malignant acinar formation, malignant melanoma, and myeloid sarcoma.
- IHC phenotype:
  - (+) CD19, CD20 (Pan-B markers) and (+) EBER (EBV in the case of PTLD) and (−) Cyclin D1
  - (+) CD10 and Bcl6, and CD23 (−), CD43 (−)
  - T(14;18) ⇒ *IGH/BCL2*

- Heterogeneity: B cells (50–60%), T cells (5–15%), histiocytic (5%), no cell markers (30–40%)
- *TRT*: Chemotherapy.
- *PGN*: Rapid progression and poor PGN, if untreated, but good results with aggressive therapy.

Extranodal Marginal Zone B-Cell Lymphoma of MALT (ENMZL/MALT)
- *DEF*: Postgerminal center B-cell NHL (aka MALT-oma) strongly associated with *H. pylori* infection and autoimmune disorders and composed of morphologically heterogeneous small B cells with an indolent course harboring the tendency to remain localized for long periods before progressing. It can also transform into a large cell lymphoma, which behaves much more aggressively. Most common EN sites are salivary glands and GI tract.
- *CLM*: Early lesions are Ag-driven (e.g., *H. pylori* in the stomach), while later, the neoplastic process is Ag-independent.
- IHC phenotype:
  - (+) CD19, CD20 (Pan-B markers) and (−) EBER and (−) CD5 and Cyclin D1
  - (−) CD10 and Bcl6, and CD23− but CD43±
  - (±) EMA

- t(1;14)(p22;q32) ⇒ *BCL10/IGH*
- t(14;18)(q32;q21) ⇒ *IGH/MALT1*
- t(11;18)(q21;q21) ⇒ *API2/MALT1*

- *DDX*: Gastric MALT lymphoma ≠ gastritis. Distinguishing features are lymphoepithelial lesions and reactive follicles with germinal centers. There are diffuse and nodular forms, where neoplastic cells have initially a perifollicular distribution and then may infiltrate as lymphocytic cuff (mantle zone pattern) or colonize germinal centers (follicular colonization). Plasma cellular differentiation with Ig light chain restriction and plasma cellular overgrowth characteristically located in the tips of the lamina propria of the mucosa by ascending CD20 (+) tumor cells may also help to delineate the differential diagnosis.
- *PGN*: Two scoring systems are needed to be mentioned here. They include the Wotherspoon scoring of the lymphocytic infiltrates and the Copie-Bergman scoring for post-therapy grading. The prognosis is variable. On univariate analysis, the longer event-free survival (EFS) and overall survival (OS) times were observed in single MALT site involvement (OS), no nodal disease (EFS and OS), skin and orbit lymphoma (OS), localized disease (EFS and OS), and stage IV disease without bone marrow involvement (OS). However, both bone marrow and nodal involvement have been indelibly associated with the shorter OS on multivariate analysis.

Immunoproliferative Small Intestine Disease (IPSID)
- *DEF*: B-NHL (aka Mediterranean lymphoma or α heavy chain disease), which is uncommon and included under the MALT lymphoma in the current WHO classification. It is a form of MALT lymphoma developing in MALT areas of the small intestine and a there is a more or less frank correlation with *C. jejuni* infection. It arises in the intestinal mucosa as a morphologically benign-appearing infiltrate with a dense plasma cell proliferation (either lymphoplasmacytic or large cell, immunoblastic plasmacytic with IgA heavy chain restriction).
- *CLI*: Youth up to 40–45 years, malabsorption, duodenum/proximal jejunum involvement, villous atrophy (Marsh classification!), and plasma cell infiltration of the adjacent intestine with > serum IgA.
- *TRT*: Antibiotics (early) and DLBCL chemotherapy (late), when IPSID evolves to large cell lymphoma.

Gastrointestinal T-NHL
Gastrointestinal T-NHL includes the enteropathy-associated T-cell lymphoma (EATL) essentially.

Enteropathy-Associated T-Cell Lymphoma (EATL)
T-NHL arising as clonal transformation of inflammatory intestinal intraepithelial T cells and

occurring with *wide age range* (20–70 years) and a *wide interval of history of malabsorption* (2 months to >5 years), mostly *celiac disease* (even in the setting of a strict compliance to gluten-free diet) and harboring a highly aggressive course with poor outcome (5-YSR: 8–20%). Of note, only 2–5% of patients with celiac disease develop NHL. The refractory celiac disease is defined as "persistent clinical enteropathy and histological features of villous atrophy with intraepithelial lymphocytosis, despite strict adherence to a gluten-free diet for at least 12 months following a prior resolution."

EATL usually involves the *proximal small bowel* (jejunum) presenting as an ulcerating mucosal mass that circumferentially invades the wall but may also be multifocal with multiple ulcers ("ulcerative jejunitis"). The presentation includes abdominal pain, diarrhea, weight loss, and anemia, which may portend to a diagnostic workup for recurrent or refractory celiac disease.

- *EATL-I:* Medium- to large-sized cells with vesicular nuclei and prominent nucleoli and large pale-staining cytoplasm and accompanied by inflammatory cells (often eosinophils).
- *EATL-II:* Small- to medium-sized monomorphic cells with hyperchromatic nuclei and a scant rim of cytoplasm.

IHC panel is useful in distinguishing two types of EATL: EATL-I is typically associated with refractory celiac disease, but the refractory celiac disease may also precede (not always occasionally) EATL-II.

- IHC phenotype:
  - EATL-I: CD3+, CD5−, CD8−, *CD4+, CD56−, TCR−β±, HLA-DQ2/-DQ8 (>90%)*
  - EATL-II: CD3+, CD5−, *CD8+,* CD4−, *CD56+, TCR-β+, HLA-DQ2/-DQ8 (40%)*

Both types are CD30+, which is present in large, bizarre multinucleated cells mimicking HL. CD5 helps to distinguish this entity from ALK lymphoma, which may share a similar morphology with EATL-I. Moreover, both types express the cytolytic granule-associated proteins. Molecular biology studies evidence *TCR* gene rearrangement.

- *TRT*: Initially surgery is the first choice where possible. Aggressive chemotherapy may be an option, while radiotherapy has no role due to the frequent dissemination of the neoplasm.

Nonnasal NK/T-Cell NHL

NK/T-cell lymphoma of East Asian and Hispanic population (fourth to sixth decade, but younger patients have been reported, ♂>♀), EBV associated with poor PGN (median survival: 10 months, reduced to 4 months in advanced disease), and targeting multiple locations. Its occurrence in the bowel is accompanied by the usual broad cytologic spectrum of the nasal type, ranging from small to intermediate size to large cells with marked nuclear pleomorphism. There is no villous atrophy of the adjacent mucosa (Marsh classification), but the lymphoma shows ulceration, angiocentricity, and angioinvasion with both fibrinoid and coagulative necrosis. IHC pattern includes cytoplasmic, but not surface CD3 expression and (+) TIA1 and (+) CD56 like EATL-II. Moreover, NK/T-cell NHL is EBV (+), but CD8(−) and βF1(−), unlike EATL-II, which is EBV (−) and CD8(+) and βF1(+). The βF1 or β Framework 1 antibody is key. The βF1 recognizes a common epitope on the β chain of the TCR for antigen which is expressed by thymocytes and peripheral T lymphocytes following the treatment of lymphoid cells with permeabilization reagents. The βF1 does not react with γδ TCR-bearing T cells.

- *CMB*: del(6)(q21:q25), i(6)(p10), TCR rearrangement have been identified.
- *TRT*: Chemotherapy and radiation are the mainstay. In advanced or relapsed disease, it has been suggested combination chemotherapy containing L-asparaginase, such as the SMILE regimen. (Steroid/dexamethasone-Methotrexate with leucovorin-Ifosfamide with mesna-L- asparaginase-Etoposide).

Peripheral T-NHL NOS can also involve the GI tract but is a rare occurrence in childhood or youth. The Ann Arbor staging system is quite well validated for nodal involvement, and some modifications are present in the literature for extranodal involvement. There are some staging systems for primary gastrointestinal lymphomas, but the

"Lugano Staging System" seems to be commonly accepted and used (Rohatiner et al. 1994). Another option is that according to Mussoff (1977), which seems quite often used in some countries. Later, an international prognostic index was proposed and accepted. Please consult the current classification and staging of tumors of the gastrointestinal lymphatic and hematopoietic system in the "blue books" of the World Health Organization.

Pseudolymphoma

It is a localized lymphoid hyperplasia, which may be observed, for instance, in the stomach, intestines, and rectum. Lymphoid proliferation may be follicular (nodular) in the terminal ileum and appendix or diffuse, such as in the stomach, proximal small intestine, and rectum. Distinguishing features are polymorphous infiltrate, well-formed reactive lymphoid follicles, the proliferation of blood vessels, dense collagenous fibrosis, and ulceration, although this latter (Caveat!) may be a malignant feature as well.

3.3.5.5 Stromal Tumors and Smooth Muscle-Differentiating Tumors

GIST: Gastrointestinal Stromal Tumor

- *DEF*: It is the most common mesenchymal neoplasm of the GIT arising from the interstitial cells of Cajal harboring-activating mutations of c-Kit or *PDGFRA* (CD117), which is tyrosine kinase receptor involved in mitogenic signals and quite resistant to standard chemotherapy and radiation.
- *EPI*: In pediatric patients, the GIST median age is 12 years, ♂:♀ = 1:3, although it remains a rare occurrence in childhood ($6.8–15/10^6$).
- *ETP*: Genetic syndromes: Carney triad (GIST, paraganglioma, pulmonary chondroma), germline *KIT* or *PDGFRA* gene mutations, and NF-1.

Gain-of-function mutation ⇒ constitutive activation!

PDGFRα type III tyrosine kinase and DOG1 are ion trafficking involving transmembrane proteins.

~80% C-Kit (tyrosine kinase): oncogenic, gain-of-function mutations.

~10% PDGFRA (a related tyrosine kinase), c-Kit being the receptor for stem cell factor.

- *CLI*: Abdominal pain, bloating, melena, anemia, or bowel obstruction, which may occasionally be dramatic and discovered in the emergency departments.
- *GRO*: Mass with tan-white, fleshy cut surface and potential foci of cystic degeneration, hemorrhage, or necrosis.
- *CLM*: Three main histologic subtypes: spindle cell type, epithelioid type, and mixed type. Histologically, there is a uniform, monotonous appearance with minimal cytologic atypia, occasional cellular pleomorphism, and low mitotic activity. Compared with the adult GIST, which is mostly spindle cell type, the pediatric GIST is epithelioid primarily with a more indolent natural history than the adult counterpart (median survival: 20 months!).

CD117 is also expressed in MZ and other hematopoietic precursor cells and neoplasms, such as melanoma, renal cell carcinoma, and seminoma. Thus, many words of caution should be expressed.

GIST with "dot-like" IHC pattern is probably extraintestinal or exhibiting an epithelioid morphology.

- *DDX*: <u>GIST</u>, which is (+) c-kit, PDGFRα, DOG1, (±) CD34 (2/3), SMA (1/3), S-100 (1/20), DES (rare), Leiomyosarcoma (<u>LMS</u>), which is (+) SMA (~3/4), (±) DES, CK, (±) CD34 (1/10), (−) CD117/c-kit, (±) S-100 (rare), <u>Schwannoma</u>, which is (±) CD34, (−) CD117/c-kit, <u>Desmoid-type Fibromatosis</u>, which is (+) β-CatN, and Solitary Fibrous Tumor (<u>SFT</u>), which is (+) CD34, (−) CD117/c-kit, (±) β-CatN.
- *PGN*: Following resection, which is the first option for treatment, the outcome depends on staging.

Smooth muscle actin (SMA) indicates smooth muscle differentiation but also identifies myofibroblasts and myoepithelial cells, while desmin suggests myogenic differentiation (cardiac, smooth, and striated muscle cells), but also myofibroblasts and may be present in desmoid-type fibromatosis. Finally, S100 is a calcium-binding protein and is positive in schwannian, melanocytic, and chondrocyte differentiation.

Smooth Muscle Tumors

Most smooth muscle tumors are benign and have unusual characteristics (extreme cellularity, infrequent large cells with eccentric hyperchromatic nuclei, marked diffuse vascularity, palisading of nuclei, clear cytoplasm) but still benign. MI is low. Finally, the solitary fibrous tumor (SFT) can occur and is distinct by (1) "patternless pattern" of growth, (2) thick collagen bands, and (3) hemangiopericytomatous/staghorn-like blood vessels.

3.3.5.6 Nonstroma/Smooth Muscle-Differentiating Tumors

These tumors include glomus tumor, lipoma, granular cell tumor, neurofibroma, and schwannoma, but they are rarities in pediatrics and youth.

3.3.5.7 Secondary Tumors

Metastases to the stomach may occur in childhood or youth, but they may be restricted to small round blue cell tumors almost exclusively.

3.4 Small Intestine

3.4.1 Small Intestinal Anomalies

In both groups of congenital and acquired anomalies, atresia and stricture following a necro-inflammatory process, such as necrotizing enterocolitis, may play a significant role in pediatric surgery theaters (Box 3.19) (Figs. 3.16, 3.17, 3.18, 3.19, 3.20, 3.21, 3.22, 3.23, 3.24, 3.25, 3.26, 3.27, 3.28, and 3.29).

Box 3.19 Small Intestine Anomalies
  I. Congenital Anomalies
  II. Vascular Changes
  III. Inflammation and Malabsorption
  IV. Metabolic and Degenerative Diseases
  V. Tumors

3.4.1.1 Infantile Colic

Most of the conditions involving an infant crying during the first few months of life are *not* organic in nature. It is estimated that up to 1/3 of the infants experience colicky pain and in 95% of cases this pain should be considered functional. However, an excessive crying baby can be dangerous for both parents and physicians. Infantile colic is still defined according to Wessel et al. (1954) as a crying infant >3 h/day, >3 days/week, and >3 consecutive weeks. Typical aspects of a developing child during the first 3 months of life usually comprise fussing and crying, and during this time, babies cry an average of 2.2 h per day, peaking at 6 weeks of age and then gradually showing a decreasing tendency. Colicky infants have attacks of screaming in the evening associated with abnormal motor behavior, including facial flush, furrowed brow, and clenched fists, pulling up of the legs to the abdomen, and emission of a piercing, formidably high-pitched scream. The etiology is variable and there are most and less common conditions. Most common conditions include alimentary disorders, constipation, anal/rectal fissure, gastroesophageal reflux, cow's milk protein intolerance, otitis media, urinary tract infections, and viral illness. Less common conditions, of which organic causes are underlying disorders in <5% of colicky infants, involve several organs and systems. To remind are lactose intolerance (deficiency of disaccharidase), renal pathologies (e.g., ureteropelvic obstruction), biliary tract diseases, intussusception, volvulus, incarcerated hernia, unrecognized bony fracture, hair tourniquet syndrome, neurological disorders (e.g., Arnold-Chiari malformation, infantile migraine, and subdural hematoma), foreign body in the eye or corneal abrasions, meningitis, trauma, abuse, and maternal consumption of drugs (both illicit and prescription drugs).

Congenital anomalies of the small bowel anatomy may be found quite often in a pathologist practicing in the neonatal period. It is essential to be aware of it to avoid missing the "colic pain or crisis" of the neonate with organic nature that needs appropriate surgical treatment.

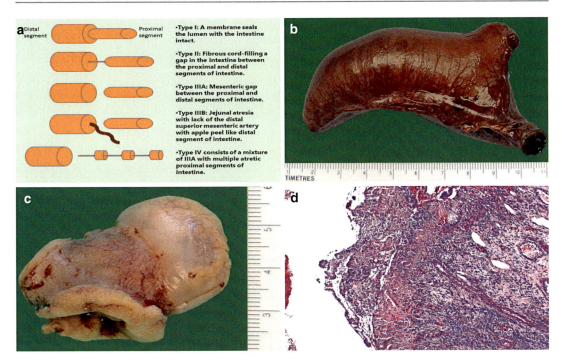

Fig. 3.16 In (**a**) is shown a schema of small bowel atresia. The gross photographs (**b**, **c**) show an atresia (**b**) and a Meckel diverticulum (**c**), respectively. The microphotograph (**d**) shows an ulcer with gastric heterotopia in a Meckel diverticulum

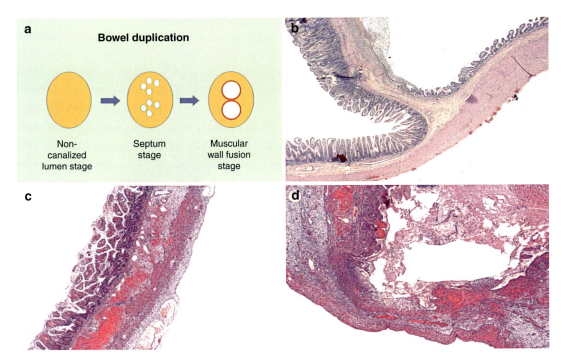

Fig. 3.17 The pathogenesis of bowel duplication is shown in (**a**), while (**b**) shows the histologic correlate of a bowel duplication. In the microphotographs (**c**, **d**) is shown a spontaneous intestinal perforation (SIP)

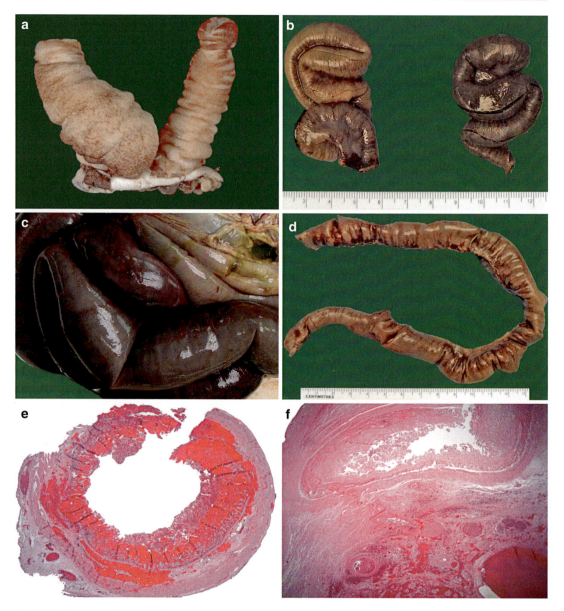

Fig. 3.18 The gross photograph in (**a**) represents a stoma prolapse (or prolapsing stoma), which is a significant complication with associated morbidity, multiple admissions, and various surgical procedures. Neonatal volvulus loops are shown in (**b**), while (**c**) shows an allograft vasculopathy-related bowel mural necrosis following solid organ transplantation in a child. The gross photograph of ischemic necrosis is shown in (**d**), while (**e**) shows bowel necrosis with transmural extension of a different pediatric patient. In (**f**) is shown a strangulated gastroschisis with transmural necrosis

3.4.2 Abdominal Wall Defects (Median-Paramedian)

3.4.2.1 Omphalocele

- *DEF*: Midline persistence of the physiologic intestinal hernia (amnion enclosed extra-abdominal sac containing intestines with or without a liver) into the umbilical cord located proximally to the fetus or infant. It may be associated with supra- and infraumbilical defects. Variable failure of infolding and fusion of the edges of the three-layered embry-

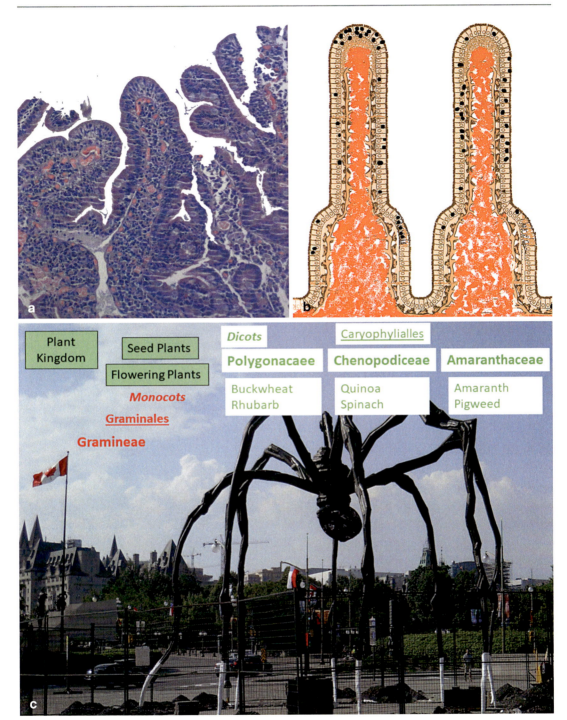

Fig. 3.19 In (**a**) is an impressive intraepithelial lymphocytosis of a patient with celiac disease. In (**b**) is presented a schema of villi showing the distribution of the lymphocytic infiltration of the surface epithelium with up in the celiac disease, laterally in inflammatory bowel disease, mainly Crohn disease, but also ulcerative colitis, and down in case of graft-versus-host disease. In (**c**) is the taxonomy of the cereals linked to celiac disease and the background is the "Maman" sculpture created by the artist Louise Bourgeois in Ottawa, ON, Canada. In 1999, this bronze, stainless steel, and marble sculpture depicts a spider of over 30 ft high and over 33 ft wide and in our context symbolizes the intricate connections of foods containing gluten

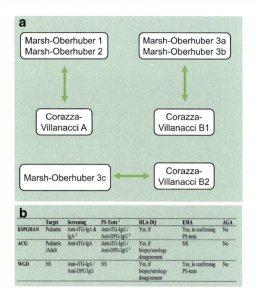

Fig. 3.20 In (**a**) are shown the two major classifications for GSE (Marsh-Oberhuber and Corazza-Villanacci), while in (**b**) is shown a comparison of the criteria used from three major gastroenterological societies

onic disc determines which organs are externally placed. The most superior abdominal defects allow exteriorization of heart with or without the liver (*pentalogy of Cantrell*), while the minor defects leave the inferior portion of the primordial gut open with *bladder or cloacal exstrophy* depending on whether or not the completion of the cloacal stage has occurred.

- Omphalocele: 1:4,000–1:5,000 live births
- Cloacal exstrophy: 1:200,000–1:400,000 live births
- *EPG*: Chromosomal aberrations (trisomies 13, 18, and 21, triploidy as well as Turner syndrome, and Klinefelter syndromes), genetic syndromes and teratogenetic complexes, such as Beckwith-Wiedemann syndrome and OEIS defects complex (omphalocele-exstrophy-imperforate anus-spinal defects). RF: Advanced maternal age.

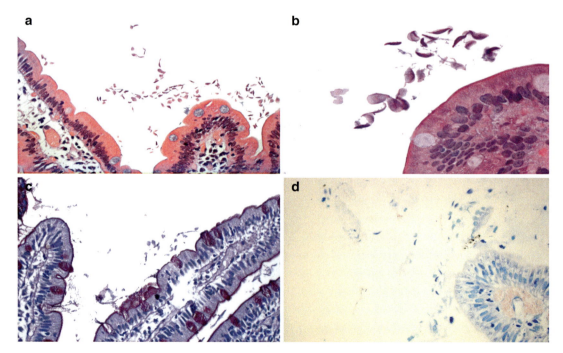

Fig. 3.21 *Giardia lamblia* (*G. intestinalis*) is a flagellated protozoan with the trophozoites firmly attach to the duodenal and proximal jejunal mucosa. It may be asymptomatic or cause chronic malabsorption and failure to thrive. Eight genetic groups of *G. lamblia* are known (two infect humans and animals; the others infect only animals). In (**a–h**) are microphotographs of the protozoan and the reaction of the intestinal mucosa. Giardiasis is a gluten-sensitive enteropathy (celiac disease) mimicker. In (**c**) the parasite is highlighted by PAS stain, while (**d**) shows a Giemsa stain-based detection of the parasites. The immunostaining of the intestinal mucosa with antibodies against several cytokines for lymphocyte subpopulations reveals CD3 (**e**), CD4 (**f**), CD8 (**g**), and CD20 (**h**) with marked CD8-based intraepithelial lymphocytosis and hyperplastic lymphoid follicle of the lamina propria (CD20 immunostain)

3.4 Small Intestine

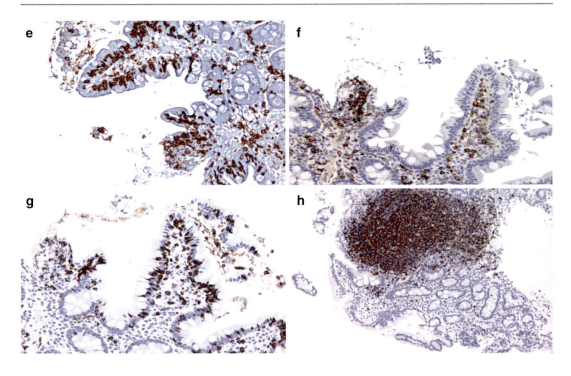

Fig. 3.21 (continued)

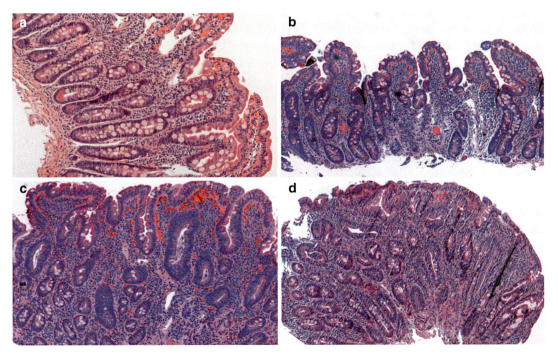

Fig. 3.22 Intraepithelial lymphocytes and Marsh classification showing (**a**) Marsh I (H&E stain, × 100), (**b**) Marsh IIIA (H&E stain, × 100), (**c**) Marsh IIIB (H&E stain, × 100), (**d**) Marsh IIIC (H&E stain, × 100). Marsh-Oberhuber classification is often shortened as Marsh

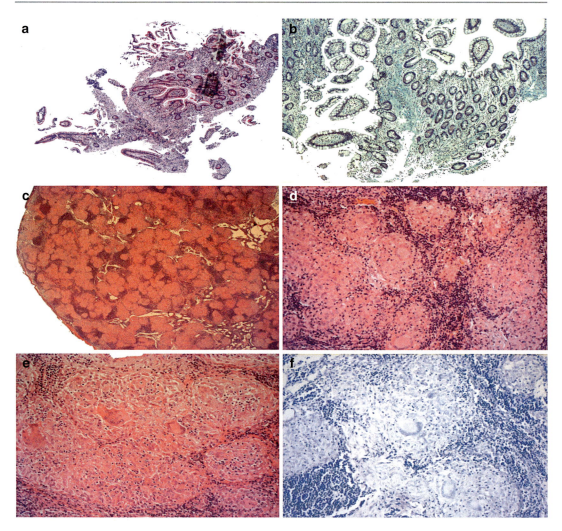

Fig. 3.23 Sarcoidosis is a multisystem, granulomatous disease of unknown etiology and may involve the gastrointestinal tract other than the respiratory tract and other extra-lymphatic organs. Sarcoid ileitis (terminal ileitis) is shown in **a** (H&E) and **b** (Masson Goldner stain) and inguinal granulomatous lymphadenitis without caseous necrosis (**c**–**e**). The microphotograph (**f**) shows Langhans-type cells (Giemsa stain)

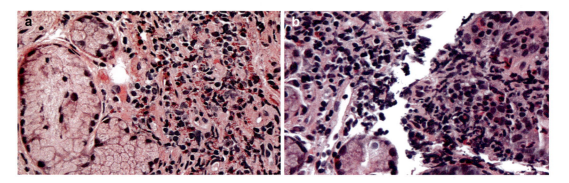

Fig. 3.24 GSE mimickers. (**a**) eosinophilic duodenitis (H&E, × 400); (**b**) peptic duodenitis (H&E, × 400). GSE, gluten-sensitive enteropathy

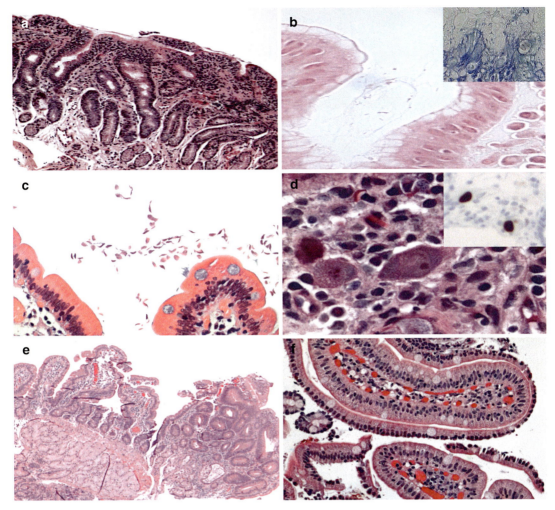

Fig. 3.25 Gluten-sensitive enteropathy mimickers. (**a**) lymphocytic gastritis with involvement of the duodenum (HE, × 100); (**b**) *H. pylori* gastritis (inset, Giemsa staining) (HE and Giemsa × 630); (**c**) giardiasis (HE, × 400); (**d**) cytomegalovirus (CMV) infection (HE, × 630) and inset showing anti-CMV antibody reacting against viral proteins using an avidin-biotin complex immunoperoxidase immunohistochemical detection (× 100); (**e**) focal adenomatous change in duodenum (HE, × 50); (**f**) sickle cell disease-related duodenitis (HE, × 200)

- *CLM*: The inner membrane of fetal parietal peritoneum and an outer layer of amnion with skin defect, which is at the basis of the increase of the maternal serum AFP (α-feto-protein) levels.
- *DDX*: Simple umbilical hernia (normal skin covering over abdominal contents, normal maternal serum AFP levels).
- *PGN*: 50% mortality.

3.4.2.2 Gastroschisis
- *DEF*: Nonenclosed paraumbilical abdominal wall defect, usually positioned to the right of a standard umbilical cord, with externalization of mostly intestine only (rarely stomach or liver) and caused by a vascular deficiency or injury leading to loss of the lateral aspect of the right (or left) lateral fold of the embryo. There is no membrane covering as seen in omphalocele, and the protruding herniated loops float freely and are continuously dangerously exposed to amniotic fluid. There is also, in this case, an increase in the maternal serum of the AFP levels. It occurs in 1:6,000 live births; in up to 1/3 of cases, there are associated anomalies; early sequelae include stenosis or "atre-

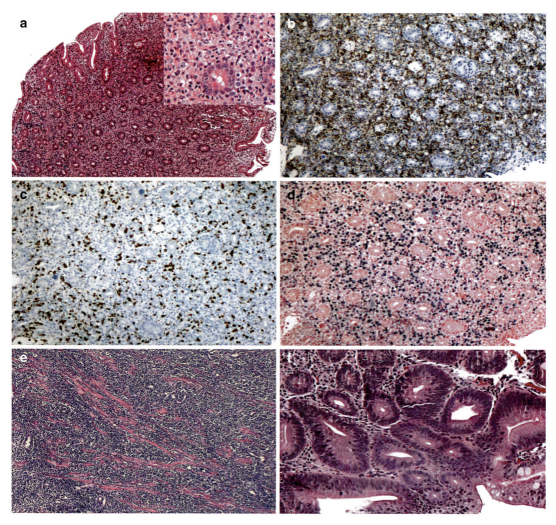

Fig. 3.26 Gluten-sensitive enteropathy mimickers. (**a–d**) Posttransplant lymphoproliferative disorder, polymorphic type (**a**, HE, × 50, and inset HE, × 200) showing mononuclear epithelial and stromal infiltration with blasts and high CD20 (**b**, × 100) on CD3 (**c**, × 100) lymphocytes and high Epstein-Barr virus replication (in situ hybridization or Epstein-Barr encoding region, × 100); (**e**) Burkitt lymphoma of the duodenum (HE, × 100); (**f**) tubular adenoma of the duodenum of a patient with familiar adenomatous polyposis (HE, × 200)

sia" of some or all of the protruded loops; and late sequelae of surgically corrected herniated loops involve dysmotility and high potential to develop necrotizing enterocolitis (NEC), which is a medical condition where a portion of the bowel becomes necrotic. Typically, NEC occurs in newborns that are either premature or otherwise unwell (e.g., congenital heart disease).

3.4.3 Continuity Defects of the Intestinal Lumen

3.4.3.1 Atresia

DEF: Failure to canalize the first endodermal tube due mostly to an in utero mechanical injury and occurring in up to nine of ten cases as solitary atresia and presenting as an acute abdomen with the characteristic "string of pearls" sign on imag-

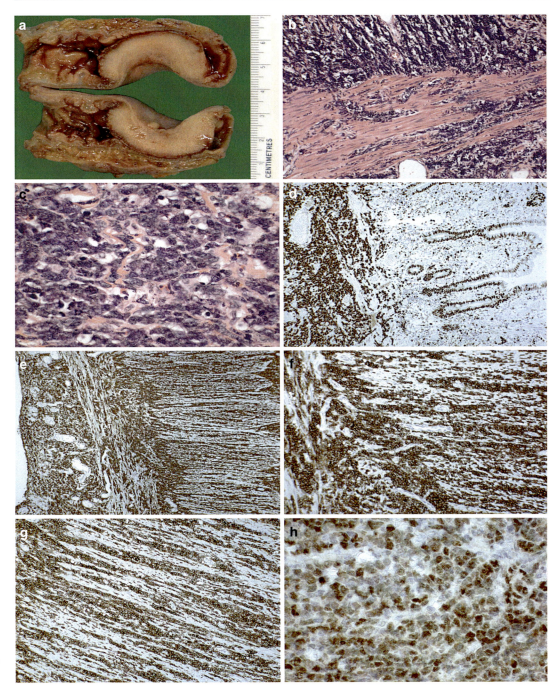

Fig. 3.27 Burkitt lymphoma presenting as intussusception at the terminal ileum (**a**). The microphotographs (**b**, **c**) reveal small- to medium-sized cells infiltrating the wall of the intestine. The microphotographs to label the infiltrate, by immunohistochemistry, are (**d–h**) and show positivity for the Ki67 proliferation antigen (**d**), CD20 (**e**), CD79a (**f**), CD10 (**g**), and BCL6 (**h**)

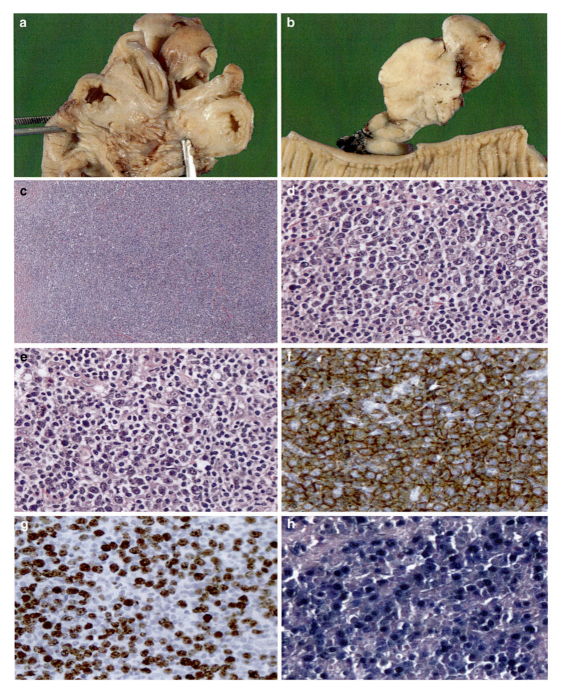

Fig. 3.28 In the ileum is also possible to recognize a posttransplant lymphoproliferative disorder (PTLD) with monomorphic type as shown in this case (**a**, **b**), which discloses a diffuse large B-cell lymphoma (**c–e**) supported by immunohistochemical stains positive for CD20 (**f**) other than other B markers (not shown). In (**g**) is shown the high proliferation activity of this lymphoma (anti-Ki67/MIB1 immunostaining). In (**h**) is shown the EBER i.e., the results of the in situ hybridization for Epstein-Barr virus (EBV)

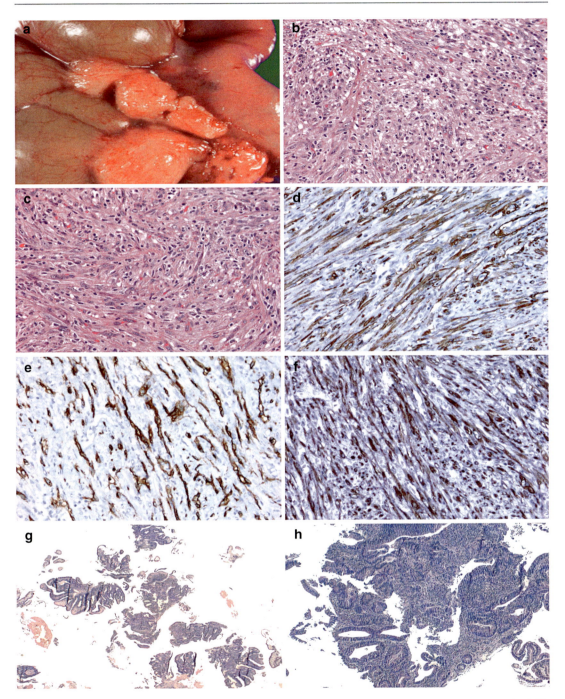

Fig. 3.29 In (**a**) is shown a lymphangioma of the mesenterial roots of the intestinal loops. The microphotographs (**b**, **c**) show the routine histology for an inflammatory myofibroblastic tumor of the small bowel. The immunohistochemistry in (**d**–**f**) shows positivity of the inflammatory myofibroblastic tumor for actin, CD34, and ALK1, respectively. The microphotographs in (**g**, **h**) show a very rare occurrence of ampullary carcinoma (ampulla of Vater) in a young male with predisposing cancerous genetics

ing. Atresia may be classified according to the occluded segment following the original Santulli and Blanc (1961) classification. The acronym BLCI may be used to group the conditions associated with bowel luminal continuity interruption. The type I (25% of cases) is defined as BLCI by a transluminal mucosal septum only (intact mesentery). A fibrous cord between proximal and distal ends (entire mesentery) defines the type II (15% of cases). The type IIIa (15% of cases) is defined by unattached blind proximal and distal ends (whole mesentery). The type IIIb (20% of cases) is defined by unattached blind proximal and shortened/spiralized distal end, while the type IV (25% of cases) is a BLCI by unattached "sausage"-like proximal end and intact distal end. The type IIIb is also called "apple-peel" type because of the spiralized distal end. The percentages should be considered approximate values derived from several statistics of the literature.

The duodenum is the most typical site of the defects involving the intestinal luminal continuity (*duodenal atresia*) and results by a persistent obliteration of the lumen that occurs between the fifth and sixth weeks of physiological development of the embryo as the epithelium proliferates fast. In the duodenum, the most common site is the ampulla of Vater, which is also the reason for associated abnormalities of the pancreas and extrahepatic biliary tract. Up to 1/5 of patients with duodenal atresia there is an annular pancreas, which is a condition characterized by the second part of the duodenum surrounded by a ring of pancreatic tissue obviously continuous with the head of the pancreas. Duodenal atresia is observed in trisomy 21 syndrome as well as VATER association. *Jejunal* and/or *ileal atresia* (1:3,000 live births) is caused by an interruption of blood flow to the affected intestinal segment. Jejunal atresias are usually multiple, and the jejunum has a good compliance that allows a dilatation before perforation occurs, while ileal atresia is generally solitary, and perforation occurs early because ileum is not prone to dilate easily. The suspicion of BLCI is further evaluated with a contrast enema using a hydro-soluble agent to differentiate ileal atresia from Hirschsprung disease or aganglionosis, the functional immaturity of the colon (meconium plug syndrome), and meconium ileus.

3.4.3.2 Stenosis

It is defined as a narrowing of a previously adequately developed and correctly canalized lumen of the bowel due mostly to fibrosis and/or stricture. A condition called *megacystis-microcolon-intestinal hypoperistalsis syndrome* (MMIH syndrome) or Berdon syndrome will be discussed in detail under the "Large Intestine" heading. Briefly, it is an autosomal recessive inherited disorder affecting newborns. It is more prevalent in female babies and presents with constipation and urinary retention, microcolon, giant bladder (megacystis), intestinal hypoperistalsis, hydronephrosis, and dilated small bowel. Ganglion cells are present in both enlarged and narrow areas of the intestine. In his first description, this syndrome was identified in five female infants, two of whom were sisters. All patients had marked dilatation of the urinary bladder, and some had hydronephrosis of the kidneys as well as the external appearance of "prune belly." The babies also had microcolon and dilatation of the small intestine.

3.4.3.3 Meconium Plug Syndrome

After atresia and malrotation, meconium ileus is the third most common cause of neonatal intestinal obstruction, and it occurs when abnormally thick, inspissated meconium occluded the distal ileum determining a low intestinal obstruction or ileus. It is almost exclusively observed in newborns with cystic fibrosis with approximately 15% as incidence in this setting. Meconium ileus may complicate with atresia, volvulus, perforation, or peritoneal pseudocyst.

3.4.4 Intestinal Muscular Wall Defects

3.4.4.1 Diverticula

Mesenteric Diverticulum
Congenital defects in the muscular wall, usually in the duodenum (pseudodiverticula show no muscle wall in outpouching).

Anti-mesenteric (Meckel's) Diverticulum
Solitary, antimesenteric outpouching due to the persistence of omphalomesenteric (vitelline)

duct (connection between GI tract and umbilicus present at 4 weeks and which usually becomes obliterated to form a fibrous band, which is subsequently absorbed). It is typically ileal (~30 cm from cecum), 2% of the population, ♂>♀, 50% with ectopic gastric mucosa (± peptic ulceration) ± pancreas, usually in the adjacent intestinal mucosa. Complications include perforation, ulceration, bleeding, and intussusception, which is the invagination of one portion of a tubular anatomical (in our case, the gastro-intestinal tract) structure within the next (Latin *intus* "within" and *susceptio* "catching, lifting up", *vide infra*).

3.4.4.2 Segmental Dilatation
(Congenital) Limited dilatation of the intestine with an abrupt transition between normal and dilated intestinal segments without intrinsic or extrinsic barrier distally to the dilatation.

3.4.5 Small Intestinal Dystopias

Dystopias or abnormal location of the intestines include intussusception, volvulus, and nonrotation or malrotation of the bowel.

3.4.5.1 Intussusception
It is generated by a telescoping phenomenon of some intestinal loops with one segment of small intestine (the *intussusceptum*) becoming "telescoped" into the immediately distal segment (the *intussuscipiens*). This situation promotes continuous propulsion of the proximal portion further inward by peristalsis, taking the mesentery (including blood vessels) with it. The consequence of involving blood vessels is an infarction. In older children, *lymphoid hyperplasia* may become the leading point of traction, while in adolescents or adults, intussusception may occur in the setting of a mass lesion, such as *Burkitt lymphoma*, which forms the leading edge.

3.4.5.2 Volvulus
Condition when a loop of the bowel twists around itself and the mesentery that supports it, resulting in a bowel obstruction or discontinuity of the lumen. Clinically, there are abdominal pain, abdominal bloating, vomiting, progressive bowel obstruction, constipation, and bloody stool.

3.4.5.3 Nonrotation and Malrotation
Intestinal nonrotation or malrotation does occur if there is no mesenteric fixation of the intestine. No fixation means an increased risk of displacement and subsequent ischemia of the abdominal segments involved because of twisting of mesenteric blood vessels. Nonrotation and malrotation can occur both in fetal and postnatal life. Acute midgut volvulus with bilious emesis and classic "corkscrew" imaging is a typical presentation of infants with malrotation.

3.4.6 Composition Abnormalities of the Intestinal Wall

Heterotopias or abnormal intestinal wall composition may include tissue that belongs to a part of the primitive stage of the primordial gut. This tissue is differentiated in a wrong location, but it is still primordially gut-linked. Thus, a heterotopia is usually a hamartia (Gr. ἁμαρτία, or errore). Choristiae are defined as heterotopic tissue parts that do not belong to the primordial gut embryologically. Hamartomas and choristomas may arise from these parts. In his *Poetics*, Aristotle first used this term to refer to the protagonist's error of a tragedy.

3.4.6.1 Pancreas Heterotopia
1–2 cm mucosal elevation, ducts, and acini, usually without islets, variable location in the small intestine, but most commonly in the periampullary region of the duodenum. It may serve as a lead point for intussusception.

3.4.6.2 Gastric Heterotopia
Discrete small nodules or sessile polyps in the duodenum or Meckel's diverticulum. It is composed of fundic-type mucosa with the chief and parietal cells.

3.4.6.3 Endometrial Heterotopia (Endometriosis)
Müllerianosis is an organoid structure of embryonic origin, i.e., a choristoma composed of

Müllerian rests – normal endometrium, normal endosalpinx, and normal endocervix – singly or in combination, incorporated within other normal organs during organogenesis.

Sampson then divided "misplaced endometrial or Müllerian tissue" into "four or possibly five groups, according to the manner in which this tissue reached its ectopic location" (1925). Sampson's classification of heterotopic or misplaced endometrial tissue is essentially based on pathogenesis. Sampson recognizes (1) "direct/primary endometriosis" [adenomyosis] (similar to the condition, which occurs in the wall of the tube [endosalpingiosis]), (2) "peritoneal or implantation endometriosis," (3) "transplantation endometriosis," (4) "metastatic endometriosis," and (5) "developmentally misplaced endometrial tissue."

3.4.7 Small Intestinal Vascular Changes

3.4.7.1 Non-NEC Vascular Necrosis

In the neonatal period, thromboemboli secondary to the use of intravascular catheters may represent an important etiology for bowel infarction. This entity is, however, different from neonatal necrotizing enterocolitis (NEC), a *multifactorial* syndrome of acute necrotizing inflammation of the *small and large intestine*. NEC is not linked to thromboembolic and needs to be kept nosologically distinct from vascular changes (infarction) of older children. Both NEC and infarction have coagulative necrosis as a crucial morphological feature, but *NEC is an inflammatory process, and the initiating site has been repeatedly identified in a venule*, although some controversy does exist. Conversely, bowel *infarction* may also have an accompanying inflammatory process, but the initiating site is an *artery*. Indeed *intestinal pneumatosis,* a characteristic finding of NEC, is not seen in infarctions. Other causes of bowel infarctions are rare and include abnormalities of intestinal blood vascular supply. It is quite convenient to classify bowel necrosis secondary to infarction according to the extent of the wall involvement (mucosal-mural-transmural).

Acute Intestinal Ischemia

EPG: *Occlusive* (arterial thrombosis and embolism, atherosclerosis, arteritis, small-vessel disease, and mesenteric venous thrombosis), *nonocclusive* (vasoconstriction-related due to decreased cardiac output, hypotension, hypovolemia, dehydration, or drugs, such as digitalis and cocaine), or dissecting aneurysm involving SMA (superior mesenteric artery), arteriopathy with intimal hypertrophy in systemic hypertension, radiation injury, or fibroelastosis adjacent to neuroendocrine tumors.

SMA is much more involved than IMA (inferior mesenteric artery) because the IMA has a more oblique take-off from the aorta. Occlusive causes of arteritis may include PAN, RA, SLE, and Takayasu disease, whereas occlusive causes of small-vessel disease may consist of HSP, leukocytoclastic vasculitis, and DIC.

Grossly hemorrhagic, whether arterial or venous in origin, arterial occlusions tend to have sharper margins, while venous occlusions tend to fade gradually into regular bowel segments. There is severe congestion, subserosal and submucosal hemorrhages, edema, and later blood in the lumen. Mesenteric venous thrombosis is associated with portal hypertension, hypercoagulable status (e.g., polycythemia), abdominal surgery or trauma, intra-abdominal sepsis, and presence of a mass ("*tumor quia tumet*"). The term "hemorrhagic gastro-enteropathy" is also used for nonocclusive, vasoconstriction-related hypovolemic conditions.

Chronic Intestinal Ischemia

It is the outcome of either acute mural infarction or chronic mesenteric insufficiency, e.g., HSP, RA, and SLE. PAN usually involves the small intestine, while phlebitis characteristically engages the colon. The ischemic bowel disease may have a gross picture that can mimic Crohn disease. There may be solitary or multiple ulcers with mucosal or transmural necrosis. The obstruction of the main branch of the mesenteric artery determines an extensive infarction, while secondary branches usually do not affect, because of numerous collaterals. Terminal branch obstruction causes a localized infarction. Fibrosis is most pronounced in the mucosa, and a fibrotic narrowing can also occur (intestinal stricture or strictura intestinalis). In scleroderma, visceral involvement may precede or occur in the absence of skin changes. The GIT is rigid, fibrotic, and often dilated, and there is atrophy of the tunica muscularis propria with the disappearance and fibrous replacement of inner circular muscular layer, concentric intimal thickening, medial fibrosis, and fibrin deposition in medium- and small-sized arteries, as well as some nonspecific mucosal chronic or active inflammation.

3.4.8 Inflammation and Malabsorption

Inflammation and malabsorption are in most cases strictly interconnected, and treatment of all of these conditions in one chapter seems a good option. We can also differentiate infective and noninfective disorders and disorders where both etiologies may play a role, but an underlying immunological problem may worsen the clinical scenario, such as inflammatory bowel diseases and specifically Crohn disease. Thus, inflammation and malabsorption are very intertwined. Malabsorption is the abnormal absorption of fat, fat-soluble vitamins, or other vitamins, proteins, carbohydrates, and minerals, usually with abnormal fecal excretion of fat (steatorrhea: >6 g/day) and due to a long list of possible conditions. Pathogenesis relies on defective intraluminal hydrolysis, mucosal abnormality, lymphatic obstruction, infection, by selecting only some of the numerous conditions associated with this disturbance.

Malabsorption is also associated with inflammation, and inflammation can be the first condition seen in premature babies. Malabsorption may be due to impaired digestion (due to liver, pancreas, gastric surgery), small bowel mucosal damage (due to celiac disease, tropical sprue, Whipple disease, parasitic disease, radiotherapy/chemotherapy, lymphoma), small bowel flora alteration (e.g., blind loop syndrome), biochemical dysregulation (e.g., disaccharidase deficiency), endocrinopathies (e.g., carcinoid syndrome, Zollinger-Ellison syndrome), and circulatory disorders (e.g., mesenteric ischemia, low-output syndrome/shock, sepsis). The pathology named *necrotizing enterocolitis* seems to be the first condition to start with because of the infantile age and involving both inflammation (and necrosis) and malabsorption.

3.4.8.1 Necrotizing Enterocolitis

- *DEF*: Etiologically multifactorial-based, homogeneous syndrome of acute *necrotizing* inflammation of *small* (*enteritis*) *and large intestine* (*colitis*) (TI, cecum, ascending colon, i.e., a "watershed" area) with quite *uniform* morphology (1/2 of cases: small and large bowel with continuous and discontinuous involvement) and occurring mostly in *premature babies* on 2nd to 3rd day of oral feeding (formula fed-risk > breastfed-risk). NEC has specific risk groups, including ~10% of all premature (4/5 of cases) newborn babies or VLBW infants (<1500 g) and ~5–10% full-term (1/5 of cases) newborn babies with CHD (right and left congenital heart disease), birth asphyxia, polycythemia, exchange transfusion, and intrauterine FGR (fetal growth restriction).

- *EPG*: Imbalance between an insult (e.g., ischemia), the degree of maturity of the intestine (morphologically and/or physiologically), and the microbiome of the gut. The baseline of intestinal vascular resistance (IVR) is high in the fetus and decreases after birth for intestinal maturation and somatic growth. IVR is dependent on the balance between dilator (NO, nitric oxide) and constrictor (endothelin) with a myogenic response. Endothelin (ET-1) is a 21 amino acid peptide with crucial regulatory function for the diameter of the blood vessels. ET-1 is produced by the endothelial cells from a 39 amino acid precursor, the so-called "big ET-1", through the actions of an endothelin converting enzyme (ECE), which is localized in the endothelial cell membrane. Angiotensin II (AII), antidiuretic hormone (ADH), thrombin, cytokines, reactive oxygen species (ROS), and shearing forces acting on the vascular endothelium stimulate ET-1 formation and release. ET-1 release is, however, inhibited by prostacyclin, atrial natriuretic peptide, and NO.

An alteration of some vasoactive mediators is present in NEC, and studies using experimental animals have evidenced that young animals have defective pressure-flow autoregulation if compared with 1-month-old animals. Moreover, PDA (patent ductus arteriosus Botalli) and L → R shunt result in compromised blood flow to the intestine during diastole, and PDA closure reduces NEC rate. Preterm intestine has an increased permeability allowing Igs, proteins, and carbohydrates to traverse into the systemic circulation, has an abnormal superficial mucus with low bacteriostatic proteins (e.g., defensins, trefoil factor), has a low IEL/EC ratio (B and T cells) reaching adult levels only at 3–4 weeks of life, and has a low sIgA. Potential enteral feeding mechanisms involve osmolar stress, by-product synthesis of toxic short-chain fatty acids, intestinal distension following an increase of feeding volume underbalanced by an abnormal vascular regulation, gut hormone activation, and altered bile acid metabolism. Activation of a pro-inflammatory cascade by binding of LPS (Gram-negative bacteria) on Toll-like receptors of inflammatory and immune cells elicits a series of signal transduction events with the final result of nuclear translocation of NF-kB and sudden increased transcription of IL-1, IL8, TNF, iNOS (inducible nitric oxide synthase), and phospholipase A2. In NEC, the anti-inflammatory signaling is inadequate. The etiopathogenetic sequences may substantially implicate five significant steps (Box 3.20). The acronym "*IBRAHIM*" may serve to remember the five major steps in the etiopathogenesis of NEC.

The trefoil factor family (TFF) comprises polypeptides that are secreted by mucus-producing epithelial cells, predominantly in the gut. In mammalians, there are three TFF members. TTF1 is expressed by surface and pit mucus cells in the stomach, TTF2 is expressed by mucus neck and glandular mucus cells of the stomach and Brunner glands of the duodenal bulb, and TTF3 is expressed by goblet cells of the small and large intestine. These peptides are considered to be enterocyte protective and promote healing in response to epithelial damage.

- *CLI*: Several stages may be crucial to assess before the neonatologist prepare the patient to send to the pediatric surgeon. Variable picture from the mild tenderness of the abdomen to sepsis, which can be life-threatening (MOF, multi-organ failure) (50% of NEC require a

---

Box 3.20 Five Major Steps in NEC Etiopathogenesis (*IBRAHIM*)

(1) Initial *INSULT* on a substrate of tissue weakness (*stress, hypoxia, prematurity*) ⇒

(2) *Bacterial colonization* (RhoA-GTPase activation, Na⁺ proton exchange pump dysregulation, pro-inflammatory COX-2 release, and NO⁻ production) ⇒

(3) ↑ *ROS* and *apoptosis* in enterocytes => Altered *health* of enterocytes

(4) *Impairment* of intestinal barrier (*mucosal*) repair evidenced by ↓TTF1 and TTF2 mRNA expression due to ROS-dependent cell toxicity, which induces DNA damage, PUFA oxidation (lipid peroxidation), oxidations of amino acids in proteins, and inactivation of specific enzymes by oxidation of cofactors ⇒

(5) Vasoconstriction-related *Maladaptation* through an *ongoing I/R-I* ⇒ NECROSIS!

Notes: I/R-I, ischemia/reperfusion – injury; ROS, reactive O_2 species; RhoA GTPase is a key intracellular regulator of actomyosin cellular dynamics as well as other functions, including adhesion, proliferation, and survival; COX-2 or cyclooxygenase-2 is an enzyme responsible for inflammation and pain; PUFA or polyunsaturated fatty acids are fatty acids that contain more than one double bond in their backbone and play a major role in inflammation. However, PUFA in culinary oils undergo oxidative deterioration at temperatures of 150 °C (302 °F) with a free radical chain reaction, able to oxidize the PUFAs into hydroperoxide and several secondary products.

surgical procedure, and the mortality rate remains still substantial with values of 25–30%). Medical imaging may show the diagnostic *pneumatosis intestinalis* (air in bowel wall) (Gr. πνευματικός "relating to wind or air" arises from πνεῦμα "air, breath, spirit, wind" and πνέω or "I breath, blow"). NEC clinical stages include *Stage I* ("Suspected") characterized by abdominal distension, ↑ pregavage residual, blood in stool, lethargy, apnea, and bradycardia; *Stage II* ("Definite"), which is the same as that for I + *pneumatosis intestinalis* or portal vein gas; and *Stage III* ("Advanced"), which is I/II + shock, DIC, ↑ pH blood, ↓ platelets, and intestinal perforation.

- *GRO*: Distended, lusterless, and gray-greenish bowel loop or segment, which is incredibly soft and fragile and potentially prone to perforation, if the process is transmural.
- *CLM*: Mucosal coagulative/ischemic necrosis with a prominent acute inflammatory infiltrate, fibrinous exudate ± pseudomembrane formation, and *pneumatosis intestinalis* are observed. Tips and pitfalls of NEC diagnostics are provided in Box 3.21.

Box 3.21 NEC Pearls and Pitfalls
- *GA and NEC*: NEC incidence, severity, and complications are inversely proportional to the GA. Advances in the supportive care of premature babies, such as a surfactant, new techniques of mechanical ventilation, and better skilled and trained PICU personnel, have enabled to increase the survival of VLBW infants determining the increase of "involuntary" NEC rate.
- *Risk factors*: Prematurity (<28 weeks), low APGAR, enteral feeding, FGR, chorioamnionitis, maternal hypertensive disease of pregnancy, placental abruption, maternal use of cocaine, absent or reversed end diastolic flow velocity, umbilical catheters, and packed cell transfusion.
- *NEC susceptibility factors*:
  1. *M*echanical factors (↓peristalsis, ↓ mucus layer, abnormal lipid composition)
  2. *M*icrobiological factors (abnormal bacterial colonization with potentially a scarcity of anaerobic bacteria)
  3. *M*iscellaneous (↓gastric acid, lactase, and abnormal, BA-deficient bile micelles)
- *Mucosal homeostasis critical factors (MHCF)*: Lactoferrin, glutamine, EGF, HB-EGF, EPO, IGF, TGF, gastric acid, PUFAs, and nucleotides.
- *Genetics*: Concordance of NEC in monozygotic vs. dizygotic twins.
- Up to 1/2 of NEC will develop an "advanced" stage, which will require surgical treatment. *Surgery* indications include *clinical deterioration*, *ileus*, *perforation*, and *intestinal stricture*.
  - Perforation: single or multiple, usually at necrotic areas or necrotic healthy boundaries.
  - Intestinal strictures: fibrosis of "NEC intestine", which occurs after 2–3 weeks to months/years after recovery.
- *SBS* (aka *intestinal failure*) occurs when small intestine (<25–40 cm) remains following NEC ± surgery.
- *Spontaneous intestinal perforation (SIP)* occurs in VLBW infants during the first 1–2 weeks of life before starting feeding and has a relatively better prognosis than NEC. Microscopically, there is neither villous necrosis nor pneumatosis in contrast to NEC.
- *Neurodevelopmental delay (NDD)* involves cerebral palsy and/or mental retardation.
- *Prevention*: Probiotic supplementation seems to ↓ the risk of NEC in VLBW infants.

Note: BA, bile acids; EGF, epithelial growth factor; HB-EGF, Heparin-binding EGF-like growth factor; IGF, insulin growth factor; EPO, FGR, fetal growth restriction; GA, gestational age; PUFA, polyunsaturated fatty acids; TGF, transforming growth factor; VLBW, very low birth weight.

PGN: In consideration of the complications, it is essential to state that NEC is the leading cause of death and long-term disability from GI disease in premature babies with *intestinal stricture*, *short bowel syndrome* (SBS), and *neurodevelopmental delay* as outmost complications. The pathogenesis of neurodevelopmental delay or damage of infants following NEC is probably multifactorial, with both nutritional and non-nutritional factors at play (Hickey et al. 2018).

Animal Models

Animal models of NEC need to have some characteristics to be labeled truly NEC models. They need to mimic the pattern of disease and show histopathology, which resembles the human condition. One of the earliest descriptions of an NEC model is that it is developed in rat pups by Barlow and Santulli in the 1970s (Sodhi et al. 2008). The development of experimental NEC was induced through the combined use of formula gavage and intermittent episodes of either cold or hypoxic stress, and microscopic investigation validated NEC. This investigation and numerous additional studies that followed it with all limitations have indicated the critical role that the rat model has played to advance our understanding of the pathogenesis of this intriguing disease. NEC mouse models have not been very much successful, mainly because the gavage-fed neonatal mice are challenging due to the smaller size of mice pups if compared to rat pups. TLR4-mutant mice have been demonstrated to be protected from the development of NEC (Jilling et al. 2006). A role of PI3K signaling was identified in NEC-induced Swiss Webster mice (Baregamian et al. 2007), while IL-18 involvement was demonstrated by Halpern et al. (2008). More NEC models pointed to the role of IFN-gamma KO mice and iNOS KO mice (Leaphart et al. 2007; Cetin et al. 2007). It remains ambiguous the underlying mechanism of how self-renewing cells are controlled in regenerating tissues and cancer. The Proliferating Cell Nuclear Antigen (PCNA)-associated factor (PAF, aka p15/KIAA0101/NS5ATP9/OEACT-1) modulates DNA repair and hyperactivates Wnt/β-catenin signaling. PAF is expressed in stem and progenitor cells (ISCs and IPCs) of the intestine and markedly upregulated during both intestinal cell regeneration and oncogenesis (Kim et al. 2018). The T-cell-mediated mucosal immunity, ROS, rectosigmoid intramural pH, and PAF have been successful using piglets. The piglet model for its simplicity is on the top of the list due to the only administration of formula to preterm piglets but does not have the potential involvement of the hypoxic insult.

3.4.8.2 Gluten-Sensitive Enteropathy

- *DEF*: Gluten-associated (sensitive) enteropathy (GSE) or celiac disease of autoimmune origin targeting peculiarly the duodenum and proximal jejunum with sparing of the ileum with serologically positive antibodies and microscopic evidence of small bowel villous atrophy of variable degree, intraepithelial lymphocytosis/lymphocytes (IEL), inflammatory infiltration of the lamina propria, and crypt hyperplasia. Serologic tests include three antibodies common in celiac disease, i.e., anti-tissue transglutaminase (tTG) antibodies endomysial antibodies (EMA), and deamidated gliadin peptide (DGP) antibodies with the most sensitive antibody tests being that of the IgA class, although IgG tests may have a role in individuals with IgA deficiency (*vide infra*).

- *EPI*: ~1:100 (1%) to 1:250 among individuals of Western European heritage (Europeans and North Americas) and two major incidence peaks for age (<3 years and third to fifth decades).

- *RF*: Some diseases are also associated with this disease. They include DM type I, autoimmune thyroiditis, Sjögren syndrome, microscopic colitis, isolated IgA deficiency (~2% of the patients), trisomy 21 syndrome (Down syndrome), dermatitis herpetiformis (Brocq-Duhring disease), and epilepsy. Familial risk is estimated to be 10–14%, but other authors report a higher incidence rate.

- *EPG*: After washing dough, the viscoelastic mass, which remains, is wheat gluten. Wheat gluten is a mixture of alcohol-soluble monomers (gliadin) and alcohol-insoluble, large polymers (glutenins). More than 200 genes are encoding gluten proteins in bread-mak-

(a) *Asymptomatic*
(b) *Mild symptoms* (≥1 of abdominal pain, flatulence, diarrhea, belching, tiredness, joint pain)
(c) *Moderate symptoms* (persistence of the previous symptoms and disturbed normal life)
(d) *Severe symptoms* (daily symptoms that restrict normal life OR excessive weight loss)

ing wheat (*Triticum aestivum*). Two other cereals, which are toxic in celiac disease, are barley and rye. Both grains have proteins structurally related to wheat gluten, which are called *hordeins* (in barley) and *secalins* (in rye). *Prolamins* are used to group *gluten*, *hordeins*, *secalins*, and *avenins*. The last proteins are the storage proteins responsible for the occasional toxicity of oats in patients with celiac disease. It has been suggested that environmental factors are highly responsible for altering the risk that HLA-predisposed subjects will develop GSE by shaping the immunologic context in which gluten is presented. Food additive may also trigger celiac disease, although more in-depth studies are necessary to enlist some factors and exculpate others.

- *CLI*: Four clinical categories, including:

Although *antiendomysial antibody* (IgA EMA) and *anti-tissue transglutaminase* (IgA TTG) antibody measure practically the same antigen, the results may vary between different laboratories and the PPV (positive predictive value) of tTG-Ab for celiac disease may be quite low. Conventional gliadin antibodies are no longer recommended. They have low sensitivity and specificity, although the new generation of *deamidated gliadin peptide antibodies* have improved diagnostic accuracy. Sites: duodenum > ileum. The PPV is defined as the number of true positives over the sum of true and false positives (positive calls) with the ideal value of the PPV, with a perfect test, being 1 (100%). Conversely, the worst possible PPV would be zero.

- *GRO*: Endoscopic features include scalloping, fissures, mosaic appearance, and a decrease of the mucosal folds, but they are neither sensitive nor specific findings for GSE.
- *CLM*: The histology of the small bowel biopsy remains the gold standard for the diagnosis of GSE, although it has been indicated that this may not be the case for some pediatric scenarios. Small bowel biopsy should be performed in the seronegative subject with signs and symptoms highly suspicious of GSE because no serologic test has a sensitivity of 100%! AGA Guidelines state that four to six specimens should be submitted to the pathologist, who should be aware of mild forms. Failure to biopsy the duodenal bulb may affect the diagnosis rate of GSE. The diagnosis of GSE should be performed according to the North American and European Societies for Gastroenterology, Hepatology, and Nutrition criteria, and the histological lesions of the intestinal mucosa are defined according to Marsh and Marsh-Oberhuber Classifications. Marsh classification, the worldwide most used system, is based on increased IELs (>40/100 enterocytes); partial, subtotal, or total villous atrophy; and elongation of the crypts with a decreased villous-to-crypt ratio. The range of the histologic patterns includes pre-infiltrative/normal (type 0), infiltrative lesion with increase of IELs (type 1), hyperplastic lesion with crypt hyperplasia (type 2), destructive lesion with variable degree of villous atrophy (type 3), and the hypoplastic lesion with total villous atrophy and crypt hypoplasia (type 4). The Corazza-Villanacci has substituted the Marsh-Oberhuber classification in some countries (Corazza and Villanacci 2005, Sergi et al. 2017).
- *DDX*: IEL mimickers.

GSE mimickers are diseases simulating GSE and leaving the patients to wrong clinical management, sometimes for a long period. IELs alone are not diagnostic of GSE, because there are numerous GSE mimickers (Sergi et al. 2017). In some situations, the location of IELs may help. GSE-associated IELs are more likely to be located at the villous tip, while non-GSE-associated IELs are laterally located, patchy distributed IEL or low down cryptically located. It is important to remember that >1 PMN defines focal acute inflammation in the lamina propria or epithelium and more than >1 focus in a tissue biopsy. Acute duodenitis (focal) is not a sensitive feature in Crohn disease but has high specificity (92%) and high predictive value (93–95%). The interobserver rate of interpreting duodenal biopsies may show different κ-factors depending on the institution but also from the gastrointestinal histopathological education of the pathologist. Lymphocyte spreading seems to be decidedly sensitive, but it may require additional training in the correct interpretation of the histology of the upper GI tract. The diagnosis of GSE may remain challenging because no single test shows 100% sensitivity and 100% specificity in every patient. The most common GSE mimickers are *Helicobacter pylori (H. pylori) infection*; *drugs*, especially NSAIDs or proton-pump inhibitors (PPIs); and inflammatory bowel disease (*IBD*). *H. pylori* infection is linked to chronic active gastritis, ulcer disease, and chronic active duodenitis-bulbitis. Nonspecific duodenitis and peptic duodenitis are situations with an acid injury. *H. pylori* infection is classically associated with duodenal gastric metaplasia, which is characterized by gastric-type mucus-secreting cells (foci) interspersed between duodenal enterocytes, which may be easily recognized by the PAS (+) staining of the cells having neutral mucin and no brush border. NSAIDs have been associated in a few cases of duodenal IEL. Brunner gland hyperplasia does not seem to be relevant. Other drugs include colchicine, mycophenolate mofetil, ipilimumab, and several chemotherapy agents or radio-/chemotherapy protocols. These drugs can produce villous architectural changes. Both Crohn disease and ulcerative colitis may have IELs in the duodenum. Duodenal granulomas are a useful finding in confirming the diagnosis of Crohn disease, but they are seen in <50% of the patients. In patients with the classical presentation of Crohn disease, villous shortening accompanied by an inflammation rich in neutrophils and edema in the lamina propria accompanied by crypt abscesses may be found. Crucial is the correlation with findings arising from other sites because isolated duodenal Crohn disease is sporadic. In up to 1/4 of patients with ulcerative colitis, variable villous blunting, inflammatory plasma cell expansion of the lamina propria, and active (neutrophilic) inflammation are also seen. Food allergy, cow's milk protein-sensitive enteropathy (CME), non-*H. pylori* infection (e.g., *Yersinia enterocolitis* and *Salmonella* spp.), small intestine bacterial overgrowth, tropical sprue, sickle cell anemia-duodenitis, and various immunological or autoimmune disorders need to be taken always into consideration. Infections include giardiasis, cryptosporidiosis, microsporidiosis, cyclosporiasis, isosporiasis, Whipple disease, *Mycobacterium avium intracellulare*, visceral leishmaniasis, cryptococcosis, and cytomegalovirus. Common variable immunodeficiency (CVID) may also present with gastrointestinal symptoms. CVID is the second most common primary immunodeficiency, and its diagnosis relies on recurrent infections, ↓ IgG levels at least two standard deviations below normal with at least ↓levels of ≥1 Ig subclass, exclusion of other causes of immunodeficiency, and a failure to mount a response to vaccination. In CVID, the duodenal biopsy may show IEL with or without villous architectural changes and scarcity or lack of plasma cells with prominent apoptosis of the crypts in CVID. GVHD and allograft bowel rejection (AGBR) may be ruled out in clinical settings. In fact, in an appropriate clinical setting, a decrescendo from base to apical villi and the finding of epithelial apoptosis deeply in the crypts, with or without some degree of architectural disturbance, may help to address the diagnosis of GVHD. Collagenous sprue (COS) is also a GSE mimicker and shares several aspects of GSE, including villous structural abnormalities, IEL, and crypt hyperplasia. An irregularly thickened layer of type 1 collagen just subjacent to the surface epithelium is useful for distinguishing COS from GSE. A monotypic, truncated Ig α heavy chain lacking an associated light chain secreted by plasma cells infiltrating the bowel wall characterizes a condition called immunoproliferative small intestine disease (IPSID), which is a MALT lymphoma. IPSID is mostly reported in individuals from the Middle East, North/South Africa, and the Far East. Several immune-related disorders, including Hashimoto thyroiditis, Graves' disease, rheumatoid arthritis, psoriasis, and systemic lupus erythematosus, may cause IEL. Finally, the enteropathy-type intestinal T-cell lymphoma (EITL) may be considered a complication of longstanding GSE and is also a GSE mimicker. EITL is frequently multifocal with ulcerative lesions. There is a tendency to perforate either at clinical presentation or during the protocols of chemotherapy. Histologically, there is a pleomorphic medium- to large-sized cell population with CD3 and lack of expression of both CD4 and CD8 as well as a small and monomorphic lymphocytic population characterized by the presence of CD3, CD8, and CD56 and no CD4. CD30 is always present in the tumor cells and may be detectable in the adjacent villi of the lymphoma lesions. It is considered an ominous marker for prognosis (*vide infra*).

- *PGN*: GSE is a lifelong disorder with a significant impact on daily life. Each GSE patient should be referred to an experienced dietitian and a celiac support group as well as carefully educated about the physiology of the disease, importance of diet, and potential consequences of inattention to gluten-free diet (GFD), such as anemia, osteopenia, and increased risk of malignant and nonmalignant complications. Lethal complications include EATL, adenocarcinoma of the small intestine, esophageal cancer, and melanoma. Nonmalignant complications include ulcers, small bowel obstruction, bleeding, and recurrent acute pancreatitis as well as nongastrointestinal complications, which are usually the consequence of malnutrition or specific deficiencies. Neurological problems include ataxia, peripheral neuropathy, or dementia that may have autoimmune pathogenesis. Other implications of GSE belong to the atypical nongastrointestinal presentations (*vide supra*). However, we found out that there are more clinical subtypes of GSE than only typical and atypical. Thus, *typical* is fully expressed GSE with intestinal symptoms of malabsorption; *atypical* is fully expressed GSE with *extraintestinal* involvement (Fe deficiency anemia, short stature, osteoporosis, infertility); *silent* is fully expressed GSE, but usually *symptom-free* (+ *family history*); *potential* is normal mucosa or minimal changes and characteristic serologic abnormalities (+ *HLA-DQ2 ± family history*); *latent* is normal mucosa, but villous atrophy develops later on, following some *events*, including tumor, environmental triggers, stress, surgery, trauma, or pregnancy; and *refractory* is severe, symptomatic, enteric atrophy *not responding* to >6 months of a strict GFD. Moreover, Catassi-Fasano five-point scoring system has been proposed for patients who have signs and symptoms of GSE but have borderline histology or who refuse an endoscopy. The presence of four out of the five criteria (or three out of four, if gene testing is not performed) would meet diagnostic criteria for GSE. Failure to respond to GFD may be due to some clinical settings, including no GSE, pancreatic insufficiency, bacterial overgrowth, microscopic colitis, irritable bowel syndrome, anal sphincter incompetence, dietary intolerance (lactose, fructose, soy, and milk protein), collagenous sprue, ulcerative jejunoileitis (UJI), and EATL. A misdiagnosis should also be considered. The differential diagnosis of villous atrophy includes acute viral enteritis, cow's milk or soy protein intolerance, lambliasis (*G. lamblia*) and other parasites, acquired immunodeficiency syndrome, hypogammaglobulinemia, Whipple disease, connective tissue disorders, radiation or drug enteritis, and tropical sprue. There are a few different etiologic factors of malabsorption with a lesion-like GSE, including refractory sprue, tropical sprue, infectious gastroenteritis, stasis syndrome, and kwashiorkor. Tropical sprue responds to folic acid and broad-spectrum antibiotics, but not to GFD, while in the stasis syndrome, lesions are usually patchy and rarely severe, and there is a response to antibiotics. Finally, kwashiorkor is a severe diffuse lesion due to protein deficiency and is not related to gluten. In celiac disease, the involvement of the gastric mucosa is often described, mainly in the form of chronic superficial gastritis and lymphocytic gastritis. Nenna et al. (2012) emphasized that mucosal participation is not only confined to the duodenum. In their series of children affected with GSE, about 1 in 5 and about 1 in 20 patients showed chronic superficial gastritis and lymphocytic gastritis, respectively. Although *Helicobacter pylori* (Hp) infection was also found in very few patients, this accompanying infection is quite debated in the literature. Gastric lesions are classified according to the Updated Sydney System that comprises several groups of diseases grouped in three categories (acute, chronic, and unique or distinctive) and inflammation degree in three grades (grade 1 for mild lesions, 2 for moderate injuries, and 3 for severe injuries) and rapid urease testing and histology diagnose Hp infection. An association has occasionally been observed in childhood and youth of celiac disease with sarcoidosis and a recent systematic review and meta-analysis found a significantly higher risk of sarcoidosis among patients with celiac disease (Wijarnpreecha et al. 2019). In patients with sarcoidosis, there is an altered gastrointestinal mucosal immune response, which is accompanied in about 40% of patients by specific sensitization to wheat protein. Gastrointestinal symptoms of GSE may appear at the same time or precede those of sarcoidosis. GSE complications occur usually in the adult age. They include *EATL-I/II* (see lymphoproliferative processes); *chronic ulcerative jejunoileitis*, which shows recurrent episodes of small intestinal ulcerations and formation of strictures; *gastrointestinal adenocarcinoma* of the usual jejunum; *collagenous disease* with collagen just under the surface epithelium with a poor prognosis; and *metabolic disorders* (lactose intolerance, ↓ [Ca] level, osteoporosis, and short stature).

- *DDX*: The differential diagnosis of GSE may be quite broad and include some differentials that may not occur in childhood (Box 3.22).

Box 3.22 Most Common Differential Diagnoses with Pathology of Malabsorption
- *Abetalipoproteinemia*: Normal villous morphology with surface epithelium showing vacuolated enterocytes filled by lipid particles (ORO+ on frozen section).
- *Eosinophilic gastroenteritis*: Villous abnormality with disproportionate eosinophils (the pathological diagnostic criteria of an eosinophilic gastrointestinal disorder are as follows: esophagus, eosinophil count ≥ 15/HPF; stomach and duodenum, eosinophil count ≥ 25/HPF; colon and rectum, eosinophil count ≥ 30/HPF according to Lucendo and Arias 2012).
- *Giardiasis (lambliasis)*: Villous abnormality of variable severity, usually mild and few to numerous parasites on the surface coat of the surface epithelium (Leung et al. 2019).
- *Intestinal lymphoma*: Villous abnormality of variable severity and atypical lymphoid infiltrate localized in the lamina propria (IHC is crucial).
- *Lymphangiectasis*: Villous abnormality of variable severity that is determined by the patchy or diffuse presence of dilated lymphatics (D2-40/podoplanin + by IHC).
- *Strongyloidiasis*: Villous abnormality of variable severity with intracryptic worms and larvae (Egg: 55 x 30 μm and oval; adult: parasitic or free-living male ~ 0.7 mm in length, while parasitic female ~ 2.2 mm in length, and free living female ~ 1 mm in length) (Beknazarova et al. 2016).
- *Whipple disease*: Villous abnormality and lamina propria filled with PAS+ macrophages with *Tropheryma whippelii* (detectable on TEM/0.5–1 μm semithin sections during active disease).

Note: IHC, immunohistochemistry; TEM, transmission electron microscopy

The small intestinal abnormalities in addition to pathognomonic lesions in other organs are summarized in Box 3.23. Here, we show diseases with impaired epithelial replacement, which can be patchy. The proximal small bowel is more damaged than the distal portions because there is significant mucosal exposure of the proximal regions to the highest concentration of gluten if we consider the untreated patient. Remarkable results came out from several experimental studies. When the distal portion is exposed to a high concentration of gluten, this portion of the small bowel will become like the proximal parts.

In case the patient does not respond to a GFD, some causes need to be taken into account, including poor compliance, T-cell lymphoma, carcinoma, and collagenous sprue.

Box 3.24 provides some tips in the diagnostics of GSE.

The ESPGHAN and NASPGHAN combined criteria are illustrated in Box 3.25. The ESPGHAN targets children exclusively, while ACG and WGO target both children and adults.

The report of GSE should contain paramount and clinically-relevant data as reported in Box 3.26.

Box 3.27 provides some pitfalls in the diagnostics of GSE.

Box 3.23 Small Intestine and Pathognomonic Lesions in Other Organs

- *Acrodermatitis enteropathica (Danbolt-Closs syndrome, Brandt syndrome)*: AR-inherited metabolic disorder affecting the Zn uptake through the mucosa of the bowel and characterized by periorificial and acral dermatitis, alopecia, and diarrhea.
- *Amyloidosis*: Amyloid in blood vessels detected in rectal biopsy using Congo red and apple-green birefringence with polarized light.
- *Capillariasis*: Parasitic infection caused by two species of nematodes, *Capillaria hepatica*, which induces hepatic capillariasis, and *C. philippinensis*, which causes intestinal capillariasis.
- *Chronic granulomatous disease*: Clumps of vacuolated pigmented macrophages in lamina propria and normal villous morphology.
- *CMV*: Variable and patchy intestinal lesions with enlarged cells with typical large intranuclear oval inclusions surrounded by a halo.
- *Crohn disease*: Transmural inflammation, ulcerations, lymphoid aggregates, and non-caseating granulomas with or without giant cells.
- *Gastrinoma* (ZES): Patchy acute inflammation with microulceration that disappears after gastrectomy and diagnosis based on excessive gastric acid secretion and ↑serum gastrin level.
- *Histoplasmosis*: Villous morphology showing club-shaped villi with lamina propria stuffed with macrophages containing characteristic encapsulated fungi (*H. capsulatum*).
- *Lipid storage disorders*: Vacuolated macrophages and ganglion cells in autonomous plexus using semithin sections, histochemistry, IHC (anti-CD68 monoclonal antibody), and TEM.
- *Macroglobulinemia*: Plasma cell dyscrasia with increased levels of macroglobulins in the circulating blood showing extracellular amorphous hyaline in lamina propria.
- *Schistosomiasis*: Ova of *S. mansoni* or *S. japonicum* in the stool on a simple smear (1–2 mg of fecal material) using the Kato-Katz technique (20–50 mg of fecal material) or the Ritchie technique.
- *Soy protein reaction*: Lesions severe, non-specific, soy protein related, but very rare.

Note: AR, autosomal recessive; CMV, cytomegalovirus; ZES, Zollinger-Ellison syndrome; TEM, transmission-electron microscopy

Box 3.24 GSE Pearls and Pitfalls
- Gluten-sensitive enteropathy (GSE) should be considered as a continuum with findings ranging from early mucosal changes through overt disease with total villous atrophy.
- The grade of Marsh classification is indicated by the most severe change in a duodenal biopsy at one time, although this grading has been controversially discussed and most authors report IIIA as partial villous blunting in a mixture of normal and subtotal villous blunting.
- Small bowel mucosal IELs without atrophy and crypt hyperplasia (Marsh I) or with crypt hyperplasia (Marsh II) and lab-based positive endomysial antibodies belong to the spectrum of genetic gluten intolerance, and patients benefit absolutely from a GFD.
- CD3+ IELs may be stained with monoclonal antibody Leu-4, $\alpha\beta$+ IELs with the monoclonal antibody BetaF1 antibody, and $\gamma\delta$+ IELs with the TCR-γ antibody.
- Villous height/crypt depth ratio (Vh/CrD) should be measured as a mean of at least five well-oriented villous-crypt pairs, and a ratio of <2 is regarded as Marsh III.
- IELs were originally counted with a 100× flat field, LM objective (10 × 10) in a minimum of 30 fields measuring 1.6 mm in epithelial length and expressed as cells/mm of the epithelium. However, a new method is typically applied and consists of counting the IELs in the 20 uppermost enterocytes in five randomly chosen villous tips (Goldstein and Underhill 2001; Biagi et al. 2004).
- Catassi-Fasano five-point scoring system has been proposed for patients who have signs and symptoms of the GSE but have borderline histology or who refuse an endoscopy (Catassi and Fasano 2010).
- Subjects with a family history positive for GSE, AID with high comorbidity with GSE (e.g., type 1 diabetes, Hashimoto's thyroiditis), and genetic diseases with high comorbidity with GSE (e.g., Down syndrome) should receive endoscopy with small bowel biopsy evaluation as early as possible.
- The tissue specimen needs to be appropriately oriented on a filter paper with the villi uppermost and measurement of the villus height/crypt depth (VH:CD) ratio, IEL, and surface enterocyte cell height (SECH) should not be used for research purposes only, but it may be convenient to report regularly, and image-analyzing software has facilitated this approach nowadays.
- Failure to respond to a GFD should raise the possibility of either poor dietary compliance or an incorrect diagnosis by the pathologist.
- Once used for diagnosis, the gluten challenge (40 g of gluten or four slices of regular bread per day for ≥2 weeks for adults and 6 weeks for children) is now reserved exclusively for occasional cases where there is doubt concerning the initial diagnosis and no serology or HLA studies have been able to demonstrate GSE.

3.4.8.3 Non-GSE Causes of Malabsorption

Giardiasis/Lambliasis
- *DEF*: An acquired form of malabsorption due to ingestion of water contaminated with infective cysts and with localization of the trophozoite residing in duodenum close to the epithelial surface, which looks like in most cases undamaged, although intraepithelial lymphocytosis can be seen.
- *CLI*: Chronic diarrhea.
- *CLM*: Villous abnormalities with parasites on the surface of the villi ± inflammation.

Comparative studies between symptomatic and asymptomatic *Giardia* infestation in humans showed no significant difference between these

Box 3.25 North American-European Divergences Across Oceans in GSE

	Screening	PS-Tests[a]	HLA-DQ	EMA	AGA
ESPGHAN	Anti-tTG-IgA and IgA[b]	Anti-tTG-IgG/ anti-DPG-IgG[c]	Yes, if ↑EMA/anti-tTG	Yes, in confirming PS tests	No
ACG	Anti-tTG-IgA	Anti-tTG-IgG/ anti-DPG-IgG[c]	Yes, if biopsy/serology disagreement	NS	No
WGO	Anti-tTG-IgA/ anti-DPG IgG	NS	Yes, if biopsy/serology disagreement	Yes, in confirming PS tests	No

Notes: *ESPGHAN* European Society for Paediatric Gastroenterology, Hepatology, and Nutrition, *ACG* American College of Gastroenterology, *WGO* World Gastroenterology Organization, *anti-tTG* antibodies anti-transglutaminase
NS, not specified
[a]postscreening tests
[b]total serum IgA
[c]anti-DPG, antideamidated gliadin peptide

Box 3.26 Synoptic Report for Gluten-Sensitive Enteropathy and Gluten-Sensitive Enteropathy Mimickers

Name: .... Case#:

Location: ....

Distal Duodenum Biopsies #:   Villous/Crypt Ratio #:   :1

Bulb Biopsies #:

Lamina Propria Inflammation #:   Lymphocytes   Plasma cells   Neutrophils

IELs #/100 epithelial cells (Top):   Enterocyte Damage:

IELs #/100 epithelial cells (Side):   Brush Border Thickness (µm):

IELs #/100 epithelial cells (Down):   Erosion:

Crypt Hyperplasia   Micro-organisms: ....

Benign (B) / Dysplasia (D) / Malignancy (M)   ....

Box 3.27 GSE Pitfalls
- Overdiagnosis of villous atrophy may occur in the setting of misinterpretation of histologic sections, which are prepared suboptimally, when the samples are too small, are sent in poor condition (no formalin or another non-conventional fixative), or have been cut tangentially and when gastric pyloric mucosa has been submitted labeled as "duodenal biopsy."
- The normal VH:CD ratio is invalid when the villi overlie Brunner glands of the duodenal cap. If proximal duodenal specimens need to be evaluated, because they can show villous atrophy, it is mandatory to receive biopsies appropriately also from the distal duodenum as well.

two groups about the CD4 and CD8 lymphocytic infiltration of patients living in Dakahlia Governorate, Egypt. Some emerging health conditions have been seen recently increased in the Western world, and most of them are linked to some change in the habit of our society.

Tropical Sprue
- *DEF*: Unrelated to gluten ingestion, malabsorption syndrome, which presents with similar clinical and pathological features to celiac disease, almost exclusively limited to tropics (living in or visiting) that may be related to enterotoxigenic *E. coli* and that responds to folic acid, vitamin B_{12}, and tetracycline.
- *CLM*: Numerous lymphocytes and occasional eosinophils efface partial villous atrophic architecture in most cases, mostly in the distal small bowel.

Whipple Disease
- *DEF*: Intestinal and systemic lipodystrophy with large macrophages (stuffed with diastase-resistant PAS-positive "bacilliform bodies," Gram+ *Tropheryma whippelii*) packing the lamina propria and distorting the villi, alternating with dilated lymphatic channels and empty spaces containing neutral lipids and occasionally with giant cells (lipogranulomas), affecting both children and adults (Fenollar et al. 2016).
- *EPI*: Usually whites, ♂ : ♀ = 10:1, 10s–40s.
- *CLI*: Sites: GI tract, LNs, heart, lung, liver, spleen, adrenal glands, and the nervous system.

Other causes of malabsorption include biliary obstruction with bile acid deficiency, chronic pancreatitis, amyloidosis, reduced small bowel length after surgical resection, infections, and lymphatic obstruction.

Anisakiasis
Human anisakiasis has been observed in children, adolescents, and young adults and is mostly based on an eosinophilic granuloma. This infection needs to be taken into account in the differential diagnosis of acute abdomen in individuals who have ingested raw fish or squid a few hours or a few days before the onset of symptoms. Anisakis is a larva (1.7–2.8 cm) of nematodes of the family *Anisakidae* (Ascaridoidea), which is the agent responsible for the parasitic zoonosis, called anisakiasis. These parasites habit adult marine mammals, and their larvae utilize crustaceans, cephalopods, and fish as intermediate and/or paratenic hosts. The infection is strictly linked to eating habits where large amounts of raw fish are widely consumed and countries, such as Japan, may harbor an increased risk for this disease. In fact, humans can be infected by eating such hosts, either fresh or inadequately treated (pickled, smoked, or salted). Once the infected food is ingested, the nematode enters the digestive tract by penetrating the gastric or intestinal wall and can sometimes reach the peritoneum, mesentery or omentum, or other viscera, causing acute and chronic inflammatory reactions that may simulate other abdominal disorders. It seems that the pathogenesis is both associated with allergic Arthus-type/anaphylaxis reactions and the physical action of perforation by the larvae.

3.4.8.4 Intractable Diarrhea of Infancy

Some conditions have been identified in childhood, and the etiology has been successfully tracked in many of them, who may affect adult life. A few diseases have received particular attention, and genetic and biochemical findings have been clarified, but their diagnosis remains delayed or idiopathic. The evidence is that some of these babies affected by intractable diarrhea of infancy (IDI) often require total parenteral nutrition (TPN) with inevitable liver sequel and transplantation to survive. There are two major categories of chronic IDI that can be distinguished one from each other on several grounds.

The first category is constituted by *congenital infantile enteropathies* characterized by intrinsic defects of the enterocytes, including *microvillous inclusion disease* (MVID), *intestinal epithelial dysplasia* or tufting enteropathy (IED-TE), and *syndromic diarrhea of infancy* (SDI). These disorders are characteristically detected for the high degree of consanguinity in the affected families. The inheritance seems to follow an autosomal recessive pattern in most of the cases, although the genes or the precise etiology remains unknown for the majority of cases.

The second category is labeled *immuno-inflammatory enteropathies* and includes *autoimmune enteropathy* (AIE) and *IPEX syndrome* (immune dysregulation, polyendocrinopathy, enteropathy, X-linkage). The identification of the molecular basis of the IDI may pinpoint new treatment strategies avoiding the early and late consequences of the actual therapy.

Microvillous Inclusion Disease (MVID)

- *DEF*: IDI characterized morphologically by *absent or abnormal microvilli* (remnants) with denudation of the brush border of the enterocytes containing *subluminal vesicles* as well as *autophagolysosomes with ectopic microvilli-forming inclusions* (osmiophilic bodies) and, in some settings (populations), genetically related to mutations of the *MYO5B* gene (Navajo populations), causing a *disruption of the epithelial cell polarity* (Iancu and Manov, 2010). Myosin Vb uncoupling from RAB8A and RAB11A (GTPases) elicits MVID (Knowles et al. 2014). Light microscopy of a jejunal biopsy shows partial villous atrophy, and characteristic accumulations of periodic acid Shiff-positive material in the apical cytoplasm of epithelial cells and ultrastructural images show absent or extensive damage of the apical microvilli with increased numbers of intracytoplasmic secretory granules and microvillous inclusions. This disturbed exocytosis of subluminal brush border vesicles has also been observed in some genetic syndromes, such as autosomal dominant hypochondroplasia (Heinz-Erian et al. 1999) with the gene defect localized, on chromosome 4p16.3, to the proximal tyrosine kinase domain of the fibroblast growth factor receptor-3 (*FGFR3*) gene.

Intestinal Epithelial Dysplasia/Tufting Enteropathy (IED-TE)

- *DEF*: IDI characterized morphologically by a variable degree of *microvillous irregularity* (size shortening, glycocalyx incompleteness, and abnormalities of the core rootlet) giving rise to *partial villous blunting with diffuse tufting change of the luminal epithelial cells* ("micropapillary or teardrop" appearance) and crypt hyperplasia. Contrary to MVID, in IED-TE, a *defect of basement membrane anchoring of enterocytes* is suspected, and, indeed, elongated desmosomes can be found ultrastructurally (TEM). Most conveniently, immunostaining more easily identifies them with antibodies directed against desmoglein. The desmogleins are a family of desmosomal cadherins that functions to maintain tissue integrity and facilitates enormously cell–cell communication. Commonly, desmogleins are the target antigens in epidermal blistering diseases such as pemphigus, which is caused by autoantibody-mediated acantholysis determining a disruption of keratinocyte adhesion. The microvillous irregularity is at the basis of this peculiar morphology (Reifen et al. 1994; Goulet et al. 2007; Goulet et al. 2008).

Syndromic Diarrhea of Infancy (SDI)

- *DEF*: IDI associated with phenotypic anomalies, aka phenotypic diarrhea or tricho-hepato-enteric syndrome, presents in *SGA* (<10th percentile) babies. SDI is related to *nonspecific (subtotal to total) villous atrophy* with a mild or very mild inflammatory mononuclear cell infiltration of the lamina propria. Moreover, there is a *dysmorphic face* (prominent forehead and cheeks with wooly hairs, which are easy to remove and poorly pigmented, broad nasal root, and hypertelorism), *immunological disorders* (T-cell immune deficiency with defective antibody production), and, in some infants, early onset of *liver fibrosis* quite progressively complicated to pigmented (Fe-loaded) cirrhosis (Goulet et al. 2008). There are mental retardation and disturbances of the fine motor activity. The investigation of the hair reveals braided hair, aniso- and poikilotrichosis (Gr. ποικίλος "variegated, spotted"), trichorrhexis (Gr. θρίξ "hair" and ῥῆξις "rupture") nodosa and longitudinal breaks, and trichothiodystrophy ("sulphur-deficient, short, brittle hair") (all of them are nonspecific abnormalities). Trichorrhexis nodosa (TN) is a quite common defect of the hair shaft with focal loss of cuticles, causing fraying of the cortical fibers and, consequently, hair breakage. Biochemically, there may be an alteration of the amino acid metabolism. TEM does not put in evidence any anomalies of the MVID or IED-TE. Prognosis is reduced, and death is often linked to end-stage liver disease (TPN complicates the first involvement of the liver, especially if continuous, lipid-free, or associated with infections) and/or sepsis.

Autoimmune Enteropathy (AIE)
- *DEF:* Severe and persistent diarrhea (GSE mimicker) with villous atrophy unresponsive to a GFD. A lack of any triggering food protein, antienterocyte antibodies, persistent diarrhea after prolonged fasting, and the presence of organ-specific serum antibodies are essential for the diagnosis of AIE.
- *MIC:* Variable degrees of architectural changes, including regular to total villous atrophy and a CD8-predominant immunophenotype of IELs. Remarkably, the lymphocytes harboring γδ immunophenotype is standard in both surface epithelium and lamina propria and help distinguishing AIE from GSE. AIE may produce subtotal villous blunting (Marsh IIIB) simulating the appearance of GSE. Helpful clues are the lack of goblet cells and Paneth cells and prominent crypt apoptosis.

IPEX Syndrome (Immune dysregulation, Polyendocrinopathy, Enteropathy, X-Linkage)
Severe and persistent diarrhea due to a rare monogenic primary immunodeficiency linked due to mutations of *FOXP3*, a key transcription factor for regulatory T (Treg) cells. The dysfunction of these cells leads to the multi-organ autoimmunity.

The diagnosis of intractable diarrhea of infancy remains a constant challenge in pediatrics, and pitfalls are common (Box 3.28).

Box 3.28 IDI: Pearls and Pitfalls
- Villous atrophy not accompanied by an inflammatory infiltrate should let you think of MVID. Immunohistochemical stains may help to address the diagnosis accurately.
- TN is an essential abnormality of SDI, and a noninvasive investigation of the hair may be a useful diagnostic tool. However, TN may be observed in several syndromes, including argininosuccinic aciduria, citrullinemia, Menkes syndrome, Netherton disease, and several syndromes associated with trichothiodystrophy hair shaft defect.

Systemic Disease with Enteric Manifestations
Congenital Disorders of Glycosylation: Group of diseases associated with a defect in the synthesis of glycans and the attachment of glycans to other molecules, presenting with neurodevelopmental abnormalities, cardiac, GI, and hepatic abnormalities. Biochemically, transferrin IEF is key, and some of the more than 18 different diseases have been genetically identified. There is partial villous atrophy, mild lymphangiectasis, and protein-losing enteropathy as well as dilatation of SER at the top of the villi and inclusions in the lamina propria. Liver biopsy shows fatty change.

Food Protein-Induced Enterocolitis Syndrome (FPIES)
- *DEF:* Non-IgE-mediated gastrointestinal food hypersensitivity representing the severe end of the spectrum of food protein-induced GI diseases in infants.
- *EPI:* FPIES has up to 40% of cow milk protein hypersensitivity in infants and young children.
- *EPG*: Cow milk or soy protein, although other foods can be triggers. Immunologic reactions may be classified as IgE-mediated, non-IgE-mediated (T cells), or mixed (IgE and T-cell mediated). FPIES has a few features that overlap with other non-IgE-mediated GI allergic

disorders, food protein-induced enteropathy, and proctocolitis. Food allergens induce local T-cell-mediated inflammation, leading to ↑ permeability, and fluid shift, but baseline intestinal absorption remain normal by oral food challenge (OFC). Activated peripheral blood mononuclear cells, ↑TNF-α, and ↓TGF-β receptors in the intestinal mucosa play a crucial role in the intestinal inflammation.

- *CLI:* Profuse, repetitive vomiting ± diarrhea, leading to dehydration and lethargy (acute form), or weight loss and failure to thrive (chronic type). FPIES usually begin within 1–4 weeks following the introduction of the trigger, rarely within the first days of life. A constant presentation is also seen. Some FPIES patients have mixed immunologic reactions to the food(s). This interaction causes their FPIES symptoms. Up to 1/4 of infants and children who fulfill the diagnostic criteria for FPIES develop IgE antibodies to the trigger food (atypical FPIES) and tend to have a more protracted course of FPIES with the potential for also developing symptoms of IgE-mediated allergy (e.g., anaphylaxis) with a severe and systemic allergic reaction with tracheal constriction, breathing difficulties, and shock (Gr. ἀνά- "thoroughly, systemic" and φύλαξις "guarding, reaction").
- *CTS/MRI*: Air-fluid levels, nonspecific narrowing, and thumbprinting of the rectum and sigmoid and thickening of the *plicae circulares* in the duodenum and jejunum with quite excessive luminal fluid.
- *LAB*: Anemia, hypoalbuminemia, ↑WBC count with a left shift, eosinophilia, thrombocytosis (>500 × 10^9/L), ↑ peripheral blood PMNs, metabolic acidosis, and methemoglobinemia (methemoglobin is a hemoglobin in the form of metalloprotein, in which the iron in the heme group is in the ferric (Fe^{3+}) state (oxidized form), not the ferrous (Fe^{2+}) of normal hemoglobin), which may be caused by severe inflammation and ↓catalase activity resulting in ↑nitrites. In chronic FPIIES, occult blood, PMNs, eosinophils, Charcot-Leyden crystals, and reducing substances are observed, while stool containing frank or occult blood, mucus, sheets of PMNs and eosinophils, and ↑carbohydrate content are more often to be seen in the acute form of FPIES. Gastric juice, which is obtained by placing a nasogastric tube into the patient and aspirating gastric fluid, may show ↑PMNs (>10/HPF).
- *GRO:* Friable mucosa with mucosal ulceration (often rectum) and bleeding.
- *CLM*: Variable degrees of villous atrophy, edema of the lamina propria, crypt abscesses, and an inflammatory cell infiltration with ↑lymphocytes, eosinophils, and mast cells as well as plasma cells displaying IgM and IgA by IHC.
- *DGN*: Consistent clinical features with improvement following the withdrawal of the suspected causal protein. An OFC is sometimes performed to confirm the diagnosis or to determine the resolution of the food allergy.
- *DDX*: Allergic food disorders, infectious diseases, intestinal obstruction due to anatomic or functional etiologies, severe GERD, and metabolic, neurologic, and cardiac conditions should be kept in the differential diagnosis. Apart from other forms of malabsorption, NEC should also be considered in the acute form of FPIES. GSE should be excluded explicitly by antibody testing (IgA-TTG) and histologic features (villous blunting, intraepithelial lymphocytes, and crypt hyperplasia), which are typically less severe of the GSE.
- *TRT*: Elimination of the food trigger(s) from the diet with symptoms resolution 2–4 h with standard management (e.g., intravenous fluids with or without intravenous glucocorticoids).
- *PGN*: Cow's milk and soy FPIES resolve in a majority of patients by the age of 3 years, although infants with solid-food FPIES and/or those with concomitant detectable food-specific IgE may show a more protracted course. Oral food challenges (OFCs) may be proposed every 18–24 months.

3.4.8.5 Crohn Disease

- *DEF*: Idiopathic, chronic, and recurrent *Th1/Th17*-driven CIBD (regional ileitis) with variable distribution in the GI tract and possible related extra-GI lesions (skin, bone, muscle, lung) showing the involvement of bowel at any level characterized by:

> 1. *Transmural inflammation*
> 2. *Ulceration of the mucosa/submucosa*
> 3. *Lymphoid aggregates*
> 4. *Granulomas of the noncaseating type* as well as *IL-12* and *IL-23* production

- *EPI*: Onset approximately 2nd-3rd decades, +++ in the USA, the UK, and Scandinavia, white>blacks, ♂<♀.
- *CLI*: Intermittent, often physical, and/or emotional stress-linked attacks of diarrhea, fever, abdominal pain with endoscopic abnormality involving terminal ileum (65–75%) and/or colon (50–70%) ± colon alone (20–30%) with "skip lesions" and intervening unaffected segments. Extraintestinal manifestations are also important (Box 3.29).

> **Box 3.29 IBD-Extraintestinal Manifestations:** *"LESS Is More"*
> - *Liver*: PSC, pericholangitis, cholangiocarcinoma
> - *Eye*: Uveitis
> - *Skeletal*: Sacroileitis, ankylosing spondylitis, and peripheral arthritis
> - *Skin*: *Erythema nodosum, pyoderma gangrenosum*

- *GRO*: Endoscopic score of the distal ileum for Crohn disease may also be performed and Laghi et al. (2003) produced one including erythema (1), a granular pattern with aphthoid ulcers (2), cobblestone pattern with serpiginous ulcers (3), and confluent serpiginous ulcers and strictures (4).
- *CLM*: *Transmural inflammation, ulceration, lymphoid aggregates, noncaseating poorly formed granulomas*, as well as dilation/sclerosis of lymphatic channels and ganglion-neuronitis. The inflammation is initially edematous but then becomes fibrous dense with longitudinal mucosal ulcerations, creeping fat, luminal narrowing, fissure, and fistulous tracts. Disease activity in patients with Crohn disease may be measured using the Pediatric Crohn Disease Activity Index (PCDAI) according to D'Haens et al. (1998) and modified by Laghi et al. (2003). According to the Cochrane Database Syst Rev 2017 (Novak et al. 2017), the most used scores, worldwide, for separate grading of both ileal and colonic specimens are the Global Histologic Disease Activity Score (GHAS) (D'Haens et al. 1999), and the Naini and Cortina Score (Naini and Cortina 2012). We consider the presence of a single neutrophil in the lamina propria as indicative of a pathologic process, possibly suggesting inflammation. Erosions are defined by the limitation of the tissue defect until the submucosa, while ulcers are more profound and go beyond the submucosa. Granulomas refer as a reaction showing epithelioid cell reaction and need to be distinguished from the mucin-related foreign body giant cell granulomas of the ulcerative colitis due to crypt rupture and diffusion of the mucin content of the goblet cells. The scoring system targets the tissue architecture, erosions and/or ulcers, epithelium, intraepithelial neutrophils, neutrophilic and lymphocytic infiltration of the lamina propria, and granuloma-poiesis.
- *DDX*: The differential diagnosis of terminal ileal stricture includes Crohn disease, diffuse enteropathy, ischemia, lymphoma, neuroendocrine tumor (carcinoid) infiltration, pelvic inflammation, radiation enteritis, and tuberculosis. Ischemic strictures of the small intestine have been described in northern Nigeria with features suggesting tubercular disease. Hypotensive drugs may also be behind ischemia at the TI. The differential diagnosis of inflammatory bowel disease (Crohn disease and ulcerative colitis) should include infections and toxicities. Recently, Wegener granulomatosis was found mimicking Crohn disease

in a child presenting with abdominal pain and weight loss and harboring gastric, duodenal, and colonic erosions at endoscopy (Radhakrishnan et al. 2006). Histology and anti-*Saccharomyces cerevisiae* (ASCA) positivity suggested Crohn disease. However, the subsequent respiratory and upper airway pathology (dyspnea and hemoptysis), as well as c-antineutrophilic cytoplasmic antibodies (c-ANCA), pointed to the right direction of Wegener granulomatosis, which is a chronic idiopathic granulomatous vasculitis with characteristic involvement of the upper and lower airways system as well as the renal system. It infrequently occurs in children compared to adults, but both Crohn disease and Wegener granulomatosis may involve multiple organ systems, including the alimentary and respiratory tracts. In Wegener granulomatosis, there is an autoimmune process going on with mainly neutrophilic inflammation in early lesions and increasing titers of c-ANCA directed against proteinase-3 (PR-3). PR-3 is a neutral serine protease localized in the azurophilic granules of the neutrophilic granulocytes and myeloperoxidase-positive monocytes and is involved in the enzymatic degradation of elastin, type IV collagen, fibronectin, laminin, and hemoglobin. Inflammatory bowel disease can also present with renal signs and symptoms including hematuria, mostly nephrolithiasis-related or rarely glomerulonephritis-related. On the other hand, the specificity of c-ANCA in Wegener granulomatosis is very high varying from 80% to 100%. ANCA may be increased in cystic fibrosis (c-ANCA), juvenile idiopathic arthritis, ulcerative colitis (p-ANCA), and Crohn disease (p-ANCA). The p-ANCA are antineutrophilic cytoplasmic antibodies against myeloperoxidase mainly. ASCA recognizes mannose sequences in the cell of the microorganism *S. cerevisiae* and is a nonautoantigen in IBD, but the reason why patients with Crohn disease develop ASCA is not known; this antibody seems to be highly specific for Crohn disease, although it may also occur in GSE, AIH, and, even, in healthy relatives of patients with Crohn disease. Finally, one differential diagnosis that should be considered is colitis of Behçet syndrome, which is an unusual disease, initially defined by the occurrence of recurrent oral and genital ulcers and relapsing iritis that is widely recognized as a multisystem disorder with involvement of skin, joints, central nervous system, kidneys, large blood vessels, lungs, and gastrointestinal tract. The histopathologic lesion, which is the primary denominator, is vasculitis, which is usually a non-necrotizing venulitis associated with a prominent perivascular inflammatory infiltrate. Gastrointestinal involvement in the setting of Behçet syndrome is rare but has been reported, particularly in the youth of some geographic areas (Yüksel et al. 2007). The colectomy specimen shows extensive mucosal ulceration with variable longitudinal, fissuring, and aphthoid configurations that are typically occurring within a normal or focally inflamed colonic mucosa and associated with a lymphocytic vasculitis involving generally submucosal veins. The differential diagnosis between Crohn colitis and colitis in Behçet syndrome is also important. As emphasized by Lee (1986), the distribution pattern, including frequent right colonic predominance, discontinuous involvement, and infrequent rectal sparing are crucial.

- *PGN*: Sequelae include fibrous strictures, fistulas, abscesses, protein-losing enteropathy, vitamin B_{12} malabsorption, and extraintestinal manifestations ("*LESS*"). There is also an increased risk of carcinoma up to 3% in some centers.

3.4.8.6 Peptic Ulceration-Associated Diseases

Peptic ulceration may occur in quite different settings, including acid hypersecretion, NSAIDs-related environment, Hp gastritis, Zollinger-Ellison syndrome, Cushing disease, systemic mastocytosis (mucosal atrophy with submucosal edema and infiltration of mast cells in the lamina propria with evidence of foveolar destruction and in the small bowel of glandular or crypt destruction), polycythemia, hyperparathyroidism, cystic fibrosis, and liver cirrhosis.

Radiation Enteritis

Grossly, there may be remarkable thickening of bowel wall due to fibrosis, particularly in the submucosa. Microscopically, there is mucosal ulceration with ↑ mucus production, nuclear changes in the lining epithelium, and, later, submucosal edema and then fibrosis and ulceration and subendothelial accumulation of lipid-laden MΦ in blood vessels, with thrombosis and calcification.

Nonsteroidal Anti-inflammatory Drug-Induced Gut Lesions

Enteritis-like histologically and patient's history is the key.

3.4.9 Short Bowel Syndrome/Intestinal Failure

The small bowel in a healthy term baby is 200–300 cm in length. One of the most common causes of short bowel syndrome (SBS) or intestinal failure (IF) is resection of the intestine following NEC, midgut volvulus, or congenital anomalies such as jejunal atresia or ileal atresia and abdominal wall defects. IF is a new term that has outmoded the traditional name, "short bowel syndrome" (SBS). Typically, newborns have a short gut characterized by loss of ileum and ileocecal valve (ICV), and several investigations have indicated that the minimum of remaining bowel for survival and adaptation should be 25 cm with an intact ICV and 42 cm without an ICV (Galea et al. 1992). However, the actual criteria to label a gut as short are functional and are based on the continued need for TPN (>6 weeks or >90 days). SBS is common in VLBW babies, and almost all of them follow NEC. In SBS, the absorptive capacity of the intestine is inadequate to meet nutrition, hydration, electrolyte balance, and growth requirements. There are three categories: traditional SBS, resulting from the loss of enough intestine to create IF; malabsorptive states, such as MVID; and motility disorders, such as intestinal pseudo-obstruction.

3.4.10 Small Intestinal Transplantation

Small intestinal transplantation (SITx) and multivisceral transplantation (MVTx) represent indeed today a reality. The hope became real. In the 1960s and 1970s the first experimental investigations were put in place. SITx and MVTx are a viable alternative to parenteral feeding for treatment of pediatric and adult intractable failure of the gastrointestinal tract. This situation has been possible because numerous factors came to the place, such as improvement of the transplant medical management and posttransplant monitoring, suitable animal models, superior immunosuppression therapies, enhanced surgical techniques, and the growth of knowledge in gut physiology and mucosal immunology among others. Several national insurance programs have approved SITx worldwide, although not in all countries this option is available. In children, the most frequent current indications for SITx are short gut following surgical resection for gastroschisis, volvulus, NEC, Hirschsprung disease, pseudo-obstruction, intestinal atresia, microvillous inclusion disease, and tufting enteropathy. In adolescents and youth, the most frequent current indications for SITx are ischemia, Crohn disease, dysmotility, trauma, volvulus, mesenteric neoplastic processes (e.g., desmoid), and intestinal graft failure. Mostly, there are three types of SITx, including the isolated small intestinal allograft (ITx), with or without colon; the composite liver-SI-allograft (LITx) with transplantation of the liver, duodenum, pancreas, and of course, small intestine, with or without colon; and the multivisceral graft (MVTx) with transplantation of the liver in addition to the same organs seen in the liver-small intestine allograft. The colon is considered optional but is frequently used to increase absorptive capacity in specific settings (e.g., Hirschsprung disease). Once the transplantation takes place, the gastrointestinal allograft (graft) presents the recipient (host) with a sizable alloantigen hematological and parenchymal cellular load harboring the ability to initiate and maintain a forceful and diverse host effector immune response. This response underlies the immune-based complications that are under continuous monitoring from the medical management team including acute and chronic rejection (host versus graft) as well as direct toxicity from continual immunosuppressive therapy (e.g., renal and neurological) and complications associated with a sustained state of immunosuppression (e.g., multisystemic infections and posttransplant malignancies). The hematolymphoid cell population introduced with the graft is also capable of initiating a graft immune response against the host (i.e., graft-versus-host disease). These immune-based and the numerous nonimmune-based complications may also be influenced by several variables such as host innate immunogenetic polymorphisms, immune subpopulations, and the amount of alloantigen disparity between the host and donor. The transplant pathologist uses biopsies and other ancillary markers to identify acute and chronic rejections as well as infections, GVHD, and malignancies. The pathologist plays a significant and/or critical role in the evaluation of patients following SITx, and intense communication with the gastroenterologist and surgeon is vital. The pathologist should know the patient's history and endoscopic findings, including the location of biopsies, native organ versus graft, and appearance of the mucosa (erosions, ulcers, and irregular mucosa). Protocol allograft biopsies are 2–3 times/week for the 1st month and 1–2 times/week for the following 2 months. At least three biopsies per location are needed, because rejection may be focal. The pathologist should also evaluate the adequacy of the biopsies. Each biopsy should include surface epithelium, muscularis mucosae, and superficial submucosa to allow the evaluation of a substantial number of crypts. Inadequate biopsies include biopsies close to the stoma (reactive changes) among others. Routine H&E sections should contain 6 levels per tissue slides to allow the contiguous evaluation of the crypts. If antibody-mediated rejection (AMR) is in the differential, a separate biopsy should be sent fresh or frozen and analyzed by immunofluorescence for C4d. In several laboratories, an immunohistochemical protocol to identify C4d in formalin-fixed and paraffin-embedded tissue sections has been established. However, C4d immunohistochemistry on formalin-fixed and paraffin-embedded tissue is, however, challenging and some laboratories still prefer to run the immunofluorescence protocols on fresh tissue. Immunostaining and in situ hybridization should also be performed for cytomegalovirus (CMV) and Epstein-Barr virus-encoded RNA (EBER), respectively. Adenovirus may be evaluated either by immunostaining or electron microscopy demonstrating the classic crystalline lattice of 72–82 nm viral particles. Early fibrosis is also assessed, and trichrome staining is performed as well. Of course, native tissues (e.g., esophagus, rectum, and skin) and biopsies from allografts (e.g., stomach, small bowel, and large bowel) are often simultaneous procured and often necessary for a complete approach. Native and allograft tissue-based approaches help to distinguish whether changes are exclusive to alloimmune response (i.e., rejection), global and indiscriminate disorders (e.g., infections, PTLD), or graft-originated modifications (e.g., GvHD).

Pathologic findings in intestinal grafts include preservation injury, acute cellular rejection, antibody-mediated rejection, chronic rejection, infections, and posttransplant lymphoproliferative disorders (PTLDs).

Preservation injuries are seen during the first few days following TX and resolve within a week and include mild ischemic-type injury of the superficial mucosa with congestion and superficial wound or drop out of the surface epithelium, partial villous atrophy with regenerative changes in crypts, and minimal or no usually neutrophilic-driven inflammation.

Acute cellular rejection (ACR) is the leading cause of graft failure in the first 2 months following TX and occurs between 1 and 9 weeks post-TX. Clinically, patients may experience fever, nausea, vomiting, diarrhea, abdominal pain, and abdominal distension with an increase of the stomal effluxion. ACR is based on the apoptotic body count (ABC), which is defined as the ratio between the numbers of apoptotic bodies per ten consecutive crypt cross sections. There have been some techniques to pick up apoptosis in the tissue slides, but morphology using a conventional H&E remains the gold standard. Apoptotic bodies are fragments of nuclei and cytoplasm with variable morphology from classic "exploding crypt cells" to "intraepithelial clusters of basophilic material." ACR is a complication that can occur in GI allografts at any time post-Tx and can be related to factors of new onset (e.g., viral infections), loss of compliance with immunosuppressive therapy, but may also have no identifiable inducing factor. The Lee et al. grading system seems to be the current evaluation method (Lee et al. 1996) with some updates according to Ruiz et al. (2004). An indeterminate grade is inserted between 0 and 1 if ABC <6/10 (often 4–5/10 in contrast to degree 0 where ABC is usually <2/10), there is minimal focal inflammation, and the surface epithelium is intact. Inflammation is mixed. The score is as follows:

- 0 = No ACR, if ABC <6/10, no inflammation, intact surface epithelium
- 1 = Mild ACR, if ABC ≥6/10, mild to focal inflammation, intact surface epithelium
- 2 = Moderate ACR, if confluent apoptosis and focal crypt dropout, moderate/severe inflammation, and focal epithelial erosion are seen
- 3 = Severe ACR, if confluent apoptosis and extensive crypt dropout, moderate/severe inflammation, and extensive erosions ± ulceration

Reactive crypt epithelial changes, including mucin depletion, nuclear enlargement, nuclear hyperchromasia, and cytoplasmic basophilia, should be kept into account to avoid to upgrade ACR. It is essential to compare graft to native biopsies to distinguish or favor enteritis over graft rejection. In severe ACR, there may be sloughed mucosa and granulation tissue. It may become arduous to find high levels of ABC.

Intestinal AMR is usually diagnosed during the first 2 weeks following transplantation in the presence of circulating IgG. The optimal use of C4d staining remains to be validated.

Chronic rejection is the principal cause of late intestinal graft failure. Clinically, patients harboring rejection of chronic type have persistent diarrhea with mucosal ulcers, which are not healing, and repeated episodes of acute rejection.

Endoscopic findings include effacement of the mucosal folds and firm appearance of the intestinal surface. Fibrosis may be found on histologic sections. Also, there may be mild ischemic changes and low-grade apoptosis. In the differential algorithms, secondary causes of fibrosis should be taken into account, including previous nonspecific episodes of rejection, ischemic injury, previous infections, drug-induced injury, and prior biopsy site.

Infections include bacteria, viruses, and fungi. Bacterial infections occur first and early, while viruses are often seen within the first 6 months with a median of about 3 months and fungi later with an average of 6 months. In the bacterial infections, extraintestinal sites are often involved, such as vascular lines, abdominal cavity, wounds, and upper and lower urinary and respiratory tracts. The most frequent viruses are CMV, EBV, adenovirus, rotavirus, and calicivirus. The most common fungus is *Candida albicans*.

The WHO classification of PTLDs includes four categories: (1) *early-type nondestructive PTLD* sectioned into *plasmacytic hyperplasia* and *IM-like (infectious mononucleosis-like) PTLD*; (2) *polymorphic PTLD*, which comprises a heterogeneous population of lymphocytes and plasma cells, without a predominance of transformed cells; (3) *monomorphic PTLD*, which recalls one of the NHLs observed in immunocompetent patients, such as diffuse large B-cell lymphoma, Burkitt lymphoma, plasma cell myeloma, or T-cell/NK-cell lymphoma; and (4) *classic Hodgkin-type PTLD*. In SITx patients, the clinical presentation of PTLD is variable and nonspecific, and this needs to be taken into consideration in evaluating biopsies taken for rejection assessment. Ulcerations and/or strictures may also be seen. However, some features may favor PTLD over reactive lymphoid aggregates, such as expansile nodules, mass lesion, the predominance of transformed cells, lymphoid atypia, numerous B-cell or plasma cell infiltrate, and some necrosis, which has been labeled as "serpiginous", in the infiltrate. Risk factors to develop PTLD include EBV seronegative blood of the recipient with primary EBV infection, duration and type of immunosuppression, and kind of allograft being the risk higher in small intestinal, cardiac, and lung transplantation. Treatment includes reduction of immunosuppression, chemotherapy, as well as rituximab (anti-CD20 monoclonal antibody), which is used as a treatment in specific settings. Of note, PTLD in SITx may occur in concomitance with rejection conversely from other organ transplants, and biopsies are assessed weekly to monitor rejection and EBV status. The recognition of well-distinct forms of rejection and their intensity are paramount to address correctly or change some modules in the therapy and are accomplished by clinical symptoms and pathological findings in endoscopically derived mucosal biopsies of the allograft. Grading of ACR is performed according to Ruiz et al. (2004), and the immune-based (T-cell or antibody-mediated) inflammatory process has to be distinguished from other inflammatory conditions in the small intestine. Acute rejection in the stomach and colon also uses criteria that are specific for each organ. There are laboratory and biomarker assays such as photo-fluorographic analysis of peripheral immune cell population cytokine profiling, quantitation of distinct gene set changes, and citrulline levels in the blood. However, these techniques are supportive only providing auxiliary backing to a morphologic diagnosis, which is utmost. Thus, the transplant pathologist is a critical member of the transplant team, and clinically correlated morphology remains the gold standard in this setting.

3.4.11 Graft-Versus-Host Disease (GVHD) of the Gut

- *DEF*: Post-BMT or organ transplantation immunologic disorder due to a response of the immunocompetent cells of the donor or graft (immunologic requirement) to the histocompatibility antigens of the recipient or host (genetic imperative). However, a third essential and indispensable requirement is that the host is unable to reject the graft. Other settings include maternal-fetal cell transfer in immunodeficient children, transfusion of nonirradiated cells and blood products, as well as malignant thymoma, which occurs rarely in childhood.

- *EPG*: Lymphokines released by the activated T lymphocytes stimulate other cells, including NK cells, which destroy the tissue. Ultrastructural studies have indicated that effector T cells extend broad pseudopods that indent but do not breach the cellular membrane that appears to be necrotic, and immunological investigations indicated that specific antibodies (anti-asialo-GM1) are a quite specific marker for NK cells in mice, can prevent GVHD.

In the esophagus, the upper 1/3 is the most affected region to be involved, and lesions may be focal or diffuse with degree variable from single-cell necrosis in the basal cell layer (apoptosis, acidophilic bodies) to prominent desquamation and inflammation of the mucosa in the most severe cases. It is extremely important to remember that infections, which are an essential differential diagnosis, may coexist with GVHD. In the stomach, the mucosa may appear congested and atrophic, and, microscopically, apoptotic cells are mainly located in the mucous neck region. Differential diagnosis includes CMV infection and Hp gastritis.

Clinical stages and histology are paramount in the diagnosis of GVHD with an acute GVHD (7–100 days, mostly) and a chronic one (>100 days up to 400 days, mostly). In the *GvHD grading*, target organs are the skin, GI tract, biliary tree, BM, and lymphoid tissues.

- *DDX*: Lymphoma/leukemia, Hodgkin lymphoma, SCID, and severe T-cell deficiency. Characteristically, the crypt basis shows vacuolated cells with karyolitic debris in empty cells giving the aspect of "popcorn lesions." GVHD histologic mimickers: *Salmonella* infection (neutrophilic clusters at the bases of the crypts), among others (*vide infra* for the differentials of GvHD in the GI tract).

Chronic GVHD is a syndrome with similarities to abnormal collagen deposition diseases, such as Sjögren disease, polymyositis, lichen planus, scleroderma, and PBC of the liver with or without a previous acute GvHD state (de novo is up to 1/3 of the chronic GVHD!).

- *CLM*: *Chronic GVHD – Histologic features* include:
  1. *Skin*:
     Early: epidermal hypertrophy and hyperkeratosis, lichenoid infiltrate in the dermis with basement membrane thickening, inflammation to atrophy of the pilar units, sweat glands (eccrine, probably) with squamous metaplasia of the ducts, and subcutaneous fat.
     Late: Epidermal atrophy (thin epidermis) with keratin-plugged hair follicles and replacement of the dermal collagen with thick, coarse fibers entrapping the sweat glands mimicking scleroderma.
  2. *Gastrointestinal*:
     Similar features to the skin may occur in the esophagus with submucosal fibrosis, which may complicate as a stricture. In the liver, PBC-like changes may be observed. Fibrosis of the salivary glands and lacrimal ducts may occur as well.
  3. *Respiratory*:
     Edema, squamous metaplasia of the bronchiolar epithelium with accompanying lymphocytic infiltrate, and fibrosis that may evolve into obstructive airways disease.
- *DDX*: CMV, cryptosporidium, mycophenolate mofetil, proton-pump inhibitors, and conditioning regimen.
- *PGN*: HLA disparity, age, germ-free environmental conditions, cytotoxic treatment of the recipient, complete chimerism (complete chimeras have a higher rate of GVHD than mixed chimeras, suggesting that the persistence of host cells may suppress the alloreactive cells), and sex mismatching. It has been suggested that a dysfunctional thymus, which may be age-dependent, may harbor some relevance for the interindividual variability.

3.4.12 Small Intestinal Neoplasms

Neoplasms of the small intestine account for less than 5% of the tumors of the GI tract, and malignancies are more than a half of benignity. Most tumors are located in the ileum. Primary GI tract malignancies account for about 5% of pediatric malignancies, although some other centers have shown that alimentary tract primary malignancies account for as few as 1.2% of pediatric cancers. There is an approximately equal distribution of lesions in male and female patients, which is in contrast to older studies, some of which have described males to have an increased risk of developing malignancies. Most of the patients are older than 10 years of age, with 15- to 19-year-olds comprising the largest group. The large percentage of GI tract malignancies are malignant lymphomas, followed by sarcomas and carcinomas. Analyses show that sarcomas before and after the year 2000 have different subset histology. Sarcomas identified previously in 2000 were mostly labeled as leiomyosarcomas (smooth muscle neoplasms), while after 2000, most sarcomas were classified as the gastrointestinal stromal tumor (GIST). In 2002, after c-kit (CD117) staining became almost universally available. Thus, about of 4/5 of GI mesenchymal tumors have been classified as GIST, with proportional reductions in the number of smooth muscle tumors. Regarding treatment modalities, across all age groups and regardless of tumor location, many patients received surgical resection for their disease. The *en bloc* resection of the neoplasm and wide nodal excision has traditionally been the definitive treatment. Review of past literature demonstrates that primary gastric adenocarcinoma in childhood and adolescence is, in general, associated with poor clinical outcome with some studies reporting the median survival of approximately 5 months, while others identified the value of 9.7 months, with a range of 5–20 months. However, if the tumor is unresected the median survival is even shorter at about 4.5 months. The mainstay of therapy is surgery with lymph node dissection. There is also some evidence that neoadjuvant and adjuvant therapy of GI adenocarcinoma, despite their resectability. In the USA, the SEER Program of the National Cancer Institute (NCI) is the essential cancer registry inclusive of incidence and survival. In 2008, Curtis et al. reported that curative resection in the absence of nodal involvement has a 5-YSR of approximately 50%, whereas nodal disease decreases 5-year survival to only 10%.

3.4.12.1 Adenoma

It is most often seen in the duodenum and ileum with similar features as present in the colon (*vide infra*), although large villous adenomas are prone to undergo malignant transformation. Brunner gland adenoma (aka polypoid hamartoma, brunneroma) is a nodular proliferation of all elements of normal Brunner glands (ducts, stroma) and usually considered as hyperplasia. Another polyp is the hamartomatous type, which is typical of Peutz-Jeghers syndrome (*vide infra*).

3.4.12.2 Adenocarcinoma

It can show a "napkin ring" or polypoid fashion by gross examination and it is often located in the duodenum. A similar histology as observed in the colon is a characteristic feature apart of its rarity. Staging is comparable to that in the colon with the exception that only N0 and N1 for regional nodes uninvolved/involved and Tis if the tumor is intraepithelial just and T1 if it is intramucosal.

3.4.12.3 Carcinoma of the Ampulla of Vater

Ampulla of Vater centered carcinoma arising from ampullary intestinal mucosa easily distinguished at an early stage, while indistinguishable from distal common bile duct or pancreas carcinoma at a late stage. The ampulla of Vater is a complex anatomic structure made up of the distal, intraduodenal portions of both the common bile duct and significant pancreatic duct, which typically join to form a common channel. On the duodenal surface, the ampulla is viewed as papilla and is lined by small intestinal-type epithelium, while a pancreatobiliary-type epithelium covers both common bile duct and pancreatic duct.

Five percent of GI malignancies is the rate of the ampulla of Vater centered carcinoma. It may originate from villous adenoma or villoglandular polyp of the ampulla (often adenomatous change in villi but malignant cells at the base) ± polyposis syndromes and neurofibromatosis, 7th decade, ♂>♀

Sites include intra-ampullary, periampullary, and mixed forms. Clinically, there may be jaundice, abdominal pain, and pancreatitis. Grossly, there is a tumor bulging into the duodenal lumen.

- *CLM*: *Adenocarcinoma*, usually poorly differentiated (G3) with either intestinal (duodenal papillary origin) or pancreaticobiliary morphology, which is essential to distinguish from bile duct or pancreatic carcinoma, because the latter two neoplasms have a worse prognosis.
  1. *Pancreatico-biliary* morphology is characterized histologically by small glands or tubules with desmoplastic stroma and specific immunohistochemical phenotype.
  2. *Intestinal* morphology is characterized histologically by mostly elongated tubules and specific immunohistochemical phenotype.

The specificity of the immunohistochemistry of the carcinoma of the ampulla of Vater is reported in Box 3.30.

DPC4-/K-ras indicate genetic heterogeneity from primary pancreatic tumors. In the differentiation, it is essential to distinguish several grades according to the amount of glands; thus, the well-differentiated one, is called if glands >95%; moderately differentiated, if glands 50–95%; poorly differentiated, if glands 5–50%; and undifferentiated, if glands <5%. Variants include colloid, hepatoid, mixed acinar-endocrine, neuroendocrine, Paneth, and signet ring cell. Moreover, the presence or absence of *pancreatic intraepithelial neoplasia* is crucial (± PanIN3), particularly in the Whipple specimen or at the resection margins. During an intraoperative frozen section consultation, check LSD as essential criteria for malignancy: incomplete ductal lumina, nuclear Ø variation of ≥4:1 between ductal epithelial cells, and disorganized ductal distribution.

Box 3.30 IHC Patterns for Carcinoma of the Ampulla of Vater

Pancreatico-biliary phenotype	MUC2−, CDX2−, MUC1+, MUC5AC+, CK7+, CK20−
Intestinal phenotype:	MUC2+, CDX2+, MUC1−, MUC5AC−, CK7−, CK20+

- *DDX*: In the differential diagnosis, it is important to remember the *periampullary myoepithelial hamartoma/adenomyoma*, which shows dilated glands lined by flattened epithelium and surrounded by smooth glands, but no cytologic atypia and no desmoplastic stroma reaction (DSSR). Moreover, other lesions include Peutz-Jeghers (PJ) polyp, juvenile polyp, pyogenic granuloma, pancreatic heterotopia, fundic mucosa heterotopia, endocrine tumors (carcinoid, somatostatinoma, gangliocytic paraganglioma, gastrinoma), mesenchymal tumors (GIST, leiomyoma, lipoma, hemangioma, neurofibroma, granular cell tumor, ganglioneuroma, schwannoma, Kaposi sarcoma), lymphoid/lymphoproliferative lesions, and metastases. The PJ polyp shows arborizing (branching) core of smooth muscle with overlying nondysplastic epithelium. Some authors favor the designation of epithelium with low malignant potential, although this latter aspect is still a matter of debate. The juvenile polyp shows a prominent lamina propria with edema, inflammatory cells, and cystically dilated glands lined by cuboidal to columnar surface epithelium with reactive changes. There may be dysplasia but the dysplastic epithelium remains an option and is not necessary for the diagnosis.
- *PGN:* 5-YSR: ~50%, but 80% if LNs are tumor-free. The tumor is metastatic at diagnosis (regional LNs) in ~1/3 of the patients. MTX: liver, lung, peritoneum, and pleura. PGN factors include tumor *size* (>2.5 cm), *LN* involvement, stage, *LVI, perineural invasion, Oddi's sphincter muscular invasion* (staging), *signet ring* histology (typing), *G3* (grading), and *positive margins* (R-classification). Histologic follow-up of ampullary adenomas in patients with Familial Adenomatous Polyposis (FAP) or other polyposis syndromes is needed because villous adenoma of the papilla of young children with FAP can transform in a malignancy later in life and genetic counseling may be appropriate.

Neuroendocrine Neoplasm
See the previous section "Neuroendocrine Tumors of the Stomach." It is one of the most common tumors of the small bowel with ~1/3 of patients presenting with concurrent malignancies (intestinal or extraintestinal).

Other carcinomas include adenosquamous carcinoma and anaplastic (sarcomatoid) carcinoma that is practically inexistent in childhood and youth.

3.4.12.4 Gangliocytic Paraganglioma
- *DEF*: Rare neuroendocrine tumor (aka non-chromaffin paraganglioma, para-ganglioneuroma), usually benign tumor of the second portion of the duodenum (almost exclusively and near ampulla of Vater), and wide age range with no gender preference (♂=♀).
- *CLI*: They can present with abdominal pain, gastrointestinal bleeding, or obstruction and may be an incidental finding at endoscopy or autopsy.
- *GRO/CLM*: Small, submucosal, pedunculated, frequently ulcerated nodule with a tendency to bleed (nonfunctional) and made up of a combination of three cell types: endocrine cells, ganglion cells, and spindle-shaped S-100-positive Schwann-like cells.
- *PGN*: Rare metastases and careful assessment before local excision are necessary with pancreaticoduodenectomy if the lesion is large.

3.4.12.5 Other Tumors
Other tumors include smooth muscle tumors, which are much more likely to be malignant than esophageal or gastric type, and malignant lymphoma (EATL, IPSID, and *de novo* lymphomas). Also, other tumors include lipoma, hemangioma, neurofibroma, and granulocytic sarcoma. The most common tumor in the small bowel is by far the metastasis that can often present as multiple polypoid lesions. Pediatric primaries include myofibroblastic tumor foremostly. Metastatic tumors are the most common neoplastic processes in the small intestine. It can often be involved as numerous polypoid lesions. Most common primaries in childhood and youth include rhabdomyosarcoma, malignant melanoma, and germ cell tumors.

There are some diseases and conditions that need a Whipple surgical operation, including ampullary carcinoma, cholangiocellular carcinoma, and complications post-Whipple surgery include pancreatic fistula and gastroparesis (early) and malabsorption, alteration in diet and loss of weight (late), malignancy of the duodenum, malignancy of the head of the pancreas, and nonmalignant conditions (e.g., hyperinsulinism).

3.5 Appendix

The *appendix vermiformis* of the intestine is found in the cecum between the small and large bowel. It is tremendously rich in lymphoid tissue in young individuals but undergoes atrophy during a lifetime. The appendix may also result in early fibrous obliteration, and immunostaining against nerve fibers reveals hypertrophy suggesting that appendix may show a schwannoma-like degeneration during lifetime, at least in some individuals (Box 3.31) (Figs. 3.30, 3.31, 3.32, 3.33, 3.34, and 3.35).

3.5.1 Appendiceal Anomalies

The primordium of the cecum and appendix appears in the 6th week as a swelling on the antimesenteric border of the caudal side of the midgut loop. A tube-shaped cecum is seen more during the fetal period, while a saccule-shaped asymmetric-type cecum is seen in adults more at the end of the fetal period. Congenital anomalies of the appendix anatomy are rare, but it is imperative to be aware of them. There is the possible confusion with other acquired entities, and they may be the basis for an additional pathology. Congenital anomalies of the appendix include agenesis, hypoplasia, duplication, and diverticula. In some genetic syndromes, congenital defects of the appendix can occur, and the pediatric pathologist should know them, in the case of surgical resection or autopsy. Appendix duplication is an extremely rare anomaly (0.004–0.009% of appendectomy specimens) that was first classified in 1936 according to their anatomic location by Cave and later modified by Wallbridge in 1963. *Agenesis of the appendix* is a rare anatomic finding; Morgagni reported the first case in 1719. Collins found 1 example in 104,066 appendectomies or an incidence of 0.0009%. The cause of a missing appendix is assumed to be secondary to an intrauterine vascular accident. An arrest of development may occur at any stage and give rise to the absence of cecum and appendix (type 1), blunt conical cecum without appendix (type 2), longitudinal symmetrical cecum with longitudinal muscle bands converging toward its apex but without appendix (type 3), and asymmetric cecum without appendix (type 4). The diagnosis of agenesis should not be made except the ileocecal area, and retrocecal space is methodically explored. Associated mesenteric lymphadenitis is noted during laparoscopy or laparotomy for congenital absence of appendix. However, an auto-amputated appendix may also be the focus for inflammation. In other cases, no apparent cause for the patient's symptoms is found. Congenital absence of the appendix should be a diagnosis of exclusion at laparotomy.

Congenital diverticula of the appendix are to be found on the antimesenteric border. It is important to remember that agenesis, duplication, and congenital diverticula may occur in the setting of other congenital anomalies, such as either associations or genetic syndromes. Acquired diverticula of the appendix are not uncommon and consist of an extension or protrusion of the mucosa through the muscularis propria and may differentiate from congenital diverticula because the natural ones have the typical layers of the appendix.

3.5.2 Appendiceal Vascular Changes

Hyperemia is a classic feature in the case of "normal"-appearing appendix without evidence of inflammation. Appendiceal necrosis is rarely seen. In newborns, a systemic hypoxic state, including extracorporeal membrane oxygenation (ECMO) and non-ECMO-related CHD or ischemia following NEC or postsurgical procedures, may be the underlying conditions.

Box 3.31 Appendix
  I. Congenital Anomalies
 II. Vascular Changes
III. Inflammation
IV. Metabolic and Degenerative Diseases
 V. Tumors (Treacherous Entities)

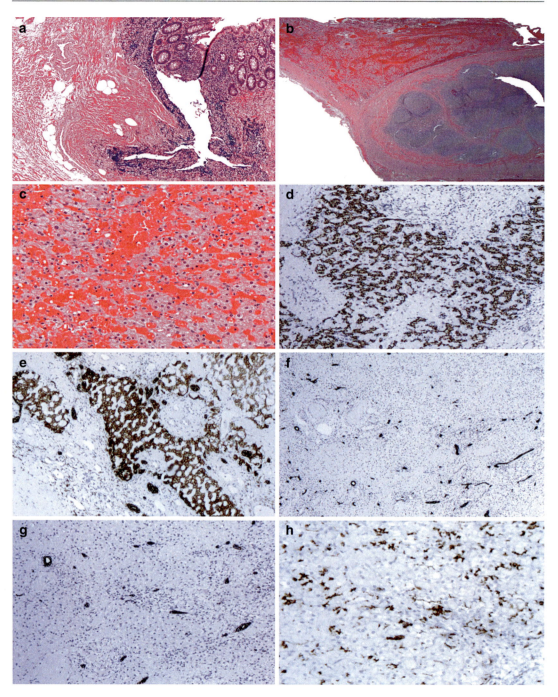

Fig. 3.30 Congenital anomalies of the appendix are shown in this panel illustrating a congenital diverticulum (**a**) in a not inflamed appendix and ectopic liver (**b**, **c**). The immunostains for Hep-Par1 (**d**), pan-keratins (**e**), CK7 (**f**), CK19 (**g**), and CEA (**h**) confirm the histologic diagnosis of hepatic tissue. CK7 and Ck19 highlight the interlobular bile ducts, while CEA detects the bile canaliculi

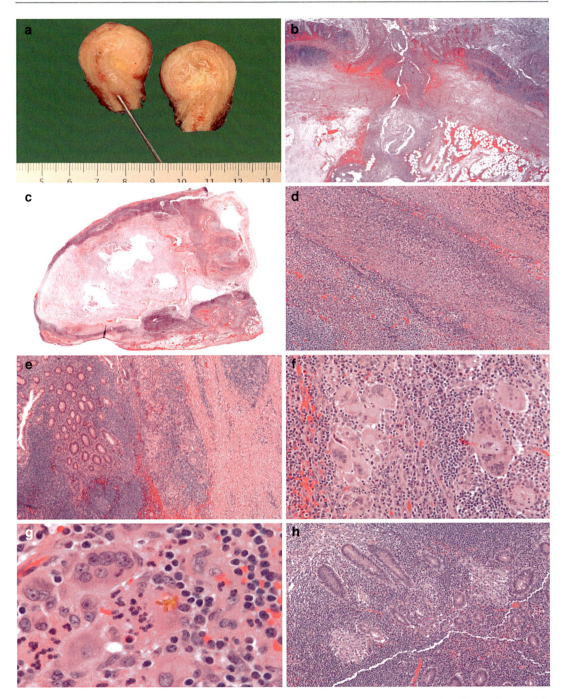

Fig. 3.31 Appendicitis with prominent inflammatory myofibroblastic component (**a**). In (**b**) is shown an appendicitis with histologic proof of a perforation. In (**c**, **d**) are shown microphotographs of a gangrenous appendicitis. In (**e–g**) are shown microphotographs of interval appendicitis with foreign body giant cells, while (**h**) represents a Crohn appendicitis with non-caseous granulomas

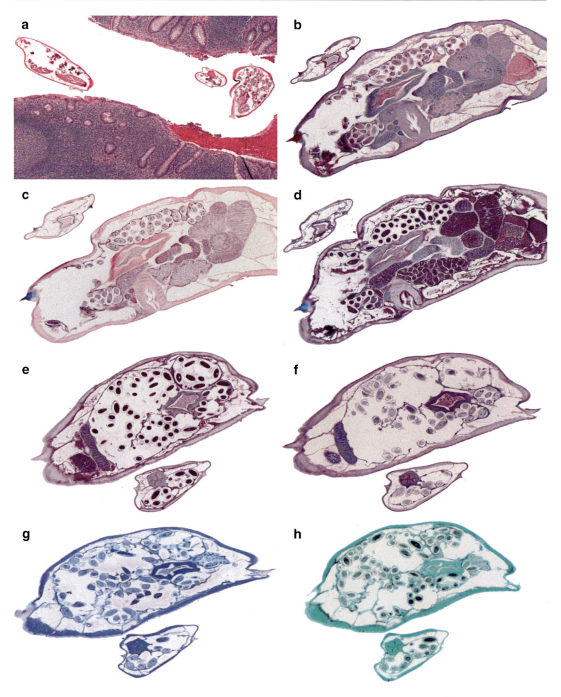

Fig. 3.32 *Enterobius vermicularis* (**a**) (H&E stain, x50) highlighted in a particularly suggestive panel with different special stains, including mucin (**b**), Alcian blue pH2.5 (**c**), Alcian blue/PAS (**d**), PAS (**e**), PASD (**f**), Giemsa (**g**), and Grocott Methenamine silver (**h**) (**b–h**, x100)

3.5 Appendix

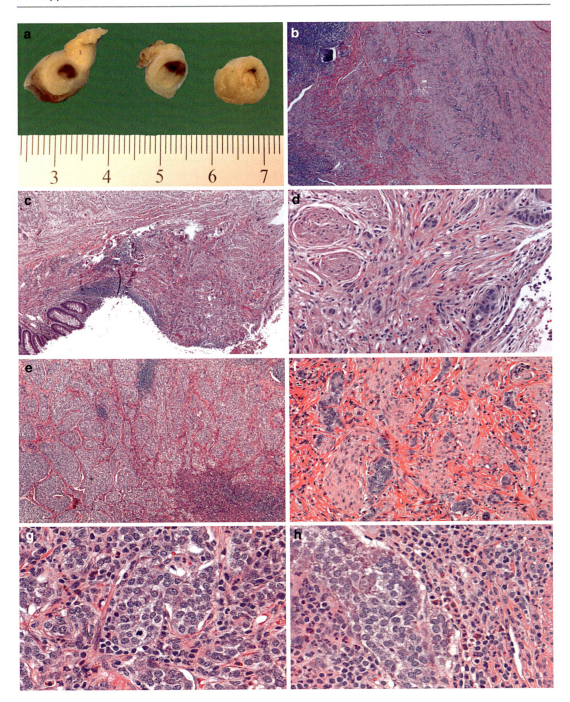

Fig. 3.33 Carcinoid in a gross photograph (**a**) and several H&E-stained histological sections (**b**–**h**)

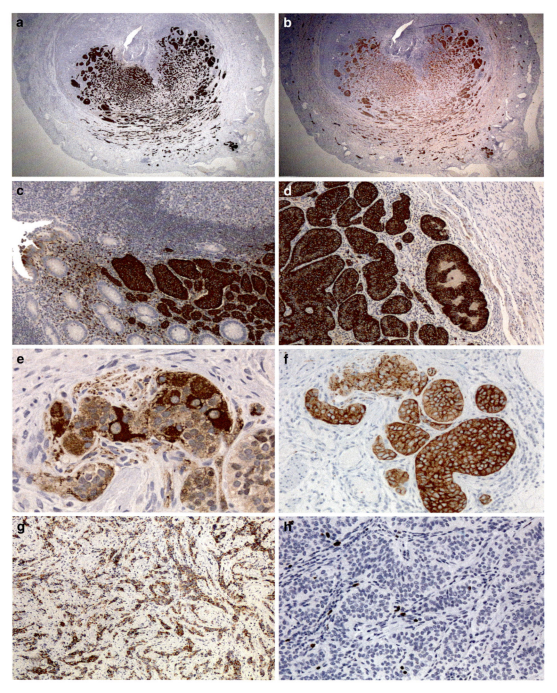

Fig. 3.34 Immunohistochemistry of an appendix carcinoid showing positive stains for chromogranin A (CGA), neuron-specific enolase (NSE), CGA, CGA, NSE, synaptophysin, AE1-3, and Ki67 proliferation antigen in (**a–h**), respectively

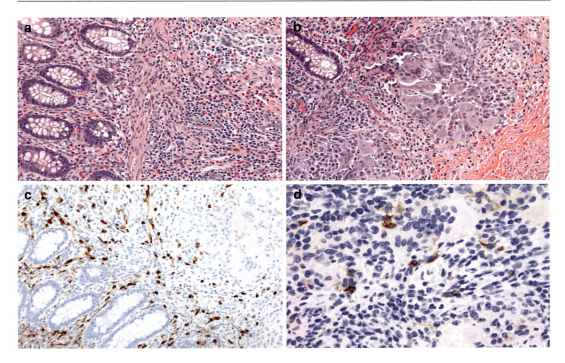

Fig. 3.35 Langerhans cell histiocytosis of the appendix is a very rare condition with very few case reports. Here is the H&E of one pediatric case (**a–b**) confirmed by S100- (**c**) and CD1a-positive (**d**) immunostains

3.5.3 Appendicitis

- *DEF*: Inflammation (acute/subacute/chronic) of the appendix.
- *EPI*: 10–30 years, ♂<♀ (probably due to a variety of gynecologic conditions). The rate of appendectomy without pathology abnormalities should be less 15% in any pediatric surgery department. If the percentage is more or equal to 15, it should be considered unusual. In this setting, the *ad hoc* review of quality assurance policies needs to put in place.
- *EPG*: A luminal obstruction, such as a fecalith or virally induced lymphoid hyperplasia, leads to luminal distension with ischemic sequelae of the wall and bacterial invasion. Alternatively, a primary viral infection or an ischemic event can induce a break of the mucosal barrier (e.g., microulcerations) representing the way for bacteria to invade and proliferate the organ. Bacteria commonly identified are *E. coli*, *P. aeruginosa*, and *Streptococcus* spp.
- *GRO*: Dull, discolored, and shaggy appendix. In nontransmural necrosis or inflammation, the appendix may show only some mild serosal hyperemia or be unremarkable.
- *CLM*: Minimal degree of neutrophilic infiltration for inflamed appendix is controversially debated with authorities ranging from mucosal to full transmural extension and involvement of the muscularis propria. This criterion is essential to avoid anomalous rates of unconfirmed surgical appendices. No confusion exists for luminal neutrophilic infiltration only, in which case, there is no appendicitis. A morphologic classification includes acute suppurative with neutrophils invading the wall, gangrenous with neutrophils and necrosis of the wall, and perforated appendicitis with a macroscopic and microscopic identification of a tissue defect of the wall (Box 3.32).

Granulomatous appendicitis is an enigmatic disease, because several etiologies may be behind it. In the differential diagnosis needs to be listed interval appendicitis (*vide infra*) or foreign body reactions, Crohn disease, sarcoidosis, and infectious agents, such as *Yersinia enterocolitica* or *Yersinia pseudotuberculosis*, although most cases

> **Box 3.32 Specific Infections of the Appendix**
> - *Adenovirus* (smudgy nuclei with intranuclear inclusions)
> - *Campylobacter jejuni* (acute self-limited inflammation with neutrophils in the lamina propria, cryptitis, and crypt abscesses, but neither architectural changes nor basal lymphoplasmacytosis)
> - *Measles* (Warthin-Finkeldey cells and lymphoid tissue hyperplasia)
> - *Oxuriasis* (pinworm infestation with intraluminal *Enterobius vermicularis*)
> - *Rotavirus* (clinical history and virology results, no apparent change at light microscopy)
> - *Salmonella spp.* (concomitant ileocecal inflammation with mildly inflamed appendix)
> - *Yersinia enterocolitica* (lymphoid hyperplasia, cryptitis, and crypt abscesses as well as granulomas of the suppurative type with central abscess formation, no architectural changes)

remain idiopathic. In *Yersinia* spp. infection of the appendix, epithelioid granulomas with lymphoid cuffing, transmural inflammation with lymphoid aggregates, mucosal ulceration, and cryptitis have been observed.

Eosinophilic appendicitis is a diffuse eosinophilic infiltration often with granulomas that has been correlated with the presence of *Strongyloides stercoralis* in the stool.

Interval appendicitis (*Interval Appendectomy*) is a specific form of appendicitis which reveals an appendicitis 4–8 weeks after the primary complaints of severe inflammation of the appendix and use of antibiotics. This delayed surgery after a course of antibiotics produces a very variable range of "post-appendicitis", including unremarkable appendices or appendices with mild serosal fibrosis to appendices showing mural thickening, fibrosis, transmural lymphoid aggregates, granulomatous features, xanthogranulomatous (florid) inflammation, as well as mucosal architectural distortion.

Appendix intussusception: It occurs with the presence of tumor, endometriosis, cystic fibrosis, and virally induced lymphoid hyperplasia.

Fibrous obliteration of the appendix tip is rare in children and has no clinical significance.

Appendix neuroma: It represents a mural disorganization of the nerve tissue with a spindle cell proliferation that progressively replaces the appendiceal mucosa and lymphoid tissue and, consequently, resulting in the obliteration of the lumen of the appendix.

The diagnosis of *chronic appendicitis* should be avoided, because it is difficult to substantiate and has several implications, probably also forensic. There are several classifications, and there is usually a difficulty in receiving an excellent intra-individual and/or inter-individual kappa score. One diagnosis that can be used in the anatomic-pathology report could be:

"Appendix, Appendectomy:
No acute inflammatory features."

Several studies have been performed to identify children at low risk of appendicitis avoiding unnecessary surgery. Some clinical judgment needs to be used. However, using the pathologist's report for children undergoing surgery and parental report exactly two weeks after discharge from the emergency department as well as reviewing nearly 2400 medical records, some pediatricians create a decision model (LOE = 1b; LOE, level of evidence), which seems working quite excellent and accurate in identifying children at low risk of having appendicitis. In diagnosis, LOE equals 1b refers as to a validating cohort study implying good reference standards, alternatively, a clinical decision rule, which has been tested within one clinical center. The model comprises two independent clinical and laboratory-based assessments 30–60 min separated one from the other. Children at low risk of appendicitis are patients if one of the following criteria is met: (1) absolute neutrophil count ≤6750/µL and no right lower quadrant tenderness OR (2) total neutrophil count ≤6750/µL with maximal tenderness in the right lower quadrant, but no abdominal pain with walking, jumping, or coughing. It seems that this model is 98% sensitive and 24%

specific, i.e., it is an excellent model to pick up children at low risk of appendicitis (Bundy et al. 2007).

Low-risk appendicitis rule is defined as a "clinical prediction rule to identify which children with acute abdominal pain are at low risk for appendicitis" (Kharbanda et al. 2012). In patients identified at low risk level, alternative strategies, such as observation or ultrasounds, rather than computed tomographic imaging or surgery have been proposed, although the clinical experience still plays a major role. Kharbanda et al.'s (2012) low-risk appendicitis rule includes an absolute neutrophil count of $\leq 6.75 \times 10^3/\mu L$ AND no maximal tenderness in the right lower quadrant (RLQ) or a total neutrophil count of $\leq 6.75 \times 10^3/\mu L$ AND maximal tenderness in RLQ, but no abdominal pain with walking/jumping or coughing. IBD (CD and UC) and granulomatous appendicitis support the precaution to look first for interval appendectomy before rendering an infection or IBD-related disorder. These aspects need to be taken into consideration carefully (Box 3.33).

3.5.4 Appendiceal Metabolic and Degenerative Changes

Metabolic and degenerative changes comprise a series of different conditions, such as appendiceal mucosal melanosis, which seems to be unrelated to laxatives as seen in the elderly. Occasionally, degenerative changes may include submesothelial deposits of decidualized cells ± endometriosis, which is a finding rarely observed in patients affected with cystic fibrosis among other conditions. An ectopic decidual reaction is the presence of heterotopic deposition of decidualized stromal cells of uterine origin. Immunohistochemistry may be used to identify the decidua cells, which are positive for VIM, DES, progesteron receptor, CD10, and alpha-1-antitrypsin. Decidualized cells show a variable expression of PLAP (Placental-like alkaline phosphatase), β-hCG (human chorionic gonadotropin), and are, usually, negative for (cyto-) keratins.

3.5.5 Appendiceal Neoplasms

Appendiceal tumors are quite rare in the pediatric age group but are more frequent in the youth. Neuroendocrine tumors of low grade or carcinoids represent most tumors. However, diagnosis may be a challenge, and treacherous entities are possible pitfalls if not all phenotypes are kept in mind. Moreover, colonic-type (non-mucinous) epithelial neoplasms of the appendix include adenomas and adenocarcinomas, tubular adenomas, traditional serrated adenomas, and sessile serrated adenomas which may be observed in children and youth with predisposing genetic syndromes (e.g., TP53 deletion syndrome, familial adenomatous polyposis).

3.5.5.1 Mucinous Cystadenoma
- *DEF*: Diffuse globular enlargement of the appendix by a significant amount of tenacious mucus into the lumen, which is lined by atypical mucinous epithelium with at least focal papillary growths.
- *SYN*: Mucocele (some pathologists reserve the term mucocele for mucinous distention secondary to non-neoplastic obstruction).
- *CLI*: In 1/5 of cases, it is associated with appendiceal perforation and implantation of

---

Box 3.33 Appendix Pearls and Pitfalls
- In scoping an appendix, ulceration of the mucosa, neutrophilic infiltration of the wall, granulomas, oxyurids, ganglia cells, and carcinoid need to be considered at any time!
- Although the appendix can be involved by idiopathic IBD (both Crohn disease and ulcerative colitis), IBD-associated granulomatous appendicitis as an isolated finding is a quite rare event. More often, the pattern is linked to an interval appendectomy.
- PCR analysis is an excellent technique to identify *Yersinia* spp. in FFPE material.

mucin within the peritoneum, although malignant cells will not be found in the mucin. Coexistence with a mucinous cystadenoma of the ovary with identical histology needs to be ruled out.

3.5.5.2 Mucinous Cystadenocarcinoma

Similar to benign counterpart except that cells are more atypical and often invade bowel wall and viable (not necessarily atypical!) tumor cells can be found in the mucin lakes.

When mucin is present in the peritoneal cavity and associated with malignant cells, the condition called *pseudomyxoma peritonei* occurs.

3.5.5.3 Adenocarcinoma

A cecal carcinoma may secondarily occur with a histology same as for colonic tumor, and a variant of signet ring cells exists.

3.5.5.4 Carcinoid Tumor

- *DEF*: Well-differentiated neuroendocrine tumor thought to arise from the subepithelial endocrine cells or Kultschitsky cells at the base of Lieberkuhn glands with a location at the tip of the appendix (mucosa and/or submucosa and/or muscularis propria) in about 70% of cases.
- *EPI*: 1:1000 children of a children's hospital, first through fourth decades of life, ♂<♀, but it may have a relationship with a higher number of incidental appendectomies due to cholecystectomy, hystero-oophorectomy, and cesarean section.
- *CLI*: Carcinoid syndrome (±).
- *CLM*: Neuroendocrine cells in the lamina propria and submucosa in the form of clusters or in single cells alone or forming glands with the tendency to invade the muscularis propria and the peritoneal surface.
- *Classic Type*: Solid nests of small monotonous cells (insular growth pattern) with often invasion of nerves, muscle, lymphatics, and even serosa.
- *Tubular Adenocarcinoid* with glandular formation without solid nests, usually tip of the appendix and usually lacks serotonin but often positive for glucagon and often misdiagnosed as adenocarcinoma.
- *Clear Cell Carcinoid*
  - Cells have clear cytoplasm but lack mucin and glycogen with behavior like the classic type.
- *Mucinous Carcinoid*
  - Goblet cell carcinoid, microglandular carcinoma, and crypt cell carcinoma (Ramani et al. 1999; van Eeden et al. 2007; Chetty 2008).
  - Unique to the appendix with tumor cells harboring a signet ring configuration.
  - Intrasubmucosal concentric growth.
  - More aggressive than the other types (tubular and clear cell) with 10–20% metastasizing, often to bilateral ovaries.
- *IHC*: (+) SYN, CGA, NSE, CD56, CDX2, and (+) CK20 (15% of cases) and concomitant (−) CK7 (−) TTF-1. Synaptophysin stain is positive in nearly all cases, while chromogranin A is positive in many cases, but not all cases. Argyrophil stain is positive in nearly all cases, while argentaffin stain is positive in most. NSE, PGP 9.5, and CD56 are also sensitive but are considered not specific immunohistochemical markers.
- *TRT*: If <1 cm, appendectomy only; 1–2 cm, right hemicolectomy if the tumor invades the meso-appendiceal fat or if the angiolymphatic invasion is detected; and >2 cm, right hemicolectomy.
- *PGN*: Outcome was excellent, with a 100% disease-free survival rate.

Although carcinoid tumors may be found throughout the gastrointestinal tract, most tumors of this kind are located within the tip of the appendix, small bowel, and rectum. Neuroendocrine tumors arise in most organs of the body and share many common pathologic features. Uniform nomenclature, grading, and staging systems have been broadly established for purposes of quality assurance and risk management in primary to tertiary healthcare institutions. Although we would prefer that anecdotal experience should not change the standards of practice, we think that a more careful evaluation should be reserved for young patients harboring a carcinoid tumor. This setting, indeed, may show a genetic cancer syndrome, and genetic counseling in this direction should not increase the hospital expenses, but it can have enormous consequences for the child's prognosis and family counseling. The guidelines of follow-up for this kind of tumor are variable. We suggested that in patients with carcinoid Ø <2 cm and no associated risk factors, metastasis, clinical and imaging follow-up may be adequate, and there may be no need of neuroendocrine markers in this subgroup of patients of appendiceal carcinoid. Current standards of practice indicate that right hemicolectomy is recommended for pediatric patients with Ø >2 cm, particularly when the mesoappendix is involved or in cases of residual tumor at the margin of resection, and simple appendectomy is justified when the tumor Ø ≤2 cm. Due to the rare occurrence of recurrent disease, follow-up is strongly recommended. Follow-up time should be reported in pediatric series, and a very variable range from 10 to 26 years has been recorded in the literature. There are rare cases of reported recurrence, including a patient who died after 33 years of follow-up for colon cancer. Thus, a 30-years follow up is a good choice. In the Spunt et al. (2000) series, it was indicated that one in four children had associated adenocarcinoma of the colon. An association between carcinoid tumors and colonic adenocarcinoma has been reported, although such an association seems extremely rare in childhood. The neural proliferation of the appendix includes neural tumors in the setting of the von Recklinghausen disease and mucosal and axial neuromas that may be considered to progress to the fibrous obliteration of the appendix. Mesenchymal tumors of the appendix include those of smooth muscle type, typically leiomyoma and rarely leiomyosarcoma and nonmyogenic tumors such as GIST, granular cell tumor, and Kaposi sarcoma among the most various entities. Pediatric lymphomas are mostly Burkitt lymphoma, while youth lymphomas are large B-cell lymphomas and, rarely, low-grade B-cell lymphomas. Occasionally, the involvement of the appendix by leukemia as secondary involvement has been reported. Secondary involvement of the appendix may occur in the setting of carcinomas of the female genital tract, such as ovary. Peritoneal endosalpingiosis most often involves the appendiceal serosa and, occasionally, the wall, but no clinical manifestations contrarily from endometriosis. Finally, the risk of complications of right hemicolectomy need to be weighed (Sánchez-Guillén et al. 2019). Complications include excessive bleeding, internal (organ) injury, infection, inability to perform the suture with obligate colostomy, and anastomotic leak, which harbors a fatality risk within 60 days of surgery of 3.1%.

3.6 Large Intestine

3.6.1 Large Intestinal Anomalies

Congenital anomalies of the large intestine include absence (agenesis), defects of canalization (atresia), muscular integrity (segmental dilation), neural crest migration (aganglionosis), persistence to yolk sac structures (Meckel's diverticulum), malrotation, duplication, and imperforate anus complete the spectrum (Box 3.34). Figures 3.36, 3.37, 3.38, 3.39, 3.40, 3.41, 3.42, 3.43, 3.44, 3.45, 3.46, 3.47, 3.48, and 3.49 illustrate some aspects of congenital and acquired pathology.

3.6.1.1 Congenital Atresia: Stenosis of the Colon
- *DEF*: Failure to have a lumen or an appropriate lumen in the large intestine.

Congenital colonic atresia is rare, but congenital colonic stenosis is exceptional with very few cases reported in the literature. Up to 1/5 of patients harboring colonic atresia also have proximal small bowel atresia, and distally, these patients may also present with imperforate anus and aganglionosis (~2%). Therefore, a suction rectal biopsy is mandatory before re-establishing

> **Box 3.34 Colon and Rectum Pathologies**
> I. Congenital Anomalies
> II. Vascular Changes
> III. Inflammation
> IV. Metabolic and Degenerative Diseases
> V. Tumors

continuity of the intestine in every patient with colonic atresia.

Malformations associated with colonic atresia include myoskeletal anomalies, such as syndactyly, polydactyly, clubfoot, absent radius, as well as ocular, cardiac, and abdominal wall defects. Some genetic syndromes have also been reported with colonic atresia and include Stickler syndrome, which is a chondrodystrophy with congenital alteration of type II collagen among others.

- *RF*: Maternal use of cocaine, amphetamines, nicotine, and decongestants.

Colonic atresias are classified similarly to the small intestinal atresia using the system of Bland-Sutton and Louw, including type I, intraluminal obstructive webs; type II, blind proximal and distal segments connected by a fibrous cord; and type III, separated ends of the bowel with an intervening V-shaped mesenteric defect. The originally 3-tier classification system by Bland-Sutton (1889) and Louw and Barnard (1955) was later expanded to five by Zerella and Martin (1976) as well as Grosfeld et al. (1979).

3.6.1.2 Acquired "Pseudo-Atresia" of the Colon

It follows various inflammatory, infectious, traumatic, and neoplastic processes in infants and children. Of all of them, NEC is probably the most common disease evolving potentially in

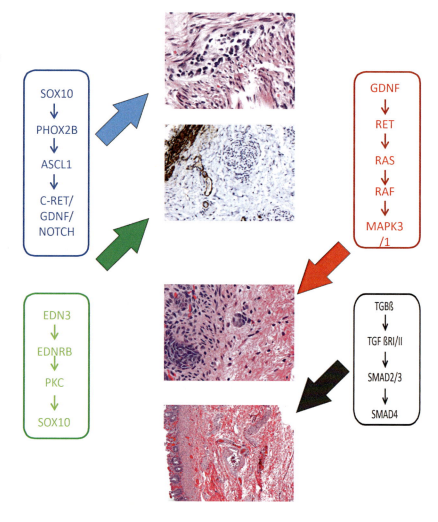

Fig. 3.36 Involvement of several genes in the cranio-caudal migration of the ganglion cells into the gut

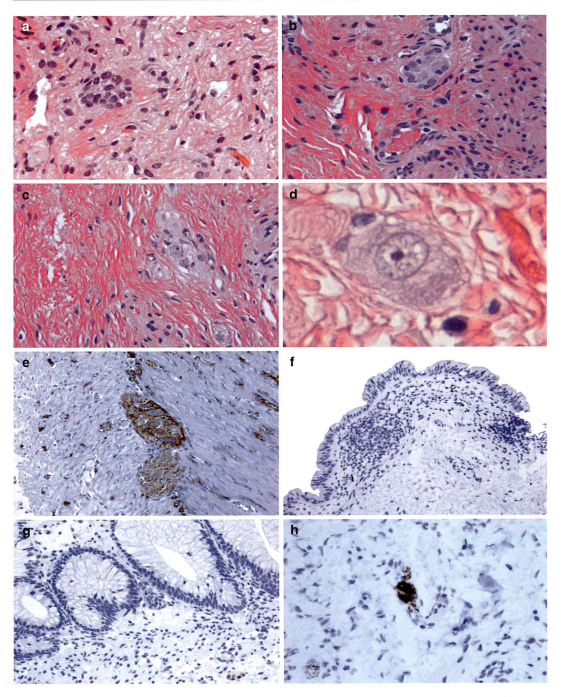

Fig. 3.37 Ganglion cells from immature (**a**, **b**) to mature (**c**, **d**) types. Synaptophysin is shown in (**e**), while (**f–h**) show a calretinin staining (**a–d**, H&E stain; **a–h**, x400, x400, x400, x630, x200, x200, x200, x400)

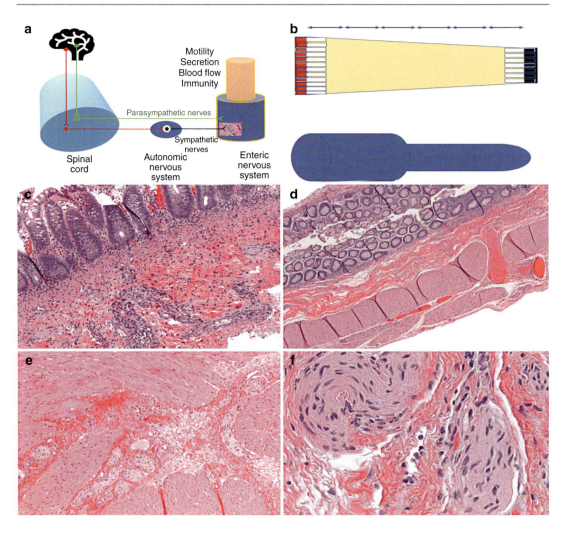

Fig. 3.38 In (**a**, **b**) are a schema of the innervation of gut and a suggested grossing protocol of the colectomy specimens with Hirschsprung's disease. The microphotographs (**c–f**) show aganglionosis and hypertrophy of the nerve fibers

acquired "pseudo-atresia" (or better stricture) of the bowel. In fact, up to 1/3 of patients affected with NEC will develop intestinal stenotic events and about 3/4 of these reside in the colon.

CLM: Submucosal fibrosis and variable maturation degrees of abdominal scar in the resected stenotic pathologic specimen. The sigmoid colon is the most common gastrointestinal site of NEC-associated stricture of the large intestine. Other acquired etiologies associated with stenotic events of the colon include complication of the hemolytic uremic syndrome, methicillin-resistant *St. aureus* enterocolitis, angio-invasive *Candida* in HIV patients, and TB infection. In adolescents and young adults, chronic nonsteroidal anti-inflammatory drug (NSAID) use/misuse has been indicated as a critical acquired etiology. Finally, neoplasia may be at the basis of acquired lesions, e.g., intraluminal lesions (polyps, neurofibromas, and adenocarcinoma in a setting of ulcerative colitis and familial adenomatous polyposis), intramural infiltrative lesions (tuberous sclerosis and neurofibromatosis), and extrinsic compressing lesions (lymphoma, retroperitoneal sarcoma, and teratomas).

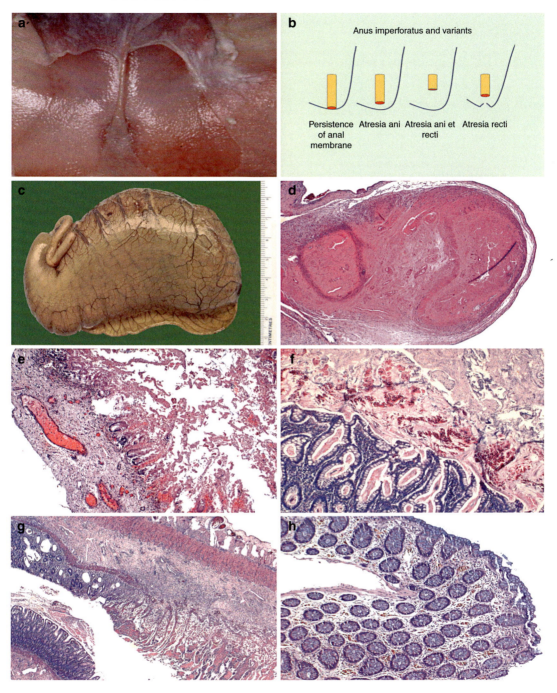

Fig. 3.39 In (**a**) is shown a patient with anal atresia, while in (**b**) a cartoon illustrates the anus imperforatus and its variants. In (**c**, **d**) are two microphotographs with colonic atresia (H&E stain, x40). In (**e**, **f**) are shown a preterm-related bowel necrosis (H&E, x100) and meconium ileus-related changes (H&E stain, x100). In (**g**) is shown a milk inspissated ileus (H&E stain, x40), while in (**h**) is shown a (pseudo-)melanosis coli in an adolescent suffering from anorexia and abusing laxatives (H&E stain, x100)

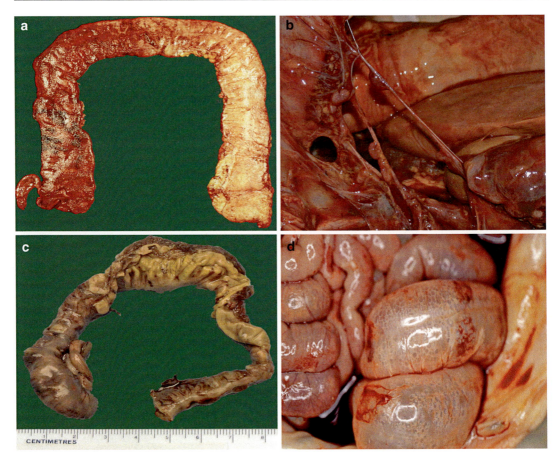

Fig. 3.40 In (**a**) is presented an ulcerative colitis, while (**b**) shows Candida-Pseudomonas infection with diffuse alveolar damage of the lungs and adhesive ileus. In (**c**) and (**d**) is shown necrotizing enterocolitis with pneumatosis intestinalis

3.6.1.3 Cloacal Membrane Teratogenesis

Robinson and Tross have proposed a probable teratogenic cause for the agenesis of the cloacal membrane. Embryonic exposure to doxylamine succinate within the first 50 years of the pregnancy was identified as "sure" in three and "probable" in two out of five pregnancies investigating a single community during a 54-year period between 1926 and 1980. Doxylamine is a first-generation antihistaminic drug, which is mainly used as the succinic acid salt (i.e., doxylamine succinate). This over-the-counter medication is often sold in Commonwealth countries in combination with paracetamol (acetaminophen) and codeine as therapy for tension headache and other types of pain, and is a quite popular "sleep-aid" in several countries (e.g., South Africa, France, Russian Federation, Spain, Portugal, and Australia). Doxylamine succinate and pyridoxine (Vitamin B6) have been approved by the FDA (U.S.A.) and Canada as the only drug approved for 'morning sickness' with a class A safety rating for pregnancy with no evidence of risk. Its "A" rating is also reported in Briggs' Reference Guide to Fetal and Neonatal Risk (Briggs et al. 2008; Slaughter et al. 2014). The International Association for Research on Cancer (IARC) inserted this chemical compound in the Group 3, i.e., "not classifiable as to its carcinogenicity to humans" indicating that there is inadequate evidence for the carcinogenicity of doxylamine

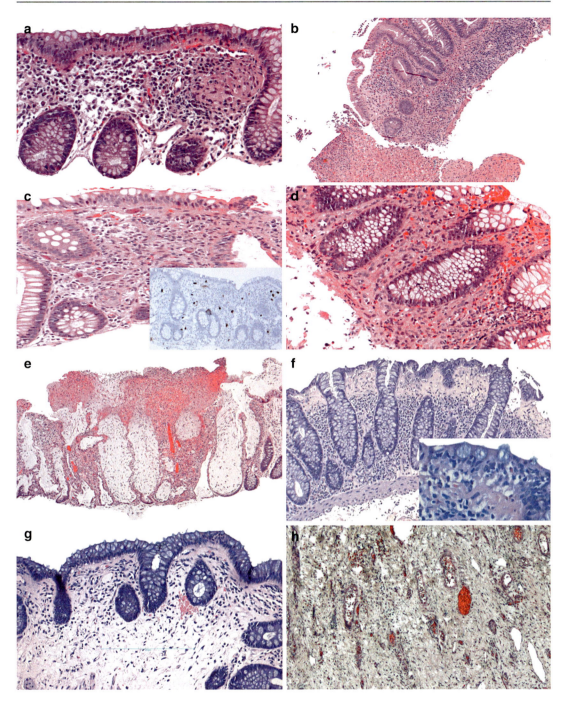

Fig. 3.41 In (**a**) is Crohn colitis, in (**b**) is ulcerative colitis, in (**c**) is Cytomegalovirus (CMV) colitis with CMV immunohistochemistry (inset). In (**d**) is protein allergic colitis with numerous eosinophils in an infant with diarrhea. In (**e**) is pseudomembranous colitis. In (**f**) is collagenous microscopic colitis with subepithelial collagenous band (inset). In (**g**) is radiation proctitis, while (**h**) is a colitis associated to a fibrosis-related stricture in a patient post-necrotizing enterocolitis (NEC) (Movat's pentachrome stain, 50×)

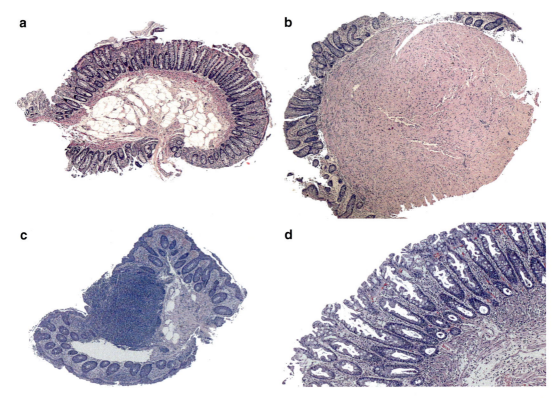

Fig. 3.42 In (**a**, **b**) are microphotographs of a submucosal lipoma (identified endoscopically as polyp), and a leiomyoma, respectively. In (**c**) is a hyperplastic lymphoid follicle and in (**d**) a hyperplastic polyp (**a–d**, H&E stain, x40, x40, x40, x100 as original magnification, respectively)

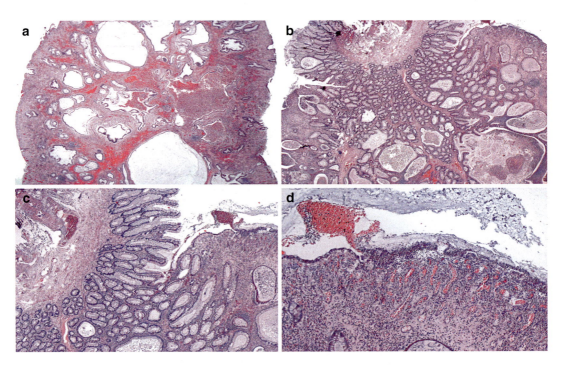

Fig. 3.43 In (**a–f**) are shown several microphotographs of a juvenile polyp (**a–f**, H&E stain, x12.5, x20, x40, x100, x100, x100 as original magnification, respectively)

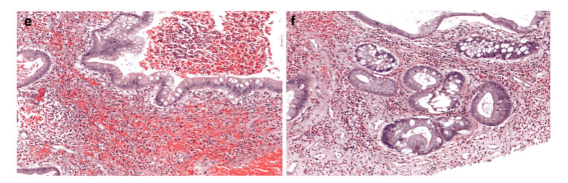

Fig. 3.43 (continued)

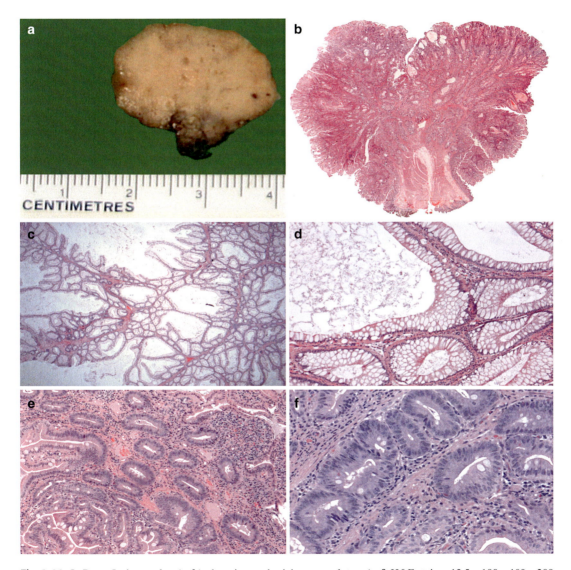

Fig. 3.44 In Peutz-Jeghers polyp (**a–b**), there is an arborizing musculature (**c–f**, H&E stain, x12.5, x100, x100, x200 as original magnification, respectively)

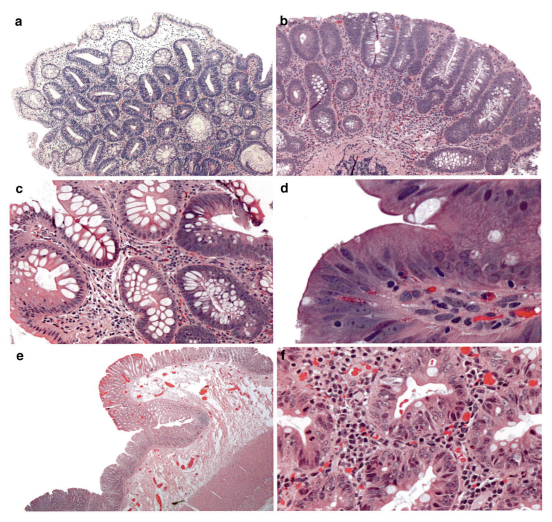

Fig. 3.45 Tubular adenomas (low-grade dysplasia) are shown in the microphotographs (**a–e**). In (**f**) is shown a tubular adenoma with high-grade dysplasia (**a–f**, H&E stain, x100, x100, x200, x630, x12.5, x400 as original magnification, respectively)

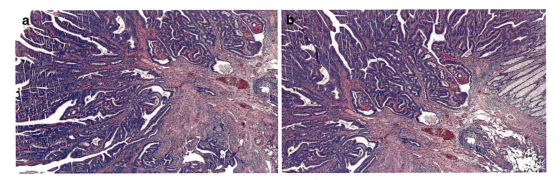

Fig. 3.46 Microinvasive carcinoma of the large bowel in an extremely rare case of youth with cancer predisposition syndrome (**a–f**, H&E stain, x40, x40, x200, x200, x200, x200 as original magnification, respectively)

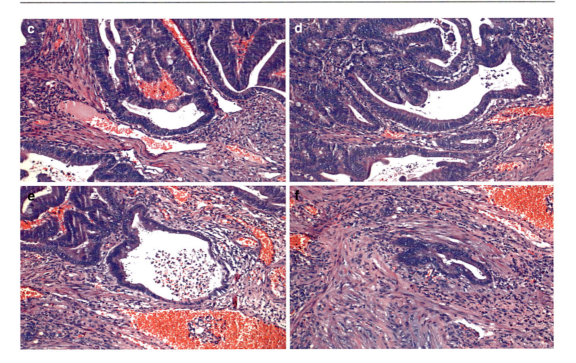

Fig. 3.46 (continued)

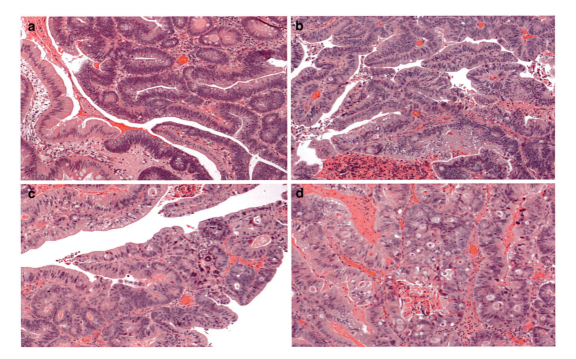

Fig. 3.47 Large bowel carcinoma (**a–d**) with high proliferation rate (**e**, anti-Ki67/MIB1 immunostaining) and *TP53* gene deletion (**f**, anti-p53 immunostaining) in a child (**a–d**, H&E stain, x100; **e–f**, Avidin-Biotin-based immunohistochemistry, x100)

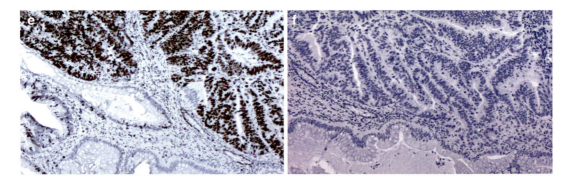

Fig. 3.47 (continued)

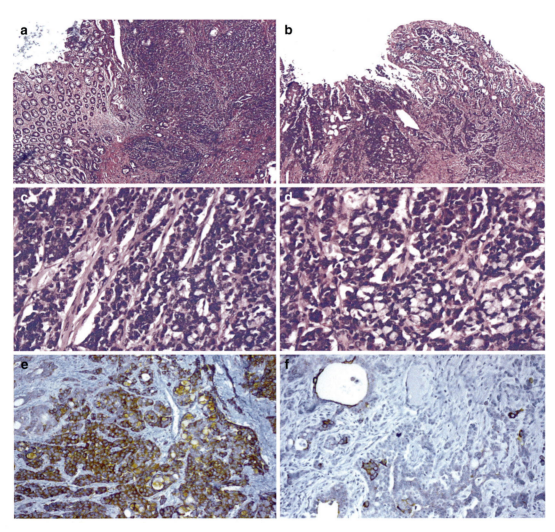

Fig. 3.48 Neuroendocrine carcinoma with classic morphology (**a–d**) (hematoxylin and eosin staining) and immunohistochemistry with small cell carcinoma of the rectum showing synaptophysin (**e**) and cytokeratin 7 (**f**) (immunostaining)

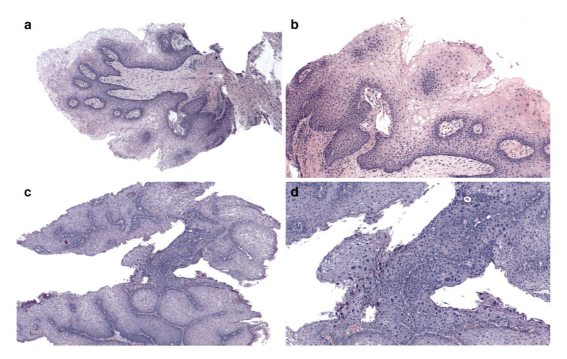

Fig. 3.49 Molluscum contagiosum in (**a**, **b**), while (**c**, **d**) show an anal intraepithelial neoplasia condyloma. Both lesions may be bring some confusion and particular care to the cytologic detail should be given

succinate in humans and limited evidence for the carcinogenicity of doxylamine succinate in experimental animals (IARC 2001).

3.6.1.4 Caudal Dysgenesis Syndrome (CDS)

- *DEF*: Malformation complex that affects the caudal spinal cord and vertebral spine, the urogenital system, the hindgut, and the lower extremities with phenotypes varying from mild (only most caudally derived structures, e.g., coccyx and lower sacrum and perineal area) to severe phenotype (lumbar and lower limb defects). Sirenomelia may be considered part of the spectrum or as a separate entity according to the classification schemes, although it is usually considered a severe degree of caudal dysgenesis in most fetal medicine centers. Moreover, cloacal/bladder exstrophy may also be part of the same spectrum of defects.
- *SYN*: Caudal regression syndrome (CRS), sirenomelia.
- *EPI*: 1:2,000–100,000.
- *EPG*: Maternal insulin-dependent diabetes mellitus (IDDM) is the most commonly recognized causal factor. In animals, some genetic factors have been identified to be etiologically related to abnormal development, including *Brachyury* (mouse, frog, zebrafish, ascidian), *Wnt3-a*, and *Danforth short tail* (*Sd*). Brachyury is located on chromosome 17 (mouse) and encodes a transcription factor that is expressed in the primitive streak and its developed mesoderm and notochord. Brachyury is also a crucial diagnostic marker for chordoma in human pathology. In experimental animal models, caudal dysgenesis can be induced using different environmental factors, including hyperglycemia, retinoids, amphetamines, trypan blue, lead, electrical fields, and ochratoxin A, which is a fungal toxin found in *Aspergillum*- or *Penicillium*-contaminated food and able to induce craniofacial, body wall, and caudal defects in mice. Some concerns may arise in considering that trypan blue is an azo dye widely used for cotton textiles in the industry (Hunger et al.

2005). Gut anomalies can also be seen in a mouse model of retinoid acid-induced caudal dysgenesis. Stage-specific retinoic acid treatment both in vitro and in vivo affects looping and rotation of the intestine and growth of asymmetrical structures with the delay in the differentiation of the intestinal mesoderm and the postcecal colon-rectum by neural crest-derived enteric neuronal precursors, potentially making an experimental model for Hirschsprung disease (Box 3.35).

> **Box 3.35 GI Biopsy Pearls and Pitfalls**
> - In examining a lower GI biopsy, we need to address a few questions that would be helpful in making the final diagnosis. In particular, we need to answer if the mucosa is inflamed, if there are features of CIIBD, if these features point to CD or UC, if acute self-limiting colitis is present, or if elements are leading to another form of inflammation. CIIBD is the generic term for referring to two very important inflammatory processes of the gastro-intestinal tract known as Crohn disease and ulcerative colitis. A schematic approach of the four compartments is required: (1) mucosal architecture, (2) cellularity of the lamina propria, (3) neutrophilic infiltration, and (4) epithelial change.
> - A normal mucosa has a crypt density of more than 6 crypts/mm, which are tightly packed, undistorted architecture. Moreover, there is an intact, columnar surface epithelial layer with standard mucin content of goblet cells, occasional branching <1/mm and no crypt shortening in a well-oriented specimen, almost flat mucosal surface, a 2:1 superficial/basal cellular density of the lamina propria with variable cell population, no granulomas, and no giant cells. No neutrophilic infiltration (≤1 PMN in an occasional crypt is allowed by some authors) (Neutrophils are the most copious type of granulocytes and that more often encountered as type of white blood cells in most mammals. Neutrophils, basophils, and eosinophils belong to the family of polymorphonuclear or PMN cellular family.

3.6.1.5 Recto-colonic Dysmotility Syndromes

It is essential to summarize that evidence-based medicine is a platform that is at the basis of modern medicine. One of the most controversially debated topics in pediatric pathology is the abnormal innervation of the GI tract and most specifically the large intestine. Since his first description in 1886, many publications, meetings, and workshops have substantiated Hirschsprung disease. The anal canal for about 2 cm above the pectinate line comprises fewer ganglion cells, more nerve trunks, and an increase of adrenergic fibers in muscle coats (Sergi 2015; Sergi et al. 2019).

Aganlionosis (HSCR, *HirSChspRung* Disease)

- *DEF:* Lack of parasympathetic ganglion cells of the submucosal (Meissner) and myenteric (Auerbach) plexuses with loss of the tonic neural inhibition and simultaneous hypertrophy of nerve fibers (sympathetic) of the distal GI tract showing a persistent contraction of the affected segment and manifesting with obstruction and dilatation ("Congenital Megacolon").
- *SYN*: Hirschsprung disease.
- *EPI:* 1:5000 births, ♂>♀, familiar cases ±.
- *ETP:* Cranio-caudal migration deficiency of (vagal) neural crest cells, which are the progenitors of ganglion cells, at the 5th to 12th weeks of gestation.

Several loci and genes have been detected to be linked with HSCR (Box 3.36).

There are a few underlying predisposing conditions with an increased risk of HSCR:

- *Syndromic HSCR*: Trisomy 21 syndrome, Waardenburg-Shah syndrome, Mowat-Wilson,

Box 3.36 Genetics of Hirschsprung's Disease

Location	Genes
10q11.21	HSCR1
13q22	HSCR2
5p13.1-p12	HSCR3
20q13	HSCR4
9q31	HSCR5
3p21	HSCR6
19q12	HSCR7
16q23	HSCR8
4q31-q32	HSCR9

Note: HSCR HirSChspRung

Goldberg-Shpritzen megacolon syndrome, primary central hypoventilation syndrome or Ondine course, cartilage-hair hypoplasia syndrome, and Smith-Lemli-Opitz syndrome.
- *Nonsyndromic HSCR*: Congenital Heart Disease (CHD), GU anomalies, congenital deafness, MEN2A, and MEN2B.

Patients are usually diagnosed with the short-segment form (S-HSCR), which occurs in approximately 80% of cases. In this form, the aganglionic segment does not extend beyond the upper sigmoid (rectosigmoid). The long-segment form (L-HSCR) (~15%) is when aganglionosis extends proximal to the sigmoid, while total colonic aganglionosis (Zuelzer-Wilson disease) (~5%) and total intestinal HSCR (exceptional) are characterized by a full deficiency of migration of neural crest cells either in all colon-rectal segments (i.e., the whole colon and rectum are aganglionic) or all bowel loops. There is a short or even an ultrashort segment, which is considered the anal form or the most distal portion, but its accurate recognition has been debated, indicating that it should be an anal manometry-based diagnosis only.

Neural Crest Migration

Neural crest cells (NCCs) are transient progenitor cells of highly motile phenotype with multipotential behavior that give rise to different cell types, including neural, endocrine, pigment, craniofacial, and cardiac cells, at the conotruncal location. According to the specific differentiation, there is a diverse array of cell types from chromaffin cells of the adrenal medulla to melanocytes of the skin. *Xenopus* and zebrafish (*Danio rerio*) embryos are two typical and attractive experimental models to study the migration. Four functionally overlapping categories can be delineated by the axial level at which neural crest cells arise (neural crest cell derivatives), including cranial neural crest cells (dorsolateral migration pathways) determining the craniofacial cells, trunk neural crest cells (anterior sclerotome pathways) defining the cutaneous melanocytes, cardiac neural crest cells identifying cardiac neurons, and vagal neural crest cells shaping the parasympathetic ganglia. The neural plate arises from neuroepithelial cells with at the edges of the plate formation of thickening to converge and the establishment of a U-shaped structure. Then, the sides fuse to form the neural tube, while the establishment of the epidermis and neural crest cells identifies the end of primary neurulation. Neural crest cell migration occurs from the neural tube and is both spatially and temporally regulated. In particular, bone morphogenetic protein, Wnt, slug, and a snail with noggin are antagonist factors decreasing in expression caudo-rostrally. Tyrosine kinase ephrin receptors and their ligand ephrinB2 provide migration in a more detailed way ensuring that neural crest cells migrate along the correct path to reach the appropriate target destination. A broad spectrum of disorders, collectively known as neurocristopathies, arises by the impaired development of these progenitor cells. Hirschsprung disease and neuroblastoma are neurocristopathies of the sympathetic and enteric nervous systems (ENSs). Although neuroblastoma is a neoplasm and Hirschsprung disease is a loss of neural crest derivatives, both disorders may occasionally co-occur, particularly in congenital central hypoventilation syndrome (CCHS), which is also referred to as Ondine's curse. Of cultural note, the name of this condition is associated with a novella of 1811 (*Undine*). In this novel as well as in a 1938 play by Jean Giraudoux, Hans and Ondine are the two major characters and Hans' deception is penalized with his death. CCHS etiology relies on a failure of the autonomic control of breathing and is characterized by the lack of the arousal of breathing under hypercapnic or hypoxic conditions. Recently, heterozygous mutations of the Paired-like homeobox 2b (*PHOX2B*) gene have been found in patients with CCHS and co-association of Hirschsprung disease and neuroblastoma. Interestingly, PHOX2B is a paired homeodomain transcription factor, which is crucial for the development of cells constituting the autonomic neural circuits and introducing non-polyalanine repeat expansion mutation of the *PHOX2B* into the mouse *Phox2b* locus reiterates the clinical features of the CCHS associated with both Hirschsprung disease and neuroblastoma. Nagashimada et al. identified that non-polyalanine repeat expansion mutation of PHOX2B is both a dominant-negative and gain-of-function mutation and *Sox10* (*s*ex-determining region Y b*ox 10*) regulation by PHOX2B is pivotal for the development and pathogenesis of the autonomic ganglia. As indicated above, other genetic diseases may be associated with Hirschsprung disease, other than Down syndrome and CCHS, and include Mowat-Wilson and other brain-related syndromes, as well as MEN2, Waardenburg syndrome and other dominant sensorineural deafness syndromes, and autosomal recessive syndromes such as Shah-Waardenburg syndrome, Bardet-Biedl syndrome, cartilage-hair hypoplasia, Goldberg-Shprintzen syndrome, and other syndromes related to cholesterol and fat metabolism (Sergi et al. 2017).

- *CLI:* Infants may present in the neonatal period or the first months of life delay in the passage of meconium, distal intestinal obstruction, abdominal distension, necrotizing enterocolitis, and persistent constipation ("*Congenital Megacolon*"). Necrotizing colitis is a life-threatening complication and should be kept in mind for both clinicians and pathologists. Anal manometry is a valuable tool for the diagnosis of aganglionosis (Takawira et al. 2015).
- *IMG:* Barium enema evidences a narrow segment of rectum/colon-rectum distally with a dilated proximal portion.
- *GRO:* Two surgical biopsies supply for the diagnosis of HSCR, including *mucosal suction* biopsies and *full-thickness* biopsies (e.g., 1 cm × 0.5 cm × 2 cm) above the pectinate line. The "pull-through" specimen (Box 3.37) is characterized by a dilated portion proximally and a stenotic portion distally due to the imbalance with the cholinergic and adrenergic innervation. During the pull-through procedure, the pathologist is required to perform intraoperative frozen sections to evaluate ganglion cells at the site where the surgeon is going to perform the colostomy.

Box 3.37 Primary Pathology Hirschsprung (Pull Through, Resection: Grossing)

- To be submitted in toto; if it is a very large specimen, consult with the pediatric pathologist.
- Note if the bowel is inside out.
- Indicate any areas of dilatation.
- Identify proximal and distal ends (there may be multiple black sutures at the distal end) and document photographically.
- Ink proximal margin blue and distal black.
- Map and submit the bowel in toto as follows:
  - Shave the proximal and distal margins, ensuring entire circumference is captured, and submit in separate cassettes.
  - Serially section remaining intestine at 1.5–2.0 cm intervals (just long enough to fit into a cassette).
  - Keep carefully these sections in order, from proximal to distal.
  - Mark each section's distal end with black ink, and blot dry and consider to document photographically.
  - Longitudinally, radially section each part so that the slices can fit, on edge in the cassette.
  - Submit the specimen in toto, from proximal to distal, sequentially (if the area of bowel is large/dilated, one section may eventually require more than one cassette).

- *CLM:* Adequacy criteria include a correct site for biopsy, which is 2–3 cm above the mucocutaneous junction, and an adequate amount of submucosa for an accurate assessment of the neural crest cell migration. Identifying ganglion cells and their maturity is crucial. Newborns often have a significant number of immature ganglion cells, characterized by diffuse chromatin, lack of nucleolus, and less cytoplasm than a mature ganglion cell. It is also essential to avoid mislabel as ganglion cells the following structures, including endothelial cells (+CD31), smooth muscle cells (+DES), lymphocytes (+CD45), and histiocytes (+CD68). Apparently, in the practice routine, there is no need to perform immunohistochemistry, but it may be needed occasionally.

The diagnosis of HSCR relies on the absence of ganglion cells of the submucosal (Meissner) and myenteric plexuses with simultaneous hyperplasia and hypertrophy of nerve fibers and some diseases should be considered in the differential diagnosis (Sergi et al. 2019). In HSCR, nerve bundles are maloriented and constituted all long way by Schwann cells, collagen, and perineural sheath. There is also an increase of adrenergic nerves in muscle layers and an increase of cholinergic fibers, especially distally. The transition zone is characterized by the termination of submucosal ganglion cells, which parallels myenteric ganglion cells (former more proximally localized), ganglion cells are found more distally along the antimesenteric border, and there is a ↓ of the number of ganglion cells with an ↑ of nerve trunks with the regular neural matrix.

The roles of special stains and IHC in HSCR diagnostics are quite controversial. An assessment of the acetylcholinesterase activity on the frozen material is performed in most laboratories of children's hospitals to identify increased cholinergic fibers in the lamina propria and muscularis mucosae, which is indicative of HSCR. It may be decreased in the transition zone. The Karnovsky-Roots method or one of its modifications is usually used. Calretinin immunostaining has been proposed as a useful alternative to the acetylcholinesterase special stain, which may take up to 8 h. Other antibodies used for immunostaining are NSE and S100. NSE stains ganglion cells and nerve fibers, while S100 stains the nerve fibers.

> **Box 3.38 Pearls and Pitfalls**
> - Pearls: correct orientation of the specimen, adequate technological support, a sufficient number of slides examined (≥50 serial sections), adequate submucosa, avoidance of lymphoid aggregates.
> - Pitfalls: Biopsy of the hypoganglionic zone, biopsy of the transition zone, hypoganglionosis (single ganglion cells and increased nerve fibers), and skip lesions; long aganglionic segments of newborns and infants do not have the classic barium enema picture. Zonal aganglionosis or skip-segment HSCR is a controversial entity indicating the presence of ganglionic segments between aganglionic segments, whereas hypoganglionosis is defined as a tenfold decrease of ganglion cell number and is seen in the transition zone or belongs to the group of NH recto-colonic dysganglionosis.
> - The presence of some inflammation (HSCR patients are prone to colitis) may be a hint for diagnosis, but this aspect needs to be carefully evaluated and never overemphasized.

- *DDX*: Non-Hirschsprung recto-colonic dysganglionosis (NHRCD).
- *PGN*: Following the initial diagnosis of HSCR, lavages are performed until a stoma is performed.

The HSCR diagnostics tips and pitfalls are crucial (Box 3.38).

3.6.1.6 Non-Hirschsprung Recto-colonic Dysmotility Syndromes

Some diseases should be differentiated from Hirschsprung disease, although they are probably quite rare. Since the diagnosis is not completely satisfied and criteria are considered variable by some authors, aspects about epidemiology and some clinical features may not be currently clari-

Intestinal neuronal dysplasia (IND) is a disorder of the myenteric nervous system with a clinical presentation similar to HSCR. There is still controversy over nomenclature and treatment, although geographic areas reach some consensus worldwide. Intestinal neuronal dysplasia (IND) type A is characterized by a decrease of the sympathetic innervation of the intestine with a moderate increase of the parasympathetic nerve fibers. IND type B shows an increase of ganglion cells in the submucosal ganglia with >8 ganglion cells/ganglion. Ganglion cells are smaller than normal and are grouped into spherical structures ("giant ganglia"). It has been stated that at least 20% of giant ganglia in 30 serial sections must be found to make a diagnosis of IND type B. Other intestinal dysmotility syndromes include *Ogilvie syndrome* (acute distension of the caecum and right colon following often the use of certain medications, particularly narcotics, that impair intestinal motility and possibly complicated by ischemic necrosis), *chronic idiopathic intestinal pseudo-obstruction*, *abnormalities of the intestinal smooth muscle (myopathy)*, *solitary neuropathy of the myenteric plexus*, and the *megacystis-microcolon-intestinal hypoperistalsis syndrome*. Moreover, *meconium plug syndrome*, *small left colon syndrome* (dysfunctional, diminutive left colon causing transient obstructive symptoms with diabetic mothers in about 50% of patients and pathogenesis is linked to hypoglycemia-related release of glucagon in the newborn), *cecal or sigmoid volvulus*, *neoplastic lesions*, *anomalous congenital peritoneal bands*, *hindgut duplication*, and *congenital segmental dilatation of the colon* have also been considered.

fied. The simplistic concept of the presence or absence of ganglion cells for HSCR or – better – as the only definitive histologic criterion for abnormal function of the hindgut is probably inadequate and should be considered misleading. It can also be related to the lack of adequate reproducible methods other than lack of satisfactory information.

Under the umbrella of non-Hirschsprung recto-colonic dysganglionosis, we should mainly list the following entities hypoganglionosis, neuronal intestinal dysplasia, type A (NID-A), neuronal intestinal dysplasia, type B (NID-B), ganglion cell immaturity (GCI), and ganglion cell heterotopia/hypogenesis.

Rectal Biopsy as Gold Standard

Rectal biopsies are a challenge, particularly in the neonatal age with or without prematurity, and a pediatric pathologist or a gastroenterological pathologist competent in reading the histologic slides is essential and should not be taken for granted. The suction biopsy of the mucosa and submucosa of the rectum is probably the best diagnostic procedure (gold standard) to rule out HSCR focusing on submucosal plexus. A full-thickness biopsy may become indispensable if the suction biopsy is nondiagnostic or if the pathologist has some concerns about material adequacy to make this diagnosis on a suction biopsy specimen. Since the full-thickness biopsy is quite tricky in the young infant, it has been widely replaced by the suction biopsy, and this occurs in most centers today. In practical terms, all newborns (<4 weeks of age), premature or not, who have delayed passage of meconium associated with a suspicious contrast enema should receive a suction biopsy.

3.6.1.7 Constipation and Chronic Intestinal Pseudo-obstruction

However, apart from Hirschsprung disease and non-HD-RCD, it may be necessary to stress that in many situations, childhood constipation has no organic defect. Childhood constipation is a multifactorial problem, and dietary review should always be part of the management package. A child can become constipated for some reasons, but it is essential to recall that a cyclical pattern can erupt. The constipation cycle starts with poor diet and ↓ fluid intake, which determines stool hardening and anal pain during the passage. The consequence is the "withholding effect", which determines constipation and abdominal discomfort. Ultimately, the cycle repeats with decreased intake of nutrients and fluid. The treatment and the challenge for the pediatrician are to break the cycle. The eating patterns in the preschool years of the child can be very unstable, and particularly important is to have a good understanding of the family situation (e.g., job, immigration status, parental relationship). There are phases of eating very little of anything, eating from a very narrow range of foods, grazing on foods, and at the same time casual meals and sudden likes and dislikes. All of the factors above are contributory factors. It is essential to promote an appropriate fiber and fluid intake during these phases since early childhood. Dietary fibers are the undigested remains of plant material, now known as nonstarch polysaccharides. These polysaccharides may be divided into fibers, which are soluble (50% of fiber intake). The source is fruits, vegetables, oats, and pulses and insoluble fibers (50% of fiber intake). These fibers have different properties and are usually found in different proportions in fiber-containing foods. Therefore, it is crucial that a variety of fiber-containing foods are eaten. Both fibers (soluble and insoluble) are essential in the diet, because the first slow the rise in blood glucose and help to decrease blood cholesterol levels, while the second act as a bulking agent and help to keep the colon healthy and draw water into it. The average intake of fiber is around 12 g per day in an adolescent and young adult, although recommendations may suggest increasing the input to 18 g per day, or even more.

It is known that children need proportionally less fibers than adults, although may be essential to teach a nutritional habit, which may be of great value later in life. There is no recommendation of minimal intake of fiber in the UK, although the USA recommends the sum of the age of the child and 5–10 g per day. Thus, a 5-year-old child should have 10–15 g per day (Williams et al. 1995; Williams 2006). For infants and preschool children, the introduction of fibers should be gradual. An excessive amount of fibers can increase the bulk of the diet, and they become easily and quickly full before they have eaten sufficient food to satisfy their need for essential vitamins, minerals, and energy. It is imperative that the amount of fiber in the diet does not compromise energy intake. Fluid intake is of the utmost importance in talking about fiber intake. Fibers draw fluid into the large bowel to get the formation of a stool, which becomes soft and bulky. The fluid intake should be around 6–8 cups daily. In many countries, children drink less than half of this recommended amount and this may jeopardize their health progressively. The type of fluid does not matter, although sugars that are not absorbed (e.g., fructose, sorbitol) increase stool frequency, because they cause high fecal water content.

Pediatric chronic intestinal pseudo-obstruction (PCIP), defined as a potentially disabling GI disorder characterized by abnormalities affecting the involuntary, coordinated muscular contractions (peristalsis) of the GI tract, may be subdivided into several categories (Box 3.39).

Box 3.39 Pediatric Chronic Intestinal Pseudo-obstruction (PCIP)

Organic colorectal diseases
  Pseudotumors (e.g., rectocele)
  Neoplasms (benign/malignant)
  Strictures (e.g., post-inflammation, post-ischemic, GvHD)
Dysmetabolism
  Thyroid hormone imbalance (hypothyroidism)
  Calcium imbalance (hypercalcemia)
  Potassium imbalance (hypokalemia)
  Diabetes mellitus (autonomic neuropathy)
Autonomic enteric neuro-myopathies
  Scleroderma or systemic sclerosis
  Amyloidosis or familial Mediterranean fever
  Sarcoidosis
Central nervous system: pathologies
  Cerebral palsy
  Spinal cord injury
  Demyelinating disease of youth
Iatrogenic PCIP (drugs-related) (e.g., Alu (antacids), anticholinergics, Ca^{2+}-channel blockers, antidepressants, antipsychotics, and narcotics)

Many scores have been designed to evaluate the severity of constipation, especially in obstructed defecation syndrome. Three different constipation scoring systems (Wexner, Knowles-Eccersley-Scott symptom (KESS), and Mansoura scoring systems) are frequently in use. The Mansoura Numeroalphabetic Constipation Score (MNCS) seems to be a simple, modified version of Wexner and KESS scores and has the advantage of accurate, rapid, and straightforward assessment of obstructed defecation, differentiating the constipated patient from the usual subject effectively.

Animal Models and *What's Going on in Research?*
An alignment of the human RET protein sequence with the orthologous sequences of 12 nonhuman vertebrates, including their comparative analysis and the evolutionary topology of the RET protein, was already performed building the basis for some genotype/phenotype correlation (Kashuk et al. 2005). There are a few animal models described in the literature. Mice with piebald trait show aganglionosis (Bielschowsky & Schofield 1962). Ileocolonic aganglionosis was described in the horse in the early 1980s, when affected foals, almost all white, die following intestinal obstruction in the first few days of life (Hultgren 1982). In both animals, the disorder is caused by a mutation in the EDNRB gene. Knockout mice experiments involving a Hox11 gene, Ncx/Hox11.L1, show animals that develop megacolon by age 3–5 weeks with some hyperinnervation of the enteric neurons in the narrow segment of the colon, some neuronal degeneration, and cell death change similar to neuronal intestinal dysplasia (Hatano et al. 1997). The gene was inactivated in embryonic stem cells by homologous recombination.

3.6.1.8 Colon Diverticulosis
- *DEF:* Outpouchings of mucosa and submucosa through a weakened area in the muscularis propria, often with hypertrophy of adjacent muscularis parts.
- *EPG:* Multifactorial pathophysiology including diet, colonic wall structure, intestinal motility, and underlying genetic aspects with some similarity with chronic IBD (focal weakness in the muscularis propria and ↑ intraluminal pressure).
- *EPI:* Acute diverticulitis is rare in patients <40 years old, representing only 2–5% of all cases. However, Williams-Beuren syndrome (WBS), Marfan syndrome (MS), Ehler-Danlos syndrome (EDS), and cystic fibrosis (CF) seem to predispose the gut to the development of diverticular disease earlier in life. Diverticulosis and diverticulitis of the left colon have been reported with a frequency of 8–12% among young adults with WBS.
- *CLI:* Symptomatic in only ~1/5 with symptoms of lower abdominal discomfort ± cramping pains, which may be present, independent of diverticulitis (left > right colon with >95% diverticula are located in the sigmoid colon).
- *IMG:* CT scan as the gold standard to diagnose the acute disease with high predictive value.
- *GRO:* When inflamed (acute and chronic), there is induration ±abscess or fistula.
- *CLM:* Acute and/or chronic inflammation features (PMNs, lymphocytes, plasma cells).
- *DDX:* Crohn disease. Pearls and pitfalls of colon diverticular disease are displayed in Box 3.40.
- *PGN:* Diverticulitis has a substantial rate of recurrence, and long-term treatment should be managed in a multidisciplinary team with clinical geneticists, pediatric gastroenterologists, and pediatric surgeons playing a significant role. Surgery does not seem to guarantee complete vanishing of symptoms, and specific medications (e.g., rifaximin, mesalamine, probiotics) are the best approach.

Box 3.40 Colon Diverticular Disease
Pearls and pitfalls: When you see a patient with WBS, MS, or CF and chronic recurrent pain in the left lower abdominal quadrant, do not forget the possibility of colonic diverticulosis or acute diverticulitis.

3.6.2 Large Intestinal Vascular Changes

There are very few conditions that may involve the large intestine apart from the shock-related changes. Necrotizing enterocolitis has been described above in the small intestine, but this condition may extend to the large intestine and determine the worrisome outcome of these patients. Moreover, ischemic colitis, although rare, may be encountered in children or young adults with congenital heart disease or abnormalities of the vascular supply to the large intestine. Ischemic colitis is not only a pathology of elderly and needs to be taken into account, particularly in children or young adults affected with genetic syndromes and congenital heart disease or that need to undergo surgery (Box 3.41).

3.6.2.1 Ischemic Colitis
- *DEF*: Ischemia of the colonic segment of the gastrointestinal tract with the most common involvement of the splenic flexure, descending colon, and sigmoid colon.
- *EPI*: Rare condition in neonatology and pediatrics with an etiology related to vascular occlusion and non-occlusive mesenteric ischemia (NOMI), which is induced by intestinal vasospasm without thromboembolic occlusion or by decreasing the mesenteric blood flow.
- *GRO*: Ulcerated bowel segment with a discrete or serpiginous pattern or cobblestone (Crohn disease-like) pattern or pseudopolyps-rich pattern resembling ulcerative colitis with a hemorrhagic appearance due to blood reflow.
- *CLM*: Edema, hemorrhage, later ulcerations, and fibrosis are often seen and potentially accompanied by pseudomembranes, pseudopolyps with not uncommonly adjacent normal mucosa. Special stain (Perls' Prussian blue) may show abundant hemosiderin according to the timing of the ischemic insult and how long it has protracted. Importantly, lymphoid follicles and granulomas are absent. Perls' Prussian Blue, which is named from the German pathologist Max Perls (1843–1881), does not involve the use of a dye. This technique determines the pigment "Prussian blue" to arise directly within the tissue. PPB targets mostly iron in the ferric state (e.g., ferritin and hemosiderin), rather than iron in the ferrous state. As a chemical (stoichiometric) note, ferrous (less stable) is the +2 oxidation state of iron, while ferric (more stable) is the +3 oxidation state of iron. Thus, Fe^{+2} can be formed through the reduction of Fe^{+3} ions, while Fe^{+3} are formed through the oxidation of Fe^{+2}. PPB relies on the following formula: $4FeCl3 + 3K4Fe(CN)6 \rightarrow Fe4[Fe(CN)6]3 + 12KCl$, i.e., (ferric iron) + (potassium ferrocyanide) $\rightarrow$ (ferric ferrocyanide or Prussian blue) + (potassium chloride) (Drury & Wallington 1980). The PPB is usually visualized microscopically as blue deposits.
- *DDX*: Crohn disease, ulcerative colitis, *E. coli* O157:H7 colitis, and pseudomembranous colitis.
- *PGN*: It depends on the extension of the lesion and the rapidity of the surgical intervention.

3.6.2.2 Angiodysplasia and Heyde Syndrome

Heyde syndrome is a syndrome of aortic valve stenosis associated with GI bleeding from colonic angiodysplasia. Aortic valve stenosis may be minimal and passed unobserved or cause of heart failure and several other cardiac problems as well as a bleeding tendency resembling that of platelet-derived coagulation disorders. Colonoscopy ± angiography usually identifies angiodysplasia. The underlying disorder is probably related to a subtle form of von Willebrand disease present in patients affected with Heyde syndrome, which resolves rapidly after heart valve replacement of the stenotic aortic valve. The coagulation abnormality may be caused by the increased breakdown of the very large von Willebrand factor

---

Box 3.41 Vascular Changes of the Large Intestine

3.6.2.1. Ischemic Colitis
3.6.2.2. Angiodysplasia and Heyde Syndrome

molecule by its natural catabolic enzyme (*ADAMTS13*) under conditions of high shear stress around the valve.

There is dilatation and increased tortuosity of the submucosal veins in the cecum and occasionally ascending colon and develops in the cecum because this GI portion has the largest diameter, and wall tension is the greatest, compressing the veins in the muscularis propria and consequently shunting more blood through the submucosal veins.

3.6.3 Large Intestinal Inflammatory Disorders

About necrotizing enterocolitis (see SMALL BOWEL), but this section will be reserved for ulcerative colitis and other colitides. Crohn colitis is described in Crohn disease, and see the chapter of the small bowel (Box 3.42).

3.6.3.1 Ulcerative Colitis

- *DEF*: Idiopathic chronic and acute inflammatory bowel disease characterized by relapsing and remitting course, variable endoscopic and fluctuating histologic activity, and evidence of continuous mucosal inflammation of the colon and rectum.
- *EPI*: 1:40,000 children (Northern Europe, North America), 4–6:100,000 in the USA; whites > black; ♂<♀; peak onset 20–25 years. Westernized lifestyle factors and UC:
  1. Cigarette smoking has been associated with mild disease, ↓ hospitalization, and ↓ use of IBD-specific medications.
  2. Early appendectomy in life has been associated with a decrease in incidence of UC.

> **Box 3.42 Large Bowel Inflammatory Disorders**
> 3.6.3.1. Ulcerative Colitis
> 3.6.3.2. Acute Self-Limiting (Infectious) Colitides
> 3.6.3.3. Microscopic/Collagenous Colitis
> 3.6.3.4. Others

Appendectomy with the resection of the appendix, a predominant "helper" organ, might influence the balance between "ileo-colonic helper and suppressor function" protecting against UC, although it does not seem affecting the UC course. (Rutgeerts et a. 1994; Selby et al. 2002).

- *EPG*: UC appears to be as *genetically heterogeneous* as CD, and several pathways have been mostly identified, including loci implicated in the dysfunction of the epithelial barrier (*ECM1*, *HNF4A*, *CDH1*, and *LAMB1*), apoptosis and autophagy (*DAP*), transcriptional regulation (*PRDM1*, *IRF5*, and *NKX2-3*), and IL-23 signaling pathway (*IL23R*, *JAK2*, *STAT3*, *IL12B*, *PTPN2*), which shares with CD. Other immunologic risk loci may also be involved, including *HLA-DR*, *Th1*, and *Th17* differentiation genes, such as *IL10*, *IL7R*, *IL23*, and *IFN-γ*. There is an association between HLAB27 and BW35. Microbiologic abnormalities, which may influence the human gut microbiota, and mucosal immune response, as well as alterations of epithelial cells and autoimmunity, may play a substantial role.
- *CLI*: An endoscopic grading system subdivides the UC colonoscopy into four quite distinct degrees or Baron Classification, which includes normal mucosa (grade 0), congestive edema and loss of vascular pattern of the mucosa (grade 1), mucosal friability and bleeding by contact (grade 2), mucosal bleeding spontaneously (grade 3), and mucosal ulceration (grade 4). Endoscopically/grossly, in pancolitis, the inflammation is limited to the colon-rectum and stops at the ileocecal valve, except backwash ileitis. Conversely, the CD is characterized by discontinuous inflammation with aphthae/ulcers, skip lesions, a cobblestone appearance, and usual rectal sparing.
- *CLM*: There is a continuous and diffuse inflammation limited to the colon-rectum and stops at the ileocecal valve (pancolitis form), but milder forms are also possible, and backwash ileitis is diagnosed (Box 3.43). As substantiated above, the UC inflammation is

> **Box 3.43 Ulcerative Colitis: Salient Features**
> - Mucosal damage of large bowel almost always affecting the rectum
> - Mucosal damage spreading proximally
> - Mucosal damage stopping at the ileocecal valve (exception: backwash ileitis)
> - Mucosal damage characterized by granular pattern, linear ulceration, and inflammatory nondysplastic polyps

typically continuous and diffuse with almost constant rectal involvement (untreated patients) without "skip lesions", deep fissural ulcers, transmural sinus tracts, and transmural lymphoid aggregates or granulomas. In children, chronicity signs (21%), relative rectal sparing (23%), and initial pancolitis (42%) are often to be recognized (Glickman et al. 2004).

A patchy involvement may be observed following therapy. Conversely, the CD is characterized by initial discontinuous inflammation with aphthous ulcers, skip lesions, cobblestone appearance of the mucosa, and usual rectal sparing. Mucosal inflammation with crypt architecture distortion diminished crypt density as well as inflammatory infiltrates of lymphocytes, plasma cells, and neutrophils that infiltrate crypts giving rise to cryptitis and crypt abscesses. There may also be goblet cell depletion and Paneth cell metaplasia. Erosion and ulcerations may also be present (severe inflammation). Epithelioid granulomas are not usually present, although some foreign body giant cell reaction may be present in the case of mucin dispersion. Inflammation grading and scoring are not an easy task for the practicing pathologist. Some methods have been proposed with variable levels of intra- and interobserver agreement and k values. Most probably, one of the most reproducible gradings is that ideated by Dr. Noam Harpaz at the Mount Sinai Hospital in 1988 (Gupta et al. 2007). The *histologic activity index* (HAI) is scored as 0 (inactive/quiescent), if no epithelial infiltration by neutrophils is found; 1 (mildly active), if neutrophilic infiltration is present and <50% of sampled crypts or cross sections are involved; 2 (moderately active), if ≥50% of sampled crypts or cross sections are involved; and 3 (strongly active), if erosions or ulcerations are seen. HIA is an important record in the UC patient's history because histologic inflammation is a risk factor for progression to colorectal neoplasia. Substantially, it is important to stress on the basal inflammation of the mucosa with limited extension to the submucosa characterized by crypt architecture distortion (branching and irregular glands), crypt density decrease, cryptitis and crypt abscesses (inflammatory infiltrates of lymphocytes, plasma cells, and neutrophils of the crypt epithelium and lumen), and regenerative/sparing features, including Paneth cell metaplasia, adipose islands, and pseudopolyps. Erosion and ulcerations may also be present indicating severe inflammation. Epithelioid granulomas are not usually present, although some foreign body giant cell reaction may be present in the case of intramucosal mucin dispersion. Clues for ulcerative colitis are abnormalities of the architecture, atrophy of the crypts, plasmacytosis of the basal mucosal stroma, Paneth cell metaplasia distally of the right flexure, depletion of the mucin close to ulcerations, continuous and diffuse (transmucosal) infiltration of the mucosa by lymphocytes and plasma cells, continuous distribution of the crypt atrophy or disturbance of the crypt architecture, and continuous distribution of the mucin depletion (active colitis). The variants of the morphologic picture of the UC include UC with discontinuous pattern with or without sparing of the rectum, pancolitis ulcers with or without marginal sparing of the rectum, *Pancolitis ulcerosa* with or without backwash ileitis, and *Proctitis ulcerosa*. ESR measurement (± blood counts), fecal lactoferrin, or calprotectin has been associated with the severity of inflammation. Calprotectin is a protein released by neutrophils and is used as an indirect measurement of the inflammatory activity of the gut by determining the amount of this protein present in the stools of patients with diarrhea. However, there is a lot of unknown around UC. Complications may be life-threatening, and surgical consult is often required for in-patients. It is important to stress that a c ombination of architectural, epithelial, and inflammatory changes is used to put in place to distinguish between CIIBD, acute infective-type colitis or self-limiting colitis, and other rare causes of colorectal inflammation in the majority of cases. Both architectural changes and distribution of the lamina propria cellularity are the most important differentiating features. Architectural abnormalities should be absent in self-limiting colitis, and inflammation of the lamina propria should be less marked and more superficial in self-limiting colitis than in CIIBD. Moreover, cryptitis is more common in UC but is also observed in Crohn disease. Some mucin depletion may be seen in any colorectal inflammation, but severe mucin, or even total, mucin depletion and erosions are more characteristically seen in UC than in Crohn disease. The infective type of colitis that a pathologist can come across during his or her practice is that related to infections of *Salmonella*, *Campylobacter*, *Shigella*, and certain strains of *E. coli*. Some immunodeficient conditions may complicate the picture and need to be considered. After 2–3 weeks of prolonged diarrhea, it is advisable to perform the biopsy; however, many features classically associated with infective-type colitis are only observed in the early acute phase. Features suggesting an infectious etiology include retention of normal architecture, superficial increase in lamina propria cellularity, neutrophilic infiltration (initial stage), mucin depletion, discontinuous inflammation, and focal cryptitis. *Campylobacter* colitis and verotoxin-producing *E. coli* colitis may show superficial hemorrhagic necrosis, which is superimposed on acute diffuse inflammation. In *C. difficile*-associated pseudomembranous colitis, superficial mucosal and crypt damage may be marked, and in the early stage, "volcano-like" lesions may be observed. In a subsequent stage, superficial necrosis involving upper crypt epithelium leading to a characteristic appearance of crypt "withering" with overlying inflammatory sloughing may mimic ischemic colitis. Some pitfalls in the setting of infectious etiology remain chronic shigellosis that may tremendously mimic CIIBD and the granulomatous inflammation of schistosomiasis, which may be confused with Crohn disease if characteristic ova are not identified. Clinical history and travel history are imperative. A clinical history of immunosuppression may suggest the diagnosis of rarer types of inflammation. It is important to list or screen for bone marrow transplantation, HIV infection, anal-receptive (child abuse-related) sexual intercourse, diversion of the gastrointestinal fecal stream, drugs (e.g., NSAIDs), chemotherapy and radiotherapy, gold-based therapy, penicillamine, and prolonged use of antibiotics. Specific features may be encountered for each of these rarer types of colonic inflammation; none of them is pathognomonic and the perusal of clinical chart, either manually or electronically; and discussion at MDT meetings or rounds are mandatory. Finally, sampling needs to be appropriate. Thus, a single biopsy is not enough to give the correct or suggestive diagnosis in up to 30% of cases of UC and up to 60% of cases of Crohn disease.

- *TRT:* It is intended to induce and then maintain remission. However, it depends upon the extent and severity of inflammation. Current medical therapy includes the following drugs: steroids, thiopurines, salicylates, and anti-Tumor Necrosis Factor (TNF) agents, such as infliximab (IFX). Inhibition of TNF effects, which is part of the inflammatory response, is key in some UC patients. TNF inhibition can be attained with monoclonal antibodies (e.g., adalimumab, certolizumab pegol, golimumab other than infliximab) or introducing a circulating receptor fusion protein such as etanercept fusing the TNF receptor to the constant end of the IgG1 antibody. Unresponsiveness to medical treatment with severe and extensive inflammation or neoplastic transformation prompts to surgery, and it is estimated that about 1/4 of patients of all ages eventually require total proctocolectomy with ileal pouch-anal anastomosis (IPAA). Other less common indications include toxic dilatation (cross diameter >6 cm), perforation, and bleeding. About neoplastic transformation, which occurs in about 5% of UC patients, studies using meta-analysis identified a transformation rate of 2% at 10 years, 8% at

20 years, and about 20% at 30 years. Dysplasia is classified as either low- or high-grade dysplasia in consideration of architectural and cytologic abnormalities. Moreover, adenoma-like lesions or masses (ALMs) and dysplasia-associated lesions or masses (DALMs) need to be biopsied for microscopic investigation. Endoscopically, ALMs are usually pedunculated or sessile lesions, while DALMs include plaques, velvety patches, thickened nodules, and broad-based masses. Indications for proctocolectomy following the diagnosis of dysplasia are traditionally (1) incomplete excision and (2) multifocality. In about 1/3 of patients with DALM, there is an underlying malignancy, and HGD is an indication for proctocolectomy because the risk is ~40%. Subjects with distal LGD had a significantly shorter time to progression to colorectal neoplasia than those with proximal LGD, and LGD is not usually an indication for proctocolectomy, but institutions have different management guidelines. IPAA consists of creating a neo-reservoir stapled to the perineum (where the rectum was originally located) by pulling down and folding loops of small intestine and stitching them together as well as removing the internal walls. IPAA is the standard of care for UC with the surgical indication. Two other techniques include (1) proctocolectomy and end ileostomy without a permanent stoma and (2) subtotal colectomy and ileorectal anastomosis when the rectum is minimally inflamed. IPAA complications are divided in acute, including acute sepsis and hemorrhage, and chronic, including pouchitis, "cuffs," small bowel obstruction, chronic pelvic sepsis, sexual dysfunction, pregnancy and delivery complication, and "pouch failure." *Pouchitis* is an acute and chronic inflammation of the pouch with ulcerations and neutrophilic infiltration on a background of chronic inflammation. Pouchitis seems to be UC specific, because FAP patients, who also require IPAA, do not usually present with it. The incidence of pouchitis is ~20% at 1 year, ~30% at 5 years, and ~40% at 10 years. *Cuffitis* is recurrent UC within the columnar cuff above the ATZ and occurs in 9–22% of patients. Pouch failure is defined as pouch excision or vague retention of the low-functioning stoma and occurs with a frequency of 5–10% in the long-term follow-up. Preoperative IFX use has been associated with more postoperative complications than without IFX use, and colectomy with end ileostomy should probably use within 8 weeks of colectomy in consideration that by 8 weeks the minimal amount of this anti-TNF agent remains in the body.

Substantially, inflammation grading and scoring is not an easy task, and some methods have been proposed with variable levels of intra- and interobserver agreement and κ-values. Most probably, one of the most reproducible gradings is that ideated by Dr. N. Harpaz in 1988. HIA is an essential record in the UC patient's history because histologic inflammation is a risk factor (promoter) for progression to colon-rectal neoplasia. ESR measurement (± blood counts), fecal lactoferrin, and calprotectin have been associated with the severity of inflammation. Another critical score is the Truelove-Richards score, which includes 0 for no neutrophil infiltration in the lamina propria (LP) of the mucosa, 1 for less than 10 neutrophils/HPF in the LP of the mucosa, 2 for 10–50 neutrophils/HPF in the LP of mucosa with >50% crypt involvement, 3 for more than 50 neutrophils/HPF with crypt abscesses, and 4 for acute inflammation and ulceration.

- *DDX*: CD and infectious colitis need to remain on the top of the differential diagnoses.

Although at least once a year has been suggested to perform a lower GI endoscopy, it is often performed when the patient is symptomatic or does not respond well to therapy. The endoscope also has the goal to identify early lesions that may be able to transform into malignancy.

- *PGN:* Complications are subdivided in non-neoplastic (intestinal and extraintestinal) and neoplastic. *Intestinal* (bleeding, toxic or fulminant colitis, toxic megacolon, stricture, dysplasia, and neoplasia) and *extraintestinal* (musculoskeletal system-related peripheral and axial arthritis; osteoporosis, osteopenia, and osteonecrosis; and bony fractures; skin-related erythema nodosum, pyoderma gangrenosum, oral ulcers; hepatobiliary PSC (25% are P-ANCA (perinuclear anti-neutrophil cytoplasmic antibodies) positive: more likely to develop primary sclerosing cholangitis or PSC), steatosis hepatis, autoimmune live disease; ocular episcleritis, scleritis, uveitis, iritis, conjunctivitis; hematopoietic-related anemia, clotting abnormalities abnormal fibrinolysis, thrombocytosis, endothelial abnormalities, and thromboembolism). Carcinoma: 1% at 10 years, 3.5% at 15 years, 10–15% at 20 years, and 30% at 30 years duration (the incidence is directly proportional to the extension of the inflammatory involvement).

Colon-Rectal Dysplasia = Intraepithelial Neoplasia

According to Riddell et al. (1983), it is "unequivocal neoplastic alteration of the intestinal epithelium that remains restricted within the basal membrane within which it originated." Thus, we have three aspects to define IEN, including nuclear abnormalities (atypia), cytoplasmic abnormalities, and growth pattern abnormalities.

1. *Nuclear Abnormalities*: nucleomegaly (↑N/C), crowding (↑ICD), hyperchromasia
2. *Cytoplasmic Abnormalities*: altered differentiation (goblet cell depletion) and clonality
3. *Growth Pattern Abnormalities*: faulty control of cellular proliferation, including glandular crowding, tubular or villous architectural change, and lack of standard of epithelial base-surface maturation

"Dysplasia" definitions include negative for dysplasia (NFD), indefinite for dysplasia (IFD), low-grade dysplasia (LGD), high-grade dysplasia (HGD). *Dysplasia*-Associated Lesion on Mucosa (DALM) is subtyped in (a) *Adenoma*-like, which may be indistinguishable from sporadic adenoma, and (b) *Non-Adenoma*-like, which may present as sessile irregular masses/nodules/strictures. Surroundings about dysplasia may be necessary to report as well. Normal or colitis may be found in NFD and sporadic adenoma; otherwise, in all other events, colitis is found. In DALM, an increased architectural disarray, abnormal villous architecture, and inflammation are often seen. Polyps with "top-down" dysplasia have a better outcome (favorable histology) than polyps with "bottom-up" dysplasia or full-thickness dysplasia. LGD needs to be carefully separated from HGD. LGD are single columnar/cuboidal epithelia with nuclei, which are visible crowded, of elliptical shape, hyperchromatic, but with smooth, delicate nuclear membranes, containing inconspicuous nucleoli, and typical mitotic figures. HGD are single, stratified, or cribriform epithelia with nuclei, which are visibly stratified haphazardly or skewed, enlarged (nucleomegaly), pleomorphic, hyperchromatic, or vesicular, containing prominent nucleoli, and atypical mitotic figures. The deceitful features of IFD are inflammation and regeneration of severe degrees, the technical inadequacy of specimen (i.e., very small, very superficial, misoriented, fragmented, poorly fixed), reactive crypts harboring unusual stratification or hyper-basophilia of nuclei, reactive crypts sheltering superficially villiform mucosa with an inadequate sampling of the basal crypts, and hyperserrated dilated crypts. The relevant immunohistochemistry-supported key features of dysplasia are as follows.

1. Flat dysplasia may be at the *base of the polyp stalk* or *surrounding mucosa*.
2. *DALM* has an *irregular architecture* and is $(+)_n p53$, $(+)_m \beta$-Cat, and $(-)_c Bcl2$.
3. *Adenoma* has a *regular architecture* and is $(-)_n p53$, $(+)_{n/m} \beta$-Cat, and $(+)_c Bcl2$.
4. *Past or present histology (+) for chronic idiopathic IBD* is required for DALM.
5. DALM does not require proctocolectomy, but *endoscopic mucosal resection* is feasible.

The rationale for these key features is that *APC gene mutations are rare in DALM*, which do not show any nuclear staining for β-catenin. Conversely, *APC gene mutations are often seen in sporadic adenomas*, which leads to loss or a reduction of the protein. APC is a protein that forms part of a complex, which degrades some proteins, including phosphorylated β-catenin. If APC function is lost, there is a reduced breakdown of β-catenin, which accumulates and migrates to the nucleus. Here, it upregulates the transcription of several cell proliferation genes. This situation is demonstrated by IHC with strong nuclear staining for β-catenin in adenomas in addition to the usual membranous staining as seen in DALM.

Also, *P53 gene mutations are common in DALM* with an accumulation of the protein p53 in the nucleus, which can be demonstrated by IHC. Conversely, *P53 gene mutations are rare in sporadic adenomas* with no avidly collection of P53 protein into the nucleus. Consequently, BCL-2 anti-apoptotic protein is typically overexpressed by epithelial cells in sporadic adenomas as a downstream effect of mutation of the *K-ras* gene as observed by IHC. Conversely, *K-ras* gene mutations are rare in DALM, and no (or very faint) BCL-2 staining is seen. Risk and clinical management of dysplasia in UC are also paramount.

In sporadic adenoma, the absence of surrounding colitis should be confirmed microscopically, and caution should be used for the pedicle, because it may be dysplastic. Caution should also be spent if the patient is younger than 40 years or lesion recurs after polypectomy. In adenoma-like dysplastic polyp, no flat dysplasia in the adjacent mucosa or elsewhere in colon and caution should be spent if the patient is younger than 40 years or lesion recurs after polypectomy. In adenoma-like dysplastic polyp, it seems that the presence of HGD should not change management. In DALM, flat LGD, and flat HGD, the confirmation of the diagnosis by a second pathologist is mandatory. In flat LGD, it seems that unifocal LGD has a similar risk as well as multifocal LGD (Box 3.44).

> **Box 3.44 Simple Features Favoring Dysplastic vs. Reactive Epithelium**
> - Diffuse nuclear changes (e.g., hyperchromatic nuclei, macronucleoli, atypical mitoses)
> - Intraluminal necrosis ("dirty necrosis")
> - Lack of base-to-surface epithelial maturation gradient
> - Loss of cellular polarity

Most recent literature has highlighted that the molecular pathogeneses of colon-rectal dysplasia and cancer in IBD have some common signals (e.g., TNF-α, IL-6, IL-23), some mediators (e.g., TLR4, NF-κB, cytosine deaminase), and numerous targets (e.g., P53, KRAS, APC, p16/p14, DCC, src) and mechanisms include oxidative damage, chromosomal instability, microsatellite instability, aberrant methylation, TSG mutational silencing, telomere shortening, clonal expansion, growth factors, epithelial-stromal interactions, and epithelial-inflammatory cell interactions. Gene targets associated with CRC in IBD include *APC* (5q21), *DCC/DPC4* (18q21), *TP53* (17p13), *CDKN2A* (9p21), *CDKN1B* (12p13), *E-cadherin*, *K-RAS*, *C-SRC*, *TGFBR2 RII*, *MSH2*, and *MLH1*. The first six are tumor suppressor genes, the subsequent two are oncogenes, and the last two are DNA mismatch repair genes, while *TGFBR2 RII* is a TGF-β1 receptor gene.

Preoperative IFX use has been associated with more postoperative complications than without IFX use, and colectomy with end ileostomy should probably use within 8 weeks of colectomy in consideration that by 8 weeks, the minimal amount of this anti-TNF agent remains in the body.

Backwash Ileitis

The term "Backwash Ileitis" in UC was used initially in 1936 from Crohn and Rosenak to describe the pathologic features and clinical course of 9 patients with a combination of ileitis and colitis, although White first described ileal involvement in 5 of 11 UC patients in 1988 at the Guy Hospital in London. Haskell et al. (2005) have reviewed the pathologic features and clinical significance of backwash ileitis in UC. Ileal changes showed a variable phenotype, including villous atrophy, crypt regeneration without increased inflammation, increased neutrophilic and mononuclear inflammation in the lamina propria, patchy cryptitis, and crypt abscesses, and focal superficial surface erosions, some with pyloric metaplasia. They also emphasized that the inflammation at this location does not affect the prevalence of pouch complications, alternatively the occurrence of dysplasia or carcinoma.

3.6.3.2 Infectious Colitides

- *DEF*: Acute self-limited colitis, whose etiology is one or more of the following etiologic agents with edema, inflammation, hyperemia, and hemorrhage.
- *ETP*: Most commonly, *Campylobacter*, *Salmonella*, *Shigella*, and *E. coli* can cause acute hemorrhagic colitis. Other infectious colitides are amebic with a preference for cecum and ascending colon and flask-shaped ulceration with minimal inflammation. The tubercular disease is most commonly ileocecal, but a tubercular mass (tuberculoma) may be found in ~1/2 of cases. CMV is most widely ileocecal with extensive ulceration, and inclusion bodies are prominent. Cryptosporidiosis represents the most common cause of severe watery diarrhea in patients who have AIDS.

3.6.3.3 "Microscopic Colitis"

Microscopic colitis is defined as a chronic or episodic watery diarrhea with radiographically and endoscopically normal bowel and "microscopic" inflammation of the colon. Patients are often >30 years of age, but children and adolescents have been reported; ♂:♀ = 1:6; patients may have an autoimmune disease. Surface epithelium is somewhat flattened with loss of mucin and cytoplasmic vacuolization and is infiltrated by lymphocytes, neutrophils, and eosinophils. Collagenous colitis: Lymphocytic colitis with deposition of ≥10-μm-thick hypocellular collagenous band beneath surface epithelium. Matta et al. (2018) reviewed 12 pediatric patients affected with collagenous gastritis, of which three had associated collagenous colitis. They found that the most common clinical presentation was iron deficiency anemia and a histologic improvement was only identified in one patient who had received oral corticosteroids and azathioprine.

3.6.3.4 Others

This group includes Crohn colitis, indeterminate colitis, diversion colitis, radiation colitis, colitis cystica profunda, and (pseudo-)melanosis coli and the reader should be referred to specific textbooks of gastrointestinal pathology.

3.6.4 Colon-Rectum Neoplasms

Pseudopolyps are seen in ulcerative colitis. Lymphoid polyps are localized but may also be diffuse. Juvenile polyps show standard components, but they have arranged abnormally cystic glands with normal or inflamed epithelium, usually in the rectum of young children; Peutz-Jeghers polyps have an arborizing structure and may occur throughout the gastrointestinal tract, adenomas are all dysplastic, but pre-malignant. The malignant potential is proportional to size, villosity, and degree of dysplasia and may be associated with abundant mucus and hypokalemia (Box 3.45).

3.6.4.1 Adenomatous and Serrated Polyps

Adenomatous and serrated polyps need to be differentiated, because of the clinical implication and therapy. Adenomatous polyps are subclassified as *tubular*, *tubulovillous*, and *villous* adenoma according to the morphology (villous component). Conventional adenomas have at least low-grade dysplasia, because it is inside of the definition of *adenomatous change*, including hyperchromasia of the nuclei of the neoplastic enterocytes, depletion of goblet cells, and initial pseudostratification of the nuclei without loss of nuclear polarity or vesicular nuclei with prominent nucleoli among other features, which identify high-grade dysplasia (*vide infra*). In the pathological report, it is important to emphasize this *two-tiered grading system* indicating the absence of high-grade dysplastic or atypical proliferating epithelium with invasion – invasive adenocarcinoma ("negative for high-grade dysplasia and malignancy") to avoid over therapy, in addition to the amount of *villosity* (villous morphology),

---

Box 3.45 Polyps: Classification

1. Inflammatory (pseudopolyp, benign lymphoid polyp)
2. Hamartomatous (juvenile polyp, Peutz-Jeghers)
3. Neoplastic (adenoma, adenocarcinoma)
4. Others (hyperplastic, lipoma, leiomyoma)

and status of the polyp *margins*. High-risk adenomas is a term used in screening programs for tubulovillous or villous adenomas or any adenomas with high-grade dysplasia or ≥1 cm on the size and need short surveillance intervals. The terms "carcinoma in situ" and "intraepithelial carcinoma" are synonymous with high-grade dysplasia and should be avoided to avoid confusion.

Sessile serrated polyps or lesions have been suggested mimicking other polyps characterized by crypt serration with or without low-/high-grade dysplasia (Torlakovic et al. 2008). Serrated polyps may be subclassified as *sessile serrated adenomas* (SSA); *traditional serrated adenoma* (TSA); *serrated polyp, unclassified* (USP); and *hyperplastic polyp* (HP). It should also be reported by the pathologist, if dysplasia is present or not, because SSA with dysplasia should be considered a lesion with an increased tendency to transform to malignancy and complete removal should be part of the standard of care, including either short-term re-endoscopy and excisional biopsy or surgical resection (Snover 2011).

Villosity: if <20–25% ⇒ *Tubular* adenoma; if ~25–75% ⇒ *Tubulo-Villous* adenoma; and if >75–80% ⇒ *Villous* adenoma. The term "*at least tubulovillous*" is recommended, if the polyp is known to be large and at least one villus is observed in a biopsy that results to be small or fragmented. Villi have been classified as classic, palmate, or foreshortened. Classic villi are long, slender upgrowths showing minimal branching and thin stromal cores. However, it may be encountered broader, branching structures, which characterize palmate villi, and leaf-like upgrowths. In addition, foreshortened villi are slender outgrowths with no branching and thin stromal core. On the other hand, villosity should be distinguished from pseudovillous morphology, such as axially sectioned crypts.

High-grade dysplasia is defined on both architectural AND cytologic features, including cribriform pattern with "back-to-back" glands, prominent glandular budding, and intraluminal papillary tufting as well as loss of cell polarity; nuclear stratification through the entire height of the surface epithelium; nuclei with open, vesicular chromatin pattern and prominent nucleoli; atypical mitoses; dystrophic (if any) goblet cells; and prominent karyorrhexis (apoptosis). As stated above, an adenoma with lamina propria invasion should be reported as high-grade dysplasia rather than intramucosal carcinoma. Since there are no lymphatics in the lamina propria, there is a negligible risk of a spread to regional lymph nodes. Intramucosal atypical glandular proliferations are not typically accompanied by stromal desmoplasia.

Invasive adenocarcinoma is defined as atypical glandular proliferation through the muscularis mucosae into the submucosa (pT1) with usually stromal desmoplasia. The pathologic report should also state if any amount of poorly differentiated (*G3*) portions, any *vascular invasion* (blood vessels and lymphatic channels), as well as any positivity (distance in micrometers) of the *marginal status* are present. It has also been recommended that a distance of the invasive component ≤1000 μm should be reported as a positive margin.

Additional optional features may be important and could be part of a requirement in the next future. They include *tumor budding* (presence of single infiltrating tumor cells or small clusters of tumor cells at the invasive front of the tumor). The *Haggitt levels* and *Kikuchi levels* are often difficult to assess, because they require correctly oriented specimens and the muscularis propria, respectively.

Serration is defined by the sawtooth luminal aspect of the crypts and identifies four kinds of polyps, including sessile serrated adenomas (SSA); traditional serrated adenoma (TSA); serrated polyp, unclassified (USP); and hyperplastic polyp (HP). *Hyperplastic polyp* (HP) is characterized by prominent serrations in the luminal halves of crypts with straight and narrow crypt bases and is usually found in the distal colon and rectum. HPs can be further subclassified as microvesicular (mucin), goblet cell-rich, and mucin-poor types. Microvesicular HPs are often found throughout the colon, while the goblet cell-rich type is predominantly seen in the left colon and rectum. *Sessile serrated adenoma* (SSA) is characterized by a frequently *right colon*-sided, flat, ill-defined outgrowth on the crests of mucosal folds with architectural and cytologic abnormalities. These features include *deep crypt serrations* (serrated architecture is observed throughout the full length of the glands) made up of abnormally located, differentiated cells (goblet or gastric-type) with *horizontal spreading* and *crypt dilation* ("boot- or anchor-shaped" crypt bases) and may present dysplasia with identical features as seen in adenomatous polyps. *Traditional serrated adenoma* (TSA) is characterized by a tubulovillous or villous-like outgrowth with prominent and rigid serrations and ectopic budding crypts, which appear to sprout into the underlying lamina propria. The cells are slender with thin, elongated "penicillate" nuclei and large eosinophilic cytoplasm. LGD/HGD with identical features as seen in adenomatous polyps may be present. *Serrated polyp, unclassified* (USP) is a lesion with aspects that are felt indeterminate between one type and another (e.g., hyperplastic polyp and SSA or SSA and TSA).

Specimen handling and processing are vital for the patient, clinician, and pathologist. Each polyp arising from a different area of the GI tract should be sent to the pathologist in a separate container with adequate fixative and correctly labeled. If multiple outgrowths are from the same area, a unique container can be used, and the electronic or paper-based requisition form should state if the specimen is polyp biopsy (incisional biopsy) or a polypectomy specimen (excisional biopsy). In the pathology laboratory, the following information should be recorded and inserted in the grossing description: number and size of polyps or tissue fragments, as well as the presence of stalks in intact polyps and their length and diameters. Polyps or tissue fragments should be grossed once they are properly fixed, and it depends on the size of the tissue specimen.

Formaldehyde penetration and fixation is a slow process, requiring at least 15 h for routine processing of pathology samples, different from monolayers of cultured cells that can typically be adequately fixed in 15–30 min. It is important to remember that a fixative labeled as 10% buffered formalin is only a 4% solution of formaldehyde because 10% buffered formalin is made from a stock bottle of 37–40% formaldehyde or more accurately a 3.7–4% solution of formaldehyde. The penetration rate of formaldehyde is mostly irrelevant to tissue up to 1.5 cm thick, although it is not 2 mm per hour, it is variable. Considering the Fick law and the solid state chemistry the depth of penetration of a specified concentration is proportional to the square root of the time of diffusion. In our context the *coefficient of diffusibility* is about 3.6. Thus, it has been determined that a 4 mm tissue block is thoroughly penetrated in less than 1 h. Although this is useful for standard H&E tissue slides, additional studies need appropriate binding time. A minimal binding time is 24 h at 22°C and 18 h at 37°C. Thus, an overnight fixation has been suggested to be appropriate in most of the cases. In the follow-up:

(a) Re-recto-colonoscopy (RCS) if >3 hyperplastic polyps proximal to the RSC or any hyperplastic polyp >1 cm
(b) 5-YSR-Re-RCS for conventional tubular adenomas <3 without HGD and size <1 cm
(c) 3-YSR-Re-RCS for ≥3 adenomas or any outgrowth >1 cm, tubulovillous/villous, any outgrowth with HGD
(d) 5 YSR-Re-endoscopy for SSA/TSA <3 and size <1 cm
(e) 3 YSR-Re-endoscopy for SSA/TSA ≥3 and any outgrowth with size >1 cm

Potential pitfalls could be "over-reliance on cytologic abnormalities" (e.g., focality or insufficient material), "over-calling architectural complexity" (e.g., focal budding), "over-calling surface changes" (e.g., following trauma, erosion, or prolapse), and "the insufficient extent of abnormalities" (e.g., less than three crypts).

3.6.4.2 Non-neoplastic, Nonserrated Polyps

Juvenile (Retention) Polyps
Most frequent polyp in children and usually located in the rectosigmoid region of the large bowel. It is granular and red due to the granulation tissue and frank inflammation, which may be disclosed on the microscopic examination. The stroma is inflamed, and there is cystic dilatation (retention cysts). Dysplasia may accompany a juvenile polyp and need to be ruled out by the pathologist.

Hamartomatous (Peutz-Jeghers-Type) Polyps
Single or multiple (in the Peutz-Jeghers syndrome) polyp of hamartomatous origin with glands maintained by broad branching bands of smooth muscle

(arborizing architecture) which extend upward from the muscularis mucosa into the lamina propria. The microscopic examination reveals both columnar and goblet cells on the surface as well as Paneth and endocrine cells near the base. There is minimal to no atypia, but dysplasia needs to be ruled out carefully.

Pseudopolyps

Islands of residual mucosa surrounded by ulceration, often seen in inflammatory bowel disease of the ulcerative colitis type.

Inflammatory Fibroid Polyps

Broad-based, a polyp with ulcerated surface and composed histologically by fibroblasts, inflammatory cells including eosinophils embedded in a collagenous or myxoid stroma.

Solitary Rectal Ulcer

Solitary ulceration of the rectal mucosa with polypoid character 4–18 cm from anal margin and associated with mucosal prolapse. Microscopically, an obliteration of lamina propria by fibrosis, smooth muscle proliferation within and extending up from the muscularis mucosa, is seen and is accompanied by variable numbers of lymphocytes. The *inflammatory cloacogenic polyp* and *mucosal prolapse syndrome* are probably variations of the solitary rectal ulcer and may be described in a spectrum.

Lymphoid Polyps

Mucosal protrusion secondary to lymphoid hyperplasia, most often found in the rectum.

3.6.4.3 Polyposis Syndromes

Polyposis syndromes are a frequent topic of the GI pathology rounds, of the routine diagnostics at a tertiary center, and of the examination for the Royal College of Pathologists as well as other colleges. Box 3.46 summarizes the hereditary syndromes, genes involved, and the GI and extra-GI features.

Box 3.46 Polyposis Syndromes (Hereditary)

Syndromes	Gene	GI/extra-GI features
FAP, classic	APC	Hypertrophy of retinal epithelium, a cribriform-morular variant of PTC, hepatocellular adenoma/carcinoma, fundic gland polyp, juvenile nasopharyngeal angiofibroma
FAP, attenuated	APC	3-99 CR adenomas
MYH-associated polyposis (MAP)	MYH	3-99 CR adenomas
FAP, variant Gardner syndrome	APC/MYH	*C*OPD: *C*ysts, epidermal, keratinous; *O*steomas, craniofacial; *P*olyposis; *D*esmoid tumors/abdominal fibromatosis
FAP, variant Turcot syndrome	APC/MYH	CNS tumors
Peutz-Jeghers polyposis	LKB1/STK11	Mucocutaneous hyperpigmentation, pancreas, breast, lung, ovary, and uterus tumors
Cowden syndrome	PTEN	"*B*TEN": *B*reast ca., follicular *T*hyroid ca., *EN*dometrial ca. type I, and facial trichilemmomas
Bannayan-Ruvalcaba-Riley syndrome	PTEN	Macrocephaly, speckled penis, developmental delay, hemangiomas, lipomas + Cowden syndrome features
Juvenile polyposis	SMAD4/BMPR1A	Digital clubbing Pulmonary AV malformations
Tuberous sclerosis	TSC1/TSC2	Rectal polyps/subependymal nodules (SEN) and subependymal giant cell astrocytomas (SEGA) of the brain, angiomyolipomas and renal cell carcinomas of the kidney, cardiac rhabdomyomas, facial angiofibromas, and ocular phakomas

Note: *GI* gastrointestinal, *FAP* familial adenomatous polyposis, *SB* small bowel, *CT* computed tomography, *MR* magnetic resonance, *CTPA* computed tomographic pulmonary angiography, *BX* biopsy, *PTC* papillary thyroid carcinoma, *US* ultrasound

FAP
Autosomal dominant, 5q21 (APC gene) with the early development of numerous polyps (usually >100, teens to 20s) constituted by tubular adenomas throughout GI tract (including small intestine and stomach) and high incidence of malignant transformation (~100% by early 30s).

Cowden Syndrome
Multiple hamartomatous polyps, but not Peutz-Jeghers type, but showing disorganization and proliferation of muscularis mucosa instead, with AD inheritance pattern and associated with facial trichilemmomas, acral keratoses, oral mucosal papillomas, and increased incidence of malignancy in various sites (e.g., breast, thyroid).

Gardner Syndrome
AD inherited, colonic polyposis in association with multiple osteomas of the skull and mandible, keratinous cysts of the skin, soft tissue tumors (particularly fibromatosis), and risk of colonic carcinoma as high as for FAP.

Juvenile Polyposis Syndrome (JPS)
AD inherited, colonic polyposis with the associated risk of development of adenomatous polyps and adenocarcinoma. The updated criteria for JPS are >5 juvenile polyps of the colorectum, multiple juvenile polyps throughout the GI tract, and any number of juvenile polyps + family history of juvenile polyposis.

Peutz-Jeghers Syndrome (PJS)
AD inherited, polyposis with multiple characteristic hamartomatous (arborizing fashion with glands supported by broad bands of smooth muscle) polyps in the colon (30%), small bowel (100%), and stomach (25%); mucocutaneous melanotic pigmentation around the lips, mouth, face, genitalia, and palmar surfaces of hands and harboring an increased risk of developing carcinomas of the pancreas, breast, lung, ovary, and uterus. PJP diagnostic criteria (≥1 criterion) are (1) ≥3 PJPs, (2) ≥1 PJP and +FHx, (3) mucocutaneous pigmentation (MCP) and +FHx, and (4) ≥1 PJP and +FHx.

Turcot Syndrome
AR inherited, polyposis with colonic adenomatous polyps seen in association with brain tumors, usually glioblastomas (brain MRI!).

Cronkhite-Canada Syndrome
Nonhereditary GI polyposis syndrome showing gastric and colonic polyps, which are sessile and show hyperplastic, cystically dilated glands with edema and quite characteristic eosinophilic inflammation of the lamina propria and associated with alopecia, nail atrophy, and hyperpigmentation.

Spiegelman Classification is widely used to guide surveillance and therapeutic interventions of upper GI polyps and is characterized by four increasing risk profiles based on number, size, architecture, and degree of dysplasia of the adenomas observed during endoscopy.

MYH protein has the critical duty to repair DNA by removing adenine residue that is mispaired with 8-oxoguanine during cellular replication of oxidized DNA. It is important to remind that deficiency of this protein leads to a *Base Excision Repair Defect* and therefore to *Somatic Mutations* in APC. This explains the clinical similarities and pathologic features between FAP and APC.

3.6.4.4 Familial vs. Sporadic Colorectal Neoplasia

Familial Adenomatous Polyposis (FAP), APC/WNT-related (~2/3)
Familial Adenomatous Polyposis (FAP), MMR-related (~2/3)
Hereditary Nonpolyposis Colorectal Cancer (HNPCC)
Sporadic Colorectal Cancer (SCRC), APC/WNT-related (~4/5)
Sporadic Colorectal Cancer (SCRC), MMR-related (<1/5)

Familial and sporadic colorectal cancer is displayed in Box 3.47.

Defective *mismatch repair (MMR) system* is featured by the *microsatellite instability (MSI)*, which typically occurs because of a germline mutation in one of the MMR genes, consisting of several proteins including *MLH1*, *MSH2*, *MSH6*,

> **Box 3.47 Familial vs. Sporadic Colorectal Neoplasia**
>
> - *FAP, APC/WNT-rel.*: AD, *APC*, no major site, CRC (tubular/tubule-villous/villous A/ACA)
> - *FAP, APC, MMT-rel.*: AR/none, *MUTYH*, no major site, CRC (SSA, Muc-ACA)
> - *HNPCC/LYNCH S.*: AD, *MLH1* and *MSH2*, R > L colon, CRC (SSA, Muc-ACA)
> - *SCRC, APC/WNT-rel.*: None, *APC*, R < L, CRC (tubular/tubule-villous/villous A/ACA)
> - *SCRC, MMR-rel.*: None, *MLH1* and *MSH2*, R > L, CRC (SSA, Muc-ACA)
>
> Note: HNPCC is also called Lynch syndrome; AD, autosomal recessive pattern of inheritance; AR, autosomal recessive pattern of inheritance

and *PMS2*, or methylation of the MLH1 promoter. Microsatellites are defined as recurring nucleotide sequences that are distributed throughout the genome. Microsatellites consist of mono-, di-, or higher-order nucleotide repeats, which are more frequently copied incorrectly when DNA polymerases cannot bind them efficiently.

Interestingly, nature has ideated a reparative mechanism, which is indeed the MMR system that is responsible for the surveillance and correction of these replication errors. MSI may be detected by either PCR comparing the length of nucleotide repeats in tumor cells and healthy cells (+MSI if the length of repeat sequences from tumor vs. normal cells differs in >30%) or immunohistochemistry. MMR defects may result from a germline mutation in one of the MMR genes followed by a variation on the second allele of that gene (somatic inactivation of the wild-type allele) or methylation of the promoter of an MMR gene (usually MLH-1). Consequently, there are a loss of protein function and lack of detection by immunohistochemistry. MMR status is critical to assess for *prognosis* (MSI-associated tumors have a more favorable outcome and less nodal and distant metastases), *response to 5-FU*, and *irinotecan therapy*. Irinotecan is a semisynthetic derivative of *camptothecin*, which is a cytotoxic alkaloid extracted from some plants (e.g., Camptotheca acuminata). Irinotecan and its active metabolite (SN-38) target the DNA replication enzyme *topoisomerase* I, which determines reversible single-strand breaks in DNA during its replication. It seems that stage II MSI tumors do not benefit and might be harmed by 5-FU therapy whereas may be more responsive to irinotecan than microsatellite-stable (MSS) tumors and *detection of HNPCC or Lynch syndrome*.

In adult CRC, the mechanisms of tumorigenesis include three pathways: (1) *chromosomal instability neoplasia* (CIN), (2) *CpG island methylator phenotype* (CIMP), and (3) *microsatellite instability* (MSI). The most common pathway is the CIN pathway, which occurs in 70–80% of CRC, whereas the CIMP pathway is the second common pathway to CRC and accounts for about 15% of CRC. Both CIN and CIMP pathways are typically used for the oncogenesis of sporadic CRC in adults. It has been determined that probably approximately 5% of CRC (i.e., 1 in 20 cancers of colon-rectum) develops via the MSI pathway in adults. In childhood CRC, the mechanism of tumorigenesis is not well investigated compared to adult CRC, and a suggested mechanism is the inheritance of biallelic germline mutations of the MMR genes. MSI test and/or IHC are recommended when:

1. CRC is diagnosed in a patient ≤50 years.
2. Syn-metachronous CRC (or another HNPCC-related neoplasm) is found.
3. CRC in a patient ≤60 years exhibiting tumor-infiltrating lymphocytes, Crohn disease-like lymphocytic reaction, mucinous/signet ring differentiation, or medullary growth pattern.
4. (+) FH according to the revised Bethesda guidelines.

If the tumor shows MSI-high and protein loss by IHC, MMR genetic testing is recommended according to the IHC patterns.

Box 3.48 Interpretation of IHC Stains for DNA MMR in MSI-H Tumors

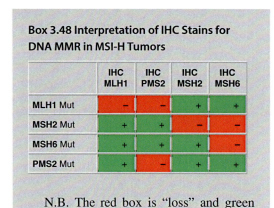

N.B. The red box is "loss" and green boxes are "maintained" nuclear staining. This box has been gathered from Sepulveda Recommendations and Table.

IHC principle: Pathogenic mutations in an MMR gene lead to the absence of any detectable nuclear staining of the gene product by IHC.

In CRC with MSI and absent MLH1 expression, BRAF analysis and/or MLH1 promoter methylation studies are recommended (Box 3.48).

HNPCC: Virtually 100% negative for BRAF mutation!

SCRC: 40–80% positive for BRAF mutation!

When a mutation occurs in an MMR gene complex, the ability to repair other random mutations is compromised, and this results in an accumulation of mutations (MSI). The occurrence of these mutations leads to inactivation TSGs and subsequent penetrance of CRC. Both IHC and mutational analysis for point mutations and large genomic arrangements (duplications and deletions) are instrumental.

HNPCC – Lynch Syndrome is defined as mutations in one of the four *MMR genes*, AD, ~5% of CRC with other tumors at *B*reast, *O*vary, *U*rinary tract, *G*I and liver, *E*ndometrium, *S*kin ("Chateau de *BOUGES*" mnemonics). The *Amsterdam Criteria II* include three cases of Lynch-associated cancer AND ≥2 successive generations affected of which first-degree relatives of the other two, AND ≥1 with diagnosis <50 years old, AND lack of FAP. Moreover, *Muir-Torre syndrome* is an HNPCC variant with keratoacanthomas and sebaceous tumors. Management: Full colonoscopy every 1–3 years ≥20 years, annual screening for ECA ≥30 years, yearly urinalysis + cytology ≥30 years, annual skin surveillance, upper GI endoscopy for families with gastric carcinoma, and prophylactic subtotal colectomy and hysterectomy and oophorectomy after childbearing age to reduce the risk.

3.6.4.5 Adenocarcinoma

Embryology plays a significant role in colonic carcinogenesis because there are two colons, two different pathways, and two different adenocarcinomas. The proximal colon, i.e., the colonic segment proximal to the splenic flexure, and the distal colon, i.e., the colon after the splenic flexure, are embryologically of different origins. The former develops from the midgut and the latter from the hindgut. Carcinomas that arise in these two colonic segments have been shown to have distinct clinical and pathological features, and their carcinogenetic genetic pathways are also suggested to be different.

- *DEF*: Malignant epithelial tumor showing atypical glandular proliferation.
- *EPI*: It is the second leading carcinoma associated with death in the USA accounting for approximately 15% of all cancer deaths and about 98% of all cancers of the large intestine. 50–70% are localized in the rectosigmoid segment, <20% in patients <40–50 years of age. CRCs in childhood are sporadic.
- *RF*: "IDAHO" mnemonics: inflammatory bowel disease; diet such as high lipid, red meat intake, and low fiber content; age such as older age and risk which increases with age; and heritage such as geographic, familiarity, susceptibility genes, and obesity/others. Dietary factors predisposing to carcinoma include the low content of absorbable vegetable fiber, the high material of refined carbohydrates, and high-fat content.
- *DGN*: Barium enema examination and lower GI endoscopy with flexible proctosigmoidoscopy and colonoscopy) are critical in the evaluation for detecting colorectal carcinoma. Barium enema examination may show a polypoid or flat lesion or an ulcerated mass or as a mass causing obstruction or as an annular con-

stricting lesion (*apple core appearance*). Lower GI may demonstrate an ulcerated mass, an exophytic or villiform tumor, or a mass completely obstructing the large bowel lumen. Cecum and ascending colon in low incidence areas and more frequently rectum and sigmoid in higher incidence areas. On the left side, tend to grow as annular encircling lesions with early obstruction; on the right, polypoid, fungating masses, generally without obstruction, and diagnosed later, usually well to moderately differentiated adenocarcinoma (G1 or G2), often with T-cell inflammatory response at the leading edge.

MSI-H CRC, both sporadic and HNPCC-associated, differ from MSI-L/MSS CRC in several pathological features. MSI-H CRCs are mainly located in the proximal colon and have a large size, degree of poor differentiation, mucinous or medullary histology, pushing margins of expansion, intense peritumoral lymphoid reaction (Crohn disease-like), and intratumoral lymphocytic infiltration. Adenocarcinoma of the colon-rectum is staged, almost universally, according to the TNM classification. Since the colon has no lymphatics in the lamina propria, if the tumor invades this subepithelial part of the mucosa, the adenocarcinoma is still considered in situ, different from other body regions.

- *PGN*: The most robust clinicopathological factors associated with poor prognosis are *pT4*, *high histologic grade*, *number of nodes assessed <12*, *LVI* and *perineural invasion*, *emergency surgery* (both due to obstruction and perforation), high preoperative *CEA level*, *MSS/MSI-L* (compared to MSI-H), *BRAF mutations* (stage II and III colon cancer), and probably a number of *RNA gene signatures*. In addition, poorly differentiated adenocarcinoma, mucinous adenocarcinoma, and signet ring cell carcinoma (*Por/Muc/Sig cancers*) have been associated with more advanced stage in the TNM classification and showed worse disease-specific survival than Wel/Mod carcinomas, but in Por/Muc/Sig neoplasms, but not in Wel/Mod cancers, proximal cancers showed significantly better disease-specific survival than distal tumors. Moreover, the MLH1 loss seems associated with a very low risk of distant metastases, while high protein levels of CD133 and β-catenin in combination with preserved MLH1 expression seem linked to advanced tumor stage in right-sided neoplasm. Particular histologic types include mucinous (collections of tumor cells suspended in lakes of extracellular mucin at >50%), signet ring (signet ring cells with diffuse infiltration and thickening of the wall), squamous (more common in cecal neoplasms, but usually adenosquamous), small cell undifferentiated carcinomas (SCUC, rare, aggressive with early spread and rapid deterioration and morphologically like those of the lung and pancreas with neuroendocrine differentiation identified by IHC and/or TEM), clear cell change (clear cells >50%, basaloid or formerly cloacogenic), and, finally, choriocarcinomatous.

Ulcerative colitis (or better IBD, including also Crohn disease) is at risk of CRC and needs periodic surveillance. Other risk factors of the history include sporadic adenomas, previous colorectal cancer, and breast, ovarian, or endometrial carcinoma.

Carcinoid Tumor

Most commonly in the rectum, but sometimes encountered in the distal sigmoid region.

Other neoplasms include lipoma, lipomatosis of the ileocecal valve, smooth muscle tumors, neural tumors, and Kaposi sarcoma.

3.7 Anus

The pathology of the anus is displayed in five categories (Box 3.49).

Box 3.49 Anus Pathologies
3.7.1. Anal Anomalies
3.7.2. Inflammatory Anal Diseases
3.7.3. Benign Anal Tumors and Non-neoplastic Anal Lesions (Pseudotumors)
3.7.4. Anal Pre- and Malignant Lesions

3.7.1 Anal Anomalies

Congenital anomalies of the appendiceal anatomy are rare.

3.7.1.1 Anorectal Defects

Anorectal Agenesis (high/supralevator anomalies, 40%): Absence of anal canal with rectum ending above the *musculus levator ani* determining obstruction and frequently associated with abnormalities of the spine and urinary tract, a defect of the innervation of pelvic muscles, and recto-cystic, recto-urethral, and recto-vaginal fistula. PGN depends on the surgery, which is quite complicated and requires high competency.

Anal Agenesis (intermediate anomalies, 15%), including anal agenesis associated with Larsen syndrome as well as anorectal stenosis and anorectal membrane. Larsen syndrome is an AD inherited genetic disorder characterized by osteochondrodysplasia, congenital dislocations of large-joints, and cranio-facial defects. Mutations in the gene that encodes the connective tissue protein, filamin B (*FLNB*) are etiologic for Larsen syndrome. PGN depends on the outcome of the surgery, which is less complicated than that performed in supra-elevator anomalies.

Low Anal Defects (translevator anal anomalies, 40%), including ectopic (perineal, vestibular, or vulvar) anus, anal stenosis, and imperforate anus derived from a failure of the cloacal diaphragm to rupture. PGN depends on a simple surgery, which is curative, being absent, or rarely reported associated anomalies and no evidence of pelvic innervation disturbances.

Anal Defects, Miscellaneous (5%), include a perineal groove, persistent anal membrane, and persistence of cloaca, which is a single narrow channel opening onto perineum with a small orifice and derived from combined bladder, genital tract, and bowel emptying into it. The exstrophy of the cloacal membrane has been reported in this condition.

VATER/VACTERL association (Vertebral defects, Anal atresia, Cardiac defects, Tracheo-Esophageal fistula, Renal defects, and Limb deformities), cloaca exstrophy sequence, and OEIS-complex (Omphalocele, Exstrophy of the cloaca, Anus imperforatus, and Spinal defects), cloacal dysgenesis sequence (urorectal septum malformation/sequence, prune belly sequence II) and sirenomelia are just a few of the numerous associations and sequences with anorectal defects. The chromosomal aberrations with anorectal defects are also numerous; among others are prevalent partial trisomy 22 ("Cat-eye" syndrome), trisomy 8 mosaicism, and trisomy 18 syndromes. Anorectal malformations may also be encountered in monogenic syndromes including Baller-Gerold syndrome, Currarino triad, Johannson-Blizzard syndrome, Pallister-Hall syndrome, and Townes-Brocks syndrome. Anorectal defects have been described in products of conception arising in pregnancies affected with probable or certain teratogenic insults, including cocaine, diazepam, maternal DM, retinol derivates, and thalidomide (Frías et al. 2007; Yu et al. 2003).

Surgery in the anorectal area needs to be performed considering the great importance of the rectogenital septum, which is known in the clinical literature as Denonvilliers' fascia. This septum forms an incomplete separation between the rectum and the urogenital organs in both genders. The rectogenital septum is composed of collagenous and elastic fibers and smooth muscle, which are intermingled with nerve fibers emerging from the autonomic inferior hypogastric plexus. Although the etiology of postoperative incontinence is multifactorial, it has been hypothesized that, following surgery such as restorative proctectomy, a surgical procedure where the ventral rectal wall is extensively dissected in its anterocaudal part, the dramatic event of incontinence may eventuate simply by disruption of the rectogenital septal anchoring mechanism. The rectogenital septum is formed at the 9th week of gestation within a local condensation of collagenous fibers. The anatomy of the perineal body is still poorly defined, but recent studies of Dr. Fritsch's group have been crucial to increasing the understanding of this very complex region. They found that the rectogenital septum and its smooth muscle components share the same innervation as the longitudinal rectal muscle layer. Therefore, it has been suggested a safer dissection of the anterior rectal wall during rectal resection limiting functional disturbance and preventing neural damage (Aigner et al. 2004; Lindsey et al. 2005; Zhai et al. 2009).

3.7.2 Inflammatory Anal Diseases

3.7.2.1 Crohn Disease
- *DEF*: Anal canal involvement in about 1/4–3/4 of the patients with Crohn disease with several signs and symptoms, including fissures, fistulas, ulcers, abscesses, and skin tags.
- *LM*: Small, tight nonnecrotic granulomas close to the mucosa ± MNGC.
- *DDX*: TB (caseating granulomas, AFB+), FB granulomas, sexual transmitting diseases (STDs), including *granuloma inguinale* or *C. granulomatis* infection, lymphogranuloma venereum or *C. trachomatis* infection, and HIV infection.

3.7.2.2 Ulcerative Colitis
In about one in ten cases with ulcerative colitis, there are some anal lesions, including midline dorsal fissures, skin excoriation, perianal or ischiorectal abscess, and recto-vaginal fistula, and the major differential diagnosis is Crohn disease other than STDs.

3.7.2.3 Lymphoid Polyp
Enlarged lymphoid follicles covered by columnar or transitional epithelium appearing as a small polypoid lesion with superficial erosions in young women or female adolescents, although male adolescents have also been observed presenting with a it.

3.7.3 Benign Anal Tumors and Non-neoplastic Anal Lesions (Pseudotumors)

Benign tumors and non-neoplastic lesions (pseudotumors) include several entities, including *granular cell tumor* (florid pseudoepitheliomatous hyperplasia, as seen in other locations, may mimic squamous cell carcinoma, particularly in the immunodeficient or posttransplantation), *hemorrhoids* (unusually in children and youth and due to prolonged constipation, cirrhosis, pregnancy, and rectal neoplasms and possibly showing pagetoid dyskeratosis with pale cells in epidermis with premature keratinization, resembling Paget cells, DDX: anorectal arterio-venous malformations), *chronic hyperproliferative plaque/mass* in HIV-infected patients, *anal skin tags* or fibroepithelial polyps (squamous epithelium with central core of inflamed, edematous, myxoid, or fibrovascular stroma with thin-walled vessels; large, multinucleated, CD34+ stellate cells; and often mast cells), *inflammatory cloacogenic polyp* (polypous lesion associated with solitary rectal ulcer syndrome or mucosal prolapse syndrome with hyperplastic or regenerative character and features are thickened muscularis mucosa with irregular strands penetrating lamina propria and lack of dysplastic changes, desmoplasia, or invasion, also observed in children and adolescents), *internal anal sphincter achalasia* (marked reduction or absence of interstitial cells of Cajal in internal anal sphincter), *leiomyoma*, *lipoma*, *radiation proctitis with anal involvement*, and *tailgut cyst* or retrorectal hamartoma (presacral hamartoma arising from remnants of postanal gut or tailgut, DDX: teratoma, dermoid cyst, duplication enterogenous cyst, anal gland cyst, and anterior sacral meningocele) (Box 3.50).

Box 3.50 Pearls and Pitfalls

- Distinguishing fissures from fistula is crucial, because of the clinical implication. The fissure is a single linear separation of tissue extending usually through mucosa and quite nonspecific, while the fistula is a chronic inflammatory process (granulation tissue) with a canal that opens typically at or above the dentate line and ends blindly in perianal soft tissue or reaches the skin and generally associated with Crohn disease, ulcerative colitis, actinomycosis, and tuberculosis.
- External, inferior hemorrhoidal plexus is below the anorectal line; internal, superior hemorrhoidal plexus is above the anorectal line.
- In the diagnostics of Hirschsprung disease, the ganglion cells are usually absent or sparse in the anorectal wall for 1–2 cm above the dentate line, but any marked reduction or absence of interstitial cells of Cajal in internal anal sphincter in these patients needs to be assessed.

3.7.4 Anal Pre- and Malignant Lesions

Irradiation and chemotherapy as a combined-modality treatment for SqCC of the anal region of the last decade have been successful with an impressive 80% of 5-YSR when compared to melanoma of the same area with grim results. AIN and anal region cancers are usually neoplastic conditions with predilections for middle-aged individuals. However, viral infection and solid organ transplantation have reduced the age of presentation, and occasionally, there are case reports in children or adolescents (Gutman et al. 1994; De Góis et al. 2005; Zaramella et al. 2013).

3.7.4.1 Condyloma Acuminatum
- *DEF*: Benign low-grade lesion characterized by papillary excrescences and usually associated with HPV 6, 11, 16, 18 and other HPV lesions (anal dysplasia/CIS, verrucous carcinoma, squamous cell carcinoma/basaloid carcinoma).
- *CLM*: Papillary squamous epithelium with hyperkeratosis is present. Koilocytotic atypia and dysplastic changes are variable.
- *DDX*: AIN, verrucous carcinoma, squamous cell carcinoma, and basaloid carcinoma.

Dysplasia (Anal Intraepithelial Neoplasia, AIN)

It is the precursor lesion for anal carcinoma, including low-grade/condyloma, intermediate-grade, and high-grade/CIS lesions, respectively. AIN may occur as single, but more often multiple in the perianal skin or anal canal in the flat mucosa or with *condyloma acuminatum* or in hemorrhoid specimens. HPV is commonly detected in these lesions (usually HPV 16, 18) and particularly in HIV patients. The risk of both prevalence and incidence of high-grade AIN increases as CD4 counting falls. Cytology thin-preps and smears may be helpful for detection. The sensitivity of the cytology is around 85% and the specificity practically half of this figure when compared with histology. Anal cytology by the Palefsky method is simple to perform, has a sensitivity and specificity comparable with cervical cytology, and may be used for individuals at high risk of developing anal carcinoma (Palefsky et al. 1997; Fox et al. 2005).

3.7.4.2 Anal Carcinoma
- *DEF*: A malignant epithelial tumor of the anal canal/anorectum and the perianal region showing several histologic types of differentiation (e.g., atypical glandular proliferation) (*vide infra*). The WHO classification of anal carcinomas includes squamous cell adenocarcinoma (rectal type, of anal glands, or within anorectal fistulae), mucinous adenocarcinoma, small cell carcinoma, and undifferentiated carcinoma.
- *EPI*: 15–90 years, but the annual incidence increases after the 2nd decade of life, ♂>♀.
- *EPG*: RFs include AIN, male anal intercourse, STD (e.g., *lymphogranuloma venereum* and *condyloma acuminatum*), virus infection (HPV and HIV), Crohn disease, heavy smoking, immunosuppression, Hx. anal fistulas, and Hx. genital warts. About HPV, E6 oncoprotein of HPV inactivates TP53, and about HIV, HIV-positive patients have a risk lower than males with a history of receptive anal intercourse. Moreover, 11q chromosomal abnormalities have been detected in the anal canal cancer.
- *CLI*: Anal bleeding, pain, itching, mass, or asymptomatic (1/4). In case of a mass, proctologic examination, with ano- and proctoscopy, and transrectal ultrasound are useful (pitfall: flat lesion).
- *PGN:* It depends from the stage (unfavorable: Ø > 2 cm, deep depth of invasion, N+, M+), nonmarginal location, ploidy, grade, and histology. Anal margin tumors harbor a better outcome. The cloacogenic carcinoma has been considered a carcinoma with a slightly better PGN, but this term is now obsolete. Worse PGN is found in nonkeratinizing basaloid and small cell carcinoma.

Tumors that arise from the columnar epithelium of proximal zone of the anal canal are considered rectal tumors because they are usually CK7−/CK20+ as found in colorectal carcinomas, but mucinous adenocarcinoma may also arise from anal glands or congenital anorectal duplications.

- *IHC:* (+) CK, EMA, CEA, p53, (−) ER, PR.

Anal Carcinoma Histologic Variants
- *Squamous Cell Carcinoma:* Nodular, ulcerated, invading deeply and spreading both proximally and distally showing squamous cell differentiation similar to tumors of the skin or upper aerodigestive tract (keratinizing/well-differentiated) ± basaloid (worse PGN) and mucoepidermoid differentiation.
- *Verrucous Carcinoma:* Condyloma-like carcinoma with very well-differentiated squamous epithelium with minimal atypia and pushing borders that invade stroma similarly to upper respiratory tract and lower female genital tract lesions (better PGN).
- *Basal Cell Carcinoma:* Mostly of nodular type, this tumor is about 0.2% among all anorectal cancers and etiopathogenetic factors include chronic trauma, chronic dermatitis, nevoid BCC syndrome, *TP53* mutation, immunodeficiency, sexually transmitted disease, and arsenic exposure.
- *Adenocarcinoma:* Ulcerated mass with gelatinous consistency showing mucinous epithelium with atypical ductal proliferation and dilated, mucin-filled cysts containing free-floating tumor cells (colloid adenocarcinoma) ± granulomatous reaction to mucin. There is Hx. of perianal fistulae and/or vaginal cysts, and the course is indolent with gradual progression.
- *Clear Cell Carcinoma:* Clear cell-dominant anal carcinoma.
- *Small Cell Carcinoma:* Very aggressive variant showing solid nests of small cells with little or minimal cytoplasm, hyperchromatic nuclei with molding, and central necrosis with IHC phenotype (+) CGA, SYN, NSE, CD57.

- *DDX*: Basal cell carcinoma of the anal region needs to be distinguished from basaloid SqCC because of the therapeutic and prognostic implications. Basal cell carcinoma is usually located in the perianal area, while basaloid SqCC originates from the anal canal/anorectum. Basal cell carcinoma is distinguished from the basaloid SqCC, because the former shows a particular retraction artifact, uncommon atypical mitoses, no *in situ* component, and IHC expression of Ber-EP4 and BCL2. SqCC usually has no retraction artifact, common atypical mitoses, in situ component, and IHC expression of CDKN2A and SOX2.

3.7.4.3 Paget Disease
- *DEF*: Erythematous, ulcerated, or eczematous perianal lesion, which may be associated with an underlying local malignancy showing single, occasionally nests or gland-like formations of large, pale-staining to clear intraepidermal carcinoma cells with abundant mucin. Other associations of PD include squamous hyperplasia, fibroepithelioma-like hyperplasia, and papillomatous hyperplasia.
- *IHC:* Mucin (mucicarmine, PAS), CK7, CEA, EMA; ± CK20.
- *DDX:* Gross cystic disease fluid protein 15 (GCDFP-15) is key. PD associated with the primary cutaneous tumor with sweat gland differentiation are GCDFP-15+ and CK20−, while PD associated with an underlying rectal ACA are GCDFP-15−, CK20+, and MUC2+. GCDFP-15 is regulated by the androgen receptor (AR) and is a crucial diagnostic marker for mammary differentiation in anatomic pathology.

Animal studies have been performed in mice, in whom anal carcinoma may be induced by chemical carcinogens that can be potentiated using growth factors. 1,2-Dimethylhydrazine (DMH) or symmetrical DMH is one of the two isomers of DMH acting as a DNA alkylating agent whose action may be potentiated using EGF (Zaafar et al. 2014). Moreover, in preserving the anal continence, the surgeon reflects on two essential components, including the internal and external sphincters. If patients need excision of both muscles of not more than half of the circumference of the anus, anal continence is preserved. Finally, in case an extensive surgery is required, an abdominoperineal resection (APR) is performed.

3.7.4.4 Nonanal Carcinoma Neoplasms

Carcinoid tumor (1/3 have a second tumor, often CRC)

Embryonal rhabdomyosarcoma (infants and children, botryoid/diffuse growth pattern, IHC, and TEM evidence of muscle differentiation)

GIST (small asymptomatic intramural nodules to large painful masses with + CD117, CD34, DOG1 highly cellular spindle cells and occasional epithelioid morphology; poor PGN if Ø > 5 cm with five mitotic figures/50 HPF)

Histiocytic sarcoma (extremely rare in childhood and adolescence)

Leiomyosarcoma (most common stromal tumor of the anus, forming a polypoid intraluminal mass and showing – CD117, +SMA spindle cells with oval to moderately elongated, often blunt-ended nuclei and distinct eosinophilic cytoplasm and numerous mitoses)

Lymphoma (AIDS and AIDS-associated, EBV ±)

Melanoma (Tumor with relatively rare occurrence being as 1/10 as common as squamous cell carcinoma. It is possibly mistaken as hemorrhoid with ~10% 5-YSR. Grossly, there is a single or multiple, polypoid mass(es), which are pigmented and exhibit a growth near the dentate line with histologic similarities to other mucosal melanomas with nests of pigmented epithelioid or spindle cells ± junctional component with lentiginous appearance, crypts invasion, and potential DSR. IHC: (+) S100, HMB-45, MART-1/Melan-A. PGN is related to tumor size and depth of invasion (≤2 mm: better outcome).

3.7.4.5 Secondary Tumors

Metastases to the anus or perianal region are rarities, but rectal carcinomas are most common. Rare cases have been reported from other sites (lung, breast, and kidney). In childhood and youth, embryonal rhabdomyosarcoma and malignant lymphoma may probably be the most frequent types.

3.7.4.6 TNM Staging and CAP Guidelines

The following staging is valid for anal canal tumors; however, perianal tumors, melanoma, carcinoid tumors, and sarcoma need to be excluded. Moreover, stage tumors overlapping the anorectal junction should be considered as anal tumors, if epicenter (not hypocenter) is distal to ≤2 cm proximal from dentate line, and as rectal tumors, if epicenter (not hypocenter) is >2 cm proximal to the dentate line. CIS, carcinoma in situ (Bowen disease, HSIL, AIN II-III); the adjacent organs are vagina, urethra, and urinary bladder, but the direct invasion of the rectal wall, perirectal skin, subcutaneous tissue, or sphincter muscle is not classified as pT4. Moreover, TX, NX, and MX refer to as primary site, lymph node, or distant sites cannot be assessed, respectively. Regional lymph nodes (LNs) are perirectal (anorectal, perirectal, lateral sacral), internal iliac (hypogastric), and inguinal (superficial). An important aspect emphasized in some reports, although not part of TNM, is the identification of single cells or small clusters of cells ≤0.2 mm (i.e., ≤200 μm), so-called *"isolated tumor cells"* (ITC), found by routine H&E, IHC, PCR, or another method. LNs or distant sites with ITC are classified as pN0 or pM0.

In a pathology report, it is mandatory to report the following features: *polyp size* (at least one dimension), *typing* or histologic type (squamous cell carcinoma, adenocarcinoma, mucinous adenocarcinoma, small cell carcinoma, undifferentiated carcinoma, other, or undetermined carcinoma), *grading* or histologic grade (well, moderately, or poorly differentiated, undifferentiated, undetermined grade), *depth/extent of invasion* (no invasion, indeterminate, lamina propria, muscularis mucosa, submucosa), *resection margin* (not assessable, positive/negative for tumor, and if negative, closest tumor to mucosal margin is mm as well as information on carcinoma in situ absent/present at mucosal margin), and *angiolymphatic invasion* (absent, present for large/small vessels, or indeterminate), *perineural invasion*, *pTNM*, and *stage*. Moreover, the diagnosis of adenocarcinoma is based on % of tumor that forms glands: well, >95%; moderate, 50–95%; poor, 5–49%; and undifferentiated, <5%. It has also been suggested to provide some data about HPV status, polyp configuration, and presence or absence of colitis as well as Crohn disease, fistula, condyloma, dysplasia, and Paget disease.

In the case of local excision (transanal disc excision), it is mandatory to report specimen type (intact, fragmented, or other), tumor size, typing, grading, depth/extent of invasion, resection margin, angiolymphatic invasion, pTNM, and stage. In the case of anus resection, it is mandatory to report the specimen type (APR, other, NOS), tumor site (e.g., anterior wall), tumor size, typing, grading, depth/extent of invasion, resection margin (proximal, distal, radial, or soft tissue closest to deepest tumor penetration), angiolymphatic invasion, invasion of other structures, nodal involvement, pTNM, and stage.

3.8 Peritoneum

The peritoneum pathology may be subdivided in cytology, non-neoplastic pathology, and neoplastic pathology (Box 3.51).

3.8.1 Cytology

The mesoderm lines the normal peritoneum-derived keratin and calretinin-positive mesothelium. Calretinin, which is encoded by the *CALB2* gene, is a calcium-binding protein involved in calcium signaling. Subserosal cells are vimentin positive, fibroblast-like, and considered to be pluripotent cells. Cytology of peritoneal fluid is often routine in both pediatrics and medicine, and it is beneficial for both diagnostic and staging purposes. The diagnostic portion helps to orientate the surgical option in most cases. Indications for peritoneal washing include staging malignant tumors of the ovary, fallopian tube, and endometrium (gynecological malignancies) and malignant tumors of the stomach and pancreas (nongynecological malignancies), exclude occult carcinoma, identify mimickers of malignancy, and assess the response to treatment. In benign peritoneal washing, it is easy to identify mesothelial cells in sheets and histiocytes, but they may be accompanied by collagen balls, adipose tissue, and skeletal muscle. Mesothelial cells are observed in flat sheets, and the cells are evenly spaced with the cytoplasm of moderate amount. Nuclear membranes are thin and contain pale chromatin, which is evenly dispersed. Commonly, mesothelial cells may exhibit small nucleoli. Endosalpingiosis is the presence of ectopic (outside the fallopian tube) cystic glands that are lined with fallopian tube-type ciliated epithelium and occurring in pelvic organs, including ovaries, fallopian tube serosa, uterine serosa, uterine myometrium, or pelvic peritoneum. Endosalpingiosis is characterized cytologically by cuboidal cells with minimal to mild atypia harboring cilia occasionally and psammoma bodies. Endometriosis is defined by the presence of ectopic endometrial glands and/or stroma. At least two out of three histologic features need to be fulfilled for the diagnosis of endometriosis: (1) endometrial-type glands, (2) endometrial-type stroma, and (3) evidence of chronic hemorrhage (hemosiderin-laden macrophages). The glandular epithelium may harbor metaplastic changes (tubal, mucinous, squamous, hobnail). Cytologically, endometriosis may disclose endometrial glandular cells, stromal cells, hemosiderin-laden macrophages, and tissue fragments containing endometrial glands and stroma. In the case of serous adenocarcinoma, small or large clusters of atypical cells and isolated atypical cells are seen. These cells disclose marked variation in nuclear size, hyperchromatic nuclei, prominence of nucleoli, some mitotic figures, and vacuolated cytoplasm. In borderline serous tumors, small or large clusters of minimal or mild atypical cells are seen. Cytoplasmic vacuoles and psammoma bodies may be seen. In endometrial cancer colonizing the peritoneal washing, there are isolated cells and clusters of atypical cells, which have enlarged nuclei, coarse chromatin pattern, nuclear pleomorphism, and scant or vacuolated cytoplasm. Mesothelial cell atypia may be encountered as a result of previous chemother-

> **Box 3.51 Peritoneal Pathology**
> 3.8.1. Cytology (Fig. 3.50)
> 3.8.2. Non-neoplastic Peritoneal Pathology (Fig. 3.51)
> 3.8.3. Neoplastic Peritoneal Pathology (Figs. 3.52, 3.53, 3.54, and 3.55)

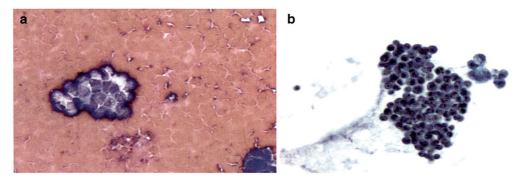

Fig. 3.50 Peritoneal fluid positive for neuroblastoma cells

apy and may be considered a mimicker of a malignancy. There are often multinucleation and marked anisonucleosis with nuclear size variable among cells and prominence nucleoli in large nuclei, keeping a strict normochromatic pattern. There is no hyperchromasia as seen in true malignancies. Secondary tumors, e.g., neuroblastoma, may also colonize the peritoneal fluid. In this case, small- to medium-sized cells in small clusters are seen (Fig. 3.50). The tumor cells show a high nucleus-to-cytoplasm ratio and hyperchromasia. Mitotic figures are not frequently encountered, but it depends on the Mitosis-Karyorrhexis index of the tumor. Thus, it may be important to evaluate the original tumor, if it is available.

3.8.2 Non-neoplastic Peritoneal Pathology

We may encounter peritonitis, but it may fully resolve, become walled off (abscess), or heal as fibrous adhesions. Peritonitis may be due to chemical injuries (bile, pancreatic juice, gastric juice, barium, meconium), bacterial injuries (primary or systemic, perforation of viscus), and foreign body-related (granulomatous). Peritoneal cysts are lined by flattened cuboidal mesothelium and contain a clear fluid, while Müllerian cysts have a fallopian tube (ciliated) epithelium. Pseudocysts have no lining. Also, the multicystic benign mesothelioma is constituted by numerous microscopic cysts, although single cysts may be larger up to 15 cm, lined by cuboidal mesothelium, and probably represent a reactive process. Hyperplasia and meta-plasia can occur in the peritoneum and be a consequence of irritation and have the architecture of nodules or papillae, and psammoma bodies may be encountered. Müllerian structures may also be observed and include endosalpingiosis, endometriosis, and ectopic decidual reaction (*vide supra*) (Fig. 3.51).

3.8.3 Neoplastic Peritoneal Pathology

It includes teratoma (Fig. 3.52), the desmoplastic small round cell tumor (DSRCT) (Fig. 3.53), malignant mesothelioma, as well as hematologic malignancies, such as posttransplant lymphoproliferative disorders (PTLD) in the form of diffuse large B-cell lymphoma (DLBCL) (Figs. 3.54 and 3.55). The teratoma may contain mature and immature elements of the three germ cell layers (ectoderm, mesoderm, endoderm) as well as malignant components, such as yolk sac tumor, embryonal carcinoma, and choriocarcinoma. The DSRCT is a rare and aggressive malignant mesenchymal neoplasm that typically occurs in adolescence and youth. There is a male preponderance ($\male:\female = 5:1$). The DSRCT is characterized by clusters of poorly differentiated small round hyperchromatic cells lying within an abundant fibrosclerotic stroma. There is a co-expression of epithelial, mesenchymal, myogenic, and neural markers in the tumor cells. Typically, the cytogenetic testing for *EWSR1-WT1* rearrangement is positive showing the chromosomal t(11;22)(p13;q12) resulting in the fusion of the Ewing

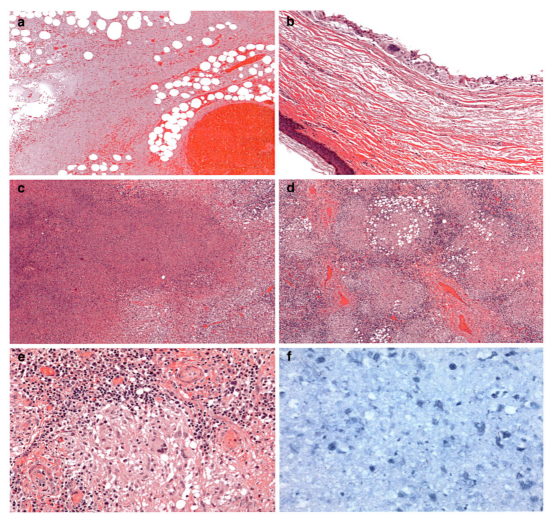

Fig. 3.51 In (**a** and **b**) omentum with nodule-mesothelial cell hyperplasia, calretinin-positive (immunostaining not shown), (**c**–**d**) omental hemorrhagic necrosis due to torsion of an inflamed epiploic appendix (H&E stain, 50×), (**e**) umbilical epidermoid cyst with foreign body giant cell reaction (H&E stain, 100×), (**f**) omental tuberculosis-necrotizing granulomatous peritonitis with acid-fast bacilli positivity (Ziehl-Nielsen staining, 630×)

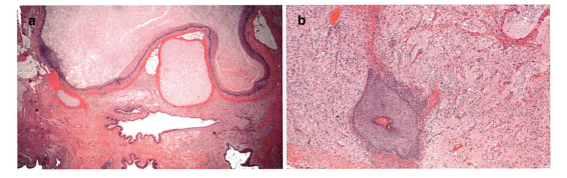

Fig. 3.52 Mature teratoma with ectodermal, mesodermal, and endodermal differentiation (hematoxylin and eosin staining)

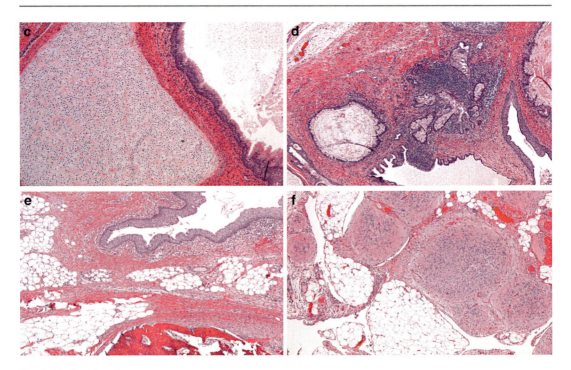

Fig. 3.52 (continued)

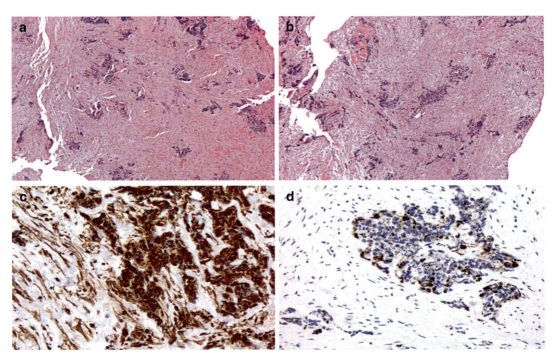

Fig. 3.53 Desmoplastic small round cell tumor with solid nests of round or oval cells surrounded by cellular desmoplastic stroma (**a**, **b**, hematoxylin and eosin staining), which is confirmed by immunohistochemistry (**c–h**) with positivity for vimentin, WT1, EMA, neuron-specific enolase, desmin, and CD99, respectively

3.8 Peritoneum

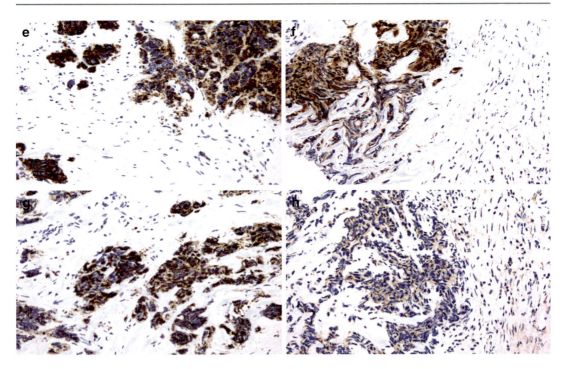

Fig. 3.53 (continued)

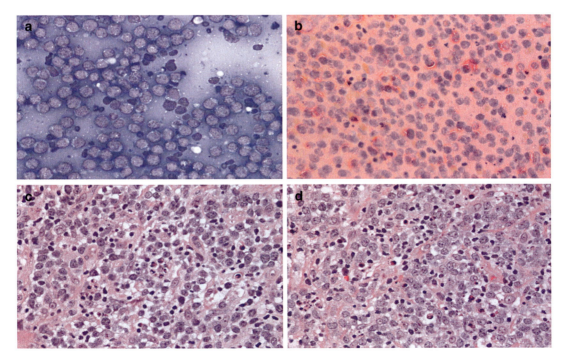

Fig. 3.54 PTLD of monomorphic type with diffuse large B-cell lymphoma in a child who underwent organ transplantation. The monomorphism and cytologic details point to a lymphoma that was confirmed by flow cytometry and immunohistochemistry

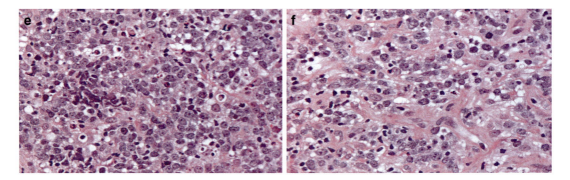

Fig. 3.54 (continued)

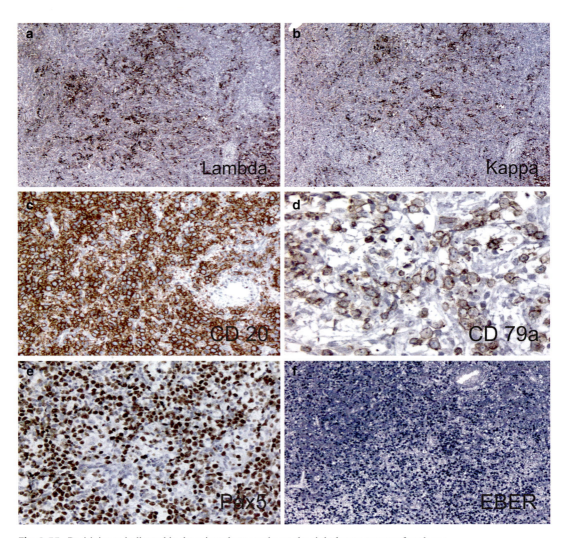

Fig. 3.55 Positivity as indicated in the microphotographs on the right lower corner of each one

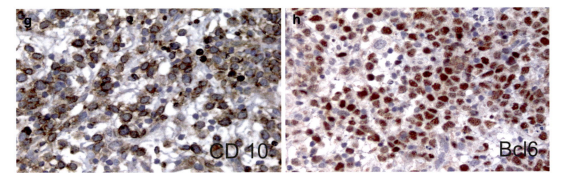

Fig. 3.55 (continued)

sarcoma (EWSR1) and Wilms' tumor (WT1) genes. Commonly, DSRCT develops in the abdomen exhibiting peritoneal spread, with subsequent metastasis to distant lymph nodes, liver, and lungs. Malignant mesothelioma of the peritoneum (MPM) may arise from submesothelial connective tissue and presents with ascites. MPM is divided into three subtypes, epithelioid, sarcomatoid, and biphasic/mixed, according to the World Health Organization (WHO). The epithelioid subtype is composed of cells that look like normal mesothelial cells in a tubule-papillary or trabecular pattern, while the sarcomatoid subtype is composed of tightly packed spindle cells with the occasional presence of osteoid, chondroid, or muscular components. The mixed or biphasic subtype is defined as containing both epithelioid and sarcomatoid components, with each contributing ≥10% of the overall histology. Commonly, lymphoma originates in lymph nodes, but it can originate in other organs outside of the lymph nodes. The infiltration of lymphomatous cells in the organs other than lymph nodes is termed as extranodal lymphoma. Although any organ can be affected, the most frequently involved system is the gastrointestinal tract with the stomach being the commonest site, which is followed by Waldeyer ring, lung, liver, spleen, bone, and skin. In the abdomen, the peritoneum can be a site of extranodal lymphoma. Precisely, patients with immunosuppression for organ transplantation may develop the PTLD. PTLD is a lethal complication that can be difficult to treat. According to the 2008 WHO classification, PTLD is categorized as one of the four major forms: early lesion, polymorphic PTLD, monomorphic PTLD, and classical Hodgkin lymphoma-type PTLD (CHL). The main risk factors for developing PTLD are age, Epstein-Barr virus (EBV) status, immunosuppression, and the type of organ transplanted. Children with PTLD tend to be EBV-positive, while most adults with PTLD are EBV-negative. As a patient receives higher doses of immunosuppression, the rate of developing PTLD and specifically monomorphic PTLD increases. PTLD has a strong association with EBV, particularly in B-cell proliferation, and rituximab, a genetically engineered chimeric murine/human monoclonal antibody (IgG1 κ immunoglobulin) targeting the surface CD20 antigen of normal and malignant B lymphocytes, is effective for CD20(+)-PTLD but is only effective in about half of these patients.

Multiple Choice Questions and Answers

- GIT-1 In esophageal atresia (EA), the upper esophagus is not connected to the lower esophagus (atretic), and in about nine out of ten newborns, there is a tracheoesophageal fistula (TEF) as well, in which the trachea and the esophagus are connected. An isolated EA/TEF occurs in about 40% of affected individuals, while the remaining have EA/TEF associated with other congenital disabilities or as

part of a multiple congenital anomaly or genetic syndrome. Esophageal atresia is NOT associated with which of the following genetic syndromes?
(a) CHARGE syndrome
(b) Trisomy 18 syndrome
(c) VACTERL association
(d) Neurofibromatosis type I

- GIT-2 Wolf-Hirschhorn syndrome (WHS) is a genetic syndrome with diaphragmatic defect and typical craniofacial features in infancy consisting of "Greek warrior helmet" appearance of the nose, microcephaly, high anterior hairline with prominent glabella, hypertelorism, epicanthus, highly arched eyebrows, micrognathia, short philtrum, and downturned corners of the mouth, as well as poorly formed ears with pits/tags. There is prenatal-onset growth deficiency with postnatal growth retardation and hypotonia. Internal findings reveal structural brain abnormalities, congenital heart defects, hearing loss, gut, and urinary tract malformations, as well as skeletal anomalies. Which chromosomal aberration is associated with WHS?

  (a) Heterozygous deletion of the WHS critical region (WHSCR) on chromosome 14p3
  (b) Heterozygous deletion of the WHS critical region (WHSCR) on chromosome 4p16.3
  (c) Heterozygous deletion of the WHS critical region (WHSCR) on chromosome 21p12.3
  (d) Heterozygous deletion of the WHS critical region (WHSCR) on chromosome 3p16.3
  (e) Heterozygous deletion of the WHS critical region (WHSCR) on chromosome 4p12.3

- GIT-3 Necrotizing enterocolitis (NEC) is a life-threatening condition that usually affects premature babies and can have clinically significant long- and short-term consequences. Which of the following statements is NOT correct?
  (a) The modified Bell staging system is used for scoring NEC.
  (b) NEC presentation can include lethargy, temperature volatility, episodes of apnea, bradycardia/hypotension, and reduced glucose homeostasis.
  (c) NEC presentation can be confused with sepsis.
  (d) Blood culture results are positive in about 90% of patients at the onset of NEC.
  (e) Typical bacteria seen are *Clostridium perfringens*, *Escherichia coli*, *Enterobacter cloacae*, *Klebsiella pneumoniae*, *Pseudomonas aeruginosa*, and *Proteus mirabilis*.
  (f) The presence of *Candida* spp. in these cultures can indicate more severe disease.

- GIT-4 A 2-week-old infant presents with constipation and the diagnosis of Hirschsprung's disease is suspected by an abnormal result of the anal manometry. A rectal suction biopsy is sent to the pathologist. What are the criteria to diagnose Hirschsprung's disease on hematoxylin and eosin staining of a rectal suction biopsy?
  (a) Presence of parasympathetic ganglion cells in the submucosal plexus and hypertrophy of the nerve fibers
  (b) Absence of the nerve fibers that innervate the submucosa of the rectum
  (c) Absence of parasympathetic ganglion cells in the submucosal plexus and myenteric plexus in addition to hypertrophy of the nerve fibers
  (d) Absence of parasympathetic ganglion cells in the submucosal plexus and hypertrophy of the nerve fibers
  (e) Atrophy of the mucosa with the presence of parasympathetic ganglion cells

- GIT-5 A 15-year-old boy presents with a 4-week history of bloody diarrhea, periumbilical pain, low fever, and weight loss. The family physician investigates stool culture for bacteria, ova, and parasites, but they were negative. According to the boy and his family, the child had these symptoms once before. The family physician also notes some perianal involvement. The child is referred to the pediatric gastroenterologist, who evidences some

patchy areas of ulcerations in the large bowel and difficulty to pass the ileocecal valve. Also, radiologic barium films showed a "string sign" in the terminal ileum. If your suspicion of diagnosis is correct, which statement is NOT valid for your diagnosis?
  (a) Patchy lesions are usually seen on a biopsy taken during the endoscopy.
  (b) Granulomas may be present at the histologic examination of the gastrointestinal biopsies.
  (c) The annual incidence of this disease in North America is reported to be 3.1–20.2 per 10^5 with a prevalence of 201 per 10^5 population.
  (d) Abnormalities of *NOD2*, *IL23R*, and *ATG16L1* genes have been found associated with this disease.
  (e) The Montreal classification is based on the age at diagnosis, disease location, the disease behavior, and the genotype.
- GIT-6 Ischemic colitis is not frequent in childhood or adolescence/youth, but it can occur and needs to be considered in the differential diagnosis if the clinical setting is appropriate. Which of the following statements is correct?
  (a) The most common symptoms are periumbilical pain and bright red rectal bleeding.
  (b) In most cases, the occlusion of the major mesenteric vessels is responsible for creating the ischemia.
  (c) The most vulnerable regions are the splenic flexure and descending colon.
  (d) Nonoperative management is not justified.
  (e) Large genomic studies have identified an autosomal recessive inheritance pattern.
- GIT-7 Colorectal polyps are not rare in childhood and adolescence in a tertiary healthcare institution, which of the following structures is/are considered premalignant/precancerous?
  (a) Tubular adenoma
  (b) Tubulovillous adenoma
  (c) Villous adenoma
  (d) Hyperplastic polyp
  (e) [b] and [c]
  (f) [a], [b], and [c]
  (g) All of the above
- GIT-8 Which TNM corresponds to the following definition of carcinoid of the appendix: "Tumor more than 2 cm but not more than 4 cm or with extension to the cecum"?
  (a) pT2
  (b) pT1b
  (c) pT3
  (d) pT4
- GIT-9 Which of the following statements is NOT correct regarding the TP53 gene?
  (a) The human *TP53* gene is located on chromosome 17p with 11 exons and 10 introns.
  (b) In the order from N- to C-terminus, the functional domains include the transactivation domain, proline-rich domain, tetramerization domain, and the primary domain.
  (c) Once activated, p53 upregulates its negative regulator, MDM2 (murine/human double minute 2) and MDM2 functions as an E3 ubiquitin-ligase, to regulate the ubiquitination of p53, which leads to its degradation.
  (d) Depending on the specific context, p53 can induce cell cycle arrest, apoptosis, or senescence, in the presence of cellular stress (e.g., DNA damage, hypoxia, oncogene activation).
  (e) Activation of p53 can trigger the mitochondrial (intrinsic), but not the death receptor-induced (extrinsic) apoptotic pathways.
- GIT-10 An 18-year-old boy presents with a presacral mass in the setting of a Currarino triad. What are the three components of the Currarino triad?
  (a) The anomaly of the *MNX1* gene, presacral mass, and anorectal malformation
  (b) Presacral mass, Wilms' tumor, and sacrococcygeal tumor
  (c) Presacral mass, Wilms' tumor, and sacrococcygeal tumor
  (d) Partial sacral agenesis, presacral mass, and anorectal malformation
  (e) Pelvis agenesis, post-sacral mass, and anorectal malformation

References and Recommended Readings

Abdull Gaffar B. Granulomatous diseases and granulomas of the appendix. Int J Surg Pathol. 2010;18(1):14–20. Review. PubMed PMID: 20106828.

Abuetabh Y, Tivari S, Chiu B, Sergi C. Semaphorins and their significance in cancer. Austin J Clin Pathol. 2014;1:1–5.

Agaram NP, Shia J, Tang LH, Klimstra DS. DNA mismatch repair deficiency in ampullary carcinoma: a morphologic and immunohistochemical study of 54 cases. Am J Clin Pathol. 2010;133(5):772–80. PubMed PMID: 20395525.

Agnes A, Estrella JS, Badgwell B. The significance of a nineteenth century definition in the era of genomics: linitis plastica. World J Surg Oncol. 2017;15(1):123. https://doi.org/10.1186/s12957-017-1187-3. PMID: 28679451; PMCID: PMC5498981.

Aigner F, Zbar AP, Ludwikowski B, Kreczy A, Kovacs P, Fritsch H. The rectogenital septum: morphology, function, and clinical relevance. Dis Colon Rectum. 2004;47(2):131–40. PubMed PMID: 15043282.

Akbulut S, Tas M, Sogutcu N, Arikanoglu Z, Basbug M, Ulku A, Semur H, Yagmur Y. Unusual histopathological findings in appendectomy specimens: a retrospective analysis and literature review. World J Gastroenterol. 2011;17(15):1961–70. PubMed PMID: 21528073; PubMed Central PMCID: PMC3082748.

Alberti L, Carniti C, Miranda C, Roccato E, Pierotti MA. RET and NTRK1 proto-oncogenes in human diseases. J Cell Physiol. 2003;195:168–86.

Alevizos L, Gomatos IP, Smparounis S, Konstadoulakis MM, Zografos G. Review of the molecular profile and modern prognostic markers for gastric lymphoma: how do they affect clinical practice? Can J Surg. 2012;55(2):117–24. Review. PubMed PMID: 22564515; PubMed Central PMCID: PMC3310767.

Allison MC, Hamilton-Dutoit SJ, Dhillon AP, Pounder RE. The value of rectal biopsy in distinguishing self-limited colitis from early inflammatory bowel disease. Q J Med. 1987;65(248):985–95. PubMed PMID: 3455554.

Alves MM, Sribudiani Y, Brouwer RW, et al. Contribution of rare and common variants determine complex diseases-Hirschsprung disease as a model. Dev Biol. 2013;382:320–9.

Ambroze WL Jr, Pemberton JH, Dozois RR, Carpenter HA, O'Rourke JS, Ilstrup DM. The histological pattern and pathological involvement of the anal transition zone in patients with ulcerative colitis. Gastroenterology. 1993;104(2):514–8. PubMed PMID: 8425694.

American Academy of Pediatrics. Committee on genetics. Health care supervision for children with Williams syndrome. Pediatrics. 2001;107:1192–204.

Amiel J, Lyonnet S. Hirschsprung disease, associated syndromes, and genetics: a review. J Med Genet. 2001;38:729–39.

Amiel J, Sproat-Emison E, Garcia-Barcelo M, et al. Hirschsprung disease, associated syndromes and genetics: a review. J Med Genet. 2008;45:1–14.

Annenkov A, Nishikura K, Domori K, Ajioka Y. Alpha-methylacyl-coenzyme A racemase expression in neuroendocrine neoplasms of the stomach. Virchows Arch. 2012;461(2):169–75. https://doi.org/10.1007/s00428-012-1272-5. Epub 2012 Jul 11. PubMed PMID: 22782380; PubMed Central PMCID: PMC3421105.

Armstrong AE, Weese-Mayer DE, Mian A, Maris JM, Batra V, Gosiengfiao Y, Reichek J, Madonna MB, Bush JW, Shore RM, Walterhouse DO. Treatment of neuroblastoma in congenital central hypoventilation syndrome with a PHOX2B polyalanine repeat expansion mutation: new twist on a neurocristopathy syndrome. Pediatr Blood Cancer. 2015;62(11):2007–10. https://doi.org/10.1002/pbc.25572. Epub 2015 May 22. PubMed PMID: 26011159.

Arnold S, Pelet A, Amiel J, et al. Interaction between a chromosome 10 RET enhancer and chromosome 21 in the Down syndrome-Hirschsprung disease association. Hum Mutat. 2009;30:771–5.

Asherson N. David Bayford. His syndrome and sign of dysphagia lusoria. Ann R Coll Surg Engl. 1979;61(1):63–7. PubMed PMID: 369446; PubMed Central PMCID: PMC2494476.

Asim A, Kumar A, Muthuswamy S, Jain S, Agarwal S. Down syndrome: an insight of the disease. J Biomed Sci. 2015;22:41.

Aslanian H, Chander B, Robert M, Cooper D, Proctor D, Seropian S, Jain D. Prospective evaluation of acute graft-versus-host disease. Dig Dis Sci. 2012;57(3):720–5. Epub 2011 Oct 20. PubMed PMID: 22011927.

Audicana MT, Kennedy MW. Anisakis simplex: from obscure infectious worm to inducer of immune hypersensitivity. Clin Microbiol Rev. 2008;21(2):360–79, table of contents. https://doi.org/10.1128/CMR.00012-07. Review. PubMed PMID: 18400801; PubMed Central PMCID: PMC2292572.

Azzoni C, Bottarelli L, Campanini N, Di Cola G, Bader G, Mazzeo A, Salvemini C, Morari S, Di Mauro D, Donadei E, Roncoroni L, Bordi C, Sarli L. Distinct molecular patterns based on proximal and distal sporadic colorectal cancer: arguments for different mechanisms in the tumorigenesis. Int J Color Dis. 2007;22:115–26. https://doi.org/10.1007/s00384-006-0093-x.

Bach SP, Mortensen NJ. Ileal pouch surgery for ulcerative colitis. World J Gastroenterol. 2007;13(24):3288–300. Review. PubMed PMID: 17659667.

Badner JA, Chakravarti A. Waardenburg syndrome and Hirschsprung disease: evidence for pleiotropic effects of a single dominant gene. Am J Med Genet. 1990;35:100–4.

Badner JA, Sieber WK, Garver KL, Chakravarti A. A genetic study of Hirschsprung disease. Am J Hum Genet. 1990;46:568–80.

Bălănescu NR, Topor L, Malureanu D, Stoica I. Ileocolic intussusception due to Burkitt lymphoma: a case report. J Med Life. 2013;6(1):61–4. Epub 2013 Mar 25. PubMed PMID: 23599821; PubMed Central PMCID: PMC3624648.

Ballard ET. Ultrashort segment Hirschsprung's disease: a case report. Pediatr Pathol Lab Med. 1996;16:319–25.

Banno K, Kisu I, Yanokura M, Tsuji K, Masuda K, Ueki A, Kobayashi Y, Yamagami W, Nomura H, Tominaga E, Susumu N, Aoki D. Epimutation and cancer: a new carcinogenic mechanism of Lynch syndrome (Review). Int J Oncol. 2012;41(3):793–7. https://doi.org/10.3892/ijo.2012.1528. Epub 2012 Jun 25. PubMed PMID: 22735547.

Barauskas G, Gulbinas A, Pranys D, Dambrauskas Z, Pundzius J. Tumor-related factors and patient's age influence survival after resection for ampullary adenocarcinoma. J Hepato-Biliary-Pancreat Surg. 2008;15(4):423–8. https://doi.org/10.1007/s00534-007-1313-7. Epub 2008 Aug 1. PubMed PMID: 18670845.

Baregamian N, Rychahou PG, Hawkins HK, Evers BM, Chung DH. Phosphatidylinositol 3-kinase pathway regulates hypoxia-inducible factor-1 to protect from intestinal injury during necrotizing enterocolitis. Surgery. 2007;142(2):295–302. PubMed PMID: 17689699; PubMed Central PMCID: PMC2613757.

Barlow B, Santulli TV. Importance of multiple episodes of hypoxia or cold stress on the development of enterocolitis in an animal model. Surgery. 1975;77(5):687–90. PubMed PMID: 1173200.

Barlow A, de Graaff E, Pachnis V. Enteric nervous system progenitors are coordinately controlled by the G protein-coupled receptor EDNRB and the receptor tyrosine kinase RET. Neuron. 2003;40:905–16.

Barnewolt CE. Congenital abnormalities of the gastrointestinal tract. Semin Roentgenol. 2004 Apr;39(2):263–81. https://doi.org/10.1053/j.ro.2004.02.002. PMID: 15143689.

Barr RG. Colic and crying syndromes in infants. Pediatrics. 1998;102(5 suppl E):1283.

Batt RE, Smith RA, Buck Louis GM, Martin DC, Chapron C, Koninckx PR, Yeh J. Müllerianosis. Histol Histopathol. 2007;22(10):1161–6. https://doi.org/10.14670/HH-22.1161. Review. PubMed PMID: 17616942.

Bayford D. An account of a singular case of deglutition. Memoirs Med Soc London. 1794;2:275–86.

Bedoya A, Garay J, Sanzón F, Bravo LE, Bravo JC, Correa H, Craver R, Fontham E, Du JX, Correa P. Histopathology of gastritis in Helicobacter pylori-infected children from populations at high and low gastric cancer risk. Hum Pathol. 2003;34(3):206–13. PubMed PMID: 12673553.

Been JV, Lievense S, Zimmermann LJ, Kramer BW, Wolfs TG. Chorioamnionitis as a risk factor for necrotizing enterocolitis: a systematic review and meta-analysis. J Pediatr. 2013;162(2):236–42.e2. PubMed PMID: 22920508.

Beknazarova M, Whiley H, Ross K. Strongyloidiasis: a disease of socioeconomic disadvantage. Int J Environ Res Public Health. 2016;13(5):pii: E517. https://doi.org/10.3390/ijerph13050517. Review. PubMed PMID: 27213420; PubMed Central PMCID: PMC4881142.

Ben Brahim M, Belghith M, Mekki M, Jouini R, Sahnoun L, Maazoun K, Krichene I, Golli M, Monastiri K, Nouri A. Segmental dilatation of the intestine. J Pediatr Surg. 2006;41(6):1130–3. PubMed PMID: 16769347.

Bentley E, Jenkins D, Campbell F, Warren B. How could pathologists improve the initial diagnosis of colitis? Evidence from an international workshop. J Clin Pathol. 2002;55(12):955–60. Review. PubMed PMID: 12461067; PubMed Central PMCID: PMC1769831.

Berrocal T, Torres I, Gutiérrez J, Prieto C, del Hoyo ML, Lamas M. Congenital anomalies of the upper gastrointestinal tract. Radiographics. 1999;19(4):855–72. Review. PubMed PMID: 10464795.

Berry-Kravis EM, Zhou L, Rand CM, Weese-Mayer DE. Congenital central hypoventilation syndrome: PHOX2B mutations and phenotype. Am J Respir Crit Care Med. 2006;174:1139–44.

Bethel CA, Bhattacharyya N, Hutchinson C, et al. Alimentary tract malignancies in children. J Pediatr Surg. 1997;32:1004–8.. [discussion 1008-9].

Biagi F, Luinetti O, Campanella J, Klersy C, Zambelli C, Villanacci V, Lanzini A, Corazza GR. Intraepithelial lymphocytes in the villous tip: do they indicate potential coeliac disease? J Clin Pathol. 2004;57(8):835–9. PubMed PMID: 15280404; PubMed Central PMCID: PMC1770380.

Bianchi A. Total esophagogastric dissociation: an alternative approach. J Pediatr Surg. 1997;32(9):1291–4. PubMed PMID: 9314245.

Bielschowsky M, Schofield GC. Studies on megacolon in piebald mice. Aust J Exp Biol Med Sci. 1962;40:395–403.

Bishop PR, Nowicki MJ, Subramony C, Parker PH. The inflammatory polyp-fold complex in children. J Clin Gastroenterol. 2002;34(3):229–32. PubMed PMID: 11873101.

Blereau RP. Eosinophilic granuloma of the appendix. Pediatr Emerg Care. 1985;1(2):85–6. PubMed PMID: 3843439.

Blund-Sutton J. Imperforate ileum. Am J Med Sci (1827–1924). 1889;98(5):457–62.

Bolk S, Pelet A, Hofstra RM, Angrist M, Salomon R, Croaker D, Buys CH, Lyonnet S, Chakravarti A. A human model for multigenic inheritance: phenotypic expression in Hirschsprung disease requires both the RET gene and a new 9q31 locus. Proc Natl Acad Sci U S A. 2000;97(1):268–73. PubMed PMID: 10618407; PubMed Central PMCID: PMC26652.

Bondurand N, Sham MH. The role of SOX10 during enteric nervous system development. Dev Biol. 2013;382:330–43.

Bondurand N, Pingault V, Goerich DE, Lemort N, Sock E, Le Caignec C, Wegner M, Goossens M. Interaction among SOX10, PAX3 and MITF, three genes altered in Waardenburg syndrome. Hum Mol Genet. 2000;9(13):1907–17. PubMed PMID: 10942418.

Bondurand N, Fouquet V, Baral V, Lecerf L, Loundon N, Goossens M, Duriez B, Labrune P, Pingault V. Alu-mediated deletion of SOX10 regulatory elements in Waardenburg syndrome type 4. Eur J Hum Genet. 2012;20(9):990–4. https://doi.org/10.1038/ejhg.2012.29. Epub 2012 Feb 29. PubMed PMID: 22378281; PubMed Central PMCID: PMC3421117.

Boot H. Diagnosis and staging in gastrointestinal lymphoma. Best Pract Res Clin Gastroenterol. 2010;24(1):3–12. Review. PubMed PMID: 20206103

Borrego S, Ruiz A, Saez ME, Gimm O, Gao X, López-Alonso M, Hernández A, Wright FA, Antiñolo G, Eng C. RET genotypes comprising specific haplotypes of polymorphic variants predispose to isolated Hirschsprung disease. J Med Genet. 2000;37(8):572–8. PubMed PMID: 10922382; PubMed Central PMCID: PMC1734658.

Borrego S, Ruiz-Ferrer M, Fernandez RM, Antinolo G. Hirschsprung's disease as a model of complex genetic etiology. Histol Histopathol. 2013;28:1117–36.

Bosman FT, Carneiro F, Hruban RH, Theise ND. WHO classification of tumours of the digestive system, WHO classification of tumours, vol. 3. 4th ed. Lyon: IARC; 2010. ISBN-13: 9789283224327 ISBN-10: 9283224329. 417 p.

Boulos BP. Complicated diverticulosis. Best Pract Res Clin Gastroenterol. 2002;16:649–62.

Bremer IL. Diverticula and duplication of intestinal tract. Arch Pathol Lab Med. 1944;38:132–40.

Brenner J, Sordillo PP, Magill GB. Malignant mesothelioma in children: report of seven cases and review of the literature. Med Pediatr Oncol. 1981;9(4):367–73. Review. PubMed PMID: 6790917.

Brierley D, Martin S. Oxidative stress and the DNA mismatch repair pathway. Antioxid Redox Signal. 2013;18(18):2420–8. PubMed PMID: 23121537.

Briggs GG, Freeman RK, Yaffe SJ. Drugs in pregnancy and lactation: a reference guide to fetal and neonatal risk. Obstet Med. 2008;2:89. https://doi.org/10.1258/om.2009.090002. Philadelphia: Lippincott Williams & Wilkins. ISBN 978-0-7817-7876-3. PMC 4989726.

Bristowe JS. Clinical lecture on the consequences of long continued constipation. Br Med J. 1885;1:1085–8. https://doi.org/10.1136/bmj.1.1274.1085. PMID: 20751277.

Broide DH. Immunologic and inflammatory mechanisms that drive asthma progression to remodeling. J Allergy Clin Immunol. 2008;121(3):560–70; quiz 571–2. https://doi.org/10.1016/j.jaci.2008.01.031. PMID: 18328887; PMCID: PMC2386668.

Bronner MP. Granulomatous appendicitis and the appendix in idiopathic inflammatory bowel disease. Semin Diagn Pathol. 2004;21(2):98–107. Review. PubMed PMID: 15807470.

Brosens LA, Montgomery EA, Bhagavan BS, Offerhaus GJ, Giardiello FM. Mucosal prolapse syndrome presenting as rectal polyposis. J Clin Pathol. 2009;62(11):1034–6. PubMed PMID: 19861563; PubMed Central PMCID: PMC2853932.

Bundy DG, Byerley JS, Liles EA, Perrin EM, Katznelson J, Rice HE. Does this child have appendicitis? JAMA. 2007;298(4):438–51. Review. PubMed PMID: 17652298; PubMed Central PMCID: PMC2703737.

Buraniqi E, Moodley M. ZEB2 gene mutation and duplication of 22q11.23 in Mowat-Wilson syndrome. J Child Neurol. 2015;30:32–6.

Buratti S, Kamenwa R, Dohil R, Collins D, Lavine JE. Esophagogastric disconnection following failed fundoplication for the treatment of gastroesophageal reflux disease (GERD) in children with severe neurological impairment. Pediatr Surg Int. 2004;20(10):786–90. PubMed PMID: 15138781.

Burke JS. Lymphoproliferative disorders of the gastrointestinal tract: a review and pragmatic guide to diagnosis. Arch Pathol Lab Med. 2011;135(10):1283–97. Review. PubMed PMID: 21970484

Burn J, Mathers J, Bishop DT. Genetics, inheritance and strategies for prevention in populations at high risk of colorectal cancer (CRC). Recent Results Cancer Res. 2013;191:157–83. https://doi.org/10.1007/978-3-642-30331-9_9. Review. PubMed PMID: 22893205.

Burns AJ, Goldstein AM, Newgreen DF, et al. White paper on guidelines concerning enteric nervous system stem cell therapy for enteric neuropathies. Dev Biol. 2016;417(2):229–51. https://doi.org/10.1016/j.ydbio.2016.04.001.

Butler Tjaden NE, Trainor PA. The developmental etiology and pathogenesis of Hirschsprung disease. Transl Res. 2013;162:1–15.

Cabane J, Bahnini A, Lebas J, Wattiaux MJ, Imbert JC. Sarcoidose révélée par l'atteinte de l'intestin grele. [Sarcoidosis disclosed by involvement of the small intestine]. Ann Med Interne (Paris). 1988;139(4):284–5. French. PubMed PMID: 3190074.

Cacciaguerra S, Miano AE, Di Benedetto A, Vasquez E, Magro G, Fraggetta F. Gastric carcinoma with ovarian metastases in an adolescent. Pediatr Surg Int. 1998;14(1–2):98–9. PubMed PMID: 9880711.

Caldwell JM, Collins MH, Stucke EM, Putnam PE, Franciosi JP, Kushner JP, Abonia JP, Rothenberg ME. Histologic eosinophilic gastritis is a systemic disorder associated with blood and extragastric eosinophilia, TH2 immunity, and a unique gastric transcriptome. J Allergy Clin Immunol. 2014;134(5):1114–24. https://doi.org/10.1016/j.jaci.2014.07.026. Epub 2014 Sep 15. PubMed PMID: 25234644; PubMed Central PMCID: PMC4254306.

Camp ER, Hochwald SN, Liu C. FAP with concurrent duodenal adenomatous polyposis and carcinoid tumor. J Surg Oncol. 2004;87:187–90.

Caplan M. Necrotizing enterocolitis and short bowel syndrome. Chapter 73. In: Gleason CA, Devaskar SU, editors. Avery's diseases of the newborn. 9th ed. Philadelphia: Saunders; 2012. p. 1022–9.

Carrasquillo MM, McCallion AS, Puffenberger EG, Kashuk CS, Nouri N, Chakravarti A. Genome-wide association study and mouse model identify interaction between RET and EDNRB pathways in Hirschsprung disease. Nat Genet. 2002;32:237–44.

References and Recommended Readings

Carroll RE, Benedetti E, Schowalter JP, Buchman AL. Management and complications of short bowel syndrome: an updated review. Curr Gastroenterol Rep. 2016;18(7):40. https://doi.org/10.1007/s11894-016-0511-3. Review. PubMed PMID: 27324885.

Carter TC, Kay DM, Browne ML, Liu A, Romitti PA, Kuehn D, Conley MR, Caggana M, Druschel CM, Brody LC, Mills JL. Hirschsprung's disease and variants in genes that regulate enteric neural crest cell proliferation, migration and differentiation. J Hum Genet. 2012;57(8):485–93. https://doi.org/10.1038/jhg.2012.54. Epub 2012 May 31. PubMed PMID: 22648184; PubMed Central PMCID: PMC3503526.

Cass D. Hirschsprung's disease: an historical review. Prog Pediatr Surg. 1986;20:199–214. https://doi.org/10.1007/978-3-642-70825-1_15. PMID: 3095874.

Castle SL, Zmora O, Papillon S, Levin D, Stein JE. Management of Complicated Gastric Bezoars in Children and Adolescents. Isr Med Assoc J. 2015;17(9):541–4. PubMed PMID: 26625542.

Catassi C, Fasano A. Celiac disease diagnosis: simple rules are better than complicated algorithms. Am J Med. 2010;123(8):691–3. PubMed PMID: 20670718.

CEBM, Centre for Evidence Based Medicine. Available at http://www.cebm.net

Cecchini S, Marchesi F, Caruana P, Tartamella F, Mita MT, Rubichi F, Roncoroni L. Cyst of the gastric wall arising from heterotopic pancreas: report of a case. Acta Biomed. 2016;87(2):215–9. PubMed PMID: 27649007.

Cetin S, Leaphart CL, Li J, Ischenko I, Hayman M, Upperman J, Zamora R, Watkins S, Ford HR, Wang J, Hackam DJ. Nitric oxide inhibits enterocyte migration through activation of RhoA-GTPase in a SHP-2-dependent manner. Am J Physiol Gastrointest Liver Physiol. 2007;292(5):G1347–58. Epub 2007 Feb 1. PubMed PMID: 17272518.

Cheluvappa R. Identification of new potential therapies for colitis amelioration using an appendicitis-appendectomy model. Inflamm Bowel Dis. 2019;25(3):436–44. https://doi.org/10.1093/ibd/izy332. PubMed PMID: 30329049.

Chen YY, Wang HH, Antonioli DA, Spechler SJ, Zeroogian JM, Goyal R, Shahsafaei A, Odze RD. Significance of acid-mucin-positive nongoblet columnar cells in the distal esophagus and gastroesophageal junction. Hum Pathol. 1999;30(12):1488–95. PubMed PMID: 10667428.

Cheng WW, Tang CS, Gui HS, So MT, Lui VC, Tam PK, Garcia-Barcelo MM. Depletion of the IKBKAP ortholog in zebrafish leads to Hirschsprung disease-like phenotype. World J Gastroenterol. 2015;21(7):2040–6. https://doi.org/10.3748/wjg.v21.i7.2040. PubMed PMID: 25717236; PubMed Central PMCID: PMC4326138.

Chetty R. Requiem for the term 'carcinoid tumour' in the gastrointestinal tract? Can J Gastroenterol. 2008a;22(4):357–8. PubMed PMID: 18414708; PubMed Central PMCID: PMC2662891.

Chetty R. Goblet cell carcinoid tumours of the appendix: a unique neuroendocrine tumour. Histopathology. 2008b;52(6):770–1. PubMed PMID: 18439158.

Chiang JM, Changchien CR, Chen JR. Solitary rectal ulcer syndrome: an endoscopic and histological presentation and literature review. Int J Color Dis. 2006;21(4):348–56. Epub 2005 Aug 17. Review. PubMed PMID: 16133006.

Choristoma. Stedman's medical dictionary. 28th ed. Philadelphia: Lippincott Williams & Wilkins; 2006. p. 371.

Coe A, Avansino JR, Kapur RP. Distal rectal skip-segment Hirschsprung disease and the potential for false-negative diagnosis. Pediatr Dev Pathol. 2016;19:123–31.

College of American Pathologists (CAP). POET REPORT. Perspectives on emerging technology. Prognostic uses of MSI testing. CAP Report. 2011; 1–7.

Concetti L. About some innate colon malformation causing habitual constipation of children. Arch Kinderheilk. 1899;27:319–53.

Cook MG, Dixon MF. An analysis of the reliability of detection and diagnostic value of various pathological features in Crohn's disease and ulcerative colitis. Gut. 1973;14(4):255–62. PubMed PMID: 4706906; PubMed Central PMCID: PMC1412589.

Corazza GR, Villanacci V. Coeliac disease. J Clin Pathol. 2005;58(6):573–4. Review. PubMed PMID: 15917404; PubMed Central PMCID: PMC1770677.

Corpron CA, Black CT, Herzog CE, Sellin RV, Lally KP, Andrassy RJ. A half century of experience with carcinoid tumors in children. Am J Surg. 1995;170:606–8.

Crohn BB, Rosenak BD. A combined form of ileitis and colitis. JAMA. 1936;106:1–7.

Cross SS, Vergani P, Stephenson TJ, Varghese M, Sidhu R, Paul HD. Dysplasia-associated mass or lesion (DALM) and sporadic adenomas in patients with chronic idiopathic inflammatory bowel disease. Diagn Histopathol. 2008;14(2):110–5. https://doi.org/10.1016/j.mpdhp.2007.12.006.

Cruz-Correa M, Poonawala A, Abraham SC, Wu TT, Zahurak M, Vogelsang G, et al. Endoscopic findings predict the histologic diagnosis in gastrointestinal graft-versus-host disease. Endoscopy. 2002;34:808–13.

Curtis JL, Burns RC, Wang L, Mahour GH, Ford HR. Primary gastric tumors of infancy and childhood: 54-year experience at a single institution. J Pediatr Surg. 2008;43(8):1487–93. https://doi.org/10.1016/j.jpedsurg.2007.11.016. PubMed PMID: 18675640.

D'Aleo C, Lazzareschi I, Ruggiero A, Riccardi R. Carcinoid tumors of the appendix in children: two case reports and review of the literature. Pediatr Hematol Oncol. 2001;18:347–51.

D'Haens GR, Geboes K, Peeters M, Baert F, Penninckx F, Rutgeerts P. Early lesions of recurrent Crohn's disease caused by infusion of intestinal contents in excluded ileum. Gastroenterology. 1998;114(2):262–7. PubMed PMID: 9453485.

Daher P, Ghanimeh J, Riachy E, Zeidan S, Eid B. Congenital segmental dilatation of the small bowel (CSD). Eur J Pediatr Surg. 2007;17(4):289–91. PubMed PMID: 17806030.

Dall'Igna P, Ferrari A, Luzzatto C, Bisogno G, Casanova M, Alaggio R, Terenziani M, Cecchetto G. Carcinoid tumor of the appendix in childhood: the experience of two Italian institutions. J Pediatr Gastroenterol Nutr. 2005;40(2):216–9. PubMed PMID: 15699700.

Dalton BG, Thomas PG, Sharp NE, Manalang MA, Fisher JE, Moir CR, St Peter SD, Iqbal CW. Inflammatory myofibroblastic tumors in children. J Pediatr Surg. 2016;51(4):541–4. https://doi.org/10.1016/j.jpedsurg.2015.11.015. Epub 2015 Nov 27. PubMed PMID: 26732283.

Danese S, Fiocchi C. Ulcerative colitis. N Engl J Med. 2011;365(18):1713–25. Review. PubMed PMID: 22047562.

Davidson J, Healy C, Blackburn SC, Curry J. Laparoscopic enteropexy for prolapsing stoma: a case series describing a novel technique. J Laparoendosc Adv Surg Tech A. 2018;28(9):1135–8. https://doi.org/10.1089/lap.2017.0730. PubMed PMID: 29624469.

de Góis NM, Costa RR, Kesselring F, de Freitas VG, Ribalta JC, Kobata MP, Taha NS. Grade 3 vulvar and anal intraepithelial neoplasia in a HIV seropositive child--therapeutic result: case report. Clin Exp Obstet Gynecol. 2005;32(2):138–40. PubMed PMID: 16108402.

De Lagausie P, Bonnard A, Schultz A, Van den Abbeel T, Bellaiche M, Hartmann JF, Cezard JP, Aigrain Y. Reflux in esophageal atresia, tracheoesophageal cleft, and esophagocoloplasty: Bianchi's procedure as an alternative approach. J Pediatr Surg. 2005;40(4):666–9. PubMed PMID: 15852275.

de Pontual L, Pelet A, Clement-Ziza M, et al. Epistatic interactions with a common hypomorphic RET allele in syndromic Hirschsprung disease. Hum Mutat. 2007;28:790–6.

DeBrosse CW, Case JW, Putnam PE, et al. Quantity and distribution of eosinophils in the gastrointestinal tract of children. Pediatr Dev Pathol. 2006;9(3):210–8.

Delves PJ, Roitt IM. The immune system. First of two parts. N Engl J Med. 2000a;343(1):37–49. Review. PubMed PMID: 10882768.

Delves PJ, Roitt IM. The immune system. Second of two parts. N Engl J Med. 2000b;343(2):108–17. Review. PubMed PMID: 10891520.

Derby LE, Jick H. Appendectomy protects against ulcerative colitis. Epidemiology. 1998;9(2):205–7. PubMed PMID: 9504292.

Derkay CS, Wiatrak B. Recurrent respiratory papillomatosis: a review. Laryngoscope. 2008;118(7):1236–47. Review. PubMed PMID: 18496162.

Dicken BJ, Sergi C, Rescorla FJ, Breckler F, Sigalet D. Medical management of motility disorders in patients with intestinal failure: a focus on necrotizing enterocolitis, gastroschisis, and intestinal atresia. J Pediatr Surg. 2011;46:1618–30.

Dixon MF, Genta RM, Yardley JH, Correa P. Classification and grading of gastritis. The updated Sydney system. International workshop on the histopathology of gastritis, Houston 1994. Am J Surg Pathol. 1996;20(10):1161–81. Review. PubMed PMID: 8827022.

Do MY, Myung SJ, Park HJ, Chung JW, Kim IW, Lee SM, Yu CS, Lee HK, Lee JK, Park YS, Jang SJ, Kim HJ, Ye BD, Byeon JS, Yang SK, Kim JH. Novel classification and pathogenetic analysis of hypoganglionosis and adult-onset Hirschsprung's disease. Dig Dis Sci. 2011;56(6):1818–27. https://doi.org/10.1007/s10620-010-1522-9. Epub 2011 Jan 11. PubMed PMID: 21222160.

Dokucu AI, Oztürk H, Kilinç N, Onen A, Bukte Y, Soker M. Primary gastric adenocarcinoma in a 2.5-year-old girl. Gastric Cancer. 2002;5(4):237–9. PubMed PMID: 12491083.

Doray B, Salomon R, Amiel J, et al. Mutation of the RET ligand, neurturin, supports multigenic inheritance in Hirschsprung disease. Hum Mol Genet. 1998;7:1449–52.

Doubaj Y, Pingault V, Elalaoui SC, Ratbi I, Azouz M, Zerhouni H, Ettayebi F, Sefiani A. A novel mutation in the Endothelin B receptor gene in a Moroccan family with Shah-Waardenburg syndrome. Mol Syndromol. 2015;6(1):44–9. https://doi.org/10.1159/000371590. Epub 2015 Jan 28. PubMed PMID: 25852447; PubMed Central PMCID: PMC4369116.

Douglas JG, Gillon J, Logan RF, Grant IW, Crompton GK. Sarcoidosis and coeliac disease: an association? Lancet. 1984;2(8393):13–5. PubMed PMID: 6145934.

Drévillon L, Megarbane A, Demeer B, Matar C, Benit P, Briand-Suleau A, Bodereau V, Ghoumid J, Nasser M, Decrouy X, Doco-Fenzy M, Rustin P, Gaillard D, Goossens M, Giurgea I. KBP-cytoskeleton interactions underlie developmental anomalies in Goldberg-Shprintzen syndrome. Hum Mol Genet. 2013;22(12):2387–99. https://doi.org/10.1093/hmg/ddt083. Epub 2013 Feb 19. PubMed PMID: 23427148.

Drury RAB, Wallington EA. Carleton's histological technique. 5th ed. Oxford University Press; 1980. ISBN 0-19-261310-3.

Drut R. Multivisceral dysplastic lesions in a patient with tuberous sclerosis and Langerhans cell histiocytosis. Pediatr Pathol. 1990;10(4):633–9. Review. PubMed PMID: 2196548.

Drut R, Drut RM. Lymphocytic gastritis in pediatric celiac disease – immunohistochemical study of the intraepithelial lymphocytic component. Med Sci Monit. 2004;10(1):CR38–42. PubMed PMID: 14704635.

Duggan EM, Marshall AP, Weaver KL, St Peter SD, Tice J, Wang L, Choi L, Blakely ML. A systematic review and individual patient data meta-analysis of published randomized clinical trials comparing early versus interval appendectomy for children with perforated appendicitis. Pediatr Surg Int. 2016;32(7):649–55. https://doi.org/10.1007/s00383-016-3897-y. Epub 2016 May 9. Review. PubMed PMID: 27161128.

Dunphy L, Clark Z, Raja MH. *Enterobius vermicularis* (pinworm) infestation in a child presenting with symptoms of acute appendicitis: a wriggly tale! BMJ Case Rep. 2017; pii: bcr-2017-220473. https://

doi.org/10.1136/bcr-2017-220473. PubMed PMID: 28988188.

Eaden J, Abrams K, McKay H, Denley H, Mayberry J. Inter-observer variation between general and specialist gastrointestinal pathologists when grading dysplasia in ulcerative colitis. J Pathol. 2001;194(2):152–7. PubMed PMID: 11400142.

Ebers G. History of a rare case of ileus. Hufland's J. 1836;83(2):62–89.

Edden Y, Shih SS, Wexner SD. Solitary rectal ulcer syndrome and stercoral ulcers. Gastroenterol Clin N Am. 2009;38(3):541–5. Review. PubMed PMID:19699413.

Ehrenpreis TH. Megacolon in the newborn. A clinical and roentgenological study with special regard to the pathogenesis. Acta Chir Scand. 1946;94(Suppl):112.

Emison ES, McCallion AS, Kashuk CS, Bush RT, Grice E, Lin S, Portnoy ME, Cutler DJ, Green ED, Chakravarti A. A common sex-dependent mutation in a RET enhancer underlies Hirschsprung disease risk. Nature. 2005;434(7035):857–63. PubMed PMID: 15829955.

Emison ES, Garcia-Barcelo M, Grice EA, Lantieri F, Amiel J, Burzynski G, Fernandez RM, Hao L, Kashuk C, West K, Miao X, Tam PK, Griseri P, Ceccherini I, Pelet A, Jannot AS, de Pontual L, Henrion-Caude A, Lyonnet S, Verheij JB, Hofstra RM, Antiñolo G, Borrego S, McCallion AS, Chakravarti A. Differential contributions of rare and common, coding and noncoding Ret mutations to multifactorial Hirschsprung disease liability. Am J Hum Genet. 2010;87(1):60–74. https://doi.org/10.1016/j.ajhg.2010.06.007. PubMed PMID: 20598273; PubMed Central PMCID: PMC2896767.

Encinas JL, Avila LF, García-Cabeza MA, Luis A, Hernández F, Martínez L, Fernández A, Olivares P, Tovar JA. Tumor carcinoide bronquial y apendicular [Bronchial and appendiceal carcinoid tumors]. An Pediatr (Barc). 2006;64(5):474–7. Spanish. PubMed PMID: 16756890.

England RJ, Pillay K, Davidson A, Numanoglu A, Millar AJ. Intussusception as a presenting feature of Burkitt lymphoma: implications for management and outcome. Pediatr Surg Int. 2012;28(3):267–70. https://doi.org/10.1007/s00383-011-2982-5. Epub 2011 Oct 4. PubMed PMID: 21969235.

Erickson RP, Larson-Thome K, Valenzuela RK, Whitaker SE, Shub MD. Navajo microvillous inclusion disease is due to a mutation in MYO5B. Am J Med Genet A. 2008;146A:3117–9.

Ernst PB, Takaishi H, Crowe SE. Helicobacter pylori infection as a model for gastrointestinal immunity and chronic inflammatory diseases. Dig Dis. 2001;19(2):104–11. Review. PubMed PMID: 11549818.

Erten EE, Çavuşoğlu YH, Arda N, Karaman A, Afşarlar ÇE, Karaman I, Özgüner IF. A rare case of multiple skip segment Hirschsprung's disease in the ileum and colon. Pediatr Surg Int. 2014;30(3):349–51. https://doi.org/10.1007/s00383-013-3428-z. Epub 2013 Nov 1. PubMed PMID: 24178302.

Falk S. Lymphomas of the upper GI tract: the role of radiotherapy. Clin Oncol (R Coll Radiol). 2012;24(5):352–7. https://doi.org/10.1016/j.clon.2012.02.003. Epub 2012 Mar 2. Review. PubMed PMID: 22386892.

Fenollar F, Minodier P, Boutin A, Laporte R, Brémond V, Noël G, Miramont S, Richet H, Benkouiten S, Lagier JC, Gaudart J, Jouve JL, Raoult D. Tropheryma whipplei associated with diarrhoea in young children. Clin Microbiol Infect. 2016;22(10):869–74. https://doi.org/10.1016/j.cmi.2016.07.005. Epub 2016 Jul 9. PubMed PMID: 27404363.

Fernandez RM, Ruiz-Ferrer M, Lopez-Alonso M, Antinolo G, Borrego S. Polymorphisms in the genes encoding the 4 RET ligands, GDNF, NTN, ARTN, PSPN, and susceptibility to Hirschsprung disease. J Pediatr Surg. 2008;43:2042–7.

Fernández RM, Bleda M, Luzón-Toro B, García-Alonso L, Arnold S, Sribudiani Y, Besmond C, Lantieri F, Doan B, Ceccherini I, Lyonnet S, Hofstra RM, Chakravarti A, Antiñolo G, Dopazo J, Borrego S. Pathways systematically associated to Hirschsprung's disease. Orphanet J Rare Dis. 2013;8:187. https://doi.org/10.1186/1750-1172-8-187. PubMed PMID: 24289864; PubMed Central PMCID: PMC3879038.

Ferris Villanueva E, Guerrero Bautista R, Chica Marchal A. Hirschsprung disease associated with Mowat-Wilson syndrome: report of a case. Nutr Hosp. 2015;31:1882–4.

Finney MT. Congenital idiopathic dilatation of the colon. Surg Gynecol Obstet. 1908;6:624–43.

Fiori MG. Domenico Battini and his description of congenital megacolon: a detailed case report one century before Hirschsprung. J Peripher Nerv Syst. 1998;3:197–206. PMID: 10959250.

Fox PA. Human papillomavirus and anal intraepithelial neoplasia. Curr Opin Infect Dis. 2006;19(1):62–6. Review. PubMed PMID: 16374220.

Fox PA, Seet JE, Stebbing J, Francis N, Barton SE, Strauss S, Allen-Mersh TG, Gazzard BG, Bower M. The value of anal cytology and human papillomavirus typing in the detection of anal intraepithelial neoplasia: a review of cases from an anoscopy clinic. Sex Transm Infect. 2005;81(2):142–6. PubMed PMID: 15800092; PubMed Central PMCID: PMC1764665.

Frei JV, Morson BC. Medical audit of rectal biopsy diagnosis of inflammatory bowel disease. J Clin Pathol. 1982;35(3):341–4. PubMed PMID: 7068926; PubMed Central PMCID: PMC497543.

Frías JL, Frías JP, Frías PA, Martínez-Frías ML. Infrequently studied congenital anomalies as clues to the diagnosis of maternal diabetes mellitus. Am J Med Genet A. 2007;143A(24):2904–9. PubMed PMID: 18000913.

Friedmacher F, Puri P. Classification and diagnostic criteria of variants of Hirschsprung's disease. Pediatr Surg Int. 2013;29:855–72.

Fu M, Tam PK, Sham MH, Lui VC. Embryonic development of the ganglion plexuses and the concentric layer structure of human gut: a topographical study. Anat Embryol (Berl). 2004;208:33–41.

Fujiyoshi Y, Nakamura Y, Cho T, Nishimura T, Morimatsu M, Shirouzu K, Fukuda S, Kimura T, Nakashima H,

Hashimoto T. Exstrophy of the cloacal membrane. A pathologic study of four cases. Arch Pathol Lab Med. 1987;111(2):157–60. PubMed PMID: 3813831.

Gabriel SB, Salomon R, Pelet A, et al. Segregation at three loci explains familial and population risk in Hirschsprung disease. Nat Genet. 2002;31:89–93.

Galea MH, Holliday H, Carachi R, Kapila L. Short-bowel syndrome: a collective review. J Pediatr Surg. 1992;27(5):592–6. Review. PubMed PMID: 1625129.

Gao H, Chen D, Liu X, Wu M, Mi J, Wang W. Polymorphisms and expression of the WNT8A gene in Hirschsprung's disease. Int J Mol Med. 2013;32:647–52. https://doi.org/10.3892/ijmm.2013.1433. PMID: 23836442.

Garcia M, Delacruz V, Ortiz R, Bagni A, Weppler D, Kato T, Tzakis A, Ruiz P. Acute cellular rejection grading scheme for human gastric allografts. Hum Pathol. 2004;35(3):343–9. PubMed PMID: 15017591.

Garcia-Barcelo MM, Tang CS, Ngan ES, Lui VC, Chen Y, So MT, Leon TY, Miao XP, Shum CK, Liu FQ, Yeung MY, Yuan ZW, Guo WH, Liu L, Sun XB, Huang LM, Tou JF, Song YQ, Chan D, Cheung KM, Wong KK, Cherny SS, Sham PC, Tam PK. Genome-wide association study identifies NRG1 as a susceptibility locus for Hirschsprung's disease. Proc Natl Acad Sci U S A. 2009;106(8):2694–9. https://doi.org/10.1073/pnas.0809630105. Epub 2009 Feb 5. PubMed PMID: 19196962; PubMed Central PMCID: PMC2650328.

Gariepy CE, Mousa H. Clinical management of motility disorders in children. Semin Pediatr Surg. 2009;18(4):224–38. https://doi.org/10.1053/j.sempedsurg.2009.07.004. Review. PubMed PMID: 19782304.

Geboes K, Desmet V, Vantrappen G, Mebis J. Vascular changes in the esophageal mucosa. An early histologic sign of esophagitis. Gastrointest Endosc. 1980;26(2):29–32.PubMed PMID: 7390102.

Geboes K, Haot J, Mebis J, Desmet VJ. The histopathology of reflux esophagitis. Acta Chir Belg. 1983;83(6):444–8. PubMed PMID: 6659822.

Geboes K, Riddell R, Ost A, Jensfelt B, Persson T, Löfberg R. A reproducible grading scale for histological assessment of inflammation in ulcerative colitis. Gut. 2000;47(3):404–9. PubMed PMID: 10940279; PubMed Central PMCID: PMC1728046.

Gee SJ. Treatment of intestinal obstruction. A discussion at the Liverpool Medical Institution. Lancet. 1884;3197:1019–20.

Genersich G. Concerning a congenital dilatation and hypertrophy of the colon. Jb Kinderheilk. 1894;37:91–100.

Gershon MD. The second brain: a groundbreaking new understanding of nervous disorders of the stomach and intestine. New York: Harper Perennial; 1999.

Gershon MD. Transplanting the enteric nervous system: a step closer to treatment for aganglionosis. Gut. 2007;56:459–61.

Gershon MD. NPARM in PHOX2B: why some things just should not be expanded. J Clin Invest. 2012;122:3056–8.

Gershon MD, Ratcliffe EM. Developmental biology of the enteric nervous system: pathogenesis of Hirschsprung's disease and other congenital dysmotilities. Semin Pediatr Surg. 2004;13:224–35.

Gervaz P, Bucher P, Morel P. Two colons-two cancers: paradigm shift and clinical implications. J Surg Oncol. 2004;88:261–6. https://doi.org/10.1002/jso.20156.

Glickman JN, Bousvaros A, Farraye FA, Zholudev A, Friedman S, Wang HH, Leichtner AM, Odze RD. Pediatric patients with untreated ulcerative colitis may present initially with unusual morphologic findings. Am J Surg Pathol. 2004;28(2):190–7. PubMed PMID: 15043308.

Goddard MJ, Lonsdale RN. The histogenesis of appendiceal carcinoid tumours. Histopathology. 1992;20:345–9.

Goedde TA, Rodriguez-Bigas MA, Herrera L, Petrelli NJ. Gastroduodenal polyps in familial adenomatous polyposis. Surg Oncol. 1992;1(5):357–61. PubMed PMID: 1341271.

Goldman L, Schafer AI. Goldman-Cecil Medicine, 2-Volume Set. 25th ed; 2016, ISBN-13: 978-1-4557-5017-7. 3024 p

Goldstein NS, Underhill J. Morphologic features suggestive of gluten sensitivity in architecturally normal duodenal biopsy specimens. Am J Clin Pathol. 2001;116(1):63–71. PubMed PMID: 11447753.

Goldstone R, Itzkowitz S, Harpaz N, Ullman T. Progression of low-grade dysplasia in ulcerative colitis: effect of colonic location. Gastrointest Endosc. 2011;74(5):1087–93. Epub 2011 Sep 10. PubMed PMID: 21907984.

Goldthorn JF, Canizaro PC. Gastrointestinal malignancy in infancy, childhood, and adolescence. Surg Clin North Am. 1986;66:845.

Gonzalez-Crussi F, Sotelo-Avila C, deMello DE. Primary peritoneal, omental, and mesenteric tumors in childhood. Semin Diagn Pathol. 1986;3(2):122–37. PubMed PMID: 3616217.

Gorter RR, Kneepkens CM, Mattens EC, Aronson DC, Heij HA. Management of trichobezoar: case report and literature review. Pediatr Surg Int. 2010;26(5):457–63. https://doi.org/10.1007/s00383-010-2570-0. Epub 2010 Mar 6. Review. PubMed PMID: 20213124; PubMed Central PMCID: PMC2856853.

Goulet O, Salomon J, Ruemmele F, de Serres NP, Brousse N. Intestinal epithelial dysplasia (tufting enteropathy). Orphanet J Rare Dis. 2007;2:20. Review. PubMed PMID: 17448233; PubMed Central PMCID: PMC1878471.

Goulet O, Vinson C, Roquelaure B, Brousse N, Bodemer C, Cézard JP. Syndromic (phenotypic) diarrhea in early infancy. Orphanet J Rare Dis. 2008;3:6. https://doi.org/10.1186/1750-1172-3-6. Review. PubMed PMID: 18304370; PubMed Central PMCID: PMC2279108.

Grabowski P, Schönfelder J, Ahnert-Hilger G, Foss HD, Heine B, Schindler I, Stein H, Berger G, Zeitz M, Scherübl H. Expression of neuroendocrine markers: a signature of human undifferentiated carcinoma of the colon and rectum. Virchows Arch. 2002;441(3):256–63. Epub 2002 Jun 7. PubMed PMID: 12242522.

Gramlich T, Petras RE. Pathology of inflammatory bowel disease. Semin Pediatr Surg. 2007;16(3):154–63. Review. PubMed PMID: 17602970.

Greco L, Torre P. Genetica della malattia celiaca. Riv Ital Ped. 2000;26:632–3.

Griffith AM, Buller MB. Inflammatory bowel dis- ease. In: Walker WA, Durie PR, Hamilton JR, Walker-Smith JA, Watkins JB, editors. Pediatric gastrointestinal disease. Pathophysiology, diagnosis, management. 3rd ed. Hamilton: BC Decker; 2000. p. 613–51.

Griseri P, Vos Y, Giorda R, Gimelli S, Beri S, Santamaria G, Mognato G, Hofstra RM, Gimelli G, Ceccherini I. Complex pathogenesis of Hirschsprung's disease in a patient with hydrocephalus, vesico-ureteral reflux and a balanced translocation t(3;17)(p12;q11). Eur J Hum Genet. 2009;17(4):483–90. https://doi.org/10.1038/ejhg.2008.191. Epub 2008 Oct 29. PubMed PMID: 19300444; PubMed Central PMCID: PMC2986215.

Groisman GM, Polak-Charcon S. Fibroepithelial polyps of the anus: a histologic, immunohistochemical, and ultrastructural study, including comparison with the normal anal subepithelial layer. Am J Surg Pathol. 1998;22(1):70–6. PubMed PMID: 9422318.

Grosfeld JL, Ballantine TV, Shoemaker R. Operative mangement of intestinal atresia and stenosis based on pathologic findings. J Pediatr Surg. 1979;14(3):368–75. PubMed PMID: 480102.

Gui H, Tang WK, So MT, et al. RET and NRG1 interplay in Hirschsprung disease. Hum Genet. 2013;132:591–600.

Guindi M, Riddell RH. Indeterminate colitis. J Clin Pathol. 2004;57(12):1233–44. Review. PubMed PMID: 15563659; PubMed Central PMCID: PMC1770507.

Gupta RB, Harpaz N, Itzkowitz S, Hossain S, Matula S, Kornbluth A, Bodian C, Ullman T. Histologic inflammation is a risk factor for progression to colorectal neoplasia in ulcerative colitis: a cohort study. Gastroenterology. 2007;133(4):1099–105; quiz 1340-1. https://doi.org/10.1053/j.gastro.2007.08.001. Epub 2007 Aug 2. PMID: 17919486; PMCID: PMC2175077.

Gupta V, Lal A, Sinha SK, Nada R, Gupta NM. Leiomyomatosis of the esophagus: experience over a decade. J Gastrointest Surg. 2009;13(2):206–11. Epub 2008 Sep 26. PubMed PMID: 18818979.

Gutierrez JC, De Oliveira LO, Perez EA, Rocha-Lima C, Livingstone AS, Koniaris LG. Optimizing diagnosis, staging, and management of gastrointestinal stromal tumors. J Am Coll Surg. 2007;205(3):479–91 (Quiz 524). Epub 2007 Jul 16. Review. PubMed PMID: 17765165.

Gutman LT, St Claire KK, Everett VD, Ingram DL, Soper J, Johnston WW, Mulvaney GG, Phelps WC. Cervical-vaginal and intraanal human papillomavirus infection of young girls with external genital warts. J Infect Dis. 1994;170(2):339–44. PubMed PMID: 8035020.

Hackam DJ, Upperman JS, Grishin A, Ford HR. Disordered enterocyte signaling and intestinal barrier dysfunction in the pathogenesis of necrotizing enterocolitis. Semin Pediatr Surg. 2005;14(1):49–57. Review. PubMed PMID: 15770588.

Haller JA. A child with dysphagia lusoria. Am J Dis Child. 1987;141(5):480. https://doi.org/10.1001/archpedi.1987.04460050022020.

Halpern MD, Khailova L, Molla-Hosseini D, Arganbright K, Reynolds C, Yajima M, Hoshiba J, Dvorak B. Decreased development of necrotizing enterocolitis in IL-18-deficient mice. Am J Physiol Gastrointest Liver Physiol. 2008;294(1):G20–6. Epub 2007 Oct 18. PubMed PMID: 17947451; PubMed Central PMCID: PMC3086795.

Handousa AE, Azab MS, El-Beshbishi SN, El-Nahas HA, Abd El-Hamid MA. Comparative study between immunohistochemical grading and giardia genotyping among symptomatic and asymptomatic humans. Egypt J Immunol. 2007;14(2):63–72. PubMed PMID: 20306658.

Hansford JR, Mulligan LM. Multiple endocrine neoplasia type 2 and RET: from neoplasia to neurogenesis. J Med Genet. 2000;37:817–27.

Haot J, Hamichi L, Wallez L, Mainguet P. Lymphocytic gastritis: a newly described entity: a retrospective endoscopic and histological study. Gut. 1988;29(9):1258–64. PubMed PMID: 3198002; PubMed Central PMCID: PMC1434343.

Harting MT, Blakely ML, Herzog CE, Lally KP, Ajani JA, Andrassy RJ. Treatment issues in pediatric gastric adenocarcinoma. J Pediatr Surg. 2004;39(8):e8–10. PubMed PMID: 15300556.

Haskell H, Andrews CW Jr, Reddy SI, Dendrinos K, Farraye FA, Stucchi AF, Becker JM, Odze RD. Pathologic features and clinical significance of "backwash" ileitis in ulcerative colitis. Am J Surg Pathol. 2005;29(11):1472–81. PubMed PMID: 16224214.

Hatano M, Aoki T, Dezawa M, Yusa S, Iitsuka Y, Koseki H, Taniguchi M, Tokuhisa T. A novel pathogenesis of megacolon in Ncx/Hox11L.1 deficient mice. J Clin Invest. 1997;100:795–801.

Hatzipantelis E, Panagopoulou P, Sidi-Fragandrea V, Fragandrea I, Koliouskas DE. Carcinoid tumors of the appendix in children: experience from a tertiary center in northern Greece. J Pediatr Gastroenterol Nutr. 2010;51:622–5.

Hawkes EA, Wotherspoon A, Cunningham D. Diagnosis and management of rare gastrointestinal lymphomas. Leuk Lymphoma. 2012;53(12):2341–50. https://doi.org/10.3109/10428194.2012.695780. Epub 2012 Jun 21. PubMed PMID: 22616672.

Heanue TA, Pachnis V. Enteric nervous system development and Hirschsprung's disease: advances in genetic and stem cell studies. Nat Rev Neurosci. 2007;8:466–79.

Heide S, Masliah-Planchon J, Isidor B, Guimier A, Bodet D, Coze C, Deville A, Thebault E, Pasquier CJ, Cassagnau E, Pierron G, Clément N, Schleiermacher G, Amiel J, Delattre O, Peuchmaur M, Bourdeaut F. Oncologic phenotype of peripheral neuroblastic tumors associated with PHOX2B non-Polyalanine repeat expansion mutations. Pediatr Blood Cancer. 2016;63(1):71–7. https://doi.org/10.1002/pbc.25723. Epub 2015 Sep 16. PubMed PMID: 26375764.

Heinz-Erian P, Schmidt H, Le Merrer M, Phillips AD, Kiess W, Hadorn HB. Congenital microvillus atrophy in a girl with autosomal dominant hypochondroplasia. J Pediatr Gastroenterol Nutr. 1999;28(2):203–5. PubMed PMID: 9932857.

Henderson L, Nour S, Dagash H. Heterotopic pancreas: a rare cause of gastrointestinal bleeding in children. Dig Dis Sci. 2018;63(5):1363–5. https://doi.org/10.1007/s10620-018-4981-z. Epub 2018 Feb 22. PubMed PMID: 29468375.

Hickey M, Georgieff M, Ramel S. Neurodevelopmental outcomes following necrotizing enterocolitis. Semin Fetal Neonatal Med. 2018;23(6):426–32. https://doi.org/10.1016/j.siny.2018.08.005. Epub 2018 Aug 17. Review. PubMed PMID: 30145060.

Hirschsprung H. Constipation of newborns as a result of dilatation and hypertrophy of the colon. Jhrb f Kinderh. 1888;27:1–7.

Hsueh W, Caplan MS, Qu XW, Tan XD, De Plaen IG, Gonzalez-Crussi F. Neonatal necrotizing enterocolitis: clinical considerations and pathogenetic concepts. Pediatr Dev Pathol. 2003;6(1):6–23. Epub 2002 Nov 11. Review. PubMed PMID: 12424605.

Hultgren BD. Ileocolonic aganglionosis in white progeny of overo spotted horses. J Am Vet Med Assoc. 1982;180:289–92.

Hummel M, Oeschger S, Barth TF, Loddenkemper C, Cogliatti SB, Marx A, Wacker HH, Feller AC, Bernd HW, Hansmann ML, Stein H, Möller P. Wotherspoon criteria combined with B cell clonality analysis by advanced polymerase chain reaction technology discriminates covert gastric marginal zone lymphoma from chronic gastritis. Gut. 2006;55(6):782–7. Epub 2006 Jan 19. Review. PubMed PMID: 16423889; PubMed Central PMCID: PMC1856242.

Hung FC, Kuo CM, Chuah SK, Kuo CH, Chen YS, Lu SN, Chang Chien CS. Clinical analysis of primary duodenal adenocarcinoma: an 11-year experience. J Gastroenterol Hepatol. 2007;22(5):724–8. PubMed PMID: 17444863.

Hunger K, Mischke P, Rieper W, Raue R, Kunde K, Engel A. Azo dyes. In: Ullmann's encyclopedia of industrial chemistry. Ullmann's encyclopedia of industrial chemistry. Weinheim: Wiley-VCH; 2005. https://doi.org/10.1002/14356007.a03_245. ISBN 978-3527306732.

Husby S, Koletzko S, Korponay-Szabó IR, Mearin ML, Phillips A, Shamir R, Troncone R, Giersiepen K, Branski D, Catassi C, Lelgeman M, Mäki M, Ribes-Koninckx C, Ventura A, Zimmer KP, ESPGHAN Working Group on Coeliac Disease Diagnosis, ESPGHAN Gastroenterology Committee, European Society for Pediatric Gastroenterology, Hepatology, and Nutrition. European Society for Pediatric Gastroenterology, Hepatology, and Nutrition guidelines for the diagnosis of coeliac disease. J Pediatr Gastroenterol Nutr. 2012;54(1):136–60. https://doi.org/10.1097/MPG.0b013e31821a23d0. Erratum in: J Pediatr Gastroenterol Nutr. 2012 Apr;54(4):572. PubMed PMID: 22197856.

Hyams JS, Ferry GD, Mandel FS, Gryboski JD, Kibort PM, Kirschner BS, Griffiths AM, Katz AJ, Grand RJ, Boyle JT, et al. Development and validation of a pediatric Crohn's disease activity index. J Pediatr Gastroenterol Nutr. 1991;12(4):439–47. PubMed PMID: 1678008.

Iancu TC, Manov I. Ultrastructural aspects of enterocyte defects in infancy and childhood. Ultrastruct Pathol. 2010;34(3):117–25. https://doi.org/10.3109/01913121003648410. Review. PubMed PMID: 20455660.

Ihtiyar E, Algin C, Isiksoy S, Ates E. Small cell carcinoma of rectum: a case report. World J Gastroenterol. 2005;11(20):3156–8. PubMed PMID: 15918209.

International Agency for Research on Cancer. Digestive system tumours: WHO classification of tumours, 5th ed., vol. 1. WHO classification of tumours editorial board. 2019. ISBN-13 (Print Book): 978-92-832-4499-8.

International Association for Research on Cancer (IARC). Doxylamine succinate. In: Some thyrotropic agents. IARC working group on the evaluation of carcinogenic risks to humans, vol. 79. Lyon: IARC; 2001. p. 145–60.

Ishihara S, Watanabe T, Akahane T, Shimada R, Horiuchi A, Shibuya H, Hayama T, Yamada H, Nozawa K, Matsuda K, Maeda K, Sugihara K. Tumor location is a prognostic factor in poorly differentiated adenocarcinoma, mucinous adenocarcinoma, and signet-ring cell carcinoma of the colon. Int J Color Dis. 2012;27(3):371–9. Epub 2011 Nov 4. PubMed PMID: 22052041.

Isidor B, Lefebvre T, Le Vaillant C, Caillaud G, Faivre L, Jossic F, Joubert M, Winer N, Le Caignec C, Borck G, Pelet A, Amiel J, Toutain A, Ronce N, Raynaud M, Verloes A, David A. Blepharophimosis, short humeri, developmental delay and Hirschsprung disease: expanding the phenotypic spectrum of MED12 mutations. Am J Med Genet A. 2014;164A(7):1821–5. https://doi.org/10.1002/ajmg.a.36539. Epub 2014 Apr 8. PubMed PMID: 24715367.

Issaivanan M, Redner A, Weinstein T, Soffer S, Glassman L, Edelman M, Levy CF. Esophageal carcinoma in children and adolescents. J Pediatr Hematol Oncol. 2012;34(1):63–7. PubMed PMID: 22052168

Jacobi A. On some important causes of constipation in infants. Am J Obstet. 1869;2:96.

Jacobson S, Marcus EM. Neuroanatomy for the neuroscientist. 2nd ed. Boston: Springer; 2011.

Jain M, Solanki SL, Bhatnagar A, Jain PK. An unusual case report of rapunzel syndrome trichobezoar in a 3-year-old boy. Int J Trichol. 2011;3(2):102–4. https://doi.org/10.4103/0974-7753.90820. PubMed PMID: 22223971; PubMed Central PMCID: PMC3250004.

Jay V. Legacy of Harald Hirschsprung. Pediatr Dev Pathol. 2001;4:203–4. https://doi.org/10.1007/s100240010144. PMID: 11178638.

Jayanthi V, Girija R, Mayberry JF. Terminal ileal stricture. Postgrad Med J. 2002;78(924):627–31. PubMed PMID: 12415093; PubMed Central PMCID: PMC1742518.

Jenkins D, Balsitis M, Gallivan S, Dixon MF, Gilmour HM, Shepherd NA, Theodossi A, Williams GT. Guidelines for the initial biopsy diagnosis of suspected chronic idiopathic inflammatory bowel disease. The British Society of Gastroenterology Initiative. J Clin Pathol. 1997;50(2):93–105. PubMed PMID: 9155688; PubMed Central PMCID: PMC499731.

Jiang Q, Turner T, Sosa MX, Rakha A, Arnold S, Chakravarti A. Rapid and efficient human mutation detection using a bench-top next-generation DNA sequencer. Hum Mutat. 2012;33:281–9.

Jiang Q, Arnold S, Heanue T, Kilambi KP, Doan B, Kapoor A, Ling AY, Sosa MX, Guy M, Jiang Q, Burzynski G, West K, Bessling S, Griseri P, Amiel J, Fernandez RM, Verheij JB, Hofstra RM, Borrego S, Lyonnet S, Ceccherini I, Gray JJ, Pachnis V, McCallion AS, Chakravarti A. Functional loss of semaphorin 3C and/or semaphorin 3D and their epistatic interaction with ret are critical to Hirschsprung disease liability. Am J Hum Genet. 2015;96(4):581–96. https://doi.org/10.1016/j.ajhg.2015.02.014. PubMed PMID: 25839327; PubMed Central PMCID: PMC4385176.

Jilling T, Simon D, Lu J, Meng FJ, Li D, Schy R, Thomson RB, Soliman A, Arditi M, Caplan MS. The roles of bacteria and TLR4 in rat and murine models of necrotizing enterocolitis. J Immunol. 2006;177(5):3273–82. PubMed PMID: 16920968; PubMed Central PMCID: PMC2697969.

Jonsson T, Johannsson JH, Hallgrimsson JG. Carcinoid tumors of the appendix in children younger than 16 years. A retrospective clinical and pathologic study. Acta Chir Scand. 1989;155:113–6.

Kalcheim C, Langley K, Unsicker K. From the neural crest to chromaffin cells: introduction to a session on chromaffin cell development. Ann N Y Acad Sci. 2002;971:544–6.

Kam MK, Cheung MC, Zhu JJ, Cheng WW, Sat EW, Tam PK, Lui VC. Perturbation of Hoxb5 signaling in vagal and trunk neural crest cells causes apoptosis and neurocristopathies in mice. Cell Death Differ. 2014;21(2):278–89. https://doi.org/10.1038/cdd.2013.142. Epub 2013 Oct 18. PubMed PMID: 24141719; PubMed Central PMCID: PMC3890950.

Karell K, Louka AS, Moodie SJ, Ascher H, Clot F, Greco L, Ciclitira PJ, Sollid LM, Partanen J, European Genetics Cluster on Celiac Disease. HLA types in celiac disease patients not carrying the DQA1*05-DQB1*02 (DQ2) heterodimer: results from the European genetics cluster on celiac disease. Hum Immunol. 2003;64(4):469–77. PubMed PMID: 12651074.

Kashuk CS, Stone EA, Grice EA, Portnoy ME, Green ED, Sidow A, Chakravarti A, McCallion AS. Phenotype-genotype correlation in Hirschsprung disease is illuminated by comparative analysis of the RET protein sequence. Proc Natl Acad Sci U S A. 2005;102(25):8949–54. Epub 2005 Jun 13. PubMed PMID: 15956201; PubMed Central PMCID: PMC1157046.

Kasprzak J, Szaładzińska B, Smoguła M, Ziuziakowski M. Intestinal parasites in stool samples and peri- anal swabs examined by The Voivodeship Sanitary-Epidemiological Station in Bydgoszcz between 2000–2014. Przegl Epidemiol. 2017;71(1):45–54. PubMed PMID: 28654741.

Kassira N, Pedroso FE, Cheung MC, Koniaris LG, Sola JE. Primary gastrointestinal tract lymphoma in the pediatric patient: review of 265 patients from the SEER registry. J Pediatr Surg. 2011;46(10):1956–64. Review. PubMed PMID: 22008334

Kaurah P, Huntsman DG. Hereditary diffuse gastric cancer. In: Adam MP, Ardinger HH, Pagon RA, Wallace SE, Bean LJH, Stephens K, Amemiya A, editors. GeneReviews® [Internet]. Seattle: University of Washington; 2002 Nov 4 [updated 2018 Mar 22]. p. 1993–2018. Available from http://www-ncbi-nlm-nih-gov.login.ezproxy.library.ualberta.ca/books/NBK1139/. PubMed PMID: 20301318.

Kawai H, Satomi K, Morishita Y, Murata Y, Sugano M, Nakano N, Noguchi M. Developmental markers of ganglion cells in the enteric nervous system and their application for evaluation of Hirschsprung disease. Pathol Int. 2014;64(9):432–42. https://doi.org/10.1111/pin.12191. Epub 2014 Aug 22. PubMed PMID: 25146344.

Kelly PJ, Lauwers GY. Gastric polyps and dysplasia. Diagn Histopathol. 2011;17(2):50–61.

Kennedy R, Potter DD, Moir C, Zarroug AE, Faubion W, Tung J. Pediatric chronic ulcerative colitis: does infliximab increase post-ileal pouch anal anastomosis complications? J Pediatr Surg. 2012;47(1):199–203. PubMed PMID: 22244417.

Keyzer-Dekker CM, Sloots CE, Linschoten IK, Biermann K, Meeussen C, Doukas M. Effectiveness of rectal suction biopsy in diagnosing Hirschsprung disease. Eur J Pediatr Surg. 2016;26:100–5.

Kharbanda AB, Dudley NC, Bajaj L, Pediatric Emergency Medicine Collaborative Research Committee of the American Academy of Pediatrics, et al. Validation and refinement of a prediction rule to identify children at low risk for acute appendicitis. Arch Pediatr Adolesc Med. 2012;166(8):738–44.

Khwaja MS, Subbuswamy SG. Ischaemic strictures of the small intestine in northern Nigeria. Trop Gastroenterol. 1984;5(1):41–8. PubMed PMID: 6740759.

Kidd M, Gustafsson BI. Management of gastric carcinoids (neuroendocrine neoplasms). Curr Gastroenterol Rep. 2012;14(6):467–72. https://doi.org/10.1007/s11894-012-0289-x. PubMed PMID: 22976575.

Kilic E, Cetinkaya A, Utine GE, Boduroglu K. A diagnosis to consider in intellectual disability: Mowat-Wilson syndrome. J Child Neurol. 2016;31(7):913–7. https://doi.org/10.1177/0883073815627884. Epub 2016 Jan 25

Kim HS. Lymphomas. In: Encyclopedia of Gastroenterology. USA: Elsevier; 2004.

Kim G, Baik SH, Lee KY, Hur H, Min BS, Lyu CJ, Kim NK. Colon carcinoma in childhood: review of the literature with four case reports. Int J Color Dis. 2013;28(2):157–64. PubMed PMID: 23099637.

Kim J, Bhagwandin S, Labow DM. Malignant peritoneal mesothelioma: a review. Ann Transl

Med. 2017;5(11):236. https://doi.org/10.21037/atm.2017.03.96. Review. PubMed PMID: 28706904; PubMed Central PMCID: PMC5497105.

Kim MJ, Xia B, Suh HN, Lee SH, Jun S, Lien EM, Zhang J, Chen K, Park JI. PAF-Myc-controlled cell stemness is required for intestinal regeneration and tumorigenesis. Dev Cell. 2018;44(5):582–596.e4. https://doi.org/10.1016/j.devcel.2018.02.010. PubMed PMID: 29533773; PubMed Central PMCID: PMC5854208.

Klenn PJ, Iozzo RV. Larsen's syndrome with novel congenital anomalies. Hum Pathol. 1991;22(10):1055–7. PubMed PMID: 1842379.

Klimstra DS, Modlin IR, Coppola D, Lloyd RV, Suster S. The pathologic classification of neuroendocrine tumors: a review of nomenclature, grading, and staging systems. Pancreas. 2010;39(6):707–12. Review. PubMed PMID: 20664470.

Klöppel G, Rindi G, Perren A, Komminoth P, Klimstra DS. The ENETS and AJCC/UICC TNM classifications of the neuroendocrine tumors of the gastrointestinal tract and the pancreas: a statement. Virchows Arch. 2010;456(6):595–7. Epub 2010 Apr 27. PubMed PMID: 20422210.

Knowles BC, Roland JT, Krishnan M, Tyska MJ, Lapierre LA, Dickman PS, Goldenring JR, Shub MD. Myosin Vb uncoupling from RAB8A and RAB11A elicits microvillus inclusion disease. J Clin Invest. 2014;124(7):2947–62. https://doi.org/10.1172/JCI71651. Epub 2014 Jun 2. PubMed PMID: 24892806; PubMed Central PMCID: PMC4071383.

Kovach SJ, Fischer AC, Katzman PJ, Salloum RM, Ettinghausen SE, Madeb R, Koniaris LG. Inflammatory myofibroblastic tumors. J Surg Oncol. 2006;94(5):385–91. Review. PubMed PMID: 16967468.

Kovesi T, Rubin S. Long-term complications of congenital esophageal atresia and/or tracheoesophageal fistula. Chest. 2004;126(3):915–25. Review. PubMed PMID: 15364774.

Kulkarni KP, Sergi C. Appendix carcinoids in childhood: long-term experience at a single institution in Western Canada and systematic review. Pediatr Int. 2013;55(2):157–62. https://doi.org/10.1111/ped.12047. Review. PubMed PMID: 23279208.

Kurppa K, Collin P, Viljamaa M, Haimila K, Saavalainen P, Partanen J, Laurila K, Huhtala H, Paasikivi K, Mäki M, Kaukinen K. Diagnosing mild enteropathy celiac disease: a randomized, controlled clinical study. Gastroenterology. 2009;136(3):816–23. Epub 2008 Nov 24. PubMed PMID: 19111551.

Ladd AP, Grosfeld JL. Gastrointestinal tumors in children and adolescents. Semin Pediatr Surg. 2006;15:37.

Laghi A, Borrelli O, Paolantonio P, Dito L, Buena de Mesquita M, Falconieri P, Passariello R, Cucchiara S. Contrast enhanced magnetic resonance imaging of the terminal ileum in children with Crohn's disease. Gut. 2003;52(3):393–7. PubMed PMID: 12584222; PubMed Central PMCID: PMC1773565.

Lake JI, Heuckeroth RO. Enteric nervous system development: migration, differentiation, and disease. Am J Physiol Gastrointest Liver Physiol. 2013;305:G1–24.

Lam-Himlin D, Montgomery EA. The neoplastic appendix: a practical approach. Diagn Histopathol. 2011;17(9):395–403.

Lamps LW, Madhusudhan KT, Greenson JK, Pierce RH, Massoll NA, Chiles MC, Dean PJ, Scott MA. The role of Yersinia enterocolitica and Yersinia pseudotuberculosis in granulomatous appendicitis: a histologic and molecular study. Am J Surg Pathol. 2001;25(4):508–15. PubMed PMID: 11257626.

Lanza G, Gafà R, Maestri I, Santini A, Matteuzzi M, Cavazzini L. Immunohistochemical pattern of MLH1/MSH2 expression is related to clinical and pathological features in colorectal adenocarcinomas with microsatellite instability. Mod Pathol. 2002;15(7):741–9. PubMed PMID: 12118112.

Larsen Haidle J, Howe JR. Juvenile polyposis syndrome. In: Adam MP, Ardinger HH, Pagon RA, Wallace SE, Bean LJH, Stephens K, Amemiya A, editors. GeneReviews® [Internet]. Seattle: University of Washington; 2003 May 13 [updated 2017 Mar 9]. p. 1993–2018. Available from http://www-ncbi-nlm-nih-gov.login.ezproxy.library.ualberta.ca/books/NBK1469/. PubMed PMID: 20301642.

Lauren P. The two histological main types of gastric carcinoma: diffuse and so-called intestinal-type carcinoma. An attempt at a histo-clinical classification. Acta Pathol Microbiol Scand. 1965;64:31–49. PubMed PMID: 14320675.

Leaphart CL, Cavallo J, Gribar SC, Cetin S, Li J, Branca MF, Dubowski TD, Sodhi CP, Hackam DJ. A critical role for TLR4 in the pathogenesis of necrotizing enterocolitis by modulating intestinal injury and repair. J Immunol. 2007a;179(7):4808–20. PubMed PMID: 17878380.

Leaphart CL, Qureshi F, Cetin S, Li J, Dubowski T, Baty C, Beer-Stolz D, Guo F, Murray SA, Hackam DJ. Interferon-gamma inhibits intestinal restitution by preventing gap junction communication between enterocytes. Gastroenterology. 2007b;132(7):2395–411. Epub 2007 Mar 21. Erratum in: Gastroenterology. 2007 Nov;133(5):1746. Batey, Catherine [corrected to Baty, Catherine]. PubMed PMID: 17570214.

Lebwohl B, Rubio-Tapia A, Assiri A, Newland C, Guandalini S. Diagnosis of celiac disease, gastrointestinal endoscopy clinics of North America. Available online 20 August 2012, ISSN 1052-5157. https://doi.org/10.1016/j.giec.2012.07.004.

Lecerf L, Kavo A, Ruiz-Ferrer M, et al. An impairment of long distance SOX10 regulatory elements underlies isolated Hirschsprung disease. Hum Mutat. 2014;35:303–7.

Lee RG. The colitis of Behçet's syndrome. Am J Surg Pathol. 1986;10(12):888–93. PubMed PMID: 3789253.

Lee RG, Nakamura K, Tsamandas AC, Abu-Elmagd K, Furukawa H, Hutson WR, Reyes J, Tabasco-Minguillan JS, Todo S, Demetris AJ. Pathology of human intestinal transplantation. Gastroenterology. 1996;110(6):1820–34. https://doi.org/10.1053/gast.1996.v110.pm8964408. PMID: 8964408.

References and Recommended Readings

Lefrancois L, Puddington L. Basic aspects of intraepithelial lymphocytic immunobiology. In: Ogra PL, Mestecky J, Lamm ME, editors. Mucosal immunology. San Diego: Academic Press; 1990. p. 413–28.

Lei H, Tang J, Li H, Zhang H, Lu C, Chen H, Li W, Xia Y, Tang W. MiR-195 affects cell migration and cell proliferation by down-regulating DIEXF in Hirschsprung's disease. BMC Gastroenterol. 2014;14:123. https://doi.org/10.1186/1471-230X-14-123. PubMed PMID: 25007945; PubMed Central PMCID: PMC4099404.

Lessells AM, Beck JS, Burnett RA, Howatson SR, Lee FD, McLaren KM, Moss SM, Robertson AJ, Simpson JG, Smith GD, et al. Observer variability in the istopathological reporting of abnormal rectal biopsy specimens. J Clin Pathol. 1994;47(1):48–52. PubMed PMID: 8132809; PubMed Central PMCID: PMC501756.

Lesueur F, Corbex M, McKay JD, Lima J, Soares P, Griseri P, Burgess J, Ceccherini I, Landolfi S, Papotti M, Amorim A, Goldgar DE, Romeo G. Specific haplotypes of the RET proto-oncogene are over-represented in patients with sporadic papillary thyroid carcinoma. J Med Genet. 2002;39(4):260–5. PubMedPMID: 11950855; PubMed Central PMCID: PMC1735081.

Leung AK, Lemay JF. Infantile colic: a review. J R Soc Promot Heal. 2004;124(4):162–6. PubMed PMID: 15301313.

Leung AKC, Leung AAM, Wong AHC, Sergi CM, Kam JKM. Giardiasis: an overview. Recent Patents Inflamm Allergy Drug Discov. 2019;13(2):134–43. https://doi.org/10.2174/1872213X13666190618124901. PubMed PMID: 31210116.

Levine AD, Fiocchi C. Immunology of inflammatory bowel disease. Curr Opin Gastroenterol. 2000;16(4):306–9. PubMed PMID: 17031093.

Levine TS, Tzardi M, Mitchell S, Sowter C, Price AB. Diagnostic difficulty arising from rectal recovery in ulcerative colitis. J Clin Pathol. 1996;49(4):319–23. PubMed PMID: 8655709; PubMed Central PMCID: PMC500459.

Li H, Tang J, Lei H, Cai P, Zhu H, Li B, Xu X, Xia Y, Tang W. Decreased MiR-200a/141 suppress cell migration and proliferation by targeting PTEN in Hirschsprung's disease. Cell Physiol Biochem. 2014;34(2):543–53. https://doi.org/10.1159/000363021. Epub 2014 Aug 8. PubMed PMID: 25116353.

Li Y, Kido T, Garcia-Barcelo MM, Tam PK, Tabatabai ZL, Lau YF. SRY interference of normal regulation of the RET gene suggests a potential role of the Y-chromosome gene in sexual dimorphism in Hirschsprung disease. Hum Mol Genet. 2015;24:685–97.

Lim CH, Dixon MF, Vail A, Forman D, Lynch DA, Axon AT. Ten year follow up of ulcerative colitis patients with and without low grade dysplasia. Gut. 2003;52(8):1127–32. PubMed PMID: 12865270; PubMed Central PMCID: PMC1773763.

Lindsey I, Warren BF, Mortensen NJ. Denonvilliers' fascia lies anterior to the fascia propria and rectal dissection plane in total mesorectal excision. Dis Colon Rectum. 2005;48(1):37–42. PubMed PMID: 15690655.

Liu Y, Lee YF, Ng MK. SNP and gene networks construction and analysis from classification of copy number variations data. BMC Bioinformatics. 2011;12(Suppl 5):S4.

Louw JH. Resection and end-to-end anastomosis in the management of atresia and stenosis of the small bowel. Surgery. 1967;62(5):940–50. PubMed PMID: 4861374.

Louw JH, Barnard CN. Congenital intestinal atresia; observations on its origin. Lancet. 1955;269(6899):1065–7. PubMed PMID: 13272331.

Lucendo AJ, Arias A. Eosinophilic gastroenteritis: an update. Expert Rev Gastroenterol Hepatol. 2012;6(5):591–601. https://doi.org/10.1586/egh.12.42. Review. PubMed PMID: 23061710.

Ludeman L, Shephard NA. Problem areas in the pathology of chronic inflammatory bowel disease. Curr Diagn Pathol. 2006;12:248–60.

Luzón-Toro B, Fernández RM, Torroglosa A, de Agustín JC, Méndez-Vidal C, Segura DI, Antiñolo G, Borrego S. Mutational spectrum of semaphorin 3A and semaphorin 3D genes in Spanish Hirschsprung patients. PLoS One. 2013;8(1):e54800. https://doi.org/10.1371/journal.pone.0054800. Epub 2013 Jan 23. PMID: 23372769; PMCID: PMC3553056.

Luzon-Toro B, Espino-Paisan L, Fernandez RM, Martin-Sanchez M, Antinolo G, Borrego S. Next-generation-based targeted sequencing as an efficient tool for the study of the genetic background in Hirschsprung patients. BMC Med Genet. 2015;16:89.

Lwin T, Melton SD, Genta RM. Eosinophilic gastritis: histopathological characterization and quantification of the normal gastric eosinophil content. Mod Pathol. 2011;24(4):556–63.

Lyle HH. VIII. Linitis plastica (cirrhosis of stomach): with a report of a case cured by gastro-jejunostomy. Ann Surg. 1911;54(5):625–68. PubMed PMID: 17862763; PubMed Central PMCID: PMC1406341.

Lynch KA, Feola PG, Guenther E. Gastric trichobezoar: an important cause of abdominal pain presenting to the pediatric emergency department. Pediatr Emerg Care. 2003;19(5):343–7. Review. PubMed PMID: 14578835.

Magrath I. Malignant non-Hodgkin's lymphoma in children. Principles and practice of pediatric oncology. 4th ed. Pholadelphia: Lippincott, Williams and Wilkins; 2002. p. 661–705.

Mahadevaiah SA, Panjwani P, Kini U, Mohanty S, Das K. Segmental dilatation of sigmoid colon in a neonate: atypical presentation and histology. J Pediatr Surg. 2011;46(3):e1–4. PubMed PMID: 21376178.

Mahmoudi A, Rami M, Khattala K, Elmadi A, Afifi MA, Youssef B. Shah-Waardenburg syndrome. Pan Afr Med J. 2013;14:60.

Majumdar S, Wood P. Congenital central hypoventilation syndrome (CCHS) with Hirschsprung disease (Haddad syndrome): an unusual cause of reduced baseline variability of the fetal heart rate. J Obstet Gynaecol. 2009;29:152–3.

Mansour-Ghanaei F, Herfatkar M, Sedigh-Rahimabadi M, Lebani-Motlagh M, Joukar F. Huge simultaneous trichobezoars causing gastric and small-bowel obstruction. J Res Med Sci. 2011;16(Suppl 1):S447–52. PubMed PMID: 22247733; PubMed Central PMCID: PMC3252776.

Markowitz J, Kahn E, Grancher K, Hyams J, Treem W, Daum F. Atypical rectosigmoid histology in children with newly diagnosed ulcerative colitis. Am J Gastroenterol. 1993;88(12):2034–7. PubMed PMID: 8249970.

Martin LW, Zerella JT. Jejunoileal atresia: a proposed classification. J Pediatr Surg. 1976;11(3):399–403. PubMed PMID: 957064.

Martucciello G, Ceccherini I, Lerone M, Jasonni V. Pathogenesis of Hirschsprung's disease. J Pediatr Surg. 2000;35:1017–25.

Mathur P, Mogra N, Surana SS, Bordia S. Congenital segmental dilatation of the colon with anorectal malformation. J Pediatr Surg. 2004;39(8):e18–20. PubMed PMID: 15300559.

Matta J, Alex G, Cameron DJS, Chow CW, Hardikar W, Heine RG. Pediatric collagenous gastritis and colitis: a case series and review of the literature. J Pediatr Gastroenterol Nutr. 2018;67(3):328–34. https://doi.org/10.1097/MPG.0000000000001975. Review. PubMed PMID: 29601434.

Mattiucci S, Paoletti M, Borrini F, Palumbo M. First molecular identification of the zoonotic parasite Anisakis pegreffii (Nematoda: Anisakidae) in a paraffin-embedded granuloma taken from a case of human intestinal anisakiasis in Italy. BMC Infect Dis. 2011;11(1):82.

McCallion AS, Emison ES, Kashuk CS, et al. Genomic variation in multigenic traits: Hirschsprung disease. Cold Spring Harb Symp Quant Biol. 2003;68:373–81.

McCormick PA, Feighery C, Dolan C, O'Farrelly C, Kelliher P, Graeme-Cook F, Finch A, Ward K, Fitzgerald MX, O'Donoghue DP, et al. Altered gastrointestinal immune response in sarcoidosis. Gut. 1988;29(12):1628–31. PubMed PMID: 3265402; PubMed Central PMCID: PMC1434106.

McDonald GB, Sale GE. The human gastrointestinal tract after allogeneic marrow transplantation. In: Sale GI, Shulman HM, editors. The Pathology of Bone Marrow Transplantation. New York: Masson; 1984. p. 83.

McGill TW, Downey EC, Westbrook J, Wade D, de la Garza J. Gastric carcinoma in children. J Pediatr Surg. 1993;28(12):1620–1. Review. PubMed PMID: 8301513.

McHugh JB, Gopal P, Greenson JK. The clinical significance of focally enhanced gastritis in children. Am J Surg Pathol. 2013;37(2):295–9. https://doi.org/10.1097/PAS.0b013e31826b2a94. PubMed PMID: 23108022.

McLetchie NG, Purves JK, Saunders RL. The genesis of gastric and certain intestinal diverticula and enterogenous cysts. Surg Gynecol Obstet. 1954;99(2):135–41. PubMed PMID: 13187191.

Mellott DO, Burke RD. Divergent roles for Eph and Ephrin in avian cranial neural crest. BMC Dev Biol. 2008;8:56. https://doi.org/10.1186/1471-213X-8-56. PubMed PMID: 18495033; PubMed Central PMCID: PMC2405773.

Mi J, Chen D, Wu M, Wang W, Gao H. Study of the effect of miR124 and the SOX9 target gene in Hirschsprung's disease. Mol Med Rep. 2014;9:1839–43.

Misdraji J, Graeme-Cook FM. Miscellaneous conditions of the appendix. Semin Diagn Pathol. 2004;21(2):151–63.

Miyamoto H, Kurita N, Nishioka M, Ando T, Tashiro T, Hirokawa M, Shimada M. Poorly differentiated neuroendocrine cell carcinoma of the rectum: report of a case and literal review. J Med Investig. 2006;53(3–4):317–20. Review. PubMed PMID: 16953071.

Modi BP, Jaksic T. Pediatric intestinal failure and vascular access. Surg Clin North Am. 2012;92(3):729–43, x. Review. PubMed PMID: 22595718.

Moertel CG, Dockerty MB, Judd ES. Carcinoid tumors of the vermiform appendix. Cancer. 1968;21:270–8.

Moertel CG, Weiland LH, Nagorney DM, Dockerty MB. Carcinoid tumor of the appendix: treatment and prognosis. N Engl J Med. 1987;317:1699–701.

Moertel CL, Weiland LH, Telander RL. Carcinoid tumor of the appendix in the first two decades of life. J Pediatr Surg. 1990;25:1073–5.

Moore SW. Chromosomal and related Mendelian syndromes associated with Hirschsprung's disease. Pediatr Surg Int. 2012;28(11):1045–58.

Moore SW. Total colonic aganglionosis and Hirschsprung's disease: a review. Pediatr Surg Int. 2015;31:1–9.

Moore SW, Zaahl MG. Multiple endocrine neoplasia syndromes, children, Hirschsprung's disease and RET. Pediatr Surg Int. 2008;24:521–30.

Moore SW, Sidler D, Schubert PA. Segmental aganglionosis (zonal aganglionosis or "skip" lesions) in Hirschsprungs disease: a report of 2 unusual cases. Pediatr Surg Int. 2013;29:495–500.

Morini S, Perrone G, Borzomati D, Vincenzi B, Rabitti C, Righi D, Castri F, Manazza AD, Santini D, Tonini G, Coppola R, Onetti Muda A. Carcinoma of the ampulla of Vater: morphological and immunophenotypical classification predicts overall survival. Pancreas. 2012;42(1):60–6. PubMed PMID: 22889982.

Morris CA, Leonard CO, Dilts C, Demsey SA. Adults with Williams syndrome. Am J Med Genet Suppl. 1990;6:102–7.

Morsi A, Abd El-Ghani Ael-G, El-Shafiey M, Fawzy M, Ismail H, Monir M. Clinico-pathological features and outcome of management of pediatric gastrointestinal lymphoma. J Egypt Natl Canc Inst. 2005;17(4):251–9. PubMed PMID: 17102814.

Moscicki AB, Darragh TM, Berry-Lawhorn JM, Roberts JM, Khan MJ, Boardman LA, Chiao E, Einstein MH, Goldstone SE, Jay N, Likes WM, Stier EA, Welton ML, Wiley DJ, Palefsky JM. Screening for anal cancer in women. J Low Genit Tract Dis. 2015;19(3 Suppl 1):S27–42. https://doi.org/10.1097/LGT.0000000000000117. Review. PubMed PMID: 26103446; PubMed Central PMCID: PMC4479419.

Murase M, Miyazawa T, Taki M, Sakurai M, Miura F, Mizuno K, Itabashi K, Toki A. Development of fatty acid calcium stone ileus after initiation of human milk fortifier. Pediatr Int. 2013;55(1):114–6. https://

doi.org/10.1111/j.1442-200X.2012.03630.x. PubMed PMID: 23409991.

Murray SB, Spangler BB, Helm BM, Vergano SS. Polymicrogyria in a 10-month-old boy with Mowat-Wilson syndrome. Am J Med Genet A. 2015;167A:2402–5.

Musshoff K. Klinische Stadieneinteilung der Nicht-Hodgkin-Lymphome [Clinical staging classification of non-Hodgkin's lymphomas (author's transl)]. Strahlentherapie. 1977;153(4):218–21. German. PubMed PMID: 857349.

Mya G. Two observations of congenital dilatation and hypertrophy of the colon. Sperimentale. 1894;48:215–31.

Nadal SR, Calore EE, Manzione CR, Horta SC, Ferreira AF, Almeida LV. Hypertrophic herpes simplex simulating anal neoplasia in AIDS patients: report of five cases. Dis Colon Rectum. 2005;48(12):2289–93. PubMed PMID: 16228826.

Nagashimada M, Ohta H, Li C, Nakao K, Uesaka T, Brunet JF, Amiel J, Trochet D, Wakayama T, Enomoto H. Autonomic neurocristopathy-associated mutations in PHOX2B dysregulate Sox10 expression. J Clin Invest. 2012;122(9):3145–58. https://doi.org/10.1172/JCI63401. Epub 2012 Aug 27. PubMed PMID: 22922260; PubMed Central PMCID: PMC3428093.

Nagy N, Goldstein AM. Endothelin-3 regulates neural crest cell proliferation and differentiation in the hindgut enteric nervous system. Dev Biol. 2006;293:203–17.

Nassar H, Albores-Saavedra J, Klimstra DS. High-grade neuroendocrine carcinoma of the ampulla of vater: a clinicopathologic and immunohistochemical analysis of 14 cases. Am J Surg Pathol. 2005;29(5):588–94. PubMed PMID: 15832081.

Nawa T, Kato J, Kawamoto H, Okada H, Yamamoto H, Kohno H, Endo H, Shiratori Y. Differences between right- and leftsided colon cancer in patient characteristics, cancer morphology and histology. J Gastroenterol Hepatol. 2008;23:418–23. https://doi.org/10.1111/j.1440-1746.2007.04923.x.

Neilson IR, Yazbeck S. Ultrashort Hirschsprung's disease: myth or reality. J Pediatr Surg. 1990;25:1135–8.

Nelson H, Petrelli N, Carlin A, Couture J, Fleshman J, Guillem J, Miedema B, Ota D, Sargent D. National Cancer Institute expert panel. Guidelines 2000 for colon and rectal cancer surgery. J Natl Cancer Inst. 2001;93(8):583–96. PubMed PMID: 11309435.

Nenna R, Magliocca FM, Tiberti C, Mastrogiorgio G, Petrarca L, Mennini M, Lucantoni F, Luparia RP, Bonamico M. Endoscopic and histological gastric lesions in children with celiac disease: mucosal involvement is not only confined to the duodenum. J Pediatr Gastroenterol Nutr. 2012;55(6):728–32. https://doi.org/10.1097/MPG.0b013e318266aa9e. PubMed PMID: 22773062.

Neumann J, Horst D, Kriegl L, Maatz S, Engel J, Jung A, Kirchner T. A simple immunohistochemical algorithm predicts the risk of distant metastases in right-sided colon cancer. Histopathology. 2012;60(3):416–26. https://doi.org/10.1111/j.1365-2559.2011.04126.x. PubMed PMID: 22276605.

Neves GR, Chapchap P, Sredni ST, Viana CR, Mendes WL. Childhood carcinoid tumors: description of a case series in a Brazilian cancer center. Sao Paulo Med J. 2006;124:21–5.

Nguyen T, Park JY, Scudiere JR, Montgomery E. Mycophenolic acid (cellcept and myofortic) induced injury of the upper GI tract. Am J Surg Pathol. 2009;33(9):1355–63. PubMed PMID: 19542873.

Noda Y, Watanabe H, Iida M, Narisawa R, Kurosaki I, Iwafuchi M, Satoh M, Ajioka Y. Histologic follow-up of ampullary adenomas in patients with familial adenomatosis coli. Cancer. 1992;70(7):1847–56. PubMed PMID: 1326395.

Noël JM, Katona IM, Piñeiro-Carrero VM. Sarcoidosis resulting in duodenal obstruction in an adolescent. J Pediatr Gastroenterol Nutr. 1997;24(5):594–8. PubMed PMID: 9161957.

Nostrant TT, Kumar NB, Appelman HD. Histopathology differentiates acute self-limited colitis from ulcerative colitis. Gastroenterology. 1987;92(2):318–28. PubMed PMID: 3792768.

Novak G, Parker CE, Pai RK, MacDonald JK, Feagan BG, Sandborn WJ, D'Haens G, Jairath V, Khanna R. Histologic scoring indices for evaluation of disease activity in Crohn's disease. Cochrane Database Syst Rev. 2017;7(7):CD012351. https://doi.org/10.1002/14651858.CD012351.pub2. PubMed PMID: 28731502; PubMed Central PMCID: PMC6483549.

O'Donnell AM, Puri P. Skip segment Hirschsprung's disease: a systematic review. Pediatr Surg Int. 2010;26:1065–9.

Oberhuber G, Granditsch G, Vogelsang H. The histopathology of coeliac disease: time for a standardized report scheme for pathologists. Eur J Gastroenterol Hepatol. 1999;11(10):1185–94. Review. PubMed PMID: 10524652.

Odze RD, Bines J, Leichtner AM, Goldman H, Antonioli DA. Allergic proctocolitis in infants: a prospective clinicopathologic biopsy study. Hum Pathol. 1993a;24(6):668–74. PubMed PMID: 8505043.

Odze R, Antonioli D, Peppercorn M, Goldman H. Effect of topical 5-aminosalicylic acid (5-ASA) therapy on rectal mucosal biopsy morphology in chronic ulcerative colitis. Am J Surg Pathol. 1993b;17(9):869–75. PubMed PMID: 8352372.

Odze RD, Goldblum J, Noffsinger A, Alsaigh N, Rybicki LA, Fogt F. Interobserver variability in the diagnosis of ulcerative colitis-associated dysplasia by telepathology. Mod Pathol. 2002;15(4):379–86. PubMed PMID: 11950911.

Oster JR, Materson BJ, Rogers AI. Laxative abuse syndrome. Am J Gastroenterol. 1980;74(5):451–8. PubMed PMID: 7234824.

Palefsky JM. Practising high-resolution anoscopy. Sex Health. 2012;9(6):580–6. https://doi.org/10.1071/SH12045. PubMed PMID: 23380236.

Palefsky JM, Holly EA, Ralston ML, Arthur SP, Hogeboom CJ, Darragh TM. Anal cytological abnormalities and anal HPV infection in men with centers for disease control group IV HIV disease. Genitourin

Med. 1997;73(3):174–80. PubMed PMID: 9306896; PubMed Central PMCID: PMC1195816.

Palefsky JM, Holly EA, Efirdc JT, Da Costa M, Jay N, Berry JM, Darragh TM. Anal intraepithelial neoplasia in the highly active antiretroviral therapy era among HIV-positive men who have sex with men. AIDS. 2005;19(13):1407–14. PubMed PMID: 16103772.

Pampiglione S, Rivasi F, Criscuolo M, De Benedittis A, Gentile A, Russo S, Testini M, Villan M. Human anisakiasis in Italy: a report of eleven new cases. Pathol Res Pract. 2002;198(6):429–34. PubMed PMID: 12166901.

Parkes SE, Muir KR, al Sheyyab M, et al. Carcinoid tumours of the appendix in children 1957–1986: incidence, treatment and outcome. Br J Surg. 1993;80:502–4.

Partsch CJ, Siebert R, Caliebe A, Gosch A, Wessel A, Pankau R. Sigmoid diverticulitis in patients with Williams–Beuren syndrome: relatively high prevalence and high complication rate in young adults with the syndrome. Am J Med Genet A. 2005;137:52–4.

Pashankar DS, Israel DM. Gastric polyps and nodules in children receiving long-term omeprazole therapy. J Pediatr Gastroenterol Nutr. 2002;35(5):658–62. PubMed PMID: 12454582.

Pasini B, Ceccherini I, Romeo G. RET mutations in human disease. Trends Genet. 1996;12:138–44.

Patel SG, Ahnen DJ. Familial colon cancer syndromes: an update of a rapidly evolving field. Curr Gastroenterol Rep. 2012;14(5):428–38. https://doi.org/10.1007/s11894-012-0280-6. PubMed PMID: 22864806; PubMed Central PMCID: PMC3448005.

Patel BK, Shah JS. Necrotizing enterocolitis in very low birth weight infants: a systemic review. ISRN Gastroenterol. 2012;2012:562594. PubMed PMID: 22997587.

Patil DT, Goldblum JR, Billings SD. Clinicopathological analysis of basal cell carcinoma of the anal region and its distinction from basaloid squamous cell carcinoma. Mod Pathol. 2013;26(10):1382–9. https://doi.org/10.1038/modpathol.2013.75. Epub 2013 Apr 19. PubMed PMID: 23599161.

Pelet A, Attie T, Goulet O, Eng C, Ponder BA, Munnich A, Lyonnet S. De-novo mutations of the RET proto-oncogene in Hirschsprung's disease. Lancet. 1994;344(8939–8940):1769–70. PubMed PMID: 7997019.

Pelet A, Geneste O, Edery P, Pasini A, Chappuis S, Atti T, Munnich A, Lenoir G, Lyonnet S, Billaud M. Various mechanisms cause RET-mediated signaling defects in Hirschsprung's disease. J Clin Invest. 1998;101(6):1415–23. PubMed PMID: 9502784; PubMed Central PMCID: PMC508697.

Pelet A, de Pontual L, Clément-Ziza M, Salomon R, Mugnier C, Matsuda F, Lathrop M, Munnich A, Feingold J, Lyonnet S, Abel L, Amiel J. Homozygosity for a frequent and weakly penetrant predisposing allele at the RET locus in sporadic Hirschsprung disease. J Med Genet. 2005;42(3):e18. PubMed PMID: 15744028; PubMed Central PMCID: PMC1736014.

Pentiuk S, Putnam PE, Collins MH, Rothenberg ME. Dissociation between symptoms and histological severity in pediatric eosinophilic esophagitis. J Pediatr Gastroenterol Nutr. 2009;48(2):152–60. https://doi.org/10.1097/MPG.0b013e31817f0197. PubMed PMID: 19179876; PubMed Central PMCID: PMC2699182

Perez EA, Gutierrez JC, Jin X, Lee DJ, Rocha-Lima C, Livingstone AS, Franceschi D, Koniaris LG. Surgical outcomes of gastrointestinal sarcoma including gastrointestinal stromal tumors: a population-based examination. J Gastrointest Surg. 2007;11(1):114–25. PubMed PMID: 17390197.

Perito ER, Mileti E, Dalal DH, Cho SJ, Ferrell LD, McCracken M, Heyman MB. Solitary rectal ulcer syndrome in children and adolescents. J Pediatr Gastroenterol Nutr. 2012;54(2):266–70. PubMed PMID: 22094902.

Persad R, Huynh HQ, Hao L, Ha JR, Sergi C, Srivastava R, Persad S. Angiogenic remodeling in pediatric EoE is associated with increased levels of VEGF-A, angiogenin, IL-8, and activation of the TNF-α-NFκB pathway. J Pediatr Gastroenterol Nutr. 2012;55(3):251–60. https://doi.org/10.1097/MPG.0b013e31824b6391. PubMed PMID: 22331014

Phusantisampan T, Sangkhathat S, Phongdara A, Chiengkriwate P, Patrapinyokul S, Mahasirimongkol S. Association of genetic polymorphisms in the RET-protooncogene and NRG1 with Hirschsprung disease in Thai patients. J Hum Genet. 2012;57:286–93.

Pickett LK, Briggs HC. Cancer of the gastrointestinal tract in childhood. Pediatr Clin N Am. 1967;14:223.

Pingault V, Ente D, Dastot-Le Moal F, Goossens M, Marlin S, Bondurand N. Review and update of mutations causing Waardenburg syndrome. Hum Mutat. 2010;31:391–406.

Piotrowska AP, Solari V, Puri P. Distribution of interstitial cells of Cajal in the internal anal sphincter of patients with internal anal sphincter achalasia and Hirschsprung disease. Arch Pathol Lab Med. 2003;127(9):1192–5. PubMed PMID: 12946224.

Pitera JE, Smith VV, Woolf AS, Milla PJ. Embryonic gut anomalies in a mouse model of retinoic acid-induced caudal regression syndrome: delayed gut looping, rudimentary cecum, and anorectal anomalies. Am J Pathol. 2001;159(6):2321–9. PubMed PMID: 11733381; PubMed Central PMCID: PMC1850584.

Plaskett J, Chinnery G, Thomson D, Thomson S, Dedekind B, Jonas E. Rapunzel syndrome: a South African variety. S Afr Med J. 2018;108(7):559–62. https://doi.org/10.7196/SAMJ.2018.v108i7.13115. PubMed PMID: 30004342.

Poddar U. Diagnostic and therapeutic approach to upper gastrointestinal bleeding. Paediatr Int Child Health. 2019;39(1):18–22. https://doi.org/10.1080/20469047.2018.1500226. Epub 2018 Jul 30. Review. Erratum in: Paediatr Int Child Health. 2019 Feb;39(1):80. PubMed PMID: 30058470.

Ponz de Leon M, Sacchetti C, Sassatelli R, Zanghieri G, Roncucci L, Scalmati A. Evidence for the existence of different types of large bowel tumor: suggestions from the clinical data of a population-based registry. J Surg Oncol. 1990;44:35–43.

Prasad KK, Thapa BR, Lal S, Sharma AK, Nain CK, Singh K. Lymphocytic gastritis and celiac disease in Indian children: evidence of a positive relation. J Pediatr Gastroenterol Nutr. 2008;47(5):568–72. PubMed PMID: 18979579.

Prommegger R, Obrist P, Ensinger C, Profanter C, Mittermair R, Hager J. Retrospective evaluation of carcinoid tumors of the appendix in children. World J Surg. 2002;26:1489–92.

Puiman P, Stoll B. Animal models to study neonatal nutrition in humans. Curr Opin Clin Nutr Metab Care. 2008;11(5):601–6. Review. PubMed PMID: 18685456.

Pusch CM, Sasiadek MM, Blin N. Hirschsprung, RET-SOX and beyond: the challenge of examining non-mendelian traits (Review). Int J Mol Med. 2002;10:367–70.

Qualia CM, Katzman PJ, Brown MR, Kooros K. A report of two children with Helicobacter heilmannii gastritis and review of the literature. Pediatr Dev Pathol. 2007;10(5):391–4. PubMed PMID: 17929990.

Rabah R. Pathology of the appendix in children: an institutional experience and review of the literature. Pediatr Radiol. 2007;37(1):15–20. Epub 2006 Oct 10. Review. PubMed PMID: 17031635.

Radhakrishnan KR, Kay M, Wyllie R, Hashkes PJ. Wegener granulomatosis mimicking inflammatory bowel disease in a pediatric patient. J Pediatr Gastroenterol Nutr. 2006;43(3):391–4. PubMed PMID: 16954966.

Raghunath BV, Shankar G, Babu MN, Kini U, Ramesh S, Jadhav V, Aravind KL. Skip segment Hirschsprung's disease: a case report and novel management technique. Pediatr Surg Int. 2014;30(1):119–22. https://doi.org/10.1007/s00383-013-3367-8. Epub 2013 Aug 15. PubMed PMID: 23948815.

Ramani DM, Wistuba II, Behrens C, Gazdar AF, Sobin LH, Albores-Saavedra J. K-ras and p53 mutations in the pathogenesis of classical and goblet cell carcinoids of the appendix. Cancer. 1999;86:14–21.

Raveenthiran V. Knowledge of ancient Hindu surgeons on Hirschsprung disease: evidence from Sushruta Samhita of circa 1200–600 BC. J Pediatr Surg. 2011;46(11):2204–8. https://doi.org/10.1016/j.jpedsurg.2011.07.007. PubMed PMID: 22075360.

Reifen RM, Cutz E, Griffiths AM, Ngan BY, Sherman PM. Tufting enteropathy: a newly recognized clinicopathological entity associated with refractory diarrhea in infants. J Pediatr Gastroenterol Nutr. 1994;18(3):379–85. PubMed PMID: 8057225.

Reissmann M, Ludwig A. Pleiotropic effects of coat colour-associated mutations in humans, mice and other mammals. Semin Cell Dev Biol. 2013;24:576–86.

Reumaux D, Sendid B, Poulain D, Duthilleul P, Dewit O, Colombel JF. Serological markers in inflammatory bowel diseases. Best Pract Res Clin Gastroenterol. 2003;17(1):19–35. Review. PubMed PMID: 12617880.

Rezvani M, Menias C, Sandrasegaran K, Olpin JD, Elsayes KM, Shaaban AM. Heterotopic pancreas: histopathologic features, imaging findings, and complications. Radiographics. 2017;37(2):484–99. https://doi.org/10.1148/rg.2017160091. Review. PubMed PMID: 28287935.

Riccabona M, Rossipal E. Sonographic findings in celiac disease. J Pediatr Gastroenterol Nutr. 1993;17(2):198–200. PubMed PMID: 8229548.

Riddell RH, Goldman H, Ransohoff DF, Appelman HD, Fenoglio CM, Haggitt RC, Ahren C, Correa P, Hamilton SR, Morson BC, Sommers SC, Yardley JH. Dysplasia in inflammatory bowel disease: standardized classification with provisional clinical applications. Hum Pathol. 1983;14(11):931–68. PubMed PMID: 6629368.

Rindi G, Klöppel G, Alhman H, Caplin M, Couvelard A, de Herder WW, Erikssson B, Falchetti A, Falconi M, Komminoth P, Körner M, Lopes JM, McNicol AM, Nilsson O, Perren A, Scarpa A, Scoazec JY, Wiedenmann B, all other Frascati Consensus Conference participants, European Neuroendocrine Tumor Society (ENETS). TNM staging of foregut (neuro)endocrine tumors: a consensus proposal including a grading system. Virchows Arch. 2006;449(4):395–401. Epub 2006 Sep 12. PubMed PMID: 16967267; PubMed Central PMCID: PMC1888719.

Rindi G, Arnold R, Bosman FT, Capella C, Klimstra DS, Klöppel G, Komminoth P, Solcia E. Nomenclature and classification of neuroendocrine neoplasms of the digestive system. In: WHO classification of tumours of the digestive system. 4th ed. Lyon: International Agency for Research on Cancer (IARC); 2010a. p. 13–4.

Rindi G, Inzani F, Solcia E. Pathology of gastrointestinal disorders. Endocrinol Metab Clin N Am. 2010b;39(4):713–27. Review. PubMed PMID: 21095540.

Rivadeneira DE, Tuckson WB, Naab T. Increased incidence of second primary malignancy in patients with carcinoid tumors: case report and literature review. J Natl Med Assoc. 1996;88:310–2.

Roberts DM, Ostapchuk M, O'Brien JG. Infantile colic. Am Fam Physician. 2004;70(4):735–40. PubMed PMID: 15338787.

Roberts RR, Bornstein JC, Bergner AJ, Young HM. Disturbances of colonic motility in mouse models of Hirschsprung's disease. Am J Physiol Gastrointest Liver Physiol. 2008;294:G996–G1008.

Robertson S. FLNB-related disorders. In: Adam MP, Ardinger HH, Pagon RA, Wallace SE, LJH B, Stephens K, Amemiya A, editors. GeneReviews® [Internet]. Seattle: University of Washington; 1993–2020. 2008 Oct 9 [updated 2013 Oct 17]. Available from http://www.ncbi.nlm.nih.gov/books/NBK2534/. PubMed PMID: 20301736.

Robinson HB Jr, Tross K. Agenesis of the cloacal membrane. A probable teratogenic anomaly. Perspect Pediatr Pathol. 1984;8(1):79–96. PubMed PMID: 6701078.

Roed-Petersen K, Erichsen G. The Danish pediatrician Harald Hirschsprung. Surg Gynecol Obstet. 1988;166:181–5. PMID: 3276016.

Rogler LE, Kosmyna B, Moskowitz D, Bebawee R, Rahimzadeh J, Kutchko K, Laederach A, Notarangelo LD, Giliani S, Bouhassira E, Frenette P, Roy-Chowdhury J, Rogler CE. Small RNAs derived from lncRNA RNase MRP have gene-silencing activity relevant to human cartilage-hair hypoplasia. Hum Mol Genet. 2014;23(2):368–82. https://doi.org/10.1093/hmg/ddt427. Epub 2013 Sep 5. PubMed PMID: 24009312; PubMed Central PMCID: PMC3869355.

Rohatiner A, d'Amore F, Coiffier B, Crowther D, Gospodarowicz M, Isaacson P, Lister TA, Norton A, Salem P, Shipp M, et al. Report on a workshop convened to discuss the pathological and staging classifications of gastrointestinal tract lymphoma. Ann Oncol. 1994;5(5):397–400. PubMed PMID: 8075046.

Roka K, Roma E, Stefanaki K, Panayotou I, Kopsidas G, Chouliaras G. The value of focally enhanced gastritis in the diagnosis of pediatric inflammatory bowel diseases. J Crohns Colitis. 2013;7(10):797–802. https://doi.org/10.1016/j.crohns.2012.11.003. Epub 2012 Nov 30. PubMed PMID: 23207168.

Romano C, Oliva S, Martellossi S, Miele E, Arrigo S, Graziani MG, Cardile S, Gaiani F, de'Angelis GL, Torroni F. Pediatric gastrointestinal bleeding: perspectives from the Italian Society of Pediatric Gastroenterology. World J Gastroenterol. 2017;23(8):1328–37. https://doi.org/10.3748/wjg.v23.i8.1328. Review. PubMed PMID: 28293079; PubMed Central PMCID: PMC5330817.

Romeo G, Ronchetto P, Luo Y, Barone V, Seri M, Ceccherini I, Pasini B, Bocciardi R, Lerone M, Kääriäinen H, et al. Point mutations affecting the tyrosine kinase domain of the RET proto-oncogene in Hirschsprung's disease. Nature. 1994;367(6461):377–8. PubMed PMID: 8114938.

Rosen R, Vandenplas Y, Singendonk M, Cabana M, DiLorenzo C, Gottrand F, Gupta S, Langendam M, Staiano A, Thapar N, Tipnis N, Tabbers M. Pediatric gastroesophageal reflux clinical practice guidelines: joint recommendations of the North American Society for Pediatric Gastroenterology, Hepatology, and Nutrition and the European Society for Pediatric Gastroenterology, Hepatology, and Nutrition. J Pediatr Gastroenterol Nutr. 2018;66(3):516–54. https://doi.org/10.1097/MPG.0000000000001889. PubMed PMID: 29470322; PubMed Central PMCID: PMC5958910.

Ruemmele P, Dietmaier W, Terracciano L, Tornillo L, Bataille F, Kaiser A, Wuensch PH, Heinmoeller E, Homayounfar K, Luettges J, Kloeppel G, Sessa F, Edmonston TB, Schneider-Stock R, Klinkhammer-Schalke M, Pauer A, Schick S, Hofstaedter F, Baumhoer D, Hartmann A. Histopathologic features and microsatellite instability of cancers of the papilla of Vater and their precursor lesions. Am J Surg Pathol. 2009;33(5):691–704. PubMed PMID: 19252434.

Ruiz P. How can pathologists help to diagnose late complications in small bowel and multivisceral transplantation? Curr Opin Organ Transplant. 2012;17(3):273–9.

Ruiz P, Bagni A, Brown R, Cortina G, Harpaz N, Magid MS, Reyes J. Histological criteria for the identification of acute cellular rejection in human small bowel allografts: results of the pathology workshop at the VIII international small bowel transplant symposium. Transplant Proc. 2004;36(2):335–7. https://doi.org/10.1016/j.transproceed.2004.01.079. PubMed PMID:15050150.

Ruiz P, Takahashi H, Delacruz V, Island E, Selvaggi G, Nishida S, Moon J, Smith L, Asaoka T, Levi D, Tekin A, Tzakis AG. International grading scheme for acute cellular rejection in small-bowel transplantation: single-center experience. Transplant Proc. 2010;42(1):47–53. https://doi.org/10.1016/j.transproceed.2009.12.026. PubMed PMID: 20172279.

Rusconi B, Good M, Warner BB. The microbiome and biomarkers for necrotizing enterocolitis: are we any closer to prediction? J Pediatr. 2017;189:40–47.e2. https://doi.org/10.1016/j.jpeds.2017.05.075. Epub 2017 Jun 29. PubMed PMID: 28669607; PubMed Central PMCID: PMC5614810.

Rutgeerts P, D'Haens G, Hiele M, Geboes K, Vantrappen G. Appendectomy protects against ulcerative colitis. Gastroenterology. 1994;106(5):1251–3. PubMed PMID: 8174886.

Ruysch F. Observationum anatomico-chirurgicarum centuria, accredit catalogus rariorum quae in museo Ruyschiano asservantur. Amsterdam: Hendrik & Theodoor Boom; 1691.

Rybalov S, Kotler DP. Gastric carcinoids in a patient with pernicious anemia and familial adenomatous polyposis. J Clin Gastroenterol. 2002;35:249–52.

Ryden SE, Drake RM, Franciosi RA. Carcinoid tumors of the appendix in children. Cancer. 1975;36:1538–42.

Saeed A, Barreto L, Neogii SG, Loos A, McFarlane I, Aslam A. Identification of novel genes in Hirschsprung disease pathway using whole genome expression study. J Pediatr Surg. 2012;47:303–7. https://doi.org/10.1016/j.jpedsurg.2011.11.017. PMID: 22325380.

Salomon R, Attie T, Pelet A, et al. Germline mutations of the RET ligand GDNF are not sufficient to cause Hirschsprung disease. Nat Genet. 1996;14:345–7.

Sampson JA. Heterotopic or misplaced endometrial tissue. Am J Obstet Gynecol. 1925;10:649–64.

San Vicente B, Bardaji C, Rigol S, Obiols P, Melo M, Bella R. Retrospective evaluation of carcinoid tumors of the appendix in children. Cir Pediatr. 2009;22:97–9.

Sánchez-Guillén L, Frasson M, García-Granero Á, Pellino G, Flor-Lorente B, Álvarez-Sarrado E, García-Granero E. Risk factors for leak, complications and mortality after ileocolic anastomosis: comparison of two anastomotic techniques. Ann R Coll Surg Engl. 2019;101(8):571–8. https://doi.org/10.1308/rcsann.2019.0098. Epub 2019 Sep 6. PubMed PMID: 31672036; PubMed Central PMCID: PMC6818057.

Sanchez-Mejias A, Fernandez RM, Lopez-Alonso M, Antinolo G, Borrego S. Contribution of RET, NTRK3 and EDN3 to the expression of Hirschsprung disease in a multiplex family. J Med Genet. 2009;46:862–4.

Santin BJ, Prasad V, Caniano DA. Colonic diverticulitis in adolescents: an index case and associated syndromes. Pediatr Surg Int. 2009;25:901–5.

Santos C, López-Doriga A, Navarro M, Mateo J, Biondo S, Martínez Villacampa M, Soler G, Sanjuan X, Paules MJ, Laquente B, Guinó E, Kreisler E, Frago R, Germà JR, Moreno V, Salazar R. Clinico-pathological risk factors of stage II colon cancer: results of a prospective study. Color Dis. 2013;15(4):414–22. https://doi.org/10.1111/codi.12028. PubMed PMID: 22974322.

Santulli TV, Blanc WA. Congenital atresia of the intestine: pathogenesis and treatment. Ann Surg. 1961;154:939–48. PubMed PMID: 14497096; PubMed Central PMCID: PMC1465932.

Sasaki H, Sasano H, Ohi R, Imaizumi M, Shineha R, Nakamura M, Shibuya D, Hayashi Y. Adenocarcinoma at the esophageal gastric junction arising in an 11-year-old girl. Pathol Int. 1999;49(12):1109–13. Review. PubMed PMID: 10632934.

Sato K, Yokouchi Y, Saida Y, Ito S, Kitagawa T, Maetani I. A small cell neuroendocrine carcinoma of the rectum diagnosed by colorectal endoscopic submucosal dissection. J Gastrointestin Liver Dis. 2012;21(2):128. PubMed PMID: 22720298.

Savino F. Focus on infantile colic. Acta Paediatr. 2007;96(9):1259–64. PubMed PMID: 17718777.

Savino F, Tarasco V. New treatments for infant colic. Curr Opin Pediatr. 2010;22(6):791–7. PubMed PMID: 20859207.

Schäppi MG, Staiano A, Milla PJ, Smith VV, Dias JA, Heuschkel R, Husby S, Mearin ML, Papadopoulou A, Ruemmele FM, Vandenplas Y, Koletzko S. A practical guide for the diagnosis of primary enteric nervous system disorders. J Pediatr Gastroenterol Nutr. 2013;57(5):677–86. https://doi.org/10.1097/MPG.0b013e3182a8bb50. Review. PubMed PMID: 24177787.

Schill EM, Lake JI, Tusheva OA, Nagy N, Bery SK, Foster L, Avetisyan M, Johnson SL, Stenson WF, Goldstein AM, Heuckeroth RO. Ibuprofen slows migration and inhibits bowel colonization by enteric nervous system precursors in zebrafish, chick and mouse. Dev Biol. 2016;409(2):473–88. https://doi.org/10.1016/j.ydbio.2015.09.023. Epub 2015 Nov 14. PubMed PMID: 26586201; PubMed Central PMCID: PMC4862364.

Schmocker RK, Lidor AO. Management of non-neoplastic gastric lesions. Surg Clin North Am. 2017;97(2):387–403. https://doi.org/10.1016/j.suc.2016.11.011. Review. PubMed PMID: 28325193.

Schriemer D, Sribudiani Y, IJpma A, Natarajan D, MacKenzie KC, Metzger M, Binder E, Burns AJ, Thapar N, Hofstra RMW, Eggen BJL. Regulators of gene expression in enteric neural crest cells are putative Hirschsprung disease genes. Dev Biol. 2016;416(1):255–65. https://doi.org/10.1016/j.ydbio.2016.06.004. Epub 2016 Jun 4. PubMed PMID: 27266404.

Schulz C, Schütte K, Malfertheiner P. Rare neoplasia of the stomach. Gastrointest Tumors. 2015;2(2):52–60. https://doi.org/10.1159/000435899. Epub 2015 Aug 7. Review. PubMed PMID: 26674659; PubMed Central PMCID: PMC4668797.

Schumacher G, Kollberg B, Sandstedt B. A prospective study of first attacks of inflammatory bowel disease and infectious colitis. Histologic course during the 1st year after presentation. Scand J Gastroenterol. 1994;29(4):318–32. PubMed PMID: 8047806.

Schwartz MG, Sgaglione NA. Gastric carcinoma in the young: overview of the literature. Mt Sinai J Med. 1984;51(6):720–3. PubMed PMID: 6335567.

Sehgal VN, Srivastava G. Trichotillomania +/− trichobezoar: revisited. J Eur Acad Dermatol Venereol. 2006;20(8):911–5. Review. PubMed PMID: 16922936.

Selby WS, Griffin S, Abraham N, Solomon MJ. Appendectomy protects against the development of ulcerative colitis but does not affect its course. Am J Gastroenterol. 2002;97(11):2834–8. PubMed PMID: 12425556.

Seldenrijk CA, Morson BC, Meuwissen SG, Schipper NW, Lindeman J, Meijer CJ. Histopathological evaluation of colonic mucosal biopsy specimens in chronic inflammatory bowel disease: diagnostic implications. Gut. 1991;32(12):1514–20. PubMed PMID: 1773958; PubMed Central PMCID: PMC1379253.

Sepulveda AR. Medscape pathology. The importance of Microsatellite Instability in Colonic Neoplasms (review). http://cme.medscape.com/viewarticle/571610

Sergi C. Hirschsprung's disease: historical notes and pathological diagnosis on the occasion of the 100(th) anniversary of Dr. Harald Hirschsprung's death. World J Clin Pediatr. 2015a;4(4):120–5. https://doi.org/10.5409/wjcp.v4.i4.120. eCollection 2015 Nov 8. Review. PubMed PMID: 26566484; PubMed Central PMCID: PMC4637802.

Sergi C. Eosinophilic esophagitis. J Pediatr Gastroenterol Nutr. 2015b;61(5):529–30. https://doi.org/10.1097/MPG.0000000000000960. PubMed PMID: 26308315.

Sergi C, Shen F, Bouma G. Intraepithelial lymphocytes, scores, mimickers and challenges in diagnosing gluten-sensitive enteropathy (celiac disease). World J Gastroenterol. 2017a;23(4):573–89. https://doi.org/10.3748/wjg.v23.i4.573. Review. PubMed PMID: 28216964; PubMed Central PMCID: PMC5292331.

Sergi CM, Caluseriu O, McColl H, Eisenstat DD. Hirschsprung's disease: clinical dysmorphology, genes, micro-RNAs, and future perspectives. Pediatr Res. 2017b;81(1–2):177–91. https://doi.org/10.1038/pr.2016.202. Epub 2016 Sep 28. Review. PubMed PMID: 27682968.

Sergi C, Hager T, Hager J. Congenital segmental intestinal dilatation: a 25-year review with long-term follow-up at the medical University of Innsbruck. Austria AJP Rep. 2019;9(3):e218–25. https://doi.org/10.1055/s-0039-1693164. Epub 2019 Jul 11.

PubMed PMID: 31304051; PubMed Central PMCID: PMC6624109.

Seri M, Yin L, Barone V, Bolino A, Celli I, Bocciardi R, Pasini B, Ceccherini I, Lerone M, Kristoffersson U, Larsson LT, Casasa JM, Cass DT, Abramowicz MJ, Vanderwinden JM, Kravcenkiene I, Baric I, Silengo M, Martucciello G, Romeo G. Frequency of RET mutations in long- and short-segment Hirschsprung disease. Hum Mutat. 1997;9(3):243–9. PubMed PMID: 9090527.

Seshadri D, Karagiorgos N, Hyser MJ. A case of cronkhite-Canada syndrome and a review of gastrointestinal polyposis syndromes. Gastroenterol Hepatol (N Y). 2012;8(3):197–201. PubMed PMID: 22675284; PubMed Central PMCID: PMC3365525.

Shahar E, Shinawi M. Neurocristopathies presenting with neurologic abnormalities associated with Hirschsprung's disease. Pediatr Neurol. 2003;28:385–91.

Sharan A, Zhu H, Xie H, Li H, Tang J, Tang W, Zhang H, Xia Y. Down-regulation of miR-206 is associated with Hirschsprung disease and suppresses cell migration and proliferation in cell models. Sci Rep. 2015;5:9302. https://doi.org/10.1038/srep09302. Erratum in: Sci Rep. 2016;6:17666. PubMed PMID: 25792468; PubMed Central PMCID: PMC4366810.

Sherman PM, Hassall E, Fagundes-Neto U, Gold BD, Kato S, Koletzko S, Orenstein S, Rudolph C, Vakil N, Vandenplas Y. A global, evidence-based consensus on the definition of gastroesophageal reflux disease in the pediatric population. Am J Gastroenterol. 2009;104(5):1278–95; quiz 1296. doi: 10.1038/ajg.2009.129. Epub 2009 Apr 7. Review. PubMed PMID: 19352345.

Shia J, Klimstra DS, Bagci P, Basturk O, Adsay NV. TNM staging of colorectal carcinoma: issues and caveats. Semin Diagn Pathol. 2012;29(3):142–53. https://doi.org/10.1053/j.semdp.2012.02.001. PubMed PMID: 23062421.

Shidrawi RG, Przemioslo R, Davies DR, Tighe MR, Ciclitira PJ. Pitfalls in diagnosing coeliac disease. J Clin Pathol. 1994;47(8):693–4. PubMed PMID: 7962617; PubMed Central PMCID: PMC502137.

Shimokaze T, Sasaki A, Meguro T, Hasegawa H, Hiraku Y, Yoshikawa T, Kishikawa Y, Hayasaka K. Genotype-phenotype relationship in Japanese patients with congenital central hypoventilation syndrome. J Hum Genet. 2015;60(9):473–7. https://doi.org/10.1038/jhg.2015.65. Epub 2015 Jun 11. PubMed PMID: 26063465.

Shing HP, Wu TT, Hwang B, Chin TW, Wei CF, Tasy SH. Malignant epithelial neoplasm consistent with primitive cystic hepatic neoplasm with mesothelial differentiation: a case report. Zhonghua Yi Xue Za Zhi (Taipei). 1997;59(4):265–8. PubMed PMID: 9216124.

Shulman HM, Kleiner D, Lee SJ, Morton T, Pavletic SZ, Farmer E, Moresi JM, Greenson J, Janin A, Martin PJ, McDonald G, Flowers ME, Turner M, Atkinson J, Lefkowitch J, Washington MK, Prieto VG, Kim SK, Argenyi Z, Diwan AH, Rashid A, Hiatt K, Couriel D, Schultz K, Hymes S, Vogelsang GB. Histopathologic diagnosis of chronic graft-versus-host disease: National Institutes of Health Consensus Development Project on Criteria for Clinical Trials in Chronic Graft-versus-Host Disease: II. Pathology working group report. Biol Blood Marrow Transplant. 2006;12(1):31–47. PubMed PMID: 16399567.

Sieben NL, Macropoulos P, Roemen GM, Kolkman-Uljee SM, Jan Fleuren G, Houmadi R, Diss T, Warren B, Al Adnani M, De Goeij AP, Krausz T, Flanagan AM. In ovarian neoplasms, BRAF, but not KRAS, mutations are restricted to low-grade serous tumours. J Pathol. 2004;202(3):336–40. PubMed PMID: 14991899.

Simpson J, Sundler F, Humes DJ, Jenkins D, Scholefield JH, Spiller RC. Post-inflammatory damage to the enteric nervous system in diverticular disease and its relationship to symptoms. Neurogastroenterol Motil. 2009;21(8):847–e58. Epub 2009 May 14. PubMed PMID: 19453515.

Singh B, Mortensen NJ, Warren BF. Histopathological mimicry in mucosal prolapse. Histopathology. 2007;50(1):97–102. Review. PubMed PMID: 17204024.

Skaba R. Historic milestones of Hirschsprung's disease (commemorating the 90th anniversary of Professor Harald Hirschsprung's death). J Pediatr Surg. 2007;42:249–51. https://doi.org/10.1016/j.jpedsurg.2006.09.024. PMID: 17208575.

Slaughter SR, Hearns-Stokes R, van der Vlugt T, Joffe HV. FDA approval of doxylamine-pyridoxine therapy for use in pregnancy. N Engl J Med. 2014;370(12):1081–3. https://doi.org/10.1056/NEJMp1316042.

Snover DC. Sessile serrated adenoma/polyp of the large intestine: a potentially aggressive lesion in need of a new screening strategy. Dis Colon Rectum. 2011a;54(10):1205–6. https://doi.org/10.1097/DCR.0b013e318228f8bc. PubMed PMID: 21904132.

Snover DC. Update on the serrated pathway to colorectal carcinoma. Hum Pathol. 2011b;42(1):1–10. https://doi.org/10.1016/j.humpath.2010.06.002. Epub 2010 Sep 24. Review. PubMed PMID: 20869746.

Sodhi C, Richardson W, Gribar S, Hackam DJ. The development of animal models for the study of necrotizing enterocolitis. Dis Model Mech. 2008;1(2–3):94–8. https://doi.org/10.1242/dmm.000315. Review. PubMed PMID: 19048070; PubMed Central PMCID: PMC2562191.

Soga J. The term "carcinoid" is a misnomer: the evidence based on local invasion. J Exp Clin Cancer Res. 2009;10(28):15. Review. PubMed PMID: 19208248; PubMed Central PMCID: PMC2657123.

Sonnenberg A, Amorosi SL, Lacey MJ, Lieberman DA. Patterns of endoscopy in the United States: analysis of data from the Centers for Medicare and Medicaid Services and the National Endoscopic Database. Gastrointest Endosc. 2008;67(3):489–96. https://doi.org/10.1016/j.gie.2007.08.041. Epub 2008 Jan 7. PubMed PMID: 18179793.

Soyer T, Talim B, Karnak İ, Ekinci S, Andiran F, Çiftçi AÖ, Orhan D, Akyüz C, Tanyel FC. Surgical treatment of childhood inflammatory myofibroblastic tumors. Eur J Pediatr Surg. 2017;27(4):319–23. https://doi.org/10.1055/s-0036-1593380. Epub 2016 Oct 3. PubMed PMID: 27699733.

Spunt SL, Pratt CB, Rao BN, Pritchard M, Jenkins JJ, Hill DA, Cain AM, Pappo AS. Childhood carcinoid tumors: the St Jude Children's Research Hospital experience. J Pediatr Surg. 2000;35(9):1282–6. PubMed PMID: 10999679.

Stanley AJ, Laine L, Dalton HR, Ngu JH, Schultz M, Abazi R, Zakko L, Thornton S, Wilkinson K, Khor CJ, Murray IA, Laursen SB. International gastrointestinal bleeding consortium. Comparison of risk scoring systems for patients presenting with upper gastrointestinal bleeding: international multicentre prospective study. BMJ. 2017;356:i6432. https://doi.org/10.1136/bmj.i6432. PubMed PMID: 28053181; PubMed Central PMCID: PMC5217768.

Stewart DR, von Allmen D. The genetics of Hirschsprung disease. Gastroenterol Clin N Am. 2003;32(3):819–37, vi. Review. PubMed PMID: 14562576.

Stobdan T, Zhou D, Ao-Ieong E, Ortiz D, Ronen R, Hartley I, Gan Z, McCulloch AD, Bafna V, Cabrales P, Haddad GG. Endothelin receptor B, a candidate gene from human studies at high altitude, improves cardiac tolerance to hypoxia in genetically engineered heterozygote mice. Proc Natl Acad Sci U S A. 2015;112(33):10425–30. https://doi.org/10.1073/pnas.1507486112. Epub 2015 Aug 3. PubMed PMID: 26240367; PubMed Central PMCID: PMC4547246.

Stringer MD. Acute appendicitis. J Paediatr Child Health. 2017;53(11):1071–6. https://doi.org/10.1111/jpc.13737. Epub 2017 Oct 17. Review. PubMed PMID: 29044790.

Strober W, Fuss IJ. Proinflammatory cytokines in the pathogenesis of inflammatory bowel diseases. Gastroenterology. 2011;140(6):1756–67. Review. PubMed PMID: 21530742.

Sullivan JS, Sundaram SS. Gastroesophageal reflux. Pediatr Rev. 2012;33(6):243–53; quiz 254. https://doi.org/10.1542/pir.33-6-243. Review. PubMed PMID: 22659255.

Sullivan JS, Sundaram SS, Pan Z, Sokol RJ. Parenteral nutrition supplementation in biliary atresia patients listed for liver transplantation. Liver Transpl. 2012;18(1):120–8. https://doi.org/10.1002/lt.22444. PubMed PMID: 21987426; PubMed Central PMCID: PMC3245380.

Surawicz CM, Belic L. Rectal biopsy helps to distinguish acute self-limited colitis from idiopathic inflammatory bowel disease. Gastroenterology. 1984;86(1):104–13. PubMed PMID: 6689653.

Szymońska I, Borgenvik TL, Karlsvik TM, Halsen A, Malecki BK, Saetre SE, Jagła M, Kruczek P, Talowska AM, Drabik G, Zasada M, Malecki M. Novel mutation-deletion in the PHOX2B gene of the patient diagnosed with neuroblastoma, Hirschsprung's disease, and congenital central hypoventilation syndrome (NB-HSCR-CCHS) cluster. J Genet Syndr Gene Ther. 2015;6(3):pii: 269. Epub 2015 Sep 7. PubMed PMID: 26798564; PubMed Central PMCID: PMC4718609.

Takawira C, D'Agostini S, Shenouda S, Persad R, Sergi C. Laboratory procedures update on Hirschsprung disease. J Pediatr Gastroenterol Nutr. 2015;60:598–605. https://doi.org/10.1097/MPG.0000000000000679. Review. PubMed PMID: 25564805.

Tam PK. Hirschsprung's disease: a bridge for science and surgery. J Pediatr Surg. 2016;51:18–22.

Tam PK, Garcia-Barcelo M. Genetic basis of Hirschsprung's disease. Pediatr Surg Int. 2009;25:543–58.

Tan H, Wang B, Xiao H, Lian Y, Gao J. Radiologic and clinicopathologic findings of inflammatory myofibroblastic tumor. J Comput Assist Tomogr. 2017;41(1):90–7. https://doi.org/10.1097/RCT.0000000000000444. PubMed PMID: 27224222.

Tanaka M, Riddell RH, Saito H, Soma Y, Hidaka H, Kudo H. Morphologic criteria applicable to biopsy specimens for effective distinction of inflammatory bowel disease from other forms of colitis and of Crohn's disease from ulcerative colitis. Scand J Gastroenterol. 1999;34(1):55–67. PubMed PMID: 10048734.

Tang CS, Sribudiani Y, Miao XP, de Vries AR, Burzynski G, So MT, Leon YY, Yip BH, Osinga J, Hui KJ, Verheij JB, Cherny SS, Tam PK, Sham PC, Hofstra RM, Garcia-Barceló MM. Fine mapping of the 9q31 Hirschsprung's disease locus. Hum Genet. 2010;127(6):675–83. https://doi.org/10.1007/s00439-010-0813-8. Epub 2010 Apr 2. PubMed PMID: 20361209; PubMed Central PMCID: PMC2871095.

Tang CS, Tang WK, So MT, Miao XP, Leung BM, Yip BH, Leon TY, Ngan ES, Lui VC, Chen Y, Chan IH, Chung PH, Liu XL, Wu XZ, Wong KK, Sham PC, Cherny SS, Tam PK, Garcia-Barceló MM. Fine mapping of the NRG1 Hirschsprung's disease locus. PLoS One. 2011;6(1):e16181. https://doi.org/10.1371/journal.pone.0016181. PubMed PMID: 21283760; PubMed Central PMCID: PMC3024406.

Tang W, Qin J, Tang J, Zhang H, Zhou Z, Li B, Geng Q, Wu W, Xia Y, Xu X. Aberrant reduction of MiR-141 increased CD47/CUL3 in Hirschsprung's disease. Cell Physiol Biochem. 2013;32(6):1655–67. https://doi.org/10.1159/000356601. Epub 2013 Dec 5. Erratum in: Cell Physiol Biochem. 2018;48(3):1398. PubMed PMID: 24334875.

Tang W, Li H, Tang J, Wu W, Qin J, Lei H, Cai P, Huo W, Li B, Rehan V, Xu X, Geng Q, Zhang H, Xia Y. Specific serum microRNA profile in the molecular diagnosis of Hirschsprung's disease. J Cell Mol Med. 2014;18(8):1580–7. https://doi.org/10.1111/jcmm.12348. Epub 2014 Jun 28. PubMed PMID: 24974861; PubMed Central PMCID: PMC4190904.

Tang W, Tang J, He J, Zhou Z, Qin Y, Qin J, Li B, Xu X, Geng Q, Jiang W, Wu W, Wang X, Xia Y. SLIT2/ROBO1-miR-218-1-RET/PLAG1: a new disease pathway involved in Hirschsprung's disease. J Cell Mol Med. 2015;19(6):1197–207. https://doi.org/10.1111/

jcmm.12454. Epub 2015 Mar 19. PubMed PMID: 25786906; PubMed Central PMCID: PMC4459835.

Tang W, Cai P, Huo W, Li H, Tang J, Zhu D, Xie H, Chen P, Hang B, Wang S, Xia Y. Suppressive action of miRNAs to ARP2/3 complex reduces cell migration and proliferation via RAC isoforms in Hirschsprung disease. J Cell Mol Med. 2016;20(7):1266–75. https://doi.org/10.1111/jcmm.12799. Epub 2016 Mar 16. Erratum in: J Cell Mol Med. 2018 Oct;22(10):5170. PubMed PMID: 26991540; PubMed Central PMCID: PMC4929290.

Thakkar K, Fishman DS, Gilger MA. Colorectal polyps in childhood. Curr Opin Pediatr. 2012;24(5):632–7. https://doi.org/10.1097/MOP.0b013e328357419f. PubMed PMID: 22890064.

The International Agency for Research on Cancer et al. WHO classification of tumours of the digestive system (IARC WHO classification of tumours). 4th ed. Geneva: World Health Organization; 2010. p. 13–4.

Theodossi A, Spiegelhalter DJ, Jass J, Firth J, Dixon M, Leader M, Levison DA, Lindley R, Filipe I, Price A, et al. Observer variation and discriminatory value of biopsy features in inflammatory bowel disease. Gut. 1994;35(7):961–8. PubMed PMID: 8063225; PubMed Central PMCID: PMC1374845.

Tiwari C, Sandlas G, Jayaswal S, Shah H. Spontaneous intestinal perforation in neonates. J Neonatal Surg. 2015;4(2):14. eCollection 2015 Apr-Jun. PubMed PMID: 26034708; PubMed Central PMCID: PMC4447467.

Torlakovic EE, Gomez JD, Driman DK, Parfitt JR, Wang C, Benerjee T, Snover DC. Sessile serrated adenoma (SSA) vs. traditional serrated adenoma (TSA). Am J Surg Pathol. 2008;32(1):21–9. Erratum in: Am J Surg Pathol. 2008 Mar;32(3):491. PubMed PMID: 18162766.

Tripathy BB. Congenital absence of appendix: a Surgeon's dilemma during surgery for acute appendicitis. J Indian Assoc Pediatr Surg. 2016;21(4):199–201. https://doi.org/10.4103/0971-9261.186555. PMCID: PMC4980886. PMID: 27695217.

Trochet D, O'Brien LM, Gozal D, Trang H, Nordenskjöld A, Laudier B, Svensson PJ, Uhrig S, Cole T, Niemann S, Munnich A, Gaultier C, Lyonnet S, Amiel J. PHOX2B genotype allows for prediction of tumor risk in congenital central hypoventilation syndrome. Am J Hum Genet. 2005;76(3):421–6. Epub 2005 Jan 18. Erratum in: Am J Hum Genet. 2005 Apr;76(4):715. Niemann, Stephan [added]. PubMed PMID: 15657873; PubMed Central PMCID: PMC1196394.

Tye-Din J, Anderson R. Immunopathogenesis of celiac disease. Current Gastroenterol Rep. 2008;10(5):458–65.

Ubeira FM, Anadón AM, Salgado A, Carvajal A, Ortega S, Aguirre C, López-Goikoetxea MJ, Ibanez L, Figueiras A. Synergism between prior Anisakis simplex infections and intake of NSAIDs, on the risk of upper digestive bleeding: a case-control study. PLoS Negl Trop Dis. 2011;5(6):e1214.

Uesaka T, Enomoto H. Neural precursor death is central to the pathogenesis of intestinal aganglionosis in Ret hypomorphic mice. J Neurosci. 2010;30:5211–8.

Umar A, Boland CR, Terdiman JP, Syngal S, Adl C, Ruschoff J, Fishel R, Lindor NM, Burgart LJ, Hamelin R, Hamilton SR, Hiatt RA, Jass J, Lindblom A, Lynch HT, Peltomaki P, Ramsey SD, Rodriguez-Bigas MA, Vasen HFA, Hawk ET, Barrett JC, Freedman AN, Srivastava S. Revised Bethesda guidelines for hereditary nonpolyposis colorectal cancer (Lynch syndrome) and microsatellite instability. J Natl Cancer Inst. 2004;96:261–8.

Ushiku T, Moran CJ, Lauwers GY. Focally enhanced gastritis in newly diagnosed pediatric inflammatory bowel disease. Am J Surg Pathol. 2013;37(12):1882–8. https://doi.org/10.1097/PAS.0b013e31829f03ee. PubMed PMID: 24121177; PubMed Central PMCID: PMC4333144.

Vakil N, van Zanten SV, Kahrilas P, Dent J, Jones R. Global Consensus Group. The Montreal definition and classification of gastroesophageal reflux disease: a global evidence-based consensus. Am J Gastroenterol. 2006;101(8):1900–20; quiz 1943. PubMed PMID: 16928254.

van der Putte SC. Normal and abnormal development of the anorectum. J Pediatr Surg. 1986;21(5):434–40. PubMed PMID: 3712197.

van Eeden S, Offerhaus GJ, Hart AA, Boerrigter L, Nederlof PM, Porter E, van Velthuysen ML. Goblet cell carcinoid of the appendix: a specific type of carcinoma. Histopathology. 2007;51(6):763–73. PubMed PMID: 18042066.

Vandenplas Y, Rudolph CD, Di Lorenzo C, Hassall E, Liptak G, Mazur L, Sondheimer J, Staiano A, Thomson M, Veereman-Wauters G, Wenzl TG. North American Society for Pediatric Gastroenterology Hepatology and Nutrition, European Society for pediatric gastroenterology hepatology and nutrition. Pediatric gastroesophageal reflux clinical practice guidelines: joint recommendations of the North American Society for Pediatric Gastroenterology, Hepatology, and Nutrition (NASPGHAN) and the European Society for Pediatric Gastroenterology, Hepatology, and Nutrition (ESPGHAN). J Pediatr Gastroenterol Nutr. 2009;49(4):498–547. https://doi.org/10.1097/MPG.0b013e3181b7f563. PubMed PMID: 19745761.

Vergouwe FWT, IJsselstijn H, Biermann K, Erler NS, Wijnen RMH, Bruno MJ, Spaander MCW. High prevalence of Barrett's esophagus and esophageal squamous cell carcinoma after repair of esophageal atresia. Clin Gastroenterol Hepatol. 2018;16(4):513–521.e6. https://doi.org/10.1016/j.cgh.2017.11.008. Epub 2017 Nov 11. PubMed PMID: 29133255.

Vignati PV, Welch JP, Cohen JL. Long-term management of diverticulitis in young patients. Dis Colon Rectum. 1995;38:627–9.

Villanacci V, Bassotti G, Liserre B, Lanzini A, Lanzarotto F, Genta RM. Helicobacter pylori infection in patients with celiac disease. Am J Gastroenterol. 2006;101(8):1880–5. Epub 2006 Jun 16. PubMed PMID: 16780559.

Vitale G, Dicitore A, Messina E, Sciammarella C, Faggiano A, Colao A. Epigenetics in medullary thyroid cancer: from pathogenesis to targeted therapy.

Recent Pat Anticancer Drug Discov. 2016;11(3):275–82. Review. PMID: 27306881.

Volante M, Brizzi MP, Faggiano A, La Rosa S, Rapa I, Ferrero A, Mansueto G, Righi L, Garancini S, Capella C, De Rosa G, Dogliotti L, Colao A, Papotti M. Somatostatin receptor type 2A immunohistochemistry in neuroendocrine tumors: a proposal of scoring system correlated with somatostatin receptor scintigraphy. Mod Pathol. 2007;20(11):1172–82. Epub 2007 Sep 14. PubMed PMID: 17873898.

Voltaggio L, Montgomery EA. Histopathology of Barrett esophagus. Diagn Histopathol. 2010;17(2):41–9.

von Herbay A, Herfarth C, Otto HF. Cancer and dysplasia in ulcerative colitis: a histologic study of 301 surgical specimen. Z Gastroenterol. 1994;32(7):382–8. PubMed PMID: 7975773.

von Herbay A, Maiwald M, Ditton HJ, Otto HF. Histology of intestinal Whipple's disease revisited. A study of 48 patients. Virchows Arch. 1996;429(6):335–43. PubMed PMID: 8982377.

Vongbhavit K, Underwood MA. Intestinal perforation in the premature infant. J Neonatal-Perinatal Med. 2017;10(3):281–9. https://doi.org/10.3233/NPM-16148. PubMed PMID: 28854518.

Vyas M, Yang X, Zhang X. Gastric hamartomatous polyps-review and update. Clin Med Insights Gastroenterol. 2016;9:3–10. https://doi.org/10.4137/CGast.S38452. eCollection 2016. Review. PubMed PMID: 27081323; PubMed Central PMCID: PMC4825775.

Walker TJ. Congenital dilatation and hypertrophy of the colon fatal at the age of 11 years. Br Med J. 1893;2:230–1. https://doi.org/10.1136/bmj.2.1700.230. PMID: 20754384.

Walsh MC, Kliegman RM, Fanaroff AA. Necrotizing enterocolitis: a practitioner's perspective. Pediatr Rev. 1988;9(7):219–26. Review. PubMed PMID: 3141910.

Wang KK, Sampliner RE. Practice guidelines updated guidelines 2008 for the diagnosis, surveillance and therapy of Barrett's esophagus. Am J Gastroenterol. 2008;103(3):788–97.

Wang LC, Lee HC, Yeung CY, Chan WT, Jiang CB. Gastrointestinal polyps in children. Pediatr Neonatol. 2009;50(5):196–201. https://doi.org/10.1016/S1875-9572(09)60063-2. PubMed PMID: 19856862.

Wang LL, Fan Y, Zhou FH, Li H, Zhang Y, Miao JN, Gu H, Huang TC, Yuan ZW. Semaphorin 3A expression in the colon of Hirschsprung disease. Birth Defects Res A Clin Mol Teratol. 2011;91(9):842–7. https://doi.org/10.1002/bdra.20837. Epub 2011 Jun 8. PubMed PMID: 21656899.

Wang LL, Zhang Y, Fan Y, Li H, Zhou FH, Miao JN, Gu H, Huang TC, Yuan ZW. SEMA3A rs7804122 polymorphism is associated with Hirschsprung disease in the Northeastern region of China. Birth Defects Res A Clin Mol Teratol. 2012;94(2):91–5. https://doi.org/10.1002/bdra.22866. Epub 2011 Dec 20. PubMed PMID: 22184102.

Wang Y, Wang J, Pan W, Zhou Y, Xiao Y, Zhou K, Wen J, Yu T, Cai W. Common genetic variations in Patched1 (PTCH1) gene and risk of Hirschsprung disease in the Han Chinese population. PLoS One. 2013;8:e75407. https://doi.org/10.1371/journal.pone.0075407. PMID: 24073265.

Washington K, Jagasia M. Pathology of graft-versus-host disease in the gastrointestinal tract. Hum Pathol. 2009;40(7):909–17. https://doi.org/10.1016/j.humpath.2009.04.001. Review. PubMed PMID: 19524102.

Weese-Mayer DE, Berry-Kravis EM, Ceccherini I, Rand CM. Congenital central hypoventilation syndrome (CCHS) and sudden infant death syndrome (SIDS): kindred disorders of autonomic regulation. Respir Physiol Neurobiol. 2008;164:38–48.

Weese-Mayer DE, Rand CM, Berry-Kravis EM, Jennings LJ, Loghmanee DA, Patwari PP, Ceccherini I. Congenital central hypoventilation syndrome from past to future: model for translational and transitional autonomic medicine. Pediatr Pulmonol. 2009;44(6):521–35. https://doi.org/10.1002/ppul.21045. Review. PubMed PMID: 19422034.

Wei X, Sulik KK. Pathogenesis of caudal dysgenesis/sirenomelia induced by ochratoxin A in chick embryos. Teratology. 1996;53(6):378–91.

Weiss SA. Drama in the modern world: plays & essays. Lexington, MA: D.C. Heath & Company; 1974. ISBN 0669831212.

Wessel MA, Cobb JC, Jackson EB, Harris GS Jr, Detwiler AC. Paroxysmal fussing in infancy, sometimes called colic. Pediatrics. 1954;14:421–35.

White WH. On simple ulcerative colitis and other rare intestinal ulcers. Guy Hosp Rep. 1888;30:131–62.

Wijarnpreecha K, Panjawatanan P, Corral JE, Lukens FJ, Ungprasert P. Celiac disease and risk of sarcoidosis: a systematic review and meta-analysis. J Evid Based Med. 2019;12(3):194–9. https://doi.org/10.1111/jebm.12355. Epub 2019 Jun 20. PubMed PMID: 31218829.

Williams CL. Dietary fiber in childhood. J Pediatr. 2006;149:S121–30.

Williams CL, Bollella M, Wynder EL. A new recommendation for dietary fiber in childhood. Pediatrics. 1995;96(5 Pt 2):985–8. PubMed PMID: 7494677.

Wolff-Bar M, Dujovny T, Vlodavsky E, Postovsky S, Morgenstern S, Braslavsky D, Nissan A, Steinberg R, Feinmesser M. An 8-year-old child with malignant deciduoid mesothelioma of the abdomen: report of a case and review of the literature. Pediatr Dev Pathol. 2015;18(4):327–30. https://doi.org/10.2350/14-06-1511-CR.1. Epub 2015 Apr 9. Review. PubMed PMID: 25856259.

Wong HH, Hatcher HM, Benson C, Al-Muderis O, Horan G, Fisher C, Earl HM, Judson I. Desmoplastic small round cell tumour: characteristics and prognostic factors of 41 patients and review of the literature. Clin Sarcoma Res. 2013;3(1):14. https://doi.org/10.1186/2045-3329-3-14. PubMed PMID: 24280007; PubMed Central PMCID: PMC4176496.

Xu C, Chen P, Xie H, Zhu H, Zhu D, Cai P, Huo W, Qin Y, Li H, Xia Y, Tang W. Associations between CYP2B6 rs707265, rs1042389, rs2054675, and

Hirschsprung disease in a Chinese population. Dig Dis Sci. 2015;60(5):1232–5. https://doi.org/10.1007/s10620-014-3450-6. Epub 2014 Nov 26. PubMed PMID: 25424204; PubMed Central PMCID: PMC4427616.

Xu Q, Wu N, Cui L, Wu Z, Qiu G, Filamin B. The next hotspot in skeletal research? J Genet Genomics. 2017;44(7):335–42. https://doi.org/10.1016/j.jgg.2017.04.007. Epub 2017 Jul 6. Review. PubMed PMID: 28739045.

Yanagisawa H, Yanagisawa M, Kapur RP, Richardson JA, Williams SC, Clouthier DE, de Wit D, Emoto N, Hammer RE. Dual genetic pathways of endothelin-mediated intercellular signaling revealed by targeted disruption of endothelin converting enzyme-1 gene. Development. 1998;125(5):825–36. PubMed PMID: 9449665.

Yannopoulos K, Stout AP. Smooth muscle tumors in children. Cancer. 1962;15:958–71. PubMed PMID: 14008985.

Yantiss RK, Farraye FA, O'Brien MJ, Fruin AB, Stucchi AF, Becker JM, Reddy SI, Odze RD. Prognostic significance of superficial fissuring ulceration in patients with severe "indeterminate" colitis. Am J Surg Pathol. 2006;30(2):165–70. PubMed PMID: 16434889.

Young JL Jr, Miller RW. Incidence of malignant tumors in U. S children. J Pediatr. 1975;86(2):254–8. PubMed PMID: 1111694.

Young RH, Gilks CB, Scully RE. Mucinous tumors of the appendix associated with mucinous tumors of the ovary and pseudomyxoma peritonei. A clinicopathological analysis of 22 cases supporting an origin in the appendix. Am J Surg Pathol. 1991;15(5):415–29. PubMed PMID: 2035736.

Yu J, Gonzalez S, Martinez L, Diez-Pardo JA, Tovar JA. Effects of retinoic acid on the neural crest-controlled organs of fetal rats. Pediatr Surg Int. 2003;19(5):355–8. Epub 2003 Jul 24. PubMed PMID: 12898162.

Yüksel Z, Schweizer JJ, Mourad-Baars PE, Sukhai RN, Mearin LM. A toddler with recurrent oral and genital ulcers. Clin Rheumatol. 2007;26(6):969–70. Epub 2006 May 24. PubMed PMID: 16721495.

Zaafar DK, Zaitone SA, Moustafa YM. Role of metformin in suppressing 1,2-dimethylhydrazine-induced colon cancer in diabetic and non-diabetic mice: effect on tumor angiogenesis and cell proliferation. PLoS One. 2014;9(6):e100562. https://doi.org/10.1371/journal.pone.0100562. PMID: 24971882; PMCID: PMC4074064.

Zaramella M, Parigi GB, Rosso R, Maccabruni A. Grade 3 anal intraepithelial neoplasia in an HIV-infected African girl. Pediatr Infect Dis J. 2013;32(3):254–6. https://doi.org/10.1097/INF.0b013e318288f912. PubMed PMID: 23376940.

Zerella JT, Martin LW. Jejunal atresia with absent mesentery and a helical ileum. Surgery. 1976;80(5):550–3. PubMed PMID: 824753.

Zhai LD, Liu J, Li YS, Yuan W, He L. Denonvilliers' fascia in women and its relationship with the fascia propria of the rectum examined by successive slices of celloidin-embedded pelvic viscera. Dis Colon Rectum. 2009;52(9):1564–71. https://doi.org/10.1007/DCR.0b013e3181a8f75c. PubMed PMID: 19690483.

Zhang XN, Zhou MN, Qiu YQ, Ding SP, Qi M, Li JC. Genetic analysis of RET, EDNRB, and EDN3 genes and three SNPs in MCS + 9.7 in Chinese Patients with isolated Hirschsprung disease. Biochem Genet. 2007;45:523–7.

Zhang Z, Jiang Q, Li Q, Cheng W, Qiao G, Xiao P, Gan L, Su L, Miao C, Li L. Genotyping analysis of 3 RET polymorphisms demonstrates low somatic mutation rate in Chinese Hirschsprung disease patients. Int J Clin Exp Pathol. 2015;8(5):5528–34. eCollection 2015. PubMed PMID: 26191260; PubMed Central PMCID: PMC4503131.

Zhou Z, Qin J, Tang J, Li B, Geng Q, Jiang W, Wu W, Rehan V, Tang W, Xu X, Xia Y. Down-regulation of MeCP2 in Hirschsprung's disease. J Pediatr Surg. 2013;48(10):2099–105. https://doi.org/10.1016/j.jpedsurg.2013.07.011. Erratum in: J Pediatr Surg. 2019 Jul;54(7):1516. PubMed PMID: 24094964.

Zhu H, Cai P, Zhu D, Xu C, Li H, Tang J, Xie H, Qin Y, Sharan A, Tang W, Xia Y. A common polymorphism in pre-miR-146a underlies Hirschsprung disease risk in Han Chinese. Exp Mol Pathol. 2014;97(3):511–4. https://doi.org/10.1016/j.yexmp.2014.11.004. Epub 2014 Nov 8. PubMed PMID: 25445498.

Zhu D, Xie H, Li H, Cai P, Zhu H, Xu C, Chen P, Sharan A, Xia Y, Tang W. Nidogen-1 is a common target of microRNAs MiR-192/215 in the pathogenesis of Hirschsprung's disease. J Neurochem. 2015;134(1):39–46. https://doi.org/10.1111/jnc.13118. Epub 2015 May 4. PubMed PMID: 25857602.

Zhuge Y, Cheung MC, Yang R, Koniaris LG. Diagnosing gastrointestinal stromal tumors before the year 2000. Cancer Epidemiol Biomark Prev. 2009;18(3):1013–4; author reply 1014-5. https://doi.org/10.1158/1055-9965.EPI-08-0865. PubMed PMID: 19273490.

Parenchymal GI Glands: Liver

Contents

- 4.1 **Development and Genetics** 426
- 4.2 **Hepatobiliary Anomalies** 430
  - 4.2.1 Ductal Plate Malformation (DPM) 431
  - 4.2.2 Congenital Hepatic Fibrosis 435
  - 4.2.3 Biliary Hamartoma (von Meyenburg Complex) 435
  - 4.2.4 Caroli Disease/Syndrome 435
  - 4.2.5 ADPKD-Related Liver Cysts 435
- 4.3 **Hyperbilirubinemia and Cholestasis** 436
  - 4.3.1 Hyperbilirubinemia 436
  - 4.3.2 Decreased Bilirubin Conjugation and Unconjugated Hyperbilirubinemia 437
  - 4.3.3 Conjugated Hyperbilirubinemia 438
- 4.4 **Infantile/Pediatric/Youth Cholangiopathies** 439
  - 4.4.1 Biliary Atresia 440
  - 4.4.2 Non-BA Infantile Obstructive Cholangiopathies (NBAIOC) 443
  - 4.4.3 The Paucity of the Intrahepatic Biliary Ducts (PIBD) 444
  - 4.4.4 Neonatal Hepatitis Group (NAG) 444
  - 4.4.5 Primary Sclerosing Cholangitis (PSC) 456
  - 4.4.6 Primary Biliary Cirrhosis (PBC) 458
  - 4.4.7 Pregnancy-Related Liver Disease (PLD) 459
- 4.5 **Genetic and Metabolic Liver Disease** 460
  - 4.5.1 Endoplasmic Reticulum Storage Diseases (ERSDs) 461
  - 4.5.2 Congenital Dysregulation of Carbohydrate Metabolism 464
  - 4.5.3 Lipid/Glycolipid and Lipoprotein Metabolism Disorders 467
  - 4.5.4 Amino Acid Metabolism Disorders 471
  - 4.5.5 Mitochondrial Hepatopathies 471
  - 4.5.6 Peroxisomal Disorders 477
  - 4.5.7 Iron Metabolism Dysregulation 479
  - 4.5.8 Copper Metabolism Dysregulation 483
  - 4.5.9 Porphyria-Related Hepatopathies 485
  - 4.5.10 Shwachman-Diamond Syndrome (SDS) 485
  - 4.5.11 Chronic Granulomatous Disease (CGD) 485
  - 4.5.12 Albinism-Related Liver Diseases 486
- 4.6 **Viral and AI Hepatitis, Chemical Injury, and Allograft Rejection** 486
  - 4.6.1 Acute Viral Hepatitis 486

© Springer-Verlag GmbH Germany, part of Springer Nature 2020
C. M. Sergi, *Pathology of Childhood and Adolescence*,
https://doi.org/10.1007/978-3-662-59169-7_4

4.6.2	Chronic Viral Hepatitis	488
4.6.3	Autoimmune Hepatitis	489
4.6.4	Drug-Induced Liver Disease (Chemical Injury) and TPN	490
4.6.5	Granulomatous Liver Disease	495
4.6.6	Alcoholic Liver Disease	496
4.6.7	Non-alcoholic Steatohepatitis	498
4.6.8	Acute and Chronic Rejection Post-Liver Transplantation	498
4.7	**Hepatic Vascular Disorders**	503
4.7.1	Acute and Chronic Passive Liver Congestion	503
4.7.2	Ischemic Hepatocellular Necrosis	505
4.7.3	Shock-Related Cholestasis	505
4.8	**Liver Failure and Liver Cirrhosis**	505
4.9	**Portal Hypertension**	508
4.10	**Bacterial and Parasitic Liver Infections**	509
4.10.1	Pyogenic Abscess	509
4.10.2	Helminthiasis	510
4.11	**Liver Tumors**	510
4.11.1	Benign Tumors	510
4.11.2	Malignant Tumors	517
4.11.3	Metastatic Tumors	534
Multiple Choice Questions and Answers		538
References and Recommended Readings		540

4.1 Development and Genetics
(Figs. 4.1 and 4.2)

The liver is the most important gland of the organism. The liver has an enormous reserve and regenerative capacity (to recall the Prometheus legend), but the regenerated livers often have abnormal biliary connections. Hepatocytic nuclear size has a various range with ploidy ranging up to octoploid (8n) with significant anisonucleosis. We distinguish three types of lobules. Classical lobule with a hexagonal shape with a central vein in the middle, portal lobule with a triangular shape with a portal tract in the middle + three central veins, and an acinar (Rappaport) lobule, which has a diamond shape with two portal tracts + two central veins. The Rappaport zones are Zone 1 (Rappaport), periportal (portal tract) ⇒ toxin-/drug-sensitive; Zone 2 (Rappaport), intermediate; and zone 3 (Rappaport), pericentral (central vein) ⇒ ischemia-sensitive. Portal tract/portal space contains the portal vein, portal hepatic artery, and interlobular bile duct (1 IBD per artery) embedded in connective tissue with lymphatics and nerves. The limiting plate is the periportal rim of hepatocytes surrounding the portal tract. Acinar cords thickness (θ): two cells in newborns and regenerating liver and one cell in adults. The histochemical special stain reticulin is useful to distinguish the thickness of the hepatocytes (*muralium*). This continuous system or *muralium* of anastomosing hepatic plates (*laminae hepatis*) and sinusoidal spaces ("*labyrinthus hepatis*") of Elias is an important milestone in the anatomical structure of the liver. It has given way to an organization of well-demarcated areas of liver parenchyma tightly associated with supply venules and draining biliary channels that follow a regular branching pattern as indicated briefly above. Sinusoids: Intercordal and lined by discontinuous, fenestrated endothelium. Lymphatics: Portal tract with flow starting from perisinusoidal space (Disse space), periportal space (Mall space), and portal tract lymphatics up to larger lymphatic ducts. Bile flow: Canaliculi → canals of Herring → portal bile ducts, left and right hepatic ducts, common hepatic bile duct → joining with cystic duct → choledocus → ampulla of Vater draining in the duodenum. Fe accumulation is mainly periportal, while lipofuscin is primarily pericentral.

4.1 Development and Genetics

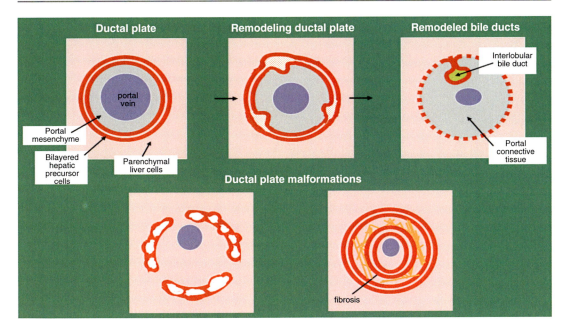

Fig. 4.1 In the upper row are shown the three different phases of the primitive intrahepatic biliary system with ductal plate, including ductal plate, remodeling ductal plate, and remodeled bile ducts. In the lower row are shown the two main types of ductal plate malformation (DPM), including type I and type II, respectively. Type I is characterized by impressive dilated embryonic peripheral biliary structures and no or mild fibrosis, while type II is characterized by one or two circles of the primitive ductal plate without obvious dilatation of the embryonic peripheral structures and marked fibrosis. The biliary proliferation of the type I-DPM differs usually from the ductal biliary proliferations seen in obstructive cholangiopathies, such as biliary atresia

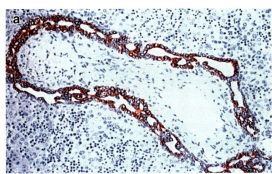
Type I-DPM

Type II-DPM

Fig. 4.2 In (**a**) and (**b**) are shown the two main types of ductal plate malformation (DPM), including type I and type II, respectively. Type I is characterized by impressive dilated embryonic peripheral biliary structures and no or mild fibrosis, while type II is characterized by one or two circles of the primitive ductal plate without obvious dilatation of the embryonic peripheral structures and marked fibrosis. The biliary proliferation of the type I-DPM differs usually from the ductal biliary proliferations seen in obstructive cholangiopathies, such as biliary atresia

The canals of Herring are among cells intermediate between hepatocytes and biliary cells. Bile flow is energy-dependent with active transport. Kupffer cells belong to the monocyte/phagocyte system in sinusoids and are highlighted by the CAM5.6 stain. Perisinusoidal Ito cells are fat-containing mesenchymal cells located in the space of Disse with the function to store vitamin A. In Box 4.1 is indicated the most common immunophenotype in a liver (child and adult). Box 4.2 classify the type of liver biopsy.

Intravital evaluation of the liver may be carried out using noninvasive techniques such as

> **Box 4.1 ICH Panel**
> - CK7: Late (fetal) biliary cell marker and permanent cholangiocyte marker
> - CK19: Early (fetal) biliary cell marker and permanent cholangiocyte marker
> - CK 8: Hepatocyte marker (fetal, child, adult)
> - CK18: Hepatocyte marker (fetal, child, adult)
> - CD68 (KP-1) → Kupffer cells
> - CAM5.2 (LMW-CK) → interlobular bile ducts
> - pCEA / CD10 → bile canalicular marker
>
> Notes: CAM5.2 represents primary immunoreactivity with CK-8 and CK-18 as well as weaker but distinct immunoreactivity with CK-7.

> **Box 4.2 Type of Biopsy**
> - Percutaneous vs. wedge (indications and risks) liver biopsy

ultrasound, computed tomography, or magnetic resonance imaging or invasive procedures such as biopsy.

During the perinatal period, the liver is still immature, and maturation changes occur in childhood. Liver diseases of acquired type, as seen in adults, are rare in children but may occur in youth. Understanding the development of the biliary system is crucial to interpreting categories of neonatal and infantile cholangiopathies.

Bile Duct Development
To properly understand several congenital and infantile disorders of the biliary system it is opportune applying a common terminology and system to analyze the fetal liver systematically. In this way, three stages summarize the development of the biliary system. Intrahepatic bile duct formation initiates at the liver hilum, where it proceeds toward the periphery accompanying the portal vein ramification between 6th and 9th post-ovulation week. Bipotential progenitor cells, supposed to be able of differentiation into either hepatocytes or bile duct cell types, latter through contact with the portal vein surrounding mesenchyme, contribute to the formation of the first stage, so-called *ductal plate* (DP). The ductal plate is formed out of a previously mono- and then double-layered cell cylinder with a slit-like lumen. In a sequence of development, the continuous migration of biliary cells from the ductal plate, into the mesenchyme toward the portal vein, characterizes the stage of *remodeling ductal plate* (RDP) in the 13th to 17th gestational week. Through gradual apoptosis of non-migrating cells, the disappearance of the ductal plate is granted, and the final stage of regular *remodeled bile ducts* (RBD) can be reached. The development of the intrahepatic bile ducts proceeds from the hilum to peripheral portions, and ≥2 are often present in the same specimen. Thus, an experienced fetal liver pathologist is key in identifying properly the stage of development of the intrahepatic biliary system. Ductal plate malformation is the result of the developmental disruption, which manifests itself through ductal plate persistence or incomplete disappearance; the excessive ramification of the leading portal vein, so-called *"pollard willow"* pattern, has been investigated as an etiological factor in the past. Two groups of congenital diseases of the intrahepatic bile ducts should draw our attention at this time: (I) illnesses defined by necro-inflammatory destruction of intrahepatic bile ducts and fibrotic changes, including biliary atresia and the lack of the intrahepatic bile ducts, and (II) conditions with characteristic ectasia of the intrahepatic bile ducts accompanied by fibrosis, both in variable degrees. These can also be called fibrocystic disorders, which can be typified with Meckel-Gruber syndrome, Caroli disease, and congenital hepatic fibrosis. The extrahepatic biliary tree includes the common hepatic duct emerging from the combined right and left hepatic ducts and continuing in the common hepatic duct after the fusion with the cystic duct. In the gallbladder, herniation of the mucosa into the smooth muscle or subserosa is quite common and takes the name of Rokitansky-Aschoff sinuses. Serosa covers the adventitia on the ventral side. On the hepatic side, bile ducts (aka *Luschka ducts*) are identified in the connective tissue of the subse-

rosa. A single layer of columnar epithelial cells similar to the gallbladder lines the extrahepatic biliary tree. The epithelium invaginates into the stroma forming saccular structures, aka *sacculi of Beale*, which are surrounded by mucinous glands. The dense subepithelial stroma of the common bile duct shows, commonly, no muscle fibers apart of the intrahepatic and intraduodenal portion. The patterns of keratin 7 (K7, CK7) expressing biliary structures of liver biopsies of infants ≤1 year with follow-up data have been investigated retrospectively. Although the term "keratins" has been proposed internationally, the term "cyto-keratins" is still in use in pathology. Thus, this book use both terms interchangeably. There are specific immunohistochemical patterns in biliary atresia, neonatal hepatitis (a category of inflammation of the liver probably including several conditions showing lobular disarray of the hepatocytes and giant cell transformation of the liver cells as well as the presence of extramedullary hematopoiesis), and the scarcity of the PIBD that have been characterized (Sergi et al. 2008) (*vide infra*). Keratins are intermediate filaments of the cytoskeleton. Ductal plate remnants have been verified to recapitulate the primitive stages of the intrahepatic biliary system. We found that the lack of intrahepatic interlobular bile ducts in infants aged <1 year is an adverse prognostic factor, which was independent of the etiology of neonatal liver disease (Sergi et al. 2008). This data has been reinforced by the expression of fibrocystin (or polyductin), the gene product of the ARPKD. There are numerous RBDs in cholangiopathies keen to be surgically corrected. Our understanding of the physiology of bile secretion showed progress in the understanding of the infantile cholangiopathies. Defects of the intrahepatic biliary system are not only a perinatal and infantile problem but have been associated with the development of cholangiocellular carcinoma (CCA), a malignant epithelial tumor of the liver, in the long term. There is indeed a 15% risk of malignant transformation approximately after the second decade of life, at an average age of 25 years, although pediatricians and pediatric gastroenterologists have occasionally observed teenagers with biliary cancer. The overall incidence of CCA in patients with untreated cysts is up to ¼ of individuals harboring this developmental anomaly. The mechanism of carcinogenesis may rely on biliary stasis, reflux of pancreatic juice causing chronic inflammation, activation of bile acids, and abnormal deconjugation of carcinogens. In fact, single nucleotide polymorphisms (SNPs) of the bile salt transporter protein in *BSEP*, *FIC1*, and *MDR3* genes can lead to unstable bile content and, then, to deconjugation of xenobiotics, which are previously conjugated in the liver. It is important to emphasize that in the background of congenital bile duct abnormalities, a genetic cross-check is crucial. The background may lead to the development of malignancy at an early stage in life and this information needs to be communicated between the pediatric gastroenterologist and the adult gastroenterologist, who is going to follow-up the patient when he or she is out of the Children's Hospital. In fact, it has been documented that individuals who are heterozygous for bile salt transporter polymorphisms have an increased predisposition to CCA as adults. The interacting exposure to cofactors that result in chronic inflammation of the biliary tree is still unknown (Box 4.3).

Box 4.3 Pearls and Pitfalls!

To highlight biliary structures and differentiate them from hepatocytes, two monoclonal antibodies may be used, including one against *CK7* (e.g., clone OV-TL 12/30, 1:50, Dako Corporation, Hamburg, Germany) and *CK19* (e.g., clone RCK 108, 1:100, Dako Corporation, Hamburg, Germany). The mixture of cytokeratin epitopes is often used to highlight both biliary cells and hepatocytes (*AE1 + AE3*). AE1 reacts with specific "Group A acidic cytokeratins" with a molecular weight of 40,000–50,000 (CK10, CK14, CK15, CK16, CK19), while AE3 recognizes all eight "Group B basic cytokeratins" with a molecular weight of 58,000–67,000 (CK1–CK8). Although the term "keratins" has been proposed internationally, the term "cyto-keratins" is still in use in pathology. Thus, this book use both terms interchangeably.

Regarding quantitative analysis, peripheral tubular structures (PTS) may be defined as ductular structures of biliary cells migrating into the intraportal mesenchyme but located at the periphery of the portal tract and partially in contact with the ductal plate. Remodeled bile ducts may be defined as mainly round, centrally located ductular structures in the portal tracts without any contact with the ductal plate. In the past, we studied portal tract parameters of the liver development during fetal age (Sergi et al. 2000). We targeted the portal tract perimeter, cross-sectional area, longest axis, and several parameters of the remodeling ductal plate. Through an orderly process of selection and deletion, the first fetal biliary structure or ductal plate is remodeled during intrauterine life into the system of bile ducts. Our morphometric data indicate that a period of slow-down of the progressive ramification of the intrahepatic biliary tree occurs between the 21st week and the 32nd week of human gestation. The *interim* hematopoietic function of the liver and, in a unique way, the intraportal granulopoiesis may play a role. Liver erythropoiesis dominates between 12 weeks of gestation until the beginning of the third trimester (25th week) when 50% of the blood cells are formed in the liver and 50% in the bone marrow. However, until 32 weeks of pregnancy, the liver plays a significant role in both the hematopoiesis and the maternofetal exchange. Only at this time do the hematopoietic cells in the liver begin to form islands out of a previously diffuse distribution. After birth, the bone marrow becomes the primary site of production of both red and white cells series. Sohn et al. (1993) investigated the eosinophilic granulopoiesis in human fetal liver from 5 to 34 weeks of gestation. Eosinophilic granulopoiesis preceded both erythropoiesis and megakaryopoiesis in the embryonic liver. Eosinophilic granulopoiesis develops more actively in the portal areas than in the hepatic laminae mainly between the 20th week and the 32nd week of gestation. Costa et al. (1998) suggested similarly a period of slow-down of the development of bile ducts occurs between the 20th week and 32nd week of gestation using semiquantitative analysis only. Our both morphometric and quantitative study support that in this period effectively the maturation of the intrahepatic biliary system does slow. Many granulopoietic cells were observed in the portal tract close to the limiting plate. Moreover, a BD/PT ratio equal to or a value higher than 0.5 was seen in 9 of 11 subjects, who were at least 35 weeks gestation old. Three individuals had a BD/PT ratio equal to or a value higher than 0.9 at the 38th, 39th, and 40th week of pregnancy. A diagnosis of scarcity of interlobular bile ducts should be made with extreme caution and always considering the gestational age of the subject. Blind evaluation of liver biopsy specimens showed that BA occurs more frequently in categories B) BD/PT = 0 and C) 0.1< BD/PT <0.5. The ductular proliferation in infants with BA is commonly encompassed in the enlarged and inflamed portal tracts where ongoing bile duct destruction may take place. BA affects the development of both the intra- and extrahepatic biliary tree and results in the progressive fibrotic obstruction of the preformed bile ducts. Analysis of our original data produced a median survival rate of 525 days without liver transplantation. If the categories and not the original diagnoses are taken into consideration, a poor median survival rate in infants with a bile duct to portal tract ratio of 0 (category B) is consistently found (252 days). This aspect is probably the most remarkable data following the identification of the ductal plate malformation. Although it may be considered reasonable, the double blind evaluation consolidated this data. In several institutions the null biopsies at time of the liver transplantation is crucial and the pathological report including the BD/PT ratio may have striking importance for the post-transplantation follow-up. To date, the sensitivity of liver biopsy in diagnosing BA is higher than 90%, and the specificity approaches 80%. The variety of presentation and the associations with other congenital disabilities has suggested the BA is a heterogeneous disorder.

4.2 Hepatobiliary Anomalies

Hepatobiliary and pancreatic aberrations are central in the dysontogenesis of the human body, and the identification of some animal models

4.2 Hepatobiliary Anomalies

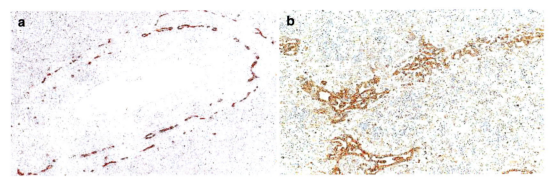

Fig. 4.3 In (**a**) ductal plate malformation of type 2 with more fibrosis and less dilatation of the biliary structures, while (**b**) shows cystic biliary structures at the periphery of the portal tract and less fibrosis, which correspond to the type 1 of DPM. In (**b**), BCL2 is highly expressed allowing the structure to stay vital and inhibit the apoptosis

and the progress of molecular biology techniques allowed us to better understand the dysregulation of these processes (Box 4.4) (Figs. 4.3, 4.4, 4.5, 4.6, and 4.7).

> **Box 4.4 Hepatic, Biliary, and Pancreatic Structural Anomalies**
> 4.2.1. *Ductal Plate Malformation* (DPM)
> 4.2.2. *Congenital Hepatic Fibrosis* (CHF)
> 4.2.3. *Biliary Hamartoma (von Meyenburg Complex)*
> 4.2.4. *Caroli Disease/Syndrome*
> 4.2.5. *ADPKD-Related Liver Cysts*

4.2.1 Ductal Plate Malformation (DPM)

We should distinguish several congenital conditions of abnormality of the biliary system because they have different outcomes. Ductal plate malformation (DPM or better ductal plate malformations), aka hepatic fibrocystic diseases, is a group of congenital disorders resulting from abnormal embryogenesis of the biliary ductal system. The abnormalities include *syndromic DPM*, *congenital hepatic fibrosis* (CHF), *choledochal cyst*, *Caroli disease* or *Caroli syndrome*, *cysts associated with the autosomal dominant polycystic liver disease* (ADPKD), and *biliary hamartoma* or von Meyenburg complex. The hepatic lesions can be associated with renal anomalies such as autosomal recessive polycystic kidney disease (ARPKD), medullary sponge kidney, and nephronophthisis. Understanding the embryology and pathogenesis of the ductal plate is central to diagnose accurately ductal plate malformations, which in turn is relevant for both clinical management and genetic counseling.

Ductal plate malformations (DPM) Morphogenesis and pathology of DPM have been intensely researched preciously in the field of congenital diseases. Since DPMs are linked to embryologic intrahepatic bile duct development, it was of great importance to achieve greater knowledge about the maturation of these structures, which has been subjecting of many previous thorough studies. Meckel-Gruber syndrome (MGS) or Meckel syndrome follows an autosomal recessive inheritance pattern and is physically expressed by occipital encephalocele, postaxial polydactyly, diffuse cystic renal dysplasia, and malformation of the ductal plate of the liver (Sergi et al. 2000). Other than MGS, there are several syndromes associated with DPM. Three underlying mechanisms may result to be behind the DPM according to Raynaud (2011). They include (1) abnormal differentiation of hepatoblasts to ductal plate cells with perturbation of cell polarization and abnormal lumen formation, (2) abnormal duct expansion after a correct ductal plate cell differentiation and maturation of primitive ductal structures, and (3) failure of maturation of the nearly remodeled interlobular bile ducts.

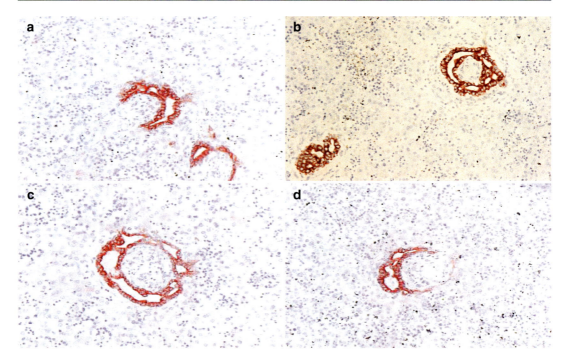

Fig. 4.4 In (**a**) through (**d**) are shown several forms of DPM, including glomeruloid (**a, b**) and polypoid lesions (**c, d**). An immunohistochemical procedure was used to reach these results. We used antibodies against the cytokeratins CK7, CK19, and AE1–AE3 (pan-keratins), although the antibody against keratin 19 is highly reliable to highlight the protoforms of the intrahepatic biliary system

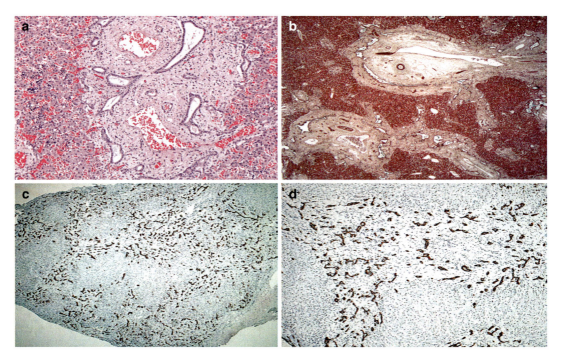

Fig. 4.5 In this panel are shown the morphological patterns of ARPKD, congenital hepatic fibrosis (**a-b**), and the CK7-abnormal patterns encountered during the microscopic evaluation of the portal tracts (**c-f**). The microphotographs (**a-b**) are H&E (x), while (**c-f**) are anti-cytokeratin 7 immunostained slides (x)

4.2 Hepatobiliary Anomalies

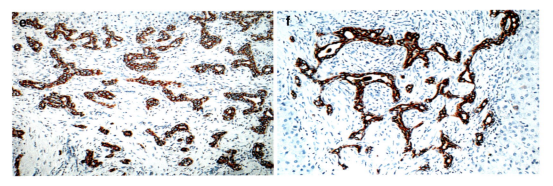

Fig. 4.5 (continued)

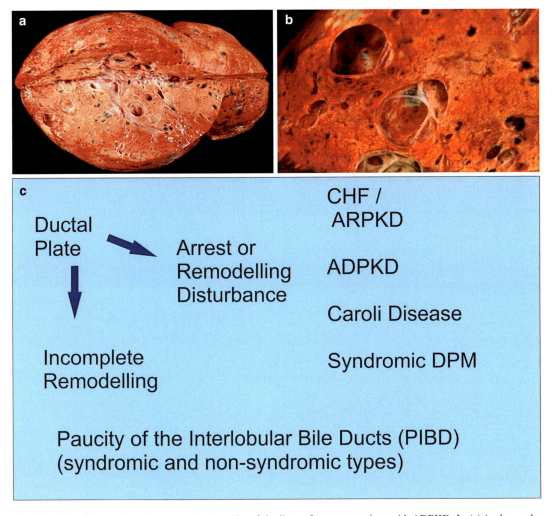

Fig. 4.6 In (**a**, **b**) are shown the gross photographs of the liver of a young patient with ADPKD. In (**c**) is shown the simplified etiopathogenesis of the ductal plate malformations

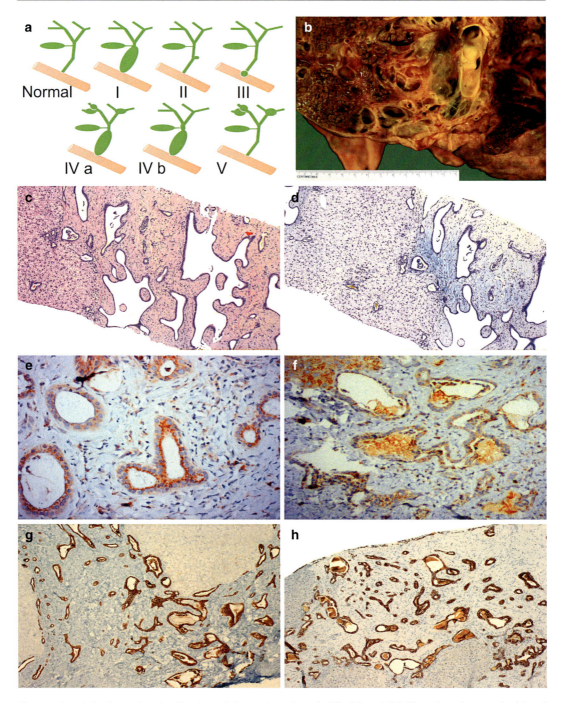

Fig. 4.7 In (**a**) is shown the classification of the cystic abnormalities of the extrahepatic biliary tract. In (**b**) is shown the liver of a patient with Caroli disease, while the histology of Caroli disease is shown in the microphotographs (**c**) through (**h**). The antibody against cytokeratin 7 revealed a cystic dilation of the segmental bile ducts in (**d**), (**g**), and (**h**). The microphotographs (**e**) and (**f**) show the intrahepatic biliary system harboring an overexpression of FP2 an antigen linked to ARPKD (fibrocystin). In this book as well as in most of the authors in the scientific literature cytokeratin 7 is interchangeable with keratin 7

In practice, the diameter of malformed ducts may determine the type of DPM. The consequence of it is, thus, that von Meyenburg complexes or congenital hepatic fibrosis (CHF)/autosomal recessive polycystic kidney disease (ARPKD) occurs if small primitive biliary structures are involved. Conversely, liver cysts of the ADPKD occur if medium-sized biliary structures are affected, while Caroli disease takes place when the dilatation involves large intrahepatic bile ducts. Both ADPKD and Caroli are usually present in adulthood, while ARPKD occurs in infancy and childhood. Von Meyenburg is an isolated phenomenon and may be encountered at different ages of life.

4.2.2 Congenital Hepatic Fibrosis

Congenital hepatic fibrosis (CHF) is characterized by diffuse periportal fibrosis and malformation of interlobular bile ducts. Progressive increase of portal fibrous tissue and degeneration of bile ducts may lead to portal hypertension, cholestasis, cholangitis, and eventually biliary cirrhosis. This autosomal recessive disease is found to be often associated with Caroli syndrome and polycystic kidney disease of autosomal recessive type (ARPKD) showing renal cysts distributed throughout the cortex and medulla of both kidneys with uncountable 1/2 mm cysts on the outer surface and dilated channels at right angles to the cortex, replacing the cortex and medulla substantially and fully.

4.2.3 Biliary Hamartoma (von Meyenburg Complex)

Benign tumorlike malformation of the liver, which is classically associated with polycystic liver disease but may present as incidental finding as well. It should be considered a focal (localized) ductal plate malformation.

4.2.4 Caroli Disease/Syndrome

Caroli disease differs itself from Caroli syndrome by showing neither inheritance pattern nor associations with renal disorders. Main properties are congenital ectasia of larger intrahepatic bile ducts due to the total or partial arrest of the remodeling of the ductal plate. The syndromic form of Caroli follows an autosomal dominant inheritance and can be connected to congenital hepatic fibrosis (CHF).

- *CLI*: It may present with cholestasis, cholangitis, cholelithiasis, and portal hypertension.
- *PGN*: Cholangitis, cirrhosis, and cholangiocarcinoma may complicate Caroli disease, and this may be considered when considering a close follow-up schedule of these patients.

4.2.5 ADPKD-Related Liver Cysts

ADPKD-associated cysts are usually large. The number of hepatic cysts can increase with the age of the patient, but the biliary malformations and hepatic fibrosis that are characteristic of ARPKD are virtually absent in most cases of patients with ADPKD. Defects in two genes have been implicated in ADPKD. Mutations in *PKD1* restricted at chromosome 16p13,23 appear to cause a more frequent and severe form of ADPKD comparing with *PKD2*, which is located at chromosome 4q21-q23. *PKD1* encodes an integral membrane glycoprotein, Polycystin-1, that is implicated in cell-cell or cell-matrix interactions. Polycystin-1 comprises multiple transmembrane domains and an N-terminal extracellular region that binds ligands in the extracellular compartment. In the C-terminal cytoplasmic region of Polycystin-1, there are phosphorylation sites and consensus sequences for several signaling molecules. This conformation suggests a role in intracellular signal transduction for this protein. The *PKD2* gene product, or Polycystin-2, shows some appealing qualities. There is a significant homology to a voltage-activated Ca^{2+} channel in the intracellular C-terminal domain. Both Polycystin 1 and Polycystin-2 seem to interact to form a heterodimeric ion channel at the plasma membrane able to regulate renal tubular morphology and function.

4.3 Hyperbilirubinemia and Cholestasis

The term hyperbilirubinemia is not interchangeable with cholestasis, because hyperbilirubinemia refers to a general increase of bilirubin in the blood, while cholestasis is associated with a clinicopathological condition with organic substrates. Hyperbilirubinemia can manifest as cholestasis, but cholestasis is a more liver-associated term than bilirubinemia that can also be due to clinical settings such as hemolytic anemia. Cholestasis involves a disturbance in the biliary secretory mechanisms, which is accompanied by accumulation in the liver and blood of substances usually secreted in the bile.

4.3.1 Hyperbilirubinemia

Bile is a complex mixture of substances produced by the hepatocytes and includes bile acids, bilirubin, phospholipids, cholesterol, organic ions, and water. It is crucial for the absorption of lipids and lipid-soluble nutrients and the excretion of endogenous by-products and toxins of exogenous origin. There are molecules located on two different sites of the hepatocytes and cholangiocytes. The *basolateral membrane* molecules aim to uptake bile salts and other particles from the portal venous system, whereas the *apical membrane* molecules translocate bile constituents from the intracellular environment of the hepatocytes to the extracellular environment of the bile canaliculi, which are the first section of the road of the intrahepatic biliary system. Biochemically, it is important to remember the difference between scramblase, flippase, and floppies. A scramblase, which is usually located in the ER only, migrates from one part to the other (in both directions). Essential for the bile excretion are two specific membranous proteins, which are of flippase and floppase type. A flippase only moves phospholipids from the extracellular leaflet of the bilayer to the cytoplasmic leaflet (outer→inner), whereas a floppase moves phospholipids from the cytoplasmic leaflet to the extracellular leaflet (inner→outer). Also, scramblases move in either direction. In other words, flippase flips the bilayer lipid from extracellular to intracellular, and floppase is vice versa. The intrahepatic biliary system is further composed of cholangioles of Hering, interlobular bile ducts, and interlobar and ductus choledocus or common bile duct. Most of the intracellular transfer protocol occurs using energy, an actin-myosin web, and an osmotic gradient that allows the attraction of water into bile. Mutations of the genes specific for the translocases and transporters of the hepatocyte membrane, enzymes involved in bile acid and bilirubin metabolism, proteins involved in maintaining the integrity of the tight junctions of the bile canaliculi, as well as the intrahepatic biliary system development are considered to be at the basis of the genetic disorders of intrahepatic cholestasis.

Hyperbilirubinemia Steps
There are seven steps of the metabolism of bilirubin that may be singly or multiply altered and determine an increase of the serum bilirubin level in the patient.

4.3.1.1 The Increase of Bilirubin Production
Causes: Hemolysis (Rh incompatibility, ABO incompatibility, other minor blood group incompatibility), RBC fragmentation (e.g., congenital spherocytosis, hereditary elliptocytosis, polycythemia, G6PD, PK, HK), or enclosed hematomas

4.3.1.2 The Decrease of Bilirubin Uptake by the Hepatocytes
Causes include hypothyroidism or gestational hormones, hypoalbuminemia, generalized hypoproteinemia, and drug-protein displacement/competition (e.g., sulfonamides, salicylates, heparin, and caffeine). A dreadful complication of unconjugated hyperbilirubinemia is bilirubin encephalopathy, which may present as acute, chronic, or subtle encephalopathy if unconjugated bilirubin (BR) is >30 mg/dL. In acute bilirubin elevation (ABE), there is a critical state of hyperbilirubinemia in the CNS and a range of different signs and symptoms (lethargy, decreased feeding, hypo-/hypertonia, high-pitched cry, spasmodic torticol-

lis, opisthotonos or severe hyperextension and spasticity - from ὄπισθεν 'behind' and τόνος 'tension' -, setting-sun sign, fever, seizures, and death). In the "yellow" opistothonos the complete "bridging" or "arching" position assumes a characteristic skin and mucosa discoloration, which has been reported in famous books of pediatrics of the last century. In chronic bilirubin elevation (CBE), there is a clinico-pathological tetrad constituted by extrapyramidal movement disorder, auditory dysfunction, oculomotor impairments, and dental enamel hypoplasia of the deciduous teeth due to bilirubin deposition in the basal ganglia as well as auditory nuclei, and oculomotor nuclei of the brain stem. Finally, subtle (or subacute or subchronic) bilirubin encephalopathy (SBE) is a chronic state of mild/very mild bilirubin-induced neurological dysfunction evolving with neurologic, cognitive, learning, and movement disorders, isolated hearing loss, and auditory dysfunction. Cholestasis is associated with increased serum alkaline phosphatase. The increase of 5′-nucleotidase and leucine aminopeptidase (LAP) parallel alkaline phosphatase and confirm the hepatic source of alkaline phosphatase. The rise of γ-glutamyl transpeptidase (or γ-glutamyltransferase, GGT) may also be found in many other conditions. In cholestasis, there is also the increase of serum cholesterol and phospholipids but not triglycerides. However, in Byler syndrome or progressive familial intrahepatic cholestasis (PFIC), both γ-glutamyl transpeptidase and cholesterol levels are low. Moreover, there is increased fasting serum bile acid (>1.5 μg/mL) with a ratio of cholic acid: chenodeoxycholic acid >1 in primary biliary cirrhosis and many intrahepatic cholestatic conditions but <1 in most chronic hepatocellular conditions (e.g., Laennec cirrhosis, chronic active hepatitis). Cholestasis may occur without hyperbilirubinemia. Therapy for unconjugated hyperbilirubinemia includes double-volume exchange transfusion, phototherapy, and drug-driven microsomal enzymatic hyperinduction (e.g., phenobarbital). Targeting the surgically correctable cholestasis means surgery (e.g., Kasai portoenterostomy), ursodeoxycholic acid, vitamin supplements, and ultimately, although not necessarily liver transplantation.

4.3.2 Decreased Bilirubin Conjugation and Unconjugated Hyperbilirubinemia

In this section, both abnormal intracellular binding or storage of bilirubin and the inefficiency of the conjugation system are two aspects that will be considered.

4.3.2.1 Abnormal Intracellular (Intrahepatocytic) Binding or Storage of Bilirubin

It is a deficit or alteration of glutathione S-transferase, which is the primary intracellular binding protein for bilirubin.

4.3.2.2 The Inefficiency of the (Intrahepatocyte) Conjugation System

Decrease or absence of bilirubin UDP-glucuronosyltransferase, which characterizes *Gilbert syndrome* or *Crigler-Najjar syndrome*, respectively. Liver damage can also be the cause for hyperbilirubinemia owing to hepatocyte damage.

Crigler-Najjar Syndrome

UDP-glucuronosyltransferase family one member A1 (UGT1A1) is an enzyme produced from the *UGT1A1* gene and is the only enzyme that glucuronidates bilirubin by converting its toxic molecule (unconjugated bilirubin) to its nontoxic molecule, which is the conjugated bilirubin. This process makes it able to be dissolved and removed from the body. We recognized two types of Crigler-Najjar, including type I CN, a sporadic, AR-inherited disorder in which patients have no UGT1A1 activity, and type II CN, a commoner AD-inherited disorder in which patients have some UGT1A1 activity. Type 1 CN is fatal in the neonatal period unless the baby gets a liver transplant, while jaundice characterizes type II CN due to a ↓ UGT1A1 activity (<30% usually). Severe unconjugated hyperbilirubinemia can lead to kernicterus, which is a form of brain damage caused by the accumulation of unconjugated bilirubin in the basal ganglia of the brain. Currently, kernicterus, which was first described by Christian Georg Schmorl (1861-1932), a German pathologist, in 1904, using the German stems Kern or "nucleus" and ikterus or "jaundice", is a rare condition. Schmorl identified a symmetrical yellow discoloration of globus pallidus, putamen, thalamus, subthalamic nuclei, and hippocampus (CA2-CA3 regions). MRI shows hyperintensity and then hypointensity in T1, while normal to hyperintensity in T2.

Gilbert Syndrome

- DEF: AR-inherited syndrome of hyperbilirubinemia with a ↑ level of unconjugated bilirubin in the bloodstream due to a ↓ UGT1A1 activity level (~30% usually) but without severe consequences.
- EPI: 5–10% of the general population.
- CLI: Mild jaundice may appear under conditions of exertion, stress, fasting, and infections, but the disease is otherwise usually asymptomatic.

Dubin-Johnson Syndrome

- DEF: AR-inherited disorder in which patients have ↑ conjugated bilirubin in the blood due to a defect in the secretion of bilirubin glucuronides across the canalicular membrane.
- CLI: Hyperbilirubinemia and jaundice.
- GRO/CLM: Darkly pigmented liver because of coarse granules present throughout within the hepatocyte cytoplasm.

Rotor Syndrome

- DEF: AR-inherited disorder in which patients have ↑ conjugated bilirubin in the blood due to multiple defects in hepatocyte uptake and excretion of bilirubin pigments.
- CLI: Hyperbilirubinemia and jaundice.
- GRO/CLM: The liver looks normal differently from the appearance observed in patients with Dubin-Johnson syndrome.

4.3.3 Conjugated Hyperbilirubinemia

In this section are considered for discussion the abnormal excretion of the conjugated bilirubin into the intrahepatic biliary tract, the structural abnormalities of the intra- and extrahepatic biliary system, and abnormalities of the enterohepatic circulation.

4.3.3.1 Abnormal Excretion of the Conjugated Bilirubin into the Intrahepatic Biliary Tract

Decrease/absence of carrier protein (MRP2) responsible for the discharge of bilirubin into the bile canaliculus, which characterizes *Dubin-Johnson syndrome* with no morbidity or mortality, but an increase of urinary coproporphyrin I levels and melanin-like intracellular (intralysosomal) black pigment of the hepatocytes. MRP2 is multidrug resistance-associated protein 2 and aliases are ABC30, CMOAT, DJS, cMRP, while the gene name is the ATP binding cassette subfamily C member 2 (*ABCC2*). The absence (or minor) of morbidity/mortality in DJS is probably due to the compensatory upregulation of MRP3 (multidrug resistance-associated protein 3) at the lateral membrane of the hepatocytes, which may mediate the reflux of conjugated bilirubin as well as other MRP2 substrates into the bloodstream.

4.3.3.2 Structural Abnormalities of the Intra- and Extrahepatic Biliary System

The etiology may be surgical or medical. Surgical conditions include mostly biliary atresia, which is a progressive destructive process of the extra- and intrahepatic biliary system usually occurring between 2 and 6 weeks of age, as a time frame, and conveying to portal fibrosis as well as cholestasis and periportal biliary proliferation in a syndromic (aka embryonal type) or non-syndromic (aka fetal type) setting. Other surgical conditions include a choledochal cyst and other surgically correctable abnormalities of the extrahepatic biliary system. Medical conditions include syndromic (e.g., Alagille syndrome) or non-syndromic paucity of the interlobular bile ducts (PIBD), progressive familial intrahepatic cholestasis (PFIC) 1-3, etc.

4.3.3.3 Abnormalities of the Enterohepatic Circulation

Intestinal obstruction, such as in Hirschsprung disease, duodenal atresia, etc., or alterations in the bacterial flora using antibiotics may be at the basis of an increase in reabsorption of bilirubin from the intestine. Impaired secretion of sitosterol (and cholesterol) from the enterocytes back into the gut lumen and the liver into the bile characterizes a rare inherited disease called sitosterolemia with mutations of either one of the half-transporters ABCG5 (ATP-binding cassette sub-family G member 5) or ABCG8 (ATP-binding cassette sub-family G member 8) leading to building up of the plant sterol sitosterol (24-ethyl-cholesterol). The ATP-binding cassette (*ABC*) transporters genes are divided into seven distinct subfamilies (ABC1, MDR/TAP, MRP, ALD, OABP, GCN20, and White). Clinically, there is atherosclerosis of the youth, tendon xanthomas (Greek ξανθός 'yellow' refers as to a deposition of yellowish material), hemolytic episodes, and arthralgias. A practical approach to hyperbilirubinemia is the following: If bilirubin is >270 μmol/L, perform direct Coombs test. If the direct Coombs test is negative, it is direct bilirubin. In the case of direct bilirubin, there is a cutoff of 25 μmol/L to take into consideration. If direct bilirubin is >25 μmol/L, then think of an infective etiology (sepsis, TORCH, hepatitis) and metabolic (galactosemia, AATD, CF) and cholestasis syndromes (BA, PFIC, bile salt dysmetabolism). If direct bilirubin is <25 μmol/L, then think of delayed umbilical cord clumping, MFT, FFT, AB 0 sensitivity, spherocytosis, elliptocytosis, thalassemia, and erythrocyte defects. If the hematocrit is healthy, then think of prematurity, breast milk icterus, hematoma resorption, hypothyroidism, diabetic fetopathy, Crigler-Najjar syndrome, and Lucey-Driscoll syndrome (transient familial hyperbilirubinemia).

4.4 Infantile/Pediatric/Youth Cholangiopathies

During the process to clarify neonatal and infantile cholestasis, some conditions should be taken into account. Cholestasis in the newborn age is a characteristic feature for different disorders of the prenatal or perinatal period. These include several metabolic, toxic, hereditary, anatomic, and infectious causes (Oliveira et al.). Nevertheless, three specific disorders compromise up to 80% of all neonatal cholestasis cases. These are *biliary atresia* (BA) and *non-BA infantile obstructive cholangiopathies* (NBAIOC), *neonatal hepatitis* (NH) or neonatal hepatitis group (NHG), and the *paucity of intrahepatic bile ducts*, which can be syndromic (aka Alagille syndrome) or non-syndromic. Significant symptoms of neonatal cholestasis consist of jaundice, hepatomegaly, acholic stool, and dark urine (Box 4.5). BA and NBAIOC will be discussed in Sects. 4.4.1 and 4.4.2, respectively. Section 4.4.3 will be devoted to PIBD. Metabolic Dysregulations will be part of the division of Genetic and Metabolic Liver Diseases in Chap. 6.

In primis, it is essential to differentiate between ballooning and feathery degeneration of the hepatocytes. Ballooning of the hepatocytes describe 2-3X larger liver cells with a central nucleus and a cytoplasm cleared with "whisps" or cobbwebs, while feathery degeneration refers as to a cholate stasis-associated, usually periportal flocculation of the hepatocytes, which are also larger than normal (2-3X). Conversely, steatosis is fatty change without ballooning degeneration (ballooning is often indicative of steatohepatitis). In many settings, it may be necessary to have some liver tissue for metabolic investigations of the living patient and one of the most often questions to the pediatric pathologist is how many cores are needed to be done for that amount of tissue. To keep in mind are the following equivalences, i.e., in case of laparoscopic cup biopsies ⇒ 45 mg of liver tissue, in case of 14G true-cut biopsy needle ⇒ 15–20 mg of liver tissue, while in case of 18G needle biopsy ⇒ 3–5 mg of liver tissue.

> **Box 4.5 Infantile/Pediatric/Youth Cholangiopathies**
>
> 4.4.1. *Biliary Atresia* (BA)
> 4.4.2. *Non-BA Infantile Obstructive Cholangiopathies* (NBAIOC)
> 4.4.3. *The Lack or Paucity of the Intrahepatic Biliary Ducts* (PIBD) (Syndromic and Non-syndromic)
> 4.4.4. *Neonatal Hepatitis Group* (NHG)
>   4.4.4.1. Bile Acid Synthesis Disorders (BASD)
>   4.4.4.2. Defects of the Intracellular Transport of Conjugated Bilirubin
>   – Progressive Familial Intrahepatic Cholestasis Types I–III
>   – Cystic Fibrosis
>   4.4.4.3. Infection-Related Neonatal Hepatitis
>   4.4.4.4. Non-BASD Metabolic Dysregulation-Related Neonatal Hepatitis
>   4.4.4.5. Toxin-Related Neonatal Hepatitis
>   4.4.4.6. Immunology-Based Neonatal Hepatitis
>   4.4.4.7. Endocrine Gland Dysregulation-Based Neonatal Hepatitis
>   4.4.4.8. Multi-etiologic Chromosomal Abnormality-Associated Cholangiopathy
>   4.4.4.9. Miscellaneous/Idiopathic NHG
> 4.4.5. *Primary Sclerosing Cholangitis* (PSC)
> 4.4.6. *Primary Biliary Cirrhosis* (PBC)
> 4.4.7. *Pregnancy-Related Liver Disease* (PLD)

Box 4.6 shows the etiology and clinical-radiologic-biochemical investigations for neonatal/infantile cholestasis (10 Major Categories).

4.4.1 Biliary Atresia

Biliary atresia is defined as the complete obliteration of both intra- and extrahepatic biliary tracts, although mainly extrahepatic bile ducts with some genetic homogeneity (Sergi, 2019) (Fig. 4.8). With an incidence of 1:8000 to 1:14,000, BA covers up to 30% of the neonatal cholestasis cases. Progressive inflammatory sclerosis of the biliary tree is postulated to be caused by infectious agents (CMV and rotavirus, among others), abnormal bile acid composition, immune-mediated reactions, physical or chemical insults, or abnormal cilia or vascular insufficiency. Some authors consider ontogenetic alteration in the remodeling of the ductal plate. The extrahepatic tract is mostly obliterated or atretic and shows a shrunken or missing gallbladder, whereas intrahepatic ducts form irregular, distorted bile ducts. Portal fibrosis and ductular proliferation are typical features in portal tracts. Since the progression of the process, condition worsens up to liver cirrhosis and its consequences. Treatments are, if indicated, Kasai procedure (hepatic porto-enterostomy) and liver transplantation. In the neonatal period, the immaturity of the liver to eliminate the unconjugated hyperbilirubinemia reflects the clinical phenomenon of physiologic jaundice, which exhibits a mild elevation of serum bilirubin, its gradual rise to a maximum of 8–9 mg/dL, and its fall to typical values in the 2nd week of life. This phenomenon seems to be more accentuated in premature infants with high peak serum bilirubin levels and long duration. It seems to run inversely proportional to the age of gestation. In case of jaundice at early onset (<24 h of birth), with very high serum bilirubin levels (>11–12 mg/dL in a formula-fed infant or >14–15 mg/dL in breastfed infant), with a conjugated component (>2 mg/dL), and persistent (>2 weeks), further investigation to determine the basis for the elevated bilirubin levels and prompt management are indicated. The *protracted conjugated hyperbilirubinemia* or *neonatal cholestasis* is a

4.4 Infantile/Pediatric/Youth Cholangiopathies

Box 4.6 Etiology and Clinical-Radiologic-Biochemical Investigations for Neonatal/Infantile Cholestasis (10 Major Categories)

1. *Obstructive cholangiopathies* (BA and NBAIOC)
   - Biliary atresia (*GGT, ultrasound, HIDA scan, ERCP, liver biopsy*)
   - Choledochal cyst (*GGT, ultrasound, HIDA scan, ERCP, liver biopsy*)
   - Caroli disease and syndrome (*GGT, ultrasound, HIDA scan, ERCP, liver biopsy*)
   - Neonatal sclerosing cholangitis (*GGT, ultrasound, HIDA scan, ERCP, liver biopsy*)
   - Idiopathic biliary system perforation (*GGT, ultrasound, HIDA scan, ERCP, liver biopsy*)
2. *The paucity of intrahepatic biliary ducts* (syndromic/non-syndromic Alagille syndrome) (*Extrahepatic anomalies, hypercholesterolemia, genetics*)
3. *Defects of the intracellular transport of conjugated bilirubin*
   - Progressive familial intrahepatic cholestasis (PFIC) (*GGT, IHC, genetics*)
   - Cystic fibrosis (CF) (*GGT, sweat test, genetics*)
4. *Infections* (TORCH) (*TORCH screen, HA/B/C/V, and HIV*)
   - Viruses/parasites: CMV, rubella, Reo-3, Parvovirus B19, coxsackie, HHV-6, HSV, *T. gondii*
   - Bacteria: Sepsis, UTI, *L. monocytogenes*, late-onset *Streptococcus* infection, TB, and *T. pallidum*
5. *Metabolic dysregulations*
   - Alpha-1-Antitrypsin Deficiency (AATD) (*AAT, IEF, genetics*)
   - Carbohydrate-related disorders (galactosemia, hereditary fructosemia, CGD)
   - Bile acid synthesis disorders (BASD)
   - Amino acid metabolism disorders (e.g., tyrosinemia) (*succinylacetone, urinary organic acids*)
   - Urea cycle disorders (UCD) (e.g., arginase deficiency)
   - Mitochondrial hepatopathies
   - Lipid metabolism disorders and lipoprotein defects (*mass spectrometry, liver biopsy*)
   - Total parenteral nutrition (TPN) (*chart review, GGT, MDT meetings/rounds*)
   - Peroxisomal disorders
6. *Toxin-related disorders*
   - Fetal alcohol syndrome (FAS) (*chart review, MDT meetings/rounds*)
   - Breast milk-related cholestasis (*chart review, MDT meetings/rounds*)
7. *Immunological disorders*
   - Neonatal lupus
   - Neonatal hemochromatosis (congenital alloimmune hepatitis)
   - ABO blood group incompatibility-linked inspissated bile syndrome
   - Autoimmune hemolytic anemia
8. *Endocrine gland dysfunctions*
   - Hypothyroidism and hypoparathyroidism
   - Hypopituitarism and diabetes insipidus
   - Hypoadrenalism
9. *Chromosomal abnormalities* (*chart review, MDT meetings/rounds, karyotype*)
   - Trisomy 13, 18, 21 syndromes
   - Non-trisomic chromosomal abnormalities (Turner syndrome, cat's eye syndrome)
10. *Miscellaneous/idiopathic*
    - Non-biliary atresia – "high GGT intrahepatic cholestasis" and idiopathic
    - Hemophagocytic lymphohistiocytosis (HLH) (ferritin, BMB)
    - ARC syndrome (arthrogryposis, renal tubular dysfunction, and cholestasis)

Notes: *BMB*, bone marrow biopsy; *CMV*, cytomegalovirus; *ERCP*, endoscopic retrograde cholangiopancreatography; *GGT*, gamma-glutamyl transpeptidase; *HHV-*, human herpesvirus 6; *HSV*, herpessimplex virus; *HIDA*, hepatobiliary iminodiacetic acid; *IEF*, isoelectrofocusing; *MDT*, multidisciplinary team; *TB*, tuberculosis; *TORCH*, Toxoplasma gondii, other viruses (HIV, measles, etc.), rubella (German measles), cytomegalovirus, and herpes simplex; *UTI*, urinary tract infections

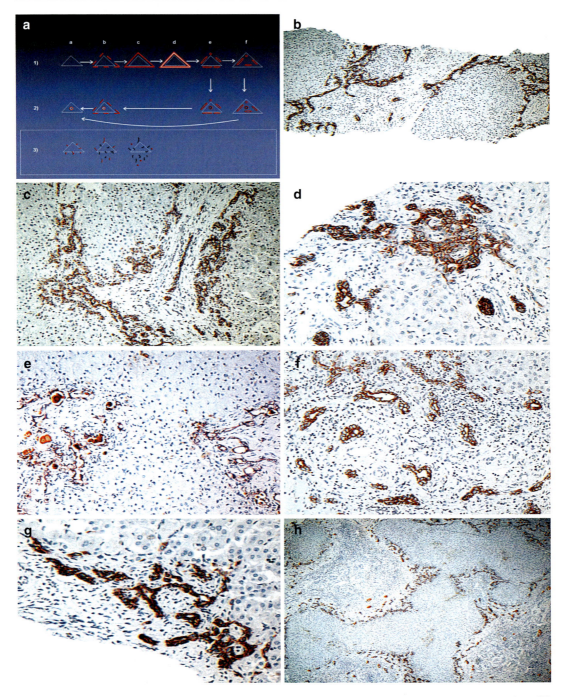

Fig. 4.8 In (**a**) is demonstrated a cartoon showing the development of the intrahepatic biliary system in a schematic way to evaluate the number of abnormal structures in infantile cholangiopathies. Using antibodies against CK7 the microphotographs (**b**) through (**h**) show the abnormalities encountered in case of biliary atresia at different days of liver biopsy. It has been stressed in the literature that a Kasai porto-enterostomy may not be efficacious, but actually deleterious in the advanced stages of biliary atresia

heterogeneous group of diseases characterized clinically by an increase of bile acids and alkaline phosphatase other than an increased serum concentration of conjugated bilirubin and pathologically by stasis of bile pigment or feathery degeneration of hepatocytes resulting from stagnation of bile acids (cholatostasis). Neonatal cholestasis represents a challenge for the pathologist in evaluating liver tissue biopsies from affected infants because the list of potential causes of neonatal cholestasis is very extensive. Different categories have been identified, including infectious agents, metabolic disorders, toxins or drugs, genetic or chromosomal diseases, anatomic abnormalities, and different conditions with either mechanical obstruction to bile flow or functional impairment of the hepatic excretory function. Substantially, it is essential to divide neonates with cholestasis into two major categories, i.e., "correctable" and "non-correctable forms." Conceptually, it is critical to keep separate the class including newborns with primary intrahepatic disease (e.g., an inborn error of bile acid metabolism, syndromic and non-syndromic form of the paucity of the intrahepatic bile ducts) from the category including newborns with primary extrahepatic disease (e.g., obliterative cholangiopathy of the common bile duct). However, there are practically multiple areas of potential overlap on the clinical, histogical, or biochemical basis. The role of the pathologist and the diagnostic accuracy of percutaneous (needle) liver biopsy in the diagnosis of surgery-correctable forms has been an argument of intense discussion, but remains crucial in the 21st century. Although first studies emphasized the usefulness to perform a liver biopsy in picking up features characteristic of surgery-correctable cholestatic forms, later studies objected the reliability of the histopathologic diagnosis because divergent opinions arose from several authorities with a failure rate of 33%. Subsequently, a percutaneous liver biopsy was re-evaluated and was seen to correlate reasonably well with open surgery liver biopsy. More recently, the diagnostic accuracy of percutaneous liver biopsy has been reported to be up to 95%. Probably, the problem resides in the impact the multi-etiology of many surgery-correctable cholestatic forms and their diverse morphology have on the developing perinatal liver. In analyzing the changes observed in the hepatocytes and intrahepatic bile ducts in both surgery correctable and non-correctable forms of neonatal cholestasis, overlap situations can still occur. Liver of infants showing extrahepatic bile duct obstruction and the paucity of intrahepatic bile ducts (replacing the previous misnomer of intrahepatic biliary atresia) has been described. Similarly, extrahepatic bile duct obstruction has been found associated with hyperplasia of intrahepatic bile ducts or congenital hepatic fibrosis. To complicate the categorization of the liver histology of infants with neonatal cholestasis, Schweizer postulated the existence of (at least) two forms of biliary atresia with or without associated malformations (e.g., *situs inversus*, malrotation, polysplenia). The Han ethnics of Chinese population is particularly useful for Genomic-Wide Association Studies. (GWAS) because of the large size and homogeneity of this ethnics. Recently, GWAS found a potential susceptibility locus for BA between the genes *ADD3* and *XPNPEP1* located with the chromosomal localization of 10q25.1 with replication in independent Chinese and Thai specimens and in a Zebrafish model (Tang et al. 2016). Adducins are heteromeric proteins constituted of different subunits, which are involved in the assembly of the spectrin-actin network in red blood cells and at sites of cell-cell contact in epithelial tissues.

4.4.2 Non-BA Infantile Obstructive Cholangiopathies (NBAIOC)

This spectrum of disorders includes the agenesis and the aplasia of the extrahepatic biliary system and the intrinsic and the extrinsic obstruction of the extrahepatic biliary system. Inherent etiology consists of inflammation with edema of the wall of the extrahepatic biliary tree, the presence of obstruction due to stone or abnormal bile composition. The extrinsic obstructive cholangiopathies include inflammatory or a neoplastic process compressing the extrahepatic biliary system. At neonatal age, one of the most common neoplasms is the neuroblastoma IVS with the presence of neuroblastoma localized in the liver. Neuroblastoma IVS is able to compress the extrahepatic biliary tree extrinsically.

4.4.3 The Paucity of the Intrahepatic Biliary Ducts (PIBD)

It is the lack of the scarcity of interlobular bile ducts per portal tract in the liver. PIBD can be divided in a syndromic and a non-syndromic form. The syndromic form also called Alagille syndrome or arterio-hepatic dysplasia reveals hepatic ductopenia, accompanied by cardiac, ocular, facial, and skeletal anomalies. It is an autosomal dominant inherited disease with its gene locus on 20p12 (jagged-1) (*JAG1*). It can also be associated with other congenital disorders like α1-antitrypsin deficiency, Turner syndrome (Monosomy X0), and Down syndrome (trisomy 21 syndrome). Major hepatic features are chronic cholestasis, portal fibrosis, and progressive decrease or lacking bile ducts in the portal tracts. The bile duct to portal area ratio lies between 0 and 0.4 in comparison with 0.9 and 1.8 in healthy children, according to Alagille definition. Cardiac and renal complications may have severe consequences on the patients' health, too. A non-syndromic form of bile ducts paucity has not shown a definite association with any genetic, congenital, or metabolic disorders but could be observed in relation with Byler disease, Zellweger syndrome, Niemann-Pick disease, cytomegalovirus infection, or Alpers disease, to cite just a few of diseases. Bruguera et al. (1992) have also suggested a relationship to idiopathic adulthood ductopenia. Moreover, Ishak (2002), Witzleben (1982), as well as other studies have been pillar for the investigation of this very complex and complicated topic. *Alagille syndrome* is a genetic disorder involving the correct development of the intrahepatic biliary system which includes first the condition named Alagille syndrome or syndromic paucity of intrahepatic interlobular bile ducts, which is due to either human jagged-1 (*JAG1*) gene (ALGS1 disease) or Notch homolog 2 (*NOTCH2*) gene (ALGS2 disease). Some illnesses that not resemble Alagille are grouped under the umbrella of "non-syndromic paucity of interlobular bile ducts" and include several syndromes (Fig. 4.9).

4.4.4 Neonatal Hepatitis Group (NAG)

NH is an intrahepatic cholestatic nonobstructive disorder of viral (HSV, CMV, rubella virus, etc.), metabolic (α1-antitrypsin deficiency, CFTR, galactosemia, etc.), or idiopathic nature (more than 50%) (Box 4.7). Besides cholestasis, lobular inflammation, focal fibrosis and a hepatocellular transformation such as giant-cell -transformation are current main findings. Progressive forms may result in massive fibrosis and liver cirrhosis (Figs. 4.10, 4.11, and 4.12).

Box 4.7 Neonatal Hepatitis Group (NAG)

4.4.4. Neonatal Hepatitis Group (NAG)
    4.4.4.1. Bile Acid Synthesis Disorders (BASD)
    4.4.4.2. Defects of the Intracellular Transport of Conjugated Bilirubin
       – Progressive Familial Intrahepatic Cholestasis Types I–III
       – Cystic Fibrosis
    4.4.4.3. Infection-Related Neonatal Hepatitis
    4.4.4.4. Non-BASD Metabolic Dysregulation-Related Neonatal Hepatitis
    4.4.4.5. Toxin-Related Neonatal Hepatitis
    4.4.4.6. Immunology-Based Neonatal Hepatitis
    4.4.4.7. Endocrine Gland Dysregulation-Based Neonatal Hepatitis
    4.4.4.8. Multi-etiologic Chromosomal Abnormality-Associated Cholangiopathy
    4.4.4.9. Miscellaneous/Idiopathic NAG

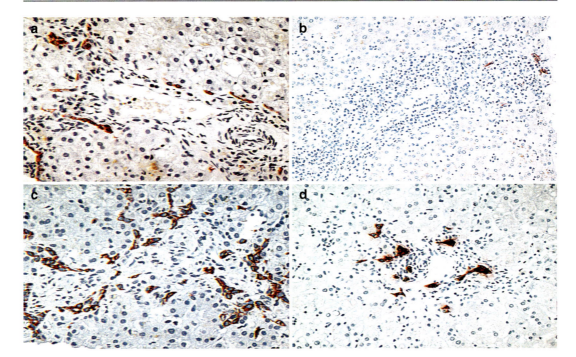

Fig. 4.9 PIBD CK7 patterns highlight the very few reactive biliary structures located at the periphery with virtual absence of any interlobular bile duct. The microphotographs (a–d) show these different patterns in different children with PIBD. It is important to remember that an interlobular bile duct should be usually present within three diameters of the adjacent portal artery (anti-CK7 immunostaining)

4.4.4.1 Bile Acid Synthesis Disorders (BASD)

When "neonatal hepatitis" ("giant cell transformation of the lobular hepatocytes") is prominent, and bile canaliculi show an immunoreactivity using the anti-BSEP and anti-GGT antibodies, bile acid synthesis defects (BASD) become highly likely. In particular, if average or low serum bile acid concentrations, normal or minimally elevated GGT, and absence of pruritus occur in a patient with jaundice and conjugated hyperbilirubinemia, the suspicion for BASD should be high and a clinical-pathological conference or multidisciplinary team meeting should be promptly called. Bile acid synthesis defects are crucial for not only the pediatrician (and pediatric pathologist) but also for the internist and neurologist. It is now demonstrated that BAS defects may present as cerebrotendinous xanthomatosis, myocardial infarction of the young/middle-aged individual, spastic paraparesis, and early dementia, other than cholestasis/liver disease at infancy. Synthesis of bile acids from cholesterol requires modifications to the nucleus of the cholesterol molecule and oxidation of the cholesterol side chain. There are two pathways used by the human organism to produce bile acids. They are called "neutral" pathway, which is confined to the liver and usually responsible for the symptomatology in children and adults, and the "acidic" or alternative pathway, which is responsible for the remotion of cholesterol and production of bile acids in extrahepatic organs, mostly the brain and organs rich in macrophages, other than the liver. This latter seems to be usually responsible for the symptomatology in neonates and infants. Bile acids are needed as detergents in the intestinal lumen for the digestion and absorption of fat, and the pumping of bile acids drives fat-soluble vitamins and the secretion of bile by the liver into the canalicular system. Failure to synthesize the standard bile acids leads to inadequate bile flow, which determines the clinical and histologic pictures of cholestasis by retention of compounds generally

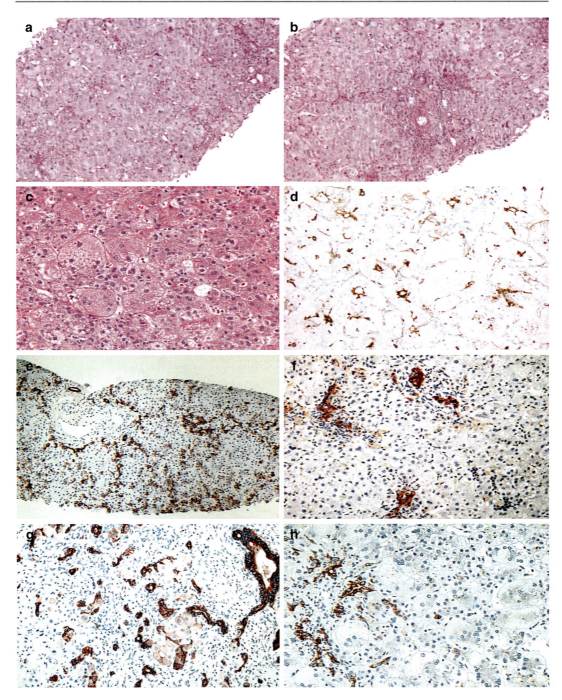

Fig. 4.10 In this panel are shown microphotographs of neonatal hepatitis (**a–c**) with numerous giant cell transformation of the hepatocytes, abnormal expression of the polyclonal antibody against carcinoembryonic antigen (CEA) for bile canaliculus (**d**), and abnormalities of biliary tract (**e–h**) with numerous biliary duct proliferation units, pseudoacinar transformation of the hepatocytes ("pseudorosettes"), and immunohistochemical phenotype change with acquisition of the biliary phenotype by some hepatocytes. The microphotographs (**a–c**) are Periodic acid Schiff with diastase digestion (PASD), x100; PASD, x100; and H&E, x200. The microphotographs (**d**) through (**h**) are pCEA, x200; anti-CK7, x50, x100, x100, respectively. The immunohistochemical slides have been performed using either an antibody against the carcinoembryonic antigen (CEA, polyclonal) or against the (cyto-)keratin 7 (CK7, monoclonal) and Avidin-Biotin-Complex (ABC)

4.4 Infantile/Pediatric/Youth Cholangiopathies

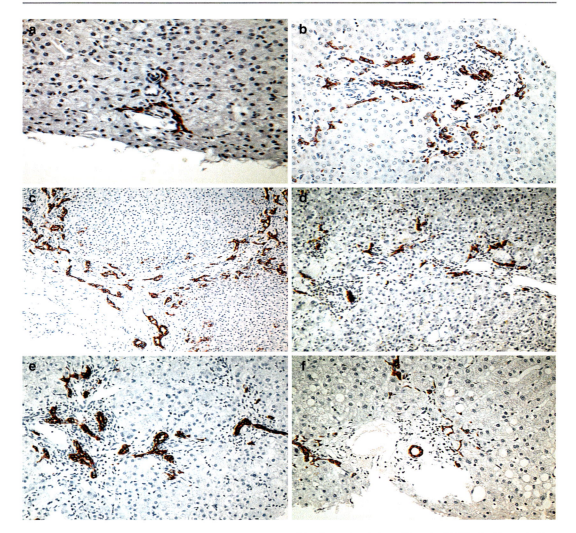

Fig. 4.11 Several patterns of CK7 expression in infantile cholangiopathies. In (**a**) is shown a ductopenia not otherwise specified (NOS) (paucity of interlobular bile ducts) in an infant without congenital defects or metabolic storage disorder (anti-CK7 immunostaining, ABC method, x200 original magnification); in (**b**) is shown some abnormal reaction patterns identified in a patient with Langerhans cell histiocytosis and hepatic involvement (S-100, CD1a positive, not shown) (anti-CK7 immunostaining, ABC method, x200 original magnification); and in microphotograph (**c**) the same patient with Langerhans cell histiocytosis-associated liver cirrhosis with nodules and pseudonodules (anti-CK7 immunostaining, ABC method, x100 original magnification). In the (**d**) microphotograph is shown the liver with abnormal reaction patterns, cholestasis, and hyperbilirubinemia in a newborn with necrotizing enterocolitis with hepatic disease (anti-CK7 immunostaining, ABC method, x100 original magnification); in the (**e**) microphotograph is shown the liver histology of an infant affected with Wissler-Fanconi allergic subsepsis syndrome (WFASS) with cholestasis and hyperbilirubinemia (anti-CK7 immunostaining, ABC method, x160 original magnification). WFASS is a rare rheumatic syndrome that was first described by Wissler in 1944 and Fanconi in 1946 with similar presentation to sepsis and is characterized by polymorphous exanthemas, recurrent high fever, leukocytosis, and arthralgia. In the (**f**) microphotograph is shown the liver histology of a patient with cholestasis and hyperbilirubinemia and fatty change NOS (anti-CK7 immunostaining, ABC method, x200 original magnification). In the (**g**) microphotograph is shown the histology of an infant with hereditary fructose intolerance (HFI) (anti-CK7 immunostaining, ABC method, x320 original magnification); while the (**h**) microphotograph exhibits the liver histology of a patient with glycogenosis 1b (anti-CK7 immunostaining, ABC method, x320 original magnification)

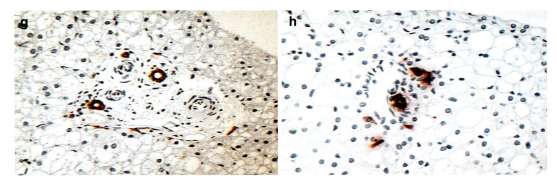

Fig. 4.11 (continued)

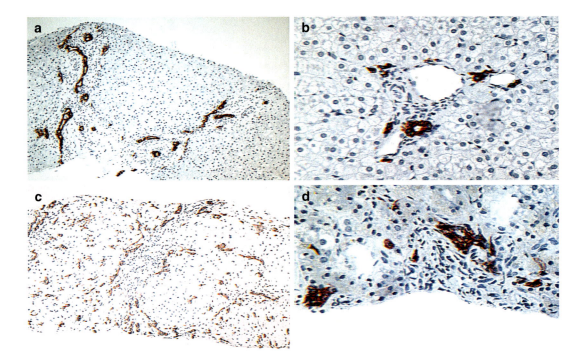

Fig. 4.12 Several patterns of CK7 expression in infantile cholangiopathies. In the (**a**) microphotograph is shown the liver of an infant with neonatal cholestasis, not otherwise specified (NOS) (anti-CK7 immunostaining, Avidin-Biotin-Complex [ABC] method, x100 original magnification); in the (**b**) microphotograph is shown the histology of a patient with neonatal cholestasis and congenital lung disease with pulmonary hypertension not otherwise specified (anti-CK7 immunostaining, ABC method, x320 original magnification); in the (**c**) microphotograph is shown the liver of an infant with cholestasis and Alpers disease (anti-CK7 immunostaining, ABC method, x100 original magnification) (Alpers disease/syndrome is a progressive neurologic disorder that begins during infancy/childhood and is complicated by serious liver disease and is due to mutations of the *POLG* gene); in the (**d**) microphotograph is shown the liver histology of a newborn with neonatal cholestasis and gestation affected with HELLP syndrome (a complication of pregnancy characterized by hemolysis, elevated liver enzymes, and a low platelet count) (anti-CK7 immunostaining, ABC method, x200 original magnification). In the microphotograph (**e**) is shown the liver histology of an infant with sepsis-related liver disease with diffuse pseudo-acinar transformation (anti-CK7 immunostaining, ABC method, x320 original magnification); in the microphotograph (**f**) is shown the liver histology of a newborn with biliary atresia (extrahepatic type) (anti-CK7 immunostaining, ABC method, x62.5 original magnification)

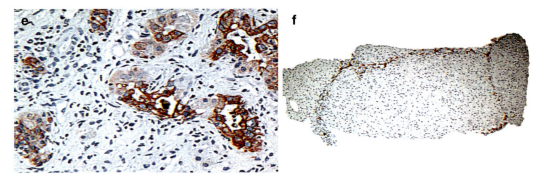

Fig. 4.12 (continued)

excreted in bile such as conjugated bilirubin. The deficiency of bile acid synthesis is also responsible for the lack of elevation of GGT, unlike other forms of cholestasis. This situation is due to the deficient detachment of GGT from the canalicular membrane, which is mediated by bile acids. This feature is essential for early recognition of this condition in the setting of neonatal/infantile cholestasis, although amidation defects do show GGT increase. Bile acids are the final products of cholesterol utilization and represent the major pathway of cholesterol catabolism in mammals. Four primary physiologic functions have been identified, including (1) elimination of excess cholesterol, (2) anti-crystallization of cholesterol in the gallbladder, (3) facilitation of the digestion of dietary triacylglycerols (triglycerides) by emulsifying them and making them accessible to pancreatic lipases, and (4) facilitation of the absorption of fat-soluble vitamins. The liver is the one organ where their full synthesis can occur, although some of the enzymes involved in the biogenesis may be expressed in several cell types. Bile acids are essential for the solubilization of dietary cholesterol, lipids, fat-soluble vitamins (A, D, E, K), and some other important nutrients. Bile acid synthesis needs the interaction of 17 enzymes occurring in multiple intracellular compartments, including endoplasmic reticulum, mitochondria, peroxisomes, and cytosol. In human bile, the most abundant bile acids are chenodeoxycholic acid or CDCA (45%) and cholic acid or CA (31%), which are referred to as the primary bile acids. There are two pathways for the biosynthesis of the bile acids, including the classic or neutral and the alternative or acidic pathway. In rodents, the acidic pathway can account for up to 1/4 of total bile acid synthesis, while in humans it accounts for about 1/20 of total bile acid synthesis. The reaction catalyzed by the 7-α-hydroxylase (CYP7A1) is the rate-limiting step of the total bile acid synthesis hydroxylating the molecule of cholesterol at the position 7. The 7-hydroxycholesterol is changed into CDCA and CA using several steps. Another important limiting step is due to the action of 3β-hydroxy-Δ5-C27-steroid oxidoreductase, which is encoded by *HSD3B7* gene. It initiates the epimerization (a stereochemistry process where an epimer is produced to transform into its chiral counterpart) of the cholesterol molecule at the three positions when the hydroxyl group in the β-orientation is converted in α-orientation. CA is distinct from CDCA because CA derives from intermediates of the *HSD3B7* gene product on molecules where CYP8B1 can act, while CDCA escapes the action of CYP8B1. The enzymatic activity of CYP8B1 is crucial in determining the CA/CDCA ratio. CYP8B1 is regulated by the end products (CA and CDCA) using their regulation of the nuclear receptor FXR. The mitochondrial enzyme sterol 27-hydroxylase (CYP27A1) generates bile acid intermediates, which are subsequently hydroxylated on the position seven by oxysterol 7α-hydroxylase (CYP7B1) (alternative pathway). The alternative or acidic pathway appears to play a major role in infancy and as a means to control cholesterol levels in extrahepatic tissues, mainly the nervous system. The secretion of the bile acids may occur following their conjugation via an amide bond at the terminal carboxyl group using either glycine or taurine, which are two amino acids. The glycol- and tauro-conjugates are essential to increase the amphipathic nature of these molecules making them more able to be secreted. The conjugated bile acids are the major solutes in human bile which are converted to salts in the presence of sodium (bile salts). Bile salts are, then, secreted from hepatocytes into the bile canaliculi using a bile salt export protein (BSEP; an ATP-binding cassette B11, ABCB11). MDR3, aka *ABCB4*, is necessary for the transport of phospholipids into the bile canaliculi. MDR3 deficiency is a very confusing topic because the terminology is a difficult topic. In fact, MDR3 deficiency applies to several disorders including not only progressive familial intrahepatic cholestasis (PFIC) type 3, benign recurrent intrahepatic cholestasis (BRIC), low phospholipid associated cholelithiasis (LPAC) syndrome, transient neonatal cholestasis (TNC), but also adult biliary fibrosis/cirrhosis, and some cases of intrahepatic cholestasis of pregnancy (ICP) and drug induced cholestasis (DIC). All these disorders are determined by mutations of the *ABCB4* gene. Bile salts, phospholipids, and cholesterol are transported from the bile canalicular system into the interlobular bile ducts, segmental ducts into the gallbladder, where they are concentrated to form bile. In addition to these compounds, electrolytes, minerals, some proteins, bilirubin, and biliverdin pigments account to complete the mixture of bile. The bile salts primary role is to solubilize cholesterol preventing cholesterol crystallization and subsequent gallstone formation (Admirand and Small, 1968). Lipid consumption induces the secretion of cholecystokinin (CCK) from duodenal enteroendocrine I cells. CCK binds to an epithelial receptor of the gallbladder. This process promotes the contraction of smooth muscle cells of the gallbladder and, subsequently, the relaxation of the sphincter of Oddi. The gallbladder-arising intraduodenal micelles promote the absorption of fat-soluble vitamins as well as the digestion of lipids by pancreatic enzymes. The enterohepatic cycle is complete when bile salts are reabsorbed using an apical Na^+-dependent bile transporter (ASBT). ASBT is located in the brush border of the intestinal surface cell or enterocyte. About 95% of bile salts are reabsorbed into the distal ileum. The intra-enterocytic transport from the cytosol of the basolateral membrane occurs using the ileal bile acid-binding protein (IBABP, aka fatty acid-binding protein subclass 6 or FABP6). The transfer from the basolateral membrane into the bloodstream occurs via a heterodimeric transporter of organic solutes (OSTα/OSTβ). Some small percentage of bile salts stay intraluminal and undergo deconjugation by intestinal bacteria before being either adsorbed or converted into secondary bile acids. These secondary deconjugation products are deoxycholate (DCA) and ursodeoxycholate (UDCA) from CA and lithocholate (LCA) from CDCA. These compounds are either excreted in the stools or passively adsorbed from the colon. Both primary and secondary bile acids are transported back to the hepatocytes via the bloodstream using a Na-Na (Na^+)-taurocholate cotransporting polypeptide (NTCP/SCL10A1) and organic anion transporters (OATP) like OAT1B2. Regulation of bile acid homeostasis occurs through the action of some important receptors, the farnesoid X receptors (FXRs), which belong to the superfamily of nuclear receptors including the steroid/thyroid hormone receptor family, liver X receptors (LXRs), retinoid X receptors (RXRs), and the peroxisome proliferator-activated receptors (PPARs). Bile acids are physiologic FXR ligands, being CDA the most potent activator of human FXR. FXR targets mostly the small heterodimer protein (*SHP*) gene, which represses the expression of *CYP7A1*, but also *SREBP-1c* or sterol 12α-hydroxylase, and solute carrier family 10 (or sodium/bile acid cotransporter family), member 1, which is identified as the Na+-taurocholate cotransporting polypeptide (NTCP). FXR also interacts with BSEP, MDR3, and MRP2. Interestingly, gene mutations altering the steric structure of BSEP and MDR3 are associated with PFIC types 2 and 3, respectively, while gene mutations affecting MRP2 are associated with Dubin-Johnson syndrome.

Inborn Errors of Bile Acid Synthesis

The combined use of mass spectroscopy and gas chromatography on blood, urine, or bile has been able to identify nine defects in the synthetic pathway. The ESI-MS or electrospray-ionization tandem mass spectrometry is used to screen BASD analyzing urinary cholanoids (bile acids and bile alcohols). The infants usually present with a neonatal hepatitis-like syndrome (lobular involvement > portal tract involvement) and absent or low bile acids and direct hyperbilirubinemia. An essential biochemical clue may be the average serum level of gamma-glutamyl transferase (GGT) found in patients affected with BASD, as seen in PFIC-1 and PFIC-2. GGT is an enzyme synthesized in the hepatocytes and secreted into the bile and then reabsorbed by damaged bile duct epithelial layer. Bile acids promote the detachment of GGT from the canalicular membrane. Thus, in the absence or low bile acids, there is no plasmatic elevation of GGT. The liver is damaged in BASD due to the lack of usual choleretic influence of CA and DCA and due to the intrahepatocytic accumulation of metabolic intermediates. Metabolic intermediates are constituted by hydrophobic bile acids, which are hepatocytic- and cholangiocyte-toxic. In comparing classic NH with BASD, some histologic clues may point to BASD, although clinical presentation and course may be quite variable with the life-threatening infantile cholestatic liver disease to pediatric/adult progressive neurological disease with signs of upper motor neuron damage (spastic paraparesis). Histologic clues of BASD are more common giant cell necrosis than NH and cytotoxic interface hepatitis with cholate stasis (swollen and necrotic cholangiocytes in the smallest Herring ductules) accompanied by normal interlobular bile ducts.

CYP7A1 (Cholesterol 7α-hydroxylase)

Marked elevation of total cholesterol and LDL, cholelithiasis, premature coronary and peripheral vascular disease, statin-resistant hypercholesterolemia, no liver dysfunction are the features.

CYP27A1 (Sterol 27-hydroxylase) or Cerebrotendinous Xanthomatosis

- *LAB*: ↑ bile alcohol glucuronides in serum and ↑ of plasma cholestanol: cholesterol ratio in the urine.
  1. Life-threatening neonatal cholestasis
  2. Infantile chronic diarrhea with developmental delay/regression and accompanied by bilateral cataracts (progressive neurological dysfunction)
  3. Adult upper motor neuron disease with spastic paraparesis, IQ fall/early dementia, ataxia, and dysarthria with seizures and/or peripheral neuropathy, xanthomata, premature atherosclerosis with myocardial infarction in the 4th decade, and osteoporosis leading to pathological skeletal fractures.

CYP7B1 (Oxysterol 7α-Hydroxylase)

Neonatal cholestasis, hepatosplenomegaly, cirrhosis, liver failure, marked an increase of serum ALT and AST with or without associated neurologic co-morbidity (AR-inherited spastic paraplegia 5A or SPG5A), being these patients more prone to have mutations of the gene encoding for CYP7B1.

- *LAB*: Marked increase of serum BA, mainly 3β-Δ_5-mono-hydroxy bile acids and an increase of sulfate and glycosulfate conjugates of 3β-Δ_5-mono-hydroxy bile acids and no primary bile acids in the urine.
- *CLM*: Giant cell hepatitis, marked fibrosis/cirrhosis, and steatosis of micro- and macrovesicular type.

HSD3B7 (3β-δ5-C27-steroid Oxidoreductase)

- *CLI*: Hepatomegaly with jaundice, pruritus, steatorrhea, and fat-soluble vitamin deficiency (e.g., "nutritional" rickets).
- *LAB*: Decrease or absent primary bile acids in serum and increase of dihydroxy- and trihydroxy cholenoic acids as well as a decrease of or missing primary bile acids in the urine.

- *CLM*: Giant cell change of the hepatocytes and hepatocyte disarray with cholestasis and cholatestasis as well as bridging fibrosis and late-onset liver disease with malabsorption.
- *PGN*: Good response to bile acid replacement therapy.

AKR1D1 (or SRD5B1) (Δ4-3-Oxosteroid-5β-reductase)

HSD3B7 deficiency like presentation, but earlier with the more severe liver disease, neonatal cholestatic jaundice, evidence of fat-soluble vitamin malabsorption, prolonged prothrombin, and progressive cirrhosis.

- *LAB*: ↑ 3-oxo-Δ4 bile acids, increase of all bile acids and a decrease of primary bile acids in serum and increase 3-oxo-Δ4 bile acids, an increase of all bile acids, and a reduction of primary bile acids in the urine.
- *CLM*: Giant cell hepatitis with steatosis and extramedullary hematopoiesis of the liver as well as "neonatal hemochromatosis"-type iron accumulation in several tissues.
- *PGN*: Bile acid supplementation is efficient, but treatment should be started before the liver is irreversibly damaged.

AMACR (2-Methyl acyl-CoA Racemase)

Neonatal cholestatic liver disease with severe fat-soluble vitamin deficiencies, hematochezia, as well as adult-onset peripheral motor neuropathy of sensory type.

- *LAB*: ↑ C27 trihydroxycholestanoic and pristanic acid, ↓ primary bile acids with normal LCFA and phytanic acid in serum, and ↓ primary bile acids in the urine.

Bile Acid Conjugation Defects

- *CLI*: Transient neonatal cholestasis and fat soluble vitamin deficiencies.
- *LAB*: Increase of unconjugated CA and DCA and absence of CDCA in serum and absence of glycine and taurine conjugates and presence of cholic glucuronidated and sulfates and unconjugated cholic acid in the urine.

4.4.4.2 Defects of the Intracellular Transport of Conjugated Bilirubin

Substantially, progressive familial intrahepatic cholestasis I, II, and III types and cystic fibrosis (mucoviscidosis) play a significant role in this subgroup.

Progressive Familial Intrahepatic Cholestasis (PFIC-1-3)

The hepatic uptake of bile salts, organic anions, and cations are regulated by the sodium-dependent taurocholate cotransporter protein (NTCP), organic anion transporters (OATP1–2), and organic cation transporter (OCT1) at the basolateral membrane of the hepatocyte. These transporters are not directly ATP-dependent for their function. ATP-dependent function is mostly vital for transporters mediating the secretion from the liver to bile drainage system (ATP-binding cassette proteins or ABC) with several members including the P-glycoprotein (MDR), the multidrug resistance proteins (MRPs), the cystic fibrosis transmembrane regulator, the transporter associated with antigen presentation (TAP) proteins, and a peroxisomal long-chain fatty acid transporter. Three proteins are at the basis of PFIC-1-3. It is important to remember that phospholipids are required in bile for the protection of the canalicular membrane and of the cholangiocytes from the toxic effects of high concentrations of bile salts, and this protection is achieved by using mixed micellar formation.

Translocases and Transporters of the Hepatocyte Membrane

ATP8B1 (18q21-22) encodes for *FIC1* located in the canalicular membrane of hepatocytes, cholangiocytes, enterocytes, pancreas, among other tissues/organs (P-type ATP-dependent aminophospholipid *flippase* translocating *phosphatidylserine* from outer to inner leaflet of bile canalicular membrane). Gene mutations, inherited in an AR fashion, can cause *PFIC-1* (Byler disease), *ICP*, and *BRIC* according to the presence of modifier and modulator genes. Clinically there is a serum increase of bile salts, normal GGT, decrease bile acid secretion, and consequent pruritus, jaundice, and diarrhea. Therapy includes partial external biliary diversion, rifampicin, and 3-hydroxy-3-methyl-glutaryl-coenzyme A reductase (HMGCR) inhibitors, such as simvastatin. In the case of milder functional defects, BRIC is diagnosed, which manifests with recurrent cholestatic/icteric attacks but no progression toward liver cirrhosis. Greenland familial cholestasis (GFC) also is associated with *PFIC-1* gene mutations. This gene, which is on chromosome 18, is also mutated in the milder phenotype, benign recurrent intrahepatic cholestasis type 1 (BRIC1). Byler disease (BD), also called progressive familial intrahepatic cholestasis (PFIC), is of autosomal recessive inheritance. Two specific gene defects could be identified: PFIC-1 located on 18q21.22 and PFIC-2 situated at band 2q24. Low serum levels of GGT activity are linked to these defects. Progressive cholestasis is accompanied by histomorphology changes including the lack of the intrahepatic bile ducts, giant cell transformation of the hepatocytes, and intensifying fibrosis. Final cirrhosis leads to death in late infancy or early childhood.

ABCB11 (2q24) encodes for *BSEP* located in the canalicular membrane of hepatocytes (ATP-dependent *bile salt exporter protein* toward the lumen of the bile canaliculus), and gene mutations (AR inheritance) can cause *PFIC-2, ICP,* and *BRIC* as well as *mixed cholelithiasis.*

ABCB4 (7q21) encodes for *multidrug resistance protein 3* or *MDR3* located in the canalicular membrane of hepatocytes, macrophages, cerebral vascular endothelial cells, among other tissues/organs (ATP-dependent phospholipid *floppase* translocating *phosphatidylcholine* from inner to outer leaflet of bile canalicular membrane), and gene mutations (AR inheritance) can cause *PFIC-3, ICP, drug-induced cholestasis, cholesterol cholelithiasis,* and *"cryptogenic" cholangiopathic cirrhosis.* In PFIC-3 jaundice and pruritus are associated with hepatosplenomegaly and portal hypertension. Therapy: ± UDCA. Even though ICP is a limited disease, which resolves after delivery, it may be related to stillbirth, premature births, or other complications of the fetus (Dixon et al. 2017). In summary, clinically and biochemically, PFIC-3 is differentiated from PFIC1-2 due to high GGT, while PFIC-1 is differentiated from PFIC-2 by the occurrence of extrahepatic symptoms, such as diarrhea or pancreatic insufficiency, which are characteristics of PFIC-1, but not of PFIC-2. Histologically, PFIC-2 can be diagnosed by the absence of BSEP-2 immunohistochemical reactivity from the canaliculi using BSEP-specific antibodies (Box 4.8). A rabbit polyclonal and a mouse monoclonal antibody are commercially available. Most often, the immunogen is the recombinant fragment corresponding to human ABCB11/BSEP between the amino acids at position 474 and 745 (internal sequence).

Box 4.8 PFIC Group of the "Three Musketeers"

PFIC-1: *ATP8B1* on 18q21-22 ⇒ ¬ FIC1 ⇒ ↓ Bile secretion into bile with normal GGT

PFIC-2: *ABCB11* on 2q24 ⇒ ¬ BSEP ⇒ ↓ Bile secretion into bile with normal GGT

PFIC-3: ABCB4 on 7q21 ⇒ ¬ MDR3 ⇒ No secretion of PC into bile with ↑ GGT

Notes: The use of the term three musketeers is linked to the close liaison that all three diseases may have in the dysregulation of the hepatocytic bile flow. *GGT* gamma-glutamyl transpeptidase, *PC* phosphatidylcholine, ¬ abnormal

---

Cystic Fibrosis

An AR-inherited disease characterized by the accumulation of thick, sticky mucus that can damage several organs. The most affected organ is the lung, but other organs are also affected, among them the pancreas and liver. The features of CF and its severity vary among affected individuals. Mutations in the *CFTR* (cystic fibrosis transmembrane conductance regulator, an ATP-binding cassette subfamily C, member 7) gene on 7q31.2 cause cystic fibrosis.

The gene product is a protein making a channel that transports negatively charged chloride ions into and out of cells. Mutations in the CFTR gene interrupt the correct functioning of the chloride channels, preventing them from regulating the flow of chloride ions and, of course, water across cell membranes. At the cholangiocellular level, *CFTR* gene mutations may determine cholestasis in the liver, and hepatobiliary complications may occur with evolution to secondary biliary cirrhosis. CF diagnosis relies on >1 characteristic phenotypic feature and one out of three laboratory findings: an elevated sweat chloride concentration on two occasions, one of the two most frequent *CFTR* gene mutations (ΔF508, i.e., deletion of the three nucleotides that comprise the codon for phenylalanine at position 508), and the demonstration of abnormal nasal epithelial ion transport (Box 4.9).

Box 4.9 CF Liver Patterns

Pattern	Rate	Comments
"Transaminitis"	10–35%	Asymptomatic patients (↑ liver enzymes only)
Neonatal cholestasis	<1%	Broad DDX
Steatosis, microvesicles	30–50%	Minimal to massive fatty change
Focal biliary fibrosis	10–70%	Focal change with portal fibrosis and biliary proliferation
Cirrhosis, micronodules	5–10%	Micronodular liver

CF may affect the liver with some histologic patterns, including steatosis, focal biliary fibrosis, and cirrhosis. In infancy or even at neonatal age, about 1/10 of CF individuals have a focal biliary disease with occasionally PIBD. The receptor protein is mainly expressed in a copious amount in the branching ducts of the pancreas, intestinal epithelium, testicular parenchyma, and bronchial glands. The receptor is not represented in the common hepatic bile duct or the hepatocytes. Conversely, the receptor is found in the apical membranes of the epithelium (epithelial cells) of branching interlobular bile ducts and gallbladder (Fig. 4.13).

Steatosis has been associated with malnutrition and essential fatty acid deficiency. Liver disease, although a component of morbidity for CF individuals, has been observed to be related to death in only about 1 in 20 cases, while 4/5 of CF individuals have lung disease-related deaths. Sclerotherapy or banding therapy has been used and reasonably substantially successful in treating bleeding esophageal varices, leaving TIPS (transjugular intrahepatic portosystemic shunt) procedure to be performed now only occasionally.

4.4.4.3 Infection-Related Neonatal Hepatitis

Sepsis and severe bacterial infections may result in a quite severe degree of cholestasis in both infant and children as well as youth. An isolated defect of excretion of conjugated bilirubin has been suggested due to the presence of conjugated hyperbilirubinemia in the presence of low ALP and low cholesterol plasma levels.

- *CLM*: Perivenular bilirubinostasis, fatty cell change of the hepatocytes, and portal inflammation ± reversible cytopathic biliary cells (swelling, pyknosis, focal bile duct damage). There is an optional ↑ # of bile ductules and Hering canals at periportal location with or without intraluminal inspissated bilirubin concrements (PAS+) and ductular neutrophilia without involvement of segmental bile ducts (DDX, bile duct obstruction with ascending cholangitis associated neutrophils in ductules + segmental bile ducts other than cholestasis).
- *DDX*: "Inspissated bile syndrome" is not an entity and has been misinterpreted as such but defines a cholestatic state with complex etiopathogenesis and often seen in the livers of infants or children who deceased within a month of birth following some illness.

4.4.4.4 Non-BASD Metabolic Dysregulation-Related Neonatal Hepatitis

Cholestasis may also be seen in metabolic disorders, which mostly are cystic fibrosis, galactosemia, fructose intolerance, tyrosinemia, and erythropoietic porphyria. Since cholestasis is

4.4 Infantile/Pediatric/Youth Cholangiopathies

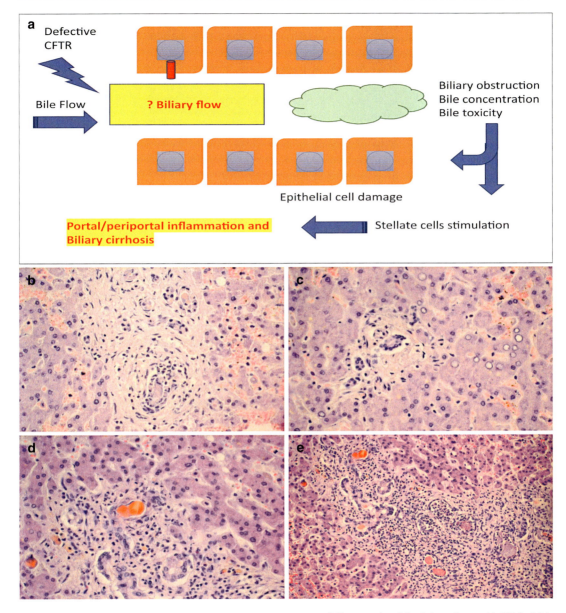

Fig. 4.13 Cystic fibrosis (CF) is a common AR-inherited disorder due to abnormal CFTR gene. Liver abnormalities in CF vary in nature ranging from defects attributable to the underlying CFTR gene defect to those related to systemic disease and malnutrition. Currently, hepatobiliary complications of CF represent the third most frequent cause of disease-related death in patients with CF. In (**a**) is shown a cartoon summarizing the pathogenesis of hepatobiliary disturbance in CF. The microphotographs (**b–e**) show the abnormalities that can be observed in the portal tracts with orangiophilic concretions of bile, biliary proliferations, and portal mononuclear inflammatory infiltrates

only part of the metabolic disorders, these syndromes are discussed in Sect. 4.6. Likewise, several diseases may also present with intraacinar cholestasis, although this would be part of a more complex picture and are discussed in other specific sections.

4.4.4.5 Endocrine Dysregulation-Based Neonatal Hepatitis (EDNH)

- *DEF*: Intraacinar cholestasis with simple bilirubinostasis associated with endocrine imbalances (e.g., hypothyroidism, hyperthyroidism, and panhypopituitarism).

Hypothyroidism and hypoparathyroidism, hypopituitarism, and diabetes insipidus, as well as hypoadrenalism, may intriguingly alter the metabolism of the bilirubin. Prematurity and/or small for gestational age have been suggested to be considered risk factors.

4.4.4.6 Miscellaneous/Idiopathic NAG

Cholestatic jaundice may also be seen in amyloidosis, which is an extremely rare event in childhood and youth. Light-chain deposit disease is due to amyloid deposition in the space of Disse. Consequently, hepatocytic atrophy and bile canaliculus compression can be seen. Moreover, sarcoidosis, malignant infiltration by neuroblastoma stage IV or IVS in infants, as well as malignant infiltration of other tumors in older children and youth (Hodgkin lymphoma, non-Hodgkin lymphomas, sarcoma, or, more rarely, carcinoma) with diffuse portal infiltration with or without bile duct obstruction are also present with cholestatic jaundice (Chiu et al. 2019).

- *DDX*: Stauffer syndrome, which is a nephrogenic (renal carcinoma) and non-nephrogenic (e.g., soft tissue sarcoma) hepatosplenomegaly with non-malignancy-related hepatic dysfunction (reactive hepatitis with zone 3 bilirubinostasis).

Trisomy 13, 18, 21 syndromes are known causes of neonatal cholestasis, but their phenotype is quite characteristic. The eclectic NAG group involves three additional heterogeneous subcategories, including non-biliary atresia-high GGT intrahepatic cholestasis, hemophagocytic lymphohistiocytosis (HLH), and ARC syndrome (arthrogryposis, renal tubular dysfunction, and cholestasis). Non-PFIC-1–3/non-arthrogryposis syndrome-low GGT intrahepatic cholestasis is quite broad. Several diseases may play a significant role in this group, but the list remains incomplete and will be updated in the nearest future. Alpers disease is an infantile diffuse cerebral degeneration of the gray matter accompanied by hepatic cirrhosis. It is an AR inherited mutation of the mitochondrial DNA polymerase (pol γ) located on the *POLG* gene.

4.4.5 Primary Sclerosing Cholangitis (PSC)

- *DEF*: It is the inflammatory destruction of the extrahepatic (early) and intrahepatic (later) biliary system with a substantial autoimmune basis and highlighting a progressive nature (HLA-B8, DR3) leading ineluctably to biliary cirrhosis of the liver (10% intrahepatic only) with or without potentially associated IgG4 diseases (Riedel thyroiditis, Ormond retroperitoneal fibrosis, and orbital pseudotumor).
- *EPI*: Worldwide, ♂:♀ = 2:1, 3rd–4th decades of life (also neonatal sclerosing cholangitis and juvenile forms), ± ulcerative colitis (UC) (60–90%) with PSC more common in p-ANCA (+) UC and UC usually developing first.
- *CLI*: Progressive fatigue, pruritus, and jaundice.
- *LAB*: "Beading"-ERCP appearance, (+) ANNA, (+) atypical p-ANCA recognizing both β isoform 5 of tubulin and bacterial cell division protein Ftsz.
- *CLM*: Liver biopsy with early and late lesions (Fig. 4.14).
- Early lesions: Large bile ducts surrounded by lymphocytes and plasma cells with a tendency to infiltrate the biliary epithelium and lumina filled with cellular debris (inflammation + ulceration + necrosis).
- Late lesions are subdivided in primary disease-related changes and chronic substantial "large duct"-like obstruction-associated changes.
  - Primary disease-related changes:
    o Portal and periportal inflammation with eosinophils and plasma cells (± "spillover") + pericholangitis (lymphocytic infiltration of the interlobular bile ducts with epithelial cell damage consisting of cytoplasmic vacuolation, nuclear pyknosis, and nuclear loss) with the possibility to evolve in a residual scar following the ductal destruction and "onion skin" fibrosis.

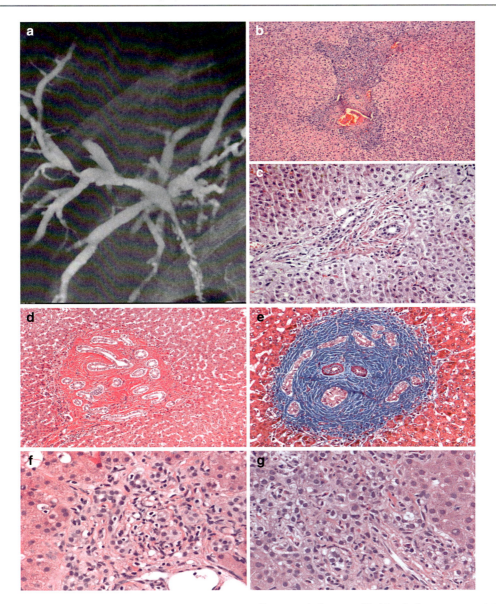

Fig. 4.14 Progressive sclerosing cholangitis gives a clear-cut radiologic picture in (**a**) and liver histopathology (**b**–**g**) with circumferential sclerosis of the bile ducts and hepatocytic cholestasis. The 11-year-old child suffered from Crohn disease. At 11 years, he presented with blood-stained stools and increased liver parameters. A colonoscopy performed was successful in identifying Crohn disease and therapy with Mesalazine was started. This drug, which is aka mesalamine or 5-aminosalicylic acid is a medication used to treat inflammatory bowel disease, including ulcerative colitis and Crohn disease. Three years later, a relapse occurred and a liver biopsy was performed following abdominal pain, jaundice, generalized itching, hepatomegaly, and splenomegaly

- Chronic substantial ("large duct-like") obstruction-related changes:
    ○ Portal tract inflammation with lymphocytes and fibrosis, bile ductular proliferation, canalicular/ductal cholestasis similar to chronic large duct obstruction, although the degree of lymphocytic infiltration appears to be quite greater with clearcut piecemeal focal necrosis.

- *DDX*:
    - Large duct obstruction (LDO) (*vide supra*) (PSC favoring clues are eosinophils, inflammatory damage to the small interlobular bile ducts, and tissue scars).
    - Primary biliary cirrhosis (PBC) (eosinophils and bile duct damage in an appropriate epidemiological setting help in the direction of PBC).
    - Chronic active hepatitis (CAH) (no obstructive features, ± PMNs/Eos, ± germinal centers, and prominent piecemeal necrosis in an appropriate virological/microbiological setting are key).
    - Other differential diagnoses to keep in mind are schistosomiasis (it involves, however, veins and not ducts), CMV infection, and cholangiocarcinoma. These diagnoses are not anymore restricted to some geographical areas because population migrations and traveling have re-shaped the medical approach to diagnosis.
- *PGN*: There is a substantial risk to develop cholelithiasis, cholangitis, strictures, and cholangiocellular carcinoma (up to 10%) and staging is key involving four grades: (1) portal, (2) periportal, (3) septal, and (4) cirrhosis.

4.4.6 Primary Biliary Cirrhosis (PBC)

- *DEF*: Lymphocytic inflammatory destruction of the bile ducts of progressive nature.
- *SYN*: Chronic nonsuppurative destructive cholangitis.
- *EPI*: Worldwide, ♂:♀ = 1:9, middle age, but youth cases have been described.
- *CLI*: ↑ ALP with normal ALT and AST, polyclonal IgM peak in serum, (+) H2-AMA (E2 pyruvate dehydrogenase subunit on inner membrane), which recognizes a mitochondrial Ag (PDC-E2) expressed in IBD, and (+) HLA-DR8.
- *CLM* – Characteristic features of PBC include:
    1. Lymphocytic inflammatory infiltration of the portal tracts with *patchiness* in the liver biopsy (different degree in several portal tracts) and within individual portal tracts (the inflammation is usually centered on the interlobular bile ducts) = *intraportal & interportal patchiness*.
    2. PBC type of bile duct damage with hypereosinophilia of the biliary epithelial cytoplasm with pseudostratification (≠ hepatitis type of bile duct damage with cytoplasmic vacuolation and loss of cells without stratification).
    3. Phenotype variability of piecemeal necrosis includes:
        - *Lymphocytic* type (lymphocytes destroy hepatocytes at the limiting plate).
        - *Fibrotic* type (irregular portal tract edge with fibrous tissue insinuating subtly between hepatocytes).
        - *Ductular* type (inexorable proliferation of bile ductules with prominent portal expansion).
        - *Biliary* type (loss of the limiting plate with bile pigment accumulation and macrophages recruitment).
            o Being the fibrotic, ductular, and biliary types of piecemeal necrosis much more characteristic of PBC than the lymphocytic type.
    4. Epithelial histiocytes, incipient granulomas, and well-formed granulomas are additional microscopic clues.
    5. Periportal cholestasis + periportal Mallory hyaline + hepatocellular Cu deposition are often observed.
    6. PMNs are usually in the regions of bile ductular proliferation and not within the lymphoid aggregates or around the interlobular bile ducts.
- *DDX*: CAH, Wilson disease, drug-induced liver disease/reaction, PSC, and large duct obstruction (LDO).
- *PGN*: Severe, if the liver function is compromised.

The bile duct damage in PSC and PBC is different and this aspect needs to be continuously stressed specially to pathologists not involved in the diagnostic of numerous liver biopsies. There is obvious cytoplasmic vacuolation and loss of individual cells and lymphocytes without stratification in PSC and cytoplasmic eosinophilia and pseudostratification in PBC. The Ludwig staging of PBC (which is also useful for PSC) includes (1) portal stage ("portitis" without piecemeal necrosis or fibrosis), (2) periportal stage ("portitis" with piecemeal necrosis or fibrosis but lack of bridging), (3) septal stage (+bridging necrosis or fibrosis), and (4) cirrhosis. The Scheuer staging of PBC (also useful for PSC) includes (1) "florid duct lesion" with "portitis" and ductal damage; (2) bile ductular proliferation with triad expansion, neo-cholangioles, and piecemeal necrosis; (3) scarring with the fibrous septal formation and ↓ of the inflammatory component, and (4) cirrhosis.

The Mallory hyaline is a cluster of keratin intermediate filaments showing as irregular eosinophilic intracytoplasmic material ("rop-like") sharply delimited from the surrounding cytoplasm and (+) keratins of both "bile duct type" and "hepatocyte type." It may be found in (1) *alcoholic liver disease* (ALD), centrilobular + neutrophilic satellitosis + surrounding fatty change; (2) *nonalcoholic steatohepatitis* (NASH), periportal/centrilobular ± neutrophilia + surrounding fatty change; (3) *drug-induced liver disease* (DILD), periportal/centrilobular + cholestasis + drug-related changes; (4) *PBC*, periportal, neutrophilic satellitosis, and surrounding fatty change; (5) *Wilson disease*, periportal ± neutrophilic satellitosis ± surrounding fatty change; and (6) long-standing *LDO*, periportal + cholestasis + bile ductular proliferation, but usually neither neutrophilic satellitosis nor surrounding fatty change.

In Box 4.10 are described some pearls and pitfalls for properly assessing the bile duct damage.

Pediatric liver cirrhosis is a final common pathway for many progressive and, ultimately, chronic diseases that may start even in the unborn child at the intrauterine fetal stage of the liver development. It is paramount to establish careful and thorough examination of the medical records of both mother and child since early infancy. Chronic cholate stasis refers to bile acid retention in periportal hepatocytes with pseudo-xanthomatous change, copper deposition, and potential Mallory hyaline formation. Children with PBC should be considered a rarity with only two reports in the scientific literature.

> **Box 4.10 Pearls and Pitfalls**
> - Hepatitis-type bile duct damage: Vacuolation of the biliary epithelial cytoplasm, loss of cells without stratification, and lymphocytic infiltrate.
> - PBC-type bile duct damage: Hypereosinophilia of the biliary epithelial cytoplasm, cellular pseudostratification, and lymphocytic infiltrate.
> - In PBC there is a mild lobular inflammation with granulomas, but the lobular changes *never* predominate unless there is a condition called "overlap syndrome" with PBC + AIH!
> - In liver cirrhosis, ductular proliferation is frequently observed regardless of etiology and should not be used as a histological finding pointing to PBC or any other specific cholangiopathy!

4.4.7 Pregnancy-Related Liver Disease (PLD)

- *DEF*: The liver function of a pregnant woman can be dysregulated and have an increase of transaminases, hyperbilirubinemia, and cholestasis but also hepatomegaly and develop seizures that can be life-threatening for herself and her baby.

PAFL:	Pregnancy-related acute fatty liver
PIC:	Pregnancy-related intrahepatic cholestasis
PPE:	Pregnancy-associated preeclampsia
HELLP:	Hemolysis, elevated liver enzymes, associated with low platelets

Acute Fatty Liver of Pregnancy

- Third trimester, ± AR inherited Fatty acid oxidation defects (FAOD)
- Diffuse microvesicular steatosis ± lymphocytes

Intrahepatic Cholestasis of Pregnancy

- Third trimester ± *MDR3* gene mutations
- "Bland cholestasis" with hepatocellular and canalicular cholestasis, which are most prominent in the centrilobular region of the liver acinus

Preeclampsia

- Third trimester, vasospasm, and disseminated intravascular coagulation (DIC) leading pregnancy complication characterized by proteinuria and hypertension that dominate the clinical picture.
- Hemorrhagic/ischemic necrosis, periportal fibrin deposition along sinusoids in the periportal areas, and fibrin thrombi in portal veins
- Glomerular endotheliosis is present because the kidney is the target organ.

HELLP Syndrome

- Complication of pregnancy characterized biochemically by *H*emolysis, *E*levated *L*iver enzymes, and a *L*ow *P*latelet count and histologically by nonspecific inflammation, glycogenated hepatocytic nuclei, and periportal fibrin/necrosis

4.5 Genetic and Metabolic Liver Disease

Hepatomegaly may have several and different underlying cytologic features and etiologic factors. We distinguish and include cellular hyperplasia/hypertrophy (hepatocytes, Kupffer cells, inflammatory cells such as in viral hepatitis and AI hepatitis), fibrosis (such as congenital hepatic fibrosis), venous congestion (such as chronic right heart failure or hepatic outflow blood vessel obstruction or Budd-Chiari syndrome), biliary tract dilatation, accumulation of metabolic substances and metals (e.g., fat, cholesterol, glycogen, sphingolipid, sphingomyelin, acute phase proteins, copper, and iron), as well as atypical cellular proliferation. Fat can usually accumulate under three main conditions, including obesity, diabetes mellitus (e.g., Mauriac syndrome), and FAOD (MCAD and LCHAD) due to the inability to use fat, which accumulates in the liver (low ratio of ketone bodies to dicarboxylic acid and low total serum carnitine). Galactosemia, oxidative phosphorylation defects (e.g., Alpers syndrome), and Reye syndrome show triglycerides (TG) accumulation and hypertrophy of SER, while cholesterol accumulation may be observed in cholesterol ester storage disease (CESD) and Wolman disease, Glycogen Storage Disease (GSD) I and IV, Gaucher disease, Niemann Pick disease (NPD), alpha-1-antitrypsin deficiency (AATD), Wilson disease, and Indian childhood cirrhosis or malnutrition. Wilson disease occurs between the second and fourth decade of life and is diagnosed if serum ceruloplasmin level is <20 mg/dL, liver copper >250 μg/g, and urinary copper >100 μg/day. Primary tumor infiltration may be an embryonal sarcoma, hepatoblastoma, hepatocellular carcinoma, rhabdomyosarcoma, teratoma, and hemangioendothelioma. Secondarily, the liver may be infiltrated by neoplasms of adjacent organs such as the adrenal glands/paraganglia (neuroblastoma), kidney (Wilms tumor), and lymph nodes (Hodgkin lymphoma and NHL). These neoplastic processes interfere with the complex metabolic processes that run in the liver determining a dysregulation at different time of the neoplastic infiltration. A standard feature of many metabolic disorders is represented by pericellular fibrosis that may exquisitely help to distinguish these disorders from other more common causes of fibrosis, such as chronic active hepatitis. Chronic active hepatitis is usually characterized by a more confluent portal fibrosis. On the other hand, prominent microvesicular steatosis in the liver of a child with hepatomegaly and with or without subsequent portal hypertension does not necessarily point to a primary dysregulation of the lipid utilization pathway. Hepatic fatty change may coexist as a confounding variable in several and different metabolic defects, either of poor metastasis intake or because the primary deficiency intrinsically interferes with lipid processing or lipid utilization.

Acute liver disease etiologies include viral hepatitis A and B viruses, Reye syndrome, and cystic liver diseases. Systemic diseases involving the liver include cystic fibrosis, sickle cell disease, obesity, TPN, celiac disease (AIH, PSC, PBC, "transaminitis" an increase of liver enzymes), PSC, IBD, Langerhans cell histiocytosis (LCH), hemophagocytic lymphohistiocytosis (HLH), muscular dystrophies, and inborn errors of glycosylation (carbohydrate-deficient glycoprotein syndromes).

4.5.1 Endoplasmic Reticulum Storage Diseases (ERSDs)

Endoplasmic reticulum storage diseases are a group of metabolic disorders with an accumulation of undigested material or inappropriately stored content in the endoplasmic reticulum of a cell (Box 4.11).

> **Box 4.11 ERSD(s)**
> 4.5.1.1. Alpha-1-antitrypsin deficiency (AATD)
> 4.5.1.2. Non-AATD endoplasmic reticulum storage diseases

4.5.1.1 Alpha-1-antitrypsin Deficiency (AATD)

Alpha-1-antitrypsin (AAT) is a serum acute phase protein with essential roles in the inflammation and other processes of the body and is synthesized in the ER and glycosylated in the Golgi complex. AAT deficiency (AATD) is probably doubtless the prototype for an ERSD along with diseases such as Alzheimer disease and Parkinson disease. AATD may also be considered a disorder of protein structure and degradation affecting a selected group of proteins called "serine proteases." This group of diseases has also been collected together under the name of serpinopathies, and the gene encoding AAT is on 14q31-32.3 and has labeled *SERPINA1*. The gene comprises four expressing regions or coding exons (II, III, IV, V) and three additional exons (IA, IB, IC) on 5'. Promoters, responsible for the steady state of the protein, and enhancers, accountable for expression during inflammatory states, are distinct. They are specifically regulated through different mechanisms in hepatocytes and monocytes, which are the two critical cells that produce AAT. In hepatocytes, there are discrete binding sites that provide well distinct pathways able to regulate the acute phase response through different cytokines interaction (Fregonese and Stolk, 2008). Among others, the control of the local transcription is associated with sites located upstream of *HNF-1a* and *HNF-4*, which are two binding sites. Probably, the humoral regulation is mostly mediated by IL-6 and oncostatin-M. In monocytes and other cell types, the promoter for *SERPINA1* is regulated by a distinct group of transcription factors aka "SP1 family." Alternative transcripts and stability of the gene product are also physiologically significant. RNA stability explicitly influences gene expression, and some gene transcripts seem to be quite more stable than others. This situation may be particularly relevant for therapy purposes as well. A defective gene in a cell type, e.g., monocyte, may be corrected by transfection of small DNA fragments of the unaffected gene to result in increased protein secretion. More than 100 genetic variants of *SERPINA1* are known, although the most common and clinically essential alleles leading to AATD are the Z and S alleles, while the normal allele is M. The Z allele is higher in Scandinavian countries, while the S allele is highest in Spain and Portugal. Alphabet letters are derived from the migration of the protein through a 5% polyacrylamide gel electrophoresis (PAGE) (pH 4–5) with isoelectric focusing (IEF) usually performed at 1400 volts or 40–50 mA for 1.5 h on an electrophoresis unit. Each protease inhibitor (Pi) type shows a characteristic set of bands from top to bottom on PAGE and is reported as Pi phenotype. Each chromosome is responsible for its gene transcription. Thus, AATD is an autosomal co-dominant inherited disease.

- *EPI*: The Z allele frequency varies in different geographic regions, being the highest in Northern Europe with a value of 0.0140 among Caucasians. In North America, the Z allele is 0.0092 and in Australia and New Zealand is 0.0151, whereas values of 0.0002 and 0.0061 have been reported in Japan and South Korea. In the remaining parts of Asia, Z allele varies from 0.0036 in Southeast Asia to 0.0056 in the Middle East.

- *CLI*: Clinically significant disease states occur mainly with PiZZ, PiSZ, and PiNull (or PiQ0), the latter defined as no protein is produced due to a stop codon. Clinically, individuals affected with AATD may have a widely variable spectrum of disease, ranging from asymptomatic individuals to patients with fatal liver or lung disease. PiZZ and PiSZ are well-known risk factors for the development of respiratory symptoms, such as dyspnea, coughing, early-onset emphysema, and airflow obstruction in youth. The 52 kDa AAT protects in the pulmonary parenchyma the healthy and fragile alveolar tissue from proteolytic damage by enzymes like elastase released by neutrophils during inflammatory states. Neonatal hepatic syndrome with prolonged jaundice after birth with conjugated hyperbilirubinemia and abnormal liver enzymes can also be seen. In about 10% of neonates, however, there is frank jaundice.
- In AATD, the Z allele is the most important genetic defect. It corresponds to Glu → Lys at position 342 of DNA. The Z allele-derived protein has an abnormal polymerization in the hepatocytes with aggregation in the ER and consequently reduced secretion and naturally decreased serum AAT concentration. The Glu to Lys amino acid substitution disrupts a crucial salt bridge in a beta-pleated sheet region of AAT. It allows the reactive loop from one molecule to insert itself into the beta sheet of an adjacent molecule. The PiS phenotype corresponds to Glu264Val, which has no dramatic consequences as observed in the PiZ phenotype.
- *CLM*: AATD liver disease is due to continuous polymerization and consequent accumulation of the protein and formation of light microscopically identifiable cytoplasmic inclusions, which are more evident in chronic inflammatory states in childhood or more often at adult age. AATD neonatal type shows, histologically, the characteristic triad of cholestasis, the giant-cell transformation of the hepatocytes, and PIBD, in the absence of the above mentioned cytoplasmic inclusions (Fig. 4.15). Thus, a pitfall may be to diagnose AATD by the simple identification of inclusions. The fate of AATD liver disease is liver cirrhosis with or without neoplastic transformation (HCC and CCA). In AATD, the inclusions are PAS+ and diastase resistant, round to oval cytoplasmic globules and located in the periportal regions of the liver. Small AAT globules may be difficult to distinguish from lipofuscin, iron, or fibrinogen or even giant mitochondria. Thus, the wise use of special stains and immunohistochemistry using an antibody against the AAT molecule or, also, the Glu-Lys protein (or protein from Z allele) is of enormous help (Callea et al. 1991). Moreover, nonspecific inclusions are associated with congestion and mostly centrilobular rather than periportal. Giant mitochondria are usually not periportal and are PAS negative. Mallory hyalin is irregular in shape rather than round and located more often in the centrilobular areas rather than periportal regions with the exception of the Mallory hyalin occurring in cholate stasis. On the other hand ("unwise use"), immunohistochemistry may lead to false-positive results, because a diffuse granular cytoplasmic staining without granulopoiesis may be detected in several disease states, including various regeneration and degenerate conditions. Finally, PiMM phenotype may occasionally be observed also with an accumulation of AAT. Genotyping of these patients characterizes a special group, called "M-like" AAT molecules. These AAT molecules harbor peculiar mutation leading to normal Pi phenotype on usual IEF but are detectable with IEF using a restricted range of isoelectric pH or molecular techniques, such as DNA sequencing. A few AAT molecules have been characterized and include M-Duarte, M-Malton, or M-Cagliari. Electron microscopy investigation shows a characteristic accumulation of finely granular material, which is located in dilated

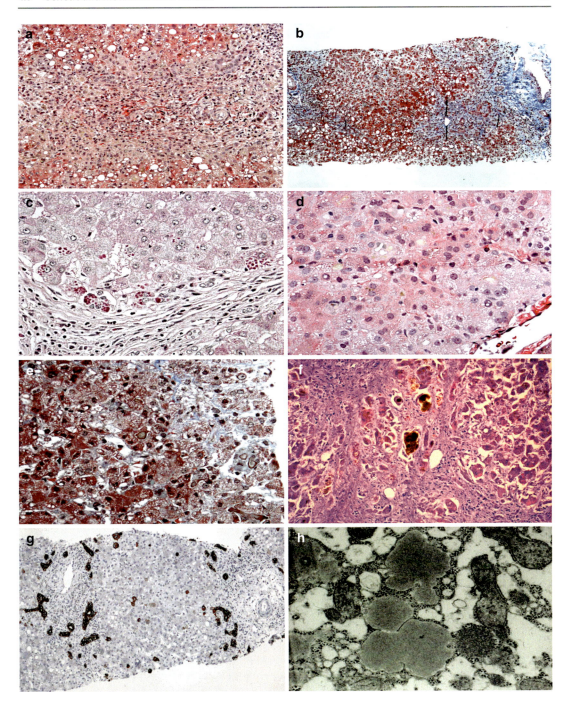

Fig. 4.15 Alpha-1-antitrypsin deficiency is associated with fibrosis and cirrhosis and the presence of eosinophilic, PAS-positive, and diastase-resistant periportal globules that are expression of an accumulation of alpha-1-antitrypin (**a–e**). Some patients may show an impressive and fast developing perihepatocytic (peri-sinusoidal) collagen fiber deposition that may points to a rapid deterioration of the liver with need for the patient to be considered for liver transplantation in the near future. In fact, the liver is unable to excrete this protein because a mutation alters the 3D structure allowing the protein to keep stacked in the endoplasmic reticulum. As expression of the endoplasmic reticulum storage disorder, there is accumulation of granular material in the endoplasmic reticulum of the hepatocytes as observed in the transmission electron microscopy (**h**). Cholestasis is present (**f**) and proliferating biliary structures (**g**) are seen at the periportal areas

cisterns of the rough endoplasmic reticulum and is moderately electron-dense. The ultrastructural evidence is particularly useful in the evaluation of neonatal or infantile cholestasis with AATD, because of the infrequent histologic evidence of periportal globules.

AATD lung disease develops in some individuals harboring AATD deficiency. In the lung, AAT that is released into the serum by the hepatocytes enters by diffusion, where it inhibits neutrophil elastase. The inhibition of neutrophil elastase is fascinating because of the proteolytic cleavage of the reactive loop of AAT by elastase. The division of this peptide bond determines the N-terminal portion of the circle to insert into the main beta sheet structure of the AAT molecule. Thus, the elastase becomes trapped against the inhibitor, which leads to its internalization and following degradation of the protease/inhibitor complex by lysosomes. Interestingly, AAT plays a role of competition/inhibition with leukocytic elastase. Reaction kinetics favors elastase complexing with AAT rather than with the elastin molecule, which is its substrate.

If transplantation of a liver from a donor with M-type genotype is useful for reverse AATD, treatment of lung disease includes intravenous AAT augmentation therapy, annual flu vaccination, and a pneumococcal vaccine every 5 years.

Non-hepatic/non-pulmonary AATD-related diseases include panniculitis and glomerulonephritis among other complications. In the UK study, four categories are recognized, which are listed in Box 4.12.

> **Box 4.12 AATD Liver Disease Categories**
> 1. Gradual improvement and normal liver by the age of 3–10 years
> 2. "Transaminitis" (e.g., persistent, abnormal serum AST)
> 3. Cirrhosis with liver transplantation by the age of 17 years
> 4. Hepatomegaly and portal hypertension with serum AST increase, but no jaundice

Pediatricians who have neonates in the ward with bleeding disorders or unexplained jaundice have a long list of differential diagnosis, such as virus infection. Unlikely to be associated with AATD would be Wilson disease, which usually occurs after 6 years of age, and AIH, which is also characteristic of older children. Hemochromatosis of neonatal type is a heterogeneous disorder and needs to be taken into consideration.

Phenotyping by IEF and genotyping by molecular biology techniques should be performed in case of a serum level of AAT < 11 µM or 50 mg/dL by radial immune diffusion or nephelometry, although these values may be falsely normal during any period of acute inflammation. Thus, IEP and genotyping are very important tools in the evaluation of neonatal cholestasis syndrome, probably independently from the AAT values in serum.

- *PGN*: ZZ subjects (1 in 2–4000) have 10–15% average level of the protein. Nevertheless, 2/3 of ZZ subjects are going to develop liver disease, and up to 1/20 of them may die by 4 years following the onset of the symptomatology if no treatment is started.

4.5.1.2 Non-AATD Endoplasmic Reticulum Storage Diseases

Other ERSDs include α-antichymotrypsin, C1 inhibitor, a/hypofibrinogenemia, and antithrombin III deficiency. These disorders are rare and the reader is advised to consult specific textbooks of hepatology.

4.5.2 Congenital Dysregulation of Carbohydrate Metabolism

Congenital dysregulation affecting the metabolism of the carbohydrate is a heterogeneous group of metabolic diseases that may be life-threatening if they are not recognized on time (Box 4.13).

4.5.2.1 Glycogen Storage Diseases

Congenital dysregulation of the metabolism of carbohydrates is a very complex and heterogeneous group, including several diseases; some

> **Box 4.13 Congenital Dysregulation of Carbohydrate Metabolism**
> 4.5.2.1. Glycogen Storage Diseases
> 4.5.2.2. Lafora Progressive Myoclonic Epilepsy
> 4.5.2.3. Galactosemia
> 4.5.2.4. Hereditary Fructose Intolerance
> 4.5.2.5. Congenital Disorders of Glycosylation

of them are particularly important in childhood and will be selectively highlighted in this textbook. The group of congenital carbohydrate metabolism dysregulation diseases (CCMDD) includes glycogenosis, galactosemia, tyrosinemia, and hereditary fructose intolerance.

4.5.2.2 Lafora Progressive Myoclonic Epilepsy

- *DEF*: AR-inherited genetic disease, characterized by mutations in one of two genes, either *EPM2A* on chromosome 6q24 or *NHLRC1* (or *EPM2B*) on 6p22.3, whose gene products are laforin and malin, respectively. Laforin is a protein tyrosine phosphatase with a carbohydrate-binding domain in the N-terminus, while malin is a Zn finger-containing protein of the RING-type in the N-terminal and six "NHL-repeat" domains in the C-terminal direction. A RING (*Really Interesting New Gene*) finger domain is a protein structural domain of Zn finger type with C_3HC_4 amino acid motif which binds two Zn cations, while the *NHL-repeat* ("*Ncl-1, HT2A and Lin-41*"), is an amino acid sequence found extensively in numerous eukaryotic and prokaryotic proteins. Laforin and malin possess an inhibitory activity to polyglucosan accumulation through their phosphorylase activity and binding/inactivation of glycogen synthase, respectively. Both laforin and malin deficiency leads to an accumulation of glycogen molecules that are insoluble, because of branching insufficiency. Insoluble glycogen molecules lead progressively and inexorably to polyglucosan or Lafora bodies.
- *EPI*: Worldwide (more common in Mediterranean countries, Central Asia, India, Pakistan, North Africa, and the Middle East) with variable age of onset (6-19 years).
- *CLI*: Myoclonus (episodes of sudden, involuntary muscle jerking or twitching affecting part of the body or the entire body), seizures (generalized tonic-clonic episodes), and progressive dementia with onset in late childhood or adolescence.
- *CLM*: Lafora bodies are the hallmark and represent an abnormal form of glycogen that cannot be degraded to be used in the cells, and neurons are particularly vulnerable. Lafora bodies are to be found in many body tissues, but signs of this disease are limited to the nervous system. Biopsy of the axillary skin is the most used procedure to show characteristic inclusion bodies in duct cells of eccrine sweat glands. The liver may be affected, and slight periportal fibrosis has been noted, although the clinical liver disease is not usually a problem in Lafora progressive myoclonus epilepsy. In the brain, Lafora bodies are found in the substantia nigra, globus pallidum, dentate nucleus, and parts of the reticular system and cerebral cortex. The inclusions are sharply demarcated, homogeneous, pink eosinophilic that displace the nucleus to the periphery of the cell and show an artefactual halo from the surrounding cytoplasm. Special stains, including PAS, DPAS, Best carmine, Lugol iodine, methenamine silver nitrate, and colloidal iron, are positive. The pivotal treatment of the histologic slides with alpha-amylase, amyloglucosidase, and pectinase digests the inclusion bodies.
- *TEM*: Non-membrane bound clusters of interwoven 6–10 nm filaments.
- *DDX*: Lafora bodies need to be differentiated from ground glass change of HBV-infected hepatocytes, which are Victoria blue, orcein, and aldehyde fuchsin positive, changes of hepatocytes following treatment with cyanamide or disulfiram or 6-thioguanine therapy for leukemia as well as glycogenosis type IV. Van Hoof-Hageman-Bol Lafora bodies classification includes type I, PAS → granular

and red (most frequently seen); type II, PAS → dense peripheral zone; and type III, PAS → homogeneously bright red (Van Hoof and Hageman-Bal, 1967).

Lafora progressive myoclonus epilepsy has a quite distinctive pattern of neurological features. The neurology of Lafora progressive myoclonus epilepsy includes behavioral changes, confusion, dementia, depression, dysarthria, generalized tonic-clonic seizures ("*Grand mal*" seizures), intellectual function decline, myoclonus, and status epilepticus. Patients affected with Lafora progressive myoclonus epilepsy lose "progressively" the ability to perform daily living activities in the middle of the second decade and generally survive only 10 years following the appearance of the symptoms.

4.5.2.3 Galactosemia

It is an AR-inherited genetic disease with a triad of fatty change and pericellular fibrosis of hepatocytes with cholestasis, similarly to acute tyrosinemia. Cholestasis is also associated with the pseudoacinar transformation of the hepatocytes as well as giant cell transformation of the primary autochthonous liver cells. Perisinusoidal fibrosis may be associated with the fibrotic expansion of the portal tracts with eventual development of cirrhosis macroregenerative or dysplastic nodules.

- Pentad: Steatosis + cholestasis + hepatocytic disarray + pseudoacini + pericellular fibrosis.
- *TEM*: Bile, lipid droplets, ↑ ER, and abnormal mitochondria are seen.

4.5.2.4 Hereditary Fructose Intolerance

It is a neonatal hepatitis (NH) with giant cell transformation, steatosis, fibrosis, and cirrhosis similar to galactosemia (see above) and tyrosinemia (*vide infra*) and needs to be rapidly diagnosed.

- *LM* Pentad: Steatosis + cholestasis + hepatocytic disarray + pseudoacini + pericellular fibrosis

- *TEM*: Concentric membranous arrays of the SER are a prominent feature and typically associated with rarefaction of the cytosol.

4.5.2.5 Congenital Disorders of Glycosylation (CDG)

- *DEF*: Group of metabolic diseases affecting the glycoprotein biogenesis with at least 18 different genetic types following almost all an AR inheritance pattern and consisting in the absence or substantial structural alteration of N-glycan chains due to either deficiency of the biosynthesis of oligosaccharide precursors or deficiencies residing in specific steps of N-glycan assembly.
- CDG-I: Defects in the assembly of the lipid-linked oligosaccharides and their transfer to the protein either located in the cytosol (CDG-Ia and CDG-Ib) or the ER (CDG-Ic, CDG-d, CDG-Ie).
- CDG-II: Defects in the processing of the protein-bound glycans, located either in the ER (CDG-IIb) or in the Golgi complex (CDG-IIa and CDG-IIc).
- *LAB*: Serum isoelectrofocusing (IEF).
- *CLI*: Variable phenotype, including the dysmorphic facies, failure to thrive, neurological abnormalities, hepatosplenomegaly, and cardiomyopathy delineating often, but not always, two clinical subtypes:
  – *Neurological* form of CDG: psychomotor retardation with *strabismus* underlying cerebellar hypoplasia and *retinitis pigmentosa*.
  – *Multivisceral* form of CDG: neurological and extra-neurological features with the involvement of the liver, kidney, heart, and gastrointestinal tract.
- *CLM*: Steatosis and fibrosis, but no specific findings.
- *TEM*: Findings consistent with phospholipidosis, including lysosomal myeloid bodies or *Ghadially myelinosomes* in the hepatocytes only with an aspect similar to inclusions found in several states of prolonged cholestasis, lysosomal storage disease (e.g., Niemann-Pick), as well as phospholipidosis following drug toxicity. However, in cholestasis autolysosomes are seen predominantly in the pericanalicular region, while

myelinosomes are often found in both hepatocytes and Kupffer cells in several lysosomal storage diseases.

4.5.3 Lipid/Glycolipid and Lipoprotein Metabolism Disorders

Metabolic defects of the lipid or glycolipid metabolism and lipoprotein deficiencies are summarized in Box 4.14.

> **Box 4.14 Lipid/Glycolipid and Lipoprotein Metabolism Disorders**
> Mannosidosis
> Fucosidosis
> Aspartylglycosaminuria
> Niemann-Pick Disease
> Gaucher Disease
> Mucolipidoses (Fig. 4.16)
> Mucopolysaccharidoses
> Disorders of Lipoprotein
> A/Hypobetaliproteineamia
> Tangier Disease
> Familial Hypercholesterolemia
> Wolman Disease
> Cholesterol Esther Storage Disease
> Gangliosidoses
> Fabry Disease
> Metachromatic Leucodystrophy
> CTX
> Ceramidase Deficiency

Before starting with the single diseases, it may be useful to recall the classification and terminology of lipids.

Sphingosine and Sphingolipids

Sphingosine is an essential component in biochemistry, and many pediatric disorders are linked to it or its derivates. Both sphingolipids and phospholipids have a polar hydrophilic head group and two nonpolar hydrophobic tails. Sphingosine is a long-chain amino alcohol and is, substantially, the core of sphingolipids, which includes sphingomyelins and glycosphingolipids. Glycosphingolipids are cerebrosides, sulfatides, globosides, and gangliosides. Palmitoyl-CoA and serine condensate to form the sphingoid bases (sphingosine, dihydrosphingosine, and ceramides on the cytosolic portion of the endoplasmic reticulum through the enzymatic action of serine palmitoyltransferase (SPT)). Sphingomyelins, are sphingolipids that are also phospholipids which are essential structural components of nerve cell membranes. Chemically sphingomyelins are structured as having palmitic or stearic acid N-acylated at carbon 2 of sphingosine. The pathway for the synthesis of sphingosine is the following: Ser → 3-ketosphinganine → dihydrosphingosine → dihydroceramide → ceramide → sphingosine and the enzymes involved are serine palmitoyltransferase, 3-ketosphinganine reductase, dihydroceramide desaturase, and ceramide synthase/ceramidase. The transfer of phosphorylcholine from phosphatidylcholine synthesizes the sphingomyelins to a ceramide under the enzymatic action of sphingomyelin synthases (SMS) with two genes, i.e., *SMS1* located in the trans-Golgi apparatus and *SMS2* located in the plasmatic membrane. Sphingomyelins are degraded by sphingomyelinases, which is encoded, in humans, by the sphingomyelin phosphodiesterase-1 gene (*SMPD1*) located on 11p15.1–11p15.4. *SMPD1* gene defects cause Niemann-Pick disease (NPD). Ceramide is important for many epithelial cells, and the biological significance has been highlighted by experimental oncology studies, particularly head and neck cancer. The glycosphingolipids or glycolipids are structurally formed by a backbone of ceramide associated with several and different carbohydrate groups attached to carbon 1 of sphingosine. Four main classes need to be kept in mind in pediatrics, including cerebrosides, sulfatides, globosides, and gangliosides. Cerebrosides contain a glycol group (galactose or glucose) linked to ceramide forming galactocerebrosides and glucocerebrosides. The first cerebrosides are an essential component of neuronal cell membranes, while the second are intermediates in the synthesis and degradation of very complex glycosphingolipids and not usually found in neuronal membranes.

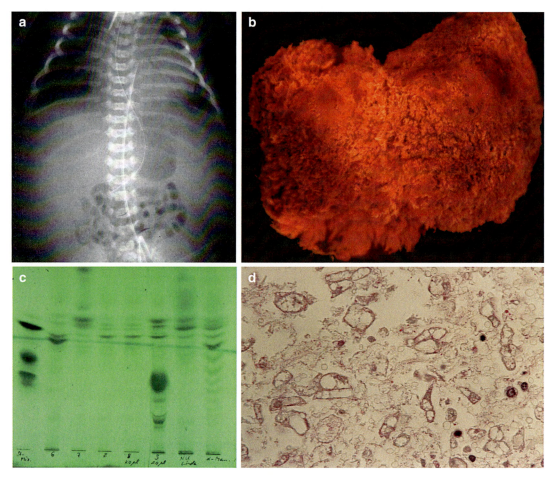

Fig. 4.16 Sialidosis is a rare metabolic disorder with ascites (**a**) identified in this babygram, hepatomegaly with liver failure (**b**), urine chromatographic pattern (**c**), and massive vacuolation of the liver cells (**d**)

Ceramide and UDP-galactose are used to form galactocerebrosides. A deficiency in the catabolism of the complex glycosphingolipids gives rise to lysosomal accumulation of glucocerebrosides as observed in diseases such as Gaucher disease. Sulfatides are the sulfuric acid esters of galactocerebrosides and are synthesized from galactocerebrosides and activated sulfate, which is 3′-phosphoadenosine 5′-phosphosulfate (PAPS). Sulfatides are found in excess primarily in metachromatic leukodystrophy, which is also known as sulfatide lipidosis. Globosides are cerebrosides with additional glycol-groups, e.g., galactose, glucose, or GalNAc. In Fabry disease, globotriaosylceramide (aka ceramide trihexoside) containing glucose and two molecules of galactose (three exposes with six atoms of carbon) accumulates in tissues, mainly in the kidneys. The gangliosides are similar to globosides with the exception that they contain NANA in variable amounts. The terminology of gangliosides is quite straightforward. The letter G refers to ganglioside, while the subscripts M, D, T, and Q point to the number of molecules of sialic acid (i.e., M = monosialic acid, D = disialic acid, T = trisialic acid, Q = quater/tetrasialic acid). The additional subscript numbers 1, 2, or 3 refer to the carbohydrate sequence, which is attached to the ceramide. Regarding the chemical formula and chemical structure, G = ganglioside + subscript letters (M/D/T/Q) + subscript number (1/2/3), ceramide (M = monosialic acid, D = disialic acid, T = trisialic acid, Q = quater/tetrasialic acid), 1 = GalGalNAcGalGlc-ceramide, 2 = GalNAcGalGlc-ceramide, and

3 = GalGlc-ceramide (see sources at http://themedicalbiochemistrypage.org).

Sphingolipidoses are incredibly devastating for the affected child, family, and pediatricians and represent one of the most complex disorders because of the accumulation in several and different organs. Lipids are the best molecules for energy storage because the cleavage of fatty acids produces both energy and water, which is essential for many metabolic processes. Acetyl-CoA is essential in many biochemical processes, but primarily it is converted from molecules of pyruvate formed in glycolysis and entering the mitochondria. In this complex series of biochemical reactions, pyruvate undergoes oxidative decarboxylation.

Sphingomyelins: N-acyl-sphingosine + phosphorylcholine, being the sphingosine a particular insature amino alcohol with a long chain represents the definition of sphingomyelin. The ceramides or N-acyl-sphingosine, through the primary alcoholic group, may be bound to hexoses or oligo-hexosidic chains giving rise to glycolipids or glycosphingosides. Lipidoses have nonspecific light microscopic findings, but the ultrastructural findings may be diagnostic. Lipidoses include Niemann-Pick disease, Fabry disease, Wolman disease, cholesterol ester storage disease, and gangliosidoses.

- *CLM*: Foamy change of the cytoplasm in both Kupffer cells and hepatocytes.
- *TEM*: See Box 4.15.
- *DDX*: Carnitine deficiency, toxic and metabolic etiologies of microvesicular steatosis such as Reye syndrome and drug reactions → vacuolation only in the hepatocytes.

4.5.3.1 Niemann-Pick Disease (NPD)

Lipid storage disorder, due to deficient activity of sphingomyelinase. Accumulation of lipids in macrophages and deposition in the nervous system lead to neurodegenerative symptomatic and can be accompanied by systemic manifestations involving the lung, liver, and spleen. Portal hypertension, cirrhosis, and ascites are common hepatic manifestations.

> **Box 4.15 TEM Features of Lipidoses**
> - *Fabry Disease*: Concentrically laminated + parallel-laminated inclusions
> - *GM1 Gangliosidosis*: Flocculent material in Kupffer cells + lamellar myelin figures + lipofuscin
> - *GM2 Gangliosidosis*: Parallel-laminated intralysosomal inclusions ("Zebra bodies")
> - *Niemann-Pick Disease*: concentrically laminated intralysosomal inclusions of sphingomyelin in both hepatocytes and Kupffer cells
> - *Wolman Disease/Cholesteryl Ester Storage Disease*: Needle-like clefts of cholesterol in Kupffer cells + neutral lipid in the hepatocytes

4.5.3.2 Gaucher Disease

In Gaucher disease, there is a portal tract infiltration of macrophages and enlarged Kupffer cells with "crinkled paper" cytoplasm ("Gaucher cells") with linear amphophilic quality on H&E, (+) DPAS, (+) Pearls. Ret stain is useful in highlighting the intrasinusoidal location of the Gaucher cells. There may be pericellular/centrilobular fibrosis leading to portal hypertension and occasionally cirrhosis. TEM shows intralysosomal tubular inclusions of accumulated glucocerebroside.

4.5.3.3 Mucolipidoses

Mucolipidoses are a group of metabolic diseases that are close to mucopolysaccharidoses and sphingolipidoses. Lowden and O'Brien classification indicates essentially two types of sialidosis, sialidosis, type I, 6p21.33, OMIM 256550, *NEU1*, mild form ("normosomatic" type or "cherry red spot myoclonus syndrome"), and sialidosis, type II, 6p21.33, OMIM 256550, *NEU1*, severe form ("dysmorphic" type), and is commonly subdivided into infantile and juvenile forms. Mucolipidoses are congenital lysosomal disorders, in which several cells and tissues show infiltrations of material stainable with Sudan black of mucolipidosis nature, of mucopolysaccharides, and, in some cases, of gangliosides (Fig. 4.16).

> **Box 4.16 "Metabolic Placenta"**
> - Gangliosidosis type 1, GM1
> - Morquio Disease
> - Mucolipidosis type I
> - Mucolipidosis type II
> - Mucolipidosis type IV
> - Niemann-Pick disease type C
> - Pompe disease

Sialidosis and not sialuria refers to progressive accumulation of sialidated glycopeptides and oligosaccharides into the lysosomal compartment due to a deficiency of the enzyme neuraminidase with subsequent accumulation and excretion of N-acetylneuraminic acid (sialic acid) bound to several and different oligosaccharides and glycoproteins. In humans, two clinically similar neurodegenerative lysosomal storage disorders, sialidosis and galactosialidosis, may link to a primary or secondary deficiency of neuraminidase. Moreover, it is important to differentiate between sialidosis and sialuria. If sialidosis is an accumulation and excretion of bound (i.e., covalently linked) sialic acid to oligosaccharides and glycoproteins, silauria is defined by the accumulation and excretion of free (not covalently linked) sialic acid (neuraminidase activity is normal or elevated, e.g., Salla disease). Mucolipidosis type II or I cell disease shows PAS (+) lysosomal inclusions in preferentially renal and mesenchymal cells of children with Hurler syndrome-like coarse features with the involvement of the placenta as well. In the examination of an infant, it is essential that the pathologist liaises with the clinician to receive the placenta and investigate this vital organ for the determination of the vacuolation of the cyto- and syncytiotrophoblast (Box 4.16). The practical workup in case of suspicious metabolic disease and potential placenta infiltration is shown in Box 4.17.

> **Box 4.17 Practical Workup in Case of a Suspicion of Metabolic Disease-Affected Placenta**
> 1. Samples for cryostat sections (1.5 × 1.5 × 0.4 cm) to stain for ORO
> 2. Samples (2× 1 mm × 1 mm × 1 mm tissue) for glutaraldehyde and epoxy embedding for TEM
> 3. Samples for biochemical, enzymatic activity (store at 4 degrees in the fridge using the appropriate transport medium)
> 4. Samples for genetic studies (3× 2 ml cryo-vials to freeze in liquid N_2-cooled isopentane) and stored at −80 C.
>
> Notes: *ORO*, oil red O; *TEM*, transmission electron microscopy

In the urea cycle environment platform, there are some essential reactions, including glutamate dehydrogenase reaction and glutamine synthetase reaction, seen among others.

Kwashiorkor: Syndrome of malnutrition seen in some underdeveloped countries where children are fed with low protein. In particular, the absence of lysine in low-grade cereal proteins leads to an inability to synthesize protein, due to the missing essential amino acids. The essential amino acids are arginine, histidine, isoleucine, leucine, lysine, methionine, phenylalanine, threonine, tryptophan, and valine, while the nonessential amino acids are alanine, asparagine, aspartate, cysteine, glutamate, glutamine, glycine, proline, serine, and tyrosine. There are three amino acids, including arginine, methionine, and phenylalanine, which are classified as essential for nonchemical reasons. Arginine is needed at a high rate because of the growing needs of the body but is synthesized by mammalian cells at a low rate. Methionine and phenylalanine are considered essential because they are needed in large amounts to form cysteine and tyrosine, respectively, if the latter two amino acids are not adequately supplied in the diet.

4.5.4 Amino Acid Metabolism Disorders

Amino acids are organic compounds with carbon (C), hydrogen (H), oxygen (O), and nitrogen (N) as critical elements and contain amine (-NH$_2$) and carboxyl (-COOH) functional groups. The amino acids also contain a side chain (R group), which is specific to each amino acid (Box 4.18).

Box 4.18 Amino Acid Metabolism Disorders
- Cystinosis
- Homocystinuria
- Tyrosinemia
- Cystathionine β-synthase deficiency

4.5.4.1 Tyrosinemia
- *DEF*: It is a life-threatening genetic disorder of the amino acid metabolism with disruption of the multistep process that breaks down the tyrosine with an accumulation of tyrosine and its by-products in tissues and organs.
    - Acute/hyperacute form with massive hepatic necrosis, cholestasis, generalized hepatocellular disarray, pseudoacinar transformation, and pericellular fibrosis
    - Chronic form: cirrhosis + steatosis
- *CLM*: Cirrhosis + patchy fatty change.
- *TEM*: Nonspecific findings, including fat accumulation, cholestasis, pleomorphism of mitochondria and peroxisomes, as well as collagen deposition.
- *DDX*: FHF of other causes (viruses, toxic or drug exposure, galactosemia, fructosemia, and neonatal hemochromatosis).
- *PGN*: Severe, if untreated.

4.5.5 Mitochondrial Hepatopathies

Mitochondrial diseases may involve not only the liver but also extrahepatic organs. The liver is mainly affected by mitochondrial disorders because the hepatocytes contain a relatively large number of mitochondria when compared with other cells of the body. Disorders targeting the mitochondrial oxidative phosphorylation (OXPHOS) and the metabolism of the hepatocytes have the consequence to influence fatty acid oxidation, which is the basis for bile flow impairment and fatty change as well as cell death and fibrogenesis. A liver manifestation of mitochondrial disease may give rise to a variable range of signs and symptoms from cholestasis and steatosis to chronic liver disease or neonatal liver failure with or without associated neuromuscular symptoms. Sokol and Treem classification includes primary and secondary mitochondrial hepatopathies, the latter being caused by a secondary insult to mitochondria by either a genetic defect or exogenous. Primary mitochondrial disorders may be further classified according to Leonard and Shapiro in diseases caused by mutations affecting mtDNA (class 1a) and conditions caused by mutations in nuclear genes encoding mitochondrial respiratory chain proteins or cofactors (class 1b) (Box 4.19) (Fig. 4.17).

Box 4.19 Mitochondrial Hepatopathies

4.5.5.1. *Electron Transport (Respiratory Chain) Defects*
4.5.5.2. *Fatty Acid Oxidation and Transport Defects*
4.5.5.3. *Disorders of Mitochondrial Translation Process*
4.5.5.4. *Urea Cycle Enzyme Defects*
4.5.5.5. *Mitochondrial Phosphoenolpyruvate Carboxykinase Deficiency*
4.5.5.6. *Secondary Mitochondrial Hepatopathies*

4.5.5.1 Electron Transport (Respiratory Chain) Defects

This group includes defects with neonatal liver failure due to complex I–III deficiencies (single or multiple), mitochondrial DNA depletion syn-

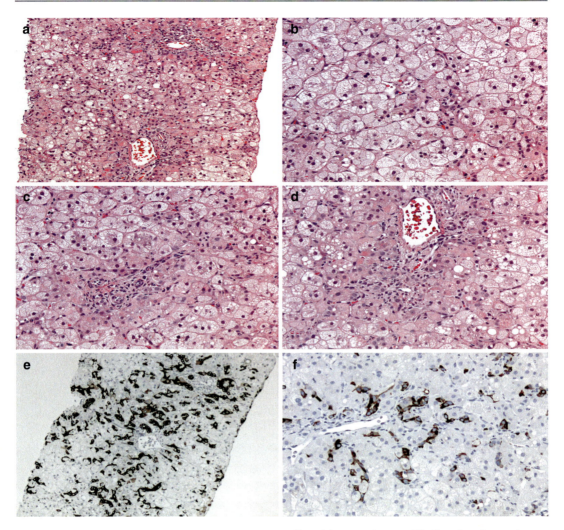

Fig. 4.17 Mitochondrial DNA (mtDNA) depletion syndromes (MDS) are a genetically and clinically heterogeneous group of AR-inherited disorders that are characterized by a severe reduction in mtDNA content. This process leads to impaired energy production in affected tissues and organs. The liver show foamy changes of the hepatocytes (**a–d**) and biliary proliferations (**e, f**) as well as reduction of cristae and swelling of the mitochondria ultrastructurally (not shown)

drome, Alpers-Huttenlocher syndrome with delayed-onset liver failure, Pearson syndrome, mitochondrial neuro-gastrointestinal encephalomyopathy, villous atrophy syndrome, Navajo neuro-hepatopathy, and electron-transfer flavoprotein (ETF) and ETF dehydrogenase deficiencies. The laboratory tools useful to investigate the respiratory chain defects include (1) plasma lactate, >2.5 mmol/L and persistently increased; (2) molar ratio of plasma lactate/pyruvate, >20.1 mol/mol; (3) molar ketone body ratio arterial 3-hydroxybutyrate/acetoacetate, >2.1 mol/mol; (4) postprandial plasma ketone bodies or lactate, paradoxical increase; (5) OGTT (2 g/kg) assessing plasma lactate/pyruvate ratio every 15 min six times for 90 min, elevated lactate/pyruvate ratio; and (6) urinary levels of lactate, succinate, fumarate, malate, 3-methylglutaconic, and 3-methyl glutaric by gas chromatography-mass spectroscopy. The oral glucose tolerance test (OGTT) is usually used to diagnose an intolerance to glucose/glycids. It involves the oral administration of a glucose drink to challenge the insulin-glucose response. All pregnant women not known to have gestational diabetes should be screened for glucose intolerance at 24th–28th weeks' gestation if they are at low or medium risk (an early screening is performed if the risk of gestational diabetes is higher). The original OGTT is performed assessing the glucose level 2 h after a 75 g glucose drink for nonpregnant women/adults and three blood samples (fasting or before the drink, 1 h, 2 h) after a 75 g glucose drink.

Neonatal Liver Failure (NLF)

- *DEF*: Complex I-, III-, and/or IV-related NLF with acute onset in the early newborn age with major clinical features, including transient hypoglycemia, neurologic involvement with severe hypotonia, apnea, lethargy, myoclonus epilepsy, psychomotor retardation, ↑ plasma lactate/pyruvate ratio, and β-hydroxybutyrate and arterial ketone body ratio of β-hydroxybutyrate to acetoacetate and rapid fatal downhill. Associated features may be low birth weight (about 1/4 of cases), polyhydramnios, hypertrophic cardiomyopathy, ventricular septal defects, cardiac arrhythmias, and hydronephrosis. The intersections between cardiac pathology and liver pathology are astonishing when investigating pediatric patients.

Mitochondrial DNA Depletion Syndrome (mtDNA-DS)

- *DEF*: Congenital multisystemic acute/chronic reduction in mtDNA copy number leading to insufficient synthesis of complexes I, III, IV, and V of mtDNA as compared to nuclear DNA, nDNA (<10% mtDNA/nDNA) with a normal mtDNA genome sequence manifesting in the early neonatal age with two phenotypes (myopathic and hepato-cerebral with hepatomegaly and progressive liver failure) with substantial heterogeneity. Three genes are involved, *POLG*, *DGUOK*, and *MPV17*.
    1. *POLG* ⇒ MtDNA polymerase-gamma (pol-γ), in which pol-γ activity is essential for mtDNA replication and repair leading to mtDNA depletion in case of defective protein.
    2. *DGUOK* ⇒Mt nucleotide salvaging dGK that together with TK2 acts to maintain a supply of dNTPs for mtDNA synthesis.
    3. *MPV17* ⇒ Mt inner membrane protein, which functions in preserving mtDNA and OXPHOS regulation leading to mtDNA-DS due to OXPHOS failure and mtDNA depletion.
- *CLI*: FTT, poor feeding psychomotor delay with seizures, nystagmus, pyramidal signs, cardiomegaly, congenital heart defects, and tubulopathy.
- *LAB*: ↑ of plasma lactate and plasma lactate/pyruvate concentration ratio.
- *CLM*: Macro- and microvesicular steatosis, hepatocytic and canalicular cholestasis, iron deposition in hepatocytes and sinusoidal cells, as well as fibrosis (Fig. 4.17).
- *TEM*: Small vesicles of lipid and mitochondrial distortion and pleomorphism.
- *PGN*: Less rapid than NLF of respiratory chain disease.

Alpers-Huttenlocher Syndrome

Delayed-onset liver disease (2 months–8 years) associated with *POLG* alteration and NADH oxidoreductase (complex I) deficiency in hepatic and muscular mitochondria and characterized by hepatomegaly, jaundice, progressive organ failure in the liver to cirrhosis, FTT, ataxia, refractory partial motor epilepsy or multifocal myoclonus, and characteristic EEG (high-amplitude slow activity with polyspikes) as extrahepatic manifestations. Lab values show normal plasmatic lactate and lactate/pyruvate ratio.

Pearson Syndrome

Infantile, insidious primary mitochondrial hemato-hepatopathy with mtDNA deletion altering complex I and III synthesis and characterized clinically by refractory sideroblastic anemia, vacuolization of bone marrow precursors, neutropenia, thrombocytopenia and hepatomegaly, cholestasis, and progressive liver failure. Other organs involved may be exocrine pancreas (pancreatic fibrosis and consequent insufficiency), endocrine pancreas (diabetes mellitus), kidneys (renal tubular disease or Fanconi's syndrome), small intestine (partial villous atrophy), skin (patchy erythematous lesions and photosensitivity), eyes (visual impairment and pigmentary retinopathy), and hydrops fetalis.

Villous Atrophy Syndrome
- *DEF*: Infantile/early childhood-onset chronic mitochondrial hepatopathy due to involvement of complex III of mtDNA rearrangement (complex III deficiency) and presenting as hepatomegaly and raised liver enzymes with vomiting, anorexia, chronic diarrhea, villous atrophy, diabetes mellitus, cerebellar ataxia, sensorineural deafness, retinitis pigmentosa, proximal muscle weakness, and seizures due to the involvement of the pyramidal tract of the CNS.
- *CLM*: Partial villous atrophy with eosinophilic infiltration in the small bowel and steatosis on liver biopsy.
- *LAB*: Complex III deficiency in skeletal muscle.

Navajo Neuro-hepatopathy
- *DEF*: Acute/chronic mitochondrial hepatopathy at onset in infancy or childhood with hepatic involvement (jaundice, ascites, Reye's syndrome-like episodes, progressive cirrhosis, and liver failure) and extrahepatic features (sensorimotor neuropathy, corneal anesthesia, acral mutilation, and progressive CNS white matter lesions with severe loss of myelinated fibers) due to MPV17 gene mutations affecting the respiratory chain complexes I, III, and IV.
- Three forms are known, including (1) infantile Navajo neuro-hepatopathy, FTT, jaundice ± neuropathology, death <2 years; (2) childhood Navajo neuro-hepatopathy, rapid development of liver failure: and (3) classic Navajo neuro-hepatopathy, progressive neurologic >> liver symptoms.
- *CLM*: Portal fibrosis with extension to micronodular cirrhosis, steatosis of micro- and macrovesicular type, pseudoacinar formation, MNGC formations, cholestasis, and periportal inflammation.

4.5.5.2 Fatty Acid Oxidation and Transport Defects

Fatty acid metabolism in the liver cells involves complex processes with the involvement of both cytoplasmic and mitochondrial compartments with probably more than a dozen defects described in the past two decades. Fatty acid oxidation defects (FAOD) include (1) long-chain hydroxy acyl CoA dehydrogenase deficiency, (2) acute fatty liver of pregnancy (AFLP) (harboring LCHAD enzyme mutations), (3) carnitine palmitoyltransferase I and II deficiencies, (4) carnitine-acylcarnitine translocase deficiencies, and (5) fatty acid transport defects. Long-chain acyl-CoA esters are carried into mitochondria using a carnitine-dependent mechanism. Once inside of the mitochondria, the long-chain acyl-CoA esters are metabolized by β-oxidation at the inner mitochondrial membrane and mitochondrial matrix. FAOD exit in decreased energy production interfering with the formation of ketone bodies and the alternate pathway of the output of dicarboxylic acid becomes dominant.

- *CLI*: Reye syndrome-like symptoms and liver disease occur in defects in carnitine handling (uptake) involving OCTN2, carnitine palmitoyltransferase (CPT1, CPT2) defects, medium chain acyl-CoA dehydrogenase (MCAD) deficiency, and LCHAD deficiency. Some FAO defects, including MCAD deficiency, LCACD deficiency, carnitine transporter defect, and carnitine translocase defect, have been indicated as etiologic factors in some SUDI and SUD in childhood (Bove KE, Histologic Pattern of Metabolic Disease, chapter 10). At autopsy, urine collection for acyl and acylcarnitine compounds, skin/tendon biopsy to establish a fibroblast culture, and freezing of six tissue samples should be considered essential. Clinical presentation is also variable with the metabolic crisis with non-ketotic hypoglycemia in the immediate neonatal period to progressive liver fibrosis with or without cardiomyopathy. FAOD determine the accumulation of neutral lipid in FAO-dependent organs such as the liver, heart, proximal renal tubule, and type 1 skeletal muscle fibers. The neutral lipid accumulation mani-

fests histologically as microvesicular steatosis, which may be transitory as seen in MCAD when homeostasis is restricted.
- *TEM*: Generic mitochondrial findings, which are not distinctive compared with abnormalities of the mitochondria seen in primary disorders of electron transport, mtDNA depletion syndrome, and RS. Thus, the diagnosis is practically based on FAB-MS measurement of plasma acylcarnitine profile utilizing Guthrie card blood spots and of urinary acyl-carnitine and organic acid profiles as well as investigations carried out on cultured fibroblasts.

Carnitine Deficiency
- *DEF*: It is a genetic disorder condition in which the individuum does not have the carnitine shuttle, and this deficit prevents the body from using some fats for energy.
- *CLI*:
  1. Reye's syndrome-like form.
  2. Severe liver involvement form (hepatomegaly, hypoglycemia, hyperammonemia, and CNS dysfunction).
- *CLM*: Variable phenotype from normal or slight "portitis" to steatosis of microvesicular and macrovesicular types.
- *TEM*: Intracellular lipids.
- *DDX*: Reye's syndrome among other metabolic defects presenting with steatosis.

4.5.5.3 Urea Cycle Enzyme Defects

The urea cycle has two aims, which are the (1) de novo biosynthesis and degradation of *arginine* and (2) serves to incorporate the surplus of nitrogen into *urea* that is considered a waste nitrogen product. The urea cycle consists of well-delineated and orchestrated enzymatic steps, in which the nitrogen arising from ammonia and aspartate is transferred to urea, which is found in the urine of mammals, amphibians, and fishes. The urea cycle enzymes are partly localized in the mitochondria and partially localized within the cytosol. Carbamyl phosphate synthetase (CPS) and ornithine transcarbamylase (OTC) are within the mitochondria, whereas argininosuccinate synthetase (ASS), argininosuccinate lyase (ASL), and arginase (ARG) are in the cytosol.

Urea cycle defects (UCD) include the deficiency of the respective enzymes:

1. Carbamyl phosphate synthetase deficiency (CPSD)
2. Ornithine transcarbamylase deficiency (OTCD)
3. Argininosuccinate synthetase deficiency (ASSD), aka citrullinemia
4. Argininosuccinate lyase deficiency (ASLD), aka argininosuccinic aciduria
5. Arginase deficiency also called hyperargininemia

ORNT1 or ornithine transporter is a carrier protein that allows the transfer of ornithine from the cytosol to the mitochondrion, and CITRIN is a hepatic mitochondrial aspartate glutamate carrier protein, and NAGS is N-acetyl glutamate synthetase which adds acetyl CoA to glutamate to form NAG that with NH_3 and HCO_{3-} is transformed to carbamyl phosphate that enters the urea cycle. Deficiency of NAGS, CPS, and OTC in the mitochondrial matrix impairs the subsequent synthesis of $CO(NH_2)_2$ or carbamide (urea) by reducing the production of the mitochondrial citrulline.

Most of the urea cycle metabolic capacity is located in the liver, and the genes for the respective enzymes are well known. CPS has a gene located on 2q35, OTC on Xp21.1, ASS on 9q34, ASL on 7cen-q11.2, and arginase on 6q23.

In case of a defect of a urea cycle enzyme, there are two consequences, including:

1. The transformation of the amino acid arginine from non-essential to essential (except hyperargininemia only)
2. The progressive accumulation of nitrogen in several molecules, particularly ammonia

Hyperammonemia is a toxic state for the brain because it interferes with the metabolism of neurotransmitters influencing the glutamine-glutamate-GABA balance and with energy production. However, differently from liver failure where ammonia is one of the multiple toxins, in UCD ammonia is the only toxin except hyperargininemia, where a synergistic effect may also play a role. Remarkably, the brain MRI of patients affected with UCD demonstrates asymmetric focal lesions, while brain MRI of patients with liver failure generalized and symmetric lesions are typically seen.

- *EPI*: 1:8200 (prevalence, USA), but the overall incidence of defects presenting is ~1:45,000 live births.
- *CLI*: Two typical presentations: neonatal and pediatric.
- A lethargic and hypotonic baby characterizes the neonatal form with the progressive development of vomiting, seizures, hypothermia, and hyperventilation and signs of increased intracranial pressure with bulging fontanel and macrocephaly and death within a few days.
- The *pediatric* form (older child) is characterized by either chronic and episodic course with lethargy, headache irritability, agitation, confusion, hallucinations, vomiting, hypotonia, ataxia, dysarthria, or stroke-like episodes with focal dysfunction of the brain. Coarse and friable hair, also called trichorrhexis nodosa, is seen in ASLD. Other organ involvements are ocular (*scotomas*), circulatory (*shock*), renal (*renal failure*), the lung (*hemorrhage*), and the liver (*hepatomegaly*). All UCD except hyperargininemia is similar. In hyperargininemia, there is a slow progression of growth failure, psychomotor retardation, progressive spastic tetraplegia, tremor, ataxia, choreoathetosis, epilepsy, and hyperactivity with or without episodes of lethargy, vomiting, and, finally, coma.
- *CLM*:
  1. Carbamyl phosphate synthetase deficiency (CPSD): Neuropathology
  2. *Ornithine transcarbamylase deficiency* (OTCD): Neuropathology
     ± liver pathology (occasional mitochondrial and peroxisomal changes)
  3. Citrullinemia: Neuropathology
  4. *Argininosuccinic aciduria* (ASLD): Neuropathology and hepatomegaly with ALT increase and macrovesicular steatosis to severe fibrosis on LM and dilation of rough and smooth endoplasmic retuculum, megamitochondria, mitochondrial polymorphism with shortened cristae and paracrystalline matrix inclusions admixed to normal mitochondria, and excess of glycogen on TEM
- *PGN*: Regarding outcome and genetic counseling of hyperargininemia, the lifespan of this disorder is longer than in other UCDs. Prenatal diagnosis can carry out on the amniotic fluid by measuring abnormal metabolites, on chorionic villi or amniocytes for DNA analysis, or on cultured amniocytes or in utero liver biopsy samples. The detection of the carrier in OTCD is performed using protein loading, alanine loading, and allopurinol challenge to induce orotic aciduria. However, a negative test does not typically rule out the carrier status. Importantly, DNA techniques can be successfully used for carrier detection if the mutation in the affected patient and parents are known.

Citrin deficiency is due to an abnormal or absent mitochondrial aspartate-glutamate carrier protein in the hepatocytes owing to gene mutations causing type II citrullinemia in adults and cholestasis in infants, mostly of ethnic East Asian origin. Microscopically, neonatal citrin deficiency manifests with NH features with macrovesicular steatosis, accompanied by lobular cholestasis, the prominent acinar transformation of the hepatocytes (pseudorosettes of cholestatic origin), and progressive pericellular/perisinusoidal fibrosis, which is unusual in NH of newborns with the ethnic European-American background. The control of the metabolism of glutamate, glutamine, and ammonia within the CNS is referred to as the glutamate-glutamine cycle. Synaptic glutamate is removed using three distinct mechanisms:

1. Absorption from the postsynaptic glutamatergic neuron
2. Reuptake into the presynaptic neuron
3. Uptake from astrocytes, which is mediated by Na+-independent and Na+-dependent systems, the later having a high affinity for gluta-

mate and subdivided into excitatory amino acid transporter 1 (EAAT1, aka GLAST) and excitatory amino acid transporter 2 (EAAT2, aka GLT1). The glutamate is then converted back to glutamine within the astrocytes.

4.5.5.4 Secondary Mitochondrial Hepatopathies

Secondary mitochondrial hepatopathies are hepatopathies with the involvement of the mitochondrial cytologic subunit secondary to several and different etiologies. We include mostly Reye syndrome, hepatic copper overload (Wilson disease, Indian childhood cirrhosis, idiopathic infantile copper toxicosis, cholestasis), hepatic iron overload (genetic hemochromatosis, juvenile hemochromatosis, GALD/neonatal hemochromatosis, tyrosinemia, type I, Zellweger syndrome), drug administration (amiodarone, barbiturates, chloramphenicol, nucleoside analogs, salicylic acid, tetracycline), bacterial toxins (cereulide, *Bacillus cereus* emetic toxin), ekiri, chemical toxins (Fe, ethanol, cyanide, antimycin A, rotenone), mitochondrial lipid peroxidation (cholestasis, hydrophobic bile acid toxicity, non-alcoholic steatohepatitis (NASH), total parenteral nutrition (TPN), bacterial contamination of small bowel, jejunoileal bypass, or idiopathic), and liver cirrhosis. Some of these disorders are found in other paragraphs of this chapter as well as other sections of this book.

4.5.6 Peroxisomal Disorders

Peroxisomes are organelles that starting their life in an individual sub-compartment of the ER are essential for the metabolism of ether-phospholipids, fatty acid beta-oxidation, and fatty acid alpha-oxidation. Peroxin 3 (PEX3) with different peroxisomal membrane proteins (PMPs) plays significant roles for the initial formation of the peroxisomes and PEX5 and PEX 7 being responsible for targeting the peroxisomal matrix proteins into the peroxisomes. Ether-phospholipids are different from the diacyl-phospholipids because they have an alkyl or alkenyl group instead of an acyl group at the sn-1 of the glycerol molecule. Ether-phospholipid biosynthesis begins in the peroxisomal environment, because the first two enzymes, i.e., dihydroxyacetone phosphate acyl-transferase (DHAPAT) and alkyl-DHAP synthase (ADHAPS), encoded by *GNPAT* gene and *AGPS*, respectively, are located in the peroxisomes, while the other enzymes are located in the endoplasmic reticulum. Peroxisomal disorders are a genetic disease that can be identified by an impairment in 1 or 1+ functions of peroxisomes and are classified into three subgroups with several subgroups (Box 4.20).

Like mitochondria, peroxisomes contain a fatty acid β-oxidation system, which harbors four enzymatic reactions and targets the chain shortening of some fatty acid substrates, of which three most important are C26:0, pristanic acid, and di-/

Box 4.20 Groups and Subgroups of Peroxisomal Disorders

Peroxisomal Biogenesis Disorders (PBDs)
  Zellweger Spectrum Disorders (ZSDs)
    Zellweger Syndrome (ZS)
    Neonatal Adrenoleukodystrophy (NALD)
    Infantile Refsum Disease (IRD)
  Rhizomelic Chondrodysplasia Punctata Type 1 (RCDP1)
Single Peroxisomal Enzyme Deficiencies (SPEDs)
  Acyl-CoA Oxidase 1 (ACOX1) Deficiency
  D-Bifunctional Protein (DBP) Deficiency
  2-methyl Acyl-CoA Racemase Deficiency
  Sterol Carrier Protein X Deficiency (SCPX)
  Phytanoyl-CoA Hydroxylase (Adult Refsum Disease) Deficiency
  Acyl-CoA-Dihydroxyacetonephosphate Acyltransferase (RCDP2) Deficiency
  Alkyl-Dihydroxyacetonephosphate Synthase (RDCP3) Deficiency
Single Peroxisomal Substrate Transport Deficiency (SPSTD)
  X-Linked Adrenoleukodystrophy (X-ALD)

trihydroxycholestanoic acid. To the success of the four subsequent enzymatic reactions, there are two distinct acyl-CoA oxidases (ACOX1 and ACOX2), two bi-functional proteins (LBP and DBP), and two different thiolases (pTH1 and pTH2), of which one (aka SCPX) is important for the last step. ACOX1, DBP, and SCPX deficiencies have been identified in humans. A fourth defect of the SPED group is constituted by the deficit of phytanoyl-CoA hydroxylase (adult Refsum disease). This enzyme is essential for fatty acids with a methyl group at the 3-position, such as phytanic acid or 3,7,11,15-tetramethyl hexadecanoic acid, which needs to undergo α-oxidation before being β-oxidized. Peroxisome biogenesis disorders consist of heterogeneous groups and can be subdivided into 13 complementation groups. Complementation group 8 contains three clinical phenotypes: Zellweger syndrome, infantile Refsum disease, and neonatal adrenoleukodystrophy. *PEX26*, the pathogenic gene for this group of peroxisomal biogenesis disorders, has recently been identified.

Zellweger Syndrome and PEX Disorders
- *DEF*: ZS is the severe form of peroxisomal disorders and expresses itself through craniofacial dysmorphism and neurological, ocular, renal, and hepatic anomalies and usually leads to death within the 1st year of life.
- *SYN*: Cerebro-hepato-renal syndrome. Zellweger syndrome is named after Hans Zellweger (1909–1990). Dr. Zellweger was a Swiss-American pediatrician, professor of Pediatrics and Genetics at the University of Iowa who identified and researched on this disorder.
- *EPI*: 1:10^5 births
- *EPG*: It is an AR-inherited disease due to mutations of several and different genes involved in peroxisomal biogenesis and characterized by multiple enzymatic defects affecting β-oxidation, plasmalogen synthesis, and phytanic acid oxidation. Mutations in *PEX1, PEX2, PEX3, PEX5, PEX6, PEX10, PEX12, PEX13, PEX14, PEX16, PEX19*, and *PEX26* genes, which encode peroxins and are responsible for the normal assembly of peroxisomes, are at the basis of this genetic disease.
- *CLI* – Cerebro-hepato-renal syndrome, including:
  – Severe and generalized hypotonia and psychomotor retardation (brain involvement)
  – Hepatomegaly ± icterus (liver involvement)
  – Renal cortical cysts (kidney involvement)

Patients harboring this syndrome evidence an abnormal shape of the head and unusual facial characteristics.

The face is characterized by a high forehead, large anterior fontanel, hypoplastic supraorbital ridges, bilateral epicanthus, and abnormal ears. The CNS examination shows a severe truncal and peripheral hypotonia, lack of visual and auditory responses, seizures, glaucoma, and retinitis pigmentosa. Moreover, cortical dysplasia with polymicrogyria and pachygyria in the perisylvian area, neuronal heterotopia, migration defects of Purkinje cells, and discontinuation in the principal nucleus of the inferior olivary complex as well as dysmyelination of the cerebral white matter have been described.

- *LAB*: ↑ Pipecolicacidemia, ↑ trihydroxycoprostanic acidemia (THCA), ↑ [VLCFA]urine (>C22).
- *PGN*: Poor outcome with death within 6–12 months of life.

X-ALD is an X-linked genetic disease due to *ABCD1* gene mutations without genotype-phenotype correlation. The *ABCD1* gene encodes ABCD1 (or ALDP), which is a peroxisomal transmembrane protein with the structure of an ATP-binding cassette transporter. There is a rate of 1:42000 (or 1:20000 males) and is the most frequent monogenetically demyelinating disease as well as the most frequent peroxisomal disorder. Clinically, there are two distinct phenotypes, including (1) adreno-myeloneuropathy (AMN) with the involvement of the adrenal gland (adrenal insufficiency or Addison disease) and spinal cord involvement with progressive stiffness, weakness of the legs, impairment of the vibration sense, sphincter disturbances, and impotence, and (2) X-ALD of cerebral demyelinating type (CDA), showing cerebral demyelination. There is an increase of VLCFAs. In NALD, there is a severe truncal peripheral hypotonia, lack of visual and auditory responses, and seizures as well as retinitis pigmentosa. Moreover, hepatomegaly is seen, and the CNS shows polymicrogyria only ± dysmyelination for infants older than 2 years. In IRD, there are some ZS-like external features but no neuronal migration disorders and no progressive white matter involvement. Psychomotor retardation, retinitis pigmentosa, and peripheral deafness are seen in surviving children. In RDCP1, the inheritance is AR, and there is rhizomelic dwarfism with growth retardation, microcephaly, characteristic face (prominent forehead and midface hypoplasia with anteverted nares and long philtrum), congenital cataracts, and neurological features, including truncal hypertonia, spastic tetraplegia, and epilepsy. X-ray shows rhizomelic dwarfism with severe shortening of humeral bones and femora, metaphyseal cupping or splaying punctate calcifications, and malformations of the spine with characteristic coronal clefts of the vertebral bodies. Acyl-CoA Oxidase 1 (ACOX1) deficiency is an AR-inherited disease with severe neonatal hypotonia, inadequate response to visual and auditory stimuli, retinitis pigmentosa, hepatomegaly, and seizures but no facial dysmorphism. X-ray is mute, but brain imaging shows progressive white matter demyelination with Blood-Brain Barrier (BBB) disruption like X-ALD. D-Bifunctional Protein (DBP) deficiency is an AR-inherited genetic disease with ZS-like facial dysmorphism, neonatal hypotonia, seizures, failure of the visual system (nystagmus, strabismus, inability to properly fixate objects, retinitis pigmentosa) and thermoregulatory system, and peripheral neuropathy. Brain imaging delay of maturation of the white mater and demyelination of cerebral hemispheres and progressive cerebral atrophy. Sterol Carrier Protein X (SCPX) deficiency is an AR genetic disease with a stutter in childhood accompanied later by dystonic head tremor and spasmodic torticollis with respective hyperintense signals in the thalamus using magnetic resonance imaging, so-called butterfly-like lesions in the pons, and injuries in the occipital areas. Other diseases include the Adult Refsum Disease (ARD), deficiency of acyl-CoA-dihydroxyacetone phosphate acyltransferase (DHAP-AT), and scarcity of alkyldihydroxyacetone phosphate (ADHAPS).

4.5.7 Iron Metabolism Dysregulation

Hepcidin and Regulation of Fe Metabolism
Liver cells have the *HAMP* gene, which encodes for hepcidin, a master regulator of iron metabolism. In response to infection and iron overload states, hepcidin is upregulated, and this upregulation is also supported by hemojuvelin (50%), HFE protein (25%), and TFR2 (25%), which increase hepcidin production. The *TFR2* gene provides instructions for building the protein called transferrin receptor 2. The release of hepcidin by hepatocytes slows down the passage of Fe^{3+} ions through the intestinal enterocytes and, simultaneously, the release of Fe^{2+} from macrophages at various locations determining a decrease of iron in serum. In MΦ, Fe^{2+} comes from the degradation of RBC, and it seems that hepcidin acts on MΦ by internalizing ferroportin (FP-1) in these cells (Figs. 4.18, 4.19, and 4.20). Box 4.21 lists the disorders of the regulation of the iron metabolism that will be part of this section.

Neonatal hemochromatosis has been a mystery for a long while, and it seems that recently some aspects of neonatal iron accumulation have clarified. As early as 1865, Trousseau described a collection of disorders including the association of diabetes, cirrhosis of the liver, fibrosis of the pan-

---

Box 4.21 Iron Metabolism Dysregulation

- Neonatal hemochromatosis
- Genetic hemochromatosis, adult type
- Hereditary hemochromatosis, juvenile type

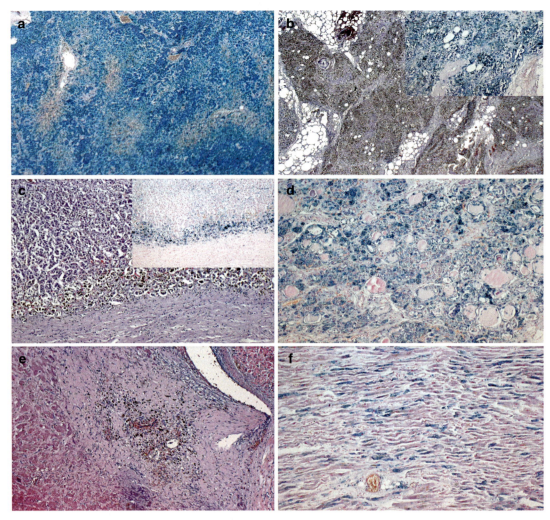

Fig. 4.18 Hereditary hemochromatosis (HFE) is a disorder that causes the body to hyperabsorb iron from the diet and store the excess iron in the body's tissues and organs, particularly the skin, heart, liver, pancreas, and joints leading to early failure of several organs and deaths of these patients. Most patients are homozygous for a C282Y mutation in the HFE gene, which is frequent in northern Europe, where 1:5–10 individuals are carriers. In (**a**) liver cirrhosis of homozygous HFE young male is seen. The accumulation is massively represented in the trabecular hepatocytes (**a**). Multiple organs are involved by iron accumulation (**b–f**), such as pancreas (**b**), adrenal gland (**c**), thyroid gland (**d**), and heart (**e–f**). The Pearls' special stain highlights the iron deposition identifying the iron as blue granules

creas, and pigmentation of the skin. Twenty-four years later, on the *Versammlung Deutscher Ärzte* at Heidelberg, Germany, von Recklinghausen labeling the disease *Haemochromatosis* reported 12 more similar cases with brown pigmentation in almost all the organs. Noteworthy, Straeter was the first author to state tissue siderosis at infancy. He described two infants, one of them probably affected by an infection showing iron accumulation in the spleen, liver, pancreas, and lymph nodes at autopsy, the other one apparently concerned with a severe hemolytic anemia showing iron accumulation in the endothelial layer of the vessels, liver, spleen, lymph nodes, lungs, thymus, kidney, and pancreas. Strictly related to the prominent role of the organ at fetal age, some liver iron storage is physiologic in the term infant. The diagnosis of neonatal iron storage disease relies on multi-organ iron accumulation with sparing of the reticuloendothelial system. Cottier

4.5 Genetic and Metabolic Liver Disease

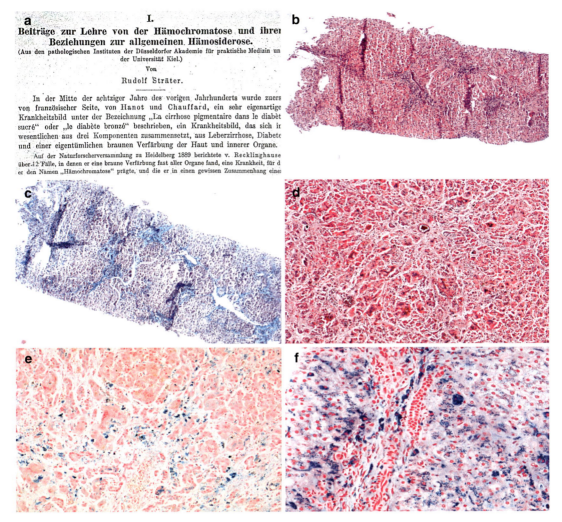

Fig. 4.19 Gestational alloimmune liver disease (GALD) is a maternofetal alloimmune disorder directed at the fetal liver, often manifesting as neonatal liver failure and neonatal iron storage disease of various extrahepatic tissues. The mechanism of iron overload in GALD relates to severe depressed hepatic expression of hepcidin (HAMP) in the injured fetal liver, which leads to dysregulation of placental iron flux. In (**a**) is shown the first demonstration of a neonatal hemochromatosis in the German literature. In (**b**) is shown the neonatal liver failure with cirrhotic nodules full or iron (**b**). Early features of iron accumulation can be discovered in infants before liver failure (**c, e**). The iron accumulation is often seen in the biliary proliferation, a common marker of inherited Fe storage disorders (**f**)

was the first author to recognize in infants the phenotype of advanced liver disease together with extrahepatic parenchymal siderosis as mimicking late-stage hemochromatosis in adults. The tissue distribution of hemochromatosis-type iron is different from the secondary siderosis, e.g., transfusional siderosis. The hemochromatosis siderosis involves, in approximate order of severity, periportal hepatocytes (the first site of iron deposition following adsorption into the portal venous plasma from chyme), the remaining hepatic lobule, exocrine pancreas, myocardium, and some endocrine epithelia. Conversely, the reticuloendothelial elements in the spleen, lymph nodes, and bone marrow and along hepatic sinusoids represent the primitive iron deposition sites in secondary siderosis. Excess iron storage in tissues led to organ dysfunction. Both in hemochro-

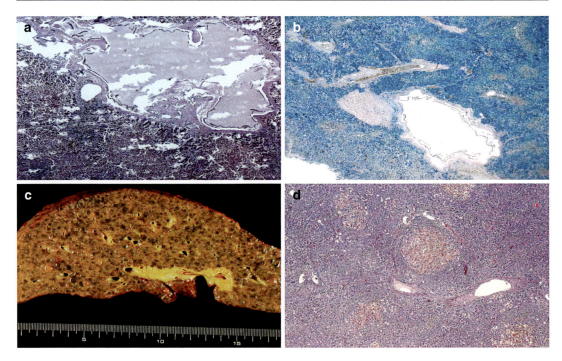

Fig. 4.20 In (**a**, **b**) are shown the microphotographs of a unique young patient harboring HFE and Caroli disease showing dilated segmental ducts and iron accumulation (**a**, H&E; **b**, Pearls' Prussian Blue stain). In (**c**) and (**d**) the gross photograph and the histopathology of liver failure of a young patient with Wilson disease are presented

matosis siderosis and in secondary siderosis, the failure of the myocardial and endocrine systems results in the death of the affected individuals. Almost certainly, neonatal iron storage disease is not one specific entity but a clinicopathologic phenotype. Currently, neonatal iron storage disorder is considered a GALD, i.e., gestational alloimmune liver disease, which is a maternofetal alloimmune disorder directed at the fetal liver, often producing neonatal liver failure with the most common presentation being the neonatal hemochromatosis. GALD is the result of neonatal complement-mediated severe liver injury. The hepatic damage is mediated by maternal *allo*-antibodies, which are detected by immunohistochemistry staining using the complement C5b-9 complex (Melin-Aldana et al. 2015). The initiation of antenatal therapy with high-dose intravenous immunoglobulin (IVIG) in women with an affected offspring to prevent the reappearance of the disease in subsequent pregnancies has given excellent results demonstrating the effectiveness of this approach. Newborns affected with neonatal hemochromatosis present clinically with liver disease-related complications: hypoglycemia (reduced capacity to store glycogen and release glucose), hypoalbuminemia (reduced synthesis of albumin), hemorrhagic diathesis (reduced synthesis of clotting factors and thrombocytopenia), and icterus (impaired hepatic synthesis and excretion of bilirubin). Although transaminase activities and blood concentrations of alpha-1-antitrypsin, ceruloplasmin, and transferrin are low, alpha-fetoprotein may be high reflecting the attempted liver regeneration due to reduced hepatocellular mass. Abnormal bile acid metabolism may represent a subgroup of this disorder. The increase in ferritinemia may be nonspecific and not diagnostic of neonatal hemochromatosis. Hypotransferrinemia reflects severe liver disease with reduction of the synthesis function of the hepatocytes, and transferrin saturation should be determined when neonatal hemochromatosis is being considered. Although liver biopsy is a procedure to diagnose neonatal iron accumulation, it is difficult to carry out safely. This situation is due to the hemorrhagic diathesis associated with this disorder. MRI can provide evidence of siderosis of the pancreas and myocardium and exclude splenic siderosis. However, anesthesia or sedation may worsen the clinical situation of an ill infant. A 3-mm punch biopsy specimen of the lower lip mucosa may be more feasible. The lower lip mucosa biopsy has been used to demonstrate extrahepatic parenchymal siderosis and to support the diagnosis of neonatal iron accumulation. Formalin-fixed and paraffin-embedded tissue obtained from the mucosa lower lip a few millimeters above the more inferior oro-alveolar sulcus generally suffices to prove siderosis without examining repeated frozen sections. Salivary gland biopsies have allowed the diagnosis to be made quite early. During the clinical approach, some conditions can be mistaken for neonatal hemochromatosis, including liver-predominant mitochondriopathies, hepatic infarction, hemophagocytic lymphohistiocytosis, and Finnish lethal neonatal metabolic syndrome. In the case of poor prognosis, orthotopic liver transplantation (OLT) is mandatory. In a 10-year retrospective analysis, Sigurdsson et al. (1998) found 14 infants with neonatal hemochromatosis, 8 patients of whom had been diagnosed after 1993 and had received the antioxidant-chelation cocktail (including deferoxamine, vitamin E, N-acetylcysteine, selenium, and prostaglandin-E1).

Genetic (or hereditary) hemochromatosis is a relatively common, AR-inherited inborn error of metabolism leading to massive iron deposits in hepatic and extrahepatic (the pancreas and other endocrine organs, heart, kidney, and salivary glands, are the first organs irrorated with a conspicuous amount of blood) tissues (often up to 100 times the average level). Mutations in five genes, including *HFE*, *HAMP*, *HJV*, *SLC40A1*, and *TFR2*, can cause genetic or hereditary hemochromatosis. *HFE* gene mutations cause type 1 hemochromatosis, while type 2 hemochromatosis results from mutations in either the *HAMP* or *HJV* gene. Mutations in the *TFR2* gene cause the type 3 of genetic hemochromatosis, while mutations in the *SLC40A1* gene are responsible for the type 4 hemochromatosis. *HFE* gene is on 6p22.2 with some relationships to HLA-A3, HLA-B7, and HLA-B14. HFE homozygosity is 1:400 to 1:200 individuals in the general population, although about 1/5 of the homozygous individuals are progressing to symptomatic disease. Heterozygosity is 1/10. The most frequent gene mutation is C282Y leading to the substitution of tyrosine for cysteine at amino acid 282 in the HFE protein. Pathogenetically, there is an impairment of the enterocytes to regulate iron absorption, which leads to a saturation of circulating transferrin with iron. Saturated transferrin reaches hepatic and extrahepatic tissues, where it is first as ferritin and then as hemosiderin. Fe toxicity is due to the generation of free radicals that cause peroxidation of organelle membranes and cell death-leading intracellular changes. Simultaneously, there is a paracrine stimulation of collagen-producing cells leading to tissue fibrosis. Clinically, the classic triad which is cirrhosis, skin pigmentation, and diabetes mellitus ("bronze disease") is rarely seen, if adequate treatment is started. Biochemically, hepatic iron index (HII) is mole Fe/grams dry weight liver/age. The HII is a quite useful parameter, although the use of clinical judgment to discern severity is paramount. Healthy subjects have HII of 0.1–1, while homozygotes and heterozygotes have >2 and 0.2–2, respectively. GH needs to be differentiated from other states with iron tissue deposition, such as multiple transfusions, Porphyria Cutanea Tarda (PCT), and chronic dietary Fe overload. In secondary siderosis, iron overload is without associated tissue damage, and iron preferentially accumulates in the Kupffer cells. Brown discoloration of most of the organs is seen during grossing (gross morphology). This aspect corresponds to the increased hemosiderin storage in most of the organs.

4.5.8 Copper Metabolism Dysregulation

Wilson disease and Menkes syndrome are metabolic conditions with altered copper metabolism (Box 4.22).

> **Box 4.22 Copper Metabolism Dysregulation**
> 4.5.8.1. Wilson Disease (Hepato-Lenticular Degeneration)
> 4.5.8.2. Menkes Syndrome

4.5.8.1 Wilson Disease

- *DEF*: AR-inherited metabolic disorder of the copper metabolism due to mutations of *ATP7B* on 13q14 leading to a ↓ Cu-transporting ATPase 2 and inability to excrete Cu into the bile.
- *SYN*: Hepato-lenticular degeneration.
- *EPG*: The gene product of *ATP7B* is a P-type ATPase that is exclusively found in the liver (intracellular trans-Golgi site) and is required for both copper incorporation into ceruloplasmin and biliary excretion.
- *EPI*: $33/10^6$ births.
- *LAB*: Cu level measured by atomic absorption spectroscopy (AAS) is useful for the initial differentiation of the homozygous patients. There is Cu > 250µg/1 g dry weight of liver in patients affected with Wilson disease.
- *CLI*: Clinically liver disease (fulminant hepatitis, fatty change or liver cirrhosis) may be associated with degeneration of the basal ganglia (= hepato-lenticular degeneration as is also known Wilson disease in the older literature) and Cu accumulation in the Descemet membrane of ocular limbus (Kayser-Fleischer rings on the cornea).
- *GRO*: Diffuse areas of necrosis in the case of fulminant hepatitis with massive hepatic necrosis, yellow hepatomegaly in case of steatosis, and no gross changes (CAH) are the grossing milestones to remember.
- *CLM* – It includes three features:
  1. Acute/chronic hepatitis (CAH)
  2. Fulminant hepatic failure (FHF)
  3. Cirrhosis

In (1) there is steatosis (micro- and macrovesicular type) and an increase of glycogen-rich nuclei in the acute phase, and the addition of a portal infiltrate characterizes the prolonged period. In (2) there is massive hepatic necrosis with a pattern indistinct from other causes of FHF, although the preexistence of fibrosis or cirrhosis at the time of onset of FHF is quite indicative of Wilson disease. Also, Mallory hyaline, Cu, and lipofuscin are seen. Mallory hyalin is periportal, does not attract neutrophils, and is not associated with other histologic features of different diseases showing Mallory hyalin such as alcoholic liver disease. Finally, in (3) there is cirrhosis with Mallory hyaline, Cu, and lipofuscin with location variability, (+) orcein staining, and Victoria blue staining for Cu-associated proteins and (+) rhodanine staining and rubeanic acid staining for elementary Cu, which is initially intracytoplasmic rather than lysosomal giving rise to a diffuse light staining of the cytoplasm.

- *TEM*: It includes abnormalities of peroxisomes and mitochondria. Peroxisomes show an enlargement and matrix flocculency, while mitochondria exhibit increased matrix density, separation of inner and outer layers of the mitochondrial limiting membrane, and bulbous dilatation of the tips of the cristae.
- *DDX* – Cu accumulation states (other than Wilson disease) include:
  - Primary biliary cirrhosis
  - Primary sclerosing cholangitis
  - Long-standing extrahepatic biliary obstruction
  - Indian childhood cirrhosis
- *PGN*: Poor outcome if the patient does not receive some treatment.

4.5.8.2 Menkes Syndrome

It is a disorder that affects Cu levels in the body and shows several findings, including sparse, kinky hair, failure to thrive, and deterioration of the CNS. Patients with Menkes syndrome usually begin to develop symptoms during infancy and survival is minimal after 3 years of age. A milder form starts in early to middle childhood (occipital horn syndrome). Mutations in the *ATP7A* gene cause Menkes syndrome. Cu accumulates mainly in the small intestine and kidneys. The decreased Cu supply can induce a decrease of the activity of numerous copper-containing enzymes that are crucial for the structure and function of the skeletal system, other than skin, the nervous system, and the vascular system. For more details, please refer to genetic books.

4.5.9 Porphyria-Related Hepatopathies

Both *porphyria cutanea tarda* (PCT) and erythropoietic protoporphyria (EPP) affect the liver showing characteristic histologic and ultrastructural changes. In PCT, there are intrahepatocytic needle-shaped inclusions, which are autofluorescent, birefringent, and hydrosoluble. There are also focal hepatocellular necrosis, portal lymphocytic infiltrate, Fe deposition, and fatty change. A regeneration (thickened liver cell plates) can also be appreciated. The ultrastructural study of PCT reveals intrahepatocytic, elongated crystalline electron-lucent bodies surrounded by electron-dense material revealing alternating areas of differing electron density using transmission electron microscopy (TEM). In EPP, there is centrilobular canalicular cholestasis with dark brown bile with red "Maltese cross" configuration when polarized → ductal cholestasis with pericellular and portal fibrosis/cirrhosis. The ultrastructural study of EPP reveals "starburst" crystalline material in hepatocytes ductal epithelium and Kupffer cells. PCT evolution comprises liver fibrosis, cirrhosis, and neoplastic degeneration (HCC), while the outcome of EPP includes liver fibrosis and cirrhosis (Box 4.23).

Box 4.23 Liver "Highlights" of Porphyria Cutanea Tarda (PCT)

1. Focal hepatocellular necrosis
2. Portal lymphocytic infiltration
3. Fe deposition
4. Diffuse fatty change

Note: 4 "F's" to recall these findings include "Foci" of necrosis, "Fullness" of lymphocytes in the portal tracts, "Fe" deposition, and "Fatty" change of diffuse type.

4.5.10 Shwachman-Diamond Syndrome (SDS)

- *DEF*: AR-inherited disease, determined by mutation of the *SBDS* gene on 7q11 and characterized by pancreatic exocrine insufficiency, BM hypoplasia (fat replacement of myeloid lineages and myeloid maturation arrest), and bony abnormalities (metaphyseal dysostosis) as well as hepatomegaly with a variable pattern from steatosis to cirrhosis.
- *CLI*: Failure to thrive and hematological abnormalities, including anemia, neutropenia, and thrombocytopenia, are seen in infancy and childhood of affected individuals.
- *PGN*: ↑ risk to develop myelodysplastic syndrome and leukemia.

Other than SDS, there is a growing list of possible metabolic defects or genetic syndromes showing NASH or early cirrhosis in life, and OMIM or other search engines may be able to create an updated list. However, some conditions should be kept in mind, because patients may be seen either under the lens or at autopsy. It is essential to remember partial or congenital lipodystrophy, Bardet-Biedl syndrome, Alstrom syndrome, Bloom syndrome, Dorfman-Chanarin syndrome, Aarskog syndrome, Donohue syndrome, or leprechaunism with fatty change associated or not with hepatocellular and reticuloendothelial siderosis.

4.5.11 Chronic Granulomatous Disease (CGD)

- *DEF*: Eclectic group of X-linked diseases in which individual cells of the immune system have difficulty forming the ROS compounds (e.g., superoxide radical due to defective phagocyte NADPH oxidase) used to kill certain pathogens following phagocytosis by the MΦ.
- *SYN*: Bridges-Good syndrome, chronic granulomatous disorder, and Quie syndrome.
- *EPI*: 1:200,000 people in the USA, with about 20 new patients/year.
- *EPG*: Deletions, frameshift, nonsense, and missense mutations in the *CYBA*, *CYBB*, *NCF1*, *NCF2*, or *NCF4* genes can cause CGD.

- *CLM*: Granulomatous inflammation and diffuse accumulation of lipofuscin pigment, Fontana (+) in RES cells. Confluent granulomas (DDX, Sarcoidosis!) may progress to necrosis and suppuration of some areas. Pigmented histiocytes are light tan in color and are PAS+ and argentophilic. On frozen section, pigmented histiocytes are sudanophilic and autofluorescent. This lobular activity is associated with portal inflammation by lymphocytes, and plasma cells with mild-moderate periportal or perisinusoidal fibrosis and occasionally steatosis are seen.

4.5.12 Albinism-Related Liver Diseases

Chediak-Higashi syndrome (CHS) and Hermansky-Pudlak syndrome (HPS) are two albinism-related liver diseases. Thus, in ACHS, hepatosplenomegaly with hemophagocytosis associate with fever, lymphadenopathy, and recurrent infections, due to pancytopenia, progressive neuropathy mostly of the cranial and peripheral nerves, and hyperlipidemia. Conversely, in AHPS (or oculo-cutaneous albinism), widespread deposition of ceroid pigment (granules intermingled with lipid and delimitated by a single membrane, ultrastructurally) in RES cells is associated with mild bleeding diathesis due to a storage pool of defects of platelets. Individuals affected with AHPS may develop lung fibrosis, cardiomyopathy, renal failure, and granulomatous colitis.

4.6 Viral and AI Hepatitis, Chemical Injury, and Allograft Rejection

Either viruses or autoimmune as well as drugs or toxins may start an inflammation of the liver, which may be in different areas and may suggest some etiologies more than others. Box 4.24 summarizes the topics that are going to be the target of this subchapter.

The clinic-pathological syndromes following a viral exposure are listed in Box 4.25.

Box 4.24 Viral and AI Hepatitis, Chemical Injury, and Allograft Rejection

4.6.1. *Acute Viral Hepatitis*
4.6.2. *Chronic Viral Hepatitis*
4.6.3. *Autoimmune Hepatitis*
4.6.4. *Drug-/Toxic-Induced Liver Disease (DILD/TILD, chemical injury)*
4.6.5. *Granulomatous Liver Disease*
4.6.6. *Alcoholic Liver Disease*
4.6.7. *Non-alcoholic Steatohepatitis (NASH)*
4.6.8. *Liver Allograft Rejection*

Box 4.25 Clinicopathologic Syndromes Following Viral Exposure

1. Acute asymptomatic infection with subsequent recovery
2. Acute symptomatic infection (hepatitis of anicteric/icteric type) with subsequent recovery
3. Chronic hepatitis (>6 months) with potential progression to liver cirrhosis
4. Fulminant hepatitis with submassive to massive parenchymal necrosis

---

4.6.1 Acute Viral Hepatitis

Hepatitis A Nonenveloped ssRNA, icosahedral, 27 nm, picornavirus without direct cytopathic effect (incubation period, about 1 month, antibodies 2 weeks later), but with portal inflammation ("portitis") with periportal involvement (conversely, Hep B and Hep C are mostly pan-lobular) and abundant plasma cells (like AIH).

- *LAB*: IgM → acute infection; IgG → previous HAV exposure.
- *PGN*: No chronic carrier state, no chronic hepatitis, no HCC, death <5/1000.

Hepatitis B Enveloped dsDNA, hexagonal, 42 nm, hepadnavirus with no direct cytopathic effect by integrating into host DNA and causing injury through inflammation (HBsAg as first, followed by HBeAg, which markedly increases early and then disappears with IgM-anti-HBc and anti-HBe, being IgM-anti-HBc positive at the time of symptoms). HBsAg drops after about 2 months, and before anti-HBs appear there is a 1–4 month "window period" (Fig. 4.21).

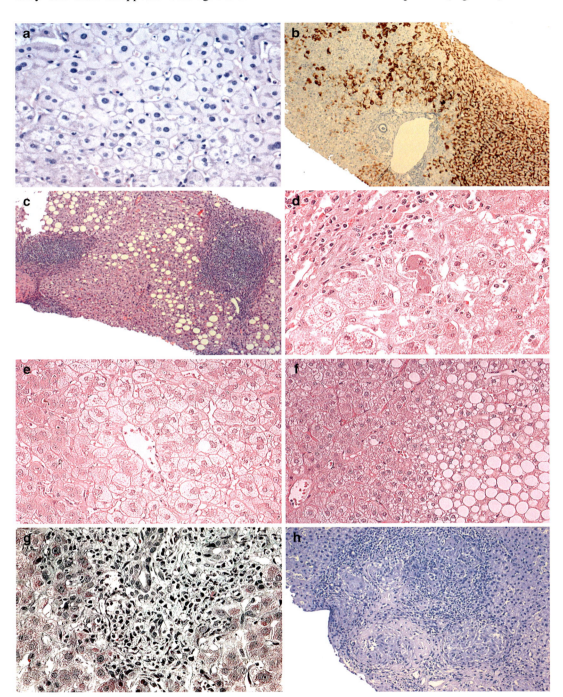

Fig. 4.21 HBV hepatitis can show ground-glass change of the hepatocytes (**a**, H&E, ×400 as original magnification), and positive HBsAg by immunohistochemistry (**b**). HCV hepatitis shows lymphocytic infiltration of the portal tracts with occasional lymphoid follicles, steatosis, and occasional vacuolar change of the hepatocytes (**c–g**). In (**h**) there is sarcoidosis of the liver with confluent noncaseating granulomas

- *CLI*: Jaundice with or without constitutional symptoms.
- *CLM*: Ground-glass hepatocytes.
- *TEM*: Smooth endoplasmic reticulum (SER) distension with virions.
- *PGN*: HBeAg persistence indicates continued active infection and likely progression to chronic hepatitis, which is more common in immunocompromised patients. In 10% with symptomatic disease, there is acute fulminant hepatitis or chronic liver disease (cirrhosis and HCC; Anti-HBsAg does confer long-term immunity!).

Hepatitis C Enveloped ssRNA, 55–65 nm, flavivirus with direct cytopathic effect.

- *CLM*: Portal inflammation with interlobular bile duct involvement, lymphoid aggregates, and follicles ± germinal centers, extension into the lobules, macro- and microvesicular steatosis, eosinophilic changes of the hepatocytes.
- *LAB*: Anti-HCV does not confer immunity.
- *PGN*: Insidious progression with chronic hepatitis in about half of the patients, tendency to progress to cirrhosis and HCC even without prominent piecemeal necrosis.

Hepatitis delta (δ) Defective RNA virus, small, 1.7 bp, requiring concurrent HBV infection, being HBsAg (envelope protein) for infectivity with or without direct cytopathic effect and tendency to cause more severe hepatitis with marked intralobular involvement characterized by microvesicular steatosis and multiple acidophilic bodies.

- LAB: IgM anti-D.

Hepatitis E ssRNA, calicivirus, cholestasis-associated virus with poor outcome in pregnancy but usually self-limited.

- LAB: HepE Ag in the hepatocytes.

The hepatitis histology is described and summarized in hepatitis bullets (Box 4.26).

Box 4.26 Hepatitis Bullets
- Acidophilic (Councilman) bodies: Eosinophilic apoptotic degeneration of hepatocytes
- Ballooning degeneration: Swelling of the hepatocytes with rarefaction of the cytoplasm
- Bridging necrosis: Collapse of reticulin network due to coalescence of the trabecular hepatocytic *muralium* due to liver cell necrosis
- Interphase hepatitis: Inflammation of the periportal area with the erosion of the limiting plate
- Lobular disarray: Disorganization of the trabecular *muralium*
- Lobular hepatitis: Patchy inflammation within the lobules (lobular more extensive than spotty)
- Piecemeal necrosis: Scattered hepatocytic necrosis of the periportal area
- Portal inflammation ("portitis"): Chronic lymphocytic inflammation of the portal tracts
- Spotty necrosis: Focal necrosis of hepatocytes ± perinecrotic lymphocytic recruitment

---

4.6.2 Chronic Viral Hepatitis

The immune system of some patients harboring hepatitis virus infection is unable to eradicate the viruses for no identified etiology. These patients will have viruses that cause chronic inflammation, which over time can lead to liver cirrhosis, liver failure, and liver carcinoma. Necro-inflammatory lesions include focal necrosis (or spotty necrosis), piecemeal necrosis, and confluent necrosis. Focal or spotty necrosis refers to an intralobular accumulation of a few mononuclear cells, mainly lymphocytes, potentially around acidophilic bodies with or without pyknotic nuclear fragment, which correspond to remnants of disintegrating hepatocytes that underwent cell destruction by cell-mediated immunological mechanisms (aka apoptosis). Areas of focal inflammation display hepatocytes with characteristic positive immunosignal on nuclei and cytoplasm for HbcAg and often membranous staining for HBsAg. HbcAg may also be detectable in the cell membrane of hepatocytes. Expression of HLA also characterizes focal necrosis-class I antigens and HLA-class II and surface IC adhesion molecules, such as ICAM-1 and leukocyte function-associated antigen 3 (LFA3). The hepatocytes show expression of γ-IFN receptor, and CD8+ is seen in the T-cell subset of the surrounding lymphocytes. Besides the lymphocytes, the focal necrosis includes MAC387+ cells, which are MΦ or dendritic cells with immunoregulatory function. Piecemeal necrosis is the hallmark of the interface hepatitis or the activity, which is characterized by lymphocytic infiltration and destruction of the hepatocytes at the connective tissue parenchymal "interface" at the periportal region and along the fibrous septa. In confluent necrosis, there is a more severe injury. Necrosis may become confluent when it involves more than one zone within the lobule and/or extend from one lobule to another adjacent lobule at a zone level. This necrosis is termed "bridging necrosis" and can occur as central-central or portal-portal level. Grading and staging histopathology of chronic hepatitis are several, including the HAI score, the Scheuer system, the Metavir system, and the Ishak modified HAI among others. These scores have easily described in several texts of pathology and can be found on the Internet at no cost. Scores can become very useful and considered perfect, but the more complicated they get, the less they are used or are used in the wrong way. Thus, most probably, there is no other organ, like the liver, harboring a babelization of several scores. A simplified score may be far from perfection but useful in reaching an agreement, such as the κ factor, and easy to memorize and discuss at multidisciplinary team meetings. A simplified Scheuer system (SSS) is probably a suggestion of this fascinating but also very complicated field, where several immunologic mechanisms, including Th_1 and Th_2 responses, may overlap at some point.

Needle Stick Injury

In the case of an occupational needle stick injury, and this may not be overall infrequent in both pathologists and residents or pathology assistants, there is a compulsory report to the lab manager to fill in following prompt medical intervention. Although rare, transmission of some disease from a percutaneous injury may theoretically lead to life-threatening illness. In case of deep trauma, the incident usually occurs with a device visibly contaminated with some patient's blood. Postexposure steps include (1) clean the wound with soap and water, (2) retrieve the source for details and check infection status, (3) report to the lab manager, (4) identify the protocol of prophylaxis in use in your institution, and (5) educate your healthcare co-workers.

4.6.3 Autoimmune Hepatitis

- *DEF*: Autoimmune (AI)-driven acute hepatitis disease with characteristics of progression and chronicity, if no treatment is undertaken, and variable intensity of hepatitis disease (acuteness, 40%) and potentially associated with other AI disorders, such as RA, AT/HT, and SS (rheumatoid arthritis, autoimmune thyroiditis/Hashimoto thyroiditis, Sjögren syndrome). AIH is considered a T-cell-mediated AI disorder (CD4 and CD8 T-cell enhancement with IFN-γ hyperproduction), and suggested triggers are viruses, drugs, and herbals (Box 4.27) (Fig. 4.22).
- *EPI*: Worldwide, ♂ < ♀ (70%), HLA-DR3 (+).
- *CLI*: Jaundice.
- *LAB*: Hyper-γ-globulinemia, hyper-IgG, auto-Abs, but no positivity for viral markers (Boxes 4.27 and 4.28).
- *GRO*: The liver failure is a more common diagnosis of the twenty-first century.
- *CLM*: L/PC (PC > L) PT inflammation with interphase activity and lobulitis ± mild fibrosis (L/PC, lymphocytes/plasma cell ratio).
- *PGN*: Cirrhosis (40% for type I and 80% for type II), although a good response to prednisone ± azathioprine is critical. 10-YSR (post-LTX): 75% and recurrence post-LTX: up to 40%.

In Box 4.29 are summarized some pearls and pitfalls during the diagnostic procedure of AIH.

Box 4.27 AIH in Childhood and Youth

Type	Age	Lab	Associations
Type I	Infancy*	ASMA, ANA, pANCA ↑↑↑ HGG	AIT, GD, UC
Type II	2–14 years	LKM-1, ↑ HGG	DM, AIT, APECED

Notes: *Type I is also seen in elderly. *AIT*, autoimmune thyroiditis; *APECED*, autoimmune polyendocrinopathy candidiasis ectodermal dystrophy; *ASMA*, antismooth muscle antibodies; *DM*, diabetes mellitus; *GD*, Graves disease; *HGG*, hyper-γ-globulinemia; *LKM-1*, antiliver-kidney microsome type 1; *pANCA*, Perinuclear antineutrophil cytoplasmic antibodies; *UC*, ulcerative colitis.

4.6.4 Drug-Induced Liver Disease (Chemical Injury) and TPN

A broad range of different chemical agents, e.g., drugs, industrial toxins, and anesthetics, can produce a restricted number of patterns of liver cell

Box 4.28 AIH – Biochemistry
AIH I: (+) ANA, SMA, AAA, anti-SLA/LP, HLA-DR3
　AIH II: (+) ALKM-1, ACL1
　Notes: *AAA* anti-actin autoantibodies, *ANA* antinuclear antibodies, *SMA* smooth muscle antibodies, *anti-SLA/LP* antibodies against soluble liver antigen/liver-pancreas, *ALKM* anti-liver-kidney microsome antibodies; ACL1 is the ATP-citrate synthase subunit 1 (ATP-citrate (pro-S-)-lyase 1), which catalyzes the formation of cytosolic acetyl-CoA, which is mainly used for the biosynthesis of fatty acids and sterols.

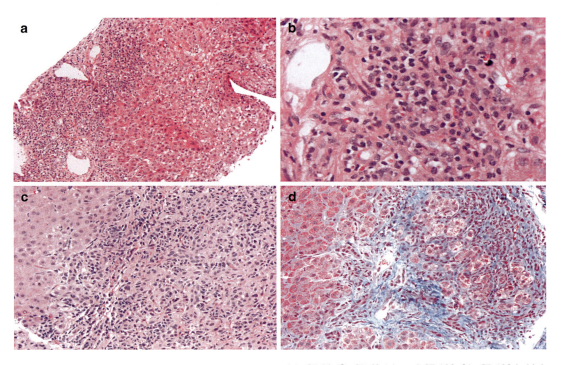

Fig. 4.22 Autoimmune hepatopathy shows marked lymphoplasmacellular infiltration with interface hepatitis (**a–d**). The immunohistochemistry (**e–h**) shows positive CD3 (**e**), CD20 (**f**), CD68 (**g**), and CD138 (**h**). CD138 is high specific for plasma cells

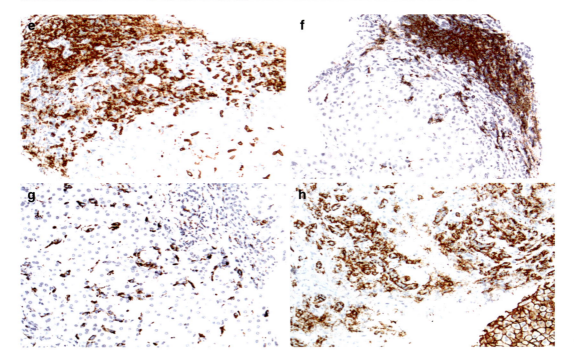

Fig. 4.22 (continued)

Box 4.29 Pearls and Pitfalls
- In AIH, AMA is not seen in high titers, unless there is a condition also known with the term "overlap syndrome," which combines AIH and PBC (AIH/PBC).
- "Overlap syndrome" shows liver histology of both AIH and PBC other than combined positivity for both AMA and ANA, with a picture of lobular hepatitis, interface hepatitis, and plasma cell infiltration more than the usual degree expected for a PBC!

damage, when ingested, inhaled, or administered parenterally. The range of liver disease varies from fatty change with an accumulation of triglycerides within the hepatocytes through cholestasis and hepatitis to extensive hepatic necrosis requiring liver transplantation. Chemical-induced damage is the etiology in about ¼ of cases of fulminant hepatic failure (FHF) and is a significant healthcare problem worldwide because of the limited number of available allografts (Figs. 4.23 and 4.24).

Hepatotoxicity may be subdivided in predictable (intrinsic) and unpredictable (idiosyncratic) forms. The predictable or inherent hepatotoxicity results from the exposure of liver to chemical agents that are intrinsically hepatotoxic injuring probably everyone having contact with it. The degree of cell damage is dose-dependent and is experimentally reproducible and is usually zonal necrosis (e.g., inorganic phosphorus targeting zone 1, whereas acetaminophen, carbon tetrachloride, and halothane determining damage in zone 3). Predictable or intrinsic hepatotoxins may be subclassified as direct or indirect toxins-related. Direct hepatotoxins or their metabolic products act on liver cells and their organelles by an immediate, nonselective effect of either physical or chemical nature distorting and destroying the basilar framework of the liver cells. The cellular injury is usually mediated by peroxidation of lipids of the membranes and leads to metabolic defects with resulting steatosis and tissue necrosis. Direct hepatotoxicity due to yellow phosphorus determines steatosis and zone 1 necrosis, whereas

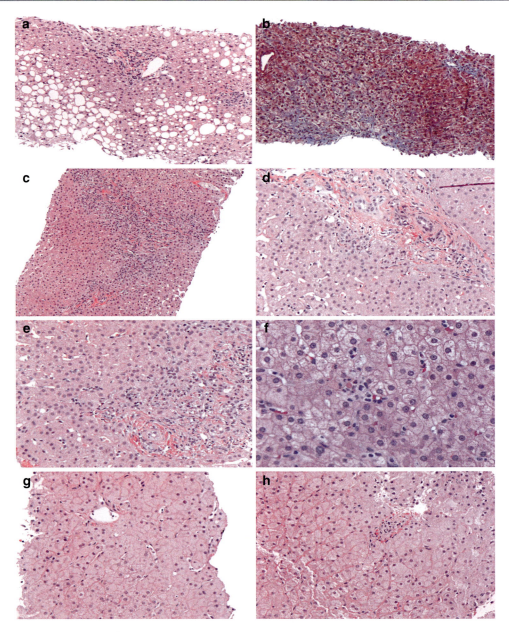

Fig. 4.23 Drug-induced liver disease can assume several histopathology patterns with steatosis, inflammation, cholestasis, eosinophilic recruitment, and ground glass-like change of the hepatocytes (**a–h**)

direct hepatotoxicity by CCl4 is steatosis and centrilobular (zone 3) necrosis acting on cytochrome P450. Indirect hepatotoxicity is due to a selective interference with a specific metabolic pathway or structural process. The consequence of the cell damage is some tissue changes, including steatosis, cholestasis, and necrosis. A few characteristic examples are targeting peculiar systems of the cell framework. Tetracycline, IV administered, can lead to microvesicular steatosis; acetaminophen determines the production of large amounts of toxic metabolites by the mixed-function oxidase (MFO) system outpacing the detoxifying abilities of thiol-rich glutathione. Thus, the substance may covalently bind to tissue macromolecules (e.g., N-acetyl-p-benzoquinone)

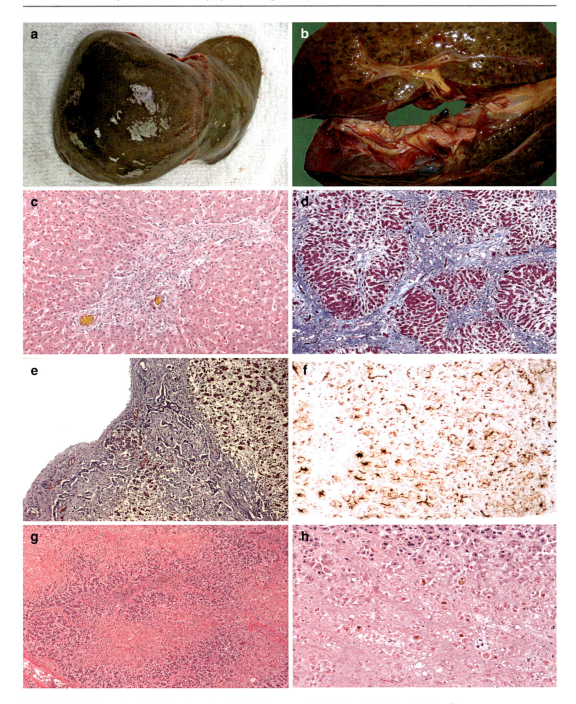

Fig. 4.24 Total parenteral nutrition hepatopathy is shown grossly (**a–b**) and histologically (**c–h**)

and determines zone 1 necrosis of liver tissue. Outpacing the detoxifying properties of the MFO system may also be seen with a limited amount of the drug in a setting of chronic alcoholism and MFO system stimulants (e.g., phenobarbital). Amanita phalloides may lead to steatosis and zone 1 necrosis. Erythromycin and anabolic or contraceptive steroids are able, instead, to induce cholestasis. Chronic alcoholism is also part of ethanol toxicity, which is linked to acetaldehyde production (*vide infra*). Unpredictable (idiosyncratic) hepatotoxic agents may act either through a hypersensitivity-related injury or an aberration of metabolic pathways. In unpredictable hepatotoxicity, liver cell damage is observed in only a very few or few individuals exposed to this agent. There is no dose-dependency, and most of the hepatotoxic drug reactions are of an idiosyncratic type. Liver tissue damage includes nonzonal necrosis to massive hepatic necrosis. In hypersensitivity-related injuries, such as the following phenytoin, chlorpromazine, sulfonamide, and halothane, the exposure to the chemical agent occurred 1–5 weeks earlier than the diagnosis of liver tissue damage. There is a recurrence of the liver tissue damage by re-exposure. Liver tissue eosinophilia and/or granulomas are seen, and systemic symptoms (fever, rash, peripheral eosinophilia) are present. In dysregulated metabolism injury, such as the following isoniazid, the hepatotoxic reaction is the consequence of an aberrant metabolic pathway due to an accumulation of toxic metabolites (weeks to several months) in a particularly susceptible patient. No systemic symptoms are seen, but a liver biopsy may be a challenge in sorting out other etiologies, such as acute viral hepatitis. Clinically, chemical hepatitis of affected individuals may show jaundice, fever, arthralgia, nausea, vomiting, dark urine, and abdominal pain. CCl4 intoxication may also determine a headache, drowsiness, and dizziness. The CIOMS (Council for International Organization of Medical Sciences) criteria for DILD are as follows. It is designated hepatocellular injury, if ↑ ALT is >2 than normal (isolated) OR ALT/ALP ≥ 5 is seen, cholestatic injury if ↑ ALP >2 than normal (isolated) OR ALT/ALP ≤ 2, mixed injury if ↑ ALT + ALP and 2 < ALT/ALP < 5, acute injury if changes as described above lasting less than or equal to 3 months, and chronic if changes as described above lasting more than 3 months.

Chemical agent's exposure may be life-threatening, and allograft transplantation may be required. However, a range of sequelae has been described, including chronic hepatitis, cirrhosis, vascular abnormalities, and hyperplastic/neoplastic nodules. A detailed pharmacologic history is crucial. We see often *acute liver failure with inflammation-free hepatocellular necrosis*, following acetaminophen, cocaine, MDMA (methylenedioxy-metamphetamine), (ecstasy), CCl_4; *acute liver failure with inflammation-associated hepatocellular necrosis*, following isoniazid, monoamine oxidase inhibitors (MAOI), anticonvulsants (phenytoin, valproate), antimicrobials (sulfonamides, cotrimoxazole, ketoconazole); *acute liver failure with inflammation-free microvesicular steatosis* after exposure to tetracycline and nucleoside analogs; *inflammation-free cholestasis* following anabolic/androgenic steroids, estrogenic steroids, and NSAIDs (nimesulide, piroxicam); *cholestatic hepatitis* after chlorpromazine and clarithromycin; *granulomatous liver disease*, after isoniazid, IFN, phenytoin, and allopurinol; moreover, *steatosis, microvesicular type*, after cocaine, tetracycline, valproic acid, and zidovudine; *steatosis, macrovesicular type*, after alcohol, steroids, total parenteral nutrition (TPN), gold, chlorinated hydrocarbons, and chemotherapeutic agents (5-fluorouracil); and *steatohepatitis* after amiodarone, chemotherapeutic agents (irinotecan), and perhexiline; an *AI chronic hepatitis* after lisinopril, sulfonamides, trazodone, uracil tegafur, tamoxifen, and methotrexate; *DILD-related AIH* following minocycline, nitrofurantoin, methyldopa, and clometacin; and *vascular abnormalities (sinusoidal obstruction syndrome)* in case of exposure to oxaliplatin, pyrrolizidine alkaloids, and chemotherapy for treatment of ALL.

It is important to remember that any pattern of primary liver disease can mimic a drug-induced liver injury. Clinical history and a systematic literature search are critical to identifying the culprit compound. The clinical history should contemplate the exposure to herbal and over-the-counter agents and toxins as well. The most common histological pattern of DILD remains acute hepatitis. In North America drugs such as acetaminophen are the most common cause of ALF. A chronic DILD is rare, although both minocycline used in youth and methotrexate used in both childhood and adolescence are two common drugs with this disorder showing either fibrosis or cirrhosis. DILD associated with a cholestatic injury will eventually determine ductopenia, i.e., lack of the interlobular bile ducts. Amiodarone, a cardiovascular drug, and irinotecan, an anticancer drug, may be associated with drug-induced steatohepatitis. Irinotecan is a topoisomerase inhibitor and works by blocking topoisomerase one which results in DNA damage and cell death. Individuals with two copies of the *UGT1A1*28* gene variant are at higher risk for side effects. Other medications can trigger or exacerbate steatohepatitis and include tamoxifen, estrogens, and nifedipine. In pattern five cholestatic hepatitises may also have a quite noticeable acute hepatocellular inflammatory damage, and some herbal products with known hepatotoxicity have been identified.

Total Parenteral Nutrition
- *DEF*: Multifactorial (toxicity of TPN solutions, gut-derived hepatotoxins, and absence of gastrointestinal stimuli for digestive secretions due to the interruption of oral intake).
- *CLM*: Hepatocellular and canalicular bilirubinostasis ± portal/periportal inflammation, extramedullary hematopoiesis, focal steatosis (Fig. 4.24).
- Newborns on long-term parenteral nutrition (PN) are prone to develop the PN-associated liver disease (PNALD). Aluminum (Al) is a known contaminant of infant PN, and a randomized control trial using a Yucatan piglet PN model was recently performed. The canalicular space was smaller, and the microvilli were shorter in the high-dose Al group than in the low-dose Al group. Hall et al. 2018 concluded that structural changes occur in the hepatocytic canalicular structure despite unaltered serum bile acids. High Al in PN is associated with the short microvillous profile of the lumen. It could decrease the functional excretion area of the hepatocytes and impair bile flow.

Postoperative Cholestasis
- *DEF*: Postoperative hepatic dysfunction ranging from a rise in conjugated serum bilirubin associated with increase of transaminases only ("benign postoperative intrahepatic cholestasis") to extrahepatic obstruction and liver failure and probably multi-etiologic in nature (e.g., bilirubin overload by hemolysis, halothane hepatotoxicity, drug-induced liver disease, viral hepatitis, shock or prolonged hypotension, bacterial infections, bile-duct injury, choledoco-lithiasis, cholecystitis, and pancreatitis).
- Benign postoperative intrahepatic cholestasis occurs within 24–48 h after major surgery and lasts 1–2 weeks. Increased bilirubin is due to either hemolysis of transfused blood cells or resorption of hematomas, which may be overamplified in patients harboring a congenital defect of the erythrocyte (e.g., sickle cell anemia). Postoperatively, the liver may also be overwhelmed due to a reduction in the normal secretory function, which may be interpreted as an acquired disorder of hepatic bilirubin transport.
- *CLM*: Bilirubinostasis in acinar zone 3 ± Kupffer cell siderosis, usually no portal inflammation.

4.6.5 Granulomatous Liver Disease

A multi-etiologic liver disease characterized by the presence of one or more hepatic granulomas. The granulomas consist of activated macrophages (epithelioid MΦ) accompanied by T lymphocytes (CD4+ T-cells) and multinucleated giant cells, which infiltrate the hepatic tissue as nodular lesions through IFN-γ and IL-2 secretion from CD4+ T-cells, and MΦ cells T-cell pool expansion in reaction to either foreign (indigestible) material or as a hypersensitivity reaction and potentially accompanied by additional hepatic inflammation and fibrosis.

Th_1 cell-type response with allied cytokines → mycobacterial granulomas

Th_1 cytokines: TNF-α, IFN-γ, IL-12, iNOS, IL-2

Th_2 cell-type response with allied cytokines → schistosomal granulomas

Th_2 cytokines: IL-4, IL-5, IL-10, IL-13

- *EPI*: 3–15% of liver biopsies, typically subclinical.

- *EPG*: See Boxes 4.30 and 4.31.
- *CLI*: Variable symptomatology including abdominal pain, weight loss, fatigue, chills, hepatosplenomegaly, lymphadenopathy, fever of unknown origin (FUO).
- *CLM*: Granulomas may be located in the lobules (TB, sarcoidosis, drugs), portal/periportal (sarcoidosis), periductal (PBC), perivenous (mineral oil lipogranulomas), and periarterial/intra-arterial (phenytoin). Fibrin strands may be highlighted using phosphotungstic acid-hematoxylin (PTAH) or Lendrum methods. Fibrin-ring granuloma, although initially considered specific for Q fever, they are now regarded as nonspecific, because of the association with multiple conditions including Hodgkin lymphoma, allopurinol ingestion, CMV, EBV, leishmaniasis, toxoplasmosis, hepatitis A, SLE, giant cell arteritis, staphylococcal infection, *boutonneuse* fever (*Rickettsia conorii*).
- *LAB*: ↑ ALP, but regular or mildly ↑ AST, ↑ ACE (sarcoidosis, PBC, silicosis, and asbestosis), ↑ serum globulins (sarcoidosis, berylliosis, and chronic granulomatous disease of childhood), peripheral blood eosinophilia (drug- or parasite-related granuloma).
- *TNT*: It should be directed toward the etiologic agent when known and use of corticosteroids for sarcoidosis.
- *PGN*: It depends on the underlying disorder.

Box 4.30 Etiology of Granulomatous Liver Disease
- Collagen vascular diseases
- Drugs
- Hodgkin lymphoma
- Infections, bacterial (e.g., TB, mycobacteria, brucellosis, tularemia, Q fever)
- Infections, fungal (e.g., histoplasmosis, cryptococcosis, blastomycosis)
- Infections, viral (e.g., EBV, CMV)
- Infestations, parasitic (e.g., schistosomiasis, toxoplasmosis, toxocariasis)
- Primary biliary cirrhosis (periportal granulomas ≠ "lipogranulomas")
- Primary sclerosing cholangitis
- Sarcoidosis (typically 2/3 of patients)

Notes: In case an etiology is not found, the granulomatous liver disease is labeled as idiopathic

---

Box 4.31 Most Common Types of Granulomas (Typology and Etiology)

Central necrosis with peripheral MΦ ± MNGC (Caseating) ⇒ TB

Cluster(s) of MΦ ± MNGC without central necrosis (noncaseating) ⇒ sarcoidosis, drugs

Lipid vacuole(s) surrounded by MΦ and lymphocytes (lipogranuloma) ⇒ fatty liver, mineral oil

Central lipid vacuole or space with MΦ, lymphocytes, and a ring of fibrin (fibrin-ring or doughnut granuloma) ⇒ Q fever, allopurinol, Hodgkin lymphoma

---

4.6.6 Alcoholic Liver Disease

It is difficult to imagine talking about alcoholic liver disease in a pediatric textbook, but the use or abuse of this substance alone or in conjunction with other substances has been increasing over time. Although the sale of these substances is forbidden in individuals younger than 18 years, many teenagers reach these substances efficiently using several methods. Ethanol exposure may induce *fatty change* (steatosis), *alcoholic steatohepatitis* (ASH), and *liver cirrhosis* (Box 4.32). Epidemiology is an extremely delicate situation, because of the substantial commercial interests of many countries, and is controversially debated due to discordant religious believes and low-social or economic drug in both developing and well-developed nations. In North America, death/illness linked to chronic alcoholism is the third largest health problem after heart disease and cancer. In some states, acute alcoholic intoxication is a particular and severe problem in youth. Typically, alcoholic liver disease develops a long-

> **Box 4.32 Alcoholic Liver Disease → Alcoholic Steatohepatitis (ASH)**
> - Fibrosis, mild to severe
> - Hepatocyte swelling and necrosis
> - Mallory hyaline body or damaged intermediate filaments of the hepatocytic cytoskeleton
> - Neutrophilic recruitment following the dying hepatocytes

term ingestion >80 g/day of ethanol (CH_3CH_2OH), which is equivalent to eight 12-ounce beers, 1 liter of wine, or half-pint of whiskey. In some studies, and, possibly in some populations, the limit of 40 g/day in males and 20 g/day in females has been emphasized. It means ¼ of wine every day in women may lead to alcoholic liver disease! Notwithstanding its cultural tolerance in industrialized or Western countries and no current prohibition campaigns, alcohol is and remains hepatotoxic and can cause liver disease even in the absence of the malnourishing, which is observed in chronic alcoholism. As early as in 1985, Novick et al. found a high rate of liver cirrhosis in youth in a hospital for addictive diseases in New York City, NY, USA. These authors reported liver cirrhosis in 53 patients under the age of 35 years. The liver cirrhosis was diagnosed clinically and biopsy-proven. Alcohol abuse was found in 96% of patients, and parenteral heroin abuse was observed in 98% of the families. In about half of the patients, the duration of alcohol abuse was 7 or fewer years. Theoretically, if a child starts drinking at age 18 (legal age in some state) or earlier, there is a risk of developing liver cirrhosis when the young adult reaches the age of 25 years.

Pathogenetically, there are a few factors that need to be taken into consideration, including the degradation system, cytotoxic products of the catabolism on membranes and cytoskeleton, and paracrine stimulation.

1. The degradation system is played by the alcoholic dehydrogenase (ADH) system for 90% of ethanol metabolism and microsomal ethanol oxidizing system (MEOS) for nearly all of the remaining 10% of ethanol metabolism.
2. Cytotoxic effects are mediated by acetaldehyde and free radicals acting on the physical state and lipid composition of the cell membranes, impairment of ER function and protido-synthesis, as well as intracellular cytoskeleton-supported molecular trafficking, determining the fatty change (steatosis) and steatohepatitis (*vide infra*).
3. Paracrine stimulation of Ito cells (aka hepatic stellate cells or perisinusoidal cells) to produce collagen has been linked to alcoholic liver disease as well and the development of liver cirrhosis. Gold chloride is used to selectively stain Ito cells. The distinguishing feature of Ito cells in routine H&E slides is the presence of multiple lipid droplets in their cytoplasm.

Clinically, there is a clinical spectrum from hepatomegaly in steatosis to viral hepatitis-like symptoms (e.g., nausea, vomiting, jaundice, hepatomegaly) in steatohepatitis with AST:ALT>2:1 and possible hepatic encephalopathy (e.g., confusion, lethargy, hallucination, asterixis, and coma) in late stages.

- *CLM*: There is a zone 3 (perivenular)-starting macrovesicular prevalent steatosis with focal lipogranulomas and retrograde restructuring of the liver following 2–4 of ethanol abstinence. Moreover, there are ASH showing a zone 3-starting steatosis with liver cell damage (ballooning degeneration, Mallory hyaline, and neutrophil-rich inflammatory infiltrate ± L/PC infiltrates) and perivenular fibrosis (phlebosclerosis) to chicken wire-like fibrosis along the spaces of Disse to central hyaline sclerosis leading ultimately to liver cirrhosis (*vide infra*). Other microscopic features are not specific and may be present in other forms of hepatitis and include siderosis, cholangiolitis, cholestasis, portal fibrosis, and interface hepatitis.

Hepatocellular steatosis (fatty change) results from three mechanisms that may act simultaneously or in tandem/subsequently like in the "National Football League."

1. *N*ADH+H+ in excess due to the action of two significant enzymes of CH_3CH_2OH metabolism, ADH, and acetaldehyde dehydrogenase inducing shunting of standard substrates away from catabolism and toward lipid biosynthesis
2. Increased *F*at catabolism at the periphery
3. Impairment of *L*ipoprotein assembly and secretion

Notes: N=NADH+H+ in excess, fat catabolism increased, and lipoprotein malfunction

- *PGN*: COD (cause of death) in ALD is due to hepatic coma, more frequently seen in teenagers. Other COD include hepatorenal syndrome developing to multi-organ failure, intercurrent infection evolving to sepsis, massive GI bleeding, and neoplastic transformation (HCC) with metastasis. In a study from Leipzig (Germany), Schoeberl et al. (2008) identified that during the years 1998–2004, the rate of alcohol-intoxicated teenagers increased at about 171.4%. A total number of 173 patients with an average age of 14.5 years was admitted to the university children hospital. They found severe symptoms in many of them, including 62 were unconscious, 2 were in a coma, and at least 3 patients had to be ventilated. Hepatic coma and death are impressively increasing in the adolescence and youth.

4.6.7 Non-alcoholic Steatohepatitis

Hyperlipidemic diet is quite typical in the Western world, but it became a problem in some other regions as well, such as the Mediterranean areas. Obesity is a reality in our countries. Dietary intake, particularly items such as juice and sugar-sweetened beverages, decreased physical activity, and sedentary behaviors, such as screen time, are quite diffuse. The body mass index (BMI) is used to determine whether a child is obese and is calculated as body weight (kg) divided by length or height (m^2). World Health Organization (WHO) growth curves are available and allow to determine the progression of children (Fig. 4.25). Obesity is influenced by primary genetic or endocrine factors and nongenetic endocrine ones, including the environment, family, and social. Primary genetic or endocrine etiologies are rare but need to be taken into consideration, if the child has dysmorphism, severe hyperphagia, developmental delay, clinical signs of Cushing impressively increasing in or Cushingoid syndromes, and hypothyroidism. Risk factors for obesity also need to be considered and include a family history of obesity, substantial pregnancy weight gain, and gestational diabetes mellitus. Health risks for the child include hyperlipidemia, hypertension, insulin resistance, low quality of life (*quoad valetudinem*), and sleep-disordered breathing and poor school performance. The identification of *acanthosis nigricans*, a sign of insulin resistance, has been proposed by some authorities. AAP guidelines recommend a fasting lipid profile for children older than 2 year of age affected with obesity, and the Canadian Diabetes Association 2013 recommends screening prepubertal children for T2DM every 2 years, if some factors are met, including obesity, high-risk ethnic group, family history of type 2 diabetes mellitus, in utero exposure to hyperglycemia, and signs of insulin resistance. NASH grading has been indicated to harbor an extreme prognostic value. There is a growing list of possible metabolic defects or genetic syndromes showing NASH or early cirrhosis in life, and OMIM or other search engines may be able to create an updated list.

Box 4.33 shows some tips and pitfalls in diagnosing of NASH/ASH.

4.6.8 Acute and Chronic Rejection Post-Liver Transplantation

Acute rejection in liver transplant recipients is an important clinical event, which may be life-threatening. Despite improvements in immunosuppressive therapy, hepatic allograft rejection remains an important cause of morbidity. Moreover, it is a critical etiology for late graft loss in patients undergoing liver transplantation.

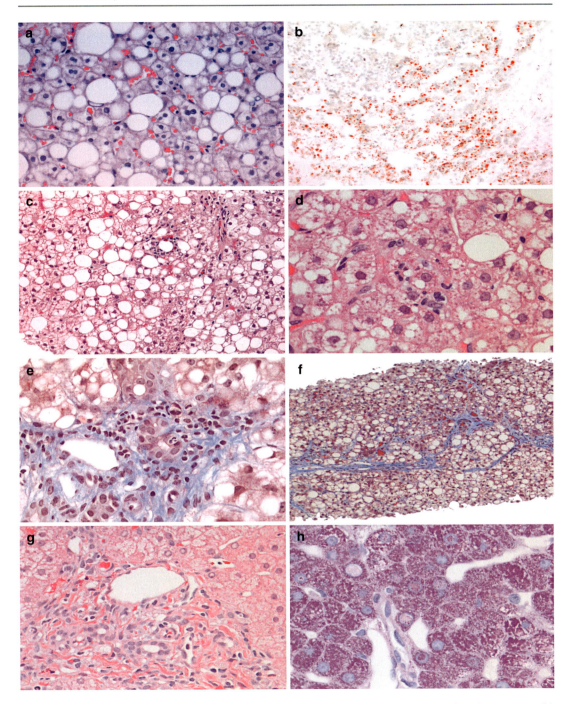

Fig. 4.25 Alcoholic steatohepatitis and NASH can be recognized by multiple vacuolation of the hepatocytes with fibrosis and neutrophilic damage (**a–h**, H&E, Masson's trichromic, periodic acid Shiff, and Oil Red O stains)

In about 5–10% of liver recipients who develop acute cellular rejection, there is a progression to severe ductopenic rejection despite antirejection therapy. Histologic features portend a terrible outcome include arteritis, ductopenia (PIBD), and/or simultaneous hepatocellular ballooning and hepatocellular dropout and necrosis. Bile duct damage in the diagnosis of acute rejection should include a predominantly lymphocytic infiltrate and a damaged bile duct. Useful features of bile duct damage include variation in biliary epithelial nuclear size, vacuolation of the cytoplasm, missing nuclei, and, naturally, intraepithelial lymphocytes. The importance of ductopenia depends mainly on the degree and persistence of the scarcity (Figs. 4.26, 4.27, and 4.28). See Box 4.34 for pearls and pitfalls on liver TX and the DDX.

Posttransplantation lymphoproliferative disorders may be a complication of the immunosuppression used in patients to avoid rejection or GvHD.

Box 4.33 NASH/ASH Pearls and Pitfalls! (Fig. 4.25)

- NASH is *Non*-ASH and includes a series of several etiologies, including obesity, DM, drug toxicity (e.g., amiodarone, perhexiline maleate, estrogens), or post-jejunoileal bypass surgery or gastropexy for adiposity. The differential diagnosis requires a clinicopathologic correlation!
- Mallory hyaline (aka Mallory-Denk body) can be highlighted by histochemistry and immunohistochemistry. Masson trichrome stain is useful to detect the Mallory-Denk bodies (MDBs). Immunohistochemistry for p62, ubiquitin, or CK8/CK18 may also be positive in MDBs.
- Mallory hyaline is also seen in Wilson disease, Indian childhood cirrhosis, some liver neoplasms, and some disorders associated with chronic cholestasis (e.g., PBC).

Notes: *CK8* cytokeratin 8, *CK18* cytokeratin 18

Box 4.34 Pearls and Pitfalls on Liver TX-DDX

- If you see spotty necrosis with neutrophilic microabscesses, then think of CMV!
- If you see spotty necrosis with portal mononuclear infiltrates ± lobular mononuclear infiltrates, then think of HBV and HCV.
- If you see spotty necrosis only, then think of hepatitis NOS.
- If you see focal lobular mononuclear clusters only, then think of a nonspecific finding!
- If you see the biliary ductular reaction with bile plugs, then think of sepsis or preservation injury (time and lab notes are essential)!
- If you see biliary ductular reaction without bile plugs, then think of severe early tissue preservation injury, ACR, LDO and severe viral hepatitis, and the list may not be complete (time and lab notes are essential)!
- If you see a mixed inflammatory infiltrate of the portal tracts with bile duct damage and/or inflammation (mononuclear, eosinophils, neutrophils), then think of ACR!
- If you see periductal neutrophilic infiltrates and duct damage, then resolving rejection or LDO are on the top of the list (clinicopathologic correlation is essential!).
- If you see cholestasis only, then tissue preservation injury, LDO, or chronic rejection should be considered (crucial is the interlobular bile ducts/portal tract ratio)!
- If you see diffuse hepatocyte ballooning, then think of ischemia and/or hepatitis.
- If you see centrilobular hepatocyte necrosis and/or swelling, minimal inflammation with or without congestion and/or fibrosis, then think of tissue preservation injury, ischemia, or vascular rejection.
- If you see parenchymal necrosis, then think of local ischemia/another injury.

Notes: *ACR* acute cellular rejection, *LDO* large duct obstruction

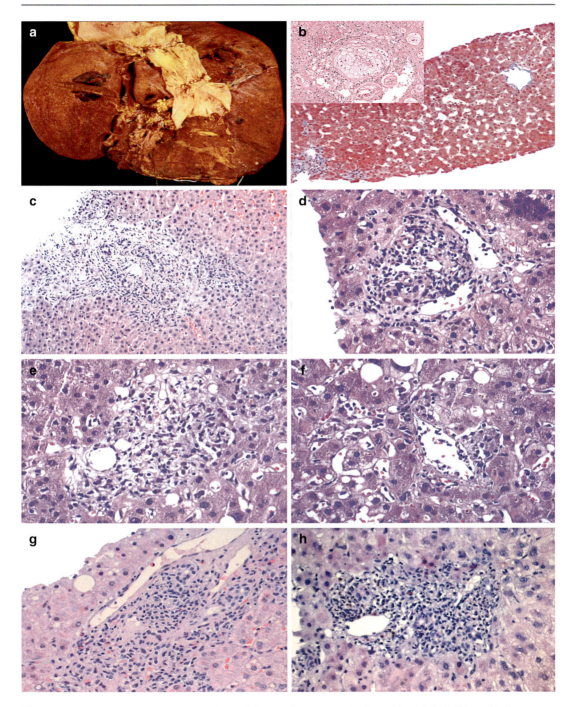

Fig. 4.26 TX hepatopathy shows hemorrhages (**a**) as well as acute rejection with endothelialitis and inflammatory infiltrate of the portal tract (**b**–**h**)

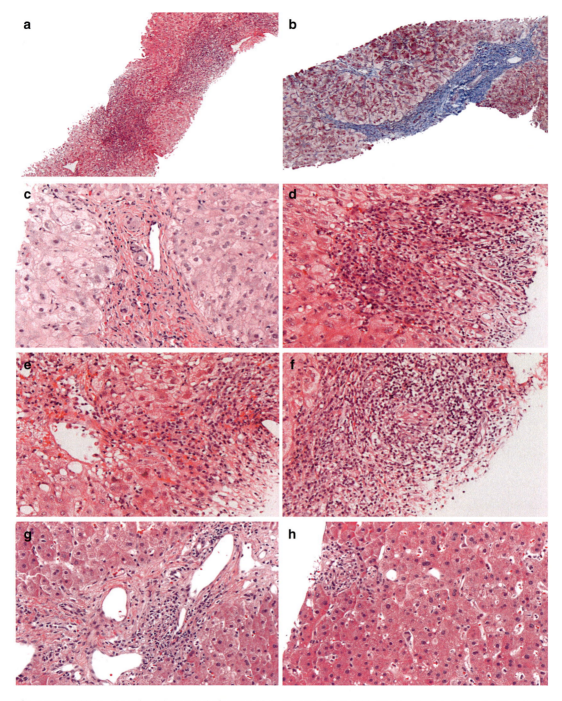

Fig. 4.27 In this panel a quite unique panel of progressive sclerosing cholangitis (de novo)-related transplant hepatopathy is shown (**a–h**)

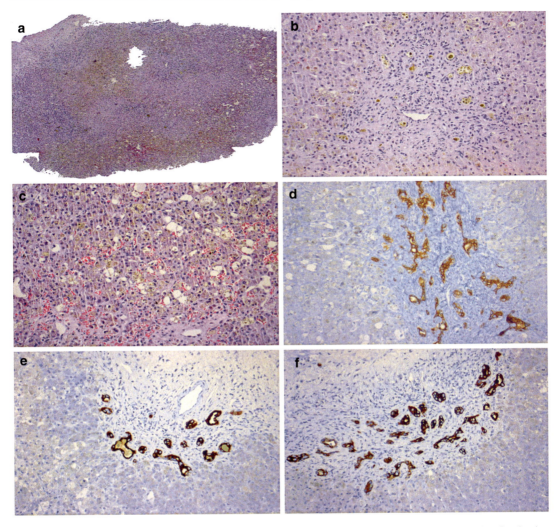

Fig. 4.28 In this panel is shown chronic graft-versus-host disease with paucity of the interlobular bile ducts in an adolescent. Note the biliary proliferation of peripherally located bile ductules, some of them showing lumina with inspissated bile (**a–c**: H&E stain; **d–f**: anti-CK7 immunostain)

4.7 Hepatic Vascular Disorders

Hepatic vascular disorders are a common coming in the pathology practice (Boxes 4.35 and 4.36).

Box 4.35 Circulatory Disorders
1. ↓ blood flow into the liver
2. ↓ blood flow through the liver
3. ↓ blood flow out of the liver

4.7.1 Acute and Chronic Passive Liver Congestion

Cholestasis, as well as liver cell damage, may also be encountered in ischemic or hemorrhagic processes involving the structure of the liver.

Acute and Chronic Passive Liver Congestion
Clinical (hepatomegaly) and pathologic (gross and histologic passive congestion) are observed in right-sided heart failure and, often, seen with children with congenital heart defects affecting

> **Box 4.36 Hepatic Vascular Disorders**
> 4.7.1. Acute and Chronic Passive Liver Congestion
> 4.7.2. Ischemic Hepatocellular Necrosis
> 4.7.3. Shock-Related Cholestasis

the functionality of the right heart. The etiology of hepatic congestion is increased pressure within hepatic venules and sinusoids due to severe right-sided heart failure and constrictive pericarditis, thrombosis of large hepatic veins (Budd-Chiari syndrome), and veno-occlusive disease (VOD). Budd-Chiari syndrome may be observed in neoplastic or non-neoplastic myeloproliferative states, hypercoagulability, malignancies of the liver, kidney, and adrenal gland, postradiation exposure, and following contraceptive use administered orally. VOD has been associated with herbal teas-containing pyrrolizidine alkaloids, radiotherapy, and chemotherapy (azathioprine and cytotoxic agents in bone marrow transplant (BMT) and kidney transplant (KTX) recipients). Clinically, hepatomegaly with or without dysregulation of liver enzymes is seen. Grossly, the classic nutmeg liver with diffuse parenchymal mottling with reddish-purple perivenular areas surrounded by paler regions is observed. Microscopically, a variable range of changes is seen, including sinusoidal congestion, atrophy and disappearance of the hepatocyte plates with the collapse of the reticulin framework and fatty change of zone 2, and reverse lobulation with pericentral fibrosis and portocentral septa. Budd-Chiari syndrome may also show fibrin thrombi, while VOD is a fibro-obliteration of the terminal venulae.

Overall, ischemia is the consequence of shock or other systemic hypotensive states or other states leading to congestion (e.g., some cirrhotic nodules in patients with hypodynamic heart function). Grossly, there are no right ischemic infarcts due to the liver's dual blood supply. Zahn infarct is a wedge-shaped area due to sinusoidal dilation due to an occlusion of intrahepatic portal vein branches. Liver changes due to either obstruction of blood inflow (hepatic artery or portal vein) or blood outflow (central vein or hepatic vein/s). In infarction, there is a sudden obstruction of the hepatic artery. Liver circulatory changes are summarized in Box 4.37. The etiology of Budd-Chiari syndrome includes *polycythemia vera*, pregnancy, oral contraceptives, hepatocellular carcinoma, and intra-abdominal malignancies.

> **Box 4.37 Liver Circulatory/Vascular Changes**
> - BCS: Partial/sub-occlusive/occlusive thrombosis and/or fibrous obliteration of the major hepatic veins or inferior vena cava ± membranous webs
> - CLN: prolonged congestive heart failure
> - CPC: Right heart failure (nutmeg appearance with centrilobular congested areas and relative pallor of the periportal areas with distension of central veins and perivenular sinusoids of the zone 3 of Rappaport)
> - Infarction: Thrombus
> - PVT: Abdominal malignancies, peritoneal sepsis, pancreatitis, post-surgery, cirrhosis, metastatic carcinoma; clinically, abdominal pain and ascites
> - VOD: Pyrrolizidine alkaloids, azathioprine, thioguanine, hepatic radiation, BMT with GvHD; sclerosis-related occlusion of the central veins with associated sinusoidal congestion of all three zones
>
> Notes: *BCS* Budd-Chiari syndrome, *CLN* centrilobular necrosis, *CPC* chronic passive congestion *PVT* portal vein thrombosis, *VOD* veno-occlusive disease

Peliosis Hepatis

Rochalimaea henselae/Rochalimaea quintana-related vascular change of the liver with cystic pools of blood lined by hepatocytes (CD31-, no endothelial cells) due to the progressive death of infected hepatocytes (skin→ bacillary angiomatosis) with non-infective mimickers, including anabolic steroid use, exposure to vinyl chloride, post renal TX, and hematologic disorders.

4.7.2 Ischemic Hepatocellular Necrosis

The patterns of hepatic injury with necrosis and apoptosis can be subdivided according to the distribution and degree of involvement. According to the distribution, we may have centrilobular necrosis, which is most common and is immediately around the terminal hepatic vein. According to the degree of involvement, we can have focal or spotty necrosis (only scattered cells within the hepatic lobulus), interface hepatitis (between periportal parenchyma and inflamed portal tracts), bridging necrosis, spanning adjacent hepatic lobules, submassive necrosis involving entire lobules, and massive necrosis, comprising most of the hepatic parenchyma.

4.7.3 Shock-Related Cholestasis

- *DEF*: Shock-related bilirubinostasis due to some etiologies that may act singly or in combination, as shown below:
  1. Inhibition of the hepatocellular Na^+-K^+-ATPase at the basolateral membrane due to endotoxin, hypoxia-related damage, and/or DILD
  2. Hyperthermia/fever, and dehydration
  3. Bile ductular proliferation with some obstruction
  4. Bacterial β-glucuronidase-related production of bile concretion
  5. Effect(s) of inflammatory substances (e.g., leukotrienes) on interlobular bile ducts and/or bile ductules.
- *CLM*: Perivenular necrosis of hepatocytes with cholestasis is due to the reduction of the O_2 supply to the liver, particularly to hepatocytes located in the acinar zone 3, which are the peripheral territory of the hepatic acini.
- *DDX*: Necrotizing hepatitis, TPN, BA, decompensated liver cirrhosis.

4.8 Liver Failure and Liver Cirrhosis

Liver failure is referred to an acute liver illness associated with encephalopathy and subdivided according to the onset of brain involvement (vide infra) (Box 4.38) (Figs. 4.29 and 4.30).

> **Box 4.38 Liver Failure**
> – Acute liver failure (ALF)
> – Chronic liver disease (CLD).
> – Hepatic metabolic dysfunction (HMD) without overt necrosis (e.g., tetracycline toxicity)

Acute liver illness is associated with encephalopathy and subdivided according to the onset of brain involvement:

- Fulminant, if <2 weeks after the initial diagnosis
- Acute, if <6 months after the initial diagnosis
- Subacute, if <3 months after the initial diagnosis

Three liver failure-related complications are mostly highly relevant to clinics:

1. Hepatic *encephalopathy*: Neurotransmission disorder due to >>> NH_4^+ ⇒ edema
2. Hepato-*renal syndrome*: Renal failure in CLD (intrinsic morphologic/functional injury)
3. Hepato-*pulmonary syndrome*: Hypoxemia and IPVD (intrapulmonary vascular dilatation) in CLD

Liver cirrhosis refers to the end stage of several and different progressive and chronic diseases leading to diffuse and irreversible parenchymal damage. Three processes of regeneration are described in the original German

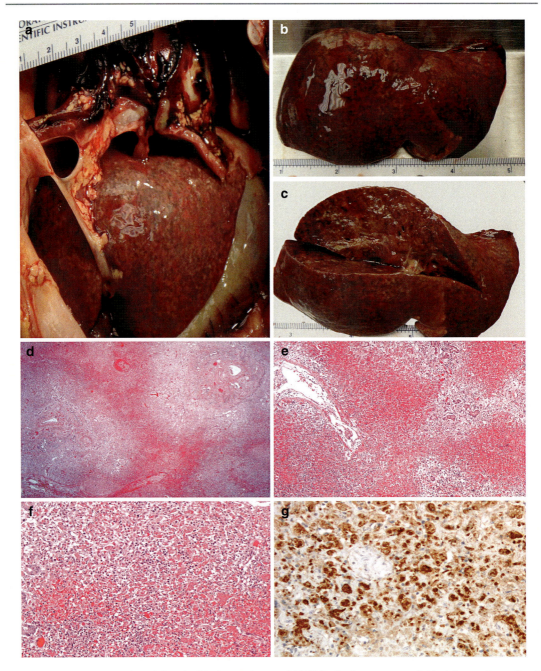

Fig. 4.29 Neonatal HSV-liver failure is life-threatening and shows liver failure both grossly (**a–c**) and histologically (**d–f**). The microphotograph (**g**) shows the detection of HSV in the hepatocytes of this infants who suffered from HSV-induced liver failure

literature of pathology schools: "Abbau-Aufbau-Umbau." In liver cirrhosis, there is simultaneous presence of all three processes, which include dismantlement of the normal liver architecture with portal triads and centrolubar zones (*Abbau*), increase of collagen and collagen-producing cells through paracrine activity (*Aufbau*), and conversion of the normal hepatic architecture into structurally abnormal nodules (*Umbau*). These three events, although simultaneous, may be incon-

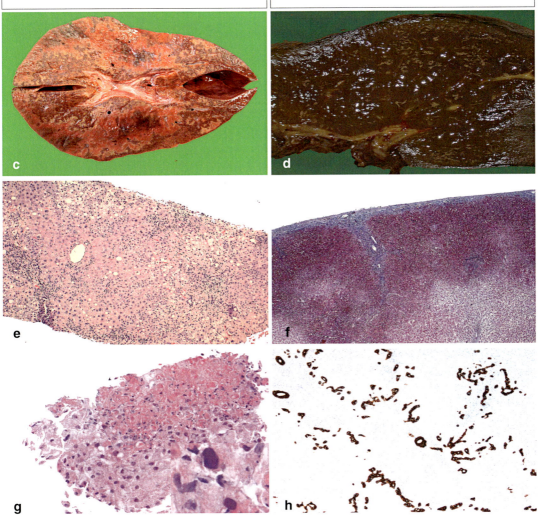

Fig. 4.30 In this panel are shown the species of the *Amanita* genus mushroom that each summer is culprit of some deaths worldwide and the mechanism of damage due to amatoxin and phallotoxin (**a–b**). In (**c**) and (**d**) are two gross photographs of common mushroom poisoning liver failure. The histopathology of liver failure can be different, because the etiology of fulminant liver failure is variable (**e–h**). In (**g**) adenovirus-related liver disease is seen (H&E stain). The microphotograph in (**h**) shows the biliary proliferation encountered in cases of subacute liver failure when the liver has some time to regenerate and rebuild some excretory biliary ways to drain the bile (anti-CK7 immunostain)

spicuous in the early stages (at least one or two out of three) and become more prominent as the disease progresses leading to the abnormal vascular relationship (AV shunts) and portal hypertension with obvious consequences for the affected individuals (Box 4.39).

> **Box 4.39 Chronic Liver Disease (CLD) End Stage**
> Cirrhosis – End stage of CLD is characterized by:
>
> - Loss of hepatocytes (*Abbau* of the German literature)
> - ECM deposition (*Anbau* of the German literature)
> - Vascular reorganization (*Umbau* of the German literature) with sinusoidal capillarization in Disse space
>
> Evolving to:
>
> - Bridging fibrous septs
> - New parenchymal nodules
> - Liver architecture disruption

Some stimuli can bring forth the activation of stellate cells, which are at the basis of the liver fibrosis. They are cyto-/chemokine release by KC, endothelial cells, hepatocytes, and biliary cells. ECM disruption, inflammation of chronic type (⇒ inflammatory cytokines, such as tumor necrosis factor (TNF), lymphotoxin, IL1-β, and lipid peroxidation products), and toxin-/drug-related direct stimulation of stellate cells (stellate cells → myofibroblasts (MFB) ⇒ collagen I and III through the expression of platelets growth factor receptor beta (PDGFR-β), transforming growth factor beta (TGF-β), matrix metalloproteinases (MMP2), and tissue inhibitor of metalloproteinases (TIMP1-2)) are important key factors in liver cirrhosis.

The easiest way to classify the liver cirrhosis is according to the size of the nodules, i.e., micronodular or macronodular or mixed. Classifications based on etiology may also be used, but the size of the nodules will direct the pathologist to the most critical criterion and then to some etiology list. The cutoff for micro- or macronodular cirrhosis is 3 mm. If more than 3 mm, the cirrhosis is macronodular, while if less or equal to 3 mm, the cirrhosis is classified as micronodular. However, in many cases the liver may show a combination of features of both macro- and micronodules. The micronodular is classically the Laennec or portal cirrhosis, while the macronodular is usually the post-necrotic or post-hepatic liver cirrhosis. Causes of death (COD) of patients with liver cirrhosis are (1) progressive liver failure, (2) portal hypertension-related complications, and (3) hepatocellular carcinoma.

4.9 Portal Hypertension

Portal venous pressure in physiological conditions is 0–10 mm Hg and is slightly above the venous pressure of the inferior vena cava. However, the hepatic venous pressure (HVP) gradient is usually ~ 5mmHg. If it rises over 6mmHg portal hypertension is considered to be present, but if it is over over 10mmHg clinical features of portal hypertension can develop and a HVP gradient greater than 12 mmHg is accompanied by varices and other complications. The gradient between the two pressure venous systems is low because:

- High compliance of the portal vein and sinusoids
- The low resistance of the portal vein and sinusoids
- The outflow of the large suprahepatic veins

Also, there is an adverse pump effect resulting from direct drainage into the right atrium and the perfect location at the thoraco-abdominal interface.

Portal hypertension: It is defined by a pressure that is higher than typical values. The etiology is dual. Portal hypertension may be due to (1) either an increased resistance to portal flow at the level of the sinusoids (2) or a partial venous blood flow produced from a hyperdynamic circulation.

There are four consequences of portal hypertension, including (1) ascites (sinusoidal hypertension, hepatic lymph percolation, splanchnic circulation vasodilation, and hyperdynamic circulation), (2) portal system venous shunts, (3) congestive splenomegaly and hypersplenism, and (4) hepatic encephalopathy. A common classification of portal hypertension includes a prehepatic (e.g., thrombosis of the portal vein and "cavernomatous" change, splenic vein thrombosis, splenic arteriovenous fistula, and idiopathic tropical splenomegaly), hepatic (pre-sinusoidal, e.g., hepatic fibrosis due to, for instance, schistosomiasis, but also hepatic neoplasms, viral hepatitis, and infiltrative disease such as tuberculosis, sarcoidosis, hemochromatosis, and rarely amyloidosis; sinusoidal, cirrhosis, for instance, in biliary atresia, but also acute viral hepatitis, NASH/ASH, and fatty liver of pregnancy; and post-sinusoidal for instance of veno-occlusive disease and in case of hyaline sclerosis of central veins), and post-hepatic (e.g., Budd-Chiar syndrome, but also inflammatory and/or neoplastic infiltration, caval inferior occlusion due to thrombus or neoplasm, right ventricular failure, tricuspid regurgitation, and constrictive pericarditis).

Arterio-Portal Hypertension

It is an excessive and constrained arterial blood inflow into the portal venous system, which is typically a closed vascular system that is limited by volume with a single outflow. Typical underlying condition: arterio-portal fistula from a vascular malformation located in the liver or the splanchnic area. The arterio-portal fistula is defined as simple or complex according to single or multiple arteries ending in a common venous channel or an aneurysm. Management of portal hypertension may be very challenging, especially in childhood (De Ville de Goyet et al. 2012). Surgical management may depend not only on parenchymal damage and hepatic dysfunction but also about the secondary injury and fibrosis occurring in patients with post-sinusoidal portal hypertension. Surgical intervention was rarely a common practice in the past assuming that blood vessels may become more stable with age and the chance of success increase. However, several studies have evidenced excellent technical outcomes even in early childhood, and the management is changing worldwide. This information is relevant to liaise correctly with the pediatric surgical team at multidisciplinary team meetings or clinical rounds and to evaluate precisely failed procedures associated with death at the time of postmortem investigations or the court for medicolegal cases. Complications following shunt and bypass may occur even in best centers. The incidence of complications is variable because some factors should be taken into consideration, mostly the underlying clinical condition(s) of the patient.

4.10 Bacterial and Parasitic Liver Infections

4.10.1 Pyogenic Abscess

It is defined as a collection of pus (purulent abscess) that has built up within the liver in a newly formed cavity (the condition in which pus and fluid from infected tissue accumulates in a pre-existing body cavity is called empyema, e.g. pleural empyema). Both abscess and empyema have Latin and Greek origin (Abscessus from abscēdō "depart" and ἐμπύημα "suppuration").

There are four significant ways in which pyogenic microorganisms invade the liver.

1. The microorganisms may travel through the portal vein from regions drained by it, such as acute suppurative appendicitis, pyophlebitis, gastric and intestinal ulcers, as well as infectious (pyogenic) diseases of the rectum, spleen, and pancreas.
2. The microorganism belongs to blood-borne infection and is transmitted through the hepatic artery. Etiologies include osteomyelitis, acute infections of the urinary tract, or pyemia from any source.
3. The microorganism gets access to the liver through a direct extension from a contiguous infection (e.g., subphrenic abscess, empyema, nephritic abscess, perinephritic abscess).
4. The microorganisms can invade the liver directly through trauma, and this may be a

penetrating injury (accidental, criminal, iatrogenic) or a subcutaneous injury (mostly, random).

4.10.2 Helminthiasis

Helminthiasis group includes echinococcosis, schistosomiasis, and ascariasis. Worldwide, helminths remain probably the primary cause of wildlife diseases and, unfortunately, represent a significant burden for the patients, healthcare, and the often-economic crisis in the livestock industry. Human socioeconomic problems are a considerable task in developing countries as well and added to the burden. Helminths often live in the gastrointestinal tract of their hosts with the option to burrow into other organs inducing relevant damages. Liver disease is derived from any macroparasitic illness of humans and animals in which the liver is infected with parasitic worms, called helminths (Greek: ἕλμινς, ἕλμινθος "intestinal worm").

4.11 Liver Tumors

4.11.1 Benign Tumors (Figs. 4.31, 4.32, and 4.33)

Benign tumors are essential to differentiate from malignant, but benign tumors also have a quite complicate or variable list of names that may confounder either the pathologist or the clinician. It is essential to differentiate between reactive (or regenerative) changes and true neoplasm (benign or malignant). In the liver, it is useful first to distinguish a *nodular regenerative group* from *liver cell adenoma*. The concept of *hepatocellular dysplasia* is presented, although is controversially discussed in the scientific literature. Benign bile duct tumors (i.e., non-hepatocytic but biliary cell-differentiated benign tumors) are listed in Box 4.40.

Finally, the remaining benign tumors include non-hepatocytic, non-biliary cell-differentiated tumors, which are a *cavernous hemangioma*, *infantile hemangioendothelioma*, *peliosis hepa-*

> **Box 4.40 Benign Bile Duct Tumors**
> - Bile duct hamartoma
> - Bile duct adenoma
> - Biliary cystadenoma
> - Biliary papillomatosis

tis, and *hepatic mesenchymal hamartoma*. The most common genetic syndromes associated with pediatric liver tumors are illustrated in Box 4.41.

The entities collected in the group of regenerative nodularity are particularly important for the differential diagnosis of hepatocellular tumors.

4.11.1.1 Nodular Regenerative Hyperplasia (NRH)

Nodular regenerative hyperplasia (aka nodular transformation) is constituted by the formation of cirrhosis-like nodules ("compensatory hyperplasia") separated by the compressed atrophic liver with atrophic features without intervening fibrosis and associated with myelo-proliferative disorders, RA, chronic venous congestion, and drugs (e.g., steroids, chemotherapy protocols). Ischemia or chemical insults may determine the hyperplasia of the hepatic *muralium* (two-cell thick indicating regeneration) and generally involving the entire liver.

4.11.1.2 Partial Nodular Transformation (PNH)

The early stage of NRH with similar changes but confined to the area of the hilum.

4.11.1.3 Focal Nodular Hyperplasia (FNH)

Well-circumscribed, hyperplastic and hypervascular, *asymptomatic* lesion (4/5 single, 1/5 multiple) due probably to a pre-existing arterial malfunction with specific arteriographic finding (hypervascularity with centrifugal filling and a particular blush of dense capillarity). Grossly, solid, gray-white, unencapsulated (usually Ø < 5 cm), subcapsular with central star-shaped scar with radiating fibrous "rays." Microscopically, there is a hyperplastic hepatocytic nodule with central fibrosis and radiating

Box 4.41 Legitimate Genetic Syndromes Associated with Pediatric Liver Tumors

Disease	Tumor(s)	Chromosome	Gene
Aicardi syndrome	HB	?X	Unknown
Alagille syndrome	HCC	20p12	*JAG1*
Ataxia-telangiectasia	HCC	11q22–11q23	*ATM*
BWS	HB, HHA	11p15.5	*P57KIP2*, others
Edwards syndrome	HB	(47, XX or 47, XY) +18	Unknown
Fanconi anemia	HCA, HCC, FL-HCC	1q42, 3p, 20q, 13.2–13.3	*FAA*, *FAC*, others
FAP	HB, HCA, HCC, HBA	5q21.22	*APC*
Genetic hemochromatosis	HCC	6.22.2	*HFE*
GSD Ia	HCA, HCC, HB	17q21.31	*G6PC*
Hereditary tyrosinemia	HCC	15q25.1	*FAH*
LFS	Hepatoblastoma, undifferentiated sarcoma	17p13	*TP53*, others
Neurofibromatosis	HCC, HAS, MPNST	17q11.2	*NF-1*
Non-Alagille familial cholestatic syndromes	HCC	18q21–18q22, 2q24	*FIC1*, *BSEP*
Simpson-Golabi-Behmel syndrome	HB, NB, WT	Xq26.2	*GPC3*
Tuberous sclerosis	Angiomyolipoma	9q34, 16p13	*TSC1*, *TSC2*

Notes: *FAP* familial adenomatous polyposis, *LFS* Li-Fraumeni syndrome, *GSD* Ia, glycogen storage disease type Ia, *HB* hepatoblastoma, *HBA* hepatic biliary adenoma, *HHA* hepatic hemangioendothelioma, *HAS* hepatic angiosarcoma, *FAH* fumarylacetoacetate hydrolase, *G6PC* glucose-6-phosphatase, catalytic, *MPNST* malignant peripheral nerve sheath tumor

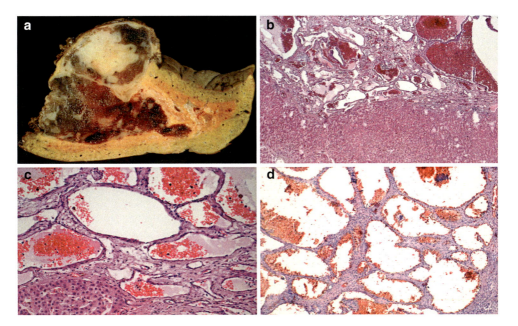

Fig. 4.31 In this gross (**a**, **e**) and microphotographs are shown hemangioma (**a**–**d**) and infantile hemangioendothelioma (**e**–**h**)

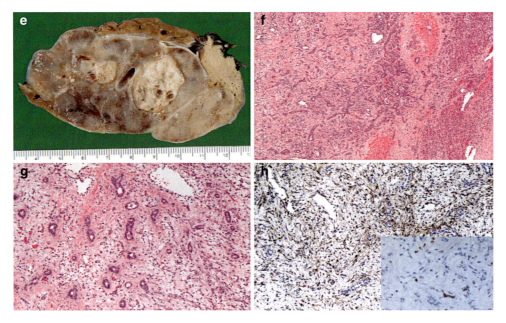

Fig. 4.31 (continued)

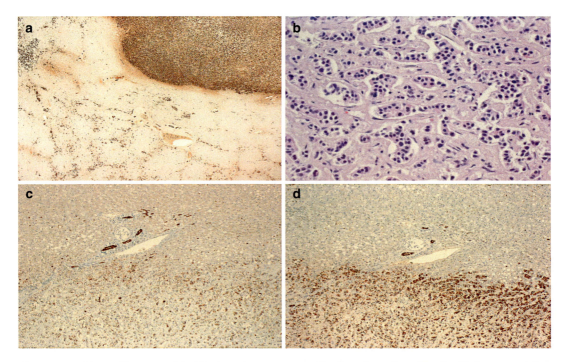

Fig. 4.32 In this panel is presented the bile duct adenoma with its characteristic morphologic features, including a tight network of simple noncystic tubular ducts harboring small or indistinct lumina and epithelial cells lacking atypias or mitotic figures but having abundant cytoplasm and pale nuclei (**a–d**: **a**, AE1–AE3 immunostaining or pan-cytokeratin markers, ×200 original magnification; **b**, hematoxylin and eosin staining, ×400 original magnification; **c**, anti-cytokeratin 7 immunostaining, ×100 original magnification; **d**, anti-cytokeratin 19 immunostaining, ×100 original magnification). All immunohistochemistry procedures have been performed using avidin-biotin complex. In figure (**a**) is particularly evident the contrast between the adenoma and the healthy liver. The tight network as better highlighted in (**b**), while the cytokeratin 7 and 19 immunostainings reveal the persistence of cytokeratin 19 over 7 prevalence of the primitive bile duct structures

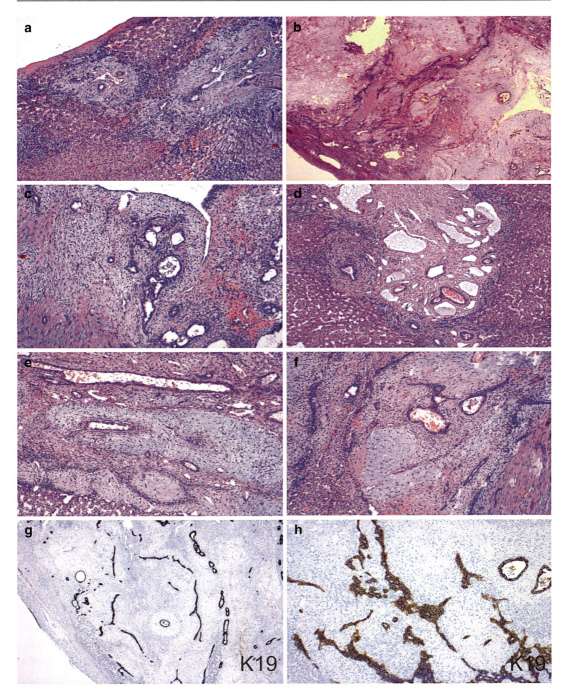

Fig. 4.33 In this panel is presented the rare occurrence of a hepatic mesenchymal hamartoma (HMH), which contains epithelial and mesenchymal elements with branching ducts of biliary epithelium without atypia. The branching ducts are embedded in loose, myxoid stroma with myofibroblast-like cells (**a**, hematoxylin and eosin staining, ×100 original magnification; **b**, hematoxylin and eosin staining, ×200 original magnification; **c**, hematoxylin and eosin staining, ×100 original magnification; **d**, hematoxylin and eosin staining, ×100 original magnification; **e**, hematoxylin and eosin staining, ×100 original magnification; **f**, hematoxylin and eosin staining, ×100 original magnification). Two immunostaining are also seen in Figures (**g**) and (**h**) using antibodies against cytokeratin 19 (**g**, ×100 original magnification) and cytokeratin 7 (**h**, ×100 original magnification). Both immunohistochemistry procedures have been performed using the avidin-biotin complex.

fibrous bands with portal tract-like areas and eccentrically thickened blood vessels in the septs.

4.11.1.4 Macro-Regenerative Nodule (or Nodularity) (MRN)

Benign, non-adenomatous nodule, usually Ø > 1 cm, occurring in the setting of cirrhosis with a thin fibrous rim, similar to surrounding smaller nodules. DDX: LCA, dysplasia.

It is paramount to mention the definitions and aspects of nodular regenerative hyperplasia (NRH) and partial nodular transformation (PNT). In NRH, there are regenerating hepatocytes with thickened cell plates and compression of adjacent areas, which present with thinning or atrophy of the cell plates in the absence of fibrosis (≠ cirrhosis) separating the several nodules. In PNT, there is an NRH morphology but limited to the perihilar region of the liver and caused by a vascular insult. Thus, location is crucial (diffuse vs. focal-perihilar) and then evaluate the presence of fibrosis. If no fibrosis, it is NRH/PNT, while in a state of fibrosis, the diagnosis portends to cirrhosis and Masson trichrome stain may be crucial. The trichromic stain is essential in differentiating NRH/PNT from cirrhosis.

4.11.1.5 Liver Cell Adenoma (LCA)

- *DEF*: A benign epithelial tumor of hepatocellular origin, which may be associated with oral contraceptives, anabolic steroids, antiestrogen therapy, Klinefelter syndrome, GSD Ia/III, Fanconi anemia, FAP, DM, mucopolysaccharidosis Type IH (MPS-IH or Hurler syndrome), tyrosinemia and classified in four types with the acronym "FAT" to remember the three most important types (Box 4.42) and presenting as a well-circumscribed, hyperplastic but hypovascular, often symptomatic (bleeding ¼ of cases) lesion with a specific arteriographic finding (hypovascularity).
- *EPI*: Worldwide, 2nd–4th decades, 2–4% liver tumors in childhood, ♂:♀ = 1:9.
- *CLI*: Mostly solitary, right lobe, usually >10 at presentation (in case of liver cell adenomatosis, Flejou et al. 1985), and symptomatic (hemorrhage 1/4 of cases) and hypovascular-

Box 4.42 Typing of Liver Cell Adenomas (Four Types)

Type	Etiology	Histology/IHC
Fatty LCA	*HNF1α*	LFABP (−)
Atypia/adenomatous LCA	*CTNNB1*	Inflammation, β-cat (+)$_{n/c}$
Telangiectatic LCA	*GP130*	SAA (+), CRP (+)
LCA, NOS	Unknown	

Notes: *CRP* C-reactive protein, *CTNNB1* beta-catenin gene, *GP130* glycoprotein 130, *HNF1α* hepatocyte nuclear factor 1 alpha, *SAA* serum amyloid-A

ity on arteriography (imaging). Flejou et al. (1985) definition of liver cell adenomatosis include the following criteria (a) arbitrarily >10 hepatic nodules, (b) equal sex distribution, (c) no association with oral contraceptives, (d) ↑ levels of serum ALP and GGT. Chiche et al. (2000) distinguish two patterns of liver cell adenomatosis, including the *massive* type presenting with hepatomegaly, a deformation of the liver contour, with nodules of 2–10 cm (Ø), and the *multifocal* type containing many adenomas up to 4 cm (Ø) but lacking a deformation of the liver contour. A rapidly progressive course is usually restricted to the massive type.
- *GRO*: Partially or unencapsulated, mostly round, yellow-tan, subcapsular and without necrosis.
- *CLM*: Well-differentiated hepatocytes with abundant cytoplasm and growing in two-cell thick trabeculae with solitary arterial blood vessels without evidence of portal triads, central veins, fibrosis, bile ducts or connection with the biliary system ("unpaired arteriae," NTBV – non-triadic blood vessels), intact reticulin framework, no mitoses, and no invasiveness. The acronym "MAGIC" is useful to remember some important criteria (MI = 0, Atypia = 0, Growth = two cell thick with unpaired AA and no BD, Invasiveness = 0, C = Clinical History, often +). The blood vessels tend to be localized at the periphery of the

tumor showing intimal thickening and smooth muscle proliferation.
- *HSS/IHC*: Ret (+), GPC-3 (−), CD34 (±) + subtyping markers (LFABP, β-cat, SAA, CRP) and E/PR (+), p53 (−).
- *DDX*: Hepatocellular nodular lesions, including FNH, NRH (even a large regenerative nodule), PNT, FLC, and HCC. The normal liver always has portal tracts, and nontriadal blood vessels are never seen in other lesions other than LCA.
- *TNT*: Resection.
- *PGN*: In case of oral contraceptive/anabolic steroid-induced LCA, the tumor often regresses when these drugs/supplements discontinued. The condition "multiple hepatocellular adenomatosis" represents an incredible challenge may be a very well-differentiated HCC and expert pathology consultation is strongly recommended.

Box 4.42 highlights the distinctive phenotyping of liver cell adenomas as we know from the most current classification.

4.11.1.6 Hepatocellular Dysplasia
The concept of HCD is controversially debated. It has been suggested that even when correctly diagnosed, no clear indications for therapy exist.

Large cell dysplasia refers to isolated large cells with enlarged nuclei but standard N/C ratio and normal nuclear contours. It seems to be often seen in some livers with cirrhosis. There is no clear-cut evidence to assign this condition a level of premalignant condition.

Small cell dysplasia refers to clusters of normal-sized hepatocytes with visible larger nuclei and distinct increased N/C ratio as well as irregular nuclear contours. The architecture seems, however, unaffected. Flow cytometry studies evidence an increase in the number of peaks, mostly 2n and 4n. MIB1/Ki67 and PCNA are increased, although not so high as much we can detect in an HCC. There is a broad platform of consensus, although not univocal, that "small cell dysplasia" may be considered a premalignant condition.

4.11.1.7 Bile Duct Hamartoma (BDH)
BDH is a ductal plate malformation or anomaly and is aka von Meyenburg complex or Moschowitz complex. Typically, there are multiple small white nodules scattered throughout the liver showing, microscopically, a mostly periportal anarchic collection of biliary structures, some of them with dilation and bile pigment in lumens embedded in an abundant fibrous stroma. There is an ischemic origin raised from some authors, although some of them are, apparently, of malformative origin and associated with ADPKD.

4.11.1.8 Bile Duct Adenoma (BDA)
Subcapsular, firm, white, small (<1 cm), mostly single (4/5 of the cases) nodule is showing, microscopically, a collection of multiple small ducts with small lumens lined by low cuboidal biliary epithelium without bile pigment and embedded in a generally scant connective tissue stroma (Fig. 4.31).

4.11.1.9 Biliary Cystadenoma (BCA)
BCA is a multilocular cystic lesion, which is, typically, seen in the right lobe with tremendous size (25–30 cm) and female preference. Microscopically, multilocular cysts filled with mucinous (mucin/AB+) or clear fluid lined by a simple cuboidal to columnar epithelium are embedded in a densely cellular ovarian stroma-like connective tissue like mucinous cystic neoplasms of the pancreas. IHC shows (+) CK7, CK19, VIM, SMA, INA, and ER/PR. Cellular pleomorphism, atypia, stromal infiltration, as well as malignant transformation have been described.

4.11.1.10 Biliary Papillomatosis
This unique biliary tumor is the biliary counterpart of IPMN of the pancreas (*vide infra*) and is constituted by dilated intra- and extrahepatic biliary epithelium-lined ductal structures with exophytic papillary proliferations of duct lining cells on fibrovascular cores. The outcome has been considered uncertain because the epithelium is cytologically benign without atypia or pleomorphism, but the lesion is frequently recurrent showing progression, which can be life-threatening.

Finally, the remaining benign tumors include non-hepatocytic, non-biliary cell-differentiated tumors, which are a *cavernous hemangioma*, *infantile hemangioendothelioma*, *peliosis hepatis*, and *hepatic mesenchymal hamartoma*.

4.11.1.11 Cavernous Hemangioma

It is probably the most common benign tumor, and almost always or exclusively of cavernous type. There is a female preference and may enlarge during pregnancy. It is a hemangioma with thick septs separating the blood-filled lacunae (Fig. 4.32).

4.11.1.12 Infantile Hemangioendothelioma

IHE may be solitary or multiple. There are two histologic types, including IHE type 1, characterized by small, well-formed blood vessels with plump endothelial lining cells (resembling a capillary hemangioma with regard to the lining), and IHE type 2, characterized by marked nuclear pleomorphism, increased MI, some papillary architecture with endothelial cells projecting into lumens of cystic spaces. IHE is associated with an increased rate of morbidity and mortality, which are linked to hepatic failure or congestive heart failure.

4.11.1.13 Peliosis Hepatis

Peliosis hepatis (PH) (Greek πελιός meaning 'livid', 'discoloured by extravasated blood') is a rare vascular condition of the liver characterized by the presence of multiple blood-filled cystic cavities, which are distributed randomly throughout the parenchyma. PH may occur in a variety of settings with similar features of "peliosis" in the spleen and bone marrow. Patients are usually asymptomatic, and PH is often found either incidentally or at autopsy. However, bleeding, hepatomegaly, and liver impairment due to mass effect have been described. Various etiologies and associations are known. These include toxins (As, polyvinyl chloride, thorium oxide), drugs (anabolics, corticosteroids, oral contraceptive pill (OCP) azathioprine, diethylstilbestrol (DES), Ig therapy, methotrexate, tamoxifen, 6-thioguanine, 6-mercaptopurine), chronic illness (HCC, Tb, leprosy, celiac disease, DM, necrotizing vasculitis, hematologic malignancies, such as Hodgkin lymphoma and multiple myeloma), infection in AIDS (*Bartonella henselae*, *B. quintana*, and *Rochalimaea henselae*), and renal or heart TX. PH is idiopathic in up to 50% of cases. PH is probably due to the breakdown of the sinusoidal edging, hepatic outflow obstruction, and subsequent dilation of the central vein of the hepatic lobule. Microscopically, there are multiple blood-filled cystic spaces within the liver with associated sinusoidal dilatation. Two forms are disntinguished, the minor and major forms. In the former, the size of the blood filled cavities are small, and the lesions involve only a part of the hepatic lobules. In the latter, the blood-filled cavities are large and exhibit a tendency to confluence. In the major form, the lesions involve a large portion of the hepatic lobules and the hepatocytic plates lining the small cavities are normal. However, those lining the large cavities are often thin and interrupted. PGN is associated to the possibility to drain this tumor avoiding hemorrhage and hypovolemic shock correctly. In the setting of HIV/AIDS, antibiotic therapy is advised. DDX includes cavernous hemangioma (globular contrast enhancement is usually centripetal rather than centrifugal), HCC (other histologic patterns are often present), hepatic abscess (inflammation with neutrophils), FNH (central scar, homogeneous arterial enhancement and star-shaped septs), LCA (various histology, but may contain fat), hypervascular metastases (patient's history), and hepatic sinusoidal dilation (isolated finding, no mass effect).

4.11.1.14 Hepatic Mesenchymal Hamartoma

HMH is an infantile, solitary, spherical, reddish tumor with variable size (Ø: 5–20 cm) with or without "peduncle" (i.e., a stalk-like part by which the neoformation is attached to hepatic tissue) and some cystic accumulation of fluid in the stroma. Microscopically, there is a combined dual structure of well-vascularized loose connective tissue quite haphazardly intermixed with irregular, elongated, branching biliary structures potentially reminiscent of the embryological ductal plate type

II more than type I, blood vessels, islands of hepatocytes, and foci of hematopoiesis (Fig. 4.33). Since the neoformation is hamartomatous in origin, HMH is considered potentially benign, with very rare chances of recurrence. The hamartomatous and heterotopic liver tissue should be classified either an accessory liver, when it is connected with a thin stalk ("peduncle") to the main liver or as a true ectopic liver, when no such relationship can be identified. Occasionally, ectopic livers may occur on the serosal surface of the gallbladder.

4.11.1.15 Teratoma

True liver teratomas are, probably, exceptional or unique tumors, with only 25 cases reported in the scientific literature in 2008. Most examples are either intra- or retroperitoneal and may be an incidental finding at CT or autopsy. Microscopically, there is a three-layered germ cell neoplasm without evidence of any malignant germ cell component.

4.11.1.16 Angiomyolipoma

It is usually solitary with a quite large range in size (Ø: 1–20 cm) and characteristic tan-yellow color but well-circumscribed and constituted by the three components as seen in other locations and hematopoietic islands in 2/3 of cases.

4.11.2 Malignant Tumors (Figs. 4.34, 4.35, 4.36, 4.37, 4.38, 4.39, 4.40, and 4.41)

4.11.2.1 Hepatoblastoma

An embryonal tumor of the liver with age <3 years, congenital in some cases and potential positive family history. Grossly, it is a heterogeneous lobulated tan-red-green intrahepatic mass with viable and necrotic areas that are single and usually unencapsulated.

- *CLM*: Hepatoblastoma can be epithelial (~3/4), mesenchymal, or mixed (~1/4). Epithelial hepatoblastomas can subtype in embryonal, fetal, SCUD, and macrotrabecular (Figs. 4.34, 4.35, and 4.36).
- *IHC*: (+) Hepar1, AFP, EMA, but also (+) HCG, VIM, β-Cat, and (±) NSE, CGA, and S100.

- The epithelial, fetal type shows small monotonous malignant hepatocytes with an irregular two-cell thick cord arrangement with sinusoids, which recalls fetal liver and foci of EMH are common. The epithelial, embryonal type has a mainly solid growth pattern with ribbons, rosettes, and papillary clusters with small cells, which show some clearcut immaturity, high N/C ratios, and high MI, and may exhibit some ductular differentiation. The macrotrabecular type looks like HCC. The mixed type shows primitive mesenchymal components with osteoid or chondroid (± striated muscle and neural tissue) associated with an epithelial component, usually of fetal type (the mesenchymal component may also be CK+, which may harbor metaplastic features and represent not true mesenchyme). The rare anaplastic microscopic phenotype is unique and shows undifferentiated small blue cells growing in sheets and forming loosely cohesive clusters. Manivel et al. (1986) first described the entity of teratoid hepatoblastoma. This category of hepatoblastomas reveals classic fetal and embryonal hepatocytic differentiation mixed with a mesenchymal component including primitive spindle cell areas and osteoid in addition to teratoid features that included neuroepithelial, neuroendocrine, cholangioblastic differentiation and basaloid/squamoid islands and melanin-containing foci. Probably, some mixed hepatoblastoma and teratoma as combined tumors described in the literature may represent teratoid hepatoblastomas. This impressive divergent differentiation of teratoid hepatoblastoma is thought to be related to the pluripotential stem cell precursors, which harbor the ability to differentiate into these different components. By immunohistochemistry, AFP, HepPar1, and GPC3 are positive in the fetal and embryonal epithelial hepatocytic areas, but negative in the basaloid squamous and glandular areas. AE1/3 and Anti-CK19 are strongly positive in the glandular and squamous areas and focally positive in the more differentiated fetal and embryonal hepatoblastoma areas. High-molecular-weight cyto-

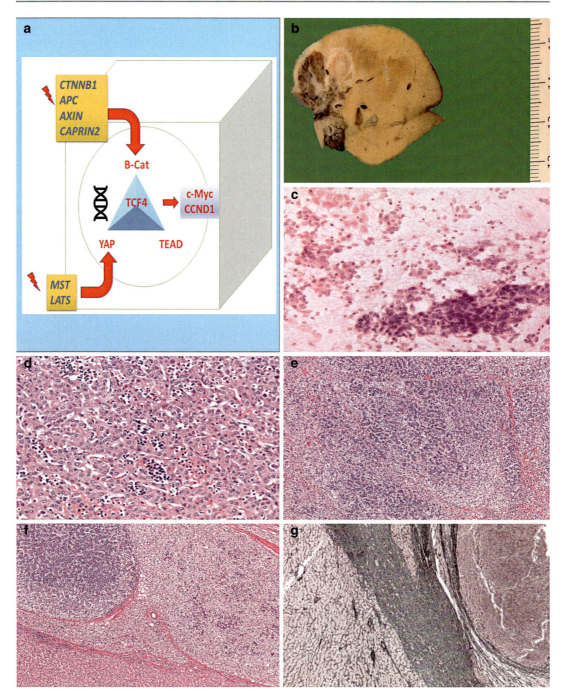

Fig. 4.34 In Figure (**a**) is presented a schema of the pathogenesis of hepatoblastoma. Figure (**b**) discloses a gross image of a liver with a hepatoblastoma, which is relatively homogeneous and brown discolored. Figure (**c**) shows the cytology features of a hepatoblastoma (**c**, hematoxylin and eosin staining, ×200 original magnification). Figures (**d–f**) show the morphologic features of hepatoblastoma, including epithelial elements in varying rates and at variable stages of differentiation (**d**, hematoxylin and eosin staining, ×200 original magnification; **e**, hematoxylin and eosin staining, ×50 original magnification; **f**, hematoxylin and eosin staining, ×50 original magnification). The reticulin staining (**g**) allows to demarcate the tumor from the healthy liver and the thickness of the capsule (Reticulin special stain, ×50 original magnification)

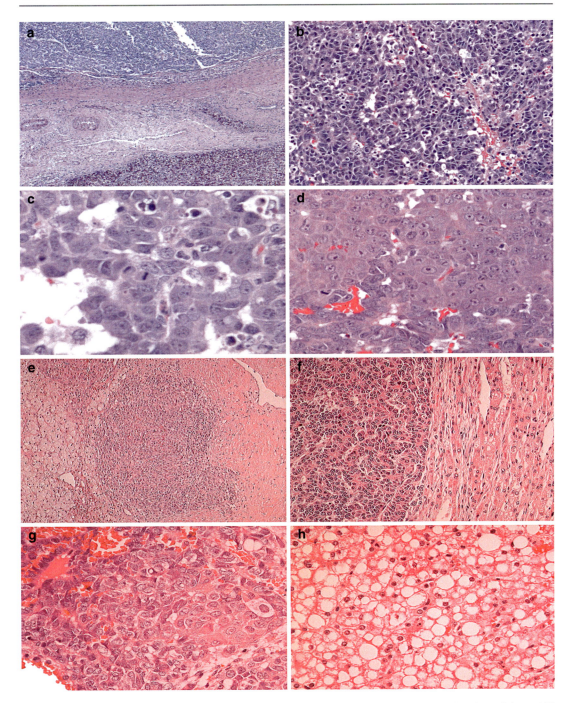

Fig. 4.35 The hepatoblastoma can disclose varying degrees of hepatocytic differentiation (**a**, periodic acid-Schiff staining, ×50 original magnification; **b**, hematoxylin and eosin staining, ×200 original magnification; **c**, hematoxylin and eosin staining, ×630 original magnification; **d**, hematoxylin and eosin staining, ×200 original magnification; **e**, hematoxylin and eosin staining, ×200 original magnification; **f**, hematoxylin and eosin staining, ×200 original magnification; **g**, hematoxylin and eosin staining, ×200 original magnification; **h**, hematoxylin and eosin staining, ×200 original magnification). The embryonal component is particularly evident in Fig. (**c**)

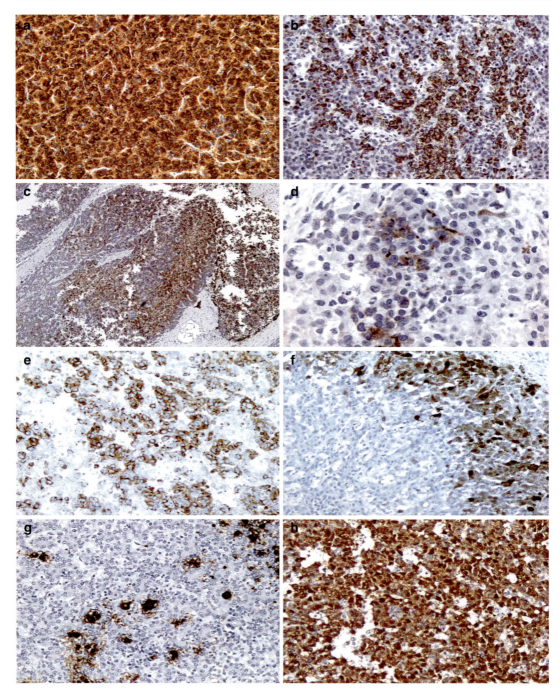

Fig. 4.36 The immunohistochemistry plays a major role in the diagnosis of hepatoblastoma. Figures (**a**–**h**) show the immunohistochemical expression of HepPar1 in the fetal component of hepatoblastoma (**a**, HepPar1 immunostaining, ×200 original magnification), of HepPar1 in the embryonal component of hepatoblastoma (**b**, HepPar1 immunostaining, ×200 original magnification), of cytokeratin 19 (**c**, anti-CK19 immunostaining, ×50 original magnification), of epithelial membrane antigen (EMA) (**d**, EMA immunostaining, ×400 original magnification), of alpha-fetoprotein (AFP) (**e**, AFP immunostaining, ×200 original magnification), of S100 (**f**, S100 immunostaining, ×200 original magnification), and of beta-catenin (**h**, anti-beta-catenin immunostaining, ×200 original magnification). Figure (**g**) shows the immunohistochemical detection of CD61-stained megakaryocytes of the extramedullary hematopoiesis (**g**, anti-CD61 immunostaining, ×200 original magnification). All immunohistochemical procedures have been performed using the avidin-biotin complex

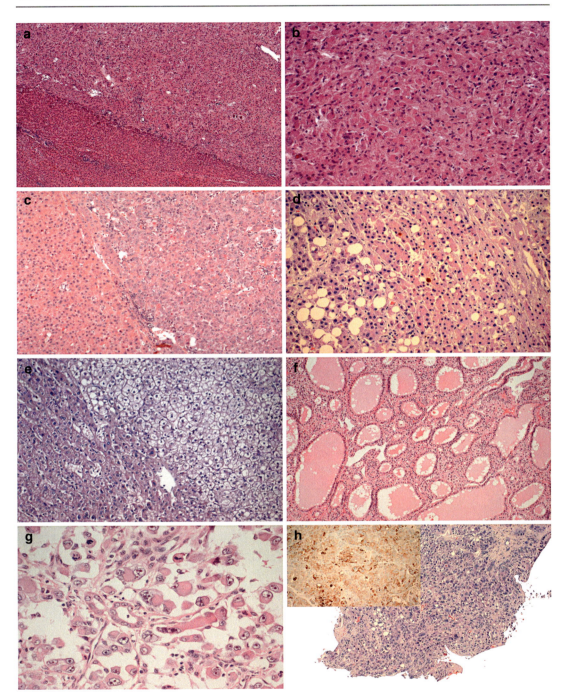

Fig. 4.37 This panel shows morphologic characteristics of hepatocellular carcinoma (HCC), which is rarer than hepatoblastoma in childhood. Figure (**a**) shows the separation of HCC (upper right) from the healthy liver (bottom left) (**a**, hematoxylin and eosin staining, ×40 original magnification). A high power magnification shows the tightly packed atypical cells with a moderate amount of cytoplasm (**b**, hematoxylin and eosin staining, ×200 original magnification). Another HCC is shown in Figure (**c**) (**c**, hematoxylin and eosin staining, ×100 original magnification), while Figure (**d**) shows a HCC with fatty cell change of some neoplastic hepatocytes (**d**, hematoxylin and eosin staining, ×200 original magnification). A clear cell pattern is shown in Figure (**e**) (hematoxylin and eosin staining, ×200 original magnification). Figure (**f**) shows a pseudoglandular differentiation of the HCC (**f**, hematoxylin and eosin staining, ×100 original magnification). Figure (**g**) shows …Figure (**h**) shows …The inset of Figure (**h**) discloses…

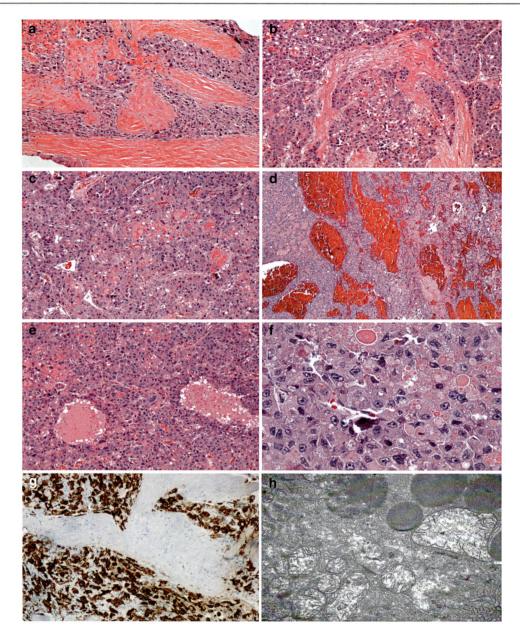

Fig. 4.38 This panel illustrates the fibrolamellar type of hepatocellular carcinoma (HCC-FL). Figures (**a**) through (**f**) show the histologic characteristics of this tumor. This tumor is constituted by trabeculae, sheets, and cords of cells with abundant oncocytic cytoplasm cells embedded in a background of dense collagen bundles arranged in a parallel fashion. The tumor cells have a polygonal shape with well-defined cell borders, copious granular and eosinophilic cytoplasm with pink bodies (**f**), vesicular nuclei with prominent nucleoli (**a–c, e**, hematoxylin and eosin staining, ×100 original magnification; **d**, hematoxylin and eosin staining, ×50 original magnification; **f**, hematoxylin and eosin staining, ×400 original magnification). Figure (**g**) shows the positivity of the tumor cells for cytokeratin 7 using a monoclonal antibody and avidin-biotin complex (anti-cytokeratin 7 immunostaining, ×100 original magnification). Figure (**h**) shows an electron microphotograph of a tumor cell showing many mitochondria and copious endoplasmic reticulum membranes with concentric whorls. The recently described internalized canaliculi lined by microvilli is not shown (x??? original magnification)

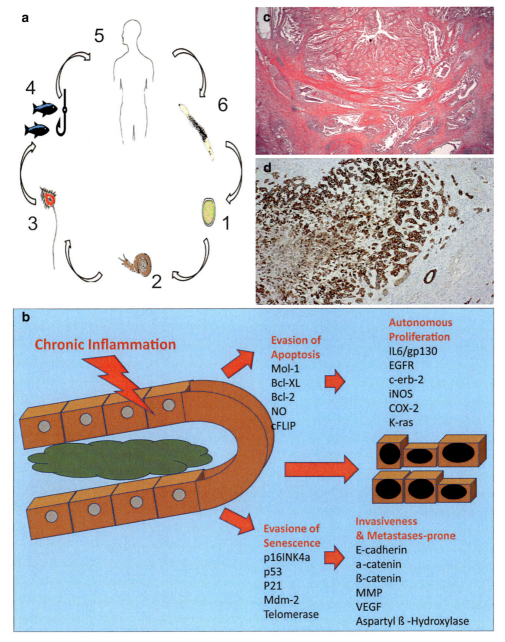

Fig. 4.39 The life cycle of *O. viverrini* is shown in Figure (**a**). Humans are infected with the *O. viverrini* by eating the raw fish which has the encysted resting or maturing stage of a trematode parasite in the tissues of an intermediate host. The metacercariae (intermediate host), then, enter into the duodenum and migrate to the biliary tree. In the biliary tree, the metacercariae mature locally, and their eggs are discharged into the human intestine by the biliary fluid. The eggs exit with the stool. In the freshwater, the eggs get eaten by freshwater snails where the eggs hatched and develop to cercariae. The cercariae invade a fish (second intermediate host) by penetrating its skin. Humans consume raw fish and are the final host where the metacercariae grow and continue the cycle by exiting through human stools. The encysted maturing stage of a trematode in its intermediate host prior to transfer to the definitive host, usually represents the organism's infectious stage. Figure (**b**) shows the intricate relationship of chronic inflammation with oncogenesis in cholangiocarcinoma. Figures (**c**) and (**d**) show a cholangiocellular carcinoma (CCC) in a young male with infiltrating atypical ductular proliferations with a desmoplastic reaction (**c**, hematoxylin and eosin staining, ×12.5 original magnification). Figure (**d**) shows the cytokeratin 19 expression by the tumor cells using a monoclonal antibody against the cytokeratin 19 and the avidin-biotin complex (anti-CK19 immmunostaining, ×100 original magnification)

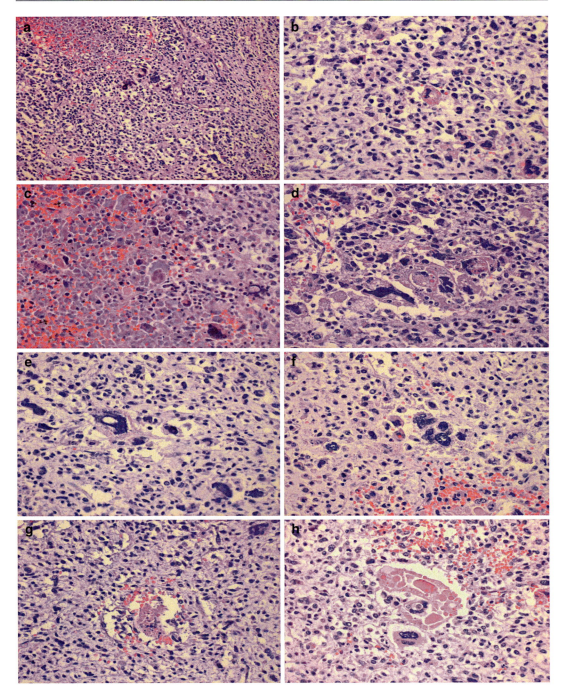

Fig. 4.40 This panel of only histologic microphotographs illustrates the characteristic morphology of the embryonal sarcoma of the liver. There are tumor cells with embryonic mesenchymal differentiation and lack of epithelial characteristics. Most often, the tumor cells exhibit fusiform and star shapes with undetermined cell boundaries. Clear pleomorphism is found (**a**, hematoxylin and eosin staining, ×200 original magnification; **b**, hematoxylin and eosin staining, ×400 original magnification; **c**, hematoxylin and eosin staining, ×400 original magnification; **d**, hematoxylin and eosin staining, ×400 original magnification; **e**, hematoxylin and eosin staining, ×400 original magnification; **f**, hematoxylin and eosin staining, ×400 original magnification; **g**, hematoxylin and eosin staining, ×400 original magnification; **h**, hematoxylin and eosin staining, ×200 original magnification)

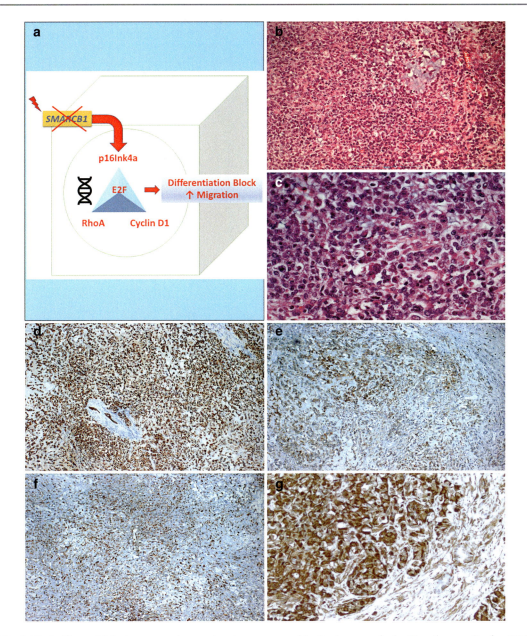

Fig. 4.41 In Figure (**a**) is shown a pathogenetic schema of the rhabdoid tumor of the liver. Figures (**b**) and (**c**) show the characteristic features of a rhabdoid tumor (**b**, hematoxylin and eosin staining, ×200 original magnification; **c**, hematoxylin and eosin staining, ×400 original magnification). In the (malignant) rhabdoid tumor, the tumor cells are eosinophilic with eccentric vesicular nuclei, prominent nucleoli, abundant cytoplasm, and periodic acid-Schiff-positive cytoplasmic inclusions. Immunohistochemistry (**d**–**g**) shows positivity for AE1–AE3 (**d**), CD99 (**e**), and vimentin at low (**f**) and high (**g**) magnification in different areas of the tumor (immunostaining, avidin-biotin complex; **d**–**f**, ×100 original magnification; **g**, ×200 original magnification), respectively. Negative results for INI1/BAF47 are not shown. The hyaline-like globoid bodies show a weakly positive PAS reaction with diastase pretreatment. These inclusions are also stained light blue by Masson trichrome method, but Fontana–Masson for melanin and Alcian blue for mucin are negative. Transmission electron microscopic studies shows bundles of cytoplasmic filaments, often occupying a paranuclear area. These filaments are approximately 10 nm in diameter and are considered to be intermediate filaments of the cytoskeleton. Some filamentous inclusions may contain few entrapped elements of rough endoplasmic reticulum, mitochondria, glycogen, and lipid droplets

keratins (keratins 5 and 6) are positive in the basaloid elements, while CGA is strongly positive in several cords and nests, which may be located adjacent to the classic fetal/embryonal hepatoblastoma, in glandular structures, and in the basaloid squamous areas, indicating neuroendocrine differentiation. CD56 reveals some of the undifferentiated glandular structures, while P63 stains some of the peripheral basaloid cells. β-catenin shows cytoplasmic and focal nuclear staining in the basaloid, squamous, and glandular structures. Typically, neither DES nor MYF4 are observed.

- PGN: Poor factors include age <1 year, Ø (large vs. small), the involvement of vital structures, macrotrabecular type, the anaplastic type, and the presence of aneuploidy. There are also several, relatively new poor prognostic markers in hepatoblastoma. These markers have been the center of a multicenter investigation with results reported by Children's Hepatic tumors International Collaboration (CHIC) (Wu et al. 2017). Currently, all pediatric oncology groups are using the PRE-Treatment EXTent of tumor (PRETEXT) grouping system as part of the risk stratification in patients harboring hepatoblastoma. Useful prognosis markers include diagnostic age (≧8 years in PRETEXT I-III, and ≧ 3 years in PRETEXT IV), initial AFP level (≤1000 ng/mL in PRETEXT I-III, and ≤100 ng/mL in PRETEXT IV). The imaging grouping systems is used to define the extent of liver involvement by the tumor. In PRETEXT (PRE-Treatment EXTent of disease), the extent of liver involvement is defined before therapy, while in POSTTEXT (POST-Treatment EXTent of disease), the extent of liver involvement is defined in response to therapy. PRETEXT is based on the Couinaud 8-segment anatomic architecture of the liver using cross-sectional imaging. In the PRETEXT system the liver is divided into four parts, called sections. Thus, the left lobe of the liver includes a lateral section (Couinaud segments I, II, and III) and a medial section (segment IV), while the right lobe of the liver includes an anterior section (segments V and VIII) and a posterior section (segments VI and VII).

4.11.2.2 Hepatocellular Carcinoma (HCC)

- DEF: Hepatocytic-differentiated malignant epithelial liver tumor
- EPI: Worldwide (Asia>Western), any age (usually >3 years), ♂ > ♀ (especially if +cirrhosis).
- CLI: Abdominal pain, ascites, hepatomegaly, and ↑ AFP in serum.
- RFs: Liver cirrhosis, HBV and HCV infections, xenobiotics, thorotrast (thorium dioxide), steroids (anabolic or progestational steroids), alcohol, radiation, metabolic diseases (PFIC-2, AATD, tyrosinemia, GH), aflatoxins (*Aspergillus flavus*), schistosomiasis, and ataxia-telangiectasia syndrome (Knisely et al. 2006).

A xenobiotic (Greek ξένος = foreigner, stranger, and βίος = life, plus the Greek suffix for adjectives -τικός, -ή, -ό) is a chemical molecule that is found in an organism, although it is not usually produced or expected to be present in it. Examples of xenobiotics may be dioxins and polychlorinated biphenyls that may alter our metabolism and initiate the cascade of cancer, either as initiators or as promoters. Our food may become contaminated by the direct use of artificial hormones on animals or plants or involuntarily by using animals or plants living downstream of sewage treatment plant outfalls or other plants where pesticides or other harmful chemical substances are used. The International Agency for Research on Cancer, an Agency of the World Health Organization, is leading in the classification of carcinogenic substances, such as glyphosate (Guyton et al. 2015; Portier et al. 2016).

- GRO: Typically, single large mass or multiple discrete nodules or, unusually, numerous small nodules diffusely scattered throughout the liver. The tumor can be encapsulated and pedunculated and may show necrosis or hemorrhage.

- *CLM*: Trabecular, solid, acinar (pseudoglandular), peliod, giant cell, sarcomatoid, and clear cell are the histologic growth patterns. A network of sinusoidal blood vessels surrounds the neoplastic liver cells, and there is usually scant stroma. Hepatocytes may show a well-differentiated to very bizarre phenotype. Angioinvasion is a standard feature, and portal vein thrombosis is often found. Other standard features include intranuclear pseudoinclusions, Mallory hyaline, bile pigment, and clear cell change (Figs. 4.37 and 4.38).
- *IHC*: (+) AAT, transferrin, CAM 5.2, but (−) for CEA, AE1 and EMA and (+) HepPar1, GPC-3, and AFP (all three highly relevant!). CAM 5.2 is a monoclonal antibody often cited to be reacting with keratins 8, 18, and 19, although it seems to be specific for keratin 8 only (Makin et al. 1984; Smedts et al. 1990).
- *PGN* – Poor with 5-YSR: 10%, but a better outcome may be observed, if low stage, encapsulation, single lesion, no cirrhosis, and OCP-Hx are (+). No effect on outcome is seen about tumor size, age, sex, and HBV status.

The sclerosing variant of HCC that needs to be differentiated from HCC, fibrolamellar type, represents 3% of all primary hepatic tumors with ♂ = ♀, usually in middle-aged individuals, and 2/3 of patients show hypercalcemia and hypophosphatemia without bony metastatic disease and half of the patients have liver cirrhosis.

- *CLM*: Histologically, there are cords of tumor cells in a dense fibrotic stroma, and in most cases, they show characteristic hepatocellular histology.
- *PGN*: Poor with the survival of 6 months only. Staging criteria for HCC include the direct invasion of adjacent organs other than gallbladder, the involvement of the principal division of the portal vein or hepatic veins, the presence or not of the involvement of visceral epithelium, the size (Ø) less or more than 5 cm, the number (solitary or multiple), and the vascular invasion at microscopic level.

Additional pathologic findings that may be relevant for the oncologist are fibrosis score (none to moderate fibrosis or Ishak score 0–4 versus severe fibrosis/cirrhosis or Ishak score 5–6), the presence or not of hepatocellular dysplasia (low-grade versus high-grade dysplastic nodule), steatosis, iron overload, and chronic hepatitis. In North America, about 2/3 of the primary malignant liver tumors are hepatocellular carcinomas. Histologically, multiple architectural growth patterns can be encountered. Numerous growth patterns may be found in the same tumor. It seems that only the fibrolamellar variant has prognostic significance. It has, however, argued that the longer survival might relate to the age of the affected patients (young) and the absence of cirrhosis. The correct labeling as fibrolamellar carcinoma (FLC) of the neoplastic formation found in the liver requires the lack of any other usual growth patterns of hepatocellular carcinoma. The occurrence of only one additional pattern makes the diagnosis of FLC incompatible. Histologic grading of a hepatocellular carcinoma follows the original grading system of Edmondston and Steiner (1954). However, it has been observed that this grading is easily underestimated in liver core biopsies, and interrater disagreement between pathologists can be astonishingly significant unless consensus meetings are regularly held (Pirisi et al. 2010). Briefly, grade I includes minor differentiation between tumor cells and hyperplastic liver cells and the diagnosis of malignancy is made by more aggressive tumor growth patterns elsewhere in the neoformation. Grade II includes tumor cells with close likeness to normal liver cells but nuclei are larger than those observed in hepatocytes and exhibit more hyperchromasia. Sharp, clear-cut borders and abundant and acidophilic cytoplasm are characteristics of the tumor cells and acini can be found and have variable size. Moreover, protein precipitate or bile are commonly seen filling the lumina of neoplastic acini. Grade III is characterized by larger tumor cells harboring more hyperchromatic nuclei with a higher N/C ratio. The cytoplasm of the tumor cells is granular and acidophilic. Acini are less frequently encountered than Grade II. Similarly, protein or bile precipitates are less commonly seen. A single cell growth in vascular channels is more often seen than Grade II. Finally, grade 4 is characterized by a very large cell volume and nuclei are intensely hyperchromatic. Cytoplasmic granularity is mild or minimal and acini are rarely seen. Medullary growth pattern and tumor cell scattering in vascular channels without cohesion are also seen. Intriguingly, some spindle cell areas or short, plump cell forms similar to lung carcinoma, small-cell variant, may be encountered.

Attention in childhood and youth should be reserved for the *HCC, fibrolamellar variant*. The fibrolamellar variant of HCC is peculiarly distinct from other malignant tumors. It is paramount to separate the FL-HCC or FLC from HCC by its unique clinicopathologic features. Edmondson initially described this tumor in 1956 as *eosinophilic hepatocellular carcinoma with lamellar fibrosis*. Its distinction from HCC took a long time. Although the World Health Organization (WHO) Classification of Tumors recognized FLC as having a unique histological pattern, only in 2010, FLC acquired its own WHO classification number (Bosman 2010). Clinically, FLC occurs in *adolescents and young adults* (no sex preference) with *unknown etiology* and is unrelated to liver cirrhosis, HBV infection, or metabolic abnormalities but symptomatology similar to classic HCC and tendency to metastasize to the abdominal lymph nodes, peritoneum, and lung. X-ray shows a central scar raising the differential diagnosis with FNH, although FLC shows some calcification, which is uncommon in FNH. Interestingly, an increased serum AFP is present in only 1/10 of the patients, as compared with more than half of the patients with HCC showing this as an important distinguishing tumor marker clue. Epidemiologically, FLC is quite rare in East Asia, which shows a high incidence of conventional HCC.

- *GRO*: Large, hard, scirrhous, well-circumscribed, bulging, brown or white-brown tumor with a central stellate scar and numerous fibrous bands throughout and preference for the *left lobe*.
- *CLM*: FLC architecture exhibits nests, sheets, and cords of large oncocytic polygonal hepatoid cells with low N/C ratio embedded in a background of dense, paucicellular collagen bundles arranged in a parallel fashion. At high power, the cells show well-defined cell borders, copious granular and eosinophilic cytoplasm with pale bodies or PAS+ hyaline globules, vesicular nuclei, and prominent nucleoli. There may be necrosis, angioinvasion, and often calcification with FBGC reaction. The surrounding liver is unremarkable. Rare patterns include trabecular, adenoid, and pelioid patterns.
- *HSS/IHC*: (+) HepPar, CK7, fibrinogen (pale bodies), Cu, Cu-binding protein, bile, AAT, pCEA and CAM5/2; (−) mucin, AFP.
- *CMB*: Diploid (2n) tumor with fewer chromosomal aberrations as compared with conventional HCC (*vide supra*).
- *TEM*: Mitochondria-rich (→ oncocytic) and RER-rich cell with fibrinogen (→ pale bodies), intracytoplasmic luminal/bile canaliculi (→ bile) without neurosecretory granules.
- *FNA*: Air-dried, Diff-Quick-stained slides, and alcohol-fixed smears stained with Papanicolau show mild-to-moderate cellular smears (dishesive cells) with occasional aggregates of fibrous tissue (collagen strands) and individual large cells with low N/C ratios, abundant granular cytoplasm, sharply outlined pale cytoplasmic vacuoles, as well as intracytoplasmic bile pigment. The nuclei are large and have prominent nucleoli, and naked nuclei with prominent nucleoli are often seen. Features associated with conventional HCC, such as vessels traversing groups of hepatocytes and "basket" pattern of endothelial cells surrounding clusters of hepatocytes, are useful for the differential diagnosis with the most frequent type of HCC. In case of a mediastinal mass showing the above-described features, some more common lesions need to be included in the differential diagnosis, such as thymic carcinoma, paraganglioma, ganglioneuroma, germ cell tumor, and metastatic melanoma.
- *DDX*: Adenosquamous carcinoma with sclerosis, CCA, FNH, HCC, sclerosing variant, metastatic carcinoma with sclerotic stroma, paraganglioma, and NE tumors.

- *TRT*: Management of HCC remains challenging since complete surgical resection is crucial for a cure. Nevertheless, in pediatric HCC, <1/5 of the patients are considered eligible for initial resection. Various studies have been conducted using different combinations of chemotherapeutic regimens to reduce tumor load. In the past, pediatric HCC patients have been treated with the same protocols as pediatric patients with hepatoblastoma. Thus, chiefly, cisplatin, doxorubicin, carboplatin, 5-fluorouracil, and vincristine have been used. Currently, PLADO (cisplatin and doxorubicin) in combination with sorafenib is endorsed with the aim of achieving operability status. Alternatively, gemcitabine and oxaliplatin in combination with sorafenib have been used (Schmid and von Schweinitz 2017).
- *PGN*: FLC has a better outcome than usual HCC, but substantially it may depend on the suitability to the surgical resection.

4.11.2.3 Cholangiocellular Carcinoma (CCA)

- *DEF*: Malignant epithelial neoplasm of the biliary system of the liver, which can be localized in the extra- and/or intrahepatic biliary network.
- *AKA*: Cholangiocellular carcinoma or cholangiocarcinoma (CCA) or bile duct/biliary carcinoma, although the term cholangiocarcinoma is only used for the intrahepatic form.
- *EPI*: CCA has been described as a "silent killer." The name is due to its relatively silent clinical progression and the consequent difficulty in diagnosis before an advanced stage. Lifestyle and environmental pollutants have been associated with the rise of neoplastic diseases. In some geographic areas of East Asia, CCA seems to be mainly liver fluke-related. Its rate appears to be increasing, and additional factors (endemic and/or epidemic) may be the cause of this increase. Genetics may also play a critical role in the etiopathogenesis of CCA. Klatskin tumor is defined as a CCA arising at the confluence of left and right hepatic ducts. The chemical compound, 1,2-dichloropropane (1,2-DCP) was classified as carcinogenic to humans (Group 1), on the basis of sufficient evidence in humans. There is relevant literature highlighting the connection between exposure to 1,2-DCP and cholangiocarcinoma. It seems that 1,2-DCP causes cholangiocarcinoma and the most important human evidence regarding such carcinogenicity arises from studies of workers in a small off set printing plant in Osaka, Japan, where a very high rate of cholangiocarcinoma was reported. Later, additional cases were identified from several other printing plants in Japan (Benbrahim-Tallaa et al. 2014). An intriguing relationship with the bile microbiome has been recently highlighted in the scientific literature (Di Carlo et al. 2019)
- *EPG*: In Box 4.43, the known risk factors are summarized (Al-Bahrani et al. 2013; Bahitham et al. 2014).

Box 4.43 CCA Risk Factors

Risk factors	Category	Geographic area
Clonorchis sinensis	Parasitic infection	Asia (Korea, Taiwan, Japan, China)
Opisthorchis viverrini	Parasitic infection	Thailand
Ascaris lumbricoides	Parasitic infection	India, Latin America, China
HBV HCV	Viral infection	Worldwide
PSC	Inflammation (chronic)	Worldwide
Hepatolithiasis	Gallstone formation	Taiwan, Japan
Anastomosis	Surgery	No preference
Thorium dioxide	Radioactive	Germany, USA, Japan
Plutonium	Radioactive	Russia

Notes: An additional risk factor is the ductal place malformation which has been reported occasionally and the exposure to 1,2-DCP (*vide supra*).

The adult trematodes deposit fully developed eggs that are passed in the feces and after ingestion by a suitable snail, which is the first intermediate host; the eggs inside of the intermediate host release *miracidia*, which undergo several developmental stages. The *cercariae* are released from the snail. They penetrate freshwater fish, which is the second intermediate host. The *cercariae* encyst as *metacercariae* in the musculature or under the scales of the fish. The definitive host is a mammalian (cats, dogs, and humans) that becomes infected by ingesting undercooked fish containing *metacercariae*. In the definitive host, the *metacercariae* arrived in the duodenum target the biliary system ascending through the ampulla of Vater. In the intrahepatic biliary system, the *metacercariae* attach and develop into adult trematodes, which lay eggs periodically after 3–4 weeks of attachment.

CCA accounts for 5–30% of all primary malignancies of the liver considering all ages, although it is quite rare in the youth and almost inexistent in childhood. The relationship of CCA in individuals affected with abnormalities of the intra- or extrahepatic biliary system (Caroli disease and Caroli syndrome) is part of an intense investigation, particularly for the role of the ductal plate malformation in youth and older age carried out by several groups worldwide, mainly in Japan and United States of America.

- *GRO*: In the WHO, CCA is classified as extra- or intrahepatic with the former type being suborganized into three types. This subdivision is according to the location of the tumor concerning the hilum of the liver: perihilar tumor (Klatskin tumor), middle, and distal extrahepatic.
- *CLM*: Histologically, most bile duct cancers are adenocarcinomas. Rarely, CCA variants occur. They include adenosquamous and squamous carcinoma, *cholangiolo*-cellular carcinoma, mucinous carcinoma, signet ring-cell carcinoma, sarcomatous carcinoma, lymphoepithelioma-like carcinoma, clear-cell carcinoma, mucoepidermoid carcinoma, bile duct cysto-adenocarcinoma, and combined HCC-CCA (Fig. 4.39). No precise data on CCA prognosis seems to be available due to their rare occurrence. Importantly, from the microscopic point of view, there is marked heterogeneity of neoplastic cells within the same gland (compare pancreatic adenocarcinoma – *vide infra*) and tendency of the neoplastic cells to spread between the hepatic *muralium* and along the intrahepatic biliary system and perineural as well as invading the lymphatic drainage system inciting a quite marked desmoplastic tissue response. The combined HCC-CCA is known but is practically never seen in childhood or youth.

HSS shows (+) Mucin, and IHC shows (+) EMA but (−) HepPar1, GPC-3, and AFP (useful for the DDX with HCC). Moreover, CCA IHC shows (+) CK7, CK19, CAM5.2, AE1, pCEA, and (−) CK20 and (±) CDX2 (useful for the DDX. with metastases from a carcinoma of the colon-rectum). Since CCA may be a PSC-related lesion, c-Erb-B2 is positive. CCA does not show bile production as seen in HCC, and there is a brush-border pattern for pCEA or Villin unlike the canalicular staining observed in HCC.

Clinical symptoms and signs of cholestasis often raise suspicion of a malignancy arising in the extrahepatic biliary system. There may be painless jaundice and pruritus with ↑ levels of serum enzyme. The preoperative diagnosis of these cancers remains a challenge. Endoscopic retrograde cholangiopancreatography (ERCP) with direct visualization is the gold standard in the evaluation of the extrahepatic biliary system and cytology in evaluating biliary ductal epithelium by brushings, and direct examination of bile and therapeutic biliary stenting to relieve symptoms of obstruction is essential. ERCP is decisive in the assessment of various inflammatory, infectious, and neoplastic conditions and complements the available imaging techniques.

In cytology, benign groups show epithelial cells arranged in flat sheets with the absence of nuclear overlapping. There are low N/C ratios, round to oval nuclei with smooth nuclear membranes, and evenly dispersed fine chromatin. Reactive cells show some variability in nuclear size and inconspicuous nucleoli. Atypical smears display cells in clusters and groups with increased N/C ratio, nucleomegaly with the irregular nuclear membrane, coarse chromatin pattern, and prominence of nucleoli with lack of bile pigment. Malignant features contain 3D clusters and papillary-like cellular fragments with overlapping of nuclei other than atypical features. Nuclei can be vesicular or hyperchromatic. Necrosis is common in half of the cases with inflammatory cells and blood cellular components as well as lack of bile pigment.

Box 4.44 indicates some tips on the diagnosis of HCC.

4.11.2.4 Hepatic Angiosarcoma (HAS)
- *DEF*: It is a malignant mesenchymal tumor of the liver showing anastomosing vascular channels with varying degrees of differentiation, probably equivalent to type 2 of Dehner-Ishak classification of hemangioendothelioma. HAS harbors, in about 1/3 of cases, an association to some environmental factors (anabolic steroids, arsenic, inorganic copper, thorium dioxide, vinyl chloride) and familial syndromes (BRCA1 gene mutation, BRCA2 gene mutation, ICF syndrome (immunodeficiency centromeric region instability and facial anomalies), Klippel-Trenaunay syndrome, Maffucci syndrome, and Neurofibromatosis type 1).
- *SYN*: Malignant hemangioendothelioma.
- *EPI*: Worldwide, ♂ > ♀, 6th–7th decades (adults) and 2–7 years (children)
- *EPG*: Vinyl chloride is a colorless organ-chloride gas with the formula $H_2C=CHCl$ with a sweet odor (aka vinyl chloride monomer, VCM or chloroethene), which is used to make the polymer polyvinyl chloride and vinyl products (e.g., plastic industry). Vinyl chloride is part of numerous products used in building and construction, industrial and household equipment, automotive industry, electrical wire insulation and cables, piping, and medical supplies.
- *CLI*: In children, there are three possible presentations:
  1. Therapy refractory "infantile multinodular hemangioma"
  2. Relapse of an infantile multinodular hemangioma, the following regression in the

Box 4.44 Pearls and Pitfalls!
- Clusters of small acini (aka periluminal sacculi of Beale) may be quite often encountered in the extrahepatic biliary system and could represent a pitfall for CCA. They should not be considered invasive CCA!
- Perineural invasion in CCA is particularly useful in the intraoperative frozen section!
- Precursor lesions for CCA are two, including biliary intraepithelial neoplasia, which is analog to PanIN (see below), and papillary intraepithelial neoplasm of the biliary tract, which is analog to IPMN of the pancreas (*vide infra*).
- CCA shows no bile production, while bile production is seen in HCC!
- The pseudoglandular or acinar variant of HCC contains colloid-like material or bile but no mucin, unlike CCC!
- CCA glandular differentiation: (+) mCEA, mucin, and DPAS, different from HCC that is (−) mCEA, mucin, and DPAS (apart the combined HCC-CCA or mixed type)!
- (+) EMA, (−) HepPar1, GPC-3, AFP (CCA) ≠ (−) EMA, (+) HepPar1, GPC-3, AFP (HCC)!
- (+) CK7, CK19 (CCA) ≠ (±) CK7, (−) CK19 (HCC)
- (±) CDX2 (CCA) ≠ (+) (colon-rectum carcinoma)

early infancy either spontaneously or after conventional therapy
3. Single/multiple liver mass in a patient older than 1 year of age
- In youth and adults, but also in younger patients, the most classical presentation is, thus, an abdominal mass. A number of signs and symptoms can occur and include jaundice, abdominal pain, vomiting, fever tachy- and dyspnea, and anemia as well as high-output cardiac failure, hemorrhagic ascites, DIC, intraperitoneal bleeding, and Kasabach-Merritt syndrome (hemangioma-thrombocytopenia syndrome or vascular tumor-thrombocytopenia syndrome).
- *CLM*: Mixed histology with some growth patterns such as sinusoidal, solid, papillary, and cavernous. In most cases, sinusoidal and solid are the encountered patterns. In the sinusoidal model, there are flat or hobnailed atypical cells sitting on both sides of the cell plate in a "scaffold-like" distribution, while spindle cells constitute the solid pattern. Atrophy and disruption of the cell plate are found among intermixed hepatocytes, although pseudoacinar formation can occur and should be considered a pitfall (see DDX). In the papillary pattern, the endothelial cell lining becomes multilayered and form papillary-like projections, while in the cavernous type areas of hemorrhage and necrosis are seen.
- *HSS/IHC*: (+) CD31, CD34, FVIIIa, *Ulex europaeus* agglutinin I, and VEGF (at least one of the first three should be positive).
- *TEM*: Weibel-Palade bodies.
- *DDX*: The rare occurrence of this neoplasm in children and adolescents makes, however, the diagnosis of angiosarcoma in a patient younger than 21 years a quite difficult topic at pediatric tumor boards. In Box 4.45 are summarized the diagnoses that need to be taken into consideration. Other less common differential diagnoses may include hepatic mesenchymal hamartoma and undifferentiated

> **Box 4.45 HAS Differential Diagnosis**
> 1. Epithelioid hemangioendothelioma (EHE): Epithelioid cytology (potential pitfall, epithelioid variant of angiosarcoma, although in this latter more spindled areas are seen), less destructive pattern (less necrosis, low MI, less atypia)
> 2. Kaposi sarcoma (KS): HIV(+), HHV-8(+)
> 3. Infantile multinodular hemangioma (IMNH): No aggressive feature imaging, no tumor necrosis, no mitoses or very low MI, no atypia
> 4. Hepatocellular carcinoma (HCC): Hx(+) risk factors (*vide supra*), hepatocytic morphology of tumor cells (grade IV may be challenging), (+) Hepar1, GPC-3, pCEA/CD10, AFP, AE1–AE3, CD34, and (−) Ret
> 5. Cholangiocarcinoma (CCA): Hx(+) PSC, choledochal cysts, lithiasis, worm infestations

embryonal sarcoma, which may be seen at pediatric age. In differentiating vascular tumors of the breast, c-MYC amplification has been used, but no current data are present for vascular tumors of the liver.
- *TNT*: Treatment may be quite challenging, and the prognosis is dismal in most of the cases. Liver TX is contraindicated in both children and adults, because of the high incidence of metastases, but this topic has no full consensus. Non-TX treatment strategies include vascular ablation, chemotherapy, and surgical exploration to perform right or left hepatectomy according to the location of the tumor.
- *PGN*: Dismal outcome and liver TX is controversially discussed, even no metastases are found. Most patients die within 6–12 months of diagnosis and two significant COD are liver

failure and peritoneal hemorrhage. In the literature, the mean survival noted is 10 months to 2 years.

4.11.2.5 Undifferentiated Embryonal Liver Sarcoma (UELS)

- *DEF*: It is a malignant mesenchymal tumor of the liver showing no lines of differentiation (apart from some focally postulated angio-differentiated areas) that may occur with hepatic mesenchymal hamartoma or vaginal embryonal rhabdomyosarcoma.
- *SYN*: Malignant mesenchymoma, mesenchymal sarcoma.
- *EPI*: Worldwide, children>adults, ♂ = ♀.
- *CLI*: Abdominal mass, abdominal pain, fever, and normal serum AFP.
- *GRO*: Solitary, well-delimited tumor with a soft consistency showing at places an appearance of cystic, gelatinous, hemorrhagic, and partly necrotic aspects.
- *CLM*: Anaplastic spindle/oval cells with intracytoplasmic hyaline globules embedded in a myxoid stroma with many thin-walled blood vessels and accompanied by foci of extramedullary hematopoiesis and margination of trapped hepatocytes and biliary structures at the periphery (Fig. 4.40).
- *HSS/IHC*: DPAS+ hyaline globules, (+) VIM, high MI as well as (+) GPC-3, CD56, paranuclear dot-like CKs, Bcl-2, AAT, AACT, and some myo-markers, but (−) AFP and myogenin.
- *TEM*: Hyaline globules → lysosomes?/apoptotic bodies?
- *DDX*: It includes *echinococcosis* (endemic areas, different imaging, but sometimes confusing), *ERMS* of the biliary system (myxoid mass, rhabdomyoblastic differentiation, cambium layer, no hyaline globules, no diffuse anaplasia, (+) DES, myogenin, and MYF4(+). *GIST*: Adults, (+) CD34, CD117, DOG1; *HCC, sarcomatoid variant* (rare in children, but not in youth); *HCC, sclerosing variant* (rare in children, intracellular bile, Mallory bodies, (+) HepPar1); *Hepatoblastoma*, mixed form (β-Cat+); and *hepatic mesenchymal hamartoma*, bland cystic tumor with DPM-like structures showing irregular branching biliary structures and no anaplasia.
- *TNT*: *In sano* surgical resection (negative margins) + CHT. Alternative TNT → LTX.
- *PGN*: Good outcome, although size may be a life-threatening factor.

4.11.2.6 Hepatic Leiomyosarcoma (HLMS) and Hepatic Rhabdomyosarcoma (HRMS)

Hepatic Leiomyosarcoma (HLMS) and Hepatic Rhabdomyosarcoma (HRMS) are rare and need to be kept in mind, although hepatic rhabdomyosarcoma needs to be differentiated from a rhabdoid tumor of the liver and from the undifferentiated pleomorphic sarcoma, which may require different chemotherapy protocols.

4.11.2.7 Rhabdoid Tumor of the Liver

- *DEF*: Very aggressive mesenchymal tumor with merging features of epithelioid sarcoma, intra-abdominal desmoplastic round cell tumor, rhabdomyosarcoma, melanoma, and carcinoma with new metastatic potential and inadequate response to therapy.
- *EPG*: *SMARCB1* loss.
- *CLI*: History of intermittent fever, abdominal pain, and an enlarged abdomen.
- *IMG*: Heterogeneously solid mass (CT) with prominent lobulation (MRI).
- *GRO*: Solid mass.
- *CLM*: Solid sheets of medium-sized cells with eosinophilic to the deep eosinophilic cytoplasm, nucleus displacement, and prominence of single nucleoli admixed with myxoid, hyalinized, and pseudoalveolar areas. The tumor infiltrates and replaces the hepatic parenchyma, sparing interlobular and segmental bile ducts, and growing into hepatic veins (Fig. 4.41).

- *IHC*: (+) AE1/AE3, VIM, WT1, (±) EMA, but (−) INI1, DES, CDS (Caldesm.), S100, GFAP.
- *TEM*: Prominent intermediate filaments of the cytoskeleton.
- *CMB*: *SMARCB1* gene loss.
- *PGN*: High rate of death within 6 months of the diagnosis.
- *TRT*: Three cycles of chemotherapy at 3-week intervals.

The first member of an ATPase chromatin remodeling complex to be involved in carcinogenesis was, indeed, *SMARCB1*, which is the abbreviation for SWI/SNF-related, matrix-associated, actin-dependent regulator of chromatin, subfamily B, member 1, although it is also known as INI1 and BAF47. *SMARCB1* provides instructions for making a protein that builds a part of several different SWI/SNF protein complexes, which regulate the gene expression by chromatin remodeling. SWI/SNF complexes are crucial in development because they are involved in many processes, including repairing damaged DNA, replicating DNA strands, and controlling the growth, division, and differentiation of cells. Both SMARCB1 protein and other SWI/SNF subunits are considered tumor suppressors. The loss of *SMARCB1* causes cell cycle progression partially through the downregulation of *p16INK4a* and upregulation of *E2Fs* and *Cyclin D1*. Also, *SMARCB1* loss triggers cell cycle checkpoints causing arrest and apoptosis. Overall, there is a histologic pattern characterized by numerous mitotic and apoptotic figures. The oncogenic synergy seems to be a result, at least partially, of the disruption of checkpoints via inactivation of TP53 in vivo with cancer formation. Moreover, there is an involvement of the cytoskeleton because of the role of *SMARCB1* in controlling the actin cytoskeleton network. Transcriptionally, *SMARCB1* loss enhances RhoA signaling conferring boosted migratory potential upon several cell lines. The electron microscopy of a rhabdoid cell with *SMARCB1* loss is particularly in syntonic clang with this finding showing prominent intermediate filaments. The cytoskeleton (microfilaments, microtubules, and intermediate filaments), unique to eukaryotic cells, acts as both muscle and skeleton, for movement and stability. Alteration of the cytoskeleton is considered necessary for the metastatic potential of tumor cells.

4.11.3 Metastatic Tumors

Liver metastases are exceptionally rare in children as compared with the adult. However, exceptions do exist. An example is the presence of neuroblastoma in the liver, which might be considered abnormal localization rather than metastasis. The *neuroblastoma IVS* refers to a special group of neuroblastomas. The infant is younger than 1 year old, and the malignancy is on one side (unilateral) of the body. It might have spread to lymph nodes on the same side of the body (ipsilateral or homolateral) but not to contralateral nodes. The neuroblastoma has spread to the skin, liver, and/or the bone marrow, but ≤10% of marrow cells are malignant, and bone is not involved as investigated by imaging tests such as an aralkylguanidine, such as MIBG or lobenguane. An MIBG scan is a nuclear medicine imaging test, which combines a small amount of radioactive material with *meta*-iodobenzylguanidine (MIBG) to find certain types of tumors (e.g., pheochromocytoma and neuroblastoma) in the body (Figs. 4.42, 4.43, 4.44, and 4.45).

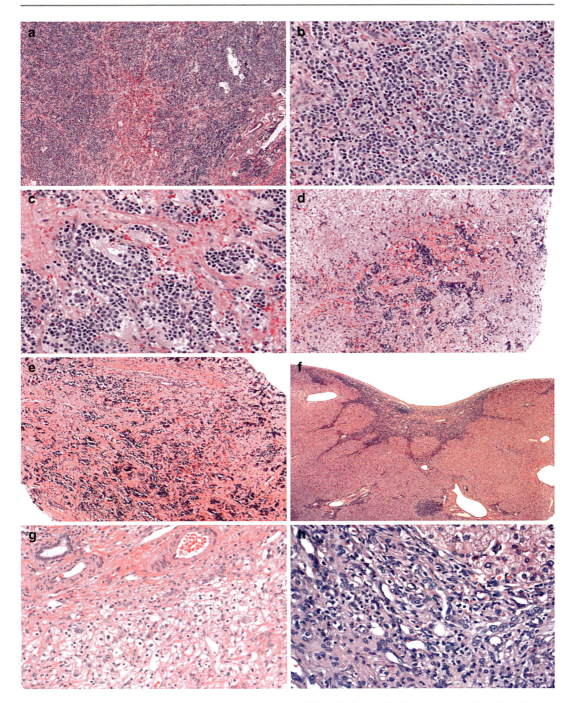

Fig. 4.42 Histological panel with several examples of neuroblastoma characterized by small round blue cells infiltrating the liver (**a**, hematoxylin and eosin staining, ×50 original magnification; **b**, hematoxylin and eosin staining, ×200 original magnification; **c**, hematoxylin and eosin staining, ×200 original magnification; **d** hematoxylin and eosin staining, ×100 original magnification; **e**, hematoxylin and eosin staining, ×100 original magnification; **f**, hematoxylin and eosin staining, ×20 original magnification; **g**, hematoxylin and eosin staining, ×200 original magnification; **h**, hematoxylin and eosin staining, ×200 original magnification). Figures (**g**) and (**h**) show post-chemotherapy regressive changes

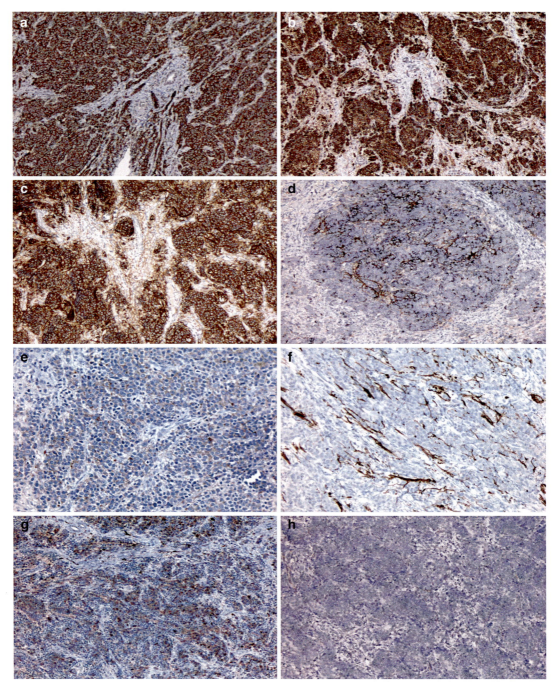

Fig. 4.43 Immunohistochemical panel of neuroblastoma showing a positivity for synaptophysin (**a**, anti-synaptophysin immunostaining, ×100 original magnification), neuron-specific enolase (NSE) (**b**, anti-NSE immunostaining, ×100 original magnification), CD56 (**c**, anti-CD56 immunostaining, ×200 original magnification), CD57 (**d**, anti-CD57 immunostaining, ×100 original magnification), chromogranin A (CGA) (**e**, anti-CGA immunostaining, ×200 original magnification), neurofilaments (**f**, anti-neurofilaments immunostaining, ×100 original magnification), ... ??? (**g**, anti-??? immunostaining, x??? original magnification), and lack of expression for CD99 (**h**, anti-CD99 immunostaining, ×100 original magnification). In CD99, the internal positive control is constituted by the intratumoral blood vessels

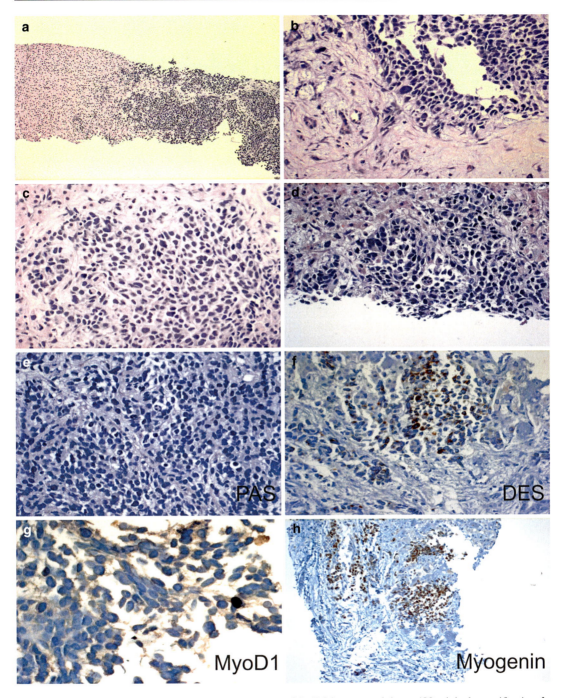

Fig. 4.44 Histologic and immunohistochemical panel of a metastatic rhabdomyosarcoma to the liver with infiltration of small round blue cell tumors at low and high power magnification (**a**, hematoxylin and eosin staining, ×100 original magnification; **b–d**, hematoxylin and eosin staining, ×400 original magnification) and positivity for desmin, MyoD1, and myogenin (**f**, anti-desmin immunostaining, ×400 original magnification; **g**, anti-MyoD1 immunostaining, ×400 original magnification; **h**, anti-myogenin immunostaining, ×200 original magnification). All immunostaining were performed using the avidin-biotin complex. Figure (**e**) shows the absence of glycogen in the tumor cells, a useful characteristic feature to distinguish from other small round blue cell tumors that may have intracytoplasmic glycogen (**e**, periodic acid-Schiff special stain, ×400 original magnification)

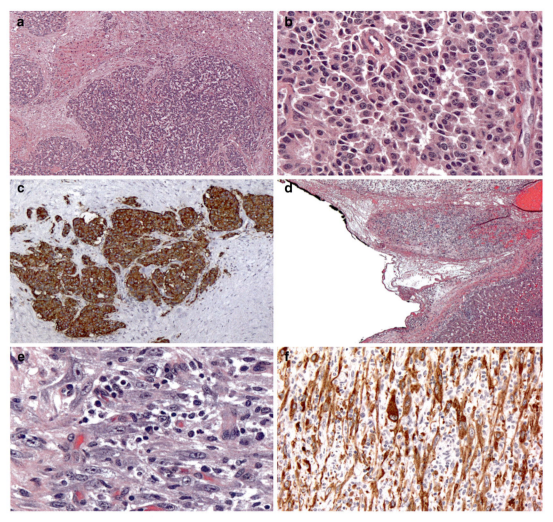

Fig. 4.45 Figures (**a–c**) show a metastatic neuroendocrine tumor of the pancreas (**a**, hematoxylin and eosin staining, ×100 original magnification; **b**, hematoxylin and eosin staining, ×400 original magnification; **c**, anti-synaptophysin immunostaining, ×100 original magnification). Figures (**d–f**) show a metastatic inflammatory myofibroblastic tumor (**d**, hematoxylin and eosin staining, ×50 original magnification; **e**, hematoxylin and eosin staining, ×200 original magnification; **f**, anti-anaplastic lymphoma kinase 1 immunostaining, ×200 original magnification). All immunostaining have been carried out using the avidin-biotin complex method

Multiple Choice Questions and Answers

- HEP-1 Cytoskeleton is made up of microtubules, actin filaments, and intermediate filaments and plays a significant role in the cell. These structures frame the cell shape and help organize the cell compartments as well as provide a basis for cellular movement and cell division. Cytokeratins are an essential component of intermediate filaments of epithelial tissue. Which of the following statement is TRUE?
  (a) Epithelial cells forming bile ducts express CK-8 and CK-18 in addition to CK-7 and CK-20, the latter two being also positive in normal adult hepatocytes.
  (b) Epithelial cells forming bile ducts express CK-7 and CK-20 in addition to CK-8 and CK-18, the latter two being also positive in normal adult hepatocytes.

- (c) Epithelial cells forming bile ducts express CK-7 and CK-19 in addition to CK-8 and CK-18, the latter two being also positive in normal adult hepatocytes.
- (d) Epithelial cells forming bile ducts express CK-7 and CK-20 in addition to CK-5 and CK-6, the latter two being also positive in normal adult hepatocytes.

- HEP-2 Which of the following statements regarding fibrocystin is NOT correct?
  - (a) The *PKHD1* gene provides a protein called fibrocystin (aka polyductin).
  - (b) More than 270 mutations in the PKHD1 gene have been identified in individuals with polycystic kidney disease, which is autosomal dominant inherited.
  - (c) The *PKHD1* gene is located on 6p12.3-p12.2.
  - (d) PKHD1 is present in fetal and adult kidney cells, the liver, and the pancreas.
  - (e) Fibrocystin may act as a receptor, interacting with extracellular molecules.
  - (f) Fibrocystin is also found in primary cilia.

- HEP-3 The paucity of the interlobular bile ducts (PIBD) can be part of a familial syndrome of cholestasis named Alagille syndrome (AGS) or occurs as non-syndromic PIBD. Which of the following conditions is NOT presenting as PIBD, non-syndromic type?
  - (a) Kartagener syndrome
  - (b) Progressive intrahepatic familial cholestasis (PFIC)
  - (c) Maternal use of progesterone during pregnancy
  - (d) Hemophagocytic lymphohistiocytosis
  - (e) Congenital pancreatic hypoplasia

- HEP-4 Which material is NOT required in the pediatric pathologist's liver biopsy cart?
  - (a) Upright light microscope
  - (b) Inverted light microscope
  - (c) Containers of 10% buffered formalin, EM grade 2.5% glutaraldehyde in 0.1M buffer glutaraldehyde, and Hank's Balanced Salt Solution with a sterile urine specimen cup containing two drops of Hank's medium
  - (d) Sterile scalpel blades, aluminum foil, the plastic mold containing OCT (optimal cutting temperature) compound, and Dewar flask of liquid nitrogen
  - (e) Glass slides, Coplin jars of Diff-Quik staining solutions, and large-caliber squeeze-bulb soft plastic pipettes

- HEP-5 What is the most common cause of neonatal cholestasis?
  - (a) Neonatal sclerosing cholangitis
  - (b) The paucity of the interlobular bile ducts
  - (c) Biliary atresia
  - (d) Progressive familial intrahepatic cholestasis
  - (e) Cystic fibrosis-related hepatopathy

- HEP-6 Which of the following etiologies of conjugated hyperbilirubinemia is NOT correct?
  - (a) Cholelithiasis
  - (b) Primary biliary cirrhosis
  - (c) Hepatitis B Virus (HBV) hepatitis
  - (d) Rotor syndrome
  - (e) Crigler-Najjar syndrome type I

- HEP-7 Autoimmune hepatitis (AIH) is a relapsing and remitting immune-mediated liver disease that most commonly affects females in their 20s and 40s, but it may be seen in children and adolescents as well. It should be diagnosed following exclusion of other etiologies (e.g., viral hepatitis, drug reaction). What are the histologic hallmarks of AIH?
  - (a) Hepatitis dominated by plasma cells and interface activity potentially accompanied by emperipolesis and pseudorosettes
  - (b) Hepatitis with pericellular fibrosis and lobular disarray
  - (c) The paucity of interlobular bile ducts with spotty necrosis in the liver lobules
  - (d) Hepatitis with bile duct proliferation with emperipolesis and pseudorosettes

- HEP-8 Computed tomography scan of the abdomen in a young patient with a liver neoplasm shows a central scar inside of the mass. In addition to a fibrolamellar carcinoma of the liver, which condition should be considered in the differential diagnosis?

(a) Mesenchymal hamartoma of the liver
(b) Focal nodular hyperplasia
(c) Nodular regenerative hyperplasia
(d) Scirrhous variant of conventional hepatocellular carcinoma
(e) Metastatic rhabdomyosarcoma

- HEP-9 Aflatoxins are a family of toxins produced by certain fungi, mostly *Aspergillus flavus* and *Aspergillus parasiticus*, that are found on farming crops such as maize (corn), peanuts, cottonseed, and tree nuts. Which neoplasm is associated with exposure to aflatoxins?
  (a) Mesenchymal hamartoma of the liver
  (b) Cholangiocellular carcinoma
  (c) Hepatocellular carcinoma
  (d) Embryonal sarcoma of the liver
  (e) Hepatoblastoma

- HEP-10 Which molecular biological abnormality is associated with the fibrolamellar carcinoma of the liver?
  (a) *MYCN* amplification
  (b) *DNAJB1–PRKACA* fusion transcript
  (c) *BRCA2* amplification
  (d) t(2;5)(p23;q35)
  (e) *NPM-ALK* fusion protein

References and Recommended Readings

Abdualmjid RJ, Sergi C. Hepatotoxic botanicals – an evidence-based systematic review. J Pharm Pharm Sci. 2013;16(3):376–404. Review. PubMed PMID: 24021288.

Abdullah A, Patel Y, Lewis TJ, Elsamaloty H, Strobel S. Extrarenal malignant rhabdoid tumors: radiologic findings with histopathologic correlation. Cancer Imaging. 2010;10:97–101. https://doi.org/10.1102/1470-7330.2010.0010. PubMed PMID: 20299301; PubMed Central PMCID: PMC2842174.

Abuetabh Y, Persad S, Nagamori S, Huggins J, Al-Bahrani R, Sergi C. Expression of E-cadherin and β-catenin in two cholangiocarcinoma cell lines (OZ and HuCCT1) with different degree of invasiveness of the primary tumor. Ann Clin Lab Sci. 2011;41(3):217–23. PubMed PMID: 22075503.

Ackermann O, Fabre M, Franchi S, Pariente D, Debray D, Jacquemin E, Gauthier F, Bernard O. Widening spectrum of liver angiosarcoma in children. Journal of Pediatric Gastroenterology and Nutrition. 2011;53(6):615–9. PMID:21832953

Admirand WH, Small DM. The physicochemical basis of cholesterol gallstone formation in man. J Clin Invest. 1968;47(5):1043–52. PubMed PMID: 5645851; PubMed Central PMCID: PMC297257

Al-Bahrani R, Abuetabh Y, Zeitouni N, Sergi C. Cholangiocarcinoma: risk factors, environmental influences and oncogenesis. Ann Clin Lab Sci. 2013;43(2):195–210. Review. PubMed PMID: 23694797.

Al-Bahrani R, Nagamori S, Leng R, Petryk A, Sergi C. Differential expression of sonic hedgehog protein in human hepatocellular carcinoma and intrahepatic cholangiocarcinoma. Pathol Oncol Res. 2015a;21(4):901–8. https://doi.org/10.1007/s12253-015-9918-7. Epub 2015 Mar 5. PubMed PMID: 25740074.

Al-Bahrani R, Tuertcher D, Zailaie S, Abuetabh Y, Nagamori S, Zetouni N, Bahitham W, Sergi C. Differential SIRT1 expression in hepatocellular carcinomas and cholangiocarcinoma of the liver. Ann Clin Lab Sci. 2015b;45(1):3–9. Erratum in: Ann Clin Lab Sci. 2015 Fall;45(6):720. PubMed PMID: 25696003.

Ariza CR, Frati AC, Sierra I. Hypothyroidism-associated cholestasis. JAMA. 1984;252(17):2392. PubMed PMID: 6481926

Aronson DC, Meyers RL. Malignant tumors of the liver in children. Semin Pediatr Surg. 2016;25:265–75.

Auron A, Brophy PD. Hyperammonemia in review: pathophysiology, diagnosis, and treatment. Pediatr Nephrol. 2012;27(2):207–22. Epub 2011 Mar 23. Review. PubMed PMID: 21431427.

Bachmann C. Inherited hyperammonimieas. In: Blau N, Duran M, Blaskovics ME, Gibson KM, editors. Physician's guide to the laboratory diagnosis of metabolic diseases. Berlin/Heidelberg: Springer; 2003.

Bahitham W, Liao X, Peng F, Bamforth F, Chan A, Mason A, Stone B, Stothard P, Sergi C. Mitochondriome and cholangiocellular carcinoma. PLoS One. 2014;9(8):e104694. https://doi.org/10.1371/journal.pone.0104694. eCollection 2014. PubMed PMID: 25137133; PubMed Central PMCID: PMC4138114.

Bakdounes K, Jhala N, Jhala D. Diagnostic usefulness and challenges in the diagnosis of mesothelioma by endoscopic ultrasound guided fine needle aspiration. Diagn Cytopathol. 2008;36(7):503–7. https://doi.org/10.1002/dc.20811. PubMed PMID: 18528879.

Bale PM, Kann AE, Dorney SFA. Renal proximal tubular dysgenesis associated with severe neonatal hemosiderotic liver disease. Pediatr Pathol. 1994;14:479–89.

Bankowski Z, Bruppacher R, Crusius I, Kremer G, Venulet J, Council for International Organizations of Medical Sciences (CIOMS). Reporting adverse drug reactions: definitions of terms and criteria for their use. Geneva: CIOMS; 1999. ISBN: 9290360712

Beiler HA, Sergi C, Wagner G, Zachariou Z. Accessory liver in an infant with congenital diaphragmatic hernia. J Pediatr Surg. 2001;36(6):E7. PubMed PMID: 11381450.

Benbrahim-Tallaa L, Lauby-Secretan B, Loomis D, Guyton KZ, Grosse Y, El Ghissassi F, Bouvard V,

Guha N, Mattock H, Straif K. International Agency for Research on Cancer Monograph Working Group. Carcinogenicity of perfluorooctanoic acid, tetrafluoroethylene, dichloromethane, 1,2-dichloropropane, and 1,3-propane sultone. Lancet Oncol. 2014;15(9):924–5. PubMed PMID: 25225686

Beuers U, Richter WO, Ritter MM, Wiebecke B, Schwandt P. Klinefelter's syndrome and liver adenoma. J Clin Gastroenterol. 1991;13(2):214–6. PubMed PMID: 1851773

Bioulac-Sage P, Laumonier H, Laurent C, Blanc JF, Balabaud C. Benign and malignant vascular tumors of the liver in adults. Semin Liver Dis. 2008;28(3):302–14. https://doi.org/10.1055/s-0028-1085098. Epub 2008 Sep 23. Review. PubMed PMID: 18814083

Birken C, Hamilton J. Obesity in a young child. CMAJ. 2014;186(6):443–4. https://doi.org/10.1503/cmaj.130238. Epub 2014 Jan 13. Review. PubMed PMID: 24418984; PubMed Central PMCID: PMC3971030

Blisard KS, Bartow SA. Neonatal hemochromatosis. Hum Pathol. 1986;17:376–83.

Bonkovsky HL, Banner BF, Lambrecht RW, Rubin RB. Iron in liver diseases other than hemochromatosis. Semin Liver Dis. 1996;16:65–82.

Bosman FT, World Health Organization. WHO classification of Tumours of the digestive system. 4th ed. Lyon: International Agency for Research on Cancer; 2010.

Bove KE. Chapter 10: Histologic patterns of metabolic disease. In: Saxena R, editor. Practical hepatic pathology: a diagnostic approach, A volume in the Pattern recognition series. 2nd ed. Milton, ON: Elsevier Canada; 2011. eBook ISBN: 9780323442862, 9780323442855; Hardcover ISBN: 9780323428736.

Bove KE, Wong R, Kagen H, Balistreri W, Tabor MW. Exogenous iron overload in perinatal hemochromatosis: a case report. Pediatr Pathol. 1991;11:389–97.

Brisigotti M, Fabbretti G, Pesce F, Gatti R, Cohen A, Parenti G, Callea F. Congenital bilateral juvenile granulosa cell tumor of the ovary in leprechaunism: a case report. Pediatr Pathol. 1993;13:549–58.

Bruguera M, Llach J, Rodés J. Nonsyndromic paucity of intrahepatic bile ducts in infancy and idiopathic ductopenia in adulthood: the same syndrome? Hepatology. 1992;15(5):830–4. Review. PubMed PMID: 1568724

Buccoliero AM, Castiglione F, Maio V, Moncini D. Teratoid hepatoblastoma. Fetal Pediatr Pathol. 2008;27(6):274–81.

Callea F, Brisigotti M, Faa G, Lucini L, Eriksson S. Identification of PiZ gene products in liver tissue by a monoclonal antibody specific for the Z mutant of alpha 1-antitrypsin. J Hepatol. 1991;12(3):372–6. PubMed PMID: 1940268

Callea F, Sergi C, Medicina D, Pizzorni S, Brisigotti M, Fabbretti G, Bonino F. From immunohistochemistry to in situ hybridization. Liver. 1992;12(4 Pt 2):290–5. Review. PubMed PMID: 1447961

Callea F, Sergi C, Fabbretti G, Brisigotti M, Cozzutto C, Medicina D. Precancerous lesions of the biliary tree. J Surg Oncol Suppl. 1993;3:131–3. Review. PubMed PMID: 8389160.

Cave D, Ross DB, Bahitham W, Chan A, Sergi C, Adatia I. Mitochondrial DNA depletion syndrome-an unusual reason for interstage attrition after the modified stage 1 Norwood operation. Congenit Heart Dis. 2013;8(1):E20–3. https://doi.org/10.1111/j.1747-0803.2011.00569.x. Epub 2011 Oct 20. PubMed PMID: 22011012.

Chandramouleeswari K, Anita S, Shivali B. Mesenchymal hamartoma of the liver: a case report. J Clin Diagn Res. 2012;6(9):1552–4. https://doi.org/10.7860/JCDR/2012/4151.2558. PMID: 23285455; PMCID: PMC3527795

Cheng G, Tang CS, Wong EH, Cheng WW, So MT, Miao X, et al. Common genetic variants regulating ADD3 gene expression alter biliary atresia risk. J Hepatol. 2013;59(6):1285–91. https://doi.org/10.1016/j.jhep.2013.07.021.

Cheung PC, Ng WF, Chan AK. Neonatal haemochromatosis associated with down syndrome. J Paediatr Child Health. 1995;31:249–52.

Chiche L, Dao T, Salamé E, Galais MP, Bouvard N, Schmutz G, Rousselot P, Bioulac-Sage P, Ségol P, Gignoux M. Liver adenomatosis: reappraisal, diagnosis, and surgical management: eight new cases and review of the literature. Ann Surg. 2000;231(1):74–81. Review. PubMed PMID: 10636105; PubMed Central PMCID: PMC1420968

Chiu B, Chan J, Das S, Alshamma Z, Sergi C. Pediatric sarcoidosis: a review with emphasis on early onset and high-risk sarcoidosis and diagnostic challenges. Diagnostics (Basel). 2019;9(4):pii: E160. https://doi.org/10.3390/diagnostics9040160. Review. PubMed PMID: 31731423

Clayton PT. Delta 4-3-oxosteroid 5 beta-reductase deficiency and neonatal hemochromatosis [letter]. J Pediatr. 1994;125:845–6.

Clayton PT. Disorders of bile acid synthesis. J Inherit Metab Dis. 2011;34(3):593–604.

Conrad RJ, Gribbin D, Walker NI, Ong TH. Combined cystic teratoma and hepatoblastoma of the liver: probable divergent differentiation of an uncommitted hepatic precursor cell. Cancer. 1993;72(10):2910–3.

Costa AM, Pegado CS, Pôrto LC. Quantification of the intrahepatic biliary tree during human fetal development. Anat Rec. 1998;251(3):297–302. PubMed PMID: 9669756

Craig JM. Sequences in the development of cirrhosis of the liver in cases of erythroblastosis fetalis. Arch Pathol. 1950;49:665–86.

Crowe A, Knight CS, Jhala D, Bynon SJ, Jhala NC. Diagnosis of metastatic fibrolamellar hepatocellular carcinoma by endoscopic ultrasound-guided fine needle aspiration. CytoJournal. 2011;8:2. https://doi.org/10.4103/1742-6413.76495. PubMed PMID: 21369523; PubMed Central PMCID: PMC3045764.

Czauderna P, Haeberle B, Hiyama E, Rangaswami A, Krailo M, Maibach R, et al. The Children's hepatic tumors international collaboration (CHIC): novel global rare tumor database yields new prognostic factors in hepatoblastoma and becomes a research model. Eur J Cancer. 2016;52:92–101.

D'Antiga L. Medical management of esophageal varices and portal hypertension in children. Semin Pediatr Surg. 2012;21(3):211–8. https://doi.org/10.1053/j.sempedsurg.2012.05.004. Review. PubMed PMID: 22800974

Danks DM, Tippett P, Adams C, Campbell P. Cerebro-hepato-renal syndrome of Zellweger. A report of eight cases with comments upon the incidence, the liver lesion, and a fault in pipecolic acid metabolism. J Pediatr. 1975;86:382–7.

Davison S. Assessment of liver disease in cystic fibrosis. Paediatr Respir Rev. 2018; https://doi.org/10.1016/j.prrv.2018.05.010. pii: S1526-0542(18)30071-X. [Epub ahead of print] Review. PubMed PMID: 29933897.

De Boissieu D, Knisely AS. Neonatal hemochromatosis. In: Suchy FJ, Sokol RJ, Balistreri WF, editors. Liver disease in children. 2nd ed. Philadelphia: Lippincott Williams & Wilkins; 2001. p. 641–7.

de Ville de Goyet J, D'Ambrosio G, Grimaldi C. Surgical management of portal hypertension in children. Semin Pediatr Surg. 2012;21(3):219–32. https://doi.org/10.1053/j.sempedsurg.2012.05.005. Review. PubMed PMID: 22800975

Degott C, Rueff B, Kreis H, Duboust A, Potet F, Benhamou JP. Peliosis hepatis in recipients of renal transplants. Gut. 1978;19(8):748–53. PubMed PMID: 355072; PubMed Central PMCID: PMC1412137

Desmet VJ. Cholestasis: extrahepatic obstrction and secondary biliary cirrhosis. Chapter 11, p. 425–6.

Di Carlo P, Serra N, Gulotta G, Giammanco A, Colomba C, Melfa G, Fasciana T, Sergi C. Bactibilia in diseases of the biliary tract and pancreatic gland in patients older than 80 years: a STROBE-retrospective cohort study in a teaching hospital in Italy. Eur J Clin Microbiol Infect Dis. 2018;37(5):953–8. https://doi.org/10.1007/s10096-018-3213-y. Epub 2018 Feb 27. PubMed PMID: 29484561.

Di Carlo P, Serra N, D'Arpa F, Agrusa A, Gulotta G, Fasciana T, Rodolico V, Giammanco A, Sergi C. The microbiota of the bilio-pancreatic system: a cohort, STROBE-compliant study. Infect Drug Resist. 2019;12:1513–27. https://doi.org/10.2147/IDR.S200378. eCollection 2019. PubMed PMID: 31354308; PubMed Central PMCID: PMC6578573

Dinakaran D, Sergi CM. Co-ingestion of aspirin and acetaminophen promoting fulminant liver failure: a critical review of Reye syndrome in the current perspective at the dawn of the 21st century. Clin Exp Pharmacol Physiol. 2018;45(2):117–21. https://doi.org/10.1111/1440-1681.12861. Epub 2017 Dec 4. Review. PubMed PMID: 28945927.

Dinakaran D, Bristow E, Armanious H, Garros D, Yap J, Noga M, Sergi C. Co-ingestion of willow bark tea and acetaminophen associated with fatal infantile fulminant liver failure. Pediatr Int. 2017;59(6):743–5. https://doi.org/10.1111/ped.13262. Epub 2017 Apr 24. PubMed PMID: 28436611.

Dixon PH, Sambrotta M, Chambers J, Taylor-Harris P, Syngelaki A, Nicolaides K, Knisely AS, Thompson RJ, Williamson C. An expanded role for heterozygous mutations of ABCB4, ABCB11, ATP8B1, ABCC2 and TJP2 in intrahepatic cholestasis of pregnancy. Sci Rep. 2017;7(1):11823. https://doi.org/10.1038/s41598-017-11626-x. PubMed PMID: 28924228; PubMed Central PMCID: PMC5603585

Donohue WL, Uchida IA. Leprechaunism: a euphemism for a rare familial disorder. J Pediatr. 1954;45:505–19.

Dorn L, Menezes LF, Mikuz G, Otto HF, Onuchic LF, Sergi C. Immunohistochemical detection of polyductin and co-localization with liver progenitor cell markers during normal and abnormal development of the intrahepatic biliary system and in adult hepatobiliary carcinomas. J Cell Mol Med. 2009;13(7):1279–90. https://doi.org/10.1111/j.1582-4934.2008.00519.x. Epub 2008 Oct 6. PubMed PMID: 19292732.

Driscoll SG, Hayes AM, Levy HL. Neonatal hemochromatosis: evidence for autosomal recessive transmission. Am J Hum Genet. 1988;43:A232.

Edmondson HA. Differential diagnosis of tumors and tumor-like lesions of liver in infancy and childhood. AMA J Dis Child. 1956;91(2):168–86. PubMed PMID: 13282629

Edmondson HA, Steiner PE. Primary carcinoma of the liver: a study of 100 cases among 48,900 necropsies. Cancer. 1954;7(3):462–503.

Edwards JH, Harnden DG, Cameron AH, Crosse VM, Wolff OH. A new trisomic syndrome. Lancet, London. 1960;1:787–90.

Egawa H, Berquist W, Garcia-Kennedy R, Cox K, Knisely AS, Esquivel CO. Rapid development of hepatocellular siderosis after liver transplantation for neonatal hemochromatosis. Transplantation. 1996;62:1511–3.

Elias H. A re-examination of the structure of the mammalian liver 11: the hepatic lobule and its relation to the vascular and biliary systems. Am J Anat. 1949;85:379–456.

Elsayes KM, Narra VR, Yin Y, et al. Focal hepatic lesions: diagnostic value of enhancement pattern approach with contrast-enhanced 3D gradient-echo MR imaging. Radiographics. 25(5):1299–320. https://doi.org/10.1148/rg.255045180.

Erez A, Shchelochkov OA, Plon SE, Scaglia F, Lee B. Insights into the pathogenesis and treatment of cancer from inborn errors of metabolism. Am J Hum Genet. 2011;88(4):402–21. https://doi.org/10.1016/j.ajhg.2011.03.005. Review. PubMed PMID: 21473982; PubMed Central PMCID: PMC3071916

Faa G, Sciot R, Farci AM, Callea F, Ambu R, Congiu T, van Eyken P, Cappai G, Marras A, Costa V, Desmet VJ. Iron concentration and distribution in the newborn liver. Liver. 1994;14:193–9.

Fabbretti G, Sergi C, Consalez G, Faa G, Brisigotti M, Romeo G, Callea F. Genetic variants of alpha-1-antitrypsin (AAT). Liver. 1992;12(4 Pt 2):296–301. Review. Erratum in: Liver. 2013 Jan;33(1):164.

Consales, G [corrected to Consalez, G]. PubMed PMID: 1447962.

Fanburg-Smith JC, Hengge M, Hengge UR, Smith JS Jr, Miettinen M. Extrarenal rhabdoid tumors of soft tissue: a clinicopathologic and immunohistochemical study of 18 cases. Ann Diagn Pathol. 1998;2(6):351–62. PubMed PMID: 9930572.

Fargion S, Bissoli F, Fracanzani AL, Suigo E, Sergi C, Taioli E, Ceriani R, Dimasi V, Piperno A, Sampietro M, Fiorelli G. No association between genetic hemochromatosis and alpha1-antitrypsin deficiency. Hepatology. 1996a;24(5):1161–4. PubMed PMID: 8903392.

Fargion S, Sergi C, Bissoli F, Fracanzani AL, Suigo E, Carazzone A, Roberto C, Cappellini MD, Fiorelli G. Lack of association between porphyria cutanea tarda and alpha 1-antitrypsin deficiency. Eur J Gastroenterol Hepatol. 1996b;8(4):387–91. PubMed PMID: 8781910.

Fellman V, Rapola J, Pihko H, Varilo T, Raivio KO. Iron-overload disease in infants involving fetal growth retardation, lactic acidosis, liver haemosiderosis, and aminoaciduria. Lancet. 1998;351:490–3.

Ferenci P, Zollner G, Trauner M. Hepatic transport systems. J Gastroenterol Hepatol. 2002;17(Suppl):S105–12. Review. PubMed PMID: 12000597.

Flejou JF, Barge J, Menu Y, Degott C, Bismuth H, Potet F, Benhamou JP. Liver adenomatosis. An entity distinct from liver adenoma? Gastroenterology. 1985;89(5):1132–8. PubMed PMID: 2412930

Fregonese L, Stolk J. Hereditary alpha-1-antitrypsin deficiency and its clinical consequences. Orphanet J Rare Dis. 2008;3:16. https://doi.org/10.1186/1750-1172-3-16. PMID: 18565211; PMCID: PMC2441617

Frijters CM, Ottenhoff R, van Wijland MJ, van Nieuwkerk CM, Groen AK, Oude Elferink RP. Regulation of mdr2 P-glycoprotein expression by bile salts. Biochem J. 1997;321(Pt 2):389–95.

Geramizadeh B, Safari A, Bahador A, Nikeghbalian S, Salahi H, Kazemi K, Dehghani SM, Malek-Hosseini SA. Hepatic angiosarcoma of childhood: a case report and review of literature. J Pediatr Surg. 2011;46(1):e9–11. https://doi.org/10.1016/j.jpedsurg.2010.09.005. Review. PubMed PMID: 21238632

Gilchrist KW, Gilbert EF, Goldfarb S, Goll U, Spranger JW, Opitz JM. Studies of malformation syndromes of man XIB: the cerebro-hepato-renal syndrome of Zellweger: comparative pathology. Eur J Pediatr. 1976;121:99–118.

Griffiths W, Cox T. Haemochromatosis: novel gene discovery and the molecular pathophysiology of iron metabolism. Hum Mol Genet. 2000;9:2377–82.

Gupta M, Pai RR, Dileep D, Gopal S, Shenoy S. Role of biliary tract cytology in the evaluation of extrahepatic cholestatic jaundice. J Cytol. 2013;30(3):162–8. https://doi.org/10.4103/0970-9371.117657. PubMed PMID: 24130407; PubMed Central PMCID: PMC3793352.

Guyton KZ, Loomis D, Grosse Y, El Ghissassi F, Benbrahim-Tallaa L, Guha N, Scoccianti C, Mattock H. Straif K; International Agency for Research on Cancer Monograph Working Group, IARC, Lyon, France. Carcinogenicity of tetrachlorvinphos, parathion, malathion, diazinon, and glyphosate. Lancet Oncol. 2015a;16(5):490–1. https://doi.org/10.1016/S1470-2045(15)70134-8. Epub 2015 Mar 20. PubMed PMID: 25801782

Guyton KZ, Loomis D, Grosse Y, El Ghissassi F, Benbrahim-Tallaa L, Guha N, Scoccianti C, Mattock H, Straif K. International Agency for Research on Cancer monograph working group, IARC, Lyon, France. Carcinogenicity of tetrachlorvinphos, parathion, malathion, diazinon, and glyphosate. Lancet Oncol. 2015b;16(5):490–1. https://doi.org/10.1016/S1470-2045(15)70134-8. Epub 2015 Mar 20. PubMed PMID: 25801782

Hall AR, Le H, Arnold C, Brunton J, Bertolo R, Miller GG, Zello GA, Sergi C. Aluminum exposure from parenteral nutrition: early bile canaliculus changes of the hepatocyte. Nutrients. 2018;10(6):pii: E723. https://doi.org/10.3390/nu10060723. PubMed PMID: 29867048

Herrmann T, Muckenthaler M, van der Hoeven F, Brennan K, Gehrke SG, Hubert N, Sergi C, Gröne HJ, Kaiser I, Gosch I, Volkmann M, Riedel HD, Hentze MW, Stewart AF, Stremmel W. Iron overload in adult Hfe-deficient mice independent of changes in the steady-state expression of the duodenal iron transporters DMT1 and Ireg1/ferroportin. J Mol Med (Berl). 2004;82(1):39–48. Epub 2003 Nov 15. PubMed PMID: 14618243.

Herrmann U, Dockter G, Lammert F. Cystic fibrosis-associated liver disease. Best Pract Res Clin Gastroenterol. 2010;24(5):585–92. https://doi.org/10.1016/j.bpg.2010.08.003. Review. PubMed PMID: 20955961.

Hicks J, Mani H, Stocker JT. The liver, gallbladder, and biliary tract. In: Stocker JT, Dehner LP, editors. Pediatric pathology. 3rd ed. Philadelphia: Lippincott Williams & Wilkins; 2011. p. 640–742.

Hsu SM, Raine L, Fanger HJ. Use of avidin-biotin-peroxidase complex (ABC) in immunoperoxidase techniques: a comparison between ABC and unlabeled antibody (PAP) procedures. Histochem Cytochem. 1981;29:577–80.

Hua Z, Sergi C, Nation PN, Wizzard PR, Ball RO, Pencharz PB, Turner JM, Wales PW. Hepatic ultrastructure in a neonatal piglet model of intestinal failure-associated liver disease (IFALD). J Electron Microsc. 2012;61(3):179–86. https://doi.org/10.1093/jmicro/dfs035. Epub 2012 Feb 26. PubMed PMID: 22366032.

Hultcrantz R, Glaumann H. Studies on the rat liver following iron overload: biochemical studies after iron mobilization. Lab Investig. 1982;46:383–92.

Iancu TC. Biological and ultrastructural aspects of iron overload: an overview. Pediatr Pathol. 1990;10:281–96.

Iannaccone R, Federle MP, Brancatelli G, et al. Peliosis hepatis: spectrum of imaging findings. AJR Am J Roentgenol. 2006;187(1):W43–52. https://doi.org/10.2214/AJR.05.0167.

Ishak KG. Inherited metabolic diseases of the liver. Clin Liver Dis. 2002;6(2):455–79, viii. Review. PubMed PMID: 12122865

Jääskeläinen J, Martikainen A, Vornanen M, Heinonen K. Neonatal haemochromatosis combined with duodenal atresia [letter]. Eur J Pediatr. 1995;154:247–8.

Jansen PL, Müller M. Genetic cholestasis: lessons from the molecular physiology of bile formation. Can J Gastroenterol. 2000;14(3):233–8. Review. PubMed PMID: 10758420

Jayaram A, Finegold MJ, Parham DM, Jasty R. Successful management of rhabdoid tumor of the liver. J Pediatr Hematol Oncol. 2007;29(6):406–8. PubMed PMID: 17551403.

JDC Y, Gross JB, Ludwig J, Purnell DC. Cholestatic jaundice in hyperthyroidism. Am J Med. 1989;86:619–20.

Jiménez-Heffernan JA, López-Ferrer P, Burgos E, Viguer JM. Pathological case of the month. Primary hepatic malignant tumor with rhabdoid features. Arch Pediatr Adolesc Med. 1998;152(5):509–10. PubMed PMID: 9605039.

Johal JS, Thorp JW, Oyer CE. Neonatal hemochromatosis, renal tubular dysgenesis, and hypocalvaria in a neonate. Pediatr Dev Pathol. 1998;1:433–7.

Johnson CA, Gissen P, Sergi C. Molecular pathology and genetics of congenital hepatorenal fibrocystic syndromes. J Med Genet. 2003;40(5):311–9. Review. PubMed PMID: 12746391; PubMed Central PMCID: PMC1735460.

Johnston J, Al-Bahrani R, Abuetabh Y, Chiu B, Forsman CL, Nagamori S, Leng R, Petryk A, Sergi C. Twisted gastrulation expression in cholangiocellular and hepatocellular carcinoma. J Clin Pathol. 2012;65(10):945–8. Epub 2012 May 25. PubMed PMID: 22639408.

Josephson J, Turner JM, Field CJ, Wizzard PR, Nation PN, Sergi C, Ball RO, Pencharz PB, Wales PW. Parenteral soy oil and fish oil emulsions: impact of dose restriction on bile flow and brain size of parenteral nutrition-fed neonatal piglets. JPEN J Parenter Enteral Nutr. 2015;39(6):677–87. https://doi.org/10.1177/0148607114556494. Epub 2014 Oct 17. PubMed PMID: 25326097.

Kallo IL, Lakatos I, Szijarto L. Leprechaunism (Donohue's syndrome). J Pediatr. 1965;66:372–9.

Kamath BM, Piccoli DA. Heritable disorders of the bile ducts. Gastroenterol Clin N Am. 2003;32(3):857–75, vi. Review. PubMed PMID: 14562578.

Keller M, Scholl-Buergi S, Sergi C, Theurl I, Weiss G, Unsinn KM, Trawöger R. An unusual case of intrauterine symptomatic neonatal liver failure. Klin Padiatr. 2008;220(1):32–6. https://doi.org/10.1055/s-2007-970591. PubMed PMID: 18172830.

Kerem B, Rommens JM, Buchanan JA, Markiewicz D, Cox TK, Chakravarti A, Buchwald M, Tsui LC. Identification of the cystic fibrosis gene: genetic analysis. Science. 1989;245:1073–80.

Kershisnik MM, Knisely AS, Sun CC, Andrews JM, Wittwer CT. Cytomegalovirus infection, fetal liver disease, and neonatal hemochromatosis. Hum Pathol. 1992;23:1075–80.

Kim L, Park YN, Kim SE, Noh TW, Park C. Teratoid hepatoblastoma: multidirectional differentiation of stem cell of the liver. Yonsei Med J. 2001;42(4):431–5.

King MW. Sphingolipid metabolism and the ceramides. In: The medical biochemistry page. http://themedicalbiochemistrypage.org. Accessed 01 Jan 2020.

Knisely AS. Iron and pediatric liver disease. Semin Liver Dis. 1994;14:229–35.

Knisely AS, Magid MS, Dische MR, Cutz E. Neonatal hemochromatosis. Birth Defects Orig Artic Ser. 1987;23:75–102.

Knisely AS, Strautnieks SS, Meier Y, Stieger B, Byrne JA, Portmann BC, Bull LN, Pawlikowska L, Bilezikçi B, Ozçay F, László A, Tiszlavicz L, Moore L, Raftos J, Arnell H, Fischler B, Németh A, Papadogiannakis N, Cielecka-Kuszyk J, Jankowska I, Pawłowska J, Melín-Aldana H, Emerick KM, Whitington PF, Mieli-Vergani G, Thompson RJ. Hepatocellular carcinoma in ten children under five years of age with bile salt export pump deficiency. Hepatology. 2006;44(2):478–86. PubMed PMID: 16871584

Kok NF, Terkivatan T, Ijzermans JN. Regarding 'liver cell adenoma and liver cell adenomatosis' by Ludger Barthelmes and Iain S. Tait. HPB (Oxford). 2006;8(1):71–2. https://doi.org/10.1080/13651820500537879. PubMed PMID: 18333245; PubMed Central PMCID: PMC2131360

Kubitz R, Keitel V, Häussinger D. Inborn errors of biliary canalicular transport systems. Methods Enzymol. 2005;400:558–69. Review. PubMed PMID: 16399370

Kubo S, Takemura S, Tanaka S, Shinkawa H, Kinoshita M, Hamano G, Ito T, Koda M, Aota T. Occupational cholangiocarcinoma caused by exposure to 1,2-dichloropropane and/or dichloromethane. Ann Gastroenterol Surg. 2017;2(2):99–105. https://doi.org/10.1002/ags3.12051. eCollection 2018 Mar. Review. PubMed PMID: 29863124; PubMed Central PMCID: PMC5881298

Lafaro KJ, Pawlik TM. Fibrolamellar hepatocellular carcinoma: current clinical perspectives. J Hepatocell Carcinoma. 2015;2:151–7. https://doi.org/10.2147/JHC.S75153. PMID: 27508204; PMCID: PMC4918295

Lee WS, Sokol RJ. Liver disease in mitochondrial disorders. Semin Liver Dis. 2007;27(3):259–73.

Leeuwen L, Fitzgerald DA, Gaskin KJ. Liver disease in cystic fibrosis. Paediatr Respir Rev. 2014;15(1):69–74. https://doi.org/10.1016/j.prrv.2013.05.001. Epub 2013 Jun 14. Review. PubMed PMID: 23769887.

Leonard JV, AHV S. Mitochondrial respiratory chain disorders I: mitochondrial DNA defects. Lancet. 2000;355:299–304.

Lim DW, Wales PW, Josephson JK, Nation PN, Wizzard PR, Sergi CM, Field CJ, Sigalet DL, Turner JM. Glucagon-like peptide 2 improves cholestasis in parenteral nutrition – associated liver disease. JPEN J Parenter Enteral Nutr. 2016;40(1):14–21. https://doi.org/10.1177/0148607114551968. Epub 2014 Oct 3. PubMed PMID: 25280755.

Litten JB, Tomlinson GE. Liver tumors in children. Oncologist. 2008;13(7):812–20. https://doi.org/10.1634/theoncologist.2008-0011. Epub 2008 Jul 21. Review. PubMed PMID: 18644850.

Lotz G, Simon S, Patonai A, Sótonyi P, Nemes B, Sergi C, Glasz T, Füle T, Nashan B, Schaff Z. Detection of chlamydia pneumoniae in liver transplant patients with chronic allograft rejection. Transplantation. 2004;77(10):1522–8. PubMed PMID: 15239615.

Lund DP, Lillehei CW, Kevy S, Perez Atayde A, Maller E, Treacy S, Vacanti JP. Liver transplantation in newborn liver failure: treatment for neonatal hemochromatosis. Transplant Proc. 1993;25:1068–71.

Makin CA, Bobrow LG, Bodmer WF. Monoclonal antibody to cytokeratin for use in routine histopathology. J Clin Pathol. 1984;37(9):975–83. PubMed PMID: 6206100; PubMed Central PMCID: PMC498911

Manivel C, Wick MR, Abenoza P, Dehner LP. Teratoid hepatoblastoma: the nosologic dilemma of solid embryonic neoplasms of childhood. Cancer. 1986;57(11):2168–74.

McKillop SJ, Belletrutti MJ, Lee BE, Yap JY, Noga ML, Desai SJ, Sergi C. Adenovirus necrotizing hepatitis complicating atypical teratoid rhabdoid tumor. Pediatr Int. 2015;57(5):974–7. https://doi.org/10.1111/ped.12674. Epub 2015 Aug 19. PubMed PMID: 26508178.

McKusick VA. Mendelian inheritance in man. A catalog of human genes and genetic disorders. 12th ed. Baltimore: Johns Hopkins University Press; 1998.

Melin-Aldana H, Park C, Pan X, Fritsch M, Malladi P, Whitington P. Gestational autoimmune disease in newborns with an indeterminate cause of death following a complete autopsy. J Neonatal Perinatal Med. 2015. [Epub ahead of print] PubMed PMID: 25766200

Metzman R, Anand A, DeGiulio PA, Knisely AS. Hepatic insufficiency and fibrosis associated with intrauterine parvovirus B19 infection in a newborn premature infant. J Pediatr Gastroenterol Nutr. 1989;9:112–4.

Meyers RL. Tumors of the liver in children. Surg Oncol. 2007;16(3):195–203. Epub 2007 Aug 21. Review. PubMed PMID: 17714939.

Meyers RL, Maibach R, Hiyama E, Häberle B, Krailo M, Rangaswami A, et al. Risk-stratified staging in paediatric hepatoblastoma: a unified analysis from the Children's hepatic tumors international collaboration. Lancet Oncol. 2017;18:122–31.

Mostoufizadeh M, Lack EE, Gang DL, Perez Atayde AR, Driscoll SG. Postmortem manifestations of echovirus 11 sepsis in five newborn infants. Hum Pathol. 1983;14:818–23.

Muiesan P, Rela M, Kane P, Dawan A, Baker A, Ball C, Mowat AP, Williams R, Heaton ND. Liver transplantation for neonatal haemochromatosis. Arch Dis Child Fetal Neonatal Ed. 1995;73:F178–80.

Nemes K, Bens S, Bourdeaut F, Hasselblatt M, Kool M, Johann P, Kordes U, Schneppenheim R, Siebert R, Frühwald MC. Rhabdoid tumor predisposition syndrome. In: Adam MP, Ardinger HH, Pagon RA, Wallace SE, Bean LJH, Stephens K, Amemiya A, editors. GeneReviews® [Internet]. Seattle: University of Washington; 2017. 1993–2018. Available from http://www-ncbi-nlm-nih-gov.login.ezproxy.library.ualberta.ca/books/NBK469816/

Nicolle D, Fabre M, Simon-Coma M, Gorse A, Kappler R, Nonell L, Mallo M, Haidar H, Déas O, Mussini C, Guettier C, Redon MJ, Brugières L, Ghigna MR, Fadel E, Galmiche-Rolland L, Chardot C, Judde JG, Armengol C, Branchereau S, Cairo S. Patient-derived mouse xenografts from pediatric liver cancer predict tumor recurrence and advise clinical management. Hepatology. 2016;64(4):1121–35. https://doi.org/10.1002/hep.28621. Epub 2016 Jun 16. PubMed PMID: 27115099.

Novick DM, Kreek MJ, Arns PA, Lau LL, Yancovitz SR, Gelb AM. Effect of severe alcoholic liver disease on the disposition of methadone in maintenance patients. Alcohol Clin Exp Res. 1985a;9(4):349–54. PubMed PMID: 3901806

Novick DM, Enlow RW, Gelb AM, Stenger RJ, Fotino M, Winter JW, Yancovitz SR, Schoenberg MD, Kreek MJ. Hepatic cirrhosis in young adults: association with adolescent onset of alcohol and parenteral heroin abuse. Gut. 1985b;26(1):8–13. PubMed PMID: 3855296; PubMed Central PMCID: PMC1432410

Oddone M, Bellini C, Bonacci W, Bartocci M, Toma P, Serra G. Diagnosis of neonatal hemochromatosis with MR imaging and duplex Doppler sonography. Eur Radiol. 1999;9:1882–5.

Oliveira NL, Kanawaty FR, Costa SC, Hessel G. Infection by cytomegalovirus in patients with neonatal cholestasis. Arq Gastroenterol. 2002;39(2):132–6. Epub 2003 Feb 19. PubMed PMID: 12612719

Oppenheimer EH, Esterly JR. Pathology of cystic fibrosis review of the literature and comparison with 146 autopsied cases. Perspect Pediatr Pathol. 1975;2:241–78.

Parashari UC, Singh R, Yadav R, Aga P. Changes in the globus pallidus in chronic kernicterus. J Pediatr Neurosci. 2009;4(2):117–9. https://doi.org/10.4103/1817-1745.57333. PMID: 21887193; PMCID: PMC3162777

Parizhskaya M, Reyes J, Jaffe R. Hemophagocytic syndrome presenting as acute hepatic failure in two infants: clinical overlap with neonatal hemochromatosis. Pediatr Dev Pathol. 1999;2:360–6.

Park RW, Grand RJ. Gastrointestinal manifestations of cystic fibrosis: a review. Gastroenterology. 1981;81:1143–61.

Patton RG, Christie DL, Smith DW, Beckwith JB. Cerebro-hepato-renal syndrome of Zellweger. Two patients with islet cell hyperplasia, hypoglycemia, and thymic anomalies, and comments on iron metabolism. Am J Dis Child. 1972;124:840–4.

PDQ® Pediatric Treatment Editorial Board. PDQ childhood liver cancer treatment. Bethesda: National Cancer Institute. Available at: https://www.cancer.gov/types/liver/hp/child-liver-treatment-pdq. Accessed 24 Nov 2019 [PMID: 26389232]

Perry TL. Tyrosinemia associated with hypermethioninemia and islet cell hyperplasia. Can Med Assoc J. 1965;97:1067–75.

Pirisi M, Leutner M, Pinato DJ, Avellini C, Carsana L, Toniutto P, Fabris C, Boldorini R. Reliability and reproducibility of the edmondson grading of hepatocellular carcinoma using paired core biopsy and surgical resection specimens. Arch Pathol Lab Med. 2010;134(12):1818–22. https://doi.org/10.1043/2009-0551-OAR1.1. PubMed PMID: 21128781

Porta EA, Stein AA, Patterson D. Ultrastructural changes of the pancreas and liver in cystic fibrosis. Am J Clin Pathol. 1964;41:451–65.

Portier CJ, Armstrong BK, Baguley BC, Baur X, Belyaev I, Bellé R, Belpoggi F, Biggeri A, Bosland MC, Bruzzi P, Budnik LT, Bugge MD, Burns K, Calaf GM, Carpenter DO, Carpenter HM, López-Carrillo L, Clapp R, Cocco P, Consonni D, Comba P, Craft E, Dalvie MA, Davis D, Demers PA, De Roos AJ, DeWitt J, Forastiere F, Freedman JH, Fritschi L, Gaus C, Gohlke JM, Goldberg M, Greiser E, Hansen J, Hardell L, Hauptmann M, Huang W, Huff J, James MO, Jameson CW, Kortenkamp A, Kopp-Schneider A, Kromhout H, Larramendy ML, Landrigan PJ, Lash LH, Leszczynski D, Lynch CF, Magnani C, Mandrioli D, Martin FL, Merler E, Michelozzi P, Miligi L, Miller AB, Mirabelli D, Mirer FE, Naidoo S, Perry MJ, Petronio MG, Pirastu R, Portier RJ, Ramos KS, Robertson LW, Rodriguez T, Röösli M, Ross MK, Roy D, Rusyn I, Saldiva P, Sass J, Savolainen K, Scheepers PT, Sergi C, Silbergeld EK, Smith MT, Stewart BW, Sutton P, Tateo F, Terracini B, Thielmann HW, Thomas DB, Vainio H, Vena JE, Vineis P, Weiderpass E, Weisenburger DD, Woodruff TJ, Yorifuji T, Yu IJ, Zambon P, Zeeb H, Zhou SF. Differences in the carcinogenic evaluation of glyphosate between the International Agency for Research on Cancer (IARC) and the European Food Safety Authority (EFSA). J Epidemiol Community Health. 2016;70(8):741–5. https://doi.org/10.1136/jech-2015-207005. Epub 2016 Mar 3. PubMed PMID: 26941213; PubMed Central PMCID: PMC4975799

Prelog M, Bergmann C, Ausserlechner MJ, Fischer H, Margreiter R, Gassner I, Brunner A, Jungraithmayr TC, Zerres K, Sergi C, Zimmerhackl LB. Successful transplantation in a child with rapid progression of autosomal recessive polycystic kidney disease associated with a novel mutation. Pediatr Transplant. 2006;10(3):362–6. Erratum in: Pediatr Transplant. 2008 Mar;12(2):256. Sergi, E Consolato [corrected to Sergi, Consolato]. PubMed PMID: 16677362.

Rabah R. Teratoid hepatoblastoma with abundant neuroendocrine and squamous differentiation with extensive parenchymal metastasis. Arch Pathol Lab Med. 2012;136(8):911–4. https://doi.org/10.5858/arpa.2012-0212-CR. PubMed PMID: 22849740

Ramachandran R, Kakar S. Histological patterns in drug-induced liver disease. J Clin Pathol. 2009;62(6):481–92. https://doi.org/10.1136/jcp.2008.058248. Review. Erratum in: J Clin Pathol. 2010 Dec;63(12):1126. PubMed PMID: 19474352

Ramsay AD, Bates AW, Williams S, Sebire NJ. Variable antigen expression in hepatoblastomas. Appl Immunohistochem Mol Morphol. 2008;16(2):140–7.

Rand EB, McClenathan DT, Whitington PF. Neonatal hemochromatosis: report of successful orthotopic liver transplantation. J Pediatr Gastroenterol Nutr. 1992;15:325–9.

Resnick MB, Kozakewich HP, Perez-Atayde AR. Hepatic adenoma in the pediatric age group. Clinicopathological observations and assessment of cell proliferative activity. Am J Surg Pathol. 1995;19(10):1181–90. PubMed PMID: 7573676

Rinaldo P, Yoon HR, Yu C, Raymond K, Tiozzo C, Giordano G. Sudden and unexpected neonatal death: a protocol for the postmortem diagnosis of fatty acid oxidation disorders. Semin Perinatol. 1999;23(2):204–10. Review. PubMed PMID: 10331471

Roberts CW, Biegel JA. The role of SMARCB1/INI1 in development of rhabdoid tumor. Cancer Biol Ther. 2009;8(5):412–6. Epub 2009 Mar 29. Review. PubMed PMID: 19305156; PubMed Central PMCID: PMC2709499.

Roels F, Espeel M, De Craemer D. Liver pathology and immunocytochemistry in congenital peroxisomal diseases: a review. J Inherit Metab Dis. 1991;14:853–75.

Rogers DR. Leprechaunism (Donohue's syndrome). A possible case, with emphasis on changes in the adenohypophysis. Am J Clin Pathol. 1966;45:614–9.

Rosenstein BJ, Cutting GR. The diagnosis of cystic fibrosis: a consensus statement. J Pediatr. 1998;132:589–95.

Roulet M, Laurini R, Rivier L, Calame A. Hepatic veno-occlusive disease in newborn infant of a woman drinking herbal tea. J Pediatr. 1988;112:433–6.

Ruchelli ED, Uri A, Dimmick JE, Bove KE, Huff DS, Duncan LM, Jennings JB, Witzleben CL. Severe perinatal liver disease and down syndrome: an apparent relationship. Hum Pathol. 1991;22:1274–80.

Savastano S, San Bortolo O, Velo E, et al. Pseudotumoral appearance of peliosis hepatis. AJR Am J Roentgenol. 2005;185(2):558–9.

Schiesser M, Sergi C, Enders M, Maul H, Schnitzler P. Discordant outcomes in a case of parvovirus b19 transmission into both dichorionic twins. Twin Res Hum Genet. 2009;12(2):175–9. https://doi.org/10.1375/twin.12.2.175. PubMed PMID: 19335188.

Schmid I, von Schweinitz D. Pediatric hepatocellular carcinoma: challenges and solutions. J Hepatocell Carcinoma. 2017;4:15–21. https://doi.org/10.2147/JHC.S94008. PMID: 28144610; PMCID: PMC5248979

Schmorl G, Zur Kenntnis d. Ikterus neonatorum, insbesondere der dabei auftretenden Gehirnveranderungen. Verh Dtsch Ges Pathol. 1904;6:109–15.

Schöberl S, Nickel P, Schmutzer G, Siekmeyer W, Kiess W. Alkoholintoxikation bei Kindern und Jugendlichen. Eine retrospektive analyse von 173 an einer Universitatskinderklinik betreuten Patienten. [Acute ethanol intoxication among children and adolescents. A retrospective analysis of

173 patients admitted to a university children hospital]. Klin Padiatr. 2008;220(4):253–8. https://doi.org/10.1055/s-2007-984367. Epub 2008 Feb 12. German. PubMed PMID: 18270881

Schoenlebe J, Buyon JP, Zitelli BJ, Friedman D, Greco MA, Knisely AS. Neonatal hemochromatosis associated with maternal autoantibodies against Ro/SS-A and La/SS-B ribonucleoproteins. Am J Dis Child. 1993;47:1072–5.

Schranz M, Talasz H, Graziadei I, Winder T, Sergi C, Bogner K, Vogel W, Zoller H. Diagnosis of hepatic iron overload: a family study illustrating pitfalls in diagnosing hemochromatosis. Diagn Mol Pathol. 2009;18(1):53–60. https://doi.org/10.1097/PDM.0b013e31817cfd4b. PubMed PMID: 19214108.

Schweizer P, Kirschner HJ, Schittenhelm C. Anatomy of the porta hepatis (PH) as rational basis for the hepatoporto-enterostomy (HPE). Eur J Pediatr Surg. 1999;9:13–8.

Sergi CM. Hepatocellular carcinoma, fibrolamellar variant: diagnostic pathologic criteria and molecular pathology update. A primer. Diagnostics (Basel). 2015;6(1):pii: E3. https://doi.org/10.3390/diagnostics6010003. PubMed PMID:26838800; PubMed Central PMCID: PMC4808818

Sergi CM. Genetics of biliary atresia: a work in progress for a disease with an unavoidable sequela into liver cirrhosis following failure of hepatic portoenterostomy. Liver Cirrhosis – Debates and Current Challenges, Georgios Tsoulfas, IntechOpen. https://doi.org/10.5772/intechopen.85071. Available from: https://www.intechopen.com/books/liver-cirrhosis-debates-and-current-challenges/genetics-of-biliary-atresia-a-work-in-progress-for-a-disease-with-an-unavoidable-sequela-into-liver-. March 23, 2019.

Sergi C, Consalez GG, Fabbretti G, Brisigotti M, Faa G, Costa V, Romeo G, Callea F. Immunohistochemical and genetic characterization of the M Cagliari alpha-1-antitrypsin molecule (M-like alpha-1-antitrypsin deficiency). Lab Investig. 1994;70(1):130–3. PubMed PMID: 8302013.

Sergi C, Goeser T, Otto G, Otto HF, Hofmann WJ. A rapid and highly specific technique to detect hepatitis C RNA in frozen sections of liver. J Clin Pathol. 1996;49(5):369–72. PubMed PMID: 8707948; PubMed Central PMCID: PMC500473.

Sergi C, Jundt K, Seipp S, Goeser T, Theilmann L, Otto G, Otto HF, Hofmann WJ. The distribution of HBV, HCV and HGV among livers with fulminant hepatic failure of different etiology. J Hepatol. 1998a;29:861–71.

Sergi C, Jundt K, Seipp S, Goeser T, Theilmann L, Otto G, Otto HF, Hofmann WJ. The distribution of HBV, HCV and HGV among livers with fulminant hepatic failure of different aetiology. J Hepatol. 1998b;29(6):861–71. PubMed PMID: 9875631.

Sergi C, Beedgen B, Kopitz J, Zilow E, Zoubaa S, Otto HF, Cantz M, Linderkamp O. Refractory congenital ascites as a manifestation of neonatal sialidosis: clinical, biochemical and morphological studies in a newborn Syrian male infant. Am J Perinatol. 1999a;16(3):133–41. PubMed PMID: 10438195.

Sergi C, Beedgen B, Linderkamp O, Hofmann WJ. Fatal course of veno-occlusive disease of the liver (endophlebitis hepatica obliterans) in a preterm infant. Pathol Res Pract. 1999b;195(12):847–51. PubMed PMID: 10631721.

Sergi C, Kahl P, Otto HF. Contribution of apoptosis and apoptosis-related proteins to the malformation of the primitive intrahepatic biliary system in Meckel syndrome. Am J Pathol. 2000a;156:1589–98.

Sergi C, Adam S, Kahl P, Otto HF. The remodeling of the primitive human biliary system. Early Hum Dev. 2000b;58:167–78.

Sergi C, Adam S, Kahl P, Otto HF. Study of the malformation of ductal plate of the liver in Meckel syndrome and review of other syndromes presenting with this anomaly. Pediatr Dev Pathol. 2000c;3:568–83.

Sergi C, Mornet E, Troeger J, Voigtlaender T. Perinatal hypophosphatasia: radiology, pathology and molecular biology studies in a family harboring a splicing mutation (648+1A) and a novel missense mutation (N400S) in the tissue-nonspecific alkaline phosphatase (TNSALP) gene. Am J Med Genet. 2001a;103(3):235–40. PubMed PMID: 11745997.

Sergi C, Penzel R, Uhl J, Zoubaa S, Dietrich H, Decker N, Rieger P, Kopitz J, Otto HF, Kiessling M, Cantz M. Prenatal diagnosis and fetal pathology in a Turkish family harboring a novel nonsense mutation in the lysosomal alpha-N-acetyl-neuraminidase (sialidase) gene. Hum Genet. 2001b;109(4):421–8. PubMed PMID: 11702224.

Sergi C, Himbert U, Weinhardt F, Heilmann W, Meyer P, Beedgen B, Zilow E, Hofmann WJ, Linderkamp O, Otto HF. Hepatic failure with neonatal tissue siderosis of hemochromatotic type in an infant presenting with meconium ileus. Case report and differential diagnosis of the perinatal iron storage disorders. Pathol Res Pract. 2001c;197(10):699–709; discussion 711–3. PubMed PMID: 11700892.

Sergi C, Arnold JC, Rau W, Otto HF, Hofmann WJ. Single nucleotide insertion in the 5′-untranslated region of hepatitis C virus with clearance of the viral RNA in a liver transplant recipient during acute hepatitis B virus superinfection. Liver. 2002;22(1):79–82. PubMed PMID: 11906622.

Sergi C, Gross W, Mory M, Schaefer M, Gebhard MM. Biliary-type cytokeratin pattern in a canine isolated perfused liver transplantation model. J Surg Res. 2008a;146(2):164–71. Epub 2007 Jul 13. PubMed PMID: 17631899.

Sergi C, Bentsz J, Feist D, Nutzenadel W, Otto HF, Hofmann WJ. Bile duct to portal space ratio and ductal plate remnants in liver disease of infants aged less than 1 year. Pathology. 2008b;40(3):260–7. https://doi.org/10.1080/00313020801911538.

Sergi C, Abdualmjid R, Abuetabh Y. Canine liver transplantation model and the intermediate filaments of the cytoskeleton of the hepatocytes. J Biomed Biotechnol. 2012;2012:131324. https://

doi.org/10.1155/2012/131324. Epub 2012 Mar 28. Review. PubMed PMID: 22536013; PubMed Central PMCID: PMC3321507.

Sergi C, Shen F, Lim DW, Liu W, Zhang M, Chiu B, Anand V, Sun Z. Cardiovascular dysfunction in sepsis at the dawn of emerging mediators. Biomed Pharmacother. 2017;95:153–60. https://doi.org/10.1016/j.biopha.2017.08.066. Epub 2017 Sep 12. Review. PubMed PMID: 28841455.

Shneider BL, Setchell KD, Whitington PF, Neilson KA, Suchy FJ. Delta 4-3-oxosteroid 5 beta-reductase deficiency causing neonatal liver failure and hemochromatosis. J Pediatr. 1994;124:234–8.

Siafakas CG, Jonas MM, Perez-Atayde AR. Abnormal bile acid metabolism and neonatal hemochromatosis: a subset with poor prognosis. J Pediatr Gastroenterol Nutr. 1997;25:321–6.

Sigurdsson L, Reyes J, Kocoshis SA, Hansen TW, Rosh J, Knisely AS. Neonatal hemochromatosis: outcomes of pharmacologic and surgical therapies. J Pediatr Gastroenterol Nutr. 1998;26:85–9.

Silver MM, Valberg LS, Cutz E, Lines LD, Phillips MJ. Hepatic morphology and iron quantitation in perinatal hemochromatosis. Comparison with a large perinatal control population, including cases with chronic liver disease. Am J Pathol. 1993;143:1312–25.

Smedts F, Ramaekers F, Robben H, Pruszczynski M, van Muijen G, Lane B, Leigh I, Vooijs P. Changing patterns of keratin expression during progression of cervical intraepithelial neoplasia. Am J Pathol. 1990;136(3):657–68. PubMed PMID: 1690513; PubMed Central PMCID: PMC1877502

Snover DC, Freese DK, Sharp HL, Bloomer JR, Najarian JS, Ascher NL. Liver allograft rejection. An analysis of the use of biopsy in determining outcome of rejection. Am J Surg Pathol. 1987;11(1):1–10. PubMed PMID: 3538917

Sohn DS, Kim KY, Lee WB, Kim DC. Eosinophilic granulopoiesis in human fetal liver. Anat Rec. 1993;235(3):453–60. PubMed PMID: 8430915

Sokol RJ, Treem WR. Mitochondria and childhood liver diseases. J Pediatr Gastroenterol Nutr. 1999;28(1):4–16. Review. PubMed PMID: 9890461

Stocker JT. Hepatic tumors in children. Clin Liver Dis. 2001;5(1):259–81, viii–ix. Review. PubMed PMID: 11218918.

Sträter R. Beiträge zur Lehre von der Hämochromatose und ihren Beziehungen zur allgemeinen Hämosiderose. Virchows Arch. 1914;218:1–301.

Sundaram SS, Sokol RJ. The multiple facets of ABCB4 (MDR3) deficiency. Curr Treat Options Gastroenterol. 2007;10(6):495–503. https://doi.org/10.1007/s11938-007-0049-4. PMID: 18221610; PMCID: PMC3888315

Sundaram SS, Bove KE, Lovell MA, Sokol RJ. Mechanisms of disease: inborn errors of bile acid synthesis. Nat Clin Pract Gastroenterol Hepatol. 2008;5(8):456–68. https://doi.org/10.1038/ncpgasthep1179. Epub 2008 Jun 24. Review. PubMed PMID: 18577977; PubMed Central PMCID: PMC3888787

Tang V, Cofer ZC, Cui S, Sapp V, Loomes KM, Matthews RP. Loss of a candidate biliary atresia susceptibility gene, add3a, causes biliary developmental defects in zebrafish. J Pediatr Gastroenterol Nutr. 2016;63(5):524–30. https://doi.org/10.1097/MPG.0000000000001375.

Taylor SA, Kelly S, Alonso EM, Whitington PF. The effects of gestational Alloimmune liver disease on fetal and infant morbidity and mortality. J Pediatr. 2018;196:123–128.e1. https://doi.org/10.1016/j.jpeds.2017.12.054. Epub 2018 Feb 27. PubMed PMID: 29499991.

Tõnisson M, Tillmann V, Kuudeberg A, Lepik D, Väli M. Acute alcohol intoxication characteristics in children. Alcohol Alcohol. 2013;48(4):390–5. https://doi.org/10.1093/alcalc/agt036. Epub 2013 Apr 30. PubMed PMID: 23632804

Tsai EA, Grochowski CM, Loomes KM, Bessho K, Hakonarson H, Bezerra JA, et al. Replication of a GWAS signal in a Caucasian population implicates ADD3 in susceptibility to biliary atresia. Hum Genet. 2014;133(2):235–43. https://doi.org/10.1007/s00439-013-1368-2.

Turner JM, Wales PW, Nation PN, Wizzard P, Pendlebury C, Sergi C, Ball RO, Pencharz PB. Novel neonatal piglet models of surgical short bowel syndrome with intestinal failure. J Pediatr Gastroenterol Nutr. 2011;52(1):9–16. https://doi.org/10.1097/MPG.0b013e3181f18ca0.

van den Brand M, Flucke UE, Bult P, Weemaes CM, van Deuren M. Angiosarcoma in a patient with immunodeficiency, centromeric region instability, facial anomalies (ICF) syndrome. Am J Med Genet A. 2011;155A(3):622–5. https://doi.org/10.1002/ajmg.a.33831. Epub 2011 Feb 18. PubMed PMID: 21337690

Van Hoof F, Hageman-Bal M. Progressive familial myoclonic epilepsy with Lafora bodies. Electron microscopic and histochemical study of a cerebral biopsy. Acta Neuropathol. 1967;7(4):315–36. PubMed PMID: 4166286

Verloes A, Lombet J, Lambert Y, Hubert AF, Deprez M, Fridman V, Gosseye S, Rigo J, Sokal E. Tricho-hepato-enteric syndrome: further delineation of a distinct syndrome with neonatal hemochromatosis phenotype, intractable diarrhea, and hair anomalies. Am J Med Genet. 1997;68:391–5.

Vitola BE, Balistreri WF. Liver disease in systemic disorders. In: Behrman RE, Kliegman RM, Jenson HB, editors. Nelson's textbook of pediatrics. 21st ed. Philadelphia: Elsevier.

Vohra P, Haller C, Emre S, Magid M, Holzman I, Ye MQ, Iofel E, Shneider BL. Neonatal hemochromatosis: the importance of early recognition of liver failure. J Pediatr. 2000;136:537–41.

Voigt M, Schneider KT, Jahrig K. Analyse des Geburtsgutes des Jahrgangs 1992 der Bundesrepublik Deutschland. Teil 1: Neue Perzentilwerte für die Körpermaße von Neugeborenen. Geburtshilfe Frauenheilkd. 1996;56:550–8.

References and Recommended Readings

Vokuhl C, Oyen F, Häberle B, von Schweinitz D, Schneppenheim R, Leuschner I. Small cell undifferentiated (SCUD) hepatoblastomas: all malignant rhabdoid tumors? Genes Chromosomes Cancer. 2016;55(12):925–31. https://doi.org/10.1002/gcc.22390. Epub 2016 Jul 29. PubMed PMID: 27356182.

Von Recklinghausen FD. Über Haemochromatose. Tageblatt der 62. Versammlung deutscher Naturforscher und Aerzte in Heidelberg. 1889; 324–5.

Vyberg M, Poulsen H. Virchows Arch. 1984.

Wiedemann HR. Hans-Ulrich Zellweger (1909–1990). Eur J Pediatr. 1991;150(7):451. https://doi.org/10.1007/BF01958418. PMID 1915492

Witzleben CL. Bile duct paucity ("intrahepatic atresia"). Perspect Pediatr Pathol. 1982;7:185–201. PubMed PMID: 6981794

Witzleben CL, Uri A. Perinatal hemochromatosis: entity or end result? Hum Pathol. 1989;20:335–40.

Wu X, Dagar V, Algar E, Muscat A, Bandopadhayay P, Ashley D, Wo Chow C. Rhabdoid tumour: a malignancy of early childhood with variable primary site, histology and clinical behaviour. Pathology. 2008;40(7):664–70. https://doi.org/10.1080/00313020802436451. PubMed PMID: 18985520.

Wu JF, Chang HH, Lu MY, Jou ST, Chang KC, Ni YH, Chang MH. Prognostic roles of pathology markers immunoexpression and clinical parameters in Hepatoblastoma. J Biomed Sci. 2017;24(1):62. https://doi.org/10.1186/s12929-017-0369-1. PMID: 28851352; PMCID: PMC5574230

Zeman MV, Hirschfield GM. Autoantibodies and liver disease: uses and abuses. Can J Gastroenterol. 2010;24(4):225–31. https://doi.org/10.1155/2010/431913. PMID: 20431809; PMCID: PMC2864616

Ziegler MM. Meconium Ileus. Curr Probl Surg. 1994;31:731–77.

Parenchymal GI Glands (Gallbladder, Biliary Tract, and Pancreas)

Contents

5.1	**Development and Genetics**	551
5.2	**Biliary and Pancreatic Structural Anomalies**	553
5.2.1	Choledochal Cysts	553
5.2.2	Gallbladder Congenital Abnormalities	553
5.2.3	Pancreas Congenital Anomalies	553
5.3	**Gallbladder and Extrahepatic Biliary Tract**	556
5.3.1	Cholesterolosis and Cholelithiasis	556
5.3.2	Acute and Chronic Cholecystitis	557
5.3.3	Gallbladder Proliferative Processes and Neoplasms	558
5.3.4	Extrahepatic Bile Duct Tumors and Cholangiocellular Carcinoma	561
5.4	**Pancreas Pathology**	561
5.4.1	Congenital Hyperinsulinism and Nesidioblastosis	561
5.4.2	Acute and Chronic Pancreatitis	563
5.4.3	Pancreatoblastoma and Acinar Cell Carcinoma in Childhood and Youth	566
5.4.4	Cysts and Cystic Neoplastic Processes of the Pancreas	567
5.4.5	PanIN and Solid Pancreas Ductal Carcinoma	571
5.4.6	Other Tumors	571
5.4.7	Degenerative Changes and Transplant Pathology	571
	Multiple Choice Questions and Answers	574
	References and Recommended Readings	575

5.1 Development and Genetics

Before starting to list the congenital anomalies of the pancreatic gland, it is essential to recall four critical steps and times in its embryologic development.

1. Derivation from two diverticula (dorsal first and ventral later) of the primitive foregut just distally to the stomach at the *3rd–4th week* of gestation, following the development of the hepatic primordium.
2. Dorsal and ventral buds develop out as body-tail and the head-uncinate process of the pancreas, respectively.
3. The rapid growth of the dorsal primordium extending into the dorsal mesentery and simultaneous elongation of the hepatic primordium (the future common bile duct) with

rotation along the ventral diverticulum to a more dorsal position.

4. At the end of the *6th week*, there is the *fusion* of the two primordial and their ductal systems with the formation of the Wirsung duct from the ventral bud and distal portions of the dorsal ductular structure and the Santorini duct from the proximal portion of the dorsal lumen.

Sinusoids are intercordal/between the acinar cords and lined by discontinuous, fenestrated endothelium. Lymphatics are embedded in the portal tract, but lymphatic flow starts from perisinusoidal space (Disse space) through periportal space (Mall area), portal tract lymphatics, and up to larger lymphatic ducts. The bile flow is as follows: hepatocytic bile canaliculi → canals of Hering → portal bile ducts → left and right hepatic ducts → common hepatic bile duct → joining with the cystic duct of the gallbladder → ductus choledochus → ampulla of Vater draining in the duodenum.

Notes: Iron accumulation is mainly periportal, while lipofuscin is mostly pericentral.

The canals of Hering are among cells intermediate between hepatocytes and biliary cells. Bile flow is energy dependent with active transport. Kupffer cells belong to the monocyte/phagocyte system in sinusoids and are highlighted by the CAM5.6 stain. Perisinusoidal Ito cells are fat-containing mesenchymal cells located in the space of Disse with the function to store vitamin A.

Immunohistochemical markers that are useful to characterize the phenotype of the biliary and hepatic epithelium are CK7 (late fetal biliary cell marker and permanent cholangiocyte marker of child and adult), CK19 (early fetal biliary cell marker and permanent cholangiocyte marker of child and adult), CK8 (hepatocyte marker of fetal, child and adult), CK18 (hepatocyte marker of fetal, child and adult), CD68 (KP-1, Kupffer cells), CAM5.2 (LMW-CK, interlobular bile ducts), and pCEA/CD10 (bile canaliculi of the hepatocytes) are useful markers in pediatric pathology. CAM5.2 reacts against cytokeratins 7 and 8, but not keratins 18 and 19. In several textbooks, and also in this book, CK (cytokeratin) is interchangeable with K (keratin).

Intravital biliary excretion evaluation of the liver may be carried out using noninvasive techniques such as ultrasound, computed tomography, or magnetic resonance imaging or invasive procedures such as biopsy.

The innervation of the pancreas is intimately linked to the development of the endocrine component (Fig. 5.1). Amella et al. (2008) studied the

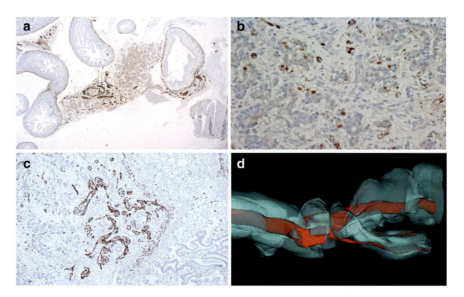

Fig. 5.1 The ductulo-insular complexes highlighted by a study performed on fetal pancreas using an antibody against S-100 are shown in (**a**–**c**) (anti-S100 immunohistochemistry, Avidin-Biotin-Complex, ×20 original magnification for **a**, ×400 original magnification for **b**, ×100 original magnification for **c**) with ductulo-insular complexes in the pancreas as a whole (**a**), in the parenchyma (**b**) and in proximity of the *papilla Vateri* (**c**), while (**d**) shows the 3-D representation at the level of the ramification of the post-ganglionic nerve bundles in the dorsal part of the pancreas

nerve structures by automatic immunostaining techniques using a polyclonal antibody against two S-100 proteins. The antibody reacts strongly with human S100A and S100B that are detected in Schwann cells. These authors found characteristic immunoreactivity patterns in the parenchyma of the head, body, and tail of the pancreas. The relative density was decreasing from head to tail. Also, they reconstruct some nerves associated with the superior mesenteric plexus. They found that the perimeter and the width of the nerve fibers seem to increase at a continuous rate up to term in all three regions of the pancreas. The study performed using spatial and temporal co-analysis identified that the head of the pancreas had a two-peak growth increase at the 14th and 22nd week of gestation about the area, perimeter, and width of the nerve structures. The body and tail regions showed a single peak at 20 weeks of pregnancy. Understanding the factors that may alter the pancreas nerve innervation can be of significance for the development of therapies in pancreatic disorders of a child and adulthood.

5.2 Biliary and Pancreatic Structural Anomalies

Hepatobiliary and pancreatic aberrations are central in the dysontogenesis of the human body, and the identification of some animal models and the progress of molecular biology techniques allowed us to understand the dysregulation of these processes better.

5.2.1 Choledochal Cysts

- *DEF*: Cystic dilations of the choledochus, including cystic dilation of the entire extrahepatic duct (Alonso-Lej A), partial saccular dilation of the duct (Alonso-Lej B), and cystic dilation of the intraduodenal portion of the duct (Alonso-Lej C or choledococele) (Alonso-Lej F). Choledochal cysts need to be distinguished by Caroli disease, which is the intrahepatic dilation of the segmental ducts. The Todani modification of the Alonso-Lej classification is the most used classification. Type III often shows a classic presentation of choledococele with the dilatation of the extrahepatic bile duct limited to the wall of the duodenum. In type III the intramural component is apparently in the duodenum. Imaging classically evidences type IVB by the "string of beads" aspect. Type V of congenital biliary cysts is represented by multiple sites of saccular or cystic dilatation of only intrahepatic biliary tree and is also known as Caroli disease or "communicating cavernous ectasias."
- *CLI*: Abnormalities of the choledochal system may present with pain, jaundice, and abdominal mass, symptoms of biliary obstruction, cholangitis-associated sepsis, or peritonitis linked to a rupture of the cyst.

5.2.2 Gallbladder Congenital Abnormalities

Gallbladder and biliary tract abnormalities may become part of the spectrum of the fetal type of biliary atresia (*vide infra*) (Fig. 5.2).

5.2.3 Pancreas Congenital Anomalies

Pancreas anomalies include agenesis/aplasia/hypoplasia, restricted exocrine hypoplasia, ductal abnormalities, choledochal cysts (vide supra), heterotopia, annular pancreas, cystic dysplasia, and congenital cysts of the pancreas (Fig. 5.2).

5.2.3.1 Agenesis/Aplasia/Hypoplasia (Both Exocrine and Endocrine Components)

Rare events detectable in nonviable fetuses and infants are due probably to the absent or low fetal insulin necessary for the pancreas development in the setting of fetal growth restriction. It may be linked with the lack of gallbladder.

5.2.3.2 Restricted Exocrine Hypoplasia (aka "lipomatous pseudohypertrophy of the pancreas")

Enlarged gland with a standard shape and fibrofatty replacement of the acinar tissue, presumably associated to intrauterine or genetic

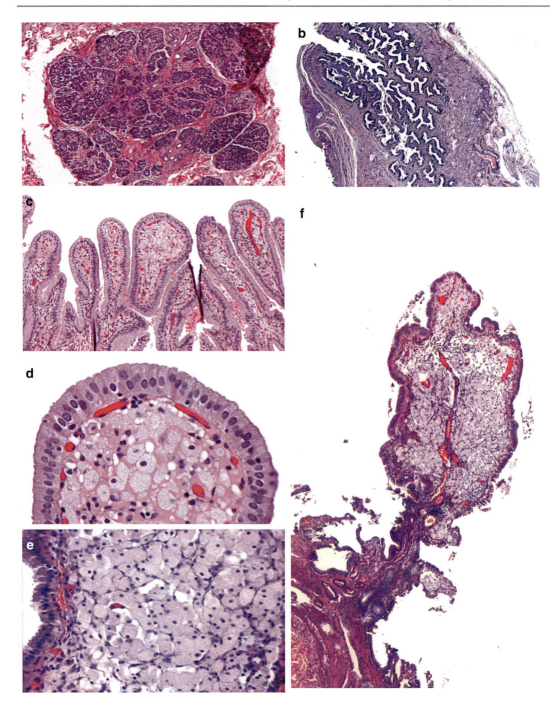

Fig. 5.2 Development and congenital anomalies of the pancreas and extrahepatic biliary tract. In (**a**) is shown a pancreatic tissue heterotopia in the gallbladder (hematoxylin and eosin staining, ×50 as original magnification). In (**b**) is shown a quite fibrotic gallbladder in an infant with biliary atresia (hematoxylin and eosin staining, ×20 as original magnification). Note the fibrosis with almost obliteration which may resemble a chronic cholecystitis of the adult. In (**c**) through (**f**) there are different phenotypic aspects of cholesterolosis, which may also assume the form of a polyp (**c–f**, hematoxylin and eosin staining, ×100, ×400, ×400, ×40 as original magnification, respectively)

or idiopathic infection, and similar to the ARVCM of the heart of young adults or to diffuse *lipomatosis cordis* of the elderly. REH (or LPP) may occur from infants to adults and may be associated with Johanson-Blizzard syndrome, which is recognized by the abnormal development of the pancreas, nose, and scalp. This syndrome, which is accompanied by mental retardation, hearing loss, and growth failure, may be life-threatening.

5.2.3.3 Ductal Abnormalities

Frequent (up to 30% of healthy individuals) anomalies of the interaction between the ventral anlage, which is responsible for the Wirsung or main duct, and the dorsal anlage, which is responsible for the Santorini or accessory duct. Ductal communication variations are many.

1. A joint ventral duct and common bile at the ampulla of Vater are seen with complete regression of the dorsal duct.
2. A crossroads of ventral and common bile ducts at ampulla and persistence of the dorsal duct.
3. Persistence of both dorsal and ventral ducts without evidence of communication (*pancreas divisum*).
4. Common channel with ventral duct communicating with the common bile duct about 5–15 mm from the ampulla.
5. Separate communication of ventral duct into the duodenum with persistence of dorsal duct. Junction of the pancreaticobiliary system and ampulla of Vater are also seen and include a complete separation of the openings of both the common bile duct and the Wirsung duct or joint opening of the two ducts surrounded by some muscular bundles ("sphincter of Oddi"). The pathogenesis of choledochal cysts has been associated with a small entrance of the CBD into the main pancreatic duct less than 2 cm proximal to the duodenal wall. Since 1945 a long common channel (up to 1/5 of healthy individuals) has been formerly since associated to pancreatitis, as well as stenosis of the common distal duct and proximal strictures of the conduit. In *pancreas divisum*, which is a quite frequent anomaly (up to 1/4 of patients with "idiopathic" pancreatitis) of the 3rd–4th decades, but also identified in children, there is a separation of both ventral and dorsal pancreatic ducts. It may present as epigastric pain radiating to the back following fatty food ingestion.

5.2.3.4 Heterotopia Pancreatitis (Gr. ἕτερος or "Other, Different" and τόπος "Place")

Aberrant presence of pancreatic tissue located outside both intra- and extra-abdominal (stomach, duodenum, jejunum, liver, gallbladder, small intestine, colon, appendix, omentum, Meckel diverticulum, bronchogenic cyst, pulmonary sequestration, umbilicus) of the standard anatomic site (up to 15% of autopsy specimens).

- *GRO*: The heterotopic pancreatic tissue is usually recognized as a discrete, firm, yellow, submucosal nodule and microscopically as pancreatic tissue with acini, islets, and ducts. Several theories have been proposed, including irregular sequestration during migration, abnormal differentiation of Brunner glands, abnormal adhesion of normal developing primordia with nearby structures, and abnormal differentiation of multipotent endodermal stem cells. A broad spectrum of symptoms may occur, and occasionally neoplastic transformation has been reported.

5.2.3.5 Annular Pancreas

Abnormality constituted by usually a flat band of pancreatic tissue completely encircling the 2nd portion of the duodenum with or without duodenal atresia or stenosis and presenting from newborns to adults. AP may usually occur in two critical clinical settings, including trisomy 21 syndrome and intestinal malrotation. In neonates, it is a surgical emergency (duodenal atresia) with persistent bilious vomiting and failure to thrive, upper abdominal pain or epigastric pain with the "double bubble" radiologic sign. However, the presentation in older children and teenagers, it is different and characterized by nausea, vomiting, postprandial fullness, bloating, upper GI bleeding, and weight loss. Three theories have been proposed and include 1) hypertrophy of both ventral and dorsal ducts resulting in a ring adherence of the ventral primordium to the duodenum before rotating and 2) overgrowth or 3) adhesion of the left diverticulum of the primitive foregut of a paired ventral primordium. Prognosis may be uncertain in the early neonatal period if a surgical bypass is not promptly performed.

5.2.3.6 Cystic Dysplasia

It is not a properly congenital anomaly but usually the consequence of inflammation and obstruction with retention in the setting of cystic fibrosis, although few "true" dysplastic cases of cystic dysplasia have been reported in some genetic syndromes.

5.2.3.7 Congenital Cysts of the Pancreas

In contract to pseudocysts (no epithelial lining), true cysts are characterized by a cuboidal or stratified squamous epithelial lining and may be classified as solitary or multiple. Single cysts are rare and sometimes are associated with some additional anomalies and may present with polyhydramnios to gastroduodenal compression. Multiple cysts are a relatively more frequent event and occur in the setting of von Hippel-Lindau syndrome, Ivemark syndrome, and Meckel-Gruber syndrome.

5.3 Gallbladder and Extrahepatic Biliary Tract

The gallbladder has not frequently been a topic of discussion in many pediatric pathology meetings, although extrahepatic BA is the most common cause of obstructive cholangiopathy in a newborn with prolonged jaundice. In this section, we will target our discussion on congenital abnormalities, inflammatory conditions, degenerative changes, as well as neoplasms of the gallbladder and extrahepatic biliary tract (Box 5.1).

5.3.1 Cholesterolosis and Cholelithiasis

In 2/3 of individuals, there is a common channel with common bile duct joining the primary pancreatic duct and emptying into the lumen of the

> **Box 5.1 Gallbladder and Extrahepatic Biliary Tract Pathology**
> 5.3.1. Cholesterolosis and Cholelithiasis
> 5.3.2. Acute and Chronic Cholecystitis
> 5.3.3. Gallbladder Proliferative Processes and Neoplasms
> 5.3.4. Extrahepatic Bile Duct Tumors and Cholangiocellular Carcinoma

duodenum at a single site, while in 1/3 of individuals, there are separate orifices. The gallbladder has neither muscularis mucosae nor submucosa but exclusively mucosa, lamina propria, muscularis propria, and connective tissue. This particular structure is specific for this bile receptacle.

5.3.1.1 Cholelithiasis
- 10% Pure
- 90% Mixed
  - Pure cholesterol stones: pale yellow, round/ovoid, granular surface, crystalline core
  - Pure Ca bilirubinate stones (hemoglobinopathies, hemolysis, cirrhosis, pigment stones): black oval, smooth surface

Most stones of mixed type are constituted by a variable combination of cholesterol Ca bilirubinate and Ca carbonate and appear to be multiple and laminated.

Some authors classify stone as cholesterol stones if cholesterol is >70% and pigment stones if cholesterol is <30%. In about 1/5 of the cases of cholelithiasis, radiopacity is reached, because the stones have enough calcium to be radiopaque.

Bile salts supersaturated with cholesterol are responsible for lithogenesis. In the triangle of Admirand-Small, it is easy to check using the triangular coordinates if a bile sample falls below or above the line of saturation of cholesterol. Bile salts and less lecithin increase the solubility of cholesterol in bile. Thus, an excess of the ratio cholesterol/bile salts, followed by nucleation precipitation, and growth of the nidus supported by mucus hypersecretion in the gallbladder are the events of the lithogenesis.

F Factors: Female (± multiple pregnancies), fat (obesity), and forty (probably in the past, but now *fourteen* considering the high-caloric diet of some teenagers) are valid, but not anymore for the age, because the hypercholesterolemic diet of children and adolescents has evidenced an increase of cholesterolosis and cholelithiasis in young individuals. Moreover, there is a more frequent association with some drugs in children and youth. Some prescriptions are lithogenic because increase cholesterol excretion or decrease bile salts levels are promoting stone formation.

5.3.2 Acute and Chronic Cholecystitis

5.3.2.1 Cholecystitis

Acute
90%: stone-associated ("calculus cholecystitis") and often caused by an impaction of the gallstone in the neck of the gallbladder or cystic duct, determining a surgical emergency clinically. In 4/5 of cases, there is an infective concause because bile juice culture shows bacilli.

10%: acalculous cholecystitis is often seen in males and following previous surgery, trauma, bacteremia (e.g., endocarditis), systemic arteritis, DM, TPN, or intravenous hyperalimentation.

- *GRO*: There are enlargement and tension of the gallbladder that may be bright red and gray on the surface due to fibrin covering. By opening the gallbladder, fibrin, pus and thick bilious juice may be seen and thickening of the wall by observing the gallbladder using a magnification lens.
- *CLM*: There may be neutrophilic infiltration of the wall, edematous wall, and fibrin on the surface (Fig. 5.3).

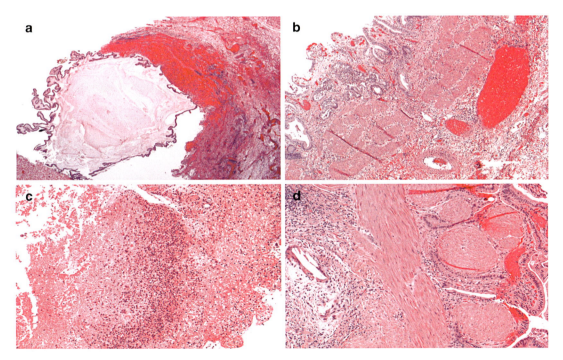

Fig. 5.3 Several patterns of cholecystitis in childhood and youth. Several patterns of cholecystitis in childhood and youth (**a–g**), hematoxylin and eosin staining; h, Luna staining; (**a**) ×12.5 original magnification; (**b**) ×50 original magnification; (**c**) ×100 original magnification; (**d**) ×100 original magnification; (**e**) ×100 original magnification; (**f**) ×50 original magnification; (**g**) ×200 original magnification; (**h**) ×200 original magnification). In (**a**) is shown a microphotograph of a hydrops of a gallbadder. The (**b**) and (**c**) microphotographs show an acute hemorrhagic cholecystitis, while the (**d**) and (**e**) microphotographs represent a chronic cholecystitis with foreign body giant cell granuloma with cholesterin crystals. The microphotographs (**f**) through (**h**) show an eosinophilic cholecystitis with numerous eosinophils highlighted by Luna staining, which is optimal to effectively demonstrate eosinophils in formalin-fixed and paraffin-embedded tissue sections

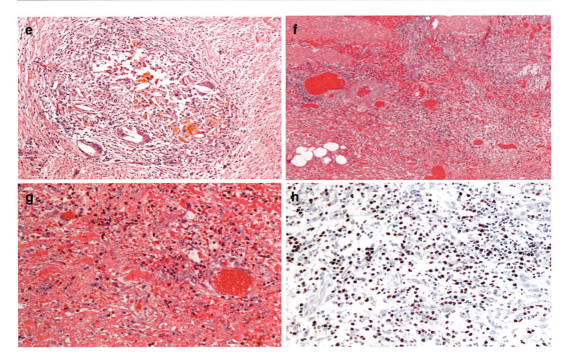

Fig. 5.3 (continued)

5.3.2.2 Chronic Cholecystitis

It is defined as a cholelithiasis-related chronic inflammation of the gallbladder with a female preference resulting from repeated mild acute inflammatory events. Chronic cholecystitis is, occasionally, in the absence of a positive history for colic pain at the right upper quadrant of the abdomen. Grossly, thickening of the wall is present, and microscopically, there is some degree of intramural chronic inflammation with entrapment of some epithelial crypts (aka *cholecystitis glandularis*), dystrophic calcification, subserosal fibrosis, as well as some overlapping features of acute cholecystitis.

Variants

- *Follicular cholecystitis*: If lymphoid follicles are present in the wall and transmurally. It would be interesting to compare with the follicular cystitis of the urinary bladder.
- *Eosinophilic cholecystitis* refers to the presence of many eosinophils with the potential to be linked to a parasitic infestation.
- *Xanthogranulomatous cholecystitis* refers to a pseudotumoral inflammatory condition, which is probably due to repeated events-linked extravasation of bile into the gallbladder wall that may be on its turn due to mucosal ulcers (neither muscularis mucosae nor submucosa are present in the gallbladder) or rupture of Rokitansky-Aschoff sinuses. Microscopically, there is a variable intramural pattern constituted by foamy macrophages, lymphocytes, plasma cells, neutrophils, foreign body multinucleated giant cells, and some fibroblasts according to the chronicity of the disease.

5.3.3 Gallbladder Proliferative Processes and Neoplasms

- *Cholesterolosis* and *cholesterol polyps* are parts of the same phenomenon and constituted by the accumulation of lipid-laden macrophages located in the lamina propria at the tips of the mucosal folds determining the gross finding of yellow flecks studding mucosal surface ("strawberry gallbladder"). There is no clinical significance if cholelithiasis and cholecystitis are not associated.

- *Inflammatory polyps* are sessile mucosal projections constituted by a fibrous stroma lined by a usually intact epithelium and stroma infiltration of chronic inflammatory cells (lymphocytes and plasma cells).
- *Mucocele* or *hydrops* is an event due to an obstruction of the outflow of the bile by a stone in the absence of acute inflammation. Grossly, the gallbladder is distended, and by opening the lumen, clear, watery, mucinous secretion is recognized, and the wall is stretched and atrophic.

Adenomyoma (localized, segmental, diffuse)/adenomyomatosis represents single or multiple proliferations of both epithelial (glandular) and smooth muscle elements with marked extension and deepening of the Rokitansky-Aschoff sinuses into the wall. Microscopically, there frequently are profoundly penetrating dilated glands surrounded by hyperplastic smooth muscle cells. In adenomyomatosis, there is a characteristic intraluminal trapping of crystals of cholesterol within the Rokitansky-Aschoff sinuses.

5.3.3.1 Metaplasia
- *Gastric metaplasia*: Pyloric mucosa type, antral mucosa type
- (size matters: pyloric gland metaplasia is $\varnothing < 5$ mm, while pyloric gland adenoma is $\varnothing \geq 5$ mm!)
- *Intestinal metaplasia*: From goblet cells mainly prevalent at the tips to a variable amount of goblet cells, columnar cells, Paneth cells, and endocrine cells
- *Squamous metaplasia*: Squamous epithelium (there is pseudostratification, which is basal only, and no significant atypia, otherwise there is dysplasia vide infra)

5.3.3.2 Dysplasia
Dysplasia is the atypical disordered proliferation of cuboidal/columnar epithelium showing loss of polarity and pseudostratification. Dysplasia is differentiated in low-grade (LGD) and high-grade dysplasia (HGD). In LGD, there is mild nuclear enlargement, mild nuclear pseudostratification, and only minimal nuclear irregularities. In HGD, there is a marked nuclear enlargement, prominent nucleoli, mitoses, nuclear hyperchromasia, and loss of polarity as marked nuclear changes. Complex architecture with ductal fusion and cribriform change would point to the diagnosis of HGD/CIS (*Carcinoma in situ*).

5.3.3.3 Neoplasms
Benign neoplasms are either papillomas or adenomas, and in both cases, they constituted localized benign overgrowths of lining epithelium. *Papillomas* have a stalk-like connection (pedunculated projections), while *adenomas* have a broad base.

Adenomas are classified as tubular, papillary, and tubular-papillary. Grossly, tubular adenomas are lobular in contour, while papillary adenomas are cauliflower-like. Another classification used is to subdivide adenomas in pyloric gland type, intestinal type, and biliary type. By definition, pyloric gland-type change with a size $\varnothing > 0.5$ cm should be classified as adenoma.

Other benign lesions, evidently a rarity in children or youth, are *paraganglioma*, *neurofibromas*, and *granular cell tumor*. Of all three, neurofibromas should be taken into consideration as a possible cause of symptomatology of the right upper quadrant of the abdominal topography in the setting of a syndromic condition known as neurofibromatosis. Finally, three lesions need to be further mentioned, because they are precancerous lesions, including *flat dysplasia*, *papillomatosis*, and biliary intraductal papillary mucinous neoplasms (IPMN) (higher potential for malignant transformation) (Fig. 5.4).

5.3.3.4 Gallbladder Carcinoma
- *DEF/EPI*: Malignant epithelial tumor of the gallbladder with 90% of patients older than 50 years, but rare cases have been described in youth.
- *EPG*: Cholelithiasis, porcelain gallbladder, cholecyst-enteric fistula, abnormalities of pancreatic biliary duct anastomosis, ulcerative colitis, and FAP as risk factors.
- *CLI*: Symptoms of gallbladder carcinoma include chronic abdominal pain, anorexia, and weight loss, and signs are palpable mass, hepatomegaly, and jaundice.
- *GRO*: The gallbladder carcinoma can be either diffuse (2/3 of cases) or papillary/polypoid (1/3 of cases) with cholelithiasis present in

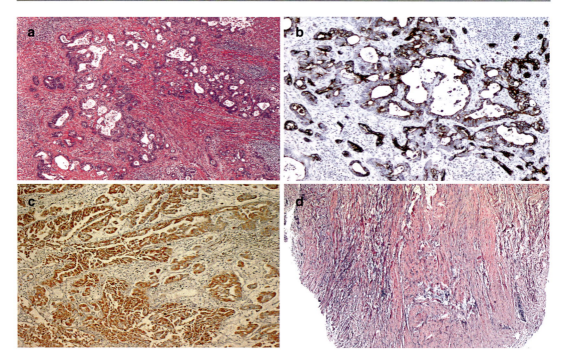

Fig. 5.4 Gallbladder carcinoma and Pseudotumor/Fibromatosis in adolescence and youth. In (**a**) is shown an atypical ductal proliferation of the gallbladder (hematoxylin and eosin staining, ×50 original magnification), which is highlighted by CK7 (Anti-CK7 immunostaining, Avidin-Biotin Complex, ×100 original magnification). The (**c**) microphotograph shows the atypical ductal proliferations, which are stained with an antibody against fibrocystin (anti-FP2 immunostaining, Avidin-Biotin Complex, ×100 original magnification), while (**d**) shows a fibromatosis of the gallbladder in a pediatric patient (hematoxylin and eosin staining, ×40 original magnification). Most neoplasms of the pancreas and bile ducts are in the category of CK7+/CK20− tumors, while carcinomas of the gall bladder show a coordinate expression of the two cytokeratins in about 1/4 of cases (Duval et al. 2000)

more than 4/5 of cases. However, 1% of cases constitute an incidentaloma as surprising finding during cholecystectomy for cholelithiasis. The diffuse form is practically indistinguishable grossly from chronic cholecystitis, while the papillary/polypoid form is identifiable, although it may show an extremely invasive basis in the wall or the liver bed.

- *CLM*: Histologically, two major types are the pancreaticobiliary type and the intestinal type of carcinomas that recall the classification of the ampullary carcinomas. Other phenotypes of carcinomas of the gallbladder are mucinous, clear cell, signet-ring cell squamous-adenosquamous, lymphoepithelioma (LE)-like carcinoma, and small and undifferentiated cell. Cytologically, they may be poorly differentiated. There are three ways of lymphatic drainage, including cholecysto-retro-pancreatic, cholecyst-celiac, and cholecyst-mesenteric.

Muduly et al. (2012) reported a case of carcinoma of the gallbladder in a 15-year-old boy, who presented with a history of pain in the right abdominal quadrant since the last 3 months and jaundice since the last 1 month with occasional fever, chills, and rigor. CT scan of the abdomen and pelvis revealed a mass lesion in the fundus and body of the gallbladder with the invasion of segment IV and V of the liver and multiple metastases throughout the liver parenchyma with the invasion of second part of the duodenum. Muduly et al. (2012) could identify just two case reports available in English literature and very few in non-English literature.

The diagnosis of gallbladder carcinoma is crucial, and this may be important in childhood as well. In gallbladder carcinoma, CIS can occur and extend into both the cystic duct and the Rokitanski-Aschoff sinuses and need to be carefully differentiated from invasive carcinoma

(→ 2nd opinion!). When dysplasia is found in gallbladder specimen, extensive/in toto embedding/sampling is required. In specific settings, jaundice is mostly due to a malignant obstruction of the biliary tree rather than a sign of a metastatic process or presence of a concurrent choledocholithiasis.

5.3.4 Extrahepatic Bile Duct Tumors and Cholangiocellular Carcinoma

5.3.4.1 Carcinoma of the Extrahepatic Bile Ducts

Cholangiocarcinoma/cholangiocellular carcinoma (CCA) is extremely rare in childhood. CCA has been reported in association with underlying disorders, mostly fluke infestation. The survival and prognosis are miserable, mainly in patients with unresectable or advanced stage CCA. Recently, Monsereenusorn et al. (2018) reported a Thai patient with CCA. The 11-years-old boy presented with progressive distension of the abdomen, which may suggest celiac disease in Western countries. However, the child had a history positive for eating uncooked fish, meat, and beef. Cisplatin and gemcitabine regimen for 1–2 cycles was started and his clinical conditions improved. Repeated CT scan of the abdomen demonstrated a shrunken mass, but the tumor recurred after the completion of 6 cycles of chemotherapy. The child's overall survival was 15 months.

5.3.4.2 Embryonal Rhabdomyosarcoma

Malignant mesenchymal tumor of the gallbladder/extra-hepatic biliary system (EHBS) that typically occurs in children younger than 5 years of age and arises in extrahepatic biliary tract (3/4 of cases in common bile duct) patients shows obstruction and jaundice. Grossly, a botryoid character is seen. Microscopically there is a botryoid morphology with characteristic *cambium layer* underneath the biliary epithelium of the biliary duct.

5.4 Pancreas Pathology

Pancreas pathology is a little extraordinary for the routine of a pediatric pathologist. In my experience, acute fulminant pancreatitis with hemorrhagic necrosis that I came across a few times in my pathology practice is one of the side effects of the "rave party" culture that find many adolescents as adepts today. Congenital hyperinsulinism may have an organic pathology that in the past was labeled as nesidioblastosis, and one of the most common pediatric pancreatic tumors is the entity known formerly as infantile pancreatic carcinoma, although the most commonly used term for this condition is pancreatoblastoma (Box 5.2). Other typical pediatric tumors may be solid and cystic. Thus, some of the following information should be very close to pediatric pathology program fellowships.

5.4.1 Congenital Hyperinsulinism and Nesidioblastosis

Congenital hyperinsulinism (*CHI*) is a significant cause of neonatal and infantile hypoglycemia with eight different genes leading to a life-threatening phenotype characterized by increased, dysregulated insulin secretion, coma, and severe brain

---

Box 5.2 Pancreas Pathology
5.4.1. Congenital Hyperinsulinism and Nesidioblastosis
5.4.2. Acute and Chronic Pancreatitis
5.4.3. Pancreatoblastoma, Acinar Cell Carcinoma, and Pancreas Ductal Carcinoma
5.4.4. Cysts and Cystic Neoplastic Processes of the Pancreas
5.4.5. PanIN and Solid Pancreas Ductal Carcinoma in Children and Youth
5.4.6. Other Tumors
5.4.7. Degenerative Changes and Transplant Pathology

Notes: *PanIN* pancreatic intraepithelial neoplasia

damage. The previous term, *nesidioblastosis*, which should not be used anymore, has been first identified by Laidlaw in 1938. The genetics of CHI and nesidioblastosis is also important, and nine genes are to mention *ABCC8, KCNJ11, GLUD1, GSK, HADH, SLC16, SLC16A1, HNF4A*, and *UCP2*. The CHI forms associated with *GLUD1* (encoding glutamate dehydrogenase), *GCK* (encoding glucokinase), *HADH* (encoding for L-3-hydroxy acyl-CoA dehydrogenase), *SLC16A1* (encoding the monocarboxylate transporter 1), *HNF4A* (encoding hepatocyte nuclear factor 4α), or *UCP2* (encoding mitochondrial uncoupling protein 2) constitutes rare forms of CHI.

- *EPI:* 1–1.4:50,000 live births (80–120 new cases/year in the USA) to 1:2500 live births in the Arabian Peninsula.
- *CLI*: The diagnosis relies on:
  1. Fasting and postprandial hypoglycemia (glucose <2.5–3 mmol/L) with associated unsuppressed insulin and c-peptide levels (insulin >1–2.0 mU/L)
  2. A positive response to the subcutaneous or IM administration of glucagon (glucose increase to 2–3 mmol/L)
  3. Harmful ketone bodies (acetone, acetoacetic acid, and β-hydroxybutyric acid) in urine or blood (plasmatic β-hydroxybutyrate <2.0 mmol/L and FFA <1.5 mmol/L)

Preoperatively, [^{18}F]-L-DOPA PET CT scan should be performed in all cases. Interventional radiology studies also include arterial stimulation with venous sampling (ASVS) or hepatic portal venous catheterization and selective sampling of the pancreatic veins (THPVS).

GRO/CLM: Two forms, including (*focal*) *nodular* form and *diffuse* form, are known. A nodule of proliferative islet cells that incorporate haphazardly or push exocrine elements aside characterizes the focal structure. Differently, from insulinoma, the protuberance of CHI retains the lobular architecture, and some exocrine elements are still present within the lesion. Abnormal islets containing 5–10% of cells with atypical (enlarged) nuclei throughout the pancreas characterize the diffuse form. Jaffe, Hashida, and Yunis (1980) identified four groups: Islet cell dysplasia (~70%), diffuse adenomatosis (<20%), no clinical or surgical pathology detectable (10%), and adenoma (very rarely). Although it has been reported a very rare occurrence, the adenoma needs to be kept in mind. Islet cell dysplasia (focal/diffuse) is characterized by the loss of the standard centrilobular arrangement of islets with scattered small aggregates of endocrine cells arranged haphazardly within the lobule and irregular borders of the islands, as well as the budding of small endocrine cell aggregates within acini and/or intralobular ducts as well as nuclear pleomorphism. In adenoma, there is one cell type and an encapsulated islet cell tissue. Adenomatosis is characterized by more than one cell type and no encapsulation of islet cell tissue. Finally, ductule-insular complexes or tubular-islet proliferations are seen and are constituted by extensive proliferation of intermingled ductular and endocrine cell clusters.

A most recent form of CHI has been labeled as mosaic form, because of the coexistence of two types of islets: There may be large islets with cytoplasm-rich β-cells and very few enlarged nuclei and shrunken islets with β-cells showing little cytoplasm and small nuclei. In particular, in small islands, β-cells had copious insulin content but little Golgi proinsulin molecules. Conversely, large islets showed low insulin storage and high proinsulin production and were mostly confined to a few lobules. The patients with mosaic histology exhibit some clinical features similar to infants presenting a loss-of-function mutation of the *HADH* gene. The specific and correct identification of changes of CHI β-cells relies on two critical parameters, including the size of the β-cell nuclei (mean nuclear radius) and the nucleus-cytoplasm ratio measured as the areas of β-cells' nuclear crowding. In the diffuse form, there are numerous abnormal β-cells with these two characteristics. Conversely, immense aggregations of islet cells, separated by thin rims of acinar cells or strands of connective tissue, may be observed. Typically, islet cells may also show large and hypertrophic nuclei.

- *DDX*: Infant of a mother with poorly controlled DM, Beckwith-Wiedemann syndrome (BWS), a glucose-based inborn error of metabolism (such as gluconeogenesis, glycogen storage disorder), birth asphyxia, hypopituitarism, and factitious

hyperinsulinemia (Munchhausen syndrome by proxy) are crucial differential diagnoses.
- *PGN*: Outcome depends from prompt diagnosis and successful therapy, including complete resection of the nodule with negative margins for the focal form or near-total (95–98%) pancreatectomy with preservation of the common bile duct for the diffuse form. Postoperative care is paramount and require insulin, glucose, and pancreatic enzyme replacement in case of pancreatic exocrine insufficiency. Annual investigations include assessment of residual insulin secretion, HbA1c values, and OGTT (oral glucose tolerance test). A hemoglobin A1c (HbA1c) test measures the amount of glucose attached to hemoglobin. HbA1c is given in percentages. A normal values is when HbA1c is below 5.7%, while a value of 6.5% or higher indicates diabetes. A pre-diabetic condition is considered when HbA1c is between 5.7% and 6.4%. Intensive prevention of hypoglycemia at birth prevents neurologic sequelae and this aspect harbors key forensic issues. Some tips and pitfalls of CHI are presented in Box 5.3.

5.4.2 Acute and Chronic Pancreatitis

5.4.2.1 Acute Pancreatitis
- *DEF*: Acute inflammation of the pancreatic gland.
- *EPI*: There is an incidence of 3.6–13.2 cases of acute pancreatitis per 100,000 pediatric individuals per year. Such value approaches the incidence of pancreatitis in adults. The mean age of patients with the disease is 9.2 ± 2.4 (SD) years ($\male:\female = 1:2$). It has also been estimated that 2–13 new cases occur annually per 10^5 children and about 1/4 of children with acute pancreatitis develop a severe complication with a mortality rate up to 10% despite substantial advances in the treatment of this dreadful inflammatory disease (Restrepo et al. 2016).
- *EPG*: The non-idiopathic etiology comprises gallstones, CH_3CH_2OH, trauma, mumps (paramyxovirus) and other viruses (EBV, CMV); autoimmune (PAN, SLE), scorpion sting and

> **Box 5.3 Pearls and Pitfalls!**
> - When you see hypoglycemia in the chart, then think of CHI.
> - If the surgeon does not find a suspected focal lesion using palpation and 3.5× magnification, intraoperative frozen sections of biopsies of the head, body, and tail are usually performed.
> - If the surgeon finds a suspected focal lesion, a partial pancreatectomy is performed with tissue sent to the pathologist for intraoperative frozen section confirmation of negative margins and to rule out tentacle-like extensions of the tumoral nodule.
> - The focal nodular form is associated with hemi- or homozygosity due to an inherited paternal mutation of the *SUR1* gene or *KIR6.2* gene and LOH for the maternal allele specifically in the hyperplastic islet cells. 11p15 LOH (maternal) leads to an unbalanced expression of 11p15 imprinted genes, such as growth factors and tumor suppressor genes. The focal form seems to be sporadic in most cases. Conversely, the diffuse type is a heterogeneous disease with involvement of *SUR1* or *KIR6.2*, which is usually recessively inherited or more rarely a dominantly inherited CHI. Dominant forms of CHI can also be due to mutations in the *GCK* gene, *GLUD1* gene, and *SLC16A1* and in the *HNF4A* when MODY is observed.

snake bites, as well as brown recluse spider bites; hypercalcemia {such as in patients with parathyroid adenoma/carcinoma}, hyperlipidemia, hypertriglyceridemia, and hypothermia; and iatrogenic, including ERCP and drugs (steroids, sulfonamides, azathioprine, NSAIDs, furosemide, thiazides, and estrogens). In case of a gallstone, a stone in the common bile duct may provoke obstruction of the excretion of the pancreatic gland secretions determining a backflow with digestive juice leading to lysis of pancreatic cells and secondary recruitment of inflammatory cells:

1. Duct obstruction in the outflow of pancreatic juices from the pancreas into the duodenum (e.g., stone, tumor) causing a backflow, which is cell lytic due to the presence of powerful enzymes
2. Acinar cell injury (e.g., EtOH, viruses, ischemia)
3. Defective intracellular transport and activation of enzymes

Acinar cell injury induces activation of proteolytic enzymes, subsequent tissue destruction (aka "acinar cell homogenization"), as well as interstitial edema and inflammation, proteolysis (→ necrosis), fat dissolution (fat necrosis → saponification), and hemorrhage.

In teenagers or youth going to "rave" parties, there is not only a risk of the alcoholic coma but also that of acute hemorrhagic pancreatitis or acute pancreatic necrosis due to the excess of alcohol drinking and causing acute or hyperacute acinar cell injury. "Rave" parties with the illicit consumption of heroin, cocaine, ecstasy, marijuana with or without combination of alcohol pose serious concerns to teenagers and youth in terms of hepatotoxicity and pancreatic toxicity (Pontes et al. 2008; Figurasin & Maguire 2020).

- *GRO*: There are edema and enlargement of the pancreas, and fat necrosis is recognized at the time of surgery or autopsy as gray white-yellow small areas (pinpoint) ± hemorrhagic parenchymal disease according to the extension of the tissue damage.
- *CLM*: It is easy to separate a mild from severe damage, although most of the acute pancreatitis-related deaths in ICU will show extensive severe damage at autopsy (an autopsy may be particularly relevant to perform for insurance issues and to deal with life insurance policies) (Figs. 5.5 and 5.6).
- Mild damage: Interstitial edema + fat necrosis, superficial.
- Severe damage: Extensive hemorrhagic necrosis involving both pancreatic tissue and extrapancreatic, periglandular, and omental fat tissue.
- Non-mortal or "nearly deadly" complications include systemic and local complications.

- Systemic: Single to multiorgan failure (MOF), ARDS, DIC, and ARF (mortality: 20%).
- Local: Pancreatic abscess, pancreatic pseudo-abscesses, duodenal obstruction, and chronic pancreatitis. The evolution from an acute to a constant process is not usual in the youth but it has occasionally been reported.

5.4.2.2 Chronic Pancreatitis

It is a constant process, usually not associated with acute pancreatitis and showing ductal system and islets that are spared initially. Three major types are as follows:

- *Chronic obstructive pancreatitis* characterized by narrowing or occlusion of the duct (e.g., cholelithiasis or carcinoma) and pancreatic damage (milder than regular calcifying variety) with acinar damage relatively uniformly throughout the exocrine gland.
- *Chronic calcifying pancreatitis* characterized by irregular and patchy damage (e.g., EtOH) with ductal system dilatation, ductal squamous metaplasia, intraluminal protein plugs with or without calcification, acinar dilatation, acinar sub-atrophy and atrophy, interlobular fibrosis as well as often pseudocystic change.
- *PRSS1-related hereditary pancreatitis* is acute pancreatitis that progresses from acute (sudden onset; duration <6 months) to recurrent acute (>1 episode of acute pancreatitis) to chronic (duration >6 months) and is associated with a genetic variation of the *PRSS1* gene. Acute pancreatitis occurs by age 10, while a chronic inflammatory process takes place by age 20.
- Pancreatitis complications include widespread, lipase-associated fat tissue necrosis of mostly subcutaneous tissue, mediastinum, pleura, pericardium, and liver, erythema nodosum-like panniculitis, islet cell loss following an initial proliferation of islet cells, and avascular bone necrosis.

Pancreatitis Variants/Special Types

Pancreatitis, Lymphoplasmacytic Sclerosing Type
♂ > ♀, ± AI (PSC, Sjogren disease), ± multifocal inflammatory fibrosclerosis of the pancreas char-

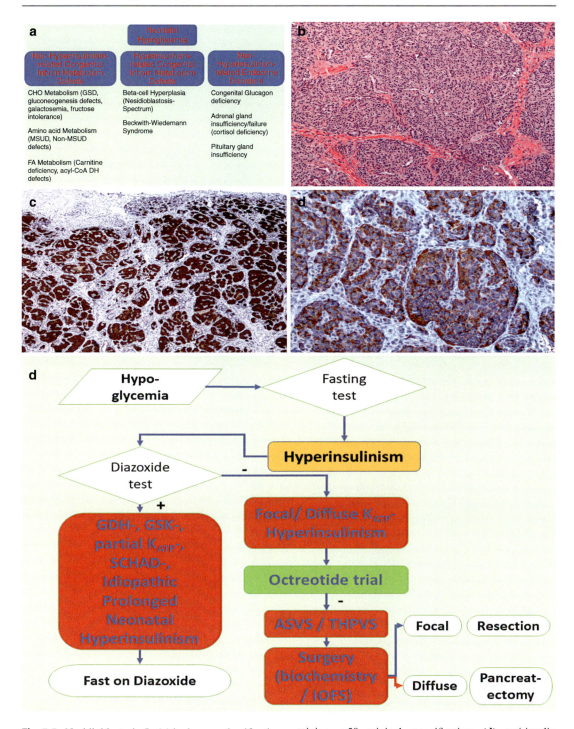

Fig. 5.5 Nesidioblastosis. In (**a**) is shown a classification of neonatal hypoglycemia. A microphotograph of nesidioblastosis is shown in (**b**) and immunohistochemical positivity for synaptophysin and insulin are shown in (**c**) and (**d**), respectively (**c**–**d**: **c**, anti-synaptophysin immunostaining, ×50 original magnification; (**d**) anti-insulin immunostaining, ×200 original magnification). In (**e**) is shown the flow chart to manage a patient with neonatal hypoglycemia

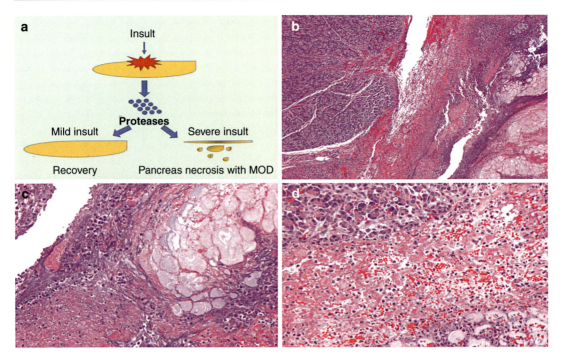

Fig. 5.6 Pancreatitis. In (**a**) is shown a cartoon with the etio-pathogenesis of acute pancreatitis, while the microphotographs (**b**) through (**d**) show the necrosis of the pancreas in the event of an acute pancreatitis (b-d, hematoxylin and eosin staining, ×50, ×200, and ×200, respectvely). In b) through d) is shown the histology of a trauma-related acute pancreatitis in a pediatric patient. There is evidence of digestion of the parenchyma, including hemorrhage, necrosis, fat necrosis, and neutrophils

acterized by one or more of the following processes such as chronic sclerosing sialoadenitis, Riedel thyroiditis, primary sclerosing cholangitis, Sjogren's disease, orbital pseudotumor, mediastinal and retroperitoneal fibrosis, an increase of serum IgG4, as well as (+) steroid-responsiveness.

- *CLM*: "Ducto-centric" dense L/PC inflammation (IgG4 PC >10/HPF) + periductal fibrous tissue expansion + *peri*-phlebitis/obliterative venulitis.

Pancreatitis, Granulocytic Epithelial Lesion Type
♂ = ♀, ± UC, normal IgG4 (it is *not* an IgG4-related AI pancreatitis)

- *CLM*: There is intraepithelial neutrophilia ± eosinophilia (i.e., intraepithelial inflammation constituted by neutrophils with or without eosinophils) extending into the ductules, acini, and interstitial tissue.

- *Paraduodenal Pancreatitis* (Cystic Dystrophy of Heterotopic Pancreas, CDHP) = Cystic Dystrophy of Duodenal Wall. It occurs in a male with a Hx (+) for EtOH abuse, waxing and waning of severe upper abdominal pain, and unwarrantable postprandial vomiting. Microscopically, there is dense myoid stromal proliferation with intervening rounded lobules of pancreatic tissue, characterizing the lesion as "myo-adenomatosis" + Brunner gland hyperplasia + small cysts lined by cellular stroma preferentially located in the area of the *papilla minor*.

5.4.3 Pancreatoblastoma and Acinar Cell Carcinoma in Childhood and Youth

The most frequent pediatric tumors in childhood and youth are pancreatoblastoma and pseudopapillary tumor of the pancreas. Rarely, some other entities have been reported, including the acinar cell carcinoma.

5.4.3.1 Pancreatoblastoma
- *DEF*: A pancreatic tumor closely related to acinar cell carcinoma (ACC) and associated with BWS and showing cell differentiation toward all three lineages (acinar, ductal, and neuroendocrine), which also emphasizes the multi-phenotypic distinction of this tumor.
- *SYN*: Infantile pancreatic carcinoma.
- *EPG*: There is an activation of APC/β-Cat pathway and chromosome 11p alterations with abnormal (N/C) immunolabeling for β-Cat and overexpression of Cyclin-D1, which is more strikingly expressed in the squamoid morulae.
- *GRO*: Partially encapsulated and lobulated tumor with grey cut surface.
- *CLM*: There are well-formed glands (ACC-like) arranged around solid nests of cells with squamoid features (aka "squamoid morulae or corpuscles") looking like an islet cell tumor or carcinoid but devoid of granules by TEM (Fig. 5.7).
- *IHC*: (+) Trypsin (acinar differentiation), (+) CK (ductal differentiation), and (+) (CGA, SYN) (focal neuroendocrine differentiation), (+) VIM, AAT (at least focally) but (−) CD10, PR (DDX: Solid Pseudo-Papillary Tumor SPPT).
- *PGN*: Favorable, although size, direct extension in neighboring organs and metastases may be encountered and should be discussed at the pediatric tumor boards

5.4.3.2 Acinar Cell Carcinoma in Childhood and Youth

Acinar cell carcinomas (ACCs) of the pancreas are rare pancreatic neoplasms accounting for about 15% in children and adolescents. Grossly, they are large at the time of presentation, well circumscribed, fleshy, and pink to tan and typically contain some areas of necrosis and/or hemorrhage. Microscopically, ACC may have different histological characteristics. There is an acinar architectural pattern, characterized by cells forming structures resembling normal pancreatic acini; a glandular pattern, characterized by proliferation of cells, sometimes arranged in multiple layers, forming glandular structures; and a trabecular pattern showing ribbons of cells actively mirroring the morphology of pancreatic neuroendocrine tumors. Also, there is a solid pattern, which shows large sheets of poorly differentiated cells without lumens. In two children, no allelic loss of chromosome 11p and no mutations in the APC/β-catenin pathway were identified. This situation may be different from the adult population (Abraham et al. 2002; Bergmann et al. 2014; La Rosa et al. 2015).

5.4.4 Cysts and Cystic Neoplastic Processes of the Pancreas

In this section, we will deal with cysts, pseudocysts, and cystic neoplasms of the pancreas. The differential diagnosis of cysts and cystic neoplastic processes of the pancreas includes *intraductal papillary mucinous neoplasms* (IPMN), *mucinous cystic neoplasms* (MCN), *serous microcystic adenoma* (PSMA), *pancreatic intraepithelial neoplasia* (PanIN), *cystic pancreatic ductal adenocarcinoma* (CPDA), and *pancreatic pseudocysts* (PPC). These entities are not a topic in pediatric pathology but the environmental burden has become massive and pollution overwhelming in some countries. Thus, some adult entities are more and more appearing in pediatric, adolescent, or young age. Moreover, the knowledge of family cancer syndromes and knowledge of a specific syndrome be relevant for genetic counseling. In fact, the cystic form of a classic pancreatic ductal adenocarcinoma is practically inexistent in childhood or youth, but it may occur in young individuals with a predisposing genetic condition and a pancreatic mass.

5.4.4.1 Pancreas Pseudocysts (PPC)
- *DEF*: Non-epithelial-lined cavities (cysts are true cysts with epithelial lining!) without connection with the ductal system and accounting for >75% of all cystic lesions in the pancreas with ♂ > ♀ preference in the setting of pancreatitis (a complication of chronic pancreatitis),

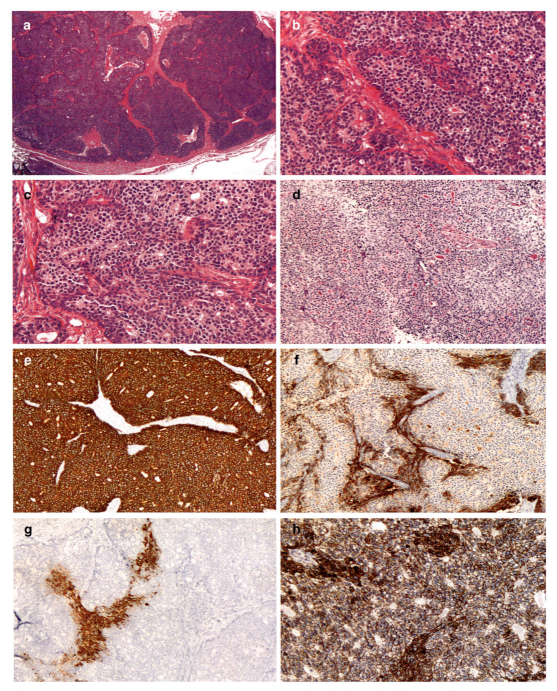

Fig. 5.7 Pancreatoblastoma. A small round blue cell tumor with pseudorosettes is seen in the event of a pancreatoblastoma (**a–c**, hematoxylin and eosin staining, (**d**) periodic acid Schiff staining; ×12.5, ×200, ×100, ×100 as original magnification, respectively). The immunohistochemistry (**e–h**) with pankeratins, carcinoembryonic antigen (CEA), neuron specific enolase (NSE), and β-catenin confirms the diagnosis of pancreatoblastoma (**e**) anti-pankeratin immunostaining, Avidin-Biotin complex, ×100 original magnification; (**f**) anti-CEA immunostaining, Avidin-Biotin complex, ×100 original magnification; (**g**) anti-NSE immunostaining, Avidin-Biotin complex, ×100 original magnification; (**h**) anti-β-catenin immunostaining, ×200 original magnification)

trauma, and occasionally neoplastic duct obstruction (Box 5.4).
- *EPI*: Childhood and youth.
- *EPG*: Two phases:
  1. Intraluminal necrotic debris, precipitated acinar secretions, and granulation tissue
  2. Stroma proliferation with cellular fibrosis (the cellularity may be quite intense and be a challenge at frozen section, and cellularity may also mimic ovarian-type stroma)

Pseudocystic fluid is necrotic but also rich in amylase.

± Extra-pancreatic extension (pancreatic pseudocysts may have an extra-pancreatic extension with peripancreatic hemorrhagic fat necrosis).

- *PGN*: Rupture, erosion (→ spleen), bleeding, shock.

As stated above, both mucinous cystic neoplasms (MCN) and intraductal papillary mucinous neoplasm (IPMN) are adult neoplasms, but occasionally they can be found in adolescent individuals and youth. Linings are serous, mucinous, squamous, acinar, endothelial, and degenerated. Pseudocysts are localized collections of pancreatic secretions, lack of an epithelial lining evidencing a lining of macrophages or fibrosis, arise practically after one or more episodes of acute or chronic pancreatitis, measure up to several centimeters, harbor no risk for malignant degeneration, but are critical mimickers of pancreatic neoplasms. Ductal Lineage Markers include mucin, mucin-related glycoproteins and oncoproteins, MUCs, CA19-9, CEA, B72.3, and DUPAN-2. PanIn1a/b-2-3 may be considered precancerous lesions for invasive ductal carcinoma and should be defined as tiny (Ø ≤ 0.5 cm) epithelial proliferations within the smaller pancreatic ductal system with short and stubby papillae showing lobocentric atrophy and scarring in the lobular units, which are located downstream of PanIN lesions due to obstruction and lack of (forward)flow (DDX: ACA, which shows desmoplastic tissue reaction). If there is a loss of DPC4 (SMAD4), there is a high probability of ACA!

IPMN has two primary variants identified as branch duct type and as main duct type. Useful criteria to distinguish between MCN and IPMN are illustrated in Box 5.5.

Polycystic disease and VHL are different from serous cystadenoma in VHL. Individual cysts are practically indistinguishable from serous tumors. They are more widely spread in the pancreas; they are often multifocal; and there is an apparent association with cysts in other organs.

Box 5.4 The Distinctiveness of Some Cystic Neoplasms of the Pancreas

	IPMN	MCN	SMAP
Age	>50	18–95	40–70
Gender	♂ ≥ ♀	♂ << ♀	♂ < ♀
Site	Head	Tail/body	Body/tail
U/M	U/M	U/M	OC/PC
DC	+	−	−
CLI	ACA	None	None
MIC	L→HGD	L→HGD	Gly-Cyto

Notes: *IPMN* intraductal papillary mucinous neoplasm, *MCN* mucinous cystic neoplasm, *SMA* serous microcystic adenoma of pancreas, *PAS* periodic acid – Schiff, *DC* ductal communication (either main pancreatic duct or a branch of the pancreatic ductal system), *L > HGD* Low-grade dysplasia > Borderline > High-grade dysplasia, *CLI* clinical features, *MIC* microscopy, *U/M* unifocal/multifocal, *OC/PC* oligocystic/polycystic, *Gly-Cyto* glycogen-rich clear cell cytology with distinct cytoplasmic borders, and round, uniform nuclei with dense homogenous chromatin

Box 5.5 Useful Criteria for the Distinction of MCN from IPMN

Features	MCN	IPMN
Gender	♂ << ♀	♂ ~ ♀
Endoscopy	None	Mucin extrusion
Radiology	Thick walled	Cystically dilated duct
Location	Tail	Head
Ductal communication	No	Yes
MIC	Ovarian-like stroma	Cystically dilated ducts
PGN		

In the differential diagnosis, one should also include endocrine neoplasms such as insulinoma, gastrinoma, glucagonoma, and nonfunctioning endocrine tumors. *Rare exocrine neoplasms* include solid papillary epithelial neoplasm and giant cell tumor. *Non-epithelial neoplasms* also include sarcoma and metastases to the pancreas. Para-pancreatic neoplasms should also be taken into consideration. Box 5.6 lists the criteria to differentiate between pseudocysts and cystic neoplasms of the pancreas.

The *solid (and cystic) pseudopapillary tumor of the pancreas* has a typical peak in female adolescents and youth.

- *DEF*: Low-grade malignant neoplasm arising from centroacinar cells (DOG1+) with characteristic involvement of young women, showing poorly cohesive monomorphic epithelial cells forming fragile solid and pseudopapillary structures with a tendency to bleeding and relatively good PGN.
- *SYN*: Frantz tumor, Hamoudi tumor.
- *EPI*: 1–2% of non-endocrine tumors of the pancreas occur in the 3rd–5th decades with female preference (♂ < ♀) and no associated clinical syndrome.

Box 5.6 Differentiation Between Pseudocysts from Cystic Neoplasms of the Pancreas

Criterion	Pancreatic Pseudocysts	Pancreatic Cystic Neoplasms
History	Previous pancreatitis, trauma	None
Locules	Single cyst, non-loculated	Multiloculate
Wall thickness	≤0.5 cm	>0.5
Duct-cyst connection	>2/3	None
Fluid	Low-viscosity with ↑amylase	High-viscosity with ↓amylase
Cytology	Inflammatory type	Atypical type
Cyst wall	Fibrosis	Malignant cells

- *FNA*: Cellular smear with both single cells and small loose clusters of scattered papillary structures with delicate fibrovascular cores, which may contain metachromatic material and hyaline globules. Fine chromatin and grooved nuclei are often seen.
- *CLM*: PEN or CNS-ependymoma-like with pseudopapillae with hyalinized fibrovascular cores lined by several layers of epithelial cells with bland cytology and imparting a delicate and discohesive appearance (abnormality of the E-Cad-β-Cat complex). Tumor cells have round/oval nuclei with finely stippled chromatin, nuclear grooves, indistinct nucleoli, and no/low MI. Moreover, they have clear to the eosinophilic cytoplasm with intracytoplasmic PAS + hyaline globules, and may exhibit a mucinous change within the core. Other cell types include foamy cells and FBGC surrounding lipid/cholesterol crystals. Of note is the lack of any stromal reaction surrounding the infiltrations of tumor cells, which is usually a characteristic of malignancy (Fig. 5.8).
- *IHC*: (+) VIM, CD10, CD56, CK (patchy) and (+) ER/PR and (+) β-Cat (N/C), E-Cad, CCND1, CD99 (paranuclear) and (+) claudin 2 c, claudin 5 m (−) CGA, CEA, acinar and ductal markers (Fig. 5.9).
- *TEM*: Large ED granules with multiple internal membranous and granular inclusions as well as a collection of AAT-granular material in ER (AAT, alpha-1-antitrypsin; ED, electron-dense; ER, endoplasmic reticulum).
- *CMB*: Mutations in exon 3 of β-Cat gene.
- *DDX*. AcCC: Acinar formations, prominent nucleoli, brisk MI, (+) trypsin, chymotrypsin, lipase.
- Adrenal cortical neoplasms: (+) CK, IN-A.
- Pan-END-T: Lack of the SPT-features (pseudopapillae, intracytoplasmic hyaline globules, longitudinal grooves) and (−) CD10.
- β-Cat(n) Pan-Pseudocyst: No epithelial cells lining the pseudocystic structures.
- *PGN*: Low-malignant neoplasm, which is usually treated with wide excision. Factors associated with poor outcome (MTX in 1/10),

angioinvasion, G3, and necrosis, although patients typically survive even with metastases.

5.4.5 PanIN and Solid Pancreas Ductal Carcinoma

Pancreatic carcinoma harbors a poor prognosis, which is attributed to failure to diagnose cancer early, while the tumor may be able to be resected to its propensity to disseminate quickly and to its resistance to systemic therapy. Although CA19-9 is a useful marker for pancreatic cancer, there are some caveats. CA19-9 is seen as increased in several benign diseases, including acute and chronic pancreatitis, liver cirrhosis, cholangitis, and obstructive jaundice, in multiple types of adenocarcinoma, no expression in subjects harboring Lewis a- b-genotype, no or low sensitivity for early or small diameter pancreatic cancer, and poorly differentiated pancreatic carcinomas. Thus, the histologic workup is mandatory to confirm a diagnostic suspicion of imaging. Pancreatic intraepithelial neoplasia can develop in youth with predisposing defects of tumor suppressor genes. PanIN lesions show alterations in *K-RAS, P16, DPC4,* and *TP53*. PanIN1a/PanIN1b lesions are typically associated with *K-RAS* gene mutations; PanIN2 lesions are, conversely, associated with *p16/CDKN2A* inactivation; and PanIN3 lesions are, alternatively, associated with *DPC4 (SMAD4)* and *TP53* loss.

The most essential serum markers of pancreatic cancer are CA242, CEA (carcinoembryonic antigen), TPA (tissue polypeptide antigen), TPS (tissue polypeptide-specific antigen), M2-pyruvate kinase, MIC-1 (macrophage inhibitory cytokine), IGFBP-1 (insulin growth factor binding protein 1), Du-Pan, haptoglobin, and serum amyloid A. The most crucial tissue markers of pancreatic cancer are K-ras, p-53, MUC-1, MUC2, MUC5AC, p21, SMAD4, and Bcl2. The features identified by electron microscopy in pancreatic neoplasms are presented in Box 5.7.

5.4.6 Other Tumors

Other tumors include pancreas endocrine neoplasms (Fig. 5.10) as well as mesenchymal, hematolymphoid, and melanocytic tumors.

5.4.7 Degenerative Changes and Transplant Pathology

5.4.7.1 Diabetes Mellitus

- *Diabetic pancreatic changes*: Decreased number and size of islets, insulitis, and amyloid
- *Diabetic macrovascular disease*: Accelerated atherosclerosis with early gangrene of the limbs and acute myocardial infarction
- *Hyaline arteriolosclerosis*
- *Diabetic microangiopathy*

There is diffuse thickening of the basal membrane (BM) in several organs and tissues, includ-

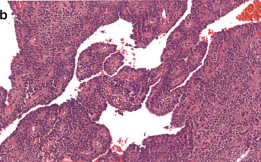

Fig. 5.8 This locally aggressive neoplasm of the pancreas may be encountered in female adolescents and can replace part of the pancreas as shown in the gross photograph in (**a**). The solid and pseudopapillary nature is seen in the microphotograph (**b**) (hematoxylin and eosin staining, ×100 original magnification)

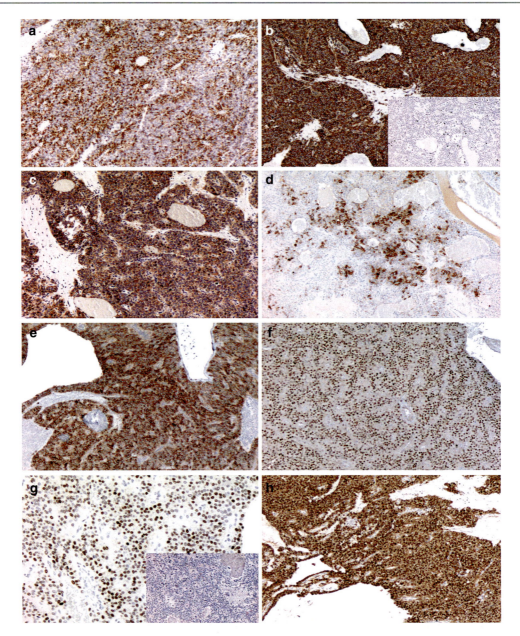

Fig. 5.9 The solid pseudopapillary tumor of the pancreas needs to be confirmed by immunohistochemistry (**a–h**) using cytokeratins, vimentin, neuron-specific enolase (NSE), alpha-1-antitrypsin (AAT), CD10, Cyclin D1, progesteron receptor, and β-catenin (**a**) anti-AE1-3 immunostaining, Avidin-Biotin Complex, ×100 original magnification; (**b**) anti-Vimentin immunostaining, Avidin-Biotin Complex, ×100 original magnification; (**c**) anti-NSE immunostaining, Avidin-Biotin Complex, ×100; (**d**) anti-AAT immunostaining, Avidin-Biotin Complex, ×100 original magnification; (**e**) anti-CD10 immunostaining, Avidin-Biotin Complex, ×100 original magnification; (**f**) anti-cyclin-D1 immunostaining, Avidin-Biotin Complex, ×100 original magnification; (**g**) anti-progesteron receptor immunostaining, ×200 original magnification; (**h**) anti-beta-catenin immunostaining, ×100 original magnification). The insets in microphotographs (**b**) and (**g**) show the low proliferation rate of this tumor (**b**) inset; anti-Ki67 immunostaining, Avidin-Biotin Complex, ×100 original magnification) and the absence of expression of estrogen receptor (**g**) inset; anto-estrogen receptor immunostaining, Avidin-Biotin Complex, ×original magnification)

ing the skin, skeletal muscle, retina, renal glomeruli, and renal medulla.

1. *Nephropathy*: Glomerular lesions (capillary BM thickening, diffuse mesangial sclerosis, nodular glomerulosclerosis), hyaline arteriolosclerosis (afferent and efferent arterioles), and medullary lesions (necrotizing papillitis or papillary necrosis and pyelonephritis)

Box 5.7 Ultrastructural Features of Pancreatic Neoplasms

Differentiation	Electron microscopy findings
Acinar	Large zymogenic granules (125 nm–1 µm) and RER prominence
Ductal	Microvillous differentiation with lumen and RER prominence
Endocrine	Neurosecretory granules

Note: RER, rough endoplasmic reticulum

2. *Neuropathy*: Distal symmetric sensory/sensory motoric neuropathy, autonomic neuropathy, and focal/multifocal asymmetric neuropathy
3. *Retinopathy*: Nonproliferative (exudative) and proliferative forms

COD of patients with DM: Acute myocardial infarction or renal failure

IgG4-related AI diseases: AI pancreatitis, lymphoplasmacytic sclerosing type, chronic sclerosing sialadenitis (Kuettner tumor), PSC, IBD, retroperitoneal fibrosis, and Riedel thyroiditis

5.4.7.2 Pancreas Transplant

Pitfalls are scattered chronic inflammatory cells in the form of a sparse lymphocytic infiltrate that is always seen and should not raise any concern. Acute rejection features include patchy inflammation in acinar tissue and ductal system, cell dropout, interstitial fibrosis, cellular atypia, and tissue

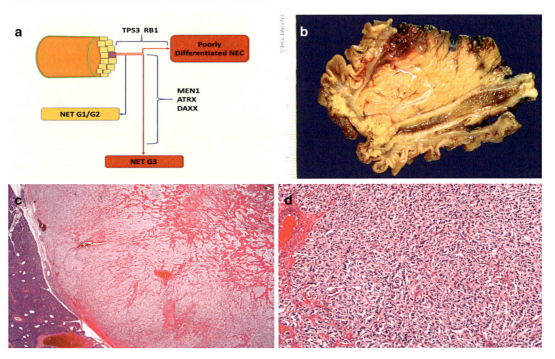

Fig. 5.10 Pancreatic Neuroendocrine Tumor. The signaling pathways of the pancreas neuroendocrine tumor (NET) are illustrated in (**a**). The gross photograph (**b**) shows a NET with infiltration of the ductus pancreaticus major (or duct of Wisung), while the microphotographs of this pancreatic NET are shown in (**c**)–(**d**) (**c–d**, hematoxylin and eosin staining, ×12.5 and ×100 as original magnifications). In (**d**) the tumor cells are arranged as ribbons and trabeculae. They are medium-sized cells with amphophilic (or eosinophilic) cytoplasm and uniform, central, oval, with finely stippled chromatin. The trabeculae and ribbons evidence a rich vascular network

capillaritis. Chronic rejection features include endarteritis with foam cell accumulation.

Multiple Choice Questions and Answers

- PBL-1 The current classification of choledochal cyst lists five types. Match each type with the correct definition.
  (a) Type 1
  (b) Type 2
  (c) Type 3
  (d) Type 4
  (e) Type 5
  1. One or more cysts of intrahepatic ducts (Caroli disease)
  2. Segmental or diffuse fusiform dilatation of common bile duct
  3. Dilatation of the intraduodenal common bile duct (choledochocele)
  4. Diverticulum of common bile duct
  5. Multiple cysts of extrahepatic bile ducts with or without cysts of intrahepatic ducts
     (a)(2) (b)(4) (c)(3) (d)(5) (e)(1)
- PBL-2 An eosinophilic cholecystitis is an infrequent condition. Apart from the idiosyncratic reaction to biliary contents, it is less commonly observed due to other conditions. Which of the following conditions is NOT associated with eosinophilic cholecystitis?
  (a) Atopy
  (b) Hydrocortisone
  (c) Cephalosporin
  (d) Erythromycin
  (e) Hypereosinophilia syndrome
  (f) Parasitic infestation
- PBL-3 A 15-year-old obese girl arrives at the emergency department of the children's hospital with acute onset of severe abdominal pain following her dinner. Her abdominal pain is in the right upper quadrant. Which of the following symptoms is most specific for the diagnosis that you are considering for this patient?
  (a) Hemoccult-positive stool
  (b) Right upper quadrant pain with a wavy pattern, which last over hours
  (c) Abdominal wall, which is rigid and shows a diffuse rebound tenderness
  (d) Obesity-related epigastric post-prandial pain
- PBL-4 Which immunohistochemical marker can be used to study the pancreas nerve innervation during the intrauterine development?
  (a) S100
  (b) NSE
  (c) Neurofilament
  (d) CD56
  (e) CD57
- PBL-5 Which of the following is NOT a complication of primary sclerosing cholangitis?
  (a) Fat-soluble vitamin deficiencies
  (b) Metabolic bone disease (osteoporosis)
  (c) Cholangiocellular carcinoma
  (d) Gallbladder carcinoma
  (e) Dominant biliary strictures
  (f) Cholangitis
  (g) Cholelithiasis
  (h) Portal hypertension
  (i) Pancreas carcinoma
  (j) Hepatocellular carcinoma in patients with liver cirrhosis
  (k) Colon-rectal carcinoma in patients with concomitant ulcerative colitis
- PBL-6 In the last couple of decades, binge drinking became the most common, costly, and probably deadly pattern of excessive alcohol use in the USA and Canada. Acute pancreatitis is a dreadful disease in children and adolescents and is not a rare condition anymore. All of the following conditions are associated with acute pancreatitis except one.
  (a) Alcohol ingestion
  (b) External trauma (e.g., car accident)
  (c) Internal trauma (e.g., abdominal surgery)
  (d) Diseases of the biliary tract
  (e) Alpha-1-antitrypsin deficiency
- PBL-7 Which of the following characteristic pathologic features of the cystic fibrosis of the pancreas does NOT belong to this disease?
  (a) Accumulation of secretion
  (b) Acute pancreatitis
  (c) Dysplasia of the ductal epithelium
  (d) Exocrine pancreatic gland atrophy

(e) Exocrine pancreatic gland atrophy with lipomatosis
(f) Fibrosis with total obliteration of the exocrine glands and ducts.
- PBL-8 The following statements correspond to pancreatoblastoma apart from which one?
  (a) Mixtures of acini and "squamoid corpuscles" with or without endocrine or ductal features
  (b) Highly cellular tumor with uniform epithelial cells in sheets and nests with acini/ducts
  (c) Hypercellular stroma
  (d) Bone/cartilage islands
  (e) Neuroblasts
- PBL-9 Which of the following parameters is associated with a favorable prognosis for gallbladder carcinoma?
  (a) Papillary growth pattern
  (b) The small cell growth pattern
  (c) Angiolymphatic invasion
  (d) Tumor budding
  (e) Dedifferentiation
- PBL-10 A 16-year-old girl presents with vague abdominal discomfort undergoes imaging, which reveals a pancreatic tumor. Histologic section of the resected mass shows solid areas with uniform cells surrounding capillary cores with the unique pseudopapillary formation in addition to necrotic/hemorrhagic cystic areas. The tumor cells are loosely cohesive, round to oval, and intermingled with eosinophilic hyaline globules. An immunohistochemical investigation reveals positivity for β-catenin expression (nuclear and cytoplasmic), loss of membrane staining (with nuclear localization) with E-cadherin, and lack of trypsin or chymotrypsin. Which of the following is the most likely diagnosis?
  (a) Pancreatoblastoma
  (b) Solid pseudopapillary tumor
  (c) Acinar cell carcinoma
  (d) Medullary carcinoma variant of pancreatic ductal adenocarcinoma
  (e) Hepatoid carcinoma variant of pancreatic ductal adenocarcinoma

References and Recommended Readings

Abraham SC, Wu TT, Hruban RH, Lee JH, Yeo CJ, Conlon K, et al. Genetic and immunohistochemical analysis of pancreatic acinar cell carcinoma. Frequent allelic loss on chromosome 11p and alterations in the APC/β-catenin pathway. Am J Pathol (2002) 160:953–62.10.1016/S0002-9440(10)64917-6.

Adzick NS. Congenital hyperinsulinism. Chapter 79. p. 611–616. In: Mattei, P., Nichol, P.F., Rollins, II, M.D., Muratore, C.S. (Eds.). Fundamentals of pediatric surgery. Springer Publ. 2017.

Adzick NS. Surgical treatment of congenital hyperinsulinism. Semin Pediatr Surg. 2020;29(3):150924. https://doi.org/10.1016/j.sempedsurg.2020.150924. Epub 2020 May 17. PMID: 32571515.

Amella C, Cappello F, Kahl P, Fritsch H, Lozanoff S, Sergi C. Spatial and temporal dynamics of innervation during the development of fetal human pancreas. Neuroscience. 2008;154(4):1477–87. https://doi.org/10.1016/j.neuroscience.2008.04.050.

Alagille D. Intrahepatic biliary atresia (hepatic ductular hypoplasia). In: Berenberg SR, editor. Liver diseases in infancy and childhood. The Hague: Martinus Nijhoff Med. Division; 1976. p. 129–42.

Alonso-Lej F, Rever WB, Pessagno DJ. Congenital choledocal cysts, with a report of 2, and analysis of 94 cases. Int Abstr Surg. 1959;108:1–30.

Arnoux JB, de Lonlay P, Ribeiro MJ, Hussain K, Blankenstein O, Mohnike K, Valayannopoulos V, Robert JJ, Rahier J, Sempoux C, Bellanné C, Verkarre V, Aigrain Y, Jaubert F, Brunelle F, Nihoul-Fékété C. Congenital hyperinsulinism. Early Hum Dev. 2010;86(5):287–94. Epub 2010 Jun 13. Review.

Bale PM, Kann AE, Dorney SFA. Renal proximal tubular dysgenesis associated with severe neonatal hemosiderotic liver disease. Pediatr Pathol. 1994;14:479–89.

Bergmann F, Aulmann S, Sipos B, et al. Acinar cell carcinomas of the pancreas: a molecular analysis in a series of 57 cases. Virchows Arch. 2014;465(6):661-672. https://doi.org/10.1007/s00428-014-1657-8.

Beresford OD, Owen TK. Lipomatous pseudohypertrophy of the pancreas. J Clin Pathol. 1957;10:63–6.

Berman LG, Prior JT, Abramow SM, Ziegler DD. A study of the pancreatic duct system in man by the use of vinyl acetate casts of postmortem preparations. Surg Gynecol Obstet. 1960;110:391–403.

Bresson JL, Schmitz J, Saudubray JM, Lesec G, Hummel JA, Rey J. Le syndrome de Johanson-Blizzard. Une autre cause de lipomatose pancréatique [Johanson-Blizzard's syndrome: another cause of pancreatic lipomatosis (author's transl)]. Arch Fr Pediatr. 1980;37(1):21–4. French. PubMed PMID: 7469679.

Chandra RS. Liver, gallbladder, and biliary tract. In: Gilbert-Barness E, editor. Potter's pathology of the fetus and infant. Philadelphia: Mosby Year Book, Inc. 1997. p. 823–62.

Cooper D, Schermer A, Sun TT. Classification of human epithelia and their neoplasms using monoclonal antibodies to keratins: strategies, applications, and limitations. Lab Investig. 1985;52:243–56.

Costa AM, Pegado CS, Porto LC. Quantification of the intrahepatic biliary tree during human fetal development. Anat Rec. 1998;251:297–302.

Curry CA, Eng J, Horton KM, Urban B, Siegelman S, Kuszyk BS, Fishman EK. CT of primary cystic pancreatic neoplasms: can CT be used for patient triage and treatment? AJR Am J Roentgenol. 2000;175(1):99–103. PubMed PMID: 10882255.

Demetris AJ, Seaberg EC, Wennerberg A, Ionellie J, Michalopoulos G. Ductular reaction after submassive necrosis in humans. Special emphasis on analysis of ductular hepatocytes. Am J Pathol. 1996;149:439–48.

Desmet VJ. Intrahepatic bile ducts under the lens. J Hepatol. 1985;1:545–59.

De Lange C, Janssen TAE. Large solitary pancreatic cyst and other developmental errors in a premature infant. Am J Dis Child. 1948;75:587–94.

Duffy MJ, Sturgeon C, Lamerz R, Haglund C, Holubec VL, Klapdor R, Nicolini A, Topolcan O, Heinemann V. Tumor markers in pancreatic cancer: a European Group on Tumor Markers (EGTM) status report. Ann Oncol. 2010;21(3):441–7. https://doi.org/10.1093/annonc/mdp332. Epub 2009 Aug 18. Review. PubMed PMID: 19690057.

Duval JV, Savas L, Banner BF. Expression of cytokeratins 7 and 20 in carcinomas of the extrahepatic biliary tract, pancreas, and gallbladder. Arch Pathol Lab Med. 2000;124(8):1196-1200.https://doi.org/10.1043/0003-9985(2000)124<1196:EOCAIC>2.0.CO;2.

Flanagan SE, Kapoor RR, Hussain K. Genetics of congenital hyperinsulinemic hypoglycemia. Semin Pediatr Surg. 2011;20(1):13–7. Review.

Feldman M, Weinberg T. Aberrant pancreas: a cause of duodenal syndrome. JAMA. 1952;148:893–8.

Figurasin R, Maguire NJ. 3,4-Methylenedioxy-Methamphetamine (MDMA, Ecstasy, Molly) Toxicity. [Updated 2020 May 24]. In: StatPearls [Internet]. Treasure Island (FL): StatPearls Publishing; 2020 Jan-. Available from: https://www-ncbi-nlm-nih-gov.login.ezproxy.library.ualberta.ca/books/NBK538482/.

Gray SW, Skandalakis JE. Embryology for surgeons: the embryological basis for the treatment of congenital defects. Philadelphia: WB Saunders Co; 1972. p. 263–82.

Hammar JA. Über die erste Entstehung der nicht kapillaren intrahepatischen allengänge beim Menschen. Z Mikrosk Anat Forsch. 1926;5:59–89.

Hamilton PW, Allen DC. Morphometry in histopathology. The Journal of Pathology. 1995;175:369–79.

Hori A, Orthner H, Kohlschütter A, Schott KM, Hirabayashi K, Shimokawa K. CNS dysplasia in dysencephalia splanchnocystica (Gruber's syndrome). A case report. Acta Neuropathol. 1980;51(2):93–7. PubMed PMID: 7435150.

Hsu SM, Raine L, Fanger H. Use of avidin-biotin-peroxidase complex (ABC) in immunoperoxidase techniques: a comparison between ABC and unlabeled antibody (PAP) procedures. J Histochem Cytochem. 1981;29:577–80.

Husen Y, Saeed MA, Siddiqui S. Acinar Cell Carcinoma of Pancreas in a Child: A Radiological Perspective. J Ayub Med Coll Abbottabad. 2015;27(4):936-937.

Jaffe R, Hashida Y, Yunis EJ. Pancreatic pathology in hyperinsulinemic hypoglycemia of infancy. Lab Invest. 1980;42(3):356–365.

James C, Kapoor RR, Ismail D, Hussain K. The genetic basis of congenital hyperinsulinism. J Med Genet. 2009;46(5):289–99. Epub 2009 Mar 1. Review.

Jones KL. Smith's recognizable patterns of human malformation. 5th ed. Philadelphia: W.B. Saunders Company; 1997.

Jones S, Zhang X, Parsons DW, Lin JC, Leary RJ, Angenendt P, Mankoo P, Carter H, Kamiyama H, Jimeno A, Hong SM, Fu B, Lin MT, Calhoun ES, Kamiyama M, Walter K, Nikolskaya T, Nikolsky Y, Hartigan J, Smith DR, Hidalgo M, Leach SD, Klein AP, Jaffee EM, Goggins M, Maitra A, Iacobuzio-Donahue C, Eshleman JR, Kern SE, Hruban RH, Karchin R, Papadopoulos N, Parmigiani G, Vogelstein B, Velculescu VE, Kinzler KW. Core signaling pathways in human pancreatic cancers revealed by global genomic analyses. Science. 2008;321(5897):1801–6. https://doi.org/10.1126/science.1164368. Epub 2008 Sep 4. PubMed PMID: 18772397; PubMed Central PMCID: PMC2848990.

Jørgensen MJ. The ductal plate malformation. A study of the intrahepatic bile duct lesion in infantile polycystic disease and congenital hepatic fibrosis. Acta Pathol Microbiol Immunol Scand. 1977;257(Suppl):1–87.

Kamath BM, Piccoli DA. Heritable disorders of the bile ducts. Gastroenterol Clin N Am. 2003;32(3):857–75, vi. Review.

Kahn E, Markowitz J, Aiges H, Daum F. Human ontogeny of the bile duct to portal space ratio. Hepatology. 1989;10:21–3.

La Rosa S, Sessa F, Capella C. Acinar Cell Carcinoma of the Pancreas: Overview of Clinicopathologic Features and Insights into the Molecular Pathology. Front Med (Lausanne). 2015;2:41. Published 2015 Jun 15. https://doi.org/10.3389/fmed.2015.00041.

Ligneau B, Lombard-Bohas C, Partensky C, Valette PJ, Calender A, Dumortier J, Gouysse G, Boulez J, Napoleon B, Berger F, Chayvialle JA, Scoazec JY. Cystic endocrine tumors of the pancreas: clinical, radiologic, and histopathologic features in 13 cases. Am J Surg Pathol. 2001;25(6):752–60. PubMed PMID: 11395552.

Marquard J, Palladino AA, Stanley CA, Mayatepek E, Meissner T. Rare forms of congenital hyperinsulinism. Semin Pediatr Surg. 2011;20(1):38–44. Review.

Meissner T, Mayatepek E. Clinical and genetic heterogeneity in congenital hyperinsulinism. Eur J Pediatr. 2002;161(1):6–20. Review.

Moll R, Franke WW, Schiller DL, Geiger B, Krepler R. The catalog of human cytokeratins: patterns of expression in normal epithelia, tumors and cultured cells. Cell. 1982;31:11–24.

Moloney WC, McPherson L, Fliegelmann L. Esterase activity in leukocytes demonstrated by the use of naphthol ASD chloroacetate substrate. J Histochem Cytochem. 1960;8:200–7.

Monsereenusorn C, Satayasoontorn K, Rujkijyanont P, Traivaree C. Cholangiocarcinoma in a Child with Progressive Abdominal Distension and Secondary Hypercalcemia. Case Rep Pediatr. 2018;2018:6828037. Published 2018 Apr 23. https://doi.org/10.1155/2018/6828037

Muduly DK, Satyanarayana Deo SV, Shukla NK, Kallianpur AA, Prakash R, Jayakrishnan T. Gall bladder cancer in a child: a rare occurrence. J Cancer Res Ther. 2012;8(4):653–54. https://doi.org/10.4103/0973-1482.106593.

Nakanuma Y, Hoso M, Sanzen T, Sasaki M. Microstructure and development of the normal and pathologic biliary tract in humans, including blood supply. Microsc Res Tech. 1997;38:552–70.

Online Mendelian inheritance in man. McKusick VA, Johns Hopkins University (Baltimore, MD) and NCBI, National Library of Medicine (Bethesda, MD). http//www3.ncbi.nlm.nih.gov/omim/. 1998.

Pontes H, Santos-Marques MJ, Fernandes E, et al. Effect of chronic ethanol exposure on the hepatotoxicity of ecstasy in mice: an ex vivo study. Toxicol In Vitro. 2008;22(4):910-920. https://doi.org/10.1016/j.tiv.2008.01.010.

Rahier J, Guiot Y, Sempoux C. Morphologic analysis of focal and diffuse forms of congenital hyperinsulinism. Semin Pediatr Surg. 2011;20(1):3–12. Review.

Restrepo R, Hagerott HE, Kulkarni S, Yasrebi M, Lee EY. Acute Pancreatitis in Pediatric Patients: Demographics, Etiology, and Diagnostic Imaging. AJR Am J Roentgenol. 2016;206(3):632–44. https://doi.org/10.2214/AJR.14.14223.

Rienhoff WF, Pickrell KL. Pancreatitis: an anatomic study of the pancreatic and extrahepatic biliary systems. Arch Surg. 1945;51:205–19.

Sergi C, Adam S, Kahl P, Otto HF. Study of the malformation of ductal plate of the liver in Meckel syndrome and review of other syndromes presenting with this anomaly. Pediatr Dev Pathol. 2000;3(6):568–83. PubMed PMID: 11000335.

Sharma S, Agarwal S, Nagendla MK, Gupta DK. Omental acinar cell carcinoma of pancreatic origin in a child: a clinicopathological rarity. Pediatr Surg Int. 2016;32(3):307-311. https://doi.org/10.1007/s00383-015-3850-5.

Sergi C, Jundt K, Seipp S, Goeser T, Theilmann L, Otto G, Otto HF, Hofmann WJ. The distribution of HBV, HCV and HGV among livers with fulminant hepatic failure of different aetiology. J Hepatol. 1998;29:861–71.

Sempoux C, Capito C, Bellanné-Chantelot C, Verkarre V, de Lonlay P, Aigrain Y, Fekete C, Guiot Y, Rahier J. Morphological mosaicism of the pancreatic islets: a novel anatomopathological form of persistent hyperinsulinemic hypoglycemia of infancy. J Clin Endocrinol Metab. 2011;96(12):3785–93. Epub 2011 Sept 28.

Sohn DS, Kim KY, Lee WB, Kim DC. Eosinophilic granulopoiesis in human fetal liver. Anat Rec. 1993;235:453–60.

Song JS, Yoo CW, Kwon Y, Hong EK. Endoscopic ultrasound-guided fine needle aspiration cytology diagnosis of solid pseudopapillary neoplasm: three case reports with review of literature. Korean J Pathol. 2012;46(4):399–406. https://doi.org/10.4132/KoreanJPathol.2012.46.4.399. Epub 2012 Aug 23. PubMed PMID: 23110037; PubMed Central PMCID: PMC3479817.

Strayer DS, Kissane JM. Dysplasia of the kidneys, liver, and pancreas: report of a variant of Ivemark's syndrome. Hum Pathol. 1979;10(2):228–34. PubMed PMID: 422192.

Streeter GL. Weight, sitting height, head size, foot length and menstrual age of human embryo. Contr Embryol Carnegie Institution. 1920;11:143–70.

Tavassoli M. Embryonic and fetal hemopoiesis: an overview. Blood Cells. 1991;1:269–81.

Thompson ED, Wood LD. Pancreatic Neoplasms With Acinar Differentiation: A Review of Pathologic and Molecular Features [published online ahead of print, 2019 Dec 23]. Arch Pathol Lab Med. 2019;10.5858/arpa.2019-0472-RA. https://doi.org/10.5858/arpa.2019-0472-RA.

Treem WR, Krzymowski GA, Cartun RW, Pedersen CA, Hyams JS, Berman M. Cytokeratin immunohistochemical examination of liver biopsies in infants with Alagille syndrome and biliary atresia. J Pediatr Gastroenterol Nutr. 1992;15:73–80.

Van Eyken P, Sciot R, Callea F, Van der Steen K, Moerman P, Desmet VJ. The development of the intrahepatic bile ducts in man: a keratin-immunohistochemical study. Hepatology. 1988;8:1586–95.

Vogel M. Atlas der morphologischen Plazentadiagnostik. 2nd ed. Berlin Heidelberg/New York: Springer Verlag; 1996.

Wood LD, Klimstra DS. Pathology and genetics of pancreatic neoplasms with acinar differentiation. Semin Diagn Pathol. 2014;31(6):491-497. https://doi.org/10.1053/j.semdp.2014.08.003.

Kidney, Pelvis, and Ureter

Contents

6.1	**Development and Genetics**	581
6.2	**Non-cystic Congenital Anomalies**	589
6.2.1	Disorders of Number	589
6.2.2	Disorders of Rotation	591
6.2.3	Disorders of Position	591
6.2.4	Disorders of Separation	591
6.3	**Cystic Renal Diseases**	591
6.3.1	Classifications	591
6.3.2	Autosomal Dominant Polycystic Kidney Disease	591
6.3.3	Autosomal Recessive Polycystic Kidney Disease	596
6.3.4	Medullary Sponge Kidney	597
6.3.5	Multicystic Dysplastic Kidney (MCDK)	597
6.3.6	Hydronephrosis/Hydroureteronephrosis	597
6.3.7	Simple Renal Cysts	598
6.3.8	Acquired Cystic Kidney Disease (CRD-Related)	598
6.3.9	Genetic Syndromes and Cystic Renal Disease	598
6.4	**Primary Glomerular Diseases**	599
6.4.1	Hypersensitivity Reactions and Major Clinical Syndromes of Glomerular Disease	599
6.4.2	Post-infectious Glomerulonephritis	600
6.4.3	Rapidly Progressive Glomerulonephritis	602
6.4.4	Minimal Change Disease	604
6.4.5	(Diffuse) Mesangial Proliferative GN	604
6.4.6	Focal and Segmental Glomerulosclerosis	604
6.4.7	Membranous Glomerulonephritis	606
6.4.8	Membranoproliferative (Membrane-Capillary) Glomerulonephritis	609
6.5	**Secondary Glomerular Diseases**	610
6.5.1	SLE/Lupus Nephritis	610
6.5.2	Henoch-Schönlein Purpura	611
6.5.3	Amyloidosis	613
6.5.4	Light Chain Disease	614
6.5.5	Cryoglobulinemia	614
6.5.6	Diabetic Nephropathy	614

6.6	**Hereditary/Familial Nephropathies**	615
6.6.1	Fabry Nephropathy	615
6.6.2	Alport Syndrome	615
6.6.3	Nail-Patella Syndrome	617
6.6.4	Congenital Nephrotic Syndrome	617
6.6.5	Thin Glomerular Basement Membrane Nephropathy (TBMN)	617
6.7	**Tubulointerstitial Diseases**	618
6.7.1	Acute Tubulointerstitial Nephritis (ATIN)	618
6.7.2	Chronic Tubulointerstitial Nephritis (CTIN)	622
6.7.3	Acute Tubular Necrosis	623
6.7.4	Chronic Renal Failure (CRF)	623
6.7.5	Nephrolithiasis and Nephrocalcinosis	624
6.7.6	Osmotic Nephrosis and Hyaline Change	625
6.7.7	Hypokalemic Nephropathy	625
6.7.8	Urate Nephropathy	625
6.7.9	Cholemic Nephropathy/Jaundice-Linked Acute Kidney Injury	625
6.7.10	Myeloma Kidney	625
6.7.11	Radiation Nephropathy	626
6.7.12	Tubulointerstitial Nephritis and Uveitis (TINU)	626
6.8	**Vascular Diseases**	626
6.8.1	Benign Nephrosclerosis	626
6.8.2	Malignant Nephrosclerosis	626
6.8.3	Renal Artery Stenosis	626
6.8.4	Infarcts	627
6.8.5	Vasculitis	627
6.8.6	Hemolytic Uremic Syndrome	628
6.8.7	Thrombotic Thrombocytopenic Purpura	630
6.9	**Renal Transplantation**	630
6.9.1	Preservation Injury	630
6.9.2	Hyperacute Rejection	630
6.9.3	Acute Rejection	631
6.9.4	Chronic Rejection	632
6.9.5	Humoral (Acute/Chronic) Rejection	632
6.9.6	Antirejection Drug Toxicity	633
6.9.7	Recurrence of Primary Disorder	633
6.10	**Hereditary Cancer Syndromes Associated with Renal Tumors**	633
6.10.1	Beckwith-Wiedemann Syndrome	633
6.10.2	WAGR Syndrome	633
6.10.3	Denys-Drash Syndrome	634
6.10.4	Non-WT1/Non-WT2 Pediatric Syndromes	634
6.10.5	Von Hippel Lindau Syndrome	634
6.10.6	Tuberous Sclerosis Syndrome	634
6.10.7	Hereditary Papillary Renal Carcinoma Syndrome	634
6.10.8	Hereditary Leiomyoma Renal Carcinoma Syndrome	634
6.10.9	Birt-Hogg-Dube Syndrome	634
6.11	**Pediatric Tumors (Embryonal)**	634
6.11.1	Wilms Tumor (Nephroblastoma)	634
6.11.2	Cystic Partially Differentiated Nephroblastoma (CPDN) and (Pediatric) Cystic Nephroma	645
6.11.3	Congenital Mesoblastic Nephroma	646
6.11.4	Clear Cell Sarcoma	646
6.11.5	Rhabdoid Tumor	647

6.11.6	Metanephric Tumors	647
6.11.7	XP11 Translocation Carcinoma	647
6.11.8	Ossifying Renal Tumor of Infancy	650
6.12	**Non-embryonal Tumors of the Young**	**650**
6.12.1	Clear Cell Renal Cell Carcinoma	650
6.12.2	Chromophobe Renal Cell Carcinoma	653
6.12.3	Papillary Adenoma and Renal Cell Carcinoma	653
6.12.4	Collecting Duct Carcinoma	656
6.12.5	Renal Medullary Carcinoma	656
6.12.6	Angiomyolipoma	656
6.12.7	Oncocytoma	656
6.12.8	Other Epithelial and Mesenchymal Tumors	659
6.13	**Non-neoplastic Pathology of the Pelvis and Ureter**	**661**
6.13.1	Anatomy and Physiology Notes	661
6.13.2	Congenital Pelvic-Ureteral Anomalies	661
6.13.3	Congenital Ureteric Anomalies	661
6.13.4	Lower Urinary Tract Abnormalities	662
6.14	**Tumors of the Pelvis and Ureter**	**666**
6.14.1	Neoplasms	666
6.14.2	Genetic Syndromes	666
Multiple Choice Questions and Answers		**667**
References and Recommended Readings		**668**

6.1 Development and Genetics

Spatiotemporal coordination of cell behavior and mesenchymal-epithelial interactions are critical for the growth and patterning that develop by branching morphogenesis in the ontogenesis of the urinary system (Figs. 6.1 and 6.2). Some abbreviations of this chapter are common to other chapters of this book. However, the numerous abbreviations of genes and proteins that play a crucial role in renal ontogenesis are collected at the end of the chapter before references and recommended reading for clarity. Although multiple and different steps and mechanisms have been delineated during organogenesis, the complex and full set of kidney development is poorly understood. Renal morphogenesis starts when the Wolffian duct (WD) forms from the intermediate mesoderm. The caudal growth of this structure induces transient organs that play a role in lower vertebrates only and definitive and functional organs that are present in higher vertebrates. There is an intricate interaction of numerous growth factors, including members of the FGF (fibroblast growth factor), TGF-β (transforming growth factor beta), and Wnt (wingless) signaling that have been implicated in the control of branching of the ureteral bud and subsequent morphogenesis. These growth factors also interact with other cell surface and ECM (extracellular) proteins such as integrins, heparan sulfate proteoglycans, and matrix metalloproteases, which appear to play specific roles in branching morphogenesis in the kidney as well as the lung, pancreas, and salivary gland. The TGF-β superfamily has been intensely investigated in the last two decades and glial cell line-derived neurotrophic factor (GDNF), neurturin (NTN), artemin, and persephin, as distant members of the TGF-β superfamily, all function as neurotrophic factors. In particular, GDNF and NTN are two structurally related, extremely potent neurotrophic factors that play crucial roles in the control of survival and differentiation of neurons. The bioactivities of GDNF, NTN, artemin, and persephin are mediated by a receptor complex composed of the non-ligand-binding signaling subunit (c-Ret receptor tyrosine kinase). Four ligand-binding subunits, designated GFRα-1 (10q26.11) through 4, are known. They are cyste-

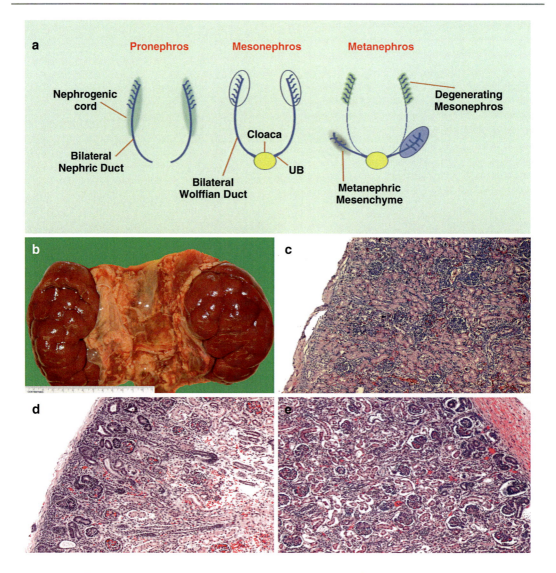

Fig. 6.1 In figure (**a**), the different stages of kidney development (pronephros, mesonephros, and metanephros) are presented. Figure (**b**) shows a case of fetal lobulation in an infant, while figures (**c**–**e**) show different gradual developments of the glomerulogenic zone in comparison with the definitive zone (**c**–**e**, hematoxylin and eosin staining, ×100 original magnification). In (**c**) is shown the histology of a kidney from an 11-month-old female infant, while in (**d**) there is a focal dilation of the forming collecting ducts, which was related to a mild outflow obstruction (posterior urethral valves) in a male baby. This patient was a male baby of 21 weeks gestation following a spontaneous miscarriage. Figure (**e**) shows the histopathology of a kidney from a premature baby, at 32 + 5 weeks of gestation with lung immaturity (saccular stage) and amnion remnants

ine-rich glycosylphosphatidylinositol (GPI)-linked cell surface proteins. One of these ligands acts on both GDNF and NTN. Moreover, multiple alternatively spliced transcript variants have been described for RET tyrosine kinase receptor. Interestingly, heterozygous *RET* mutations determining loss of function in humans cause Hirschsprung disease, a developmental defect of the autonomic nervous system of the gut that is associated sometimes with renal defects. Conversely, *Ret* heterozygous mice are mostly healthy. Homozygous mutations of *Ret* in mice induce severe congenital anomalies of the kidney and enteric nervous system.

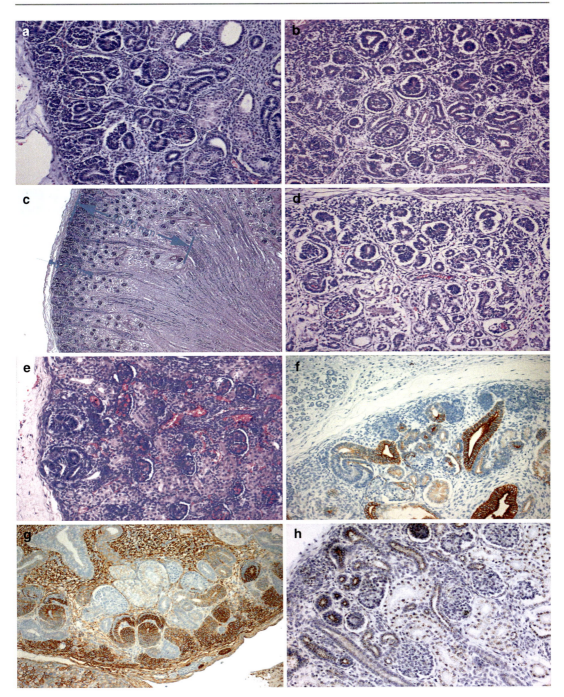

Fig. 6.2 Figure (**a**) shows the kidney histology of a female 15-week-old gestation fetus with trisomy 18 (hematoxylin and eosin staining, ×200, original magnification). Figure (**b**) shows the kidney histology of a male 14 weeks gestation old fetus with trisomy 21 (hematoxylin and eosin staining, ×200, original magnification). Figure (**c**) shows the low-power kidney histology of a premature baby (hematoxylin and eosin staining, ×40, original magnification). Figure (**d**) shows the kidney histology of a 16-week-old gestation female fetus without anomalies but chorioamnionitis of the placenta (hematoxylin and eosin staining, ×200, original magnification). Figure (**e**) shows the kidney histology of a female premature baby of 32 + 2 weeks gestation with non-renal external anomalies (hematoxylin and eosin staining, ×200, original magnification). Figures (**f**) and (**g**) show the immunohistochemical detection in early second trimester of pregnancy of the forming tubular system and blastema in the fetal kidney using an antibody against CK19 and CD56, respectively (avidin-biotin complex, ×200, original magnification). Figure (**h**) shows the immunohistochemical expression of SHH in the first trimester of pregnancy (hematoxylin and eosin staining, ×200, original magnification)

The urogenital tract arises from the intermediate mesoderm of the early embryo. There are three significant intriguing aspects considering the development of the urogenital system:

- First, the interconnection of the urinary and genital ontogeny, in which the components of both systems interact until the late phases of urinary development and gradually model the sexual differentiation.
- Second, the ontogeny of the kidney with organ isoforms that play a role in the formation of proto-kidneys in some vertebrates.
- Third, the extraordinary cross talk between epithelium and mesenchyme with mutual induction, which may be the basis for further investigations of developmental and molecular biology.

The intermediate mesoderm generates the urogenital system, including kidneys, gonads, and excretory systems. Before any gonadal development is evident, the urinary system starts first with the formation of an elongated pair of excretory organs, which will be terminal in lower vertebrates only, but specific components are retained for the development of some parts of the urogenital system. In mammals, the adult kidney is the final product of three embryonic organs, i.e., the pronephros, the mesonephros, and the metanephros (Faa et al. 2012). The first two stages are transient, while the last persists as a functional kidney. The pronephros starts on day 22 post-ovulation in humans and day 8 in mice. The pronephric duct arises in the intermediate mesoderm ventrally to anterior somites following induction from the nephrogenic mesenchyme and migrates caudally. During the migration, the anterior region of the pronephric duct induces the adjacent mesenchyme to form the tubules of this initial renal isoform, which is functioning in fish and amphibian larvae. In mammals, the pronephric tubules and the anterior regions of the pronephric duct degenerate, while the more caudal regions persist and serve as the basis for the formation of the mesonephros, which starts on day 25 in humans. Then, mesonephric tubules are formed along a craniocaudal gradient and empty separately into the continuation of the primary pronephric duct, i.e., the mesonephric (Wolffian) duct. The first 4–6 pairs of mesonephric tubules arise similar to the pronephric tubules as outgrowths from the central nephric ducts. However, the other pairs of mesonephric tubules occur separately in the intermediate mesoderm. Approximately on day 28, two mesonephric ducts with a lumen throughout each attach to the cloaca. As more tubules are induced caudally, the anterior tubules disappear progressively by apoptosis in humans. In mice, the anterior tubules remain, while the posterior tubules regress. The mesonephros plays essential roles in ontogeny, including hematopoiesis and excretory structures of the male genital system (vas deferens and efferent ducts of the testes). The metanephros represents the final kidney and starts early in the 5th week. Substantially, it originates very near the attachment of the mesonephric duct to the cloaca when an epithelial outgrowth called *ureteric bud* takes place. The successful and continuous interaction between this structure and the intermediate mesoderm is at the basis of a correct formation of the definitive kidney (*nephrogenesis*). Three genes seem to play a significant role in the pronephros and mesonephros stages. Knockout experiments have suggested that the homeobox domain genes, *Lim-1* and *Pax-2*, are essential in the earliest phases of the renal ontogenesis. The knockout experiments involving the *WT-1* gene have satisfactorily demonstrated the absence of posterior mesonephric tubules. The stages of the definitive kidney development are five and include (i) the primary *ureteric bud*, (ii) the *metanephrogenic cap mesenchyme*, (iii) the *mesenchymal-epithelial transition*, (iv) *glomeruli and tubulogenesis*, and (v) *interstitial maturation*. Fundamentally, two intermediate mesodermal tissues, the primary ureteric bud and the metanephrogenic mesenchyme, interact closely and reciprocally inducing each other to form the kidney. The metanephrogenic mesenchyme promotes the elongation and branching of the ureteric bud or metanephric *diverticulum*. The tips of these branches induce the mesenchymal cells to be aggregated and form epithelial clusters of about 20 cells (*metanephric blastema*), which will proliferate and differentiate into the structure of the final nephron with characteristic shapes. A crucial step is the breakdown of the basal lamina of the ureteric bud ducts. It takes place by the growing epithelial nodules with subsequent fusion, which creates a connection between the ureteric bud and the differentiated metanephrogenic mesenchyme. This setting enables material to pass through the excretory system linked to the cloaca. The branches of the ureteric bud are indeed the basis for the formation of the collecting system, including collecting ducts, bilateral renal pelvis, ureters, and median bladder trigone. Several gene regulatory networks play a role in the metanephric stage of development. Inducers are GDNF, C-RET, WNT11, AT-II, AT1R, AT2R, COFILIN I, DESTRIN, ETv4, ETv5, FGFR1, FGFR2, FGF8, MMP-9, SOX8, SOX9, and TP53, whereas inhibitors are BMP4, FOXC1, FOXC2, ROBO2, SEMA3a, SLIT2, and SPRY1. In this chapter there are numerous abbreviations of genes and proteins that play a crucial role in renal ontogenesis and most of them are explained, but the reader is invited to consult molecular biology books for some that are not explained here in detail. The GDNF/c-Ret/Wnt1 pathway is now considered the major positive regulator of the ureteric bud development. In fact, the outgrowth of the ureteric bud from the mesonephric duct is a response to the secretion of GDNF by the mesenchymal cells of the metanephrogenic blastema. This signal is closely bound by C-RET, a member of the TKR superfamily, which is located in the plasmatic membranes of the epithelial cells of the early ureteric *diverticulum*. Interestingly, *Gdnf-/Ret*-deficient experimental animals show severe branching abnormalities, and *Ret* knockout mice display bilateral renal agenesis. The lack of Gdnf blocks the development of the ureteric bud, whereas a balanced action of Ret tyrosine kinase on distal ureter maturation controls the connection between the ureter and urinary bladder. *Pax2* and *Pax8* are necessary for the morphogenesis and topometry of the primary ureteric bud with a major role of *Pax2* regulating the formation of Gdnf in the metanephric mesenchyme. A decrease of *Pax2* expression during kidney development, due to its gene mutation, determines an excessive amount of apoptosis in the tips of the ureteric bud with subsequent loss or scarcity of the ureteric bud branches. Thus, essentially, there are eight sets of signals operating in the mutual induction of the metanephric kidney as indicated in Box 6.1. Although some of these studies are related to lower vertebrates, most of the genes play a substantial role in higher species. Thus, the terminology used with capitols or not is interchangeable most of the times. Specific embryology books and publications dealing with renal morphogenesis are essential for gathering completely this chapter of human embryology.

The activity of the Gdnf/c-Ret/Wnt1 pathway is mainly inhibited by Spry1 and Sema3a. Spry1 is a critical regulator of Gdnf-/Ret-mediated kidney induction, while class 3 semaphorins are involved in ureteric bud branching. Class 3 semaphorins are a family of guidance proteins that play an essential role in branching morphogenesis of epithelial structures. Sema3a acts inhibiting the growth and branching of the ureteric bud downregulating Gdnf signaling, inhibiting Vegf-a, and decreasing the activity of the Akt sur-

Box 6.1 Induction Keys of the Metanephric Kidney
1. WT1-induced formation of the metanephrogenic mesenchyme
2. GDNF and HGF secretion of the metanephrogenic mesenchyme to promote and direct the ureteric bud
3. FGF2 and BMP7 secretion of the ureteric bud is able to prevent mesenchymal apoptosis
4. LIF secretion from the ureteric bud induces the mesenchymal cells to be aggregated and secrete WNT4
5. Conversion of the mesenchymal-epithelial transdifferentiated clusters into nephrons
6. Mesenchymal signals induce the progressive and sequentially multimeric branching of the ureteric bud
7. Induction of mesenchymal stem cells by the ureteric bud
8. Differentiation of the nephron and progressive growth of the ureteric bud

vival pathway. Development, growth, and branching of the ureteric diverticulum require, finally, several intracellular changes of the cytoskeleton. At this time, investigations using transmission electron microscopy have been crucial in identifying the enormous role of the cytoskeleton in glomerular and renal morphogenesis. Cofilin I and Destrin are actin-depolymerizing factors useful for shape changes of the cells.

The metanephrogenic cap mesenchyme is constituted by self-renewing progenitor cells, which condensate and aggregate around the tips of epithelial branches of the primary ureteric bud and transform themselves uniquely and brilliantly into the cap mesenchyme cells. Subsequently, these cells undergo mesenchymal-epithelial transdifferentiation forming most of the epithelia of the definitive nephron. Genes involved in the metanephrogenic mesenchyme include *CD24*, Lim-type homeobox gene *LhxI*, *EyaI*, and *SixI* as part of a transcription factor complex among others, although Pax2 and Wnt4 are the two gene products mainly involved in the process of metanephric mesenchymal differentiation toward the cap mesenchyme. The mesenchymal-to-epithelial transition is paralleled by the activation of genes linked to epithelial differentiation (e.g., keratins, desmosomal components, *adherens* junctions and tight junctions, basement membrane collagen, laminin) and inactivation of genes related to mesenchymal differentiation (e.g., vimentin, collagen). The pretubular clustering and the epithelial transformation are probably the most critical phases during nephrogenesis. The formation of the renal vesicle, which is essentially a simple tubule undergoing extensive growth, segmentation, and differentiation, is characterized by an increase of expression of *LhxI*, *Fgf8,* and *Wnt4* and a decrease of expression of *Six2* and *Cited2*. Cap mesenchymal cells respond to Fgf8/Wnt4 expression by differentiating into the renal vesicle. It is important to remember that the renal vesicle is the first epithelial structure originating from the metanephrogenic cap mesenchyme. The mature nephrons, ranging from 300,000 to 1,800,000 per kidney, are endowed with 36 weeks of gestation. Nephron differentiation includes a series of steps such as segmentation and patterning of the renal vesicle, fusion of the vesicle with

> **Box 6.2. Nephron Differentiation Steps**
> 1. Renal vesicle
> 2. Comma-shaped transformed vesicle
> 3. S-shaped transformed vesicle
> 4. Capillary loop stage
> 5. Maturing nephron

the ureteric component of the fetal kidney to create the patency of the excretory system. These steps are also defined in Box 6.2.

Progression of the mesenchymal-epithelial transition characterizes the first stage. There is the transformation of the robust pretubular mesenchymal clusters into round or ovoid epithelial aggregates with a central lumen. This "lumenification" is crucial to many structures in the human organisms as well as in lower vertebrates (e.g., intra- and extrahepatic biliary system of the liver). In the proximal renal vesicles, there is a high expression of *Tmem100* and *Wt1*, whereas in the distal renal vesicles, there is elevated expression of *Dkk I*, *Papss2*, *Greb I*, *DII*, *Pcsk9*, *LhxI*, *BMP2*, and *Pou3f3*. The study of the appearance of intercellular adhesion molecules such as cadherin-6 and E-cadherin is crucial in the second stage when clear segmentation of the renal vesicle determines the "*poiesis*" of comma- and S-shaped epithelial structures. In the second stage, Notch1 and Notch2 contribute to the process of segmentation of the maturing epithelial structures. The third stage is constituted by the appearance of the first loop of Henle and its crucial localization and development in the renal cortex. This stage is characterized by the expression of vascular markers such as CD31 and CD34. Finally, the differentiation of the main components of the renal corpuscles and the tubular segments and the development of the juxtaglomerular complex, *macula densa*, *mesangium*, and part of the afferent arterioles characterize stage IV. CD10 appears to be highly expressed in the undifferentiated metanephric mesenchyme and during the differentiation of podocytes and Bowman capsule cells. During this stage, PAX2 and other transcription factors useful in the early stages of the metanephric ontogenesis are progressively downregulated and appear before tubule formation. TBX18, SOX9, and HOX10 seem to play a critical role in the human development of the stromal cell compartment and the regulation of patterning and differentiation and integration of different stromal cell types. However, the role of PAX2 in the early stages of distinction as a repressor of the interstitial cell fate needs to be further emphasized. Intriguingly, PAX2 is a developmental boundary factor located between the nephron and non-nephron lineages favoring the connection of numerous transcription factor families.

The Role of Localized Gdnf/Ret Signaling in Ureter Formation

The critical part of Gdnf and its receptors in ureteral and renal development was initially revealed by experiments of knockout mice (Costantini and Shakya 2006). If any of the three main actors such as Gdnf, Ret, and Gfr-α1 working in the playground of kidney morphogenesis is missing or malfunctioning, excretory system defects ranging from renal agenesis to blind-ending ureters with no renal tissue or blastema to extremely hypoplastic kidney rudiments are typically observed. Gdnf signaling is crucial to induce the outgrowth of the ureteral bud from the Wolffian duct and to promote its continued growth of the early stages and its branching. In the absence of Gdnf, Ret or Gfra-1 signaling failure of ureteric bud formation is seen. It is now clear that the outgrowth of the bud is not a simple morphogenetic event, but a quite complex one. It requires an intricate network of positive and negative regulatory mechanisms. Before the outgrowth of the ureteric bud, Ret is present along the Wolffian duct, while Gdnf is localized in a broad region of the adjacent mesenchyme. As the ureteric bud grows out into the metanephric blastema and starts to branch, the expression of *Ret* gene is restricted to the distal tips of the ureteral bud, while the expression of *Gdnf* is limited to the undifferentiated mesenchyme at the periphery of the kidney surrounding the tips of the ureteral bud and corresponding to the glomerulogenic zone. Although Gdnf and Ret remain crucial, mainly *Ret−/−* mice show rudimentary organ and ureteral formation pointing out to additional factors, including Fgf7, Fgf10, and Hgf. This phenomenon is known in developmental biology as redundancy. The Gdnf expression domain is restricted by the transcription factors FoxC1 and by the action of the combined Slit2/Robo2 signaling. In fact, in the absence of both FoxC1 and Slit2/Robo2, Gdnf is present also anteriorly, which results in ectopic ureteric buds. A negative regulation on Gdnf has been demonstrated with the expression of Spry1 gene in the Wolffian duct. Sprouty1, a negative regulator of Ras/Erk Map kinase signaling, is highly expressed in the posterior part of the Wolffian duct. Bmp4 is also essential to suppress the Gdnf response and is itself suppressed by Gremlin. Both Bmp4 and FoxC1 on one side and Slit2/Robo2 and Spry1 on the other side are crucial to the correct outgrowth of the ureteric bud and formation of a single bud. *Spry1−/−* animals show, indeed, many ureteric buds leading to numerous ureters and numerous nephric parenchymas. An additional network of transcription factors that promote the expression of the *Gdnf* gene in the nephrogenic cord and metanephric mesenchyme include *Pax2, Eya1, Sall1,* and *Hox11* paralogs.

Epithelial Branching Regulation

The use of Hoxb7/Gfp transgenic mice and cell chimeras has been pivotal in gathering more details about the branching morphogenesis of the kidney. After the outgrowth of the ureteric bud in the metanephric mesenchyme, branching of this structure occurs. It continues to elongate for about 11 generations of branching. The tips of the ureteral bud are more involved in the branching than lateral portions, differently from the lung epithelium where the lateral branching is prominent. It seems clear that GDNF signaling also plays a significant role in the human epithelial branching regulation, and this is not a wonder considering that GDNF is also a neurotrophic factor, which is useful for the neuritic branching as well. The network constituted by GDNF, RET, and GFR-α1 is present only in the peripheral region of the developing kidney. This region is, indeed, the area of the renal parenchyma, which shows most of the ureteral bud growth and branching. It represents the glomerulogenic zone of the late fetal kidney. It is difficult to separate outgrowth from branching. Both processes are coupled and separate them would be artificial. In addition to Gfr-α1 and Ret gene expressions, other genes are involved in this delicate process of elongation and branching. Some of these genes are mainly or, as indicated by some embryologists and developmental biologists, exclusively expressed in the tip domain, but not in the trunks. They include Wnt11, Crif1, Cxcl14, among others. Conversely, some other genes are represented only in the trunks such as Wnt7b and collagen XVIII. To the best of our current knowledge, the site of GDNF synthesis does not provide positional information to ascertain the pattern of ureteric bud branching morphogenesis.

Mesenchymal-to-Epithelial Transdifferentiation (MET)

If GDNF/RET signaling is central to outgrow and branch the ureteral bud in humans and experimental animals, WNT signaling plays a crucial role in mesenchymal-to-epithelial transdifferentiation with both WNT4 and WNT9b being involved in MET. Other important factors involved in MET include FGF8, WT1, and ODD1. While epithelial cords originate from the ureteric bud and go through the processes of elongation and branching, self-renewing progenitor cells go through a process of condensation and aggregation around the ureteral tips and transform into the mesenchymal cap cells, which can transdifferentiate and give rise to most of the epithelial components of the nephron. Renal vesicles are formed at E12.5 from the mesenchymal cap cells, whereas S-shaped bodies are formed just subsequently. S-shaped bodies then fuse with the collecting duct epithelium, and by E13.5 the S-shaped bodies become infiltrated by endothelial precursors to form the glomerular tuft, which is key for the excretory function of the kidney. This complex structure includes capillary loops, mesangium, glomerular basement membrane, and podocytes. Cre-Lox mice have been essential to indicate that Six2 and Cited1 genes are expressed by the nephron-committed, multipotent, self-renewing progenitor cells, which are thus responsible for the above-described structures of the glomerular tuft. The master control gene that regulates the expression of other genes linked to renal development remains probably WT1. It is now clear-cut that WT1 plays a central role in the regulation of the MET process and, even, in the podocyte maturation. WT1 is present in the cap mesenchyme progenitor cells, and one of the most important target genes of WT1 is Bmp7, which is expressed in the early kidney of embryonic life in the renal progenitor cells, as well as in the ureteric bud cells and podocytes. Nephron induction and maintenance of the renal progenitor cell population in the cap mesenchyme are associated with Fgf8, which is also required for cell survival and anti-apoptosis machinery and mechanisms at distinct stages of nephrogenesis. It is also central for regulating the gene expression in primitive nephrons. Fgf8 is expressed earlier than Wnt4 and is necessary for both Wnt4 and Lim1. Finally, Eya1, Pax2, and Hox11 positively regulate the expression of both Gdnf and Six2.

Molecular Patterning and Vascularization of the Glomerulus

Wt1 is limited to the proximal S-shaped body and is contained in the podocyte cell population throughout all stages of nephrogenesis. Also, Wnt9b initially activates the expression of Fgf8, Wnt4, and Pax8 in the pretubular aggregates, and, subsequently, Wnt4 is vital to maintaining the appearance of these genes and induces Lim1. Another essential gene involved in the patterning of the proximal nephron is Notch2. Megalin, Umo, integrin $\alpha 3/\beta 1$, and Nphs play also critical roles in the distinct patterning of three nephron areas, including the loop of Henle, distal tubule, and the formation of functional filters. Finally, the vascularization of the glomerulus is under the control of VEGF-A, which is produced in the forming podocytes, simultaneously with the spatiotemporal expression of Flt1 and Flk1 produced by angioblasts. VEGF is essential in the development of the human kidney by promoting endothelial cell differentiation, capillary formation, and proliferation of tubular epithelia. PDGFβ and PDGFRβ are necessary for the expansion of the mesangium and configuration of the glomerular capillary network. Moreover, the correct expression of Bmp4 during podocyte differentiation is also critical to the development of the glomerular tuft. Bmp4 is counterbalanced by the appearance of Noggin, a Bmp antagonist. Finally, the proteins Glepp1 and Kreisler are involved in the differentiation and final maturation of the podocytes. Intriguingly, more roles have also been observed for Nphs1,

Nphs2, integrin α3/β1, dystroglycan, and Col4. These factors play an essential role in the glomerular basement membrane. Of note, gene mutations of *Nphs1* (slit diaphragm nephrin) and *Nphs2* (slit diaphragm podocin) cause the congenital nephrotic syndrome.

Renal and Ureteral Congenital Defects

The most critical step in the morphogenesis of the ureter and kidney is the initial outgrowth of the ureteric bud. Failure of this crucial step determines as a consequence the absence of the ureter and kidney to develop, which is called renal agenesis. In humans, renal agenesis is found to be associated with *RET* gene mutations in about 40%, while it is associated with *GDNF* gene mutations in approximately 1/20 to 1/10 of the cases. If the ureteric bud and the kidney have improper connections, the fate is abnormalities of the urinary system including megaureter, vesicoureteral reflux, and ureteroceles. These defects are quite frequent in humans. Hypo-/dysplastic kidneys result, in vivo, from hypomorphic mutations in *RET* or from heterozygosity for *GDNF* null mutations. Oligonephronia is observed in *bmp7* null embryos in experimental animals, which show the poor development of the kidney due primarily to the premature loss of the progenitor cell population. The histological picture of oligonephronia is characterized by a reduced number of nephrons and by hypertrophy of all the cytologic elements of these nephrons. Importantly, dysplastic lesions are never observed. Glomerulovascular defects are associated with lack of expression of *VEGF-A* in the podocytes, while overexpression of *VEGF-A* gene in podocytes is seen in collapsing glomerulopathy. Glomerular microaneurysms and collapsed glomerular tufts are detectable if BMP4 is lost in the podocytes, while overexpression of *BMP4* gene results in clumps without endothelia. Fetal growth restriction (FGR) produces oligonephronia with secondary development of arterial hypertension and renal disease. It has also been demonstrated that oligonephronia is produced independently by FGR and by unrelated hypertension and other renal diseases.

The kidney receives 20–25% of the cardiac output. The arteries include renal, interlobar, arcuate, interlobular, afferent, glomerular, and efferent arteries. The renal lobules demarcated by interlobular arteries contain collecting ducts. The renal lobe is constituted by the Malpighian pyramid plus an overlying cortex (6–18 lobes/kidney) and 10–25 papillary ducts (of Bellini) per papilla. The juxtaglomerular cells are adapted smooth muscle cells located within afferent arterioles, while the *macula densa* is a specialized region of the distal convoluted tubule. It is located where it adjoins its derived glomerulus. Lacis cells (aka Polkissen or Goormaghtigh cells) are extraglomerular mesangial cells situated between afferent arteriole, *macula densa*, and glomerulus. Lacis cells are considered light-staining pericytes. The glomerulus contains capillaries with fenestrated endothelium, and the GBM is ~3.5 nm and contains polyanions and acidic glycoproteins (negatively charged). The ultrafiltration is 180 L filtrate/day and ~ 1 L off final excretion per day. The reabsorption occurs for about 2/3 of the filtrate in the first third (proximal convoluted tubules) where the output is isotonic (glucose, amino acids, HCO_3- reabsorption); 15% occurs in the middle 1/3 (loop of Henle) where the output is hypotonic (low salt permeability and high H_2O in the descending portion of the loop, high salt permeability and low water in the ascending portion of the loop).

6.2 Non-cystic Congenital Anomalies

6.2.1 Disorders of Number

Disorders of number include several anomalies ranging from renal agenesis to hypoplasia. Renal agenesis: consists in the lack of one or both kidneys (Fig. 6.3).

Bilateral renal agenesis: Rare, associated with severe oligohydramnios, early death.

Unilateral renal agenesis: 1/1000 with compensatory hypertrophy of the other kidney.

Hypoplasia renal agenesis: Usually unilateral and due to insufficient development with ↓# of lobules (<5) that needs to be kept separated from vascular/infectious insult-related hypoplasia showing no decreased of lobules (# of lobules >10).

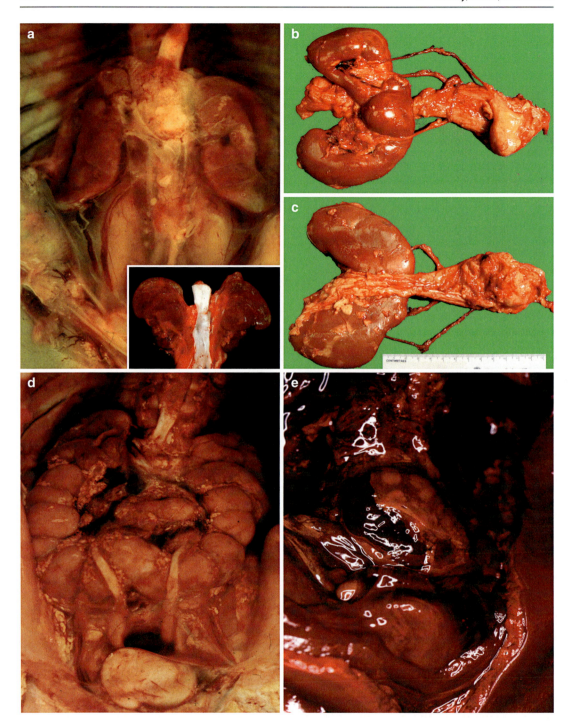

Fig. 6.3 Non-cystic congenital anomalies of the kidney showing (**a**) two cases of renal agenesis, while in (**b**) and (**c**), the anterior and posterior view of a horseshoe kidney with duplicitas of the ureters. Figure (**d**) shows a horseshoe kidney in *in situ* preparation, while figure (**e**) shows an abnormal low location of the left kidney (renal ptosis)

> **Box 6.3 Renal Dystopia (Unilateral/Bilateral)**
> - *Subdiaphragmatic Renal Dystopia* (the renal arteries depart at the level of the T12 vertebra with abdominal or thoracic kidney)
> - *Lumbar Renal Dystopia* (the renal arteries extend from the aorta at the level of L2 to the aortic bifurcation)
> - *Ivy Renal Dystopia* (the renal arteries depart from the common iliac arteries, and both kidneys are located in the ileum)
> - *Pelvic Renal Dystopia* (the renal arteries depart from the internal iliac artery so that the kidney can lodge in the sacrum or between the urinary bladder and the rectum in male and the Douglas space in female)

6.2.2 Disorders of Rotation

Disorders of rotation include the renal pelvis malrotation and the duplications of the upper urinary tract.

6.2.3 Disorders of Position

The left kidney is located higher than the right due to the presence of the liver, but renal ptosis can occur at the time of the ontogenesis (*ectopia renalis sive renis/renum*) (Box 6.3).

6.2.4 Disorders of Separation

Fusion of both kidneys (1/600 individuals) with the formation, typically, of a horseshoe kidney mostly of the lower poles (90%) and rarely of the upper poles (10%) constitute the disorders of separation.

6.3 Cystic Renal Diseases

6.3.1 Classifications

Renal cystic diseases are a very heterogeneous group of disorders with the potential to become life-threatening conditions leading to chronic renal insufficiency, dialysis, and death unless renal transplantation is not successfully performed (Figs. 6.4, 6.5, 6.6, 6.7, and 6.8). All renal cysts have mostly in common only the formation of fluid and are filled with it. Cystic change may be uni- or bilateral, may affect part or all the kidney, and may involve the cortex, medulla, or both. A renal cyst may be defined as "a fluid-filled sac due to dilation in any part of the nephron unit or collecting duct." There are quite a few modalities to approach renal cysts, and there is no perfect classification useful for both congenital and acquired cystic diseases. A classic concept of renal cystic diseases is to classify them according to the embryologic disorder, although it is not always known. A primary disorder of renal development or secondary changes in portions of the nephrons or collecting ducts that initially differentiated may constitute a mode to distinguish two essential categories. The classic system of Osathanondh and Potter was based on detailed microdissection studies with no or few clinical or genetic correlations. In addition to this system, the Welling-Grantham and the Bernstein-Gilbert classifications are used (Box 6.4).

The term *renal dysplasia* implies any developmental abnormality resulting from anomalous metanephric differentiation. The anlage of the kidney, including the *ureteric bud* and the *metanephric blastema*, have both formed embryologically. However, both subsequently have failed to interact correctly and develop in a standard way. According to the time of occurrence of the injury, such as physical, chemical, or genetic, on the particularly sensitive system that involves the interaction of *ureteric bud* and *nephrogenic blastema*, a spectrum of abnormalities can be observed, including renal agenesis or aplasia to hydronephrosis.

6.3.2 Autosomal Dominant Polycystic Kidney Disease

It is the most common inherited form ("adult"; Potter type III, relatively familiar with an incidence of 1:500), which is usually bilateral but eventually unilateral or, even, focal. ADPKD

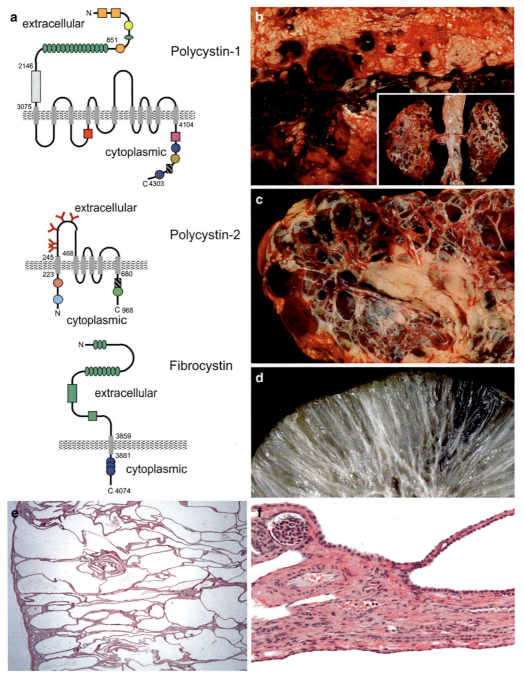

Fig. 6.4 Autosomal dominant polycystic kidney disease and autosomal recessive polycystic kidney disease. Figure (**a**) shows three proteins, the abnormal composition of which can be linked to renal cystic disease. These are the schematic diagrams of the protein domain structures of polycystin-1, polycystin-2, and fibrocystin. The colored boxes or circles on the main backbone of the proteins (identified as thick black line) indicate conserved domains or regions that have a high degree of homology with known proteins. Annotation of domains and regions is based on the evaluation of SwissProt and Genbank databases for the following accession numbers: NP_000287 (polycystin-1), Q13563 (polycystin-2), and AAL74290 (fibrocystin). Also, literature data has been used. The boundaries of selected regions are given as the number of amino acid residues from the N-terminus of the protein (N). A wavy horizontal line corresponds to the cell membane, while the transmembrane domains are shown by gray rectangles. These rectangles cross the cell membrane. Figures (**b**) and (**c**) show the gross photographs of kidneys with autosomal dominant polycystic kidney disease (ADPKD), while figures (**d-f**) show the gross photograph and the microphotographs of autosomal recessive polycystic kidney disease

6.3 Cystic Renal Diseases

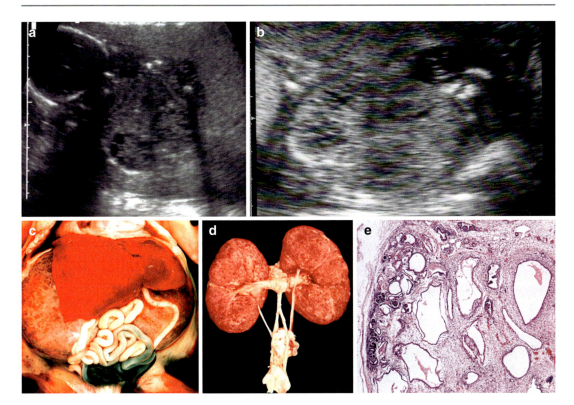

Fig. 6.5 Ultrasound and anatomic-pathologic panel of a fetus affected with Meckel-Gruber syndrome in the second trimester of pregnancy. Figure (**a**) shows the encephalocele on ultrasound, while figure (**b**) shows the cystic renal changes. Figures (**c**) and (**d**) show the nephromegaly of both kidneys, and figure (**e**) shows the histology of cystic renal dysplasia, which is characteristic of Meckel-Gruber syndrome

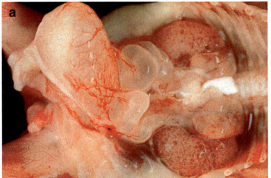 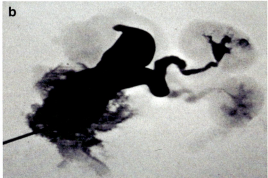

Fig. 6.6 Figures (**a**) and (**b**) show the anatomopathologic view and the postmortem radiologic view using contrast medium of multicystic kidney disease due to an obstruction of the urethra (urethral valves) with dilatation of the ureters and cystic changes of the kidney cortex bilaterally (**c**). Figure (**e**) represented the schematic diagram of nephrocystin. Nephrocystin-1 is a protein that in humans is encoded by the *NPHP1* gene, which encodes a protein with src homology domain 3 (SH3) patterns. *NPHP1* gene mutations cause familial juvenile nephronophthisis as shown in a histology macro-slide of the kidney (**d**)

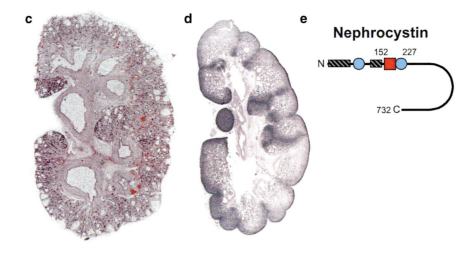

Fig. 6.6 (continued)

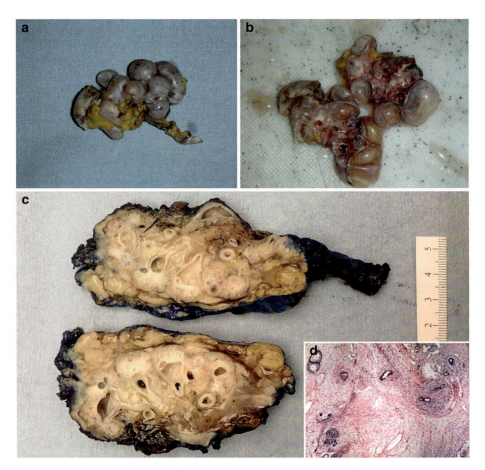

Fig. 6.7 Figures (**a**) through (**d**) show multicystic kidney dysplasia, which is often unilateral and very rarely may be accompanied by tumors (e.g., Wilms tumor or renal cell carcinoma). In the inset of (**c**), histology of the multicystic kidney dysplasia (**d**) is presented

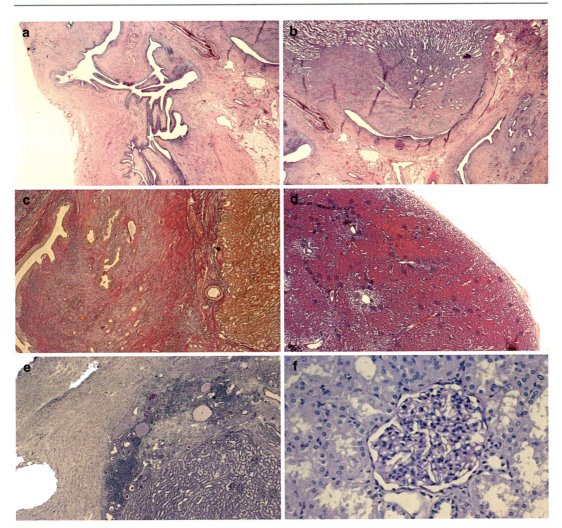

Fig. 6.8 Segmental renal dysplasia with abnormal structures identified on histology (**a–f**) Renal dysplasia can be uni- or bilateral, segmental, and of variable severity, with renal aplasia identifying the extreme dysplasia. The segmental type is highlighted by a dysplastic area involving a portion only of the kidney. In the focal type there is an admixture of normal and aberrantly formed nephrons (a, hematoxylin and eosin staining, x40; b, hematoxylin and eosin staining, x40; c, elastica Van Gieson staining, x50; d, hematoxylin and eosin staining, x40; e, periodic acid Schiff staining, x40; f, periodic acid Schiff staining, x400)

accounts for 90% of all polycystic kidney disease, which manifests around age 30 and is under the influence of mutations in at least three different genes (*PKD1*, *PKD2*, and *PKD3*). *PKD1* is the gene responsible in approximately 85% of the patients, is located on chromosome 16 (16p13.3, 46 exons, 4302 amino acids), and encodes polycystin 1 (PKD1), which is an integral membrane glycoprotein. *PKD2* is the gene involved in the clear majority of the remaining cases and is located on chromosome 4 (4q21–23) and encodes for polycystin 2 (PKD2), which is an integral membrane glycoprotein. PKD1 is restricted to epithelial cells and shows high levels of expression in cortical tubules and moderate levels of expression in loops of Henle and collecting ducts. PKD2 is a transmembrane protein with amino acid similarity with PKD1 and the family of voltage-activated Ca^{2+} (and Na^+) channels. PKD1 and PKD2 function together as part of a multicomponent membrane-spanning complex involved exquisitely in cell-cell or cell-matrix interactions. Finally, a further subset seems to be linked to *PKD3* gene, although it is still under intense investigation. Larger cysts that form over time and grow in size constitute

Box 6.4 Three Classifications of Renal Cysts at a Glance

Morphologic patterns of renal cystic disease	Potter	Welling-Grantham	Bernstein-Gilbert
Renal cysts distributed *throughout* the cortex and *medulla* of both kidneys with uncountable 1/2 mm cysts on the outer surface and dilated channels at right angles to the cortex, replacing the cortex and medulla substantially and fully	*Type I*	ARPKD	–
Large dysplastic kidneys with peripherally located cysts and *ureteral atresia*	*Type II*	–	*Multicystic dysplasia*
Barely recognizable dysplastic kidneys, minimal/no cystic at all with *ureteral atresia*	*Type II*	–	*Aplastic dysplasia*
Small dysplastic kidneys with *patency of dilated ureters*	*Type II*	–	*Hypoplastic dysplasia*
Large, diffusely dysplastic kidneys with numerous thick-walled collecting ducts, no lobar development or nephronic differentiation and *patency of ureters*	*Type II*	–	*Diffuse cystic dysplasia*
Renal cysts distributed *throughout* the cortex and *medulla of both kidneys* with numerous, spherical fluid-filled cysts with Ø from few mm to cm	*Type III*	ADPKD	–
Cystic changes in the peripheral cortex only with variable dysplastic medullary development	*Type IV*	–	*Obstructive dysplasia*
Isolated bars of metaplastic cartilage or rudimentary lobules in an otherwise apparently normally developed kidney ± ureteral duplication	(*Type IV*)	–	*Segmental dysplasia*

the pathology. Kidneys show a diffuse hyperechogenicity with enlarged kidneys and bilateral cysts. ADPKD is characterized by cystic dilation in all parts of the nephron, including the Bowman's space, in a patient of age <30 years with at least two cysts in one kidney. Kidneys show areas of normal and abnormal nephron differentiation. Prenatal ultrasounds show normal or enlarged and echogenic kidneys. Interestingly, cysts are also present in the liver, pancreas, spleen, and lungs. Affected children and adolescents show gross or microscopic hematuria, hypertension, cyst infection, and renal insufficiency. Urine concentrating defect is an early marker. The floppy mitral valve of the heart is commonly seen in 1/6 of cases, while Berry aneurysms in the brain are present in 1/3 of ADPKD patients, who died from renal failure, and in 1/3 of ADPKD patients from complications related to arterial hypertension.

6.3.3 Autosomal Recessive Polycystic Kidney Disease

It is a rare genetic disorder with small cysts ("infantile"; Potter type I, perinatal, juvenile renal cysts) of the collecting ducts. Symptoms begin in utero at 19th–20th week of gestation (hyperechogenic and enlarged kidneys, pulmonary hypoplasia, and oligohydramnios) or in the postnatal period. The incidence is 1 in 40,000 and typically presents in infancy with hyponatremia during the first weeks of the neonatal period, reduced the renal concentrating ability, and decreased urinary acidification capacity, metabolic acidosis, and chronic pyuria. Arterial hypertension is quite common and occurs early in the disease course. In older children, hepatic disease predominates, and ¼ of cases show variceal bleeding. End-stage renal disease (ESRD) usually occurs after the age of 15 years. Morphologically, there is a nonobstructive dilatation/ectasia of the collecting ducts in the renal medulla with longitudinal cysts that run parallel from the cortex to the central medulla starting at a right angle from the cortex or the renal capsule. There is a broad range of severity proportional to the % of nephrons affected by cysts. The hepatic involvement shows cysts, fibrosis, and portal hypertension and is called "congenital hepatic fibrosis (CHF)."

6.3.4 Medullary Sponge Kidney

It is also considered a form of pseudocyst formation of the kidney in the papillary areas (aka familial juvenile nephronophthisis, hereditary tubulointerstitial nephritis). There is congenital dilatation of collecting ducts ranging from ectasia, which may appear as linear papillary striations on urography, to cystic pools. This latter pattern is the origin of the name "spongelike" appearance on a section of the kidney. MSK is often hereditary (65%, juvenile onset, AR; 15%, adult onset, AD). The onset in childhood is progressive (5–10 years to renal failure). Renal insufficiency results from tubulointerstitial damage. There is an estimated prevalence of 1 in 5000. There is a predisposition to nephrolithiasis in the dilated ducts with clustering of calcifications in the papillary areas on plain films. Medullary sponge kidney may be associated with other congenital and inherited disorders, including Beckwith-Wiedemann syndrome, ADPKD, CHF, and Caroli disease. Histologically, there are contracted, granular, medullary cysts, most prominent at the corticalmedullary junction and associated with tubular atrophy, thickening of basement membranes, and interstitial fibrosis.

6.3.5 Multicystic Dysplastic Kidney (MCDK)

It is a sporadic, non-hereditary form of cystic renal disease with developmentally abnormal kidney ("multicystic kidney"; Potter types II and IV). It consists of an ureteral bud abnormality leading to the atresia or absence of the ureter, persistence within the organ of undifferentiated mesenchyme, immature and cystically dilated collecting ductules (primitive ducts), and, tremendously important from a diagnostic point of view, primitive cartilaginous islands (20% pathognomonic). MCDK can be unilateral (60–70%) or bilateral (30–40%) and occurs in 1:3000 with ♂ > ♀ prevalence. On the US, the affected kidney is grossly enlarged. There is loss of the reniform shape. There is an atretic or absent ureter with multiple variably sized noncommunicating cysts separated by little or no echogenic parenchyma. Technetium-99m dimercaptosuccinic acid or DMSA is a radioisotope that is injected into the body is rapidly adsorbed and excreted by the kidneys. A gamma camera, which is a special apparatus, is used to gather photographs of the kidney (DMSA renal scan). This diagnostic imaging exam is able to evaluate the function, size, shape, and position of both kidneys detecting precisely tissue scarring caused by congenital defects or frequent infections. Voiding cystourethrography (VCUG) is a technique able to detect the urethra and urinary bladder at time of urination. VCUG requires the filling of the urinary bladder by a radiocontrast agent (e.g., diatrizoic acid). DMSA shows an absence of function in the affected kidney, VCUG is useful in evaluating vesico-urethral refulx (VUR), and US should be repeated every 6–12 months. Morphologically, there are enlarged usually cystic organs often showing an irregular contour, with disorganized parenchyma and difficulty to distinguish the cortex from the medulla. In 90% of cases, there is outflow obstruction from ureteral atresia or posterior urethral valves. MCDK has been reported in a variety of syndromes, including Beckwith-Wiedemann syndrome, trisomy 18 syndrome, and VACTERL among others. The role of nephrectomy is controversial. It has been recommended to treat it in order to prevent urinary tract infection (UTI), arterial hypertension (HTN), or renal malignancy (e.g., Wilms tumor or renal carcinoma).

6.3.6 Hydronephrosis/ Hydroureteronephrosis

It represents an obstruction of urinary outflow leading to cystic dilatation of the ureter, pelvis, and renal calyces. If the blockage occurs early in gestation, it invariably produces some degree of cystic renal dysplasia in the setting of the multicystic dysplastic kidney group. In young adults, there is progressive atrophy of renal cortex until only a thin rim remains, and this depends on the time and degree of obstruction.

6.3.7 Simple Renal Cysts

Renal cysts may be single or multiple. They are commonly noted on X-ray, and their incidence increases with age. The lining is smooth, and the lumen is filled with clear fluid. There is, usually, no impact on renal function. In some cases, bleeding can occur causing flank pain. The most critical management issue is to differentiate the cysts both radiologically and histologically from a neoplasm.

6.3.8 Acquired Cystic Kidney Disease (CRD-Related)

In the setting of chronic uremia, the development of many cysts characterizes this form of polycystic kidney disease (chronic renal disease-related). The diagnosis is based on detecting at least three cysts in each kidney in a patient with chronic renal failure. This form of polycystic kidney disease averages 10% at the onset of dialysis treatment. Later, it increases, to reach 60% and 90% of the renal parenchyma at 5 and 10 years, respectively. Grossly, kidneys are small or even shrunken in the early stage, and cysts are usually smaller than 0.5 cm. Subsequently, cystic numbers and kidney volume increase with time.

6.3.9 Genetic Syndromes and Cystic Renal Disease

This group collects a large and heterogenous pool of cystic renal disease in the setting of genetic syndromes. There is an obstruction of urinary outflow leading to cystic dilatation of the ureter, pelvis, and kidney. Genetic syndromes associated with cystic renal disease are numerous. Wolf-Hirschhorn syndrome (WHS) is a genetic disease that affects several organs and compartments of the body. The significant features of WHS include a dysmorphic face, delayed growth and development, mental retardation, hypotonia, and seizures. Other features may include skeletal abnormalities, congenital heart defects, diaphragmatic defects, hearing loss, urinary tract malformations, and brain abnormalities. In WHS, there is the deletion of genetic material close to the end of the short arm of chromosome 4. Although most patients with WHS are not inherited with parents showing a numerically and structurally normal karyotype, some individuals with WHS are inherited, and the deletion originates intriguingly from a parent who does not have WHS. The infants show bradycardia and difficulty to breath spontaneously at birth. Chest X-ray shows a diaphragmatic defect, bilateral lung hypoplasia, and a shift of the mediastinum to the right. The examination of the face of patients with WHS shows microcephaly, prominent glabella, broad nasal bridge, hypertelorism, poorly differentiated and low-set ears, a bilateral palatoschisis, and micrognathia. Midline closure defects of the cervical spine bodies, lower jaw, and skull base may also be identified. Cytogenetics confirms the clinical diagnosis of WHS identifying the 4p deletion. As indicated above, there is often bilateral lung hypoplasia and bilateral renal hypoplasia with renal cortex harboring primitive cystic ducts and reduced nephrogenesis, and the medulla contains an ↑ amount of connective tissue. In cortical cysts, a *stenosis of the ureteropelvic junction* can be postulated, and the cystic renal disease may correspond to a somewhat late injury affecting the nephrogenic zone more severely. It leads to the formation of primitive tubules and nephronic elements in the outer cortex (Potter type IV). In *renal dysplasia*, the concept of *altered metanephric differentiation* is basilar. It can be recognized in at least one of the following components is present. An altered metanephric differentiation reveals *primitive ducts* (ducts or small cysts lined with undifferentiated columnar epithelium and surrounded by fibromuscular collars), *metaplastic cartilage* (island of hyaline cartilage), and the *lobar disorganization*, i.e., the lack of a traditional architectonics of the kidney. The Potter microdissection studies have demonstrated that in Potter type II, the cysts occur most often at the *terminal ends of short primitive ducts*, while in Potter type IV, the cysts occur *at the ends or along the course of collecting tubules and terminal duct branches*. This situation suggests that the obstruction alters the differentiation of late generations of nephrons after that the early generations of nephrons had begun to secrete urine. Conversely, no contact of the ureteral bud with the metanephric blastema produces renal agenesis or aplasia, while an obstruction in wholly concluded development of the kidney results in hydronephrosis.

6.4 Primary Glomerular Diseases

6.4.1 Hypersensitivity Reactions and Major Clinical Syndromes of Glomerular Disease

Multi-etiologic GN (aka acute diffuse intracapillary proliferative GN) is presenting with an abrupt onset. Glomerulonephritis is defined as a renal disease in which the significant pathologic changes are confined to the glomeruli. It is crucial to distinguish diffuse vs. global involvement when all glomeruli vs. the entire glomerulus are involved and focal vs. segmental involvement when one area of the kidney or part of the tuft are included, respectively.

Primary pathogenic mechanisms are in Box 6.5, while the hypersensitivity reactions are listed in Box 6.6. Before examining the primary glomerular diseases, it is essential to recall the hypersensitivity responses, and a list of the major clinical syndromes of glomerular diseases will be the introduction to this section as well.

Type I is characterized by the initial introduction of Ag, which produces an Ab response, particularly an IgE (Ab) response. IgE binds very exquisitely and specifically to receptors on the surface of mast cells, which remain circulating. The reintroduction of the Ag, which interacts with IgE localized on the surface of mast cells, causes the cells to degranulate and release histamine, lipid mediators, and chemotactic factors. The chemical-related events are smooth muscle contraction, vasodilation, increased vascular permeability, bronchoconstriction, and edema (e.g., asthma).

Type II or cytotoxic hypersensitivity also involves antibody-mediated cellular reactions, but the isotype is usually IgG, and the "culprit" cells are cytotoxic lymphocytes rather than mast cells. Thus, type II is also labeled antibody-dependent cell-mediated cytotoxicity (ADCC). Of course, there may be activation of complement, which binds to the cell-bound antibody. Moreover, the antibodies can cross-react with "self" antigens determining "collateral" damage of the host tissue, as observed, for instance, in pemphigus, autoimmune hemolytic anemia, and Goodpasture syndrome.

Type III or immune complex hypersensitivity involves a circulating antibody that reacts specifically with a free circulating antigen. The Ag-Ab complexes are circulating and eventually deposit on tissues, which can start the activation of complement. The consequence is manifested in the tissue damage. Two variables of type III or IC hypersensitivity include the type of Ag and Ab and the final size of the resulting immune complex. Very small ICs remain in circulation, and huge ICs are removed by the glomerulus mesangial cells, while an intermediate size of IC may facilitate its

---

Box 6.5 Pathogenic Mechanisms
- Deposition of *preformed* IC in the glomeruli
- Formation of IC by the interaction of circulating antibodies with *Ag deposited in the glomeruli*
- Direct communication of *circulating antibodies with components of the glomerulus*

Box 6.6 Hypersensitivity Reactions
- *Type I*: Allergy-related (immediate) hypersensitivity
- *Type II*: Cytotoxic hypersensitivity
- *Type III*: Immune complex hypersensitivity
- *Type IV*: Delayed hypersensitivity

> **Box 6.7 Major Clinical Syndromes of Glomerular Disease**
> - *Nephritic Syndrome or Acute Glomerulonephritis* (abrupt onset, ± previous infection, ~ degree of hematuria/proteinuria and ↓GFR, and tendency to spontaneous recovery)
> - *Nephrotic Syndrome* (insidious onset, ± previous infection, proteinuria, edema, hypertension, lipidemia, and lipiduria)
> - *Rapidly Progressive Glomerulonephritis* (insidious onset, ± previous infection, progressive loss of renal function with oliguria, no/little tendency to spontaneous recovery)
> - *Chronic Glomerulonephritis* (insidious onset, difficult to discriminate between infection, vascular, or immunologic cause, ~ degree of hematuria/proteinuria/hypertension, relentless progressive deterioration of renal function)
> - *Mostly Asymptomatic Urinary Abnormalities* (proteinuria <3 g/day ± hematuria with few or no symptoms)

lodging and, finally, determine renal damage. An example of type III is serum sickness.

Type IV starts with the introduction of an Ag (e.g., an intracellular pathogen such as *M. tuberculosis*), which produces a cell-mediated response. Recovery requires induction of specific T-cell clones, which cause activation of macrophages and destruction.

The primary clinical syndromes of glomerular disease are in Box 6.7.

6.4.2 Post-infectious Glomerulonephritis

- *DEF*: Multi-etiologic GN presenting with an abrupt onset of the acute nephritic syndrome, ↑ASO, and ↓C3 in the plasma, ↓ GFR, salt/H$_2$O retention (ASO, antistreptolysin O; GFR, glomerular filtratio rate) following an infection.
- *SYN*: Acute diffuse intracapillary proliferative GN.
- *EPI*: Children > adults, ♂ > ♀, worldwide.
- *EPG*: Bacteria (nephritogenic type group A β-hemolytic *Streptococcus*, *Salmonella*, *Enterococcus*, *Staphylococcus*, spirochetes), viruses (HBV, EBV), protozoa (malaria, *T. gondii*) as triggering factors and *Type III hypersensitivity* (immune complex mediated) as pathogenesis. ASO is a toxic enzyme produced by group A *Streptococcus* bacteria.
- *CLI*: Abrupt onset of *acute nephritic syndrome* with darkening of urine, malaise, oliguria, edema, proteinuria, ↓ GFR with salt and H$_2$O retention ± ↑ BUN, FENa<0.5%, ↑ P-Cr, edema involving the face, eyelids, and hands (worse in the morning), circulatory congestion, HTN without retinal alterations (BUN, blood urea nitrogen; FENa, fractional excretion of sodium; P-Cr, plasmatic creatinine). It manifests 1–4 wks (~10 days) after "illness" (atypical features: encephalopathy, headache, somnolence, or seizures) ± (+) throat or skin cultures, antibody markers (ASO, antihyaluronidase, antideoxyribonuclease, and anti-nicotine adenine dinucleotidase), and ↓C3 with ~nl. C1q-C2-C4 due to the prominent involvement of the alternative pathway of complement activation.
- *GRO*: Pale renal cortex with petechial hemorrhages grossly.

> - *CLM*: Diffuse and global *endocapillary proliferative GN* with *tuft enlargement and hypercellularity* with neutrophils, monocytes, and eosinophils within the capillary lumina leading to partial capillary obliteration, filling of Bowman space and mesangial and epithelial proliferation, with or without crescents (MT or TB stains useful to highlight "gumdrop" or "domed" deposits on the epithelial side of the BM) (Fig. 6.9).

- *IFM*: (1) Granular ("*lumpy bumpy*" or "*starry sky*") deposition of *IgG-C3* in peripheral loops (glomerular capillary wall) in the first 2 weeks;

6.4 Primary Glomerular Diseases

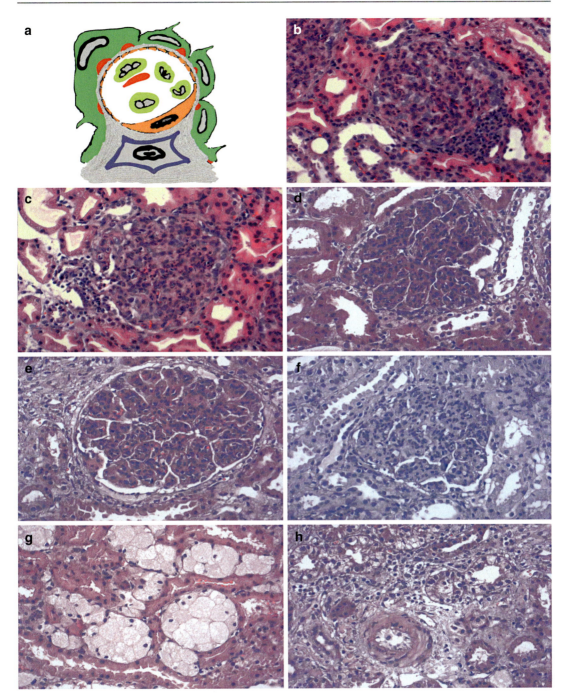

Fig. 6.9 Post-infectious glomerulonephritis with infiltration of the glomeruli by neutrophils (more than 20 per glomerulus) (**b–f**). In (**a**) is shown a cartoon of the post-infectious glomerulonephritis with neutrophilic recruitment. Figure (**g**) and figure (**h**) show some foamy cell changes of tubular epithelial cells and a thickening of the arteriole. Both findings are rarely encountered in post-infectious glomerulonephritis and may suggest an interstitial involvement. All microphotographs have been taken from H&E-stained slides apart of (**f**), which was stained with periodic acid Schiff and all photos have an original magnification of x400

(2) *mesangial IgG-C3 deposition* in the subsiding phase; (3) *garland* pattern (densely packed deposits decorating the peripheral capillary walls) in persistent proteinuria (poor PGN factor); (4) fibrinogen in a mesangial pattern.
- *TEM*: Subepithelial "humps" (dome deposits due to IgG-C3 complexes projecting outward from the epithelial side of the BM) ± small subendothelial deposits, foot process effacement over deposits and atypical humps (numerous and involving adjacent areas).
- *TRT*: H_2O/Na^+ restriction, loop-acting diuretics.
- *PGN*: 95% spontaneous recovery (if no crescents → diuresis back <1 week and histology usually unremarkable after 6–36 months), but ~½ of youth may develop RPGN. Poor PGN factors: (1) crescents ≥30% of the glomeruli, (2) garland IF pattern, and (3) atypical humps.

Crescents are due to injury to glomerular capillary wall. There may be *circulating antibodies to intrinsic or aka "planted" glomerular antigens* (e. g., classical anti-GBM Ab-mediated disease, glomerulosclerosis, type I immune complex disease) and *circulating immune-complexes-mediated disease* (e. g., toward exogenous or endogenous antigens or ICGN type II), and *cell-mediated disease* (e. g., ICGN type III). The cellular composition of the crescents depends on two factors, including the stage of disease and intactness of the Bowman capsule. An intact capsule is at the basis of an almost pure epithelial cellular component of the crescent, but a disruption of the capsule allows monocytes/macrophages to activate fibroblast migration, which may undoubtedly lead to a fibrous scar.

6.4.3 Rapidly Progressive Glomerulonephritis

- *DEF*: Multi-etiologic GN characterized by acute nephritic syndrome and poor prognosis.
- *SYN*: Diffuse crescentic GN (extracapillary proliferative GN).
- *CLI*: Acute nephritic syndrome (oliguria, azotemia, proteinuria, hematuria, HTN with eventual anuria & ESRD).

Three *Etiologic Different Subtypes (PIG-PI)*

Post-infectious-Anti-*G*BM-*P*auci-*i*mmune with *Monomorphic LM Phenotype*
Glomerular crescents (proliferating parietal epithelial cells and infiltration of Mo/Mφ following GBM/Bowman capsule disruption with fibrin and collagen deposition through the activation of fibroblasts), segmental necrosis, glomerular capillary collapse, tubular atrophy, and interstitial inflammatory infiltrate (Fig. 6.10).

Type I or anti-GBM GN (If + lung involvement = *Goodpasture syndrome*)

IFM: Diffuse linear staining of GBM with IgG > IgM > IgA, granular staining for C3, fibrinogen within glomerular capillary loops; ± linear staining of tubular BM for Ig.

TEM: Fibrin associated with GBM breaks, but no deposits

Type II or post-infectious GN (severe form/progression of PIGN but better PGN than other types)

IFM: Granular IgG-C3 IC (type III hypersensitivity)

TEM: Subepithelial "humps," mesangial deposits, and fibrin deposition with GBM breaks

Type III or pauci-immune GN: ± vasculitis, + ANCA

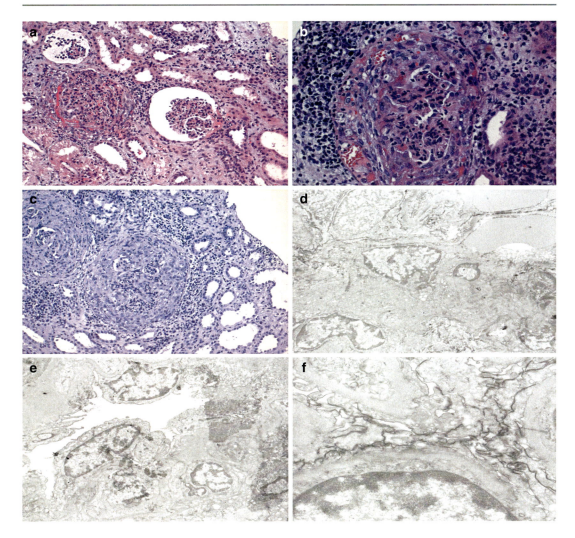

Fig. 6.10 The rapid progressive glomerulonephritis (RPGN) is a renal syndrome that is characterized by a rapid loss of renal function approaching typically a 50% decline in the glomerular filtration rate (GFR) within 3 months and glomerular crescent formation seen in 50–75% of glomeruli observed on kidney biopsies. In half of cases, RPGN is associated with an underlying disease such as Goodpasture syndrome, systemic lupus erythematosus, or granulomatosis with polyangiitis; the other half of cases are idiopathic. Histologically, there is hypercellularity of the glomerulus with crescents as seen in figures (**a**–**f**) using both light and electron microscopy

p-ANCA, MPO-ANCA, MPA (MPA, Microscopic Poly-Angitis)

c-ANCA, PR3-ANCA, WG (WG, Wegener Granulomatosis)

IFM: No Ig staining, ± C, and fibrinogen with crescents

TEM: Fibrin associated with GBM breaks, but no deposits

Acute Nephrotic Syndrome-Associated Glomerulopathies

A multi-etiologic clinical entity with increased glomerular permeability and presentation as massive proteinuria and lipiduria with edema, hypoalbuminemia, and hyperlipidemia. Proteinuria >40 mg/m²/h (or > 3.5 g/day/1.73 m² BSA or > 50 mg/kg/24 h). Non-nephrotic range of proteinuria: 4–40 mg/m²/h. Since clinical and pathologic overlap exists, minimal change disease, diffuse mesangial proliferative, and focal and segmental glomerulosclerosis may be part of a spectrum of the same disease process.

- *PGN*: Rapid, generally irreversible course and therapy (steroids, cyclophosphamide, plasmapheresis) usually do not help.

6.4.4 Minimal Change Disease

- *DEF*: Multi-etiologic GN occurring as an acute nephrotic syndrome in 2–6-year-old children with phenotypically intact glomeruli on light microscopy but fusion of foot processes on electron microscopy.
- *SYN*: Lipoid nephrosis, nil disease, foot process disease.
- *EPG*: Idiopathic (primary) nephrotic syndrome (with atopy and HLA B12 or without atopy, often with Asian-Arabian ethnics) and secondary to HL, NHL, and solid tumors such as RCC in youth compared to pancreatic carcinoma, mesothelioma, colon carcinoma, prostate carcinoma, and syphilis of older adults. Immune-mediated T-cell dysregulation-related process with selective proteinuria usually without hematuria, HTN, or loss of renal function.
- *CLI*: Proteinuria-related symptoms and signs.
- *CLM*: Unremarkable, although occasionally tubules may show hyaline droplets.
- *IFM*: Negative (occasionally, complement and fibrinogen in peripheral capillary walls).
- *TEM*: Foot process "fusion" (polyanion content change → juxtaposition of the foot processes of visceral epithelial cells and obliteration of the slit pore), microvillus change of epithelial cells.
- *TRT*: Good response to steroids, but, rarely, immunosuppression is required (e.g., cyclosporine).
- *PGN*: Relapsing/polycyclic course; 70–80% full remission; 1–2% death.

6.4.5 (Diffuse) Mesangial Proliferative GN

- *DEF*: Glomerulonephritis with diffuse cellular proliferation of the mesangium and some increase in mesangial matrix with usually IgA or IgM deposits and variable outcome.
  - With predominant IgA mesangial deposits (IgA nephropathy or Berger disease) => hematuria
  - With predominant IgM and/or C3 deposits => variable (~) presentation
  - With other patterns of Ig and C3 deposits
  - No Ig or C3 deposits
- *EPG*: Secondary to underlying disease, including resolving PIGN, SLE, RA, HSP, Alport syndrome, Goodpasture syndrome, Kimura disease, D-penicillamine. Penicillamine is an anti-rheumatic drug used to treat the active phases of rheumatoid arthritis. Penicillamine is also used for chelation of Cu (copper) in Wilson disease.
- *CLM*: Variable degree of increase in mesangial cellularity (diffuse and global) without capillary wall changes (Figs. 6.11 and 6.12).
- *IFM*: ± IgA, IgM, C3 mesangial deposits (IgG in PIGN) (*vide supra*).
- *TEM*: ± Finely granular or homogeneous mesangial deposits.

6.4.6 Focal and Segmental Glomerulosclerosis

- *DEF*: Multi-etiologic GN presenting as an acute nephrotic syndrome (nonselective proteinuria) and considered in most cases (at least with secondary causes) a pattern of response rather than a specific disease with hyalinosis and sclerosis representing entrapment of plasmatic proteins in extremely permeable glomeruli and reactive mesangial contribution.
- *EPI*: ~10% of childhood and ~15% of the adult nephrotic syndrome
- *EPG*: Primary and secondary (*DISH*), including *D*rugs (heroin), *I*nfections (HIV, VUR-associated), *S*oft tissue (obesity, obstruction, radiation, sarcoidosis, solid tumors, nephron ablation including partial nephrectomy, bilateral cortical necrosis, segmental hypoplasia,

6.4 Primary Glomerular Diseases

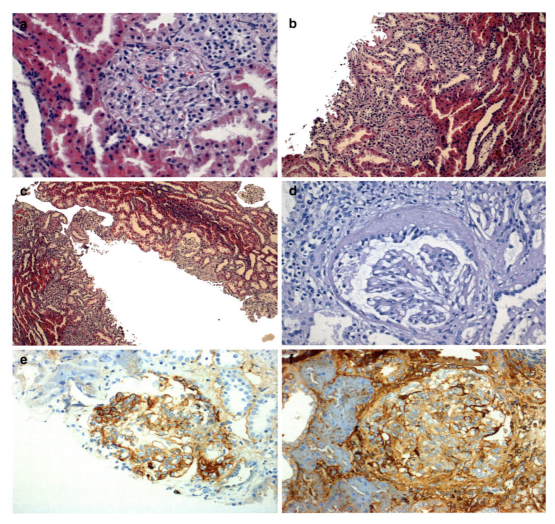

Fig. 6.11 Mesangial proliferative glomerulonephritis is a hypocomplementemic (C3) glomerulonephritis due to subendothelial and mesangial deposition of complement from dysregulation of alternative pathway. MesPGN includes dense deposit disease (membranoproliferative glomerulonephritis type 2) and proliferative glomerulonephritis with C3 deposits. There are large glomeruli with accentuation of lobules, irregular thickening of glomerular basement membrane by interposition of mesangial cells between basement membrane and endothelium (**a-c**). There is crescent formation (**d**). Ig M and IgG depositions are shown in figures (**e**) and (**f**), respectively. Both figures (**e**) and (**f**) have been performed using antibodies against IgM and IgG by immunohistochemistry (avidin-biotin complex, ×400 original magnification). In MesPGN genetic mutations or the development of autoantibodies to complement may be encountered

and unilateral agenesis), and *H*ematologic (sickle cell disease, aging).
- Primary causes include mutations in five genes that encode slit diaphragm proteins: NPHS1-*nephrin*, NPHS2-*podocin*, *Actinin-4*, TRCP6 (*transient receptor potential channel 6*), and CD2AP (*CD2-associated protein*).
- HIV-associated (AIDS) FSGS (IVDA): *FSGS-collapsing glomerulopathy* with rapid progression to RF and characterized by *hyperplasia of glomerular epithelial cells*, *tubular hyaline casts*, and *interstitial nephritis* on CLM and *tubular-reticular structures* in endothelial cells on TEM.
- *CLM*: (1) *FS-SCLEROSIS of glomeruli* (more common juxtamedullary) with hyaline subendothelial material ± lipid inclusions, synechiae, and foamy histiocytes; (2)

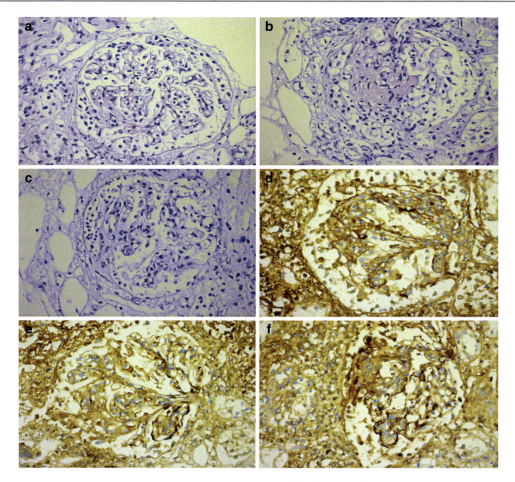

Fig. 6.12 IgA nephropathy is a type of diffuse mesangio-proliferative glomerulonephritis (**a-c**) with elevated serum IgA and IgA present in mesangium (**d-f**; antibody against IgA, avidin-biotin complex, ×400 original magnification). All microphotographs have been taken at x400 as original magnification. The microphotographs (**a**) and (**b**) are from H&E stained slides, while (**c**) is from a periodic acid Schiff-stained slide. The poor eosin staining in (**a**) and (**b**) is due to a low exposure to eosin or prolonged exposure to hematoxylin

epithelial cell detachment from GBM (clear zone or halo) (early) and hypertrophy and hyperplasia (late); (3) *TBM thickening, tubular atrophy*, and *interstitial fibrosis* (Fig. 6.13).
- *IFM*: IgM and C3 in sclerotic segments (non-specific and seldom seen).
- *TEM*: Collapse of glomerular capillary loops with foot process fusion and ↑ of mesangial matrix and sclerosis.
- *TRT*: Steroid-resistant PGN, 10-YSR, 50% (early presentation of FSGS has worse clinical course than the late presentation) as risk of recurrence in allografts.

6.4.7 Membranous Glomerulonephritis

- *DEF*: Clinicopathologic insidious and indolent entity commonly associated with *Nephrotic Syndrome* and characterized by diffuse thickening of the glomerular capillary loops due to subepithelial IC deposits (type II and type III hypersensitivity).
- *EPI*: MGN accounts for 1/20 and approximately half of nephrotic syndrome in children and adults, respectively.
- *EPG*: Idiopathic or type II related (85%) and secondary or type III related with several etiolo-

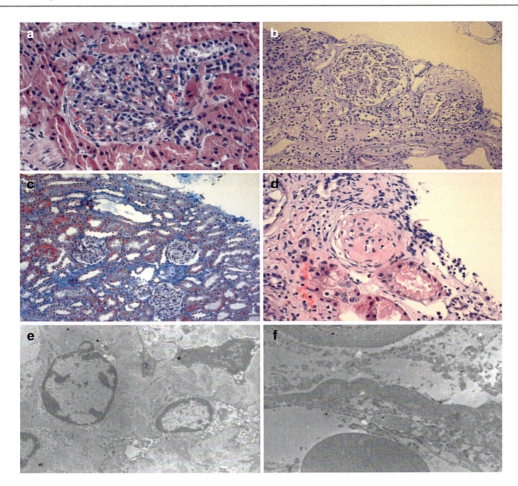

Fig. 6.13 Focal segmental glomerulosclerosis shows a sclerosis, which is focal and segmental with mesangial sclerosis in lobules that appear to adhere to Bowman's capsule, starting in corticomedullary region. Inframembranous hyaline deposits (periodic acid-Schiff positive, trichrome staining red, silver staining negative) and endocapillary foam cells or lipoid droplets in focal glomeruli can be found. On the beginning there is mild mesangial hypercellularity, but in the advanced stage, the lesions become hypocellular as shown in the microphotographs (**a–f**) using light and electron microscopy

gies ("*DISH*"), including *D*rugs and toxic metals (e.g., D-penicillamine, captopril, heroin, gold, mercury), *I*nfections (HBV, malaria, schistosomiasis, syphilis, leprosy, filariasis, hydatid disease), *S*oft tissue collagenopathies (SLE, RA, PSS, MCTD, dermatomyositis, sarcoidosis), and *H*emato-vascular and neoplastic diseases (HL, NHL, renal vein thrombosis, SCD, Hashimoto thyroiditis, lung/colon/stomach carcinoma). Moreover, it may also occur as de novo in renal allografts, Gardner-Diamond syndrome, bullous pemphigoid, Fanconi syndrome, Kimura disease, and Weber-Christian syndrome. In practice, about ¾ of MGN is secondary to D-penicillamine, gold, mercury, HBV, SLE, carcinoma, and *de novo* (1/4) in renal allografts.

- *CLM*: *Uniform and diffuse capillary wall thickening* ("stiff" glomerular loops) without any significant proliferation of endothelial, mesangial, or epithelial cells and *spikes* and *domes* of argyrophilic deposits (SS +) growing in the direction of the urinary space (Fig. 6.14).

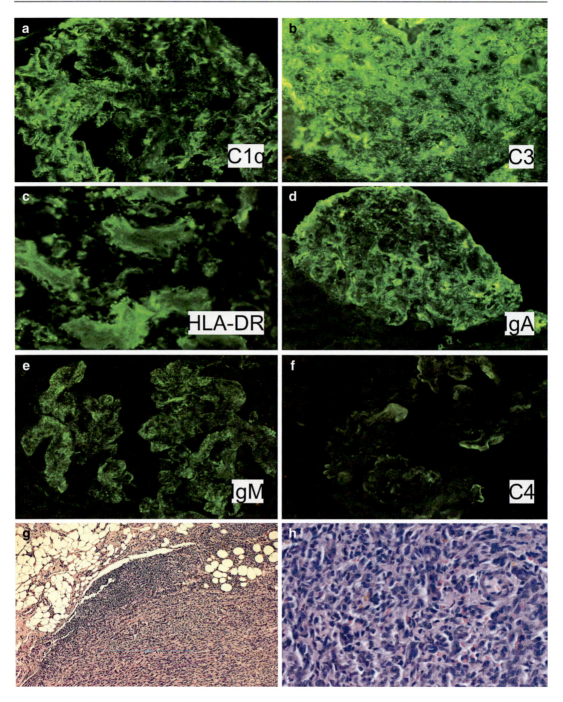

Fig. 6.14 Hereditary C4 deficiency (C4def) is a very rare inherited condition that predisposes to immune complex disease and end-stage renal failure with most of the patients suffering from systemic lupus erythematosus. Figures (**a**) through (**e**) show the normal appearance of the deposition of C1q, C3, HLA-DR, IgA, and IgM by immunofluorescence. Figure (**f**) shows no or minimal C4 deposition. In figures (**g**) and (**h**), the microphotographs of Kaposi sarcoma are illustrated with infiltration of hyperchromatic spindle cells into the interstitium and slit-like lumina with entrapped erythrocytes (hematoxylin-eosin staining; g, ×20 and ×200 as original magnifications)

- *IFM*: *Diffuse granular subepithelial* deposits (IgG > IgM > IgA; ±C3)
- *TEM*: Four stages, including stage I with scattered *subepithelial* electron-dense deposits (*domes*), stage II with more deposits and projections of BM material deposited in between (*spikes and domes*), stage III with encircling of the deposits (*chain-link*), and stage IV with dissolution of deposits but irregular persistence of the thickening of GBM (*swiss-cheese*).
- *PGN*: Excellent in children (10-YSR, 90%) but less benign in adults (10-YSR, 75%). MGN can lead to persistent proteinuria, mesangial cell proliferation, and slow progression to EDRF, but steroid therapy may help. In some cases, a transformation to anti-GBM type is possible, and the progression to ESRF can become quite rapid. The outcome is tightly linked to the presence of proliferating mesangial cells. If mesangial cell proliferation (+) → worse PGN.

6.4.8 Membranoproliferative (Membrane-Capillary) Glomerulonephritis

- *DEF*: Multi-etiologic GN, presenting as an acute nephritic syndrome, acute nephrotic syndrome, and HTN in children and young adults. It is characterized by a slowly progressive and relentless course with intermittent remissions and gradual loss of renal function as well as a tendency to relapse in renal grafts. It has been described with cannabis use.
- *SYN*: Hypocomplementemia (C3) or lobular GN.

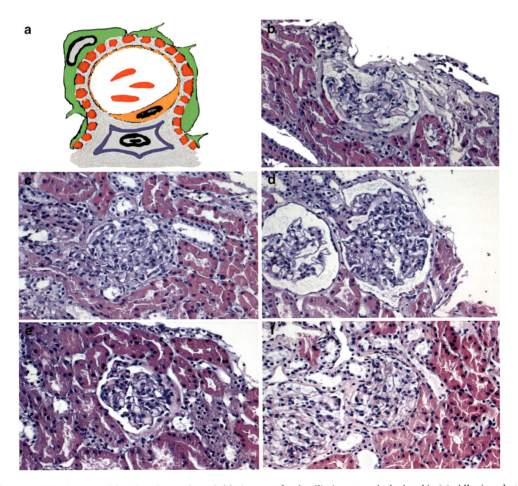

Fig. 6.15 Membranoproliferative glomerulonephritis (see text for detail). A cartoon is depicted in (**a**). All microphotographs (**b-f**) have been taken from H&E stained slides at x400 as original magnification

Three subtypes: (1) *classical/mesangiocapillary*, (2) "dense deposits," (3) *mixed* with *monomorphic LM phenotype* (mnemonic: *ELMIC*)

CLM: *E*nlargement and *L*obular accentuation of the Glomeruli, *M*esangial cell number and matrix increase, *I*rregular capillary wall thickening with interposition of mesangial cells between the BM and the endothelium giving rise to the double contour appearance or "tram track" or "reduplication of the GBM" pattern, and *C*rescents (1/5 of cases) (Fig. 6.15).

Type I – Classical, "mesangiocapillary" (2/3): C-classical pathway activation (better PGN)

- IFM: *IgG, IgM* ± *IgA* deposition and *C3, C1q*, and *C4* deposition ("lumpy bumpy") as well as granular *fibrin* deposition (in serum <C3, C1q, C4, and C2 => ↓CH50)
- TEM: *Subendothelial* and *mesangial* deposits with an increase of the mesangial matrix, which determines a "*mesangialization*" of the capillary loops, and foot process fusion

Type II – Dense deposit GN (1/3): C-alternative pathway activation (worse PGN)

- IFM: *No Ig* deposition, but extensive *C3* deposition (normal C1q, C4, C2 because C3 NeF antibody stabilizes alternative pathway (C3 convertase)) and occasional *fibrin* deposition
- TEM: *Dense deposits in lamina densa of GBM* with a long discontinuous ribbon of hazy material

Type III – Mixed (rare). It may represent probably an advanced form of type I with subendothelial and subepithelial deposits.

- *PGN*: Severity of disease – renal function (FE_{Na}), kidney size <9 cm (US), necrosis, sclerosis, and tubular and vascular fibrosis (histology).

6.5 Secondary Glomerular Diseases

6.5.1 SLE/Lupus Nephritis

- *DEF*: Mono-etiologic GN associated with systemic lupus erythematosus (SLE), which is a systemic AI disease of the connective tissue that can affect multiple organs with type III hypersensitivity-driven inflammation and tissue damage (Fig. 6.16).
- *EPI*: ♂ < ♀, 2nd-4th decades of life, non-European origin harboring an unpredictable course with periods of illness ("flares") alternating with remissions.
- *EPG*: Type III hypersensitivity-driven inflammation and tissue damage with clearance deficiency ("LE cell": Viable neutrophil engulfed with the ingested nuclear material).
- *CLI*: *S*erositis with pleuritis and pericarditis; *O*ral ulcers; *A*rthritis; *P*hotosensitivity; *B*lood dyscrasias with anemia, leucopenia, and thrombocytopenia; *R*enal involvement; *A*NA+; *I*mmunologic dysregulation with LE cell, dsDNA, anti-Sm, and false VDRL; *N*eurologic symptoms with psychosis and seizures; *M*alar rash; and *D*iscoid rash. The diagnosis is based on completion of 4/11 criteria.
- *CLM*: The World Health Organization (WHO) Classification ("*MES-FO-DI-MEM-SCL*") is used.
- *WHO Class I* No lesions, asymptomatic.
- *WHO Class II* (*Me*sangial) Proteinuria (mild to moderate), good prognosis.
- *CLM*: IIA, no significant changes; IIB, mild mesangial hypercellularity distally from vascular pole.

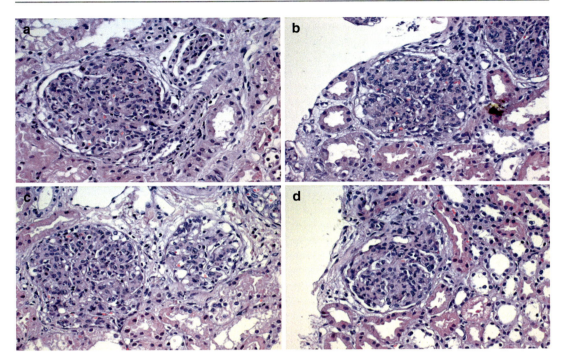

Fig. 6.16 Systemic lupus erythematosus (SLE)/lupus nephritis is an inflammation of the kidneys caused by SLE, an autoimmune disease. Figures (**a**) through (**d**) show hypercellularity, hyperlobulation, adhesion, and thickening of the capillary wall (hematoxylin-eosin staining, ×400 original magnification)

- *IFM*: Mesangial IgG and C3 and EM, *mesangial* deposits.
- WHO Class III (*Fo*cal Segmental) Proteinuria, ± aggressive course.
- *CLM*: FS necrosis/proliferation (<50% involvement of <50% glomeruli) with mild mesangial prominence, segmental capillary proliferation with lumen obliteration, hyaline wire loops, and focal crescents.
- *IFM*: Granular capillary Ig and C3 and EM, *subendothelial* deposits and *mesangial* deposits.
- WHO Class IV (*Di*ffuse Proliferative) – most often – the aggressive course is leading to RF!
- CLM: Membranoproliferative GN + mesangial proliferation ± crescents and "hematoxylin body," thickened "wire loops," and lobular accentuation.
- *IFM*: Coarsely granular *Full house* in mesangium and capillary loops (IgG + IgM > IgA > IgE).
- If + IgE − > worst PGN!
- *TEM*: "Fingerprint" *subendothelial* deposits and *mesangial, subepithelial*, and *intramembranous* deposits as well as tubular-reticular structures in endothelial cells
- WHO Class V (*Mem*branous) Indolent progression.
- *CLM/IFM/TEM*: Identical to lesions of idiopathic membranous GN!
- WHO Class VI (*Scl*erosing).
- *CLM*: Global glomerulosclerosis with fibrous crescents, interstitial fibrosis, and tubular atrophy.
- *TEM*: Irregular thickening of capillary BM with *intramembranous* deposits.
- *PGN*: Young adult SLE – renal involvement, 50–80% of patients with SLE and PGN is directly related to the extent of renal disease. *Subendothelial* deposits, the most hostile deposits, correlate well with RF ± *lymphoplasmacytic tubulointerstitial infiltrate with eosinophils.*

6.5.2 Henoch-Schönlein Purpura

- *DEF*: Systemic leukocytoclastic vasculitis of the 2nd-3rd infancy characterized by deposi-

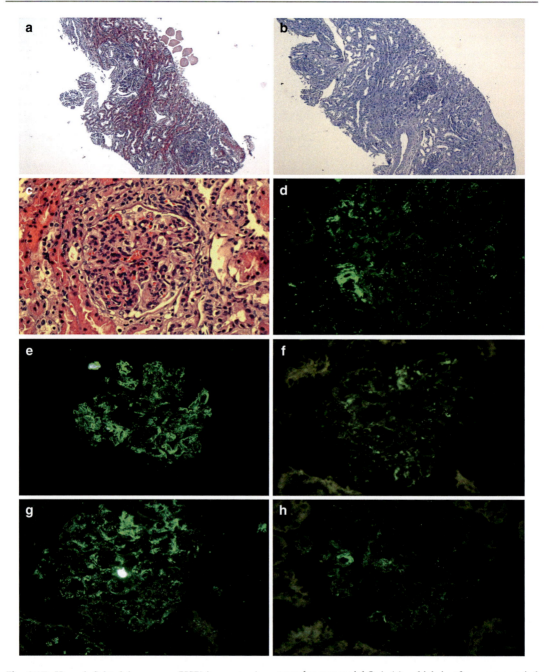

Fig. 6.17 Henoch-Schönlein purpura (HSP) is a systemic vasculitic disorder with nephropathy, including glomerular proliferation and cell infiltrates (**a** and **b**, hematoxylin-eosin staining, ×100 original magnification; (**c**) hematoxylin and eosin staining, ×400 original magnification) and granular mesangial IgA (**e**), which is often accompanied by C3 (**g**), fibrinogen, and both light chains and less frequently by IgG (**d**) and/or IgM (**f**). Figure (**h**) shows the C1q deposition. All immunofluorescence microphotographs have been taken at ×400 as original magnification

tion of IgA-immune complexes linked often to infections, drugs, or idiopathic (~1/3) with mostly *restitutio ad integrum*.
- *CLI*: HSP presents as *A*rthralgia (non-migratory), *R*enal involvement (asymptomatic hematuria/proteinuria to GN), *C*utaneous palpable purpura (subepidermal hemorrhages + necrotizing vasculitis involving the small BV of the dermis with IgA deposits), and *I*ntestinal involvement (pain, vomiting, bleeding ± intussusception) ("*ARCI*").
- *EPG*: Infectious (e.g., *β-hemolytic Streptococcus*, Lancefield group A, HBV, HSV, PV-B19, cox-

sackievirus, adenovirus, *Helicobacter pylori*, measles, mumps, rubella, *Mycoplasma* spp.), drugs linked to HSP, usually as an idiosyncratic reaction, include the antibiotics vancomycin and cefuroxime, ACE inhibitors enalapril and captopril, anti-inflammatory agent diclofenac, as well as ranitidine and streptokinase.
- *CLM*: Similar to IgA nephropathy but often crescentic GN when severe (Fig. 6.17).
- *PGN*: Typically, HSP resolves within several weeks and requires no treatment apart from symptom control but may relapse in 1/3 of the patients with potentially irreversible kidney damage (~1/100).

6.5.3 Amyloidosis

- *DEF*: Group of diseases having in common the deposition of a similar-appearing protein, the *Amyloid*, which is a pathologic proteinaceous substance, deposited in the extracellular space in various tissues and organs of the body in a wide array of clinical settings.

Most common forms of *Amyloid Proteins*:

1. *AL* (A-light chain): Ig light chain ($\lambda > \kappa$) secreted by monoclonal PC (Bence-Jones protein).
2. *AA* (A-associated): Derived by proteolysis of SAA (*Serum Amyloid Associated*), a non-Ig protein synthesized by the liver in inflammation as part of the Acute Phase Response: Secondary Amyloidosis.
3. *Aβ* (*β-protein*): Derived by proteolysis of APP (amyloid precursor protein) and is the core of cerebral plaques in Alzheimer disease and cerebral BV as well as trisomy 21 syndrome.
4. Other biochemically distinct proteins include *TTR*, which is specifically involved in familial amyloid polyneuropathies and senile systemic amyloidosis, *A-β2-microglobulin* engaged in long-term hemodialysis, and misfolded *prion* proteins.

- *EPG*: Misfolded proteins are *unstable* with an intrinsic tendency to *self-associate* = > deposition = > disruption of the normal function of tissues by alteration of the proteasome. The Ubiquitin Proteasome Pathway (UPP) is considered the main mechanism for protein catabolism in mammals. The UPP affects numerous cellular processes. Defects in the system can result in the pathogenesis of several human diseases. Its role in medicine and biology has been universally recognized in 2004 with the Nobel Prize for Chemistry which was awarded to Avram Hershko, Aaron Ciechanover and Irwin Rose. The proteasome is supposed to degrade IC misfolded proteins. Genetically determined structural abnormality in the SAA molecule renders it resistant to degradation of Mφ.
- Localized amyloidosis forms comprise senile cerebral (Alzheimer disease), which occurs early in individuals with trisomy 21 syndrome, medullary carcinoma of the thyroid, T2DM in the islets of the pancreas (AIAPP), and the isolated atrial amyloidosis (AAMI).
- *SAA-Conditions*: TB, bronchiectasis, and chronic osteomyelitis in the pre-antibiotic era. RA, ankylosing spondylitis, IBD, heroin/"skin-popping" of narcotics, HL, RCC, and FMF in the post-antibiotic era.
- *GRO*: Enlarged and firm organ with waxy appearance (iodine- > yellow +H_2SO_4 - > blue-violet), such as the kidney, spleen ("*sago spleen*" with tapioca-like granules and "*lardaceous spleen*" with map-like areas), liver, heart (restrictive CMP, congestive heart failure, arrhythmias), and tongue (macroglossia).

In pediatrics, amyloidosis is AA-type or reactive to chronic inflammatory diseases, and the most important causes are the *Familial Mediterranean Fever* (FMF) (> IL-1, pyrin, polyserositis, Middle East) and other autoinflammatory disorders (e.g., *juvenile rheumatoid arthritis*). *Tuberculosis* has practically disappeared or strongly decreased in pediatrics of developed countries but still present in underdeveloped countries.

> Amyloid: Hyaline, eosinophil, Congo-red+ and apple-green birefringence when polarized. Historically, the amyloid detection with the iodine-sulphuric acid reaction had been identified in the 19th century (Aterman 1976).

> Amyloid protein A deposits (+Congo-red/ AG-B) ⇒ massive proteinuria (12–20 g/24 h!).

> CLM: Homogeneous deposits in glomeruli, tubular BM, and BV walls.

> IFM: ±Ig, but nonspecific.

> TEM: Amyloid fibrils, β-pleated sheet, 7–12 nm, randomly arranged, non-branching in mesangium and peripheral capillary BM (≠18–24 nm θ of fibrillary GN or 30–50 nm θ of immunotactoid GN).

6.5.4 Light Chain Disease

7–10% of patients with MM

- *CLM*: Capillary wall thickening, nodular sclerosis, and AL-deposits ("AL Amyloid").
- *IFM*: Monoclonal κ or λ light chain – linear deposition.
- *TEM*: Granular deposits in glomerular and tubular BM.

6.5.5 Cryoglobulinemia

50% of patients (usually adults, rare in children) with lymphoproliferative disorders (MM, WM, CLL, B-cell NHL), infections (mainly HCV), AID, and essential. It usually presents as type I membranoproliferative (mesangiocapillary) GN.

> - *Type I* (monoclonal IgM): Waldenstrom macroglobulinemia (WM), multiple myeloma (MM)
> - *Type II* (monoclonal IgM + polyclonal IgG): WM, MM, CLL
> - *Type III* (*polyclonal* IgM and IgG): RA, SLE, PAN

>90% of mixed cryoglobulinemia → (+) HCV and 50% of (+) HCV have cryoglobulinemia.

- *CLM*: Glomerular enlargement with diffuse ↑ of mesangial matrix and # of mesangial cells accompanied by lobular accentuation, neutrophils, and intraluminal eosinophilic occlusive thrombi ± crescents.
- *IFM*: IgG and IgM deposition in peripheral capillary deposits as well as C3, C1, and C4 in a granular fashion (mixed cryoglobulinemia type II or type III).
- *TEM*: Subendothelial and mesangial deposits.

6.5.6 Diabetic Nephropathy

Diabetes is one of the most common internal diseases globally and across ages. Dramatically, the worldwide incidence of type 1 diabetes mellitus (T1DM) is rising by 3% per year. Diabetic nephropathy may occur in children and adolescents as early tubulopathy and in many countries a growing number of children and adolescents are suffering from type 2 diabetes mellitus (T2DM) with or without obesity. This increase is not more restricted to Caucasian children, but rising rates have also been scored in Asian communities worldwide. The high-fructose corn syrup is the ingredient of several pop drinks and has been considered the culprit but probably increased sugar content in the diet of children and adolescents plays a major role in several communities and cultures. Recent advances in DM treatment have been successful in decreasing morbidity and mortality from diabetes-related retinopathy (DM-Ret), nephropathy (DM-Nep), and neuropathy (DM-Neu). T1DM-Nep/T2DM-Nep is more common in early onset or poorly controlled DM and presents with recurrent proteinuria, often nephrotic, with papillary necrosis of both kidneys and slow progression to CRF.

Diffuse and Nodular Types → "ThIN"
- *CLM*: *Th*ickening, diffuse hyaline type of capillary wall, "*in*sudative" glomerular lesions (fibrin cap, capsular drop), *n*odular (Kimmelstiel-Wilson) sclerosis, arteriolar nephrosclerosis, and duration-related global sclerosis.
- *IFM*: Diffuse "*thin*" linear staining for IgG, C3, and albumin.
- *TEM*: Diffuse GBM thickening (up to 5–10× normal), ↑ mesangial matrix and subendothelial granular deposits.
- *PGN*: Diabetic children, even with normal levels of albuminuria, show ↑ urine neutrophil gelatinase-associated lipocalin (NGAL) supporting the hypothesis of a "tubular phase" of diabetic disease preceding overt diabetic nephropathy. Thus, the use of urine NGAL measurement for early evaluation of renal involvement may be beneficial.

6.6 Hereditary/Familial Nephropathies

6.6.1 Fabry Nephropathy

Although arising in late adolescence and youth, Fabry disease may play a role for genetic counseling. Fabry nephropathy is X-linked recessive lysosomal storage disease, and there are several mutations of the *GLA* gene leading to a lack or deficiency of the lysosomal enzyme α-galactosidase A. It results in progressive glycotriaosylceramide accumulation in multiple organs, of which the kidney is primarily involved. Fabry nephropathy is characterized by vacuolization of the cells in the glomeruli, tubules, interstitium, and arteries. The TEM shows myelin bodies. Although the kidney biopsy is not necessary for diagnosis, it does have a critical role in the evaluation of disease evolution and treatment efficiency with the enzyme replacement therapy (Fig. 6.18).

6.6.2 Alport Syndrome

- *DEF*: Genetic disorder, mostly X-linked dominant (*COL4A5* gene), which is characterized by *glomerulosclerosis*, *tubular atrophy*, and *interstitial foam cells*. There is a real chance to progress to ESRD, sensorineural hearing loss, abnormalities of the eyes, and male fertility disturbances. Ocular findings include anterior lenticonus, maculopathy (whitish or yellowish specks or granulations in the perimacular region), posterior polymorphous corneal dystrophy (PPMD, vesicles of the Descemet membrane and the corneal endothelium), and recurrent corneal erosion. *Lenticonus anterior* is an anterior bulging of the lens capsule and the underlying cortex diagnosed by biomicroscopic examination and is virtually pathognomonic of Alport syndrome. Conversely, the *lenticonus posterior* is not associated with systemic disease.
- *IFM*: No specific staining and no reactivity using anti-GBM from patients with Goodpasture syndrome (linear fluorescence would be present in healthy individuals or patients with TBMN). By IHC, the α5 chain of collagen IV is expressed.
- *TEM*: *"Basket-weave" pattern* of the lamina densa, which is split into multiple interwoven lamellae showing thinning, thickening, and splitting of GBM.
- Subtypes of collagen IV-related nephropathies are X-linked Alport syndrome (*XLAS*), AR-Alport syndrome (*ARAS*), AD-Alport syndrome (*ADAS*), and *TBMN* (*vide infra*). Any of these genes harboring mutations determine proteins that prevent the proper production or assembly of the network of collagen IV, which plays an essential structural element of basement membranes in some organs such as the kidney, inner ear, and eye. *COL4A5* (aka *ASLN*, *ATS*) is collagen IV, α5 and is located on Xq22. *COL4A5* translates one of the six subunits of collagen IV. Collagen IV is the primary structural component of basement membranes. *COL4A5* shares some similarities with other members of the collagen gene family. *COL4A5* is organized in a head-to-head conformation with another collagen IV gene. The fate of this phenomenon is that each gene pair shares a common promoter.

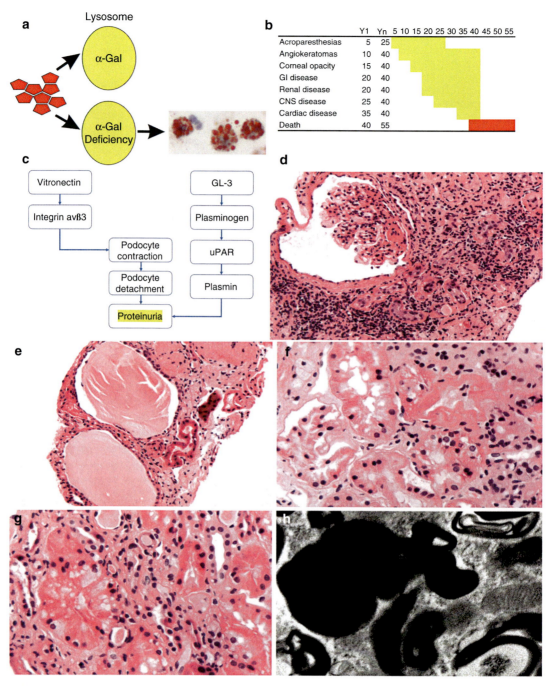

Fig. 6.18 Fabry disease or Anderson-Fabry disease (AFD) is a lysosomal storage disorder caused by genetic deficiency of the enzyme encoded by the alpha-galactosidase A (a-Gal A) enzyme, which is located on the X chromosome (Xq22.1) with the lipids highlighted by special stains for the accumulation products (**a**). This causes progressive intracellular accumulation of globotriaosylceramide and other sphingolipids throughout the body with progressive life quality impairment (**b**). Renal histology (**d–g**) shows hypertrophic glomerular podocytes distended with foamy appearing vacuoles, mesangial widening, and varying degrees of glomerular obsolescence. Vacuolation is also observed in the capillary endothelium and distal tubular epithelial cells, including the epithelial cells of Henle's loop and the collecting duct and less often in proximal tubular epithelial cells (**d** and **e**, hematoxylin and eosin staining, ×200 as original magnification; **f** and **g**, hematoxylin and eosin staining, ×400 as original magnification). Figure (**h**) shows the ultrastructural findings with enlarged secondary lysosomes, which are also known as myeloid or zebra bodies, packed with lamellated membrane structures (×1500, original magnification)

6.6.3 Nail-Patella Syndrome

- *DEF*: Classic clinical tetrad of a particular phenotype with AD inheritance involving changes of the ungueal apparatus (most constant feature), knees, elbows, and iliac bones. Nails may be absent, hypoplastic, or dystrophic. They have been described with different morphologies such as nails with longitudinal or horizontal ridges, pitted, discolored, or even separated into two halves by a longitudinal cleft or ridge of the skin. In the knees, the patellae may be small. Occasionally, the patellae are irregularly shaped, or, even, absent. Elbows may have the limitation of extension, pronation, and supination, *cubitus valgus*, and antecubital pterygia. Iliac horns are bilateral, conical, bony protuberances that project posterolaterally from pelvic iliac bones. Renal involvement may first manifest as proteinuria with or without hematuria up to half of the patients, and ESRD occurs in 1:20 patients.
- *TEM* shows thickening of GBM with collagen-like fibers within the GBM and electron-lucent areas giving a mottled "moth-eaten" appearance. Additional complications include primary open-angle glaucoma and ocular HTN (annual visual monitoring is advisable). Diagnosis is based on clinical features and *LMX1B* gene screening. LMX1B (aka NPS1) is LIM homeobox transcription factor 1, β gene with protein product located on 9q33.3. This gene belongs to the homeoboxes or LIM class gene families. There are two homologs in mouse and rat species (*Lmx1b*). This gene translates a member of LIM-homeodomain family of proteins, which contains two N-terminal zinc-binding LIM domains, a homeodomain, and a C-terminal glutamine-rich domain. The protein works as a transcription factor, which plays an essential role for the healthy development of dorsal limb structures, the GBM, the anterior segment of the eye, and for the development of dopaminergic and serotonergic neurons.

6.6.4 Congenital Nephrotic Syndrome

Nephrotic syndrome type 1 (NPHS1), aka Finnish congenital nephrosis, is due to homozygous or compound heterozygous mutation in nephrin gene (*NPHS1*; 602716), which is located on chromosome 19q13.1. The congenital nephrotic syndrome is characterized by massive proteinuria at prenatal onset and enlarged placenta (1/3 body weight of the newborn). At birth, there is a severe steroid-resistant nephrotic syndrome with rapid progression to ESRD within 1–3 years. It is essential to remind to the genetic heterogeneity of nephrotic syndrome and FSGS (*vide supra*). Both disorders represent a spectrum of hereditary renal diseases. In fact, NPHS2 (600995) is caused by mutation in the *PODOCIN* gene (604766), NPHS3 (610725) is linked to the *PLCE1* gene (608414), NPHS4 (256370) is related to the *WT1* gene (607102), NPHS5 (614199) is associated with *LAMB2* gene (150325), and NPHS6 (614196) is connected to the *PTPRO* gene (600579). FSGS1 (603278) is caused by a mutation in the *ACTN4* gene (604638), while FSGS2 (603965) is linked to the *TRPC6* gene (603652). Moreover, FSGS3 (607832) is associated with genetic variation in the *CD2AP* (604241), FSGS4 (612551) is on chromosome 22q12. FSGS5 (613237) is caused by a mutation in the *INF2* gene (610982). Renal histology of congenital nephrotic syndrome shows immature glomeruli with sclerosis (mesangial → global sclerosis), tubular cyst formation (aka microcystic disease) and atrophy, interstitial fibrosis, and medial arteriolar hypertrophy. TEM exhibits an increased mesangial matrix and an obliteration of foot processes.

6.6.5 Thin Glomerular Basement Membrane Nephropathy (TBMN)

TBMN is an AD-inherited disorder with >50% of the patients having a family history of hematuria. In about half of the families affected with TBMN, the condition co-segregates with heterozygous *COL4A3/COL4A4* gene mutations. TBMN is characterized by diffuse thinning of the lamina densa and the GBM as a whole and manifests as persistent microscopic hematuria (persistent or intermittent) with onset in childhood, no uremia or renal insufficiency, with rare extrarenal abnormalities. LM and IF are usually unremarkable, although small amounts of Ig and C3 deposited along GBM may be encountered. IF with anti-GBM antibodies from the serum of patients with Goodpasture syndrome is positive like in healthy individuals, a feature helping to differentiate from Alport syndrome. The intraglomerular variability in GBM width, observed in TEM, is small in individuals with TBMN. GBM is ~200 nm vs. 300–400 nm of GBM in healthy individuals. The GBM of persons affected with TBMN remains attenuated over time, rather than undergoing the progressive thickening and multi-lamellation identified in renal biopsies of patients affected with Alport syndrome. Individuals with TBMN exhibit standard GBM staining for the collagen $\alpha 3(IV)$ chain, $\alpha 4(IV)$ chain, and $\alpha 5(IV)$ chain.

6.7 Tubulointerstitial Diseases

6.7.1 Acute Tubulointerstitial Nephritis (ATIN)

Acute inflammation of the interstitium with often involvement of the tubules – four diseases

6.7.1.1 Acute Pyelonephritis
- *DEF*: Patchy, *wedge-shaped mostly cortical suppurative inflammation* with *edema and neutrophils* in tubular lumina and interstitium, as well as areas or necrosis/abscess in the cortex, but sparing of glomeruli and BVs.
- *CLI*: Flank pain, fever, malaise, dysuria, and pyuria with Gram-negative GIT bacilli (>85%), e.g., *E. coli*, *Proteus* spp., *Klebsiella* spp., and *Enterobacter* spp., occurring in the setting of obstruction of the lower urogenital tract (e.g., prostatic valves, pregnancy), hematogenous spread, ascending infection, and VUR (during micturition) (Figs. 6.19 and 6.20).

6.7.1.2 Pyonephrosis
Purulent inflammation of the kidney with a filling of the pelvic system and kidney caused by near complete obstruction.

6.7.1.3 Acute Hypersensitivity Nephritis (AHN)
- *DEF*: Hypersensitivity-based acute nephritis.
- *EPG*: AHN may be associated to various etiologies, including *allergic/drug-induced*, *infectious* (Hantavirus, EBV, HSV, CMV, adenovirus, HIV), and *autoimmune/systemic* (SLE, Sjögren's, RA); otherwise *idiopathic* forms of the disease also exist. Drugs constitute the most common etiology of AHN in roughly 2/3 of patients. Multiple agents from many different classes of drugs can be the underlying cause, and clinical presentation, as well as laboratory findings, may vary according to the class of drug involved. Most common drugs are aminoglycosides (e.g., gentamycin), rifampin, amphotericin, beta-lactam antibiotics, NSAIDs, PPI, diuretics, and phenytoin.
- *CLM*: Tubular and interstitial lymphoplasmacytic inflammation with macrophages and eosinophils as well as ± tubular damage and regeneration with interstitial fibrosis. There is sparing of glomeruli and BV. In NSAID-linked AHN, there is also glomerular foot process fusion. Noninvasive tests, such as gallium-67 ($_{67}$Ga) scintigraphy and eosinophiluria testing, have limited diagnostic utility, but a renal biopsy can establish a definitive diagnosis.
- *PGN*: Good, if there is timely discontinuation of the causative agent and partial recovery of kidney function with or without steroids is customarily observed. Since patients can eventually develop ESRD and RF, early recognition is crucial.

6.7.1.4 Renal Papillary Necrosis
- *DEF*: Necrosis of the renal papillae due to DM, acetaminophen intoxication (also phenacetin, aspirin, codeine, and EtOH), and vascular occlusion (e.g., SCD, leukemia) – related TIN, generally affecting the upper and lower pole renal papillae.
- Synchronized lesions in DM and temporal heterogeneity in non-DM causes.
- Analgesic nephropathy is dose-dependent ≠ nephrotoxic ATN due to NSAIDs and penicillins.

Three stages (Fig. 6.21):

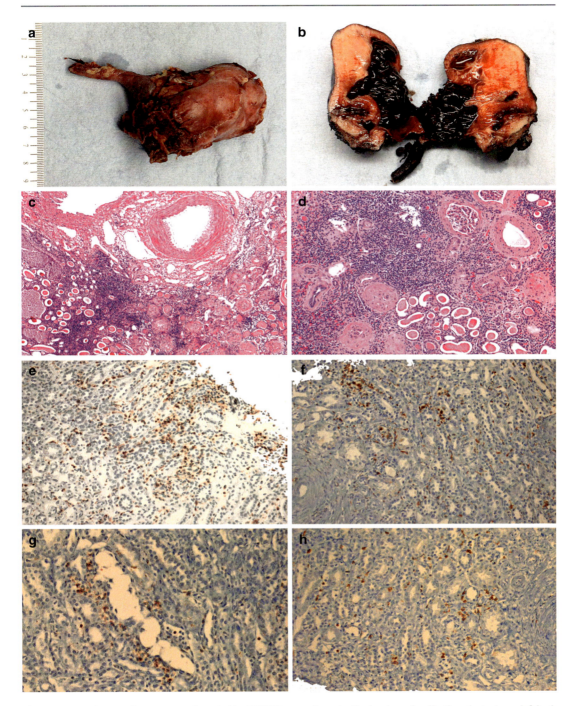

Fig. 6.19 Xanthogranulomatous pyelonephritis (XGPN) with bosselated surface of the kidney (**a**) and homogeneous gray-whitish/yellowish cut surface and hemorrhagic pelvis (**b**). Figures (**c**) and (**d**) show the histology with granulomatous mixed (lymphocytic-rich) inflammatory infiltrate with fibrosis, glomerular obsolescence, and thyroidization of the parenchyma with eosinophilic concretions in the lumina of collecting ducts (**c** and **d** both hematoxylin and eosin staining taken at ×50 and ×100 as original magnification). Figures (**e**) through (**h**) show an interstitial pyelonephritis with immunohistochemical detection of antibodies against CD45, CD68, CD4, and CD8, respectively (avidin-biotin complex, ×200 original magnification)

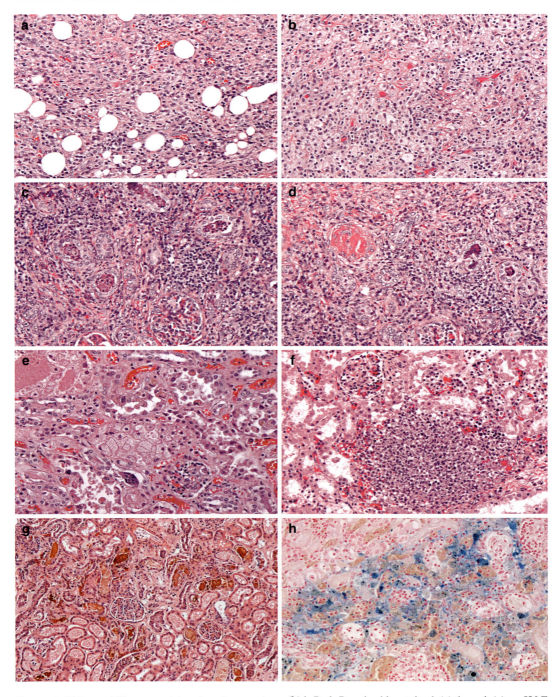

Fig. 6.20 This panel illustrates eight microphotographs (**a-h**) showing the variability seen in sepsis with bilirubin casts in case of hepatorenal failure (**a-g**). In some case, hemosiderin deposition is particularly prominent (**h**). If (**h**) is Perls Prussian blue stained, (**a**) through (**g**) are H&E stained. All microphotographs are at x200 as original magnification, apart (**a**), (**b**), and (**g**), which are at x100 as original magnification

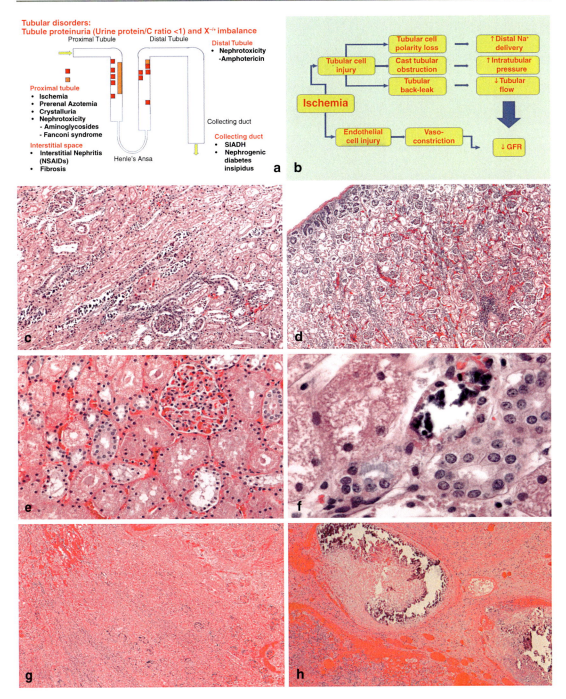

Fig. 6.21 Acute tubular necrosis with the two subsets on the top (**a-b**) showing the alterations seen in tubular disorders of ischemic and toxic type and the complications of renal ischemia. In (**c**) through (**h**), there are examples of necrosis with disappearance of the structure, hyperemia, and dystrophic calcifications of the kidney (see text for details). All microphotographs are from H&E-stained slides taken at different degree of magnification

- I: Papillae *firm with linear gray streaks* ⇒ interstitial "homogenization", BM thickening, focal cell necrosis, and fine calcification with normal cortex
- II: Papillae *shrunken (brown)* ⇒ confluent necrosis, tubular atrophy in the overlying cortex
- III: *Small kidneys* ⇒ total papillary necrosis with extensive calcification/ossification, overlying tubular atrophy, and interstitial fibrosis

6.7.2 Chronic Tubulointerstitial Nephritis (CTIN)

- *DEF*: Etiologically heterogeneous chronic renal disease with *chronic inflammation* of the interstitium with *lymphoplasmacytic infiltration* and *fibrosis* (hallmark of chronic injury) as well as involvement of the tubules showing *tubular atrophy* and *thyroidization*, initially sparing of glomeruli and then *periglomerular fibrosis* until *obsolescent glomeruli* as well as *intimal fibrosis and medial thickening of the arteries*.
- *EPG*: Etiologic heterogeneity, *obstructive* CTIN (e.g., vesico-urethral reflux or VUR) and *nonobstructive* CTIN (drugs, such as lithium, cyclosporine, and analgesics) identify four clinical settings depending on the laterality of the inflammation (uni-/bilateral) and the association or no with Urinary Tract (UT) abnormalities – (1) *bilateral chronic pyelonephritis (CPN) with UT malformations and lithiasis* (most of the cases); (2) *bilateral CPN with isolated VUR*; (3) *bacterial CPN without uropathy, lithiasis, or VUR* (extremely rare in children); and (4) CPN without urinary infection ("*a-bacterial CPN*").
- The fourth clinical setting is probably associated with an immune process resulting from persistence of bacterial antigen following a single or asymptomatic infection.
- *GRO*: *Irregular, asymmetrical scarring* involving the calyces and pelvis as well as the cortex with the finely granular appearance of the intervening cortex due to hypertensive nephrosclerosis. It is paramount to recall that scars are more extensive and flatter than infarction-related injuries. There is mostly involvement of the upper and lower lobes, where there are compound papillae, which are responsible for the drainage of 2–3 lobes and VUR is more frequent.

Specific forms or complications:

1. *Xanthogranulomatous pyelonephritis* (CTIN complication associated with *E. coli*, *Proteus* spp., and *S. aureus* that may mimic RCC and Tb. It presents grossly as friable yellow-orange nodules surrounding the dilated calyces initially. Later, there is a diffuse (scattered throughout the parenchyma), hemorrhagic purulent exudate in dilated calyces with thickened pelvic mucosa. Microscopically, XGPN shows an infiltrate of finely granular foamy macrophages with neutral fat and cholesterol esters, MNGC of foreign body type, and an admixture of lymphocytes, plasma cells, PMN, and fibroblasts as well as dense inflammatory fibrous tissue demarcating the granulomatous zone from the atrophic cortex).
2. *Malakoplakia*, which is often a complication of a chronic bacterial (*E. coli*) UTI. There is an insufficient killing of bacteria by host macrophages ("Von Hansemann histiocytes") and an accumulation of partially digested bacteria with Fe and Ca deposition (+Prussian Blu, +Von Kossa, but also +PAS). It occurs mostly in females of the 5th decade with a history of immunosuppression due to lymphoma, DM, and renal TX or because of long-term therapy with systemic corticosteroids. Malakoplakia has been reported only rarely in children, particularly in patients affected with primary immune deficiencies.
3. *Tuberculosis* is frequently unilateral and may be either miliary (numerous small tubercles scattered throughout the cortex) or isolated to the UT, showing the progressive destruction of the renal parenchyma.
4. *Pelvic lipomatosis*, which appears to be comparable to *lipomatosis pancreatis et cordis*, represents a replacement of the pelvis by adipose tissue following atrophy.

5. *Ask-Upmark kidney*, which is extensive scarring of a renal lobule/pole, ensuing in apparent focal hypoplasia.

6.7.3 Acute Tubular Necrosis

- *DEF*: Destruction of renal tubular cells from several and different causes (e.g., tubular obstruction by cellular debris), which are classified into two subgroups, *ischemic* ATN and *nephrotoxic* ATN, and aggressive medical treatment, including dialysis, is typically required.
- *CLI*: Oliguria, <400 mL urine/24 h, or anuria, <100 mL urine/24 h with four clinical phases:

> 1. Onset (<36 h)
> 2. Oliguric (days to weeks with fluid overload, uremia, hyperkalemia)
> 3. Early diuretic, characterized by a stable increase in urine volume, hypokalemia, but also $X^{+/-}$ imbalance
> 4. Vulnerability to infection and late diuretic with the recovery of function

- *GRO*: Swollen, kidneys showing a pale cortex and a dark hyperemic medulla.
- *Ischemic ATN* (aka tubulorrhexis), which involves the proximal tubules and a thick portion of the ascending limb, is most susceptible to ischemic injury (highest ATPase activity) and results from ↓ perfusion and inadequate oxygenation to the renal tubules. It ensues *S*kip lesions, *T*ubulorrhexis, and *T*amm-Horsfall protein in distal tubules (+) "STT" associated with interstitial edema, dilated *vasa recta* with leukocytes, and rhabdomyolysis pigment.
- *Nephrotoxic ATN*, which involves the proximal tubules and results from numerous agents (heavy metals, organic solvents, sulfonamides, neomycin, methicillin, and anesthetics) that directly and specifically damage the renal tubule epithelium. There are no *S*kip lesions, *T*ubulorrhexis, and *T*amm-Horsfall protein in distal tubules → (−) "STT."

The mechanisms may underlie a (1) depolarization of highly polar tubular epithelial cells, (2) persistent *pre*-preglomerular arteriolar vasoconstriction, (3) tubular backflow following destruction or alteration of the integrity of tubular cells, (4) tubular obstruction from casts of necrotic tubular cells and proteinaceous material, and (5) permeability of the glomerular tuft (direct effect).

- *CLM*: Interstitial edema, accumulation of leukocytes in *vasa recta*, and dilatation of tubules, followed by epithelial cell damage and necrosis, tubular dilatation, and cast formation. In the recovery phase, features are epithelial cell regeneration with flattened cells containing hyperchromatic nuclei and mitotic figures. The massive accumulation of calcium oxalate crystals in tubules is indicative of ethylene glycol intoxication.

6.7.4 Chronic Renal Failure (CRF)

- *DEF*: It is a slow and progressive decline of kidney function that occurs because of a complication from another serious medical condition of renal or extra-renal origin.
- *CLI*: Gradual manifestation unlike ARF leading to ESRD. In the USA, approximately 1 in 1000 people progress to ESRD with a prevalence of about 19 million individuals (children and adults) living with some CRF. In Canada, the number is proportionally the same at approximately 1.9 to 2.3 million individuals or higher according to the Provinces. Characteristic features of CRF are early dehydration, then systemic edema (*anasarca*) with loss of concentrating ability, then filtration rate changes, metabolic acidosis, Kussmaul breathing, GI bleeding, congestive heart failure, uremic pericarditis, HTN (↑ volume, renin, or both), anemia (↓erythropoietin and hemorrhage associated), and abnormal bone metabolism (↓Ca, ↑PO_4, ↑PTH) with renal osteodystrophy.

6.7.5 Nephrolithiasis and Nephrocalcinosis

Nephrolithiasis occurs primarily without renal abnormality or infection, and several kinds of stones may be observed. Secondary nephrolithiasis occurs in a setting of renal or extrarenal underlying disorder. Mucopolysaccharides are admixed with all kidney stones with calcium oxalate and calcium phosphate (>3/4) (often in sarcoidosis, hyperparathyroidism, excess intake of vitamin D, multiple myeloma). In 5–20% of cases, stones are composed of uric acid and xanthine (often in gout, Lesch-Nyhan, glycogen storage disease, excess protein intake), 10-15% are magnesium ammonium phosphate ("triple stones" or "struvite stones"), and 1–2% is composed of cysteine (often seen in areas of ulceration on the surface of papillae). Oxalate is built by the liver but is also a naturally occurring substance found in food, and some fruits, vegetables, nuts, and chocolate have high oxalate levels. Dietary products, vitamin D in high dosage, intestinal bypass surgery, and some metabolic disorders can increase the concentration of Ca^{2+} oxalate in urine (Box 6.8).

Selected causes of hypercalciuria in children include increased intestinal calcium absorption due to excessive intake of vitamin D, renal tubular dysfunction due to renal tubular phosphate leak, impaired absorption of the tubular calcium, type 1 (distal) renal tubular acidosis, Dent disease, and Bartter syndrome. Endocrine disorders may be due to hypothyroidism, an excess of adrenocortical hormones, and hyperparathyroidism. Disorders of the bone metabolism associated with nephrolithiasis include immobilization, rickets, juvenile rheumatoid arthritis, and malignant neoplasms. Other miscellaneous conditions related to pediatric nephrolithiasis include familial idiopathic hypercalciuria, UTI, glycogen storage disease, William syndrome, increased renal PGE2 production, and some drugs (diuretics, corticosteroids). Idiopathic infantile hypercalcemia (aka Williams syndrome) (IIH/WS) is a contiguous gene deletion syndrome with a chromosomal deletion on 7q11.23 that occurs hugely rarely with a phenotype of dysmorphic features (elfin facies), cardiovascular abnormalities (aortic stenosis), infantile hypercalcemia, renovascular hypertension, and intellectual disability. It is important to remember that PTH and vitamin D metabolism are standard, but the response of calcitonin to Ca infusion may be abnormal pointing together with the phenotype to suspicion of IIH/WS and starting a genetic test. Nephrolithiasis complications include kidney failure, obstruction of the urinary excretory system with acute and chronic pyelonephritis and hydronephrosis, and sepsis.

Nephrocalcinosis consists of calcification in renal parenchyma, typically on tubular BM, and in the interstitium. Nephrocalcinosis has been associated with hypercalcemia and hypercalciuria. Usually, there is a microscopic picture of tubular atrophy, interstitial fibrosis, and periglomerular fibrosis.

Primary hyperoxalurias (three types, PH 1–3) are rare inborn errors of glyoxylate metabolism characterized by the overproduction of oxalate. This substance is deposited as Ca^{2+} oxalate in various organs, being the kidney the primary target for oxalate deposition. The consequence is ESRD and the patient will eventually need a renal transplantation. PH is primarily caused by autosomal recessive enzymatic defects in glyoxylate metabolism that result in enhanced production of oxalate. PH type 1 is due to the defects in the hepatic peroxisomal enzyme alanine: glyoxylate aminotransferase (*AGT*) gene (~80% of PH cases) and PH type 2 is due to defects in the *GRHPR* gene (~10%). PH type 3 is due to mutations in the *HOGA1* gene that encodes the mito-

---

Box 6.8 Kidney Stones: Composition and Most Common Pathogenesis

Stone	%	Pathogenesis
Ca^{2+} oxalate	70	Hypercalciuria, hyper-PTH, hypocitraturia, RTA
Ca^{2+} phosphate	15	Hypercalciuria, hyper-PTH, Hypocitraturia, RTA
Uric acid	10	Hyperuricosuria
Struvite	10	UTI
Cystine	1-2	Cystinuria

Notes: *Hyper-PTH* hyperparathyroidism, *RTA* renal tubular acidosis, *UTI* urinary tract infections, *Struvite* is a magnesium phosphate mineral ($NH_4MgPO_4 \cdot 6H_2O$) which crystallizes as white to yellowish or brownish pyramidal crystals or, alternatively, in platy mica-like forms

chondrial 4-hydroxy-2-oxoglutarate aldolase enzyme and occurs in about 5% of cases.

6.7.6 Osmotic Nephrosis and Hyaline Change

It is a foamy clearing of the cytoplasm of the proximal tubular epithelium secondary to sucrose or mannitol injection. It does not seem to have a clinical significance. Hyaline change is a pattern constituted in the proximal tubular epithelium containing cytoplasmic eosinophilic PAS-positive droplets and is observed in patients with marked proteinuria. It should represent reabsorption of filtered proteins.

6.7.7 Hypokalemic Nephropathy

There is a vacuolar change of the tubular epithelial cells, which is secondary to chronic protein depletion (e.g., GI disease). Microscopically, there is coarse vacuolization of tubular epithelial cells, mainly of the proximal limb of the nephron unit. It seems to be associated with dilatation of intercellular spaces.

6.7.8 Urate Nephropathy

It is nephropathy due to precipitation of uric acid crystals in renal tubules, mainly collecting ducts, usually occurring in leukemia/lymphoma patients on chemotherapy. Distinguishing between acute and chronic urate nephropathy is important. The first is not a true form of acute TIN but rather intraluminal obstructive uropathy caused by intraluminal uric acid crystal deposition. The etiology includes tumor lysis syndrome after treatment of lymphoma, leukemia, or other myeloproliferative disorders, seizures, treatment of solid tumors, and rarely due to primary disorders of urate overproduction (hypoxanthine-guanine phosphoribosyltransferase HGPRT deficiency) or overexcretion due to decreased proximal tubule reabsorption (Fanconi-like syndromes). The chronic form is not seen in children or young adults. It is a CTIN, which is caused by deposition of Na^+ urate crystals in the medullary interstitium in individuals affected by chronic hyperuricemia. The chronic form may occur in patients with tophaceous gout.

6.7.9 Cholemic Nephropathy/ Jaundice-Linked Acute Kidney Injury

The historical term of *cholemic nephrosis* is still used in PICU but should be replaced with the new term of Jaundice-associated acute kidney injury. This injury is associated with liver dysfunction in jaundiced patients, and tubular injury by direct toxicity of bile salts and bilirubin seems the most plausible explanation. The renal function can, indeed, recover if the jaundice is managed. There are often large intratubular bile casts, which may impair the functionality of the kidney either by direct tubular toxicity, nephron obstruction, or both.

6.7.10 Myeloma Kidney

Insidious and progressive renal failure due to several factors, which have been listed as (1) myeloma cast nephropathy, (2) paraprotein deposition, (3) nephrocalcinosis, and (4) TIN changes. In the myeloma cast nephropathy, Bence-Jones protein (BJP) forms large, laminated casts, which are toxic to tubular epithelial cells inducing the recruitment of MNGC and PMN. BJP is different from Tamm-Horsfall (uromodulin) casts, which have a flat, atrophic lining, DPAS (++) but weakly (+) if diluted by paraproteins. Paraprotein deposition involves amyloid/κ light chain deposition in blood vessels, tubules, and glomeruli ("light chain nephropathy"). Nephrocalcinosis is due to calcium deposition in tubular epithelial cells and BM, two processes that induce interstitial fibrosis and chronic inflammation with or without CRF. Finally, TIN changes are constituted from plasma cell infiltrates, urate crystals (like a gout of the middle-aged individual/elderly), and nephritic effects of drugs. Grossly, there are normal to shrunken and pale kidneys. Microscopically, pink to amorphous blue casts, giant cells, tubular necrosis, interstitial inflammation ± granulomas (*vide supra*).

6.7.11 Radiation Nephropathy

It is characterized by glomerular capillary loop thickening and fusion, tubular atrophy, interstitial (stromal) fibrosis, and vascular changes, including fibrinoid necrosis.

6.7.12 Tubulointerstitial Nephritis and Uveitis (TINU)

TINU is a unique subset of adolescents and young women with interstitial nephritis with tubulointerstitial nephritis, uveitis with data supporting that modified C-reactive protein (mCRP), an autoantigen common to both the uvea, and renal tubular cells, may be involved in the pathogenesis. The inflammation in TINU syndrome is T-lymphocyte driven, although there is a paradoxical suppression of cytokine production and a decrease in peripheral immune response as shown by anergy to skin testing, similar to sarcoidosis. Clinical history of patients affected with TINU includes prior infection or the use of specific drugs, the Chinese herb, "goreisan," AI diseases like hypoparathyroidism, hyperthyroidism, IgG4-related autoimmune disease, and rheumatoid arthritis.

6.8 Vascular Diseases

6.8.1 Benign Nephrosclerosis

Renal changes that occur in the setting of mild (benign) HTN. Grossly, there is symmetric bilateral hypo-/atrophy of kidneys with fine, even granular renal surface with cortical narrowing. Microscopically, there is *hyaline arteriolosclerosis* (hyaline thickening of the walls of the small arteries and arterioles) and *fibroelastic hyperplasia*, which is characterized by reduplication of elastic lamina of interlobar and arcuate arteries and medial fibrosis, focal tubular atrophy, and interstitial fibrosis.

6.8.2 Malignant Nephrosclerosis

Renal changes associated with rapidly fatal malignant hypertension. Microscopically, there are *Hyperplastic arteriolitis*, which is characterized by intimal thickening by the proliferation of concentric smooth muscle cells ("onion skinning") and *Fibrinoid necrosis* of arterioles, often with inflammatory infiltrates in the vessel wall (necrotizing arteriolitis or "flea-bitten" kidney).To recall the microscopic features the "H & F" of benign nephrosclerosis are not capitalized, while the "H & F" of malignant nephrosclerosis are capitalized.

6.8.3 Renal Artery Stenosis

Seventy percent is caused by an atheromatous plaque at the origin (*atherosclerosis*); ♂ > ♀ and causes include lifestyle, obesity, etc. *Fibromuscular dysplasia* accounts for abundant remaining cases, especially those occurring in younger patients (♂ > ♀), and other causes in the DDX include Takayasu arteritis, EDS type IV, Marfan syndrome, Williams syndrome, and NF1. The current pathologic classification of FMD is based on the dominant arterial wall layer involved (intima, media, or adventitia). In the intimal fibroplasia, there is circumferential/eccentric collagen deposition in the intima with fragmentation/duplication of the internal elastic lamina. In medial fibroplasia, there is a discontinuous area of thickened collagen-rich fibromuscular ridges alternating with thinned media replacing smooth muscle with fragmentation/duplication of the internal elastic lamina. In perimedial fibroplasia, there is an extensive collagen deposition in the outer 1/3 of the media. In medial hyperplasia, there is smooth muscle cell hyperplasia, but no fibrosis. In the adventitial fibroplasia, there is dense collagen that substitutes the fibrous tissue of the adventitia ± extension into the surrounding tissue (± retroperitoneal fibrosis).

Atherosclerosis can be distinguished from FMD due to the age and occurrence of the process at the ostium or the proximal portion of the arteries, unlike the FMD that occurs in the middle or distal portion of the blood vessels. In differentiating with a vasculitis, typical values for ESR and CRP would point to FMD unless there is infarction of the kidney or bowel. A condition challenging to distinguish from FMD and possibly a subtype of FMD is *segmental arterial mediolysis*.

6.8.4 Infarcts

Wedge-shaped, with the base at the capsule, pale-white areas with hyperemic border (fresh) to tissue retraction (scar) by fibrous tissue replacement. They are usually embolic in origin and search for cholesterol crystals may be helpful (Figs. 6.22 and 6.23).

6.8.5 Vasculitis

- *DEF*: Hypersensitivity response to antigens (e.g., drugs) and various clinical syndromes (e.g., WG, PAN) and the distinction is based on organ involvement (Fig. 6.24).
- *CLM*: Focal and segmental necrotizing to diffuse crescentic GN ± granulomas in WG, extending through Bowman capsule into the interstitium and presence of tubulointerstitial infiltrate with eosinophils.
- *IFM*: Mesangial and subendothelial fibrin, ± IgG, IgM, C3 staining.
- *TEM*: Mesangial deposits, fibrin in capillaries, and capillary rupture.

Pauci-immune GN: ± vasculitis, + ANCA (Fig. 6.25). ANCA, antineutrophilic cytoplasmic antigens

p-ANCA, MPO-ANCA, MPA.

c-ANCA, PR3-ANCA, WG

No Ig staining, ± C, and fibrinogen with crescents.

TEM: Fibrin associated with GBM breaks, but no deposits.

P-ANCAs are seen in small vessel vasculitides, polyarteritis nodosa, rheumatoid disorders, and PSC (unlike AMA of PBC and ANA of AIH).

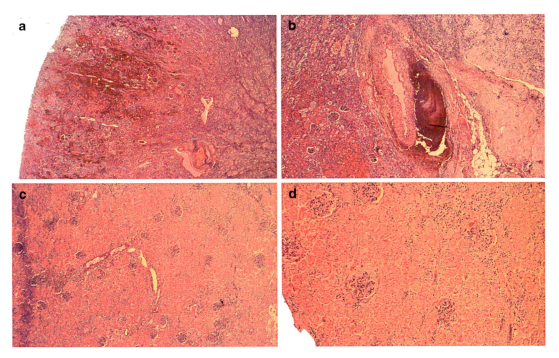

Fig. 6.22 Infarct necrosis is shown in this panel of four microphotographs taken at low and medium power of magnification. A hemorrhagic infarction is seen in both (**a**) and (**b**). Necrosis is particularly evident and highlighted better in the lower set of microphotographs by the pale color (**c** and **d**)

6.8.6 Hemolytic Uremic Syndrome

- *DEF*: Shiga-like toxin 0157, H7 *E. coli* (e.g., undercooked beef, postpartum, oral contraceptive users)-caused microangiopathic hemolytic anemia, thrombocytopenia, and ARF (1/4-HUS → CRF) with toxin damaging endothelium, promoting vasoconstriction, thrombosis and necrosis (Fig. 6.26).
- *CLM*: Thrombi and fragmented RBCs in glomeruli, focal necrosis without PMN infiltration,

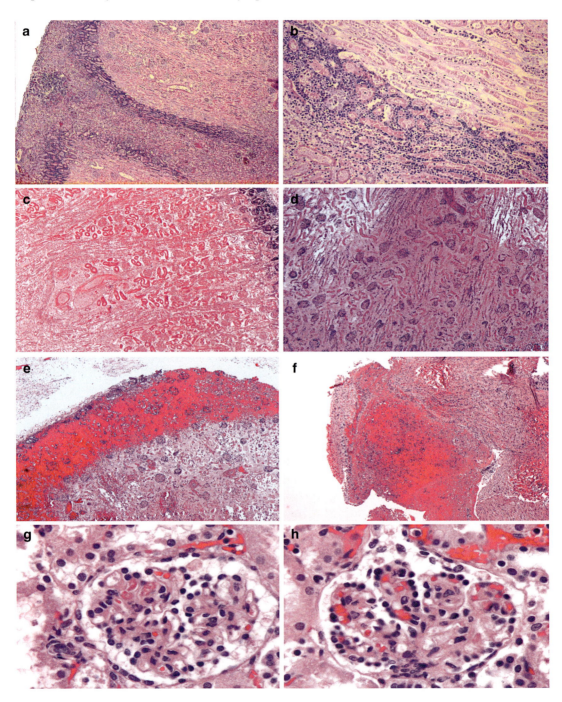

Fig. 6.23 Infarct necrosis (**a–d**) with subcortical hemorrhages (**e, f**) and thrombi in the glomeruli (**g-h**) seen at high magnification (H&E, 400×, original magnification). A lower magnification was used in the microphotographs (**a-f**)

Childhood Vasculitis

Systemic symptoms + ARCI
Arthropathy (arthralgia) & involvement
-> Renal (hematuria & red cell casts)
-> Cutaneous (purpuric skin rash)
-> Intestinal (abdominal pain)
+/- cardiac / pulmonary disease
Anemia, leukocytosis, thrombocytosis
raised ESR or CRP, ANCA +ve

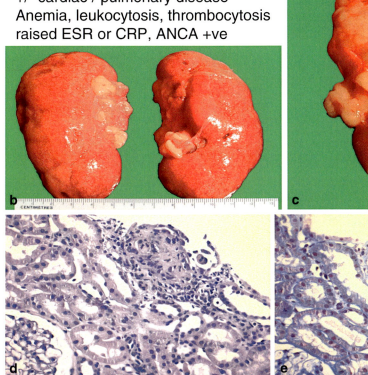

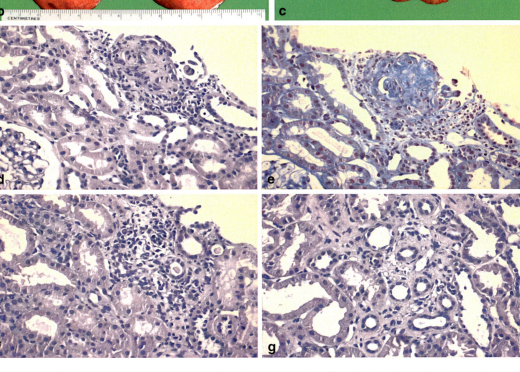

Fig. 6.24 Childhood vasculitis is often Henoch-Schonlein purpura with glomerulonephritis and extrarenal signs and symptoms (arthropathy, purpuric skin rash or palpable purpura, abdominal pain and associated or not cardiopulmonary diseases). The gross photographs are depicted in (**b**) and (**c**) with petechial hemorrhages and loss of the smooth surface of the kidney. In the microphotographs (**e–h**), glomerulonephritis with occasional sclerosis and lymphocytic infiltration of the interstitium are seen

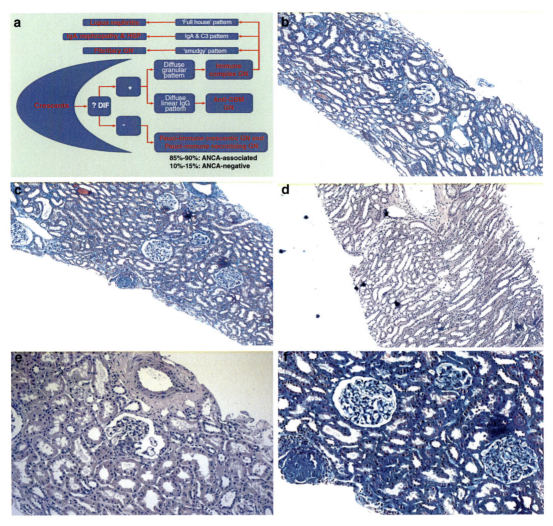

Fig. 6.25 In (**a**) is depicted the flowchart in case a crescent is seen, and the differential diagnosis knot in this flowchart is characterized by direct immunofluorescence (DIF), which is a key method in nephrology. In (**b**) through (**f**), there are microphotographs of a pauci-immune glomerulo- nephritis with cellular and fibrous crescents, fibrinoid necrosis of the glomerular tuft. Tubular atrophy/casts/ necrosis and occasional mononuclear inflammatory cells are observed in the interstitium (see text for detail)

thickening of capillary walls ("tram tracks"), and myo-intimal proliferation in small arterioles.
- *IFM*: Fibrin, occasional Igs.
- *TEM*: Separation of endothelium from BM with light granular material located in the subendothelial space.

6.8.7 Thrombotic Thrombocytopenic Purpura

Identical to HUS, with perhaps more prominent platelet-related pathology.

6.9 Renal Transplantation

6.9.1 Preservation Injury

ATN-like changes in the tubules with swelling of tubular epithelial cells with occlusion of the lumen by cytoplasmic "blebs" ± necrosis. TEM: Swelling of mitochondria and lysosomes.

6.9.2 Hyperacute Rejection

Minutes to hours after TX (<1 week) and mediated by *preformed antibodies* vs. *donor endothe-*

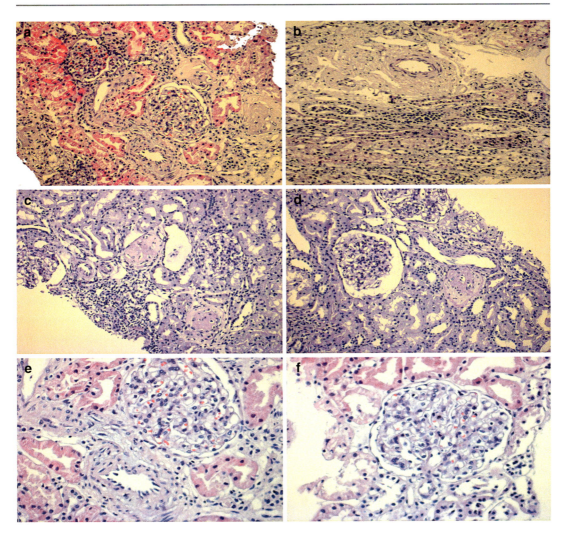

Fig. 6.26 The microphotographic panel (**a-f**) shows fibrin deposits in capillary lumina, endothelial cell proliferation in small arteries and arterioles, ischemic changes in glomeruli with endothelial swelling and capillary luminal narrowing, and focal red-colored mesangial thrombi, which can be at least better appreciated in (**a**), (**e**), and (**f**)

lium (type 2 hypersensitivity) ⇒ *thromboses* in arteries and glomeruli with consequent necrosis and interstitial tissue infarcts and hemorrhage, congested interstitial capillaries with PMN margination in peritubular BV (DDX: cold ischemia-induced graft vascular endothelial damage which induces less pronounced inflammation, systemic vasculitis, DIC, CNI.

IFM: C3 and fibrin deposits in small BVs.

6.9.3 Acute Rejection

Weeks to years after TX and subdivided into two branches (Fig. 6.27):

- *Interstitial* (cellular or type 4 hypersensitivity): Interstitial edema and chronic inflammatory cell infiltrate, initially at the corticomedullary junction with immunoblasts, lymphocytes (CD4+ and CD8+), plasma cells, and scattered neutrophils and eosinophils – tubulitis. Reversible with immunosuppression, e.g., OKT3, a monoclonal antibody to T cells
- *Vascular* (hormonal or type 2 hypersensitivity): Endothelial cell swelling and lifting with subendothelial infiltration of inflammatory cells, interstitial hemorrhage, necrotizing arteritis, and fibrinoid necrosis. Less easily reversible than the interstitial type of rejection.

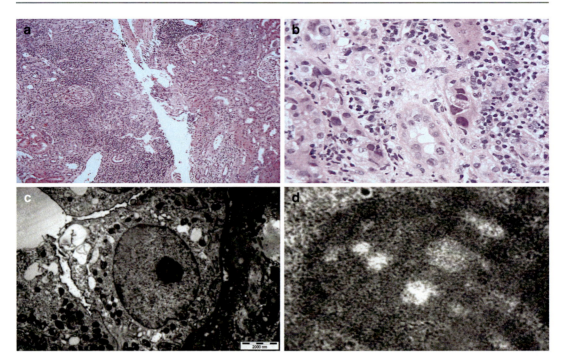

Fig. 6.27 TX SV40 (polyoma)-nephropathy with tubulointerstitial nephritis and infiltration by inflammatory cells with enlarged tubular epithelial cells and characteristic cytologic changes (viral cytopathy with smudged nuclei and basophilic chromatin) (**a**, **b**). Two electron microscopy photographs (**c** and **d**) show an enlarged epithelial cell with vacuolation (**c**) and crystalline arrays of nonenveloped, round, electron-dense particles in loose crystalline lattices (**d**). The SV40 is 45 nm in diameter

6.9.4 Chronic Rejection

Several months to years after TX and subdivided into two branches of changes, which are the vascular and the cellular changes, and the autonomous disease labeled chronic allograft glomerulopathy:

- *Vascular*: Concentric myo-intimal fibrous proliferation not attributable to other causes (e.g., donor HTN), variable luminal stenosis, fragmentation of the internal elastic lamina, foamy cell, and mononuclear cell infiltration. The reduplicative multilayering of the internal elastic lamina without foamy/mononuclear cell infiltration favors HTN.
- *Cellular:* Tubular atrophy with BM thickening or tubular diameter < 50% of normal and interstitial fibrosis with lymphocytic infiltration.
- *Chronic Allograft Glomerulopathy*: Glomerular sclerosis ± lymphocytic infiltration ± changes like TMA. Banff scoring parameters: increase of mesangial matrix and BM thickening ± double contour in capillary loops. In distinguishing from MCGN, the focality of the lesions would help to address to CAG.

6.9.5 Humoral (Acute/Chronic) Rejection

Several months to years after TX and subdivided into two branches of changes.

Early HR: Interstitial edema, neutrophilic margination and accumulation in capillaries (peritubular and glomerular), fibrin thrombi in glomeruli, and endothelial swelling, peritubular capillary congestion, fibrinoid necrosis, and ischemic tubular damage.

Late HR: Progressive peritubular and glomerular capillary damage with luminal dilation and BM thickening and reduplication with consequent tubular atrophy and interstitial fibrosis with ↑ of intravascular mononuclear cells and loss of peritubular capillaries.

IFM/IHC: +C4d is lining the peritubular and glomerular capillaries.

6.9.6 Antirejection Drug Toxicity

Cyclosporine and tacrolimus produce identical lesions, which are focal and can be overlooked, necessitating the evaluation of serial tissue sections. Acute toxicity is characterized histologically by necrosis and early hyalinosis of individual smooth muscle cells in the afferent arterioles and isometric vacuolation of the proximal straight tubules, while chronic toxicity may disclose the damaged media with smooth muscle cells in afferent arterioles replaced by beaded medial hyaline deposits that bulge into the adventitia. Thrombotic microangiopathy may be an additional element in acute toxicity. The interstitial space displays banded fibrosis and tubular atrophy. Acute toxicity is now rare because dosage has been lowered, but chronic toxicity is still present and may induce chronic allograft damage.

6.9.7 Recurrence of Primary Disorder

Most commonly seen in patients with FSGN and MPGN occurring in 10–20% and it accounts for 2% of graft failures.

6.10 Hereditary Cancer Syndromes Associated with Renal Tumors

Box 6.9 lists the genetic syndromes associated with renal tumors.

6.10.1 Beckwith-Wiedemann Syndrome

Beckwith-Wiedemann syndrome (BWS) is a genetic syndrome associated with overgrowth and cancer predisposition. BWS is character-

---

Box 6.9 Genetic Cancer Syndromes Associated with Renal Tumors

6.10.1. *Beckwith-Wiedemann Syndrome/ WT2 Associated*
6.10.2. *WAGR Syndrome/WT1 Associated*
6.10.3. *Denys-Drash Syndrome/WT1 Associated*
6.10.4. *Non-WT1/WT2 Pediatric Syndromes*
6.10.5. *Von Hippel Lindau Syndrome/VHL Associated*
6.10.6. *Tuberous Sclerosis Syndrome*
6.10.7. *Hereditary Papillary Renal Carcinoma Syndrome (HPRCS)*
6.10.8. *Hereditary Leiomyoma Renal Carcinoma Syndrome (HLRCS)*
6.10.9. *Birt-Hogg-Dube Syndrome*

---

ized by macrosomia, macroglossia, visceromegaly, hemihyperplasia, ear creases/pits, omphalocele, adrenocortical cytomegaly, neonatal hypoglycemia, renal abnormalities (e.g., medullary dysplasia, nephrocalcinosis, medullary sponge kidney, and nephromegaly), and, very importantly, embryonal tumors (e.g., Wilms tumor, hepatoblastoma, neuroblastoma, and rhabdomyosarcoma). In BWS there is abnormal regulation of gene transcription in two imprinted domains on chromosome 11p15.5. In early and late infancy, all patients with BWS and BWS spectrum should be screened for BWS-associated malignancies. Occasionally, BWS-linked cancer may be the first recognized manifestation of the syndrome. In particular, patients with BWS have an increased risk of cancer through age 7-8 years and should be screened every three months with serum and ultrasound investigations. In particular, complete abdominal ultrasonography through age 7-8 years for Wilms tumor and α-fetoprotein levels until age four years for hepatoblastoma.

6.10.2 WAGR Syndrome

WT1 gene associated (*W*ilms tumor, *a*niridia, *g*enitourinary defects, *r*etardation, mental) syndrome.

6.10.3 Denys-Drash Syndrome

WT1 gene associated (Wilms tumor, glomerulonephritis – diffuse mesangial sclerosis and intersex disorders – pure gonadal dysgenesis with male pseudohermaphroditism) syndrome.

6.10.4 Non-WT1/Non-WT2 Pediatric Syndromes

This group, which is incomplete, includes FWT, trisomy 18 syndrome, Perlman syndrome, Bloom syndrome, Fraser syndrome, and KTS (Klippel-Trenaunay syndrome or Klippel-Trenaunay-Weber syndrome). Although controversially debated, patients with KTS are frequently considered potential candidates for developing WT because they have unilateral overgrowth of the lower limb. This aspect has been argued because of the results from 95% confidence intervals for the two risks (KTS and WT) and compared with the general population risks of KTS (1 in ~47,000) and WT (1 in 10,000).

6.10.5 Von Hippel Lindau Syndrome

Clear cell RCC – chromosome 3 (3p25, *RAF-1*: pVHL inhibits the elongation step of RNA synthesis by interacting with *ELONGIN B* and *C*. To memorize the abnormalities and neoplastic findings of the von Hippel Lindau syndrome, the mnemonic acronym.
"K-PASH" can be used. It includes the kidney (CCRCC or clear cell renal cell carcinoma/renal cysts), pancreas (cysts/serous cystadenoma), adrenal (pheochromocytoma), skin (cafe au lait spots), hemangioblastoma of the retina and cerebellum (± EPO-induced polycythemia) (The acronym recalls K-PAX, the 2001 American SF movie with Kevin Spacey and Jeff Bridges).

6.10.6 Tuberous Sclerosis Syndrome

Angiomyolipoma, RCC, and chromosomes 9 and 16 changes.

6.10.7 Hereditary Papillary Renal Carcinoma Syndrome

Papillary RCC type 1 on chromosome 7 (7q31).

6.10.8 Hereditary Leiomyoma Renal Carcinoma Syndrome

Papillary RCC type 2 on chromosome 1 (1q42–43).

6.10.9 Birt-Hogg-Dube Syndrome

Ch-RCC/oncocytoma on chromosome 17 (17p11–2). It occurs in association with cutaneous lesions, including fibrofolliculomas, trichodiscomas, and acrochordons as well as pulmonary cysts and spontaneous pneumothorax.

6.11 Pediatric Tumors (Embryonal)

6.11.1 Wilms Tumor (Nephroblastoma)

- *DEF*: WT is the most common pediatric, mostly *triphasic* appearing renal tumor (epithelial, mesenchymal, and blastemal components) with *genetic heterogeneity* (±11p13 deletion involving the *WT1* gene, *WTX*, and *CTNNB1*) and arising from *nephrogenic rests*, which are an expression of the abnormal development of the *glomerulogenic zone* of the fetal kidney.
- *EPI*: 50% WT at age < 3 years, ~100% <10 years (Median age at diagnosis of WT:

Nephrogenic Rests: *Perilobar* (peripheral, sharp demarcated, often multifocal, and composed of blastema, tubules, and scant stroma) and *intralobar* (random, irregular, often unifocal, and composed of blastema, tubules, and stroma occasionally predominating).

> Perilobar rests are associated with ~1/3 of cases with synchronous WT.

> Intralobar rests are usually found in *metachronous* bilateral WT and younger age of presentation.

> Peri- and intralobar may also be further subclassified as:
>
> - *Dormant*, if there is no evolution/neoplastic transformation, usually occurring in older patients
> - *Regressing/sclerosing*, if there is evidence of maturation
> - *Obsolescent*, if there is mainly hyalinized stroma ("scar" stage)
> - *Hyperplastic*, if there is evidence of a tendency to progress to WT

> WT developing in a nephrogenic rest is characterized by (1) spherical expansile growth and (2) formation of a peritumoral fibrous pseudocapsule isolating the tumor from the nephrogenic rest and normal renal parenchyma.

> *Nephroblastomatosis*: Diffuse or multifocal dissemination of either abnormal persistence of nephrogenic rests with or without hyperplastic features or nephrogenic blastema. There is an extremely high probability of developing clonal transformation and develop multiple WTs, or as single WT with numerous neoplastic foci at presentation.

~3.5 years) with no sex predilection in unilateral cases (♂ = ♀), but female prevalence in bilateral cases (♂ : ♀=0.6:1).

- *EPG*: RF/Genetics include somatic biallelic inactivation of *WT1* gene, which is a gene encoding for a Zn finger transcription factor involved in nephron-gonadal development and is observed in 5–20% of sporadic WT. *WTX* gene encodes a protein that negatively regulates the Wnt/β-catenin signaling pathway and mediates the binding of *WT1*, and *WTX* has been reported to play a role in the WNT/beta-catenin signaling pathway. Somatic mutations of *CTNNB1*, which encodes β-catenin, have also been observed in WT. Interestingly, WTX deletion/truncation mutations appeared to be quite rare in tumors carrying exon three mutations of *CTNNB1* gene, encoding β-catenin. Approximately 1/10 WT patients have a genetic syndrome, including:
  - *WAGR syndrome* (WT, aniridia, genitourinary malformation, and mental retardation)
  - *Denys-Drash syndrome* (DDS)
  - *Beckwith-Wiedemann syndrome* (BWS)

  Individuals with WAGR syndrome have an inherited 11p13 deletion involving the *WT1* gene and 30% risk of developing WT. Mesangial sclerosis and pseudohermaphroditism constitute DDS and 90% risk to develop WT, while hemihypertrophy, macroglossia, omphaloceles, and visceromegaly characterize BWS. A gene (*WT2*) has been localized to chromosome 11p15.
  Intralobar nephrogenic rests have an increased rate of progression to WT and are more commonly associated with *WT1* mutations, DDS, and WAGR syndrome, while perilobar nephrogenic rests are seen in sporadic tumors and are related to genetic/epigenetic dysregulation at 11p15, idiopathic hemihypertrophy, and BWS.

- *CLI*: WT presents as an asymptomatic abdominal mass (rarely as hematuria, arterial hypertension, or proteinuria) in 4/5 of children. Hypertension, gross hematuria, and fever are present in 5–30% of children at presentation, while an abdominal pain or hematuria is present in 1/4 of children at presentation. UTI and varicocele are present in <1/20 of children at presentation, while the triad "hypertension, anemia, and fever" is present in <1/20 of children at pre-

sentation. Respiratory symptoms related to lung metastases remain rare at presentation.
- *LAB*: CBC, chemistry profile with kidney function tests, electrolytes (e.g., sodium and calcium), urinalysis, coagulation cascade studies, cytogenetical studies, including 1p and 16q deletion are paramount.
- *IMG*: Imaging studies will contemplate renal ultrasonography, chest radiography, abdominal and chest CT, and an abdominal MRI.
- *GRO*: WT is large, solid, gray-white, well-circumscribed with hemorrhage and necrosis.
- *CLM*: The three variably represented components include (1) blastemal component that may show or not cellular cohesiveness (nodular or serpentine pattern vs. diffuse); (2) epithelial component, which shows embryonic tubular structures (with or without glomeruli) to small, round cell rosettes; and (3) stromal component with spindle cell (fibroblast-like) possibly differentiating toward smooth or skeletal muscle (Figs. 6.28, 6.29, 6.30, 6.31, 6.32, and 6.33). Grade histology: "favorable" vs. "unfavorable" according to the presence of anaplasia. Anaplasia is defined as nuclei 3× than the size of neighbor tumor cells AND hyperchromatic nuclei AND abnormal (multipolar) mitoses. All three conditions need to be satisfied before the diagnosis of anaplasia is given. A nephroblastomatosis needs to be ruled out (Fig. 6.34).
- *IHC*: (+) VIM, WT1 (mostly, but nonspecific!) (±) AE1–AE3 (epithelium), DES, and NSE.
- *TEM:* There are ultrastructural similarities between normal metanephric blastema of the intrauterine life and WT. The ultrastructural examination of the blastema shows solid sheets of uniform cells without lumina or tubular structures and deep infoldings of the nuclear membranes, relatively scant cytoplasm containing free ribosomes but few mitochondria, granular endoplasmic reticu-

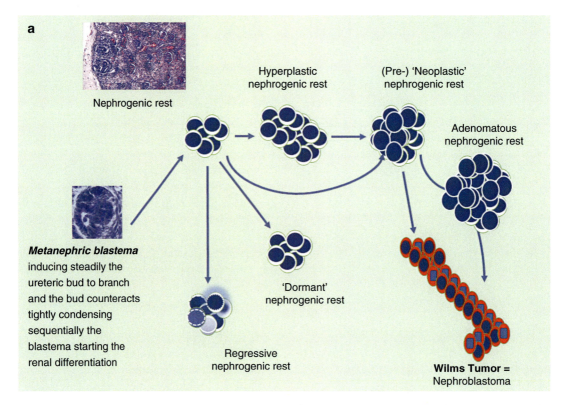

Fig. 6.28 In (**a**) is shown the evolution of metanephric blastema into nephrogenic rests (different subtypes) and Wilms tumor. In (**b**) and (**c**) are imprints of a Wilms tumor with high nucleus to cytoplasm ratio and dense chromatin. The distinction of the varying nephrogenic rests may be vague

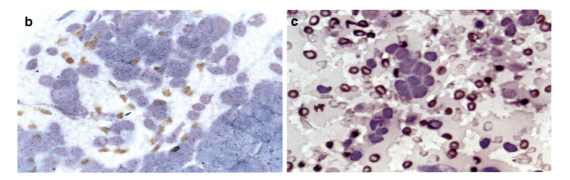

Fig. 6.28 (continued)

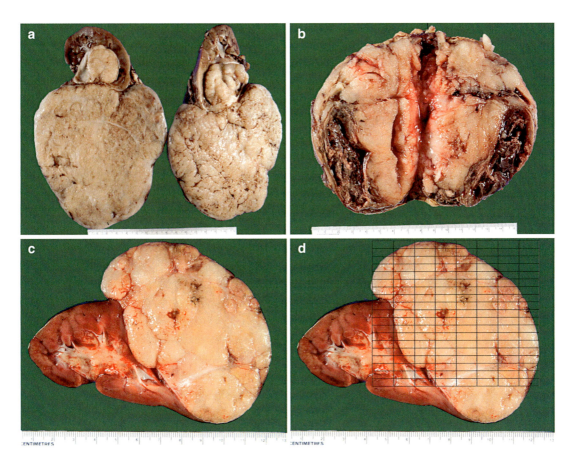

Fig. 6.29 Gross photographs of kidneys with Wilms tumor. Grossly, the tumor is large and well-circumscribed and typically solitary, although one in ten cases is bilateral or multicentric. The tumor is soft and tan-gray, but not entirely or always homogeneous displaying occasional hemorrhage, necrosis, and cysts (**a–d**). A chessboard helps for an adequate sampling with the tumor being sampled with at least one block per centimeter of at least the largest dimension

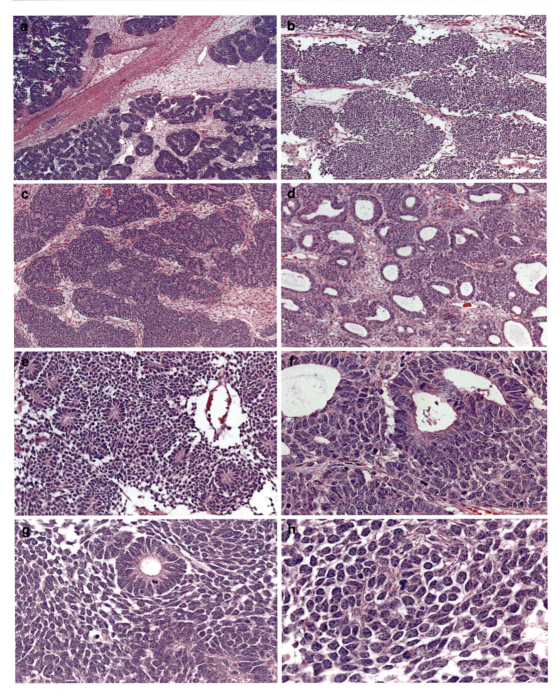

Fig. 6.30 In this panel of microphotographs, Wilms tumor shows the blastemal, the epithelial, and focally the stromal differentiation. The tubular differentiation (epithelial differentiation) is particularly striking with the presence of true rosettes different from the pseudorosettes seen in the neuroblastoma. All microphotographs have been taken from tumor without previous chemotherapy

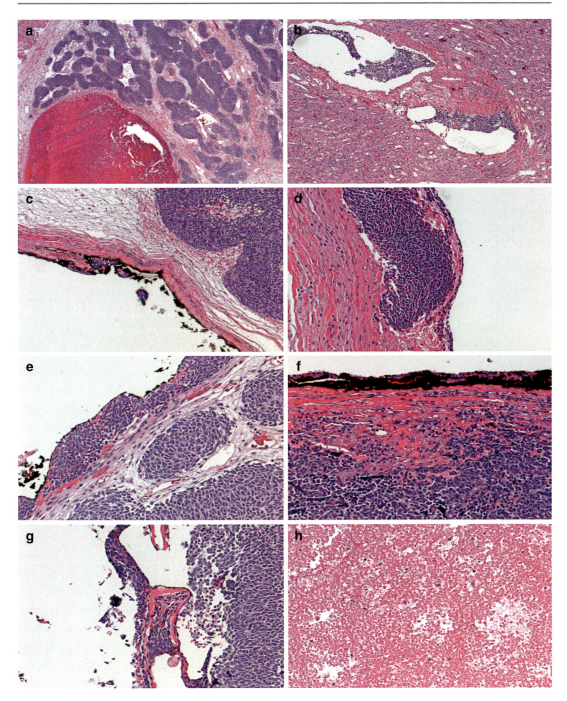

Fig. 6.31 Wilms tumor can present with substantial challenges for the pathologist, specifically if it is close to a blood vessels (**a**), inside of the lumen of a blood vessel (**b**), close to the resection margins (**c**–**g**), and when it is mostly necrotic (**h**) (all microphotographs are H&E and their magnification is x25, x50, x100, x200, x200, x200, x200, and x200, respectively)

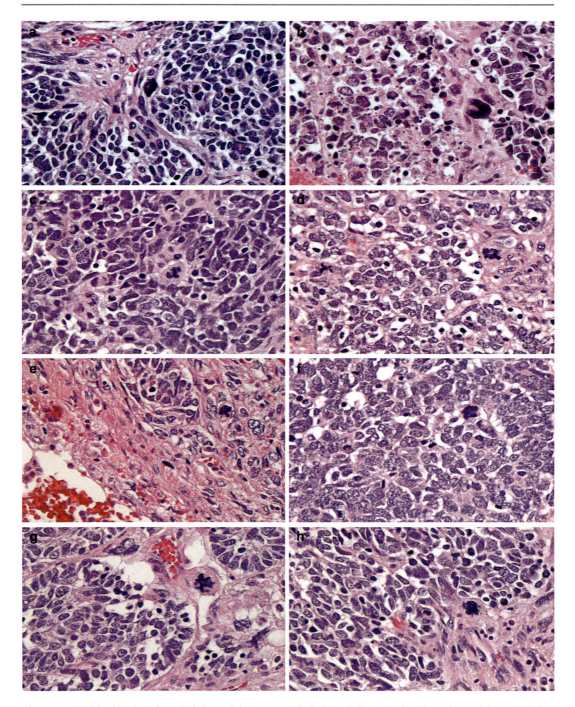

Fig. 6.32 The identification of anaplasia is crucial because it upstages the Wilms tumor to unfavorable histology, which requires a more aggressive chemotherapy protocol. The anaplasia is defined by the fulfillment of all three criteria (enlarged size more than three times of the normal size, multipolarity of the mitotic spindles, and hyperchromatism) (**a-h**). All microphotographs have been taken from H&E-stained slides at x400 as original magnification

6.11 Pediatric Tumors (Embryonal)

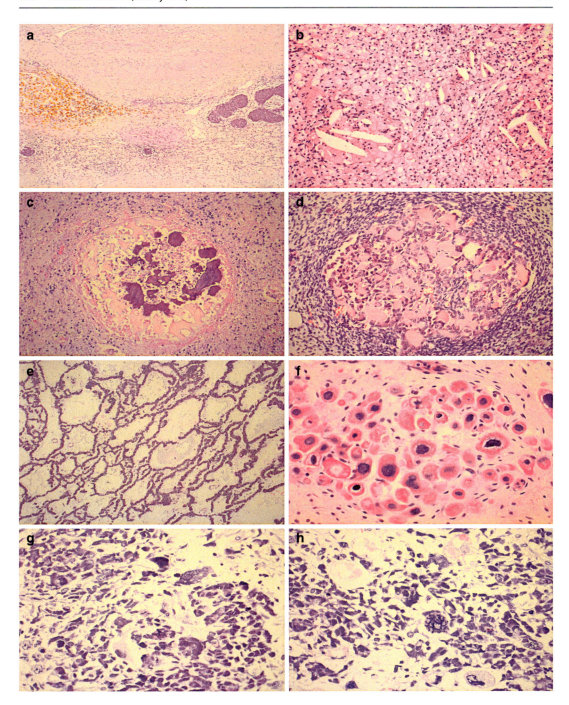

Fig. 6.33 The effects of chemotherapy on Wilms tumor are important to recognize to avoid favoring a wrong diagnosis (**a-h**). In particular, the rhabdoid differentiation is shown in (**f**) and two anaplastic Wilms tumor with persistence of blastema and anaplasia. All microphotographs (**a-h**) have been taken from H&E-stained slides at variable degree of magnification: x100, x200, x200, x200, x100, x400, x400, and x400, respectively

lum cisterns, or other organelles. The cells are strictly opposed and are only separated from each other by a narrow intercellular cleft.

- *TRT:* Immediate nephrectomy with histologic confirmation is the usual procedure in North America (the USA and Canada); also there may be an exploration of the contralateral kidney to ensure that the disease is unilateral and biopsy sampling of the regional LNs for staging purposes. In children presenting with bilateral malignancy, an immediate nephrectomy is not performed. About the chemotherapy, regimens classically comprise vincristine and dactinomycin, initially. The chemotheraputic drugs doxorubicin and then cyclophosphamide and etoposide are added for the high-risk malignant disease. In case children show loss of heterozygosity at 1p and 16q in *WT*, more aggressive chemotherapy is usually recommended and represents the subsequent management. The SIOP Risk Stratification for Post-chemotherapy Specimen is crucial. In the Post-chemotherapy Specimen, it is important to assess the rate (%) of necrosis. If necrosis is 100%, then the tumor is labeled as "completely necrotic type." If necrosis is between 2/3 and full (100%) necrosis, the tumor is of "regressive type." If necrosis is <2/3 (67%), viable tumor tissue needs to be subtyped as blastemal type, epithelial type, stromal type, or mixed type according to the % of the three components. According to the revised SIOP working classification of renal tumors of childhood (2001), for pretreated cases, low-risk tumors are congenital mesoblastic nephroma, cystic partially differentiated nephroblastoma, and the utterly necrotic nephroblastoma. Intermediate risk tumors are nephroblastoma epithelial type, nephroblastoma stromal type, nephroblastoma mixed type, nephroblastoma regressive type, and nephroblastoma focal anaplasia. High-risk tumors include nephroblastoma blastemal type, nephroblastoma diffuse anaplasia, clear cell sarcoma of the kidney, and rhabdoid tumor of the kidney. In case of primary nephrectomy cases, low-risk tumors are congenital mesoblastic nephroma and cystic partially differentiated nephroblastoma, while intermediate-risk tumors are non-anaplastic nephroblastoma and its variants, as well as nephroblastoma focal anaplasia. High-risk tumors include the nephroblastoma with diffuse anaplasia, the clear cell sarcoma of the kidney, and the rhabdoid tumor of the kidney.
- *PGN:* Up to 90% of children survive with current multimodality therapy, which has been an impressive goal reached in the last couple of decades of chemotherapy protocols and successful clinical trials. Two most important factors are established in indicating adverse PGN in WT and reveal the importance of the pathologist in assessing properly a WT specimen. The two factors include *histology* (i.e., the presence or absence of *anaplasia*) (favorable vs. unfavorable histology depending on the presence or not of anaplasia) and *tumor stage*. Spreading pattern is local to perirenal soft tissue, adrenal gland, bowel, and liver. MTX to regional LN is observed in ~15%. Distantly, WT can metastasize to the lungs, liver, and peritoneum. Chemotherapeutical regimens ± radiation induce massive necrosis of the immature component but spare the mature portions according to SIOP guidelines. Unilateral tumors have good PGN (~90%), but unfavorable histology is accompanied by bad PGN (~5%) with an increase of mortality until 1/2 cases. Following numerous observational and clinical trials studies good PGN signs include first infancy (age < 2 years), epithelial differentiation (no independent factor), favorable histology (no anaplasia), and low stage.

WT Genetic Heterogeneity

There is a prevalent model for WT, which underlines from one side a bifurcated model and on the other side a most elaborate and genetic heterogeneous background. One of the two arms of the bifurcated model is constituted by *biallelic WT1 mutations* resulting in the development of an ILNR, which is the basis for additional

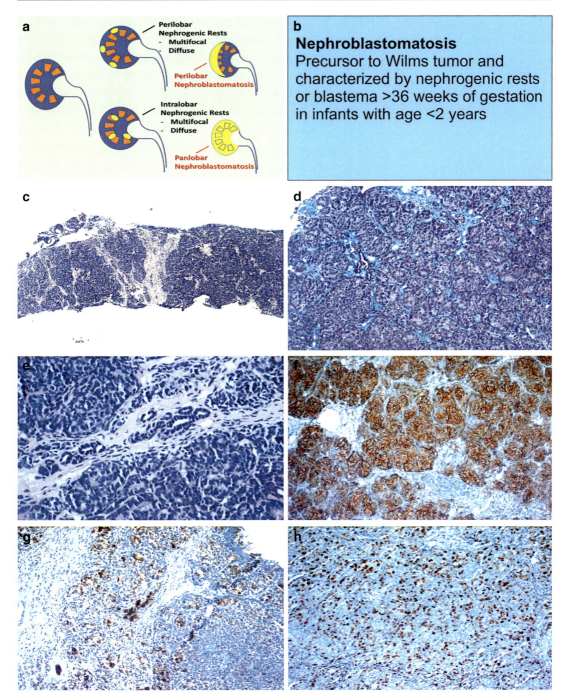

Fig. 6.34 Nephroblastomatosis can occur when numerous nephrogenic rests occur in a multiple fashion (**a-b**). In other words, it is a rare preneoplastic lesion defined as the presence of multiple nephrogenic rests, which persist beyond 36 weeks gestation and are thought to be precursors for the development of Wilms tumor (**c-e**), H&E staining, x40, x200, and x400 as original magnification, respectively. CD56 immunostaining is shown in (**f**), AE1–AE3 is shown in (**g**), and the proliferation antigen Ki67 (MIB1) is shown in (**h**). All three immunostainings were performed using avidin-biotin complex and taken at 200× as original magnification

genetic changes, such as *WNT*-activating mutations, which are unquestionably responsible for the development of neoplasia. The other arm of the bifurcated model is constituted by genetic and or epigenetic modifications resulting in *biallelic 11p15 methylation at imprint control region 1 (ICR1)*, resulting in biallelic expression of *IGF2*, and development of a PLNR, which may be the basis of additional genetic changes giving rise to the neoplasia. However, most recent data support further heterogeneity in this bifurcated WT model. Five subsets have been recently delineated using global gene expression patterns, *WT1*, *CTNNB1*, and *WTX* mutations, as well as 11p15 copy number and methylation patterns. In subset 5, biallelic methylation of *ICR1* occurring before induction results in persistent elevation of IGF2 causing preferential mesenchymal proliferation preventing the development of nephrons and resulting in ILNR. If biallelic methylation of *ICR1* happens after initiation within the foci of early nephron cells, the persistence of increased IGF2 hampers terminal differentiation of the epithelium and ends in the development of a PLNR.

New Differentiation Markers for Non-embryonal Renal Tumors

Current markers are useful in differentiating chromophobe RCC and oncocytoma from eosinophilic variants of clear cell RCC (CCRCC), papillary RCC, and other tumor subtypes. The use of microarrays has led to the development and application of several immunohistochemistry-based assays. Although partially applied in the diagnostic routine, these tests harbor a remarkable potential clinical utility. This situation is the example of glutathione S-transferase α and adipophilin for CCRCC, AMACR for papillary RCC, and c-Kit, beta-defensin 1, and parvalbumin for chromophobe RCC. Proteins of the claudin family are localized in tight epithelial junctions and are dysregulated in several carcinomas. Claudin-7 and claudin-8 are expressed in the normal distal nephron epithelium and seem to be more specific for chromophobe RCC rather than oncocytoma.

Beckwith-Wiedemann Syndrome (BWS) as Genomic Imprinting Syndromes and Cancer

Genomic imprinting is a form of epigenetic control of gene expression able to preferably express one allele of a gene according to the parent-of-origin of the allele. It means that some "imprinted" genes are derived from the mother and others from the father, but there are also imprinted genes, which show a tissue-specificity and others that show complex mechanisms of interaction according to the stage of developing an embryo. It has been widely demonstrated that alterations occurring during the "imprinting" process can be associated with the development of genetic syndromes and oncogenesis. There are several mechanisms implicated in the imprinting process that include DNA methylation, insulation, chromatin change, and expression of large noncoding RNAs. Most probably, DNA methylation plays a significant role in genomic imprinting by stopping the binding of transcription factors to the methylated gene promoter or preventing the binding of insulator proteins to regions that are differentially methylated (DMR) and inducing changes in the structure of chromatin. Mechanisms leading to disorders of imprinting include mutations or epimutations, loss of imprinting (LOI), uniparental disomy (UPD), and intragenic mutations or copy number alterations affecting the function of an imprinted gene directly. Differently, from genetic mutations, epimutations do not change the nucleotide sequence of DNA and may be reversible. Similar to genetic mutations, epimutations harbor comparable specific patterns of gene expression that are heritable through mitosis. In LOI both alleles are expressed. In BWS an alteration of the gene expression from aberrant loss or gain of methylation at an imprinting DMR is at the basis of pathogenicity for disease and cancer. An aggregation of five imprinted genes (*IGF2*, *H19*, *CDKN1C*, *KCNQ1*, and *KCNQ1OT1*) is contained in the 11p15.5 imprinted region, and BWS is a heterogeneous genetic disorder with a broad range of molecular mechanisms which can ultimately induce dysfunction of the imprinting process, particularly of *IGF2* and *CDKN1C*.

There is also the possibility of WT occurring in patients with polycycyic disease as demonstrated by Thankamony et al. (2016). These authors reported on a 17-months-old baby with bilateral polycystic kidneys and PHACE syndrome (posterior fossa anomalies, hemangioma, arterial lesions, congenital heart defects, and eye abnormalities). Cyst formation is cell signaling defect in both epithelial cell proliferation and fluid accumulation. PKD increases intracellular Ca^{2+}, and cAMP stimulates mitogen-activated protein (MAP) kinase, which induces cell proliferation. The altered calcium ions homeostasis induces an alteration of cAMP levels and, subsequently, dysregulation of ion transporters. The next step is the dysregulated expression of aquaporins and fluid accumulation.

Extrarenal WT is sporadic, and most extrarenal WTs are in the retroperitoneum. The exact origin of the extrarenal form of the disease is heavily debated with some plausible hypotheses including the origin from ectopic metanephric blastema, the origin from primitive mesodermal tissue, and the Connheim cell rest theory.

6.11.2 Cystic Partially Differentiated Nephroblastoma (CPDN) and (Pediatric) Cystic Nephroma

- *DEF*: Infantile, almost fully differentiated and cystic, renal tumor, which usually presents as a single abdominal mass with ureteral obstruction and harboring a good prognosis.
- *GRO*: Sharply demarcated, variably large, coarsely nodular, and white tumor with numerous serous cysts.
- *CLM*: There is tubular lining epithelium with hobnailing and intercystic spindle cell stroma between cysts ± islands of scattered blastemal cells (Fig. 6.35).
- The (pediatric) cystic nephroma and the cystic, partially differentiated nephroblastoma are on the benign end and polycystic WT on the malignant end of pediatric cystic renal neoplasms. Typically, cystic nephroma does not contain nephroblastomatous elements.

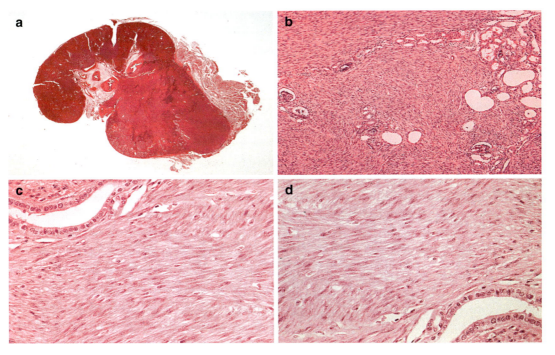

Fig. 6.35 In (**a**) is a low power microphotograph of a kidney with congenital mesoblastic nephroma. In (**b**) the tumor has similarity with infantile fibromatosis and shows fascicles and whorls of bland spindled cells and thin collagen fibers. As seen in (**b**), the tumor surrounds tubules and glomeruli and has irregular borders. In the cellular variant, there is a similarity with infantile fibrosarcoma with a sheetlike proliferation of plump, atypical spindle cells (**c–d**). Mesenchymal differentiation may occur such as the chondroid metaplasia (**e–h**). Inflammatory cells may also be seen

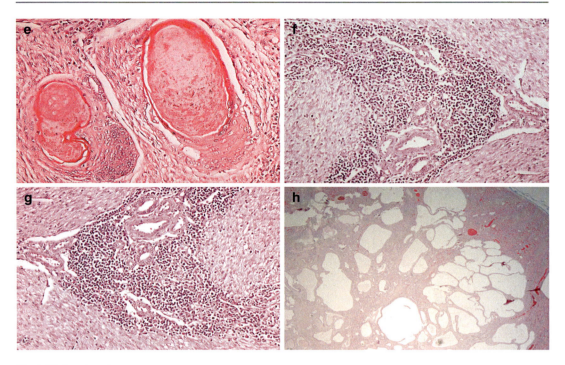

Fig. 6.35 (continued)

6.11.3 Congenital Mesoblastic Nephroma

- Infantile renal tumor, which most commonly occurs in the 1st year of life (8/million in children <15 years), presenting as a solid, yellow-gray to tan, mass with whirled configuration (*leiomyomatous hamartoma*) and *t(12;15)*(p13;q25) resulting in *ETV6-NTRK3* gene fusion (sharing with infantile fibrosarcoma). Two forms are recognized and they include "congenital" and "atypical" (or cellular) characterized by the absence/presence of a noncapsular infiltrative growth pattern (pushing vs. infiltrative borders) without hemorrhage or necrosis (Fig. 6.35). The CMN, common type, usually harbors a favorable PGN, unlike the atypical variant. CMN tends to recur if not completely excised. An atypical variant is characterized by many mitoses, infiltration of the pelvis or perirenal tissue, and tendency to metastasize.

6.11.4 Clear Cell Sarcoma

- *DEF*: Bone-metastasizing tumor (4% of pediatric renal tumors) with a peak at the end of the first infancy presenting with a well-demarcated, homogeneous tan, unilateral cystic mass with rare hemorrhage and necrosis and diagnostic arborizing vascular peritumoral cellular vascular plexus.
- *CLM*: There are small nests or cords composed of pale, rounded, or polygonal cells (clear cytoplasm in ~20%) with a delicate chromatin pattern and separated by an arborizing network of thin-walled, small blood vessels (Ret+) as well as areas with stellate and spindle cells. Focal tubular differentiation at the periphery of the tumor (small cysts lined by low cuboidal epithelium may occasionally be observed arising probably from dilated tubules). There are several growth patterns, including epithelioid, spindled, sclerosing, myxoid, palisading, cystic, pericytoma-like, and pleomorphic. The alcian blue-positive material may be seen in some areas.

- *IHC*: (+) VIM, Cyclin D1, (±) p53, BCL2, (−) WT1, CD56, EMA, CD99, DES, S-100, CEA, SYN, AE1–AE3. CD34 decorates the septal capillaries, and MIB1 (Ki67) is about 20% of the tumor cells (mild proliferation activity).

6.11.5 Rhabdoid Tumor

- *DEF*: Highly malignant early pediatric renal neoplasm with extrarenal sites of occurrence including soft tissues and the CNS ("atypical teratoid/RT") and harboring a point mutation or large deletion of the *SMARCB1/hSNF5/INI-1* gene located on chromosome 22q11.3–5, which encodes a component of the SWI/SNF chromatin remodeling complex that plays a major role in transcriptional regulation (SWItch/Sucrose Non-Fermentable is an important nucleosome remodeling complex of eukaryotes).
- *CLI*: Clinically, over 2/3 of patients with RT present with the non-localized disease, ± association with medulloblastoma.
- *GRO*: The tumor is soft and solid, with infiltrative margins.
- *CLM*: Histologically, the monomorphic proliferation of medium-sized round/oval (also spindle) cells with eccentric nuclei, eosinophilic cytoplasmic inclusions, macronucleoli, and abundant cytoplasm, the absence of muscular differentiation (≠ WT), and involving both the cortex and medulla.
- *IHC*: (+) VIM, EMA, CK, and (−) INI1 (i.e., the INI1 staining is not retained comparing the surrounding tissue and other organs).
- *TEM*: Whorled intermediate filaments in the cytoplasm.
- *TRT/PGN*: Chemotherapy alone is rarely curative (overall survival: ¼ of patients).

6.11.6 Metanephric Tumors

- *DEF*: Not exclusively pediatric, but in a wide age range occurring renal tumor group of metanephric origin, including *metanephric adenoma* (MNA), *metanephric adenofibroma* (MNAF), and *metanephric stromal tumor* (MNST) that share some aspects between each other with a favorable outcome. Occasionally, it has been reported a case of *metanephric adenosarcoma* (Figs. 6.36 and 6.37).
- *MNA*: ♂ < ♀, 2nd-4th decades of life, "incidentaloma", polycythemia, well-demarcated tumor, solid or lobulated, gray-tan-yellow unencapsulated with variable size ± H/N/C/Ca (hemorrhage/necrosis/calcification/calcium) change and characterized microscopically by *densely packed small round uniform acini* of round/ovoid cells with small, monotonous bland nuclei with no or inconspicuous nucleoli, with no or very rarely mitoses, and scant pale or pink cytoplasm in a acellular stroma ± papillary structures, glomeruloid bodies, and psammoma bodies (IHC: +WT1, CD57, pan-CK, VIM, −EMA, -CK7; *DDX*, PRCC (+CK7, EMA); WT, triphasic, blastema, nucleoli, and mitoses).
- *MNAF*: Metanephric tumor with both epithelial (*densely packed small round uniform acini*) and stromal elements (+CD34), first–fouth decades, ♂ > ♀.
- *MNST*: Metanephric tumor with common stromal components characterized by the spindle and epithelioid stromal cells with angiodysplasia and glia-epithelial complexes (+CD34, ± S100 and GFAP; -CK, -DES), first decade, ♂ > ♀, triad or extrarenal vasculopathy.

6.11.7 XP11 Translocation Carcinoma

- *DEF*: Renal tumor harboring gene fusions involving members of the microphthalmia-associated transcription (MiT) factors family, including *TFE3* and *TFEB* genes, more common in children/young adults than middle-aged adults and seniors.
- *EPI*: 3% of adult RCC and ~40% of RCC in children/young adults.
- *EPG*: Exposure to cytotoxic chemotherapy is a risk factor with overexpression of *TFE3* or *TFEB* activating multiple downstream targets.

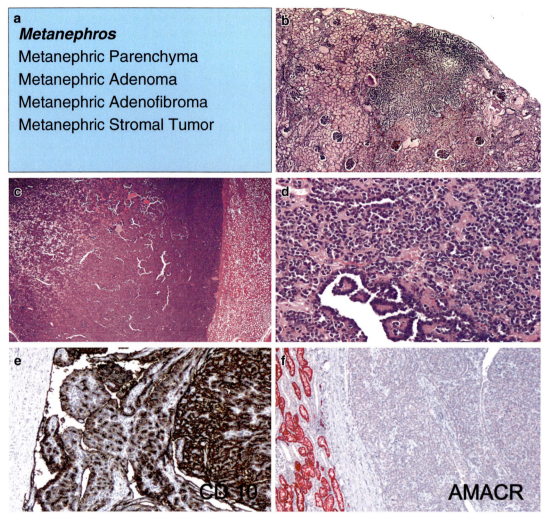

Fig. 6.36 Metanephric differentiation is pillar to understand between metanephros, metanephric parenchyma, and tumors arising from it, including metanephric adenoma, metanephric adenofibroma, and metanephric stromal tumor (**a**). In (**b**) through (**d**), a metanephric adenoma is displayed. The tumor is composed of uniform, closely packed papillae or tubules "floating" in loose stroma. The tumor cells are small with minimal cytoplasm and uniform chromatin. No atypia or mitotic activity is seen. The microphotographs (**e**) and (**f**) show the expression of CD10 and the lack of expression of AMACR or P504S (avidin-biotin complex, immunostaining, ×100 original magnification)

- *GRO*: Tan-yellow, often hemorrhagic and necrotic.
- *CLM*: Papillary, alveolar, and nested growth pattern with clear and eosinophilic cells accompanied by psammoma bodies. Clear to eosinophilic discohesive pseudostratified cells with voluminous cytoplasm and high-grade nuclei, psammoma bodies, and ± melanin pigment.
- *IHC*: (+) TFE3 or TFEB (strong nuclear staining), PAX8, Cathepsin K, CD10, AMACR, VIM, E-cadherin, ± AE1–AE3/EMA, and melanocytic markers (HMB45 and Melan-A), more commonly in t(6:11) carcinomas and infrequently in Xp11, but (−) Carbonic Anhydrase/CAIX, CD45, Calretinin, and SMA.
- *TEM*: Features of clear cell carcinoma, including cell junctions, numerous mitochondria, microvilli, BM, and abundant glycogen.
- *DDX*: Clear cell PRCC (homogeneous low-grade clear cells without eosinophilic cells, no

6.11 Pediatric Tumors (Embryonal)

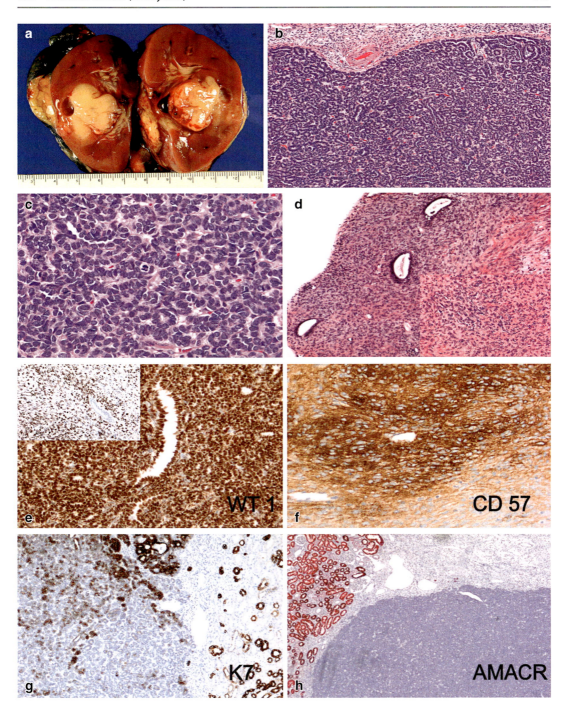

Fig. 6.37 This panel shows gross (**a**) and microphotographs of a metanephric adenofibroma (**b** through **h**). Grossly, the tumor is a solitary, nonencapsulated, yellow mass located in the medulla and pelvis and having a cyst. Microscopically (**b**–**d**), there is focal embryonal epithelium identical to metanephric adenoma and bland spindle cell stroma with concentric cuffing of entrapped tubules (**d**). There is positivity for WT1, CD57, and K7 and lack of expression for AMACR (avidin-biotin complex immunostaining, ×200 original magnification for **e** and **f** and ×100 original magnification for g and ×50 original magnification for **h**). The inset of figure **e** is another view (×100) of the diffuse expression of WT1 and spans morphologic spectrum between benign pure stromal and pure epithelial lesions. Epithelial cells are bland with no mitotic figures. Entrapped tubules may have papillary hyperplasia

calcifications, commonly branched ductular structures, and secretory cells with nuclei aligned above BM and CK7+, CAIX+, CD10-, and AMACR-), CCRCC (older patients, no true papillae, (+) AE1-AE3/EMA+, VIM+, TFE3-, diffuse CAIX+, 3p deletion present), CCSK (children, cells with indistinct cell borders, TFE3-), chromophobe RCC (diffuse CD117/c-kit), and PRCC (no nested alveolar patterns, no extensive areas of clear cells; CK7+, TFE3-, trisomy 7 and 17).
- *TRT*: Surgical resection and mTOR and tyrosine kinase inhibitors, although selective MET inhibitors appear to have modest antitumoral activity.
- *PGN*: Aggressive clinical course.

6.11.8 Ossifying Renal Tumor of Infancy

- *DEF*: Sporadic tumor showing a calcified mass in renal pelvis composed of spindle cells embedded in the partially calcified osteoid matrix. Clinically, it presents in infancy as macrohematuria or abdominal mass and seems to have good behavior. Very few case reports are described in the literature. The WHO classification of tumors of the urinary system is paramount in distinguishing this entity as well as other pediatric renal tumors from non-embryonic renal tumors of the youth.

6.12 Non-embryonal Tumors of the Young

6.12.1 Clear Cell Renal Cell Carcinoma

- *DEF*: Most frequent (~9/10) adult renal tumor (aka hypernephroma or renal adenocarcinoma) of the proximal tubular epithelium with a median age in the sixth decade, ♂ > ♀, ± #3p−/~ (RAF-1) (PVHL ⇒ ↓ HIF-1a ⇒ HRE binding), bilateral (1%) and in the setting of syndromes, including *VHLS, ADPKD, tuberous sclerosis*, and *long-standing dialysis*.
- *EPI*: RFs are summarized by the acronym *IDAHO* as the mnemonic word recalling several factors including *I*rrational living (smoking), *D*M, *A*ge and *A*cquired cystic kidney disease (long-term dialysis), *H*TN and *H*eritage (hereditary cancer syndromes), and *O*besity.
- *CLI*: Hematuria, flank pain, mass ± paraneoplastic syndromes (*polycythemia, hypercalcemia, HTN, Cushing syndrome, Stauffer syndrome or liver dysfunction, femininization/masculinization/gynecomastia*), and *systemic amyloidosis*.
- *GRO*: Well-demarcated, mostly cortex-located (bosselated) with extrarenal (early) > intrapelvic (late) extensions, a golden yellow tumor with fibrous pseudocapsule, variegated CS ± H/N/Ca/Cy change, and multiple nodularities (5%) (H/N/Ca/Cy change: hemorrhage, necrosis, calcification, and cystic change or degeneration).
- *CLM*: Tubular, papillary, solid, alveolar, or trabecular pattern of large cells with distinct cell membranes, central nuclei (Fuhrman grading using a 10× objective), and clear or granular cytoplasm ± hyaline droplets and micro-/macrocysts containing eosinophilic fluid. Variants: oncocytic, rhabdoid, and sarcomatoid (Fig. 6.38). Very rarely, RCC can occur in the setting of multicystic kidney dysplasia (Fig. 6.39). In this situation, the Ridson-Hill lesions of multicystic kidney dysplasia need to be recognized to support this pathogenetic pathway. In fact, multicystic kidney dysplasia contain progenitor cells able to disclose some autonomy and differentiate into neoplastic cells.
- *CYT*: The typical cytologic presentation of RCC consists of neoplastic cells in papillary groups and irregular clusters as well as single. Neoplastic cells are large with copious and vacuolated (sometimes granular) cytoplasm and low nuclear/cytoplasmic ratio with nuclear features depending on the grade of the RCC. Typically, nuclei look round to oval, uniform, with finely granular chromatin and a single nucleolus. Grade I is <10 μm, round, uniform nuclei (mature lymphocyte-like); grade II is 10–15 μm finely granular nuclei with open chromatin and inconspicuous nucleoli; grade III corresponds to 15–20 μm nuclei with prominent nucleoli; and grade IV corresponds to pleomorphic/multilobated nuclei, >20 μm sized ± spindle cell change. Numerous cytoplasmic vacuoles can be highlighted using a modified Giemsa (Diff-Quik) stain.

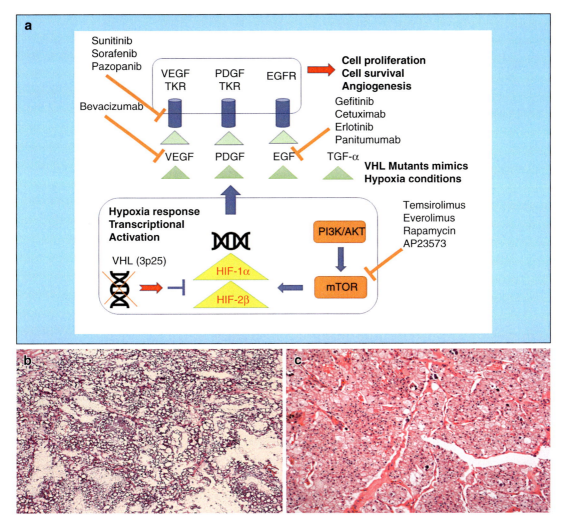

Fig. 6.38 Carcinogenetic pathways of the renal cell carcinoma with the useful chemotherapy molecules are displayed in (**a**). The microphotograph (**b**) shows a clear cell renal cell carcinoma of the young and figure (**c**) a chromophobe type (hematoxylin and eosin staining, ×100 original magnification)

- *IHC*: (−) CK7, (−) AMACR, (+) CD10, (+) VIM, AE1/AE3, LMWCK, RCCM, *PAX2*, *PAX8*, and CAIX (±EMA) and (−) HMWCK, CK7, CK20, CD117, kidney-specific cadherin, and parvalbumin.
- *TEM*: Clear cells contain fat, glycogen, and few organelles.
- *PGN*: Occasionally, this tumor regresses without treatment, but ± spread to perinephric fat or regional LN, renal vein and IVC (1/3 of cases), and distant organs (often solitary metastases) – lung, bones (pelvis, femur), adrenal gland, and liver (1/3 of cases) are typically observed.
- 5-YSR: I, 60–80%; II, 40–70%; III, 10–40%; IV, 5%. Good PGN factors: absence of metastases, <3 cm best, clear cell pattern.
- *Multilocular Cystic RCC:* RCC with wide age range and peaks in the early youth (20–76 years), ♂ > ♀, favorable PGN, and characterized by *serous/hemorrhagic cysts* lined by clear and flat cuboidal cells (no hobnail cells, unlike tubulocystic carcinoma) separated by irregular septa of variable thickness containing aggregates of clear cells, Fuhrman 1 (unlike cystic nephroma of the adult or tubulocystic carcinoma), and no nephroblastomatous elements (≠ CPDN).

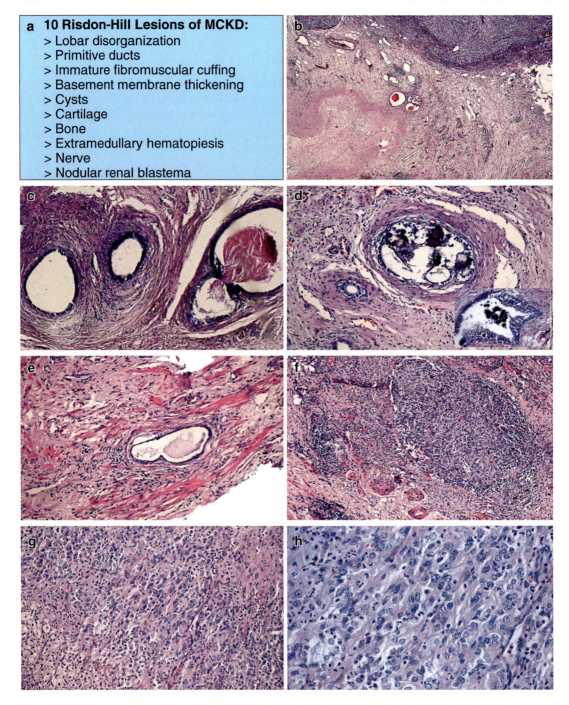

Fig. 6.39 In (**a**) are displayed the Ridson-Hill lesions of multicystic kidney dysplasia. The microphotographs (**b**) through (**h**) show the rare occurrence of a renal cell carcinoma arising on the ground of a multicystic kidney dysplasia

- *Xp11 RCC*: RCC of children and young adults defined by several different translocations involving *Xp11* with fusion at different chromosomes and involving the *TFE3* gene (*MITF*), being grossly, well-demarcated, but nonencapsulated ± irregular margins and perirenal and renal sinus extensions, yellow-tan CS ± H/N, and, microscopically, characterized by *papillary/nested/alveolar* patterns with *psammoma bodies* (often within hyaline nodules) composed of *Fuhrman G3–G4 cells with clear or granular cytoplasm* and uncertain PGN.

- IHC: −CK7, +AMACR, +CD10, − EMA, +TFE3
- DDX: CCRCC: −CK7, −AMACR, +CD10, +EMA, −TFE3
         PRCC: +CK7, +AMACR, +CD10, −EMA, −TFE3

The immunostaining using TFE3 is useful to pick up PRCC-TFE3 renal carcinomas, ASPL-TFE3 renal carcinomas, and PSF-TFE3 renal carcinomas, as well as ASPS.

6.12.2 Chromophobe Renal Cell Carcinoma

- *DEF*: Relatively favorable behavior (5-YSR: ~90%)-accompanied RCC (1/20 adult renal epithelial tumors) of the intercalated cell of cortical collecting duct with distinctive cytology occurring in youth to middle-aged patients with no gender preference (♂ = ♀) and ± *BHDS* (*FLCN* at 17p11.2).
- *GRO*: Well-circumscribed, tan-brown with geographic necrosis, micro cysts, ± multifocality/bilaterality, and no central scar (≠ renal oncocytoma).
- *CLM*: Sheets or alveolar pattern with thick-walled ± eccentrically hyalinized BVs composed of cells with cytoplasm ranging from bright to flocculent to eosinophilic (3-cytotypes) with greater cytoplasmic density at the periphery of the tumor cells and perinuclear clearing making the cytoplasmic borders thick and distinct ("plant cell appearance"). No *chicken-wire* vasculature (more fibrovascular than vascular) and no PGN value of Fuhrman grading:

- Type 1: Small cells with slightly granular eosinophilic cytoplasm
- Type 2: Medium-large cells with prominent perinuclear halo
- Type 3: Large, polygonal cells with thick cell border and abundant cytoplasm

- *HSS/IHC*: (+) Hale's colloidal iron (acid mucopolysaccharides in microvesicles), LMWCK, CK7 (diffuse and strong, compared to oncocytoma with scattered single cells), EMA/MUC1 (diffuse cytoplasmic), parvalbumin (calcium-binding protein), CD117/C-Kit (membranous), E-Cadherin, Claudin-7 (distal nephron marker), kidney-specific cadherin, ± RCC-Ma, CD10 – VIM (or weak), and N-Cadherin, there is low Ki-67 expression (MIB1) indicating a low proliferation rate.
- *TEM*: Microvesicles (Ø: 150–300 nm) and optional abundant mitochondria in the eosinophilic variant of Ch-RCC.
- *PGN*: Poor PGN factors include specifically sarcomatoid change, necrosis, high stage, and vascular invasion

6.12.3 Papillary Adenoma and Renal Cell Carcinoma

Papillary Adenoma

- *DEF*: Most common renal tubular epithelial tumor, usually occurring as an asymptomatic event in 1/10 individuals <40 years, with abnormality of the Y chromosome and the chromosomes 7 and 17 and good behavior.
- *GRO*: Well-demarcated round or wedge-shaped, encapsulated, yellow or gray, subcapsular cortical nodule (*0.1–1.5 cm*).

- *CLM*: Tubular, papillary, or tubular-papillary pattern of small cells with regular round-ovoid nuclei with inconspicuous nucleoli and scant cytoplasm ± rare mitoses (low MI), psammoma bodies, and MΦ.

Papillary Carcinoma

- *DEF*: Relatively favorable behavior (5-YSR: ~85%)-accompanied "chromophil" RCC (10–20% of adult RCC) of the proximal or distal convoluted tubule occurring in mostly males with #7+, #17+, Y-, which are like the cytogenetic (chromosomal) changes seen in papillary adenoma, but also +12, +16, +20, +3q, −X (p or q for all chromosomes). Interestingly, PRCC may show point mutation in *C-KIT/CD117* intron 1 and mutations in *MET* proto-oncogene on #7, but no evidence of *TP53* mutations or 3p-.
- *GRO*: >0.5 cm, red/brown tumor with thick capsule and reactive changes (hemorrhage) ± multifocal (80%) and characteristically "pouring out" of the kidney.
- *CLM*: *Papillary or tubulo-papillary* tumor with foamy MΦ and intracellular Fe deposition as well as columnar or cuboidal cells with G1-G2 nuclei with longitudinal nuclear grooves and finely granular cytoplasm ± glassy hyaline globules, psammoma bodies, PMNs, and necrosis (Figs. 6.40 and 6.41).
- *IHC*: (+) CK7, AMACR, CD10, LMWCK (CK 8/18 and CK19), but also (+) m-EMA/MUC1, RCC-Marker, CD117/C-Kit (cytoplasmic); (±) VIM; (−) HMWCK (34BE12: CK1, CK5/6, CK10, CK14), UEA, PVA, WT1
- *TEM*: Variable size of microvilli, small amount of intracytoplasmic lipid, but no glycogen.
- *DDX*: Papillary adenoma (similar morphology, but <0.5 cm), CDC (tubular infiltration with desmoplasia), CCRCC (-CK7, -AMACR; 3p- present, no +7 or + 17, no Y-), Xp11-RCC (+TFE3).

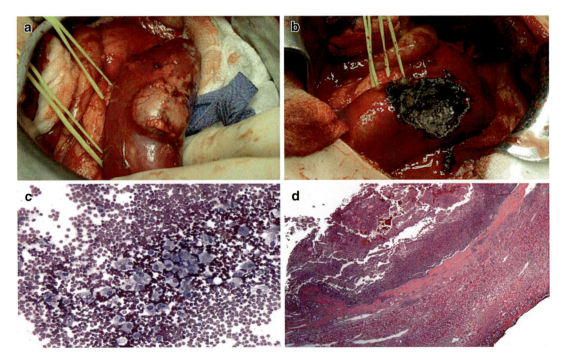

Fig. 6.40 Intraoperative photograph demonstrating exophytic lesion before (**a**) and after (**b**) resection. Imprints show very large cells with eccentric nuclei (**c**). The tumor infiltrated the tumor capsule invading the surrounding renal parenchyma (**d, e**). Papillary structures are evident in (**f**) though (**h**). Typical papillary configuration of the tumor with focal thyroid-like appearance in some papillary projections are seen

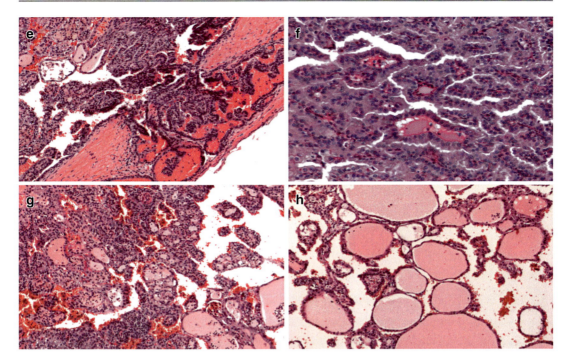

Fig. 6.40 (continued)

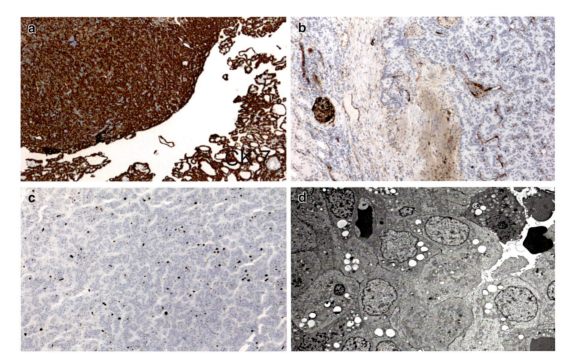

Fig. 6.41 The tumor expressed cytokeratin 7 (**a**), CD10 (**b**), and vimentin (**c**), but did not express WT1 (**d**). The proliferation activity is low (**e**). The electron microscopy shows epithelial cells with prominent luminal brush border (**f**) (immunostaining: antibody against cytokeratin 7, immunostaining, ×50; antibody against CD10, immunostaining, ×100; antibody against vimentin, immunostaining, ×100). For the proliferation activity, the antibody MIB1 was used against the antigen Ki67. The transmission electron microscopy has ×5000 as original magnification

6.12.4 Collecting Duct Carcinoma

- *DEF*: Central medullary carcinoma of the collecting ducts of Bellini occurring in a wide age range from children to elderly, mostly males (♂:♀ = 2:1), + triad (arterial hypertension, anemia, and fever) and accompanied by a poor outcome.
- *GRO*: Solid with irregular borders.
- *CLM*: Tubular or tubular-papillary histologic pattern with DPSR and PMNs, but also microcystic, solid, cord-like, and sarcomatoid patterns, and composed of G3–G4 Fuhrman cytology with eosinophilic cytoplasm ± hobnail features.
- *HSS/IHC*: (+) HMW-CK, VIM, UEA1, Peanut; (±) CD15, EMA; (−) CD10, Villin.
- *TEM*: Features of adenocarcinoma, including intracellular and extracellular lumina, well-formed cell junctions, prominent basal lamina, and short apical microvilli.
- *DDX*: PRCC, RMC, MTX-Ca, Uro-Ca.

6.12.5 Renal Medullary Carcinoma

- *DEF*: Rapidly growing tumor of the renal medulla associated with sickle cell trait (Afro-American/Hispanic/Brazilian), associated with *#11-* (monosomy, β-globin gene on 11p) and *INI1-* (loss of *INI1* similarly to rhabdoid tumor), occurring in 10–40 years, ♂ > ♀, (+) clinical triad, mild response to RT/CHT and poor PGN (even death within 15 weeks).
- *GRO*: Solid ± H/N (no cyst!).
- *CLM*: *Solid/adenoid-cystic/reticular* growth pattern with DPSR of G3–G4 cells with large vesicular nuclei and prominent nucleoli and eosinophilic cytoplasm admixed with PMNs and bounded by lymphocytes (intense inflammatory response of the "host" at the interface between tumor and adjacent renal parenchyma.
- *HSS/IHC*: (+) AE1/AE3, LMWCK, EMA, VIM, HIF, VEGF, (±) CK7, CEA, (±) HMWCK, Ulex.
- *DDX*: Uro-Ca (+CK20, adjacent CIS) and CDC (+HMWCK, Ulex).
- "Sickle cell trait" – 7 nephropathies: gross hematuria, papillary necrosis, nephrotic syndrome, renal infarction, hyposthenuria, pyelonephritis, and RMC.

6.12.6 Angiomyolipoma

- *DEF*: Benign mesenchymal cortico-/medullary neoplasm of the *perivascular epithelioid cells* (PEComa) associated with TS, VHLS, and ADPKD, occurring mostly in youth to middle-aged females if sporadic or in children/adolescents with no sex preference (♂ = ♀) if hereditary, ± LOH of *TSC1* (9q34) or *TSC2* (16p13).
- *CLI*: (±) triad, ± LAM of the lung and indolent course except for massive intratumoral hemorrhage, fat replacing-associated renal parenchyma with CRF, and malignant transformation/RCC.
- *GRO*: Intrarenal, well-circumscribed, unencapsulated, lobulated, yellow to the tan-pink tumor with hemorrhage.
- *CLM*: A variable admixture of thick-walled tortuous BVs without elastic lamina, *myo*-proliferation (smooth muscle) spinning off from the outer layers of BVs walls ("hair-on-end" appearance), and *fat* (Figs. 6.42 and 6.43).
- *IHC*: Myomelanocytic co-expression: (+) SMA, MSA, DES & HMB45, MART-1, MiTF.
- *TEM*: Dense bundles of intermediate filaments located predominantly in the periphery of the cytoplasm some of the spindle cells, well developed RER, pre-melanosomes and numerous vesicles in the edge of the cytoplasm. There is no continuous basal lamina. Premelanosomes contain longitudinal filaments that may be cross-linked with one another with an enveloping bilayered membrane.
- Variant: *Epithelioid AML* (EAML), which has a locally infiltrative growth pattern and composed of sheets of malignant-appearing epithelioid cells (DDX: sarcomatoid RCC). EAML seems to be less aggressive than believed to be at the time of the original description.

6.12.7 Oncocytoma

- *DEF*: 1/20 of all renal tumors, ♂ > ♀, III–VI decades, benign tumor of the intercalated cells

6.12 Non-embryonal Tumors of the Young

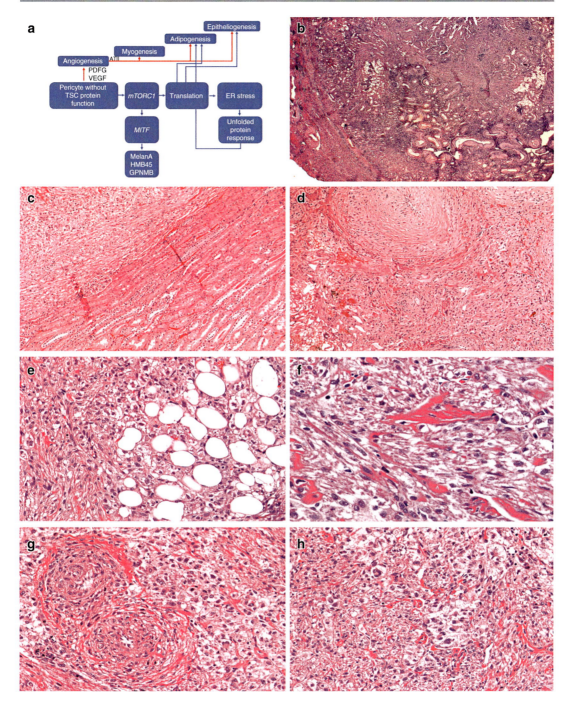

Fig. 6.42 In (**a**) is shown the carcinogenetic pathways with differentiation of an angiomyolipoma. The microphotographs (**b**) through (**h**) show an angiomyolipoma with the characteristic triphasic morphology with myoid spindle cells, islands of mature adipose tissue (**e**), and abnormally structured ("dysmorphic") thick-walled blood vessels (**g**) (hematoxylin and eosin staining). All microphotographs (**b-h**) are H&E and the original magnification is x20, x100, x100, x200, x400, x200, and x200, respectively

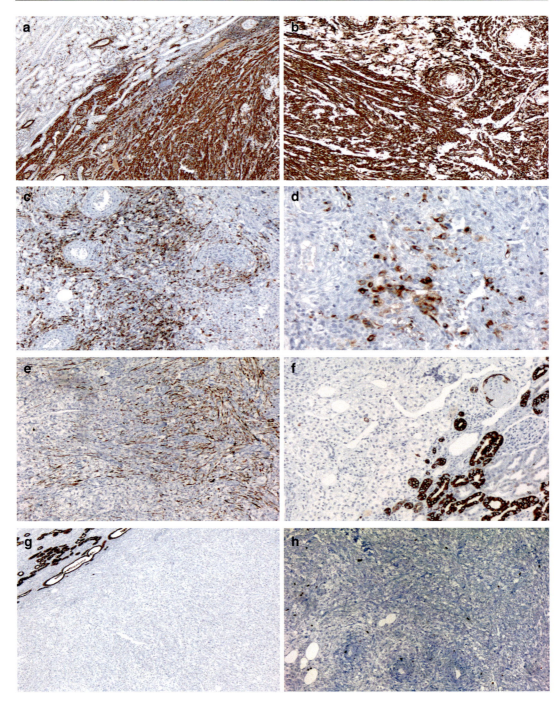

Fig. 6.43 The immunohistochemical panel of angiomyolipoma shows an expression of actin (**a**), caldesmon (**b**), HMB45 (**c, d**), and desmin (**e**), but no expression of AE1–AE3 (**f**), epithelial membrane antigen (**g**), and low proliferation activity (**h**). Avidin-biotin complex immunostaining (**a**) at ×50, (**b**) at ×100, (**c**) at ×100, (**d**) at ×200, (**e**) at ×100, (**f**) through (**h**) at ×100 as original magnification

6.12 Non-embryonal Tumors of the Young

of collecting duct, usually solitary (multiple, 1/20) either asymptomatic or with hematuria/flank pain/mass, ± BHDS, and *#Y-, #1-, #14-, 11q13~ = > CCDN1- >* cyclin D1 (MCL).
- *GRO*: Mahogany-brown, large, and solid tumor with central stellate scar (focal hemorrhage in 1/5 of cases).
- *CLM*: *"Archipelagineous" architecture* with solid sheets or nests of cells with small, round, and regular nuclei (Fuhrman 1–2) and abundant granular acidophilic cytoplasm in a loose fibrous stroma.
- *HSS/IHC*: ±*CK7*, ±AMACR, ±CD10, +*EMA*, −*VIM*, +*CD117*, and + *CK14*.
- *DDX*: Ch-RCC ±*CK7*, -AMACR, ±CD10, +*EMA*, −*VIM*, +*CD117*, and ±*CK14*, other than cytology, Hale, IHC, TEM. If Fuhrman 3–4 → degenerative change, but Fuhrman 3–4 and + MI and necrosis → *RCC, oncocytic variant*.
- *TEM*: Many mitochondria (≠ Ch-RCC that exhibits numerous cytoplasmic microvesicles of Ø: 150–300 nm).
- *PGN*: Good, but the invasion of the capsule or renal vein is possible, and debate still exists to call it as oncocytoma or regards it as RCC, oncocytic variant. Exclusion criteria: areas of clear cell carcinomas, sarcomatoid carcinoma, prominent/diffuse papillary architecture, gross/microscopic necrosis, ↑MI, and atypical mitoses.

6.12.8 Other Epithelial and Mesenchymal Tumors

- TUBULOCYSTIC CARCINOMA: LMP-RCC, an incidental and indolent tumor with an unknown cell of origin and occurring in a wide age range of mostly *males* (♂:♀ = 6:1), and no well established cytogenetics.
- *GRO*: Well-demarcated, spongy tumor with *"bubbly wrap-like"* CS.
- *CLM*: *Small-to-intermediate-sized tubules admixed with cystically dilated tubules ("spider-web" pattern) lined by a single layer of flat, "hobnail," or cuboidal to columnar cells* in a *fibrotic intervening stroma* (no clear cell aggregates!).
- *IHC*: +*CK7*, +*AMACR*, +*CD10* (*DDX*: MTS-RCC, PRCC, CDC).
- *TEM*: *Many* microvilli (like PRT) or sparse microvilli and cytoplasmic interdigitations (ICC of collecting duct-like).
- MUCINOUS TUBULAR AND SPINDLE CELL CARCINOMA (MTS-RCC): Very low-grade, incidental, and indolent tumor with epithelial and mesenchymal components, of the proximal nephron (+CK7 and AMACR), occurring in a wide range of age of mostly *females* (♂:♀ = 1:4), ± nephrolithiasis, and variable cytogenetics (losses of *#1*, 4, 6, *8*, 13, and 14, and gains of *#7*, 11, 16, and *17*).
- *GRO*: Cortical or central located, well-circumscribed occasionally encapsulated with gray-tan and glistening CS.
- *CLM*: *Tightly packed elongated tubuli of G1 cytology in a bubbly myxoid stroma* with abundant basophilic extracellular mucin and *G1 spindle cells areas*.
- *IHC*: +*CK7*, +*AMACR*, -*CD10* (DDX: CCRCC and PRCC); +LMWCK and -HMWCK.
- *TEM*: *Loop* of Henle-like morphology.
- CYSTIC NEPHROMA (OF THE ADULT): Very low-grade, incidental, and indolent tumor with epithelial and mesenchymal components of mostly *females* (♂:♀ = 1:8) older than 30 years.
- *GRO*: Solitary, unilateral, encapsulated, and well-circumscribed tumor of the *upper pole* composed of numerous cysts of variable size with uneventful adjacent renal parenchyma.
- *CLM*: *Multilocular cysts* lined by a single layer of *flat, "hobnail," or cuboidal to columnar cells* and separated by *fibrous septa* of variable thickness (no nephronic differentiation, no blastema, no clear cell aggregates!).
- MIXED EPITHELIAL AND STROMAL TUMOR: Very low-grade, incidental/symptomatic (triad), and indolent tumor with epithelial and mesenchymal components of mostly *females* (♂:♀ = 1:8) of the 4th decade. Grossly, *centrally* located, well-circumscribed, and encapsulated tumor with protrusion into the renal pelvis and microscopically, consisting of highly complex epithelial cells lined

papillae (true papillae) lined by a single layer of *flat, "hobnail," or cuboidal to columnar cells* and variable cellular stroma composed of spindle cells with plump nuclei and copious cytoplasm ± myxoid change or collagen, smooth muscle, or fatty tissue.
- *IHC*: (+) CK and VIM in the epithelial component & (+) SMA and DES in the mesenchymal component.
- *RENOMEDULLARY INTERSTITIAL CELL TUMOR* (aka "medullary fibroma"): Mostly asymptomatic, incidental mesenchymal tumor with a wide age range from adolescents to elderly originating from a prostaglandin-producing interstitial cell in renal medulla involved in the regulation of the blood pressure. Grossly, pyramid-located, well-circumscribed (Ø < 0.5 cm), gray-white nodule and, microscopically, consisting of small stellate and spindle cells embedded in a loose myxoid stroma.
- *JUXTAGLOMERULAR CELL TUMOR*: Benign renin-secreting (DDX: WT, RCC, renal oncocytoma) mesenchymal tumor with a wide age range from children to elderly (♂ < ♀) originating from of the modified smooth muscle cells of the juxtaglomerular apparatus and presenting with refractory HTN, hyper-reninemia, secondary hyperaldosteronism, and hypokalemia. Grossly, unilateral, solitary, cortical, solid, encapsulated, well-demarcated (Ø < 5 cm), gray-white/yellow-tan tumor ± hemorrhage and, microscopically, composed of solid sheets of uniform, round to polygonal and spindle cells with round to oval nuclei, low or nil MI, and copious granular eosinophilic cytoplasm as well as thin- and thick-walled prominent BVs with hemangiopericytoma-like "antler vascular pattern" in numerous fields.
- *IHC*: (+) VIM, SMA, CD34, Renin.
- *TEM*: Rhomboid crystals → Renin-specific (DDX: rhomboid crystals of ASPS!).

Finally, rhabdo- and leiomyosarcoma can also occur in the kidney (Fig. 6.44). In 2016 has been published the new W.H.O. classification of the

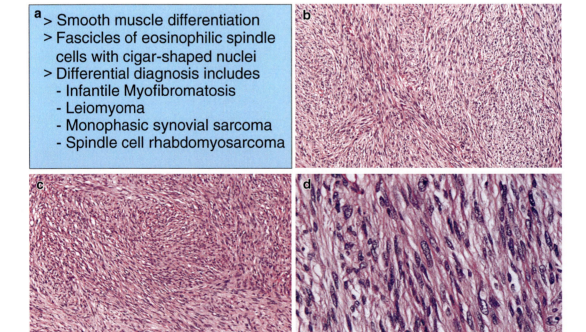

Fig. 6.44 In (**a**) are shown the characteristic features of a leiomyosarcoma that are displayed in the microphotographic panel (**b**) through (**d**). All three microphotographs (**b-d**) have been taken from H&E-stained slides and the magnification is x100, x100, and x400, respectively

urinary system. Several familial forms of RCC have been linked to the corresponding histology, which can also arise sporadically. The multilocular cystic renal cell carcinoma or multilocular clear cell renal cell carcinoma is currently described as multilocular cystic renal neoplasm of low malignant potential. The size cutoff for papillary adenoma was increased from 0.5 cm to 1.5 cm. Two forms, pediatric and adult, are collected under the term "cystic nephroma" and, as indicated above, the adult form is included along a spectrum of mixed epithelial and stromal tumors. Similar to carcinoids of other systems, renal carcinoids are re-classified into well-differentiated neuroendocrine tumors of the kidney.

6.13 Non-neoplastic Pathology of the Pelvis and Ureter

6.13.1 Anatomy and Physiology Notes

Ureter: ~30 cm in length and 0.5 cm in Ø. Three points of luminal narrowing, including the ureteropelvic junction, the crossing point with the iliac BV, and at entry into the urinary bladder, need to be kept in mind by the pediatric urologist and pathologist.

6.13.2 Congenital Pelvic-Ureteral Anomalies

1. *Pelvic-Ureteral Junction (PUJ) Stenosis/Obstruction (ab externo*: PUJ stenosis is the most common cause of collecting duct system dilation in the fetal kidney and one of the most frequent anomalies occurring in the urinary tract in children. There is an association with anomalies of the contralateral kidney, including pelvicureteral junction obstruction, renal agenesis, renal dysplasia, multicystic dysplastic kidney (MCDK), and obstructive dysplasia (Potter type IV). PUJ obstruction *ab externo* may occur in compression or kinking of the excretory system by a neoplastic or inflammatory tumor of the abdomen or pelvis. In UPJ stenosis, which usually occurs in males, there is an abnormal organization of and excess stromal deposition of collagenous tissue between smooth muscle bundles, and cells are derived from medullary interstitial cells. Surgical (urological) intervention is indicated if there is impairment of the kidney function, pyelonephritis, lithiasis, or pain.

2. *Hydrocalyx/Hydrocalycosis*: Dilation of a significant calyx may occur in the context of central obstruction, as described in infundibular stenosis, or by extrinsic obstruction by a blood vessel or non-neoplastic (e.g., parapelvic cyst) or neoplastic (e.g., desmoplastic small round blue cell tumor). This condition needs to be distinguished from megacalycosis, which is nonobstructive dysplastic lesion with dilated and over numeracy calyces with or without renal medullary hypoplasia.

3. *Calyceal Diverticula*: Cystic structures connected to an adjacent minor calyx by a narrow ductule or channel.

4. *Partial Duplication of the Renal Pelvis and Ureter*: Typically, unilateral and no clinical significance.

6.13.3 Congenital Ureteric Anomalies

Ectopia of the ureters is usually associated with complete ureteric and renal duplication, and about 10% are bilateral. Interestingly, there is a dysplastic upper pole of a *duplex kidney*, which is drained by the ectopic ureter that is connected below the common ureterovesical junction into the lower trigone or the proximal urethra. Ectopia of the ureters is much more common in females, and the insertion occurs in either vagina or vulva, with subsequent urinary incontinence. Ectopia of the ureters may also be associated with ureteroceles, which is a cystic dilation of the terminal ureter. Clinically, there is UTI, obstruction of the bladder neck, blockage of the contralateral kidney, or nephrolithiasis. Megaureters may be differentiated in primary and secondary causes. Primary ureter results from a functional obstruction of the distal ureter, which is caused by a peristaltic segment, while secondary megaureter is multi-etiologic with causes ranging from intrinsic ureteric obstruction (e.g., stone) to bladder outflow obstruction, vesicoureteral reflux (VUR), and compression *ab externo* of the distal ureter. With an incidence rate of 1–2% in children and diagnosed by a voiding cystourethrogram (VCUG) and an intravenous pyelogram (IVP), VUR is defined as a retrograde flow of urine from the urinary bladder into the ureters due to a deficiency of the functional valve-like mechanism at the ureterovesical junction. The competence of this valve depends on three factors, including the length of the intramural length of the ureter, the position of the ureteric orifice in the urinary bladder, and the integrity of the musculature of the bladder wall (tripodal competence or "competentia tripodalis"). Shortage of the intramural portion of the distal ureter and lateral ectopic position of the orifice are the most common causes of VUR in children. Genetic factors may play a substantial role, owing to a 30- to the 50-fold ↑ risk of VUR if an immediate relative is also affected by this pathology. An important note to recall is that the intramural ureter lengthens with age and subsequently VUR is progressively rarer or disappears when the child grows up. VUR may also be an associated pathology to Prune Belly syndrome and posterior urethral valves. A dramatic complication related to primary and secondary VUR is reflux nephropathy. Reflux encourages more or less infected urine from the urinary bladder to be swept up into the kidneys. Subsequently, this phenomenon leads to recurrent pyelonephritis, tubulointerstitial nephritis with "thyroidization" of the renal parenchyma, and gross scarring at the renal poles. The glomeruli become sclerotic with consequent proteinuria, arterial HTN, and progressive loss of renal function. Antibiotic therapy (sulfamethoxazole and trimethoprim) reduces recurrent infections by about 40% in children with age <18 years with prior UTI, while surgery seems to be the treatment of choice for severe grades of VUR and mainly secondary VUR associated with abnormal development of the lower urinary tract (UT-congenital anomalies associated VUR or MCA associated VUR). VUR grading divides vesicoureteral reflux according to the height of reflux up the ureters and degree of dilatation of the ureters (Figs. 6.45, 6.46 and 6.47). It is important to remember that each side may have a different degree of VUR.

6.13.4 Lower Urinary Tract Abnormalities

The pathologies that need to be considered include *Prune Belly syndrome* (PBS) (aka Eagle-Barrett syndrome or "triad" syndrome), *non-PBS, urinary bladder wall abnormalities,* and *posterior urethral valves*. PBS is almost exclusively expressed in males and seems to be due to a defect of the mesenchymal development. In PBS, there is poor differentiation of prostate and urinary bladder with prostatic gland hypoplasia and bladder wall thickening, ureteral smooth muscle aplasia with ureteral peristalsis and gross ureteral dilation, absence or deficiency of the abdominal wall musculature, bilateral undescended testes (cryptorchidism), and varying degrees of renal dysplasia similar to MCDK or Potter type IV. PBS also has extra-GU manifestations in about 3/4 of patients. The prognosis of PBS patients depends on the severity of the clinical and pathologic features such as defects of the cardiopulmonary system, gastrointestinal tract, and skeleton. The degree of the bilateralism of renal dysplasia is also a significant prognostic factor leading to a variable outcome following urodynamic management procedures. *Non-PBS abnormalities of the urinary bladder* include *bladder exstrophy* and *neuropathic or neurogenic bladder* essentially. A midline closure defect involving the lower abdominal wall, the urinary bladder, and the external genitalia underlies the bladder exstrophy it is probably due to a primary weakness of the differentiation of the cloacal membrane. Bladder exstrophy may be associated with imperforate anus and rectal atresia. In children, the neuropathic or neurogenic bladder is associated with myelomeningocele (*spina bifida*), spinal dysraphism (*spina bifida occulta*), and sacral agenesis. In young adults, the neuropathic or neurogenic bladder is due to CNS trauma, spinal trauma, multiple sclerosis, stroke, and trau-

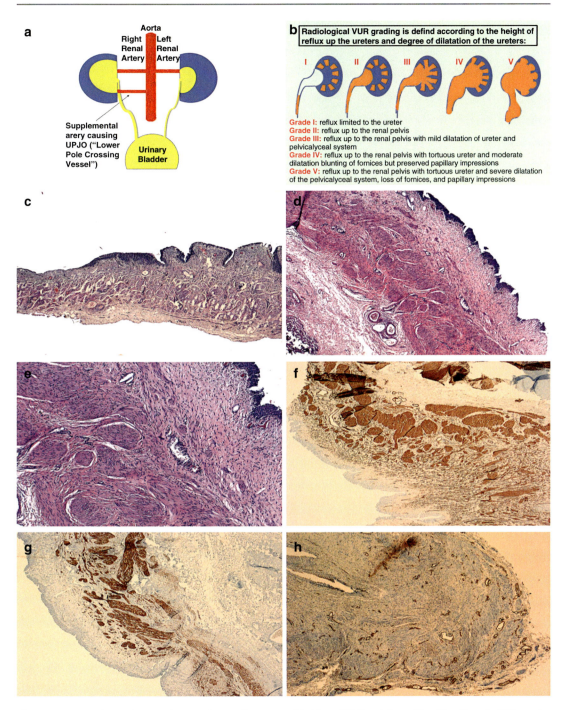

Fig. 6.45 In (**a**) is a schema of a supplemental artery causing ureteropelvic junction obstruction, while (**b**) shows the radiological grading of the vesicoureteral reflux. The microphotographs (**c**) through (**e**) show the longitudinal section of an ureteropelvic junction without stenosis (**c**) and with stenosis (**d**, **e**) Microphotographs (**c**)-(**e**) are H&E-staining with x40, x40, and x100 as original magnification, while (**f**)-(**h**) are immunostained slides (avidin-biotin complex) using antibodies against smooth muscle actin (SMA), desmin (DES), and CD31, an endothelial cell marker. The microphotographs of (**f**) through (**h**) is x40 for all three

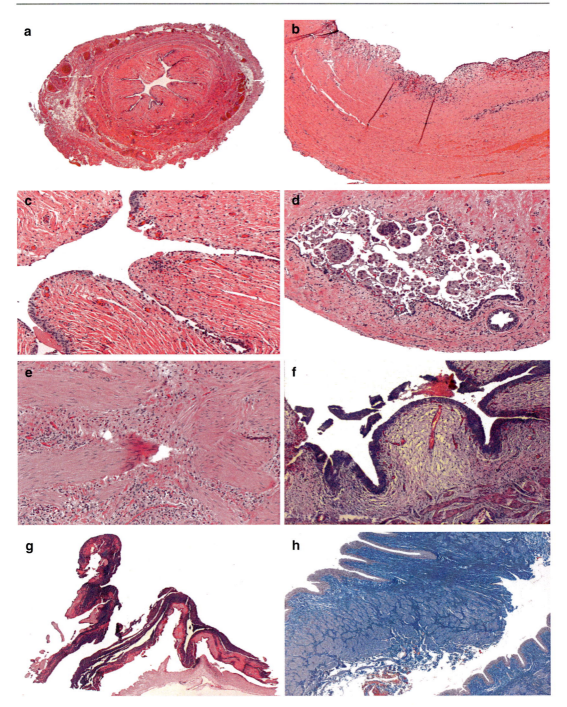

Fig. 6.46 The microphotographs (**a**) through (**d**) show a stenosis of the ureteropelvic junction accompanied by an inflammatory reaction (**c**, **d**) (H&E stain, x12.5, x50, x100, and x100 as original magnification, respectively). In (**e**) is shown a case of eosinophilia in ureteropelvic junction and some basophilic (hyperchromasia) changes of the epithelium (**f**). The microphotographs (**g**, **h**) show ureteral valve (**g**) with marked fibrosis (**f**). The microphotographs (**e**) through (**g**) are H&E and the magnification is x100 for all three, while the microphotograph (**h**) is Celestin Alcian Blue at x40 as original magnification

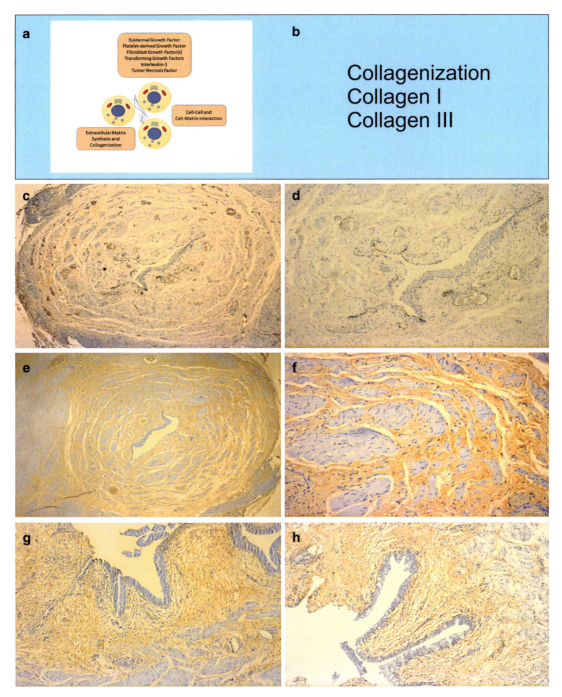

Fig. 6.47 Collagen I and collagen III play a major role in the collagenization of the stenosis seen as morphologic substrate for the ureteropelvic junction obstruction. In (**c**) and (**d**), a control without collagenization and (**e**) through (**h**) collagenization with overexpression of collagen I (**e**, **f**) and III (**g**, **h**). The original magnification of (**c**) through (**h**) is x40, x100, x40, x200, x100, and x100, respectively

matic or iatrogenic peripheral nerve damage. *Posterior urethral valves* (PUV) are the most common etiology of bladder outflow obstruction in male infants. PUV lead to bilateral megaureter and hydronephrosis, although it can be easily demonstrated only in 1/10 of patients with hydronephrosis. The underlying anatomic-pathologic abnormality is a deficiency of the reabsorption of mucosal folds in the posterior urethra, almost always just distally to the *verumontanum*. The consequence of the PUV is a dilation of the urethra located proximally, hypertrophy and trabeculation of the urinary bladder, VUR of higher grade, and renal dysplasia. Two factors suggest to opt for a surgical management. They include the age of the child and the degree of renal failure, while the severity of renal dysplasia is the most critical limiting factor for survival and long-term renal outcome.

6.14 Tumors of the Pelvis and Ureter

6.14.1 Neoplasms

Tumors of the pelvis and ureter are rare in children and young adults. However, benign entities seem to play a significant role including single or multiple fibroepithelial polyps, hematoma, teratoma, lipoma, leiomyoma, and hemangioma. Practically inexistent are carcinoid, sarcomas (various types), lymphoma, or plasmacytoma, while rarely urothelial carcinomas have been described in the pediatric patient. Urothelial or transitional cell carcinoma may occur in the setting of renal transplantation or young adults with a long-standing history of analgesic abuse and cyclophosphamide and renal papillary necrosis. Clinical presentation includes hematuria. Endoscopically, These masses or nodules are soft, white to grayish red with glistening surfaces. Microscopically, glandular or squamous metaplasia in association with a urothelial differentiation can be observed. Theoretically, these tumors can also acquire sarcomatoid appearance, and the occurrence of pools of mucin may suggest a differential diagnosis with adenocarcinoma. Multicentricity may also occur. Non-urothelial carcinomas are epidermoid carcinomas and adenocarcinomas. The latter may be found in a setting of long-standing chronic inflammation such as long-standing *ureteritis follicularis* and *ureteritis cystica*. Metastatic tumors usually bilateral may be associated with indirect colonization from WT of the kidney and desmoplastic small round blue cell tumor or other sarcomas (e.g., Ewing sarcoma) of the abdomen. Botryoid WT of the renal pelvis extending into the urinary bladder needs to be taken into account. Some pathologies may require special consideration, including florid von Brunn's nests mimicking urothelial carcinoma (DDX of the nested variant of urothelial carcinoma) and nephrogenic adenoma. These pathologies are better described in the chapter of the lower urinary tract.

6.14.2 Genetic Syndromes

Two syndromes have been described with urothelial carcinomas, including the *Costello syndrome* and *Turner syndrome*. If Turner syndrome is quite known, Costello syndrome may benefit some notes for the pathologist. Costello syndrome is a *HRAS*-associated genetic syndrome with a phenotype characterized by macrocephaly with coarse facial features (large mouth with full lips and full nasal tip), curly or sparse, beautiful hair, soft skin with deep palmar and plantar creases, as well as diffuse hypotonia and joint laxity with ulnar deviation (wrists and fingers) and tight Achilles tendons. There is a failure to thrive in infancy, short stature, and psychomotoric developmental delay. Additional findings include Chiari I malformation with associated CNS anomalies including hydrocephalus or syringomyelia, hypertrophic cardiomyopathy, stenosis of the pulmonary valve (other congenital heart defects are also possible), and arrhythmias (usually chaotic atrial rhythm/multifocal atrial tachycardia or ectopic atrial tachycardia). There is an increased frequency of papillomata of the face and perianal region and an approximately 15% lifetime risk for the development of malignancies including rhabdomyosarcoma, neuroblastoma, and urothelial carcinoma of the urinary bladder in children, adolescents, and young adults. It has been suggested an abdominal and pelvic ultrasound screening for rhabdomyosarcoma and neuroblastoma every 3–6 months until age 8–10 years and annual urinalysis for evidence of hematuria to screen for bladder malignancy starting at age of 10 years. Costello syndrome is AD inherited, although most of the patients affected with this syndrome showed *de novo* mutation. There are case reports of somatic mosaicism in one parent, and occasional recurrence in siblings has been detected because of germline mosaicism in a parent.

Multiple Choice Questions and Answers

- UUS-1 Which of the following statement is NOT correct regarding the immune complexes?
  (a) The Ag-Ab complexes are circulating and eventually deposit on tissues.
  (b) Immune complexes are involved in the pathogenesis of Henoch-Schönlein purpura.
  (c) Immune complexes have not been associated with the lupus nephritis.
  (d) Immune complexes represent a form of type III hypersensitivity.
  (e) Immune complexes are involved in the pathogenesis of post-streptococcal glomerulonephritis.
- UUS-2 Which disorder is at the basis of the hemolytic-uremic syndrome?
  (a) Anaerobic glycolysis
  (b) Disorder of the mitochondrial electron transport chain
  (c) Vascular endothelium
  (d) Hemoglobinopathy
  (e) Coagulation
- UUS-3 Glycogen storage disease type Ia is also called von Gierke disease and belongs to the group of the glycogenosis. Which of the following statements is TRUE?
  (a) There is a deficiency of the phosphorylase A enzyme.
  (b) There is a deficiency in maltase.
  (c) There is a decrease of intrahepatocyte glycogen.
  (d) There is an increase of glycogen in the cortical tubular epithelial cells.
- UUS-4 A 6-month-old infant presents with progressive hepatic and renal failure. Despite intensive therapy, the infant dies, and the parents consent to an autopsy. The pathologist identifies radial spoke arranged cysts during the gross examination of the kidneys. Which mode of inheritance may be suggested according to this finding?
  (a) Autosomal recessive
  (b) Autosomal dominant
  (c) X-linked recessive
  (d) X-linked dominant
  (e) Mitochondrial
- UUS-5 A 2-year-old boy is brought to the emergency department with generalized edema. The medical team identifies ascites and hyaline casts and lipid droplets at the urinalysis. Blood examination reveals hypoalbuminemia and decrease of total plasma protein, while there is hyperlipidemia. A biopsy is ordered, but light microscopy does not reveal any changes in the structure of the glomerulus. Which statement is CORRECT for the diagnosis of this boy?
  (a) Response to steroid therapy is poor.
  (b) The long-term outlook of this disease is meager.
  (c) There is a vascular problem with early atherosclerosis.
  (d) There is a hepatic problem with early hepatic failure.
  (e) Electron microscopy will show attenuation and flattening of the foot processes of the podocytes.
- UUS-6 What is the meaning of fractional excretion of sodium (FENa) higher than 3%?
  (a) Renal failure
  (b) Syndrome of inappropriate secretion of ADH (SIADH)
  (c) Prerenal oliguria
  (d) Diuretic-induced polyuria
  (e) Hypertension

- UUS-7 Which of the following does NOT belong to the causes of acquired nephrogenic diabetes insipidus?
  (a) Interstitial nephritis
  (b) Chronic renal inefficiency
  (c) Papillary necrosis
  (d) Hypokalemia
  (e) Sickle cell disease
  (f) Hereditary spherocytosis
- UUS-8 An 8-year-old child has three episodes of gross hematuria. At daytime, the parents bring the child to the emergency department, and the senior resident on call prescribes a series of exams. Which of the following exams has limited relevance to the current pediatric situation of this child?
  (a) Phase-contrast microscopy or Wright stain for the red cell morphology
  (b) Prothrombin time, partial thromboplastin time, and platelet count
  (c) Ultrasonography of the kidneys and urinary bladder
  (d) Urinary screening for hypercalciuria
  (e) Sulfosalicylic acid precipitation test for proteinuria
- UUS-9 Electron-dense deposits are composed of immunoglobulin and complement and are in the particular site of the glomerulus in different renal disease. Electron microscopy is an essential tool in nephrology. Match the electron-dense deposits location with the specific type of glomerulonephritis.
  (a) ED subepithelial deposits
  (b) ED subendothelial deposits
  (c) ED epimembranous deposits (similar to subepithelial deposits)
  (d) ED deposits within the basement membrane
  (e) ED deposits within the mesangial matrix
    1. Membranous glomerulonephritis
    2. Acute glomerulonephritis
    3. Type I membranoproliferative glomerulonephritis (MPGN)
    4. Type II MPGN
    5. IgA nephropathy
       (a)(2) (b)(3) (c)(1) (d)(4) (e)(5)
- UUS-10 A 4-year-old child has a history of abdominal pain and bloody diarrhea. The child develops acute glomerulonephritis, Coombs-negative hemolytic anemia, and renal failure, and renal biopsy shows fibrin deposits in capillary lumina, endothelial swelling, and capillary luminal narrowing as well as fibrinoid necrosis of larger vessels, associated with thrombosis and endothelial cell proliferation in small arteries and arterioles. Which of the following diseases is the most likely diagnosis in this child?
  (a) Lipoid nephrosis
  (b) Hemolytic-uremic syndrome
  (c) Lupus erythematosus
  (d) Infective endocarditis
  (e) Interstitial nephritis

References and Recommended Readings

Andrews PE, Kelalis PP, Haase GM. Extrarenal Wilms' tumor: results of the National Wilms' tumor study. J Pediatr Surg. 1992;27:1181–4. [PubMed].

Arora A, Saluja SS, Bhardwaj V, Singh M, Singh A, Puri AS. Malakoplakia masquerading as ulcerative colitis in an immunocompetent child. J Pediatr Gastroenterol Nutr. 2012. [Epub ahead of print] PubMed PMID: 23114470.

Arts HH, Knoers NV. Current insights into renal ciliopathies: what can genetics teach us? Pediatr Nephrol. 2013;28(6):863–74. https://doi.org/10.1007/s00467-012-2259-9. Epub 2012 Jul 25. Review. PubMed PMID: 22829176; PubMed Central PMCID: PMC3631122.

Belis JA, Post GJ, Rochman SC, Milam DF. Genitourinary leiomyomas. Urology. 1979;13(4):424–9. Review. PubMed PMID: 373208.

Bergmann C. Educational paper: ciliopathies. Eur J Pediatr. 2012;171(9):1285–300. https://doi.org/10.1007/s00431-011-1553-z. Epub 2011 Sep 7. PubMed PMID: 21898032; PubMed Central PMCID: PMC3419833.

Broeker BH, Calamone AA. Primary extrarenal Wilm's tumour in children. J Pediatr Surg. 1989;24:1283–8.

Brugmann SA, Cordero DR, Helms JA. Craniofacial ciliopathies: a new classification for craniofacial disorders. Am J Med Genet A. 2010;152A(12):2995–3006. Review. PubMed PMID: 21108387; PubMed Central PMCID: PMC3121325.

Cacalano G, Fariñas I, Wang LC, Hagler K, Forgie A, Moore M, Armanini M, Phillips H, Ryan AM, Reichardt LF, Hynes M, Davies A, Rosenthal A. GFRalpha1 is an essential receptor component for GDNF in the developing nervous system and kidney. Neuron. 1998;21(1):53–62. PubMed PMID: 9697851; PubMed Central PMCID: PMC2710137.

References and Recommended Readings

Chen YH, Liu HP, Chen HY, Tsai FJ, Chang CH, Lee YJ, Lin WY, Chen WC. Ethylene glycol induces calcium oxalate crystal deposition in Malpighian tubules: a Drosophila model for nephrolithiasis/urolithiasis. Kidney Int. 2011;80(4):369–77. https://doi.org/10.1038/ki.2011.80. Epub 2011 Mar 30. PubMed PMID: 21451462.

Clarkson MR, Giblin L, O'Connell FP, O'Kelly P, Walshe JJ, Conlon P, O'Meara Y, Dormon A, Campbell E, Donohoe J. Acute interstitial nephritis: clinical features and response to corticosteroid therapy. Nephrol Dial Transplant. 2004;19(11):2778–83. Epub 2004 Aug 31. PubMed PMID: 15340098.

Costantini F, Shakya R. GDNF/Ret signaling and the development of the kidney. Bioessays. 2006;28(2):117–27. Review. PubMed PMID: 16435290.

Dähnert W. Radiology review manual: Lippincott Williams & Wilkins; 2011. isbn:1609139437.

Divya P, Crasta JA. Pediatric malakoplakia of colon: a report of two cases. Pediatr Surg Int. 2010;26(3):323–5. Epub 2010 Jan 26. PubMed PMID: 20101508.

Enomoto H, Hughes I, Golden J, Baloh RH, Yonemura S, Heuckeroth RO, Johnson EM Jr, Milbrandt J. GFRalpha1 expression in cells lacking RET is dispensable for organogenesis and nerve regeneration. Neuron. 2004;44(4):623–36. PubMed PMID: 15541311.

Faa G, Gerosa C, Fanni D, Nemolato S, Di Felice E, Van Eyken P, Monga G, Iacovidou N, Fanos V. The role of immunohistochemistry in the study of the newborn kidney. J Matern Fetal Neonatal Med. 2012a;25(Suppl 4):135–8. https://doi.org/10.3109/14767058.2012.715018. Review. PubMed PMID: 22958045.

Faa G, Gerosa C, Fanni D, Monga G, Zaffanello M, Van Eyken P, Fanos V. Morphogenesis and molecular mechanisms involved in human kidney development. J Cell Physiol. 2012b;227(3):1257–68. https://doi.org/10.1002/jcp.22985. Review. PubMed PMID: 21830217.

Fanni D, Fanos V, Monga G, Gerosa C, Locci A, Nemolato S, Van Eyken P, Faa G. Expression of WT1 during normal human kidney development. J Matern Fetal Neonatal Med. 2011;24(Suppl 2):44–7. https://doi.org/10.3109/14767058.2011.606619. Epub 2011 Sep 2. PubMed PMID: 21888469.

Fanni D, Fanos V, Gerosa C, Senes G, Sanna A, Van Eyken P, Iacovidou N, Monga G, Faa G. CD44 immunoreactivity in the developing human kidney: a marker of renal progenitor stem cells? Ren Fail. 2013;35(7):967–70. https://doi.org/10.3109/0886022X.2013.808955. Epub 2013 Jul 5. PubMed PMID: 23826724.

Fanos V, Gerosa C, Loddo C, Faa G. State of the art on kidney development: how nephron endowment at birth can shape our susceptibility to renal dysfunction later in life. Am J Perinatol. 2019;36(S 02):S33–6. https://doi.org/10.1055/s-0039-1691798. Epub 2019 Jun 25. PubMed PMID: 31238356.

Gadd S, Huff V, Huang CC, Ruteshouser EC, Dome JS, Grundy PE, Breslow N, Jennings L, Green DM, Beckwith JB, Perlman EJ. Clinically relevant subsets identified by gene expression patterns support a revised ontogenic model of Wilms tumor: a Children's oncology group study. Neoplasia. 2012;14(8):742–56. PubMed PMID: 22952427; PubMed Central PMCID: PMC3431181.

Geavlete O, Călin C, Croitoru M, Lupescu I, Ginghină C. Fibromuscular dysplasia – a rare cause of renovascular hypertension. Case study and overview of the literature data. J Med Life. 2012;5(3):316–20. Epub 2012 Sep 25. PubMed PMID: 23049635; PubMed Central PMCID: PMC3465001.

Gillspie RS, Stapleton FB. Nephrolithiasis in children. Pediatr Rev. 2004;25(4):131–8.

Greene AK, Kieran M, Burrows PE, Mulliken JB, Kasser J, Fishman SJ. Wilms tumor screening is unnecessary in Klippel-Trenaunay syndrome. Pediatrics. 2004;113(4):e326–9. https://doi.org/10.1542/peds.113.4.e326. PMID: 15060262.

Gripp KW, Lin AE. Costello Syndrome. 2006 Aug 29 [updated 2012 Jan 12]. In: Pagon RA, Adam MP, Bird TD, Dolan CR, Fong CT, Stephens K, editors. GeneReviews™ [Internet]. Seattle: University of Washington, Seattle; 1993–2013. Available from http://www.ncbi.nlm.nih.gov/books/NBK1507/ PubMed PMID: 20301680.

Guay-Woodford LM. Hereditary Nephropathies and Developmental Abnormalities of the Urinary Tract (Chap 130). In: Gupta NP, editor. Challenging and Rare Cases in Urology. Philadelphia: Saunders Elsevier; 2011.

Gully M, Frauger É, Spadari M, Pochard L, Pauly V, Romain F, Gondouin B, Sallée M, Moussi-Frances J, Burtey S, Dussol B, Daniel L, Micallef J, Jourde-Chiche N. Effets uro-nephrologiques des produits utilises par les usagers de drogues: revue de la litterature et enquete pharmaco-epidemiologique en France et dans la region de Marseille. Nephrol Ther. 2017;13(6):429–38. https://doi.org/10.1016/j.nephro.2017.01.024. Epub 2017 Sep 27. French. PubMed PMID: 28958905.

Gunay-Aygun M. Liver and kidney disease in ciliopathies. Am J Med Genet C Semin Med Genet. 2009;151C(4):296–306. Review. PubMed PMID: 19876928; PubMed Central PMCID: PMC2919058.

Hafez MH, El-Mougy FA, Makar SH, Abd El Shaheed S. Detection of an earlier tubulopathy in diabetic nephropathy among children with normoalbuminuria. Iran J Kidney Dis. 2015;9(2):126–31. PubMed PMID: 25851291.

Hill LM, Nowak A, Hartle R, Tush B. Fetal compensatory renal hypertrophy with a unilateral functioning kidney. Ultrasound Obstet Gynecol. 2000;15(3):191–3. PubMed PMID: 10846772.

Hornsby CD, Cohen C, Amin MB, Picken MM, Lawson D, Yin-Goen Q, Young AN. Claudin-7 immunohistochemistry in renal tumors: a candidate marker for chromophobe renal cell carcinoma identified by gene expression profiling. Arch Pathol Lab Med. 2007;131(10):1541–6. PubMed PMID: 17922590.

Hovda KE, Guo C, Austin R, McMartin KE. Renal toxicity of ethylene glycol results from internalization of calcium oxalate crystals by proximal tubule cells. Toxicol Lett. 2010;192(3):365–72. Epub 2009 Nov 18. PubMed PMID: 19931368.

Kaiserling E, Kröber S, Xiao JC, Schaumburg-Lever G. Angiomyolipoma of the kidney. Immunoreactivity with HMB-45. Light- and electron-microscopic findings. Histopathology. 1994;25(1):41–8. PubMed PMID: 7959644.

Kajbafzadeh A, Baharnoori M. Renal malakoplakia simulating neoplasm in a child: successful medical management. Urol J. 2004;1(3):218–20. PubMed PMID: 17914695.

Kamath BM, Podkameni G, Hutchinson AL, Leonard LD, Gerfen J, Krantz ID, Piccoli DA, Spinner NB, Loomes KM, Meyers K. Renal anomalies in Alagille syndrome: a disease-defining feature. Am J Med Genet A. 2011; https://doi.org/10.1002/ajmg.a.34369. [Epub ahead of print] PubMed PMID: 22105858.

Kimura H, Sato O, Deguchi JO, Miyata T. Surgical treatment and long-term outcome of renovascular hypertension in children and adolescents. Eur J Vasc Endovasc Surg. 2010;39(6):731–7. Epub 2010 Apr 28. PubMed PMID: 20430657.

Kojima Y, Lambert SM, Steixner BL, Laryngakis N, Casale P. Multiple metachronous fibroepithelial polyps in children. J Urol. 2011;185(3):1053–7. https://doi.org/10.1016/j.juro.2010.10.046. Epub 2011 Jan 21. PubMed PMID: 21256525.

Lebowitz RL, Olbing H, Parkkulainen KV, Smellie JM, Tamminen-Möbius TE. International system of radiographic grading of vesicoureteric reflux. International Reflux Study in Children. Pediatr Radiol. 1985;15(2):105–9. PubMed PMID: 3975102.

Lee JH, Gleeson JG. The role of primary cilia in neuronal function. Neurobiol Dis. 2010;38(2):167–72. Epub 2010 Jan 22. Review. PubMed PMID: 20097287; PubMed Central PMCID: PMC2953617.

Li L, Parwani AV. Xanthogranulomatous pyelonephritis. Arch Pathol Lab Med. 2011;135(5):671–4. Review. PubMed PMID: 21526966.

Lim DHK, Maher ER. Genomic imprinting syndromes and Cancer. Adv Genet. 2010;70:145–75.

Liptak P, Ivanyi B. Primer: histopathology of calcineurin-inhibitor toxicity in renal allografts. Nat Clin Pract Nephrol. 2006;2(7):398–404; quiz following 404. Review. PubMed PMID: 16932468.

Lott IT. Neurological phenotypes for down syndrome across the life span. Prog Brain Res. 2012;197:101–21. Review. PubMed PMID: 22541290; PubMed Central PMCID: PMC3417824.

Moore MW, Klein RD, Fariñas I, Sauer H, Armanini M, Phillips H, Reichardt LF, Ryan AM, Carver-Moore K, Rosenthal A. Renal and neuronal abnormalities in mice lacking GDNF. Nature. 1996;382(6586):76–9. PubMed PMID: 8657308.

Naesens M, Kuypers DR, Sarwal M. Calcineurin inhibitor nephrotoxicity. Clin J Am Soc Nephrol. 2009;4(2):481–508. Review. PubMed PMID: 19218475.

Nagahara A, Kawagoe M, Matsumoto F, Tohda A, Shimada K, Yasui M, Inoue M, Kawa K, Hamana K, Nakayama M. Botryoid Wilms' tumor of the renal pelvis extending into the bladder. Urology. 2006;67(4):845.e15–7. Epub 2006 Apr 5. Review. PubMed PMID: 16600349.

Niu ZB, Yang Y, Hou Y, Chen H, Wang CL. Ureteral polyps: an etiological factor of hydronephrosis in children that should not be ignored. Pediatr Surg Int. 2007;23(4):323–6. Epub 2007 Feb 15. PubMed PMID: 17377827.

Olin JW, Sealove BA. Diagnosis, management, and future developments of fibromuscular dysplasia. J Vasc Surg. 2011;53(3):826–36.e1. Epub 2011 Jan 13. Review. PubMed PMID: 21236620.

Olin JW, Froehlich J, Gu X, Bacharach JM, Eagle K, Gray BH, Jaff MR, Kim ES, Mace P, Matsumoto AH, McBane RD, Kline-Rogers E, White CJ, Gornik HL. The United States registry for Fibromuscular dysplasia: results in the first 447 patients. Circulation. 2012;125(25):3182–90. Epub 2012 May 21. PubMed PMID: 22615343.

Olsen TS, Holm-Nielsen P. Ultrastructure of nodular renal blastema and sclerotic metanephric hamartoma. Scand J Urol Nephrol Suppl. 1987;104:25–30. PubMed PMID: 2830662.

Peltomäki P. Mutations and epimutations in the origin of cancer. Exp Cell Res. 2012;318(4):299–310. Epub 2011 Dec 13. Review. PubMed PMID: 22182599.

Penton AL, Leonard LD, Spinner NB. Notch signaling in human development and disease. Semin Cell Dev Biol. 2012;23(4):450–7. Epub 2012 Jan 28. Review. PubMed PMID: 22306179.

Perazella MA, Markowitz GS. Drug-induced acute interstitial nephritis. Nat Rev Nephrol. 2010;6(8):461–70. Epub 2010 Jun 1. Review. PubMed PMID: 20517290.

Perlman EJ. Pediatric renal tumors: practical updates for the pathologist. Pediatr Dev Pathol. 2005;8(3):320–38. Epub 2005 Jul 14. Review. PubMed PMID: 16010493.

Pichel JG, Shen L, Sheng HZ, Granholm AC, Drago J, Grinberg A, Lee EJ, Huang SP, Saarma M, Hoffer BJ, Sariola H, Westphal H. Defects in enteric innervation and kidney development in mice lacking GDNF. Nature. 1996;382(6586):73–6. PubMed PMID: 8657307.

Plouin PF, Perdu J, La Batide-Alanore A, Boutouyrie P, Gimenez-Roqueplo AP, Jeunemaitre X. Fibromuscular dysplasia. Orphanet J Rare Dis. 2007;2:28. Review. PubMed PMID: 17555581; PubMed Central PMCID: PMC1899482.

Risdon RA. Renal dysplasia. I. A clinico-pathological study of 76 cases. J Clin Pathol. 1971;24(1):57–71. PubMed PMID: 5573004; PubMed Central PMCID: PMC478026.

Risdon RA, Yeung CK, Ransley PG. Reflux nephropathy in children submitted to unilateral nephrectomy: a clinicopathological study. Clin Nephrol. 1993;40(6):308–14. PubMed PMID: 8299237.

Rostand SG. Oligonephronia, primary hypertension and renal disease: 'is the child father to the man?'. Nephrol Dial Transplant. 2003;18(8):1434–8. Review. PubMed PMID: 12897076.

Ruteshouser EC, Robinson SM, Huff V. Wilms tumor genetics: mutations in WT1, WTX, and CTNNB1 account for only about one-third of tumors. Genes Chromosomes Cancer. 2008;47(6):461–70. PubMed PMID: 18311776.

Rutigliano DN, Georges A, Wolden SL, Kayton ML, Meyers P, La Quaglia MP. Ureteral reconstruction for retroperitoneal tumors in children. J Pediatr Surg. 2007;42(2):355–8. PubMed PMID: 17270548.

Sánchez MP, Silos-Santiago I, Frisén J, He B, Lira SA, Barbacid M. Renal agenesis and the absence of enteric neurons in mice lacking GDNF. Nature. 1996;382(6586):70–3. PubMed PMID: 8657306.

Sanna A, Fanos V, Gerosa C, Vinci L, Puddu M, Loddo C, Faa G. Immunohistochemical markers of stem/progenitor cells in the developing human kidney. Acta Histochem. 2015;117(4–5):437–43. https://doi.org/10.1016/j.acthis.2015.02.014. Epub 2015 Mar 20. PubMed PMID: 25800980.

Sattar S, Gleeson JG. The ciliopathies in neuronal development: a clinical approach to investigation of Joubert syndrome and Joubert syndrome-related disorders. Dev Med Child Neurol. 2011;53(9):793–8. https://doi.org/10.1111/j.1469-8749.2011.04021.x. Epub 2011 Jun 17. Review. PubMed PMID: 21679365.

Schuchardt A, D'Agati V, Larsson-Blomberg L, Costantini F, Pachnis V. Defects in the kidney and enteric nervous system of mice lacking the tyrosine kinase receptor Ret. Nature. 1994;367(6461):380–3. PubMed PMID: 8114940.

Shakya R, Jho EH, Kotka P, Wu Z, Kholodilov N, Burke R, D'Agati V, Costantini F. The role of GDNF in patterning the excretory system. Dev Biol. 2005;283(1):70–84. PubMed PMID: 15890330.

Slavin RE, Inada K. Segmental arterial mediolysis with accompanying venous angiopathy: a clinical pathologic review, report of 3 new cases, and comments on the role of endothelin-1 in its pathogenesis. Int J Surg Pathol. 2007;15(2):121–34.

Slavin RE, Saeki K, Bhagavan B, Maas AE. Segmental arterial mediolysis: a precursor to fibromuscular dysplasia? Mod Pathol. 1995;8(3):287–94.

Stanley JC, Gewertz BL, Bove EL, Sottiurai V, Fry WJ. Arterial fibrodysplasia. Histopathologic character and current etiologic concepts. Arch Surg. 1975;110:561–6.

Stathopoulos IP, Trovas G, Lampropoulou-Adamidou K, Koromila T, Kollia P, Papaioannou NA, Lyritis G. Severe osteoporosis and mutation in NOTCH2 gene in a woman with Hajdu-Cheney syndrome. Bone. 2012; https://doi.org/10.1016/j.bone.2012.10.027. pii: S8756-3282(12)01335-X. [Epub ahead of print] PubMed PMID: 23117206.

Steffens J, Nagel R. Tumours of the renal pelvis and ureter. Observations in 170 patients. Br J Urol. 1988;61(4):277–83. Review. PubMed PMID: 3289673.

Thankamony P, Sivarajan V, Mony RP, Muraleedharan V. Wilms Tumor in a Child With Bilateral Polycystic Kidneys and PHACE Syndrome: Successful Treatment Outcome Using Partial Nephrectomy and Chemotherapy. J Pediatr Hematol Oncol. 2016;38(1):e6–9. https://doi.org/10.1097/MPH.0000000000000472. PMID: 26583622.

Tu BW, Ye WJ, Li YH. Botryoid Wilms' tumor: report of two cases. World J Pediatr. 2011;7(3):274–6. https://doi.org/10.1007/s12519-011-0310-8. Epub 2011 Aug 7. PubMed PMID: 21822996.

Tuchman M, Gahl WA, Gunay-Aygun M. Genetics of fibrocystic diseases of the liver and molecular approaches to therapy. Clinical Gastroenterology, 2010. In: Fibrocystic diseases of the liver, part 1. p. 71–102.

Van Acker KJ, Vincke H, Quatacker J, Senesael L, Van den Brande J. Congenital oligonephronic renal hypoplasia with hypertrophy of nephrons (oligonephronia). Arch Dis Child. 1971;46(247):321–6. PubMed PMID: 5090664; PubMed Central PMCID: PMC1647725.

Vinocur C, Hitzig G, Marboe C, Hensle TW. Renal pelvic tumors in childhood. Urology. 1980;16(4):393–5. PubMed PMID: 7414786.

Volmar KE, Chan TY, De Marzo AM, Epstein JI. Florid von Brunn nests mimicking urothelial carcinoma: a morphologic and immunohistochemical comparison to the nested variant of urothelial carcinoma. Am J Surg Pathol. 2003;27(9):1243–52. PubMed PMID: 12960809.

Wang XM, Jia LQ, Wang Y, Wang N. Utilizing ultrasonography in the diagnosis of pediatric fibroepithelial polyps causing ureteropelvic junction obstruction. Pediatr Radiol. 2012;42(9):1107–11. https://doi.org/10.1007/s00247-012-2404-4. Epub 2012 Jun 1. PubMed PMID: 22653534.

Waters AM, Beales PL. Ciliopathies: an expanding disease spectrum. Pediatr Nephrol. 2011;26(7):1039–56. Epub 2011 Jan 6. Review. PubMed PMID: 21210154; PubMed Central PMCID: PMC3098370.

Weissleder R, Wittenberg J, Harisinghani MG. Primer of diagnostic imaging: Mosby Inc; 2007. ISBN:0323040683.

WHO Classification of Tumours of the Urinary System and Male Genital Organs. WHO Classification of Tumours, 4th Edition, Volume 8, 2016 Edited by Moch H, Humphrey PA, Ulbright TM, Reuter VE ISBN-13 (Print Book): 978-92-832-2437-2 World Health Organization, Geneva, Switzerland.

Yang X, Zhang Y, Hu J. The expression of Cajal cells at the obstruction site of congenital pelviureteric junction obstruction and quantitative image analysis. J Pediatr Surg. 2009;44(12):2339–42. https://doi.org/10.1016/j.jpedsurg.2009.07.061. PubMed PMID: 20006022.

Zisman A, Tieder M, Alon H, Eidelman A. Benign fibroepithelial polyps of the ureter and renal pelvis in childhood. Two case reports. Scand J Urol Nephrol. 1994;28(2):191–3. PubMed PMID: 7939472.

Lower Urinary and Male Genital System

Contents

7.1	**Development and Genetics**	674
7.1.1	Urinary Bladder and Ureter	674
7.1.2	Testis, Prostate, and Penis	675
7.2	**Lower Urinary and Genital System Anomalies**	676
7.2.1	Lower Urinary System Anomalies	676
7.2.2	Male Genital System Anomalies	678
7.3	**Urinary Tract Inflammatory and Degenerative Conditions**	684
7.3.1	Cystitis, Infectious	684
7.3.2	Cystitis, Non-infectious	684
7.3.3	Malacoplakia of the Young	686
7.3.4	Urinary Tract Infections and Vesicoureteral Reflux	686
7.3.5	Megacystis, Megaureter, Hydronephrosis, and Neurogenic Bladder	687
7.3.6	Urinary Tract Endometriosis	688
7.3.7	Tumorlike Lesions (Including Caruncles)	689
7.4	**Preneoplastic and Neoplastic Conditions of the Urinary Tract**	690
7.4.1	Urothelial Hyperplasia (Flat and Papillary)	690
7.4.2	Reactive Atypia, Atypia of Unknown Significance, Dysplasia, and Carcinoma In Situ (CIS)	691
7.4.3	Noninvasive Papillary Urothelial Neoplasms	691
7.4.4	Invasive Urothelial Neoplasms	693
7.4.5	Non-urothelial Differentiated Urinary Tract Neoplasms	696
7.5	**Male Infertility-Associated Disorders**	696
7.5.1	Spermiogram and Classification	696
7.6	**Inflammatory Disorders of the Testis and Epididymis**	696
7.6.1	Acute Orchitis	697
7.6.2	Epidermoid Cysts	697
7.6.3	Hydrocele	700
7.6.4	Spermatocele	700
7.6.5	Varicocele	704
7.7	**Testicular Tumors**	704
7.7.1	Germ Cell Tumors	704
7.7.2	Tumors of Specialized Gonadal Stroma	720
7.7.3	Rhabdomyosarcoma and Rhabdoid Tumor of the Testis	727
7.7.4	Secondary Tumors	727

© Springer-Verlag GmbH Germany, part of Springer Nature 2020
C. M. Sergi, *Pathology of Childhood and Adolescence*,
https://doi.org/10.1007/978-3-662-59169-7_7

7.8	**Tumors of the Epididymis**	727
7.8.1	Adenomatoid Tumor	727
7.8.2	Papillary Cystadenoma	728
7.8.3	Rhabdomyosarcoma	728
7.8.4	Mesothelioma	728
7.9	**Inflammatory Disorders of the Prostate Gland**	728
7.9.1	Acute Prostatitis	728
7.9.2	Chronic Prostatitis	728
7.9.3	Granulomatous Prostatitis	729
7.9.4	Prostatic Malakoplakia of the Youth	729
7.10	**Prostate Gland Overgrowths**	729
7.10.1	Benign Nodular Hyperplasia of the Young and Fibromatosis	729
7.10.2	Rhabdomyosarcoma, Leiomyosarcoma, and Other Sarcomas (e.g., Ewing Sarcoma)	729
7.10.3	Prostatic Carcinoma Mimickers	730
7.10.4	Prostatic Intraepithelial Neoplasia (PIN)	732
7.10.5	Prostate Cancer of the Young	733
7.10.6	Hematological Malignancies	739
7.10.7	Secondary Tumors	739
7.11	**Inflammatory and Neoplastic Disorders of the Penis**	739
7.11.1	Infections	739
7.11.2	Non-infectious Inflammatory Diseases	740
7.11.3	Penile Cysts and Noninvasive Squamous Cell Lesions	741
7.11.4	Penile Squamous Cell Carcinoma of the Youth	741
7.11.5	Non-squamous Cell Carcinoma Neoplasms of the Penis	744
Multiple Choice Questions and Answers		744
References and Recommended Readings		746

7.1 Development and Genetics

7.1.1 Urinary Bladder and Ureter

Urogenital sinus and mesoderm contribute to the development of the urinary bladder (Box 7.1). Anatomically, only the most superior and anterior portion of the urinary bladder is covered by peritoneum, and this is important to keep in mind when bladder specimens are grossed. Urothelium is a transitional epithelium constituted by 5–8 cells thick. The urothelium is divided into three layers (superficial zone, intermediate zone, and basal zone). The external area is a single layer of large, flattened cells, aka "umbrella cells," whose characteristic is a unique trilaminar rigid membrane, which is labeled "asymmetric unit membrane" that is made up of a unique family of proteins labeled uroplakins. The intermediate zone is

> **Box 7.1 Sequential Steps of Sex Organ Development in Humans**
> (1) Germ cell formation in the 6th–10th week in the yolk sac and migration into the gonads.
> (2) Embryonic kidney-close bulging germ cell formation with sac development with a core.
> (3a) Germ cells drift to the *cortex* of the gonad and *medulla* degenerates (**46, XX**).
> (3b) Germ cells float to the *center* of the gonad and *cortex* atrophies (**46, XY**).
> (4) Bilateral Double Ductal System, i.e., Müllerian and Wolffian Ductal Systems with joining at the lower end (cloaca) at 8th–10th week of gestation.

(5) *Internal Genitalia Determination* by ± secretion of Müllerian Duct Inhibitor (MDI).

(5a) If **XX**, (−) MDI ⇒ Müllerian Ductal System ⇒ fallopian tubes, uterus, and upper vagina, while Wolffian Ductal System disappear under the influence of testosterone

(5b) In **XY**, (+) MDI ⇒ Müllerian Ductal System disappears, while Wolffian Ductal System ⇒ Epididymis, vas deferens, and sem. Vesicle under the influence of testosterone (prostate gland develops under the DHT from the conversion of T);

(6) *External Genitalia Determination* arises from median and paramedian embryonic vestigial.

(6a) In **XX**, (−) DHT ⇒ tubercle folds, and swellings transform into the clitoris, labia minora, distal vagina, and labia majora

(6b) In **XY**, (+) DHT ⇒ tubercle folds, and swellings transform into glans, penile shaft, and scrotum.

(7) *Puberty* (10–12 years in XX, 12–14 years in XY): *reproductive function*

(7a) In **XX**: FSH/LH ⇒ *ovular development*, *menarche*, *breast development*, *pelvis enlargement*, and *pubic* and *armpits hair*

(7b) In **XY**: T/DHT ⇒ *prostate growth*, *facial hair*, *hairline recession*, and *acne*

Notes: *MDI* Müllerian Duct Inhibitor, *T* testosterone, *DHT* dihydrotestosterone

7.1.2 Testis, Prostate, and Penis

The testis is a bilateral organ with two principal functions: FSH-stimulated spermatogenesis and LH-stimulated testosterone production. "Didymus" (δίδυμος, i.e., twins, testicles, double) consists of 250 lobules each with up to 4 tightly convoluted seminiferous tubules (each ~0.2 mm = 200 μm) layered by germ cells for the active spermatogenesis and Sertoli cells (standard ratio: 11:1) as support cells. In a study performed on developing human fetal testes, Bendsen et al. 2003 found the ratio between prespermatogonia and Sertoli cells of 1:11 during week 6 to week 9 post conception, while the ratio between prespermatogonia and the total number of somatic cells of 1:44 in the same period of observation. Sertoli cells form the blood-testis barrier, while the spermatogonia, the most primitive germ cells, are located on the blood side of the wall. Spermatogenesis starts with spermatogonia (two types: pale and dark), primary spermatocytes, secondary spermatocytes, spermatids, through the cytological transformation to spermatozoa. About 50% of germ cells should be in late spermatid stage in the normal testicular parenchyma. Testicular septa separate each lobule, and the spermatozoic drainage system consists of straight seminiferous tubules → rete testis → efferent ductules that drain in the deferent duct of the spermatic cord. In the interstitial tissue between the tubules, Leydig cells contain Reinke crystalloid (electron-dense hexagonal prisms with tapered ends) and produce testosterone. A synopsis of the anatomy and developmental biology of the testis is shown in Box 7.2.

Box 7.2 Testis: Useful Anatomic Notes
Newborn: 2 g
1-Year infant: 3 g
15-Year child: 18 g
Adult (20–50 years): 35 g
Capsule → 3 layers:

- *Tunica vasculosa* or internal layer
- *Tunica albuginea* or fibrous capsule
- *Tunica vaginalis* or outer serosa

---

made up of 4–5 layers when stretched. The cytological appearance changes at 6–8 sheets when contracted. The basal region is a single layer of small cells, which is in the basal position and constituted by cylindrical to flattened shape. There is no submucosa in the urinary bladder, and the lamina propria is juxtaposed to the *muscularis propria*.

The septa separate each lobule and extend from the testicular parenchyma into the tunica albuginea. Epididymis (Greek ἐπιδιδυμίς, from ἐπί (epi, "upon") and δίδυμος, is a narrow, tightly "coiled" tube connecting the posterior side of the testicles to the deferent duct of the spermatic cord and consists of three parts: head, body, and tail. The head of the epididymis is situated on the superior pole of the testis and serves as a storage system of the sperm for maturation. The body is a highly convoluted duct that connects the head to the tail of the epididymis. The rear is continuous with the deferent duct, which is the excretory duct joining the excretory duct of the ipsilateral seminal vesicle to form the ejaculatory duct.

Ontogenesis of the testicular tissue includes three phases:

1. Static (birth → age 4): Histological examination of the testis reveals seminiferous tubules compactly filled with small, hyperchromatic undifferentiated cells and many Leydig interstitial cells.
2. Growth (age 4–10 years): Histological examination of the testis reveals seminiferous tubules loosely filled with small, hyperchromatic undifferentiated cells, intratubular luminal formation, and virtual disappearance of Leydig interstitial cells.
3. Development and maturation (from the age of 10 years through puberty): Gonadotropin-driven growth of size and volume with gradual maturation of the spermatogenesis with mitoses, primary and secondary spermatocytes (age 8–11), and spermatids (age 12).

The penis is a single and median organ with two principal functions, namely, urinary and reproductive. Box 7.3 summarized the structures of the penis.

7.2 Lower Urinary and Genital System Anomalies

7.2.1 Lower Urinary System Anomalies

The anomalies of the lower urinary tract that we are going to explore in this paragraph are double ureters, the diverticula of the urinary bladder, the

> **Box 7.3 Penis: Useful Anatomic Notes**
> Erectile tissue: Two CCA and one CS containing the urethra and extending into the glans.
> CCA is covered by *tunica albuginea* and incomplete separation by the *septum penis*.
> CS is covering the urethra and is responsible for the distal formation of the glans penis.
> Superficial fascia: Dartos (The scrotal part of dartos - δαρτός "flayed" - is the dartos properly, while the penile part is referred to as the *fascia superficialis* of penis or, alternatively, the subcutaneous tissue of penis).
> Deep fascia: Buck's fascia → continuous with the fascia of the external oblique muscles.
> *Urethra*: Single distal meatus opening at the central ventral glans.
> Penis cover: Squamous epithelial layer (nonkeratinized on the glans only).
> Lymphatic drainage:
> Skin of the penis and prepuce drain to the superficial inguinal LN
> *Glans penis* and corporal bodies drain to the exterior or deep inguinal/external iliac LN.
> Notes: *CCA* corpora cavernosa, *CS* corpus spongiosum

persistence of the urachus, and the urinary bladder exstrophy (Fig. 7.1).

7.2.1.1 Double Ureters

Double ureters or duplication of the ureters, often also called renal duplication, is the most common anomaly of the urinary tract. Complete is three times less frequent than incomplete overlap. The complete duplication, which can be discovered in childhood or later life, occurs in ~1/500 individuals. Duplication of the ureters arises when a single central bud branches before it reaches the metanephric blastema during the embryogenesis of the urinary tract. There are four significant complications, including the vesicoureteral reflux (VUR), ectopic ureterocele, and ectopic ureteral insertion, and the ureteropelvic junction obstruction of the lower pole.

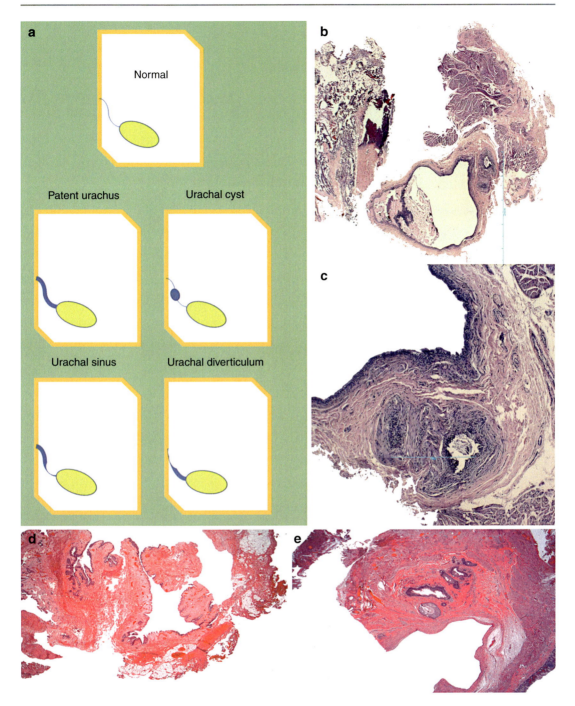

Fig. 7.1 Urinary Bladder Anomalies. In (**a**) are shown the anomalies arising from a dysgenesis of the ductus omphalo-mesentericus with a patent urachus, an urachal cyst, an urachal sinus, or an urachal diverticulum. In (**b**) and (**c**) is shown an urachus cyst (hematoxylin and eosin staining, ×20 and ×100 as original magnifications). In (**d**) and (**e**) are shown a bladder diverticulum and cloacal dystrophy, respectively (hematoxylin and eosin staining, ×12.5 for both microphotographs as original magnifications)

7.2.1.2 Urinary Bladder Diverticula

Diverticula are pouch-like evaginations of the bladder wall classified as congenital or acquired. The formers are due to a focal muscular defect, and some muscular tissue is retained within the wall, while the most recent origin follows urethral outflow obstruction, and no muscular tissue is seen (only mucosa and lamina propria in the pouch). Acquired diverticula are more common than congenital and occur most often in the posterior wall above the *trigonum vesicae urinariae*. Cystocele is a pouch created by protrusion of the urinary bladder into the vagina caused by relaxation of the pelvic support in women leading to uterine prolapse. In the setting of cystocele, an important DDX is the enlargement of the urinary bladder performed in cases of spinal cord injuries or neurologic diseases involving the spinal cord. In these cases, serious and irreversible forms of urinary bladder incontinence are treated using a technique called bladder augmentation. Bladder dysfunction and incontinence may be due to inefficient reservoir capacity of the urinary bladder. During a procedure of bladder augmentation, a skin incision to get access to the abdominal organs is followed by the exposure of the intestines and urinary bladder. A section of the ileum is removed and opened. Following sterilization of the ileal segment, it is grafted onto the urinary bladder to increase the reservoir capacity. Occasionally, the appendix and cecum may also be used, although both organs need to be inflammation-free.

7.2.1.3 Persistence of the Urachus

Urachus, of which the name derives from οὐρά ± αχός, "projecting tip," is an important embryonic structure that may persist in later life. A remnant of this embryological structure refers to the permanence of the vestigial ductal structure located between the apex of the urinary bladder and the internal side of the abdomen at the umbilicus. During the embryonal life, the *urachus* connects the bladder to the allantois (Greek: ἀλλαντοειδής or "sausage-shaped"), which is a portion of a developing amniote's conceptus (primitive embryological form) helping the embryo to exchange gases and handle liquid waste. Urachal remnants, in the form of a cord or a *ductus* and an incomplete or complete connection of the portions, are found in ~1/3 of all individuals. The lining may be transitional epithelium or columnar epithelium. The prognosis of patients with urachal remnants is linked to the potential development of malignancy. There is, indeed, an increased incidence of adenocarcinoma of the urinary bladder in patients harboring urachal remnants.

7.2.1.4 Urinary Bladder Exstrophy

Exstrophy or congenital eversion of a hollow organ refers to a developmental defect consisting in the abnormal closure of the anterior wall of the abdomen and the bladder with sequential communication of the urinary bladder with the exterior of the body through a substantial defect or an open sac. Infections most often complicate the exstrophy.

7.2.2 Male Genital System Anomalies

Anomalies of the male genital system include anorchia/anorchism, testicular regression syndrome ("vanishing testis"), cryptorchidism, polyorchidism/polyorchism, heterotopias and heterotopia-mimickers, bifid scrotum, prostate gland agenesis, prostate gland dysgenesis, phimosis and paraphimosis, urethral meatus abnormalities, and size and numerical abnormalities of the penis. Failure of pubertal maturation may include genetic syndromes, such as Klinefelter syndrome, hypogonadotropic eunuchoidism, or testicular aplasia. It is important to evaluate in a young adult or adolescent the varieties of hypogonadisms and differentiate between primary (testis) and secondary (hypothalamus-hypophysis axis). Typically, urinary FSH and LH are high in case of primary hypogonadism, while low hormonal levels are found in the involvement of the hypothalamus-hypophysis axis. The cutoff value to determine a low testosterone level varies from laboratory to laboratory, but generally, values in an adult with completed puberty <200–250 ng/dL are considered small, while values of 250–350 ng/dL may be regarded as "borderline low." Etiologies of primary hypogonadism include karyotype abnormalities (e.g., Klinefelter syndrome or 47, XXY syndrome). Other causes include toxin exposure, chemotherapy, congenital defects (e.g., anorchia, cryptorchidism), orchitis (e.g., mumps, autoimmune inflammation), testicular trauma or infarction, genetic hemochromatosis and secondary hemosiderosis, and androgen biosynthesis-inhibiting drugs (e.g., ketoconazole and increase in temperature of the testicular environment due to varicocele or a sizeable chronic inflammation). Etiologies of secondary hypogonadism causing gonadotroph suppression include alcohol or drugs (long-term therapy with opioids or corticosteroids), obesity and related conditions, aging, hemochromatosis, hyperprolactinemia, estrogen excess, anabolic steroid abuse, anorexia nervosa, acute illness and chronic medical conditions, HIV infection and AIDS, severe primary hypothyroidism, and pubertal delay. Secondary hypogonadism acquired, causing gonadotroph damage, includes sellar mass or infiltrative lesion, metastasis, trauma (head injury), radiation exposure, surgery, stalk severance, and pituitary apoplexy. The disorders of sexual development include disorders of chromosomal sex (gonadal dysgenesis, mixed gonadal dysgenesis, true hermaphroditism, Klinefelter syndrome, and XX male), diseases of gonadal sex (pure gonadal dysgenesis and absent testis syndrome), diseases of phenotypic sex of the female proband (congenital adrenal hyperplasia, congenital non-adrenal female pseudohermaphroditism, and developmental disorders of Müllerian ducts), and disorders of phenotypic sex of the male proband (abnormalities in androgen metabolism, abnormalities in androgen activity, persistent Müllerian duct syndrome, and developmental defects of male genitalia).

The postnatal-onset male hypogonadism includes *primary hypogonadism* due to *whole gonadal dysfunction* (orchitis, testicular torsion or trauma, trisomy 21 syndrome, varicocele, chronic illness, and late-onset hypogonadism), *primary hypogonadism* due to a *dissociated gonadal dysfunction* (Y chromosome deletions and *AZF* microdeletions; gene mutations of *CILD1*, *USP9Y*, etc.; chemotherapy; abdominopelvic radiotherapy; and drugs spironolactone and ketoconazole), *central hypogonadism* due to *whole gonadal dysfunction* with CNS and pituitary lesions (tumors, LCH, trauma, etc.), *primary hypogonadism* due to *functional primary hypogonadism* (including impaired general health, acromegaly, hypothyroidism, alcohol, and drug abuse), and *combined hypogonadism* due to *whole gonadal dysfunction* (cranial radiotherapy, chemotherapy, Pb intoxication, cannabis consumption, and total body irradiation) (Figs. 7.2, 7.3, and 7.4). There is some confusion about the terminology and concepts of liberalization of some drugs and/or plants for recreational use, which may have dreadful consequences for pregnant women. Cannabis is the plant itself, while hemp and marijuana are essentially specific parts of the plant. Hemp, which may be commercially available and found in grocery stores refers substantially to the stems, stalks, roots, and sterilized seeds. Hemp has less than 0.3% of tetrahydrocannabinol (THC). Marijuana refers to the viable seeds, leaves, and flowers and contains greater than 0.3% THC. THC is the principal psychoactive portion of the plant.

7.2.2.1 Anorchia/Anorchism
Rare condition, which is constituted by bilateral absence of testes in a genotypically and phenotypically normal male. Anorchia/anorchism indicates a lack of Müllerian-derived tissue, which infers that functional testicular tissue was present at the fetal stage but vanished after 14 weeks of gestation. The management includes androgens to induce secondary sex characteristics, psychological counseling, and testicular prosthesis.

7.2.2.2 Testicular Regression Syndrome
Testicular regression syndrome (TRS) or vanishing regression syndrome (VRS), also called "vanishing testis", is a form of gonadal dysgenesis. A more simplified gonadal dysgenesis includes seminiferous tubular dysgenesis (Klinefelter syndrome and 46. XX male), syndromes of gonadal dysgenesis (X0, pure gonadal dysgenesis, mixed gonadal dysgenesis, and partial gonadal dysgenesis), and bilateral vanishing testis (testicular regression syndrome). Pauci-/agonadal children with male, female, or ambiguous phenotype may harbor TRS/VRS, which may be either without gonads or testes or evidence a regression of testicular tissue with residual fibrovascular

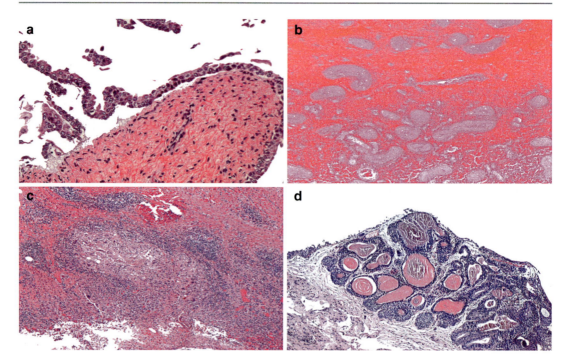

Fig. 7.2 Urinary Bladder Inflammation (Cystitis) and Pseudo-Inflammation. In (**a**) is postmortem urothelial sloughing (non-pathologic). It needs to be differentiated from true cystitis and sometimes may become a challenge. In (**b**) is hemorrhagic necrosis of the urinary bladder without frank inflammation (cystitis). In (**c**) is shown extensive granulomatous cystitis consistent with BCG effect. BCG immunotherapy is the gold-standard treatment for non-muscle-invasive urothelial carcinoma of the urinary bladder, which is at high risk of recurrence or progression. There is a strong inflammatory response to BCG that involves the attachment of BCG; its internalization of BCG into resident immune cells, normal cells, and tumor urothelial cells; the BCG-mediated induction of innate immunity, and BCG-mediated initiation of tumor-specific immunity. In (**d**) is shown an example of cystitis glandularis et cystica, which is a hyperproliferative condition where initial submucosal solid clusters of epithelial cells, termed "Brunn's nests," undergo cavitation to form fluid-filled cystic structures. It may be mistaken for a neoplasm. (**a**–**d**; hematoxylin and eosin staining, ×200, ×40, ×50, ×100 as original magnifications, respectively)

nodule (Ø: ~1 cm), punctate calcification, and hemosiderin-laden macrophages. The histological examination of the tissue may reveal rudimentary epididymis and spermatic cord. The child may show an arrest of the development of the external genitalia depending on the chronology of gonadal injury.

- *EPG*: Cryptorchidism, infarct, infection, trauma, torsion, prenatal hormone-induced atrophy and overproduction of androgens.

The pathologist needs to identify the rest of the testis (TRS/VRS) because the presence of only fat and connective tissue does not rule out an intra-abdominal testis, which may suggest the pediatric urologist may need a second look into the abdomen.

7.2.2.3 Cryptorchidism and Polyorchidism (Polyorchism)

Cryptorchidism is well-defined as a failure of one or both testes to properly descend from the abdomen to the scrotal sac. It is found in 1/10 of male newborns and is subclassified as inguinal in 80% of the cases and abdominal in 20% of the cases. Most of the cases (~90%) probably descend in the first year, and the remaining patients should undergo medical therapy or surgery by age 2 or 3, according to bilaterality or unilaterality of this condition. Typically, the testicular tissue is fertile in 100% of cases if correction is performed by 10 years of age in case of single (unilateral) disease, while fertility is probably ~50% if correction is performed by age 5 years. The impossibility or the non-correction by puberty induces a progressive transformation

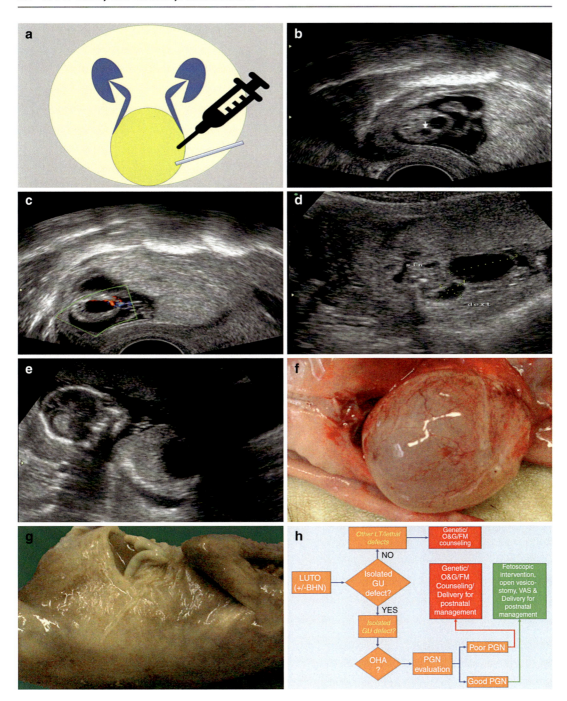

Fig. 7.3 Megacystis. A fetal megacystis is shown in the cartoon in (**a**). The syringe expresses the current management which requires the aspiration of the urinary fluid from the bladder regularly during pregnancy. In (**b**) through (**e**) are shown the examples of ultrasound taken during pregnancy of a patient with fetal megacystic. Unfortunately, the fetal megacystic is recurring and is complicated by severe outcome. In (**f**) is shown the postmortem preparation of a fetal megacystis. In (**g**) is shown the often abnormal enlargement of the abdomen, and in (**h**) is shown the current management of the lower urinary tract obstruction

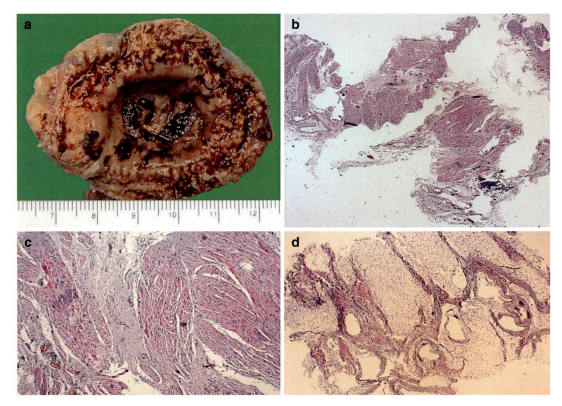

Fig. 7.4 Prune belly syndrome is a rare disorder characterized by partial or complete absence of the abdominal musculature, bilateral cryptorchidism, and/or urinary tract malformations. In (**a**) is shown the urinary bladder which is wide and hypertrophic neck with associated distal urethral obstruction. In PBS, prostatic hypoplasia is responsible for the urethral dilation, and a wide bladder neck is seen instead of the bladder neck hypertrophy that is seen in PUV. In (**b**), (**c**), and (**d**) are shown the abnormal musculature of urinary bladder (**b** and **c**) and the abnormal musculature of the abdominal wall (**d**) of a newborn affected with Prune belly syndrome. (**b–d**; hematoxylin and eosin staining; ×20, ×100, and ×20 as original magnifications, respectively)

of the testicular parenchyma with potential oncogenesis. By this age, the cryptorchid testis is small and brown grossly. Microscopically, atrophic tubules with a ↓ # of total germ cells, failure of fetal gonocytes (stem cells) to transform into dark adult spermatogonia, failure for adult spermatogonia to mature into primary spermatocytes, and a thick BM with multiple foci of Sertoli cell hyperplasia are encountered. CD99 marker by IHC may highlight the Sertoli cell hyperplasia. In fact, CD99, a marker for MIC-2, reacts typically with normal Sertoli cells and granulosa cells (Gordon et al. 1998). Moreover, there is prominent hyperplasia of the interstitial Leydig cells and in advanced cases frank testicular fibrosis. This data has been found to have a strong negative correlation with the individual boy's age at the time of orchidopexy and harbor prognostic significance regarding fertility potential. This fertility potential is especially real when confined with key clinical and laboratory findings such as hormonal measurements, age at orchidopexy, testicle size, laterality, and location of cryptorchid testes. In the past, there was a risk of malignancy assessed at 10–50 times higher in individuals with cryptorchidism than the regular rate of the non-cryptorchid population. In the testicular tissue, a seminoma can indeed develop. This potential fatal condition is much more frequent if the child has not been inspected surgically by the age of 6 years. Currently, the relative risk of testicular cancer in all patients with cryptorchidism is 2.75 to 8 with lower risk (~2–3) in boys undergoing prepubertal orchiopexy. Such decrease of the risk has been successfully reached by orchestrating new revised guidelines and increasing the level of accuracy and diagnosis in pediatric radiology. However, a higher risk does occur in boys with bilateral cryptorchidism, associated genitourinary anomalies, or late (>10–12 years) or uncorrected cryptorchidism. Polyorchidism/polyorchism refers to a very rare medical condition of harboring more than two testicles. Two types have been delineated including Type A and Type B (Bergholz and Wenke, 2009). Type A is characterized by a supernumerary reproductively functional testicle, which is attached to a *vas deferens*. It has been recognized a subtype A1 with complete duplication of the testicle, *epididymis* and *vas deferens*, a subtype A2 with the supernumerary testicle harboring its own *epididymis* and sharing a *vas deferens*, and a subtype A3 with the supernumerary testis sharing the *epididymis* and the *vas deferens* of the other testes. Also, type B is characterized by a supernumerary not reproductively functional testicle not attached to a *vas deferens*. It has been further delineated that type B contains some subtypes, including subtype B1 with the supernumerary testicle having its own *epididymis* but not roping to *a vas deferens*, and subtype B2 with the supernumerary testicle consisting exclusively of testicular tissue. Epidemiologically, subtype A3 is the most common form of polyorchidism and in more than half of the cases the supernumerary testicle is found in the left scrotal sac. Apart of the mythological folklore linked to this condition, a number of authors have tried to summarize the theories of maldevelopment. They include 1) abnormal cellular appropriation, 2) duplication of the urogenital ridge, 3) transverse division of the urogenital ridge, 4) incomplete degeneration of a part of the mesonephros, and 5) singular development of peritoneal bands.

7.2.2.4 Heterotopias and Heterotopia-Mimickers

In case of abnormal testicular tissue the pathologist needs first to rule out either malignancy or a teratoma. The category of "heterotopias and heterotopia-mimickers" includes splenic-gonadal fusion, which occurs on the left side when the spleen and testis are nearby during embryogenesis, and adrenal cortical rests. Splenic-gonadal fusion is subclassified as a continuous variant when a cord of fibrous or splenic tissue is seen and discontinuous variant when no connection is recognized.

7.2.2.5 Bifid Scrotum

It is defined as deep midline cleft in the scrotum, which is determined by incomplete fusion of the labioscrotal folds. This abnormality may accompany penoscrotal or perineal hypospadias. A variant of the bifid scrotum is the penoscrotal transposition, which is recognized as a penis lying within or beneath the scrotum.

7.2.2.6 Prostate Gland Agenesis/Dysgenesis

Prostate development starts prenatally when fetal androgens bind to androgen receptors in the mesenchyme of the urogenital sinus to stimulate prostatic ductal progenitors or buds in the epithelium of the urogenital sinus. Prostatic buds, like buds in other organs, are formed in sequential stages. In utero 2,3,7,8-tetrachlorodibenzo-p-dioxin (TCDD) exposure at time of the smooth muscle development in the urogenital sinus causes dysgenesis of the prostatic gland.

7.2.2.7 Phimosis and Paraphimosis

Phimosis (φίμωσις, "muzzling" and φῑμός, "muzzle") refers to an orifice too small to allow normal retraction, while paraphimosis (παρά, "beside" as prefix) defines a retracted prepuce but too tight to be re-extended indicating that the foreskin is trapped behind the glans. The causes of phimosis may be a congenital anomaly or a chronic infection. Paraphimosis, which is a not quite rare medical condition, shows an acute swelling and inflammation of the distal penis and glans. The constriction force acting as a tourniquet. The constriction is a medical emergency rushing the child to the emergency because the paraphimosis needs to be eliminated by either conservative manual ways or surgically.

7.2.2.8 Urethral Meatus Abnormalities

The urethral meatus may open not at the apex of the penis, but at different locations such as dorsally, a condition called epispadia (ἐπί, "on top of" ±σπαδίας and σπάω, "to break"), and ventrally, a condition called hypospadia (ὑπό "under" ±σπαδίας and σπάω, "to break"), or, ultimately, multiple accessory urethral canals may be present. Treatment consists of a surgical correction due to the recurrent infections in the urinary tract.

7.2.2.9 Size and Numerical Abnormalities of the Penis

Penis size abnormalities or abnormalities of a number of the penis are conditions usually associated with other abnormalities of the genitourinary tract. From a classification consideration of view, it is crucial to distinguish aphallia or penile agenesis, hypophilia or penile hypoplasia, hyperplasia or penile hypertrophy, and diphallia or duplications of the penis, which is a medical condition occurring in one out of every 5 to 6 million live births. Aphallia derives from the word φαλλός, "penis" and privative "a-", which in turn derives from Latin – *Alpha prīvātīvum* and Greek α στερητικόν used in medicine and sciences to express absence or negation.

7.3 Urinary Tract Inflammatory and Degenerative Conditions

7.3.1 Cystitis, Infectious

- *DEF*: Inflammation of the urinary bladder, which often precedes pyelonephritis. The most common microorganisms of cystitis are *E. coli*, *Proteus*, *Klebsiella*, *Enterobacter*, *Chlamydia*, *Mycoplasma spp.*, *M. tuberculosis*, *C. albicans*, *Cryptococcus*, *Schistosomiasis*, and *Adenovirus*. Other factors are cytotoxic chemotherapy, radiation, or trauma (Fig. 7.5).
- *CLI*: Dysuria, lower abdominal pain, and increased frequency of urination (urinary frequency).

Predisposing conditions are congenital malformations, acquired malformations, and systemic diseases such as DM and immunosuppression.

7.3.2 Cystitis, Non-infectious

Non-infectious (peculiar forms of cystitis)

- Gangrenous cystitis: Infection-related and corrosive chemical injury-related complications of inflammation of the urinary bladder particularly in some clinical settings, including DM, sepsis, and vascular disease. Microscopically, necrosis starts in the mucosa and progresses eventually to the entire wall of the urinary bladder.
- Granulomatous cystitis: Caseating or noncaseating granulomatous inflammation with overlying urothelium showing reactive atypia or ulceration and due to BCG-therapy, *M. tuberculosis/M. bovis*, sarcoidosis, Crohn disease, rheumatoid arthritis, and postsurgical granulomatous change following TUR (transurethral resection) of cancer of the urinary bladder. Bacillus Calmette–Guerin (BCG) is a form of immunotherapy used to treat some forms of bladder cancer.
- Follicular cystitis: Chronic cystitis characterized by numerous size-variable lymphoid follicles with the prominent germinal center formation in lamina propria associated with long-term UTI (urinary tract infection/s) and bladder cancer. DDX includes follicular lymphoma (see hematologic chapter for IHC).
- Emphysematous cystitis: It is a gas-filled vesicular cystitis in the lamina propria of the urinary bladder with accompanying numerous histiocytes, multinucleate giant cells, and mild perivesicular interstitial fibrosis due to gas-forming bacteria (*Aerobacter aerogenes*, *E. coli*, *Clostridium perfringens*) in both pediatric (e.g., NEC) and adult pathologic (e.g., DM) conditions.
- Encrusted cystitis: It is an inorganic salt deposition-related cystitis in bladder mucosa with a diffuse gritty appearance cystoscopically and fibrinous exudate with calcified, necrotic debris and acute and chronic inflammatory infiltrate microscopically. It is due to urea-splitting bacteria that alkalinize the urine. In healthy individuals, the normal urine pH is slightly acidic (6.0–7.5), although the normal range is 4.5–8.0. The identification of an urinary pH > 8.0 is suggestive of a urea-splitting organism (e.g., *Proteus*, *Klebsiella*, or *Ureaplasma urealyticum*). Alkaline (basic) pH suggests "struvite nephrolithiasis" or "infection" kidney stones.
- Hemorrhagic cystitis: Mucosal (lamina propria) hemorrhage-characterized cystitis due to chemotherapy (cyclophosphamide, busulfan, thiotepa), industrial products (aniline and toluidine derivatives present in dyes and insecticides), and viruses (adenovirus, type 11 and 21, polyomavirus, HSV-2).

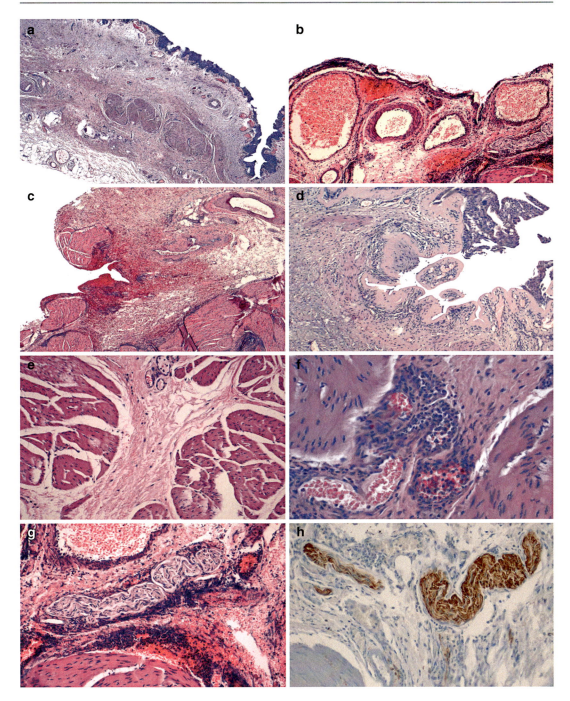

Fig. 7.5 In (**a**) through (**h**) are shown the microphotographs of a neurogenic bladder of a patient with spinal muscular atrophy type 2 who underwent cystectomy (**a–g**; hematoxylin and eosin staining; ×40, ×100, ×40, ×100, ×200, ×400, ×200 as original magnifications, respectively; **h**, anti-S100 immunostaining, Avidin-Biotin-Complex, ×200 original magnification)

- Interstitial (Hunner's) cystitis is a persistent, chronic inflammation of the urinary bladder presenting with painful hematuria, urinary frequency, urgency, clinically and edema, hemorrhage, granulation tissue, mononuclear inflammatory infiltrates (also perineural) with numerous mast cells, as well as fibrosis of all layers microscopically. A localized ulcer is often present, and mast cells are present beneath the ulcer, within the detrusor muscle, and in the bladder mucosa between the urothelial cells. The diagnosis is generally one of exclusion after the elimination of other causes (infection, malignancy). Mast cells can be highlighted by histochemistry (tryptase, Giemsa stain, Leder stain, toluidine / methylene blue) or IHC (e.g., CD117).
- Radiation cystitis may present as polypoid cystitis, which is dome/posterior wall located inflammation with polypoid features simulating neoplasm and secondary to radiation, but also indwelling catheter, bladder lithiasis, and fistulas (DDX). Microscopically, there are stromal edema, congestion, and chronic inflammation with atypical mesenchymal cells and telangiectasia of the blood vessels with hyalinization and thrombosis as well as urothelium with karyorrhexis and nuclear and cytoplasmic vacuolation but standard N/C ratio. The IHC is crucial in distinguishing radiation cystitis (e.g., a child with leukemia receiving radiation) and CIS. The IHC profile of radiation cystitis is CD44 (+) CK20 (−) p53 (−), while CD44 (−) CK20 (+) p53 (+) suggests a CIS.

Giant cell cystitis is the presence of giant cells in lamina propria due to chemotherapy toxicity or radiation but is not a distinct clinical entity. Eosinophilic cystitis is not a clinical entity and the term "eosinophilic cystitis" should be used as a descriptive term only. The inflammatory process has a polypoid aspect mimicking polypoid cystitis, rhabdomyosarcoma of botryoid type, and urothelial carcinoma. Eosinophilic cystitis is often seen in pediatric specimens from children with allergic gastroenteritis, asthma, or allergic disorders while in adults following prior transurethral biopsy, prostatic hyperplasia, and bladder carcinoma. Rarely, parasitic infection is also a critical etiology that needs to be kept in the DDX in particular settings.

7.3.3 Malacoplakia of the Young

Malakoplakia is a vesical lesion characterized by soft yellow 3–4 cm mucosal plaques or polyploidy bladder mass covered with intact mucosa clinically and constituted by accumulation of tightly packed, large, foamy histiocytes (von Hansemann cells, +CD68, +CD163) with characteristic cytoplasmic inclusions (Michaelis-Gutmann bodies, PAS+, Pearls Prussian Blue stain+, von Kossa+), occasional giant cells, and interspersed lymphocytes microscopically. Michaelis-Gutmann bodies are laminated mineralized concretions tightly located within and between histiocytes. However, in the early lesions (youth), Pearls and von Kossa stains may be negative, because mineralization occurs later in the disease process. Malakoplakia is an inadequate host response to bacterial infection (insufficient phagolysosomal activity), typically Gram-negative bacilli (*E. coli*, *Proteus*) and needs to be differentiated from granulomatous inflammation, Langerhans cell histiocytosis (LCH), small round blue cell tumors, urothelial carcinoma, and prostatic carcinoma. Obviously, granulomatous inflammation, small round blue cell tumors, and LCH play a major role in childhood and adolescence, while both urothelial carcinoma and prostatic carcinoma will play a major role later in life.

7.3.4 Urinary Tract Infections and Vesicoureteral Reflux

Infections may spread from the urinary bladder to the urinary tract along an ascendant pathway to the ureter, pelvis, and ipsilateral and contralateral kidney (UTI, urinary tract infections). In patients with vesicoureteral reflux (VUR), infections are quite common. VUR refers to a medical condition in which urine flows in a retrograde way from the urinary bladder into the ureters/kidneys. VUR grade 5 indicates UTI with or without bacteriuria/yeast urea and recurrent pyelonephritis. DMSA scan, which is a radionuclide scan that uses dimercaptosuccinic acid, assesses the renal function carefully. To date, DMSA scan is considered the most reliable test for the diagnosis of acute pyelonephritis. The VUR grading system is as follows. Grade I means ureteral dilatation, but VUR does not reach the renal pelvis. Grade II means ureteral dilatation, and VUR reaches the pelvis without dilatation of the collecting system. Grade III means ureteral dilatation ± "kinking" and dilatation of the collecting system and healthy fornices, while grade IV indicates moderate ureteral dilatation ± "kinking" and dilatation of the collecting system and blunt fornices with still visible papillary impressions. Finally, grade V means severe dilatation with "kinking" of the ureter and collecting system with the disappearance of the papillary impressions and clear-cut intraparenchymal (renal) reflux.

7.3.5 Megacystis, Megaureter, Hydronephrosis, and Neurogenic Bladder

The megacystis is relatively poorly defined and harbors a multi-etiologic spectrum. It is mostly defined as distended urinary bladder identified in utero. It may be diagnosed by ultrasound after the fetus starts producing urine at about the 10th week of gestation. If there is a megacystis, there may be a mechanical or functional bladder outlet obstruction, which can be either partial or complete. Occasionally, the urinary bladder is normal. Its etiology includes posterior urethral valves (PUV), prune belly syndrome (PBS), urethral atresia/stenosis, cloacal malformities, and megacystis-microcolon-intestinal hypoperistalsis syndrome (MMIHS).

Megaureter is a dramatic condition that applies to the ureter showing a dilation out of proportion to the rest of the urinary tract due to a vast number of elastic fibers. The following classification seems the most accepted one by most of the pediatric urologists worldwide. It separates the obstructed megaureter from the reflux and non-obstructed and non-reflux megaureter. The obstructed megaureter is further classified as primary (e.g., intrinsic obstruction, stenosis, adynamic segment) and secondary (e.g., post-urethral obstruction, post-neuropathic bladder, post-extrinsic obstruction, retroperitoneal tumor, benign or malignant). The reflux megaureter is also further classified as primary (e.g., primary reflux megaureter, prune belly syndrome) and secondary (e.g., post-urethral obstruction, post-neuropathic bladder). The non-obstructed and non-reflux-linked megaureter is subdivided in primary and secondary (e.g., polyuria, infection).

Hydronephrosis refers to a unilateral or bilateral swelling of a kidney due to a retrograde (backward) buildup of urine as consequence of a blockage or obstruction.

Ureteral dilatation may essentially be caused by VUR, obstructive disease, high urine flow from non-concentrating kidneys, and maldevelopment of the ureteral muscular layers. The fibers of the renal pelvic muscle run obliquely, and connective tissue well separates the muscle bundles. The ureteropelvic junction is an ill-defined structure showing muscle bundles with different orientation lying side by side. Consequently, the ureter consists of braided bundles of muscle fibers arranged in interlacing spirals. The distal pelvic ureter shows an outer circular muscular orientation and an inner longitudinal muscular orientation. The Waldeyer spaces (or better "*Ureterscheide*", separators) separate the loops of the outer spirals. Importantly, the muscle fibers of the roof of the intravesical ureter sweep laterally to intersect the floor, and then both continue distally at the superficial trigone (*trigonum vesicae urinariae*), which is specifically a structure separated from the detrusor muscle. In the examination of the congenital ureteral muscle abnormalities, either quantitative or qualitative or both changes may occur. Right hydronephrosis and hydroureter/megaloureter should be addressed by right ureterostomy and cystoscopy for detection of PUV, and if the diameter of the ureter is ≥1 cm, it has been advised to perform a closure ureterostomy and bilateral reimplant (new ureterostomy) after ureteral tapering.

Prune belly syndrome (Eagle-Barrett syndrome) refers to a congenital disorder characterized by partial or complete lack of the abdominal muscles, failure of both testes to descend into the scrotum (bilateral cryptorchidism), and urinary tract malformations (Fig. 7.4).

Posterior urethral valves (PUV/PUVs) are obstructing membranes, which are in the posterior male urethra as a result of abnormal intrauterine development. The Woodart classification entails PUV I–III: category I (renal dysplasia, oligohydramnios, lung hypoplasia, Potter facies, urethral atresia), category II (full triad features, minimal/unilateral renal dysplasia, no lung hypoplasia, tendency to progress to renal failure), and category III (incomplete/mild trial features, mild to moderate uropathy, no renal dysplasia, stable renal function, and absence of any form of pulmonary hypoplasia). PUV is the most common cause of bladder outlet obstruction in male newborns and occurs in ~1/8000 newborns. There are a few prognostic markers in assessing PUVs in the amniotic fluid. These biological markers include $Na^+ < 1000$ mEq/L, $Cl^- < 90$ mEq/L, osmolarity less 210 mOsm/L, β2-microglobulin <4 mg/L, and $Ca^{2+} < 8$ mg/dL and suggest a good outcome.

Neurogenic bladder refers to a urinary bladder dysfunction caused by neurologic damage with the urinary bladder being flaccid or spastic and presenting with overflow incontinence, urinary frequency, urgency, incontinence, and urine retention (Fig. 7.5).

- *GRO*: Most often, the urinary bladder is reduced in size with a nonelastic and fibrotic wall. On the internal surface, the mucosa is frequently irregularly grayish-white in appearance.
- *CLM*: Severe fibrosis in both lamina propria and muscularis propria are the main characteristics. Some areas are particularly rich in collagen fibers, which are better highlighted by Masson-Trichrome stain, while other regions show paucicellular hyalinization. There is a moderate to marked disarray of smooth muscle cells on the background of fibrosis, and in this setting some authors have suggested the term "leiomyomatous-like hyperplasia." Occasionally, lymphocytic and plasma cellular inflammatory infiltrates are scattered in the lamina propria. The normal umbrella cells cover most parts of the mucosa, but occasional squamous metaplasia is also seen. On immunohistochemistry, the umbrella cells show no cytoplasmic expression of keratin 20 on the surface, while there is a high expression of S100 protein, an antigen expressed by Schwann cells of peripheral nerves, in the superficial and deep wall layers. It exhibits marked hypertrophy and hyperplasia of the nerve fibers in the lamina propria and the muscularis propria. These findings have been explained as a reaction to the persistent mechanical strain on the bladder wall. Histologically, the bladder wall may also show paucicellular hyalinization and disarray of smooth muscle cells on a background of fibrosis, which at places may look like leiomyomatous-like hyperplasia. A clinicopathological study targeting the urothelial differentiation in patients with spinal cord injury showed in 4/5 of their patients a lack of staining with an antibody against keratin 20. This fact points to the incomplete urothelial differentiation leading to a higher risk of developing inflammation and pre-cancerogenous dysplasia.
- *PGN*: Risk of serious complications (e.g., recurrent infection, vesicoureteral reflux, autonomic dysreflexia) is high. Although the mechanism underlying the predisposition of the urothelium of the neurogenic bladder to urothelial dysplasia and transitional cell carcinoma is still unclear, these patients are at increased risk of developing some form of urothelial neoplasm and should, therefore, be followed up routinely.

7.3.6 Urinary Tract Endometriosis

Müllerian anomalies occur with a reported rate of 0.1%–3.5%, and Müllerian aplasia (class I) occur in 1 in 4000–5000 cases. The affected individuals usually have Mayer-Rokitansky-Kuster-Hauser syndrome (MRKHS), which is a disorder that occurs in females and mainly affects the reproductive system. MRKHS patients have an underdeveloped or absent vagina and uterus with normal external genitalia. Approximately 90% of affected women have some degree of Müllerian development, most often showing bilateral fibromuscular uterine remnants along the pelvic sidewall, with functional endometrium in only ~5% of these remnants. Typically, ovarian endocrine and oocyte function are healthy, and the karyotype is 46, XX. Patients affected with MRKHS present usually with primary amenorrhea, and cyclic or chronic pelvic pain secondary to *hematometra/hematosalpinx* and endometriosis can occur in those MRKHS patients with functional endometrium. Management involves removal of the functional endometrium and the potential creation of a neovagina. In 2011, Elliott et al. described the case of a 12-year-old girl with primary amenorrhea and persistent cyclic pelvic pain and imaging studies of MRKHS. It is prudent to use a combination of oral contraceptives to control endometriosis and dysmenorrhea.

7.3.7 Tumorlike Lesions (Including Caruncles)

Tumorlike lesions include *cystitis cystica et glandularis*, fibroepithelial polyp, malakoplakia (*vide supra*), Müllerian lesions, nephrogenic adenoma, papillary-polypoid cystitis, postoperative lesions (granulomas, spindle-cell nodule), prostatic-type polyps, pseudocarcinomatous hyperplasia, schistosomiasis, squamous lesions (squamous metaplasia, keratinizing and nonkeratinizing) and squamous hyperplasia, papillary and flat, as well as von Brunn's nests.

Papillary-polypoid cystitis (PPC): Non-neoplastic inflammatory lesion of the urinary bladder characterized irritable bladder symptoms and hematuria clinically, an exophytic polypoid or papillary projection endoscopically, and papillae lacking complex branching (DDX papillary urothelial neoplasia), stromal edema and fibrosis without fibrovascular cores (DDX papillary urothelial neoplasia), and reactive urothelial atypia microscopically. PPC is generally due to the indwelling catheter or vesical fistula and needs to be differentiated from papillary urothelial neoplasia and nephrogenic adenoma.

Von Brunn's nests, *cystitis cystica* and *cystitis glandularis* are all three related lesions constituting a spectrum of disease. Von Brunn's nests refer to invagination of transitional epithelium into the underlying lamina propria, while *cystitis cystica* relates to invagination that loses connection with the surface becoming cystically dilated with a flattened transitional epithelium to the cuboidal lining, and *cystitis glandularis* when columnar metaplasia takes place with mucous secreting cells in the cyst lining. Von Brunn's nests DDX should include a nested variant of urothelial carcinoma, inverted papilloma, carcinoid tumor, and paraganglioma. In the inverted papilloma, anastomosing cords or islands of urothelium originating from the superficial urothelium extend down into the underlying lamina propria (deeper than von Brunn's Nests!).

Nephrogenic adenoma is a benign epithelial lesion, which may be localized or diffuse presenting as papillary-polypoid lesion occurring in response to chronic infection, lithiasis, prolonged catheterization, and renal TX. Endoscopic view shows a papillary, polypoid, or sessile lesion; generally small and microscopic sections show papillary or tubular patterns with cuboidal and hobnail arrangement of epithelial cells. There are small tubules (microcysts) lined by cuboidal and hobnail cells without atypia, intratubular colloid-like material, and prominent BM. There is no basal cell layer (potential pitfall). IHC shows (+) AE1/3, CAM5.2 (LMW-CK), and CK7; (+) PAX2 and PAX8; (+) AMACR, PSA, and PAP; (−) CD31 and CD34; (−) HMW-CK; and (−) p63. DDX should include mesonephroid clear cell adenocarcinoma, prostatic carcinoma, and capillary hemangioma. Prostatic specific antigen (PSA) and prostatic acid phosphatase (PAP) are the tumor markers used to monitor disease progression or improvement in patients with adenocarcinoma of the prostatic gland.

Fibroepithelial polyp referred to a non-neoplastic polyp of the urinary bladder observed mostly in male children and adolescents in urethra and ureter and characterized microscopically by a polypoid lesion with variably bulbous and elongated papillae and cloverleaf-like projections with normal urothelium as lining and no / very focal scattered atypical stromal cells. DDX includes embryonal rhabdomyosarcoma (ERMS), florid *cystitis cystica et glandularis*, papillary-polypoid cystitis, urothelial papilloma, inverted papilloma, and bladder hamartoma. Notably, ERMS shows the cambium layer, rhabdomyoblastic differentiation, and (+) DES, MYOG (MYF-4), and MYOD1 (MYF-3).

Prostatic-type polyp is a non-neoplastic lesion of the prostatic urethra at the trigone (*trigonum vesicae urinariae*) manifesting as hematuria and hematospermia (blood in the semen) and constituted by a polypoid mass with variable exophytic fronds. Microscopically, there is a benign prostatic secretory urothelium on the surface, and the stroma consists of benign prostatic type-secretory glandular epithelium and urothelium. IHC is (+) PSA and PAP. DDX includes prostatic ductal adenocarcinoma, papillary urothelial neoplasm, benign prostatic hyperplasia, nephrogenic adenoma, papillary-polypoid cystitis, and *cystitis cystica et glandularis*.

Amyloidosis may be part of a systemic disease or may present as a localized nodule of the urinary bladder, the so-called amyloid tumor. Microscopically, there are AL (immunoglobulin Amyloidosis Light chain) eosinophilic amorphous deposits of amyloid ("apple green" birefringence on Congo red under polarized light, + thioflavin T fluorescence, and randomly arranged, rigid nonbranching 7–12 nm fibrils by TEM) in the interstitium of lamina propria, often penetrating the superficial muscularis propria. Some minimal inflammation may occur, and other locations described in the scientific literature include ureter, renal pelvis, and urethra.

Postoperative spindle-cell nodule (POSN) is a chronic inflammatory tumorlike lesion occurring following instrumentation (TUR) and characterized clinically by a small, friable, sessile nodule with superficial ulceration and some bleeding. Microscopically, there is a high resemblance of this lesion to both nodular fasciitis and granulation tissue observed in the soft tissue. There are fascicles or bundles of proliferative spindle cells with infiltrative margins. There is usually bland cytology and no or mild atypia.

Müllerian lesions are ectopic benign Müllerian tissue within the wall of the urinary bladder of reproductive females and subclassified as endometriosis (endometrial glands and stroma), endocervicosis (endocervical-type glands), endosalpingiosis (tubal-type glands), and Müllerianosis (a mixture of different types of Müllerian epithelium). DDX includes invasive adenocarcinoma, urothelial carcinoma with glandular differentiation, and microcystic urothelial carcinoma, which is a nested variant of urothelial carcinoma characterized by a bland cytomorphology and harbors a clinical behavior simulating the high-grade conventional urothelial carcinoma.

Squamous metaplasia refers to change of the urothelium toward squamous epithelium with or without keratinization and commonly seen in trigone (*trigonum vesicae urinariae*) of female adolescents and responsive to estrogen in case of the nonkeratinizing form secondary to trauma, diverticula, bladder lithiasis, and keratinizing form, such as following schistosomiasis.

Pseudocarcinomatous hyperplasia refers to radiation-induced cystitis and characterized by small urothelial nests with round or ovoid jagged borders and prominent cytoplasmic eosinophilia. DDX includes invasive urothelial carcinoma, a nested variant of urothelial carcinoma, and florid von Brunn's nests.

7.4 Preneoplastic and Neoplastic Conditions of the Urinary Tract

7.4.1 Urothelial Hyperplasia (Flat and Papillary)

- *Flat type*: Thickening of the mucosa with no cytological atypia (no premalignant potential!).
- *Papillary type*: Slight "tenting" with undulating or papillary growth and urothelium of average thickness and no cytological atypia (no premalignant potential!).

Papillary hyperplasia is usually asymptomatic and recognized during routinely performed follow-up cystoscopies for papillary urothelial neoplasms in both adults and children.

- *DEF*: Marked thickening of the mucosa without cytologic atypia potentially lying adjacent to low-grade papillary urothelial neoplasm but harboring no potential malignant requiring no treatment.
- *EPI*: Unknown.
- *CLI*: Asymptomatic, incidental.
- *GRO*: Marked thickening of the mucosa of the urinary bladder.
- *CLM*: Marked thickening of mucosa ≥10 cell layers with morphologic evidence of maturation from basal cells to superficial cells and neither cytologic atypia nor mitoses.
- *IHC*: CK20, CD44, and p53.
- CK20 staining is limited to the umbrella cells (differently it stains deeper cells in dysplasia), while CD44 staining is limited to the basal cells (differently stains all layers in reactive atypia, but no staining occurs in dysplasia), and p53 is negative in flat hyperplasia.
- *DDX*: Dysplasia and reactive atypia.
- *PGN*: Molecular studies (genetic alterations in Chr. 9) suggest a preneoplastic potential in flat hyperplasia concurrent with low-grade papillary tumors.

7.4.2 Reactive Atypia, Atypia of Unknown Significance, Dysplasia, and Carcinoma In Situ (CIS)

Flat lesions with atypia refer to non-papillary lesions without atypia. Nuclear abnormalities in reactive inflammatory atypia include enlargement, vesiculation or hyperchromasia, and prominence of nucleoli, all features that per se do not represent malignancy! A positive history for instrumentation, lithiasis, and therapy is usually found in the medical records. Flat lesions are divided into reactive (inflammatory) atypia, atypia of unknown significance, low-grade intra-urothelial neoplasia (dysplasia), and high-grade intra-urothelial neoplasia (carcinoma in situ, CIS). CIS may have specific patterns including scattered CIS cells in a pagetoid spread and "clinging" CIS. Since the definition of CIS relies on architectural and cytologic changes of the intraepithelial cells constituting the urothelium, an umbrella cell layer may still be present. Urothelial neoplasms are divided into invasive urothelial tumors into the lamina propria and urothelial tumors with invasion into the muscularis propria (detrusor muscle, *musculus detrusor vesicae urinariae*).

Papillary hyperplasia needs to be differentiated from other benign papillary lesions, including papilloma, condyloma, and fibroepithelial polyp (aka mesodermal stromal polyp). In papillary hyperplasia, capillaries are seen at its basis, but no central branching fibrovascular core is observed unlike papilloma, which shares with papillary hyperplasia the papillary fronds with benign unremarkable squamous lining. Thus, papilloma is a discrete papillary growth with delicate papillae and central fibrovascular core lined by urothelium of standard thickness and cytology. The inverted papilloma is a polypoid or sessile mass showing invagination of the urothelium without papillae. In inverted papilloma, few mitoses may be observed, but cell maturation is normal. In condyloma, arborizing architecture and koilocytosis are essential features, while fibroepithelial polyp has a polypoid architecture with thin squamous lining and prominent stroma with occasional large pleomorphic cells.

7.4.3 Noninvasive Papillary Urothelial Neoplasms

Malignant tumors include urothelial carcinoma, squamous cell carcinoma, adenocarcinoma, small-cell carcinoma, sarcomatoid carcinoma, rhabdomyosarcoma, and leiomyosarcoma. Entities that can occur in children and young individuals are usually restricted to rhabdomyosarcoma, papillary urothelial neoplasm of low malignant potential (PUNLMP) (*vide infra*), and urothelial carcinoma. PUNLMP is an important entity that needs to be considered especially for the follow-up. Noninvasive papillary urothelial neoplasms refer to neoplasms with papillary architecture and no invasiveness (Figs. 7.6 and 7.7). The distinguishing features of noninvasive urothelial papillary lesions fol-

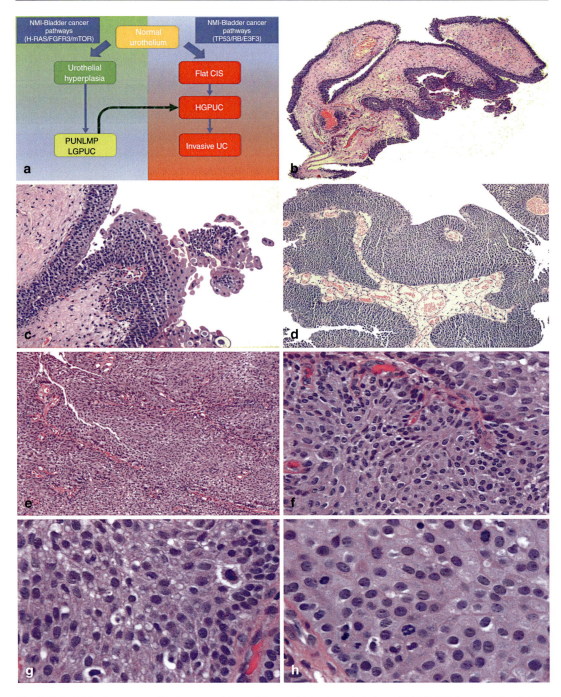

Fig. 7.6 In (**a**) is shown a graphical representation of the papillary neoplasms of the urinary bladder. The microphotographs (**b**) and (**c**) show a papilloma, while the microphotographs (**d**) through (**h**) show several features which characteristics of PUNLMP (see text for details) (**b–h**; hematoxylin and eosin staining; ×40, ×200, ×100, ×100, ×400, ×630, ×630 as original magnification, respectively)

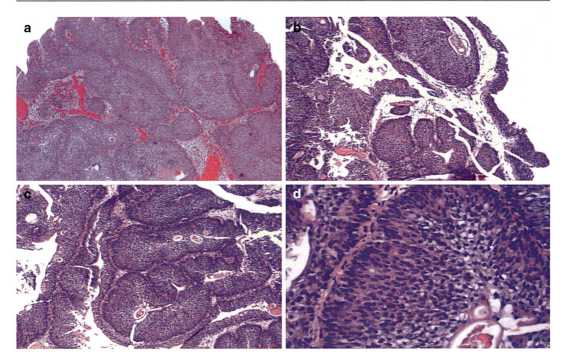

Fig. 7.7 All four microphotographs are indicative of an inverted papilloma (see text for details) (**a–d**; hematoxylin and eosin staining, ×50, ×100, ×100, ×400 as original magnifications, respectively)

low architectural and cellular indicators (*p*apillae, *o*rganization of cells, *s*ize of nuclei, *s*hape of nuclei, *u*mbrella cells, and *m*itotic index). The mnemonic word POSSUM can be used to memorize the features of this entity. Importantly, PUNLMP is characterized by delicate (+) occasional fused papillae, normal cell polarity, any thickness, cohesive cells, more or less enlarged nuclei, which are uniformly round-oval, harboring fine chromatin, and no or an inconspicuous nucleolus. There are uniform umbrella cells on the surface, and the mitotic index is nil or slightly increased at the basal layer only.

7.4.4 Invasive Urothelial Neoplasms

Although very few children and few young individuals may have a urothelial carcinoma, it enters of right in the critical task of the differential diagnosis of urothelial papillary neoplasms.

- *DEF*: Malignant epithelial tumor originating from the urothelium and harboring a poor prognosis (Fig. 7.8).
- *EPI*: Generally, urothelial carcinoma accounts for 90% of all primary tumors of the urinary bladder independent of the age of the patients and 3% of all cancer deaths, but most patients are obviously older than 50 years of age with a clear-cut male preference. In young adults, specifically with conditions predisposing the exposure of the urothelium to carcinogenic chemicals, urothelial carcinoma needs to be considered.
- *CLI*: Since its first description the most common symptom remains hematuria.
- *RF*: Exposure to aniline dyes, tobacco smoking, cyclophosphamide, nonsteroidal anti-inflammatory drugs (NSAIDs), phenacetin, and schistosomiasis. Aniline is an organic compound ($C_6H_5NH_2$), which consists of a phenyl group attached to an amino group. This compound is used in the manufacture of precursors to polyurethane and other industrial chemicals. The International Agency for Research on Cancer (I.A.R.C.), an intergovernmental agency forming part of the World Health Organization (W.H.O.) of the United Nations lists aniline in Group 3 (not classifiable

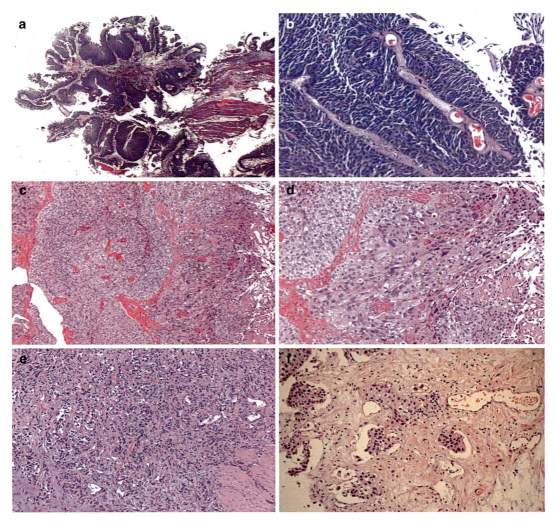

Fig. 7.8 All six microphotographs are indicative of a urothelial carcinoma. Urothelial carcinoma can arise in children and youth, although it is extremely rare (**a–f**; hematoxylin and eosin staining, ×20, ×200, ×50, ×100, ×100, ×200 as original magnifications, respectively)

as to its carcinogenicity to humans) due to the limited and contradictory data available, because the early manufacture of aniline resulted in increased incidents of bladder cancer, seem to be currently attributed to naphthylamines, not anilines. cyclophosphamide is a chemotheraputic anti-cancer agent used in oncology (leukemia, malignant lymphoma, multiple myeloma, ovarian cancer, breast cancer, small cell lung cancer, neuroblastoma, and sarcoma). Phenacetin ($C_{10}H_{13}NO_2$), a pain-relieving and fever-reducing drug, was broadly used from 1887 until 1983, when a ban was imposed by the U.S. Food and Drug Administration (F.D.A.) due to its carcinogenicity. NSAIDs include aspirin, celecoxib, diclofenac, diflunisal, etodolac, ibuprofen, indomethacin, ketoprofen, ketorolac, nabumetone, naproxen, oxaprozin, piroxicam, salsalate, sulindac, and tolmetin. Schistosomiasis ("snail fever", "bilharzia"), is a disease caused by parasitic flatworms (*Schistosoma haematobium*).

- *CGB*: LOH on 9p, involving p16, and 17p, involving p53. *TP53* is a major tumor suppressor gene very important for cells in multicellular organisms to suppress cancer. *CDKN2A* is the tumor suppressor gene responsible for the production of p16, which is an inhibitor of cyclin-dependent kinases (CDK). CDKs are the families of protein kinases that play a role in regulating the cell cycle.

- *GRO*: Origin in ¾ of cases in the trigone (*trigonum vesicae urinariae*) causing commonly partial or complete ureteral obstruction. Noninvasive forms of papillary and nonpapillary (in situ or flat lesions) frequently recur up to 2/3 of cases before showing an invasion into the *lamina propria* or more in-depth into the detrusor muscle.
- *CLM*: Malignant epithelial tumor with urothelial differentiation and grading is two-tiered (low and high grade). The Masson's trichromic stain and immunostain using an antibody against smooth muscle actin (SMA) are useful ancillary techniques to differentiate lamina propria from *muscularis propria* invasion. Both stains may highlight many muscle fibers, which are distributed throughout an extensive tumor leading to a straight diagnosis of *muscularis propria* invasion. Foci of squamous metaplasia and glandular metaplasia are also common. Urothelial carcinoma may also show a mixed differentiation (non-metaplasia!), including foci of squamous epithelial differentiation, glandular epithelial differentiation, and trophoblastic differentiation. Areas of dysplasia may be multifocal at time of the examination of the internal side of the urinary bladder harboring a urothelial carcinoma, such that a thorough study of the mucosa is paramount.
- *IHC*: (+) CK7 and CK20, (+) HMWCK and p63, and (+) CEA and CD15.

Markers altered in noninvasive forms of urothelial carcinoma include deletion of surface A, B, O, H, and Lewis, alteration of Thomsen-Friedenreich antigen, downregulation of p21 and p27, and upregulation of cyclin D1 and D3, high MIB1 (ki67) and p53, and aneuploidy.

Invasive urothelial carcinoma variants include nested, microcystic, micropapillary, lympho-epithelioma-like, lymphoma-like and plasmacytoid, sarcomatoid, clear cell (glycogen-rich), lipoid cell, undifferentiated, pleomorphic giant cell, and urothelial carcinoma with unusual stroma reactions, i.e., with pseudosarcomatous stroma and stromal osseous and cartilaginous metaplasia. The nested variant of urothelial carcinoma is particularly important to keep in mind for the differential diagnosis of von Brunn's nests and the spectrum of lesions that may derive from them. Irregularly distributed and compact clusters or "nests" of malignant cells that infiltrate through the bladder wall and may mimic von Brunn's nests characterize indeed the nested variant.

Typically, *Bacillus* Calmette-Guérin (BCG) therapy may be a standard method of choice for early cancer. Immunotherapy involving the BCG, prepared from attenuated (weakened) live bovine tuberculosis bacillus, *Mycobacterium bovis*, with no virulence in humans, may be used to treat and prevent the recurrence of superficial tumors. Although prepared to fight tuberculosis, BCG is effective in up to 2/3 of the cases at the initial stage of cancer. In randomized trials, BCG has been shown to be superior to standard chemotherapy. BCG elicits an inflammatory reaction at the local level that destroys the tumor. A radical cystectomy is a therapy option when one or more of the conditions are met. These conditions include (1) invasion of the detrusor muscle (muscularis propria) or T2, being either T2a <50% of the thickness or T2b >50% of the thickness of the muscularis propria; (2) CIS or high-grade (HG) papillary urothelial neoplasia, which are refractory to BCG therapy; and (3) CIS which is extending into prostatic urethra ± prostatic ducts.

- *PGN*: It depends on the depth of invasion, grading, typing, and multifocality. High stage (detrusor muscle invasion), LN involvement, high grade, dome or anterior wall neoplasm site, diffuse dysplasia elsewhere in the bladder, and LVI are poor outcome parameters. Young age and quick inflammatory response are good outcome factors.

7.4.5 Non-urothelial Differentiated Urinary Tract Neoplasms

Non-urothelial differentiated benign lesions are fibromas and hemangiomas (Fig. 7.9). Non-urothelial differentiated malignancy of the urinary tract accounts for about 10% of all primary tumors of the urinary bladder but is extremely rare in youth and young adults. These malignancies include squamous cell carcinoma, adenocarcinoma, small-cell carcinoma, sarcomatoid carcinoma (extremely or inexistent in youth), rhabdomyosarcoma, leiomyosarcoma (sporadic in youth), and rhabdoid tumor (Fig. 7.10), of which only the rhabdomyosarcoma and the rhabdoid tumor play a substantial role in childhood. A non-metastasizing aggressive condition that needs to be taken into account is the fibromatosis (Fig. 7.11). The rhabdomyosarcoma of the urinary tract (Figs. 7.12, 7.13, and 7.14) has a similar resemblance in all organs of the urinary tract and will be treated extensively in the prostatic gland overgrowth. The rhabdomyosarcoma of the urinary tract that most often is recognized in childhood and adolescence is the paratesticular rhabdomyosarcoma. The pure rhabdoid tumor of the urinary bladder is a very rare neoplasm (Fig. 7.10). Histologically, there are sheets of tumor cells with round or oval cells with eccentric nuclei. The nuclei have prominent nucleoli, and an eosinophilic cytoplasm is a common finding. There is a no immunohistochemical staining for SMARCB1/INI1 due to deletion and/or mutation of the *INI1* gene on chromosome 22 (22q11). This genetic change inactivates indeed the tumor suppressor gene *SMARCB1*. Extrarenal rhabdoid tumors are best managed with cystectomy, adjuvant chemotherapy, and potential additional radiotherapy. Chemotherapy drugs include combinations mostly of vincristine, cyclophosphamide, etoposide, ifosfamide with the addition of doxorubicin, actinomycin, carboplatin, and cisplatin.

7.5 Male Infertility-Associated Disorders

7.5.1 Spermiogram and Classification

Evaluation of the testicular function includes a series of investigations that start with a semen investigation and hormonal testing aiming to get FSH, LH, FAI (free-androgen index), testosterone, and prolactin. Semen delivery includes split ejaculates, pooled ejaculates, or standard spermiogram aiming to detect the main parameters of density, motility, and morphology. Other diagnostic procedures include biochemical and urine analysis; sperm examination using the DNA fragmentation index (DFI); ultrasound of the prostate and seminal glands; immunological testing, including detection of antisperm antibodies; and investigations on the interaction of sperm and cervical mucus, e.g., the postcoital test according to Sims-Huhner, or the SCMC (sperm-cervical mucus contact) contact test, and FISH sperm test. In particular, a FISH sperm test can reveal information on the genetic causes of male infertility and, at the same time, assesses the frequency of abnormalities in some chromosomes in spermatozoa. Function tests in erectile dysfunction and testicular biopsy may complete the evaluation for male infertility. The most common techniques used for sperm collection in case of an obstruction of the vas deferens, either due to vasectomy or other obstruction, include micro epididymal sperm aspiration (MESA), percutaneous epididymal sperm aspiration (PESA), and testicular sperm extraction (TESE). MESA involves the aspiration of sperm from the epididymis using a specific fine needle, TESE aims to take a small tissue fragment from the testis to isolate the sperm from the seminiferous tubule, and PESA is a technique to obtain sperm for intracytoplasmic sperm injection (ICSI). Finally, sperm chromatin structure assay (SCSA) uses the DFI to provide accurate results of sperm count. The morphologic classification of a testicular biopsy includes early maturation arrest, late maturation arrest, and tubular hyalinization.

7.6 Inflammatory Disorders of the Testis and Epididymis

Inflammatory conditions of the testis are infrequently seen due to proper vaccination against mumps. However, orchitis can be encountered by the pediatrician and needs to be differentiated from tumors and other tumor mimickers. Inflammatory or teratogenous conditions can mimic tumors and include epidermoid cysts,

7.6 Inflammatory Disorders of the Testis and Epididymis

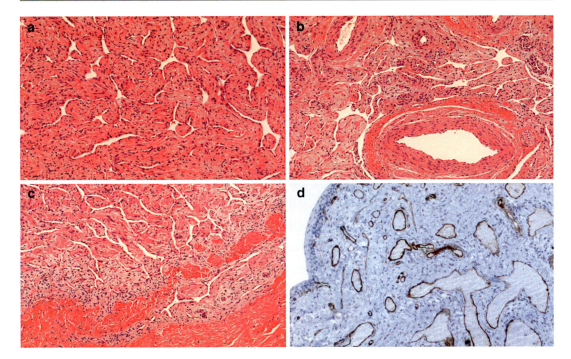

Fig. 7.9 Urinary bladder with evidence of a hemangioma with proliferating blood vessels highlighted using an antibody against endothelial cells in a child with relapsing UTI (anti-CD31 immunostain, ×100) (**a–c**; hematoxylin and eosin staining, ×100 as original magnification for all three microphotographs; **d**, anti-CD31 immunostaining, Avidin-Biotin Complex, ×100 original magnification). The hemangioma of this urinary bladder showed also positivity for GLUT1 or glucose transporter 1. GLUT1 is solute carrier family 2 and a uniporter protein that is encoded by the *SLC2A1* gene in humans. GLUT1 facilitates the transport of glucose across the plasma membranes of mammalian cells and is a marker of infantile cutaneous hemangiomas

hydrocele, spermatocele, and varicocele among others (Fig. 7.15, Fig. 7.16, Fig. 7.17, Fig. 7.18, and Fig. 7.19).

7.6.1 Acute Orchitis

Orchitis may be subdivided into acute and chronic as well as granulomatous and tuberculous orchitis mostly with epididymitis. The *vasitis nodosa* is a granulomatous condition of vas deferens that looks like spermatic granuloma of the epididymis. It is usually postvasectomy or herniorrhaphy, and occasionally it occurs associated with recanalization. Histologically, there are proliferating ductules and dilated tubules containing spermatozoa in the wall of vas deferens. There are also small bundles of hyperplastic smooth muscle, and the vas deferens may display irregular thickening. It may even resemble the *salpingitis isthmica nodosa*. An acute hemorrhagic necrosis may be due to torsion of the testis (Fig. 7.18).

7.6.2 Epidermoid Cysts

Epidermoid cysts refer to intraparenchymal cysts, which are filled with keratin and lined by squamous epithelium. They occur mostly in Caucasians and the second to fourth decades of age. It is mandatory to rule out teratoma, which unlike ovarian teratoma, can behave aggressively with possible metastases. Thus, the testis needs to be sampled thoroughly for adnexal structures. Four theories try to understand the epidermoid cysts. The first theory claims that these cysts are teratomatous in origin. They can arise from a monodermal proliferation of epidermal (ectoderm) cells. True teratomas contain at least two different embryonic germ layers, while epidermoid cysts lack both mesodermal and

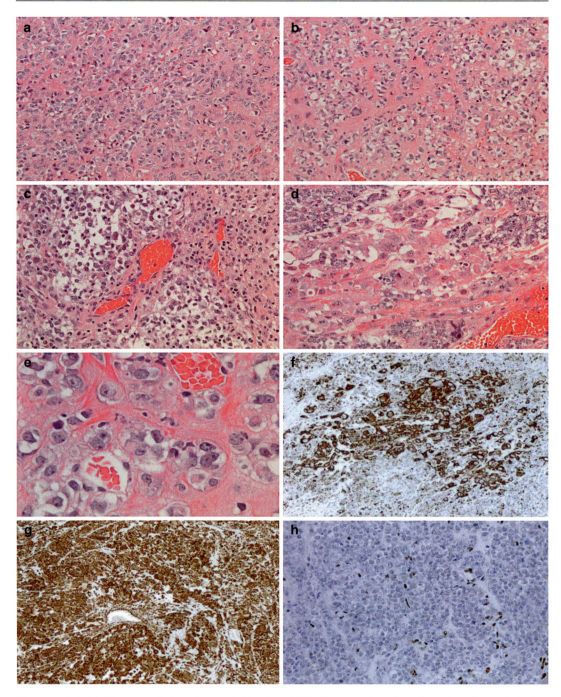

Fig. 7.10 Rhabdoid tumor of the urinary bladder is extremely rare but enters in the differential diagnosis of tumors in childhood (**a–e**; hematoxylin and eosin staining; ×200, ×200, ×200, ×200, ×630 as original magnification, respectively). The microphotographs (**a**) through (**e**) show the morphology of the tumor. The microphotographs (**f**), (**g**), and (**h**) show the expression of keratins (AE1–3), mesenchymal marker (vimentin), and the loss of INI1 (**f–h**: anti-AE1-3 immunostaining, Avidin-Biotin Complex; anti-VIM immunostaining, Avidin-Biotin-Complex; anti-INI1 immunostaining, Avidin-Biotin Complex, respectively, and ×100, ×100, ×200 as original magnification, respectively)

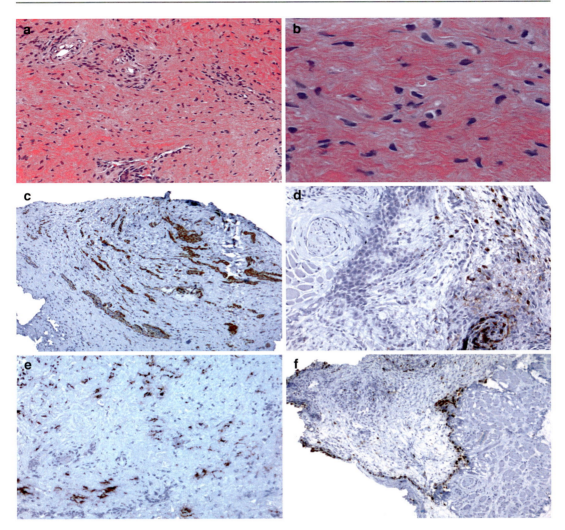

Fig. 7.11 In this panel, there is the presentation of a fibromatosis of the prostatic gland with (**a**) and (**b**) showing the characteristic morphology (**a–b**; hematoxylin and eosin staining; ×200 and ×630 original magnifcations, while (**c**) through (**f**) showing the expression of smooth muscle actin (SMA), CD3, desmin (DES), and Ki67 (MIB1), respectively (**c–f**; anti-SMA immunostaining, anti-CD3 immunostaining; anti-DES immunostaining, MIB1 immunostaining, respectively; Avidin-Biotin Complex for all immunohistochemical procedures; ×100, ×200, ×200, ×100 as original magnifications, respectively)

endodermal components. Monolayer teratomas of the ovary are the *struma ovarii* and pseudomucinous cystadenoma of the ovary. Epidermal cells are benign, do not recur, and there is no metastatic tendency. Also, epidermoid cysts do not have markers of chromosome 12p anomalies, which are often seen in germ cell tumors. Epidermoid cysts also lack an increase in alpha-fetoprotein and beta human chorionic gonadotropin. The 2nd theory targets metaplasia of epidermal cells from the rete testis secondary to obstruction, but the different sites of the cysts are incompatible with this theory. The 3rd theory uses the development of epidermoid cysts from keratinization of the rete testis as a ground to explain the epidermoid cysts. Finally, the 4th theory explains the epidermoid cysts considering the displacement of embryological derived squamous cells from the scrotal skin to the testis.

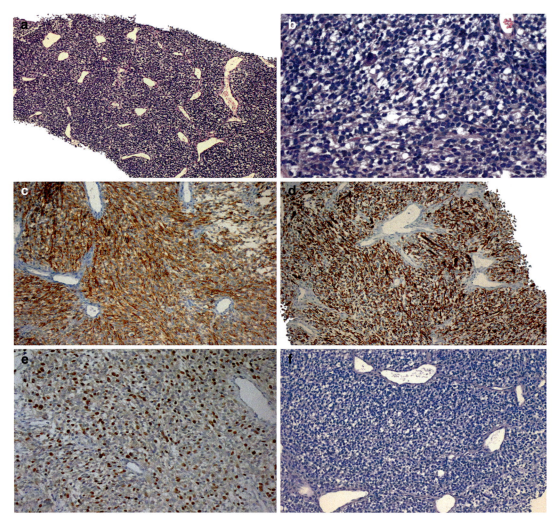

Fig. 7.12 In this panel is shown the case In this panel, is shown the case of a prostatic embryonal rhabdomyosarcoma with the microscopic morphology shown in (**a**) and (**b**) microphotographs (hematoxylin and eosin staining for both microphotographs, ×100 and ×400 as original magnification, respectively), while the immunostaining of (**c**) though (**e**) show the expression of CD56, desmin (DES), myogenin (**c–e**; anti-CD56 immunostaining, anti-DES immunostaining, and anti-myogenin immunostaining, respectively; Avidin-Biotin Complex; ×200 as original magnification for all three microphotographs). In (**f**) is shown the PAS staining of this rhabdomyosarcoma highlighting no glycogen vacuoles (×200 original magnification)

7.6.3 Hydrocele

A hydrocele (Greek: ὑδροκήλη originates from the combination of ὑδρο- 'water' and κήλη 'tumor') is a quite common benign condition and refers to an accumulation ("*tumor quia tumet*") of fluid in *tunica vaginalis*. It is a consequence of epididymitis or can follow a testicular trauma. Pathogenetically, the hydrocele is caused by fluid secreted from a remnant peritoneal piece wrapped around the *tunica vaginalis*.

7.6.4 Spermatocele

Spermatocele is a cystic dilatation of efferent ducts of the testicular parenchyma. Microscopically, the efferent ducts are lined by tall ciliated cells.

7.6 Inflammatory Disorders of the Testis and Epididymis

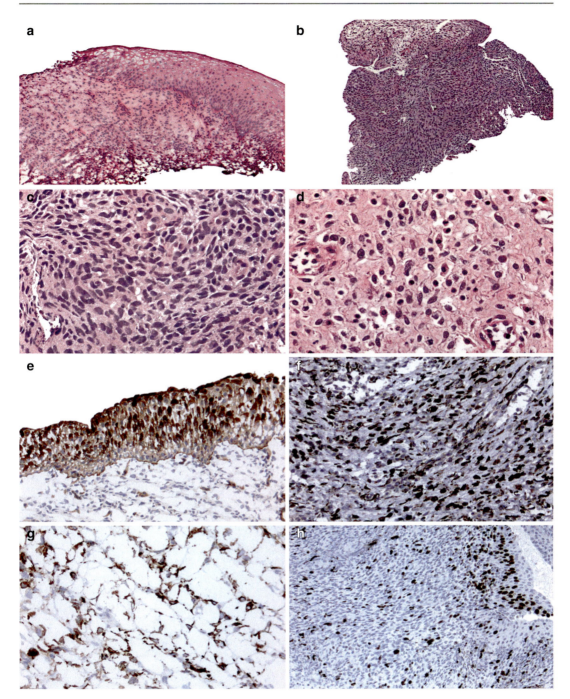

Fig. 7.13 In this panel, is shown another rhabdomyosarcoma of the prostate with the morphology shown in (**a**) through (**d**) (hematoxylin and eosin staining; ×100, ×100, ×400, and ×400 as original magnifications, respectively). The immunostains of (**e**) through (**h**) show the expression of epithelial membrane antigen (EMA) of the surface urothelium with the underlying cambium layer, the expression of desmin (DES) in the solid phase and in the cambium layer, and the Ki67 proliferation activity (MIB1), respectively (**a**, hematoxylin and eosin staining on frozen tissue as intraoperative frozen section; ×100 original magnification; **b**, hematoxylin and eosin staining, ×100 original magnification; **c–d**, hematoxylin and eosin staining, ×400 original magnification; **e**, anti-EMA immunostaining, Avidin-Biotin Complex, ×200 original magnification; **f**, anti-DES immunostaining, Avidin-Biotin Complex, ×200 original magnification; **g**, anti-DES immunostaining, Avidin-Biotin Complex, ×400 original magnification; **h**, anti-Ki67 (MIB1) immunostaining, Avidin-Biotin Complex, ×200 original magnification)

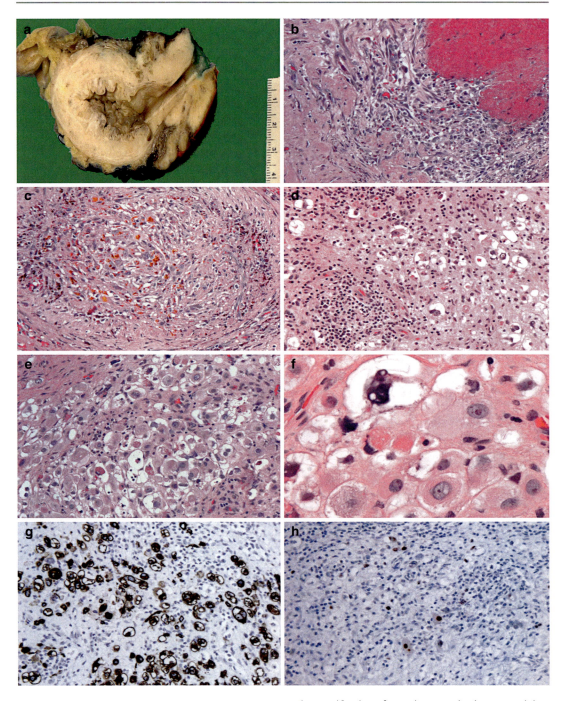

Fig. 7.14 Post-Chemotherapy changes are rarely highlighted and this gross-microscopic panel shows the changes in a rhabdomyosarcoma of the prostate gland following chemotherapy with cysto-prostatectomy (**b–f**: Hematoxylin and eosin staining, ×200, ×200, ×200, ×200, ×630 as original magnification, respectively; **g**, anti-desmin immunostaining, Avidin-Biotin Complex, ×200 original magnification; **h**, anti-myogenin immunostaining, Avidin-Biotin Complex, ×200 original magnification). Myogenin, also known as MYOG (1q32.1), is a a muscle-specific basic-helix-loop-helix transcription factor involved in the coordination of myogenesis and tissue repair. The MyoD family of transcription factors includes myogenin, MyoD, Myf5, and Mrf4

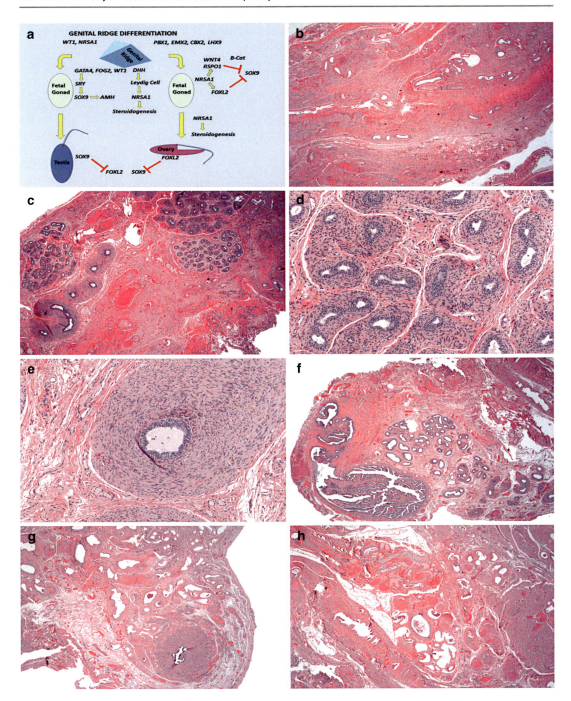

Fig. 7.15 In (**a**) is a diagram of the gonadal dysgenesis, while the microphotographs (**b**) through (**h**) show different aspects encountered in a patient undergoing excision of the gonads for gonadal dysgenesis (**b–h**: Hematoxylin and eosin staining; ×25, ×25, ×100, ×100, ×25, ×25, ×25 as original magnifications, respectively)

7.6.5 Varicocele

Varicocele is an abnormal dilatation of the venous drainage system of the testicular parenchyma in the spermatic cord with a prominent left-sided occurrence due to the different angulation of opening of the spermatic vein. Varicocele may induce decreased fertility.

7.7 Testicular Tumors

Testicular tumors, like their ovarian counterpart, can arise from the surface epithelium (*tunica vaginalis*), stroma, or germ cells. In particular, testicular neoplasms include intraepithelial or better intratubular germ cell neoplasia (ITGCN), seminomatous and non-seminomatous germ cell tumors, tumors of the specialized stroma, hematopoietic tumors, and secondary tumors. Unlike the ovary, germ cell tumors account for the most part (~ 95%) of all testicular tumors and arise from germinal epithelium, and 60% result to be of mixed type. The testicular neoplasm is the most common malignancy in males of the 3rd decade.

7.7.1 Germ Cell Tumors

Germ cell tumors derivate from the embryonal germ cell layers. The risk factors include (+) FHx, (+) history of a previous germ cell tumor, cryptorchidism, intersex syndrome, oligospermia, and infertility.

- *CLI*: "painless unilateral testicular enlargement," which may extend to the contralateral testis in up to 3% of cases with descended testis and up to 15% if bilateral testis is undescended. The clinical presentation of the contralateral testis usually follows the presentation of the first testicular enlargement, which is also termed metachronous clinical presentation.
- *EPG*: Genetically, there is an isochromosome 12 or i(12p) in many germ cell tumors, regardless of subtype. The centromere is divided on the transverse plane rather than longitudinal plane. It results in an iso-chromosome, i.e., a metacentric chromosome consisting of isologous arms that are structurally identical. Both arms contain the same genes (ἴσος, "equal").

Several fertility scores have been proposed, but the Johnsen Fertility Score (modified here) is the most appropriate for the evaluation of testis fertility to the best of my knowledge. A score of 9 means spermatogenesis complete and active with morphologically well-formed tubules. A score of 8 means seminiferous tubules containing many spermatozoa but disorganized spermatogenesis. A score of 7 means only a few spermatozoa are present. A score of 6 indicates no spermatozoa are present, but numerous spermatids. A score of 5 is given when only a few spermatids are seen. A score of 4 is given when neither spermatozoa nor spermatids are present but only many spermatocytes. A score of 3 is when only a few spermatocytes are present. A score of 2 is given when only spermatogonia are present. A score of 1 is given when no germ cells are detected, and, finally, a score of 0 means that neither germ cells nor Sertoli cells are observed.

In prepubertal testis, yolk sac tumor or endodermal sinus tumor (YST/EST) teratoma as well as Sertoli cell tumors are most often observed. Several authors report YST with ~60% and teratoma with ~40%. In postpubertal testis, seminoma and embryonal carcinomas are most often diagnosed. The lymphatic drainage of a testicular tumor may target some specific LNs with (1) periaortic and iliac (ipsilateral in 4/5 of cases) first, then (2) mediastinal and left supraclavicular LN, and finally (3) inguinal LN if the tumor has involved skin or scrotum.

Hematogenous spread: Lungs, liver, brain, and bone. The brain is a preferential site for choriocarcinoma, while skeletal colonization is seen in seminoma. The WHO biochemical evaluation of tumor markers (lactic dehydrogenase/LDH, alpha-fetoprotein/AFP, beta human chorionic gonadotropin /β-HCG) is useful for the staging of these patients.

- *TRT*: Biopsy and orchiectomy
  → If diagnostic of seminoma, then RTX of retroperitoneum (high radiosensitivity).
  → If diagnostic of NSGCT, then LN-ectomy if LVI (+) and CHT if M1a/M1b.
- *PGN*: Factors include the extension into the spermatic cord, LVI, ↑TNM stage, old age, ↑LDH, and Leydig cell hyperplasia in residual testicular tissue.

7.7.1.1 Intratubular Germ Cell Neoplasia (ITGCN)

- *DEF*: Intraepithelial neoplasia or carcinoma in situ of the seminiferous tubules of the testis (Tis). ITGCN is defined as an autonomous neoplastic proliferation of uncommitted germinal cells within the seminiferous tubules. Of note, embryonic stem cells are a type of pluripotent cell able to grow and differentiate into several and different cell types in culture, while primordial germ cells are the founder cells for the germline of the organism.
- *CLM*: Tubules with large cells harboring a centrally located nuclei with a chromatin pattern, which is evenly dispersed, and prominence of nucleoli (≥ 2) and large abundant clear to the faintly eosinophilic cytoplasm. If tumor cells fill and distend the tubules, the diagnosis of intratubular seminoma (ITS, ITGCN or GCNIS/germ cell neoplasia in situ) is appropriate.
- *HSS/IHC*: Both HSS and IHC are used to identify or confirm ITGCN exhibiting PAS (+), D-PAS(−), and (+) CD117, Oct3/4, D2–40, PLAP, as well as (+) NANOG. The same histochemical and immunohistochemical pattern can be seen in seminoma. Oct3/4 (currently relabelled as Oct4) and NANOG are two transcription factors associated with the characteristic of cellular pluripotentiality. Both markers are expressed in normal (non-tumoral) embryonic stem cells. Oct-4 (octamer-binding transcription factor 4), aka POU5F1 (POU domain, class 5, transcription factor 1) is a protein encoded by the *POU5F1* gene in humans. Oct4 participates crucially in the self-renewal of undifferentiated embryonic stem cells. NANOG is a key transcriptional factor helping embryonic stem cells to maintain pluripotency by suppressing cell determination factors. NANOG, OCT4, and SOX2 (Sex determining region Y-bOX2) function to establish the identity of the embryonic stem cells. There is a loss of expression for FHIT (fragile histidine triad protein) (3p14.2) and RNA-binding motif. *FHIT* gene, a member of the histidine triad gene family, encodes a diadenosine P1,P3-bis(5′-adenosyl)-triphosphate adenylohydrolase, which is specifically involved in purine metabolism. It is also known that FHIT acts as a tumor suppressor of HER2-driven breast carcinoma. Biologically, ITGCN can demonstrate i12p and *CKIT* (CD117) activating mutations.
- *DDX*: Normal spermatogonia, lymphoma, metastasis of carcinoma or malignant melanoma, and spermatocyte seminoma. ITGCN with microscopic invasion should be carefully ruled out.
- *ITGCN risk factors*: Cryptorchidism, gonadal dysgenesis, AIS/Androgen Insensitivity Syndrome (~ testicular feminization syndrome), (+) FHx, and prior testicular GCT. Moreover, testicular microlithiasis is an additional risk factor. Full or classical testicular microlithiasis is defined if ≥ 5 echogenic foci are found in either testes by ultrasound per view. Conversely, limited testicular microlithiasis if criteria for classic testicular microlithiasis are not met in full. In 80% of cases, both organs are affected.
- ITGCN is usually seen in 4/5 of germ cell neoplasm with invasion but is not associated with spermatocytic seminoma or pediatric teratoma and possibly yolk sac tumor. Progression to S/NSGCT is found in ~50% of ITGCN within 5 years or > 70% within 7 years post-(primary) diagnosis. ITGCN reminds another tumor pathology where the contralateral organ can be intrinsically involved such as lobular carcinoma of the breast. Two 3 mm tissue biopsies are essential to be received to evaluate contralateral testis that may be required according to international protocols (up to 95% of ipsilateral uninvolved testis with germ cell tumor and up to 5% of contralateral testis with germ cell tumor). ITGCN is identified in 1% of testicular biopsies performed for male infertility. Bouin fixation is better than formaldehyde fixation for the purpose to investigate ITGCN.
- *TRT*: Uni or bilateral orchiectomy, but RTX is also an option for bilateral cases being ITGCN radiation sensitive. Other options include "wait and see", as some authors are advocating surveillance only. ITGCN may be unclassified or separated in differentiated, with extra-tubular extension ("microinvasive"),

intratubular seminoma, intratubular embryonal carcinoma, and others.

7.7.1.2 Seminoma

- *DEF*: Tumor of neoplastic uncommitted pluripotential germinal cells and accounts for 50% of all testicular germ cell tumors (Fig. 7.20).
- *SYN*: SEMGCT (seminomatous germ cell tumor).
- *EPI*: Peculiarly, seminoma is the most likely germ cell tumor to present with a single histologic pattern and incidence rates peaks in the third decade (classic seminoma). The variant spermatocytic seminoma peaks at 65 years and is histologically and biologically different (*vide infra*).
- *CLI*: Testicular enlargement with or without pain, gynecomastia, exophthalmos, and infertility. Extratesticular locations of seminoma include retroperitoneum, mediastinum, and pineal gland.
- *GRO*: Solid, homogeneous, coarsely lobulated, fresh cream-colored or light white-yellowish ("spoiled cream") that may replace the testis entirely. Necrosis may be seen, but bleeding, frank necrosis, and cystic changes would probably suggest a non-seminomatous component. In 2% of cases, seminoma is also bilateral. It presents a gray-white mass, generally devoid of hemorrhage and necrosis. Tunica albuginea usually remains intact.
- *CLM*: Classic pattern with a compact architecture of nests of large cells separated by a delicate fibrovascular septal stroma with scattered lymphocytes. Other histologic patterns include cord-like (trabecular), sheet-like (minimal stromal trabeculae), intratubular, microcystic (edema-forming irregular microcysts or spaces surrounded by tumor cells), plasmacytoid (abundant eosinophilic cytoplasm and some atypia), and "burn-out" (rims of viable tumor cells delimitating areas of necrosis and tissue scarring following radiation therapy). On high power, tumor cells are approximately equilateral ("squared-off" cells) with copious clear to the lightly eosinophilic cytoplasm with some granular appearance, very rich in glycogen (PAS+, D-PAS-), large central nuclei with irregularly thickened nuclear membranes, granular chromatin, 1–2 prominent nucleoli, and well-delineated cell borders. In Box 7.4 the IHC markers that may be used for the testicular tumors are listed.

EMA negativity in germ cell tumors, granulosa cell tumors, as well as Sertoli-Leydig cell tumors needs to be kept in mind. This situation is particularly crucial when dealing with metastases from an unknown primary. Germ cell tumors, granulosa cell tumors, as well as Sertoli-Leydig cell tumors are positive for pan-keratin but do not express EMA. Seminoma is only weakly and focally positive for cytokeratin markers. CD143 is (+) in seminoma but (−) in spermatocytic seminoma and non-seminomatous germ cell neoplasms. Syncytiotrophoblastic cells are Langhans-type cells aka mulberry cells. The application of Masson trichromic stain + PLAP is particularly useful in case of seminoma with "burn-out" histologic pattern because the tumor cells can stand out in the necrotic areas of the neoplasm and the areas of hyalinized fibrotic or calcified tissue or tissue with calcification foci. Naturally, coarse intratubular calcifications may correspond to regression of intratubular embryonal carcinoma, and the large-cell calcifying Sertoli cell tumor needs indeed to be into account in case a similar pattern is recognized. In this setting, other histologic patterns are also seen, and the clinical and morphologic background is different in these latter two tumors. The seminoma variants are seminoma with syncytiotrophoblastic cells, seminoma, anaplastic type, and seminoma, spermatocytic type.

Seminoma with Syncytiotrophoblastic Cells
Multinucleated giant cells may be seen singly or in clusters (often perivascular) and can correspond to focal hemorrhagic foci that may be observed on gross examination. This variant of seminoma with trophoblastic cells identified as syncytiotrophoblastic cells (5–25% of cases) is NOT choriocarcinoma, because cytotrophoblasts or cytotrophoblastic epithelium is lacking. This is crucial and cannot be stressed enough. Syncytiotrophoblastic cells are pan-keratin (AE1/AE3+) and β-HCG+. Some authors have considered this kind of seminoma as a slightly more

Box 7.4 IHC Panel (Mnemonic Word: "COD-PLAP" or Code to Play) (See Text)

Marker	Name and structure	Function	(+)/(−)
CD117 (c-kit)	Type III receptor tyrosine kinase	It operates in cell signal transduction in several cell types	(+)
OCT4 (POU5F1)	Octamer-binding transcription factor 3/4, POU domain, class 5, encoded by the POU5F1 gene	Homeodomain transcription factor of the POU family, involved in the self-renewal of embryonic stem cells	(+)
D2-40	Monoclonal antibody to a 40,000 O-linked sialoglycoprotein	The sialoglycoprotein reacts with a fixation-resistant epitope in lymphatic endothelium	(+)
PLAP	An allosteric enzyme encoded by the ALPP gene (human)	Membrane-bound glycosylated dimeric protein expressed mainly in the placenta	(+)
SALL4	Stem cell marker (Sal-like protein 4) and oncofetal protein like α-fetoprotein and glypicans, which are heparan sulfate proteoglycans bound to the outer cell membrane using a glycosyl-phosphatidylinositol anchor.	Zinc-finger transcriptional factor important for embryonic development, at 20q13	(+)
VIM	Structural protein that is encoded by the VIM gene	Type III intermediate filament protein that is expressed in mesenchymal cells of all metazoan and bacteria	(+)
CD143	Type I, single chain transmembranal protein (metallopeptidase), whose cofactor is zinc	The primary targets are angiotensin I and bradykinin, acting as a blood pressure regulator	(+)
EMA	Large cell surface mucin glycoprotein expressed by glandular and ductal epithelial cells (mostly) and some hematopoietic cells	The barrier to the apical surface of epithelial cells with a protective and regulatory role	(−)
SOX2	SRY (sex-determining region Y)-box 2 is a transcription factor	It is essential for maintaining self-renewal, or pluripotency, of undifferentiated embryonic stem cells	(−) [S] (+) [NS]

Notes: S Seminomatous, NS Non-Seminomatous

aggressive variant of the classic seminoma without syncytiotrophoblastic cells, but there controversial opinion on this topic. Serum hCG and NSE are useful tumor markers following orchiectomy. Metastases of choriocarcinoma needs always to be taken into consideration. In the adjacent testicular parenchyma, there is tubular atrophy and ± ITGCN.

Anaplastic Seminoma

The variant called "anaplastic seminoma" refers to a seminoma with brisk mitotic activity (MI > 3/HPF) in addition to some marked pleomorphism of the tumor cells. Anaplasia (Greek: ἀνά "backward" and πλάσις "formation") is a biological condition of cells exhibiting poor cytological differentiation. In Wilms' tumor, anaplasia is much better defined. Criteria include nuclei with a diameter three times that of adjacent cells, hyperchromasia due to increased chromatin, and multipolar mitoses.

Spermatocytic Seminoma

It accounts for up to 1/20 of all seminomas and is probably unrelated to classic seminoma on clinical, morphologic, and cytogenetic grounds, i.e., no association with teratomatous component, no extragonadal involvement, no relationship with cryptorchidism, and no ITGCN. Microscopically, three cells are seen including small cells or lymphocyte-like, medium cells or 15–20 microns in size and large cells or 50–100 microns in size. No specific and straightforward immunocytological marker for spermatocytic seminoma is routinely used, but genes and antigens that are highly expressed in spermatogonia are usually detected in spermatocytic seminoma. They are aka "cancer-testis antigens" and are highly expressed

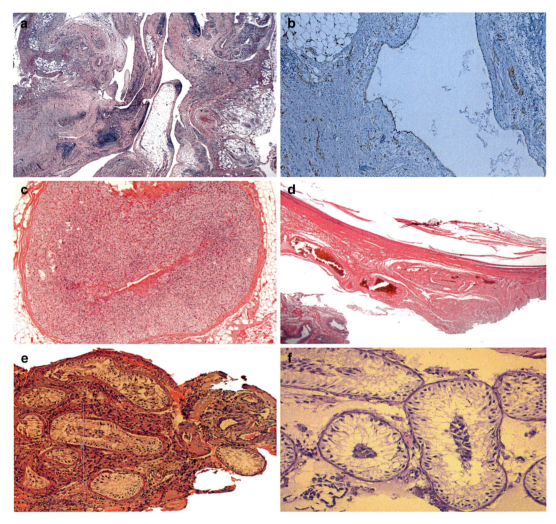

Fig. 7.16 In (**a**) and (**b**) are microphotographs of a paratesticular lymphangioma (**a**, hematoxylin and eosin staining, ×40 original magnification; **b**, anti-D2-40 or podoplanin immunostaining, ×100 original magnification), wnile (**c**) shows an ectopic adrenal gland tissue in the testis. The microphotograph (**d**) shows an epidermoid cyst of the testis (**c**, hematoxylin and eosin staining, ×50 original magnification; **d**, hematoxylin and eosin staining, ×12.5 original magnification). The microphotographs in (**e**) and (**f**) show the characteristic histology of Sertoli Cell Only Syndrome of the testis (**e**, hematoxylin and eosin staining, ×100 original magnification; **f**, hematoxylin and eosin staining, ×200 original magnification)

in spermatocytic seminomas both at the transcript and protein level. Cancer-testis antigens include SSX1, SSX-2, SSX3, SSX4 (Stoop et al. 2001; Lim et al. 2011), MAGE-A4 (Aubry et al. 2001), NY-ESO-1/CTAG1A (Satie et al. 2002), GAGE4 (Looijenga et al. 2006), SAGE1 (Looijenga et al. 2006; Lim et al. 2011), SYCP1, NSE (neuron-specific enolase), CHK2, VASA, DMRT1 (Looijenga et al. 2006), FGFR3, or HRAS (Goriely et al. 2009; Waheeb and Hofmann 2011). The above-described markers are all proteins that are expressed in spermatocytes and spermatogonia and are also most often expressed in spermatocytic seminoma. Particularly useful may be ChK2, MAGE-A4, NSE, SSX, and NY-ESO-1 according to the most recent literature. Newly, a subset of spermatocytic seminomas that express OCT2, a marker of stem spermatogonia in the testis, was also detected (Lim et al. 2011). Cytogenetically, diploid and polyploid cells, but no haploid values, have been observed in seminoma of spermatocytic

7.7 Testicular Tumors

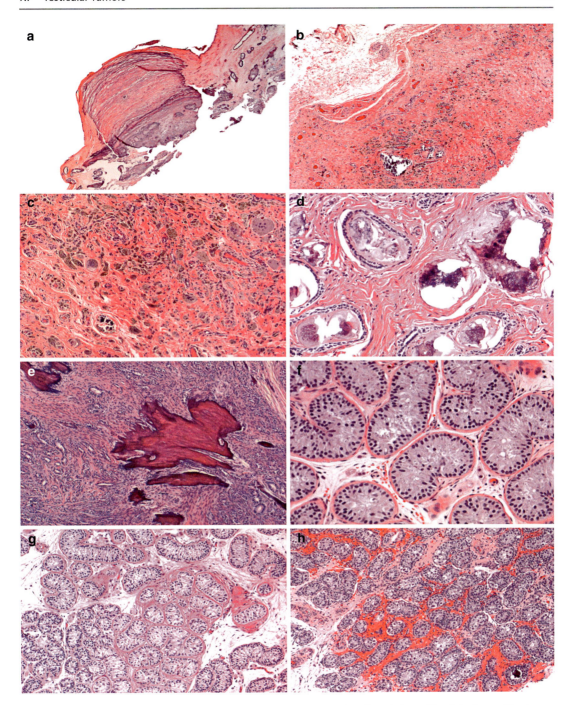

Fig. 7.17 In (**a**) is a calcified Hydatis Morgagni, in (**b**) and (**c**) are microphotographs of a testicular regression syndrome. In (**d**) is a calcified rete testis, while **e**) shows an early membranous ossification of the testis without evidence of a teratoma. The microphotographs show the histology identified in patients undergoing orchiectomy folowing the diagnosis of cryptorchidism (**a**, hematoxylin and eosin staining, ×50 original magnification; **b**, hematoxylin and eosin staining, ×50 original magnification; **c**, hematoxylin and eosin staining, ×100 original magnification; **d**, hematoxylin and eosin staining, ×200 original magnification; **e**, hematoxylin and eosin staining, ×100 original magnification; **f**, hematoxylin and eosin staining, ×200 original magnification; **g**, hematoxylin and eosin staining, ×100 original magnification; **h**, hematoxylin and eosin staining, ×100 original magnification)

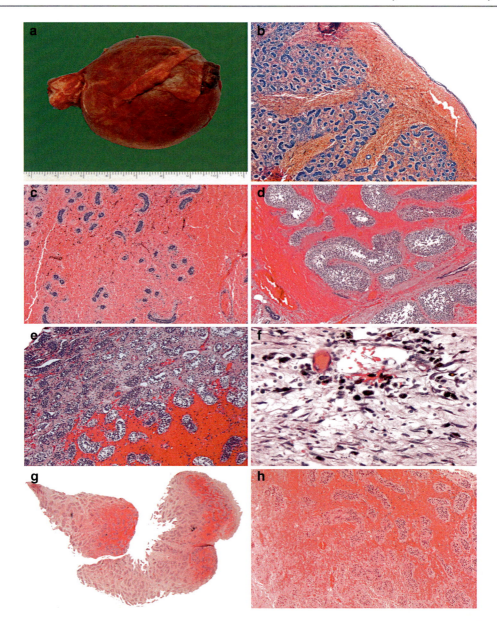

Fig. 7.18 In (**a**) through (**h**) are shown a gross photograph of orchitis (**a**) and orchitis with hemorrhagic necrosis and dystrophic calcification (In **a**) through **h**) are shown a gross photograph of orchitis and some aspects of orchitis under the lens with varying degree of destruction of testicular parenchyma (**b**, hematoxylin and eosin staining, ×40 original magnification; **c**, hematoxylin and eosin staining, ×100 original magnification; **d**, hematoxylin and eosin staining, ×50 original magnification; **e**, hematoxylin and eosin staining, ×100 original magnification; **f**, hematoxylin and eosin staining, ×400 original magnification; **g**, hematoxylin and eosin, ×12.5 original magnification; **h**, hematoxylin and eosin staining, ×100 original magnification)

type. Comparative genomic hybridization (CGH), a molecular cytogenetic method to detect unbalanced chromosomal abnormalities, shows a gain of chromosome 9, and some gains of chromosomes 1 and 20, as well as loss of chromosome 22 (Rosenberg et al. 1998). The specific gain of chromosome 9, notably 9p, suggests the involvement of DMRT1 or doublesex and mab-3 related transcription factor 1 (Looijenga et al. 2006). DMRT1 is found in a group comprising two other members of the gene family, which harbor in common a zinc finger-like DNA-binding motif.

The DNA motif domain is a conserved component of the vertebrate sex-determining pathway, which is also a key regulator of development (male) in nematodes and flies. Genetically, one consistent amplification in 9p was also reported (Looijenga et al. 2006) and uses parallel sequencing activating somatic mutations in two genes, which are physiologically connected to each other: *FGFR3* (1948A > G) and *HRAS* (181C > A and 182A > G, QK650E) (Goriely et al. 2009). The occurrence of these somatic mutations harbors an oncogenic value having been linked to other cancers such as urinary bladder malignancy. If these mutations are transmitted as germline mutations, severe/lethal skeletal osteochondrodysplasia occur such as *thanatophoric dysplasia* or *Costello syndrome*. Milder ligand-dependent mutations in *FGFR3* can give origin to achondroplasia. Spermatocytic seminoma is a definitive diagnosis of middle-aged adults, but the study of this neoplasm is particularly crucial for the relationship with genes expressed during spermatogenesis that can speed up research in pediatric urology. The differential diagnosis of seminoma is crucial for any treatment option (Box 7.5).

> **Box 7.5 Outstanding Practical Differential Diagnosis of Seminoma**
> 1. *YST*: (−) Lymphocytes, flatlining, AE1/3+, AFP+, Schiller-Duval bodies
> 2. *SCT*: Elongated cell islands with palisading nuclei at the periphery, (+) "VINCAP."
> 3. *ECA*: Glands/papillae, nuclear pleomorphism, crowding and irregularity, cytoplasm hyperdensification, indistinct cell membranes, (+) CK, CD30, OCT4, D2–40, (−) PLAP
> 4. *SSE*: Three-cell type tumor, (−) ITGCN, (−) OCT4, (−) PLAP, (+) CD117
> 5. *DLBCT*: Interstitial/intertubular NOT intratubular involvement, (−) ITGCN, (+) CD45, CD20, CD79a, PAX5, and (−) CD117, OCT4, D2–4, PLAP
>
> Notes: *YST* yolk sac tumor, *SCT* Sertoli cell tumor, *ECA* embryonal carcinoma, *SSE* spermatocytic seminoma, *DLBCL* Diffuse large B-cell lymphoma

7.7.1.3 Embryonal Carcinoma (Testicular EC or TEC)

- *DEF*: CD30 expressing non-seminomatous germ cell tumor (pure or mixed with other tumor components) with early tendency to metastasize and harboring a worse prognosis than seminoma (Fig. 7.25).
- *GRO*: TEC is heterogeneous with a solid gray appearance and multiple foci of hemorrhage and necrosis.
- *CLM*: Embryonal carcinoma may be quite polymorphic showing solid (sheets), glandular, tubular, and papillary areas or alveolar nests of undifferentiated cells with vesicular nuclei, prominent nucleoli, pleomorphism, and high MI (i.e., epithelioid features of carcinoma).
- *IHC*: (+) CK & CD30, OCT4, D2–40, and PLAP as well as (+) SALL4 and SOX2, but (−) CD117 and EMA. TEC occurs most often mixed with other histologic types, being pure in ~5% of the cases only. Thus, thorough sampling is paramount. When pure, AFP is negative.

A note should be made for CD30 in metastatic disease after chemotherapy, because this marker, especially if investigated in LNs or other possible metastatic sites, is frequently lost giving false-negative results that may influence the outcome of the patients. Chemotherapy may indeed alter the immunological phenotype. The characteristic histologic appearance may be retained forming the basis to support and emphasize the importance of H&E slides. PLAP is always a marker that needs to be ordered in investigating a testicular tumor, because, in the DDX between TEC and immunoblastic lymphoma/Ki-1 lymphoma, PLAP is determining.

- *PGN*: The cure rate for patients with TEC is excellent if the proper chemotherapeutic regimens are employed with a careful follow-up evaluation post-therapy, including radiographic evaluation and serial AFP and β-hCG determinations.

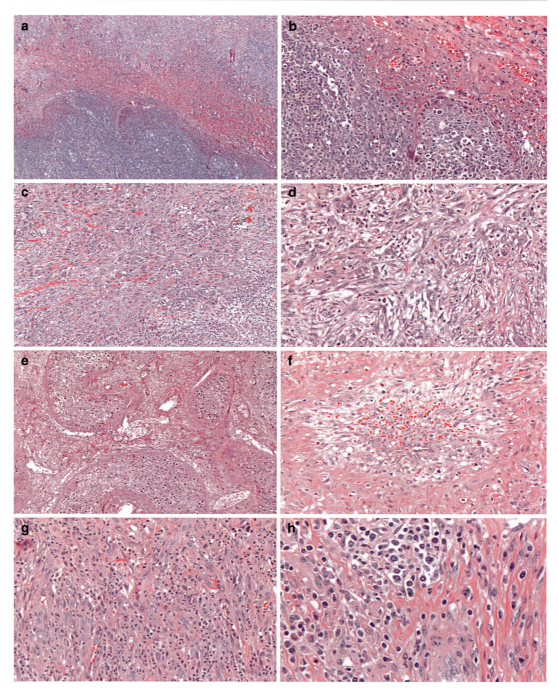

Fig. 7.19 The microphotographs (**a**) through (**h**) show an exaggerated myofibroblastic proliferation in a patient with orchi-epididymitis. Alk1 immunostain, quite specific for inflammatory myofibroblastic tumor, was negative (not shown) (**a–h**, hematoxylin and eosin staining; **a**, ×200; **b**, ×200; **c**, ×100; **d**, ×200; **e**, ×200; **f**, ×200; **g**, ×200; **h**, ×400, as original magnifications, respectively)

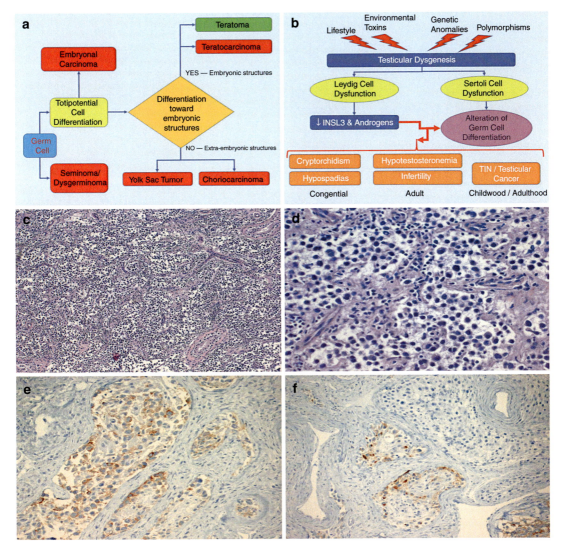

Fig. 7.20 In (**a**) and (**b**) are schemas for germ cell tumor differentiation, while (**c**) through (**f**) show seminoma and testicular intraepithelial neoplasia (germ cell neoplasia in situ or GCNIS) microphotographs (**c–d**, hematoxylin and eosin staining, **c**, ×100; **d**, ×400; **e–f**, Avidin-Biotin Complex anti-PLAP or placental alkaline phosphatase immunostaining, ×200, as original magnifications). Positive markers othern than PLAP include OCT3/4, SALL4, D2-40 (podoplanin), NANOG, SOX17, and CD117. Of note, PLAP is positive in infantile germ cells until age one and CD117 is displayed in spermatogonia as patchy membranous or cytoplasmic staining. Both PLAP and CD117 can constitute common pitfalls

7.7.1.4 Teratoma

Testicular teratomas are biologically different from the mature cystic teratoma of the ovary or dermoid cyst, being the first with low malignant potential and the latter thoroughly benign. On the other hand, it is well known that teratomas occurring in prepubertal boys are invariably benign, regardless of the composition, whereas those occurring after puberty possess a malignant potential, even when they look like histologically mature and benign Fig. 7.21 and Fig. 7.22).

- *GRO*: Heterogeneous tumors with multiloculated cysts and cartilage. Teratoma is indeed a different tumor possibly presenting as monodermal or pluridermal form. The monodermal forms are subdivided in "carcinoid" and "Primitive Neuroectodermal Tumor (PNET)."

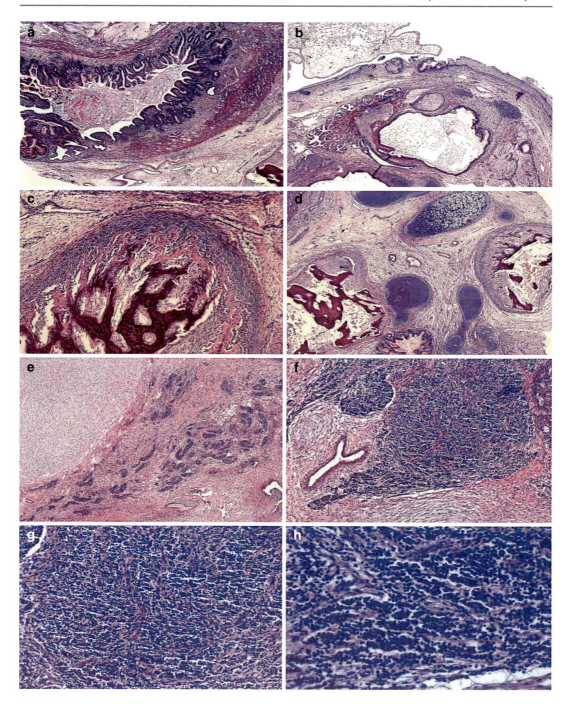

Fig. 7.21 Teratoma of the testis with the presence of all three germ cell layers (ectoderm, mesoderm, endoderm) as well as primitive neuroectodermal tissue portend to a diagnosis of a neuroblastoma, poorly differentiated (hematoxylin and eosin staining; **a**, ×40; **b**, ×40; **c**, ×40; **d**, ×40; **e**, ×40; **f**, ×100; **g**, ×200; **h**, ×400, as original magnifications, respectively)

7.7 Testicular Tumors

Box 7.6 IHC Staining Panel for Granulosa Cell Tumors

INA	Inhibin A	+c
CALR	Calretinin	+n/c
LMW-CK (CAM 5.2)	Low molecular weight keratins	+[a]
EMA	Epithelial membrane antigen	–
LCA	Leucocyte common antigen (CD45)	–
CGA	Chromogranin A	–

Notes: [a]The LMW-CK staining of GrCT is dot-like or globoid perinuclear staining

Box 7.7 IHC Staining Panel for SLCT (VINCAP)

VIM	Vimentin
INA	Inhibin A
WT1	Nephroblastoma Ag -> WT1
CALR	Calretinin and CKs
CD99	PNET Ag -> CD99
EMA	Epithelial membrane antigen
LCA	Leucocyte common antigen (CD45)
CGA	Chromogranin A

The pluridermal forms contain elements of >1 germ cell layer (ectoderm, mesoderm, endoderm).

- *CLM*: Mature and immature forms of teratomas are identified. Immature tissue involves the presence of immature neural tissue, i.e., neuroepithelium. Practically, PNET has all immature neural tissue. Various foci resembling different forms of CNS tumors, such as medulloepithelioma, ependymal neoplasms, or neuroblastoma, may be encountered considering that the malignant component of *monodermal germinal* origin may have several histological mimickers in the CNS. Cancerous part of *germinal* origin may constitute mixed tumors with teratoma and embryonal carcinoma, YST, seminoma, and choriocarcinoma. The combination of embryonal carcinoma and teratoma is designated as a mixed tumor of teratocarcinoma type. YST microfoci can also occur. The identification of microfoci of YST may be based on Heifetz et al.'s data. Recently, microfoci have been reported when one or more microfoci were seen, each occupying not more than two adjacent HPF (40X). If YST foci bigger than two adjacent HPF are found, the tumor should be classified as a malignant teratoma. The presence of a malignancy, e.g., adenocarcinoma, arises in a mature component labeled cancer as teratoma with malignant transformation (a lethal part of *somatic* origin). Finally, teratomas in children have no ITGCN, while teratomas in adults may contain ITGCN. The grading used in immature teratoma of the ovary may be applicable to teratomas of the testis, although there are some controversial issues in adopting this ovarian grading system according to O'Connor and Norris (immature tissue <1 LPF (x 4)/slide = grade 1, immature tissue 1–3 LPF/slide = grade 2, and immature tissue >3 LPF/slide = grade 3).

7.7.1.5 Yolk Sac Tumor (YST)/ Endodermal Sinus Tumor (EST)

- *DEF*: Non-seminomatous germ cell tumor recognized as a monomorphic teratoma mimicking embryonal yolk sac (Fig. 7.23).
- *GRO*: YST is homogeneous, soft, microcystic, yellow-white, and mucinous.
- *CLM*: Various patterns can be encountered explaining the organoid intermingling of epithelial and mesenchymal elements that are frequently observed. The presence of multiple histologic patterns is, on the other hand, a considerable aid in diagnosis. The histologic patterns are microcystic, endodermal sinus, solid, pseudoglandular, macrocystic, trabecular, alveolar, myxoid, sarcomatoid, polyvesicular-vitelline, hepatoid, and parietal. Two additional characteristics are of support in diagnosis, i.e. (non-glycogen-harboring, diastase resistant) intracellular hyaline globules (PAS+, D-PAS+) and Schiller-Duval bodies, which are constituted by a central vessel intimately rimmed by loose connective tissue that in turn is outlined by dysplastic epithelium (epithelial cells with atypia and abnormal

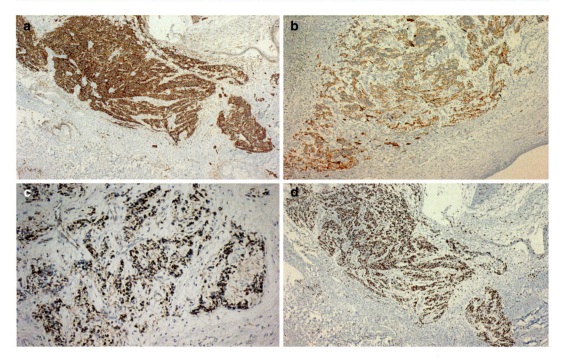

Fig. 7.22 The neuroblastoma identified in the previous teratoma panel of the testis shows positivity for CD56, synaptophysin, neuroblastoma marker, and Ki67 in (**a**), (**b**), (**c**), and (**d**), respectively. S100 was focal positive (not shown), while chromogranin A was quite diffusely positive (not shown)

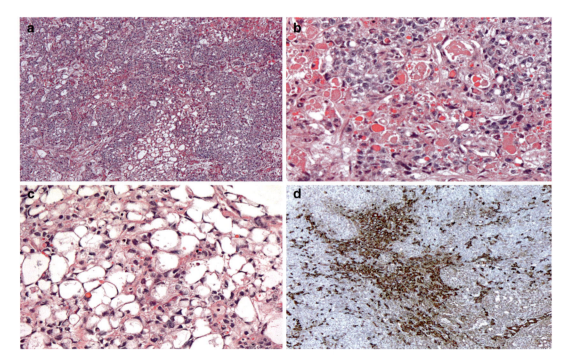

Fig. 7.23 The testis may also show yolk sac tumor (**a**, **b**, **c**) with alpha-fetoprotein positivity (**d**) (**a–c**, hematoxylin and eosin staining; **d**, anti-alpha-feto-protein Avidin-Biotin Complex immunostaining; **a**, ×50; **b**, ×200; **c**, ×200; **d**, ×50, as original magnifications, respectively)

architecture), all within a cystic space. Schiller-Duval bodies are seen in endodermal sinus pattern. Pink extracellular bands of eosinophilic BM material may also be encountered between the tumor cells.

- *IHC*: (+) CK, AFP, GPC-3, SALL4, and PLAP. GPC3 (Glypican-3) is a protein encoded by the *GPC3* gene (Xq26) in humans. *GPC3* gene mutations have been identified in Simpson-Golabi-Behmel syndrome, a multiple congenitaly anomaly syndrome characterized by craniofacial, skeletal, cardiac, and renal abnormalities.
- *PGN*: Generally, the prognosis for YST of the testis is good in infants but poor in adolescents and adults. It and depends on the size of the tumor, the stage of the tumor, the overall health of the patient, the location in the testes of the tumor, and the number of tumor masses/nodules present within the testes.

7.7.1.6 Polyembryoma

Polyembryoma is a very rare, probably quite sporadic non-seminomatous germ cell tumor showing multiple embryoid bodies, i.e., an amnion-like cavity with embryo-like cellular invagination, which overlies a yolk sac-like structure. It is seen as a mixed tumor with YST or TEC. Diffuse embryoma is termed a mixed tumor constituted by embryonal, yolk sac, and trophoblastic elements in the systematic arrangement.

7.7.1.7 Choriocarcinoma

- *DEF*: Non-seminomatous germ cell tumor accounting for 5% of all testicular neoplasms with syncytio- and cytotrophoblastic differentiation (Fig. 7.24).
- *GRO*: A small, hemorrhagic, and often partially necrotic tumor (tumor-related necrosis).
- *CLM*: Syncytiotrophoblast AND cytotrophoblasts are seen in choriocarcinoma. Both components need to be recognized. It is a biphasic, plexiform pattern with giant syncytial trophoblasts with large atypical nuclei that is intermixed with cytotrophoblasts. If the primary tumor regresses, there is necrosis and hemosiderin pigments, because of the extensive hemorrhagic component present in this tumor.

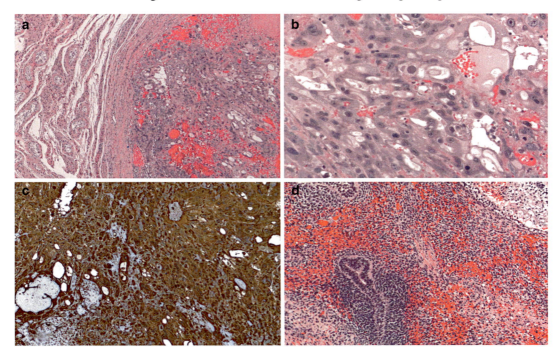

Fig. 7.24 Testis with choriocarcinoma, pure and mixed with other teratoma portions. In (**c**), HCG immunostain confirms the diagnosis of choriocarcinoma (**a–b** and **d–f**, hematoxylin and eosin staining; **c**, Avidin-Biotin Complex immunostaining; **a**, ×50, **b**, 200; **c**, ×100; **d–f**, ×100, as original magnifications, respectively)

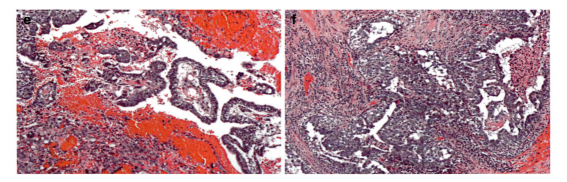

Fig. 7.24 (continued)

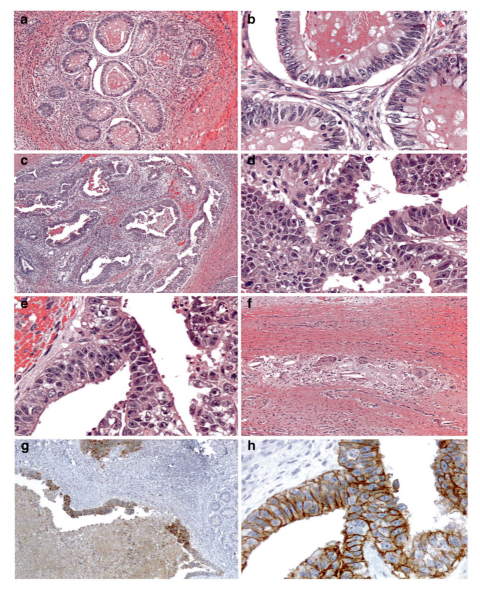

Fig. 7.25 In this panel is shown a mixed embryonal carcinoma and teratoma (**a–h**), and CD30 immunostaining characteristic for embryonal carcinoma is shown in (**g**) and (**h**) (**a–f**, hematoxylin and eosin staining; **g–h**, Avidin-Biotin Complex immunostaining; **a**, ×50; **b**, 400; **c**, ×50; **d**, ×400; **e**, ×400; **f**, ×100; **g**, ×50; **h**, ×400, as original magnifications)

7.7 Testicular Tumors

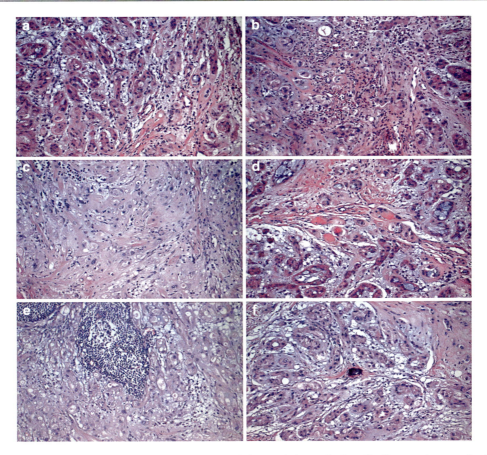

Fig. 7.26 The panel presented here shows the characteristic morphology of a Sertoli cell tumor (see text for detail) (hematoxylin and eosin staining, **a–f**, ×200, as original magnifications)

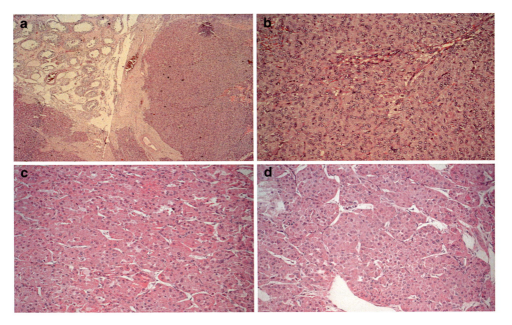

Fig. 7.27 In this panel, a Leydig cell tumor is shown with the characteristic architecture and cell morphology (see text for details) (hematoxylin and eosin staining; **a**, ×40; **b–d**, 200, as original magnifications)

- *HSS/IHC*: hCG (+) in syncytial cells; PPB (±) due to both iron deposition and the degree of regressive changes.
- *PGN*: The outcome is unfortunate because it is often disseminated at presentation.

7.7.2 Tumors of Specialized Gonadal Stroma

Sex cord-stromal tumors include Leydig cell tumor, Sertoli cell tumor, classic type and sclerosing variant, as well as large-cell calcifying Sertoli cell tumor and granulosa cell tumor of adult and juvenile type.

7.7.2.1 Leydig Cell Tumor (LeCT)
- *DEF*: Tumor of the sex cord-stromal tumor group of testicular cancers arising from Leydig cells.
- *EPI*: Leydig cell tumor (LeCT) accounts for up to 1/20 of testicular neoplasms. Epidemiological studies have revealed that this tumor in 4/5 of cases occurs in adults, while 1/5 of the patients in children (Fig. 7.27).
- *CLI*: Testicular mass with or without gynecomastia in adults, while children harboring LeCTs can manifest with endocrine symptoms, including virilization and gynecomastia.
- *GRO*: Small (~ 1–3 cm), solid, sharply demarcated, lobulated, yellow-tan, relatively soft, and mostly single nodule.
- *CLM*: The histologic patterns include solid (or sheets), trabecular (or cords), and pseudoglandular (or acinar) growth patterns of tumor cells, which are medium-sized cells with well-defined cell borders, round to oval nucleus with single prominent nucleolus, and abundant eosinophilic or clear acidophilic cytoplasm. Intervening stroma may be either

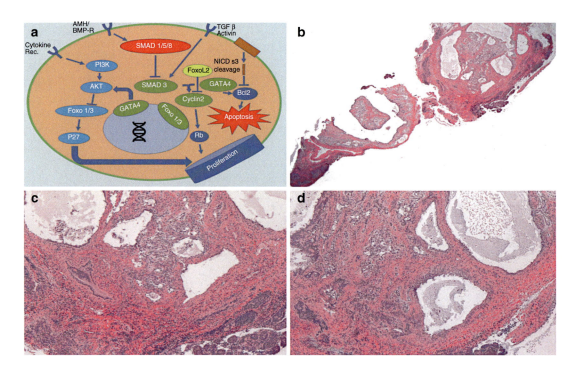

Fig. 7.28 In this panel is depicted the molecular pathogenesis of juvenile granulosa cell tumor (**a**) and shown the characteristic morphology of a juvenile granulosa cell tumor with macrocysts of the testis and granulosa cells (hematoxylin and eosin staining; **b**, ×12.5; **c**, ×50; **d**, ×50; **e**, ×50; **f**, ×50; **g**, ×200; **h**, ×200, as original magnifications)

7.7 Testicular Tumors

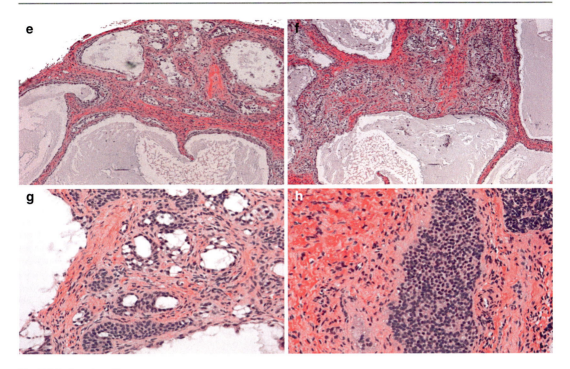

Fig. 7.28 (continued)

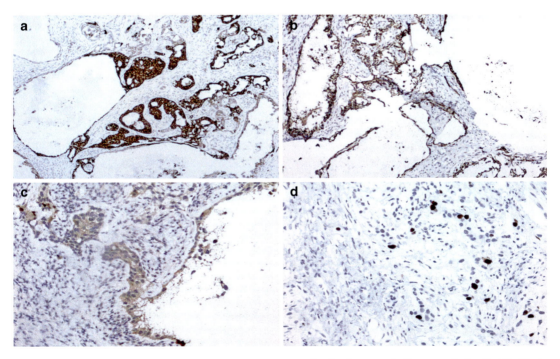

Fig. 7.29 The immunohistochemistry of the previous JGCT shows a positivity for inhibin A, CAM5.2 (low MW cytokeratins), calretinin, and the Ki67 proliferation antigen in (**a**, **b**, **c**, and **d**), respectively (Avidin-Biotin Complex immunostaining; **a**, ×100; **b**, ×100; **c**, ×200; **d**, ×200, as original magnifications)

edematous or fibrous. TEM shows mitochondria with tubular cristae, SER, and Reinke crystalloids (moderately electron dense, elongated structures with tapered ends) as well as some lipochromes. Lipochrome (Greek λίπος "fat" and χρῶμα "color") is a naturally occurring fat-soluble pigment, one of the which most known is lipofuscin, which is a fat-byproduct associated with the decomposition of cellular membranes.
- *DDX*: Nodular Leydig cell hyperplasia (multiple, each Ø < 0.5 cm, (+) cryptorchidism), testicular tumor of adrenogenital syndrome (bilateral, dark-brown, prominent fibrous intervening stroma, (−) Reinke crystalloids, and (±) INA), and Nelson syndrome (ACTH-related pituitary adenoma, multinodular testicular tumor, no effacement of the seminiferous tubules, and (−) Reinke crystalloids).
- *PGN*: Epidemiological and oncological indicate that 10% of LeCTs in adults are malignant, while all pediatric cases are usually benign. Destructive features include large size (>5 cm), necrosis, invasion of surrounding structures, MI (>5/10HPF), and LVI (+).

7.7.2.2 Sertoli Cell Tumors, Pseudotumors, and Related Lesions
- *DEF*: Sertoli cell tumors (SeCT) account for up to 1% of testicular neoplasms and in ~85% of cases occur in adults, while ~15% of the patients are typically children. Generally clinical presentation is characterized by a testicular mass with gynecomastia, which is present in 1/3 of cases, and associated with accompanying impotence. Both symptoms are due to estrogen production by the tumor cells (Fig. 7.26).
- *GRO*: SeCT are well-delimitated solid, pale yellow, or gray-yellow nodule.
- *CLM*: Solid (or sheets), trabecular (or cords), and pseudoglandular (or acinar) growth patterns of tumor cells, although most of the cases show "solid and hollow tubules" containing Sertoli cells in the paucicellular stroma. Tumor cells have a polygonal nucleus and clear to pale to eosinophilic cytoplasm.
- *TEM*: Studies evidentiate Charcot-Buettner filaments, which are perinuclear aggregates of intermediate filaments.
- *IHC*: The immunohistochemistry of the granulomas cell tumors is shown in Boxes 7.6 and 7.7.
- *PGN*: Studies indicate that 10% of SeCTs in adults are malignant, while all pediatric cases are usually benign. Similar to LeCTs destructive features include large size (>5 cm), necrosis, invasion of surrounding structures, MI (>5/10HPF), and LVI (+).
- *DDX*: Seminoma, androgen insensitivity syndrome (aka testicular feminization), Sertoli cell nodules, Sertoli cell hyperplasia (hypoplastic tubules with Sertoli cell prominence), Sertoli cell adenoma (elongated tubules lined by Sertoli-like cells and centralized BM deposits), and sex cord tumor with annular tubules (Sertoli cell adenoma-related and male counterpart of the ovarian SCTAT, usually seen only in patients with Peutz-Jeghers syndrome).

Two variants are significant to mention, because they may occur in children and adolescents and include sclerosing Sertoli cell tumor, which is characterized by solid and hollow tubules embedded in a hypo-/acellular abundant sclerotic stroma) and large-cell calcifying Sertoli cell tumor (LCCSCT), which shows sheets and cords of tumor cells with copious cytoplasm separated by abundant fibrous intervening stroma with calcifications. LCCSCT is seen in patients with Peutz-Jeghers syndrome, Carney Complex, adrenal cortical hyperplasia, Leydig cell tumors (Fig. 7.27), and pituitary tumors. Carney complex is AD-inherited genetic disorder due to mutations of *PRKAR1A* (Protein Kinase c-AMP-Dependent Type I Regulatory Subunit Alpha) on 17q23–24, which may function as a tumor suppressor gene. The protein is a critical component of type I protein kinase A (PKA), which is also the main mediator of cAMP signaling in mammals. It belongs to the class of type 1A regulatory subunits of protein kinase A. Inactivating germline mutations of PRKAR1A are seen in 2/3 of patients with Carney Complex, labeled as LAMB (lentiginosis, atrial myxoma, blue nevi) or NAME (nevi, atrial myxoma, myxoid neurofibromas, ephelides).

Other sex cord-stromal tumors include granulosa cell tumor and counterparts of ovarian surface epithelial tumors (serous, mucinous, endometrioid, clear cell, and Brenner). Granulosa cell tumor is characterized by lobular arrangements of large follicles lined by polygonal cells in an arrangement fashion, diffuse or solid pattern, and microcystic foci ± intervening fibrous stroma (Figs. 7.28 and 7.29).

Mixed germ cell and sex cord-stromal tumor include also the unique type of gonadoblastoma, which is characterized by multiple round-ovoid clusters of seminoma-like cells with dark sex cord (Sertoli-like) cells, peripheral palisading of small hyperchromatic Sertoli-like cells, and irregular deposits of eosinophilic BM material. The study of the BM material indicates that this structure is similar to that seen in Sertoli cell adenoma and Sertoli cell nodule. The gonadoblastoma should be considered definitely as premalignant, because it can progress to an invasive germ cell tumor, typically seminoma.

Other tumors include tumors of the adrenogenital syndrome, adenocarcinoma of the rete testis, and hematopoietic neoplasms (lymphoma, plasmacytoma, leukemia, and capillary hemangioma). Malignant lymphomas are rare in the youth, while they are more characteristic of the elderly. The differential diagnosis may incorporate them if the patient has been transplanted (PTLD, post-transplant lymphoproliferative disorder). Lymphomas are bilateral in 20% against 2% of bilateralism observed typically in seminomas. Malignant lymphomas are generally germinal center and post-germinal center neoplasms. Finally, about the hematopoietic malignancies, testes are a "sanctuary" site for leukemia, and half of the patients who die with leukemia have leuke-

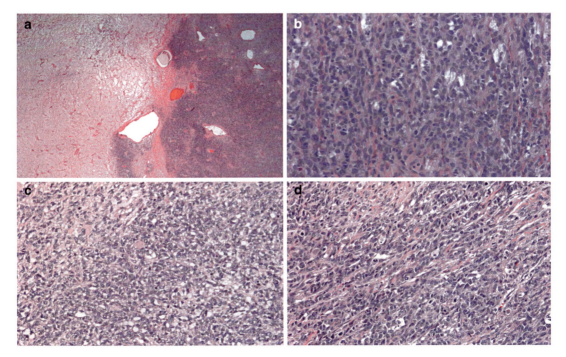

Fig. 7.30 In this panel are shown the morphology with strap cells of a classic embryonal rhabdomyosarcoma (see text for details) (Hematoxylin and eosin staining; **a**, ×12.5; **b**, ×200; **c**, ×200; **d**, ×200; **e**, ×200; **f–h**, ×630 as original magnifications)

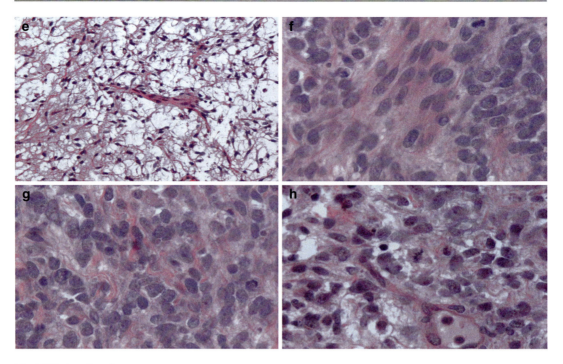

Fig. 7.30 (continued)

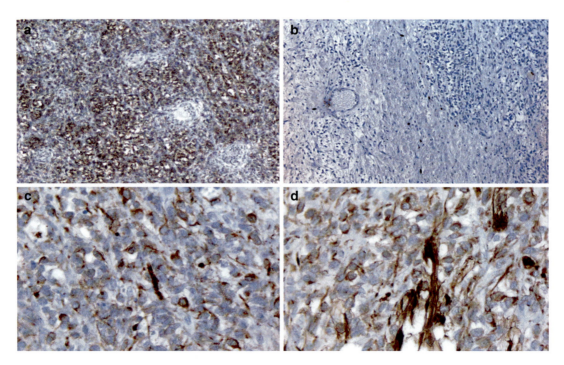

Fig. 7.31 The immunohistochemistry of the previous rhabdomyosarcoma shows a positivity for CD56 (**a**), CD57 (**b**), desmin (**c–f**), and MYF4 (**g**). The Ki67 proliferation antigen is moderately high in (**h**) (**a**, ×100; **b**, ×100; **c**, ×400; **d**, ×400; **e**, ×100; **f**, ×400; **g**, ×100; **h**, ×50)

7.7 Testicular Tumors

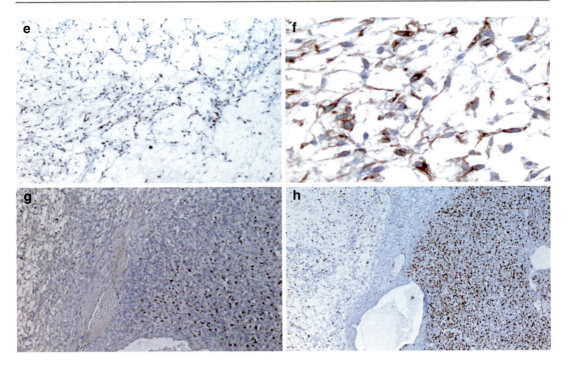

Fig. 7.31 (continued)

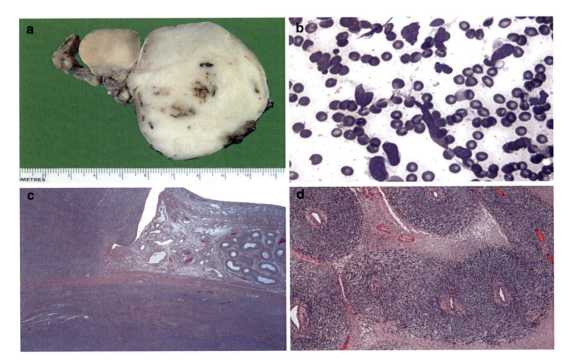

Fig. 7.32 The microphotographs show a spindle-cell rhabdomyosarcoma with gross photograph (**a**), cytology (**b**), and histology (**c–h**) Microscopically, tumor cells are arranged in lobules, small nests, microalveoli and single file arrays. They are embedded in a matrix, which may exhibit abundant hyalinization. Scattered rhabdomyoblasts can be encountered and show eccentric nuclei and copious eosinophilic cytoplasm. Mitotic figures are frequent

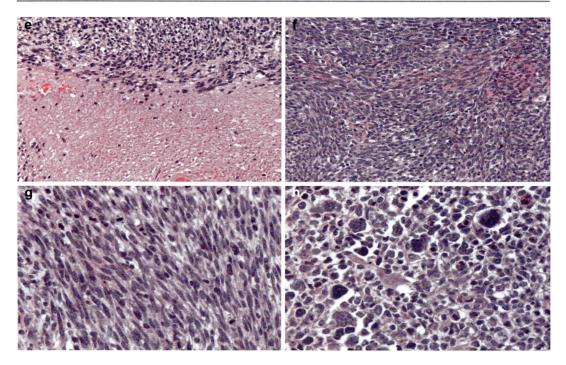

Fig. 7.32 (continued)

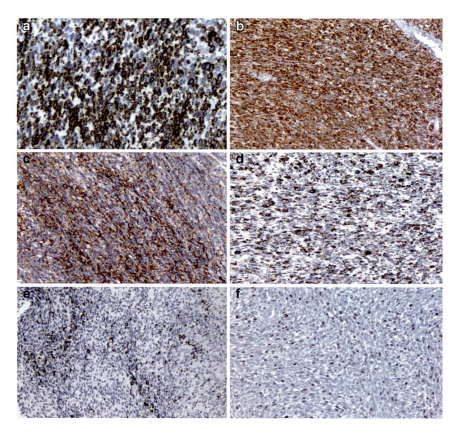

Fig. 7.33 The microphotographs of the previous spindle-cell rhabdomyosarcoma show positivity for vimentin (**a**), CD56 (**b**), CD57 ©, desmin (**d**, **e**), and MYF4 (**f**)

mic cells in the testis at the time of the autopsy. However, a recent Chinese study identified genitourinary lymphomas in children and youth as 20% of the patients studied. In this study, DLBCL was found in about 2/3 of cases, followed by mucosa-associated lymphoid tissue (MALT) lymphoma, Burkitt lymphoma, peripheral T-cell lymphoma, and plasmacytoma. PNET is an infrequent diagnosis but can occur and needs to be differentiated from germ cell tumors. PNET has been considered a variant of teratoma. It is CD99+, and neuroendocrine markers (NSE, CGA, SYN) may also be positive.

> **Box 7.8 Prostatic Carcinoma Mimickers**
> Hyperplasia-
> based: Adenosis, sclerosing, adenomatous hyperplasia, basal cell hyperplasia, BNH/PNH-small gland type, clear cell cribriform hyperplasia
> Anatomy-
> based: Cowper's glands, mesonephric gland remnants, paraganglion, seminal vesicles/ejaculatory ducts, verumontanum glands
> Atrophy-
> related: Simple, sclerotic, streaming, cystic, post-atrophic hyperplasia
> Atypia,
> Reactive: Inflammatory, ischemic, radiation-induced
> Inflammation-
> associated: Usual and granulomatous prostatitis, xanthogranulomatous prostatitis, malakoplakia
> Metaplasia-
> associated: Mucinous, nephrogenic (adenoma)

7.7.3 Rhabdomyosarcoma and Rhabdoid Tumor of the Testis

Rhabdomyosarcoma of the Testis
The most common rhabdomyosarcoma that occurs in the scrotal region in children and youth is the paratesticular spindle-cell variant of embryonal RMS, which has a better outcome than classic RMS (Figs. 7.30 and 7.31), although metastases to retroperitoneal LNs are seen in 40% of cases (Figs. 7.32 and 7.33).

Rhabdoid Tumor of the Testis
The rhabdoid tumor of the testis is similar to an extrarenal rhabdoid tumor in morphology and immunohistochemical phenotype (Figs. 7.34 and 7.35).

7.7.4 Secondary Tumors

Metastasis to testis is not a typical event in children or young individuals.

7.8 Tumors of the Epididymis

7.8.1 Adenomatoid Tumor

Adenomatoid tumor is a D2-40 (podoplanin) expressing benign epididymal tumor of the youth that can manifest as a mass with or without pain. Grossly, the adenomatoid tumor is small, solid, firm and white and contains tiny cysts. Microscopically, the adenomatoid tumor is unencapsulated with cords of flattened to cuboidal epithelium lining dilated channels and prominent stroma with smooth muscle and elastic fibers. Lining cells have vacuolated cytoplasm, and hyaluronidase-sensitive mucin staining is observed. IHC shows (+) AE1/3, CK7, CAM5.2, CALR, VIM and D2–40, as well as WT1. The occurrence of D2–40 may represent a pitfall for germ cell tumors.

7.8.2 Papillary Cystadenoma

Papillary cystadenoma is seen in individuals of families with von Hippel-Lindau syndrome (VHLS) and is small, well-delimited cystic or solid tumor grossly. Microscopically, papillary structures with infoldings lined by tumor cells with clear cytoplasm are seen. The malignant counterpart is rare. VHLS is highly suspected when an individual harbors 1) multiple hemangioblastomas of the brain, spinal cord, or eye, or 2) one hemangioblastoma and tumors including clear cell renal cell carcinoma, pancreatic cysts, pheochromocytoma, endolymphatic sac tumor, or a epididymal cyst, or, alternatively, 3) multiple bilateral clear cell renal cell carcinomas are diagnosed.

7.8.3 Rhabdomyosarcoma

The rhabdomyosarcoma (RMS) of the epididymis is a rare tumor with rhabdomyosarcomatous differentiation that recapitulates the aspects of the RMS described in the soft tissue chapter (Figs. 7.32 and 7.33). Most of these tumors are of spindle-cell type and harbor a better prognosis than the more classical type (e.g., embryonal rhabdomyosarcoma, ERMS).

7.8.4 Mesothelioma

The *tunica vaginalis* is lined by mesothelial cells and participates all but the posterior aspect of the testis. There is a visceral layer around the testis and a parietal layer against the scrotal wall. The visceral layer of the *tunica vaginalis* blends imperceptibly with the tunica albuginea. Mesothelioma is extremely rare in children or young people and arises from *tunica vaginalis* of the testis and is usually of fibrous type. DDX includes adenocarcinoma of the rete testis. In rare cases, the *tunica vaginalis* may be involved by mesothelioma. However, scrotal mesotheliomas are less common than those found in the pleural or peritoneal compartments. While the mean age of patients at presentation is 53 years, the age range is broad, with some patients presenting in adolescence and youth. Asbestos exposure has been linked to mesothelioma, but less than half of patients have such a positive history. Benign and malignant mesotheliomas have been reported. Concomitant malignant pleural or peritoneal mesothelioma may be present.

7.9 Inflammatory Disorders of the Prostate Gland

7.9.1 Acute Prostatitis

Bacterial infection of the prostate gland may represent a medical emergency. It should be distinguished from other forms of prostatitis such as chronic bacterial prostatitis and chronic pelvic pain syndrome (CPPS).

7.9.2 Chronic Prostatitis

Chronic prostatitis is a syndrome of low-pelvic pain and urinary symptoms. It occurs either with recurrent bacterial infection (chronic bacterial prostatitis [CBP]) or as pain without evidence of bacterial infection (CPPS). Sporadically, there may be positive bacterial cultures from prostatic secretions in CPPS, but there is no evidence that these are causative of the symptomology. An inflammatory "tumor" was diagnosed in a 4.5-year-old boy. The child suffered from pollakiuria, dysuria, secondary urinary incontinence, and decreased urine stream. Although pollakiuria associated with other signs and/or symptoms may be indicative of urinary tract infections (UTI), the isolated finding of pollakiuria is not associated with an underlying pathological condition. Pollakiuria can be triggered by the child's heightened awareness of the urinary bladder. The urinary bladder is constantly filled up with urine produced by the kidneys, which determines it to progressively expand. This feeling stimulates the recently toilet trained child and may alarm the parents. Ultrasonography, magnetic resonance imaging, and transrectal tumor biopsy confirmed the diagnosis of an inflammatory tumor of the prostate with a size of up to 5 cm. Under antibiotic therapy, the clinical symptoms disappeared within 6 weeks. Ultrasonography performed 3 months later exposed a healthy prostate with a residual midline cyst of 3x4mm. The inflammatory tumor can be explained by the embryologic development of the prostatic gland and the persistence of an intraprostatic cyst.

7.9.3 Granulomatous Prostatitis

Chronic granulomatous disease (CGD) is a rare hereditary disease in which phagocytes have small cytologic tools to generate the superoxide radical (O^-_2) required to kill specific pathogens. Patients affected with CGD are outstandingly susceptible to a particular set of infections and granulomatous lesions. A 15-year-old boy with X-linked CGD revealed a necrotizing granulomatous inflammation in the prostate with computed tomography showing a fluid collection in the prostatic fossa, later determined to be a prostatic abscess and granulomatous prostatitis (Agochukwu et al. 2012).

7.9.4 Prostatic Malakoplakia of the Youth

Malakoplakia is a rare inflammatory condition that is considered to develop secondary to a macrophagic inadequacy to digest the bacteria of chronic *E. coli* infection. Typically, malakoplakia affects the genitourinary tract but may also be observed in the colon, stomach, liver, lungs, uterus, bones, and skin in both children and adults (Diapera et al. 2009). Malakoplakia of the genitourinary system usually involves the urinary bladder and the prostate glands. Although malakoplakia of the prostate is mostly exclusively seen in elderly, there are occasional case reports in adolescents and youth. Malakoplakia may be clinically mistaken for prostatic benign tumors and malignancies.

7.10 Prostate Gland Overgrowths

7.10.1 Benign Nodular Hyperplasia of the Young and Fibromatosis

Benign nodular hyperplasia (BNH) or benign prostate hyperplasia (BPH) is stromal and glandular hyperplasia of the prostatic gland and usually seen in men older than 40 years. The stromal component has small blood vessels, and clinically relevant prostate cancer may be encountered in ~1.5% of resections for BPH. Occasionally, it may be observed in individuals younger than 40 years (Berry et al. 1984). The fibromatoses of the prostatic gland are rare. They constitute a group of nonmetastasizing fibrous growths with a tendency to invade surrounding tissues and ambiguous biological behavior, varying from spontaneous regression to aggressive and devastating local growth (Scholtmeijer et al. 1988) (Fig. 7.11).

7.10.2 Rhabdomyosarcoma, Leiomyosarcoma, and Other Sarcomas (e.g., Ewing Sarcoma)

Rhabdomyosarcoma (RMS) is the most common sarcoma identified in the first two decades of life (Figs. 7.12, 7.13, 7.14). Urinary bladder/prostate (BP) RMS accounts for 5% of all cases. Through efforts from multiple cooperative study groups inside of the Children's Oncology Group and the International Society for Paediatric Oncology, survival has improved significantly. Gross find-

---

Box 7.9 IHC Markers for Prostatic Carcinoma Mimickers

Basal cell markers:	34βE12, CK5/6, p63
Secretory markers:	PSA, PSAP, CD57
NE markers:	CGA, SYN, NSE, CD56, S100 (sustentacular)
WBC-macrophage markers:	CD45, CD3, CD20, CD56, CD68, MPO
Others:	AMACR

Notes: Alpha Methyl Acyl Coenzyme A Racemase (P504S) is a mitochondrial/peroxisomal enzyme identified from PrCa involved in β-oxidation of dietary branched-chain fatty acids and fatty acid derivatives.

> **Box 7.10 PIN Main Histologic Patterns**
> (1) Tufted: Nuclear enlargement, prominent nucleoli, and some residual basal cells
> (2) Micropapillary: Finger-like arrangement of the acinar epithelial cells
> (3) Flat: Pseudostratified and linear arrangement of the acinar epithelial cells
> (4) Cribriform: Interconnecting intraluminal bridges of the acinar epithelial cells
> (5) Small cell: Small cells
> (6) Hobnail: Polarization of enlarged nuclei of secretory cells toward the lumen

ings and histology are not dissimilar from other locations. Positron emission tomography–computed tomography (PET-CT) scanning remains superior to conventional metastatic workup. All cooperative oncology groups agree on surgical biopsy for diagnosis and staging of BP RMS. Patients are then grouped and risk classified before receiving chemotherapy. Other sarcomas in pediatrics include PNET or Ewing sarcoma. Leiomyosarcoma is not seen in children or youth but needs to be differentiated from the spindle-cell variant of the rhabdomyosarcoma.

7.10.3 Prostatic Carcinoma Mimickers

Mimickers of the prostatic carcinoma are crucial pitfalls. The mnemonic word "Harim" (Head of the 3 of 24 priestly departments instituted by King David) may be used (Box 7.8).

Cowper's Glands Well-circumscribed lobules (lobular architecture) of small, compact tubuloalveolar glands constituted by mucus-containing cells (+DPAS, Alcian Blue, mucicarmine) with bland cytologic features (bland cytology: small nuclei and inconspicuous nucleoli), radiating from a central excretory duct lined by pseudostratified epithelium, typically located within skeletal muscle. IHC: (+) CK, HMWCK (only ductal cells), SMA (basal cells at the periphery of the acini), (−) S100, PSAP, but (±) PSA. DDX: mucinous prostate carcinoma (lumens never almost occluded as in Cowper's glands) and foamy gland carcinoma (large atypical glands, mucin-negative).

Mesonephric Gland Remnants Mesonephric ducts arc primordia (organs or tissues in its earliest recognizable stage of embryological development) for ejaculatory ducts, and primordial ureters and remnants can pose an unembellished DDX with carcinoma. In fact, mesonephric gland remnants show a vaguely lobular and/or infiltrative proliferation of crowded, small atrophic tubules with a single cell of bland-appearing epithelial cells with scant cytoplasm with neither nuclear enlargement or prominent nucleoli, which contain dense eosinophilic colloid-like secretion (unlike loose granular eosinophilic secretion of acinar carcinoma), (±) HMWCK, (−) PSA and PSAP.

Paraganglion Solid, rounded structure with small, solid nests of cells with clear to amphophilic, finely granular cytoplasm, often with a "Zellballen" arrangement with a prominent/delicate background of the vascular capillary network, (+) CGA, SYN, S100 (sustentacular cells), (−) PSA, and PSAP. DDX: G3-Pr-Ca or grade 3 prostate carcinoma (vascular background, the granular cytoplasm of the cells, IHC).

Seminal Vesicles/Ejaculatory Ducts Paired, coiled, and tubular sex accessory glands located posterolaterally to the urinary bladder and anterior to Denonvilliers' fascia (rectoprostatic fascia) and rectum with ejaculatory ducts arising from each seminal vesicle medially surrounded by a band of loose fibrovascular connective tissue and running through the base of the prostate to the urethra. Histologically, there is an adenosis-like pattern with large dilated glands and peripherally small glands with nuclear hyperchromasia and pleomorphism, but no mitoses and the presence of prominent golden-brown lipofuscin granules. IHC: (+) HMWCK, p63, (−) PSA, PSAP. DDX: Pr-Ca, benign prostate tissue, PIN-III.

Verumontanum Mucosal Gland Hyperplasia (VMGH) Suburethral (adjacent to the posterior prostatic urethra) small proliferation of uniform, closely packed, round small acini with an intact basal cell layer (+ HMWCK), bland-appearing cytology, and inconspicuous nucleoli and prominent intraluminal corpora amylacea and distinctive brown-orange secretions and intracytoplasmic lipofuscin pigment. DDX: G1-PrCa or grade 1 prostate carcinoma (infiltrative and haphazard arrangement of the atypical glands without the peculiar intraluminal concretions characteristic of VMGH).

Atrophy A common mimicker of the adenocarcinoma of the prostate with possible occurrence adjacent to any neoplastic proliferation with a wide age range (from second decade to elderly). The anatomic prostate zones are Central (CZ), Peripheral (PZ) and Transitional (TZ). Atrophy occurs in PZ > CZ > TZ without evidence of an increased risk of harboring or developing a prostate malignancy. There are four patterns, including simple, sclerotic, streaming, and cystic as well as post-atrophic hyperplasia. There are glands arranged in multiple lobules separated by fibrotic stroma with well-formed lumens and cells with hyperchromatic nuclei, small/inconspicuous nucleoli, scant cytoplasm, and fragmented basal layer (+HMWCK, p63). DDX: Atrophic variant of Pr-Ca (prostate carcinoma). In post-atrophic hyperplasia, there is a combination of atrophic acini and acini lined by cuboidal secretory cells with abundant clear/amphophilic cytoplasm and increased N/C ratio.

Atypia (Reactive), which includes inflammatory-induced atypia, ischemic-induced atypia, and radiation-induced atypia.

Inflammation Usual and granulomatous prostatitis, XGP (xanthoma/xanthogranulomatous prostatitis), and malakoplakia.

Metaplasia

Mucinous gland metaplasia: The lobular architecture of mucinous glands lined by tall columnar cells, small basal nuclei with inconspicuous nucleoli resembling Cowper's glands (+DPAS, AB, MCM -PSA, PSAP).

Nephrogenic adenoma/metaplasia: Small, solitary proliferation of gland-like structures mostly involving the urinary bladder associated with radiation, surgery, stones, instrumentation, intravesical thiotepa, BCG, and renal TX and grossly presenting as papillary, polypoid, friable, and velvety lesion and microscopically in four histologic patterns, including tubular, cystic, polypoid-papillary, and diffuse. Microscopically, there are often small hollow tubules with low columnar cells ± hobnailing and reactive atypia, ± colloid-like secretions, and ± surrounded by a thickened hyaline sheath. IHC: ± PSA, PSAP (tubular secretions), + AMACR, + HMWCK. Thiotepa is an organophosphorus compound ("alkylating agen") with the formula $SP(NC_2H_4)_3$ and anti-cancer properties. There are several types of alkylating agents: Mustard gas derivatives (mechlorethamine, cyclophosphamide, chlorambucil, melphalan, and ifosfamide), Ethylenimines (thiotepa and hexamethylmelamine), Alkylsulfonates (busulfan), Hydrazines and Triazines (altretamine, procarbazine, dacarbazine, and temozolomide), Nitrosureas (carmustine, lomustine and streptozocin), and Metal salts (carboplatin, cisplatin, and oxaliplatin).

Hyperplasia

Adenosis, sclerosing: Circumscribed nodular proliferation of small- to medium-sized acini (± poorly formed glands and single cells) with thick eosinophilic BM separated by a cellular spindle cell-/edematous stroma (lightly basophilic), mainly in the TZ, and constituted by clear cells with abundant cytoplasm and bland-appearing nuclei with inconspicuous nucleoli (±PSA, PSAP) and a fragmented basal cell layer

(+HMWCK, p63, SMA, MSA, S100). TEM: Myoepithelial cell differentiation (abundant microfilaments with prominent dense bodies). DDX: WD-PrCA (lobular architecture and IHC cell markers).

Adenomatous, hyperplasia: Circumscribed nodular proliferation of peripherally closely packed small acini with centrally elongated, larger glands with papillary infolding set in a paucicellular stroma, mainly in the TZ, periurethral area, and apex, and constituted by clear cells with abundant cytoplasm and bland-appearing nuclei with inconspicuous nucleoli (± PSA, PSAP) and a fragmented basal cell layer (+HMWCK, p63, SMA, MSA, S100). DDX: WD-PrCA (haphazard gland architecture, often at right angles to each other, enlarged nuclei, large nucleoli, the absence of basal cell layer).

Basal cell hyperplasia: *Nodular and diffuse expansion* of uniform round (occasionally cribriform) glands set in acellular/paucicellular stroma is present. It is part of the spectrum of BPH, mainly in the TZ and PZ with inflammation, ranging from incomplete with still present residual small lumina lined by secretory cells (± PSA, PSAP) surrounded by multiple layers of basal cells (multilayering) (+HMWCK, p63) to complete forms with solid nests of cells with hyperchromatic nuclei and lack of luminal differentiation. BCH with prominent nucleoli is a BCH variant with prominent nucleoli and part of a continuum of BCH and adenoid basal cell tumor (+HMWCK).Well-formed lamellar calcifications and intracytoplasmic eosinophilic globules may be found. DDX: Ca (nodular arrangement, +NH, cellular uniformity, ± prominent nucleoli, +34βE12-multilayering).

BNH/PNH-small gland type: Very few cases of youth with benign nodular hyperplasia have been described.

Clear cell cribriform hyperplasia: It is a peculiar form of nodular hyperplasia with the cribriform arrangement, but without atypia, mainly in the TZ and after androgen deprivation therapy, and constituted by the crowded proliferation of complex glands with clear cells, cribriform and uniform lumina, evident basal cells (+HMWCK), and small uniform nuclei with inconspicuous nucleoli.

DDX: PIN-high grade (prominent nucleoli and no basal cell layer) and cribriform adenocarcinoma (lack of low-power nodularity and basal cells and the presence of significant cytologic atypia).

The immunohistochemical markers for mimickers of the prostatic carcinoma are shown in Box 7.9.

(Cyto-)keratins (CK or K) are intermediate filaments of the cytoskeleton, and each type I of acidic CK (Moll type I), CK 10, 15, 16, and 19, forms a pair with a type II of basic CK (Moll type II), CK 1–8. All epithelial cells contain at least two CK (exception is CK19, which is unpaired). AE1/3: AE1 reacts with acidic (Moll type I) CK 10, 15, 16, 19, whereas AE3 reacts with basic (Moll type II) CK 1–8. AE1/3 does not detect CK17 and CK18. 34βE12 (HMWCK): CK 1, 5/6, 10, 14 useful to stain basal cells in prostatic glands and to differentiate in situ neoplasm. CAM5.2 (LMWCK): CK 8, 18, 19, while PAN-CK entails high and low molecular weight cytokeratins.

Other tumorlike conditions include sarcoma-like nodules; exuberant stromal reaction; melanosis; urethral polyps, which occur in children and youth; and amyloid nodule.

7.10.4 Prostatic Intraepithelial Neoplasia (PIN)

Prostatic intraepithelial neoplasia (PIN) is a pre-invasive form of prostatic adenocarcinoma and usually identifiable on low power by light microscopy.

The glands have standard architecture but have the acronym "PNH":

1. Papillary projection into the lumen
2. Nuclear enlargement with overlapping and stratification
3. Hyperchromasia

The PIN is almost exclusively confined to PZ, and the hallmark for HGPIN (high-grade PIN) is the presence of prominent nucleoli. The PIN main histological patterns are shown in Box 7.10.

Other patterns include apocrine, foamy gland; mucinous, Paneth cell-like, pleomorphic, signet-ring cell; and small cell neuroendocrine. The bubbly pattern is made of pale/foamy cells with voluminous xanthoma-like cytoplasm that can build solid and cribriform patterns. The desquamating apoptotic model shows desquamating acinar cells containing apoptotic nuclear material that coalesce in the luminal gland giving rise to basophilic intraluminal masses. The mucinous model shows mucinous distension of glands. There is flat epithelial lining. The intraluminal blue-stained mucinous secretions are +DPAS and AB (periodic acid Schiff reaction with diastase treatment and alcian blue stain).

Grading: PIN I, II, or III or low (I) and high (II or III) grade with prominent nucleoli making at least PIN II. However, there is poor reproducibility for a PIN of low grade, which seems to be found in 70% of patients, classified as "normal." Conversely, HG-PIN indicates up to a 70% chance that in this prostatic gland, there may be a coexisting adenocarcinoma elsewhere.

7.10.5 Prostate Cancer of the Young

Prostate cancer in individuals with age 40 years or younger is sporadic. Reyes and Slutsky reported on two male patients on moderately differentiated prostatic adenocarcinoma aged 35 and 40 years that initially manifested as lymph node metastasis of an unknown origin and reviewed the topic in the oncologic community. A few more cases have been published in the scientific literature. It is probably unknown the actual incidence of prostatic carcinoma and PIN in this age group. However, some studies have tried to shed light on this controversial topic. In healthy men aged 20–30 years who died from trauma or homicide in Detroit, the incidence of some PIN was found 8%. PSA seems to be probably questionable in this age group. Searching in the SEER (Surveillance, Epidemiology, and End Results Program of the National Cancer Institute) database, an authoritative source for cancer statistics in the US., 1673 cases of prostate cancer in men aged 35–44 years out of 453,195 total cases have been identified, although most of them have low-grade neoplasms. Interestingly, the 10-YSR was equivalent to that of male individuals whose diagnosis was made at older ages. Only a subset of young men with high-grade malignancy behaves worse than older men with the high-grade disease. Thus, a natural consequence of it is to think that the outcome of these young men was better than older men indicating that young age group at diagnosis does not portend a poor prognosis as observed in breast cancer for young women. Overall, the approximate incidence of prostate cancer among men 40 years old or younger seems to be around 1% according to most relevant literature data, but the incidence seems to be lower if data from single hospitals or institutions are analyzed. In 1980, prostate cancer was reported in a young boy of 11 years age and summarized the clinicopathologic characteristics of prostatic carcinoma in infant and adolescents (Shimada et al. 1980). It was thought that the aggressive behavior of prostate cancer in young individuals could be related to the undifferentiated histology. In another retrospective analysis by Astigueta et al. (2010), a remarkable number of 41 patients of prostate cancer under 50 years of age were identified and carefully evaluated from 1952 to 2005. All patients had bone metastasis. Also, 20 patients had retroperitoneal metastasis, 3 patients had mediastinal lymph node metastasis, 4 had liver, 3 had lung, and 1 patient had testicular metastasis. All prostatic biopsies were reported as poorly differentiated or undifferentiated, and according to current classification, predominant Gleason score was 9. Patients received different kinds of palliative treatment like bilateral orchidectomy and adrenalectomy in the 1970s and oral or parenteral ADT (androgen deprivation therapy) in recent years. Additionally, patients received palliative radiotherapy and other symptomatic therapy. Median survival was 16.1 month, and all patients died of progressive disease. It has been controversially discussed that with every 10-years' time, the PIN will progress further to become invasive eventually and will manifest in 4/5 of men later in life. About the genetics familial clusters of prostate cancer (defined as the presence in at least two family members) seem to be seen in about 15%–25% of cases and some hereditary form of prostate malignancy compatible with a Mendelian inheritance in up to 1/10 of patients. Genetic heterogeneity is explained on the basis that in some families with recurrent prostatic malignancy, aggregation of several tumors may suggest the involvement of common predisposing genes, while in other families, the genetic component appears to be polygenic. Some single nucleotide polymorphisms (SNPs) have been associated with an increased risk of developing a prostatic malignancy.

A unique role of the pediatric pathologist is reserved for prostatic malignancy, which includes investigating PIN/cancer in the young at the autopsy, notably, if a promoter factor (e.g., inflammation for congenital GU anomalies) is added to an initiator factor (e.g., genetic alteration or abnormal SNPs). Moreover, he/she needs to provide tissue material for morphologic, epidemiological studies and NGS studies and the exclusion of the relatively most frequent tumor of the prostatic gland in childhood and adolescence (rhabdomyosarcoma).

- *DEF*: 2nd most common malignancy in men (following skin cancer) in developed countries, blacks>white and rare in Asians and no carcinoma if prepubertal castration and low incidence with hyperestrogenism (liver cirrhosis) and no association with STD, smoking, occupational exposure, diet, and BNH.
- *EPG*: Genetics: *BRCA2, AR, PI3K/AKT, PTEN, ERG/TMPRSS*.
- *CLI*: DRE (digital rectal examination), TUUS (3D trans-urethral ultrasound), and elevated PSA (> 4 or increasing over time) are useful, and PrC secretes 10x the PSA of healthy tissue, but routine screening programs using PSA with or without DRE do not significantly affect mortality overall or from prostate cancer. Consequently, the value of routine screening has been challenged and is controversially debated in medical prevention.
- *DDX*: BNH, granulomatous prostatitis, tuberculosis, infarct, and lithiasis other than prostatic carcinoma, and DDX of elevated PSA include BNH, prostatitis, infarct, trauma (biopsy, TURP), other tumors (e.g., salivary duct carcinoma).

The diagnosis of prostatic adenocarcinoma is based on biopsy and evaluation using a specific score called "Gleason score" based on histologic patterns and considering several tips arising from years of practice of pathology colleagues working with the lower uro-genital system.

1. Transrectal biopsies have been found to be more accurate than transperineal biopsies.
2. "Six-pack" model: Six samples from selected portions of the prostate, using a spring-loaded 18-gauge biopsy, have a little false-negative rate of ~10% due to sampling error and is further reduced when using 12–18 samples as used in many centers.
3. Gleason score in biopsy mostly correlates with the Gleason score of the final report coming out from prostatectomy specimens (same: ~ 60%, ± 1 unit: ~90%).
4. More errors occur with Gleason scores 5 or 6, which tend to underestimate prostatectomy Gleason score.
5. In 2005 a new consensus was reached to change some assignments of histologic pattern to a specific Gleason pattern. Consequently, Gleason score of needle biopsies <4 is challenging; if ever made, most of the cribriform patterns have been upgraded to pattern 4, because very few lesions would now satisfy the criteria for cribriform pattern 3; there are different Gleason scores used for needle biopsy and prostatectomies; the high grade of any quantity on needle biopsy should be included within the Gleason score. In particular, for the tertiary pattern on needle biopsy specimens, both the primary pattern and the highest grade should be reported and finalized, whereas, for prostatectomies, the reporting pathologist should sign out a report indicating the primary and secondary patterns, and comment on the tertiary pattern should be made. If specimens of needle biopsies show different grades in separate cores, individual Gleason scores should be

assigned to these cores, while in the eventuality of different grades observed in separate tumor nodules of prostatectomies, a separate Gleason score should be given to each of the dominant tumor nodules.

6. Tumor seeding of needle tract is a rare event in perineal needle biopsy and less common with transrectal biopsy. Most of the cases may involve poorly differentiated carcinomas.
7. 12-core TRU-guided prostate needle biopsy findings plus preoperative PSA may predict advanced local disease at prostatectomy specimen.
8. In processing core biopsy, three levels are recommended, and > 3 are needed if atypical glands are encountered.
9. Pathology review of core biopsies is strongly recommended before radical prostatectomy is performed.
10. A needle biopsy is considered unsatisfactory if neither prostatic glands nor stroma are found. The presence of stroma only may indicate a hyperplastic stromal nodule and should be considered satisfactory.
11. A single histologic level misses approximately 1/4 of total length of a core, and pre-embedding cores techniques may be used (e.g., "stretch" method) to yield more tumor/core, more cores with tumor, more cases with tumor, fewer atypical small acinar diagnoses, and fewer cases with 3 mm or less of Gleason 6 or less cancer.
12. The assessment of intraluminal, amorphous eosinophilic material should be carried out ruling out decapitation-like secretions or fractured corpora amylacea. Also, collagenous micronodules are considered accumulations of paucicellular, eosinophilic, fibrillar stroma impinging on acinar lumens. Moreover, glomerulations are defined as rounded epithelial tufts within glands reminiscent of renal glomeruli and are not observed in benign lesions.
13. In a core, features suggestive of malignancy include prominent nucleoli, marginated nucleoli, multiple nucleoli, blue-tinged mucinous secretions, intraluminal crystalloids, intraluminal amorphous eosinophilic material, collagenous micronodules, glomerulations, perineural invasion, and fatty tissue invasion.
14. HMWCK and p63 detect basal cells, which are lacking in prostatic adenocarcinoma. However, it is diagnostic of malignancy, if there is a high pretest suspicion of carcinoma, and positive control in benign glands should be demonstrated. Positive staining is useful in identifying benign mimickers of prostatic adenocarcinoma.
15. AMACR is sensitive and specific for prostate carcinoma on needle biopsies and has been recommended to use a combination of P504S and 34βE12 to diagnose limited prostatic adenocarcinoma, but negative staining with P504S in suspicious foci does not necessarily indicate a benign diagnosis. In fact, AMACR may stain hyperplastic nodules and benign glands close to TZ prostatic adenocarcinoma.
16. Minimal prostatic adenocarcinoma is a challenging diagnosis and should be reserved after careful consideration and probably approved by another urological pathologist. It relies on some common malignant features (nucleomegaly, infiltrative growth pattern, intraluminal secretions, prominent nucleoli, associated HG-PIN, amphophilic cytoplasm, hyperchromatic nuclei, and intraluminal crystalloids).
17. Uncommon features of minimal prostatic adenocarcinoma include perineural invasion, collagenous micronodules, and mitoses.
18. DDX should always include adenosis, atrophy, and high-grade PIN primarily before a diagnosis of cancer is given.

Tumor extension occurs locally through the seminal vesicles and bladder base, rarely via prostatic urethra, whereas rectal invasion is rare due to the Denonvilliers' fascia (rectoprostatic fascia). Seminal vesicle invasion occurs by three modalities including (a) direct spread along ejaculatory duct complex; (b) spread outside the prostate, through the capsule, and then into a seminal vesicle; and (c) isolated deposits of tumor in the vesicle with no apparent contiguity to the primary lesion diagnosed in the prostate.

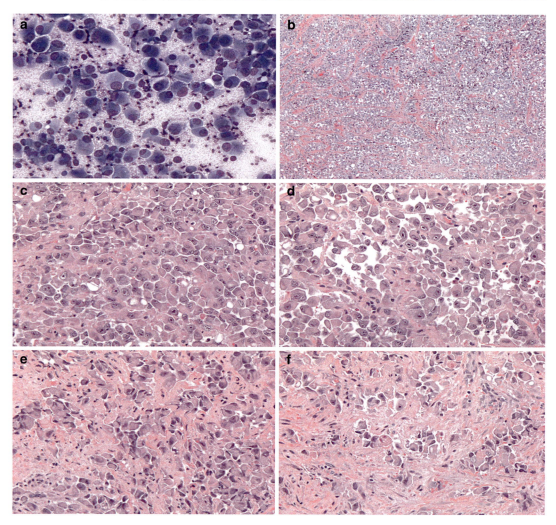

Fig. 7.34 The microphotographs of this rhabdoid tumor of the testis show a Diff-Quick cytology in (**a**) and characteristic morphology in (**b**) through (**f**). The tumor is characterized by solid sheets of large cells with eosinophilic cytoplasm, laterally displaced nucleus, and prominent nucleoli. Myxoid, hyalinized, and/or pseudoalveolar areas are frequently encountered

PrAC that is incidentally detected is considered clinically significant if total tumor volume is ≥0.5 ml, Gleason grade 4 or 5, extraprostatic extension, seminal vesicle invasion, and lymph node metastasis present. In cystoprostatectomy specimens for bladder cancer, incidental prostatic adenocarcinoma is clinically significant in about 1/2 of cases. Metastases of PrAC involve usually skeletal system (osteoblastic, upward spread: from lumbar to cervical via Batson's vertebral venous plexus (a complex valveless network of veins located near the vertebral column) and accompanied by hypocalcemia, hypophosphatemia, and increased alkaline phosphatase), lung/pleura, liver, adrenals and lymph nodes and testes (breast metastasis if estrogen therapy is performed, and the carcinoma may metastasize to male papillary breast cancer), and dura. Markers of prostatic origin include PSA, PAP, and NKX3.1 (NK3 Homeobox 1), which is recently identified marker of prostatic origin in metastatic tumors. *NKX3-1* is a protein coding gene and genetic alteration shave been found in Chromosome 8p Deletion and Rete Testis

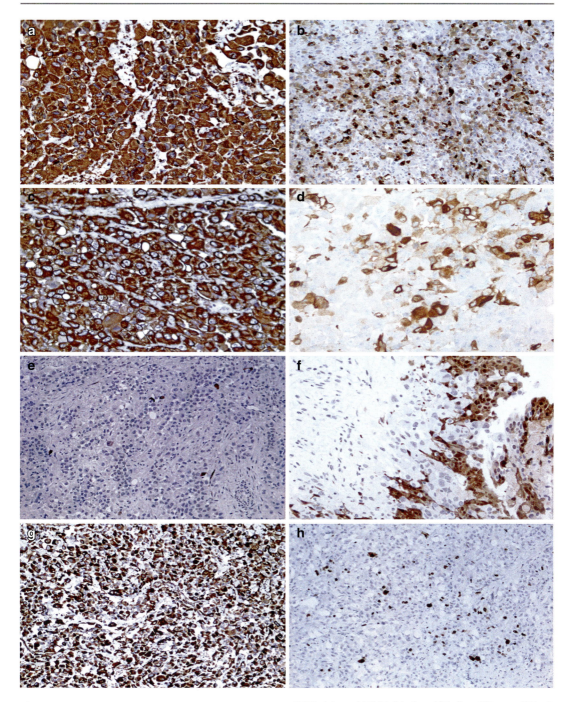

Fig. 7.35 The immunostaining microphotographs highlight the positivity of the rhabdoid tumor for vimentin (**a**), EMA (**b**), AE1–3 (**c**), synaptophysin (**d**), calretinin (**e**, **f**), WT1 (**g**), and Ki67 (**h**) (**a**, ×200; **b**, ×100; **c**, ×200; **d**, ×400; **e**, ×100; **f**, ×400; **g**, ×100; **h**, ×100 as original magnifications)

Box 7.11 Penile Noninvasive Squamous Lesions (HPV and Non-HPV Related)

Disease	Etiology	Gross	Microscopy
Condyloma acuminatum	HPV types 6 and 11	Cauliflower-like nodule at urethral meatus, fossa navicularis, or corona of glans	(+) Acanthosis with para-/hyperkeratosis and hypergranulosis (+) Koilocytosis (+) Cell maturation
Bowenoid papulosis	HPV types 16 and 18	Multiple shiny erythematous papulae that may coalesce at penis shaft and perineum	CIS similar histology (+) Cell maturation
Bowen disease	HPV types 16 and 18	Single scaly plaque at penis shaft or scrotum	CIS (full-thickness dysplasia) (−) Cell maturation
Erythroplasia of Queyrat	HPV types 16 and 18	Shiny red erythematous plaque(s) at glans and foreskin	CIS (full-thickness dysplasia) (−) Cell maturation
Pseudoepitheliomatous hyperplasia	Chronic inflammation	White-gray plaque	Irregular acanthosis of the epidermis and chronic dermal inflammation (+) Cell maturation

Adenocarcinoma. Recurrence after radical prostatectomy of PrAC is 40 months as median interval, and often there are no visible histologic features of malignancy. Subsequently, lower threshold for diagnosis should be kept in mind because atypical glands should be absent.

Prognostic factors include stage, Gleason score, surgical margins, preoperative PSA, perineurial invasion, ALVI, and size of LN metastasis. However, patients younger than 20 years usually have obstructive symptoms, advanced stage, high grade, inadequate response to treatment, and survival <1 year.

- *GRO*: The tumor is gritty and firm, gray-yellow, poorly delimited on slices of the prostatectomy specimens.
- *CLM*: The histologic patterns depend on the degree of differentiation. There may be small, medium-sized, cribriform with even or uneven borders or diffuse single cell infiltration with necrosis. The cytoplasm is usually finely granular but may be clear or foamy due to intracellular lipid (e.g., bubbly cell prostatic adenocarcinoma). The nuclei may be enlarged and hyperchromatic and have prominent nucleoli (>3 μm is quite specific for malignancy, but also >1 μm is suggestive for neoplasm). An HG-PIN is generally observed in 4/5 of carcinomas. Mitoses are not usually present, apart from in high-grade tumors. Loss of basal cells accompanies the malignant transformation of acini, and the immunohistochemistry may play a significant role. The most common pattern is infiltrative medium-sized glands (Gleason 3), but Gleason 1–2 may be found in TZ or CZ. The cribriform pattern may appear intraductal with preserved basal cell layer but is invasive, and correct grading assessment is vital for the management. Gleason 3 should be given if cribriform glands harbor smooth borders, while in Gleason 4 the cribriform glands should have uneven edges. In the experience of many urinary pathologists, this last statement regarding Gleason 4 is the most frequent occurrence in routine practice. A single cell infiltration (Gleason 5 pattern) may resemble lobular carcinoma of the breast.

Box 7.12 Pearls and Pitfalls of Penile Malignancies
- In the report for both epithelial and mesenchymal malignancies, it is essential to specify the location (glans, foreskin mucosal surface, foreskin surface, coronal sulcus, the skin of the shaft, and penile urethra).
- Surgical procedures that can be performed for penile lesions include incisional biopsy, excisional biopsy, partial penectomy, total penectomy, and circumcision.
- There are three different foreskins. These types include the short type, in which the orifice of the preputium is located behind the coronal sulcus, the medium type, in which the hole is between the corona and the meatus, and the long foreskin, when the glans is covered in full, and there is no possibility to identify the meatus unless the foreskin is not retracted.
- If >2 ipsilateral inguinal LN (+) ⇒ high probability for (+) in contralateral inguinal LN and ipsilateral pelvic LN.
- Resection margins of a penectomy specimens are three and include (1) the margin involving the proximal urethra and the surrounding periurethral cylinder (epithelium, lamina propria, corpus spongiosum, and penile fascia), (2) the margin affecting the proximal shaft (corpora cavernosa, tunica albuginea, and Buck's fascia), and (3) the cutaneous margin with underlying corporal dartos.
- Resection margins of circumcision specimens need a full histologic examination of the coronal sulcus margin and cutaneous margin.
- LVI subclassify T1 tumors in T1a and T1b.

- *TRT*: It is variable and depends on contingent factors:
  → Radical prostatectomy (not warranted if positive pelvic nodes)
  → Brachytherapy (radioactive seeds)
  → External beam radiation therapy
  → Watchful waiting (for low-grade tumors, localized tumor, or limited life expectancy)
  → Chemotherapy or hormonal treatment (LHRH analogs or luteinizing hormone-releasing hormone agonists, antiandrogens, orchiectomy, because most tumors are androgen sensitive initially) for metastatic neoplastic disease

7.10.6 Hematological Malignancies

Recurrence of acute lymphoblastic leukemia has been reported in adolescence and youth. However, leukemia can manifest or relapse in the prostate.

7.10.7 Secondary Tumors

Secondary tumors are practically inexistent in pediatric age, apart from metastases from rhabdomyosarcoma of the epididymis.

7.11 Inflammatory and Neoplastic Disorders of the Penis

7.11.1 Infections

Cystic anomalies need to be differentiated from inflammatory processes (Fig. 7.36). Infections can determine balanitis, posthitis, or balanoposthitis (Fig. 7.37).

7.11.1.1 Balanoposthitis
Balanitis refers to inflammation of the glans, while posthitis is the inflammation of the prepuce, but most commonly both conditions are associated,

and the term "balanoposthitis" is widely used. It affects about 1/10 of the uncircumcised male infants or children, possibly related to poor hygiene, aeration, irritation by smegma (Greek: σμῆγμα, "unguent, soap"). Balanoposthitis is usually a nonspecific inflammation of the glans and prepuce due to an infection with streptococci, staphylococci, *N. gonorrhoeae*, *Gardnerella vaginalis*, and *Trichomonas* particularly in uncircumcised infants and children. Balanoposthitis may be a mimicker or a form of presentation of penile cancer, which is typically a malignancy of the middle-age and elderly but can occur in patients as young as 30 years old.

7.11.1.2 Syphilis

Treponema pallidum-induced infection with three stages at level of the penis, including primary syphilis (firm, erythematous papule located on glans, which ulcerates aka "hard chancre" with a characteristic lymphocytic and plasmacytic infiltrate, endothelial cell proliferation, and capillaritis), secondary syphilis (flat, maculopapular, aka *condylomata lata*), and tertiary syphilis (penile syphilitic granulomas, aka "gummas").

7.11.1.3 Lymphogranuloma Venereum

Chlamydia trachomatis-induced short-lived painless papula or ulcer, which evolves into suppurative inflammation of the regional (inguinal) lymph nodes.

7.11.1.4 Others

Other infections of the penis include granuloma inguinale or donovanosis, which is due to an infection with *K. granulomatis* (aka *Calymmatobacterium granulomatis*) affecting the skin and mucous membranes in the genital region and resulting in nodules that evolve into beefy-red destructive ulcers, *Herpes genitalis* (HSV 1 and 2), and molluscum contagiosum (MCV, a DNA poxvirus). Moreover, pilonidal sinus of the penis is another possible but rare infection and reported to be recently caused by *Actinomyces* spp. In young individuals, granulomatous disease following intravesical BCG for urinary bladder carcinomas has also been described. A pilonidal cyst or sinus is an acquired pathology characterized by an abscess or a chronic draining sinus, which typically contains hairs and can manifest as a localized inflammation with pain, infection, and redness but also with chronic ulceration or a draining sinus or abscess formation. In the penis, it has been observed in the midline pits in the natal cleft. Dead or distorted hair pushed by mechanical forces into tiny abrasions or scars in the penile skin (coronal sulcus) form a one-ended tunnel where bacteria may collect and generate. A foreign-body giant cell reaction with multi-nucleated giant cells of foreign body cell type is typical. The biological basis for HIV infection of the circumcision has also been recently discussed. However, the discussion of circumcision on young boys is not going to settle down so smooth and is controversially addressed by several medical panels.

7.11.2 Non-infectious Inflammatory Diseases

Non-infectious inflammatory diseases include the most common balanitis xerotica obliterans, the quite obscure *balanitis circumscripta plasmacellularis*, and the Peyronie disease.

7.11.2.1 Balanitis Xerotica Obliterans

Balanitis xerotica obliterans (BXO) (*lichen sclerosus*) is inflammation with a striking equivalent to *lichen sclerosus (et atrophicus)* of the vulva of elderly. Grossly, there is a gray-white area of partial atrophy, which shows under the lens thinning of the epidermis and loss of the lamina propria structures such as diffuse fibrosis or a subepithelial hyaline band delimitated in depth by a lymphocytic infiltrate.

7.11.2.2 Balanitis Circumscripta Plasmacellularis

Balanitis circumscripta plasmacellularis (Zoon's Balanitis, BCP) is the primary differential diagnosis with erythroplasia of Queyrat and is characterized by well-circumscribed lesions, which are orange-red with a shiny appearance and multiple pinpoint red spots (aka "cayenne pepper spots") associated with pain, irritation, and discharge. Microscopically there are epidermal atrophy, loss of rete ridges, "lozenge keratinocytes," and spongiosis as well as a plasmacytic infiltrate at the subepidermal location. The "lozenge keratinocytes" are rhomboidal elongated keratinocytes. The etiology is unknown, although some argue against a distinct entity, while some other physicians recall *Mycobacterium smegmatis* as the leading case.

7.11.2.3 Peyronie Disease

Peyronie disease is a localized fibromatosis between the corpora cavernosa and tunica albuginea, which limits the movements of these structures past each other during an erection and is visualized with a curvature toward the side of the lesion. In 2/3 of cases, Peyronie disease is idiopathic, but either a blunt penile trauma or a trauma incurred during sexual intercourse has been identified in some cases, and some conditions are also associated with this fibromatosis such as diabetes mellitus, hypertension, dyslipidemia, ischemic cardiomyopathy, smoking, excessive alcohol consumption, and radical prostatectomy. A study investigating the possible development of Peyronie disease in boys following radical prostatectomy for ERMS does not seem to have been performed. Peyronie disease is probably related to other conditions, including Dupuytren contractures, Ledderhose disease, and tympanosclerosis.

7.11.3 Penile Cysts and Noninvasive Squamous Cell Lesions

Penile cysts may be distinct in congenital and acquired. The congenital cysts are usually epidermoid and typically originate from the median raphe. The site of origin is at any level from the urethral meatus to the base of the scrotum. It is thought that they occur from epithelial rest incidental inclusions to incomplete closure of urethral or genital folds. Alternatively, an epithelial split off growths after primary closure of the folds. The cysts can extend to the pelvis but have no urethral communication. Conversely, the acquired cysts are generally inclusion cysts after penile surgery (e.g., hypospadias repair or circumcision) (Fig. 7.36). STDs are the primary relationship with noninvasive squamous lesions of the penis, and HPV plays a significant role. Bowenoid papulosis occurs in patients younger than 35 years, unlike Bowen disease, which is more characteristic of patients older than 35 years. Both conditions need to be carefully discriminated against, because the malignant potential is entirely different, being the latter a squamous cell carcinoma in situ (CIS). On the other hand, Queyrat erythroplasia is a CIS of the elderly and should not be part of the central panel of differential diagnoses. Moreover, up to 1/3 of Bowen disease lesions will progress to invasive SqCC, and 1/3 of patients harboring this condition have some association with visceral malignancy (lung, GI tract, or urinary tract), unlike Queyrat erythroplasia, which needs to be differentiated by the Bowenoid papulosis. This entity is a skin lesion characterized by pigmented verrucous papules on the body of the penis. Bowenoid papulosis is associated with human papillomavirus, the etiologic agent of genital warts (Fig. 7.37).

The penile noninvasive squamous lesions are listed in Box 7.11.

Penile intraepithelial neoplasia (PeIN) refers to an intraepithelial neoplastic proliferation without invasion. PeIN is a lesion that can affect any part of the penile surface. There are different degrees of dysplasia. PeIN is classified into PeIN 1, 2, and 3, which is also known as CIS.

7.11.4 Penile Squamous Cell Carcinoma of the Youth

Penile squamous cell carcinoma of the youth (PeSqCCY) is a rare event with few cases described in the literature, and the incidence seems to have some geographic characteristics being most often reported in Central Africa (Uganda) and Brazil, an area where STD, mainly HIV, is quite prevalent. In general terms, SqCC represents 1% of all male cancers in the USA, and the incidence is lower among circumcised populations. Patients are usually 40–70 years, and the central localization is on the glans. Young patients aged 18 years have been

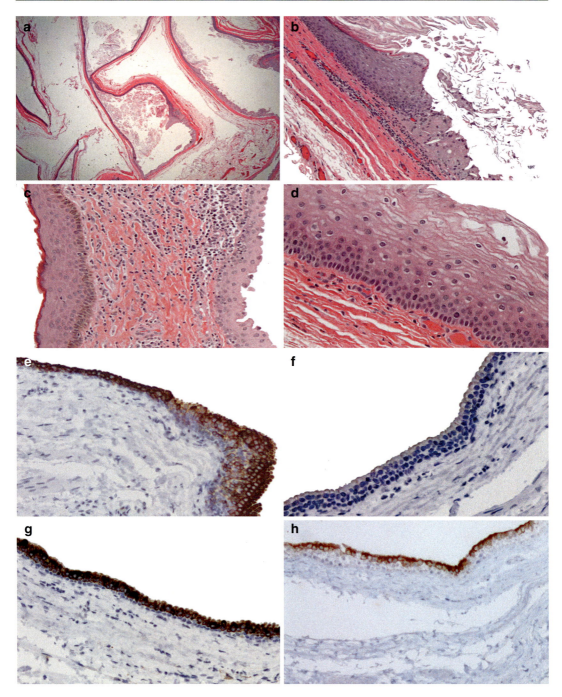

Fig. 7.36 In this panel are shown cysts of the penis with epithelial inclusion cyst (**a** and **b**) and median raphe cyst (**c** and **d**). The immunostaining of the median raphe cyst in (**e**) through (**h**) show positivity for CK7, EMA, CEA, but not for CK20 (**f**) (**a–d**, hematoxylin and eosin staining, ×12.5, ×100, ×200, ×200 as original magnifications, respectively; **e–h**: Anti-CK7 immunostaining, anti-CK20 immunostaining, anti-EMA immunostaining, and anti-CEA immunostaining, Avidin-Biotin Complex, respectively: ×200 as original magnification for all four immunohistochemical microphotographs)

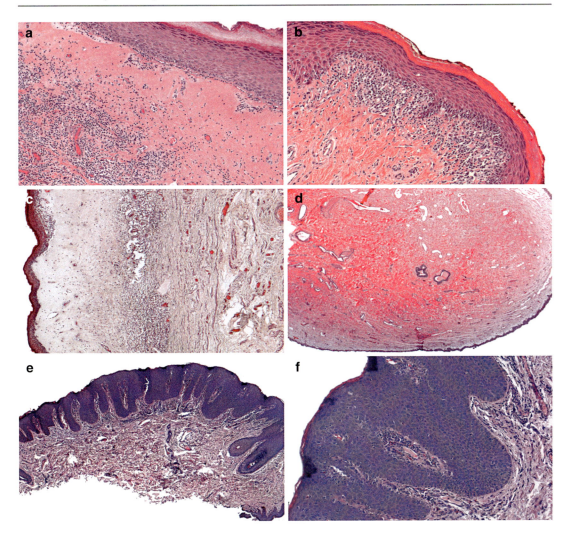

Fig. 7.37 Overgrowth in the penis in childhood and youth is rare but needs to be kept differentiated from the infections. In (**a**), (**b**), and (**c**) are shown lichen sclerosus et atrophicus (balanitis xerotica obliterans), while in (**d**) is shown a verumontanum polyp. In (**e**) and (**f**) are shown microphotographs of a Bowenoid papulosis (**a–f**; hematoxylin and eosin staining apart from the **c**) mictophotograph, which is stained with Movat pentachrome staining; ×100, ×100, ×50, ×12.5, ×40, ×200 as original magnifications, respectively)

reported (Coelho et al 2018). The risk factors for penile squamous cell carcinoma remain HPV infection (HPV types 16 and 18 but also types 11 and 30), hygiene (poor), lack of circumcision, and phimosis.

- *GRO*: Two growth patterns, including the exophytic/papillary with characteristic extensive keratinization, and the flat/ulcerating, mostly on the glans and prepuce.
- *CLM*: A slow-growing carcinoma but about 15% of penile carcinomas have already metastasis at diagnosis. Nearly half of the penile carcinoma are squamous cell carcinomas NOS. The tumor basis at the infiltration may display two different patterns. There is an infiltrating invasion pattern, which is characterized by tumor cell invasion into the stroma with small solid strands and a pushing invasion pattern, which is characterized by tumor cell invasion into the stroma with large blocks with clear-cut tumor-stroma interface. SqCC grading is conventionally subdivided in grade 1, when prominent keratinization with mini-

mal deviation from the morphology of normal or hyperplastic epithelium is seen. Grade 2 occurs when moderate keratinization with relatively high N/C ratio, evident mitotic activity, and less prominent keratinization are encountered, while grade 3 are defined as tumors showing any proportion of anaplasia in neoplastic cells.

The depth of invasion is determined as a measurement (mm) from the epithelial-stromal junction of an adjacent non-neoplastic epithelium to the deepest level of aggression if a non-verruciform neoplasm is examined and from a superficial plane underneath the keratin layer to the deepest level of invasion if a verruciform tumor is under the lens.

- *PGN*: Depends on the extension of the neoplastic disease. If the penile carcinomas are limited to the glans, there is more than 95% of survival, but survival drops to less than 50% in case of the involvement of shaft and regional LNs. Moreover, tumors invading corpora cavernosa have a higher risk of lymph nodal metastasis than tumor invading the *corpus spongiosum* only. Other factors influencing the outcome include the percentage of the poorly differentiated tumor, the depth of invasion, the size of the most extensive nodal metastasis, the extranodal and extracapsular extension, and the HPV status.

The tips and pitfalls of penile malignancies are shown in Box 7.12.

7.11.5 Non-squamous Cell Carcinoma Neoplasms of the Penis

Penile mesenchymal lesions are rare. However, these lesions may occur more often than squamous cell lesions in children and young adults. The most critical mesenchymal lesion is doubtless rhabdomyosarcoma of the penis, which has been described a few times in the literature. Single reports include clear cell sarcoma and leiomyosarcoma.

Multiple Choice Questions and Answers

- LUS-1 What are the histopathologic features of a testicular torsion after 6 hours of the torsion?
  (a) Interstitial hemorrhage and apoptosis of germ cells
  (b) Intratubular hemorrhage, germ cell atypia, and Leydig cell hyperplasia
  (c) Interstitial hemorrhage and edema, germ cell desquamation, and/or necrosis of germ cells
  (d) Interstitial hemorrhage and germ cell atypia
- LUS-2 Which features correspond to the Mikuz score 2 of testicular torsion?
  (a) Diffuse infarction of testicular parenchyma
  (b) Diffuse interstitial hemorrhage, focal necrosis of the germ cells
  (c) Interstitial edema, desquamation of germ cells
  (d) Interstitial edema, limited blood extravasation, desquamation of germ cells
- LUS-3 Which one of the following histopathologic features does NOT belong to acute Mumps orchitis?
  (a) Edema and perivascular lymphocytic exudate
  (b) Diffuse lymphocytic infiltration of the interstitial tissue with focal hemorrhage
  (c) Pronounced destruction of the germinal epithelium with plugging of the tubules by epithelial debris, fibrin, polymorphonuclear leukocytes, and giant cells
  (d) Collagenization
  (e) Calcification
- LUS-4 Gonadoblastoma is a rare lesion of that consists of germ cells that resemble those of dysgerminoma and gonadal stroma cells that look like those of a granulosa or Sertoli tumor. These neoplasms usually arise in dysgenetic gonads. Which of the following disorders is associated with the highest risk of gonadal neoplasia?
  (a) Turner syndrome
  (b) Pure gonadal dysgenesis
  (c) Mixed gonadal dysgenesis

(d) Klinefelter syndrome
  (e) WAGR (Wilms' tumor, aniridia, genital anomalies, mental retardation) syndrome
- LUS-5 A 21-year-old young man presents at his family doctor with a painless testicular mass. A testicular biopsy is performed. Histologic sections show relatively uniform polyhedral cells in sheets that are divided into poorly demarcated lobules by delicate fibrous septa. Lymphocytes and plasma cells are seen in the septa. Some granulomatous inflammation with Langhans-type multinucleated giant cells is seen at places. An immunohistochemical investigation shows the polyhedral cells to be positive for PLAP, OCT3/OCT4, CD117, D2-40, and SALL4 and negative for EMA, CD30, AFP, hCG, and glypican 3. Which is the most likely diagnosis?
  (a) Embryonal carcinoma
  (b) Yolk sac tumor
  (c) Foreign body orchitis with reactive changes
  (d) Seminoma, classic
  (e) Choriocarcinoma
- LUS-6 Octamer-binding transcription factor 4 (OCT 4) is a critical embryonic stem cell regulating gene regulating self-renewal and differentiation of embryonic stem cells and is encoded by the POU5F1 gene at 6p21.3. Which neoplasm is not expressing OCT 4 by immunohistochemistry?
  (a) Non-small cell lung carcinoma
  (b) Neuroblastoma
  (c) Seminoma/dysgerminoma
  (d) Embryonal carcinoma
  (e) Intratubular germ cell neoplasia
  (f) Teratoma-immature neuroepithelium
  (g) Spermatocytic seminoma
- LUS-7 A 15-year-old child presented with dysuria and pyuria for 2 weeks unresponsive to antibiotic treatment. She complained right flank pain and fever accompanied by hematuria for 4 days. Both past medical history and family history were unremarkable. On physical examination, lower abdominal tenderness and mild right costovertebral angle tenderness were noted. An ultrasound of the abdomen showed an ellipsoid mass located in the urinary bladder with inhomogeneous echogenicity. Transurethral resection was performed. The tumor reveals a spindle cell proliferation and an inflammatory infiltrate. There is positive immunohistochemical staining for vimentin, smooth muscle actin, and anaplastic lymphoma kinase (ALK) and negative immunohistochemical staining for desmin, myogenin, myoglobin, CD117, and S100. There is also a polyclonal kappa and lambda light chain Ig staining in plasma cells. Which of the following diagnosis corresponds to the tumor in this question?
  (a) Angiosarcoma
  (b) Synovial sarcoma
  (c) Rhabdomyosarcoma
  (d) Nodular fasciitis
  (e) Inflammatory pseudotumor
- LUS-8 Which genes with germline mutations are associated with an increased risk of developing adenocarcinoma of the prostatic gland?
  (a) *BRCA1*, *BRCA2*, and *HOXB13*
  (b) *RET*, *WNT*, and *RB*
  (c) *TP53*, *WNT*, and *AAT*
  (d) *BRCA1*, *BRCA2*, and *HOXB16*
  (e) *WNT*, *BRCA2*, and *RB*
- LUS-9 Which bacterium is a risk factor to develop adenocarcinoma of the prostate in youth?
  (a) *Moraxella catarrhalis*
  (b) *Propionibacterium acnes*
  (c) *Mycobacterium avium intracellulare*
  (d) *Methicillin-resistant Staphylococcus aureus*
  (e) *Fusobacterium nucleatum*
- LUS-10 Which of the following statements does NOT correspond to prostatic intraepithelial neoplasia (PIN)?
  (a) The intratubular neoplasia shows acinar cells with the presence of prominent nucleoli visible at a 40x magnification or lower.
  (b) There is complete or partial retention of the basal cell layer.
  (c) The most common patterns are micropapillary/cribriform, flat, and tufted.
  (d) Often multicentric in prostatectomy specimens.

References and Recommended Readings

Acién P, Acién MI. The history of female genital tract malformation classifications and proposal of an updated system. Hum Reprod Update. 2011;17(5):693–705. https://doi.org/10.1093/humupd/dmr021. Epub 2011 Jul 4. Review.

Acién P, Acién M, Sánchez-Ferrer M. Complex malformations of the female genital tract. New types and revision of classification. Hum Reprod. 2004;19(10):2377–84. Epub 2004 Aug 27.

Agochukwu NQ, Rastinehad AR, Richter LA, Barak S, Zerbe CS, Holland SM, Pinto PA. Prostatic abscess in a pediatric patient with chronic granulomatous disease: report of a unique case and review of the literature. J Pediatr Surg. 2012;47(2):400–3. https://doi.org/10.1016/j.jpedsurg.2011.11.003. Review. PubMed PMID: 22325401; PubMed Central PMCID: PMC3282836.

Agrons GA, Wagner BJ, Lonergan GJ, Dickey GE, Kaufman MS. From the archives of the AFIP. Genitourinary rhabdomyosarcoma in children: radiologic-pathologic correlation. Radiographics. 1997;17(4):919–37.

AIRTUM Working Group. Italian cancer figures, report 2013: Multiple tumours. Epidemiol Prev. 2013;37(4–5 Suppl 1):1–152. English, Italian.

Akiyama H, Kim JE, Nakashima K, Balmes G, Iwai N, Deng JM, Zhang Z, Martin JF, Behringer RR, Nakamura T, de Crombrugghe B. Osteochondroprogenitor cells are derived from Sox9 expressing precursors. Proc Natl Acad Sci U S A. 2005;102(41):14665–70. Epub 2005 Oct 3. PubMed PMID: 16203988; PubMed Central PMCID: PMC1239942.

Alam S, Goebel J, Pacheco MC, Sheldon C. Papillary urothelial neoplasm of low malignant potential in a pediatric renal transplant recipient (PUNLMP): a case report. Pediatr Transplant. 2007;11(6):680–2.

Alanee S, Shukla AR. Bladder malignancies in children aged <18 years: results from the Surveillance, Epidemiology and End Results database. BJU Int. 2010;106(4):557–60. https://doi.org/10.1111/j.1464-410X.2009.09093.x. Epub 2009 Dec 14.

Alessi E, Coggi A, Gianotti R. Review of 120 biopsies performed on the balanopreputial sac. from zoon's balanitis to the concept of a wider spectrum of inflammatory non-cicatricial balanoposthitis. Dermatology. 2004;208(2):120–4.

Amir H, Mbonde MP, Kitinya JN. Cutaneous squamous cell carcinoma in Tanzania. Cent Afr J Med. 1992;38(11):439–43.

Ander H, Dönmez Mİ, Yitgin Y, Tefik T, Ziylan O, Oktar T, Özsoy M. Urothelial carcinoma of the urinary bladder in pediatric patients: a long-term follow-up. Int Urol Nephrol. 2015;47(5):771–4. https://doi.org/10.1007/s11255-015-0950-z. Epub 2015 Mar 19.

Anton RC, Chakraborty S, Wheeler TM. The significance of intraluminal prostatic crystalloids in benign needle biopsies. Am J Surg Pathol. 1998;22(4):446–9.

Aprikian AG, Zhang ZF, Fair WR. Prostate adenocarcinoma in men younger than 50 years. A retrospective review of 151 patients. Cancer. 1994;74:1768–77. [PubMed: 8082080].

Argani P, Epstein JI. Inverted (Hobnail) high-grade prostatic intraepithelial neoplasia (PIN): report of 15 cases of a previously undescribed pattern of high-grade PIN. Am J Surg Pathol. 2001;25(12):1534–9.

Astigueta JC, Abad MA, Morante C, Pow-Sang MR, Destefano V, Montes J. Characteristics of metastatic prostate cancer occurring in patients under 50 years of age. Actas Urol Esp. 2010;34:327–32. [PubMed: 20470694].

Aubry F, Satie AP, Rioux-Leclercq N, Rajpert-De Meyts E, Spagnoli GC, Chomez P, De Backer O, Jégou B, Samson M. MAGE-A4, a germ cell specific marker, is expressed differentially in testicular tumors. Cancer. 2001;92(11):2778–85.

Barnes RD, Sarembock LA, Abratt RP, Pontin AR. Carcinoma of the penis—the Groote Schuur Hospital experience. J R Coll Surg Edinb. 1989;34(1):44–6. PubMed.

Barnholtz-Sloan JS, Maldonado JL, Pow-sang J, Giuliano AR. Incidence trends in primary malignant penile cancer. Urol Oncol. 2007;25(5):361–7.. Erratum in: Urol Oncol. 2008;26(1):112. Guiliano, Anna R [corrected to Giuliano, Anna R].

Bastianpillai C, Warner R, Beltran L, Green J. Cystitis cystica and glandularis producing large bladder masses in a 16-year-old boy. JRSM Open. 2018;9(3). https://doi.org/10.1177/2054270417746060. eCollection 2018 Mar. PubMed PMID: 29552345; PubMed Central PMCID: PMC5846953.

Bechis SK, Carroll PR, Cooperberg MR. Impact of age at diagnosis on prostate cancer treatment and survival. J Clin Oncol. 2011;29:235–41. [PMCID: PMC3058279] [PubMed: 21135285].

Becker K. Lichen sclerosus in boys. Dtsch Arztebl Int. 2011;108(4):53–8. https://doi.org/10.3238/arztebl.2011.053. Epub 2011 Jan 28. Review. PubMed PMID: 21307992; PubMed Central PMCID: PMC3036008.

Bendsen E, Byskov AG, Laursen SB, Larsen HP, Andersen CY, Westergaard LG. Number of germ cells and somatic cells in human fetal testes during the first weeks after sex differentiation. Hum Reprod. 2003;18(1):13–8. PubMed PMID: 12525434.

Bergholz R, Wenke K. Polyorchidism: a meta-analysis. J Urol. 2009;182(5):2422–7. https://doi.org/10.1016/j.juro.2009.07.063. Epub 2009 Sep 17. PubMed PMID: 19765760.

Bergholz R, Koch B, Spieker T, Lohse K. Polyorchidism: a case report and classification. J Pediatr Surg. 2007;42(11):1933–5. PubMed PMID: 18022450.

Berney DM, Shamash J, Pieroni K, Oliver RT. Loss of CD30 expression in metastatic embryonal carcinoma: the effects of chemotherapy? Histopathology. 2001;39(4):382–5. PubMed PMID: 11683938

Berney DM, Lee A, Randle SJ, Jordan S, Shamash J, Oliver RT. The frequency of intratubular embryonal carcinoma: implications for the pathogenesis of germ cell tumours. Histopathology. 2004;45(2):155–61.

Berrettini A, Castagnetti M, Salerno A, Nappo SG, Manzoni G, Rigamonti W, Caione P. Bladder urothelial neoplasms in pediatric age: experience at three tertiary centers. J Pediatr Urol. 2015;11(1):26.e1–5. https://doi.org/10.1016/j.jpurol.2014.08.008. Epub 2014 Sep 28.

Berry SJ, Coffey DS, Walsh PC, Ewing LL. The development of human benign prostatic hyperplasia with age. J Urol. 1984;132(3):474–9. PubMed PMID: 6206240.

Bismar TA, Lewis JS Jr, Vollmer RT, Humphrey PA. Multiple measures of carcinoma extent versus perineural invasion in prostate needle biopsy tissue in prediction of pathologic stage in a screening population. Am J Surg Pathol. 2003;27(4):432–40. Review.

Boltshauser E, Isler W, Bucher HU, Friderich H. Permanent flaccid paraplegia in children with thoracic spinal cord injury. Paraplegia. 1981;19(4):227–34.

Bono MJ, Reygaert WC. Urinary tract infection. [Updated 2018 Nov 15]. In: StatPearls [Internet]. Treasure Island: StatPearls Publishing; 2019. Available from: https://www.ncbi.nlm.nih.gov/books/NBK470195/

Bostwick DG, Wollan P, Adlakha K. Collagenous micronodules in prostate cancer. A specific but infrequent diagnostic finding. Arch Pathol Lab Med. 1995;119(5):444–7.

Bowen DK, Dielubanza E, Schaeffer AJ. Chronic bacterial prostatitis and chronic pelvic pain syndrome. BMJ Clin Evid. 2015;2015. pii: 1802. Review. PubMed PMID: 26313612; PubMed Central PMCID: PMC4551133.

Brawn PN, Kuhl D, Speights VO, Johnson CF 3rd, Lind M. The incidence of unsuspected metastases from clinically benign prostate glands with latent prostate carcinoma. Arch Pathol Lab Med. 1995;119(8):731–3.

Briet S, Trémeaux JC, Piard F, Striffling VL. 'adenocarcinome de la prostate de l'adolescent et l'adulte jeune. A propos d'une observation chez un homme de 20 ans [Adenocarcinoma of prostate in adolescents and young adults. Apropos of a case in a 20-year-old man]. J Urol (Paris). 1986;92(8):565–8. French. PubMed PMID: 3805751.

Bubendorf L, Schöpfer A, Wagner U, Sauter G, Moch H, Willi N, Gasser TC, Mihatsch MJ. Metastatic patterns of prostate cancer: an autopsy study of 1,589 patients. Hum Pathol. 2000;31(5):578–83.

Calonje E, Fletcher CD, Wilson-Jones E, Rosai J. Retiform hemangioendothelioma. A distinctive form of low-grade angiosarcoma delineated in a series of 15 cases. Am J Surg Pathol. 1994;18(2):115–25.

Cancel-Tassin G, Cussenot O. Prostate cancer genetics. Minerva Urol Nefrol. 2005;57(4):289–300. Review. PubMed PMID: 16247350.

Carter BS, Beaty TH, Steinberg GD, Childs B, Walsh PC. Mendelian inheritance of familial prostate cancer. Proc Natl Acad Sci U S A. 1992;89:3367–71. [PMCID: PMC48868] [PubMed: 1565627].

Cha S, Lee J, Shin JY, Kim JY, Sim SH, Keam B, Kim TM, Kim DW, Heo DS, Lee SH, Kim JI. Clinical application of genomic profiling to find druggable targets for adolescent and young adult (AYA) cancer patients with metastasis. BMC Cancer. 2016;16:170. https://doi.org/10.1186/s12885-016-2209-1. PubMed PMID: 26925973; PubMed Central PMCID: PMC4772349.

Chaumoitre K, Merrot T, Petit P, Sayegh-Martin Y, Alessandrini P, Panuel M. Voiding cystourethrography in boys. Does the presence of the catheter during voiding alter the evaluation of the urethra? J Urol. 2004;171(3):1280–1.

Cheng L, Bergstralh EJ, Cheville JC, Slezak J, Corica FA, Zincke H, Blute ML, Bostwick DG. Cancer volume of lymph node metastasis predicts progression in prostate cancer. Am J Surg Pathol. 1998;22(12):1491–500.

Chertin B, Koulikov D, Fridmans A, Farkas A. Dorsal tunica albuginea plication to correct congenital and acquired penile curvature: a long-term follow-up. BJU Int. 2004;93(3):379–81.

Chikkamuniyappa S, Scott RS, Furman J. Pilonidal sinus of the glans penis associated with actinomyces case reports and review of literature. ScientificWorldJournal. 2004;22(4):908–12.

Chiu CL, Weber DL. Prostatic carcinoma in young adults. JAMA. 1974;230(5):724–6.

Chrisofos M, Papatsoris AG, Lazaris A, Deliveliotis C. Precursor lesions of prostate cancer. Crit Rev Clin Lab Sci. 2007;44(3):243–70. Review.

Clapuyt P, Saint-Martin C, De Batselier P, Brichard B, Wese FX, Gosseye S. Urachal neuroblastoma: first case report. Pediatr Radiol. 1999;29(5):320–1.

Claros OR, Sakai AT, Consolmagno H, Nogueira MP, Testagrossa LA, Fugita OEH. Granulosa cell tumor of the testis in a newborn. Autops Case Rep. 2014;4(1):39–44. https://doi.org/10.4322/acr.2014.006. eCollection 2014 Jan-Mar. PubMed PMID: 28652991; PubMed Central PMCID: PMC5470563.

Coelho RWP, Pinho JD, Moreno JS, Garbis DVEO, do Nascimento AMT, Larges JS, Calixto JRR, Ramalho LNZ, da Silva AAM, Nogueira LR, de Moura Feitoza L, Silva GEB. Penile cancer in Maranhão, Northeast Brazil: the highest incidence globally? BMC Urol. 2018;18(1):50. https://doi.org/10.1186/s12894-018-0365-0. PubMed PMID: 29843769; PubMed Central PMCID: PMC5975591.

Cohen RJ, McNeal JE, Edgar SG, Robertson T, Dawkins HJ. Characterization of cytoplasmic secretory granules (PSG), in prostatic epithelium and their transformation-induced loss in dysplasia and adenocarcinoma. Hum Pathol. 1998;29(12):1488–94.

Cohen RJ, McNeal JE, Redmond SL, Meehan K, Thomas R, Wilce M, Dawkins HJ. Luminal contents of benign and malignant prostatic glands: correspondence to altered secretory mechanisms. Hum Pathol. 2000a;31(1):94–100.

Cohen RJ, Beales MP, McNeal JE. Prostate secretory granules in normal and neoplastic prostate glands: a diagnostic aid to needle biopsy. Hum Pathol. 2000b;31(12):1515–9.

Cohen RJ, O'Brien BA, Wheeler TM. Desquamating apoptotic variant of high-grade prostatic intraepithelial neoplasia: a possible precursor of intraductal prostatic carcinoma. Hum Pathol. 2011;42(6):892–5. https://doi.org/10.1016/j.humpath.2010.09.008. Epub 2011 Feb 1.

Copeland JN, Amin MB, Humphrey PA, Tamboli P, Ro JY, Gal AA. The morphologic spectrum of metastatic prostatic adenocarcinoma to the lung: special emphasis on histologic features overlapping with other pulmonary neoplasms. Am J Clin Pathol. 2002;117(4):552–7.

Cornejo KM, Frazier L, Lee RS, Kozakewich HP, Young RH. Yolk Sac Tumor of the Testis in Infants and Children: A Clinicopathologic Analysis of 33 Cases. Am J Surg Pathol. 2015;39(8):1121–31. https://doi.org/10.1097/PAS.0000000000000432.

D'Aprile M, Santini D, Di Cosimo S, Gravante G, Vincenzi B, Spoto S, Costantino S, Rabitti C, Tonini G. Atypical case of metastatic undifferentiated prostate carcinoma in a 36 years old man: clinical report and literature review. Clin Ter. 2000;151(5):371–4. Review. PubMed PMID: 11141722.

Dahm P, Silverstein AD, Weizer AZ, Crisci A, Vieweg J, Paulson DF. When to diagnose and how to treat prostate cancer in the "not too fit" elderly. Crit Rev Oncol Hematol. 2003;48(2):123–31. PubMed PMID: 14607375.

Dalkin B, Zaontz MR. Rhabdomyosarcoma of the penis in children. J Urol. 1989;141(4):908–9. Review.

Dangle PP, Correa A, Tennyson L, Gayed B, Reyes-Múgica M, Ost M. Current management of paratesticular rhabdomyosarcoma. Urol Oncol. 2016;34(2):84–92. https://doi.org/10.1016/j.urolonc.2015.10.004. Epub 2015 Nov 11. Review.

Dardik M, Epstein JI. Efficacy of restaining prostate needle biopsies with high-molecular weight cytokeratin. Hum Pathol. 2000;31(9):1155–61.

Davis BE, Weigel JW. Adenocarcinoma of the prostate discovered in two young patients following total prostatovesiculectomy for refractory prostatitis. J Urol. 1990;144:744–5.

Dehner LP, Smith BH. Soft tissue tumors of the penis. A clinicopathologic study of 46 cases. Cancer. 1970;25(6):1431–47.

Diapera MJ, Lozon CL, Thompson LD. Malacoplakia of the tongue: a case report and clinicopathologic review of 6 cases. Am J Otolaryngol. 2009;30(2):101–5. https://doi.org/10.1016/j.amjoto.2008.02.014. Epub 2008 Jul 22. Review. PubMed PMID: 19239951.

Dickinson SI. Premalignant and malignant prostate lesions: pathologic review. Cancer Control. 2010;17(4):214–22. Review.

Djulbegovic M, Beyth RJ, Neuberger MM, Stoffs TL, Vieweg J, Djulbegovic B, Dahm P. Screening for prostate cancer: systematic review and meta-analysis of randomised controlled trials. BMJ. 2010;341:c4543. https://doi.org/10.1136/bmj.c4543. Review. PubMed PMID: 20843937; PubMed Central PMCID: PMC2939952.

Dowling CR, Reddihough D, Smith P, Webb N, McNeill R, Clouston D. Transitional cell carcinoma in the paediatric population: be aware of unusual aetiologies. J Paediatr Child Health. 2007;43(11):773–5.

Eakins M, Crooks KK. Congenital polyp of the verumontanum. Urol Radiol. 1982;4(1):49–50.

Edwards S. Balanitis and balanoposthitis: a review. Genitourin Med. 1996;72(3):155–9. Review. PubMed PMID: 8707315; PubMed Central PMCID: PMC1195642.

El Ouakdi M, Zermani R, Boujnah H, Ayed M, Ben Jilani S, Zmerli S. Genitourinary malacoplakia. Apropos of 5 cases. A review of the literature. Ann Urol (Paris). 1989;23(3):197–200. Review. French.

Elliott JE, Abduljabar H, Morris M. Presurgical management of dysmenorrhea and endometriosis in a patient with Mayer-Rokitansky-Kuster-Hauser syndrome. Fertil Steril. 2011;96(2):e86–9. https://doi.org/10.1016/j.fertnstert.2011.06.006. Epub 2011 Jun 30.

Epsi EZ, Sultana SZ, Mannan S, Azam AS, Choudhury S, Ahmed Z, Farjan S, Kabir A, Ismatsara M, Yesmin M, Zisa RS, Khan SH. Study of prostatic volume and its variations in different age groups of Bangladeshi cadaver. Mymensingh Med J. 2016a;25(4):615–9. PubMed PMID: 27941719.

Epsi EZ, Khalil M, Sultana SZ, Zaman US, Choudhury S, Ameen S, Sultana R, Tabassum R, Nawshin N, Azam MS, Akhter SM. Histomorphological study on number of acini of the prostate gland of Bangladeshi cadaver. Mymensingh Med J. 2016b;25(2):232–6. PubMed PMID: 27277353.

Epstein JI. Gleason score 2–4 adenocarcinoma of the prostate on needle biopsy: a diagnosis that should not be made. Am J Surg Pathol. 2000;24(4):477–8. Review.

Epstein JI. Precursor lesions to prostatic adenocarcinoma. Virchows Arch. 2009;454(1):1–16. https://doi.org/10.1007/s00428-008-0707-5. Epub 2008 Dec 2. Review.

Epstein JI. An update of the Gleason grading system. J Urol. 2010;183(2):433–40. https://doi.org/10.1016/j.juro.2009.10.046. Epub 2009 Dec 14. Review.

Fernbach SK, Feinstein KA, Spencer K, Lindstrom CA. Ureteral duplication and its complications. Radiographics. 1997;17(1):109–27.

Fistarol SK, Itin PH. Diagnosis and treatment of lichen sclerosus: an update. Am J Clin Dermatol. 2013;14(1):27–47. https://doi.org/10.1007/s40257-012-0006-4. Review. PubMed PMID: 23329078; PubMed Central PMCID: PMC3691475.

Frias-Kletecka MC, MacLennan GT. Mesothelioma of the tunica vaginalis. J Urol. 2007;178(4 Pt 1):1489. Epub 2007 Aug 16.

Game X, Villers A, Malavaud B, Sarramon J. Bladder cancer arising in a spina bifida patient. Urology. 1999;54(5):923.

References and Recommended Readings

Ganesan GS, Cory D, Mitchell ME, Jones JA. Magnetic resonance imaging of penile rhabdomyosarcoma. Br J Radiol. 1992;65(770):175–8.

Garcia FU, Taylor CA, Hou JS, Rukstalis DB, Stearns ME. Increased cellularity of tumor-encased native vessels in prostate carcinoma is a marker for tumor progression. Mod Pathol. 2000;13(7):717–22.

Gatto-Weis C, Topolsky D, Sloane B, Hou JS, Qu H, Fyfe BS. Ulcerative balanoposthitis of the foreskin as a manifestation of chronic lymphocytic leukemia: case report and review of the literature. Urology. 2000;56(4):669. Review.

Ghousheh AI, Durkee CT, Groth TW. Advanced transitional cell carcinoma of the bladder in a 16-year-old girl with Hinman syndrome. Urology. 2012;80(5):1141–3. https://doi.org/10.1016/j.urology.2012.04.057. Epub 2012 Jun 27.

Giaquinto C, Del Mistro A, De Rossi A, Bertorelle R, Giacomet V, Ruga E, Minucci D. Vulvar carcinoma in a 12-year-old girl with vertically acquired human immunodeficiency virus infection. Pediatrics. 2000;106(4):E57.

Godfrey JC, Vaughan MC, Williams JV. Successful treatment of bowenoid papulosis in a 9-year-old girl with vertically acquired human immunodeficiency virus. Pediatrics. 2003;112(1 Pt 1):e73–6. Review.

Gomella LG, Liu XS, Trabulsi EJ, Kelly WK, Myers R, Showalter T, Dicker A, Wender R. Screening for prostate cancer: the current evidence and guidelines controversy. Can J Urol. 2011;18(5):5875–83. Review.

Gordon MD, Corless C, Renshaw AA, Beckstead J. CD99, keratin, and vimentin staining of sex cord-stromal tumors, normal ovary, and testis. Mod Pathol. 1998;11(8):769–73. PubMed PMID: 9720506.

Goriely A, Hansen RM, Taylor IB, Olesen IA, Jacobsen GK, McGowan SJ, Pfeifer SP, McVean GA, Rajpert-De Meyts E, Wilkie AO. Activating mutations in FGFR3 and HRAS reveal a shared genetic origin for congenital disorders and testicular tumors. Nat Genet. 2009;41(11):1247–52. https://doi.org/10.1038/ng.470. Epub 2009 Oct 25. PubMed PMID: 19855393; PubMed Central PMCID: PMC2817493.

Gray SW, Skandalakis JE. Embryology for Surgeons: The Embryological Basis for the Treatment of Congenital Defects. Philadelphia: WB Saunders Co; 1972. p. 263–82.

Greenfield SP, Williot P, Kaplan D. Gross hematuria in children: a ten-year review. Urology. 2007;69(1):166–9.

Grimbizis GF, Gordts S, Di Spiezio Sardo A, Brucker S, De Angelis C, Gergolet M, Li TC, Tanos V, Brölmann H, Gianaroli L, Campo R. The ESHRE-ESGE consensus on the classification of female genital tract congenital anomalies. Gynecol Surg. 2013a;10(3):199–212. Epub 2013 Jun 13. PubMed PMID: 23894234; PubMed Central PMCID: PMC3718988.

Grimbizis GF, Gordts S, Di Spiezio SA, Brucker S, De Angelis C, Gergolet M, Li TC, Tanos V, Brölmann H, Gianaroli L, Campo R. The ESHRE/ESGE consensus on the classification of female genital tract congenital anomalies. Hum Reprod. 2013b;28(8):2032–44. https://doi.org/10.1093/humrep/det098. Epub 2013 Jun 14. PubMed PMID: 23771171; PubMed Central PMCID: PMC3712660. Jun;49(6):944–55.

Guillou L, Wadden C, Coindre JM, Krausz T, Fletcher CD. "Proximal-type" epithelioid sarcoma, a distinctive aggressive neoplasm showing rhabdoid features. Clinicopathologic, immunohistochemical, and ultrastructural study of a series. Am J Surg Pathol. 1997;21(2):130–46. Review.

Gülpinar O, Soygür T, Baltaci S, Akand M, Kankaya D. Transitional cell carcinoma of bladder with lamina propria invasion in a 10-year-old boy. Urology. 2006;68(1):204.e1–3.

Gurel B, Ali TZ, Montgomery EA, Begum S, Hicks J, Goggins M, Eberhart CG, Clark DP, Bieberich CJ, Epstein JI, De Marzo AM. NKX3.1 as a marker of prostatic origin in metastatic tumors. Am J Surg Pathol. 2010;34(8):1097–105. https://doi.org/10.1097/PAS.0b013e3181e6cbf3. PubMed PMID: 20588175; PubMed Central PMCID: PMC3072223.

Hacker HW, Winiker H, Caduff J, Schwoebel MG. Inflammatory tumour of the prostate in a 4-year-old boy. J Pediatr Urol. 2009;5(6):516–8. https://doi.org/10.1016/j.jpurol.2009.03.023. Epub 2009 May 23.

Halabi M, Oliva E, Mazal PR, Breitenecker G, Young RH. Prostatic tissue in mature cystic teratomas of the ovary: a report of four cases, including one with features of prostatic adenocarcinoma, and cytogenetic studies. Int J Gynecol Pathol. 2002;21(3):261–7. PubMed PMID: 12068172.

Halasz C, Silvers D, Crum CP. Bowenoid papulosis in three-year-old girl. J Am Acad Dermatol. 1986;14(2 Pt 2):326–30.

Harel M, Ferrer FA, Shapiro LH, Makari JH. Future directions in risk stratification and therapy for advanced pediatric genitourinary rhabdomyosarcoma. Urol Oncol. 2016;34(2):103–15. https://doi.org/10.1016/j.urolonc.2015.09.013. Epub 2015 Oct 28. Review.

Harvey AM, Grice B, Hamilton C, Truong LD, Ro JY, Ayala AG, Zhai QJ. Diagnostic utility of P504S/p63 cocktail, prostate-specific antigen, and prostatic acid phosphatase in verifying prostatic carcinoma involvement in seminal vesicles: a study of 57 cases of radical prostatectomy specimens of pathologic stage pT3b. Arch Pathol Lab Med. 2010;134(7):983–8. https://doi.org/10.1043/2009-0277-OA.1.

Heifetz SA, Cushing B, Giller R, Shuster JJ, Stolar CJ, Vinocur CD, Hawkins EP. Immature teratomas in children: pathologic considerations: a report from the combined pediatric oncology group/Children's Cancer group. Am J Surg Pathol. 1998;22(9):1115–24. PubMed PMID: 9737245.

Hemal AK, Kumar R, Wadhwa SN. Carcinoma penis in a young boy. A case report. Indian J Cancer. 1996;33(2):108–10. Review.

Henneberry JM, Kahane H, Humphrey PA, Keetch DW, Epstein JI. The significance of intraluminal crystalloids in benign prostatic glands on needle biopsy. Am J Surg Pathol. 1997;21(6):725–8.

Herman CM, Wilcox GE, Kattan MW, Scardino PT, Wheeler TM. Lymphovascular invasion as a predictor of disease progression in prostate cancer. Am J Surg Pathol. 2000;24(6):859–63.

Hodges KB, Lopez-Beltran A, Davidson DD, Montironi R, Cheng L. Urothelial dysplasia and other flat lesions of the urinary bladder: clinicopathologic and molecular features. Hum Pathol. 2010;41(2):155–62. https://doi.org/10.1016/j.humpath.2009.07.002. Epub 2009 Sep 16. Review.

Huppmann AR, Pawel BR. Polyps and masses of the pediatric urinary bladder: a 21-year pathology review. Pediatr Dev Pathol. 2011;14(6):438–44. https://doi.org/10.2350/11-01-0958-OA.1. Epub 2011 Jul 27.

Inagaki T, Nagata M, Kaneko M, Amagai T, Iwakawa M, Watanabe T. Carcinosarcoma with rhabdoid features of the urinary bladder in a 2-year-old girl: possible histogenesis of stem cell origin. Pathol Int. 2000;50(12):973–8.

Isa SS, Almaraz R, Magovern J. Leiomyosarcoma of the penis. Case report and review of the literature. Cancer. 1984;54(5):939–42.

Isaac J, Lowichik A, Cartwright P, Rohr R. Inverted papilloma of the urinary bladder in children: case report and review of prognostic significance and biological potential behavior. J Pediatr Surg. 2000;35(10):1514–6. Review. PubMed.

James GK, Pudek M, Berean KW, Diamandis EP, Archibald BL. Salivary duct carcinoma secreting prostate-specific antigen. Am J Clin Pathol. 1996;106(2):242–7.

Janzen J, Bersch U, Pietsch-Breitfeld B, Pressler H, Michel D. Bu¨ltmann B. Urinary bladder biopsies in spinal cord injured patients. Spinal Cord. 2001;39:568–70.

Jiang Z, Wu CL, Woda BA, Dresser K, Xu J, Fanger GR, Yang XJ. P504S/alpha-methylacyl-CoA racemase: a useful marker for diagnosis of small foci of prostatic carcinoma on needle biopsy. Am J Surg Pathol. 2002;26(9):1169–74.

Johnsen SG. Testicular biopsy score countDOUBLEHYPHENa method for registration of spermatogenesis in human testes: normal values and results in 335 hypogonadal males. Hormones. 1970;1(1):2–25.

Jones MA, Young RH, Scully RE. Malignant mesothelioma of the tunica vaginalis. A clinicopathologic analysis of 11 cases with review of the literature. Am J Surg Pathol. 1995;19(7):815–25. Review.

Jones GE, Richmond AK, Navti O, Mousa HA, Abbs S, Thompson E, Mansour S, Vasudevan PC. Renal anomalies and lymphedema distichiasis syndrome. A rare association? Am J Med Genet A. 2017;173(8):2251–6. https://doi.org/10.1002/ajmg.a.38293. Epub 2017 May 23.

Kahn DG. Ossifying seminoma of the testis. Arch Pathol Lab Med. 1993;117(3):321–2.

Kao CS, Cornejo KM, Ulbright TM, Young RH. Juvenile granulosa cell tumors of the testis: a clinicopathologic study of 70 cases with emphasis on its wide morphologic spectrum. Am J Surg Pathol. 2015;39(9):1159–69. https://doi.org/10.1097/PAS.0000000000000450.

Kawada T, Muto K, Kanai T, Kuratomi Y. Prostate-specific antigen screening of workers under the age of 40 in Japan. Cancer Epidemiol. 2009;33(3–4):309–10. https://doi.org/10.1016/j.canep.2009.07.010. Epub 2009 Aug 29.

Khan MA, Puri P, Devaney D. Mesothelioma of tunica vaginalis testis in a child. J Urol. 1997;158(1):198–9.

Kieran K, Shnorhavorian M. Current standards of care in bladder and prostate rhabdomyosarcoma. Urol Oncol. 2016;34(2):93–102. Review.

Kimura T, Onozawa M, Miyazaki J, Matsuoka T, Joraku A, Kawai K, Nishiyama H, Hinotsu S, Akaza H. Prognostic impact of young age on stage IV prostate cancer treated with primary androgen deprivation therapy. Int J Urol. 2014;21(6):578–83. https://doi.org/10.1111/iju.12389. Epub 2014 Jan 9. PubMed PMID: 24405474.

Kleinschmidt-DeMasters BK. Dural metastases. A retrospective surgical and autopsy series. Arch Pathol Lab Med. 2001;125(7):880–7.

Koga S, Arakaki Y, Matsuoka M, Ohyama C. Parameatal urethral cysts of the glans penis. Br J Urol. 1990;65(1):101–3.

Komasara L, Gołębiewski A, Anzelewicz S, Czauderna P. A review on surgical techniques and organ sparing procedures in bladder/prostate rhabdomyosarcoma. Eur J Pediatr Surg. 2014;24(6):467–73. https://doi.org/10.1055/s-0034-1396424. Epub 2014 Dec 8. Review.

Korrect GS, Minevich EA, Sivan B. High-grade transitional cell carcinoma of the pediatric bladder. J Pediatr Urol. 2012;8(3):e36–8. https://doi.org/10.1016/j.jpurol.2011.10.024. Epub 2011 Nov 21.

Krishnan B, Truong LD. Prostatic adenocarcinoma diagnosed by urinary cytology. Am J Clin Pathol. 2000;113(1):29–34.

Kristiansen S, Svensson Å, Drevin L, Forslund O, Torbrand C, Bjartling C. Risk factors for penile intraepithelial neoplasia: a population-based register study in Sweden, 2000–2012. Acta Derm Venereol. 2019;99(3):315–20. https://doi.org/10.2340/00015555-3083. PubMed PMID: 30426132.

Kryvenko ON, Diaz M, Meier FA, Ramineni M, Menon M, Gupta NS. Findings in 12-core transrectal ultrasound-guided prostate needle biopsy that predict more advanced cancer at prostatectomy: analysis of 388 biopsy-prostatectomy pairs. Am J Clin Pathol. 2012;137(5):739–46. https://doi.org/10.1309/AJCPWIZ9X2DMBEBM.

Kurdgelashvili G, Dores GM, Srour SA, Chaturvedi AK, Huycke MM, Devesa SS. Incidence of potentially human papillomavirus-related neoplasms in the United States, 1978 to 2007. Cancer. 2013;119(12):2291–9. https://doi.org/10.1002/cncr.27989. Epub 2013 Apr 11.

Lane RB Jr, Lane CG, Mangold KA, Johnson MH, Allsbrook WC Jr. Needle biopsies of the prostate: what constitutes adequate histologic sampling? Arch Pathol Lab Med. 1998;122(9):833–5.

Laurichesse Delmas H, Kohler M, Doray B, Lémery D, Francannet C, Quistrebert J, Marie C, Perthus I. Congenital unilateral renal agenesis: Prevalence, prenatal diagnosis, associated anomalies. Data from two birth-defect registries. Birth Defects Res. 2017;109(15):1204–11. https://doi.org/10.1002/bdr2.1065. Epub 2017 Jul 19.

Leav I, McNeal JE, Ho SM, Jiang Z. Alpha-methylacyl-CoA racemase (P504S) expression in evolving carcinomas within benign prostatic hyperplasia and in cancers of the transition zone. Hum Pathol. 2003;34(3):228–33.

Lebowitz RL, Olbing H, Parkkulainen KV, Smellie JM, Tamminen-Möbius TE. International system of radiographic grading of vesicoureteric reflux. International reflux study in children. Pediatr Radiol. 1985;15(2):105–9. PubMed PMID: 3975102.

Lee PA, Houk CP. Cryptorchidism. Curr Opin Endocrinol Diabetes Obes. 2013;20(3):210–6. https://doi.org/10.1097/MED.0b013e32835ffc7d.

Lee TK, Miller JS, Epstein JI. Rare histological patterns of prostatic ductal adenocarcinoma. Pathology. 2010;42(4):319–24. https://doi.org/10.3109/00313021003767314.

Lerena J, Krauel L, García-Aparicio L, Vallasciani S, Suñol M, Rodó J. Transitional cell carcinoma of the bladder in children and adolescents: six-case series and review of the literature. J Pediatr Urol. 2010;6(5):481–5. https://doi.org/10.1016/j.jpurol.2009.11.006. Epub 2010 Jan 18. Review.

Leroy X, Ballereau C, Villers A, Saint F, Aubert S, Gosselin B, Porchet N, Copin MC. MUC6 is a marker of seminal vesicle-ejaculatory duct epithelium and is useful for the differential diagnosis with prostate adenocarcinoma. Am J Surg Pathol. 2003;27(4):519–21.

Lestre SI, Gameiro CD, João A, Lopes MJ. Granulomas of the penis: a rare complication of intravesical therapy with Bacillus Calmette-Guerin. An Bras Dermatol. 2011;86(4):759–62. Review. English, Portuguese.

Li R, Wheeler T, Dai H, Ayala G. Neural cell adhesion molecule is upregulated in nerves with prostate cancer invasion. Hum Pathol. 2003;34(5):457–61.

Li XB, Xing NZ, Wang Y, Hu XP, Yin H, Zhang XD. Transitional cell carcinoma in renal transplant recipients: a single center experience. Int J Urol. 2008;15(1):53–7. https://doi.org/10.1111/j.1442-2042.2007.01932.x.

Lim J, Goriely A, Turner GD, Ewen KA, Jacobsen GK, Graem N, Wilkie AO, Rajpert-De ME. OCT2, SSX and SAGE1 reveal the phenotypic heterogeneity of spermatocytic seminoma reflecting distinct subpopulations of spermatogonia. J Pathol. 2011;224(4):473–83. https://doi.org/10.1002/path.2919. Epub 2011 Jun 27. PubMed PMID: 21706474; PubMed Central PMCID: PMC3210831.

Lin DW, Porter M, Montgomery B. Treatment and survival outcomes in young men diagnosed with prostate cancer: a population-based cohort study. Cancer. 2009;115:2863–71.

Looijenga LH, Hersmus R, Gillis AJ, Pfundt R, Stoop HJ, van Gurp RJ, Veltman J, Beverloo HB, van Drunen E, van Kessel AG, Pera RR, Schneider DT, Summersgill B, Shipley J, McIntyre A, van der Spek P, Schoenmakers E, Oosterhuis JW. Genomic and expression profiling of human spermatocytic seminomas: primary spermatocyte as tumorigenic precursor and DMRT1 as candidate chromosome 9 gene. Cancer Res. 2006;66(1):290–302.

Looijenga LH, Stoop H, Hersmus R, Gillis AJ, Wolter OJ. Genomic and expression profiling of human spermatocytic seminomas: pathogenetic implications. Int J Androl. 2007;30(4):328–35; discussion 335–6. Epub 2007 Jun 15.

Lorentzen M, Rohr N. Urinary bladder tumours in children. Case report of inverted papilloma. Scand J Urol Nephrol. 1979;13(3):323–7.

Lunacek A, Schwentner C, Oswald J, Fritsch H, Sergi C, Thomas LN, Rittmaster RS, Klocker H, Neuwirt H, Bartsch G, Radmayr C. Fetal distribution of 5alpha-reductase 1 and 5alpha-reductase 2, and their input on human prostate development. J Urol. 2007a;178(2):716–21. Epub 2007 Jun 14.

Lunacek A, Oswald J, Schwentner C, Schlenck B, Horninger W, Fritsch H, Longato S, Sergi C, Bartsch G, Radmayr C. Growth curves of the fetal prostate based on three-dimensional reconstructions: a correlation with gestational age and maternal testosterone levels. BJU Int. 2007b;99(1):151–6. Epub 2006 Oct 11.

Madan R, Singh L, Haresh KP, Rath GK. Metastatic Adenocarcinoma of Prostate in a 28-Year-Old Male: The outcome is poor in young patients? Indian J Palliat Care. 2015;21(2):242–4. https://doi.org/10.4103/0973-1075.156510. PubMed PMID: 26009681; PubMed Central PMCID: PMC4441189.

Magoha GA, Ngumi ZW. Cancer of the penis at Kenyatta National Hospital. East Afr Med J. 2000;77(10):526–30.

Mallon E, Hawkins D, Dinneen M, Francics N, Fearfield L, Newson R, Bunker C. Circumcision and genital dermatoses. Arch Dermatol. 2000;136(3):350–4.

Marchalik D, Krishnan J, Verghese M, Venkatesan K. Clear cell adenocarcinoma of the bladder with intravesical cervical invasion. BMJ Case Rep. 2015;2015. https://doi.org/10.1136/bcr-2015-209893. pii: bcr2015209893. PubMed PMID: 26109625; PubMed Central PMCID: PMC4480087.

Marcus DM, Goodman M, Jani AB, Osunkoya AO, Rossi PJ. A comprehensive review of incidence and survival in patients with rare histological variants of prostate cancer in the United States from 1973 to 2008. Prostate Cancer Prostatic Dis. 2012;15(3):283–8. https://doi.org/10.1038/pcan.2012.4. Epub 2012 Feb 21. Review. PubMed PMID: 22349984.

Marinaccio A, Binazzi A, Di Marzio D, Scarselli A, Verardo M, Mirabelli D, Gennaro V, Mensi C, Merler E, De Zotti R, Mangone L, Chellini E, Pascucci C, Ascoli V, Menegozzo S, Cavone D, Cauzillo G,

Nicita C, Melis M, Iavicoli S. Incidence of extrapleural malignant mesothelioma and asbestos exposure, from the Italian national register. Occup Environ Med. 2010;67(11):760–5. https://doi.org/10.1136/oem.2009.051466. Epub 2010 Aug 25.

Marinoni F, Destro F, Selvaggio GGO, Riccipetitoni G. Urothelial carcinoma in children: A case series. Bull Cancer. 2018;105(6):556–61. https://doi.org/10.1016/j.bulcan.2018.03.002. Epub 2018 May 1. Review.

Maru N, Ohori M, Kattan MW, Scardino PT, Wheeler TM. Prognostic significance of the diameter of perineural invasion in radical prostatectomy specimens. Hum Pathol. 2001;32(8):828–33.

Mazzucchelli R, Barbisan F, Scarpelli M, Lopez-Beltran A, van der Kwast TH, Cheng L, Montironi R. Is incidentally detected prostate cancer in patients undergoing radical cystoprostatectomy clinically significant? Am J Clin Pathol. 2009;131(2):279–83. https://doi.org/10.1309/AJCP4OCYZBAN9TJU.

McDaid J, Farkash EA, Steele DJ, Martins PN, Kotton CN, Elias N, Ko DS, Colvin RB, Hertl M. Transitional cell carcinoma arising within a pediatric donor renal transplant in association with BK nephropathy. Transplantation. 2013;95(5):e28–30. https://doi.org/10.1097/TP.0b013e31828235ec.

McKenney JK. Mesenchymal tumors of the prostate. Mod Pathol. 2018;31(S1):S133–42. https://doi.org/10.1038/modpathol.2017.155.

Mehta L, Jim B. Hereditary Renal Diseases. Semin Nephrol. 2017;37(4):354–61. https://doi.org/10.1016/j.semnephrol.2017.05.007. Review.

Merrill RM, Bird JS. Effect of young age on prostate cancer survival: a population-based assessment (United States). Cancer Causes Control. 2002;13(5):435–43.

Mesia L, Georgsson S, Zuretti A. Ossified intratesticular mucinous tumor. Arch Pathol Lab Med. 1999;123(3):244–6.

Miettinen M, Fetsch JF. Reticulohistiocytoma (solitary epithelioid histiocytoma): a clinicopathologic and immunohistochemical study of 44 cases. Am J Surg Pathol. 2006;30(4):521–8.

Minkowitz G, Lee M, Minkowitz S. Pilomatricoma of the testicle. An ossifying testicular tumor with hair matrix differentiation. Arch Pathol Lab Med. 1995;119(1):96–9.

Molberg KH, Mikhail A, Vuitch F. Crystalloids in metastatic prostatic adenocarcinoma. Am J Clin Pathol. 1994;101(3):266–8.

Montgomery DA, Azmy AF. Rhabdomyosarcoma relapse in an unusual site. Pediatr Surg Int. 2005;21(7):555–6. Epub 2005 Apr 21.

Montironi R, Mazzucchelli R, Lopez-Beltran A, Scarpelli M, Cheng L. The Gleason grading system: where are we now? Diagn Histopathol. 2011;17(10):419–27. https://doi.org/10.1016/j.mpdhp.2011.06.008.

Mooney EE, Vaidya KP, Tavassoli FA. Ossifying well-differentiated Sertoli-Leydig cell tumor of the ovary. Ann Diagn Pathol. 2000;4(1):34–8.

Morris BJ, Wamai RG. Biological basis for the protective effect conferred by male circumcision against HIV infection. Int J STD AIDS. 2012;23(3):153–9. https://doi.org/10.1258/ijsa.2011.011228. Review.

Morris BJ, Wodak AD, Mindel A, Schrieber L, Duggan KA, Dilley A, Willcourt RJ, Lowy M, Cooper DA. The 2010 Royal Australasian College of Physicians' policy statement 'Circumcision of infant males' is not evidence based. Intern Med J. 2012a;42(7):822–8. https://doi.org/10.1111/j.1445-5994.2012.02823.x.

Morris BJ, Waskett JH, Banerjee J, Wamai RG, Tobian AA, Gray RH, Bailis SA, Bailey RC, Klausner JD, Willcourt RJ, Halperin DT, Wiswell TE, Mindel A. A 'snip' in time: what is the best age to circumcise? BMC Pediatr. 2012b;28(12):20. https://doi.org/10.1186/1471-2431-12-20. Review. PubMed PMID: 22373281; PubMed Central PMCID: PMC3359221.

Moul JW. Population screening for prostate cancer and emerging concepts for young men. Clin Prostate Cancer. 2003;2(2):87–97. Review.

Neogi S, Kariholu PL, Dhakre G, Gupta V, Agarwal N, Bhadani P. Malignant urothelial carcinoma of urinary bladder in a young child: a rare case report. Urology. 2013;81(4):888–90. https://doi.org/10.1016/j.urology.2012.12.016. Epub 2013 Feb 8.

Nistal M, Paniagua R. Testicular biopsy. Contemporary interpretation. Urol Clin North Am. 1999;26:555–93.

Nixon AJ, Neuberg D, Hayes DF, Gelman R, Connolly JL, Schnitt S, Abner A, Recht A, Vicini F, Harris JR. Relationship of patient age to pathologic features of the tumor and prognosis for patients with stage I or II breast cancer. J Clin Oncol. 1994;12(5):888–94. PubMed PMID: 8164038.

Novák J, Bárta J, Klézl P. Carcinoma of the penis. Cas Lek Cesk. 2007;146(10):767–70. Czech.

Oesterling JE, Epstein JI, Gearhart JP. Transitional cell carcinoma of the bladder in an adolescent with Turner's syndrome. J Urol. 1987;137(3):398–400.

Ohori M, Scardino PT, Lapin SL, Seale-Hawkins C, Link J, Wheeler TM. The mechanisms and prognostic significance of seminal vesicle involvement by prostate cancer. Am J Surg Pathol. 1993;17(12):1252–61.

Oliai BR, Kahane H, Epstein JI. Can basal cells be seen in adenocarcinoma of the prostate?: an immunohistochemical study using high molecular weight cytokeratin (clone 34betaE12) antibody. Am J Surg Pathol. 2002;26(9):1151–60.

Oosterhuis JW, Looijenga LH. Testicular germ-cell tumours in a broader perspective. Nat Rev Cancer. 2005;5(3):210–22. Review.

Oppelt P, Renner SP, Brucker S, Strissel PL, Strick R, Oppelt PG, Doerr HG, Schott GE, Hucke J, Wallwiener D, Beckmann MW. The VCUAM (Vagina Cervix Uterus Adnex-associated Malformation) classification: a new classification for genital malformations. Fertil Steril. 2005;84(5):1493–7.

Özel A, Alıcı Davutoğlu E, Erenel H, Karslı MF, Korkmaz SÖ, Madazlı R. Outcome after prenatal diagnosis of fetal urinary tract abnormalities: A tertiary center experience. J Turk Ger Gynecol Assoc. 2018.; https://doi.org/10.4274/jtgga.2017.0132. [Epub ahead of print]. PubMed Central PMCID: PMC5387761.

Pacelli A, Lopez-Beltran A, Egan AJ, Bostwick DG. Prostatic adenocarcinoma with glomeruloid features. Hum Pathol. 1998;29(5):543–6.

Pantalone KM, Faiman C. Male hypogonadism: more than just a low testosterone. Cleve Clin J Med. 2012;79(10):717–25. https://doi.org/10.3949/ccjm.79a.11174. Review.

Papantoniou N, Papoutsis D, Daskalakis G, Chatzipapas I, Sindos M, Papaspyrou I, Mesogitis S, Antsaklis A. Prenatal diagnosis of prune-belly syndrome at 13 weeks of gestation: case report and review of literature. J Matern Fetal Neonatal Med. 2010;23(10):1263–7. https://doi.org/10.3109/14767050903544777. Review. PubMed PMID: 20504067.

Patel R, Tery T, Ninan GK. Transitional cell carcinoma of the bladder in first decade of life. Pediatr Surg Int. 2008;24(11):1265–8. https://doi.org/10.1007/s00383-008-2251-4. Epub 2008 Sep 24.

Pavlakis K, Stravodimos K, Kapetanakis T, Gregorakis A, Athanassiadou S, Tzaida O, Constantinides C. Evaluation of routine application of P504S, 34betaE12 and p63 immunostaining on 250 prostate needle biopsy specimens. Int Urol Nephrol. 2010;42(2):325–30. https://doi.org/10.1007/s11255-009-9622-1. Epub 2009 Aug 5. PubMed PMID: 19655267.

Pettenati C, Ingersoll MA. Mechanisms of BCG immunotherapy and its outlook for bladder cancer. Nat Rev Urol. 2018; https://doi.org/10.1038/s41585-018-0055-4. [Epub ahead of print] Review.

Pickup M, Van der Kwast TH. My approach to intraductal lesions of the prostate gland. J Clin Pathol. 2007;60(8):856–65. Epub 2007 Jan 19. Review. PubMed PMID: 17237185; PubMed Central PMCID: PMC1994484.

Polat H, Utangac MM, Gulpinar MT, Cift A, Erdogdu IH, Turkcu G. Urothelial neoplasm of the bladder in childhood and adolescence: a rare disease. Int Braz J Urol. 2016;42(2):242–6. https://doi.org/10.1590/S1677-5538.IBJU.2015.0200. PubMed PMID: 27256177; PubMed Central PMCID: PMC4871383.

Querin G, Bertolin C, Da Re E, Volpe M, Zara G, Pegoraro E, Caretta N, Foresta C, Silvano M, Corrado D, Iafrate M, Angelini L, Sartori L, Pennuto M, Gaiani A, Bello L, Semplicini C, Pareyson D, Silani V, Ermani M, Ferlin A, Sorarù G, Italian Study Group on Kennedy's disease. Non-neural phenotype of spinal and bulbar muscular atrophy: results from a large cohort of Italian patients. J Neurol Neurosurg Psychiatry. 2016;87(8):810–6. https://doi.org/10.1136/jnnp-2015-311305. Epub 2015 Oct 26. PubMed PMID: 26503015; PubMed Central PMCID: PMC4975824.

Rajpert-De ME. Testis: Spermatocytic seminoma. Atlas Genet Cytogenet Oncol Haematol. 2013;17(6):435–40. http://AtlasGeneticsOncology.org/Tumors/SpermatSeminID5119.html

Rajpert-De Meyts E, Jacobsen GK, Bartkova J, Aubry F, Samson M, Bartek J, Skakkebaek NE. The immunohistochemical expression pattern of Chk2, p53, p19INK4d, MAGE-A4 and other selected antigens provides new evidence for the premeiotic origin of spermatocytic seminoma. Histopathology. 2003;42(3):217–26.

Ramos JZ, Pack GT. Primary embryonal rhabdomyosarcoma of the penis in 2-year-old child. J Urol. 1966;96(6):928–32.

Rando Sous A, Pérez-Utrilla Pérez M, Aguilera Bazán A, Tabernero Gomez A, Cisneros Ledo J, De la Peña Barthel J. A review of penile cancer. Adv Urol. 2009:415062. https://doi.org/10.1155/2009/415062. Epub 2010 Feb 16. PubMed PMID: 20182534; PubMed Central PMCID: PMC2825548.

Renshaw AA. Adequate tissue sampling of prostate core needle biopsies. Am J Clin Pathol. 1997;107(1):26–9.

Rey RA, Grinspon RP, Gottlieb S, Pasqualini T, Knoblovits P, Aszpis S, Pacenza N, Stewart Usher J, Bergadá I, Campo SM. Male hypogonadism: an extended classification based on a developmental, endocrine physiology-based approach. Andrology. 2013;1(1):3–16. https://doi.org/10.1111/j.2047-2927.2012.00008.x. Epub 2012 Oct 9. Review.

Reyes CV, Slutky JN. Prostate cancer in men 40 years of age and younger: report of two cases. Commun Oncol. 2009;6:425–7.

Reyes AO, Swanson PE, Carbone JM, Humphrey PA. Unusual histologic types of high-grade prostatic intraepithelial neoplasia. Am J Surg Pathol. 1997;21(10):1215–22.

Rifkin MD, Kurtz AB, Pasto ME, Goldberg BB. Polyorchidism diagnosed preoperatively by ultrasonography. J Ultrasound Med. 1983;2(2):93–4. PubMed PMID: 6842663.

Ritchey ML, Ribbeck M. Successful use of tunica vaginalis grafts for treatment of severe penile chordee in children. J Urol. 2003;170(4 Pt 2):1574–6; discussion 1576.

Roberts JT, Essenhigh DM. Adenocarcinoma of prostate in 40-year-old body builder. Lancet. 1986;2:742.

Rogatsch H, Moser P, Volgger H, Horninger W, Bartsch G, Mikuz G, Mairinger T. Diagnostic effect of an improved preembedding method of prostate needle biopsy specimens. Hum Pathol. 2000;31(9):1102–7.

Rogozinski TT, Janniger CK. Bowenoid papulosis. Am Fam Physician. 1988;38(1):161–4. Review.

Rosenberg AE. Morphology, translocations, and clinical behavior: where do we go from here? Adv Anat Pathol. 1998;5(4):235–8. PubMed PMID: 9859755.

Rosenblum S, Pal A, Reidy K. Renal development in the fetus and premature infant. Semin Fetal Neonatal Med. 2017;22(2):58–66. https://doi.org/10.1016/j.siny.2017.01.001. Epub 2017 Feb 1. Review. PubMed PMID: 28161315.

Rubenwolf P, Herrmann-Nuber J, Schreckenberger M, Stein R, Beetz R. Primary non-refluxive megaureter in children: single-center experience and follow-up of 212 patients. Int Urol Nephrol. 2016;48(11):1743–9. Epub 2016 Aug 4.

Rubin MA, Dunn R, Kambham N, Misick CP, O'Toole KM. Should a Gleason score be assigned to a minute focus of carcinoma on prostate biopsy? Am J Surg Pathol. 2000;24(12):1634–40.

Ruska KM, Partin AW, Epstein JI, Kahane H. Adenocarcinoma of the prostate in men younger than 40 years of age: diagnosis and treatment with emphasis on radical prostatectomy findings. Urology. 1999;53(6):1179–83.

Sakr WA, Haas GP, Cassin BF, Pontes JE, Crissman JD. The frequency of carcinoma and intraepithelial neoplasia of the prostate in young male patients. J Urol. 1993;150(2 Pt 1):379–85. PubMed PMID: 8326560.

Sakr WA, Grignon DJ, Haas GP, Heilbrun LK, Pontes JE, Crissman JD. Age and racial distribution of prostatic intraepithelial neoplasia. Eur Urol. 1996;30(2):138–44. Review. PubMed PMID: 8875194.

Saltzman AF, Cost NG. Current Treatment of Pediatric Bladder and Prostate Rhabdomyosarcoma. Curr Urol Rep. 2018;19(1):11. https://doi.org/10.1007/s11934-018-0761-8. Review.

Sam P, LaGrange CA. Anatomy, Pelvis, Perineum, Urogenital Triangle, Penis. [Updated 2018 Feb 3]. In: StatPearls [Internet]. Treasure Island (FL): StatPearls Publishing; 2018 Jan-. Available from: https://www.ncbi.nlm.nih.gov/books/NBK482236/

Satie AP, Rajpert-De Meyts E, Spagnoli GC, Henno S, Olivo L, Jacobsen GK, Rioux-Leclercq N, Jégou B, Samson M. The cancer-testis gene, NY-ESO-1, is expressed in normal fetal and adult testes and in spermatocytic seminomas and testicular carcinoma in situ. Lab Investig. 2002;82(6):775–80.

Saw D, Tse CH, Chan J, Watt CY, Ng CS, Poon YF. Clear cell sarcoma of the penis. Hum Pathol. 1986;17(4):423–5.

Scholtmeijer RJ, van Unnik AM, ten Kate FW. Acute urine retention in a child due to fibromatosis of the prostate. Eur Urol. 1988;14(5):412–3. https://doi.org/10.1159/000472994. PMID: 3169086.

Sebo TJ, Cheville JC, Riehle DL, Lohse CM, Pankratz VS, Myers RP, Blute ML, Zincke H. Perineural invasion and MIB-1 positivity in addition to Gleason score are significant preoperative predictors of progression after radical retropubic prostatectomy for prostate cancer. Am J Surg Pathol. 2002;26(4):431–9.

Segal RL, Burnett AL. Surgical Management for Peyronie's Disease. World J Mens Health. 2013;31(1):1–11. https://doi.org/10.5534/wjmh.2013.31.1.1. Epub 2013 Apr 23. PubMed PMID: 23658860; PubMed Central PMCID: PMC3640147.

Seydoux G, Braun RE. Pathway to totipotency: lessons from germ cells. Cell. 2006;127(5):891–904. Review. PubMed PMID: 17129777.

Shanggar K, Zulkifli MZ, Razack AH, Dublin N. Granulomatous prostatitis: a reminder to clinicians. Med J Malaysia. 2010;65(1):21–2.

Shao IH, Chen TD, Shao HT, Chen HW. Male median raphe cysts: serial retrospective analysis and histopathological classification. Diagn Pathol. 2012;7:121. https://doi.org/10.1186/1746-1596-7-121. PubMed PMID: 22978603; PubMed Central PMCID: PMC3487840.

Shelmerdine SC, Lorenzo AJ, Gupta AA, Chavhan GB. Pearls and Pitfalls in Diagnosing Pediatric Urinary Bladder Masses. Radiographics. 2017;37(6):1872–91. https://doi.org/10.1148/rg.2017170031. Review.

Shimada H, Misugi K, Sasaki Y, Iizuka A, Nishihira H. Carcinoma of the prostate in childhood and adolescence: Report of a case and review of the literature. Cancer. 1980;46:2534–42. [PubMed: 7002284].

Silber I, McGavran MH. Adenocarcinoma of the prostate in men less than 56 years old: A study of 65 cases. J Urol. 1971;105:283–5. [PubMed: 5090372].

Song HC, Sun N, Zhang WP, Huang CR. Primary Ewing's sarcoma/primitive neuroectodermal tumor of the urogenital tract in children. Chin Med J. 2012;125(5):932–6. Review.

Stamm AW, Kobashi KC, Stefanovic KB. Urologic Dermatology: a Review. Curr Urol Rep. 2017;18(8):62. https://doi.org/10.1007/s11934-017-0712-9. Review.

Stein N, Henkes D. Mesothelioma of the testicle in a child. J Urol. 1986;135(4):794.

Stoop H, van Gurp R, de Krijger R, Geurts van Kessel A, Köberle B, Oosterhuis W, Looijenga L. Reactivity of germ cell maturation stage-specific markers in spermatocytic seminoma: diagnostic and etiological implications. Lab Investig. 2001;81(7):919–28. PubMed PMID: 11454979.

Sugarman J, Anderson J, Baschat AA, Herrera Beutler J, Bienstock JL, Bunchman TE, Desai NM, Gates E, Goldberg A, Grimm PC, Henry LM, Jelin EB, Johnson E, Hertenstein CB, Mastroianni AC, Mercurio MR, Neu A, Nogee LM, Polzin WJ, Ralston SJ, Ramus RM, Singleton MK, Somers MJG, Wang KC, Boss R. Ethical Considerations Concerning Amnioinfusions for Treating Fetal Bilateral Renal Agenesis. Obstet Gynecol. 2018;131(1):130–4. https://doi.org/10.1097/AOG.0000000000002416. Review.

Sutcliffe S, Pakpahan R, Sokoll LJ, Elliott DJ, Nevin RL, Cersovsky SB, Walsh PC, Platz EA. Prostate-specific antigen concentration in young men: new estimates and review of the literature. BJU Int. 2012;110(11):1627–35. https://doi.org/10.1111/j.1464-410X.2012.11111.x. Epub 2012 Apr 13. Review.

Székely E, Törzsök P, Riesz P, Korompay A, Fintha A, Székely T, Lotz G, Nyirády P, Romics I, Tímár J, Schaff Z, Kiss A. Expression of claudins and their prognostic significance in noninvasive urothelial neoplasms of the human urinary bladder. J Histochem Cytochem. 2011;59(10):932–41. https://doi.org/10.1369/0022155411418829. Epub 2011 Aug 10. PubMed PMID: 21832144; PubMed Central PMCID: PMC3201131.

Szymanski KM, Tabib CH, Idrees MT, Cain MP. Synchronous Perivesical and Renal Malignant Rhabdoid Tumor in a 9-year-old Boy: A Case Report and Review of Literature. Urology. 2013; https://doi.org/10.1016/j.urology.2013.04.050. pii: S0090-4295(13)00583-9. [Epub ahead of print].

Taib F, Mohamad N, Mohamed Daud MA, Hassan A, Singh MS, Nasir A. Infantile fibrosarcoma of the penis in a 2-year-old boy. Urology. 2012;80(4):931–3. https://doi.org/10.1016/j.urology.2012.05.021. Epub 2012 Jul 31.

Tamsen A, Casas V, Patil UB, Elbadawi A. Inverted papilloma of the urinary bladder in a boy. J Pediatr Surg. 1993;28(12):1601–2. Review.

Tavora F, Epstein JI. High-grade prostatic intraepithelial neoplasialike ductal adenocarcinoma of the prostate: a clinicopathologic study of 28 cases. Am J Surg Pathol. 2008;32(7):1060–7. https://doi.org/10.1097/PAS.0b013e318160edaf.

Tekgül S, Riedmiller H, Hoebeke P, Kočvara R, Nijman RJ, Radmayr C, Stein R. Dogan HS; European Association of Urology. EAU guidelines on vesicoureteral reflux in children. Eur Urol. 2012;62(3):534–42. https://doi.org/10.1016/j.eururo.2012.05.059. Epub 2012 Jun 5. Review.

The American Fertility Society classifications of adnexal adhesions, distal tubal occlusion, tubal occlusion secondary to tubal ligation, tubal pregnancies, müllerian anomalies and intrauterine adhesions. Fertil Steril. 1988; 49(6):944–955. PubMed PMID: 3371491.

Thompson IM, Pauler DK, Goodman PJ, Tangen CM, Lucia MS, Parnes HL, Minasian LM, Ford LG, Lippman SM, Crawford ED, Crowley JJ, Coltman CA Jr. Prevalence of prostate cancer among men with a prostate-specific antigen level < or =4.0 ng per milliliter. N Engl J Med. 2004;350(22):2239–46. Erratum in: N Engl J Med. 2004 Sep 30;351(14):1470. PubMed PMID: 15163773.

Thorson P, Humphrey PA. Minimal adenocarcinoma in prostate needle biopsy tissue. Am J Clin Pathol. 2000;114(6):896–909.. Review

Thorson P, Vollmer RT, Arcangeli C, Keetch DW, Humphrey PA. Minimal carcinoma in prostate needle biopsy specimens: diagnostic features and radical prostatectomy follow-up. Mod Pathol. 1998;11(6):543–51.

Tokuda Y, Carlino LJ, Gopalan A, Tickoo SK, Kaag MG, Guillonneau B, Eastham JA, Scher HI, Scardino PT, Reuter VE, Fine SW. Prostate cancer topography and patterns of lymph node metastasis. Am J Surg Pathol. 2010;34(12):1862–7. https://doi.org/10.1097/PAS.0b013e3181fc679e. PubMed PMID: 21107093; PubMed Central PMCID: PMC3414911.

Trpkov K, Bartczak-McKay J, Yilmaz A. Usefulness of cytokeratin 5/6 and AMACR applied as double sequential immunostains for diagnostic assessment of problematic prostate specimens. Am J Clin Pathol. 2009;132(2):211–20; quiz 307. https://doi.org/10.1309/AJCPGFJP83IXZEUR.

Tsuzuki T, Magi-Galluzzi C, Epstein JI. ALK-1 expression in inflammatory myofibroblastic tumor of the urinary bladder. Am J Surg Pathol. 2004;28(12):1609–14. PubMed PMID: 15577680.

Ulbright TM, Young RH. Primary mucinous tumors of the testis and paratestis: a report of nine cases. Am J Surg Pathol. 2003;27(9):1221–8.

Ulbright TM, Srigley JR, Hatzianastassiou DK, Young RH. Leydig cell tumors of the testis with unusual features: adipose differentiation, calcification with ossification, and spindle-shaped tumor cells. Am J Surg Pathol. 2002;26(11):1424–33.

Urakami S, Igawa M, Shiina H, Shigeno K, Kikuno N, Yoshino T. Recurrent transitional cell carcinoma in a child with the Costello syndrome. J Urol. 2002;168(3):1133–4.

Uzoh CC, Uff JS, Okeke AA. Granulomatous prostatitis. BJU Int. 2007;99(3):510–2. Epub 2006 Nov 7. Review.

Vaidyanathan S, McDicken IW, Ikin AJ, Mansour P, Soni BM, Singh G, Sett P. A study of cytokeratin 20 immunostaining in the urothelium of neuropathic bladder of patients with spinal cord injury. BMC Urol. 2002;2:7. PubMed PMID: 12147174; PubMed Central PMCID: PMC125297.

Vargas SO, Jiroutek M, D'Amico AV, Renshaw AA. Distribution of carcinoma in radical prostatectomy specimens in the era of serum prostate-specific antigen testing. Implications for delivery of localized therapy. Am J Clin Pathol. 1999a;112(3):373–6.

Vargas SO, Jiroutek M, Welch WR, Nucci MR, D'Amico AV, Renshaw AA. Perineural invasion in prostate needle biopsy specimens. Correlation with extraprostatic extension at resection. Am J Clin Pathol. 1999b;111(2):223–8.

Varma M, Lee MW, Tamboli P, Zarbo RJ, Jimenez RE, Salles PG, Amin MB. Morphologic criteria for the diagnosis of prostatic adenocarcinoma in needle biopsy specimens. A study of 250 consecutive cases in a routine surgical pathology practice. Arch Pathol Lab Med. 2002;126(5):554–61. PubMed PMID: 11958660.

Vasconcelos MAPS, de Lima PP. Prune-belly syndrome: an autopsy case report. Autops Case Rep. 2014;4(4):35–41. https://doi.org/10.4322/acr.2014.037. eCollection 2014 Oct-Dec. PubMed PMID: 28573127; PubMed Central PMCID: PMC5443131.

Vázquez-Costa JF, Arlandis S, Hervas D, Martínez-Cuenca E, Cardona F, Pérez-Tur J, Broseta E, Sevilla T. Clinical profile of motor neuron disease patients with lower urinary tract symptoms and neurogenic bladder. J Neurol Sci. 2017;378:130–6. https://doi.org/10.1016/j.jns.2017.04.053. Epub 2017 May 2.

Vezina CM, Allgeier SH, Moore RW, Lin TM, Bemis JC, Hardin HA, Gasiewicz TA, Peterson RE. Dioxin causes ventral prostate agenesis by disrupting dorsoventral patterning in developing mouse prostate. Toxicol Sci. 2008;106(2):488–96. https://doi.org/10.1093/toxsci/kfn183. Epub 2008 Sep 8. PubMed PMID: 18779384; PubMed Central PMCID: PMC2581676.

Vohra S, Badlani G. Balanitis and balanoposthitis. Urol Clin North Am. 1992;19(1):143–7. Review.

Volmar KE, Fritsch MK, Perlman EJ, Hutchins GM. Patterns of congenital lower urinary tract obstructive uropathy: relation to abnormal prostate and bladder development and the prune belly syndrome. Pediatr Dev Pathol. 2001;4(5):467–72.

von Kopylow K, Kirchhoff C, Jezek D, Schulze W, Feig C, Primig M, Steinkraus V, Spiess AN. Screening for biomarkers of spermatogonia within the human testis: a whole genome approach. Hum Reprod. 2010;25(5):1104–12. https://doi.org/10.1093/humrep/deq053. Epub 2010 Mar 5.

Waheeb R, Hofmann MC. Human spermatogonial stem cells: a possible origin for spermatocytic seminoma. Int J Androl. 2011;34(4 Pt 2):e296–305; discussion e305. https://doi.org/10.1111/j.1365-2605.2011.01199.x. Review. PubMed PMID: 21790653; PubMed Central PMCID: PMC3146023.

Waldeyer W. Über die sogenannte Ureterscheide. Anat Anz, Verhandl d Anatom Gesellsch Wien. 1892;June:259–60.

Wang X, Guo Z, Wang HT, Si TG. 小于等于59岁前列腺癌患者的临床特点及其预后分析 [Clinical features and prognostic analysis in prostate cancer patients under 59 years of age: a report of 72 cases]. Zhonghua Yi Xue Za Zhi. 2012;92(19):1300–3. Chinese. PubMed PMID: 22883113.

Weitzner S, Sarikaya H, Furness TD. Adenocarcinoma of prostate in a twenty-seven-Year old man. Urology. 1980;16:286–8. [PubMed: 7423709].

Weitzner JM, Fields KW, Robinson MJ. Pediatric bowenoid papulosis: risks and management. Pediatr Dermatol. 1989;6(4):303–5.

Woodburne RT. The Ureter, ureterovesical junction, and vesical trigone. Anat Rec. 1965;151:243–9. PubMed PMID: 14324081.

Woods JE, Soh S, Wheeler TM. Distribution and significance of microcalcifications in the neoplastic and nonneoplastic prostate. Arch Pathol Lab Med. 1998;122(2):152–5.

Woodward PJ, Schwab CM, Sesterhenn IA. From the archives of the AFIP: extratesticular scrotal masses: radiologic-pathologic correlation. Radiographics. 2003;23(1):215–40. Review.

Yagi H, Igawa M, Shiina H, Shigeno K, Yoneda T, Wada Y, Urakami S. Inverted papilloma of the urinary bladder in a girl. Urol Int. 1999;63(4):258–60. Review.

Yamamoto S, Senzaki A, Yamagiwa K, Tanaka T, Oda T. 若年性前立腺癌の1例 [Prostatic carcinoma in a young adult: a case report]. Hinyokika Kiyo. 1990;36(5):617–22. Review. Japanese. PubMed PMID: 1698013.

Yang JB, Jeong BC, Seo SI, Jeon SS, Choi HY, Lee HM. Outcome of Prostate Biopsy in Men Younger than 40 Years of Age with High Prostate-Specific Antigen (PSA) Levels. Korean J Urol. 2010;51(1):21–4. https://doi.org/10.4111/kju.2010.51.1.21. Epub 2010 Jan 21. PubMed PMID: 20414405; PubMed Central PMCID: PMC2855470.

Yarmohammad A, Ahmadnia H, Asl Zare M. Transitional cell carcinoma in children: report of a case and review of the literature. Urol J. 2005;2(2):120–1.

Young RH, Scully RE. Pseudosarcomatous lesions of the urinary bladder, prostate gland, and urethra. A report of three cases and review of the literature. Arch Pathol Lab Med. 1987;111(4):354–8.

Zhang CF, Guo Q, Liu C, Ma J, Ma HH, Zhou HB, Shi QL. Male genitourinary system lymphoma: a clinicopathological analysis. Zhonghua Nan Ke Xue. 2012;18(1):52–7. Chinese.

Zhao J, Epstein JI. High-grade foamy gland prostatic adenocarcinoma on biopsy or transurethral resection: a morphologic study of 55 cases. Am J Surg Pathol. 2009;33(4):583–90. https://doi.org/10.1097/PAS.0b013e31818a5c6c.

Zhou M, Shah R, Shen R, Rubin MA. Basal cell cocktail (34betaE12 + p63) improves the detection of prostate basal cells. Am J Surg Pathol. 2003;27(3):365–71.